Proctor and Hughes'
CHEMICAL HAZARDS
of the WORKPLACE

Fourth Edition

PROCTOR AND HUGHES'
CHEMICAL HAZARDS
of the
WORKPLACE
Fourth Edition

Gloria J. Hathaway, Ph.D.
Nick H. Proctor, Ph.D.
James P. Hughes, M.D.

VAN NOSTRAND REINHOLD
I(T)P® A Division of International Thomson Publishing Inc.

New York • Albany • Bonn • Boston • Detroit • London • Madrid • Melbourne
Mexico City • Paris • San Francisco • Singapore • Tokyo • Toronto

I(T)P™ Van Nostrand Reinhold is a division of International Thomson Publishing, Inc.
The ITP logo is a trademark under license

Printed in the United States of America

For more information, contact:

Van Nostrand Reinhold
115 Fifth Avenue
New York, N.Y. 10003

International Thomson Publishing
GmbH
Königswinterer Strasse 418
353227 Bonn
Germany

International Thomson Publishing Europe
Berkshire House 168-173
High Holborn
London WCIV 7AA
England

International Thomson Publishing Asia
221 Henderson Road #05-10
Henderson Building
Singapore 0315

Thomas Nelson Australia
102 Dodds Street
South Melbourne, 3205
Victoria, Australia

International Thomson Publishing Japan
Hirakawacho Kyowa Building, 3F
2-2-1 Hirakawacho
Chiyoda-ku, 102 Tokyo
Japan

Nelson Canada
1120 Birchmount Road
Scarborough, Ontario
Canada M1K 5G4

International Thomson Editores
Seneca 53
Col. Polanco
11560 Mexico D.F. Mexico

1 2 3 4 5 6 7 8 9 10 HAM 02 01 00 99 98 97 96

Library of Congress Cataloging-in-Publication Data
Proctor, Nick H.
    [Chemical hazards of the workplace]
    Proctor and Hughes' Chemical hazards of the workplace / Nick H.
Proctor, James P. Hughes ; [edited by] Gloria J. Hathaway.—4th
ed.
        p. cm.
    Includes bibliographical references and indexes.
    ISBN 0-442-02050-3
    1. Industrial toxicology. 2. Industrial hygiene. I. Hughes,
James P., 1920– . II. Hathaway, Gloria J. III. Title.
RA1229.P76 1996
615.9′02—dc20   96-18126
                                              CIP

PREFACE

This volume is intended primarily for the health professional seeking a brief introductory statement on the toxicology of some 600 chemicals most likely to be encountered at work.

In reviewing the monographs the reader might keep in mind a few details. The Threshold Limit Values (TLVs) listed are those for 1995. As explained in Chapter 1, the numerical values of the TLVs do not take into account the toxicity that might result from skin absorption. However, in instances where this may occur, skin absorption is listed as a route of exposure. Obviously, all substances existing in liquid or solid form conceivably could be ingested, but ingestion is rare and of minor importance except for certain highly toxic substances such as arsenic and lead. Only in those cases has ingestion been listed as a route of exposure.

Each monograph describes the chief signs and symptoms caused by overexposure to the chemical and the clinical effects in human that are related, where data are available, to exposure levels.

If the literature does not provide sufficient data to illustrate human effects, data from animal tests are included to point out target organs that should be monitored closely in exposed workers. Sometimes no human data exist, and animal results must be relied upon entirely. In such cases, the experimental data from the literature are summarized. In many instances the odor threshold or other warning properties are identified.

In response to comments by readers of earlier editions, the format of alphabetized brief monographs supplemented by updated literature citations is maintained, providing a volume to be kept readily at hand as an initial reference. In order to accommodate over 60 additional chemical compounds within the handbook concept, it was decided to delete some of the introductory text of the earlier editions. The plethora of larger works of recent publication encouraged the authors in the decision to maintain the familiar format of "Chemical Hazards of the Workplace" in this welcome revision.

Each chemical substance listed in previous editions has been revised if new toxicological information is available. Hundreds of new references have been added.

The authors of this fourth edition, Drs. Gloria J. Hathaway and Nick Proctor, participated in the preparation of each of the previous editions and have richly fulfilled the intent of the original work.

James P. Hughes, M.D.
Piedmont, California

ACKNOWLEDGMENTS

We wish to thank once again the California Department of Health Services for the use of their excellent Occupational and Health Library. To librarian Marianne Mahoney, we extend our appreciation for her continuing help in our literature review. For technical and computer assistance, we wish to thank Barron V. Wesenberg, whose wizardry with the disks made this project possible. The first edition of this text was performed pursuant to contract No. CDC-99-74-43 under the National Institute for Occupational Safety and Health.

CONTENTS

❏

INTRODUCTION:

TOXICOLOGICAL CONCEPTS

❏

Toxicologic Concepts—Setting Exposure Limits

Nick H. Proctor, Ph.D.

DEFINITIONS

In occupational health practice, the following terms describe the states of matter in which chemical atmospheres may occur:

Gas: A formless fluid that completely occupies the space of an enclosure at 25°C and 760-torr (1-atm) pressure.

Vapor: The gaseous phase of a material that is liquid or solid at 25°C and 760-torr (1-atm) pressure.

Aerosol: A dispersion of particles of microscopic size in a gaseous medium; may be solid particles (dust, fume, smoke) or liquid particles (mist, fog).

Dust: Airborne solid particles (an aerosol) that range in size from 0.1 to 50 μ and larger in diameter. A person with normal eyesight can see dust particles as small as 50 μ in diameter. Smaller airborne particles cannot be seen unless strong light is reflected from the particles. Dust of respirable size (below 10 μ) cannot be seen without the aid of a microscope.

Fume: An aerosol of solid particles generated by condensation from the gaseous state, generally after volatilization from molten metals. The solid particles that make up a fume are extremely fine, usually less than 1.0 μ in diameter. In most cases, the volatilized solid reacts with oxygen in the air to form an oxide. A common example is cadmium oxide fume.

Smoke: An aerosol of carbon or soot particles less than 0.1 μ in diameter that results from the incomplete combustion of carbonaceous materials such as coal or oil. Smoke generally contains droplets as well as dry particles.

Mist: An aerosol of suspended liquid droplets generated by condensation from the gaseous to the liquid state or by the breaking up of a liquid into a dispersed state, such as by splashing, foaming, or atomizing. Examples are the oil mist produced during cutting and grinding operations, acid mists from electroplating, acid or alkali mists from pickling operations, and pain spray mist from spraying procedures.

Fog: A visible liquid aerosol formed by condensation.

The following terms of measurement are

commonly used in toxicologic testing and in industrial hygiene practice:

ppm: Parts of vapor or gas per million parts of air by volume.

mg/m^3: Milligrams of a substance per cubic meter of air

mg/l: Micrograms of a substance per liter of air

TOXICOLOGIC CONCEPTS

Routes of Entry of Chemicals Into The Body

In the occupational setting, inhalation is the most important route of entry of chemical agents into the body, followed by contact with skin and subsequent cutaneous absorption. Although the gastrointestinal tract is a potential site of absorption, the oral ingestion of significant amounts of chemicals is not of great significance except in cases of inhaled particles which may be transported up the mucociliary escalator and ingested.

Inhalation

The respiratory tract is exposed to chemicals in the inspired air. The two main factors that determined the tissue responses to chemicals are the functional anatomy of the respiratory tract and the physiocochemical nature of the material.[1-3]

Physicians are accustomed to considering the respiratory tract as being divided into three major regions: the nasopharyngeal (upper airways), the tracheobronchial tree (lower airways), and the pulmonary (alveoli).

The nasopharynx begins with the anterior nares and extends down to the larynx. The nasal passages are lined with vascular mucous epithelium composed of ciliated epithelium and scattered mucous glands. The nasopharynx filters out large inhaled particles and is the area where the relative humidity is increased and temperature of the air is moderated.

The airways (trachea, bronchi, and bronchioles, or tracheobronchial tree) serve as conducting passages between the nasopharynx and alveoli. They are lined with ciliated epithelium and coated with a thin layer of mucus secreted primarily by goblet cells in the upper airways and primarily by Clara cells at the bronchiolar level. This mucous covering terminates at the film covering the alveolar membrane. The surface of the airways serves as a mucociliary escalator, moving particles up to the oral cavity, where they are swallowed and excreted or expectorated.

The ciliated cells are the most vulnerable to damage. The most frequent degenerative changes in these cells are loss of cilia, necrosis, and sloughing of cells into the airway lumen. Necrosis and desquamation of nonciliated and secretory cells are less frequently observed.

After acute mild insult, the nonciliated cells proliferate and the epithelium regenerates to normal. In the airways, nonciliated basal cells are the main proliferating population. In the bronchioles, the Clara cell is the main precursor cell for regneration. Because of the delicate nature of the respiratory tract epithelium and the proximity of subepithelial blood vessels, an inflammatory response occurs to all but the mildest form of injury. Many lesions are therefore diagnosed as rhinitis, tracheitis, and bronchiitis and qualified as acute, subacute, and chronic, depending on the stage of the response.

If the insult persists, hyperplasia (cell proliferation) proceeds and leads to an abnormal epithelium. Injury produced by chronic exposure to irritants such as SO_2, NO_2, O_3, formaldehyde, and tobacco smoke includes undifferentiated basal cells (hyperplasia), squamous metaplasia, and goblet cell metaplasia. If the process of insult continues long enough, neoplasia is a possibility.

In practice, many irritants produce responses between mild and severe, and various combinations of degeneration, inflammation, and proliferation may be observed.

The lower respiratory tract (pulmonary region or alveolar ducts and sacs) is the area where gas exchange occurs. Alveolar sacs, clusters of two or more alveoli, branch from alveolar ducts. It is generally considered that there are approximately 300 million alveoli in the lungs of adult humans.[4] The total alveolar surface area in the lungs of adult humans is about 35 m^2 during expiration, 70–80 m^2 at three-fourths total lung capacity, and 100 m^2 during deep inspiration.[5]

The alveoli are lined by two main types of epithelial cells. Type I cells (squamous pneumonocytes) have flattened nuclei and thin but very extensive cytoplasm covering most of the alveolar wall. Because the cell has a very large surface area, it is very susceptible to injury.

Type II cells (granular pneumonocytes) are distributed throughout the alveoli between Type I cells. Although they are more numerous than Type I cells, they are cuboidal in shape and occupy far less of the alveolar surface area. The prime function of this cell is the production of pulmonary surfactant, and it is generally less susceptible to injury than the Type I cell.

The other main cell type in the alveoli is the alveolar macrophage, which plays an important role by phagocytizing particulates and removing them from the alveoli. Phagocytosis of toxic particulates may injure macrophages, and the discharge of their contents may cause alveolar damage. Stromal cells, such as fibroblasts, are infrequent but may increase sufficiently in number during chronic inflammatory reactions to interfere with gaseous exchange and compromise lung function.

Most direct toxins entering the alveoli primarily affect Type 1 cells and their associated capillary endothelial cells. Following acute injury, the epithelium and/or underlying capillary endothelia cells may swell, disrupt, distort, or lose their connections with others, leaving large areas of basement membrane uncovered. This allows fluid to move into the alveolar lumen from capillaries, with subsequent pulmonary edema.

The sequel to acute injury depends on the potency and concentration of the toxic agent and the duration of exposure. Potent gases produce a severe vascular reaction and alveolar flooding. The fluid prevents gaseous exchange, and death of the human or animal ensues. Following acute mild nonlethal damage, excess fluid is removed, and the resistant Type II cells proliferate and reline the alveoli. The cells subsequently differentiate into Type I cells.

If the chemical is a moderate irritant and causes significant damage to the basement membrane and stroma as well to the epithelial cells, fibroblastic repair and fibrous scarring result in the alveoli. These fibrotic alveoli are generally lined by atypical Type II cells. The lining of alveoli by Type II cells, either in the early phases of repair of mild damage or as an end stage of more severe damage, is often referred to as alveolar epithelialization. Occasionally, the alveoli may be relined by a proliferation of bronchiolar epithelium. This is termed alveolar bronchiolization. Intra-alveolar accumulation of macrophages is also a prominent feature.

Gases: The rate of removal of gases from the airstream during inhalation depends mostly on the water solubility of the gas. Highly water-soluble gases, such as ammonia, hydrogen chloride, and hydrogen fluoride, dissolve readily in the moisture associated with the mucous coating of the nasopharyngeal region, causing irritation at those sites. At high atmospheric concentrations, some of the gas will not be absorbed at the upper respiratory sites, and amounts sufficient to reach the alveoli can cause severe irritation and pulmonary edema.

Comparatively insoluble gases, such as nitrogen dioxide and phosgene, are not removed by the moisture in the upper respiratory tract and can easily reach the alveoli. Substances of intermediate solubility, such as chlorine, can cause irritation at points all along the respiratory tract.

Bronchoconstriction is one of the most common immediate responses observed upon inhalation of a number of reactive gases. The constriction may be due to a direct action on the airway smooth muscles or indirectly through the release of histamine and other mediators.

Particulates: The chief factor that determines the site of deposition of particulate matter in the respiratory tract is its size.[3,6] Particles having an aerodynamic diameter of 5–30 μ are primarily deposited in the nasopharyngeal region by impaction with nose hairs and the angular walls of the nasopharyngeal passages.

Particles with an aerodynamic diameter of 1–5 μ are deposited in the airways (tracheobronchial regions) by sedimentation under gravitational forces. As the alveolar regions are approached, the velocity of the airflow decreases significantly, allowing more time for sedimentation. The very small particles, generally less than 1 μ, that penetrate to the alveoli are deposited there, mainly by diffusion.

In extrapolating results from rodents to humans, it is important to understand the differences in deposition that occur.[7–8] Small rodents usually have lower fractional deposition of inhaled particles in the lung than humans, but rodents inhale more air per unit of lung mass or lung surface than humans do. The most important interspecies differences in deposition are associated with particles larger than about 5 μ in aerodynamic diameter since these larger particles cannot readily reach the pulmonary region in small nose-breathing rodents.

In contrast, a decreasing proportion of particles from 1 μ (100%) to 10 μ (1%) reach the pulmonary region in the human lung during normal breathing via the nose.[9] Once there, maximum pulmonary deposition occurs for particle sizes of 1–4 μ: about 25% of 1 μ, 35% of 2 μ, 30% of 3 μ, and 25% of 4 μ.[10]

Mouth breathing by humans during exertion may result in deposition that is distinctly different from that associated with nasal breathing, with increased deposition of the larger particles up to about 15 μ in both the tracheobronchial and pulmonary regions.[8]

Particle Clearance: Particles deposited in the nasopharyngeal region are moved to the pharynx by the ciliated cells and mucus and expectorated or swallowed.[10] The clearance rate is relatively rapid with a half-life of 12 to 24 hours.

Particles deposited on or in the lung parenchyma are cleared primarily by alveolar macrophages. These phagocytized particles migrate to the ciliated epithelium or to the lymphatic system at rates ranging from 2 to 6 weeks. For some materials, however, this time is longer such that half-lives of many months occur.

Certain chemicals, such as silicon dioxide, have a cytotoxic effect on the alveolar macrophage, which results in the accumulation of particles in a given area. As the macrophages lose their activity, these particles become less subject to removal, leading to the development of masses containing dead cells and particles.

Fibers: When using animal inhalation studies for assessment of the risk to human health of airborne fibers, it is critical to demonstrate that the characteristics and concentrations of the experimental fiber aerosols are comparable to those in human exposure situations.[11] NIOSH has two criteria for defining fibers: "A" rules (total fibers) and "B" rules (respirable fibers).[12]

NIOSH A rules count fibers with a length/diameter ratio ≥3:1, length ≥5 μ.

NIOSH B rules count fibers with a length/diameter ratio >5:1, length ≥5 μ, and diameter <3 μ.

Skin Contact

Skin structure varies widely in different regions, from the delicate and relatively permeable skin of the scrotum, to the rough thick covering of the palms and soles.[13] The skin of the scrotum has a relatively thin layer of keratin, whereas the palms and soles have a thick layer.

The skin consists of a thin outer layer (epidermis) and a relatively thicker inner layer (dermis). The epidermis is approximately 0.1 mm in thickness, whereas the dermis is generally 2 to 4 mm thick. Dermoepidermal ridges provide a large interface area between the epidermis and dermis. This is of great importance in that the epidermis, which is not vascularized, received all its nutrients from the blood supply of the dermis.

The epidermis consists of several types of cells. The epidermal cell type apposed to the dermis is the stratum germinativum (basal cell layer), over which are the stratum spinosum, stratum granulosum, stratum ludicum, and the outermost layer, or stratum corneum. The basal cell layer consists of one layer of columnar epithelial cells. Upon division, the basal cells are pushed up and become the stratum spinosum, which consists of several layers of cells. As these cells approach the surface of the skin, they become larger and form the stratum granulosum.

At this point, the nuclei are broken up, resulting in the death of the cell. The next layer, stratum ludicum, is ill defined except in areas of thick skin and is said to contain eleidin, a transformation product of the keratohyalin present in the stratum granulosum.

In the outermost layer, the stratum corneum, the eleidin has been converted into keratin, which represents the ultimate fate of the epidermal cell. Keratin, continuously sloughed off or worn away, is replaced by the cells beneath it. The time required for a basal cell to migrate from the stratum germinativum to the outer part of the stratum corneum is estimated at 26–28 days.

The dermis is a thick fibrous network of collagen and elastin and is composed of two layers. The outer, thinner one is the papillary layer, which has prominent papillae that merge with the thick reticular layer. The papillae are well supplied with blood by the capillaries that are prominent in them, which serve the basal cell layer in the dermis with nutrients.

The dermis contains several types of cells, including fibroblasts, fat cells, macrophages, histiocytes, mast cells, and cells associated with the blood vessels and nerves of the skin. The predominant cell is the fibroblast, which is associated with biosynthesis of the fibrous proteins and ground substances such as hyaluronic acid, chondroitin sulfates, and mucopolysaccharides.

The appendages of skin are hair follicles, sebaceous glands, eccrine and apocrine sweat glands, hair, nails and arrectores pilorum muscle.

When a substance contacts the skin, various actions are possible:

The agent may penetrate the skin, enter the blood, and act systemically (fat soluble chemicals such as chlorinated pesticides, solvent, aniline, and parathion).

The skin and its associated film of lipid and sweat may act as an effective barrier that the substance cannot penetrate.

The substance may react with the skin surface and cause primary irritation (acids, alkalies, many organic solvents).

The substance may penetrate the skin and cause allergic contact dermatitis (formaldehyde, nickel, phthalic anhydride).

In order to pass into the skin, the substance must enter through one or more of the following routes: the epidermal cells, sweat glands, sebaceous glands, or the hair follicles. The pathway through the stratum corneum and the epidermal cells is the main avenue of penetration, as this tissue constitutes the majority of the surface area of the skin.

The stratum corneum plays a critical role in determining cutaneous permeability. Absorption is faster through skin that is abraded or inflamed. Chemicals that are not normally considered hazardous may be dangerous to individuals suffering from active inflammatory dermatoses.

The skin is not only a barrier to restrict diffusion of chemicals into the body, it is also an organ that can metabolize a variety of topically applied substances before they become systemically available.[14] The skin has many of the same enzymes as the liver. The activities of several cutaneous enzymes in whole skin homogenates have been measured and compared to hepatic activity in the mouse.[15] The activities of the enzymes in the whole skin homogenates were typically 2% to 6% of the hepatic values. However, there is evidence that the enzymes are present primarily in the epidermis. Since the epidermis makes up only 2% to 3% of the total skin, the real activities may range from 80% to 240% of those in the liver. Enzyme systems present include a cytochrome P-450 system and a mixed-function oxidase system.

EXPOSURE

Exposure to chemicals in toxicological tests of animals is classified according to frequency and duration, as follows:

- *Acute* exposure is exposure for up to 24 hours.
- *Subacute* exposure is repeated exposure for 1 month or less.
- *Subchronic* exposure is repeated exposure for 1 to 3 months.
- *Chronic* exposure is repeated exposure that lasts for more than 3 months, and often for 24 months or the lifetime of rodent species.

In the occupational setting, acute human exposure generally refers to exposure that causes an effect within 24 hours, whereas chronic exposure is applied to repeated exposures over time.

Risk Assessment

Substances that are considered harmful to biological systems are generally referred to as toxins, toxicants, or exotoxins. However, many substances, whether naturally occurring in the environment or synthetic, are not readily classifiable under this definition. Substances that are harmful at high doses may be innocuous or even essential to biological systems at lower doses, illustrating the *dose-response relationship*. Substances may also have different effects on different species, being harmful to one species but not harmful to another species at the same exposure level. There maybe effects on only one organ system or a general effect on the entire organism. The effect of the substances may be immediate or delayed, even to the point of expressing itself in later generations, and it may be temporary or permanent.

Risk assessment is the name given to a complex series of steps that describe the process whereby scientific data are used to define the health effects of the exposure of individuals to various materials or situations. It usually consists of four steps:

- Hazard identification: The determination of whether or not a particular substance is causally linked to a particular health effect.
- Dose-response assessment: The determi-

nation of the relationship between the magnitude of the exposure to a substance and the probability of occurrence of the health effect in question.
- Exposure assessment: The determination of the extent of human exposure to the substance under various conditions.
- Risk characterization: The description of the nature and often the magnitude of human risk, including attendant uncertainty.

Risk management, on the other hand, is the process of setting and implementing policies that integrate the results of the risk assessment with engineering data and social, regulatory, and legal concerns.

THE STANDARDS-SETTING PROCESS

Threshold Limit Value

The American Conference of Governmental Industrial Hygienists (ACGIH) has prepared a list of the threshold limit values (TLVs) for approximately 800 substances. The following three categories of TLVs are specified in the list for 1994–1995.[16]

- *Threshold Limit Value—Time-Weighted Average (TLV-TWA)*. The time-weighted average concentration for a normal 8-hour workday and a 40-hour workweek, to which nearly all workers may be repeatedly exposed, day after day, without adverse effect.
- *Threshold Limit Value—Short-Term Exposure Limit (TLV-STEL)*. The concentration to which workers can be exposed continuously for a short period of time without suffering from (1) irritation, (2) chronic or irreversible tissue damage, or (3) narcosis of sufficient degree to increase the likelihood of accidental injury, to impair self-rescue or to materially re-

duce work efficiency, and provided that the daily TLV-TWA is not exceeded. It is not a separate independent exposure limit; rather, it supplements the time-weighted average (TWA) limit where acute effects are recognized from a substance whose toxic effects are primarily of a chronic nature. STELs are recommended only where toxic effects have been reported from high short-term exposures in either humans or animals.

A STEL is defined as a 15-minute TWA exposure that should not be exceeded at any time during a workday even if the 8-hour TWA is within the TLV-TWA. Exposures above the TLV-TWA up to the STEL should not be longer than 15 minutes and should not occur more than four times per day. There should be at least 60 minutes between successive exposures in this range. An averaging period other than 15 minutes may be recommended when this is warranted by observed biological effects.

- *Threshold Limit Value—Ceiling (TLV-C).* The concentration that should not be exceeded during any part of the working exposure.

    In the absence of a STEL, excursions in worker exposure levels may exceed 3 times the TLV-TWA for no more than a total of 30 minutes during a workday, and under no circumstances should they exceed 5 times the TLV-TWA, provided that the TLV-TWA is not exceeded.

"Skin" Notation. Substances on the list followed by the designation "Skin" refer to the potential significant contribution to the overall exposure by the skin route, including mucous membranes and the eyes, either by contact with vapors or, of probable greater significance, by direct skin contact with the substance.

TLVs are revised by the ACGIH annually as new information becomes available. Each year, additional substances of interest are added to the TLV list. Certain compounds that are proven or suspected carcinogens in humans, such as benzidine, 4-aminodiphenyl, and 4-nitrodiphenyl, have no TLV value, and human

exposure to these agents should be avoided. *Note:* For a detailed discussion of carcinogenic risks to humans, the publications of IARC should be consulted.[17]

OSHA Standards

The first occupational safety and health standards were set when, with only minor changes, the 1968 ACGIH list of nearly 400 TLVs, as well as certain stands of the American National Standards Institute (ANSI), were incorporated into the Walsh–Healey Public Contracts Act. These standards thereby became limits of exposure for employees of federal government contractors.

Subsequently, under the authority of the Occupational Safety and Health Act of 1970, these same 1968 TLVs and ANSI standards were promulgated by the Occupational Safety and Health Administration (OSHA) as the start-up Permissible Exposure Limits (PEL) for all workers covered by the act.

REFERENCES

1. Glaister JR: *Principles of Toxicological Pathology*, pp 62–74. Philadelphia, Taylor & Francis, 1986
2. West JB: *Respiratory Physiology—The Essentials.* Baltimore, MD, Williams & Wilkins, 1985
3. Gordon T, Amdur MO: Responses of the Respiratory System to Toxic Agents. In Amdur MO, Doull J, Klaasen CD (eds): *Casarett and Doull's Toxicology*, 4th ed, pp 383–406. New York, Pergamon Press, 1991
4. Charnock EL, Doershuk CF: Development aspects of the human lung. *Pediatr Clin North Am* 20:275–292, 1973
5. Weibel ER: *Morphometry of the Human Lung.* New York, Academic Press, 1963
6. Salem H: Principles of inhalation toxicology. In Salem H (ed): *Inhalation Toxicology*, pp 1–34. New York, Marcel Dekker, 1987
7. Raabe OG: Deposition and clearance of inhaled particles. In Gee JBL, Morgan WKC, Brooks SM (eds): *Occupational Lung Disease*, pp 1–38. New York, Raven Press, 1984
8. Raabe OG et al: Regional deposition of in-

haled monodisperse coarse and fine particles in small laboratory animals. *Annals Occup Hyg* 32, Supp 1:53–63, 1988

9. American Conference of Governmental Industrial Hygienists: *1994–1995 Threshold Limit Values and Biological Exposure Indices*, p 45. Cincinnati, OH, ACGIH, 1994

10. Kennedy GK Jr: Inhalation toxicology. In Hayes AW: *Principles and Methods of Toxicology*, 2d ed, pp 361–382. New York, Raven Press, 1989

11. Hesterberg TW, Hart GA: Comparison of human exposures to fiberglass with those used in a recent rat chronic inhalation study. *Reg Tax Pharmacol* 20:S35–S47, 1994

12. NIOSH (National Institute for Occupational 9434 Safety and Health): *NIOSH Manual of Analytical Methods, Method 7400*, Revision 3. Washington, DC, US Government Printing Office, 1989

13. Rongue EL: Skin structure, function, and biochemistry. In Marzulli FN, Maibach HI (eds) *Dermatotoxicology*, 3rd ed, pp 1–70. New York, Hemisphere, 1987

14. Noonan PK, Wester RC: Cutaneous biotransformations. In Marzulli FN, Maibach HI (eds): *Dermatotoxicology*, 3d ed, pp 71–94. New York, Hemisphere, 1987

15. Pohl R, Philpot R, Fouts J: Cytochrome P-450 content and mixed-function oxidase activity in microsomes isolated from mouse skin. *Drug Metab Dispos* 4:442–450, 1976

16. American Conference of Governmental Industrial Hygienists: *1994–1995 Threshold Limit Values for Chemical Substances and Physical Agents and Biological Exposure Limits*. Cincinnati, OH, ACGIH, 1994

17. IARC (International Agency for Research on Cancer): *Monographs on the Evaluation of Carcinogenic Risks to Human, Overall Evaluations of Carcinogenicity: An Updating of IARC Monographs Vols 1–42*, Suppl 7, 440 p. Lyon, France, International Agency for Research on Cancer, 1987

Part

II

❏

THE CHEMICAL HAZARDS

❏

Gloria J. Hathaway, Ph.D
Nick H. Proctor, Ph.D
James P. Hughes, M.D.

ACETALDEHYDE
CAS: 75-07-0

CH₃CHO

Synonyms: Ethanal; acetic aldehyde; ethylaldehyde; methyl formaldehyde

Physical Form.   Colorless liquid

Uses.   As a chemical intermediate in synthesis of acetic acid, pentaerythritol, and pyridine; in the production of perfumes, polyester resins, and dyes; as a food preservative and flavoring agent

Exposure.   Inhalation

Toxicology.   Acetaldehyde is an irritant of the eyes, skin, and respiratory tract; at high concentrations, it causes narcosis; it is carcinogenic in experimental animals.

Nausea, loss of consciousness, and pulmonary edema have been reported with heavy exposure.[1] At 134 ppm for 30 minutes, there was mild upper respiratory irritation, whereas 15 minutes at 50 ppm produced mild eye irritation.[2] Sensitive subjects have noted eye irritation following 15-minute exposures at 25 ppm.[3] Splashed in the eyes, the liquid causes a burning sensation, lacrimation, blurred vision, and corneal injury.[1] On the skin for a prolonged period of time, the liquid causes erythema and burns.

In animal studies, the 4-hour inhalation LC_{50} for hamsters was 17,000 ppm and 13,300 ppm for rats.[4] Exposure to 5000 ppm for 10 minutes produced a 50% decrease in respiration rate in mice; in anesthetized rats, significant increases in blood pressure were observed at 1700 ppm, and concentrations above 6000 ppm significantly increased heart rate.[5,6]

Hamsters repeatedly exposed to 4500 ppm for 3 months had growth retardation, ocular and nasal irritation, increased erythrocyte counts, and severe histopathological changes in the respiratory tract.[7]

Chronic inhalation of acetaldehyde produced tumors of the respiratory tract in rats and hamsters.[8] The incidence of laryngeal carcinomas increased in hamsters exposed for 1 year to 1500 ppm.[9] In a lifetime inhalation study (52 weeks, with recovery for 26 or 52 weeks), rats exposed at 750, 1500, or 3000 ppm had exposure-related increases in adenocarcinomas and squamous-cell carcinomas of the nasal mucosa.[10] Associated changes included growth retardation, degenerative changes of the olfactory epithelium, and metaplasia of the respiratory epithelium, frequently accompanied by keratinization.[10,11]

The IARC has determined that there is sufficient evidence for carcinogenicity of acetaldehyde to experimental animals. One limited epidemiological study provided inadequate evidence for human carcinogenicity.[4,8]

Acetaldehyde has demonstrated genotoxicity in a variety of cell culture systems.[12]

The 1995 ACGIH ceiling threshold limit value (C-TLV) for acetaldehyde is 25 ppm (45 mg/m³) with an A3 animal carcinogen designation.

REFERENCES

1. Von Burg R, Stout T: Toxicology update: acetaldehyde. *J Appl Toxicol* 11:373–76, 1991
2. Silverman L, Schulte HF, First MW: Further studies on sensory response to certain industrial solvent vapors. *J Ind Hyg Toxicol* 28:262–266, 1946
3. *Chemical Hazard Information Profile: Acetaldehyde.* Washington, DC, US Environmental Protection Agency, 1983
4. *IARC Monographs on the Evaluation of the Carcinogenic Risk of Chemicals to Humans.* Vol 36, Allyl compounds, aldehydes, epoxides and peroxides, pp 101–136. Lyon, International Agency for Research on Cancer, 1985
5. Kane LE, Dombroske R, Alaire Y: Evaluation of sensory irritation from some common industrial solvents. *Am Ind Hyg Assoc J* 41:451–455, 1980
6. Egle JL Jr: Effects of inhaled acetaldehyde and propionaldehyde on blood pressure and heart rate. *Toxicol Appl Pharmacol* 23:131–135, 1972
7. Kruysse A, Feron VJ, Til HP: Repeated exposure to acetaldehyde vapor: studies in Syrian golden hamsters. *Arch Environ Health* 30:449–452, 1975
8. *IARC Monographs on the Evaluation of Carcinogenic Risks to Humans, Overall Evaluations of Carcinogenicity: An Updating of IARC Mono-*

graphs Volumes 1 to 42, Suppl 7, pp 77–78. Lyon, International Agency for Research on Cancer, 1987

9. Feron VJ, Kruysse A, Woutersen RA: Respiratory tract tumors in hamsters exposed to acetaldehyde vapor alone or simultaneously to benzo(a)pyrene or diethylnitrosamine. *Eur J Cancer Clin Oncol* 18:13–31, 1982

10. Woutersen RA, Appleman LM, et al: Inhalation toxicity of acetaldehyde in rats. III. Carcinogenicity study. *Toxicology* 41:213–231, 1986

11. Woutersen RA, Feron VJ: Inhalation toxicity of acetaldehyde in rats. IV. Progression and regression of nasal lesions after discontinuation of exposure. *Toxicology* 47:295–305, 1987

12. Heck H: Mechanisms of aldehyde toxicity: structure activity studies. *CIIT Activities* 5(10):106, 1985

ACETAMIDE
CAS: 60-35-5

CH_3CONH_2

Synonyms: Acetic acid amide; ethanamide

Physical Form.   Deliquescent crystals

Uses.   Cryoscopy; organic synthesis; general solvent; lacquers; explosives; soldering flux; wetting agent; plasticizer

Exposure.   Ingestion; inhalation; skin absorption

Toxicology.   Acetamide is a mucous membrane irritant, a liver toxin, and a carcinogen in animals.

There are no data regarding the toxicity of acetamide to humans.

In animals, acetamide was stated to be a mild irritant to skin and eyes although experimental details were not available. Oral administration of acetamide to rodents produces lethality with doses of 1 to 7 g/kg.[1] In another report, single oral dose LD_{50} values for male rats and male mice were 10.3 and 10.1 g/kg, respec-

tively.[2] Minor changes in liver histology occur following acute exposures in rats.[1]

Oral doses of 0.3 g/kg acetamide administered on days 6 through 18 of gestation produced no toxicity or terata in rabbits. No maternal toxicity was seen at 1 g/kg although one rabbit aborted; fetal numbers and body weights were lowered, with no terata. At 3 g/kg, maternal toxicity was encountered, fetal numbers and weights were reduced, the number of dead implants was elevated, and cleft palate was seen.[1] No reproductive, embryotoxic, or teratogenic effects were observed in rats.[1]

The International Agency for Research on Cancer (IARC) has determined that there is sufficient evidence of carcinogenicity for acetamide in experimental animals. Acetamide produced benign and malignant liver tumors in rats following oral administration. In male mice, an increased incidence of malignant lymphomas also was observed.[3]

Acetamide was mutagenic in *Escherichia coli* and *Salmonella typhimurium*; this effect was independent of dose. Acetamide did produce morphological transformation in Syrian hamster embryo cells in the absence of metabolic activation. However, acetamide did not induce reversions in several *S. typhimurium* strains.[1]

ACGIH has not established a threshold limit value for acetamide.

REFERENCES

1. Kennedy GL Jr: Biological effects on acetamide, formamide, and their monomethyl and dimethyl derivatives. *CRC Crit Rev Toxicol* 17:129–182, 1986

2. *IARC Monographs on the Evaluation of the Carcinogenic Risk of Chemicals to Man*, Vol 7, *Some Anti-thyroid and related substances, nitrofurans and industrial chemicals*, pp 197–200. Lyon, International Agency for Research on Cancer, 1974

3. *IARC Monographs on the Evaluation of the Carcinogenic Risks to Humans, Overall Evaluations of Carcinogenicity: An Updating of IARC Monographs Volumes 1 to 41*, Suppl 7, pp 389–390. Lyon, International Agency for Research on Cancer, 1987

ACETIC ACID
CAS: 64-19-7

CH₃COOH

Synonyms: Ethanoic acid; ethylic acid; methane carboxylic acid; vinegar (4–6% solution in water)

Physical Form. Liquid

Uses. In the production of cellulose and vinyl acetate; dyeing; pharmaceuticals; and food processing

Exposure. Inhalation

Toxicology. Acetic acid vapor is a severe irritant of the eyes, mucous membranes, and skin.

Exposure to 50 ppm or more is intolerable to most persons and results in intensive lacrimation and irritation of the eyes, nose, and throat, with pharyngeal edema and chronic bronchitis.[1] Unacclimatized humans experience extreme eye and nasal irritation at concentrations in excess of 25 ppm; conjunctivitis from concentrations below 10 ppm has been reported.[1]

In one case report, a 37-year-old male maintenance fitter, while disconnecting a pressurized pump, was accidentally exposed to a large cloud of hot acetic acid.[2] The patient suffered first-degree burns on the hands and face and developed progressive dyspnea. At 3 months, there were persistent extensive crackles in the basal area of the lungs, widespread bronchial inflammatory changes, and diffuse moderate interstitial pneumonitis, which promptly improved following treatment with corticosteroids and bronchodilators.

In a study of five workers exposed for 7 to 12 years to concentrations of 80 to 200 ppm at peaks, the principal findings were blackening and hyperkeratosis of the skin of the hands, conjunctivits (but no corneal damage), bronchitis and pharyngitis, and erosion of the exposed teeth (incisors and canines).[3]

Digestive disorders with pyrosis and constipation have also been reported at unspecified prolonged exposures.[4]

Glacial (100%) acetic acid caused severe injury when applied to the eyes of rabbits; in humans, it has caused permanent corneal opacification.[5] A splash of vinegar (4% to 10% acetic acid solution) in the human eye causes immediate pain and conjunctival hyperemia, sometimes with injury of the corneal epithelium.[5]

On the guinea pig skin, the liquid in concentrations in excess of 80% produced severe burns; concentrations of 50% to 80% produced moderate to severe burns; solutions below 50% caused relatively mild injury; no injury was produced by 5% to 10% solutions.[3]

Although ingestion is unlikely to occur in industrial use, as little as 1.0 ml of glacial acetic acid has resulted in perforation of the esophagus.[1]

The 1995 ACGIH threshold limit value–time-weighted average (TLV-TWA) for acetic acid is 10 ppm (25 mg/m³) with a short-term exposure limit of 15 ppm (37 mg/m³).

REFERENCES

1. *AIHA Hygienic Guide Series: Acetic Acid.* Akron, OH, American Industrial Hygiene Association, 1978
2. Rajan KG, Davies BH: Reversible airways obstruction and interstitial pneumonitis due to acetic acid. *Br J Ind Med* 46:67–68, 1989
3. Guest D et al: Aliphatic carboxylic acids. In Clayton GD, Clayton FE (eds): *Patty's Industrial Hygiene and Toxicology*, 3rd ed, rev, Vol 2C, *Toxicology*, pp 4909–4911. New York, Wiley-Interscience, 1982
4. Hazard Data Bank, Sheet No 64, Acetic Acid. *Safety Practitioner*, pp 11–12, April 1985
5. Grant WM: *Toxicology of the Eye*, 2nd ed, pp 80–82, Springfield, IL, Charles C Thomas, 1974

ACETIC ANHYDRIDE
CAS: 108-24-7

(CH₃CO)₂O

$(CH_3CO)_2O$

Synonyms: Acetic oxide; acetyl oxide; ethanoic anhydrate; acetic acid anhydride

Physical Form.   Colorless liquid

Uses.   In manufacture of cellulose esters, plastics, pharmaceuticals, photographic films, cigarette filters, and magnetic tape; inorganic synthesis as an acetylating agent, bleaching agent, and dehydrating agent

Exposure.   Inhalation

Toxicology.   Acetic anhydride vapor is a severe irritant of the eyes, mucous membranes, and skin.

Humans exposed to undetermined but high vapor concentrations complained immediately of severe conjunctival and nasopharyngeal irritation, harsh cough, and dyspnea.[1] Workmen exposed to vapors from a boiling mixture complained of severe eye irritation and lacrimation.[1] The immediate effect of exposure to vapor concentrations above 5 ppm is acute irritation of the eyes and upper respiratory tract; inhalation of high vapor concentrations may produce ulceration of the nasal mucosa and, in some instances, bronchospasm.[2]

Rats exposed to 2000 ppm for 4 hours died, but 1000 ppm for 4 hours was not lethal.[3]

Both the liquid and the vapor can cause severe damage to the human eye; this is characterized by immediate burning, followed some hours later by an increasing severity of reaction with corneal and conjunctival edema.[1] Interstitial corneal opacity may develop over a period of several days as a result of progression of tissue infiltration; in mild cases, this condition is reversible, but permanent opacification with loss of vision may also occur. Workmen exposed to acetic anhydride vapor may show evidence of conjunctivitis with associated photophobia.[1]

Prolonged dermal contact with the liquid may cause the skin to redden and subsequently turn white and wrinkled but may not be painful.[4] Skin burns may appear later. Repeated skin exposure to the liquid or vapor may cause irritation.

Generalized skin reactions in guinea pigs sensitized to acetic anhydride have been demonstrated, and skin sensitization in humans occasionally occurs.[2] Although ingestion of the liquid is unlikely in ordinary industrial use, the highly corrosive nature of the substance may be expected to produce serious burns of the mouth and esophagus.

Acetic anhydride has good warning properties.

The 1995 ACGIH threshold limit value–time-weighted value is 5 ppm (21 mg/m³).

REFERENCES

1. AIHA Hygienic Guide Series: *Acetic Anhydride.* Akron, OH: American Industrial Hygiene Association, 1978
2. Fassett DW: Organic acids and related compounds. In Fassett DW, Irish DD (eds): *Patty's Industrial Hygiene and Toxicology,* 2nd ed, Vol 2, *Toxicology,* pp 1817–1818, New York, Interscience, 1963
3. Smyth HF Jr: Hygienic standards for daily inhalation. *Am Ind Hyg Assoc Q,* 17:129–185, 1956
4. Hazard Data Bank: Sheet No. 70, Acetic anhydride. *Safety Practitioner* 3:12–13, October 1985

ACETONE
CAS: 67-64-1

(CH₃)₂CO

$(CH_3)_2CO$

Synonyms: Dimethyl ketone; 2-propanone; β-ketopropane

Physical Form.   Colorless liquid

Uses.   Solvent for fats, oils, waxes, rubber, plastics; in the production of lubricating oils;

in the dyeing and celluloid industries; as a chemical intermediate; paint and varnish remover; major component of nail polish remover

Exposure.   Inhalation; skin absorption

Toxicology.   Acetone is an irritant of the eyes and mucous membranes; at very high concentrations, it is a central nervous system depressant.

Acetone is considered to be of low risk to health because few adverse effects have been reported despite widespread use for many years.[1] One early study, often quoted, reports eye, nose, and throat irritation in volunteers exposed to 500 ppm.[2]

In more recent studies, subjects exposed to 500 ppm were aware of odor but exhibited no effects.[3] Mild eye irritation occurred around 1000 ppm.[4] Higher concentrations produced headache, light-headedness, and nose and throat irritation.[4] Concentrations above 12,000 ppm depress the central nervous system, causing dizziness, weakness, and loss of consciousness.[5]

Neurobehavioral tests have found slight but statistically significant performance decrements following 4-hour exposure to 250 ppm, suggesting mild CNS depression at this level.[6]

Topical application of 1 ml acetone for 90 minutes produced reversible skin damage in humans.[7]

Acetone is metabolized mainly in the liver by three separate pathways leading to the production of glucose, with the subsequent liberation of carbon dioxide.[8] None of the intermediate metabolites appear to be toxic, with the possible exception of formate. Acetone and acetone-derived carbon dioxide are excreted in expired air and have little tendency to accumulate in the body.

In animal studies, acetone has been found to potentiate the toxicity of other solvents by altering their metabolism by induction of microsomal enzymes, particularly cytochrome P-450. Reported effects include: enhancement of the ethanol-induced loss of righting reflex in mice by reduction of the elimination rate of ethanol; increased hepatotoxicity of compounds such as carbon tetrachloride and trichloroethyl-

ene in the rat; potentiation of acrylonitrile toxicity by altering the rate at which it is metabolized to cyanide; and potentiation of the neurotoxicity of *n*-hexane by altering the toxicokinetics of its 2,4-hexanedione metabolite.[9–12]

Significant developmental toxicity, as determined by increased incidences of resorptions, occurred in mice at levels of 6600 ppm, which also caused maternal toxicity.[13] Depressed sperm motility and epididymal weight and elevated evidence of abnormal sperm were observed in male rats receiving 50,000 ppm acetone in their drinking water for 13 weeks.[14]

Acetone may be weakly genotoxic, but the majority of assays were negative.[8] It was not tumorigenic in skin-painting studies in mice.

The 1995 ACGIH threshold limit value–time-weighted average (TLV-TWA) for acetone is 750 ppm (1780 mg/m³) with a short-term excursion level 1000 ppm (2380 mg/m³).

REFERENCES

1. National Institute for Occupational Safety and Health, US Department of Health, Education and Welfare: *Criteria for a Recommended Standard . . . Occupational Exposure to Ketones.* DHEW (NIOSH) 78-173. Washington, DC, US Government Printing Office, 1978
2. Nelson KW, Ege JF Jr, Ross M, et al: Sensory response to certain industrial solvent vapors. *Am Ind Hyg Assoc J* 25:282–285, 1943
3. DiVincenzo GO, Yanno FJ, and Astill BD: Exposure of man and dog to low concentrations of acetone vapor. *Am Ind Hyg Assoc J* 34:329–336, 1973
4. Raleigh RL, McGee WA: Effects of short, high-concentration exposures to acetone as determined by observation in the work area. *J Occup Med* 14:607–610, 1972
5. Ross DS: Short communications—acute acetone intoxication involving eight male workers. *Ann Occup Hyg* 16:73–75, 1973
6. Dick RB, Setzer JV, Taylor BT, et al: Neurobehavioral effects of short duration exposures to acetone and methyl ethyl ketone. *Br J Ind Med* 46:111–121, 1989
7. Lupulescu AP, Birmingham DJ: Effect of protective agent against lipid-solvent-induced damages—ultrastructural and scanning elec-

tron microscopical study of human epidermis. *Arch Environ Health* 31:33–36, 1976

8. Agency for Toxic Substances and Disease Registry (ATSDR): *Toxicological Profile for Acetone*. US Department of Health and Human Services, Public Health Service, pp 157, 1992
9. Cunningham J, Sharkawi M, Plaa G: Pharmacological and metabolic interactions between ethanol and methyl *n*-butyl ketone, methyl isobutyl ketone, methyl ethyl ketone, or acetone in mice. *Fund Appl Toxicol* 13:102–109, 1989
10. Charbonneau M, Perreault F, Greselin E, et al: Assessment of the minimal effective dose of acetone for potentiation of the hepatoxicity induced by trichloroethylene–carbon tetrachloride mixtures. *Fund Appl Toxicol* 10:431–438, 1988
11. Freeman JJ, Hayes EP, Microsomal metabolism of acetonitrile to cyanide: Effects of acetone and other compounds. *Biochem Pharmacol* 37:1153–1160, 1988
12. Ladefoged O, Perbellini L: Acetone induced changes in the toxicokinetics of 2,5-hexanedione in rabbits. *Scand J Work Environ Health* 12:627–629, 1987
13. Mast TJ, Rommereim RL, Weigel RJ, et al: Developmental toxicity study of acetone in mice and rats. *Teratology* 39(5):468, 1989. Abstract
14. Dietz DD, Leininger JR, Rauckman EJ, et al: Toxicity studies of acetone administered in the drinking water of rodents. *Fund Appl Toxicol* 17:347–360, 1991

ACETONITRILE
CAS: 75-05-8

CH_3CN

Synonyms: Methyl cyanide; cyanomethane; ethanenitrile

Physical Form.   Colorless volatile liquid with sweetish odor

Uses.   Chemical intermediate; solvent; extractant for animal and vegetable oils

Exposure.   Inhalation; skin absorption

Toxicology.   Acetonitrile causes headache, dizziness, and nausea; at extremely high concentrations, it can cause convulsions, coma, and death.

Of 15 painters exposed to the vapor of a mixture containing 30% to 40% acetonitrile for 2 consecutive workdays, 10 developed symptoms ranging in severity from nausea, headache, and lassitude among the more lightly exposed to vomiting, respiratory depression, extreme weakness, and stupor in the more heavily exposed. Five cases required hospitalization and one died; this worker experienced the onset of chest pain 4 hours after leaving the job on the second day of exposure, followed shortly by massive hematemesis, convulsions, shock, and coma, with death occurring 14 hours after cessation of exposure.[1] At autopsy, cyanide ion concentrations (in micrograms percent) were: blood 796, urine 215, kidney 204, spleen 318, and lung 128; cyanide ion was not detected in the liver.[1]

Two human subjects inhaled 160 ppm for 4 hours; one of them experienced a slight flushing of the face 2 hours later and a slight feeling of bronchial tightness 5 hours later. A week prior to this, the same two subjects had inhaled 80 ppm with no effects.[2] Blood cyanide and urine thiocyanate levels did not correlate with exposure and, therefore, are not reliable indicators of brief exposure to low concentrations.

In male rats, the LC_{50} was 7500 ppm for a single 8-hour exposure; there was prostration, followed by convulsive seizures; at autopsy, there was pulmonary hemorrhage.[2] Rats exposed 6 hr/day, 5 days/week, for 4 weeks to concentrations greater than 600 ppm had respiratory and ocular irritation and anemia.[3] In another study, rats repeatedly exposed to 665 ppm for 7 hours daily developed pulmonary inflammation, and some animals showed minor changes in the liver and kidneys.[2]

All mice and some rats receiving 1600 ppm by inhalation 6 hr/day for up to 13 weeks died.[4] Clinical findings included hypoactivity, abnormal posture and, in rats, clonic convulsions. Male mice administered 400 ppm and females given 200 ppm, also for 13 weeks, had focal epithelial hyperplasia and ulceration of the forestomach.

In chronic studies, mice exposed 6 hr/day,

5 days/week, for 2 years to concentrations of up to 200 ppm had no increases in the incidences of neoplasms. High-dose females had a significantly increased incidence of squamous hyperplasia of the epithelium of the forestomach. In male rats receiving up to 400 ppm for the same duration, there was a slight increase in the combined incidence of hepatocellular adenoma and carcinoma. There were no exposure-related liver lesions in female rats.

No malformations related to acetonitrile exposure were observed in the offspring of rats orally exposed at maternally toxic levels.[5,6] Inhalation of 5000 or 8000 ppm for 60 minutes by pregnant hamsters on day 8 of gestation was associated with production of severe axial skeletal disorders; maternal toxicity, including irritation, respiratory difficulty, lethargy, ataxia, hypothermia, and increased mortality, was noted.[7] At lower doses, there were no signs of maternal toxicity, and offspring were normal.[7]

In the rabbit eye, a drop of the liquid caused superficial injury.[8] The liquid on the belly of a rabbit caused a faint erythema of short duration.[9]

Acetonitrile was not mutagenic in *Salmonella typhimurium* assays, with or without metabolic activation.[4]

The toxic effects of acetonitrile are attributed to the metabolic release of cyanide via hepatic metabolism; cyanide, in turn, acts by inhibiting cytochrome oxidase and thus impairs cellular respiration.[10] Evidence of the cyanide effect is supported by the reported effectiveness of specific cyanide antidotes in acetonitrile poisonings.[10]

The 1995 ACGIH threshold limit value-time-weighted average (TLV-TWA) for acetonitrile is 40 ppm ($67 mg/m^3$) with a short-term exclusion level of 60 ppm ($101 mg/m^3$).

Note. For a description of diagnostic signs, differential diagnosis, clinical laboratory tests, and specific treatment of overexposure to acetonitrile and other cyanogens, see "Cyanides."

REFERENCES

1. Amdur, ML: Accidental group exposure to acetonitrile. *J Occup Med* 1:627–633, 1959
2. Pozzani UC et al: An investigation of the mammalian toxicity of acetonitrile. *J Occup Med* 1:634–642, 1959
3. Roloff V et al: Comparison of subchronic inhalation toxicity of five aliphatic nitriles in rats. *Toxicologist* 5:30, 1985
4. National Toxicology Program: *NTP Technical Report on the Toxicology and Carcinogenesis Studies of Acetonitrile (CAS No. 75-05-8) in F344/N Rats and B6C3F₁ Mice.* NTP TR 447, NIH Pub No. 94-3363, US Department of Health and Human Services, Public Health Service, National Institutes of Health, 1994
5. Saillenfait AM, Bonnet P, Guenier JP, et al: Relative developmental toxicities of inhaled aliphatic mononitriles in rats. *Fund Appl Toxicol* 20:365–375, 1993
6. Berteau PE et al: Teratogenic evaluation of aliphatic nitriles in rats. *Toxicologist* 2:118, 1982
7. Willhite CC: Developmental toxicology of acetonitrile in the Syrian golden hamster. *Teratology* 27:313–325, 1983
8. Grant W.M.: *Toxicology of the Eye*, 3rd ed, p 52. Springfield, IL, Charles C. Thomas, 1986
9. *Toxicology Studies, Acetonitrile.* New York, Union Carbide Corporation, 1965
10. National Institute for Occupational Safety and Health, US Department of Health, Education, and Welfare: *Criteria for a Recommended Standard . . . Occupational Exposure to Nitriles.* DHEW (NIOSH) Pub 78–212, pp 155. Washington, DC, US Government Printing Office, 1978

2-ACETYLAMINOFLUORENE
CAS: 53-96-3

$C_{15}H_{13}NO$

Synonyms: N-2-Fluorenylacetamide; 2-acetaminofluorene; N-acetylaminophenathrene; AAF

Physical Form. Crystals

Uses. As a laboratory reagent for research purposes; formerly used as a pesticide

Exposure. Inhalation

Toxicology. 2-Acetylaminofluorene (AAF) is a potent carcinogen in dogs, hamsters, and rats.

There is no toxicity information on humans.[1]

Four of five dogs developed tumors of the liver and urinary bladder after ingestion of 0.6 to 1.2 g AAF/kg diet for up to 91 months.[2] Animals developing tumors received a total of 90 to 198 g AAF, whereas the animal with no tumor formation ingested 45 g; another group of 4 dogs receiving 32 to 37 g during 2.25 years did not develop tumors.[2] The extent of tumor formation was directly related to the amount of AAF consumed, being most marked in those animals that received nearly 200 g during the feeding period.[2] Liver tumors of varied types were observed. Multiple papillomas were produced in the urinary bladder and, in one dog, there was invasion of the submucosa and muscle by the tumor cells.

Intratracheal administration of 5 to 15 mg AAF 1 to 2 times per week for 17 months in hamsters (total dose 1100 mg) caused bladder tumors in 10 of 23 animals; all tumors were transitional cell carcinomas with or without focal squamous-cell carcinomas.[3]

In rats, AAF had no demonstrable acute toxicity in quantities up to 50 mg/kg subcutaneously and 1 g/kg gastrically; however, AAF was very toxic when administered in the diet.[4] Incorporation of 0.031% AAF or higher for at least 95 days led to epithelial hyperplasia of the bladder, renal pelvis, liver, pancreas, and lung; 19 of 39 rats developed malignant tumors, 16 of which were carcinomas.[4]

Animal studies have indicated that *N*-hydroxy-2-acetylaminofluorene (*N*-hydroxy-AAF) is a proximate carcinogenic metabolite of AAF.[5] AAF is not carcinogenic in the guinea pig, and no *N*-hydroxylation of AAF has been detected in vivo or in vitro in this species; however, administration of *N*-hydroxy-AAF causes tumors in guinea pigs.[5] In addition, *N*-hydroxy-AAF has proved to be a carcinogen of much greater potency than AAF in rats, mice, hamsters, and rabbits at sites of local application.[5]

AAF is classified as a cytotoxic teratogen.[1] Because of demonstrated carcinogenicity in animals, contact by all routes should be avoided.

In recent years, this compound has been used only in laboratories as a model of tumorigenic activity in animals.[6] It is of little occupational health importance.

The ACGIH has not established a threshold limit value for AAF.

REFERENCES

1. Doull J, Klaasen CD, Amdur MO (eds): Casarett and Doull's *Toxicology: The Basic Science of Poisons*, 2nd ed, p 163. New York, Macmillan 1980
2. Morris HP, Eyestone WH: Tumors of the liver and urinary bladder of the dog after ingestion of 2-acetylaminofluorene. *J Natl Cancer Inst* 13:1139–1165, 1953
3. Oyasu R, Kitajima T, Hoop ML, et al: Induction of bladder cancer in hamsters by repeated intratracheal administrations of 2-acetylaminofluorene. *J Natl Cancer Inst* 50:503–506, 1973
4. Wilson RH, Deeds F, Cox AJ Jr: The toxicity and carcinogenic activity of 2-acetaminofluorene. *Cancer Res* 1:595–608, 1941
5. Miller EC, Miller JA, Enomoto M: The comparative carcinogenicities of 1-acetylaminofluorene and its *N*-hydroxy metabolite in mice, hamsters, and guinea pigs. *Cancer Res* 24:2018–2031, 1964
6. Benya TJ, Cornish HH: Aromatic nitro and amino compounds. In Clayton GD, Clayton FE (eds): *Patty's Industrial Hygiene and Toxicology*, 4th ed, Vol IIB, *Toxicology*, pp 968–970. New York, John Wiley & Sons, 1994

ACETYLENE TETRABROMIDE
CAS: 79-27-6

CHBr₂CHBr₂

Synonyms: Tetrabromoethane; Muthmann's liquid; 1,1,2,2-tetrabromoethane

Physical Form. Colorless to yellow liquid

Uses. Gauge fluid; solvent; refractive index liquid in microscopy

Exposure. Inhalation

Toxicology. Acetylene tetrabromide is a central nervous system depressant and hepatotoxin.

A chemist working with the substance for 7.5 hours with no local exhaust ventilation developed severe, nearly fatal liver damage and was hospitalized for 9 weeks; his estimated exposure during most of the work shift prior to the onset of symptoms was 1 to 2 ppm although he had a single 10-minute period exposure to approximately 16 ppm.[1] He complained first of headache, anorexia, and nausea within hours of the exposure and, within 5 days, developed abdominal pain with bilirubinuria and a monocytosis of 17%. In this case, exposure to higher concentrations or significant skin absorption might have occurred. The similarity of the symptoms to viral hepatitis is also noted. Other workers in the same laboratory complained only of slight eye and nose irritation with headache and lassitude.

Rats exposed to a saturated atmosphere for 7 hours exhibited slight eye and nose irritation.[2] Guinea pigs exposed for 90 minutes to a saturated vapor became comatose, seemed to recover, but died after several days; the same exposure for up to 3 hours was not lethal to rats and rabbits.[3] No mortality was observed in rats, guinea pigs, rabbits, mice, and a monkey exposed 7 hr/day for 100 days to 14 ppm; findings at 14 ppm did include edema of the lungs and slight fatty degeneration of the liver in all species except guinea pigs, which showed only growth depression.[2] Repeated exposure to 4 ppm for 180 days caused slight histopathologic changes in the liver and lungs of some animals, but no effects were observed at 1 ppm.

Repeated applications of 15 mg to the skin of mice caused a statistically significant increase in the incidence of forestomach papillomas.[4]

The liquid instilled in the rabbit eye caused slight to moderate pain, conjunctival irritation, and corneal injury, which disappeared after 24 hours.[1] When bandaged onto the shaved abdomen of the rabbit for 72 hours, moderate redness, edema, and blistering were observed.[1]

Acetylene tetrabromide has a sweetish, unpleasant odor, which is readily apparent and objectionable to most persons at concentrations greater than 1 to 2 ppm.[1,2]

The 1995 ACGIH threshold limit value–time-weighted average (TLV-TWA) for acetylene tetrabromide is 1 ppm (14 mg/m³).

REFERENCES

1. Van Haaften AB: Acute tetrabromoethane (acetylene tetrabromide) intoxication in man. *Am Ind Hyg Assoc J* 30:251–256, 1969
2. Hollingsworth RL, Rowe VK, Oyen F: Toxicity of acetylene tetrabromide determined on experimental animals. *Am Ind Hyg Assoc J* 24:28–35, 1963
3. Gray MG: Effect of exposure to the vapors of tetrabromoethane (acetylene tetrabromide). *Arch Ind Hyg Occup Med* 2:407–419, 1950
4. Van Duuren BL et al: Carcinogenicity of halogenated olefinic and aliphatic hydrocarbons in mice. *J Natl Cancer Inst* 63:1433–1439, 1979

ACROLEIN
CAS: 107-02-8

C₃H₄O

Synonyms. Acrylaldehyde; 2-propenal; allyl aldehyde; propylene aldehyde; aqualin

Physical Form. Colorless or yellowish liquid

Uses. Intermediate in the manufacture of acrylic acid; herbicide; algicide; in pharmaceuticals, perfumes, food supplements, and resins; as a warning agent in methyl chloride refrigerating systems

Exposure. Inhalation

Toxicology. Acrolein is an intense irritant of the upper respiratory tract, eyes, and skin.

Exposure to high concentrations may cause tracheobronchitis and pulmonary edema.[1] The irritation threshold in man is 0.25 to 0.5 ppm, and concentrations above 1 ppm are extremely irritating to all mucous membranes within 5 minutes.[1] Fatalities have been reported at levels as low as 10 ppm, and 150 ppm was lethal after 10 minutes.[2,3] The violent irritant effect usually prevents chronic toxicity in man.[1] Skin contact causes irrritation, burns, and epidermal necrosis.[4] Eye splashes cause corneal damage, palpebral edema, blepharoconjunctivitis, and fibrinous or purulent discharge.[5]

In experimental animals, the respiratory system is a primary target of acrolein exposure following inhalation, and there is an inverse relationship between the exposure concentration and the time it takes for death to occur.[4] Inhalation LC_{50}s of 327 ppm for 10 minutes and 130 ppm for 30 minutes have been reported in rats.[4] Of 57 male rats, 32 died following exposure to 4 ppm for 6 hr/day for up to 62 days.[6] Desquamation of the respiratory epithelium, followed by airway occlusion and asphyxiation, is the primary mechanism for acrolein-induced mortality in animals.[4] Sublethal acrolein exposure in mice at 3 and 6 ppm suppressed pulmonary antibacterial defense mechanisms.[7]

Intra-amniotic administration of acrolein in rats induced a significant number of fetal malformations, whereas intravenous administration was embryolethal.[8] Pregnant rabbits given 4.0 and 6.0 mg/kg/day on days 7 through 19 of gestation had high incidences of maternal mortality, spontaneous abortion, resorptions, clinical signs, gastric ulceration, and sloughing of the gastric mucosa.[9] Acrolein did not cause statistically significant embryofetal effects at lower doses and was not considered to be a developmental toxicant at doses that did not cause severe maternal toxicity. Similar results were reported in two generations of rats administered up to 6 mg/kg; reduced pup weight occurred at levels that also produced significant maternal deaths.[10]

The carcinogenic potential of acrolein has been examined in a number of studies. Hamsters exposed to 4.0 ppm, 7 hr/day, for 52 weeks showed no evidence of respiratory tract tumors or tumors in other tissues and organs.[11] Rats exposed for 10 to 18 months to 8 ppm, 1 hr/day, also showed no evidence of a tumorigenic response.[4]

Extensive histopathological examination did not reveal any carcinogenic effects in rats,

mice, or dogs after oral exposure to 2.5, 4.5, or 2 mg/kg/day acrolein, respectively, for 12 to 24 months.[4] In the one study that reported positive findings, 20 female rats given acrolein in drinking water (625 mg/l water equivalent to daily doses approaching 40 mg/kg body weight/day) for 104 to 124 weeks had an increased incidence of adrenal cortex neoplasms as compared to controls.[12] The small numbers of animals used in this study make it unsuitable for evaluating the carcinogenic potential of acrolein. Furthermore, reevaluation of this study by an independent pathology group failed to confirm the original findings.[13] (The working group determined that the slightly elevated incidence of pheochromocytomas in the treated females was within limits for historical controls and was of no biological significance.) A recent two-year study of rats treated by daily gavage with 0, 0.05, 0.5, or 2.5 mg/kg for 102 weeks found no evidence of a neoplastic response.[13] Chronic gavage studies in mice for 18 months and capsular administration to dogs for 1 year have also revealed no indication of a carcinogenic response.[14]

Acrolein has been found to be mutagenic to bacteria and to induce sister chromatid exchanges in vitro.[15]

The 1995 threshold limit value–time-weighted average (TLV-TWA) is 0.1 ppm (0.23 mg/m^3) with a short-term excursion level of 0.3 ppm (0.69 mg/m^3).

REFERENCES

1. Beauchamp RO Jr, Andjelkovich DA, Kligerman AD, et al: A critical review of the literature on acrolein toxicity. *Crit Rev Toxicol* 14:309–380, 1985
2. Henderson Y, Haggard HW: *Noxious Gases*, p 138. New York, Reinhold, 1943
3. Prentiss AM: *Chemicals in War. A Treatise on Chemical Warfare*, pp 139–140. New York, McGraw-Hill, 1937
4. Agency for Toxic Substances and Disease Registry (ASTDR): *Toxicological Profile for Acrolein*, US Department of Health and Human Services, Public Health Service, TP-90-01, pp 145, 1990
5. Grant WM: *Toxicology of the Eye*, 3rd ed, pp 49–50. Springfield, IL, Charles C Thomas, 1986

6. Kutzman RS et al: Changes in rat lung structure and composition as a result of subchronic exposure to acrolein. *Toxicology* 34:139–151, 1985

7. Astry CL, Jakab GJ: The effects of acrolein exposure on pulmonary antibacterial defenses. *Toxicol Appl Pharmacol* 67:49–54, 1983

8. Slott VL, Hales BF: Teratogenicity and embryolethality of acrolein and structurally related compounds in rats. *Teratology* 32:65–72, 1985

9. Parent RA, Caravello HE, Christian MS, et al: Developmental toxicity of acrolein in New Zealand white rabbits. *Fund Appl Toxicol* 20:248–256, 1993

10. Parent RA, Caravello HE, Hoberman AM: Reproductive study of acrolein on two generations of rats. *Fund Appl Toxicol* 19:228–237, 1992

11. Feron VJ, Kruysse A: Effects of exposure to acrolein vapor in hamsters simultaneously treated with benzo(a)pyrene or diethylnitrosamine. *J Toxicol Environ Health* 3:379–394, 1977

12. Lijinsky W, Reuber MD: Chronic carcinogenesis studies of acrolein and related compounds. *Toxicol Ind Health* 3:337–345, 1987

13. Parent RA, Caravello HE, Long JE: Two-year toxicity and carcinogenicity study of acrolein in rats. *J Appl Toxicol* 12:131–139, 1992

14. Parent RA, Caravello HE, Balmer MF, et al: One-year toxicity of orally administered acrolein to the Beagle dog. *J Appl Toxicol* 12:311–316, 1992

15. *IARC Monographs on the Evaluation of the Carcinogenic Risks to Humans*, Suppl 7, p 78. Lyon, International Agency for Research on Cancer, 1987

ACRYLAMIDE
CAS: 79-06-1

C₃H₅NO

C_3H_5NO

Synonyms: Acrylic amide; propenamide; ethylenecarboxamide; vinyl amide

Physical Form. White crystalline powder

Uses. In the production of polyacrylamides used in water and waste treatment, paper and pulp processing, cosmetic additives, and textile processing; in adhesives and grouts; as cross-linking agents in vinyl polymers

Exposure. Inhalation; skin absorption; ingestion

Toxicology. Acrylamide causes central-peripheral axonopathy; it is carcinogenic in laboratory animals and is a suspected human carcinogen.

A variety of signs and symptoms have been described in cases of acrylamide poisoning that suggest involvement of the central, peripheral, and autonomic nervous systems.[1] Effects on the central nervous system are characterized by abnormal fatigue, memory difficulties, and dizziness,[1] with severe poisoning, confusion, disorientation, and hallucinations occuring. Truncal ataxia, nystagmus, and slurred speech have also been observed. Peripheral neuropathy symptoms can include muscular weakness; paresthesia; numbness in hands, feet, lower legs, and lower arms; unsteadiness; and difficulties in walking and standing. Clinical signs are loss of peripheral tendon reflexes, impairment of vibration sense, and muscular wasting in the extremities. Nerve biopsy shows loss of large-diameter nerve fibers as well as regenerating fibers. Autonomic nervous system involvement is indicated by excessive sweating, peripheral vasodilation, and difficulties in micturation and defecation.

Central nervous system effects predominate in acute exposures at massive doses, whereas peripheral neuropathy is more common with lower doses.[1,2] After cessation of exposure to acrylamide, most cases recover, although the course of improvement can extend over months to years and depends on the severity of exposure.[1,2] Since peripheral neurons can regenerate and central axons cannot, severely affected individuals may still experience residual ataxia, distal weakness, reflex loss, or sensory disturbance.

Because most cases of human poisoning have included skin absorption, the dose-response relationship has not been determined. On the skin, it causes local irritation characterized by blistering and desquamation of the palms and soles combined with blueness of the hands and feet.[1]

For a number of species, the oral LD_{50} was approximately 150 to 180 mg/kg body weight. In cats, a total cumulative dose of 70 to 130 mg/kg was characterized by delayed onset of ataxia.[3] Cats fed 10 mg/kg diet/day developed definite hind limb weakness after 26 days; at 3 mg/kg/day, there was twitching in the hind-quarters after 26 days and signs of hind limb weakness after 68 days.[4] The underlying lesion involves distal retrograde degeneration of long and large-diameter axons.[5]

Teratogenic effects were not observed in the offspring of rats given up to 50 mg/kg diet for 2 weeks prior to mating and for 19 days during gestation.[1] In mice, high doses produced decreased sperm count and an increase in abnormal sperm morphology.[6]

A statistically significant increase in mesothelioma of the scrotal cavity was observed in rats given drinking water formulated to provide 0.5 mg/kg bw/day for 2 years; in females, there were significant increases in the number of neoplasms of the central nervous system, thyroid, mammary gland, oral cavity, clitoral gland, and uterus.[7]

Acrylamide has also been reported to act as a skin tumor initiator in mice by three exposure routes and to increase the yield of lung adenomas in another strain of mice.[8]

In a human mortality study of 371 workers, no increase in total malignant neoplasms or in any specific cancers attributable to acrylamide exposure was found.[9] Exposure levels reached 1.0 mg/m³ prior to 1957 and were between 0.1 and 0.6 mg/m³ after 1970. This study was of such a limited sample size, however, that only large excesses could have been detected.

A much larger cohort of 8854 men from four chemical plants, 2293 of whom were exposed to acrylamide, was examined from 1925 to 1983 for mortality.[10] No statistically significant excess of all-cause or cause-specific mortality was found among acrylamide workers. Analysis by acrylamide exposure levels showed no trend of increased risk of mortality from several cancer sites. While the authors concluded that the results do not support the hypothesis that

acrylamide is a human carcinogen, this view was challenged on the basis that the comparison group included individuals from one of the four plants who had a small but significant excess of lung cancer (SMR = 1.32), which had been attributed by the authors to another occupational exposure in the production of muriatic acid.[11]

The IARC has determined that there is sufficient evidence in experimental animals for the carcinogenicity of acrylamide and inadequate evidence for carcinogenicity to humans. Overall, it is considered "probably carcinogenic to humans."[12]

Acrylamide is genotoxic in a number of test systems.[12] It induces gene mutation, structural chromosomal aberrations, sister chromatid exchange, and cell transformation. Furthermore, acrylamide forms covalent adducts with DNA in rodents and covalent adducts with hemoglobin in humans. Hemoglobin adducts have been used for biomonitoring of acrylamide. Recent studies indicate that the adducts are useful predictors of acrylamide-induced peripheral neuropathy.[13]

The 1995 ACGIH threshold limit value–time-weighted average for acrylamide is 0.03 mg/m³, with a notation for skin absorption and an A2 suspected human carcinogen designation.

REFERENCES

1. The International Programme on Chemical Safety: *Environmental Health Criteria 49: Acrylamide*, pp 1–121. World Health Organization, Geneva, 1985
2. Smith EA, Oehme FW: Acrylamide and polyacrylamide: a review of production, use, environmental fate and neurotoxicity. *Rev Environ Health* 9:215–228, 1991
3. Kuperman AS: Effects of acrylamide on the central nervous system of the cat. *J Pharmacol Exp Ther* 123:180–192, 1958
4. McCollister DD et al: Toxicology of acrylamide. *Toxicol Appl Pharmacol* 6:172–181, 1964
5. Miller MS, Spencer PS: The mechanisms of acrylamide axonopathy. *Annu Rev Pharmacol Toxicol* 25:643–666, 1985
6. Sakamoto J, Hashimoto K: Reproductive toxicity of acrylamide and related compounds in

mice—effects on fertility and sperm morphology. *Arch Toxicol* 59:201–205, 1986

7. Johnson KA et al: Chronic toxicity and onco-genicity study on acrylamide incorporated in the drinking water of Fischer 344 Rats. *Toxicol Appl Pharmacol* 85:154–168, 1986
8. Bull RJ et al: Carcinogenic effect of acryl-amide in sencar and A/J mice. *Cancer Res* 44:107–111, 1984
9. Sobel W et al: Acrylamide cohort mortality study. *Br J Ind Med* 43:785–788, 1986
10. Collins JJ et al: Mortality patterns among workers exposed to acrylamide. *J Occup Med* 31:614–617, 1989
11. Hogan KA, Scott CLS: Mortality patterns and acrylamide exposure (letters). *J Occup Med* 32:947–949, 1990
12. *IARC Monographs on the Evaluation of Carcino-genic Risks to Humans,* Vol 60, Some industrial chemicals, pp 389–429. Lyon, International Agency for Research on Cancer, 1994
13. Calleman CJ, Wu Y, He F, et al: Relation-ships between biomarkers of exposure and neurological effects in a group of workers ex-posed to acrylamide. *Toxicol Appl Pharmacol* 126:361–371, 1994

ACRYLIC ACID
CAS: 79-10-7

C₃H₄O₂

$C_3H_4O_2$

Synonyms: 2-Propenoic acid; acroleic acid; ethylenecarboxylic acid; vinylformic acid

Physical Form. Colorless, fuming liquid

Uses. Starting material for acrylates and polyacrylates used in plastics, water purifica-tion, paper and cloth coatings, and medical and dental materials

Toxicology. Acrylic acid is a severe irritant of the eyes, nose, and skin. The major route of absorption is ingestion of inhaled vapors.

Medical reports of acute human exposures (concentration unspecified) include moderate and severe skin burns, moderate eye burns, and mild inhalation effects.[1] Although acrylic acid is acutely irritating at sites of initial contact, it causes little systemic toxicity. The low systemic toxicity of acrylic acid is likely to be a conse-quence of its rapid and extensive metabolism to carbon dioxide.[2]

There is considerable variability in the re-ported values for the oral LD_{50} in rats, ranging from 350 to 3200 mg/kg.[3,4] The dermal LD_{50} in rabbits was 750 mg/kg.[5]

Rats exposed to 1500 ppm for four 6-hour periods exhibited nasal discharge, weight loss, lethargy, and kidney congestion.[6] At 300 ppm, twenty 6-hour exposures produced all but the latter effect. No toxic signs resulted from expo-sure to 80 ppm for twenty 6-hour periods.

Exposure to 0, 5, 25, or 75 ppm 6 hr/day, 5 days/week, for 13 weeks produced slight de-generative lesions of the nasal mucosa in rats at the high dose but none at 25 ppm.[7] In contrast, lesions of the nasal mucosa appeared in at least some of the mice at all dose levels but not in the control.

There were no indications of systemic tox-icity and/or carcinogenicity in rats administered 0, 120, 400, or 1200 ppm in the drinking water for over 2 years.[8]

The application of 0.1 ml of a 4% acrylic acid solution in acetone to the skin of mice 3 times per week for 13 weeks led to distinct skin irritation from 1 week on. Only minimal proliferative processes were observed when 0.1 ml of a 1% acrylic acid solution was applied.[9]

In two-week studies preliminary to a life-time dermal carcinogenicity study in mice, a concentration of 5% in acetone caused peeling and flaking of the skin.[10] A 1% solution in ace-tone applied to the skin of 40 C3H/HeJ male mice 3 days/week for 1.5 years caused no treat-ment-related tumors or effects on mortality.

A 1% solution in the eye of a rabbit caused significant injury.[5]

Intraperitoneal injection of female rats on days 5, 10, 15 of gestation with 4.7 or 8 mg/kg produced a significant increase in the number of gross abnormalities, including skeletal ab-normalities at the higher dose level.[11] In another reproductive study of rats, acrylic acid was in-jected intra-amniotically on day 13 of gestation at doses of 10, 100, or 1000 mg/fetus.[12] The highest dose level was significantly embryotoxic. One fetus at the 100-μg level was malformed, but there was no dose-response relationship.

It can be noted that these studies are of limited value in assessing teratogenic potential in the workplace because very small numbers of animals were used, the route of exposure bears no resemblance to normal exposure of mothers and embryos, and the injection process may have physically traumatized the embryo.

In an inhalation teratogenicity study, pregnant rats were exposed from day 6 to day 15 to 0, 40, 120, or 360 ppm.[13] Marked effects were observed in the dams at 360 ppm, including reduced weight gain, decreased food intake, and clinical signs of an irritant effect on mucous membranes. There were no signs, however, of embryotoxicity or teratogenicity at any of the doses tested.

Acrylic acid was not mutagenic in five strains of *Salmonella typhimurium* with or without metabolic activation by liver microsomes.[14] Results were also negative (nonmutagenic) in a number of in vivo assays in both somatic and germ cells.[15]

Alpha-, beta-diacryloxypropionic acid has been found to be a sensitizing impurity in commercial acrylic acid.[16]

The 1995 ACGIH threshold limit value–time-weighted average (TLV-TWA) for acrylic acid is 2 ppm (5.9 mg/m³) with a notation for skin.

REFERENCES

1. Sittig M: *Handbook of Toxic and Hazardous Chemicals and Carcinogens*, 2nd ed, p 43. Park Ridge, N.J. Noyes Publishing, 1985
2. Black KA, Finch L, Frederick CB: Metabolism of acrylic acid to carbon dioxide in mouse tissues. *Fund Appl Toxicol* 21:97–104, 1993
3. Carpenter CP, Weil CF, Smyth HF Jr: Range-finding toxicity data: List VIII. *Toxicol Appl Pharmacol* 28:313, 1974
4. Miller ML: Acrylic acid polymers. In Bikales NM (ed): *Encyclopedia of Polymer Science and Technology, Plastics, Resins, Rubbers, Fibers*, Vol 1, p 197. New York, Interscience, 1964
5. Toxicology studies-acrylic acid, glacial, 2 May. Union Carbide Corporation, New York, Industrial Medicine and Toxicology Department, 1977
6. Gage JC: The subacute inhalation toxicity of 109 industrial chemicals. *Br J Ind Med* 27:1, 1970
7. Miller RR, Ayres JA, Jersey GC, McKenna MJ: Inhalation toxicity of acrylic acid. *Fund Appl Toxicol* 1:271, 1981
8. Hellwig J, Deckardt K, Freisberg KO: Subchronic and chronic studies of the effects of oral administration of acrylic acid to rats. *Food Chem Toxicol* 31:1–18, 1993
9. Tegeris AS, Balmer MF, Garner FM, et al: A thirteen week skin irritation study with acrylic acid in three strains of mice. *Toxicologist* 8: 504, 1988. Abstract 504
10. DePass LR et al: Dermal oncogenicity bioassays of acrylic acid, ethyl acrylate and butyl acrylate. *J Toxicol Environ Health* 14:115, 1984
11. Singh AR, Lawrence WH, Autian J: Embryonic-fetal toxicity and teratogenic effects of a group of methacrylate esters in rats. *J Dent Res* 51: 1632, 1972
12. Slott VL, Hales BF: Teratogenicity and embryolethality of acrolein and structurally related compounds in rats. *Teratology* 32:65, 1985
13. Klimisch HJ, Hellwig J: The prenatal inhalation toxicity of acrylic acid in rats. *Fund Appl Toxicol* 16:656–666, 1991
14. Lijinsky W, Andrews AW: Mutagenicity of vinyl compounds in *Salmonella typhimurium*. *Terato Carcin Mut* 1:259, 1980
15. McCarthy KL, Thomas WC, Aardema MJ, et al: Genetic toxicology of acrylic acid. *Food Chem Toxic* 30:505–515, 1992
16. Waegemaekers THJ, van der Walle HB: Alpha, beta-diacryloxypropionic acid, a sensitizing impurity in commercial acrylic acid. *Derm Beruf Umwelt* 32:55, 1984

ACRYLONITRILE

CAS: 107-13-1

C_3H_3N

Synonyms: ACN; cyanoethylene; propenenitrile; vinyl cyanide

Physical Form. Colorless liquid

Uses. Manufacture of acrylic fibers; synthesis of rubberlike materials; pesticide fumigant

Exposure. Inhalation; skin absorption

Toxicology. Acrylonitrile is an eye, skin, and upper respiratory tract irritant; systemic effects are nonspecific but may include the central nervous, hepatic, renal, cardiovascular, and gastrointestinal systems. It is carcinogenic in experimental animals and is a suspected human carcinogen.

Most cases of intoxication from industrial exposure have been mild, with rapid onset of eye irritation, headache, sneezing, and nausea; weakness, light-headedness, and vomiting may also occur.[1] Acute exposure to higher concentrations may produce profound weakness, asphyxia, and death.[1] Acrylonitrile is metabolized to cyanide by hepatic microsomal reactions. Deaths from acute poisoning result from inhibition of mitochondrial cytochrome oxidase activity by metabolically liberated cyanide.

Prolonged skin contact with the liquid results in both systemic toxicity and the formation of large vesicles after a latent period of several hours.[1] The affected skin may resemble a second-degree thermal burn.

Administration of 65 mg/kg/day by gavage to rats on days 6 to 15 of gestation produced significant maternal toxicity and an increased incidence of malformation in the offspring.[2] Inhalation of 80 ppm 6 hr/day by the dams resulted in a significant increase of fetal malformations, including short tail, missing vertebrae, short trunk, omphalocele, and hemivertebra; maternal toxicity consisted of decreased weight gain.[2]

Oral administration of 10 mg/kg/day for 60 days to male mice induced histopathologic changes in the testis and reduced sperm counts compared with controls. These changes were not observed at a dosage level of 1 mg/kg/day.[3]

In a number of chronic bioassays in rats, administration of acrylonitrile by gavage, by inhalation, and in the drinking water produced tumors of the mammary gland, the gastrointestinal tract, the Zymbal glands, and the central nervous system.[4-6] Administration of 500 ppm in drinking water caused a statistically significant increase in microscopically detectable primary brain tumors.[7] Neurological signs were observed in 29 of 400 rats within 18 months, and brain tumors occurred in 49 of 215 animals that died or were sacrificed in the first 18 months.

In an initial study of 1345 workers potentially exposed to acrylonitrile and followed for 10 or more years, there was a greater than expected incidence of lung cancer (8 observed vs. 4.4 expected)[8] A trend toward increased risks of cancer of all sites was also observed with increased duration of exposure and with higher severity of exposure. However, in a follow-up of this cohort through 1983, the only statistically significant excess was for prostate cancer (5 observed vs. 1.9 expected)[9] An excess number of lung cancer cases remained (10 observed vs. 7.2 expected) but was not as marked.[9,10] A study of 1774 workers potentially exposed to acrylonitrile and followed for 32 years reported no significant excess of all-site or site-specific cancer mortality rates.[11] Other epidemiologic studies have reported excess lung cancer deaths but lacked statistical significance because of small cohort size.[12]

As a result of the consistent production of tumors in rats and the suspicion of cancer in humans raised by the cited studies, the ACGIH recommends that acrylonitrile be regarded as a suspected human carcinogen.[13] Some investigators suggest, however, that epidemiological studies have not provided evidence of an increased cancer risk in occupationally exposed workers.[14]

In vitro genotoxic studies have given positive results for gene mutations, chromosomal aberrations, DNA damage, and cell transformation; in vivo assays have generally been negative.[15]

Biological monitoring of acrylonitrile exposure has been proposed via measurement of hemoglobin adducts and urinary metabolites.[16]

The 1995 ACGIH threshold limit value–time-weighted average (TLV-TWA) for acrylonitrile is 2 ppm (4.3 mg/m^3) with an A2 suspected human carcinogen designation and a notation for skin absorption.

Note. For a description of diagnostic signs, differential diagnosis, clinical laboratory tests, and specific treatment of overexposure to acrylonitrile and other cyanogens, see "Cyanides."

REFERENCES

1. Willhite CC: Toxicology updates. Acrylonitrile. *J Appl Toxicol* 2:54–56, 1982
2. Murray FJ, Schwetz BA, Nitschke KD, et al: Teratogenicity of acrylonitrile given to rats by gavage or by inhalation. *Food Cosmet Toxicol* 16:547–551, 1979
3. Tandon R, Saxena DK, Chandra SV, et al: Testicular effects of acrylonitrile in mice. *Toxicol Letters* 42:55–63, 1988
4. Maltoni C, Ciliberti A, Di Maio V: Carcinogenicity bioassays on rats of acrylonitrile administered by inhalation and ingestion. *Med Lav* 68:401–411, 1977
5. *IARC Monographs on the Evaluation of the Carcinogenic Risk of Chemicals to Man.* Suppl 4, pp 25–27. Lyon, International Agency for Research on Cancer, 1982
6. Gallagher GT, Maull EA, Kovacs K, et al: Neoplasms in rats ingesting acrylonitrile for two years. *J Am College Toxicol* 7:603–615, 1988
7. Bigner DD, Bigner SH, Burger PC, et al: Primary brain tumors in Fischer 344 rats chronically exposed to acrylonitrile in their drinking water. *Food Chem Toxicol* 24:129–137, 1986
8. O'Berg MT: Epidemiologic study of workers exposed to acrylonitrile. *J Occup Med* 22:245–252, 1980
9. O'Berg MT, Chen JL, Burke CA, et al: Epidemiologic study of workers exposed to acrylonitrile: an update. *J Occup Med* 27:835–840, 1985
10. Chen JL, Walrath J, O'Berg MT, et al: Cancer incidence and mortality among workers exposed to acrylonitrile. *Am J Ind Med* 11:157–163, 1987
11. Collins JJ, Page C, Caporossi JC, et al: Mortality patterns among employees exposed to acrylonitrile. *J Occup Med* 31:368–371, 1989
12. Koerselman W, van der Graaf M: Acrylonitrile: a suspected human carcinogen. *Int Arch Occup Environ Health* 54:317–324, 1984
13. *Acrylonitrile. Documentation of the TLVs and BEIs*, 5th ed, pp 15–16. Cincinnati, OH, American Conference of Governmental Industrial Hygienists (ACGIH), 1986
14. Ward CE, Starr TB: Comparison of cancer risks projected from animal bioassays to epidemiological studies of acrylonitrile-exposed workers. *Reg Toxic Pharmacol* 18:214–232, 1993
15. Agency for Toxic Substances and Disease Registry (ATSDR): *Toxicological Profile for Acrylonitrile.* US Department of Health and Human Services, Public Health Service, pp 129, TP-90-02, 1990
16. Ivanov V, Hashimoto K, Inomata K, et al: Biological monitoring of acrylonitrile exposure through a new analytical approach to hemoglobin and plasma protein adducts and urinary metabolites in rats and humans. *Int Arch Occup Environ Health* 65:S103–S106, 1993

ALDRIN
CAS: 309-00-2

$C_{12}H_8C_{16}$

Synonyms: 1,2,3,4,10,10-Hexachloro-1,4,4α, 5,8,8α-hexahydro-1,4-endo, exo-5,8-dimethanonaphthalene; HHDN; aldrine

Physical Form. White, crystalline, odorless solid

Uses. Insecticide

Exposure. Inhalation; skin absorption; ingestion

Toxicology. Aldrin is a convulsant; its metabolite, dieldrin, causes liver cancer in mice.

In humans, early symptoms of intoxication may include headache, dizziness, nausea, vomiting, malaise, and myoclonic jerks of the limbs; clonic and tonic convulsions and, sometimes, coma follow and may occur without the premonitory symptoms.[1,2] A suicidal person who ingested 25.6 mg/kg developed convulsions within 20 minutes, which persisted recurrently until large amounts of barbiturates had been administered. Hematuria and azotemia occurred the day after ingestion and continued for 18 days. Liver function studies were within normal limits except for an elevated icterus index; an electroencephalogram revealed generalized cerebral dysrhythmia, which returned to normal after 5 months.[3]

Once aldrin is absorbed, it is rapidly metabolized to dieldrin.[4] In a study of five workers exposed to concentrations of aldrin of up to 8.5 mg/m³ who had suffered convulsive seizures or myoclonic limb movements, the probable concentration of dieldrin in the blood during intoxication ranged from 16 to 62 μg/100 g of blood; in healthy workers, the concentration of dieldrin ranged up to 22 μg/100 g of blood.[4]

Aldrin is reported to have caused erythematobullous dermatitis in a single case. Minor erythema may be observed from skin contact, but dermatitis associated with aldrin is unusual.[5]

Epidemiological studies of workers employed in the manufacture of aldrin have been inadequate to assess carcinogenicity.[6] The most recent follow-up of a cohort having mixed exposure to aldrin, dieldrin, and endrin found 9 deaths from cancer versus 12 expected. The workers had been exposed to the pesticides for a mean of 11 years and followed a mean 24 years.[7]

Aldrin induced an increased incidence of hepatocellular carcinoma at two dietary doses in male mice; the tumors showed a significant dose-response trend and were statistically significant at the high dose.[8]

In rats, follicular-cell tumors of the thyroid and adrenal cortical-cell adenomas (females only) were increased in the low-dose group but not in the high-dose group; the results could not be clearly associated with treatment.[8] The IARC has determined that there is inadequate evidence for carcinogenicity in humans and limited evidence in animals.[6]

Single high doses of aldrin (50 mg/kg) administered orally to hamsters during the period of organogenesis caused a high incidence of fetal deaths, congenital anomalies, and growth retardation.[9] No information on the health status of maternal animals was provided, but this dose is in the range of reported LD_{50}s.

Although some positive genotoxic results have been reported for aldrin, it is not thought to react directly with DNA to produce mutations.[10]

The 1995 ACGIH threshold limit value-time weighted average for aldrin is 0.25 mg/m³ with a notation for skin absorption.

REFERENCES

1. Kazantzis G, McLaughlin AIG, Prior PF: Poisoning in industrial workers by the insecticide aldrin. *Br J Ind Med* 21:46–51, 1964
2. Hoogendam I, Versteeg JPJ, DeVlieger M: Nine years toxicity control in insecticide plants. *Arch Environ Health* 10:441–448, 1965
3. Spiotta EJ: Aldrin poisoning in man. *AMA Arch Ind Hyg Occup Med* 4:560–566, 1951
4. Brown VKH, Hunter CG, Richardson A: A blood test diagnostic of exposure to aldrin and dieldrin. *Br J Ind Med* 21:283–286, 1964
5. Hayes WJ Jr: *Pesticides Studied in Man*, pp 234–237. Baltimore, MD, Williams and Wilkins, 1982
6. *IARC Monographs on the Evaluation of Carcinogenic Risks to Humans. Overall Evaluations of Carcinogenicity: An Updating of IARC Monographs Volumes 1 to 42*, Suppl 7, pp 88–89. Lyon, International Agency for Research on Cancer, 1987
7. Ribbens PH: Mortality study of industrial workers exposed to aldrin, dieldrin and endrin. *Int Arch Occup Environ Health* 56:75–79, 1985
8. National Cancer Institute: Carcinogenesis Technical Report Series No 21: *Bioassays of Aldrin and Dieldrin for Possible carcinogenicity*. DHEW (NIH) Pub No 78–821. Washington, DC, US Government Printing Office, 1978
9. Ottolenghi AD, Haseman JK, and Suggs F: Teratogenic effects of aldrin, dieldrin, and endrin in hamsters and mice. *Teratology* 9:11–16, 1974
10. Agency for Toxic Substances and Disease Registry (ATSDR): *Toxicological Profile for Aldrin/Dieldrin*. US Department of Health and Human Services, Public Health Service, pp 184, TP-92/01, 1993

ALLYL ALCOHOL
CAS: 107-18-6

C_3H_6O

Synonyms: 2-Propen-1-ol; 1-propenol-3; vinyl carbinol

Physical Form. Colorless liquid

Uses. In manufacture of allyl compounds, resins, plasticizers; fungicide and herbicide

Exposure. Inhalation; skin absorption

Toxicology. Allyl alcohol a potent lacrimator and an irritant of the mucous membranes and skin.

In humans, severe eye irritation occurs at 25 ppm, and irritation of the nose is moderate at 12.5 ppm.[1] In workers exposed to a "moderate" vapor level, there was a syndrome of lacrimation, retrobulbar pain, photophobia, and blurring of vision.[1] The symptoms persisted for up to 48 hours. Skin contact with the liquid has a delayed effect, causing aching that begins several hours after contact, followed by the formation of vesicles. Splashes of the liquid in human eyes have caused moderately severe reactions.[2]

In rats, the 1-, 4-, and 8-hour LC_{50}s were 1060, 165, and 76 ppm, respectively.[1] Signs of toxicity included lethargy, excitability, tremors, convulsions, diarrhea, coma, pulmonary and visceral congestion, and varying degrees of liver injury. Repeated 7-hr/day exposure at 60 ppm caused gasping during the first few exposures, persistent eye irritation, and death of 1 of 10 rats.[1] In several species of animals exposed to 7 ppm for 7 hr/day for 6 months, observed effects were minimal; at autopsy, findings were focal necrosis of the liver and necrosis of the convoluted tubules of the kidneys.[3]

The warning properties are thought to be adequate to prevent voluntary exposure to acutely dangerous concentrations but inadequate for chronic exposure.

The 1995 ACGIH threshold limit value–time-weighted average for allyl alcohol is 2 ppm (4.8 mg/m^3) with a short-term excursion limit of 4 ppm (9.5 mg/m^3) and a notation for skin absorption.

REFERENCES

1. Dunlap MK et al: The toxicity of allyl alcohol. *Arch Ind Health* 18:303–311, 1958
2. Grant WM: *Toxicology of the Eye*, 2nd ed, pp 105–106. Springfield, IL, Charles C Thomas, 1974.
3. Torkelson TR, Wolf MA, Oyen F, Rowe VK: Vapor toxicity of allyl alcohol as determined on laboratory animals. *Am Ind Hyg Assoc J* 20:224–229, 1959

ALLYL CHLORIDE
CAS: 107-05-1

C_3H_5Cl

Synonyms: Chlorallylene; 3-chloroprene; 1-chloro-2-propene; 3-chloropropylene

Physical Form. Liquid

Uses. Manufacture of epichlorohydrin, epoxy resin, glycerin pesticides, and sodium allyl sulphonate

Exposure. Inhalation; skin absorption

Toxicology. Allyl chloride is an irritant of the eyes, mucous membranes, and skin; chronic exposure may cause toxic polyneuropathy. In animals, it causes renal, hepatic, and pulmonary damage and, at high concentrations, central nervous system depression.

The most frequent effects in humans following overexposure have been conjunctival irritation and eye pain with photophobia; eye irritation occurs between 50 and 100 ppm.[1] Irritation of the nose occurs at levels below 25 ppm.

In one report from China, 26 factory workers exposed to allyl chloride ranging from 0.8

ppm to 2100 ppm complained of lacrimation and sneezing, which gradually diminished.[2] After 2.5 months to 5 years of exposure, most had developed weakness, paresthesia, cramping pain, and numbness in the extremities, with sensory impairment in the glove-stocking distribution as well as diminished ankle reflexes. Electroneuromyography showed neurogenic abnormalities in 10 of 19 subjects. Similar but much milder symptoms appeared in other workers exposed at 0.06 to 8 ppm for 1 to 4.5 years. Diagnostic findings suggested mild neuropathy in 13 of 27 of these subjects.

The liquid is a skin irritant and may be absorbed through the skin, causing deep-seated pain.[1] If splashed in the eye, severe irritation would be expected.

Rats survived 15 minutes at 32,000 ppm, 1 hour at 3200 ppm, or 3 hours at 320 ppm, but 0.5-, 3-, and 8-hour exposures, respectively, were lethal to all within the following 24 hours.[3] Exposure to 16,000 ppm for up to 2 hours in rats or 1 hour in guinea pigs caused eye and nose irritation, drowsiness, weakness, instability, labored breathing, and ultimately, death. Postmortem findings were severe kidney injury, alveolar hemorrhage in the lungs, and slight liver damage. No significant effects were found in rats exposed at 200 ppm for 6 hours; renal toxicity appeared at 300 ppm, but mortality was not affected until 1000 ppm was reached.[4] Several species exposed to 8 ppm for 7 hours daily for 1 month showed no apparent ill effects, but histopathologic examination revealed focal necrosis in the liver and necrosis of the convoluted tubules of the kidneys. Exposed at 3 ppm for 6 months, rats showed slight centrilobular degeneration in the liver.[5]

In other reports, rats and mice showed no effects at 20 ppm, 7 hr/day, for 90 days, but adverse effects were found following the 50-ppm regime.[4] In a limited inhalation study, rabbits exposed at 206 mg/m³, 6 hr/day, for 2 months developed unsteady gait and flaccid paralysis, whereas rabbits exposed at 17 mg/m³ for 5 months showed no evidence of toxic effects.[6]

In a recent study, rats were given 2 mM/kg allyl chloride by subcutaneous injection 5 days/week for 3 months.[7] Animals showed clinical signs of neurotoxicity after the treatment period and biochemical evidence of neurofilament protein accumulation in both the central and peripheral nervous systems. No evidence of neurofilament protein cross-linking was found, however, suggesting that allyl chloride may not share a common mechanism for the accumulation of neurofilaments with other neurotoxins such as 2,5-hexanedione.

Allyl chloride was fetotoxic to rats exposed during gestation to 300 ppm, which also caused considerable maternal toxicity in the form of kidney and liver injury.[4]

Administered by gavage for 1.5 years, allyl chloride was not carcinogenic to rats but caused a low incidence of squamous-cell carcinomas of the forestomach in mice.[8]

Although allyl chloride is detectable below 3 ppm, the warning properties are insufficient to prevent exposure to concentrations that may be hazardous with chronic exposure.[4]

The 1995 ACGIH threshold limit value–time-weighted average (TLV-TWA) for allyl chloride is 1 ppm (3 mg/m³) with a short-term excursion limit of 2 ppm (6 mg/m³).

REFERENCES

1. National Institute for Occupational Safety and Health, US Department of Health, Education, and Welfare: *Criteria for a Recommended Standard . . . Occupational Exposure to Allyl Chloride.* DHEW (NIOSH) Pub No 76-204, pp 19–38. Washington, DC, US Government Printing Office, 1976
2. He F, Zhang SL: Effects of allyl chloride on occupationally exposed subjects: *Scand J Work Environ Health* 11, (Suppl 4):43–45, 1985
3. Adams EM et al: The acute vapor toxicity of allyl chloride. *J Ind Hyg Toxicol* 22:79–86, 1940
4. Torkelson TR, Rowe VK: Halogenated aliphatic hydrocarbons. In Clayton GD and Clayton FE (eds) *Patty's Industrial Hygiene and Toxicology* 3rd rev ed, Vol 2B, *Toxicology*, pp 3568–3572. New York, Wiley-Interscience, 1981
5. Torkelson TR, Wolf MA, Oyen F, Rowe VK: Vapor toxicity of allyl chloride as determined on laboratory animals. *Am Ind Hyg Assoc J* 20:217–223, 1959
6. He F, Lu B, Zhang S, et al: Chronic allyl chloride poisoning. an epidemiology clinical, toxi-

cological and neuropathological study. *G Ital Med Lav* 7:5–15, 1985

7. Nagano M, Yamamoto H, Harada K, et al: Comparative study of modification and degradation of neurofilament proteins in rats subchronically treated with allyl chloride, acrylamide or 2,5-hexanedione. *Environ Res* 63:229–240, 1993
8. National Cancer Institute: *Carcinogenesis Technical Report Series. Bioassay of Allyl Chloride for Possible Carcinogenicity.* DHEW (NIH) Pub No 78-1323, p 53. Washington, DC, US Government Printing Office, 1978

ALLYL GLYCIDYL ETHER
CAS: 106-92-3

$C_6H_{10}O_2$

Synonyms: AGE; allyl 2,3-epoxypropyl ether

Physical Form. Liquid

Uses. Reactive diluent in epoxy resin systems; stabilizer of chlorinated compounds; manufacture of rubber

Exposure. Inhalation; skin absorption

Toxicology. Allyl glycidyl ether (AGE) causes dermatitis and eye irritation; in animals, high concentrations cause pulmonary edema and narcosis, whereas chronic low-level exposures induce nasal lesions.

Workers exposed to the vapor and/or liquid complained of dermatitis with itching, swelling, and blister formation.[1] Skin sensitization has occurred; cross-sensitization probably can occur with other epoxy agents.[2]

Three workers applying an epoxy-based waterproofing paint containing glycidyl ether inside an underground water tank died of asphyxia in the tank. Constituents of epoxy resin will displace oxygen in a confined space and may have an independent narcotic effect on exposed workers. Strict precautionary measures are recommended under these conditions.[3]

In rats, the LC_{50} for 8 hours was 670 ppm; effects were lacrimation, nasal discharge, dyspnea, and narcosis.[1] In rats repeatedly exposed to 600 ppm for 8 hours daily, effects were pronounced irritation of the eyes and respiratory tract; more than half of the rats developed corneal opacity; at necropsy, after 25 exposures, pulmonary findings were inflammation, bronchiectasis, and bronchopneumonia.[1]

Inhalation of 7 ppm for 6 hr/day caused necrosis and complete erosion of nasal mucosa after 4 days; squamous metaplasia of the respiratory epithelium and focal erosion of the olfactory epithelium, with evidence of regeneration of some epithelial surface, occurred in mice after 9 to 14 days at this exposure level.[4] Rats and mice exposed to concentrations as low as 4 ppm for 13 weeks had squamous metaplasia, hyperplasia, and inflammation of the nasal mucosa.[5]

Chronic 24-month inhalation exposure to 5 or 10 ppm AGE induced nasal lesions in rats and mice.[6] Inflammation, degeneration, regeneration, metaplasia, hyperplasia, and neoplasia were observed in the nasal mucosa. Although the incidence of primary nasal tumors was not statistically significant compared to the incidence in concurrent controls, the relative rarity of primary nasal tumors occurring spontaneously and the presence of other nonneoplastic lesions suggest that the tumors observed may be related to AGE exposure. It was concluded that there was some evidence of carcinogenicity of inhaled AGE for male mice, equivocal evidence of carcinogenicity for female mice and male rats, and no evidence of carcinogenicity for female rats.

Percutaneous absorption has been documented in rabbits.[2] The liquid dropped into the eye of a rabbit caused severe but reversible conjunctivitis, iritis, and corneal opacity.[1] Cytotoxic effects on rat bone marrow cells, with reduction in leukocyte counts, and testicular degeneration were observed after intramuscular injections at 400 mg/kg/day. AGE has shown mutagenic activity in bacteria.[7]

The 1995 ACGIH threshold limit value–time-weighted average (TLV-TWA) for allyl glycidyl ether is 5 ppm (23 mg/m³) with a short-term excursion limit of 10 ppm (47 mg/m³).

REFERENCES

1. Hine CH et al: The toxicology of glycidol and some glycidyl ethers. *AMA Arch Ind Health* 13:250–264, 1956
2. Allyl glycidyl ether (AGE). *Documentation of the TLVs and BEIs*, 6th ed, p 43–44. Cincinnati, OH, American Congress of Governmental Industrial Hygienists, 1991
3. Centers for Disease Control: Occupational fatalities associated with exposure to epoxy resin paint in an underground tank—Makati, Philippines. *MMWR* 39:373–380, June 8, 1990
4. Gagnaire F, Zissn D, Bonnet P, et al: Nasal and pulmonary toxicity of allyl glycidyl ether in mice. *Toxicol Lett* 39:139–145, 1987
5. National Toxicology Program. *Toxicology and Carcinogenesis Studies of Allyl Glycidyl Ether (CAS No. 106-92-3) in Osborne–Mendel Rats and B6C3F₁ Mice (Inhalation Studies)*. Technical Report No 376. Public Health Service, National Institutes of Health. NIH Pub No 90-2831, Research Triangle Park, NC, 1990
6. Renne RA, Brown HR, and Jokinen MP: Morphology of nasal lesions induced in Osborne–Mendel Rats and B6C3F₁ Mice by chronic inhalation of allyl glycidyl ether. *Toxicol Path* 20:416–425, 1992
7. National Institute for Occupational Safety and Health, U.S. Department of Health, Education and Welfare: *Criteria for a Recommended Standard . . . Occupational Exposure to Glycidyl Ethers*. DHEW (NIOSH) Pub No 78-166. Washington, DC, US Government Printing Office, 1978

ALLYL PROPYL DISULFIDE
CAS: 2179-59-1

$C_6H_{12}S_2$

Synonyms: Disulfide; allyl propyl, Onion oil

Physical Form. Pale, yellow oil

Source. Onions

Exposure. Inhalation

Toxicology. Allyl propyl disulfide vapor is a mucous membrane irritant.

No systemic effects have been reported from industrial exposure. At an average concentration of 3.4 ppm in an onion-dehydrating plant, there was irritation of eyes, nose, and throat in some workers.[1]

The 1995 ACGIH threshold limit value–time-weighted average (TLV-TWA) for allyl propyl disulfide is 2 ppm (12 mg/m³) with a short-term excursion limit of 3 ppm (18 mg/m³).

REFERENCES

1. Feiner B, Burke WJ, Baliff J: An industrial hygiene survey of an onion dehydrating plant. *J Ind Hyg Toxicol* 28:278–279, 1946

ALUMINUM
CAS: 7429-90-5

Al

Physical Forms. Metal dusts, pyro powders, welding fumes: When exposed to air, an aluminum surface becomes oxidized to form a thin coating of aluminum oxide, which protects against ordinary corrosion. Powder and flake aluminum are flammable and can form explosive mixtures in air, especially when treated to reduce surface oxidation (pyro powders)

Uses. Structural material in construction, automotive and aircraft industries; in the production of metal alloys; in cooling utensils, cans, food packaging, and dental materials; as pyro powders in fireworks and aluminum paint

Exposure. Inhalation

Toxicology. Inhalation of very fine aluminum powder (pyropowder) in massive concentrations may rarely be associated with pneumoconiosis in some persons. The metallic dust produced by grinding aluminum products is regarded only as inert dust.

In humans, the symptoms of long-term overexposure to only some fine powders of aluminum may include dyspnea, cough, and weakness. It has been noted that these workers are usually exposed to a number of other toxicants that could cause similar symptoms. Typically, there may be radiographic evidence of fibrosis and occasional pneumothorax. At autopsy, there is generalized interstitial fibrosis, predominantly in the upper lobes, with pleural thickening and adhesions. Particles of aluminum are found in the fibrotic tissue. A rare fatal case of pulmonary fibrosis from inhalation of a heavy concentration of fine aluminum dust was reported in a 22-year-old British worker; autopsy revealed a generalized nonnodular fibrosis and interstitial emphysema with right ventricular hypertrophy. There had been work exposure to varying concentrations of a wide range of particle sizes, but the quantity of dust in the atmosphere below 5 μ was of the order of 19 mg/m^3.[1]

Of 27 workmen with heavy exposures to aluminum powder in the same plant as the above-mentioned case, 6 were found to have evidence of pulmonary fibrosis. The finer dust was more dangerous than the coarse dust; of the 12 men exposed to fine aluminum powder, 2 died and 2 others were affected; and, of 15 men who worked exclusively with coarser powder, 2 had radiologic changes but no symptoms.[2]

Fine metallic aluminum powders inhaled by hamsters and guinea pigs caused no pulmonary fibrosis; in rats that inhaled the dust, small scars resulted from foci of lipid pneumonitis. Alveolar proteinosis developed in all three species; it resolved spontaneously, and the accumulated dust deposits cleared rapidly from the lungs after cessation of exposure. The failure of inhaled aluminum powder to cause pulmonary fibrosis in experimental animals parallels the clinical experience in the United States, where pulmonary fibrosis has not been observed in aluminum workers.[3]

It has been suggested that the explanation of pulmonary disease among powder workers in other countries may lie in the duration of exposure, the size of the particles, the density of the dust, and especially the fact that all reported cases have been associated with exposure to a submicron-sized aluminum pyrotechnic flake (powder) that has been lubricated with a nonpolar aliphatic oil rather than with the usually employed stearic acid.[2,4]

Evidence of the relatively benign nature of aluminum dust in measured concentrations lies in the 27-year experience of administration of freshly milled metal particles to workers exposed to silica as a suggested means of inhibiting the development of silicosis. Inhalation of aluminum powder of 1.2-m particle size (96%), given over 10- or 20-minute periods several times weekly, resulted in no adverse health effects among thousands of workers over several years.

The etiologic role of aluminum in neurologic disorders following chronic ingestion has been of increasing interest in recent years. However, in general, workers exposed to large amounts of aluminum dusts in factories have not been shown to develop significant neurologic deficits, and so the route of exposure may be important.[5] Several mortality studies of aluminum reduction plant workers, in which the study cohorts totaled nearly 28,000 long-term employees, recorded no excess deaths due to organic brain disorders of the dementia type, and an analysis of the occupational mortality experience of nearly 430,000 men who died in Washington State during the years 1950–1979 showed no excess deaths from this cause among the 1238 former aluminum workers included in the study.[6-8] However, three cases of a progressive neurologic disorder, characterized by incoordination, intention tremor, and cognitive deficit, in workers at an aluminum reduction plant have been reported, and the investigators postulated that they may have been related to occupational exposure to aluminum in some form.[9]

People on renal dialysis who have received high doses of aluminum in medications and in dialysate fluid for a number of years are at increased risk of developing encephalopathy or "dialysis dementia."[5] The disease is characterized by altered speech, personality changes, seizures, and motor dysfunction. Although symptoms have been reversed when aluminum exposures were lowered, aluminum has not been confirmed as the etiologic agent.

Aluminum ingestion has also been associated with Alzheimer's disease, which has clinical and histopathological features distinct from dialysis encephalopathy. Alzheimer's disease is pathologically characterized by the formation of neurofibrillary tangles and senile plaques; these tangles in the cerebral cortex and hippocampus have been reported to contain aluminum. It is not known whether aluminum is a causal agent or if the neurodegenerative disease just allows more aluminum to accumulate in the brain. It has been noted that a large number of other factors, including genetic predisposition, viral infections, and immune system dysfunction, have also been associated with Alzheimer's.[5]

One study suggested a possible link of aluminum in public water supplies with the occurrence of Alzheimer's disease in 88 county districts of England and Wales.[10] In districts where the mean aluminum concentration in water exceeded 0.11 mg/l, rates were 1.5 times higher than in districts where the mean levels were less than 0.01 mg/l. Results have been challenged on the basis of study design and on the interpretation of the relative significance of the dose of aluminum from water as a fraction of total dietary intake.[11]

Ingested aluminum is poorly absorbed, and there appears to be no retention of aluminum from nutritional sources in individuals with normal kidneys. Dusts of metallic aluminum and aluminum oxide are not significantly absorbed systemically, although fume from welding aluminum is absorbed through the lung, producing a rise in aluminum levels in plasma and urine.[12]

The biologic half-life of absorbed aluminum fume, as measured by urinary excretion, is about 9 days in workers exposed less than 1 year and increases to about 6 months after 10 years of occupational exposure.[13]

Aerosols of the soluble salts of aluminum, such as the chloride and sulfate, are irritants of little occupational importance. Although the aluminum alkyls may also be irritants, toxicity information on these compounds is inadequate.

The 1995 ACGIH threshold limit value–time-weighted average (TLV-TWA) for aluminum is 10 mg/m^3 for the metal dust, 5 mg/m^3 for pyro powders and welding fumes, as Al, and 2 mg/m^3 for the soluble salts and alkyls, as Al.

REFERENCES

1. Mitchell J: Pulmonary fibrosis in an aluminum worker. *Br J Ind Med* 16:123–125, 1959
2. Mitchell J, Manning GB, Molyneux M, Lane RE: Pulmonary fibrosis in workers exposed to finely powdered aluminum. *Br J Ind Med* 18:10–20, 1961
3. Gross P, Harley RA Jr, deTreville RTP: Pulmonary reaction to metallic aluminum powders. *Arch Environ Health* 26:277–236, 1973
4. Dinman BD: Aluminum in the lung: the pyropowder conundrum. *J Occup Med* 29:869–876, 1987
5. Agency for Toxic Substances and Disease Registry (ASTDR): *Toxicological Profile for Aluminum.* US Department of Health and Human Services, Public Health Service, p 136, TP-91/01, 1992
6. Gibbs GW: Mortality experience in eastern Canada. In Hughes JP (ed): *Health Protection in Primary Aluminium Production,* Vol 2, pp 56–69. London, International Primary Aluminium Institute, 1981
7. Rockette HE, Arena VC: Mortality studies of aluminum reduction plant workers: potroom and carbon department. *J Occ Med* 25:549–557, 1983
8. Milham S: *Occupational Mortality in Washington State, 1950–1979.* DHHS (NIOSH) Pub No 83–116, p 38. Washington, DC, US Government Printing Office, 1983
9. Longstreth WT, Rosenstock L, Heyer NJ: Potroom palsy? Neurologic disorder in three aluminum smelter workers. *Arch Intern Med* 145:1972–1975, 1985
10. Martyn CN, Osmond C, Edwardson JA, et al: Geographical relationship between Alzheimer's disease and aluminium in drinking water. *Lancet* 1:59–62, 1989
11. Schupf N, Silverman W, Zigman WB, et al: Aluminium and Alzheimer's disease. *Lancet* 4:267–269, 1989
12. Mussi I, Calzaferri G, Buratti M, et al: Behavior of plasma and urinary aluminum level in occupationally exposed subjects. *Intl Arch Occup Environ Health* 54:155–161, 1984

13. Sjogren B, Elinder CG, Lidums V, et al: Uptake and urinary excretion of aluminum
among welders. *Intl Arch Occup Environ Health*
60:77–79, 1988

ALUMINUM OXIDES

Chemical Compound: Aluminum oxide (Al_2O_3)
Mineral Name: Corundum
Synonym: α-Alumina
CAS: 1344-28-1

Chemical Compound: Aluminum oxyhydrox-
ide (AlO_2H)
Mineral Name: Boehmite; diaspore
Synonym: Alumina monohydrate
CAS: 24623-77-6

Chemical Compound: Alumina trihydroxide
($Al(OH)_3$)
Mineral Name: Gibbsite; bayerite; nords-
trandite
Synonym: Alumina trihydrate; aluminum hy-
droxide
CAS: 21645-51-2

Uses. In production of aluminum; synthetic
abrasives; refractory material

Exposure. Inhalation

Toxicology. The aluminas are considered to
be nuisance dusts; their role in fibrogenic lung
disease remains unclear.

Assessment of the toxicity of alumina has
been complicated by the chemical and physical
variants of the compound and inconsistencies
in the nomenclature used to describe them.[1]
The group of compounds referred to as aluminas is composed of various structural forms of
aluminum oxide, trihydroxide, and oxyhydrox-
ide.[2] As these aluminas are heated, dehydration
occurs, producing a variety of transitional
forms; temperatures between 200°C and 500°C
result in low-temperature-range transitional
aluminas characterized by increased catalytic
activity and larger surface area.[2] (Transitional

aluminas include chi, eta, and gamma forms
which, taken together, were formerly termed
"gamma.")[2]

Despite the problems in defining precise
exposures (in terms of structure and form), population studies of potentially exposed workers
have shown minimal evidence for pulmonary
fibrosis or pneumoconiosis.

A report from an aluminum production facility found that 7% to 8% of potentially exposed alumina workers had small, irregular
opacities as determined by chest radiograph.[2]
The prevalence of opacities was increased
among smokers and among nonsmokers with
high cumulative dust exposures. The pulmonary pathologic changes responsible for the
opacities are not clear.[2] A slight but significant
decrement in ventilatory function among nonsmoking workers was also observed in this population.[3] The findings were consistent with a
minor degree of nonspecific chronic industrial
bronchitis associated with excessive protracted
nuisance dust exposure (100 mg/year for more
than 20 years).[1]

A number of epidemiologic studies of aluminum smelter workers have confirmed either
minimal or absent fibronodular disease and no
excess mortality associated with pneumoconiosis.[4-6] A report of four subjects exposed for many
years to alumina dust found a correlation between radiographic opacities and apparent pulmonary burden of aluminum as determined by
neutron activation analysis.[7] A recent study of
nine aluminum oxide workers with abnormal
chest roentgenograms found histological evidence of interstitial lung fibrosis in three of the
most severely affected workers who underwent
lung biopsy.[8] The absence of asbestos bodies
and silicotic nodules, despite concurrent exposure to these substances, and the large number
of aluminum-containing particles in lung tissue
indicated to the investigators that aluminum
oxide was the common exposure. The role of
smoking in altering the host response in these
cases is unknown. In other case reports of lung
fibrosis, the exposure to aluminum oxide was
not well quantified, and there was concurrent
exposure to other dust and fumes.[9]

Animal experiments with alumina have
shown that the type of reaction in lung tissue
is dependent on the form of alumina and its

particle size, the species of animal used, and the route of administration. For example, intratracheal administration into rats of γ-alumina of 2 μ average size caused only a mild fibrous reaction of loose reticulin.[10] However, intratracheal administration of γ-alumina of 0.02 to 0.04 μ size into rats produced reticulin nodules that later developed into areas of dense collagenous fibrosis.[11] The latter alumina by the same route in mice and guinea pigs caused development of a reticulin network with occasional collagen whereas, in rabbits, only a slight reticulin network was observed.[10] Intratracheal administration in rats of another form of alumina, corundum of particle size less than 1 μ, caused the development of compact nodules of reticulin.

In rats, inhalation of massive levels of γ-alumina with an average particle size of 0.0005 to 0.04 μ for up to 285 days caused heavy desquamation of alveolar cells and secondary inflammation but only slight evidence of fibrosis.[12] The dust concentration in the exposure chamber was described as so high that visibility was reduced; a few breaths of the atmosphere by the investigators caused bronchial irritation and persistent cough.

A review of the animal studies concluded that a fibronodular response has resulted only from intratracheal insufflation of catalytically active, low-temperature-range transitional aluminas, and high-surface-area aluminas.[1] In general, alumina is efficiently eliminated from the lung and has a low degree of fibrogenicity.

The 1995 ACGIH threshold limit value–time-weighted average (TLV-TWA) for aluminum oxide is 10 mg/m³ for total dust containing no asbestos and less than 1% crystalline silica.

REFERENCES

1. Dinman BD: Alumina-related pulmonary disease. *J Occup Med* 30:328–335, 1988
2. Townsend MC, Sussman NB, Exterline PE, et al: Radiographic abnormalities in relation to total dust exposure at a bauxite refinery and alumina-based chemical products plant. *Am Rev Respir Dis* 138:90–95, 1988
3. Townsend MC, Enterline PE, Sussman NB, et al: Pulmonary function in relation to total dust exposure at a bauxite refinery and alumina-based chemical products plant. *Am Rev Respir Dis* 132:1174–1180, 1985
4. Saia B, Cortese S, Piazza G, et al: Chest x-ray findings among aluminum production plant workers. *Med Lav* 72:323–329, 1981
5. Chen-Yeung M, Wong R, McLean L, et al: Epidemiologic health study of workers in an aluminum smelter in British Columbia: effects on the respiratory system. *Am Rev Respir Dis* 127:465–469, 1983
6. Gibbs GW: Mortality of aluminum reduction plant workers, 1950 through 1977. *J Occup Med* 27:761–770, 1985
7. Gaffuri E: Pulmonary changes and aluminum levels following inhalation of alumina dust: a study of four exposed workers. *Med Lav* 76:222–227, 1985
8. Jederlinic PJ, Abraham JL, Churg A, et al: Pulmonary fibrosis in aluminum oxide workers; investigation of nine workers, with pathologic examination and microanalysis in three of them. *Am Rev Resp Dis* 142:1179–1184, 1990
9. Agency for Toxic Substances and Disease Registry (ATSDR): *Toxicological Profile for Aluminum.* US Department of Health and Human Services, Public Health Service, p 136, 1992
10. Stacy BD el al: Tissue changes in rats' lungs caused by hydroxides, oxides and phosphates of aluminum and iron. *J Pathol Bact* 77:417–426, 1959
11. King EJ, Harrison CV, Mohanty GP, Nagelschmidt G: The effect of various forms of alumina on the lungs of rats. *J Pathol Bact* 69:81–92, 1955.
12. Klosterkotter W: Effects of ultramicroscopic gamma-aluminum oxide on rats and mice. *AMA Arch Ind Health* 21:458, 1960.

4-AMINODIPHENYL

CAS: 92-67-1

$C_{12}H_{11}N$

Synonym: *p*-Xenylamine

Physical Form. Colorless, crystalline compound; darkens on oxidation

Uses. Formerly used as a rubber antioxidant; no longer produced on a commercial scale

Exposure. Inhalation, skin absorption

Toxicology. 4-Aminodiphenyl exposure is associated with a high incidence of bladder cancer in humans; in animals, it has produced bladder and liver tumors.

Of 171 workers exposed to 4-aminodiphenyl for 1.5 to 19 years, 11% had bladder tumors; the tumors appeared 5 to 19 years after initial exposure.[1]

In a study of 503 exposed workers, there were 35 histologically confirmed bladder carcinomas and an additional 24 men with positive cytology.[2]

Two bladder papillomas and three bladder carcinomas were observed in six dogs fed a total of 5.5 to 7 g (1.0 mg/kg, 5 days/week for life).[3] In another study, each of four dogs developed urinary bladder carcinomas with predominantly squamous differentiation in 21 to 34 months after ingestion of 0.3 g 4-aminodiphenyl three times per week (total dose 87.5 g to 144 g/dog); hematuria, salivation, loss of body weight, and vomiting were also noted, and all animals died within 13 months of the first appearance of a tumor.[4]

Rats injected subcutaneously with a total dose of 3.6 to 5.8 g/kg had an abnormally high incidence of mammary gland and intestinal tumors.[5] Nineteen of 20 newborn male mice and 6 of 23 newborn female mice developed hepatomas in 48 to 52 weeks after three subcutaneous injections of 200 μg of 4-aminodiphenyl; in control animals, 5 of 41 males and 2 of 47 females had hepatomas.[6]

The IARC has determined that there is sufficient evidence for carcinogenicity to humans and animals.[7] Furthermore, the accumulated experimental and epidemiologic evidence has demonstrated that 4-aminodiphenyl may be the most hazardous of the aromatic amines in terms of carcinogenic potential.[8] Because of demonstrated carcinogenicity, contact by all routes should be avoided.[9]

ACGIH has designated aminodiphenyl as an A1 human carcinogen with no assigned threshold limit value and a notation for skin absorption.

REFERENCES

1. Melick WF et al: The first reported cases of human bladder tumors due to a new carcinogen—xenylamine. *J Urol* 74:760–766, 1955
2. Koss LG, Myron R, Melamed MR, Kelly RE: Further cytologic and histologic studies of bladder lesions in workers exposed to para-aminodiphenyl: progress report. *J Natl Cancer Inst* 43:233, 1969
3. Deichmann WB et al: Synergism among oral carcinogens, simultaneous feeding of four bladder carcinogens to dogs. *Ind Med Surg* 34:640, 1965
4. Deichmann WB et al: The carcinogenic action of *p*-aminodiphenyl in the dog. *Ind Med Surg* 27:25–26, 1958
5. Walpole AL, Williams MHC, Roberts DC: The carcinogenic action of 4-aminodiphenyl and 3:2'-dimethyl-4-aminodiphenyl. *Br J Ind Med* 9:255–261, 1952
6. Gorrod JW, Carter RL, Roe FJC: Induction of hepatomas by 4-aminobiphenyl and three of its hydroxylated derivatives administered to newborn mice. *J Natl Cancer Inst* 41:403–410, 1968
7. *IARC Monographs on the Evaluation of Carcinogenic Risks to Humans, Overall Evaluations of Carcinogenicity: An updating of IARC Monographs Volumes 1 to 42.* Suppl 7, pp 91–92. Lyon, International Agency for Research on Cancer, 1987
8. Department of Labor: Occupational Safety and Health Standards—Carcinogens. *Federal Register* 39:3756, 3781–3784, 1974
9. 4-Aminodiphenyl. *Documentation of TLVs and BEIs*, 6th ed, pp 50–51. Cincinnati, OH, American Hygienists (ACGIH), 1991

p-AMINOPHENOL
CAS: 123-30-8

$NH_2C_6H_4OH$

Synonyms: Activol; 4-amino-1-hydroxybenzene; 4-hydroxyaniline; PAP

Physical Form. White or reddish-yellow crystals; discolors to lavender when exposed to air

Uses. Oxidative dye; developing agent for photographic processes; precursor of pharmaceuticals; used in hair dyes

Exposure. Inhalation; skin absorption

Toxicology. *p*-Aminophenol is of moderately low toxicity but has caused kidney injury and dermal sensitization in animals; the potential for producing methemoglobin is of relatively minor importance.

There are no reports of adverse health effects for human exposure.

The oral LD_{50} in rats was 671 mg/kg.[1] Effects included central nervous system depression. *p*-Aminophenol caused dermal sensitization in guinea pigs.[2] A solution of 2.5% applied to abraded skin of rabbits was a mild irritant.[1] The dermal LD_{50} in rabbits was greater than 8 g/kg, which strongly suggests that absorption through the skin is minimal.[3] Single nonlethal acute doses in rats produced proximal renal tubular necrosis of the pars recta.[4,5] Recent studies suggest that oxidative metabolism of *p*-aminophenol to a metabolite that can react with glutathione is an important step in toxicity, while mitochondria appear to be a critical target for the reactive intermediate formed.[6]

Early animal studies of *p*-aminophenol administered in the diet and topical studies of oxidative hair dyes containing *p*-aminophenol have not shown definitive carcinogenic effects.[7-9] *p*-Aminophenol has tested positive in a number of mutagenic assays, but the relevance to human exposure has not been determined.[9] Studies of the teratogenic effects of *p*-aminophenol indicated both positive and negative effects, depending on the route of administration. Hamsters given intravenous or intraperitoneal injections of *p*-aminophenol at 100 to 250 mg/kg showed significant increases in malformed fetuses and resorptions in a dose-dependent manner.[10] However, oral studies using hamsters and topical application of hair dyes containing *p*-aminophenol on rats showed no teratogenic effects.[11]

The ACGIH has not assigned a threshold limit value to aminophenol.

REFERENCES

1. Lloyd GK et al: Assessment of the acute toxicity and potential irritancy of hair dye constitutents. *Food Cosmet Toxicol* 15:607, 1977
2. Kleniewska D, Maibach H: Allergenicity of aminobenzene compounds: structure-function relationships. *Derm Beruf Umwelt* 28: 11, 1980
3. Mallinckrodt, Inc. For your information (FYI) Submission FYI-OTS-1083-0272 Supp. Seq. C. Bio/Tox data on *p*-aminophenol from 1980. Washington, DC, Office of Toxic Substances, US Environmental Protection Agency, 1983
4. Briggs D, Calder I, Woods R, Tange J: The influence of metabolic variation on analgesic nephrotoxicity: experiments with the Gunn rat. *Pathology* 14:349, 1982
5. Klos C, Koob M, Kramer C, et al: *p*-Aminophenol nephrotoxicity: biosynthesis of toxic glutathione conjugates. *Toxicol Appl Pharm* 115:98–106, 1992
6. Lock EA, Cross TJ, Schnellmann RG: Studies on the mechanism of 4-aminophenol-induced toxicity to renal proximal tubules. *Hum Exp Toxicol* 12:383–388, 1993
7. Miller JA, Miller CE: The carcinogenicity of certain derivatives of *p*-dimethylaminobenzene in the rat. *J Exp Med* 87:139–156, 1948
8. Jacobs MM, Burnett CM, Penienak AJ, et al: Evaluation of the toxicity and carcinogenicity of hair dyes in Swiss mice. *Drug Chem Toxicol* 7:573–586, 1984
9. Final Report on the safety assessment of *p*-aminophenol, m-aminophenol and o-aminophenol. *J Am Coll Toxicol* 7(3):279–333, 1988
10. Rutkowski JV, Fermn VH: Comparison of

the teratogenic effects of isomeric forms of aminophenol in the Syrian golden hamster. *Toxicol Appl Pharmacol* 63:264, 1982
11. Burnett C et al: Teratology and percutaneous toxicity studies on hair dyes. *Toxicol Environ Health* 1:1027, 1976

2-AMINOPYRIDINE

CAS: 504-29-0

$(NH_2)C_5H_4N$

Synonyms: α-aminopyridine; α-pyridylamine

Physical Form. Crystalline solid

Uses. In manufacture of pharmaceuticals, especially antihistamines

Exposure. Inhalation; skin absorption

Toxicology. 2-Aminopyridine is a convulsant.

In industrial experience, intoxication has occurred from inhalation of the dust or vapor or by skin absorption following direct contact.[1] Fatal intoxication occurred in a chemical worker who spilled a solution of 2-aminopyridine on his clothing during a distillation; he continued to work in contaminated clothing for 1.5 hours. Two hours later, he developed dizziness, headache, convulsions, and respiratory distress, which progressed to respiratory failure and death; it is probable that skin absorption was a major factor in this case.

A nonfatal intoxication from exposure to an undetermined concentration of 2-aminopyridine in air resulted in severe headache, weakness, convulsions, and a stuporous state that lasted several days. A chemical worker exposed to an estimated air concentration of 20 mg/m^3 (5.2 ppm) for approximately 5 hours developed a severe pounding headache, nausea, flushing of the extremities, and elevated blood pressure, but he recovered fully within 24 hours.

The LD_{50} in mice by intraperitoneal injection was 35 mg/kg; lethal doses in animals produced excitement, tremors, convulsions, tetany, and death.[1] Fatal doses were readily absorbed through the skin. A 0.2 M aqueous solution dropped in a rabbit's eye was only mildly irritating.[2]

The 1995 ACGIH threshold limit value–time-weighted average (TLV-TWA) for aminopyridine is 0.5 ppm (1.9 mg/m^3).

References

1. Reinhardt CF, Brittelli MR: Heterocyclic and miscellaneous nitrogen compounds. In Clayton GD, Clayton FE (eds): *Patty's Industrial Hygiene and Toxicology*, 3rd ed, rev, Vol 2, pp 2731–2832. New York, Wiley-Interscience, 1981
2. Grant WM: *Toxicology of the Eye*, 3rd ed, p 383. Springfield, IL, Charles C Thomas, 1986

AMITROLE

CAS: 61-82-5

$C_2H_4N_4$

Synonyms: Aminotriazole; 3-amino-1,24-triazole; ATA; Amizol; Azolan; Cytrol; Weedazol

Physical Form. White crystalline powder

Uses. Herbicide

Exposure. Inhalation; ingestion

Toxicology. Amitrole has low acute toxicity; in experimental animal studies, subchronic exposures were associated with changes in the thyroid, and chronic exposures were carcinogenic.

Intentional ingestion of a mixture that contained 20 mg/kg amitrole did not cause any signs of intoxication.[1] In a recently reported case study, inhalation of a large amount of amitrole containing herbicide was associated with acute toxic reaction of the lungs.[2] Lung injury was thought to be secondary to direct toxic damage to the alveolar lining cells. The remarkable absence of any other reports describing pulmonary toxicity of this herbicide was noted, in addition to the presence of other chemicals in the herbicide solution.

The LD_{50}s in animal studies are high, indicating very low acute toxicity but varying considerably according to species.[3] The oral LD_{50} in mice was 11,000 mg/kg, while 4000 mg/kg was fatal to sheep. No detectable signs of toxicity were noted in rats at 4080 mg/kg.[4] Poisoning in animals is characterized by increased intestinal peristalsis, pulmonary edema, and hemorrhages in various organs.[3]

At a level of 1000 ppm in the diet of rats, significant enlargement of the thyroid could be detected as soon as 3 days.[5] At a dietary level of 60 or 120 ppm, there was enlargement of the thyroid within 2 weeks.[6] Morphological changes were noted in the thyroid of rats fed 10 or 50 mg/kg amitrole for 11 to 13 weeks.[7] Amitrole is thought to interfere with the formation of thyroxine by inhibiting the peroxidase-dependent iodide oxidation in the thyroid.[1] Suppression of thyroid function leads to further stimulation by the pituitary, with resultant hyperplasia and tumor formation.

Like other antithyroid compounds, or like diets that are low in iodine, continuous exposure for long periods produces adenomatous changes in the thyroid glands of rats.[3] Male and female rats fed diets containing 10 or 100 mg/kg amitrole for life had marked increases in the incidence of thyroid tumors in the high-dose group. Benign thyroid tumors in males occurred at a rate of 45/75 vs. 5/75 for controls and in females the occurrence was 44/75 vs. 7/74 for controls. For malignant thyroid tumors, the incidence was 18/75 high-dose versus 3/75 for controls in males, whereas females had 28/75 versus none in controls.[8] (The high-dose female group also had an increased incidence of benign pituitary tumors.) Early studies, although limited, also found increased incidence of thyroid tumors in rats chronically fed amitrole.[3,9]

In mice, thyroid and liver tumors were produced after oral administration. Mice administered 1000 mg/kg amitrole by gavage for 4 weeks, followed by diets containing 2192 mg/kg for up to 60 weeks, had an incidence of 64/72 for thyroid tumors and 67/72 for liver tumors.[10] Among 55 male mice, 9 hepatocellular adenomas and 11 hepatocellular carcinomas were observed following a continuous diet of 500 mg/kg for 90 weeks; among the 49 females, there were 5 hepatocellular adenomas and 4 hepatocellular carcinomas. The untreated controls had one hepatocellular adenoma and no carcinomas in the males and females combined.[11] There was no indication of a carcinogenic effect in mice (or hamsters) fed up to 100 mg/kg for life.[8] No skin tumor was observed following weekly topical applications of up to 10 mg amitrole for life.[9]

Very little human data are available to assess the long-term effects of amitrole. In a small cohort study of Swedish railroad workers, there was a statistically significant excess of all cancers among those exposed to both amitrole and chlorophenoxy herbicides (6 deaths vs. 2.9 expected) but not among those exposed primarily to amitrole (5 deaths vs. 3.3 expected).[12]

Cytogenic studies have reported an increased incidence in the frequency of chromosomal aberrations and an increased frequency of sister chromatid exchanges, but these studies are of limited value because they involve occupational exposure to a number of other pesticides in addition to amitrole.[1]

The IARC has determined that there is sufficient evidence for the carcinogenicity of amitrole in experimental animals, and inadequate evidence for its carcinogenicity in humans.[1]

No effect on offspring growth or viability was observed in rats given up to 100 mg/kg in the diet for two generations; litter size and weight, as well as postnatal viability, were reduced in the offspring of breeding pairs exposed to 500 mg/kg in the diet.[3]

The 1995 ACGIH threshold limit value–time-weighted average (TLV-TWA) for amitrole is 0.2 mg/m^3.

REFERENCES

1. *IARC Monographs on the Evaluation of the Carcinogenic Risk of Chemicals to Humans*, Vol 41, pp 293–317. Lyon, International Agency for Research on Cancer, 1986
2. Balkisson R, Murray D, Hoffstein V: Alveolar

damage due to inhalation of amitrole-containing herbicide. *Chest* 101:1174–1176, 1992

3. Hayes WJ Jr: *Pesticides Studies in Man*, pp 564–566. Baltimore, MD, Williams and Wilkins, 1982
4. Gaines TB, Kimbrough RD, Linder RE: The toxicity of amitrole in the rat. *Toxicol Appl Pharm* 26:118–129, 1973
5. Mayberry WE: Antithyroid effects of 3-amino-1,2,4-triazole. *Proc Soc Exp Biol Med* 129:551–556, 1968
6. Jukes TH, Shaffer CB: Antithyroid effects of aminotriazole. *Science* 132:296–297, 1960
7. Fregly MJ: Effect of aminotriazole on thyroid function in the rat. *Toxicol Appl Pharm* 13:271–286, 1968
8. Steinhoff D, Weber H, Mohr U, et al: Evaluation of amitrole (aminotriazole) for potential carcinogenicity in orally dosed rats, mice, and golden hamsters. *Toxicol Appl Pharmacol* 69:161–169, 1983
9. Hodge HC, Maynard EA, Downs WL, et al: Tests on mice for evaluating carcinogenicity. *Toxicol Appl Pharmacol* 9:583–596, 1966
10. Innes JRM, Ulland BM, Valerio MG, et al: Bioassay of pesticides and industrial chemicals for tumorgenicity in mice; a preliminary note. *J Natl Cancer Inst* 42:1101–1114, 1969
11. Vesselinovitch SD: Perinatal hepatocarcinogenesis. *Biol Res Pregnancy Perinatol* 4:22–25, 1983
12. Axelson O, Sundell L, Andersson K, et al: Herbicide exposure and tumor mortality: an updated epidemiologic investigation on Swedish railroad workers. *Scand J Work Environ Health* 6:73–79, 1980

AMMONIA
CAS: 7664-41-7

NH_3

Synonyms: Ammonia gas

Physical Form. Colorless gas

Uses. Refrigeration; petroleum refining; blue-printing machines; manufacture of fertilizers, nitric acid, explosives, plastics, and other chemicals

Exposure. Inhalation

Toxicology. Ammonia is a severe irritant of the eyes, skin, and respiratory tract, and can be corrosive when high concentrations are inhaled.

Exposure to and inhalation of concentrations of 2500 to 6500 ppm, as might result from accidents with liquid anhydrous ammonia, cause severe corneal irritation, dyspnea, bronchospasm, chest pain, and pulmonary edema, which may be fatal. Upper airway obstruction due to laryngeal/pharyngeal edema and desquamation of mucous membranes may occur early in the course and require endotracheal intubation or tracheostomy.[1-3] Secondary effects which may complicate the clinical picture include infection and renal failure. Case reports have documented chronic airway hyperreactivity and asthma, with associated obstructive pulmonary function changes following massive ammonia exposures.[3,4]

In a human experimental study that exposed 10 subjects to various vapor concentrations for 5 minutes, 134 ppm caused irritation of the eyes, nose, and throat in most subjects and one person complained of chest irritation; at 72 ppm, several subjects reported the same symptoms; at 50 ppm, two reported nasal dryness and, at 32 ppm, only one reported nasal dryness.[2] Surveys of workers have generally found that the maximum concentration not resulting in significant complaints is 20 to 25 ppm.[2]

Tolerance to usually irritating concentrations of ammonia may be acquired by adaptation, a phenomenon frequently observed among workers who become inured to the effects of exposure; no data are available on concentrations that are irritating to workers who are regularly exposed to ammonia and who presumably have a higher irritation threshold.

In animal studies, pigs exposed at 25, 50, and 100 ppm continuously for 6 days exhibited lethargy and a concentration-related depression of body weight gain.[5] Concentrations greater than 50 ppm altered the pulmonary vascular response to endotoxins.

Liquid anhydrous ammonia in contact with the eyes may cause serious injury to the cornea and deeper structures and sometimes blindness;

on the skin, it causes first- and second-degree burns, which are often severe and, if extensive, may be fatal. With skin and mucous membrane contact, burns are of three types: cryogenic (from the liquid ammonia), thermal (from the exothermic dissociation of ammonium hydroxide), and chemical (alkaline).[3]

The 1995 ACGIH threshold limit value–time-weighted average (TLV-TWA for ammonia is 25 ppm (17 mg/m³) with a short-term excursion limit of 35 ppm (24 mg/m³).

REFERENCES

1. Department of Labor: Exposure to ammonia, proposed standard. *Federal Register* 40:54684–54693, 1975
2. National Institute for Occupational Safety and Health, US Department of Health, Education and Welfare: *Criteria for a Recommended Standard . . . Occupational Exposure to Ammonia.* (NIOSH) Pub No 74-136. Washington, DC, US Government Printing Office, 1974
3. Arwood R, Hammond J, Ward G: Ammonia inhalation. *J Trauma* 25:444–447, 1985
4. Flury K, Dines D, Rodarto J, Rodgers R: Airway obstruction due to inhalation of ammonia. *Mayo Clinic Proc* 58:389–393, 1983
5. Gustin P, Urbain B, Prouvost JF, et al: Effects of atmospheric ammonia on pulmonary hemodynamics and vascular permeability in pigs: interaction with endotoxins. *Toxicol Appl Pharmacol* 125:17–26, 1994

Exposure. Inhalation

Toxicology. Ammonium chloride fume is a cause of occupational asthma.

Two cases of occupational asthma caused by exposure to soft, corrosive soldering fluxes have been reported.[1] The first case involved a 56-year-old man who developed chest tightness and wheeze 18 months after beginning work making tins. He was using a flux containing ammonium chloride and zinc chloride when a work-related deterioration in mean daily peak expiratory flow was noted, which improved when the man was away from work.

The second case involved an 18-year-old man who had cough, wheeze, chest tightness, and sneezing while working in a small firm that made and repaired car and truck radiators. Symptoms developed 1 year after he started work at the shop. He also was using a flux containing ammonium chloride and zinc chloride.

The 1995 threshold limit value–time-weighted average (TLV-TWA) is 10 mg/m³ with a short-term excursion level of 20 mg/m³.

REFERENCES

1. Weir DC, Robertson AS, Jones S, Burge PS: Occupational asthma due to soft corrosive soldering fluxes containing zinc chloride and ammonium chloride. *Thorax* 44:220–223, 1989

AMMONIUM CHLORIDE FUME
CAS: 12125-02-9

NH₄Cl

Synonyms: Ammonium muriate

Physical Form. Fume

Uses. In manufacture of dry-cell batteries; component of fluxes in zinc and tin plating; in dyeing and printing; fertilizer; hardener for formaldehyde-based adhesives

AMMONIUM PERFLUOROOCTANOATE
CAS:3825-26-1

C₈F₁₅O₂H₄N

Synonyms: Octanoic acid; pentadecafluoroammonium salt; ammonium pentadecafluorooctanoate; ammonium perfluorocaprylate; FC-143

Physical Form. White powder

Uses. Polymerization of fluorinated monomers; surfactant

Exposure.   Inhalation

Toxicology.   Ammonium perfluorooctanoate is a hepatotoxin in rats; there are no reports of adverse effects in humans.

In rats, ammonium perfluorooctanoate induced hepatomegaly, which was more pronounced in the male than in the female.[1-4] Male rats are thought to be more sensitive to the toxic effects of ammonium perfluorooctanoate because of their slower excretion rate. The rapid excretion by female rats is due to active renal tubular secretion, which is considered to be hormonally controlled by estadiol and testosterone levels. The hepatomegaly was hypertrophic rather than hyperplastic and involved proliferation of peroxisomes.

The LC_{50} for 4 hours in male rats was 980 mg/m^3; this exposure also caused an increase in liver size and corneal opacity, which diminished over time in survivors.[5] Exposure of male rats to 8 mg/m^3, 6 hr/day, for 10 of 12 days produced reversible changes in liver weight, reversible increases in serum enzyme activities, and liver necrosis. No ocular changes occurred. No observable effects occurred at 1 mg/m^3.

In a 90-day oral study in rhesus monkeys at levels ranging from 3 to 100 mg/kg/day the gastrointestinal tract and reticuloendothelial system were the sites of toxic effects at 30 and 100 mg/kg/day.[1] Histopathological effects were seen in the gastrointestinal tract, spleen, lymph nodes, and bone marrow. Sex-related differences were not evident in the monkeys as they are in rats. No tissue changes were observed at 3 or 10 mg/kg/day.

Dermal application of 500 mg for 24 hours to rabbit skin produced mild skin irritation.[6] The dermal LD_{50} was 4300 mg/kg. Dermal application of 200 mg/kg/day to rats for 10 to 12 days caused mild decreases in body weights and increases in serum enzyme activities, indicating hepatic effects. The effects were more obvious in males than females, and all findings resolved during a 42-day recovery period.

In a teratology study, rats were exposed from days 6 through 15 of gestation by inhalation 6 hr/day to levels of 0. 0.1, 10, and 25 mg/m^3 and by gavage at 100 mg/kg/day in corn oil.[7] Maternal deaths occurred in the groups given the highest level by each route, and overt toxicity in dams was evident at 10 mg/m^3. A teratogenic response was not demonstrated.

In workers exposed to airborne levels up to 7.6 mg/m^3, blood levels of organic fluoride were higher than background, but there were no adverse health effects attributable to the exposure.[8]

Rats fed diets containing 30 or 300 ppm ammonium perfluorooctanoate for 2 years had increased liver weights with occasional necrosis and an apparent dose-dependent increase in Leydig-cell adenomas, but there was no evidence of an increased incidence of hepatocellular carcinoma.[9]

The 1995 ACGIH threshold limit value–time-weighted average (TLV-TWA) for ammonium perfluorooctanoate is 0.01 mg/m^3 with an A3 animal carcinogen designation and a notation for skin absorption.

REFERENCES

1. Griffith FD, Long JE: Animal toxicity studies with ammonium perfluorooctanoate. *Am Ind Hyg Assoc J* 41:576–583, 1980
2. Kawashima Y, Uy-Yu N, Kozuka H: Sex-related difference in the inductions by perfluorooctanoic acid of peroxisomal beta-oxidation, microsomal 1-acylglycerophosphocholine acyltransferase and cytosolic long-chain acyl-CoA hydrolase in rat liver. *Biochem J* 261:595–600, 1989
3. Hanhijarvi H, Ylinen M, Kojo A, Kosma VM: Elimination and toxicity of perfluorooctanoic acid during subchronic administration in the Wistar rat. *Pharmacol Toxicol* 6166–6168, 1987
4. Pastoor TP, Lee KP, Perri MA, Gillies PJ: Biochemical and morphological studies of ammonium perfluorooctanoate-induced hepatomegaly and peroxisome proliferation. *Exp Mol Pathol* 47:98–109, 1987
5. Kennedy GL Jr, Hall GT, Barnes JR, Chen HC: Inhalation toxicity of ammonium perfluorooctanoate. *Food Chem Toxicol* 24:1325–1329, 1986
6. Kennedy GL Jr: Dermal toxicity of ammonium perfluorooctanoate. *Toxicol Appl Pharmacol* 81:348–355, 1985
7. Staples RE, Burgess BA, Kerns WD: The embryo-fetal toxicity and teratogenic potential of

ammonium perfluorooctanoate (APFO) in the rat. *Fund Appl Toxicol* 4:429–440, 1984
8. Ubel FA, Sorenson SD, Roach DE: Health status of plant workers exposed to fluorochemicals—a preliminary report. *Am Ind Hyg Assoc J* 41:584–589, 1980
9. Cook JC, Murray SM, Frame SR, et al: Induction of Leydig Cell adenomas by ammonium perfluorooctanoate: a possible endocrine-related mechanism. *Toxicol Appl Pharm* 113: 209–217, 1992

REFERENCES

1. Ambrose AM: Studies on the physiological effects of sulfamic acid and ammonium sulfamate. *J Ind Hyg Toxicol* 25:26–28, 1943
2. Lehman AJ: Chemicals in foods: a report to the association of food and drug officials on current developments; part II. Pesticides. *Q Bull Assoc Food Drug Off US* 15:122–133, 1951

AMMONIUM SULFAMATE
CAS: 7773-06-0

$NH_4SO_3NH_2$

Synonyms: Ammate; Amicide

Physical Form. Hygroscopic crystals

Uses. In manufacture of weed-killing compounds and fire-retardant compositions

Exposure. Inhalation

Toxicology. Ammonium sulfamate is of low toxicity; there are no reports of systemic effects in humans.

Repeated application of a 4% solution to the anterior surface of one arm of each of five human subjects for 5 days caused no skin irritation.[1]

The oral LD_{50} values were 3900 mg/kg for rats and 5760 mg/kg for mice.[2]

In rats, the intraperitoneal injection of 0.8 g/kg caused the death of 6 of 10 animals; effects were stimulation of respiration and then prostration.[1]

Continuous feeding of 1% (10,000 ppm) in the diet of rats for 105 days caused no effect; 2% in the diet caused growth inhibition, but no histological effects were observed.[2]

The 1995 ACGIH threshold limit value–time-weighted average (TLV-TWA) for ammonium sulfamate is 10 mg/m³.

n-AMYL ACETATE
CAS: 628-63-7

$CH_3COOC_5H_{11}$

Synonyms: Amyl acetic ether; pentyl acetate

Physical Form. Liquid

Uses. As a solvent in laquers, paints, leather polishes, inks, adhesives, degreasers, and cosmetics

Exposure. Inhalation; minor skin absorption

Toxicology. *n*-Amyl acetate is an irritant of mucous membranes; at high concentrations, it causes narcosis in animals, and it is expected that severe exposure would produce the same effect in humans.

Several grades of technical amyl acetate are known; isoamyl acetate is the major component of some grades, whereas *n*-amyl acetate predominates in others.[1]

In humans, exposure to amyl acetate vapor for 3 to 5 minutes at 200 ppm caused mild eye and and nose irritation and severe throat irritation; at 100 ppm, slight throat discomfort has been reported.[2]

Inhalation of excessive concentrations may also cause headache, fatigue, excessive salivation, "oppression in the chest, and occasional vague nervousness."[3]

Air saturated with 5200 ppm of technical amyl acetate (*n*-amyl acetate the principal component) was fatal to 6 of 6 rats in 8 hours but caused no deaths in 4 hours.[4]

In standardized testing on rabbit eyes, amyl acetate was graded as only slightly injurious.[5] No evidence of delayed-contact hypersensitivity due to 20% amyl acetate was observed in repeat-insult skin-patch tests of 211 human subjects.[3]

Amyl acetates may be recognized at concentrations of 7 ppm by the fruitlike odor characteristic of esters; the mean olfactory detection threshold is 0.2 ppm.[1,3]

The 1995 ACGIH threshold limit value–time-weighted average (TLV-TWA) for *n*-amyl acetate is 100 ppm (532 mg/m^3).

REFERENCES

1. Hygienic Guide Series: Amyl acetate. *Am Ind Hyg Assoc J* 26:199–202, 1965
2. Nelson KW, Ege JF Jr, Ross M, et al: Sensory response to certain industrial solvent vapors. *J Ind Hyg Toxicol* 25:282–285, 1943
3. Final report on the safety assessment of amyl acetate and isoamyl acetate. *J Am College of Toxicol* 7(16):705–719, 1988
4. Smyth JF Jr, Carpenter CP, West CS, et al; Range finding toxicity data: List VI. *Am Ind Hyg Assoc J* 23:95–107, 1962
5. Grant WM: *Toxicology of the Eye*, 3rd ed, pp 97–98. Springfield, IL: Charles C Thomas, 1986

---

*sec-***AMYL ACETATE**
CAS: 626-38-0

$C_7H_{14}O_2$

Synonyms: α-Methyl butyl acetate; banana oil

Physical Form. Liquid

Uses. Manufacture of lacquers, artificial leather, photographic film, artificial glass, celluloid, artificial silk, and furniture polish

Exposure. Inhalation

Toxicology. *sec*-Amyl acetate is an irritant of the eyes, mucous membranes, and skin; high concentrations cause narcosis in animals, and severe exposure is expected to produce the same effect in humans.

In humans, exposure to 5000 to 10,000 ppm for short periods of time caused irritation of the eyes and nasal passages.[1] Exposure to 1000 ppm for 1 hour is expected to produce serious toxic effects.

In guinea pigs, 2000 ppm for 13.5 hours produced no abnormal signs except irritation of the eyes and nose. At 5000 ppm, there was lacrimation after 5 minutes; incoordination occurred within 90 minutes; at 9 hours there was narcosis, from which animals recovered. A concentration of 10,000 ppm was fatal after 5 hours.[1]

The *sec*-amyl acetates are more volatile than the primary isomers and appear to be somewhat less toxic. The odor threshold for *sec*-amyl acetate has been determined as 2 ppb in air.[2]

The 1995 ACGIH threshold limit value–time-weighted average (TLV-TWA) for *sec*-amyl acetate is 125 ppm (665 mg/m^3).

REFERENCES

1. von Oettingen WF: The aliphatic acids and their esters: toxicity and potential dangers. *AMA Arch Ind Health* 21:28–64, 1960
2. Stahl WH (ed): *Compilation of Odor and Taste Threshold Values Data*. ASTM Data Series DS48, Philadelphia, Am Soc Testing Materials, 1973

ANILINE
CAS: 62-53-3

$C_6H_5NH_2$

Synonyms: Aminobenzene; phenylamine

Physical Form. Colorless to light yellow liquid that tends to darken on exposure to air and light.

Uses. Intermediate in chemical synthesis; in manufacture of synthetic dyestuffs

Exposure. Inhalation; skin absorption

Toxicology. Aniline absorption causes anoxia due to the formation of methemoglobin.

In early studies, human exposure to vapor concentrations of 7 to 53 ppm was said to cause slight symptoms, whereas concentrations in excess of 100 to 160 ppm were associated with serious disturbances if inhaled for 1 hour.[1] Rapid absorption through the intact skin is frequently the main route of entry, either from direct contact with the liquid or the vapor.

The formation of methemoglobinemia is often insidious; following skin absorption, the onset of symptoms may be delayed for up to 4 hours.[2] Headache is commonly the first symptom and may become quite intense as the severity of methemoglobinemia progresses. Cyanosis occurs when the methemoglobin concentration is 15% or more. Blueness develops first in the lips, the nose, and the earlobes and is usually recognized by fellow workers. The individual usually feels well, has no complaints, and insists that nothing is wrong until the methemoglobin concentration approaches approximately 40%. At methomoglobin concentrations of over 40%, there usually is weakness and dizziness; up to 70% concentration, there may be ataxia dyspnea on mild exertion and tachycardia. Coma may ensure with methemoglobin levels about 70%, and the lethal level is estimated to be 85% to 90%.[3] In general, higher ambient temperatures increase susceptibility to cyanosis from exposure to methemoglobin-forming agents.[4]

The development of intravascular hemolysis and anemia due to aniline-induced methemoglobinemia has been postulated, but neither is observed often in industrial practice, despite careful and prolonged study of numerous cases. Occasional deaths from asphyxiation caused by severe aniline intoxication are said to occur. The existence of chronic aniline poisoning is controversial, but some investigators have suggested that continuous exposure to small doses of aniline may produce anemia, loss of energy, digestive disturbance, and headache.[5]

The mean lethal dose by ingestion in humans has been estimated to be between 15 and 30 g, although death has been reported after as little as 1 g.[6] A significant elevation in methemoglobin levels was reported in adult volunteers given 25 mg orally.[6]

Peak methemoglobin levels may occur some hours after exposure, and it has been postulated that metabolic transformation of aniline to phenylhydroxylamine is necessary for the production of methemoglobin.[6]

Liquid aniline is mildly irritating to the eyes and may cause corneal damage.[7]

No evidence of embryolethal or teratogenic effect was observed in the offspring of rats dosed with aniline hydrochloride during gestation.[8] Signs of maternal toxicity included methemoglobinemia, increased relative spleen weight, decreased red blood cell count, and hematological changes indicative of increased hematopoietic activity. Transient signs of toxicity were observed postnatally in the offspring through day 30.

Aniline hydrochloride was not carcinogenic to mice when administered orally.[9] In one experiment, it produced fibrosarcomas, sarcomas, and hemangiosarcomas of the spleen and body cavities in rats fed diets containing 3000 mg/kg or 6000 mg/kg for 103 weeks.

The high risk of bladder cancer observed originally in workers in the aniline dye industry has been attributed to exposure to chemicals other than aniline.[9] Studies showing significant increase in bladder cancers, such as the study of 1749 rubber antioxidant workers that found 13 cases of bladder cancer versus 3.61 expected, have involved significant exposure to chemicals such as *o*-toluidine or contaminants that are considered to be more potent carcinogens based on animal and human studies.[10,11] Epidemiological studies of workers exposed to aniline but to no other known bladder carcinogen have shown little evidence of increased risk; one study showed one death from bladder cancer versus 0.83 expected in a population of 1223 men producing or using aniline.[9] Nonetheless, NIOSH has released an alert for aniline, recommending that exposures be reduced to the lowest possible levels.[12]

The IARC has determined that evidence

for carcinogenicity is limited in animals and inadequate in humans.[9]

The 1995 ACGIH threshold limit value–time-weighted average (TLV-TWA) for aniline is 2 ppm (7.6 mg/m³).

Diagnosis. *Signs and Symptoms:* Headache; signs of anoxia, including cyanosis of lips, nose, and earlobes; eye irritation; anemia; and hematuria.

Differential Diagnosis: Other causes of cyanosis must be differentiated from methemoglobinemia due to aniline exposure. These include hypoxia due to lung disease, hypoventilation, and decreased cardiac output. Lung disease may be suspected from results of pulmonary function tests and arterial blood gas analysis. The arterial P_{O_2} may be normal in methemoglobinemia but tends to be decreased in cyanosis due to lung disease. Hypoventilation will cause elevation of arterial P_{CO_2}, which is not seen in aniline exposure. Decreased cardiac output states will cause cyanosis only when accompanied by arterial hypotension. If blood withdrawn from the vein shows the characteristic chocolate-brown coloration, the diagnosis of an abnormal pigment is almost certain, especially if the color remains after shaking the blood in air.[13]

Special Tests: These include examination of urine for blood; determination of methemoglobin concentration in the blood when aniline intoxication is suspected, and at regular intervals until the methemoglobin has been fully reduced to normal hemoglobin.[14] Methemoglobin may be differentiated from sulfhemoglobin by the addition of a few drops of 10% potassium cyanide, which results in the rapid production of bright red cyanomethemoglobin but has no effect on the color of sulfhemoglobin. Spectrophotometry is required for the precise identification of the pigment and its quantitation. Normal acid methemoglobin has a characteristic absorption spectrum with peaks at 502 and 632 nm, which disappear with the addition of cyanide, whereas sulfhemoglobin has a peak at 620 nm, which does not disappear with cyanide.[13]

Treatment. All aniline on the body must be removed. Immediately remove all clothing and wash the entire body from head to foot with soap and water. Pay special attention to the hair and scalp, fingernails and toenails, nostrils, and ear canals. Administer oxygen to alleviate the headache and general sense of weakness; confine to bed. Determine the methemoglobin concentration in the blood, and repeat every 3 to 6 hours for 18 to 24 hours. Repeat skin cleansing if the methemoglobin concentration appears to rise after 3 to 4 hours. In general, patients will return to normal within 24 hours, provided all sources of further absorption are completely eliminated.

The only justifiable use of methylene blue would be in cases of coma or stupor, usually at methemoglobin levels over 60%. In those patients who require therapy, methylene blue may be given intravenously, 1 to 2 mg/kg, over a 5-minute period as a 1% solution; if cyanosis has not disappeared within an hour, a second dose of 2 mg/kg should be administered.[13,14] The total dose should not exceed 7 mg/kg because methylene blue may cause toxic effects such as dyspnea, precordial pain, restlessness, apprehension, red cell hemolysis, and changes in the electrocardiogram (reduction in the height or even reversal of the T wave, frequently with lowering of the R wave).[14]

REFERENCES

1. Henderson Y, Haggard HW: *Noxious Gases*, 2nd ed. Reinhold, New York, 1943
2. Benya TJ, Cornish HH: Aromatic nitro and amino compounds. In Clayton GD, Clayton FE (eds): *Patty's Industrial Hygiene and Toxicology*, 4th ed, Vol IIB, *Toxicology*, pp 949–953, 982–984. New York, Wiley-Interscience, 1994
3. Chemical Safety Data Sheet, SD-21, Nitrobenzene, pp 56, 1214. Washington, DC, MCA, Inc, 1967
4. Linch AL: Biological monitoring for industrial exposure to cyanogenic aromatic nitro and amino compounds. *Am Ind Hyg Assoc* 35:426–432, 1974
5. Hazard Data Sheet: Sheet Number 78, Aniline. *The Safety Practitioner*, pp 44-45, June 3, 1986
6. Kearney TE et al: Chemically induced methemoglobinemia from aniline poisoning. *West J Med* 140:282–286, 1984
7. Chemical Safety Data Sheet SD-17, Aniline,

pp 45, 1214. Washington, DC, MCA, Inc, 1967

8. Price CJ et al: Teratologic and postnatal evaluation of aniline hydrochloride in the Fischer 344 rat. *Toxicol Appl Pharmacol* 77:465–489, 1985

9. *IARC Monographs on the Evaluation of the Carcinogenic Risk of Chemicals to Man*. Suppl 4, pp 49–50. Lyon, International Agency for Research on Cancer, 1982

10. Ward E, Carpenter A, Markowitz S, et al: Excess number of bladder cancers in workers exposed to ortho-toluidine and aniline. *J Natl Cancer Inst* 83:501–506, 1991

11. Sellers C, Markowitz S: Reevaluating the carcinogenicity of ortho-toluidine: a new conclusion and its implications. *Reg Toxicol Pharmacol* 16:301–317, 1992

12. National Institute for Occupational Safety and Health (NIOSH): Preventing bladder cancer from exposure to *o*-toluidine and aniline. *Am Ind Hyg Assoc J* 52:A260–A262, 1991

13. Rieder RF: Methemoglobinemia and sulfhemoglobinemia. In Wyngaarden JB, Smith LH (eds.): *Cecil Textbook of Medicine*, 16th ed, p 894. Philadelphia, WB Saunders, 1982

14. Mangelsdorff AF: Treatment of methemoglobinemia. *AMA Arch Ind Health* 14:148–153, 1956

ANISIDINE

CAS: 29191-52-4

$NH_2C_6H_4OCH_3$

Synonyms: Methoxyaniline; aminoanisole

Physical Form. *o*-Anisidine is a yellowish liquid that darkens on exposure to air; *p*-anisidine is a white solid

Uses. In the preparation of azo dyes

Exposure. Inhalation; skin absorption

Toxicology. Anisidine, *o*- and *p*-isomers, causes anoxia as a result of the formation of methemoglobin. *o*-Anisidine was carcinogenic in experimental animals.

Workers exposed to 0.4 ppm for 3.5 hours for 6 months did not develop anemia, but there were some cases of headache and vertigo, which may have been related to the increased levels of methemoglobin and sulfhemoglobin; erythrocytic inclusions (Heinz bodies) were observed, and absorption through the skin may have been a contributing factor.[1] Anisidine is a mild skin sensitizer, and local contact may cause dermatitis.

Mice exposed 2 hr/day at 2 to 6 ppm for a year developed anemia and reticulocytosis.

The oral LD_{50} of *o*-anisidine is reported to be 2000 mg/kg in rats, 1400 mg/kg in mice, and 870 mg/kg in rabbits. The oral LD_{50} of *p*-anisidine is 1400 mg/kg in rats, 1300 mg/kg in mice, and 2900 mg/kg in rabbits.[2] For both isomers, subacute effects included hematological changes, anemia, and nephrotoxicity.

A significant increase in transitional-cell carcinomas of the urinary bladder was found in mice and rats fed diets containing 5000 mg/kg *o*-anisidine hydrochloride for 103 weeks.[3]

The IARC has determined that there is sufficient evidence for the carcinogenicity of *o*-anisidine hydrochloride in experimental animals and, in the absence of human data, it should be regarded as if it presented a carcinogenic risk to humans.[2] Available data were inadequate to evaluate the carcinogenicity of *p*-anisidine hydrochloride.[2]

The 1995 ACGIH threshold limit value–time-weighted average (TLV-TWA) for the *o*- and *p*-isomers of anisidine is 0.1 ppm (0.5 mg/m³).

Note. For a description of diagnostic signs, differential diagnosis, and medical control, including clinical laboratory tests, as well as specific treatment of overexposure to methemoglobin-forming agents, see "Aniline."

REFERENCES

1. Pacseri I, Magos L, Batskor IA: Threshold and toxic limits of some amino and nitro compounds. *AMA Arch Ind Health* 18:1–8, 1958

2. *IARC Monographs on the Evaluation of the Carcinogenic Risk of Chemicals to Humans*, Vol 27, pp 63–77. Lyon, International Agency for Research on Cancer, 1982

3. National Cancer Institute: *Bioassay of o-Anisidine Hydrochloride for Possible Carcinogenicity, TR-89.* DHEW (NIH) Pub No 78-1339. Washington, DC, US Government Printing Office, 1978

ANTIMONY (AND COMPOUNDS)
CAS:7440-36-0

Sb

Compounds: Antimony trioxide; antimony trisulfide; antimony trichloride; antimony pentoxide; antimony pentasulfide; antimony pentachloride

Physical Form. Silvery-white soft metal

Uses. Constituent of alloy with other metals (tin, lead, copper); sulfides used in compounding of rubber and manufacture of pyrotechnics; trioxide used as a fire retardant in plastics, rubbers, textiles, and paints; chlorides used as coloring agents and as catalysts; fluorides used in organic synthesis and pottery manufacture

Exposure. Inhalation

Toxicology. Antimony is an irritant of the mucous membranes, eyes, and skin; heavy exposure to antimony trioxide and pentoxide is associated with pulmonary injury; antimony trisulfide is considered cardiotoxic. Antimony trioxide is carcinogenic in experimental animals.

Antimony poisoning was reported in 69 of 78 smelter workers during a 5-month period when antimony concentrations of breathing-zone samples in the smelter building averaged 10.07 to 11.81 mg/m^3 of air (range 0.92 to 70.7 mg/m^3); dermatitis and rhinitis were reported most frequently, but other symptoms included irritation of eyes, sore throat, headache, pain or tightness in chest, shortness of breath, metallic taste, nausea, vomiting, diarrhea, weight loss, and dysosmia.[1]

Symptomless radiographic lung changes resembling the simple pneumoconiosis of coal workers were found in 44 of 262 men exposed to antimony oxide concentrations of 0.5 to 37 mg/m^3.[2,3] In another roentgenographic study of 51 workers exposed 9 or more years to antimony oxides, numerous small opacities were densely distributed in the middle and lower lung fields.[4] There were no characteristic pulmonary function abnormalities, but chronic cough was a common symptom. Brief exposures to antimony trichloride, approximately 73 mg Sb/m^3, caused gastrointestinal symptoms, as well as irritation of the skin and respiratory tract; urinary antimony ranged up to 5 mg/l.[5]

Six sudden deaths and two deaths due to chronic heart disease occurred among 125 abrasive-wheel workers exposed to antimony trisulfide for 8 to 24 months.[6] At air concentrations averaging over 3.0 mg/m^3, 37 of 75 workers had electrocardiogram changes, and 38 had abnormalities in blood pressure. The lack of electrocardiographic changes in the oxide exposures would seem to indicate a special effect of the sulfide.

Contact of antimony compounds with the skin causes papules and pustules around sweat and sebaceous glands.[2]

Female rats exposed to 4.2 and 3.2 mg/m^3 antimony trioxide 6 hr/day, 5 days/week, for 1 year had lung tumors after an additional year of observation.[7] Similar findings were reported in another study involving heavier exposures; 27% of female rats exposed to 45 mg/m^3 antimony trioxide for 1 year and 25% of the females exposed to 38 mg/m^3 antimony ore (mainly antimony trisulfide) developed lung neoplasms.[8] No lung tumors were seen in the male rats exposed to either compound or in controls. Based on these studies, the IARC has determined that there is sufficient evidence for the carcinogenicity of antimony trioxide in animals and limited evidence for the carcinogenicity of antimony trisulfide.[9]

A recent chronic inhalation study in rats using lower exposure levels found no evidence of carcinogenicity.[10] A dose-related increase in cataracts and microscopic changes in the lungs were the primary effects noted from 12 months of exposure at 0.06, 0.51, or 4.5 mg/m^3, followed by a 12-month recovery period.

In a report from Russia, an increase in the number of spontaneous abortions was reported in women exposed to antimony in the workplace.[11,12] Exposure levels were not available. In animal studies, no effects were observed in the offspring of rats given low levels of antimony trichloride in the drinking water.

Both positive and negative results have been reported in genotoxic assays of antimony and compounds.[12]

The 1995 ACGIH threshold limit value–time-weighted average (TLV-TWA) for antimony and compounds is 0.5 mg/m³ as Sb; antimony trioxide production is given an A2 suspected human carcinogen designation with no assigned TLV.

REFERENCES

1. Renes LE: Antimony poisoning in industry. *AMA Arch Ind Hyg Occup Med* 7:99–108, 1953
2. McCallum RI: The work of an occupational hygiene service in environmental control. *Ann Occup Hyg* 6:55–64, 1963
3. McCallum RI: Detection of antimony in process workers lungs by x-radiation. *Trans Soc Occup Med* 17:134–138, 1967
4. Potkonjak V, Pavlovich M: Antimoniosis: A particular form of pneumoconiosis I. Etiology, clinical and x-ray findings. *Int Arch Occup Environ Health* 51:199–207, 1983
5. Taylor PJ: Acute intoxication from antimony trichloride. *Br J Ind Med* 23:318–321, 1966
6. Brieger H, Semisch CW, Stasney J, Piatnek DA: Industrial antimony poisoning. *Ind Med Surg* 23:521–523, 1954
7. Department of Labor: Antimony metal; antimony trioxide; and antimony sulfide response to the interagency testing committee. *Federal Register* 48:717–724, 1983
8. Groth DH, Stettler LE, Burg JR, et al: Carcinogenic effects of antimony trioxide and antimony ore concentrate in rats. *J Tox Env Health* 18:607–626, 1986
9. *IARC Monographs on the Evaluation of Carcinogenic Risks to Humans.* Vol 47, pp 291–304, *Some organic solvents, resin monomers and related compounds, pigments and occupational exposures in paint manufacture and painting.* Lyon, International Agency for Research on Cancer, 1989
10. Newton PE, Bolte HF, Daly IW, et al: Subchronic and chronic inhalation toxicity of antimony trioxide in the rat. *Fund Appl Toxicol* 22:561–576, 1994
11. National Institute for Occupational Safety and Health, US Department of Health, Education and Welfare: *Criteria for a Recommended Standard . . . Occupational Exposure to Antimony.* DHEW (NIOSH) 78-216. Washington, DC, US Government Printing Office, 1978
12. Agency for Toxic Substances and Disease Registry (ATSDR): *Toxicological Profile for Antimony.* US Department of Health and Human Services, Public Health Service, TP-91/02, pp 135, 1992

ANTU (ALPHA-NAPHTHYL-THIOUREA)
CAS: 86-88-4

$C_{11}H_{10}N_2S$

Synonyms: α-Naphthylthiourea; α-naphthyl-thiocarbamide

Physical Form.   Blue to gray powder

Uses.   Rodenticide

Exposure.   Inhalation; ingestion

Toxicology.   ANTU dust causes pulmonary edema and pleural effusion in animals.

ANTU is probably not toxic to man except in large amounts; the lethal dose by ingestion is estimated to be approximately 4 g/kg.[1] In an instance of human intoxication by ANTU, 80 g of a rat poison containing 30% ANTU was ingested, along with a considerable amount of ethanol; signs attributable to ANTU were prompt vomiting, dyspnea, cyanosis, and coarse pulmonary rales; no pleural effusion occurred, and the pulmonary signs gradually cleared.[1]

Oral administration to rats of 35 mg/kg was fatal to 60% of the animals; effects were labored respiration and muscular weakness; autopsy revealed pleural and pericardial effusion as well as mild liver damage.[2] Tolerance to the acute toxicity of ANTU has been observed following repeated administrations; intraperitoneal injection of 2.5 mg/kg produced moderate pulmonary edema and large pleural effusions, but two additional 2.5-mg/kg doses at 2-day intervals caused lesser degrees of edema and minimal pleural fluid.[3] However, daily doses of 200 mg/kg (20% of the median lethal dose) in rabbits were cumulative, causing death in 5 to 6 days but without pleural effusions.[2]

Recent studies on the mechanism of thiourea toxicity have shown that thioureas have a high degree of specificity for pulmonary endothelial cells and that the thioureas require metabolic activation before toxic effects are manifested.[4] Reduced glutathione levels have been associated with increased toxicity, but there is no evidence to suggest that the appearance of edema coincides with a decrease in glutathione. Furthermore, the induction of tolerance or resistance is not correlated with an increase in glutathione levels in rats.[4]

The 1995 threshold limit value–time-weighted average (TLV-TWA) for ANTU is 0.3 mg/m^3.

REFERENCES

1. Gosselin RE, Smith RP, Hodge HC: *Clinical Toxicology of Commercial Products*, 5th ed, Section III, pp 40–42. Baltimore, Williams and Wilkins, 1984
2. McClosky WT, Smith MI: Studies on the pharmacologic action and the pathology of alpha-naphthylthiourea (ANTU). I. Pharmacology. *Public Health Rep* 60:1101–1113, 1945
3. Sobonya RE, Kleinerman J: Recurrent pulmonary edema induced by alpha-naphthyl thiourea. *Am Rev Respir Dis* 108:926–932, 1973
4. Scott AM, Powell GM, Upshall DG, et al: Pulmonary toxicity of thioureas in the rat. *Environ Health Perspect* 85:43–50, 1990

ARSENIC AND COMPOUNDS
CAS: 7440-38-2

As

Synonyms and Compounds: Gray arsenic; metallic arsenic; arsenic trichloride; arsenic trioxide; arsenic salts

Physical Form. Metallic arsenic is a steel-gray, brittle metal; arsenic trichloride is an oily liquid; arsenic trioxide is a crystalline solid

Uses/Sources. In metallurgy for hardening copper, lead, alloys; in pigment production; in the manufacture of certain types of glass; insecticides, fungicides, rodent poison; by-product in the smelting of copper ores; dopant material in semiconductor manufacture

Exposure. Inhalation; skin absorption; ingestion

Toxicology. Arsenic compounds are irritants of the skin, mucous membranes, and eyes; arsenical dermatoses and epidermal carcinoma are reported risks of exposure to arsenic compounds, as are other forms of cancer.[1]

Acute arsenic poisoning is rare in the occupational setting and results primarily from ingestion of contaminated food and drink.[2] Initial symptoms include burning lips, constriction of the throat, and dysphagia, followed by excruciating abdominal pain, severe nausea, projectile vomiting, and profuse diarrhea.[3] Other toxic effects on the liver, blood-forming organs, central and peripheral nervous systems, and cardiovascular system may appear.[4] Convulsions, coma, and death follow within 24 hours in severe cases.[3] Levels of exposure associated with acute arsenic toxicity vary with the valency form of the element; trivalent arsenic compounds are the most toxic, presumably because of their avid binding to sulfhydryl groups. For arsenic trioxide, the reported estimated lethal dose ranges from 70 to 300 mg.[3,4]

Acute inhalation exposures have resulted in irritation of the upper respiratory tract, even leading to nasal perforations.[4] Occupational exposure to arsenic compounds results in hyperpigmentation of the skin and hyperkeratoses of palmar and plantar surfaces, as well as dermatitis of both primary irritation and sensitization types.[1] Impairment of peripheral circulation and Raynaud's phenomenon have been reported as related to long-term exposure.[5]

Chronic arsenic intoxication by ingestion is characterized by weakness, anorexia, gastrointestinal disturbances, impairment of cognitive function, peripheral neuropathy, and skin disorders. Liver damage has been observed in animals after both ingestion and inhalation of arsenic compounds, but this has not been observed with occupational exposure.[1]

Arsenic trichloride is a vesicant and can cause severe damage to the respiratory system upon inhalation; it is rapidly absorbed through the skin, and a fatal case following a spill on the skin has been reported.[6] The vapor of arsenic trichloride is highly irritating to the eyes. Some organic arsenicals, such as arsanilates, have a selective effect on the optic nerve and can cause blindness.

Teratogenic effects, including exencephaly, skeletal defects, and genitourinary system defects, of arsenic compounds administered intravenously or intraperitoneally at high doses have been demonstrated in hamsters, rats, and mice.[4] Only minimal fetal effects have been observed in studies of pregnant rats or mice exposed to lower levels via drinking water.[4] In general, the developing fetus is not considered to be especially susceptible to the effects of inorganic arsenic except at doses that are also toxic to the pregnant female.

In a large number of studies, exposure to inorganic arsenic compounds in drugs, food, and water, as well as in an occupational setting, have been causally associated with the development of cancer, primarily of the skin and lungs.[1-4] An excess mortality in respiratory cancer has been found among smelter workers and workers engaged in the production and use of arsenical pesticides. It should be noted, however, that, in a number of these studies, levels of exposure are uncertain and there is simultaneous exposure to other agents. In a follow-up of 8045 smelter workers, those with the highest estimated exposure and the longest follow-up had a ninefold increase in respiratory cancer mortality.[7]

Another large retrospective cohort study followed 3916 smelter workers and reported an overall standardized mortality ratio of 372.[8] Lung cancer mortality was related to intensity of exposure but not to duration. Histologic types of lung carcinomas were similar to those seen in smokers.

Information on the association of arsenic with skin cancer has involved primarily nonoccupational populations exposed to contaminated drinking water.[4] In more recent reports, ingestion of arsenic has also been asociated with lung, liver, bladder, and kidney cancers. Dose-response data for these cancers are available from epidemiological studies of a Taiwanese population exposed for 45 years to high levels of arsenic in the drinking water and involving more than 7000 cases of arsenical disease. For water-arsenic concentrations of 170, 470, and 800 $\mu g/l$, the corresponding mortality rate ratios for bladder cancer were 5.1, 12.1, and 28.7 for males and 11.9, 25.1, and 65.4 for females and for kidney cancer, 4.9, 11.9, and 19.6 for men and 4.0, 13.9, and 37.0 for women respectively.[9] An epidemiological study of lung cancer has shown a linear correlation between the standard mortality ratio of lung cancer and the concentration of arsenic found in the urine.

Chronic ingestion of trivalent arsenic in medicinal preparations was also associated with an increased incidence of hyperkeratosis and skin cancer.[4]

Because there is no convincing evidence of the carcinogenicity of arsenic compounds in animals, it has been suggested that the compounds are not direct carcinogens but act in some other way. One theory holds that arsenic acts as an indirect, gene-inducing carcinogen that causes cancer in man through activation of an oncogenic virus.[10] Despite the absence of suitable animal models, the IARC has determined that there is sufficient evidence for carcinogenicity to humans.[11]

The 1995 ACGIH threshold limit value–time-weighted average for arsenic, elemental and inorganic compounds (except arsine), as As is 0.01 mg/m³ with an A1 confirmed human carcinogen designation.

Diagnosis. *Signs and Symptoms:* Conjunctivitis, visual disturbances; ulceration and perforation of nasal septum; pharyngitis, pulmonary irritation; peripheral neuropathy; hyperpigmentation of skin, palmar and plantar hyperkeratoses, dermatitis, skin cancer. Arsenic may cause cancer of the lung, larynx, lymphoid system, or viscera.

Special Tests: Urinary levels of arsenic above 0.7 to 1.0 mg/l in exposed individuals may be indicative of harmful exposure, but dietary factors must be ruled out before significance is attached to any increase in the arsenic content of urine. The biologic half-life of arsenic in urine in subjects with normal renal function is 1 to 2 days.[12] Seafood, especially shellfish, is a rich source of organic arsenic compounds that are essentially nontoxic but that affect the total urinary arsenic determination. Levels of 1.35 mg/l of urine have occurred after a meal of lobster tails. In individuals without known exposure to arsenic who have not ingested a seafood meal for 2 days before sampling, the urinary total arsenic is usually below 100 μg/g creatinine. Inorganic arsenic is excreted in the urine unchanged or as monomethylarsenic or dimethylarsenic (cacodylic) acid; these compounds can be measured by hydride-generation atomic absorption spectrophotometry, yielding an index of exposure to inorganic arsenic. Levels of urinary arsenic measured by this method are generally below 20 μg/gm creatinine in unexposed subjects. One estimate is that a time-weighted average exposure of 50 μg/m³ to inorganic arsenic over several days would yield an average urinary arsenic level of 220 μg/gm creatinine by the latter method.

The Biological Exposure Indice Committee has recommended monitoring of the sum of inorganic arsenic and its methylated metabolites in urine, collected at the end of the workweek, as an indicator of exposure to inorganic arsenic compounds. A value of 50 mg/g of creatinine is recommended as a BEI.[13]

A determination of arsenic in hair and nails may be useful although its value has been questioned in industrial exposures because of the difficulty in removing all external contamination.

Treatment. Severe acute arsenic poisoning from occupational exposure is unlikely; if it should occur, administer BAL (dimercaprol) 10% in oil intramuscularly (gluteal), 3 mg/kg for each injection; first and second days, one injection every 4 hours day and night; third day, one injection every 6 hours; fourth to fourteenth day, one injection twice a day until recovery is complete.[14,15] Side effects of BAL (nausea, vomiting, hypertension, tachycardia, headache, anxiety, diaphoresis, muscle cramps, seizures, coma, and urticaria) usually respond to supportive measures and decreased dosing and are typically transient.[16] With severe acute intoxication, *d*-penicillamine may also be given orally in a dose of 25 mg/kg (to maximum 500 mg) every 6 hours for the first 10 days, in addition to BAL. BAL may also be utilized to treat chronic arsenic intoxication.[17] Meso-2,3-dimercaptosuccinic acid (DMSA) is an effective oral chelating agent. This is an "orphan drug" for which a New Drug Application has been filed with the USFDA. DMSA has few reported side effects, although experience with the drug has been limited.[18,19]

Caution: This section should not be used as a protocol for treatment of exposures to arsenic when specific exposures have occurred. Treatment protocols are constantly changing, and poison control centers and medical toxicologists should be consulted for medical advice.

REFERENCES

1. National Institute for Occupational Safety and Health, US Department of Health, Education and Welfare: *Criteria for a Recommended Standard . . . Occupational Exposure to Inorganic Arsenic, New Criteria—1975.* DHEW (NIOSH) Pub 75-149, pp 14–71. Washington, DC, US Government Printing Office, 1975
2. Landrigan PJ: Arsenic—state of the art. *Am J Ind Med* 2:5–14, 1981
3. Winship KA: Toxicity of inorganic arsenic salts. *Adv Drug React Ac Pois Rev* 3:129–160, 1984

4. Health Assessment Document for Inorganic Arsenic. Final Report. Research Triangle Park, NC, US Environmental Protection Agency, March 1984

5. Lagerkrist BA, Linderholm H, Nordberg GF: Arsenic and Raynaud's phenomenon. *Int Arch Occup Environ Health* 60:361–364, 1988

6. Hygienic Guide Series: Arsenic and its compounds (except arsine). *Am Ind Hyg Assoc J* 25:610–613, 1964

7. Lee-Feldstein A: Cumulative exposure to arsenic and its relationship to respiratory cancer among copper smelter employees. *J Occup Med* 28:296–302, 1986

8. Jarup L, Pershagen G, Wall S: Cummulative arsenic exposure and lung cancer in smelter workers: a dose-response study. *Am J Ind Med* 15:31–41, 1989

9. Smith A, Hopenhayn-Rich C, Bates MN, et al: Cancer risks from arsenic in drinking water. *Environ Health Persp* 97:259–267, 1992

10. Stohrer G: Arsenic: opportunity for risk assessment. *Arch Toxicol* 65:525–531, 1991

11. *IARC Monographs on the Evaluation of the Carcinogenic Risk of Chemicals to Humans.* Chemicals, industrial processes and industries associated with cancer in humans, Suppl 4, pp 50–51. Lyon, International Agency for Research on Cancer, 1982

12. Lauwerys RR: *Industrial Chemical Exposure: Guidelines for Biological Monitoring*, pp 12–15. Davis, California, Biomedical Publications, 1983

13. BEI Committee: notice of intended change—arsenic and its soluble inorganic compounds, including arsine. *Appl Occup Environ Hyg* 6:1049–1056, 1991

14. Martin DW Jr, Woeber KA: Arsenic poisoning. *Calif Med* 118:13, 1973

15. Arena JM: *Poisoning*, 3rd ed, pp 25–26. Springfield, IL, Charles C Thomas, 1974

16. Linden CH: Antidotes in poisoning. In Callaham ML (ed): *Current Therapy in Emergency Medicine*, pp 951–952. Toronto, BC Decker, 1987

17. Gilman AG, Goodman LS (eds.): *Goodman and Gilman's The Pharmacological Basis of Therapeutics*, 6th ed, pp 1629–1632, New York, Macmillan, 1980

18. Aposhian HV, Aposhian MM: Newer developments in arsenic toxicity. *J Am College Toxicol* 8:1297–1305, 1989

19. Gorby MS: Arsenic poisoning. *West J Med* 149:308–315, 1988

ARSINE
CAS: 7784-42-1

AsH₃

Synonyms: Arsenic hydride; arseniurretted hydrogen; arsenous hydride; hydrogen arsenide

Physical Form. Colorless, heavier-than-air gas

Uses/Sources. In the electronics industry, to manufacture gallium arsenide and gallium arsenide phosphide for semiconductors and as a dopant; produced accidentally as a result of generation of nascent hydrogen in the presence of arsenic or by the action of water on a metallic arsenide

Exposure. Inhalation

Toxicology. Arsine is a severe hemolytic agent; abdominal pain and hematuria are cardinal features of arsine poisoning and are frequently accompanied by jaundice.

Arsine is the most acutely toxic form of arsenic.[1] It binds with oxidized hemoglobin, causing profound hemolysis of sudden onset.[2] Inhalation of 250 ppm may be fatal within 30 minutes, whereas 10 to 50 ppm may cause anemia and death with more prolonged exposure. Human experience has indicated that there is usually a delay of 2 to 24 hours after exposure before the onset of headache, malaise, weakness, dizziness, and dyspnea, with abdominal pain, nausea, and vomiting.[4-7] Dark red urine is frequently noted 4 to 6 hours after exposure. This often progresses to brown urine, with jaundice appearing at 24 to 48 hours after exposure.

An unusual bronze skin color has been noted in some patients; pigmentation of the skin and mucous membranes is more often described as ordinary jaundice and is seen in most poisoning cases. Oliguria or anuria, the most serious manifestation, may become manifest before the third day. In fatal cases, death may result from renal shutdown. Kidney failure occurs as extensive lysis by-products precipitate in the tubules and/or from hypoxic damage resulting from the reduced oxygen-carrying capacity of blood.[3] Other tissues at risk from hemolysis, anemia, and sludging of red cell debris within the microcirculation are the myocardium, liver, marrow, lungs, and skeletal muscles.[3] Massive hemolysis that persists for several days may produce hyperkalemia, which can result in cardiac arrest.[8] Reticulocytosis and leukocytosis are expected.[4-7] Normal red cell fragility and a negative Coombs' test are observed. Plasma hemoglobin values greater than 2g/100 ml are reported. Symptoms of arsenic poisoning, in addition to those of arsine, may be present. In two reported cases, arsenic encephalopathy, with extreme restlessness, memory loss, agitation, and disorientation, occurred several days after an acute exposure and lasted 10 days.[6] Peripheral neuropathy appeared within a few weeks; symptoms included numbness of the hands and feet, severe muscle weakness, and photophobia.[6]

In a report of chronic arsine poisoning in workers engaged in the cyanide extraction of gold, there was severe anemia in the absence of other signs and symptoms.[9] Hemoglobin values ranged as low as 3.2 g/100 ml; marked basophilic stippling was observed. Previous exposure to trace amounts of arsine for a period of 8 months was documented. It appears that, in very small concentrations, arsine exerts a cumulative effect.[4]

Inhaled arsine is oxidized to form elemental trivalent arsenic (AS^{+3}) and arsenous oxide (As_2O_3), two human carcinogens.[10] Excess cancers from trivalent arsenic and arsenic trioxide have been associated with cumulative lifetime arsenic exposure. Exposure to arsine above 0.004 ppm is associated with increased urinary arsenic excretion, indicating exposure to arsenic. Current exposure limits may not prevent potential chronic toxicity.[10]

Animal studies have also shown that cumulative exposure to small amounts of arsine may cause deleterious effects. In rats, repeated exposure to 0.025 ppm caused significant anemia, whereas a single exposure to 0.5 caused no effects on the hematopoietic system.[11]

Arsine, at concentrations that induced maternal toxicity in rats and mice, did not affect endpoints of developmental toxicity.[12]

Arsine is nonirritating, with a garliclike odor. Warning properties of exposure to hazardous concentrations are inadequate.[1]

Diagnosis. *Signs and Symptoms:* Headache, malaise, weakness, dizziness, dyspnea; abdominal pain, nausea, vomiting; hematuria; jaundice; oliguria, anuria.

Differential Diagnosis: Distinguish hemolysis due to arsine poisoning from paroxysmal nocturnal hemoglobinuria, cold agglutinin disease, thalassemia syndromes, sickle cell anemia, congenital hemolytic icterus, and poisoning by other hemolytic agents such as stibine.[1]

Special Tests: Plasma hemoglobin; white cell count; urinalysis, urinary concentration of hemoglobin.[2] Changes in the hematocrit value may not be of sufficient magnitude to be useful in diagnosis.[2] Analysis of blood and urine for arsenic may be done but, in a medical emergency, is of less value than the above-mentioned tests. Normal blood levels of arsenic are usually below 20 µg/100 mg.[2] In arsine poisoning, the urinary arsenic level usually remains elevated for several days or until normal renal function is restored.[2] In cases of chronic exposure, indicators of hematological regeneration, such as reticulocyte counts and γ-aminolevulinic acid dehydratase activity, should be considered because packed cell volumes and complete blood counts may not be adequate for the detection of anemia due to arsine.[11]

Treatment. The treatment of choice for acute and severe arsine poisoning is exchange transfusion and, if renal failure develops, hemodialysis.[1,2] As a rough guide when there is a history of arsine exposure, replacement transfusion should be done if the serum hemoglobin reaches 1.5 g/dl or higher or if oliguria develops.[2] Renal function should be closely monitored. Alkaline diuresis should be instituted in an attempt to avoid hemoglobin precipitation in the renal tubules. Dimercaprol (BAL) therapy is of no use in arsine poisoning because it affords no protection against hemolysis.[2]

The 1995 ACGIH threshold limit value–time-weighted average (TLV-TWA) for arsine is 0.05 ppm (0.16 mg/m³).

REFERENCES

1. NIOSH: *Current Intelligence Bulletin 32, Arsine (Arsenic Hydride) Poisoning in the Workplace.* DHEW (NIOSH) Pub No 79-142. Cincinnati, OH, National Institute for Occupational Safety and Health, 1979
2. Hesdorffer CS et al: Arsine gas poisoning: the importance of exchange transfusions in severe cases. *Br J Ind Med* 43:353–355, 1986
3. Luckey TD, Venugopal B: *Metal Toxicity in Mammals,* Vol 2, p 209. New York, Plenum Press, 1977
4. Fowler BA, Weissberg JB: Arsine poisoning. *N Engl J Med* 291:1171–1174, 1974
5. Pinto SS: Arsine poisoning: evaluation of the acute phase. *J Occup Med* 18:633–635, 1976
6. Levinsky WJ, Smalley RV, Hillyer PN, Shindler RL: Arsine hemolysis. *Arch Environ Health* 20:436–440, 1970
7. Teitelbaum DT, Kier LC: Arsine poisoning. *Arch Environ Health* 19:133–143, 1969
8. Benowitz NL: Cardiotoxicity in the workplace. *Occ Med: State of the Art Reviews* 7:465–479, 1992
9. Bulmer FMR et al: Chronic arsine poisoning among workers employed in the cyanide extraction of gold: a report of fourteen cases. *J Ind Hyg Toxicol* 22:111–124, 1940
10. Landrigan PJ et al: Occupational exposure to arsine. An epidemiologic reappraisal of current standards. *Scand J Work Environ Health* 8:169–177, 1982
11. Blair PC, Thompson MB, Morrisey RE, et al: Comparative toxicity of arsine gas in B6C3F1 mice, Fischer 344 rats, and Syrian golden hamsters: system organ studies and comparison of clinical indices of exposure. *Fund Appl Toxicol* 14:776–787, 1990
12. Morrissey RE, Fowler BA, Harris MW, et al: Arsine: absence of developmental toxicity in rats and mice. *Fund Appl Toxicol* 15:350–356, 1990

ASBESTOS
CAS: 1332-21-4

Amosite—CAS: 12173-73-5
Chrysotile—CAS: 12001-29-5
Crocidolite—CAS: 12001-28-4

Synonyms: Asbestos is a generic term applied to a number of hydrated mineral silicates, including amosite, chrysotile, tremolite, actinolite, anthophyllite, and crocidolite

Physical Form. Fibers of various sizes, colors, and textures

Uses. Thermal and electrical insulation; fireproofing; cement products

Exposure. Inhalation

Toxicology. Asbestos causes chronic lung disease (asbestosis), inflammation of the pleura, mesothelioma, and cancers of the lungs.

Asbestosis is a disorder characterized by a diffuse interstitial pulmonary fibrosis, at times including pleural changes of fibrosis and calcification.[1] Chest X ray reveals a granular change, chiefly in the lower lung fields; as the condition progresses, the heart outline becomes shaggy, and irregular patches of mottled shadowing may be seen. Typically, the patient exhibits restrictive pulmonary function. Accompanying clinical changes may include fine rales, finger clubbing, dyspnea, dry cough, and cyanosis.

The onset of asbestosis is dependent on intensity of dust exposure, length of exposure, and the physical and chemical properties of the asbestos fiber.[2] In general, the grade of pulmonary fibrosis relates to the fiber burden carried by the lungs.[3] Fiber morphology is also important. Alveolar macrophages, which normally phagocytize foreign bodies deposited in the lungs, seek to engulf the asbestos fibers and remove them. The macrophages are unable to remove long fibers in this manner, which results in an ongoing focal inflammatory response. Ultimately, epithelial cells are replaced by fibrous tissue, resulting in a progressive loss of lung compliance and respiratory function. Occasionally, asbestosis may develop fully in 7 to 9 years and may cause death as early as 13 years after first exposure. Usually, however, pneumoconiosis becomes evident 20 to 40 years after the first exposure to asbestos. Once established, asbestosis progresses even after exposure has ceased.[1] Increased risk of ischemic heart disease has also been associated with asbestosis as a result of impaired lung function.[4]

The pleura may also be affected by asbestos. Often, there is thickening of the visceral pleura from extension of the parenchymal inflammation. The parietal pleura may show patches of severe thickening, particularly over the diaphragm and the lower portions of the chest wall, resulting in the so-called pleural hyaline plaques. These may be seen by X ray, especially if calcified. The health significance of pleural abnormalities is not precisely defined, but many investigators consider the pleural plaques to be essentially benign.[5] In some cases, however, pleural thickening can lead to decreased ventilatory capacity, with severe consequences.

Bronchogenic carcinoma and mesothelioma of the pleura and peritoneum are causally associated with asbestos exposure; excesses of cancer of the stomach, colon, and rectum have also been observed.[5] Among 632 asbestos workers observed from 1943 to 1967, there were 99 excess deaths (above that expected on the basis of the US white male population) for three types of malignancies: bronchogenic (63), gastrointestinal (26), and all other sites combined (10).[1]

Mesothelioma, a relatively rare and rapidly fatal neoplasm seen chiefly in crocidolite workers, may occur without radiological evidence of asbestosis at exposure levels lower than those required for prevention of radiologically evident asbestosis.[1] Mesothelioma can occur after a short, intensive exposure; cases in children under 19 years of age indicate that the latent time period for development may be shorter than at first estimated, although the disease may occur following a very limited exposure 20 to 30 years earlier.

Fiber characteristics, including durability, harshness, surface chemistry, and dimensions appear to play a role in the carcinogenic process. Width and length of fibers are important parameters in determining the carcinogenic potential of various asbestos forms, where a fiber is defined as a particle with a length-to-width ratio of at least 3:1 and a length of 5 μm or more. In animal studies, fibers longer than 8 μm and narrower than 0.25 μm were more closely linked to pleural tumors irrespective of fiber type.[6] In general, fibers with widths greater than 1 μm are not implicated in the occurrence of lung cancer or mesothelioma.[7]

Cigarette smoking is strongly implicated as a cocarcinogen among asbestos workers.[8] The incidence of lung carcinoma among nonsmoking asbestos workers is not significantly greater than that of nonasbestos workers, whereas asbestos workers who smoke have a much higher incidence. Cigarette-smoking asbestos workers have approximately 15 times the risk of developing lung cancer compared to nonsmoking asbestos workers.[9]

No obvious developmental effects were observed in animals exposed to high levels of asbestos during gestation.[4]

The 1995 ACGIH proposed threshold limit value-time-weighted average (TLV-TWA) for asbestos is 0.2 f/cc with an A1 confirmed human carcinogen designation.

REFERENCES

1. National Institute for Occupational Safety and Health, U.S. Department of Health, Education, and Welfare: *Criteria for a Recommended Standard . . . Occupational Exposure to Asbestos*. DHEW (HSM) Pub No 72-10267. Washington, DC, US Government Printing Office, 1972
2. Parkes WR: *Occupational Lung Disorders*, 2nd ed, p 255. London, Butterworths, 1982
3. Becklake MR: Pneumoconioses. In Murray JF, Nadel JA: *Textbook of Respiratory Medicine*, Vol 2, p 1577. Philadelphia, WB Saunders, 1988
4. Sanden A, Jarvholm B, Larsson S, et al: The importance of lung function, nonmalignant diseases associated with asbestos, and symptoms as predictors of ischaemic heart disease in shipyard workers exposed to asbestos. *Br J Ind Med* 50:785–790, 1993
5. Agency for Toxic Substances and Disease Registry (ATSDR): *Toxicological Profile for Asbestos.*

US Department of Health and Human Services, Public Health Service, p 141, 1993

6. Stanton MF, Layard M, Tegeris A, et al: Relation of particle dimension to carcinogenicity in amphibole asbestoses and other fibrous minerals. *J Natl Cancer Inst* 57:965–975, 1981

7. Wylie AG, Bailey KF, Kelse JW, et al: The importance of width in asbestos fiber carcinogenicity and its implications for public policy. *Am Ind Hyg Assoc J* 54:239–252, 1993

8. Selikoff IJ, Hammond EC, Churg J: Asbestos exposure, smoking and neoplasia. *JAMA*, 204:106–112, 1968

9. Selikoff IJ, Lee DHK: *Asbestos and Disease*, p 327. New York, Academic Press, 1978

ASPHALT FUMES
CAS:8052-42-4

Synonyms: Asphaltic bitumen; asphaltum; petroleum asphalt; bitumen

Physical Form. Brownish-black viscous liquid or solid composed essentially of hydrocarbons; residue from the evaporation of the lighter hydrocarbons from petroleum

Uses/Sources. Asphalt fumes arise from asphalt used for road construction, roofing, coating of construction materials, and in association with the production of asphalt from petroleum

Exposure. Inhalation; skin contact

Toxicology. Acute exposure to asphalt fumes has caused irritative effects. Certain extracts of asphalt have caused a carcinogenic skin response in experimental animals.

Following acute exposure, subjective symptoms, including abnormal fatigue, reduced appetite, and throat and eye irritation, have been reported.[1] In a study of road repair and construction workers, symptoms increased with increasing concentration of asphalt fumes and with increasing asphalt temperature. In another report of female workers in a commercial lighting factory, there was a causal association between exposure to asphalt fumes, irritative symptoms (nausea, headache, fatigue, skin rashes, and eye, nose and throat irritation), and macrothrombocytosis (enlarged platelets), which reversed with a reduction of exposure.[2]

In mice skin-painting studies, skin tumors were produced by steam-refined petroleum bitumens, an air-refined bitumen in toluene, two cracking-residue bitumens, and a pooled mixture of steam- and air-blown petroleum bitumens.[3] In contrast, standard roofing petroleum asphalts produced no tumors.

The IARC stated in 1985 that there was inadequate evidence that bitumens alone are carcinogenic to humans, but there was sufficent evidence of animal carcinogenicity of certain extracts as detailed in the preceding paragraph.[3]

Since then, additional epidemiologic studies have appeared.

In a historical cohort study of 1320 workers in the asphalt industry, there was a significant increase in brain cancer (SMR 500) but not in respiratory, bladder, or gastrointestinal cancer.[4] In a study of 679 Danish men who were heavily exposed to asphalt, significant increases occurred in the incidences of cancer of the mouth (SMR 1111), esophagus (698), rectum (318), and lung (344).[5]

A subsequent mortality study of this same cohort found significant increases for death due to lung cancer.[6] (Mortality from noncarcinogenic respiratory diseases, including bronchitis, emphysema, and asthma, also occurred in excess.)

The conflicting evidence in epidemiologic studies reflects the difficulties in establishing the exact nature of the material to which the workers are exposed and in ensuring that the exposure is to asphalt alone. Cohorts of workers such as roofers are often also exposed to coal tar pitches, which are generally considered to contain more polynuclear aromatic hydrocarbons than asphalt and to be more potent carcinogens.[7]

There was a five fold range in mutagenicity in fumes from asphalts derived from a variety of crude oils, and the asphalts were far less mutagenic than coal tar fumes.[8]

The 1995 ACGIH theshold limit value–time-weighted average (TLV-TWA) for asphalt fumes is 5 mg/m^3.

REFERENCES

1. Norseth T, Waage J, Dale I: Acute effects and exposure to organic compounds in road maintenance workers exposed to asphalt. *Am J Ind Med* 20:737–744, 1991
2. Chase RM, Liss GM, Cole DC, et al: Toxic health effects including reversible macrothrombocytosis in workers exposed to asphalt fumes. *Am J Ind Med* 25:279–289, 1994
3. *IARC Monographs on the Evaluation of the Carcinogenic Risk of Chemicals to Humans*, Vol 35, Polynuclear aromatic compounds, Part 4, Bitumens, coal-tars and derived products, shale-oils and soots, pp 38–81. Lyon, International Agency for Research on Cancer, 1985
4. Hansen ES: Cancer mortality in the asphalt industry: a ten year follow up of an occupational cohort. *Br J Ind Med* 46:582–585, 1989
5. Hansen ES: Cancer incidence in an occupational cohort exposed to bitumen fumes. *Scand J Work Environ Health* 15:101–105, 1989
6. Hansen ES: Mortality of mastic asphalt workers. *Scand J Work Environ Health* 17:20–24, 1991
7. *IARC Monographs on the Evaluation of Carcinogenic Risks to Humans. Overall Evaluations of Carcinogenicity: An Updating of IARC Monographs Volumes 1 to 42*, Suppl 7, pp 133–134. Lyon, International Agency for Research on Cancer, 1987
8. Machado ML, Beatty PW, Fetzer JC, et al: Evaluation of the relationship between PAH content and mutagenic activity of fumes from roofing and paving asphalts and coal tar pitch. *Fund Appl Toxicol* 21:492–499, 1993

ATRAZINE
CAS:1912-24-9

C$_8$H$_{14}$ClN$_5$

Synonyms: 2-Chloro-4-ethylamino-6-isopropylamino-1,3,5-triazine; 6-chloro-*N*-ethyl-*N'*-(1-methylethyl)-1,3,5-triazine-2,4-diamine

Physical Form. Colorless, crystalline solid

Uses. Herbicide

Exposure. Inhalation

Toxicology. The acute toxicity of atrazine to animals is low.

The oral LD$_{50}$ in rats was 3080 mg/kg, and the dermal LD$_{50}$ in rabbits was 7500 mg/kg.[1] There was minimal irritation on rabbit skin, and moderate irritation when atrazine was placed in the rabbit eye. No human skin or eye irritation has been reported.[2]

In a teratology study, oral doses of 0, 10, 70, or 700 mg/kg/day were given to rats on days 6 through 15 of gestation, and rabbits were given oral doses of 0, 1, 5, or 75 mg/kg/day on days 7 through 19 of gestation.[3] Maternal toxicity was seen in rats at 70 mg/kg and in rabbits at 5 mg/kg. Fetal toxicity was seen in rats at 70 mg/kg and in rabbits at 75 mg/kg. Teratogenesis was not demonstrated at any of the treatment levels.

When rats were administered atrazine in drinking water at 0.1, 0.2, or 0.5 g/l for 1 or 3 weeks, they excreted as the principal metabolite 2-chloro-4-ethylamino-6-amino-*s*-triazine.[4] Atrazine and its metabolite have been shown to alter the activity of some testosterone-metabolizing enzymes in the rat pituitary and hypothalamus and to decrease hormone-receptor binding in the prostate.[5]

Male rats fed diets containing 1000 mg/kg atrazine for 8 weeks, followed by 750 mg/kg of diet for up to 118 weeks, had a significantly increased incidence of benign mammary gland tumors; females similarly dosed had a significantly increased incidence of uterine adenocarcinomas and an increased incidence of leukemias and lymphomas.[6]

In humans, a number of case control studies have indicated an association between atrazine exposure and various cancers, including lymphomas, ovarian cancer, and colon cancer.[5]

The IARC concluded that there was limited evidence for the carcinogenicity of atrazine in animals and inadequate evidence of carcinogenicity in humans.

Atrazine was not mutagenic in bacteria and did not cause chromosomal aberrations in cultured rodent cells; it did induce DNA strand breaks in stomach, liver and kidney cells of rats treated orally.[5]

The 1995 ACGIH threshold limit value–time-weighted average (TLV-TWA) for atrazine is 5 mg/m^3.

REFERENCES

1. Atrazine Ciba-Geigy toxicology data, Agric Div, Ciba-Geigy Corp, Ardsley, NY, October 1, 1972
2. *Aatrex Herbicide Technical Bulletin*, Geigy Agric Chem Div, Ciba-Geigy Corp, Ardsley, NY, June 1971
3. Infurna R et al: Teratological evaluations of atrazine technical, a triazine herbicide, in rats and rabbits. *J Toxicol Environ Health* 24:307–319, 1988
4. Ikonen R, Kangas J, Savolainen H: Urinary atrazine metabolites as indicators for rat and human exposure to atrazine. *Toxicol Lett* 44:109–112, 1988
5. *IARC Monographs on the Evaluation of Carcinogenic Risks to Humans* Vol 53, Occupational exposures in insecticide application, and some pesticides, pp 441–465, International Agency for Research on Cancer, World Health Organization, 1991
6. Pinter A, Torok G, Borzsonyi M, et al: Long-term carcinogenicity bioassay of the herbicide atrazine in F344 rats. *Neoplasma* 37:533–544, 1990

AZINPHOS-METHYL
CAS: 86-50-0

$C_{10}H_{12}N_3O_3PS_2$

Synonyms: *O,O*-dimethyl-*S*-[4-oxo-1,2,3-benzotriazin-3(4H-yl) methyl] phosphorothioate; Guthion; Methyl Guthion; Gusathion

Physical Form. White crystalline solid

Uses. Acaricide; insecticide

Exposure. Inhalation; skin absorption; ingestion

Toxicology. Azinphos-methyl is an indirect inhibitor of cholinesterase.

Dosages given to volunteers for approximately 30 days ranged from 4.0 to 20 mg/man/day and did not produce clinical effects or a significant change in cholinesterase levels. Apparently, a level high enough to inhibit cholinesterase remains to be studied. It is regarded as of only moderate toxicity to man.[1]

In a study of eight workers engaged in the formulation of a Guthion wettable powder and exposed to concentrations of up to 9.6 mg/m^3, the lowest activity of cholinesterase in blood serum was 78% of the value before exposure, and there were no signs or symptoms of illness.[2]

In animals, azinophos-methyl has an acute oral toxicity similar to that of parathion, although the acute dermal toxicity is less than that of parathion.[1]

Rats that inhaled azinphos-methyl at 4.72 mg/m^3, 6 hr/day, 5 days/week for 12 weeks, showed significant depression of red cell and plasma cholinesterases; concentrations of 0.195 and 1.24 mg/m^3 were without effect.[3]

Rats fed azinophos-methyl for 2 years at rates of 50 ppm and later 100 ppm had normal growth rates, but plasma, red cell, and brain cholinesterase activity was depressed in the females.[4] Dietary levels of 5 ppm were without effect, and no tumorgenic activity was noted at any dosage level. Dogs receiving 300 ppm in their feed had tremors, weakness, lethargy, and some weight loss; 5 ppm administered in the feed for 2 years was without effect on cholinesterase levels.

A reproductive study of mice and rats demonstrated embryotoxicity at levels associated with significant maternal toxicity. No teratogenicity was observed at lower doses.[5]

The 1995 ACGIH threshold limit value–time-weighted average (TLV-TWA) for azinphos-methyl is 0.2 mg/m^3 with a notation for skin absorption.

Note. For a description of diagnostic signs, differential diagnosis, and medical control, including clinical laboratory tests, as well as specific treatment of overexposure to anticholinesterase insecticides, see "Parathion."

REFERENCES

1. Hayes WJ Jr: Organic phosphorus pesticides. In *Pesticides Studied in Man*, pp 358–359. Baltimore, MD, Williams and Wilkins, 1982
2. Jegier Z: Exposure to Guthion during spraying and formulating. *Arch Environ Health* 8:565–569, 1964
3. Kimmerle G: Subchronic inhalation toxicity of azinophos-methyl in rats. *Arch Toxicol* 35:83–89, 1976
4. Woden AN, Wheldon GH, Noel PRB, et al: Toxicity of Gusathion for the rat and dog. *Toxicol Appl Pharmacol* 24:405–412, 1973
5. Short RD et al: Developmental toxicity of Guthion in rats and mice. *Arch Toxicol* 43:177–186, 1980

BARIUM AND COMPOUNDS
CAS: 7440-39-3

Ba

Compounds: Soluble—barium nitrate; barium sulfide; barium chloride; barium hydroxide; barium acetate. Insoluble—barium sulfate

Physical Form. Elemental barium is a silver-white metal; many of the compounds are white powders or crystals

Uses. Catalyst for organic reactions; lubricating oil additive; rat poison; in manufacture of paper electrodes; in fireworks; in electroplating; in medicine as a radiopaque substance for X-ray diagnosis

Exposure. Inhalation; ingestion

Toxicology. Certain compounds of barium are toxic to the cardiovascular, respiratory, gastrointestinal, hepatic, and renal systems in man and animals.

The toxicity of barium compounds depends on their solubility, with the more soluble forms being more toxic than the relatively insoluble forms, which are inefficient sources of Ba^{2+} ions.[1]

Inhalation of insoluble barium-containing dusts may produce a benign pneumoconiosis, termed baritosis.[2] The condition is without clinical significance. Characteristic X-ray changes are those of small, extremely dense, circumscribed nodules, evenly distributed throughout the lung fields, reflecting the radiopacity of the barium dust. Exposure of workers to concentrations ranging to 92 mg/m³ of barium sulfate caused no abnormal signs or symptoms, including no interference with lung function or liability to develop pulmonary or bronchial infection.[3] Ingestion of insoluble barium compounds also presents no problems to health, barium sulfate being widely used as a contrast agent in radiography.[4]

The barium ion is a muscle poison causing stimulation and then paralysis. Initial symptoms are gastrointestinal, including nausea, vomiting, colic, and diarrhea, followed by myocardial and general muscular stimulation with tingling in the extremities.[2] Severe cases continue to loss of tendon reflexes, general muscular paralysis, and death from respiratory arrest or ventricular fibrillation. The threshold of a toxic dose in humans is reported to be about 0.2 to 0.5 g Ba absorbed from the gut; the lethal dose is 3 to 4 g Ba.

In animal studies, rats receiving 110 mg barium/kg body weight in the drinking water as barium chloride dihydrate for 15 days had no clinical findings of toxicity. In female mice administered 85 mg/kg/day and in male mice given 70 mg/kg/day in the drinking water for the same time period, there was no histopathological evidence of toxicity although relative liver weights of the dosed animals were significantly greater than controls.[5] In 13-week studies in mice, liver weights were lower than controls at doses above 100 mg/kg/day and, at doses of 450 mg/kg/day and 495 mg/kg/day in males and females, respectively, there was multifocal to diffuse nephropathy characterized by tubule dilation, regeneration, and atrophy. In two-year studies, there were no chemical-related increased incidences of neoplasms in mice or rats receiving up to 2500 ppm barium chloride dihydrate in the drinking water.[5] There were dose-related increased incidences of nephropathy in the mice.

Barium chloride dihydrate was not mutagenic in *Salmonella typhimurium*, nor did it induce sister chromatid exchanges or chromosomal aberrations in cultured Chinese hamster ovary cells.[5]

In a mating trial, no adverse anatomical effects were observed in the offspring of rats or mice receiving up to 4000 ppm in the drinking water, although rat pup weight was reduced. Reproductive indices in rats and mice were unaffected.[6]

The barium ion is a physical antagonist of potassium, and it appears that the symptoms of barium poisoning are attributable to Ba2+-induced hypokalemia.[2] The effect is probably due to a transfer of potassium from extracellular to intracellular compartments rather than to urinary or gastrointestinal losses. Signs and symptoms are relieved by intravenous infusion of K+.[2]

Barium hydroxide and barium oxide are strongly alkaline in aqueous solution, causing severe burns of the eye and irritation of the skin.[7]

The ACGIH threshold limit value–time-weighted average for barium, and soluble barium compounds, as Ba is 0.5 mg/m³; for barium sulfate, it is 10 mg/m³ for total dust containing no asbestos and less than 1% silica.

REFERENCES

1. Agency for Toxic Substances and Disease Registry (ATSDR): *Toxicological Profile for Barium.* US Department of Health and Human Services, Public Health Service, p 138, TP-91/03, 1992
2. Reeves AL: Barium. In Friberg L et al (eds): *Handbook on the Toxicology of Metals*, pp 321–328. New York, Elsevier North-Holland, 1979
3. Barium Sulfate. *Documentation of the TLVs and BEIs*, 5th ed, p 48. Cincinnati, OH, American Conference of Governmental Industrial Hygienists, 1986
4. Dare PRM, Hewitt PJ, Hicks R, et al: Short communication: barium in welding fume. *Ann Occup Hyg* 2:445–448, 1984
5. National Toxicology Program: *NTP Technical Report on the Toxicology and Carcinogenesis Studies of Barium Chloride Dihydrate (CAS No. 10326-27-9) in F344/N Rats and B6C3F₁ Mice Drinking Water Studies).* NTP TR 432, NIH Pub No 94-3163, US Department of Health and Human Services, Public Health Service, National Institutes of Health, Research Triangle Park, NC, 1994
6. Dietz DD, Elwell MR, Davis WE Jr, et al: Subchronic toxicity of barium chloride dihydrate administered to rats and mice in the drinking water. *Fund Appl Toxicol* 19:527–537, 1992
7. Grant WM: *Toxicology of the Eye*, 3rd ed, p 134. Springfield, IL, Charles C Thomas, 1986

BAUXITE
CAS: 1318-16-7

$Al_2O_3 \cdot 2H_2O$

Synonym: Beauxite

Physical Form. Dust (red, brown, or yellow)

Uses. Ore for production of alumina; adsorbent in oil refining

Exposure. Inhalation

Toxicology. Bauxite can be considered to be a nuisance particulate; long experience with mining and refining of bauxite has not revealed significant adverse health effects.

Nuisance particulates have little adverse effect on lungs and do not produce significant organic disease or toxic effect when exposures are kept under reasonable control.[1] When inhaled in excessive amounts, however, all dusts may be expected to evoke some cellular response. According to ACGIH, the lung-tissue reaction caused by inhalation of nuisance particulates has the following characteristics: (1) The architecture of the air spaces remains intact; (2) collagen (scar tissue) is not formed to a significant extent; and (3) the tissue reaction is potentially reversible.

In one recent case report of a 70-year-old worker exclusively exposed to the dust of raw bauxite, deposits of bauxite were found in the lungs in areas of mild pulmonary fibrosis.[2] There were no clinical symptoms, and it is not clear if the fibrosis was a response to the bauxite or if the bauxite accumulated in pre-existing fibrotic areas.

The nuisance dust aspect of bauxite is in sharp contrast to the limited industrial situation in which lung injury was reported in Canadian workers who, in the 1940s, engaged in the manufacture of alumina abrasives in the virtual absence of fume control.[3,4] Fusing of bauxite at 2000° C gave rise to a fume composed of freshly formed particles of amorphous silica and aluminum oxide. In spite of the poor choice of the term bauxite fume pneumoconiosis, sometimes used to describe the disease, scientific opinion favors the silica component as the probable toxic agent. It should be emphasized that bauxite from some sources may contain small amounts of silica.

The 1995 ACGIH threshold limit value for bauxite is 10 mg/m³.

REFERENCES

1. Particulates Not Otherwise Classified (PNOC). *Documentation of TLVs and BEIs*, 6th ed, pp 1166–1167. Cincinnati, OH, American Conference of Government Industrial Hygienists (ACGIH), 1991
2. Bellot SM, Schade Van Westrum JAFM, Wagenvoort CA, et al: Deposition of bauxite dust and pulmonary fibrosis. *Path Res Pract* 179:225–229, 1984
3. Hatch TF: Summary. In Vorwald, AJ (ed): *Pneumoconiosis, Beryllium, Bauxite Fumes*, pp 498–501. New York: Harper Brothers, 1950
4. Shaver CG, Riddel AR: Lung changes associated with the manufacture of alumina abrasives. *J Ind Hyg Toxicol* 29:145–157, 1947

BENOMYL
CAS: 17804-35-2

$C_{14}H_{18}N_4O_3$

Synonyms: Methyl 1-(butylcarbamoyl)-2-benzidimidazolecarbamate; Benlate; Benex

Physical Form.   White crystalline solid

Uses.   Fungicide; ascaracide

Exposure.   Inhalation; skin contact

Toxicology.   Benomyl is of low acute toxicity to experimental animals but is teratogenic in rats; it can cause dermatitis in workers.

The oral LD_{50} for rats was greater than 10 g/kg, and the dermal LD_{50} in rabbits was also greater than 10 g/kg.[1] There was mild irritation when benomyl was placed on the skin of the rabbit and in the rabbit eye.

In a 90-day inhalation study, equal groups of male and female rats were exposed nose-only 6 hr/day, 5 days/week, to levels of 0, 10, 50, or 200 mg/m³.[2] At 45 days, half the animals were sacrificed and necropsied. Degeneration of the olfactory epithelium was observed in all the males and 8 of 10 females at the highest dose level. Two of 10 males at the 50-mg/m³ level had less severe olfactory degeneration. After 90 days of exposure, the remainder of the animals were sacrificed, and findings were essentially the same as those seen at the end of 45 days. There were no other effects observed. In a follow-up to this study, it was determined that the olfactory epithelial damage reported following inhalation exposure is specific to the route of exposure since the nasal cavity is not a target following dietary administration of benomyl.[3] Specifically, rats fed diets containing 0, 5000, 10,000, or 15,000 ppm benomyl for 32 days had toxicity only in the form of decreased body weight gain and food consumption at the two highest dose levels.

In another study, mice were given diets containing 0, 500, 1500, 5000, or 7500 ppm benomyl for 2 years. An oncogenic response was reported in the livers of male mice dosed at 500 and 1500 ppm but not in the 5000 to 7500 group; an increase in nonmalignant liver tumors was observed in all female treatment groups.[4] At this time, there does not appear to be any conclusive evidence that benomyl is carcinogenic.

In a teratology study, female rats were administered 62.5 mg/kg, beginning at day 6 of gestation.[5] Fetuses examined at day 16 or day 20 showed a high incidence of craniocerebral anomalies, including hydrocephalus. In another study, male rats were gavaged daily with 0, 1, 5, 15, or 45 mg/kg/day.[6] After 76 to 79 days, the animals were evaluated for reproductive endpoints. At the highest dose level, minimal to moderate changes were observed, including decreased testis weight and sperm production. At 62 days, the males had been mated with females, and reproductive performance was not affected.

Contact dermatitis has been reported in Japanese women who worked in a greenhouse where benomyl had been used.[7] Eruptions on the backs of the hands and on the forearms consisted of redness and edema. Other cases of dermatitis have also been reported.[8]

The 1995 ACGIH threshold limit value-time weighted average for benomyl is 0.84 ppm (10 mg/m³).

REFERENCES

1. EI Du Pont de Nemours & Co, Inc: Technical Data Sheet, *Benomyl*, June 1974
2. Warheit DB, Kelly DP, Carakostas, MC, Singer AW: A 90-day inhalation toxicity study with benomyl in rats. *Fund Appl Toxicol* 12:333–345, 1989
3. Hurtt ME, Mebus CA, Bogdanffy MS: Investigation of the effects of benomyl on rat nasal mucosa. *Fund Appl Toxicol* 21:253–255, 1993
4. Von Burg R: Toxicology update: benomyl. *J Appl Toxicol* 13:377–381, 1993
5. Ellis WG, Semple JL, Hoogenboom JR, Kavlock RJ, Zeman FJ: Benomyl-induced craniocerebral anomalies in fetuses of adequately nourished and protein-deprived rats. *Teratogen Carcin Mutagen* 8:377–391, 1988
6. Linder RE, Rehnberg LF, Strader LF, Diggs, JP: Evaluation of reproductive parameters in adult male Wistar rats after subchronic exposure (gavage) to benomyl. *J Toxicol Environ Health* 25:285–298, 1988
7. Savitt LE: Contact dermatitis due to benomyl insecticide. *Arch Dermatol* 105:926–927, 1972
8. Hayes WJ Jr: *Pesticides Studied in Man*, pp 610–612. Baltimore, MD, Williams and Wilkins, 1982

BENZ[a]ANTHRACENE
CAS: 56-55-3

$C_{18}H_{12}$

Synonyms: BA; benzanthracene; 1,2-benz(a) anthracene; benzo(a)anthracene; 2,3-benzophenanthrene; naphthanthracene; tetraphene

Physical Form. Solid; often associated with or adsorbed onto ultrafine airborne particulate matter

Sources. A major component of the total content of polynuclear aromatic hydrocarbons, also known as polycyclic aromatic hydrocarbons; human exposure primarily through smoking of tobacco, inhalation of products of incomplete organic combustion such as automobile exhaust, and ingestion of food contaminated by combustion effluents such as those that are smoked or barbecued.

Exposure. Inhalation

Toxicology. Benz[a]anthracene (BA) is carcinogenic to experimental animals.

The IARC considers that there is "sufficient evidence" that BA is carcinogenic to experimental animals.[1] BA has produced carcinogenic results in the mouse by several routes of administration. It caused hepatomas and lung adenomas after oral administration by stomach tube of 15 doses of 1.5 mg each over a period of 5 weeks early in the lifetimes of the mice.[2]

BA undergoes metabolism in animals and

humans to intermediates responsible for its toxicity. These metabolic intermediates include arene oxides, dihydrodiols, and diol epoxides such as BA 3,4-dihydrodiol and BA 3,4-diol-1,2-epoxide.[3]

BA is a complete carcinogen for the mouse skin. A 0.2% solution of BA in dodecane three times weekly produced skin tumors in 11/21 animals with an average latent period of 61 weeks, whereas a 1% solution produced tumors in 17/22 animals with an average l atent period of 42 weeks.[4]

BA's metabolites are genotoxic in the Ames mutation test and caused unscheduled DNA synthesis in primary rat hepatocytes.[5-6] In an in vivo mutagenic assay, male CD rats (6/group) were dosed 3 times with BA at a 24-hour interval by intratracheal instillation.[7] Lung cells were enzymatically separated and used to determine the frequency of DNA adducts, sister chromatid exchanges (SCEs), and micronuclei. BA induced DNA adducts, SCEs, and micronuclei in this rat lung-cell system.

Benz(a)anthracene is designated an A2 suspected human carcinogen by ACGIH and has no assigned threshold limit value.

REFERENCES

1. *IARC Monographs on the Evaluation of the Carcinogenic Risk of Chemicals to Humans*, Vol 32, Polynuclear aromatic compounds, Part 1, Chemical, environmental and experimental data, pp 135–146. Lyon, International Agency for Research on Cancer, December 1983
2. Klein M: Susceptibility of strain B6AF1/J hybrid infant mice to tumorigenesis with 1,2-benzanthracene, deoxycholic acid, and 3-methylcholanthrene. *Cancer Res* 23:1701, 1963
3. Agency for Toxic Substances and Disease Registry: *Toxicological Profile for Benzo[a]pyrene.* ATSDR/TP-88/04, Public Health Service, Centers for Disease Control, Atlanta, GA, pp 25–29, 1990
4. Bingham E, Falk HL: Environmental carcinogens: the modifying effect of cocarcinogens on the threshold response. *Arch Environ Health* 19:779, 1969
5. McCann J, Choi E, Yamasaki E, and Ames BN: Detection of carcinogens as mutagens in the Salmonella/microsome test: assay of 300 chemicals. *Proc Natl Acad Sci*, 72:5135–5139, 1975
6. Probst GS et al: Chemically-induced unscheduled DNA synthesis in primary rat hepatocyte cultures: a comparison with bacterial mutagenicity using 218 compounds. *Environ Mutagen*, 3:11–32, 1981
7. Whong WZ, Stewart JD, Cutler D, Ong T: Induction of in vivo DNA adducts by 4 industrial by-products in the rat-lung-cell system. *Mutat Res* 312(2):165–72, 1994

BENZENE
CAS: 71-43-2

C_6H_6

Synonyms: Benzol; cyclohexatriene

Physical Form. Colorless liquid

Uses. Intermediate in the production of styrene, phenol, cyclohexane, and other organic chemicals; in manufacture of detergents, pesticides, solvents, and paint removers; in gasoline

Exposure. Inhalation; skin absorption

Toxicology. Acute benzene exposure causes central nervous system depression; chronic exposure causes bone marrow depression leading to aplastic anemia and is also associated with an increased incidence of leukemia.

Human exposure to very high concentrations, approximately 20,000 ppm, is fatal in 5 to 10 minutes.[1-3] Concentrations of 7500 ppm are dangerous to life within 30 minutes. Convulsive movements and paralysis, followed by unconsciousness, follow severe exposures. Brief exposure to concentrations in excess of 3000 ppm is irritating to the eyes and respiratory tract; continued exposure may cause euphoria, nausea, a staggering gait, and coma. Inhalation of lower concentrations (250 to 500 ppm) produces vertigo, drowsiness, headache, and nausea, whereas 25 ppm for 8 hours is without clinical effect.

The most significant toxic effect of benzene exposure is an insidious and often irreversible injury to the bone marrow. Long-term exposures to low concentrations have been observed to have an initial stimulant effect on the bone marrow, followed by aplasia and fatty degeneration.[3] Clinically, an initial increase, followed by a decrease in the erythrocytes, leukocytes, or platelets, is observed, with progression to anemia, leukopenia, and/or thrombocytopenia, respectively.[3] If pancytopenia (the depression of all three cell types) occurs and is accompanied by bone marrow necrosis, the syndrome is termed aplastic anemia. The hypocellularity varies greatly from conditions in which the marrow is completely devoid of recognizable hematopoietic precur sors to those in which the precusors of only one cell line are absent or arrested in their development.[4] Typical symptoms may include light-headedness, headache, loss of appetite, and abdominal discomfort. With more severe intoxication, there may be weakness, blurring of vision, and dyspnea on exertion; the mucous membranes and skin may appear pale, and a hemorrhagic tendency may result in petechiae, easy bruising, epistaxis, bleeding from the gums, or menorrhagia.[5] The most serious cases of aplastic anemia succumb within 3 months of diagnosis as a result of infection or hemorrhage.[6] The mechanism of benzene-induced aplastic anemia appears to involve the concerted action of several benzene metabolites, perhaps acting in concert with unmodified benzene, on early stem and progenitor cells, as well as blast cells, to inhibit maturation and amplification.[1,7] Metabolities may also inhibit stromal cells, which are necessary to support growth of differentiating and maturing marrow cells.[1]

Numerous case reports and epidemiologic studies suggest a leukemogenic action of benzene in humans—the leukemia tending to be acute and myeloblastic in type, often following aplastic changes in the bone marrow. Benzene may also induce chronic types of leukemia.[8]

One study indicated a fivefold excess of all leukemias and a tenfold excess of myelomonocytic leukemia among benzene-exposed workers compared to the US Caucasian male population.[9] Among shoemakers chronically exposed to benzene, the annual incidence of leukemia was 13.5 per 100,000, whereas the incidence in the general population was 6 per 100,000.[10] Four cases of acute leukemia were reported in shoemakers exposed to concentrations of benzene up to 210 ppm for 6 to 14 years; two of the four had aplastic anemi a prior to leukemia; three of the four cases of leukemia were of the acute myeloblastic type; the fourth patient developed thrombocythemia in the second year after an episode of aplastic anemia, and acute monocytic leukemia developed later.[11]

A retrospective cohort study in China of 28,460 benzene-exposed workers found a leukemia mortality rate of 14 per 100,000 person-years in the benzene cohort and 2 per 100,000 in the control cohort.[12] The standardized mortality ratio (SMR) was 574, and the mean latency period for induction of benzene leukemia was 11.4 years. Concentrations in the workplace where the patients had been employed were reported to range from 3 to 300 ppm but were mostly in the range of 16 to 160 ppm. The SMR in this study was simil ar to that in a study of two pliofilm manufacturing plants with 748 workers and exposures ranging from 16 to 100 ppm (SMR = 560).[13] In another report, a mortality update through 1982 for 956 employees exposed to benzene, there was a nonsignificant excess of total death from leukemia based on four observed cases; however, all four cases involved myelogenous leukemias, which represented a significant excess in this subcategory.[14]

Persons with aplastic anemia due to benzene exposure have been found to be at a much greater risk for developing leukemias. A follow-up of 51 benzene-exposed workers with pancytopenia revealed 13 cases of leukemia.[15] The cumulative incidence of leukemia among individuals with clinically ascertained benzene hemopathy has ranged from 10% to 17% in various studies.[16]

The IARC has concluded that epidemiological studies have established the relationship between benzene exposure and the development of acute myelogenous leukemia and that there is sufficient evidence that benzene is carcinogenic to humans.[16] Although a benzene-leukemia association has been made, the exact shape of the dose-response curve and/or the existence of a threshold for the response is unknown and has been the source of speculation and controversy.[17-21] Some risk assessments suggest exponential increases in relative risk (of leukemias) with increasing cumulative exposure to benzene. At low levels of exposure, however, a small increase in leukemia mortality cannot be distinguished from a no-risk situation.[1] In addition to cumulative dose, other factors such as multiple solvent exposure, familial connection, and individual susceptibility may play a role in leukemia development.[15]

A relationship between benzene exposure and the production of lymphoma and multiple myeloma is controversial. In one report, a statistically significant increase in deaths from multiple myeloma was found, although the numbers were small.[22]

An increased incidence of neoplasms at multiple sites has been found in chronic inhalation and gavage studies in rodents. Anemia, lymphocytopenia, bone marrow hyperplasia, and an increased incidence of lymphoid tumors occurred in male mice exposed at 300 ppm for life.[23] Gavage administration to rats in one study, and rats and mice in another, caused an increase in tumors; especially significant was an increase in zymbal gland tumors (tumors of the auditory sebaceous glands) in both reports.[24,25]

Although consistent findings of chromosomal aberrations (stable and unstable) in the nuclei of lymphocytes have been reported in exposed workers, the implications with respect to leukemia are not clear.[26] Data on exposure levels are limited but are said to range from 10 to a few 100 ppm.[16] In controlled rat studies, exposure to 1, 10, 100, or 1000 ppm for 6 hours caused a dose-response relationship in the percentage of cells with abnormalities and aberrations at the two highest dose levels.[27] An increase in polychromatic erythrocytes with micronuclei (thought to be broken fragments of chromosomes that are left behind) has also been observed in benzene-treated animals.[15] Mice exposed at 10 ppm for 6 hours had a significantly increased incidence of micronuclei compared to controls.[28]

In addition to tumor induction and cyto-genic damage, inhaled benzene can cause immunodepressive effects in mice at 100 ppm as manifested by reduced host resistance to a transplantable syngeneic tumor.[29]

Exposure to benzene vapor produces feto-toxicity, such as growth retardation, in mice and rats. In one study, a teratogenic potential, as evidenced by exencephaly, angulated ribs, and dilated brain ventricles, was observed in rats exposed to 500 ppm.[30] These anomalies have not been duplicated in other studies despite the use of higher be nzene levels, and the teratogenic findings have been attributed to nutritional alterations, chance events, or an underestimation of spontaneous malformations.[17]

Tests for phenol levels in urine have been used as an index of benzene exposure; urinary phenol concentrations of 200 mg/l are indicative of exposure to approximately 25 ppm of benzene in air.[31]

Direct contact with the liquid may cause erythema and vesiculation; prolonged or repeated contact has been associated with the development of a dry, scaly dermatitis or with secondary infections.[3] Some skin absorption can occur with lengthy exposure to solvents containing benzene and may contribute more to toxicity than originally believed, but the dermal route is considered only a minor source of human exposure for the general population.[32]

The proposed 1995 ACGIH threshold limit value–time-weighted average (TLV-TWA) for benzene is 0.3 ppm (0.96 mg/m^3) with an A1 confirmed human carcinogen designation and a notation for skin absorption.

REFERENCES

1. *Environmental Health Criteria 150: Benzene.* World Health Organization, Geneva, 1993
2. Gerarde HW: *Toxicology and Biochemistry of Aromatic Hydrocarbons*, pp 97–108. New York, Elsevier, 1960
3. US Department of Labor: Occupational exposure to benzene. *Federal Register* 42:22516–22529, 1977
4. Goldstein BD: Hematotoxicity in humans. In Laskin S, Goldstein BD: Benzene toxicity: a

critical evaluation. *J Toxicol Environ Health* Suppl 2:69–105, 1977

5. Committee on Toxicology of the National Research Council: *Health Effects of Benezene—a Review.* US Department of Commerce, National Technical Information Service PB-254 388, pp 1–23. Washington, DC, National Academy of Sciences, 1976

6. Rappaport JM, Nathan DG: Acquired aplastic anemias: pathophysiology and treatment. *Adv Intern Med* 27:547–590, 1982

7. Snyder R, Witz G, Goldstein BD: The toxicology of benzene. *Environ Health Persp* 100:293–306, 1993

8. Vigliani EC: Leukemia associated with benzene exposure. *Ann NY Acad Sci* 271:143–151, 1976

9. Infante PF, Rinksy RA, Wagoner JK, Young RJ: *Leukemia Among Workers Exposed to Benzene.* National Institute for Occupational Safety and Health (NIOSH), US Department of Health, Education and Welfare. Washington, DC, US Government Printing Office, April 26, 1977

10. Aksoy M et al: Leukemia in shoe-workers exposed chronically to benzene. *Blood* 44:837–841, 1974

11. Aksoy M, Dincol K, Erden S, Dincol G: Acute leukemia due to chronic exposure to benzene. *Am J Med* 52:160–165, 1972

12. Yin SN et al: Leukaemia in benzene workers: a retrospective cohort study. *Br J Ind Med* 44:124–128, 1987

13. Rinsky RA et al: Leukemia in benzene workers. *Am J Ind Med* 2:217–245, 1981

14. Bond GG et al: An update of mortality among chemical workers exposed to benzene. *Br J Ind Med* 43:685–691, 1986

15. Aksoy M: Malignancies due to occupational exposure to benzene. *Am J Ind Med* 7:395–402, 1985

16. *IARC Monographs on the Evaluation of the Carcinogenic Risk of Chemicals to Humans*, Vol 29, Some industrial chemicals and dyestuffs, pp 93–148. Lyon, International Agency for Research on Cancer, May 1982

17. Marcus WL: Chemical of current interest—benzene. *Toxicol Ind Health* 3:205–266, 1987

18. Rinsky RA et al: Benzene and leukemia. An epidemiologic risk assessment. *N Engl J Med* 316:1044–1050, 1987

19. Paustenbach DJ, Price PS, Ollison W, et al: Reevaluation of benzene exposure for the pli-

ofilm (rubberworker) cohort (1936–1976). *J Toxicol Environ Health* 36:177–231, 1992

20. Paxton MB, Chinchilli VM, Brett SM, et al: Leukemia risk associated with benzene exposure in the pliofilm cohort. II. Risk estimates. *Risk Anal* 14:155–161, 1994

21. Crump KS: Risk of benzene-induced leukemia: a sensitivity analysis of the pliofilm cohort with additional follow-up and new exposure estimates. *J Toxicol Environ Health* 42:219–242, 1994

22. Occupational Safety and Health Administration: Occupational exposure to benzene; final rule, 29 CFR Part 1910. *Federal Register* 52(176):34460, 1987

23. Snyder CA et al: The inhalation toxicology of benzene: Incidence of hematopoietic neoplasms and hematotoxicity in AKR/J and C57BL/6J Mice. *Toxicol Appl Pharmacol* 54:323–331, 1980

24. Maltoni C et al: Benzene: a multipotential carcinogen. Results of long-term bioassays performed at the Bologna Institute of Oncology. *Am J Ind Med* 4:589–630, 1983

25. National Toxicology Program: *Toxicology and Carcinogenesis Studies of Benzene in F344N Rats and B6C3F1 Mice (Gavage Studies)*, pp 1–277. National Cancer Institute, NTP Technical Report 289, 1986

26. Tough IM, Brown WM: Chromosome aberrations and exposure to ambient benzene. *Lancet* 1:684, 1965

27. Styles JA, Richardson CR: Cytogenetic effects of benzene: dosimetric studies on rats exposed to benzene vapor. *Mutat Res* 135:203–209, 1984

28. Erixson GL et al: Induction of sister chromatid exchanges and micronuclei in male DBA/2 mice by the inhalation of benzene. *Environ Mutagen* 6:408, 1984. Abstract

29. Rosenthal GJ, Synder CA: Inhaled benzene reduces aspects of cell-mediated tumor surveillance in mice. *Toxicol Appl Pharmacol* 88:35–43, 1987

30. Kuna RA, Kapp RW Jr: The embryotoxic/teratogenic potential of benzene vapor in rats. *Toxicol Appl Pharmacol* 57:1–7, 1981

31. Walkley JE, Pagnotto LD, Elkins HB: The measurement of phenol in urine as an index of benzene exposure. *Am Ind Hyg Assoc J* 22:362–367, 1961

32. Susten AS et al: Percutaneous penetration of benzene in hairless mice: an estimate of der-

mal absorption during tire building operations. *Am J Ind Med* 7:323–335, 1985

BENZIDINE
CAS: 92-87-5

$C_{12}H_{12}N_2$

Synonyms: 4,4'-Biphenyldiamine; 4,4'-diaminobiphenyl; 4,4'-bianiline

Physical Form. Colorless, crystalline compound that darkens on oxidation

Uses. In manufacture of dyestuffs; hardener for rubber; laboratory reagent

Exposure. Skin absorption; inhalation

Toxicology. Benzidine exposure is associated with a high incidence of bladder cancer in humans.

Relatively little information is available on the noncarcinogenic effects of benzidine in humans.[1] Acute dermal exposure has reportedly caused severe, recurrent eczematous dermatitis, and chronic exposure may result in sensitizatio n dermatitis.[1]

Numerous studies have reported the occurrence of bladder cancer in workers exposed to benzidine by inhalation and through skin absorption.[1]

Of 25 workers involved in benzidine manufacture, 13 developed urinary bladder tumors, and 4 renal tumors also occurred. The average duration of exposure was 13.6 years, and the average induction time from first exposure to detection of the first tumor was 16.6 years. Initial tumors made their appearance as late as 9 years after cessation of exposure. It is not known if the cancers were influenced by concurrent exposure to other chemicals in the occupational environment.[2]

In a 30-year follow-up of a cohort of 984 workers employed at a benzidine manufacturing facility, there was a significant excess of bladder tumors among men with the highest estimated level of benzidine exposure.[3] The bladder cancer risk declined in those first employed after 1950, when preventive measures were instituted.

In a plant that manufactured β-naphthylamine and benzidine, a cohort of 639 male employees with exposure from 1938 or 1939 to 1965 was studied; concentration of initial exposure, duration of exposure, and years of survival after the exposure are factors that affect the incidence of tumor formation.[4] Of all malignant neoplasms, 35% were of the bladder and kidney. The observed mortality rate for cancer of the bladder was 78 per 100,000 in the cohort, as compared to 4.4 per 100,000 expected for men of the same age. Of 42 bladder and kidney neoplasms, 16 were attributed to benzidine exposure, and 18 were attributed to combined exposure.

During a 17-year period, 83 workers in a benzidine department were examined cystoscopically; 34 workers had congestive lesions, 3 had pedunculated papillomas, 4 had sessile tumors, and carcinoma was found in 13 of the workers.[5]

The onset of occupational bladder tumors is insidious and, occasionally, the disease may be in an advanced stage before any signs or symptoms appear. In general, however, benzidine exposure may produce a variety of lesions in the urinary bladder, such as hyperemia, inflammation and papillomata, which precede malignancy.[6] The presence of blood in the urine or pain upon urination may indicate such lesions. Detection of premalignant or malignant changes may be possible through cystoscopic examination, cytological evaluation of bladder epithelial cells shed in urine, and screening for occult blood.[6] Recurrences are frequent, and tumors may recur as papillomas or carcinomas irrespective of the nature of the original lesion.[7]

Susceptibility to bladder cancer in humans has been linked to the slow acetylator phenotype of the polymorphic NAT2 *N*-acetyltransferase gene.[1] In a recent study from China, a 25-fold increase in bladder cancer incidence and a 17-fold increase in bladder cancer mortality were determined in 1972 benzidine-exposed workers.[8] In the Asian population, the slow acetylator phenotype occurs significantly less often than in Caucasian populations, but an association between those who contracted bladder cancer and the phenotype has yet to be determined for this group.

The IARC has determined that there is sufficient evidence for carcinogenicity to humans.[9]

The ACGIH has classified benzidine as an A1 confirmed human carcinogen with no assigned threshold limit value and a notation for skin absorption.

REFERENCES

1. Agency for Toxic Substances and Disease Registry (ATSDR): *Toxicological Profile for Benzidine.* US Department of Health and Human Services, Public Health Service, p 121, 1993
2. Zavon MR, Hoegg U, Bingham E: Benzidine exposure as a cause of bladder tumors. *Arch Environ Health* 27:1–7, 1973
3. Meigs JW, Marrett LD, Ulrich FU, et al: Bladder tumor incidence among workers exposed to benzidine: a thirty year follow-up. *JNCI* 76:1–8, 1986
4. Mancuso TF, El-Attar AA: Cohort study of workers exposed to beta-naphthylamine and benzidine. *J Occup Med* 9:277–285, 1967
5. Barsotti M, Vigliani EC: Bladder lesions from aromatic amines. *AMA Arch Ind Hyg Occup Med* 5:234–241, 1952
6. *IARC Monographs on the Evaluation of the Carcinogenic Risk of Chemicals to Humans*, Vol 29, pp 149–183. Lyon, International Agency for Research on Cancer, 1982
7. Scott TS, Williams MHC: The control of industrial bladder tumours. *Br J Ind Med* 14:150–163, 1957
8. Bi W, Hayes RB, Feng P, et al: Mortality and incidence of bladder cancer in benzidine-exposed workers in China. *Am J Ind Med* 21:481–489, 1992
9. *IARC Monographs on the Evaluation of Carcinogenic Risks to Humans*, Suppl 7, pp 123–125. Lyon, International Agency for Research on Cancer, 1987

2,3-BENZOFURAN
CAS: 271-89-6

C_8H_6O

Synonyms: Benzofuran; benzo(b)furan; coumarone; cumarone; 1-oxindene

Physical Form.   Liquid

Uses.   As an intermediate in the polymerization of coumarone-indene resins found in various corrosion-resistant coatings such as paints and varnishes; in water-resistant coatings for paper products and fabrics; and in adhesives for use in food containers.

Exposure.   Inhalation

Toxicology.   Benzofuran is carcinogenic in experimental animals and causes kidney damage.

In 13-week studies designed to set the dose levels for a 2-year study, benzofuran in corn oil was given by gavage to both sexes of rats and mice.[1] Based on reduced mean body weights, increased severity of nephropathy, and hepatocellular necrosis, doses selected for the 2-year studies in rats were 30 and 60 mg/kg for males and 60 and 120 mg/kg for females. Based on increased mortality and nephrosis in male mice, doses selected were 60 and 120 mg/kg for males and 120 and 240 mg/kg for females.

There was clear evidence of carcinogenic activity for male and female mice, based on an increased incidence of neoplasms of the liver, lung, and forestomach. There was no evidence of carcinogenic activity in male rats. There was some evidence of carcinogenic activity for female rats, based on an increased incidence of tubular cell adenocarcinomas of the kidney.

Exposure to benzofuran increased the severity of nephropathy in male rats, increased the incidence of nephropathy in female rats, and induced hepatocellular metaplasia in the pancreas of female rats. Nonneoplastic lesions observed in mice included syncyt ial alteration of the liver, bronchiolar epithelial hyperplasia, and epithelial hyperplasia of the forestomach.

Although no studies provide data concerning human susceptibility, it is reasonable to assume that humans with kidney or liver disease would be more susceptible to the toxic effects of 2,3-benzofuran.[2]

A threshold limit value–time-weighted average (TLV-TWA) for benzofuran has not been assigned.

REFERENCES

1. National Toxicology Program: *Toxicology and Carcinogenesis Studies of Benzofuran (CAS No. 271-89-6) in F344/N Rats and B6C3F1 Mice (Gavage Studies).* Technical Report Series No 370, October 1989. National Technical Information Service, US Department of Commerce, Springfield, VA
2. Agency for Toxic Substances and Disease Registry: *Toxicological Profile for Benzofuran.* ATSDR/TP-91/04, Public Health Service, Centers for Disease Control, Atlanta, GA, pp 5–20, 1992

BENZOIC ACID
CAS:65-85-0

C_6H_5COOH

Synonyms: Benzenecarboxylic acid; phenyl carboxylic acid; phenylformic acid

Physical Form.   White crystals or powder

Uses.   As a food preservative; in plasticizers, flavors, and perfumes; as an antifungal agent

Exposure.   Inhalation; ingestion

Toxicology.   Benzoic acid is an irritant of the eyes and respiratory system.

Although specific dose levels and durations are not available, it is assumed that exposure to the dust may be irritating to the nose and eyes.[1] At elevated temperatures, fumes may cause irritation of the eyes, respiratory system, and skin.

The systemic toxicity of benzoic acid is low. Extremely large oral doses are expected to produce gastric pain, nausea, and vomiting.[2] In one case, a 67-kg man ingested a single dose of 50 mg without ill effects. In other cases, daily intake of 4 to 6 mg caused slight gastric irritation.[3] After ingestion, benzoic acid is conjugated with glycine and excreted as hippuric acid in the urine. However, no quantitative relationship exists between benzoic acid intake and the hippuric acid excreted.

The oral LD_{50} in cats and dogs is 2 mg/kg.[2] When benzoic acid is injected in rats, tremors, convulsions, and death occur.

On human skin, intermittent exposure to 22 mg for 3 days caused moderate irritation.[4] Application to rabbit skin of 500 mg for 24 hours produced mild irritation.[5] In the eyes of rabbits, 100 mg was severely irritating.[4]

The ACGIH has not assigned a threshold limit value for benzoic acid.

REFERENCES

1. Weiss G (ed): *Hazardous Chemicals Data Book*, p 147. Park Ridge, NJ, Noyes Data Co, 1980
2. Gosselin RE, Smith RP, Hodge HC: *Clinical Toxicology of Commercial Products*, 5th ed, p II–203. Baltimore, MD, Williams and Wilkins, 1984
3. Gilman AG, Goodman LS, Rall TW, et al: *Goodman and Gilman's The Pharmacological Basis of Therapeutics*, 7th ed, p 961. New York, Macmillan, 1985
4. Material Data Safety Sheet No 402. Benzoic acid. Genium Publishing Co, Schenectady, NY. February 1987
5. NIOSH: *Registry of Toxic Effects of Chemical Substances*, Vol 1, Sweet DV (ed), p 945. US Department of Health and Human Services, Public Health Service, 1987

BENZOTRICHLORIDE
CAS: 98-07-7

$C_7H_5Cl_3$

Synonyms: Benzenyl chloride; benzoic trichloride; benzylidyne chloride; benzyl trichloride; phenylchloroform; toluene trichloride; trichlorotoluene

Physical Form. Clear, oily liquid

Uses. Chemical intermediate primarily in benzoyl chloride production; dye intermediate

Exposure. Inhalation; skin absorption

Toxicology. Benzotrichloride is an irritant and a suspected human carcinogen.

The liquid has been reported to be highly irritating to the skin and mucous membranes in humans.[1]

In rats, benzotrichloride was lethal following a 4-hour exposure at 1000 mg/m³ (125 ppm). The oral LD_{50} in rats was 6 g/kg. The 2-hour LC_{50} in rats was 150 mg/m³ (19 ppm) and was 60 mg/m³ (8 ppm) in mice. Toxic effects included central nervous system excitation, irritation of the eyes and upper respiratory tract, and slowed respiration. Hyperemia of the extremities was also observed. Motor automatism and twitching of peripheral muscles were seen at 1000 mg/m³ (125 ppm) in mice and rats, respectively. Leukopenia, mild anemia, and decreases in renal function occurred in rats following continuous inhalation exposure at 100 mg/m³ (12.5 ppm) for one month.[1]

There are no data clearly relating exposure to benzotrichloride to cancer in humans. However, an excess of respiratory cancer (six cases total) was reported in benzoyl chloride manufacturing workers who were potentially exposed to benzotrichloride.[3]

Squamous-cell carcinomas of the skin were produced in three studies following skin application of benzotrichloride to mice.[1,4] Lung carcinomas, pulmonary adenomas, and lymphomas were also observed. Intraperitoneal injection of benzotrichloride produced a significant increase in the lung tumor response in strain A/J mice within 24 weeks.[5] Administration by gastric intubation of doses ranging from 2.0 to 0.0315 µl/mouse, twice a week for 25 weeks, to female ICR mice produced forestomach tumors (squamous-cell carcinoma and papilloma), lung tumors (adenocarcinoma and adenoma), and tumors of the hematopoietic system (thymic lymphosarcoma and lymphatic leukemia) with dose-related response by 18 months.[6] It was concluded that the target organ of benzotrichloride carcinogenesis in mice is the local tissue, which is primarily exposed, and the lung and hematopoietic tissue when the dose is administered systemically.

The International Agency for Research on Cancer has determined that there is limited evidence that employment in the production of benzoyl chloride and its chlorinated toluene precursors that involves exposure to benzotrichloride represents a carcinogenic risk.[1] There is sufficient evidence that benzotrichloride is carcinogenic in mice.

The ACGIH has not established a threshold limit value for benzotrichloride.

REFERENCES

1. *IARC Monographs on the Evaluation of Carcinogenic Risk to Humans*, Vol 29, Some industrial chemicals and dyestuffs, pp 73–80. Lyon, France, International Agency for Research on Cancer, 1982
2. *Registry of Toxic Effects of Chemical Substances*, pp 26910–26911. Washington, DC, National Institute for Occupational Safety and Health, 1990
3. Sorahan T, Waterhouse JAH, Coke MA, et al: A mortality study of workers in a factory manufacturing chlorinated toluenes. *Ann Occup Hyg* 27:173–182, 1983
4. Fukuda K, Matsushita H, Sakabe H, et al: Carcinogenicity of benzyl chloride, benzal chloride, benzotrichloride and benzoyl choride in mice by skin application. *Gann* 72:655–664, 1981
5. Stoner GD, You M, Morgan MA, et al: Lung tumor induction in strain A mice with benzotrichloride. *Cancer Lett* 33:167–173, 1986
6. Fukuda K, Matsushita H, Takemoto K, et al:

Carcinogenicity of benzotrichloride administered to mice by gastric intubation. *Ind Health* 31:127–131, 1993

BENZOYL PEROXIDE
CAS: 94-36-0

$(C_6H_5CO)_2O_2$

Synonyms: Benzoyl superoxide; dibenzoyl peroxide

Physical Form. Granular, white solid

Uses. In bleaching flour and edible oils; additive in self-curing of plastics

Exposure. Inhalation

Toxicology. Benzoyl peroxide is an irritant of mucous membranes and causes both primary irritation and sensitization dermatitis.

Exposure of workers to levels of 12.2 mg/m^3 and higher has caused pronounced irritation of the nose and throat.[1]

Application to the face as lotion for acne treatment in two persons caused facial erythema and edema; patch tests with benzoyl peroxide were positive.[2] In contact with the eyes, it may produce irritation and, if allowed to remain on the skin, it may produce inflammation.[3] No systemic effects have been reported in humans. The major hazards of benzoyl peroxide are fires and explosions, which have caused serious injuries and death.[4]

Rats exposed at an atmospheric concentration of 24.3 mg/l of 78% benzoyl peroxide showed the following signs during a 4-hour exposure period: eye squint, difficulty in breathing, salivation, lacrimation, erythema, and an increase followed by a decrease in motor activity.[4] All rats appeared normal at 24 and 48 hours postexposure.

Benzoyl peroxide has been tested for carcinogenicity in mice and rats by administration in the diet and by subcutaneous injection and in mice by skin application.[5] Although no significant increases in tumor incidences were found, the IARC has determined that all the studies were inadequate for a complete evaluation of carcinogenicity in animals. Two studies indicated that benzoyl peroxide may act to promote cancer on mouse skin.[6,7]

Among a small factory population, two cases of lung cancer were found in men primarily involved in the production of benzoyl peroxide, but they were also exposed to benzoyl chloride and benzotrichloride.[8] According to the IARC,[5] no evaluation of the carcinogenicity of benzoyl peroxide to humans can be made.

The 1995 ACGIH threshold limit value–time-weighted average (TLV-TWA) for benzoyl peroxide is 5 mg/m^3.

REFERENCES

1. Benzoyl peroxide. *Documentation of the TLVs and BEIs*, 6th ed, pp 123–124. Cincinnati, OH, American Conference of Governmental Industrial Hygienists, 1991
2. Eaglstein WH: Allergic contact dermatitis to benzoyl peroxide—report of cases. *Arch Dermatol* 97:527, 1968
3. Chemical Safety Data Sheet SD-81, Benzoyl peroxide, pp 3–4, 10. Washington, DC, MCA, Inc, 1960
4. National Institute for Occupational Safety and Health: *Criteria for a Recommended Standard . . . Occupational Exposure to Benzoyl Peroxide.* DHEW (NIOSH) Pub 77–166, p 117. Washington, DC, US Government Printing Office, June 1977
5. *IARC Monographs on the Evaluation of the Carcinogenic Risks of Chemicals to Humans*, Vol 36, Allyl compounds, aldehydes, epoxides and peroxides, pp 267–283. Lyon, International Agency for Research on Cancer, 1985
6. Reiners JJ Jr et al: Murine susceptibility to two-stage skin carcinogenesis is influenced by the agent used for promotion. *Carcinogenesis* 5:301–307, 1984
7. Slaga TJ et al: Skin tumor—promoting activity of benzoyl peroxide, a widely used free radical-generating compounds. *Science* 213:1023–1025, 1981
8. Sakabe H, Fukuda K: An updating report on cancer among benzoyl chloride manufacturing workers. *Ind Health* 15:173–174, 1977

BENZO[a]PYRENE

CAS: 50-32-8

$C_{20}H_{12}$

Synonyms: B[a]P; BP; 3,4-benzopyrene; 3,4-benzpyrene

Physical Form. Yellow crystals

Sources. B[a]P is a major component of polynuclear aromatic hydrocarbons, also known as polycyclic aromatic hydrocarbons, and is usually bound to small particulate matter present in urban air, industrial and natural combustion emissions, and cigarette smoke.

Exposure: Inhalation

Toxicology. Benzo[a]pyrene (B[a]P) causes hematological and immunological effects; it is carcinogenic to experimental animals.

Systemic effects from B[a]P exposure have not been reported in humans.

Intermediate-duration oral exposure of mice has caused death due to adverse hematological effects, including aplastic anemia and panocytopenia.[1] B[a]P has been shown to inhibit the immune system markedly, especially T cell–dependent antibody production by lymphocytes exposed either in vivo or in vitro. It may also induce autoimmune responses.

B[a]P has been carcinogenic in all animal species tested to date, including mouse, rat, hamster, rabbit, guinea pig, duck, newt, dog, monkey, and fish.[2] It is metabolized to approximately 20 primary and secondary oxidized metabolites and to a variety of conjugates.[3] The most potent carcinogenic metabolite is 7,8-dihydroxy-9,10-epoxy-7,8,9,10-tetrahydrobenzo[a]pyrene. This ultimate carcinogen binds predominantly to guanine bases in DNA to form covalent adducts.

B[a]P metabolites have been shown to bind to DNA in cultured human hepatocytes and in human bladder and tracheobronchial explants.[4-5] The metabolites identified were identical to those produced in other species and differed only in the relative percentages of formation.[5] Human tissues were most active in metabolizing B[a]P and exhibited at least a threefold higher covalent binding of metabolites to DNA than hamsters, dogs, monkeys, or rats. In addition, B[a]P has been tested extensively in several bacterial and mammalian cell systems and has been chosen as a positive control for the validation of some of these systems.[6]

The IARC considers that there is "sufficient evidence" that B[a]P is carcinogenic to experimental animals.[7]

Developmental toxicity and impaired reproductive capacity were seen in two oral studies in mice.[8,9] The lowest observed adverse effect level was 10 mg/kg/day from days 7–16 of gestation.[9]

Benzo[a]pyrene is designated an A2 suspected human carcinogen by ACGIH and has no assigned threshold limit value.

REFERENCES

1. Agency for Toxic Substances and Disease Registry (ATSDR): *Toxicological Profile for Polycyclic Aromatic Hydrocarbons (PAHs).* US Department of Health and Human Services, Public Health Service, 1993
2. *IARC Monographs on the Evaluation of the Carcinogenic Risk of Chemicals to Humans,* Vol 3, Certain polycyclic aromatic hydrocarbons and heterocyclic compounds, pp 91–136. International Agency for Research on Cancer, 1973
3. Pelkonen O, Nebert DW: Metabolism of polycyclic aromatic hydrocarbons: etiologic role in carcinogenesis. *Pharmacol Rev* 34:189–222, 1982
4. Monteith DK, Novoting A, Michalopoulos G, Stron SC: Metabolism of benzo[a]pyrene in primary cultures of human hepatocytes: dose-response over a four-log range. *Carcinogenesis* 8:983–988, 1987
5. Daniel FB, Schut HAJ, Sandwisch DW, et al: Interspecies comparisons of benzo[a]pyrene metabolism and DNA-adduct formation in cultured human and animal bladder and tracheobronchial tissues. *Cancer Res* 43:4723–4729, 1983
6. Agency for Toxic Substances and Disease Registry: *Toxicological Profile for Benzo[a]pyrene.* ATSDR/TP-88/05, Public Health Service,

Centers for Disease Control, Atlanta, GA, pp 48–53, 1990
7. *IARC Monographs on the Evaluation of the Carcinogenic Risk of Chemicals to Humans*, Vol 32, Polynuclear aromatic compounds, Part 1, Chemical, environmental and experimental data, pp 211–224. Lyon, International Agency for Research on Cancer, December 1983
8. Mackenzie KM, Angevine DM: Infertility in mice exposed in utero to benzo[a]pyrene. *Biol Reproduc* 24:183–191, 1981
9. Legraverend C, Guenther TM, Nebert DW: Importance of the route of administration for genetic differences in benzo[a]pyrene-induced in utero toxicity and teratogenicity. *Teratology* 29:35–47, 1984

BENZYL CHLORIDE
CAS: 100-44-7

$C_6H_5CH_2Cl$

Synonyms: α-Chlorotoluene; (chloromethyl) benzene

Physical Form.   Colorless liquid

Uses.   In manufacture of benzyl compounds, cosmetics, dyes, and resins

Exposure.   Inhalation

Toxicology.   Benzyl chloride is a severe irritant of the eyes, mucous membranes, and skin.

Benzyl chloride is a powerful lacrimator (an immediate warning sign) and, at 31 ppm, it is unbearably irritating to the eyes and nose.[1] At 16 ppm, it is intolerable after 1 minute. Workers exposed to 2 ppm complained of weakness, irritability, and persistent headache.[2]

One author reported disturbances of liver function and mild leukopenia in some exposed workers, but this study has not been confirmed.[2]

Splashes of the liquid in the eye will produce severe irritation and corneal injury. Skin contact may produce dermatitis, and skin sensitization has been reported in guinea pigs.[3]

The LC_{50}s in mice and rats for a 2-hour inhalation exposure are 80 and 150 ppm, respectively.[4] In another investigation, it was found that all mice and rats survived 400 ppm for 1 hour.[5] Cats exposed to 100 ppm 8 hr/day for 6 days exhibited eye and respiratory tract irritation, which appeared sooner and with increasing severity each exposure day.[6]

Repeated high-dose subcutaneous injections produced local tumors in rats; skin application to a limited number of mice caused an increase in squamous-cell carcinomas of the skin, which were not statistically significant.[7,8] Administered by gavage at a dose of 50 or 100 mg/kg in mice and 15 or 30 mg/kg in rats 3 times per week for 2 years, benzyl chloride produced a statistically significant increase in the incidence of papillomas and carcinomas of the forestomach in mice and an increase in thyroid tumors in female rats.[9] The carcinogenic potential has not been determined in man.[6,7] Evidence of efficient detoxification mechanisms suggests that the risk from chronic low-level exposure is small.[6] The IARC has determined that there is inadequate evidence for carcinogenicity to humans and limited evidence for carcinogenicity to animals.[10]

Benzyl chloride was not teratogenic in rats orally administered 100 mg/kg on days 6–15 of gestation; slight fetotoxicity in the form of reduced fetal length was observed at this level.[11]

The 1995 ACGIH threshold limit value–time-weighted average (TLV-TWA) for benzyl chloride is 1 ppm (5.2 mg/m³).

REFERENCES

1. Smyth HF Jr: Hygienic standards for daily inhalation. *Am Ind Hyg Assoc Q* 17:147, 1956
2. Mikhailova TV: Benzyl chloride. In International Labour Office: *Encyclopaedia of Occupational Health and Safety*, Vol I, A–K, pp 169–170. New York, McGraw-Hill, 1971
3. Landsteiner K, Jacobs J: Studies on the sensitization of animals with simple chemical compounds—II. *J Exp Med* 64:625–639, 1936
4. Mikhailova TV: Comparative toxicity of chloride derivatives of toluene—benzyl chloride, benzal chloride and benzotrichloride. *Fed Proc* 24:T877–800, 1965

5. Back KC et al: *Reclassification of Materials Listed as Transportation Health Hazards,* Report TSA-20-72-3, pp 24–25. A-264–A-265. Washington DC, Department of Transportation, Office of Hazardous Materials, Office of Assistant Secretary for Safety and Consumer Affairs, 1972
6. National Institute for Occupational Safety and Health: *Criteria for a Recommended Standard . . . Occupational Exposure to Benzyl Chloride.* DHEW (NIOSH) Pub No 78–182, p 92. Washington, DC, US Government Printing Office, 1978
7. Preussman R: Direct alkylating agents as carcinogens. *Food Cosmet Toxicol* 6:576–577, 1968
8. Fukuda K, Matsushita H, Sakabe H, et al: Carcinogenicity of benzyl chloride, benzal chloride, benzotrichloride and benzoyl chloride in mice by skin application. *Gann* 72:655–664, 1981
9. Lijinsky W: Chronic bioassay of benzyl chloride in F344 Rats and (C57BL/6JxBALB/c) F1 Mice. *J Natl Cancer Inst* 76:1231–1236, 1986
10. *IARC Monographs on the Evaluation of the Carcinogenic Risk of Chemicals to Humans,* Suppl 7, *Overall Evaluations of Carcinogenicity: An Updating of IARC Monographs Volumes 1–42,* pp 148–149. Lyon, France, International Agency for Research on Cancer, 1987
11. Skowronski G, Abdel-Rahman MS: Teratotoxicity of benzyl chloride in the rat. *J Tox Environ Health* 17:51–56, 1986

BERYLLIUM AND COMPOUNDS

CAS: 7440-41-7

Be

Synonyms/Compounds: Glucinium; beryllium oxide; beryllium chloride; beryllium fluoride; beryllium hydroxide; beryllium phosphate; beryllium nitrate; beryllium sulfate; beryllium carbonate

Physical Form. Elemental beryllium is a gray metal

Uses. Beryllium metal sheet or wire; ceramics; hardening agent in alloys used especially in the electronics field

Exposure. Inhalation

Toxicology. Exposure to compounds of beryllium may cause dermatitis, acute pneumonitis, and chronic pulmonary granulomatosis (berylliosis) in humans. The compounds are carcinogenic in experimental animals and are suspected human carcinogens.

Acute lung disease, now chiefly of historical importance because of improved working conditions, has resulted from brief exposures to high concentrations of the oxide, phosphor mixtures, or the acid salts.[1] All segments of the respiratory tract may be involved, with rhinitis, pharyngitis, tracheobronchitis, and pneumonitis.[2] The pneumonitis may be fulminating, following high exposure levels, or less severe, with gradual onset, from lesser exposures.[3] In the majority of cases with acute beryllium pneumonitis, recovery occurs within 1 to 6 months; however, fatalities due to pulmonary edema or to spontaneous pneumothorax have been reported. The human threshold of an injurious concentration by inhalation is approximately 30 mg Be/m^3 for the high fired oxide, 1 to 3 mg Be/m^3 for the low fired oxide, and 0.1 to 0.5 mg Be/m^3 for the sulfate.[2]

Beryllium disease is regarded as chronic if it persists for a year or more, and it is usually due to granulomas in the lungs.[1,4] The onset of "berylliosis" may be insidious, with only slight cough and fatigue, which can occur as early as 1 year or as late as 25 years after exposure.[2] Progressive pulmonary insufficiency, anorexia, weight loss, weakness, chest pain, and constant hacking cough characterize the advanced disease. Cyanosis and clubbing of fingers may be seen in approximately a third of cases, and cor pulmonale is another frequent sequela.[2]

Early X-rays show a fine diffuse granularity in the lungs, whereas a diffuse reticular pattern is observed in the second stage and, finally, in the third stage, distinct nodules appear.[5]

There are many similarities between berylliosis and sarcoidosis but, in sarcoidosis, the systemic effects are much more pronounced.[6]

An immunologic basis for chronic beryllium disease has been postulated and a hyper-sensitivity phenomenon demonstrated.[4,6] Consistent with the concept of chronic berylliosis as a hypersensitivity pulmonary reaction are the following factors: Persons with berylliosis also show delayed cutaneous hypersensitivity reactions to beryllium compounds; their peripheral blood lymphocytes undergo blast transformation and release of macrophage inhibition factor after exposure to beryllium in vitro; they have depressed helper/suppressor T-cell ratios; and a dose-response relationship is lacking in chronic beryllium cases.[2,4] Hypersensitization may lead to berylliosis in people with relatively low exposures, whereas nonsensitized individuals with higher exposures may have no effects.

Skin contact with soluble beryllium salts may produce either primary irritation or sensitization dermatitis characterized by pruritis with an eruption of erythematous, papular, or papulovesicular nature; the eruption usually subsides within 2 weeks after cessation of exposure.[1] Implantation of beryllium or its compounds beneath the skin may cause necrosis of adjacent tissue and formation of an ulcer; implantation of comparatively insoluble compounds may produce a localized gran uloma, as has occurred from lacerations with old fluorescent tubes containing the phosphor.[3] Healing of ulcers and granulomas requires the surgical removal of the beryllium substance.[3] Conjunctivitis may accompany contact dermatitis resulting from exposure to soluble beryllium compounds; angioneurotic edema may be striking.[1,3]

Beryllium metal, beryllium-aluminum alloy, beryl ore, beryllium chloride, beryllium fluoride, beryllium hydroxide, beryllium sulfate, and beryllium oxide all produce lung tumors in rats exposed by inhalation or intratracheally.[7] The oxide and the sulfate produce lung tumors in monkeys after intrabronchial implantation or inhalation. A number of compounds produce osteosarcomas in rabbits following their intravenous or intramedullary administration.[7]

Although a number of epidemiologic studies have reported an increased risk of lung cancer among occupationally exposed beryllium workers, deficiencies in the studies limit any unequivocal conclusion.[7-9] Specific criticisms

include no consideration of latent effects, of smoking history, or of exposure to other potential carcinogens and underestimation of expected lung cancer deaths in comparison populations.[10,11]

A more recent study that accounted for smoking, included females, and extended the latency period has strenghtened the evidence of carcinogenicity in humans.[12] A cohort mortality study of 689 patients with beryllium disease, as determined by a case registry, found a lung cancer standardized mortality ratio (SMR) of 2.00 based on 28 observed lung cancer deaths.[13] The lung cancer excess was more pronounced in individuals with a history of acute forms of beryllium disease than among those with chronic disease. Patients with a history of acute beryllium disease and lung cancer were found to be employed by one plant in Lorain, Ohio, where exposures as high as 4700 $\mu g/m^3$ were reported during the 1940s.[14] It has been noted that this exposure level is several orders of magnitude higher than those in existence today.[14] Slight excesses in lung cancer rates were found in four of five plants operating during the 1950s, when exposures were considered to be lower than the extremely high 1940s levels.[15]

The IARC has determined that there is sufficient evidence in both humans and animals for the carcinogenicity of beryllium and beryllium compounds.[7] Based on a variety of assays, it appears that soluble beryllium compounds are weakly genotoxic.[12]

The 1995 ACGIH threshold limit value–time-weighted average (TLV-TWA) for beryllium and compounds as Be is 0.002 mg/m^3 with an A2 suspected human carcinogen designation.

REFERENCES

1. Tepper LB, Hardy HL, Chamberlin RI: *Toxicity of Beryllium Compounds*, pp 31–80. New York, Elsevier, 1961
2. Reeves AL: Beryllium. In Friberg L et al (eds): *Handbook on the Toxicology of Metals*, 2nd ed, Vol II, pp 95–116. Elsevier/North-Holland Biomedical Press, 1986
3. Hygienic Guide Series: Beryllium and its compounds. *Am Ind Hyg Assoc J* 25:614–617, 1964

4. Deodhar SD, Barna B, Van Ordstrand HS: A study of the immunologic aspects of chronic berylliosis. *Chest* 63:309–313, 1973
5. Hardy HL, Tabershaw IR: Delayed chemical pneumonitis occurring in workers exposed to beryllium compounds. *J Ind Hyg Toxicol* 28:197–211, 1946
6. *Criteria for a Recommended Standard . . . Occupational Exposure to Beryllium*, (NIOSH) Pub No 72-10268. National Institute for Occupational Safety and Health, US Department of Health, Education and Welfare. Washington, DC, US Government Printing Office, 1972
7. *IARC Monographs on the Evaluation of the Carcinogenic Risks to Humans*, Vol 58, Beryllium, cadmium, mercury and exposures in the glass manufacturing industry, pp 41–117. Lyon, International Agency for Research on Cancer, 1993
8. Mancuso TF: Mortality study of beryllium industry workers' occupational lung cancer. *Environ Res* 21:48–55, 1980
9. Wagoner JK, Infante PF, Bayliss DL: Beryllium: an etiologic agent in the induction of lung cancer, nonneoplastic respiratory disease, heart disease among industrially exposed workers. *Environ Res* 21:15–34, 1980
10. Smith RJ: Beryllium report disputed by listed author: a standing controversy over an NIOSH study is revived by one of the scientists involved. *Science* 211:556–557, 1981
11. Reeves AC: Beryllium: toxicological research of the last decade. *J Am College Toxicol* 8:1307–1312, 1989
12. Agency for Toxic Substances and Disease Registry (ATSDR): *Toxicological Profile for Beryllium*. US Department of Health and Human Services, Public Health Service, p 129, TP-92/04, 1993
13. Steenland K, Ward E: Lung cancer incidence among patients with beryllium disease: a cohort mortality study. *J Natl Cancer Inst* 83:1380–1385, 1991
14. Eisenbud M: Re: Lung cancer incidence among patients with beryllium disease [letter]. *J Natl Cancer Inst* 85:1697–1698, 1993
15. Ward E, Okun A, Ruder A, et al: A mortality study of workers at seven beryllium processing plants. *Am J Ind Med* 22:885–904, 1992

BIPHENYL
CAS: 92-52-4

$C_6H_5C_6H_5$

Synonyms: Diphenyl; phenylbenzene; bibenzene; 1,1-biphenyl; PhPh

Physical Form. Colorless to yellow solid

Uses. Heat-transfer agent; fungistat for citrus fruits; in organic synthesis

Exposure. Inhalation; skin absorption

Toxicology. Biphenyl is an irritant of the eyes and mucous membranes and may exert a toxic action on the central and peripheral nervous systems.

In a study of 33 workers in one plant with prolonged exposure to concentrations ranging up to 123 mg/m³, the most common complaints were headache, gastrointestinal symptoms (diffuse pain, nausea, indigestion), numbness and a ching of limbs, and general fatigue.[1] Neurophysiologic examination of 22 of these workers showed that 19 had changes consistent with central and/or peripheral nervous system damage. In one fatal case in this plant, exposure was high for 11 years, symptoms were as just described and, at autopsy, there was widespread liver necrosis with some cirrhotic areas, nephrotic changes, heart muscle degeneration, and edematous brain tissue.[1]

In a follow-up study, 10 of 24 workers showed electroencephalographic abnormalities, which persisted for 1 and 2 years after the initial investigation; nine workers had electromyographic abnormalities, which also persisted.[2]

Irritation to the eyes and mucous membranes has been reported in humans exposed at 3 to 4 ppm.[3] Repeated skin contact may produce sensitization or dermatitis.

Exposure of rats to biphenyl dust impregnated on diatomaceous earth at a concentration of 300 mg/m³ for 7 hr/day for 64 days caused irritation of the nasal mucosa, bronchopulmonary lesions, and slight injury to the liver and kidneys.[4]

The 1995 ACGIH threshold limit value–time-weighted average (TLV-TWA) is 0.2 ppm (1.3 mg/m³).

REFERENCES

1. Hakkinen I, Silatanen E, Hernberg S, et al: Diphenyl poisoning in fruit paper production. *Arch Environ Health* 26:70–74, 1973
2. Sappalainen AM, Hakkinen I: Electrophysiological findings in diphenyl poisoning. *J Neurol Neurosurg Psychiatry* 38:248–252, 1975
3. Sandmeyer EE: Aromatic hydrocarbons. In Clayton GD, Clayton FE (eds): *Patty's Industrial Hygiene and Toxicology*, 3rd ed, rev. Vol 2B, *Toxicology*, pp 3325–3330. New York, Wiley-Interscience, 1981
4. Deichmann WB, Kitzmiller KV, Dierker M, Witherup S: Observations on the effects of diphenyl, *o*- and *p*-aminodiphenyl, *o*- and *p*-nitrodiphenyl and dihydroxyoctachlorodiphenyl upon experimental animals. *J Ind Hyg Toxicol* 29:1–3, 1947

BISMUTH TELLURIDE
CAS: 1304-82-1

Bi_2Te_3

Synonym: Dibismuth tritelluride

Physical Form.   Gray solid

Uses.   In semiconductors; for thermoelectric cooling; power generation application; for commercial use, Bi_2Te_3 is "doped" with selenium sulfide to alter its conductivity

Exposure.   Inhalation

Toxicology.   Bismuth telluride, either alone or doped with selenium sulfide, is apparently of very low toxicity.

In limited industrial experimental work with bismuth telluride under controlled conditions (vacuum hoods), no adverse health effects were encountered other than tellurium breath.[1]

In a multispecies study, dogs, rabbits, and rats were exposed to 15 mg/m³ of bismuth telluride doped with stannous telluride for 6 hr/day, 5 days/week, for 1 year.[1] Small granulomatous lesions without fibrosis occurred in the lungs of dogs at 6 months. In dogs autopsied 4 months after an 8-month exposure, the lesions had regressed, indicating a reversible process. Rabbits showed a similar reaction but with a decreased number of pulmonary macrophages, no fibrous tissue activity, and no cellular or fibrous tissue reaction around the dust deposits in the lymph nodes. The rats exhibited no fibrosis and no lymph node reactions. The pulmonary lesions seen in the study were present in all three exposed species but were interpreted as mild and reversible and not of serious physiologic consequence.

In a similar 11-month study in which animals were exposed to undoped bismuth telluride dust of 0.04-μm diameter at 15 mg/m³, no adverse responses of any type were observed other than the pulmonary responses to the inhalation of an inert dust.

The 1995 ACGIH threshold limit value–time-weighted average (TLV-TWA) is 10 mg/m³ for undoped, and 5 mg/m³ for doped, bismuth telluride.

REFERENCES

1. Stokinger HE: The metals. In Clayton GD, Clayton, FE (eds): *Patty's Industrial Hygiene and Toxicology*, 3rd ed, rev, Vol 2A, *Toxicology*, pp 1558–1563. New York, Wiley-Interscience, 1981

BORATES, TETRA, SODIUM SALTS
CAS: 1303-96-4

$Na_2B_4O_7$: Anhydrous
$Na_2B_4O_7 \cdot 5H_2O$: Pentahydrate
$Na_2B_4O_7 \cdot 10H_2O$ Decahydrate (borax)

Synonyms/Compounds: Sodium borate; sodium pyroborate; boric acid, disodium salts

Physical Form. Anhydrous—gray solid; pentahydrate—whitesolid; decahydrate—white solid

Uses. In cleaning compounds; in fertilizers; in manufacture of glazes and enamels

Exposure. Inhalation; ingestion

Toxicology. Borates are irritants of the eyes, nose, and throat; at high concentrations, ingestion of the compounds can result in gastrointestinal irritation, kidney injury, and even death from central nervous system depression or cardiovascular collapse.

Under normal conditions of exposure, borates are primarily irritants of the skin and respiratory system.[1] Workers exposed to anhydrous sodium tetraborate complained of nasal irritation, nose bleeds, cough, shortness of breath, and dermatitis.[2] Exposure levels were not measured, but total dust levels were described as high enough to obscure visibility in production areas. In another study of borax workers, symptoms of acute respiratory irritation, including dryness of the mouth, nose, or throat; cough; nose bleeds; and shortness of breath were related to exposures of 4 mg/m³ or more.[3]

There were more frequent symptoms of respiratory tract irritation and mucous membrane irritation among workers exposed during a 7-year period to average borax concentrations of 1.5 mg/m³ compared to unexposed controls.[1] Occasional excursions to levels of 10 mg/m³ produced no functional changes in respiration, and irritation was classified as mild.

Dermal effects may be noted following either direct contact with the compounds or ingestion.[1] Erythematous rash with desquamation may develop.

Systemic toxicity may occur following chronic or multiple exposures.[1] Possible effects include gastrointestinal irritation with nausea, vomiting, and diarrhea; kidney injury such as oliguria or anuria; central nervous system depression; and vascular collapse.

In rats, the oral LD_{50} values for borates are essentially the same as for boric acid, ranging from 3.16 to 6.08 g/kg.[4] When borax was fed to dogs and rats for 2 years, 350 ppm as boron in the diet had no effect. In a three-generation feeding study in rats, 350 ppm had no effect on fertility, litter size, weight, or appearance.

Sodium borate tested negatively in the Ames bioassay but was found to be cytotoxic to cultured human fibroblasts.[5]

The 1995 ACGIH threshold limit value–time-weighted average (TLV-TWA) for anhydrous and pentahydrate borates is 1 mg/m³ and 5 mg/m³ for the decahydrate.

REFERENCES

1. Von Burg R: Toxicology update: boron, boric acid, borates and boron oxide. *J Appl Toxicol* 12:149–152, 1992
2. Birmingham DJ, Key MM: Preliminary Survey, US Borax Plant, Boron, California (Feb 20, 1963). Occupational Health Research and Training Facility, Division of Occupational Health, Public Health Service, US Department of Health, Education and Welfare, Cincinnati, OH, 1963
3. Garabrant DH, Bernstein L, Peters JM, et al: Respiratory effects of borax dust. *Br J Ind Med* 42:831–837, 1984
4. Weir RJ Jr, Fisher RS: Toxicologic studies on borax and boric acid. *Toxicol Appl Pharmacol* 23:351–364, 1972
5. Landolph JR: Cytotoxicity and negligible genotoxicity of borax and boric ores to cultured mammalian cells. *Am J Ind Med* 7:31–44, 1985

BORON OXIDE
CAS: 1303-86-2

B_2O_3

Synonyms: Boric anhydride; boron sesquioxide; boron trioxide; fused boric acid

Physical Form. Colorless crystals

Uses. In preparation of fluxes; component of enamels and glass; catalyst in organic reaction

Exposure. Inhalation

Toxicology. Boron oxide is an eye and respiratory irritant.

In 113 workers exposed to boron oxide and boric acid dusts, there were statistically significant increases in symptoms of eye irritation; dryness of the mouth, nose, and throat; sore throat; and productive cough compared with controls.[1] The mean exposure level was 4.1 mg/m^3, with a range of 1.2 to 8.5 mg/m^3. Exposures may occasionally have exceeded 10 mg/m^3. Because of mixed exposures, the study does not indicate whether boron oxide or boric acid dust is more important in causing symptoms, nor does it indicate the minimum duration of exposure necessary to produce symptoms.

Excessive absorption of boron oxide may lead to cardiovascular collapse, alterations in temperature regulation, and coma.[2]

Repeated exposure of rats to an aerosol at a concentration of 470 mg/m^3 for 10 weeks caused only mild nasal irritation; repeated exposure of rats to 77 mg/m^3 for 23 weeks resulted in elevated creatinine and boron content of the urine in addition to increased urinary volume.[3] Conjunctivitis resulted when the dust was applied to the eyes of rabbits, probably the result of the exothermic reaction of boron oxide with water to form boric acid; topical application of boron oxide dust to the clipped backs of rabbits produced erythema that persisted for 2 to 3 days.[3]

The 1995 ACGIH threshold limit value–time-weighted average (TLV-TWA) for boron oxide is 10 mg/m^3.

REFERENCES

1. Garabrant DH, Bernstein L, Peters J, et al: Respiratory and eye irritation from boron oxide and boric acid dusts. *J Occup Med* 26:584–586, 1984
2. Von Burg R: Toxicology update: boron, boric acid, borates and boron oxide. *J Appl Toxicol* 12:149–152, 1992
3. Wilding JL, Smith WJ, Yevich P, et al: The toxicity of boron oxide. *Am Ind Hyg Assoc J* 20:284–289, 1959

BORON TRIBROMIDE
CAS: 10294-33-4

BBr_3

Synonyms: Boron bromide

Physical Form. Colorless, fuming liquid with a sharp, irritating odor

Uses. Catalyst in manufacture of diborane, ultrahigh-purity boron, and semiconductors

Exposure. Inhalation

Toxicology. Boron tribromide is expected to be an irritant of the eyes, nose, and mucous membranes.

Boron tribromide reacts violently and explosively with water to yield hydrogen bromide.[1]

Effects of short-term exposure are expected to be irritation of the eyes, nose, throat, and skin. Contact with skin or the eyes may cause burns.[2]

The 1995 threshold limit value–time-weighted average (TLV-TWA) is 1 ppm (10 mg/m^3) as a ceiling limit.

REFERENCES

1. Lower LD: Boron halides. In Grayson M (ed): *Kirk–Othmer Concise Encyclopedia of Chemical*

Technology, p 180. New York, Wiley-Interscience, 1985

2. Chemical Safety Information: *Occupational Safety and Health Guidelines for Chemical Hazards*. US Department of Health and Human Services, Public Health Service, Centers for Disease Control, National Institute for Occupational Safety and Health, Division of Standards Development and Technology Transfer, 4676 Columbia Parkway, Cincinnati, OH, 1992

BORON TRIFLUORIDE
CAS: 7637-07-2

BF_3

Synonym: Boron fluoride

Physical Form. Colorless gas; forms dense white fume in moist air

Uses. In catalysis with and without promoting agents; fumigant; flux for soldering magnesium

Exposure. Inhalation

Toxicology. Boron trifluoride gas is a severe irritant of the lungs, eyes, and skin.

Examination of 13 workers with present or past occupational exposure found 8 with abnormalities of pulmonary function; chest X rays were negative, and preshift urinary fluoride concentrations did not exceed 4 mg/l.[1] Air sampling showed concentrations ranging from 0.1 to 1.8 ppm. Dryness of the nasal mucosa and epistaxis were attributed to boron trifluoride exposure in workers exposed to high concentrations for 10 to 15 years.[2]

Cotton soaked with boron trifluoride in water and placed on the skin for a day or so resulted in a typical acid burn; there was no evidence of the more severe hydrogen fluoride burn occurring because boron trifluoride has the ability to complex the fluoride ion effectively.[1]

In rats, the 4-hour LC_{50} was 436 ppm for boron trifluoride dihydrate, which is formed when boron trifluoride gas reacts with moisture. Clinical signs included gasping, excessive oral and nasal discharge, and lacrimation.[3] In a 2-week study, all animals exposed at 67 ppm, 6 hr/day, died prior to the sixth exposure, and histopathology showed necrosis and pyknosis of the proximal tubular epithelium of the kidneys; at 24 ppm and 9 ppm, signs of respiratory irritation, depression of body weight, increased lung weights, and depressed liver weights were observed. Repeated exposure for 13 weeks at 6 ppm, 6 hr/day, 5 days/week, resulted in renal toxicity in 2 of 40 rats; although clinical signs of respiratory irritation were seen, morphological examination showed no evidence of damage. The same 13-week exposure regime at 2 ppm caused elevation of urinary, serum, and bone fluoride levels but did not result in a toxic response.[3] Guinea pigs and rats showed pneumonitis and congestion in the lungs after a 6-month exposure to a calculated concentration of 3.0 ppm (1.5 ppm by analysis), and a 4-month exposure at 1.0 ppm caused reversible tracheitis and bronchitis.[1,4]

Boron trifluoride combines with atmospheric moisture to form a white mist containing hydration and hydrolysis products.[1] The odor is detectable at 3.0 ppm, but this does not serve as an adequate warning.[1]

The 1995 ACGIH threshold limit value–ceiling limit (TLV-C) for boron trifluoride is 1 ppm (2.8 mg/m³).

REFERENCES

1. National Institute for Occupational Safety and Health, US Department of Health, Education and Welfare: *Criteria for a Recommended Standard . . . Occupational Exposure to Boron Trifluoride*. DHEW (NIOSH) Pub No 77–22. Washington, DC, US Government Printing Office, 1976
2. Kasparov AA: Boron trifluoride. In International Labour Office: *Encyclopaedia of Occupational Health and Safety*, Vol I, pp 204–205. New York, McGraw-Hill, 1974

3. Rusch GM, Hoffman, GM, McConnell, RF, et al: Inhalation toxicity studies with boron trifluoride. *Toxicol Appl Pharmacol* 83:69–78, 1986
4. Torkelson TR, Sadek SE, Rowe VK: The toxicity of boron trifluoride when inhaled by laboratory animals. *Am Ind Hyg Assoc J* 22:263–270, 1961

BROMINE

CAS: 7726-95-6

Br₂

Synonyms: None

Physical Form. Dark reddish-brown, fuming, volatile liquid

Uses. In antiknock compounds for gasoline; fire retardants; sanitation preparations

Exposure. Inhalation

Toxicology. Bromine is a severe irritant of the eyes, mucous membranes, lungs, and skin.

In humans, 10 ppm is intolerable, causing severe irritation of the upper respiratory tract; lacrimation occurs at levels below 1 ppm.[1] Symptoms and signs in humans also include dizziness, headache, epistaxis, cough, followed some hours later by abdominal pain, diarrhea and, sometimes, a measleslike eruption on the face, trunk, and extremities.[2] Exposure at 40 to 60 ppm is thought to be dangerous for brief periods, and 1000 ppm may be rapidly fatal from choking caused by edema of the glottis and from pulmonary edema.[3] Pneumonia may be a late complication of severe exposures.[4] Delayed mortality after bromine exposure has been associated with peribronchiolar abscesses and is thought to be due to deep tissue penetration and damage caused by the relatively soluble bromine.

A mild degree of spermatogenic suppression and impaired reproductive performance was reported in a follow-up study of eight men accidentally exposed to bromine vapor.[6] The men were exposed for 50 to 240 minutes to unknown concentrations following a spill. Clinical manifestations, including respiratory distress and chemical skin burns, were noted at the time of the incident. Because of the small number in the cohort, a cause-result linkage cannot be established for bromine exposure and reproductive effects.

The liquid or concentrated vapor in contact with the eye will cause severe and painful burns.[4] Liquid bromine spilled on the skin causes a mild cooling sensation on first contact, followed by a burning sensation. If bromine is not removed from the skin immediately, deep surface burns result; a brown discoloration appears, leading to the development of deep-seated ulcers, which heal slowly.

Nearly 50% of mice exposed at 240 ppm for 2 hours died within 30 days; at 750 ppm, a 7-minute exposure was lethal to 40% during the same follow-up period.[7]

The 1995 ACGIH threshold limit value–time-weighted average (TLV-TWA) for bromine is 0.1 ppm (0.66 mg/m³) with a short-term excursion limit of 0.2 ppm (1.3 mg/m³).

REFERENCES

1. AIHA Hygienic Guide Series: *Bromine*. Akron, OH, American Industrial Hygiene Association, 1978.
2. Stokinger HE: The halogens and the nonmetals boron and silicon. In Clayton GD, Clayton FE (eds): *Patty's Industrial Hygiene and Toxicology*, 3rd ed, rev, Vol 2B, *Toxicology*, pp 2965–2968. New York, Wiley-Interscience, 1981
3. Henderson Y, Haggard HW: *Noxious Gases*. New York, Reinhold, 1943
4. Chemical Safety Data Sheet SD-49, Bromine, pp 5, 16–18. Washington, DC, MCA Inc, 1968
5. Kraut A, Lilis R: Chemical pneumonitis due to exposure to bromine compounds. Chest 94:208–210, 1988
6. Potashnik G, Carel R, Belmaker I, et al: Spermatogenesis and reproductive performance following human accidental exposure to bromine vapor. *Reproduc Toxicol* 6:171–174, 1992
7. Bitron MD, Aharonson EF: Delayed mortality of mice following inhalation of acute doses of

CH$_2$O,SO$_2$, CL$_2$ and Br$_2$. *Am Ind Hyg Assoc J* 39:129–138, 1978

American Conference of Governmental Industrial Hygienists (ACGIH), 1986

BROMINE PENTAFLUORIDE
CAS:7789-30-2

BrF$_5$

Synonyms: None

Physical Form. Pale yellow liquid

Uses. Oxidizer in rocket-propellant systems; fluorinating agent

Exposure. Inhalation

Toxicology. Bromine pentafluoride is an extremely reactive oxidizer and is an irritant of the eyes, mucous membranes, and lungs.

Contact of the vapor or liquid with skin or eyes is expected to cause severe burns; inhalation may cause lung injury, and lower concentrations may cause watering of the eyes and difficulty breathing.[1]

Exposure of animals to 500 ppm caused immediate gasping, swelling of eyelids, corneal opacity, lacrimation, and excessive salivation.[2] Levels of 100 ppm produced the same effects after 3 minutes; 50 ppm for 30 minutes caused deaths. Chronic exposure above 3 ppm produced severe nephrosis, marked toxic hepatosis, and severe respiratory difficulty in some of the exposed animals.

The 1995 TLV-TWA is 0.1 ppm (0.72 mg/m^3).

REFERENCES

1. Bromine pentafluoride. *Documentation of the TLVs and BEIs*, 6th ed, pp 152–153. Cincinnati, OH, American Conference of Governmental Industrial Hygienists (ACGIH), 1991
2. Bromine pentafluoride. *Documentation of the TLVs and BEIs*, 5th ed, p 66. Cincinnati, OH,

---

BROMODICHLOROMETHANE
CAS: 75-27-4

CHBrCl$_2$

Synonyms: Dichlorobromomethane; monobromodichloromethane; dichloromonobromomethane

Physical Form. Colorless liquid

Uses/Source: As a chemical intermediate for organic synthesis and as a laboratory reagent; formerly used as a solvent and flame retardant. Formed as a by-product during chlorination of water

Exposure. Ingestion; inhalation; skin absorption

Toxicology. Bromodichloromethane is a central nervous system depressant and causes damage to the liver and kidneys; it is carcinogenic in experimental animals.

No studies are available regarding health effects of bromodichloromethane in humans.

The LD$_{50}$ for a single gavage dose in both mice and rats has ranged from 450 mg/kg to 970 mg/kg.[1-3] Clinical signs associated with these exposures include piloerection, sedation, flaccid muscle tone, ataxia, and prostration; enlargement and congestion of the liver and kidneys were observed at autopsy.[3]

Subchronic exposure to bromodichloromethane in the range of 100 to 300 mg/kg/day has caused hepatic injury in mice and rats characterized by increased liver weight, pale discoloration, increased levels of hepatic enzymes, and focal areas of inflammation or degeneration.[4,5] Mild effects, including slightly increased liver weights and microscopic changes, have been noted at doses as low as 40 to 50 mg/kg/day for 2 weeks.[4]

Damage to the kidneys has also been reported at doses similar to those that affect the liver. Increased renal weights were observed in rat s receiving 200 mg/kg/day for 10 days, and increased blood urea nitrogen has been reported in mice dosed with 250 mg/kg/day for 2 weeks.[4,5]

Chronic oral studies in mice and rats show clear evidence that bromodichloromethane is carcinogenic. Male mice administered 50 mg/kg/day by gavage 5 days/week for 2 years had an increased incidence of renal carcinoma; hepatic tumors were observed in female mice similarly dosed with 75 or 150 mg/kg/day.[3] Tumors of the large intestine (intestinal carcinoma) occurred in rats at incidences of 13/50 and 45/50 in males exposed to 50 or 100 mg/kg/day, 5 days/week, for 2 years, respectively; 12/47 females were affected at the higher dose. Kidney tumors were observed in both male and female rats exposed to 100 mg/kg/day and, in another study, liver tumors occurred in females exposed to 150 mg/kg/day for 180 weeks.[3,6]

The IARC has determined that there is sufficient evidence for the carcinogenicity of bromodichloromethane in experimental animals and inadequate evidence for carcinogenicity in humans.[7]

In genotoxic assays, bromodichloromethane produced positive and negative results. It caused sister chromatid exchange in human lymphocytes but not in Chinese hamster cells; chromosomal aberrations were observed in two out of three studies; it induced mutations in some bacterial assays.[7]

Bromodichloromethane was fetotoxic at doses that also caused significant maternal toxicity. Rats exposed to doses of 50 to 200 mg/kg/day on days 6 to 15 of gestation had an increased incidence of sternebral anomalies; maternal toxicity was evidenced by a 40% reduction in body weight.[8]

The ACGIH has not established a threshold limit value for bromodichloromethane.

REFERENCES

1. Chu I, Villeneuve DC, Secours VE, et al: Toxicity of trihalomethanes: I The acute and subacute toxicity of chloroform, bromodichloromethane, chlorodibromomethane and bromoform in rats. *J Environ Sci Health* B17:205–224, 1982
2. Bowman FJ, Borzelleca JF, Munson AE: The toxicity of some halomethanes in mice. *Toxicol Appl Pharmacol* 44:213–215, 1978
3. National Toxicology Program: *Toxicology and Carcinogenesis Studies of Bromodichloromethane (CAS No. 75-27-4) in F344/N Rats and B6C3F1 Mice (gavage studies).* US Dept of Health and Human Service, NIH Pub No. 88-2537, TR-321, 1987
4. Condie LW, Smallwood CL, Laurie RD: Comparative renal and hepatotoxicity of halomethanes: bromodichloromethane, bromoform, chloroform, dibromochloromethane and methylene chloride. *Drug Chem Toxicol* 6:563–578, 1983
5. Munson AE, Sain LE, Sanders VM, et al: Toxicology of organic drinking water contaminants: trichloromethane, bromodichloromethane, dibromochloromethane and tribromomethane. *Environ Health Perspect* 46:117–126, 1982
6. Tumasonis CF, McMartin DN, and Bush B: Lifetime toxicity of chloroform and bromodichloromethane when administered over a lifetime in rats. *Ecotoxicol Environ Safety* 9:233–240, 1985
7. *IARC Monographs on the Evaluation of Carcinogenic Risks to Humans*, Vol 52, Chlorinated drinking-water; chlorination by-products; some other halogenated compounds; cobalt and cobalt compounds, pp 179–207. International Agency for Research on Cancer, World Health Organization, Lyon, 1991
8. Ruddick JA, Villeneuve DC, and Chu I: A teratological assessment of four trihalomethanes in the rat. *J Environ Sci Health* B18:333–349, 1983

BROMOFORM
CAS: 75-25-2

CHBr$_3$

Synonyms: Methyl tribromide; tribromomethane

Physical Form. Colorless liquid

Uses. As a fluid for mineral ore separation; as a laboratory reagent; in the electronics industry for quality assurance programs; formerly used as a sedative and antitussive

Exposure. Inhalation; skin absorption; ingestion

Toxicology. Bromoform is a central nervous system depressant; in experimental animals, it causes liver damage and is carcinogenic to rats.

Ingestion of the liquid has produced central nervous system depression with coma and loss of reflexes at doses in the range of 150 mg/kg; smaller doses have led to listlessness, headache, and vertigo; the approximate lethal dose in humans is considered to be 300 mg/kg.[1,2] Chronic effects have not been reported from industrial exposure.

In an early report, very high concentrations of 56,000 ppm and above were reported to cause death in dogs. The chief symptom was initial excitation, followed by deep sedation.[2]

The oral LD_{50} was 933 mg/kg body weight (bw) in rats and 707 and 1072 mg/kg bw in male and female mice, respectively.[3] Signs of acute toxicity were prostration, lachrimation, and lethargy. In 14-day studies, daily administration of 600 mg/kg bw induced lethargy, shallow breathing, and ataxia and was lethal to rats. In another two-week study, 250 mg/kg/day resulted in decreases in several indices of cellular and humoral immunity in male mice; slight liver damage, as indicated by altered liver enzymes, was also noted.[4] Mild tubular hyperplasia and glomerular degeneration were observed in the kidneys of male mice administered 289 mg/kg bw for 14 days.[5]

Hepatocellular vacuolization was observed in male rats administered up to 200 mg/kg bw for 13 weeks and in male mice dosed at 200 and 400 mg/kg bw for the same time period.[3]

In two-year carcinogenicity studies, bromoform induced adenomatous polyps and adenocarcinomas of the large intestines of rats administered 200 mg/kg/day by gavage; no increase in tumor incidence was observed in mice similarly treated with 100 mg/kg/day.[3]

The IARC has determined that there is limited evidence for the carcinogenicity of bromoform in experimental animals.[6] The evidence for carcinogenicity in humans is inadequate.

Bromoform has given positive and negative results in a variety of genotoxic assays.

An increased incidence of minor skeletal variations occurred in the offspring of rats dosed at 100 or 200 mg/kg/day on days 6 to 15 of gestation.[7] No adverse effect on fertility was found in either the parental or F_1 generation of mice treated for 18 weeks at doses of up to 200 mg/kg/day in a continuous-breeding reproductive study; a decrease in neonatal (F_1) survival was noted in the high-dose group.[6]

The undiluted liquid was moderately irritating to rabbit eyes, but healing was complete in 1 to 2 days. Repeated skin contact caused moderate irritation to rabbit skin.[8]

The 1995 ACGIH threshold limit value–time-weighted average (TLV-TWA) for bromoform is 0.5 ppm (5.2 mg/m^3) with a notation for skin absorption.

REFERENCES

1. von Oettingen WF: *The Halogenated Aliphatic, Olefinic, Cyclic, Aromatic, and Aliphatic-Aromatic Hydrocarbons, Including the Halogenated Insecticides, Their Toxicity and Potential Dangers.* US Public Health Service Pub No 414, pp 65–67. Washington, DC, US Government Printing Office, 1955
2. Agency for Toxic Substances and Disease Registry (ATSDR): *Toxicological Profile for Bromoform and Chlorodibromomethane.* US Department of Health and Human Services, Public Health Service, TP-90–05, p 121, 1990
3. National Toxicology Program: *Toxicology and Carcinogenesis Studies of Tribromomethane (Bromoform) (CAS No. 75-25-2) in F344/N Rats and B6C3F1 Mice (Gavage Studies).* Technical Report Series No 350. US Department of Health and Human Services, Public Health Service, National Institutes of Health, NIH Pub No 88-2805, 1988
4. Munson AE, Sain LE, Sanders VM, et al: Toxicology of organic drinking water contaminants: trichloromethane, bromodichloromethane, dibromochloromethane, and tribromomethane. *Environ Health Perspect* 46:117–126, 1982
5. Condie LW, Smallwood CL, Laurie RD:

Comparative renal and hepatotoxicity of halo-methanes: Bromodichloromethane, bromoform, chloroform, dibromochloromethane and methylene chloride. *Drug Chem Toxicol* 6:563–578, 1983

6. *IARC Monographs on the Evaluation of Carcinogenic Risks to Humans*, Vol 52, Chlorinated drinking-water; chlorination by-products; some other halogenated compounds: cobalt and cobalt compounds, pp 213–242, International Agency for Research on Cancer, World Health Organization, Lyon, 1991
7. Ruddick JA, Villeneuve DC, Chu I, et al: A teratological assessment of four trihalomethanes in the rat. *Environ Sci Health* B18:333–349, 1983
8. Torkelson TR, Rowe VK: Halogenated aliphatic hydrocarbons. In Clayton GD, Clayton FE (eds): *Patty's Industrial Hygiene and Toxicology*, 3rd ed, rev, Vol 2B, *Toxicology*, pp 3469–3470. New York: Wiley-Interscience, 1981

1,3-BUTADIENE
CAS: 106-99-0

C_4H_6

Synonyms: Butadiene; biethylene divinyl; erythrene

Physical Form. Colorless

Uses. In manufacture of synthetic rubber

Exposure. Inhalation

Toxicology. 1,3-Butadiene is an irritant of the eyes and mucous membranes; at extremely high concentration, it causes narcosis in animals, and severe exposure is expected to produce the same effect in humans. It is carcinogenic in experimental animals and is considered a suspected human carcinogen.

Human subjects tolerated 4000 ppm for 6 hours without apparent effect other than slight irritation of the eyes; tolerance to higher exposures appears to develop following a single exposure of 1,3-butadiene.[1] Exposure of two human volunteers to 8000 ppm for 8 hours caused eye and upper respiratory tract irritation.[1]

Dermatitis and frostbite may result from exposure to liquid and evaporating 1,3-butadiene.[2]

Deep anesthesia in rabbits was induced after 8 to 10 minutes at 200,000 to 250,000 ppm, and death occurred in 23 minutes at 250,000 ppm.[1] Recovery from brief periods of anesthesia occurred within 2 minutes of terminating exposure; no tissue changes were detectable microscopically after daily induction of anesthesia for as many as 34 times.[1] Daily exposure of rats, guinea pigs, rabbits, and dogs at 6700 ppm over 8 months resulted in no significant chronic effects.[1] In contrast, appreciable mortality occurred in mice exposed to 5000 ppm for 14 weeks.[3]

Toxicological studies have shown butadiene to be a multisite animal carcinogen with marked differences in potency across rodent species.[4]

Exposure of mice to 625 or 1250 ppm 6 hr/day, 5 days/week, caused early deaths primarily due to malignant neoplasms involving multiple organs.[3] At the end of 61 weeks, there were tumors in 20% of control males and 12% of control females compared to 80% and 94% of the exposed mice, respectively. The most common tumors were malignant lymphomas, heart hemangiosarcomas and alveolar-bronchiolar neoplasms. Nonneoplastic effects associated with these exposures included testicular and ovarian atrophy and nasal cavity lesions. A second long-term inhalation study in mice over an expanded concentration range of 6.25, 20, 62.5, 200, or 625 ppm als o caused increased lymphomas, hemangiosarcomas of the heart, and lung neoplasms.[5]

Chronic exposure of rats for 2 years, 6 hr/day, 5 days/week, to 1000 or 8000 ppm caused a significant increase in neoplasms of the mammary gland, thyroid, uterus, and Zymbal glands of exposed females and in neoplasms of the testes and pancreas (8000 ppm only) in exposed males.[6]

Research aimed at explaining the species differences in organ sites of butadiene-induced carcinogenicity has focused on the potential influence of an endogenous retrovirus in B6C3F1, mice and/or pharmacokinetic-metabolic considerations.[7] A retrovirus involvement has been suggested because the incidence of butadiene-induced thymic lymphomas was less in NIH Swiss mice compared to B6C3F$_1$ mice, and the endogenous ecotropic retrovirus was not recovered in Swiss mice tissue as it was in the B 6C3F$_1$.[7,8] Species differences in metabolic rate, detoxification, repair, and DNA adduct formation may also contribute to different carcinogenic responses.

Associations between occupational exposure to butadiene and increased risk of cancer have been examined in a number of epidemiological studies. In one study, no statistically significant excess in total or cause-specific mortality was found in 2756 styrene-butadiene rubber workers.[9] The average length of employment was 10 years, and butadiene exposures ranged from 0.11 to 174 ppm. In one plant group, there was a small excess of lymphatic and hematopoietic cancers, which was not statistically significant. In another study of nearly 14,000 workers, there was a low overall mortality rate with a slight but not significant increase in cancers of the digestive system.[10]

No significant excesses were observed for any cause of death except lymphosarcoma in a cohort of 2582 men employed at a butadiene manufacturing plant for at least 6 months between 1943 and 1979.[11]

The largest butadiene cohort study evaluated mortality between 1943 and 1982 for 12,110 workers at eight plants in the United States and Canada.[12] For the total cohort, mortality from all cancers was significantly lower than US rates, and mortality for lymphopoietic cancers was less than expected. A higher SMR was found in one subcategory; for black production workers, there was a significant excess of lymphopoietic cancers (6 observed vs. 1.2 expected) due to leukemia and other lymphatic cancers. Further analysis showed that the lymphopoietic cancers were concentrated in short-term workers, with 4 of 6 having worked less than 4 years.

Taken as a whole, the various epidemiologic studies show no consistent pattern of mortality among long-term butadiene-exposed workers but do show lymphopoietic cancer excesses among certain short-term worker subgroups.[3]

The IARC has concluded that there is sufficient evidence for carcinogenicity to animals and limited evidence of carcinogenicity to humans.[13]

Butadiene induced a variety of genotoxic effects in vivo in mice, including chromosomal aberrations, sister chromatid exchange, micronucleus formation, and sperm-head abnormalities; it did not induce micronuclei or sister chromatid exchange in rats.[13]

Pregnant rats exposed to 200, 1000, or 8000 ppm during days 6 to 15 of gestation had depressed body weight gain.[14] At the 8000 ppm exposure level, fetal growth was significantly retarded, and a significant increase in major skeletal abnormalities was observed.

The 1995 ACGIH threshold limit value–time-weighted average (TLV-TWA) for butadiene is 2 ppm (4.4 mg/m^3) with an A2 suspected human carcinogen designation.

REFERENCES

1. Carpenter CP, Shaffer CB, Weil CS, Smyth, WJ Jr: Studies on the inhalation of 1,3-butadiene; with a comparison of its narcotic effect with benzol, toluol and styrene, and a note on the elimination of styrene by the human. *J Ind Hyg Toxicol* 26:69–78, 1944
2. Parmeggiani, L (ed.) *Encyclopaedia of Occupational Health and Safety*, 3rd ed, rev, Vol 1, A–K, pp 347–348. 1,3-Butadiene. International Labour Office, Geneva, 1983.
3. Huff JE, Melnick RL, Solleveld HA, et al: Multiple organ carcinogenicity of 1,3-butadiene in B6C3F$_1$ mice after 60 weeks of inhalation exposure. *Science* 227:548–549, 1985
4. Acquavella JF: The paradox of butadiene epidemiology. *Exp Pathol* 37:114–118, 1989
5. National Toxicology Program: *Toxicology and Carcinogenesis Studies of 1,3-Butadiene (CAS No. 106-99-0) in B6C3F$_1$ Mice (Inhalation Studies)*. NTP Technical Report No 434, Research Triangle Park, NC 1994
6. Owen PE, Glaister JR, Gaunt IF, et al: Inhalation toxicity studies with 1,3-butadiene: a two-year toxicity/carcinogenicity study in rats. *Am Ind Hyg Assoc J* 48:407–413, 1987

7. Melnick RL, Huff JE, Bird MG, et al: Symposium overview: toxicology, carcinogenesis, and human health aspects of 1,3-butadiene. *Environ Health Perspect* 86:3–5, 1990
8. Melnick RL, Shackelford CC, Huff J: Carcinogenicity of 1,3-butadiene. *Environ Health Perspect* 100:227–236, 1993
9. Meinhardt TJ, Lemen RA, Crandall MS, et al: Environmental epidemiologic investigation of the styrene-butadiene rubber industry; mortality patterns with discussion of the hematopoietic and lymphatic malignancies. *Scand J Work Environ Health* 8:250–259, 1982
10. Matanoski GM, Schwartz L: Mortality of workers in the styrene-butadiene rubber polymer production. *JOM* 29:675–680, 1987
11. Divine BJ: An update on mortality among workers at a 1,3-butadiene facility–preliminary results. *Env Health Perspect* 86:119–128, 1990
12. Matanoski GM, Santos-Burgoa C, Schwartz L: Mortality of a cohort of workers in the styrene-butadiene polymer manufacturing industry (1943–1982). *Environ Health Perspect* 86:107–117, 1990
13. *IARC Monographs on the Evaluation of the Carcinogenic Risks to Humans*, Vol 54, Occupational exposures to mists and vapours from strong inorganic acids; and other industrial chemicals, pp 237–285. Lyon, International Agency for Research on Cancer, 1992
14. Owen PE, Irvine LFH: 1,3-Butadiene: Inhalation teratogenicity study in the rat. Final report. Unpublished report submitted to the International Institute of Synthetic Rubber Producers, Inc, by Hazleton Laboratories Ltd, Harrogate, England, November 1981 (cited by NIOSH *Current Intelligence Bulletin* 41, February 9, 1984)

---

n-BUTANE
CAS: 106-97-8

C_4H_{10}

Synonyms: Butane; butylhydride; methylethylmethane

Physical Form. Colorless gas

Uses. As a gasoline blending component to enhance volatility; as a constituent of liquid petroleum gas, which is usually a mixture of butane and propane and is used as a home fuel; in organic synthesis; as a solvent; as a refrigerant and aerosol propellant; as a food additive

Exposure. Inhalation

Toxicology. *n*-Butane is a central nervous system depressant at high concentrations.

In six men and women, a 10-minute exposure to butane gas at 10,000 ppm has resulted in drowsiness.[1] Narcotic effects may be accompanied by exhilaration, dizziness, and headache. There may also be loss of appetite, nausea, confusion, loss of fine motor coordination and, in extreme cases, loss of consciousness.[2]

Over a 10-year period, 38 deaths from the voluntary inhalation of butane have been identified in the United Kingdom. Possible mechanisms for the cause of death included the central respiratory and circulatory sequelae of the anesthetic properties of butane, laryngeal edema, chemical pneumonia, and the combined effects of cardiac toxicity and increased sympathetic activity.[3]

In animal studies, the 4-hour LC_{50} in rats was 278,000 ppm, and the 2-hour LC_{50} in mice was 287,000.[4] Early studies reported similar values with 270,000 ppm for 2 hours, causing death in 40% of exposed mice, and 310,000 ppm for 2 hours, causing 60% mortality.[5] In dogs, lethality was observed at concentrations of 200,000 to 250,000 ppm; anesthesia and relaxation preceded death. In animal studies, there was only a small margin of safety between anesthetic and lethal concentrations.

Several studies have indicated that *n*-butane sensitizes the myocardium to epinephrine-induced cardiac arrhythmias. In anesthetized dogs, 5000 ppm caused hemodynamic changes, such as decreases in cardiac output, left ventricular pressure and stroke volume, myocardial contractility, and aortic pressure.[6] Exposure of dogs to 1% to 20% butane for periods of 2 minutes to 2 hours hypersensitized the heart to ventricular fibrillation induced by epinephrine.[7]

Dermal penetration of butane is not expected to any large extent as skin contact would be transient because of volatility.[8] *n*-Butane did not cause respiratory or eye irritation in rabbits, but it was mildly to moderately irritating to the skin.[9] Liquefied butane may cause frostbite when applied directly to the skin.[8]

The high odor threshold does not provide adequate warning of overexposure.[1] The 1995 ACGIH threshold limit value–time-weighted average for *n*-butane is 800 ppm (1900 mg/m³), which was established because of explosivity hazards rather than toxicological concerns.

REFERENCES

1. Patty FA, Yant WP: Odor intensity and symptoms produced by commercial propane, butane, pentane, hexane and heptane vapors. US Bureau of Mines Report of Investigation No 2979, pp 1–10, 1929
2. Hamilton A, Hardy HL: *Industrial Toxicology*, 3rd ed, pp 263–264. Acton, MA, Publishing Sciences Group, Inc, 1974
3. Anderson HR, Dick B, MacNair RS, et al: An investigation of 140 deaths associated with volatile substance abuse in the United Kingdom (1971–1981). *Hum Toxicol* 1:207–221, 1982
4. Shugaev BB: Concentrations of hydrocarbons in tissues as a measure of toxicity. *Arch Environ Health* 18:878–882, 1969
5. Stoughton RW, Lamson PD: The relative anesthetic activity of the butanes and the pentanes. *J Pharmacol Exp Ther* 58:74–77, 1936
6. Aviado DM, Zakheri S, Watanabe T: *Nonfluorinated Propellants and Solvents for Aerosols*, pp 49–81. Cleveland, OH, CRC Press, 1977
7. Chenoweth MB: Ventricular fibrillation induced by hydrocarbons and epinephrine. *J Ind Hyg Toxicol* 28:151–158, 1946
8. Low LK, Meeks JR, Mackerer CR: *n*-Butane. In R Snyder (ed), *Ethel Browning's Toxicity and Metabolism of Industrial Solvents*, 2nd ed, Vol 1, *Hydrocarbons*, pp 267–272. New York, Elsevier Science Pub. Co., 1987
9. Moore AF: Final report on the safety assessment of isobutane, isopentane, *n*-butane, and propane. *J Am College Toxicol* 1:127–142, 1982

n-BUTYL ACETATE
CAS: 123-86-4

$C_6H_{12}O_2$

Synonyms: Butyl ethanoate; acetic acid, butyl ester

Physical Form. Colorless liquid

Uses. Solvent for nitrocellulose, oils, fats, resins, waxes, and camphor; in manufacture of lacquer and plastics

Exposure. Inhalation

Toxicology. *n*-Butyl acetate causes irritation of mucous membranes and eyes; at high concentrations, it causes narcosis in animals, and it is expected that severe exposure will cause the same effect in humans.

In humans, *n*-butyl acetate affected the throat at 200 ppm; severe throat irritation occurred at 300 ppm, and the majority of the subjects also complained of eye and nose irritation.[1]

Only slight eye and pulmonary irritation (as determined by lung function tests) was observed in volunteers exposed to 1400 mg/m³ for 20 min or 700 mg/m³ for 4 hours.[2]

Guinea pigs exhibited signs of eye irritation at 3300 ppm; at 7000 ppm, there was narcosis within 700 minutes but no deaths following exposure for 810 minutes; 14,000 ppm was lethal after 4 hours.[3]

Cats exposed to 4200 ppm for 6 hours for 6 days showed weakness, loss of weight, and minor blood changes.[4]

The 1995 ACGIH proposed threshold limit value–time-weighted average (TLV-TWA) for *n*-butyl acetate is 20 ppm (95 mg/m³).

REFERENCES

1. Nelson KW, Ege JF, Ross M, Woodman LE, Silverman L: Sensory response to certain in-

dustrial solvent vapors. *J Ind Hyg Toxicol* 25:282–285, 1943
2. Iregren A, Lof A, Toomingas A, et al: Irritation effects from experimental exposure to *n*-butyl acetate. *Am J Ind Med* 24:727–742, 1993
3. Sayers RR et al: Acute response of guinea pigs to vapors of some new commercial organic compounds, XII: Normal butyl acetate. *Pub Health Repts* 51:1229–1236, 1936
4. Sandmeyer EE, Kirwin CJ: Esters. In Clayton GD, Clayton FE (eds): *Patty's Industrial Hygiene and Toxicology*, 3rd ed, p 2273. New York, Wiley-Interscience, 1981

sec-BUTYL ACETATE
CAS: 105-46-4

$C_6H_{12}O_2$

Synonyms: 2-Butanol acetate; acetic acid, secondary butyl ester

Physical Form.   Colorless liquid

Uses.   In solvents, especially lacquer solvents; textile sizes and paper coatings

Exposure.   Inhalation

Toxicology.   *sec*-Butyl acetate causes irritation of the eyes and respiratory tract. By analogy with chemically similar substances, it is considered a central nervous system depressant at very high concentrations.[1]

Skin irritation may occur. *sec*-Butyl acetate has not been studied regarding its toxicity, nor are there any reports concerning harmful effects on humans.[2]

The odor of *sec*-butyl acetate is milder than *n*-butyl acetate, and it appears less irritative to the eyes and respiratory tract.[1]

The 1995 ACGIH threshold limit value–time-weighted average (TLV-TWA) for *sec*-butyl acetate is 200 ppm (950 mg/m³).

REFERENCES

1. *sec*-Butyl Acetate. *Documentation of the TLVs and BEIs*, 6th ed, p 166. Cincinnati, OH, American Conference of Governmental Industrial Hygienists (ACGIH), 1991
2. von Oettingen WF: The aliphatic acids and their esters: toxicity and potential dangers. *AMA Arch Ind Health* 21:28–65, 1960

tert-BUTYL ACETATE
CAS: 540-88-5

$CH_3COOC(CH_3)_3$

Synonyms: Acetic acid, *tert*-butyl ester

Physical Form.   Colorless liquid

Uses.   Gasoline additive; lacquer solvent

Exposure.   Inhalation

Toxicology.   By analogy to other acetate esters, *tert*-butyl acetate is expected to cause irritation of the eyes and throat. It is considered a central nervous system depressant at very high concentrations.[1]

The toxicity of *tert*-butyl acetate has not been studied, nor are there any published reports concerning harmful effects on human beings.[2] Comparative tests indicate that it is definitely less irritating to the throat than *n*-butyl acetate.[1]

The 1995 ACGIH threshold limit value–time-weighted average for *tert*-butyl acetate is 200 ppm (950 mg/m³).

REFERENCES

1. *tert*-Butyl acetate. *Documentation of the TLVs and BEIs*, 6th ed, p 167. Cincinnati, OH, American Conference of Governmental Industrial Hygienists (ACGIH), 1991
2. von Oettingen WF: The aliphatic acids and

their esters: toxicity and potential dangers. *AMA Arch Ind Health* 21:28–65, 1960

---

n-BUTYL ACRYLATE
CAS: 141-32-2

$C_7H_{12}O_2$

Synonyms: Acrylic acid butyl ester; 2-propenoic acid butyl ester

Physical Form. Colorless liquid with acrid odor; commerical form contains hydroquinone (1000 ppm) or hydroquinone methyl ether (15 or 200 ppm) to prevent polymerization

Uses. In manufacture of polymers and resins for textiles, paints, and leather finishes

Exposure. Inhalation; skin absorption

Toxicology. Butyl acrylate is an irritant of the eyes and skin.

In one report, a woman with dermatitis from the plastic nose pads of her spectacle frames was found on patch testing to react to 1% butyl acrylate but not to ethyl or methyl acrylate.[1] The sensitization was attributed to butyl acrylate, which might have been present in the plastic nose pads.

In an early range-finding study, exposure of rats to 1000 ppm for 4 hours was lethal to 5 of 6 animals.[2] In a more recent study, the LC_{50} for 4 hours in rats was 2730 ppm.[3] Behavior of the animals suggested irritation of the eyes, nose, and respiratory tract, with labored breathing. At necropsy, there were no discernible gross abnormalities of the major organs.

The rabbit dermal LD_{50} was on the order of 1800 mg/kg.[4] On the skin of rabbits, butyl acrylate was moderately irritating.[2] In the rabbit eye, the liquid produced corneal necrosis.

Hamsters and rats were exposed to an average concentration of 817 and 820 ppm, respectively, for 4 days.[5] In both animal species, there were distinct signs of toxicity; 4 of 10 hamsters

died during the exposure. The chromosome analysis carried out in the bone marrow after the exposure indicated no chromosome-damaging effects.

In a dermal carcinogenesis study, 25 µl of 1% butyl acrylate in acetone was applied 3 times weekly to mice for their lifetime.[6] No epidermal tumors were observed, and there was, therefore, no evidence of local carcinogenic activity.

No neoplastic effect was observed in rats exposed to 0, 15, 45, or 135 ppm for 6 hr/day, 5 days/week, for 2 years.[7] Dose-related changes, which included atrophy of the neurogenic epithelial cells and hyperplasia of the reserve cells, mainly affected the anterior part of the olfactory epithelium. In the high-dose group, there was opacity of the cornea. Following a 6-month postexposure period, reconstructive effects were observed in both tissues.

The IARC has determined that there is inadequate evidence for carcinogenicity of *n*-butyl acrylate to experimental animals, and no data are available on humans.[8]

In a reproductive study, inseminated rats were exposed to butyl acrylate at 0, 25, 135, and 250 ppm 6 hr/day, from day 6 to day 15 post–coitum.[9] During the inhalation period, the two high doses led to maternal toxicity, including signs of mucous membrane irritation. The same levels induced embryolethality, measured as an increased number of dead implantations. The 25-ppm level did not cause maternal toxicity or embryolethality. A teratogenic effect was not seen at any of the exposure levels.

The 1995 ACGIH threshold limit value–time-weighted average for butyl acrylate is 10 ppm (52 mg/m³).

REFERENCES

1. Hambly EM, Wilkinson DS: Contact dermatitis to butyl acrylate in spectacle frames. *Contact Dermatitis* 4:115, 1978
2. Smyth HF Jr, Carpenter CP, Weil CS: Range-finding toxicity data: List IV. *AMA Arch Ind Hyg* 4:119, 1951
3. Oberly R, Tansy MF: LC_{50} values for rats acutely exposed to vapors of acrylic and methacrylic acid esters. *J Toxicol Environ Health* 16:811, 1985

4. Carpenter CP et al: Range-finding toxicity data: List VIII. *Toxicol Appl Pharmacol* 28:313, 1974
5. Englehardt G, Klimisch HJ: *n*-Butyl acrylate monomer: Cytogenetic investigations in the bone marrow of Chinese hamsters and rats after 4-day inhalation. *Fund Appl Toxicol* 3:640, 1983
6. DePass LR et al: Acrylic acid, ethyl acrylate, and butyl acrylate: Dermal oncogenicity bioassays of acrylic acid, ethyl acrylate, and butyl acrylate. *J Toxicol Environ Health* 14:115–120, 1984
7. Reininghaus W, Koestner A, Klimisch HJ: Chronic toxicity and oncogenicity of inhaled methyl acrylate and *n*-butyl acrylate in Sprague–Dawley rats. *Food Chem Toxicol* 29:329–339, 1991
8. *IARC Monographs on the Evaluation of the Carcinogenic Risk of Chemicals to Humans*, Vol 39, Some chemicals used in plastics and elastomers, pp 67–79. Lyon, International Agency for Research on Cancer, 1986
9. Merkle J, Klimisch HJ: *n*-Butyl acrylate: prenatal inhalation toxicity in the rat. *Fund Appl Toxicol* 3:443–447, 1983

---

n-BUTYL ALCOHOL
CAS: 71-36-3

$C_4H_{10}O$

Synonyms: *n*-Butanol; butyric alcohol; propyl carbinol; butyl hydroxide; 1-butanol

Physical Form. Colorless liquid

Uses. Lacquer solvent; in manufacture of plastics and rubber cements

Exposure. Inhalation; skin absorption

Toxicology. *n*-Butyl alcohol is an irritant of the eyes and mucous membranes and may cause central nervous system depression at very high concentrations.

Chronic exposure of humans to concentrations above 50 to 200 ppm causes irritation of the eyes with lacrimation, blurring of vision, and photophobia.[1,2]

In a 10-year study of workers exposed to average concentrations of 100 ppm, no systemic effects were observed.[1] Contact dermatitis of the hands may occur as a result of a defatting action of the liquid, and toxic amounts can be absorbed through the skin.[3] Direct contact of the hands with *n*-butyl alcohol for 1 hour results in an absorbed dose that is 4 times that of inhalation of 50 ppm for 1 hour.[3]

No effects were observed in mice exposed to 3300 ppm for 7 hours, whereas exposure to 6600 ppm produced prostration within 2 hours, narcosis after 3 hours, and some deaths.[3]

Administered to pregnant rats by inhalation 7 hr/day on days 1 to 19 of gestation, 8000 ppm caused reduced fetal weights, increased incidence of skeletal malformations, and significant maternal toxicity in the form of narcosis and reduced feed consumption.[4] At 3500 ppm, for the same exposure time, there were no fetal or maternal effects.

n-Butyl alcohol was not mutagenic in the Ames *Salmonella typhimurium* assay.[3] The odor threshold is approximately 15 ppm but, following adaptation, the threshold can increase to 10,000 ppm.[3]

The 1995 ACGIH threshold value–limit ceiling value (TLV-C) for *n*-butyl alcohol is 50 ppm (152 mg/m^3) with a notation for skin absorption.

REFERENCES

1. Sterner JI, Crouch HC, Brockmyre HF, Cusack M: A ten-year old study of butyl alcohol exposure. *Am Ind Hyg Assoc Q* 10:53–59, 1949
2. Tabershaw IR, Fahy JP, Skinner JB: Industrial exposure to butanol. *J Ind Hyg Toxicol* 26:328–330, 1944
3. Rowe VK, McCollister SB: Alcohols. In Clayton GD, Clayton FE (eds): *Patty's Industrial Hygiene and Toxicology*, 3rd ed, *Toxicology*, Vol 2C, pp 4571–4578. New York, Wiley-Interscience, 1982
4. Nelson BK, Brightwell WS, Krieg EF Jr: Developmental toxicology of industrial alcohols: a summary of 13 alcohols administered by in-

halation to rats. *Toxicol Ind Health* 6:373–387, 1990

---

sec-BUTYL ALCOHOL
CAS: 78-92-2

$C_4H_{10}O$

Synonyms: 2-Butanol; ethylmethyl carbinol; butylene hydrate; 2-hydroxybutane

Physical Form. Colorless liquid

Uses. Polishes, cleaning materials, paint removers, fruit essences, perfumes, and dyestuffs; synthesis of methyl ethyl ketone; lacquer solvent

Exposure. Inhalation

Toxicology. At high concentrations *sec*-butyl alcohol causes narcosis in animals, and it is expected that severe exposure will produce the same effect in humans.

Heavy exposure reportedly causes eye, nose, and throat irritation; headache; nausea; fatigue; and dizziness.[1]

In mice, ataxia, prostration, and narcosis occurred at various times after exposure to concentrations ranging from 3300 ppm to 19,800 ppm.[1] Exposure to 16,000 ppm for 4 hours was lethal to 5 of 6 rats.[2] Mice repeatedly exposed to a concentration of 5330 ppm for a total of 117 hours were narcotized but survived.

Administered by inhalation to pregnant rats on days 1 to 19 of gestation for 7 hr/day, 7000 ppm caused an increased incidence in resorptions and reduced fetal weights as well as significant maternal toxicity in the form of narcosis, reduced feed consumption, and reduced weight gain.[3] At 3500 ppm, some maternal toxicity was observed, but there were no fetal effects.

When instilled directly into a rabbit eye, the liquid caused severe corneal injury, but it was not irritating to the skin of rabbits.[2]

sec-Butyl alcohol has an odor similar to, but less pungent than, *n*-butyl alcohol. The malodorous and irritating properties probably prevent exposure to toxic levels.

The 1995 ACGIH threshold limit value–time-weighted average (TLV-TWA) for *sec*-butyl alcohol is 100 ppm (303 mg/m³).

REFERENCES

1. Rowe VK, McCollister SB: Alcohols. In Clayton GD, Clayton FE (eds): *Patty's Industrial Hygiene and Toxicology*, 3rd ed, *Toxicology*, Vol 2C, pp 4582–4585. New York, Wiley-Interscience, 1982
2. Smyth HF Jr et al: Range-finding toxicity data: List V. *AMA Arch Ind Hyg Occup Med* 20:61–68, 1954
3. Nelson BK, Brightwell WS, Krieg EF Jr: Developmental toxicology of industrial alcohols: a summary of 13 alcohols administered by inhalation to rats. *Toxicol Ind Health* 6:373–387, 1990

---

tert-BUTYL ALCOHOL
CAS: 75-65-0

$(CH_3)_3COH$

Synonyms: 2-Methyl-2-propanol; trimethyl carbinol; *tert*-butanol

Physical Form. Colorless liquid or solid

Uses. In plastics, lacquers, cellulose esters, fruit essences, perfumes; chemical intermediates; additive to unleaded gasoline

Exposure. Inhalation

Toxicology. At high concentrations, *tert*-butyl alcohol causes narcosis in animals, and it is expected that severe exposure in humans will result in the same effect; in rodents, the urinary tract is the primary target with subchronic exposures.

In humans, heavy exposure may cause irritation of the eyes, nose, and throat; headache; nausea; fatigue; and dizziness.[1] Systemic effects have not been reported. Application of *tert*-butyl alcohol to skin causes slight erythema and hyperemia.[1]

Signs of intoxication in rats were ataxia and narcosis; the oral LD_{50} was 3.5 g/kg.[2] In a 90-day study in F344 rats and $B6C3F_1$ mice, 0, 0.25, 0.5, 1, 2, or 4% was administered in the drinking water.[3] The high dose was lethal to some animals, and clinical signs included ataxia in rats and, in mice, ataxia, hypoactivity, and abnormal posture. Gross lesions at necropsy were urinary tract calculi, renal pelvi and ureteral dilation, and thickening of the urinary bladder mucosa. In male rats, the microscopic renal changes were suggestive of a-2μ-globulin nephropathy.

Administered by inhalation for 7 hr/day on gestation days 1 to 19, 5000 ppm caused reduced fetal weights and maternal toxicity in the form of narcosis, reduced feed consumption, and reduced maternal weights.[4]

tert-Butyl alcohol was not genotoxic in the *Salmonella* assay, the mouse lymphoma assay, or Chinese hamster ovary-cell assays for increased frequencies of chromosomal aberrations and sister chromatid exchanges.[5]

The malodorous quality and irritant effects of *tert*-butyl alcohol may prevent inadvertent exposure to toxic levels.

The 1995 proposed ACGIH threshold limit value-time-weighted average (TLV-TWA) for *tert*-Butyl alcohol is 100 ppm (303 mg/m³) with an A4 not classifiable as a human carcinogen designation.

REFERENCES

1. Rowe VK, McCollister SB: Alcohols. In Clayton GD, Clayton FE (eds): *Patty's Industrial Hygiene and Toxicology*, 3rd ed, Vol 2C, *Toxicology*, pp 4585–4588. New York, Wiley-Interscience, 1982
2. Schaffarzick RW, Brown BJ: The anticonvulsant activity and toxicity of methyl parafynol (Vormison) and some other alcohols. *Science* 116:663–665, 1952
3. Lindamood C, Farnell DR, Giles HD, et al: Subchronic toxicity studies of *t*-butyl alcohol in rats and mice. *Fund Appl Toxicol* 19:91–100, 1992
4. Nelson BK, Brightwell WS, Krieg EF Jr: Developmental toxicology of industrial alcohols: a summary of 13 alcohols administered by inhalation to rats. *Toxicol Indus Health* 6:373–387, 1990
5. *tert*-Butyl alcohol. *Documentation of the TLVs and BEIs*, 6th ed, pp 174–175. Cincinnati, OH, American Conference of Governmental Industrial Hygienist (ACGIH), 1991

BUTYLAMINE
CAS: 109-73-9

$CH_3(CH_2)_3NH_2$

Synonyms: 1-Aminobutane; *n*-butylamine

Physical Form. Colorless liquid

Uses. Intermediate for pharmaceuticals, dyestuffs, rubber chemicals, emulsifying agents, insecticides, synthetic tanning agents

Exposure. Inhalation; skin absorption

Toxicology. Butylamine is an irritant of the eyes, mucous membranes, and skin.

In humans, the liquid on the skin causes severe primary irritation and second-degree burns with vesiculation.[1] Workers exposed daily at 5 to 10 ppm complained of irritation of the nose, throat, and eyes and, in some instances, headache and flushing of the skin of the face.[1,2] Concentrations of 10 to 25 ppm are unpleasant and even intolerable to some subjects for exposures of more than a few minutes' duration; daily exposures of workers to less than 5 ppm (usually 1–2 ppm) resulted in no symptoms.[1]

In rats exposed to 3000 to 5000 ppm, there was an immediate irritant response, followed by labored breathing, pulmonary edema, and death within minutes to hours.[1]

The oral LD_{50} for *n*-butylamine in rats was 372 mg/kg versus 228, 152, and 80 mg/kg for isobutylamine, *sec*-butylamine, and *tert*-butylamine, respectively.[3] Signs of toxicity included sedation, ataxia, nasal discharge, gasping, salivation, and death. Pathological examination showed pulmonary edema.

The liquid produced severe eye damage and skin burns in animals.[1]

The 1995 ACGIH threshold limit value–ceiling limit (TLV-C) for butylamine is 5 ppm (15 mg/m³) with a notation for skin absorption.

REFERENCES

1. Beard RR, Noe JT: Aliphatic and alicyclic amines. In Clayton GD, Clayton FE (eds): *Patty's Industrial Hygiene and Toxicology*, 3rd ed, rev, Vol 2B, *Toxicology*, pp 3135–3155. New York, Wiley-Interscience, 1981
2. Hygienic Guide Series: *n*-Butylamine. *Am Ind Hyg Assoc J* 21:532–533, 1960
3. Cheever KL, Richards DE, Plotnick HB: Short communication. The acute oral toxicity of isomeric monobutylamines in the adult male and female rat. *Toxicol Appl Pharmacol* 63:150–152, 1982

tert-BUTYL CHROMATE
CAS:1189-85-1

$[(CH_3)_3CO]_2CrO_2$

Synonyms: Bis(*tert*-butyl)chromate; chromic acid, di-*tert* butyl ester

Physical Form. Clear, colorless liquid

Uses. In specialty reactions as a source of chromium; in manufacture of catalysts; in polymerization of olefins; as curing agent for urethane resins

Exposure. Inhalation; skin absorption

Toxicology. *tert*-Butyl chromate is expected to cause irritation of the eyes, nose, and skin.

Skin contact with *tert*-butyl chromate caused necrosis of the skin and death of rats.[1]

NIOSH considers *tert*-butyl chromate to be an inferred carcinogen because it is a hexavalent chromium compound.[2]

The 1995 ceiling threshold limit value (C-TLV) is 0.1 mg/m³ with a notation for skin absorption.

REFERENCES

1. Roubal J, Krivucova M: Hygienic problems in the application of tertiary butyl chromate as a passivation inhibitor of metal corrosion. *Arch Gewerbepath Gewerbehyg* 17:589–596, 1960
2. *tert*-Butyl chromate. *Documentation of the TLVs and BEIs*, 6th ed, p 178. Cincinnati, OH, American Conference of Governmental Industrial Hygienists, 1991

n-BUTYL GLYCIDYL ETHER
CAS: 2426-08-6

$C_4H_9OCH_2CHOCH_2$

Synonyms: BGE; 1-*n*-butoxy-2,3-epoxypropane; 1,2-epoxy-3-butoxypropane; 2,3-epoxypropyl butyl ether

Physical Form. Colorless liquid

Uses. Viscosity-reducing agent; acid acceptor for solvents; chemical intermediate

Exposure. Inhalation

Toxicology. *n*-Butyl glycidyl ether (BGE) causes central nervous system depression and is a mild irritant of the eyes and skin in animals; it is expected that severe exposure will cause the same effects in humans.

No chronic systemic effects have been reported in humans. However, sensitization dermatitis may occur with repeated skin contact.[1]

Intragastric and intraperitoneal injection of BGE in animals produced incoordination and ataxia, followed by coma.[1] In rats exposed to graded vapor concentrations of BGE, effects were lacrimation, nasal irritation, and labored breathing. The LC$_{50}$ for an 8-hour exposure in rats was 1030 ppm, and greater than 3500 ppm for 4 hours in mice. At autopsy, pneumonitis was frequently observed. Three intramuscular injections of 400 mg/kg produced minimal toxic effects and a slight increase in leukocyte counts.[2] In male mice topically treated with 1.5 g/kg for 8 weeks and then mated, there was a significant increase in the number of fetal deaths compared with controls.[3]

BGE produced widely disparate degrees of skin irritation, ranging from very mild to severe, in tests by different investigators using similar methodology.[3] After a series of intracutaneous injections, 16 of 17 guinea pigs became sensitized.[4] The undiluted liquid in rabbit eyes caused mild eye irritation.[1]

The 1995 ACGIH threshold limit value–time-weighted average (TLV-TWA) for butyl glycidyl ether is 25 ppm (133 mg/m³).

REFERENCES

1. Hine CH et al: The toxicology of glycidol and some glycidyl ethers. *AMA Arch Ind Health* 14:250–264, 1956
2. Kodama JK et al: Some effects of epoxy compounds on the blood. *Arch Environ Health* 2:56–67, 1961
3. National Institute for Occupational Safety and Health: *Criteria for a Recommended Standard . . . Occupational Exposure to Glycidyl Ethers.* DHEW (NIOSH) Pub No 78-166. Washington, DC, US Government Printing Office, 1978
4. Weil CS et al: Experimental carcinogenicity and acute toxicity of representative epoxides. *Am Ind Hyg Assoc J* 24:305–325, 1963

***

n-BUTYL MERCAPTAN
CAS: 109-79-5

CH₃(CH)₃SH

***

Synonyms: 1-Butanethiol; thiobutyl alcohol; butyl sulfhydrate

Physical Form. Colorless liquid

Uses. Solvent; intermediate in the production of insecticides and herbicides; gas odorant

Exposure. Inhalation

Toxicology. n-Butyl mercaptan is a central nervous system depressant.

Accidental exposure of seven workers to concentrations estimated to be between 50 and 500 ppm for 1 hour caused muscular weakness and malaise; six of the workers experienced sweating, nausea, vomiting, and headache; three experienced confusion, and one lapsed into a coma for 20 minutes.[1] On admission to the hospital, all the workers had flushing of the face, increased rate of breathing, and obvious mydriasis. Six of the patients recovered within a day, but the most seriously affected patient experienced profound weakness, dizziness, vomiting, drowsiness, and depression.

In rats, the LC_{50} for 4 hours was 4020 ppm; effects were irritation of mucous membranes, increased respiration, incoordination, staggering gait, weakness, partial skeletal muscle paralysis beginning in the hind limbs, light to severe cyanosis, and mild to heavy sedation.[2] Animals that survived single near-lethal doses by the intraperitoneal and oral routes frequently had liver and kidney damage at autopsy up to 20 days post treatment. The liquid dropped in the eyes of rabbits caused slight to moderate irritation. No dermal changes were observed when 0.2 ml of a 20% solution was applied to the clipped skin of guinea pigs for 10 days.[1]

Female mice and rats exposed 6 hr/day at concentrations of 10, 68, or 152 ppm during gestation had reduced maternal weight increments at the higher doses; embryotoxic effects and increased resorptions occurred in the mice exposed at 68 and 152 ppm.[3]

The disagreeable skunklike odor is detectable at about 0.0001 to 0.001 ppm.[1]

The 1995 ACGIH threshold limit value–time-weighted average (TLV-TWA) for butyl mercaptan is 0.5 ppm (1.8 mg/m³).

REFERENCES

1. National Institute for Occupational Safety and Health: *Criteria for a Recommended Stan-

dard . . . *Occupational Exposure to* n-*Alkane Mono Thiols, Cyclohexanethiol, Benzenethiol.* DHEW (NIOSH) Pub No 78–213. Washington, DC, US Government Printing Office, 1978
2. Fairchild EJ, Stokinger HE: Toxicologic studies on organic sulfur compounds. 1. Acute toxicity of some aliphatic and aromatic thiols (mercaptans). *Am Ind Hyg Assoc J* 19:171–189, 1958
3. Thomas WC, Seckar JA, Johnson JT, et al: Inhalation teratology studies of n-butyl mercaptan in rats and mice. *Fund Appl Toxicol* 8:170–178, 1987

REFERENCES

1. *o-sec*-Butylphenol. *Documentation of the TLVs and BEIs*, 6th ed, p 185. Cincinnati, OH, American Conference of Governmental Industrial Hygienists (ACGIH), 1991
2. US Department of Health and Human Services: *Registry of Toxic Effects of Chemical Substances (RTECS)*, 1985–86 ed, Vol 4, Sweet DV (ed), p 3243. Public Health Service, Centers for Disease Control, National Institute for Occupational Safety and Health, 1987

o-sec-BUTYLPHENOL
CAS: 89-72-5

$C_{10}H_{14}O$

Synonyms: Phenol, *o-sec*-butyl; 2-(1-methylpropyl)phenol

Physical Form. Liquid

Uses. Chemical intermediate in the production of resins, plasticizers, and other products

Exposure. Inhalation; skin

Toxicology. *o-sec*-Butylphenol is a skin, eye and respiratory irritant. Acute occupational exposures have resulted in mild respiratory irritation, as well as skin burns.[1]

Rats survived a 7-hour exposure to an atmosphere saturated with the vapor.[1] The oral LD_{50} for rats is 2700 mg/kg, according to foreign literature citations.[2] In guinea pigs, the oral and skin absorption LD_{50}s ranged between 600 and 2400 mg/kg.[1]

On the skin of rabbits, 500 mg for 24 hours caused severe skin irritation, and 50 μg in the eyes of rabbits for 24 hours also produced severe irritation.[2] In another report, no corneal injury was caused by direct contact of the liquid with the eyes of guinea pigs.[1]

The 1995 ACGIH threshold limit value–time-weighted average for *o-sec*-butylphenol is 5 ppm (31 mg/m³) with a notation for skin.

p-tert-BUTYLTOLUENE
CAS: 98-51-1

$(CH_3)_3CC_6H_4CH_3$

Synonyms: *p*-Methyl-*tert*-butylbenzene; TBT; 1-methyl-4-tertiary-butylbenzene

Physical Form. Clear, colorless liquid

Uses. Solvent for resins; intermediate in organic synthesis

Exposure. Inhalation

Toxicology. *p-tert*-Butyltoluene is an irritant of the mucous membranes.

Exposure of human volunteers for 5 minutes to concentrations of 5 to 160 ppm caused complaints of irritation of the nose and throat, nausea, and metallic taste; moderate eye irritation occurred at 80 ppm.[1] Exposed workers have complained of nasal irritation, nausea, headache, malaise, and weakness. Signs and symptoms included decreased blood pressure, increased pulse rate, tremor, anxiety, and evidence of chemical irritation from skin contact. Laboratory findings suggested slight bone marrow depression.

The LD_{50} in female rats ranged from 934 ppm for 1 hour to 165 ppm for 8 hours.[1] Principal effects were irregular gait, paralysis, narcosis, and dyspnea, as well as eye irritation. At autopsy, there were pulmonary edema and severe hemorrhage in some animals. Repeated exposures of rats to 50 ppm produced liver and kidney changes and lesions in the spinal cord and brain. The liquid on the rabbit skin was only a mild irritant.

The odor is recognized by most people at 5 ppm, but tolerance may be readily acquired. The irritating property may not be sufficient to protect from hazardous concentrations.

The 1995 ACGIH threshold limit value–time-weighted average for *p-tert*-butyltoluene is 1 ppm (6.1 mg/m³).

REFERENCES

1. Hine CH et al: Toxicological studies on *p-tertiary*-butyltoluene. *Arch Ind Hyg Occup Health* 9:227–244, 1954

CADMIUM AND COMPOUNDS
CAS: 7440-43-9

Cd

Compounds: Cadmium oxide; cadmium carbonate; cadmium chloride; cadmium sulfate; cadmium sulfide

Physical Form. The metal is soft, ductile, silver-white, electropositive; cadmium oxide may take the form of a colorless amorphous powder or of red or brown crystals

Uses. The metal is used in electroplating, in solder for aluminum, as a constituent of easily fusible alloys, as a deoxidizer in nickel plating, in process engraving, in cadmium-nickel batteries, and in reactor control rods. Cadmium compounds are employed as TV phosphors, as pigments in glazes and enamels, in dyeing and printing, and in semiconductors and rectifiers

Exposure. Inhalation

Toxicology. Cadmium oxide fume is a severe pulmonary irritant; cadmium dust also is a pulmonary irritant, but it is less potent than cadmium fume because it has a larger particle size. Chronic exposure is associated with nephrotoxicity. Several inorganic cadmium compounds cause malignant tumors in animals.

Inhalation exposure to high levels of cadmium fumes or dust is intensely irritating to respiratory tissue.[1] Particle size appears to be a more important determinant of toxicity than chemical form.[1] Symptoms may include tracheobronchitis, pneumonitis, pulmonary edema, and death. However, most acute intoxications have been caused by inhalation of cadmium fume at concentrations that did not provide warning symptoms of irritation. Concentrations of fume responsible for fatalities have been 40 to 50 mg/m³ for 1 hour or 9 mg/m³ for 5 hours.[2] Nonfatal pneumonitis has been reported from concentrations of 0.5 to 2.5 mg/m³, and relatively mild cases have been attributed to even lower concentrations. Following an asymptomatic latent period of 4 to 10 hours, there is characteristically nasopharyngeal irritation, a feeling of chest constriction or substernal pain, cough, and dyspnea; there also may be headache, chills, muscle aches, nausea, vomiting, and diarrhea.[3,4] Pulmonary edema may develop rapidly, with decreased vital capacity and markedly reduced carbon monoxide–diffusing capacity.[4] In about 20% of cases, dyspnea is progressive, accompanied by wheezing or hemoptysis, and may result in death within 7 to 10 days after exposure; at autopsy, the lungs are markedly congested, and there is an intra-alveolar fibrinous exudate, as well as alveolar cell metaplasia.[3,4] Among survivors, the subsequent course is unpredictable; most cases resolve slowly, but respiratory symptoms may linger for several weeks, and impairment of pulmonary function may persist for months.[4]

Longer-term inhalation exposure at lower levels leads to decreased lung function and emphysema.[1] Early minor changes in ventilatory functions may progress, with continued exposure, to respiratory insufficiency.

Chronic exposure to cadmium results in renal damage. This damage can be identified by increased urinary levels of β_2-microglobulin, retinol-binding protein, or other low-molecular-weight proteins.[1,5] Increasing damage results in excretion of higher-molecular-weight proteins, indicating either glomerular damage or severe tubular damage.[1] The frequency of occurrence of proteinuria increases with length of exposure; in one study, persons exposed to cadmium compounds for less than 2 years had no proteinuria, whereas most of those exposed for 12 years or more had proteinuria with little other evidence of renal damage.[4] It has been estimated that overt proteinuria can occur only after 5 to 10 years of exposure to approximately 100 μg cadmium/m^3.[6] Renal damage may continue to progress even after exposure ceases.

The urinary excretion of cadmium itself bears no known relationship to the severity or duration of exposure and is only a confirmation of absorption.[3] Absorbed cadmium is retained by the body to a large extent, and excretion is very slow.[7]

Other consequences of cadmium exposure are anemia, eosinophilia, yellow discoloration of the teeth, rhinitis, occasional ulceration of the nasal septum, damage to the olfactory nerve, and anosmia.[8,9]

Chronic exposure to high levels of cadmium in food has caused bone disorders, including osteoporosis and osteomalacia.[1] Long-term ingestion, by a Japanese population, of water, rice, and beans, and rice contaminated with cadmium, was associated with a crippling condition, "itai-itai" disease. The affliction is characterized by pain in the back and joints, osteomalacia, bone fractures, and occasional renal failure, and most often affects women with multiple risk factors such as multiparity and poor nutrition.[10]

Occupational exposure to cadmium has been implicated in a significant increase in lung cancer.[11] Occupational cohort studies from the United Kingdom and Sweden have found increased mortality rates from lung cancer, but they were not necessarily related to level and duration of cadmium exposure.[11] In an American cohort, a 2.8-fold increase in lung cancer was found in the group with the highest cadmium exposure, and the dose-response trend over three exposure groups was also significant.[12] Epidemiological studies are confounded by a number of factors, such as smoking; con-comitant exposure to other carcinogens, including nickel and arsenic; small exposure populations; and limited exposure data.[11]

A number of early studies also reported an increased risk for prostate cancer, which has not been confirmed in later studies.[11]

Long-term inhalation studies in rats exposed to aerosols of cadmium chloride, cadmium sulfate, cadmium sulfide, and cadmium oxide has caused lung tumors.[13] Subcutaneous or intramuscular injection with certain cadmium salts has caused rhabdomyosarcomas and fibrosarcomas in rats; with cadmium sulfate or cadmium sulfide, there were local sarcomas and, with cadmium chloride, there were local pleomorphic sarcomas and testicular interstitial cell tumors.[14-16]

The IARC has determined that there is sufficient evidence in humans for the carcinogenicity of cadmium and cadmium compounds.[11] In animals, there is sufficient evidence for the carcinogenicity of cadmium compounds and limited evidence for cadmium metal.

Both positive and negative genotoxic results have been reported. Overall, cadmium appears to have the capability to alter genetic material, particularly chromosomes.

In rat developmental studies, fetal effects, including delayed ossification and decreased locomotor activity, occurred at doses that also caused maternal toxicity.[17] Cadmium sulfate injected into the lingual vein of female hamsters on day 8 of pregnancy caused a high incidence of resorption and malformed offspring.[18] Acute necrosis of rat testes follows large doses orally or parenterally, but testicular effects have not been reported thus far in humans.[15]

The 1995 ACGIH threshold limit value–time-weighted average (TLV-TWA) for elemental cadmium and compounds as Cd is 0.01 mg/m^3 for total particulate dust or 0.002 mg/m^3 for the respirable fraction of dust; there is an A2 suspected human carcinogen designation for both forms.

REFERENCES

1. Agency for Toxic Substances and Disease Registry (ATSDR): *Toxicological Profile for*

Cadmium. US Department of Health and Human Services, Public Health Service, pp 1-171, TP-92/06, 1993

2. Cadmium and compounds. *Documentation of TLVs and BEIs*, 5th ed, pp 87–89. Cincinnati, OH, American Conference of Governmental Industrial Hygienists (ACGIH), 1986

3. Dunphy B: Acute occupational cadmium poisoning: a critical review of the literature. *J Occup Med* 9:22–26, 1967

4. Louria DB, Joselow MM, Browder AA: The human toxicity of certain trace elements. *Ann Intern Med* 76:307–319, 1972

5. Tsuchiya K: Proteinuria of workers exposed to cadmium fume. *Arch Environ Health* 14:875–890, 1967

6. *Environmental Health Criteria 134. Cadmium*. World Health Organization, Geneva, 1992

7. Stokinger HE: The metals. In Clayton GD, Clayton FE (eds): *Patty's Industrial Hygiene and Toxicology*, 3rd ed, Vol 2, *Toxicology*, pp 1563–1583. New York, Wiley-Interscience, 1981

8. Bonnell JA: Cadmium poisoning. *Ann Occup Hyg* 8:45–49, 1965

9. Fassett DW: Cadmium: Biological effects and occurrence in the environment. *Annu Rev Pharmacol* 15:425–435, 1975

10. Emerson BT: "Ouch-ouch" disease: the osteomalacia of cadmium nephropathy. *Ann Intern Med* 73:854–855, 1970

11. *IARC Monographs on the Evaluation of Carcinogenic Risks to Humans*, Vol 58, Beryllium, cadmium, mercury, and exposures in the glass manufacturing industry, pp 119–237. Lyon, International Agency for Research on Cancer, 1993

12. Thun MJ, Schnorr TM, Smith A, et al: Mortality among a cohort of U.S. cadmium production workers–an update. *J Natl Cancer Inst* 74:325–333, 1985

13. Oldiges H, Hochrainer D, Glaser U: Long-term inhalation study with Wistar rats and four cadmium compounds. *Toxicol Environ Chem* 19:217–222, 1989

14. Heath JC, Daniel MR, Dingle JT, Webb M: Cadmium as a carcinogen. *Nature* 193:592–593, 1962

15. Haddow A, Roe FJC, Dukes CE, Mitchley BCV: Cadmium neoplasia: sarcomata at the site of injection of cadmium sulphate in rats and mice. *Br J Cancer* 18:667–673, 1964

16. Gunn SA, Gould TC, Anderson WAD: Specific response of mesenchymal tissue to cancerogenesis by cadmium. *Arch Pathol* 83:483–499, 1967

17. Baranski B: Effect of exposure of pregnant rats to cadmium on prenatal and postnatal development of the young. *J Hyg Epidemiol Microbiol Immunol* 29:253–262, 1985

18. Holmberg RE Jr, Ferm VH: Interrelationships of selenium, cadmium, and arsenic in mammalian teratogenesis. *Arch Environ Health* 18:873–877, 1969

CALCIUM CARBONATE
CAS: 1317-65-3

CaCO₃

Synonyms: Limestone; chalk; dolomite; marble

Physical Form. Odorless, tasteless powder or crystal

Uses. In manufacture of quicklime, portland cement, and paints; United States Pharmacopeia (USP) grades used in dentifrices, cosmetics, food, and pharmaceuticals such as antacids

Exposure. Inhalation

Toxicology. Calcium carbonate is considered to be a harmless dust.

No adverse effects have been reported in the literature among workers exposed to calcium carbonate.[1]

The 1995 threshold limit value–time-weighted average (TLV-TWA) is 10 mg/m³, total dust, containing no asbestos and less than 1% crystalline silica.

REFERENCES

1. Calcium carbonate. *Documentation of the TLVs and BEIs*, 6th ed, pp 195–196. Cincinnati, OH,

American Conference of Governmental Industrial Hygienists, 1991

CALCIUM CYANAMIDE
CAS: 156-62-7

$CCaN_2$

Synonyms: Cyanamide, calcium salt; calcium carbimide; cyanamide (although this synonym commonly refers to hydrogen cyanamide)

Physical Form. White or gray crystalline solid

Uses. Manufacture of calcium cyanide and dicyandiamide; formerly used as a defoliant and herbicide

Exposure. Inhalation

Toxicology. Calcium cyanamide is an irritant of the eyes, mucous membranes, and skin.

Calcium cyanamide is severely irritating to the eyes and skin and causes skin ulceration.[1] A sensitization dermatitis has been reported in 0.5% to 1% of exposed workers. Inhalation of the dust, presumably at high levels, has caused headache, tachypnea, hypotension, and pulmonary edema.[2] Calcium cyanamide does not liberate cyanide when acidified or in vivo. The lethal oral dose in humans is 40 to 50 g. In commercial form, calcium cyanamide may also contain calcium hydroxide and calcium carbonate.

Calcium cyanamide is an inhibitor of aldehyde dehydrogenase, and concurrent intake of ethanol can increase susceptibility to a transient vasomotor disturbance apparent in the face, chest and arms known as "cyanamide blush."[3] In six male alcoholic volunteers, oral administration of 0.7 mg/kg of the chemical and ingestion of ethanol produced tachycardia and decreased diastolic blood pressure.

In a carcinogenesis feeding study at levels of 2000 ppm in both sexes of mice, and at 400 ppm in female rats and 200 ppm in male rats, calcium cyanamide did not act as a carcinogen.[4]

The 1995 ACGIH threshold limit value–time-weighted average (TLV-TWA) for calcium cyanamide is (0.5 mg/m^3).

REFERENCES

1. Grant WM: *Toxicology of the Eye*, 3rd ed, p 286. Springfield, IL, Charles C Thomas, 1986
2. Arena JM: *Poisoning*, 4th ed, pp 236, 623. Springfield, IL, Charles C Thomas, 1979
3. Brien JF, Peachey JE, Rogers BJ, et al: A study of the calcium carbimide-ethanol interaction in man. *Eur J Clin Pharmacol* 14:133–141, 1978
4. National Cancer Institute: *Carcinogenesis Technical Report Series. Bioassay of Calcium Cyanamide for Possible Carcinogenicity.* DHEW (NIH) Pub No 79-1719 pp 1–112. Washington, DC, US Government Printing Office, 1979

CALCIUM HYDROXIDE
CAS: 1305-62-0

$Ca(OH)_2$

Synonyms: Slaked lime; hydrated lime; calcium hydrate

Physical Form. White microcrystalline powder

Uses. In the manufacture of mortar, plaster, whitewash, and paper pulp; in lubricants; in drilling fluids

Exposure. Inhalation

Toxicology. Calcium hydroxide is a relatively strong base and, therefore, a caustic irritant of all exposed surfaces of the body, including the respiratory tract.

Calcium hydroxide is one of the most common causes of severe chemical eye burns.[1] In almost all cases, there is a semisolid particulate paste in contact with the cornea and conjunctiva, tending to adhere and to dissolve slowly. Strongly alkaline calcium hydroxide solution is formed and causes severe injury if not removed promptly.

The oral LD_{50} for rats is between 4.8 and 11.1 g/kg.[2] Rats given tap water containing 50 or 350 mg/l had reduced food intake and were restless and aggressive at 2 months; at 3 months, they showed a loss in body weight, decreased counts for erythrocytes and phagocytes, and decreased hemoglobin.[3] Autopsy showed inflammation of the small intestine and dystrophic changes in the stomach, kidneys, and liver.

The 1995 ACGIH threshold limit value–time-weighted average (TLV-TWA) for calcium hydroxide is 5 mg/m³.

REFERENCES

1. Grant WM: *Toxicology of the Eye*, 3rd ed, pp 167–172. Springfield, IL Charles C Thomas, 1986
2. Smyth HF Jr et al: Range-finding toxicity data: List VII. *Am Ind Hyg Assoc J* 30:470, 1969
3. Wands RC: Alkaline materials. In Clayton GD, Clayton FE (eds): *Patty's Industrial Hygiene and Toxicology*, 3rd ed, rev, Vol 2B, *Toxicology*, pp 3052–3053. New York, Wiley-Interscience, 1981

CALCIUM OXIDE
CAS: 1305-78-8

CaO

Synonyms: Burnt lime; calx; lime; quicklime

Physical Form.   Crystals, white or grayish-white lumps or granular powder

Uses.   In construction materials; in manufacture of steel, aluminum, and magnesium; as a scrubbing agent to remove sulfur dioxide emissions from smokestacks; in manufacture of glass, paper, and industrial chemicals; in fungicides, insecticides, and lubricants

Exposure.   Inhalation

Toxicology.   Calcium oxide is an irritant of the eyes, mucous membranes, and skin.

The irritant effects are probably due primarily to its alkalinity, but dehydrating and thermal effects also may be contributing factors.[1] Strong nasal irritation was observed from exposure to a mixture of dusts containing calcium oxide in the range of 25 mg/m³, but levels of 9 to 10 mg/m³ produced no observable irritation.[2] Inflammation of the respiratory tract, ulceration and perforation of the nasal septum, and pneumonia have been attributed to inhalation of calcium oxide dust; severe irritation of the upper respiratory tract ordinarily causes persons to avoid serious inhalation exposure.[1,2]

Particles of calcium oxide can cause severe burns of the eyes.[3] It can also produce skin burns as well as fissuring and brittleness of the nails.[4]

The 1995 ACGIH threshold limit value–time-weighted average (TLV-TWA) for calcium oxide is 2 mg/m³.

REFERENCES

1. Calcium oxide. *Documentation of TLVs and BEIs*, 6th ed, pp 200–201. Cincinnati, OH, 1991
2. Wands RC: Alkaline materials. In Clayton GD, Clayton FE (eds): *Patty's Industrial Hygiene and Toxicology*, 3rd ed, rev, Vol 2B, *Toxicology*, pp 3053–3054. New York, Wiley-Interscience, 1981
3. Grant WM: *Toxicology of the Eye*, 3rd ed, pp 167–172. Springfield, IL Charles C Thomas, 1986
4. Fisher AA: *Contact Dermatitis*, p 17. Philadelphia, Lea & Febiger, 1973

CALCIUM SILICATE, Synthetic nonfibrous
CAS: 1344-95-2

CaSiO₃

Synonyms: Calcium hydrosilicate; calsil; Microcel; Calflo E; Florite R; Marimet 45; tobermorite (crystalline form of synthetic calcium silicate); wollastonite (naturally occurring fibrous form)

Physical Form. White powder

Uses. Anticaking agent in table salt, foods, pharmaceuticals, and agricultural pesticides; replacement for asbestos in thermal insulation

Exposure. Inhalation

Toxicology. Synthetic nonfibrous calcium silicate is considered to be a nuisance dust.

Effects of three commercially produced calcium silicate insulation materials were examined in rats by inhalation and intraperitoneal injection.[1] Exposure to 10 mg/m³ of respirable dust for 7 hr/day, 5 days/week, for 12 months had no effect on the survival of treated animals compared to controls.

Although two pulmonary neoplasms, one malignant and one benign, were found in exposed animals, neither was the cause of death, and the incidence was not significantly different from the control group, in which no tumors were found. One peritoneal mesothelioma was found in an animal from one of the inhalation groups, but this was considered to be a spontaneous tumor because none of over 100 animals injected intraperitoneally with 25 mg of calcium silicate developed that kind of tumor.

The 1995 threshold limit value–time-weighted average (TLV-TWA) is 10 mg/m³ for total dust.

REFERENCES

1. Bolton RE et al: Effects of the inhalation of dusts from calcium silicate insulation materials in laboratory rats. *Environ Res* 39:26–43, 1986

CALCIUM SULFATE
CaSO₄
CAS: 7778-18-9

CaSO₄. 2H₂O
CAS: 10104-41-4

Synonyms: Anhydrous calcium sulfate; anhydrous sulfate of lime; gypsum (CaSO₄ 2H₂O); plaster of paris (CaSO₄1/2H2O)

Physical Form. Crystal or powder

Uses. The insoluble anhydrite is used in cement formulations and as a paper filler; the soluble anhydrite is used as a drying agent; the hemihydrate is used for wall plaster and wallboard; gypsum is used in the manufacture of plaster of paris and portland cement and in water clarification

Exposure. Inhalation

Toxicology. Calcium sulfate is considered to be a nuisance dust.

There have been no reports of adverse effects in humans exposed to calcium sulfate. One report of gypsum miners attributed adverse effects to respirable quartz rather than the calcium sulfate.[1]

The 1995 threshold limit value–time-weighted average (TLV-TWA) is 10 mg/m³, total dust, containing no asbestos and less than 1% crystalline silica.

REFERENCES

1. Oakes D, Douglas R, Knight K, et al: Respiratory effects of prolonged exposure to gypsum dust. *Ann Occup Hyg* 26:833–840, 1982

CAMPHOR
CAS: 76-22-2

$C_{10}H_{16}O$

Synonyms: 2-Bornanone; 2-camphanone; 2-keto-1,7,7-trimethylnorcamphane

Physical Form. Translucent crystals with characteristic odor

Uses. Plasticizer for cellulose esters and ethers; in manufacture of plastics; in lacquers and varnishes; in explosives; in pyrotechnics; as moth repellant; as preservative in pharmaceuticals and cosmetics

Exposure. Inhalation; skin absorption

Toxicology. Camphor is an irritant of the eyes and the nose; at high concentrations, it is a convulsant.

Camphor is readily absorbed from all sites of administration, producing a feeling of coolness on the skin, and oral doses cause a sensation of warmth in the stomach.[1]

Symptoms of vapor exposure in humans are irritation of the eyes and nose and anosmia; these symptoms occur at concentrations above 2 ppm.[2] Heavy exposures cause nausea, anxiety, confusion, headache, dizziness, twitching of facial muscles, spasticity, convulsions, and coma.[1,3,4]

Most camphor poisonings in humans are due to accidental ingestion.[5] With mild poisoning, gastrointestinal tract symptoms are more common than neurological symptoms and include irritation of the mouth, throat, and stomach. Severe poisoning is characterized by convulsions.

Ingestion of 6 to 10 g of camphor by two men resulted in psychomotor agitation and hallucinations.[6] The probable lethal dose for humans is in the 50 to 500 mg/kg range.[5] Camphor may be expected to be somewhat irritating on contact with the eyes, but no serious eye injuries have been reported.[7]

The 1995 ACGIH threshold limit value– time-weighted average for camphor is 2 ppm (12 mg/m³) with a short-term excursion limit of 3 ppm (19 mg/m³).

REFERENCES

1. Gosselin RE, Smith RP, Hodge HC: *Clinical Toxicology of Commercial Products, Section III*, 5th ed, pp 84–86. Baltimore, MD, Williams and Wilkins, 1984
2. Gronka PA, Bobkoskie RL, Tomchick GJ, Rakow AB: Camphor exposures in a packaging plant. *Am Ind Hyg Assoc J* 30:276–279, 1969
3. Arnow R: Camphor poisoning. *JAMA* 235:1260, 1976
4. Ginn HE et al: Camphor intoxication treated by lipid dialysis. *JAMA* 203:164–165, 1968
5. Segal S: Camphor: Who needs it? *Pediatrics* 62:404–406, 1978
6. Koppel C et al: Camphor poisoning. Abuse of camphor as a stimulant. *Arch Toxicol* 51:101–106, 1982
7. Grant WM: *Toxicology of the Eye*, 3rd ed, p 173. Springfield, IL, Charles C Thomas, 1986

CAPROLACTAM
CAS: 105-60-2

$C_6H_{11}NO$

Synonyms: Epsilon caprolactam; 2-oxohexamethylenimine; aminocaproic lactam

Physical Form. White crystalline solid

Uses. Monomer for manufacture of polycaprolactam (nylon 6) used in carpets, textiles, clothing, and tires

Exposure. Inhalation

Toxicology. Caprolactam is an irritant of the eyes, mucous membranes, respiratory tract, and skin and, rarely, a convulsant.

Human test panel exposures to vapor levels ranging from 53 mg/m³ to 521 mg/m³ resulted in eye and throat irritation in all those exposed.[1] In a study of workers exposed to vapor over a period of 18 years at levels up to 100 ppm, there were complaints of severe discomfort from burning of the eyes, nose, and throat.[2] Eye irritation did not occur at 25 ppm, but nose and throat irritation occurred in some at 10 ppm.

An earlier study of German workers exposed to the dust at various levels (mean 61 mg/m³) reported eye, nose, and throat irritation, epistaxis, and a bitter taste in the mouth.[3] Seizures, fever, and dermatitis occurred in a worker after three days of occupational exposure to caprolactam at unmeasured levels.[4] An absence of organic central nervous system abnormalities on physical examination strongly implicated caprolactam as the cause of the seizures.

In eight workers chronically exposed to approximately 70 times the TLV, the only effects noted were peeling and/or fissuring of the skin.[5]

Doses of 350 to 600 mg/kg intraperitoneally to rats produced tremor, convulsions, and bloody eye discharge.[6] In a three-generation reproduction study, rats were given caprolactam in the diet at 0, 1000, 5000, and 10,000 ppm.[7] No teratogenic effects were observed. Caprolactam was tested for carcinogenicity in the diet of mice and rats, and no carcinogenic effect was observed.[8] The IARC evaluation concluded that there was no evidence for the carcinogenicity of caprolactam in experimental animals.[9]

The 1995 ACGIH threshold limit value–time-weighted average (TLV-TWA) for caprolactam dust is 1 mg/m³ with a short-term excursion limit of 3 mg/m³; the TLV-TWA for caprolactam vapor is 5 ppm (23 mg/m³) with a short-term excursion limit of 10 ppm (46 mg/m³).

REFERENCES

1. Caprolactam: *Documentation of the TLVs and BEIs*, 6th ed, pp 208–211. Cincinnati, OH, American Conference of Govermental Industrial Hygienists (ACGIH), 1991
2. Ferguson WS, Wheeler DD: Caprolactam vapor exposure. *Am Ind Hyg Assoc J* 34:384–389, 1973
3. Hohensee F: On pharmacological and physiological action of ε-caprolactam (Ger) *Faserforsch Textiltech* 8:299–303, 1951
4. Tuma SN, Orson F, Fossella FV, Waidhofer W: Seizures and dermatitis after exposure to caprolactam. *Arch Intern Med* 141:1544–1545, 1981
5. Kelman GR: Effects of human exposure to atmospheric epsilon–caprolactam. *Hum Toxicol* 5:57–59, 1986
6. Goldblatt MW, Farquharson ME, Bennet G, Askew BM: E-Caprolactam. *Br J Ind Med* 11:1–10, 1954
7. Serota DG, Hoberman AM, Friedman MA, Gad SC: Three-generation reproduction study with caprolactam in rats. *J Appl Toxicol* 8:285–293, 1988
8. National Toxicology Program: *Carcinogenesis Bioassay of Caprolactam (CAS No. 105-60-2) in F344 Rats and B6C3F1 Mice (Feed Study)*. Technical Report Series No 214, Research Triangle Park, NC, 1982
9. IARC Monographs on the Evaluation of the Carcinogenic Risk of Chemicals to Humans, Vol 39, Some chemicals used in plastics and elastomers, p 264. Lyon, International Agency for Research on Cancer, 1986

CARBARYL

CAS: 63-25-2

$C_{12}H_{11}NO_2$

Synonyms: 1-Naphthyl methylcarbamate; Sevin

Physical Form. Crystals of a white or grayish, odorless solid

Uses. Insecticide

Exposure. Inhalation; skin absorption; ingestion

Toxicology. Carbaryl is a short-acting anticholinesterase agent with the important characteristic of rapid reversibility of inhibition of the enzyme.

The clinical picture of carbaryl intoxication results from inactivation of cholinesterase, causing accumulation of acetylcholine at synapses in the nervous system, skeletal and smooth muscle, and secretory glands.[1-4] Signs and symptoms of overexposure may include: (1) muscarinic manifestations, such as miosis, blurred vision, lacrimation, excessive nasal discharge or salivation, sweating, abdominal cramps, nausea, vomiting, and diarrhea; (2) nicotinic manifestations, including fasciculation of fine muscles and tachycardia; and (3) central nervous system manifestations characterized by headache, dizziness, mental confusion, convulsions, coma, and depression of the respiratory center.

A single dose of 250 mg (approximately 2.8 mg/kg) ingested by an adult resulted in moderate poisoning; after 20 minutes, there was sudden onset of abdominal pain, followed by profuse sweating, lassitude, and vomiting; 1 hour after ingestion, and following administration of a total of 3 mg atropine sulfate, the person felt better and, after another hour, was completely recovered.[1] In one reported case of long-term exposure, a 75-year-old man was exposed for 8 months following repeated excessive applications of a 10% dust formulation inside his home. Signs and symptoms were compatible with cholinesterase inhibition in addition to a significant weight loss.

Workers exposed to carbaryl dust at levels that occasionally reached 40 mg/m³ had slight depression in blood cholinesterase activity but no clinical symptoms.[3] In general, cases of occupational poisoning to carbaryl are rare because mild symptoms appear long before a dangerous dose is absorbed; furthermore, rapid spontaneous recovery of inhibited cholinesterase occurs.

In a study of 59 workers exposed to concentrations ranging from 0.23 to 31 mg/m³ during a 19-month period, there were no signs or symptoms of anticholinesterase activity.[5] In the most heavily exposed workers, relatively large amounts of 1-naphthol (a metabolite of carbaryl) were excreted in the urine, and the blood cholinesterase activity was slightly depressed. It was concluded that an excretion level of total (free plus conjugated) 1-naphthol significantly above 400 μg/100 ml of urine indicates absorption and metabolism of carbaryl.

On the skin, concentrated solutions may cause irritation and systemic intoxication.[1] Allergic skin reactions are rare but have been reported.[3] Men exposed in error to 85% water-wettable powder as a dust complained of burning and irritation of the skin but recovered in a few hours without any treatment except bathing. Their blood cholinesterase levels were only slightly depressed.[2]

The possible effect of carbaryl on reproduction and/or teratogenesis has been explored in rats, mice, guinea pigs, rabbits, dogs, and other species. In general, developmental toxicity, including reduced fetal weight, fetal resorptions, and the occurrence of malformation, has occurred only at doses that cause significant maternal toxicity.[3,6] No significant changes in sperm count or fertility have been found in cohort studies of exposed workers.[3]

The 1995 ACGIH threshold limit value–time-weighted average (TLV-TWA) for carbaryl is 5 mg/m³.

Note. Treatment of anticholinesterase effects with pralidoxime (2-PAM) chloride is contraindicated.[7] In a number of reports, carbaryl poisoning appeared to be aggravated by the administration of 2-PAM Cl.[3] Recently, investigators have suggested that certain oximes, such as 2-PAM Cl, act as allosteric effectors of cholinesterases, resulting in enhanced inhibition rates and potentiation of carbaryl toxicity.[7]

REFERENCES

1. Hayes WJ Jr: *Clinical Handbook on Economic Poisons. Emergency Information for Treating Poisoning*, US Public Health Service Pub No 476, pp 44–46. Washington, DC, US Government Printing Office, 1963
2. Hayes WJ Jr: *Pesticides Studied in Man*, pp 438–447. Baltimore, MD, Williams and Wilkins, 1982
3. *Carbaryl. Environmental Health Criteria 153.* World Health Organization, pp 12–251, 1994
4. National Institute for Occupational Safety and Health: *Criteria for a Recommended Standard . . . Occupational Exposure to Carbaryl.* DHEW (NIOSH) Pub No 77-107, pp 17–96, 109–117.

Washington, DC, US Government Printing Office, 1976

5. Best EM Jr, Murray BL: Observations on workers exposed to Sevin insecticide: a preliminary report. *J Occup Med* 10:507–517, 1962
6. Mathur A, Bhatnagar, P: A teratogenic study of carbaryl in Swiss albino mice. *Food Chem Toxicol* 29:629–632, 1991
7. Lieske CN, Clark JH, Maxwell DM, et al: Studies of the amplification of carbaryl toxicity by various oximes. *Toxicol Lett*: 62:127–137, 1992

CARBON BLACK
CAS: 1333-86-4

Synonyms: Carbon; activated carbon; acetylene carbon; decolorizing carbon; actibon; channel black; furnace black; thermal black; gas black; lamp black; ultracarbon

Physical Form. Black crystal; powder that varies in particle size and degree of aggregation

Uses. In the rubber, plastic, printing, and paint industries as a reinforcing agent and a pigment

Exposure. Inhalation

Toxicology. There are no well-demonstrated health hazards to humans from acute exposure to carbon black.

Commercial carbon black is a spherical colloidal form of nearly pure carbon particles and aggregates with trace amounts of organic impurities adsorbed on the surface. Potential health effects usually are attributed to these impurities rather than to the carbon itself. Soots, by contrast, contain mixtures of particulate carbon, resins, tars, and so on, in a nonadsorbed state.[1]

A significant loss in pulmonary function was reported in a group of 125 Nigerian carbon-black workers exposed to levels of up to 34 mg/m³.[2] The most common respiratory symptom was cough with phlegm, but radio-grams were normal. Significant annual declines in FEV_1 and FVC and radiological lung changes were reported in another group of 35 workers exposed to concentrations of less than 10 mg/m³.[1] In contrast, a survey of over 500 carbon-black workers in the United States and in the United Kingdom found no statistical difference in spirometry, chest radiograph, physical examination, or reported symptoms.[3] A 1988 report on 913 men employed in the production and handling of carbon black in the United States also found no evidence of pulmonary function effects from dust exposure, as determined by spirometry.[4]

A study of over 3000 carbon-black workers employed primarily in Western Europe determined that smoking was the principal factor affecting lung function in the workers and that exposure to carbon black has no more effect than that expected from a nuisance dust. There was no evidence of any increased incidence of radiological abnormality in the workers surveyed.[5] A recent follow-up on much of this same cohort found a correlation between small opacities of the lungs (category 0/1 or greater) and cumulative dust exposure.[6] Exposure to carbon black was also associated with some increased prevalence of respiratory effects, including cough, sputum, and symptoms of chronic bronchitis, as well as small decrements in lung function tests. These results are considered to be consistent with a nonirritant effect of carbon-black dust on the airways combined with dust retention in the lungs.

A retrospective cohort study of 1200 men employed at four carbon-black plants from 1935 to 1974 found no significant increase in total mortality, mortality from heart disease, or mortality due to malignant neoplasms.[7]

Repeated inhalation by monkeys caused deposition of the dust in the lungs with minimal or no fibrous tissue proliferation.[8] Repeated exposure of the monkeys to 56.5 mg/m³ of furnace black for a total of 10,000 hours produced marked electrocardiographic changes, interpreted as right atrial and right ventricular strain, probably a result of massive deposition of the dust in the lungs.[1,8]

Carbon black itself does not appear to be a carcinogen. The major concern is the simultaneous exposure to polycyclic aromatic hydro-carbons, which are strongly adsorbed to the respirable carbon-black particles and from which PAHs may be elutriated in vivo under conditions of human exposure.[9] In a number of studies, however, attempts to elutriate PAH with biological fluids have been largely unsuccessful, and prolonged extraction with boiling aromatic solvents is required for quantitative desorption. A nonstatistically significant increase in lung cancer mortality was found among persons engaged in carbon-black production in Great Britain, which was unassociated with environmental levels or lengths of exposure.[10]

Carbon black has been implicated as a co-carcinogen in animal studies in the presence of high-fat diets and other carcinogens.[1]

The IARC has determined that there is sufficient evidence that solvent extracts of carbon black are carcinogenic, but there is inadequate evidence to evaluate the carcinogenicity of carbon black to humans.[1]

There are no human or animal reports suggesting a mutagenic or teratogenic potential for carbon black.[8]

The 1995 ACGIH threshold limit value–time-weighted average (TLV-TWA) for carbon black is 3.5 mg/m³.

REFERENCES

1. *IARC Monographs on the Evaluation of the Carcinogenic Risk of Chemicals to Humans. Polynuclear Aromatic Compounds*, Part 2, Carbon blacks, mineral oils and some nitroarsines. Vol 33, pp 35–85. Lyon, International Agency for Research on Cancer, 1984
2. Oleru UG et al: Pulmonary function and symptoms of Nigerian workers exposed to carbon black in dry cell battery and tire factories. *Environ Res* 39:161–168, 1983
3. Crosbie WA et al: Survey of respiratory disease in carbon black workers in the UK and USA. Abstract. *Am Rev Respir Dis* 119:209, 1979
4. Robertson JM, Diaz JF, Fyfe IM, Ingalls TH: A cross-sectional study of pulmonary function in carbon black workers in the United States. *Am Ind Hyg Assoc J* 49:161–166, 1988
5. Crosbie WA: The respiratory health of car-bon black workers. *Arch Environ Health* 41:346–353, 1986
6. Gardiner K, Trethowan NW, Harrington JM, et al: Respiratory health effects of carbon black: a survey of European carbon black workers. *Br J Ind Med* 50:1082–1096, 1993
7. Robertson JM, Ingalls TH: A mortality study of carbon black workers in the United States from 1935 to 1974. *Arch Environ Health* 35:181–186, 1980
8. Nau CA, Neal J, Stembridge VA, Cooley RN: Physiological effects of carbon black. IV. Inhalation. *Arch Environ Health* 4:45–61, 1962
9. National Institute for Occupational Safety and Health: *Criteria for a Recommended Standard . . . Occupational Exposure to Carbon Black*. DHEW (NIOSH) Pub No 78-204, pp 1–99. Washington, DC, US Department of Health, Education and Welfare, US Government Printing Office, 1978
10. Hodgson JT, Jones RD: A mortality study of carbon black workers employed at five United Kingdom factories between 1947 and 1980. *Arch Environ Health* 40:261–268, 1985
11. Pence BC, Buddingh F: Co-carcinogenic effect of carbon black ingestion with dietary fat on the development of colon tumors in rats. *Tox Lett* 37:177–182, 1987

CARBON DIOXIDE
CAS: 124-38-9

CO_2

Synonym: Carbonic acid gas

Physical Form. Colorless gas (solid is "dry ice")

Uses/Source. By-product of ammonia production, lime kiln operations, and fermentation; used in carbonation of beverages, as propellant in aerosols, and as dry ice for refrigeration. Exposures may occur in a variety of work settings, including farm silos, fermentation tanks, wells, shipping, mining, and firefighting, and in frozen food industries utilizing dry ice.

Exposure. Inhalation

Toxicology. Carbon dioxide is usually considered a simple asphyxiant although it is also a potent stimulus to respiration and both a depressant and an excitant of the central nervous system.

Numerous human fatalities have occurred after people entered fermentation vats, wells, and silos in which the air had been replaced largely by carbon dioxide.[1,2] In other cases, death or injuries may be caused by the toxicity of carbon dioxide alone and are not due to oxygen deprivation. At levels considered immediately dangerous to life and health, oxygen displacement by carbon dioxide may be as little as 1%.[3] The most immediate and significant effects of acute exposure at high concentrations are those on the central nervous system.[1] Concentrations of 20% to 30% (200,000 to 300,000 ppm) result in unconsciousness and convulsions within 1 minute of exposure. At concentrations of approximately 120,000 ppm, unconsciousness may be produced with longer exposures of 8 to 23 minutes. Neurological symptoms, including psychomotor agitation, myoclonic twitches, and eye flickering, have appeared after 1.5 minutes at 100,000 to 150,000 ppm.[1] Inhalation of concentrations from 60,000 to 100,000 ppm may produce dyspnea, headache, dizziness, sweating, restlessness, paresthesias, and a general feeling of discomfort; at 50,000 ppm, there may be a sensation of increased respiration, but subjects rarely experience dyspnea.[4] After several hours of exposure to 2% carbon dioxide (20,000 ppm), subjects develop headache and dyspnea on mild exertion.[5] Circulatory effects in humans exposed to carbon dioxide include increases in heart rate and cardiac output.[6]

Adaptation to low levels, 1.5% to 3.0% carbon dioxide, has occurred with chronic exposure.[1] Carbon dioxide at room temperature will not injure the skin, but frostbite may result from contact with dry ice or from the gas at low temperatures.

It is important to note that, because carbon dioxide is heavier than air, pockets of the gas may persist for some time in areas such as pits unless ventilation is provided.

Limited experimental studies in test animals have raised some concerns about the ability of carbon dioxide to harm reproductive parameters. Acute exposures to 25,000 to 100,000 ppm were reported to cause mild and reversible testicular injury in rats at all test levels.[7] Twenty-four-hour exposures of rats to 60,000 ppm on days 5 to 21 of pregnancy (1 day per cohort) caused an increase in cardiac malformations; the incidence of cardiac malformations was 23.4% in the test group versus 6.8% in the control group, with the highest incidence occurring when exposure occurred on day 10.[8]

The 1995 ACGIH threshold limit value–time-weighted average (TLV-TWA) for carbon dioxide is 5000 ppm (9000 mg/m^3) with a short-term excursion limit of 30,000 ppm (54,000 mg/m^3).

REFERENCES

1. National Institute for Occupational Safety and Health: *Criteria for a Recommended Standard . . . Occupational Exposure to Carbon Dioxide.* DHEW (NIOSH) Pub No 76-194, pp 17–105, 114–126. Washington, DC, US Government Printing Office, 1976
2. Williams HI: Carbon dioxide poisoning—report of eight cases, with two deaths. *Br Med J* 2:1012–1014, 1958
3. Jacobs DE, Smith MS: Exposures to carbon dioxide in the poultry processing industry. *Am Ind Hyg Assoc J* 49:624–629, 1988
4. Smith TC et al: The therapeutic gases. In Gilman A et al (eds): *Goodman and Gilman's The Pharmacological Basis of Therapeutics,* 7th ed, pp 333–335. New York, Macmillan, 1985
5. Schulte JH: Sealed environments in relation to health and disease. *Arch Environ Health* 8:438–452, 1964
6. Cullen DJ, Eger EI: Cardiovascular effects of carbon dioxide in man. *Anesthesiology* 41:345–349, 1974
7. Vandemark NL, Schanbacher BD, Gomes WR: Alterations in testes of rats exposed to elevated atmospheric carbon dioxide. *J Reproduc Fertil* 28:457–459, 1972
8. Haring OM: Cardiac malformations in rats induced by exposure of the mother to carbon dioxide during pregnancy. *Circ Res* 8:1218–1227, 1960

CARBON DISULFIDE
CAS: 75-15-0

CS_2

Synonyms: Carbon bisulfide; carbon disulphide

Physical Form.   Colorless liquid

Uses.   In preparation of rayon viscose fibers; solvent for lipids, sulfur, rubber, phosphorus, oils, resins, and waxes; insecticide

Exposure.   Inhalation; skin absorption

Toxicology.   Carbon disulfide causes damage to the central and peripheral nervous systems and may accelerate the development of, or worsen, coronary heart disease.

Exposure of humans to 4800 ppm for 30 minutes causes coma and may be fatal.[1] Carbon disulfide intoxication can involve all parts of the central and peripheral nervous systems, including damage to the cranial nerves and development of peripheral neuropathy with paresthesias and muscle weakness in the extremities, unsteady gait, and dysphagia.[2] A follow-up of workers with clinical and electromyographic evidence of neuropathy attributed to carbon disulfide exposure showed no significant improvement 10 years after exposure was discontinued, suggesting a permanent axonal neuropathy.[3]

In extreme cases of intoxication, a Parkinsonism-like syndrome may result, characterized by speech disturbances, muscle spasticity, tremor, memory loss, mental depression, and marked psychic symptoms; permanent disability is likely.[2] Psychosis and suicide are established risks of overexposure to carbon disulfide.[4]

Other reported effects of exposure to carbon disulfide are ocular changes (blind spot enlargement, contraction of peripheral field, corneal anesthesia, diminished pupillary reflexes, nystagmus, and microscopic aneurysms in the retina), gastrointestinal disturbances (chronic gastritis and achlorhydria), renal impairment (albuminuria, microhematuria, elevated blood urea nitrogen, and diastolic hypertension), and liver damage.[2,5] Hearing loss to high-frequency tones has also been reported.[6]

Effects commonly caused by repeated exposure to carbon disulfide vapor are exemplified by a group of workers with a time-weighted average (TWA) exposure of 11.2 ppm (range 0.9–127 ppm) who complained of headaches and dizziness; in other workers with a TWA of 186 ppm (range 23–389 ppm), complaints also included sleep disturbances, fatigue, nervousness, anorexia, and weight loss. The end-of-the-day exposure coefficient of the iodine azide test on urine was a good indicator of workers who were, or had been, symptomatic.[7]

Overexposure to carbon disulfide has been associated with an increase in coronary heart disease. In a mortality study of viscose rayon workers, 42% of deaths were certified to coronary heart disease versus 17% in unexposed workers.[8] A follow-up of this cohort showed a similar pattern with a standardized mortality ratio (SMR) for ischemic heart disease of 172 in spinning operatives.[9] This study also found that the risk declined after exposure ceased, suggesting a direct cardiotoxic or thrombotoxic effect of carbon disulfide rather than an atherogenic effect. A retrospective cohort mortality study of 10,418 men employed in the US rayon industry between 1957 and 1979 found excess deaths from arteriosclerotic heart disease among those potentially most heavily exposed (242 vs. 195.6 expected).[10] There also were excess deaths from suicide (29 vs. 18.8 expected) in one of the four plants investigated. In a Finnish cohort, removal from exposure of workers with coronary risk factors and reduction of levels to 10 ppm caused a dramatic decrease in cardiovascular mortality.[11] Other cardiovascular effects observed in workers repeatedly exposed to carbon disulfide are bradycardia, tachycardia, other arrhythmias, and electrocardiographic changes consistent with both nonspecific and ischemic wave changes.[5]

Conflicting studies have appeared regarding the ability of carbon disulfide to affect reproductive function.[6] Hypospermia, abnormal sperm morphology, menstrual cycle irregularities, increased menstrual flow and pain, and a slight increase in miscarriages have been reported in some studies, whereas other studies have not found adverse effects. A retrospective coho rt study of 265 female workers exposed 15 years prior to the study to concentrations averaging 1.7 to 14.8 mg/m³ showed no significant differences in rates of toxemia, spontaneous abortion, stillbirth, premature or over due delivery, or congenital malformation.[12] However, exposed females had a higher incidence of menstrual disturbances (primarily irregularity) than the nonexposed group. Pregnant rats and rabbits exposed at 20 and 40 ppm, 7 hr/day, showed no evidence of embryotoxicity or teratogenicity. In another report, hydrocephalia was observed in rats exposed to 32 and 64 ppm, 8 hr/day, throughout gestation.[6]

Chronic exposure of animals for periods of less than one year has not shown a carcinogenic potential for carbon disulfide.[6] Furthermore, epidemiological studies do not support a carcinogenic risk under moderate exposure conditions.[13]

Splashes of the liquid in the eye cause immediate and severe irritation; dermatitis and vesiculation may result from skin contact with the vapor or the liquid.[1,2] Although ingestion is unlikely to occur, it may cause coma and convulsions.[1,2]

Both positive and negative results have been found in genotoxic assays.[14]

Carbon disulfide is foul-smelling, but the odor is not sufficient to give adequate warning of hazardous concentrations.

The 1995 ACGIH threshold limit value–time-weighted average (TLV-TWA) for carbon disulfide is 10 ppm (31 mg/m³) with a notation for skin absorption.

REFERENCES

1. Teisinger J: Carbon disulphide. In International Labour Office: *Encyclopaedia of Occupational Health and Safety*, Vol 1, pp 252–253. New York, McGraw-Hill, 1974
2. Tolonen M: Vascular effects of carbon disulfide: a review. *Scand J Work Environ Health* 1:63, 1975
3. Corsi G et al: Chronic peripheral neuropathy in workers with previous exposure to carbon disulphide. *Br J Ind Med* 40:209–211, 1983
4. Mancuso TF, Locke BZ: Carbon disulfide as a cause of suicide—epidemiological study of viscose rayon workers. *J Occup Med* 14:595, 1972
5. Davidson M, Feinleib M: Carbon disulfide poisoning: a review. *Am Heart J* 83:100, 1972
6. Beauchamp RO Jr, Bus JS, Popp JA, et al: A critical review of the literature on carbon disulfide toxicity. *CRC Crit Rev Toxicol* 11:169–278, 1983
7. Rosensteel RE, Shama SK, Flesch JP: Occupational health case report—No 1. *J Occup Med* 16:22, 1974
8. Tiller JR, Schilling RS, Morris JM: Occupational toxic factor in mortality from coronary heart disease. *Br Med J* 4:407–411, 1968
9. Sweetnam PM, Taylor SW, Elwood PC: Exposure to carbon disulphide and ischemic heart disease in a viscose rayon factory. *Br J Ind Med* 44:220–227, 1987
10. MacMahon B, Monson R: Mortality in the US rayon industry. *J Occup Med* 30:698–705, 1988
11. Nurminen M, Hernberg S: Effects of intervention on the cardiovascular mortality of workers exposed to carbon disulphide: a 15-year follow-up. *Br J Ind Med* 42:32–35, 1985
12. Zhou SY, Liang YX, Chen ZQ, Wang YL: Effects of occupational exposure to low-level carbon disulfide (CS_2) on menstruation and pregnancy. *Ind Health* 26:203–214, 1988
13. Nurminen M, Hernberg S: Cancer mortality among carbon disulfide–exposed workers. *J Occup Med* 26:341, 1984
14. Agency for Toxic Substances and Disease Registry (ATSDR): *Toxicology Profile for Carbon Disulfide*. US Department of Health and Human Services, Public Health Service, p 172, TP-91/09, 1992

CARBON MONOXIDE
CAS: 630-08-0

CO

Synonyms: Carbonic oxide; exhaust gas; flue gas

Physical Form. Odorless, colorless, tasteless gas

Sources. Incomplete combustion of organic

fuels; vehicle exhaust; space heaters; gas and kerosene lanterns

Exposure.   Inhalation

Toxicology.   Carbon monoxide (CO) causes tissue hypoxia by preventing the blood from carrying sufficient oxygen.

Carbon monoxide combines reversibly with the oxygen-carrying sites on the hemoglobin molecule with an affinity ranging from 210 to 240 times greater than that of oxygen; the carboxyhemoglobin thus formed is unavailable to carry oxygen.[1] In addition, partial saturation of each hemoglobin molecule with carbon monoxide results in tighter binding of oxygen to hemoglobin; this shifts the oxygen-hemoglobin dissociation curve, further reducing oxygen delivery to the tissues.[2] Carbon monoxide also may exert a direct toxic effect by binding to myoglobin and cellular cytochromes, such as those contained in respiratory enzymes.

Although carbon monoxide poisoning represents a multisystem insult, the cardiac and central nervous systems are particularly sensitive to the effects of hypoxia.[1,2] Most clinical manifestations are referable to the central nervous system, but it is likely that myocardial ischemia is responsible for many carbon monoxide–induced deaths.[3]

With exposure to high concentrations (4000 ppm and above), transient weakness and dizziness may be the only premonitory warnings before coma supervenes; the most common early aftermath of severe intoxication is cerebral edema.[4,5] Severe visual disturbances also occur as a consequence of acute poisoning in which there has been a period of unconsciousness.[6] After recovery from coma, in cases with residual loss of vision, the pupils are reactive to light despite subject blindness, indicating that the damage is cortical in origin. Typically, complete recovery takes place in a few hours to a few days. Exposure to concentrations of 500 to 1000 ppm causes the development of headache, tachypnea, nausea, weakness, dizziness, mental confusion and, in some instances, hallucinations; the person is commonly cyanotic.[1-4] Because carboxyhemoglobin has a bright red color, occasionally, someone will exhibit the unusual combination of hypoxia together with a bright red color of the fingernails, mucous membranes, and skin; however, this "cherry-red cyanosis" usually is seen only at autopsy.[4]

Exposure to 50 ppm for 90 minutes may cause aggravation of angina pectoris; exposed anginal patients may show a negative inotropic effect (weakened force of myocardial contraction); 50 ppm for 120 minutes may aggravate intermittent claudication.[7]

The clinical effects of CO exposure are aggravated by heavy labor, high ambient temperature, and altitudes above 2000 ft; pregnant women are particularly susceptible to the effects of CO.[1]

The reaction to a given blood level of carboxyhemoglobin is extremely variable; some persons may be in a coma with a carboxyhemoglobin level of 38%, whereas others may maintain an apparently clear sensorium with levels as high as 55%. Levels of carboxyhemoglobin over 60% usually are fatal; 40% is associated with collapse and syncope; above 25%, there may be electrocardiographic evidence of a depression of the S-T segment; at 15% to 25%, there may be headache and nausea; levels below 15% rarely produce symptoms. The blood of cigarette smokers usually contains 2% to 10%, with the content sometimes as high as 18% carboxyhemoglobin, and nonexposed persons have an average level of 1%; heme metabolism is an endogenous source of CO.[1]

Exposure of nonsmokers to 50 ppm for 6 to 8 hours results in carboxyhemoglobin levels of 8% to 10%.[1-3] Several investigators have suggested that the results of behavioral tests such as time discrimination, visual vigilance, choice response tests, visual evoked responses, and visual discrimination threshold may be altered at levels of carboxyhemoglobin below 5%.[1]

Transient central nervous system symptoms or rapid death are not the only results of CO poisoning.[3] The occurrence of late, fatal demyelination is a rare but dreaded complication. Further, it is inappropriate to assume that, because a patient with CO poisoning shows improvement, residual mental damage may not occur.[3] A report of 63 patients studied 3 years after CO poisoning indicated that 13% showed gross neuropsychiatric damage directly attributable to their CO intoxication, 33% showed a "deterioration of personality" after poisoning, and 43% reported memory impairment.[8] A syndrome of headache, fatigue, dizziness, paresthesias, chest pains, palpitations, and visual disturbances has been associated with chronic carbon monoxide poisoning.[9]

Chronic carbon monoxide poisoning may be difficult to diagnose because carboxyhemoglobin levels correlate poorly with symptoms and symptoms may be misdiagnosed as a viral syndrome or psychological depression. Distinguishing features of chronic carbon monoxide poisoning include the absence of myalgias, fever, sore throat, and adenopathy; simultaneous illness in homebound family members and pets; and improvement with exposure to fresh air. The diagnosis can be confirmed by finding a source of carb on monoxide in the home (e.g., defective furnaces), workplace, or vehicle; negative screenings for other illnesses; abnormal carboxyhemoglobin levels; and abatement of symptoms when the CO source has been eliminated.

Occupational exposure of New York City tunnel officers to excess levels of CO was associated with a 35% excess risk of arteriosclerotic heart disease mortality.[10] The excess risk was thought to be due to repeated, short-term peak exposures on the order of 400 ppm and appeared to be reversible on cessation of exposure.

A review of 60 case reports of carbon monoxide exposure during pregnancy found fetal outcome related to maternal blood carboxyhemoglobin and maternal toxicity.[11] In cases where the mother did not become unconscious, fetal outcome was generally good. However, where the mother experienced unconsciousness or coma, fetal outcome tended to be poor (death, or survival with anatomical or functional abnormalities.) Anatomical malformations, including mongoloid-type features, missing and deformed limbs, and oral cavity anomalies, also showed a marked correlation to exposure during the first trimester.

Animal experiments support the developmental findings found in humans and suggest that prenatal exposure at maternally nontoxic levels may also damage the fetal central nervous system.[12,13] Exposure of pregnant rats to 150 ppm produced only minor reductions in pup birthweights, but evaluation of learning and memory processes suggested a functional deficit in the central nervous system that persisted into adulthood of exposed offspring.[12,13]

The 1995 ACGIH threshold limit value–time-weighted average for carbon monoxide is 25 ppm (29 mg/m³).

Treatment. The specific therapy is administration of oxygen. Breathing normal air will result in a 50% clearance of blood carbon monoxide in approximately 5 hours. Administration of 100% oxygen at sea level will achieve a 50% reduction in carboxyhemoglobin in 80 minutes, whereas oxygen at 3-atm pressure will achieve the same clearance in approximately 25 minutes. The use of 95% oxygen–5% carbon dioxide will hasten oxygenation, largely by lowering the pH (Bohr effect). In the acutely ill victim, the presence of metabolic acidosis is a contraindication to further reductions in pH by administration of carbon dioxide.[14]

REFERENCES

1. National Institute for Occupational Safety and Health, US Department of Health, Education and Welfare: *Criteria for a Recommended Standard . . . Occupational Exposure to Carbon Monoxide.* (HSM) 73-11000. Washington, DC, US Government Printing Office, 1972

2. Olson KR: Carbon monoxide poisoning: mechanisms, presentation, and controversies in management. *J Emerg Med* 1:233–243, 1984

3. Winter PM, Miller JN: Carbon monoxide poisoning. *JAMA* 236:1502–1504, 1976

4. Swinyard EA: Noxious gases and vapors. In Goodman LS, Gilman A (eds): *The Pharmacological Basis of Therapeutics*, 5th ed, pp 900–904, 910–911. New York, Macmillan, 1975

5. Beard RR: Inorganic compounds of O, N, and C. In Clayton GD, Clayton FE (eds): *Patty's Industrial Hygiene and Toxicology*, 3rd ed rev, Vol 2C, *Toxicology*, pp 4114–4124. New York, Wiley-Interscience, 1982

6. Grant WM: *Toxicology of the Eye*, 3rd ed, pp 183–186. Springfield, IL, Charles C Thomas, 1986

7. Goldsmith JR, Aronow WS: Carbon monoxide and coronary heart disease: a review. *Environ Res* 10:236–248, 1975
8. Smith J, Brandon S: Morbidity from acute carbon monoxide poisoning at a three-year follow-up. *Br Med J* 1:318–321, 1973
9. Kirkpatrick JN: Occult carbon monoxide poisoning. *West J Med* 146:52–56, 1987
10. Stern FB, Halperin WE, Hornung RW, et al: Heart disease mortality among bridge and tunnel officers exposed to carbon monoxide. *Am J Epidemiol* 128:1276–1288, 1988
11. Norman CA, Halton DM: Is carbon monoxide a workplace teratogen? A review and evaluation of the literature. *Ann Occup Hyg* 34:335–347, 1990
12. Mactutus CF, Fechter LD: Prenatal exposure to carbon monoxide: learning and memory deficits. *Science* 223:409–411, 1984
13. Mactutus CF, Fechter LD: Moderate prenatal carbon monoxide exposure produces persistent, and apparently permanent, memory deficits in rats. *Teratology* 31:1–12, 1985
14. Dinman BD: The management of acute carbon monoxide intoxication. *J Occup Med* 16:662, 1974

CARBON TETRABROMIDE
CAS: 558-13-4

CBr₄

Synonyms: Tetrabromomethane; methane tetrabromide

Physical Form. Colorless solid

Uses. To a limited extent, as an intermediate in organic synthesis

Exposure. Inhalation

Toxicology. Carbon tetrabromide is a lacrimator; high concentrations may cause upper respiratory irritation and injury to the lungs, liver, and kidneys. Chronic exposure is expected to cause liver injury.

Exposure of rats 7 hr/day, 5 days/week, for 6 months at 0.1 ppm caused no effects.[1] Exposure at higher but unstated levels caused poor growth and fatty changes in the liver. In the eyes of rabbits, the material caused severe irritation and irreversible corneal damage. The vapor is a lacrimator. On the skin of rabbits, it caused slight irritation.

An early report in the Russian literature indicated that chronic exposure of rats to 0.07 to 74 ppm for 4 months caused irritation of the eyes and respiratory tract, as well as damage to the liver.[1] A more recent study exposed rats to a single intraperitoneal injection of 25 to 125 μl/kg. Renal dysfunction rather than hepatic effects was seen in the form of oliguria, aciduria, and hypo-osmolality.[2]

An in vitro study of metabolism of carbon tetrabromide with rat liver microsomes showed that 4-bromo-2,6-dimethylphenol was a metabolite.[3]

The 1995 ACGIH threshold limit value–time-weighted average (TLV-TWA) for carbon tetrabromide is 0.1 ppm (1.4 mg/m³) with a short-term excursion limit of 0.3 ppm (4.1 mg/m³).

REFERENCES

1. Torkelson, TR, Rowe VK: Halogenated aliphatic hydrocarbons containing chlorine, bromine, and iodine. In Clayton GD, Clayton FE (eds): *Patty's Industrial Hygiene and Toxicology*, 3rd ed, Vol 2B *Toxicology*, pp 3478–3480, New York, Wiley-Interscience, 1981
2. Agarwal AK, Berndt WO, Mehendale HM: Possible nephrotoxic effect of carbon tetrabromide and its interaction with chlordecone. *Toxicol Lett* 17:557–562, 1983
3. Mico BA, Branchflower RV, Pohl LR, et al: Oxidation of carbon tetrachloride, bromotrichloromethane, and carbon tetrabromide by rat liver microsomes to electrophilic halogens. *Life Sci* 30:131–137, 1982

CARBON TETRACHLORIDE
CAS: 56-23-5

CCl₄

CCl_4

Synonyms: Carbon tet; tetrachloromethane

Physical Form. Colorless liquid

Uses. In the manufacture of chlorofluorocarbons which, in turn, are used primarily as refrigerants; formerly used widely as a solvent and as a grain fumigant and in fire extinguishers; consumer uses discontinued because of toxicity, with only industrial use remaining.

Exposure. Inhalation; skin absorption; ingestion

Toxicology. Carbon tetrachloride causes central nervous system depression and severe damage to the liver and kidneys; it is carcinogenic in experimental animals and has been classified as a potential human carcinogen.

In animals, the primary damage from intoxication is to the liver but, in humans, the majority of fatalities have been due to renal injury with secondary cardiac failure.[1,2] Human autopsy reports have confirmed renal tubular necrosis. In humans, liver damage occurs more often after ingestion of the liquid than after inhalation of the vapor.

Human fatalities from acute renal damage have occurred after exposure for 1/2 to 1 hour to concentrations of 1000 to 2000 ppm; occasional sudden deaths may have been due to ventricular fibrillation.[1] Exposure to high concentrations results in symptoms of central nervous system depression, including dizziness, vertigo, incoordination, and mental confusion; abdominal pain, nausea, vomiting, and diarrhea are frequent.[1-4] Cardiac arrhythmias and convulsions have also occurred. Polycythemia, followed by anemia and hemodilution, may occur. Within a few days, jaundice may appear, and liver injury may progress to toxic necrosis. At the same time, acute nephritis may occur, with albumin, red and white blood cells, and casts in the urine; there may be oliguria, anuria, and increased nitrogen retention, resulting in the development of uremia. The no-observed-adverse-effect level for acute human exposure is 10 ppm for a 3-hour exposure.[5]

There are several reports of adverse effects in workmen who were repeatedly exposed to concentrations of 25 to 30 ppm; nausea, vomiting, dizziness, drowsiness, and headache were frequently noted.[1] Chronic exposure has caused various abnormalities of the eyes such as reduced visual field.

Carbon tetrachloride is absorbed through the skin of humans, though much less readily than from the lung.[6] Following use as a shampoo or as a solvent for removal of adhesives from skin, a number of fatal or near-fatal cases have been reported. It has been noted that these exposures must also have involved high levels of inhalation exposure as well as dermal exposure. It has been estimated that immersion of both hands in the liquid for 30 minutes would yield an exposure equivalent to breathing 100 to 500 ppm for 30 minutes.

The liquid splashed in the eye causes pain and minimal injury to the conjunctiva. Prolonged or repeated skin contact with the liquid may result in skin irritation and blistering.[1,4]

A number of substances, including ethanol, isopropyl alcohol, polybrominated biphenyls, phenobarbital, and benzo(a)pyrene, have been shown to affect carbon tetrachloride toxicity synergistically.[4] Alcohol has been a concomitant factor in many of the human cases of poisoning, especially in cases where severe liver and kidney damage have occurred.[2] Some substances, such as chlordecone, greatly potentiate the toxicity of carbon tetrachloride at doses where both substances are not considered toxic; effects include extensive hepatoxicity characterized by total hepatic failure and greatly potentiated lethality.[7]

The mechanism of carbon tetrachloride hepatotoxicity generally is viewed as an example of lethal cleavage, where the CCl_3–Cl bond is split in the mixed-function oxidase system of the hepatocytes. Following this cleavage, damage may occur directly from the free radicals (·CCl and ·Cl) and/or from the formation of toxic metabolites such as phosgene.[4]

Animal studies demonstrate that carbon tetrachloride produces hepatocellular carcinomas in the mouse, rat, and hamster.[4] Mice administered 1250 mg/kg or 2500 mg/kg approached a 100% incidence of hepatocellular carcinomas versus 6% or less in various controls. Hamsters receiving 190 and 380 mg/kg by gavage had a 100% liver-cell carcinoma incidence for those animals surviving past the 43rd week.[8]

Sensitivity to carbon tetrachloride-induced neoplasms varied widely among 5 strains of rats receiving twice weekly subcutaneous injections of 2080 mg/kg as a 50% solution in corn oil.[9]

A number of animal studies suggest that hepatomas occur only after liver necrosis and fibrosis have occurred and, therefore, that carbon tetrachloride is not a direct liver carcinogen.[4] One early study, however, found that liver necrosis and its associated chronic regenerative state probably was not necessary for tumor induction although a correlation was found between the degree of liver necrosis and the incidence of hepatomas.[10]

In humans, cases of hepatomas have appeared years following acute exposure to carbon tetrachloride, however, none of the cases could establish a causal link between the exposure and development of neoplasms.[4] Epidemiologic studies have also given inconclusive results. A cancer mortality study of a population of rubber workers reported a significantly elevated odds ratio relating carbon tetrachloride to lymphatic leukemia, to lymphosarcoma, and to reticulum-cell carcinoma.[11,12] However, several solvents were used simultaneously, and effects should not be attributed solely to carbon tetrachloride.

IARC evaluations for carbon tetrachloride include sufficient evidence for carcinogenicity in animals, inadequate evidence for carcinogenicity in humans, and an overall evaluation that carbon tetrachloride is probably carcinogenic to humans.[13]

Carbon tetrachloride was fetotoxic to rats when administered on days 6 through 15 of gestation at 300 or 1000 ppm, 7 hr/day; an increase in skeletal anomalies due to delayed development was observed in the offspring. Signs of maternal toxicity included weight loss and hepatic damage.[14]

The sweetish odor of carbon tetrachloride does not provide satisfactory warning of exposure.

The 1995 ACGIH threshold limit value–time-weighted average (TLV-TWA) for carbon tetrachloride is 5 ppm (31 mg/m³) with a short-term excursion limit of 10 ppm (63 mg/m³), an A3 animal carcinogen designation, and a notation for skin absorption.

REFERENCES

1. National Institute for Occupational Safety and Health: *Criteria for a Recommended Standard . . . Occupational Exposure to Carbon Tetrachloride.* DHEW (NIOSH) Pub No 76-133, pp 15–68, 84–112. Washington, DC, US Government Printing Office, 1975
2. Fassett DW: Toxicology of organic compounds: a review of current problems. *Annu Rev Pharmacol* 3:267–274, 1963
3. von Oettingen WF: *The Halogenated Aliphatic, Olefinic, Cyclic, Aromatic, and Aliphatic Aromatic Hydrocarbons Including the Halogenated Insecticides, Their Toxicity and Potential Dangers,* pp 75–112. US Public Health Service Pub No 414. Washington DC, US Government Printing Office, 1955
4. *Health Assessment Document for Carbon Tetrachloride.* Cincinnati, OH, US Environmental Protection Agency, Environmental Criteria and Assessment Office, 1984
5. Stewart RD, Gay HH, Erley DS et al: Human exposure to carbon tetrachloride vapor-relationship of expired air concentrations to exposure and toxicity. *J Occup Med* 3:586–590, 1961
6. Agency for Toxic Substances and Disease Registry: *Toxicological Profile for Carbon Tetrachloride.* US Department of Health and Human Services, Atlanta, Ga, 1992
7. Mehendale HM: Potentiation of halomethane hepatotoxicity: chlordecone and carbon tetrachloride. *Fund Appl Toxicol* 4:295–308, 1984
8. Della Porta G et al: Induction with carbon tetrachloride of liver cell carcinomas in hamsters. *J Natl Cancer Inst* 26:855–863, 1961
9. Rueber MD, Glover EL: Cirrhosis and carcinoma of the liver in male rats given subcutaneous carbon tetrachloride. *J Natl Cancer Inst* 44:419–427, 1970

10. Eschenbrenner AB, Miller E: Studies on hepatomas–size and spacing of multiple doses in the induction of carbon tetrachloride hepatomas. *J Natl Cancer Inst* 4:385–388, 1943
11. Wilcosky TC et al: Cancer mortality and solvent exposures in the rubber industry. *Am Ind Hyg Assoc J* 45:809–811, 1984
12. Checkoway H et al: An evaluation of the associations of leukemia and rubber industry solvent exposures. *Am J Ind Med* 5:239–249, 1984
13. *IARC Monographs on the Evaluation of the Carcinogenic Risk of Chemicals to Man*, Vol 20, Some halogenated hydrocarbons, pp 371–399. Lyon, International Agency for Research on Cancer, 1979
14. Schwetz DW et al: Embryo- and fetotoxicity of inhaled carbon tetrachloride, 1,1-dichloroethane and methyl ethyl ketone in rats. *Toxicol Appl Pharmacol* 28:452–464, 1974

CATECHOL

CAS: 120-80-9

$C_6H_4(OH)_2$

Synonyms: 1,2-Dihydroxybenzene; pyrocatechol; 1,2-benzenediol

Physical Form. Colorless crystals

Uses. In the manufacture of rubber antioxidants and monomer inhibitors to stop radical polymerization; in dyes; as a photographic developer; in formulations for pharmaceuticals, perfumes, inks, and insecticides

Exposure. Inhalation; skin absorption

Toxicology. Catechol is a skin, eye, and respiratory tract irritant and, at high concentrations, may cause convulsions; it acts as a cocarcinogen in animal skin-painting studies.

Skin contact with catechol causes dermatitis. Absorption through the skin may give rise to symptoms similar to those seen in phenol poisoning: an increase in blood pressure and the occurrence of convulsions.[1]

A report on the health effects of Japanese factory workers exposed to catechol and phenol for 2 years found that most of the workers complained of cough and sputum, occasional sore throat, and eye irritation.[2] The respiratory disorders were not noted in the control group of workers. The incidence of skin eruptions (7/13) was also higher in the exposed workers compared to the controls (2/13). Concentrations of catechol in workroom air ranged from 8 mg/m³ up to 322 mg/m³ of air.

In rats, the single-dose oral LD_{50} was estimated to be 0.3 g/kg, based on mortality during a 14-day postexposure period. At autopsy, the rats that died during the observation period had hyperemia of the stomach and intestines.[3]

Injected intraperitoneally into female mice, a dose of 0.37 mmol/kg produced convulsions in 50% of the animals.[4]

Administered by gavage at dose levels of 150 and 300 mg/kg, 5 days/week, for 13 weeks, catechol induced lesions of the forestomach in mice and rats.[5] The higher dose was lethal to most of the animals, and histopathological examination showed acanthosis and squamous papillomas of the forestomach. Male mice also had carcinoma in situ of the forestomach, which was considered to be treatment-related.

No deaths resulted when rats inhaled 1500, 2000, or 2800 mg/m³ catechol for 8 hours; in the two higher-exposure groups, tremors appeared in 6 to 7 hours and persisted through the first postexposure day.[3] Following a 14-day holding period, the six rats exposed at 2800 mg/m³ had blackened toes and tails; some of the toes were missing, as well as the tips of the tails of all exposed animals. Similar tail loss occurred in two of six animals exposed at 2000 mg/m³. No toxic signs were seen in the 1500-mg/m³ group.

In skin-painting studies in mice, catechol increased the carcinogenic effects of benzo[a]pyrene.[6] A group of 50 mice were treated with 2 mg catechol plus 5 μg of B(a)P in 0.1 ml acetone 3 times per week for 52 weeks. The incidence of skin tumors was compared to that obtained from groups treated with B(a)P only, catechol only, or vehicle only, or from untreated controls. The catechol plus B(a)P group had incidences of 35/50 papillomas and 31/50 squamous carcinomas compared to incidences of 13/50 and 10/50 for the B(a)P-only group, respectively. No tumors occurred in the catechol-only, vehicle-only, or untreated control groups. In a later study, four dose levels of catechol in B(a)P were evaluated for carcinogenicity.[7] The catechol-only B(a)P-treated groups had the following incidences of skin tumors: 0.25 mg catechol + B(a)P was 72%, 0.1 mg catechol + B(a)P was 66%, 0.01 mg catechol + B(a)P was 18%, and 0.001 mg catechol + B(a)P was 24%. No skin tumors were observed in the vehicle-only control group, whereas 11% of the B(a)P-treated group and 21% of the catechol-only treated groups had skin tumors. It was determined that doses of 0.1 mg and above were cocarcinogenic but that the lower doses were not.[7]

Contact of 0.5 g of catechol with the intact and abraded skin of rabbits for up to 24 hours produced slight to moderate erythema and slight edema of the intact areas as well as necrosis of the abraded areas.[3] The single-dose skin penetration LD_{50} was estimated to be 0.8 g/kg. Subdermal hyperemia and edema were noted at autopsy, but there were no internal gross lesions.[3]

Application of 0.1 g into the eyes of rabbits caused moderate conjunctivitis, with exudate and corneal opacity; at 72 hours following exposure, they showed severe conjunctivitis, iritis, and diffuse corneal opacities; 14 days after exposure, all the treated eyes had pannus formation and keratoconus.[3]

The 1995 TLV-TWA is 5 ppm (23 mg/m³) with a notation for skin absorption.

REFERENCES

1. *IARC Monographs on the Evaluation of the Carcinogenic Risk of Chemicals to Man*, Vol 15, *Some fumigants, the herbicides 2,4-D and 2,4,5-T, chlorinated dibenzodioxins and miscellaneous industrial chemicals*, pp 155–175. Lyon, International Agency for Research on Cancer, August 1977
2. Hirosawa I, Asaeda G, Arizono H, et al: Effects of catechol on human subjects: a field survey. *Int Arch Occup Environ Health* 37:107–114, 1976.
3. Flickinger CW: The benzenediols: catechol, resorcinol and hydroquinone—a review of the industrial toxicology and current industrial exposure limits. *Am Ind Hyg Assoc J* 37:596–607, 1976
4. Angel A, Rogers KJ: Convulsant action of polyphenols. *Nature* 217:84–85, 1968
5. National Toxicology Program: *Report of Subchronic Test of Catechol Using Fischer-344 Rats and B6C3F₁ mice.* 1981. Cited in NTP Executive Summary of Data, Catechol, pp 1–49, 1986
6. Van Duuren BL, Katz C, Goldschmidt BM: Cocarcinogenic agents in tobacco carcinogenesis. *J Natl Cancer Inst* 51:703–705, 1973
7. Hecht SS, Carmella S, Furuya K, et al: Polynuclear aromatic hydrocarbons and catechol derivatives as potential factors in digestive tract carcinogenesis. *Environ Mutagens Carcinog, Proc Int Conf* 3:545–556, 1982

CELLULOSE AND COMPOUNDS
CAS: 9004-34-6

$(C_6H_{10}O_5)n$

Synonyms: None

Physical Form. Natural cellulose is a highly crystalline, white solid with a molecular weight varying from 300,000 to greater than 1 million

Uses/Sources. Wood contains 50% to 70% cellulose; cotton and other textile fibers of plant origin contain 65% to 95%; rayon is prepared by dissolving natural cellulose and then precipitating it from solution, with some loss of crystallinity. Cellulose is made into cellophane film and is used to form fibers, resins, coatings, and gums

Exposure. Inhalation

Toxicology. Cellulose is inert and is classified as a nuisance dust.

It has little, if any, adverse effect on the lung, and there are no reports of organic disease or toxic effect.[1] The health effects attributed to wood, cotton, flax, jute, and hemp are not attributable to their cellulose content but rather to the presence of other substances.

Cellulose fibers were found in the blood and urine of human volunteers fed dyed cellulose; there were no ill effects.[2]

In animal studies of cellulose derivatives, the only consistent effect of very high doses in the feed appears to be a reduction in the nutritional value of the feed, which manifests itself as a decrease in body weight gain or an increase in food consumption.[3] Doses up to 5000 mg/kg bw/day, or 10% in the diet, have been found to be nontoxic.

The 1995 ACGIH threshold limit value–time-weighted average (TLV-TWA) for cellulose is 10 mg/m^3.

REFERENCES

1. Cellulose. *Documentation of the TLVs and BEIs for Substances in Workroom Air*, 6th ed, pp 241–242. Cincinnati, OH, American Conference of Governmental Industrial Hygienists (ACGIH), 1991
2. Schreiber G: Ingested dyed cellulose in the blood and urine of man. *AMA Arch Environ Health* 29:39–42, 1974
3. Thomas WC, McGrath LF, Baarson KA, et al: Subchronic oral toxicity of cellulose acetate in rats. Food Chem Toxicol 29:453–458, 1991

CESIUM HYDROXIDE
CAS: 21351-79-1

CsOH

Synonym: Cesium hydrate

Physical Form. Colorless or yellow solid

Uses. As a catalyst in the polymerization of cyclic siloxanes; for electrolytes in batteries

Exposure. Inhalation, skin contact

Toxicology. Cesium hydroxide is an irritant of the eyes.

The oral LD$_{50}$ in rats was 1026 mg/kg.[1] In rabbits, a 5% solution was irritating to abraded skin and extremely irritating in the eyes. No evidence of skin sensitization was found in treated guinea pigs.

There are no reports of adverse effects in humans. By analogy to NaOH, the effects from dust or mist could be expected to vary from mild irritation of the upper respiratory tract to pneumonitis, depending on the severity of the exposure. The greatest industrial hazard is rapid tissue destruction of the eyes on contact with the solid or a concentrated solution. If cesium hydroxide is not removed from the skin, it is anticipated that burns willl occur after a period of time. Ingestion would be expected to cause corrosion of the lips, mouth, tongue, and pharynx, as well as abdominal pain.

The 1995 ACGIH threshold limit value–time-weighted average (TLV-TWA) for cesium hydroxide is 2 mg/m^3.

REFERENCES

1. Johnson GT, Lewis TR, Wagner WD: Acute toxicity of cesium and rubidium compounds. *Toxicol Appl Pharmacol* 32:239–245, 1975

CHLORDANE
CAS: 57-74-9

C$_{10}$H$_6$Cl$_8$

Synonyms: Chlordan; Vesicol 1068; CD-68; Toxichlor; Octa-Klor; 1,2,4,5,6,7,8-octachloro-2,3,3a,4,7,7a-hexahydro-4,7-methanoindene

Physical Form. Viscous amber liquid; technical-grade chlordane contains about 45 constituents, including 7% to 10% heptachlor

Uses. Insecticide; currently approved for underground termite control only

Exposure.   Skin absorption; ingestion; inhalation

Toxicology.   Chlordane is a convulsant; it is carcinogenic in experimental animals.

Established cases of chlordane poisoning have been associated with gross exposure, either by ingestion or skin contact.[1] Typically, the poisoning is characterized by onset of violent convulsions within 1/2 to 3 hours and either death or recovery within a few hours to a day. Following ingestion, nausea and vomiting may precede signs of central nervous system overactivity. Convulsions may be accompanied by confusion, incoordination, excitability, or coma. In one instance, accidental ingestion of approximately 300 ml of a 75% chlordane solution (215 g chlordane) was survived despite rapid onset of respiratory, gastrointestinal, and neurologic effects.[2] In this case, the chlordane level in whole blood was 5 mg/l at 3.5 hours post-ingestion. Kinetic analysis of blood chlordane levels with time suggested a half-life of 7 hours for distribution in the body and 34 days for elimination.[2]

Skin absorption of chlordane is rapid; a worker who spilled a 25% suspension of chlordane on his clothing, which was not removed, began having convulsions within 40 minutes and died shortly thereafter.[3]

Technical-grade chlordane is stated to be irritating to the skin and mucous membranes, but this may be truer of earlier chlordane formulations with significant hexachlorocyclopentadiene contamination.[1,3]

Mice kept in saturated vapor of technical chlordane without hexachlorocyclopentadiene for 25 days showed no symptomatic effects.[1] The oral LD_{50} values for rats range from 200 to 590 mg/kg.[3]

In experimental animals, prolonged exposure to dietary levels exceeding 3 to 5 mg/kg resulted in the induction of hepatic microsomal enzymes and, at a later stage, liver hypertrophy with histological changes.

At dosages above 30 mg/kg in the diet, chlordane interfered with reproduction in rats and mice, but this effect was reversible after exposure ceased.[3]

A dose-related increase in the incidence of hepatocellular carcinomas was found in male and female mice fed approximately 60 mg/kg chlordane for 80 weeks.[4] In rats, increases in the incidences of thyroid follicular-cell neoplasms were observed.[5]

In human case reports, chlordane exposure has been linked to neuroblastoma, aplastic anemia, and acute leukemia, but only circumstantially.[1] In a 1987 report, 25 new cases of blood dyscrasia, including leukemias, production defects, and thrombocytopenic purpura (generally following home termite treatment with chlordane/heptachlor), were reported.[6] The authors noted the rarity of many of the conditions and, hence, the difficulty of finding statistically significant results.

Epidemiologic studies have not shown a clear association between chlordane exposure and cancer mortality. No excess deaths from lung cancer were observed in termite-control workers (with particular exposure to chlordane and heptachlor) in comparison to other pesticide applicators.[7] Follow-up of 1400 men employed in the manufacture of chlordane, heptachlor, and/or endrin also showed a deficit of deaths from all cancers and a small excess of lung cancers although smoking histories were not documented.[8] A study of 800 workers employed at a chlordane production plant for 3 months or more during the period 1946–1985 showed a slightly less than expected overall death rate, and no trend with duration of employment was seen for respiratory cancer.[9]

The IARC has concluded that there is inadequate evidence for carcinogenicity of chlordane to humans and sufficient evidence for its carcinogenicity to animals.[5]

The 1995 ACGIH threshold limit value–time-weighted average (TLV-TWA) for chlordane is 0.5 mg/m^3 with a notation for skin absorption.

REFERENCES

1. Hayes WJ Jr: *Pesticides Studied in Man*, pp 229–233. Baltimore, MD, Williams and Wilkins, 1982
2. Olanoff LS et al: Acute chlordane intoxication. *J Toxicol Clin Toxicol* 20:291–306, 1983
3. *Environmental Health Criteria 34, Chlordane*, p 82. Geneva, World Health Organization, 1984

4. National Cancer Institute: *Bioassay of Chlordane for Possible Carcinogenicity*, Technical Report Series No 8. DHEW (NIH) Pub No 77-808. Washington, DC, US Government Printing Office, 1977

5. *IARC Monographs on the Evaluation of Carcinogenic Risk to Humans*, Vol 53, Occupational exposures in insecticide application, and some pesticides, pp 115–175. Lyon, International Agency for Research on Cancer, 1991

6. Epstein SS, Ozonoff D: Leukemias and blood dyscrasias following exposure to chlordane and heptachlor. *Ter Carcin Mut* 7:527–540, 1987

7. Wang HH, MacMahon B: Mortality of pesticide applicators. *J Occup Med* 21:741–744, 1979

8. Wang HH, MacMahon B: Mortality of workers employed in the manufacture of chlordane and heptachlor. *J Occup Med* 21:745–748, 1979

9. Shindell S, Ulrich S: Mortality of workers employed in the manufacture of chlordane: an update. *J Occup Med* 28:497–501, 1986

CHLORDECONE
CAS: 143-50-0

$C_{10}Cl_{10}O$

Synonym: Kepone

Physical Form. Tan to white crystalline solid

Uses. Pesticide (leaf-eating insects and fly larvae); products containing chlordecone were canceled in 1978

Exposure. Inhalation

Toxicology. Chlordecone is toxic to the nervous system, liver, and reproductive system.

The first reports of effects in humans concerned the cases of intoxication of workers of the Life Science Products Company in Hopewell, Virginia.[1,2] This was a small, improvised unit lacking control of dust generated by a process that was operated over a period of 16 months. On initial assessment, at least 9 of the 33 employees of the company were severely affected and showed memory impairment, slurred speech, tremor, opsoclonus (eye twitching), and liver damage; blood levels of the compound of up to 25 ppm were found.[2]

Subsequent examination of 117 of the 149 current or previous employees of this plant showed that 57 had present or past symptoms of intoxication, including weight loss, tremor of the upper extremities, ataxia, incoordination, arthralgia, skin rash, and abnormal liver-function tests. The incidence of illness was 67% in production workers and 16% for other employees of the plant. The wives of two workers had objective tremor; each had washed her husband's work clothes.[3] Sural-nerve biopsies from affected workers showed significant histological damage to nonmyelinated and smaller myelinated fibers, with relative sparing of larger myelinated fibers.[4]

In early animal studies, the compound reportedly caused tremor, the severity of which depended on the dosage level and duration of exposure; tremors persisted for a week or more following single exposures and cumulatively developed from daily repeated, individually ineffective doses. In male rats, the oral LD_{50} was 132 mg/kg.[5]

Reproductive studies showed that 14 pairs of mice that received 40 ppm chlordecone in the diet for 2 months before mating and during the test produced no litters, whereas 14 control pairs produced 14 first litters and 14 second litters. Further studies in mice found that infertility in chlordecone-exposed females was due to an absence of, or reduction in, the number of ovulated ocytes.[6] Chlordecone is thought to produce some of its reproductive effects by mimicking the effects of excessive estrogens. The ability to cause constant estrus and other estrogenlike effects has been repeatedly confirmed in rodents.[3]

Gestational exposure of rats and mice caused embryo/fetotoxicity and teratogenicity at doses that were severely toxic to dams.[7] Chronic exposure of mice and rats caused an increase in liver tumors.[7] Chlordecone is not considered genotoxic but may act as a tumor promoter.[7]

Chlordecone in blood is a good biomarker

of exposure because of chlordecone's association with plasma proteins and its long half-life.[7]

The ACGIH has not established a threshold limit value for chlordecone.

REFERENCES

1. Bureau of National Affairs, Inc: OSHA cites chemical manufacturer, labels Kepone exposure "catastrophe." *Occup Safety Health Rep* 5:379–380, 1975
2. Bureau of National Affairs, Inc: Allied, Hooker Chemical firms named as defendants in worker suit. *Occup Safety Health Rep* 5:516–517, 1975
3. Hayes WJ Jr, Laws ER Jr: *Handbook of Pesticide Toxicology*, Vol 2, *Classes of Pesticides*, pp 860–869. New York, Academic Press, 1991
4. Martinez AJ, Taylor JR, Dyck PJ, et al: Chlordecone intoxication in man. II. Ultrastructure of peripheral nerves and skeletal muscle. *Neurology* 28:631–635, 1978
5. Toxicological studies on decachlorooctahydro-1,3,4-metheno-2H-cyclobuta[cd]pentalen-2-one (Compound no. 1189) (Kepone). New York, Allied Chemical Corp, March 1960
6. Swartz WJ, Mall GM: Chlordecone-induced follicular toxicity in mouse ovaries. *Reproduc Toxicol* 3:203–206, 1989
7. Agency for Toxic Substances and Disease Registry (ATSDR): *Toxicological Profile for Mirex and Chlordecone*. US Department of Health and Human Services, Public Health Service, p 235, 1994

CHLORINATED DIPHENYL OXIDE
CAS: 55720-99-5

$C_6H_2Cl_3OC_6H_2Cl$ *(approximate)*

Synonyms: Chlorinated phenyl ethers; monochlorodiphenyl oxide; dichlorodiphenyl oxide, . . . , through hexachlorodiphenyl oxide

Physical Form. Varies from colorless, oily liquids to yellowish, waxy semisolids as the equivalents of chlorine increase from 1 to 6

Uses. Chemical intermediates; in the electrical industry

Exposure. Inhalation; skin absorption

Toxicology. Chlorinated diphenyl oxide causes an acneform dermatitis (chloracne)

Limited experience with humans has shown that exposure to even small amounts of the higher chlorinated derivatives, particularly hexachlorodiphenyl oxide, may result in appreciable acneform dermatitis.[1] Chloracne is usually persistent and affects the face, ears, neck, shoulders, arms, chest, and abdomen (especially around the umbilicus and on the scrotum). The most sensitive areas are below and to the outer side of the eye (malar crescent) and behind the ear.[2] The skin is frequently dry, with noninflammatory comedones and pale yellow cysts containing sebaceous matter and keratin.

No cases of systemic toxicity have been reported in humans.

In laboratory animals, cumulative liver damage has resulted from repeated intake and, in general, the toxicity increases with the degree of chlorination. Liver injury is characterized by congestion and varying degrees of fatty degeneration. In animals, these compounds cause severe skin irritation with topical application.[3] Animal experiments suggest that absorption from dermal application can result in systemic toxicity, including liver injury and weight loss. In guinea pigs, a single oral dose of 0.05 to 0.1 g/kg of material containing four or more equivalents of chlorine resulted in death 30 days after administration.[1]

The 1995 ACGIH threshold limit value–time-weighted average (TLV-TWA) for chlorinated diphenyl oxide is 0.5 mg/m³.

REFERENCES

1. Kirwin CJ Jr, Sandmeyer EE: Ethers. In Clayton GD, Clayton FE (eds): *Patty's Industrial Hygiene and Toxicology*, 3rd ed, rev, Vol 2A, *Toxicology*, pp 2546–2551. New York, Wiley-Interscience, 1981
2. Crow KD: Chloracne (halogen acne). In Marzulli FN, Maibach HI (eds): *Dermatotoxicology*,

2nd ed, pp 462–470. New York, Hemisphere Publishing, 1983
3. von Oettingen WF: *The Halogenated Aliphatic, Olefinic, Cyclic, Aromatic, and Aliphatic-Aromatic Hydrocarbons Including the Halogenated Insecticides, Their Toxicity and Potential Dangers.* US Public Health Service Pub No 414, pp 311–313. Washington DC, US Government Printing Office, 1955

CHLORINE
CAS: 7782-50-5

Cl_2

Synonyms: None

Physical Form. Greenish-yellow gas with an irritating odor

Uses. Sterilization of water supplies and swimming pools; bleaching agent; synthesis of chlorinated organic chemicals and plastics; in pulp and paper manufacturing; detinning and dezincing iron; metal fluxing

Exposure. Inhalation

Toxicology. Chlorine is a potent irritant of the eyes, mucous membranes, and skin; pulmonary effects range from respiratory irritation to edema.

Mild mucous membrane irritation may occur at 0.2 to 16 ppm; eye irritation occurs at 7 to 8 ppm, throat irritation at 15 ppm, and cough at 30 ppm.[1] A level of 1000 ppm is fatal after a few deep breaths.[1] Other studies have shown that at least some subjects develop eye irritation, headache, and cough at concentrations as low as 1 to 2 ppm.

The location and the severity of respiratory tract involvement are functions of both the concentration and the duration of exposure. With significant exposures, laryngeal edema with stridor, acute tracheobronchitis, and chemical pneumonitis have been described.[2] Death at high exposure is mainly from respiratory failure or cardiac arrest due to toxic pulmonary edema.[3] Bronchopneumonia may be a potentially lethal complication. In one accident, exposure of humans to unmeasured but high concentrations for a brief period of time caused burning of the eyes with lacrimation, burning of the nose and mouth with rhinorrhea, cough, choking sensation, and substernal pain.[4] These symptoms were frequently accompanied by nausea, vomiting, headache, dizziness and, sometimes, syncope. Of 33 victims who were hospitalized, all suffered tracheobronchitis, 23 progressed to pulmonary edema and, of those, 14 progressed to pneumonitis.[2] Respiratory distress and substernal pain generally subsided within the first 72 hours; cough increased in frequency and severity after 2 to 3 days and became productive of thick mucopurulent sputum; cough disappeared by the end of 14 days.

In another accidental exposure of 5 chlorine plant workers and 13 nonworkers, rales, dyspnea, and cyanosis were observed in the most heavily exposed, and cough was present in nearly all the patients. Pulmonary function tests 24 to 48 hours post-exposure showed airway obstruction and hypoxemia; these conditions cleared within 3 months except in 4 of the chlorine workers, who still showed reduced airway flow and mild hypoxemia after 12 to 14 months.[5]

Following acute exposures to chlorine gas, both obstructive and restrictive abnormalities on pulmonary function tests have been observed. Eighteen healthy subjects exposed after a leak from a liquid storage tank had diminished FEV_1, FEF 25% to 75%, and other flow rates within 18 hours of exposure. Follow-up studies at 1 and 2 weeks demonstrated resolution of these abnormalities in the 12 subjects, with an initial chief complaint of cough, whereas the six subjects with a chief complaint of dyspnea had persistently reduced flow rates. Repeat studies in 5 months were normal in all patients studied except for mildly reduced flow rates in two patients who were smokers.[6]

Of 19 healthy persons exposed in an accident at a pulp mill and tested within 24 hours, 10 (53%) had a reduced FEV_1 (<75%), and 13 (68%) had increased residual volumes

(>120%), suggesting obstruction with air trapping. Periodic follow-up testing over the next 700 days demonstrated gradual resolution of these abnormalities in all but three subjects tested, who had persistently reduced FEV_1. Two of these three patients were smokers.[7] In contrast, a study of four healthy patients exposed to a leak at a swimming pool showed reversible mild reductions in forced vital capacity, total lung capacity, and diffusing capacity, presumably related to mild interstitial edema. These decrements were not evident until repeat testing at 1 month, which showed significant improvements above the 100% predicted in all these parameters.[8]

In all these studies, some subjects acutely exhibited mild arterial hypoxemia, increases in alveolar-arterial oxygen tension difference, and respiratory alkalosis. Mild transient hyperchloremic metabolic acidosis, with a normal anion gap, has been described in a patient following chlorine inhalation, presumably related to systemic absorption of hydrochloric acid.[9]

Some studies of survivors of massive chlorine exposures have shown either persistent obstructive or restrictive deficits; but pre-exposure data on these patients were not available. There currently is not enough evidence to conclude that there is a potential for chronic impairment of pulmonary function following acute or chronic chlorine exposure.[10] In several cases, prolonged symptoms following chlorine exposure may be due to aggravation of pre-existing conditions, such as tuberculosis, asthma, chronic obstructive pulmonary disease, or heart disease.[10,11]

In high concentrations, chlorine irritates the skin and causes sensations of burning and prickling, inflammation, and vesicle formation.[10] Liquid chlorine causes eye and skin burns on contact.[12]

Administered in the drinking water for 2 years, 0.05 to 0.3 mM/kg/day did not cause a clear carcinogenic response in rats or mice.[13]

The range of reported odor thresholds for chlorine is 0.03 to 3.5 ppm; however, because of olfactory fatigue, odor does not always serve as an adequate warning of exposure.[1]

The 1995 ACGIH threshold limit value–time-weighted average (TLV-TWA) is 0.5 ppm (1.5 mg/m³) with a short-term excursion limit of 1 ppm (2.9 mg/m³).

REFERENCES

1. Committee on Medical and Biological Effects of Environmental Pollutants, National Research Council: *Chlorine and Hydrogen Chloride*, pp 116–123. Washington, DC, National Academy of Sciences, 1976
2. Chlorine poisoning (editorial). *Lancet*, pp 321–322. February 11, 1984
3. Baxter PJ, Davies PC, Murray V: Medical planning for toxic releases into the community: The example of chlorine gas. *Br J Ind Med* 46:277–285, 1989
4. Chasis H et al: Chlorine accident in Brooklyn. *J Occup Med* 4:152–176, 1947
5. Kaufman J, Burkons D: Clinical, roentgenological and physiological effects of acute chlorine exposure. *Arch Environ Health* 23:29–34, 1971
6. Hasan F et al: Resolution of pulmonary dysfunction following acute chlorine poisoning. *Arch Environ Health* 38:76–80, 1983
7. Charan N et al: Effects of accidental chlorine inhalation on pulmonary function. *West J Med* 143:333–336, 1985
8. Ploysonsang Y et al: Pulmonary function changes after acute inhalation of chlorine gas. *Southern Med J* 75:23–26, 1982
9. Szerlip H, Singer I: Hyperchloremic metabolic acidosis after chlorine inhalation. *Am J Med* 77:581–582, 1984
10. National Institute for Occupational Safety and Health: *Criteria for a Recommended Standard . . . Occupational Exposure to Chlorine.* DHEW (NIOSH) Pub No 76-170, pp 29, 36, 56, 84, 101. Washington, DC, US Government Printing Office, 1976
11. Das R, Blanc PD: Chlorine gas exposure and the lung: a review. *Toxicol Ind Health* 9:439–455, 1993
12. Chemical Safety Data Sheet, SD-80, Chlorine, pp 23–26. Washington, DC, MCA, Inc, 1970
13. Dunnick JK, Melnick RL: Assessment of the carcinogenic potential of chlorinated water: experimental studies of chlorine, chloramine, and trihalomethanes. *J Natl Cancer Inst* 85:817–822, 1993

CHLORINE DIOXIDE
CAS: 10049-04-4

ClO_2

Synonyms: Chlorine oxide; chlorine peroxide

Physical Form. Yellow to reddish-yellow gas

Uses. Bleaching agent in cellulose, paper pulp, flour; for purification and taste/odor control of water; oxidizing agent; bactericide and antiseptic

Exposure. Inhalation

Toxicology. Chlorine dioxide gas is a severe respiratory and eye irritant.

Exposure of a worker to 19 ppm for an unspecified time period was fatal.[1] Repeated acute exposure of workers to undetermined concentrations caused eye and throat irritation, nasal discharge, cough, wheezing, bronchitis, and delayed onset of pulmonary edema.[2] Repeated exposure may also cause chronic bronchitis.[2]

Delayed deaths occurred in animals after exposure to 150 to 200 ppm for less than 1 hour.[3] Rats exposed daily to 10 ppm died after 10 to 13 days of exposure; effects were nasal and ocular discharge and dyspnea; autopsy revealed purulent bronchitis. Another study reports that two to four 15-minute exposures to 5 ppm for 1 month did not alter the blood composition or lung histology of rats; similar exposures to 10 to 15 ppm caused bronchitis, bronchiolitis, catarrhal alveolar lesions, and peribronchial infiltration.[4] Lesions healed within 15 days posttreatment. Rats and rabbits exposed for 30 days to 5 or 10 ppm (2 hr/day) had localized bronchopneumonia, with elevated leucocyte counts; slight reversible pulmonary lesions were found after exposures of 2.5 ppm for 4 to 7 hr/day. No adverse reactions were observed in rats exposed to about 0.1 ppm for 5 hr/day for 10 weeks.[3]

Administered in the drinking water of rats daily for 9 months, 1, 10, 100, or 1000 mg/l

chlorine dioxide caused a depression in red blood cell counts, hemoglobin concentration, and packed cell volumes, along with a decrease in erythrocytic fragility. Rat body weight was decreased in all groups after 10 and 11 months of treatment.[5]

Oral gavage of rat pups with 14 mg/kg/day from postnatal day 5 through 20 caused reductions in serum thyroxine levels that correlated with depressed behavioral parameters.[6] Further studies, using the same protocol, reported decreased cell proliferation in the cerebellum and forebrain on postnatal days 11 and 21, respectively.[7] In yet another study, 14 mg/day of chlorine dioxide on postnatal days 1 to 20 was associated with some neurotoxicity (decreased forebrain weight and reduced synapse formati on on day 35), but the neurotoxicity was not correlated with any antithyroid activity of this chemical.[8]

In a multigenerational study, doses up to 10 ml/kg administered for 2.5 months prior to breeding and through the breeding and gestational periods did not cause any adverse effects in the parental generation.[9] All parameters examined in the F_1 generation except vaginal weight in female weanlings were unaffected by gestational and lactational chlorine dioxide exposure. There were no changes in thyroid hormone parameters that appeared to be attributable to chlorine dioxide treatment.

The 1995 ACGIH threshold limit value–time-weighted average (TLV-TWA) for chlorine dioxide is 0.1 ppm (0.28 mg/m³) with a short-term excursion limit of 0.3 ppm (0.83 mg/m³).

REFERENCES

1. Elkins HB: *The Chemistry of Industrial Toxicology*, 2nd ed. pp 89–90. New York, John Wiley and Sons, 1959
2. Gloemme J, Lundgren KD: Health hazards from chlorine dioxide. *AMA Arch Ind Health* 16:169, 1957
3. Dalhamn T: Chlorine dioxide-toxicity in animal experiments and industrial risks. *AMA Arch Ind Health* 15:101, 1957
4. Masschelein WJ, Rice RG: *Chlorine Dioxide Chemistry and Environmental Impact of Oxychlo-*

rine Compounds, pp 5–7. Ann Arbor Science Ann Arbor, MI, 1979

5. Abdel-Rahman MS, Couri D, Bull RJ: Toxicity of chlorine dioxide in drinking water. *J Am College Toxicol* 3:277–284, 1984
6. Orme J, Taylor DH, Laurie RD, et al: Effects of chlorine dioxide on thyroid function in neonatal rats. *J Toxicol Environ Health* 15:315–322, 1985
7. Taylor DH, Pfohl RJ: Effect of chlorine dioxide on neurobehavioral development of rats. In Bull RJ, Davis WP, Katz S, et al (eds): *Water Chlorination, Chemistry, Environmental Impact, and Health Effects*, Vol 5, pp 355–364. Chelsea, MI, Lewis, 1985
8. Toth GP, Long RE, Mills TS, et al: Effects of chlorine dioxide on the developing rat brain. *J Toxicol Environ Health* 31:29–44, 1990
9. Carlton BD, Basaran AH, Mezza LE, et al: Reproductive effects in long-Evans rats exposed to chlorine dioxide. *Environ Res* 56:170–177, 1991

CHLORINE TRIFLUORIDE
CAS: 7790-91-2

ClF$_3$

Synonym: Chlorine fluoride

Physical Form.   Colorless gas, pale green liquid, or white solid

Uses.   Fluorinating agent; incendiary; igniter and propellant for rockets; in nuclear reactor fuel processing; pyrolysis inhibitor for fluorocarbon polymers

Exposure.   Inhalation

Toxicology.   Chlorine trifluoride gas is an extremely severe irritant of the eyes, respiratory tract, and skin in animals.

The injury caused by chlorine trifluoride is in part attributed to its hydrolysis products, including chlorine, hydrogen fluoride, and chlorine dioxide. Effects in humans have not been reported but may be expected to be very severe; inhalation may cause pulmonary edema, and contact with eyes or skin may cause severe burns.[1-4]

Exposure of rats to 800 ppm for 15 minutes was fatal, but nearly all survived when exposed for 13 minutes. There was severe inflammation of all exposed mucosal surfaces, resulting in lacrimation, corneal ulceration, and burning of exposed areas of skin.[4] In another study, exposure of rats to 480 ppm for 40 minutes or to 96 ppm for 3.7 hours was fatal; in the latter group, effects were pulmonary edema and marked irritation of the bronchial mucosa. Chronic exposure of dogs and rats to about 1 ppm, 6 hr/day, for up to 6 months caused severe pulmonary irritation and some deaths.[2]

The 1995 ACGIH ceiling–threshold limit value (C-TLV) for chlorine trifluoride is 0.1 ppm (0.38 mg/m^3).

REFERENCES

1. Horn HJ, Weir RJ: Inhalation toxicology of chlorine trifluoride. I. Acute and subacute toxicity. *AMA Arch Ind Health* 12:515–521, 1955
2. Horn HJ, Weir RJ: Inhalation toxicology of chlorine trifluoride. II. Chronic toxicity. *AMA Arch Ind Health* 13:340–345, 1956
3. Boysen JE: Health hazards of selected rocket propellants. *Arch Environ Health* 7:71–75, 1963
4. Dost FN, Reed DJ, Smith VN, Wang CH: Toxic properties of chlorine trifluoride. *Toxicol Appl Pharmacol* 27:527–536, 1974

CHLOROACETALDEHYDE
CAS: 107-20-0

ClCH$_2$CHO

Synonyms: Monochloroacetaldehyde; 2-chloroacetaldehyde

Physical Form.   Colorless liquid

Uses/Sources.   In the manufacture of 2-aminothiazole; to facilitate bark removal from tree trunks; formed during the chlorination of drinking water; a metabolite of vinyl chloride

Exposure. Inhalation; skin absorption

Toxicology. Chloroacetaldehyde is a severe irritant of the eyes, mucous membranes, and skin; it is toxic and carcinogenic to the liver of male mice.

Inhalation of 5 ppm by rats caused eye and nasal irritation.[1] In rabbits, the LD_{50} for skin absorption was 0.022 ml/kg for 30% chloroacetaldehyde in water solution.[2] This solution on the skin or in the eyes of rabbits produced severe damage.

Male B6C3F$_1$ mice exposed to 0.1 g/l chloroacetaldehyde via drinking water for 104 weeks (mean ingested dose 17/mg/kg/day) had a significant increase in the prevalence of liver carcinomas (31% vs. 10% in controls). No significant changes were noted in the spleen, kidneys or testes of treated animals compared to controls.[3]

Chloroacetaldehyde has been reported to be an inhibitor of DNA synthesis and to form DNA cross-linkages; it is mutagenic in *Salmonella typhimurium* and in Chinese hamster cells.[4–6]

The 1995 ACGIH ceiling–threshold limit value (C-TLV) for chloroacetaldehyde is 1 ppm (3.2 mg/m^3).

REFERENCES

1. Chloroacetaldehyde. *Documentation of the TLVs and BEIs*, 6th ed, pp 260–261. Cincinnati, OH, American Conference of Governmental Industrial Hygienists (ACGIH), 1991
2. Lawrence WH, Dillingham EO, Turner JE, Autian J: Toxicity profile of chloroacetaldehyde. *J Pharm Sci* 61:19–25, 1972
3. Daniel FB, DeAngelo AB, Stober JA, et al: Hepatocarcinogenicity of chloral hydrate, 2-chloroacetaldehyde, and dichloroacetic acid in the male B6C3F$_1$ mouse. *Fund Appl Toxicol* 19:159–168, 1992
4. Kandala JC, Mrema JEK, DeAngelo A, et al: 2-Chloroacetaldehyde and 2-chloroacetal are potent inhibitors of DNA synthesis in animal cells. *Biochem Biophys Res Comm* 167:457–463, 1990
5. Huberman E, Bartsch H, Sachs L: Mutation induction in Chinese hamster V79 cells by two vinyl chloride metabolites, chloroethylene oxide and 2-chloroacetaldehyde. *Int J Cancer* 15:539, 1975
6. McCann J, Simmon V, Streitwieser D, et al: Mutagenicity of chloroacetaldehyde, a possible metabolic product of 1,2-dichloroethane (ethylene dichloride), chloroethanol (ethylene chlorohydrin), and cyclophosphamide. *Proc Nat Acad Sci USA* 72:3190–93, 1975

CHLOROACETONE
CAS: 78-95-5

$ClCH_2COCH_3$

Synonyms: Monochloroacetone; chloropropanone; 1-chloro-2-propanone; acetonyl chloride

Physical Form. Colorless to amber liquid, often with 5% CaCl$_2$ as a stabilizer

Uses. In manufacture of couplers for color photography; intermediate in manufacture of perfumes, antioxidants, drugs, plant growth regulators, defoliants, and herbicides

Exposure. Inhalation; skin absorption

Toxicology. Chloroacetone is a lacrimator and a severe irritant of the eyes, mucous membranes, and skin.

Chloroacetone was introduced as a war gas in 1914.[1] An airborne level of 605 ppm was found to be lethal for humans after 10 minutes, and 26 ppm was intolerable after 1 minute of exposure. Effects of exposure are immediate lacrimation, followed by irritation of the upper respiratory tract and a burning sensation on the skin. The odor is pungent and suffocating but is not considered adequate for warning.

An employee who was directly exposed to hot chloroacetone was hospitalized with irritation of the eyes and upper respiratory tract plus skin irritation. Eight hours after exposure, blisters developed on the skin. All signs and symptoms disappeared after 7 days.

The 1-hour LC_{50} in rats was 262 ppm, and the oral LD_{50} was 100 mg/kg. The dermal LD_{50} in rabbits was 141 mg/kg, indicating significant skin absorption.

The 1995 ACGIH ceiling–threshold limit value (C-TLV) for chloroacetone is 1 ppm (3.8 mg/m^3) with a notation for skin absorption.

REFERENCES

1. Sargent EV, Kirk GD, Hite M: Hazard evaluation of monochloroacetone. *Am Ind Hyg Assoc J* 47:375–378, 1986

α-CHLOROACETOPHENONE
CAS: 532-27-4

$C_6H_5COCH_2Cl$

Synonyms: 2-Chloro-1-phenylethanone; phenacyl chloride; phenyl chloromethyl ketone; tear gas; Chemical Mace

Physical Form. Crystals

Uses. Chemical warfare agent (CN); principal constituent in riot control agent Mace

Exposure. Inhalation

Toxicology. α-Chloroacetophenone is a potent lacrimating agent and an irritant of mucous membranes; it causes dermatitis of both primary irritation and the sensitization type.

In one fatal case of exposure, death occurred as a result of pulmonary edema; exposure occurred under unusual circumstances that caused inhalation of high concentrations.[1] Human volunteers exposed to levels of 200 to 340 mg/m^3 could not tolerate exposure for longer than 30 seconds.[2] Effects were lacrimation, burning of the eyes, blurred vision, tingling of the nose, rhinorrhea, and burning of the throat.[1] Less frequent symptoms included burning in the chest, dyspnea, and nausea.

Sporadic cases of dermatitis due to primary irritation by α-chloroacetophenone have been reported.[3,4] Allergic contact dermatitis to this substance in Chemical Mace has been documented by patch-test evaluation, and it is said to be a potent skin sensitizer.[3,4]

Eye splashes cause marked conjunctivitis and may result in permanent corneal damage.[5] The lacrimation threshold ranges from 0.3 mg/m^3 to 0.4 mg/m^3, and the odor threshold is 0.1 mg/m^3.[5]

The 1995 ACGIH threshold limit value–time-weighted average (TLV-TWA) for α-chloroacetophenone is 0.05 ppm (0.32 mg/m^3).

REFERENCES

1. Stein AA, Kirwan WE: Chloroacetophenone (tear gas) poisoning: clinico-pathological report. *J Foren Sci* 9:374–382, 1964
2. Punte CL, Gutentag PJ, Owens EJ, Gongwer LE: Inhalation studies with chloroacetophenone, diphenylaminoarsine and pelargonic morpholide. II. Human exposure. *J Am Ind Hyg Assoc* 23:199–202, 1962
3. Penneys NS: Contact dermatitis to chloroacetophenone. *Fed Proc* 30:96–99, 1971
4. Penneys NS, Israel RM, Indgin SM: Contact dermatitis due to 1-chloroacetophenone and Chemical Mace. *N Eng J Med* 281:413–415, 1969
5. Mackison FW et al: Occupational health guideline for α-chloroacetophenone. In NIOSH/OSHA *Occupational Health Guidelines for Chemical Hazards.* DHHS (NIOSH) Pub No 81-123. Washington, DC, US Government Printing Office, 1981

CHLOROACETYL CHLORIDE
CAS: 79-04-9

ClCH₂COCl

$ClCH_2COCl$

Synonyms: Monochloroacetyl chloride; chloroacetic acid chloride; CAC

Physical Form. Colorless liquid

Uses. Intermediate in manufacture of chloroacetophenone and various other chemicals

Exposure. Inhalation; skin absorption

Toxicology. In humans, chloroacetyl chloride (CAC) is a lacrimator; it also causes respiratory effects, including dyspnea, cyanosis, and cough, as well as skin effects such as erythema and burns.

A 44-year-old male worker experienced a large-skin-area exposure to a mixture of CAC, benzene, and xylidine.[1] The worker was put under a shower within 5 minutes of the accident but, shortly thereafter, he began to have respiratory difficulties and experienced an apparent grand mal seizure. The patient was still comatose 2 years after the accident. Burns caused by the chloroacetyl chloride were believed to have enhanced skin absorption of the other two chemicals although the relative contribution of the three chemicals to the patient's condition is unknown. The authors stated that the CAC manufacturer had provided information on two fatalities from CAC exposure, one after massive skin contact, followed by death within a few minutes. According to that manufacturer, other data indicated that CAC may promote ventricular arrhythmias.

In a similar incident, a worker was drenched by a mixture of the same three materials and sodium carbonate.[2] He suffered extensive first- and second-degree burns and pulmonary edema in spite of being placed immediately under a shower. The outcome was not reported. Other workers involved in the rescue suffered blisters on their hands and complained of chest tightness and nausea up to 2 days later.

An industrial hygienist was not able to detect odor at 0.011 ppm, found 0.023 ppm barely detectable, and considered 0.140 ppm to be "strong".[3] He experienced no eye irritation at 0.140 ppm but reported painful eye irritation and lacrimation around 1.0 ppm.

The oral LD_{50} in rats was between 187 and 229 mg/kg.[3] The 1-hour LC_{50} was 660 ppm for male rats and 750 for females; lacrimation and labored breathing occurred during exposures, and autopsy confirmed lung and nasal tissue congestion. In a 30-day inhalation study with rats, mice and hamsters, concentrations of 5 ppm or 2.5 ppm, 6 hr/day, 5 days/week, caused deaths in rats and mice but not hamsters; respiratory tract lesions were visible at necropsy, with the most severe response observed in the nasal region. Slight respiratory tract and eye irritation were observed in all species at 0.5 ppm. Applied to the skin of rabbits, the lethal dose was between 300 and 500 mg/kg.

The 1995 ACGIH threshold limit value-time weighted average (TLV-TWA) for chloroacetyl chloride is 0.05 ppm (0.23 mg/m³) with a notation for skin absorption.

REFERENCES

1. Raskin W, Canada A: Acute topical exposure to a mixture of benzene, chloroacetyl chloride and xylidine. *Vet Hum Toxicol* 23 (Suppl 1):42–44, 1981
2. Chloroacetyl chloride. *Documentation of the TLVs and BEIs for Substances in the Workroom Air,* 5th ed, p 123. Cincinnati, OH, American Conference of Governmental Industrial Hygienists (ACGIH), 1986 (1989 rev)
3. Chloroacetyl chloride. *Documentation of the TLVs and BEIs for Substances in the Workroom Air,* 6th ed, pp 268–269. Cincinnati, OH, American Conference of Governmental Industrial Hygienists (ACGIH), 1991

CHLOROBENZENE
CAS: 108-90-7

C_6H_5Cl

Synonyms: Phenylchloride; monochloroben-zene; chlorobenzol; benzene chloride

Physical Form. Colorless liquid

Uses. Manufacture of phenol, aniline, DDT; solvent for paint; color printing; dry cleaning

Exposure. Inhalation

Toxicology. Chlorobenzene is irritating to the skin and mucous membranes; it can cause central nervous system depression and liver and kidney damage.

In humans, eye and nasal irritation occur at 200 ppm and, at that level, the odor is pro-nounced and unpleasant; industrial experience indicates that occasional short exposures are not likely to produce more than minor skin irrita-tion but that prolonged or frequently repeated contact may result in skin burns.[1] In one case of accidental poisoning from ingestion of the liquid by a child, there was pallor, cyanosis, and coma, followed by complete recovery.[2]

Cats exposed to 8000 ppm showed severe narcosis after 1/2 hour and died 2 hours after removal from exposure, but tolerated 660 ppm for 1 hour.[3] Exposed animals showed eye and nose irritation, drowsiness, incoordination, and coma, followed by death from the most severe exposures. Several species of animals exposed daily to 1000 ppm for 44 days showed injury to the lungs, liver, and kidneys but, at 475 ppm, there was only slight liver damage in guinea pigs.

Leukopenia and depressed bone marrow activity were found in mice exposed at 544 ppm, 7 hr/day, for 3 weeks or at 22 ppm, 7 hr/day, for 3 months.[4] Only slight transient hematologic effects were found in rats and rabbits exposed at 250 ppm, 7 hr/day, for 24 weeks.[5] Adminis-tered to dogs in capsule form, 272.5 mg/kg/day for up to 92 days caused an increase in immature

leukocytes and some deaths.[6] Postmortem find-ings included gross and/or microscopic pathol-ogy in liver, kidneys, gastrointestinal mucosa, and hematopoietic tissue. No consistent effects were observed at 54.5 mg/kg/day.

In 91-day gavage studies, dose-dependent necrosis of the liver, degeneration or focal ne-crosis of the renal proximal tubules, and lymphoid or myeloid depletion of the spleen, bone marrow, and thymus were produced by doses of 250 mg/kg/day or greater in both sexes of rats and mice, although the incidences of lesions varied considerably by sex and species.[7,8] No toxic effects were observed at doses of 125 mg/kg/day or less. Gastric intubation of 120 mg/kg/day for 2 years produced a slight but statistically significant increase in neoplastic nodules of the liver in male rats. Increased tu-mor frequencies were not observed in female rats or in male or female mice receiving mono-chlorobenzene.

Concentrations up to 450 ppm, 7 days/week, 6 hr/day, did not adversely affect repro-ductive performance or fertility in a two-gener-ation rat study.[9] In rats and rabbits, inhalation of 590 ppm, 6 hr/day, during periods of major organogenesis did not produce structural mal-formations.[10]

Chlorobenzene was not mutagenic in a va-riety of bacterial and yeast assays. Existing data suggest that genotoxicity may not be an area of concern for chlorobenzene exposure in humans.[11]

Although the odor of chlorobenzene is pro-nounced and unpleasant, it is not sufficient to give warning of hazardous concentrations.[1]

The 1995 ACGIH threshold limit value-time weighted average (TLV-TWA) for chlo-robenzene is 10 ppm (46 mg/m^3).

REFERENCES

1. Hygienic guide series: chlorobenzene. *Am Ind Hyg Assoc J* 25:97–99, 1964
2. von Oettingen WF: *The Halogenated Aliphatic, Olefinic, Cyclic, Aromatic, and Aliphatic-Aro-matic Hydrocarbons Including the Halogenated Insecticides, Their Toxicity and Potential Dan-gers.* US Public Health Service Pub No 414,

pp 283–285. Washington, DC, U S Government Printing Office, 1955

3. Deichmann WB: Halogenated cyclic hydrocarbons. In Clayton GD, Clayton, FE (eds): *Patty's Industrial Hygiene and Toxicology*, 3rd ed, rev, Vol 2B, *Toxicology*, pp 3604–3611. New York, Wiley-Interscience, 1981

4. Zub M: Reactivity of the white blood cell system to toxic actions of benzene and its derivatives. *Acta Biol Cracoviensia* 21:163–174, 1978

5. *Toxicology Evaluation of Inhaled Chlorobenzene (Monochlorobenzene)*. NTIS PB-276-623. Cincinnati, OH, National Institute for Occupational Safety and Health (NIOSH), Division of Biomedical and Behavioral Sciences, 1977

6. Knapp WK et al: Subacute oral toxicity of monochlorobenzene in dogs and rats. *Toxicol Appl Pharmacol* 19:393, 1971

7. National Toxicology Program: *Toxicology and Carcinogenesis Studies on Chlorobenzene (CAS No. 108-90-7) in F344/N Rats and B6C3F Mice (Gavage Studies)*, Technical Report Series 261, NIH Pub No 86-2517, p 220. Washington, DC, US Department of Health and Human Services, October 1985

8. Kluwe WM, Dill G, Persings A, et al: Toxic responses to acute, subchronic and chronic oral administrations of monochlorobenzene to rodents. *J Toxicol Environ Health* 15:745–767, 1985

9. Nair RS, Barter JA, Schroeder RE, et al: A two generation reproduction study with monochlorobenzene vapor in rats. *Fund Appl Toxicol* 9:678–686, 1987

10. John JA, Hayes WC, Hanley TR Jr, et al: Inhalation teratology study on monochlorobenzene in rats and rabbits. *Toxicol Appl Pharmacol* 76:365–373, 1984

11. Agency for Toxic Substances and Disease Registry (ASTDR): *Toxicological Profile for Chlorobenzene*. US Department of Health and Human Services, TP-90-06, 1990

o-CHLOROBENZYLIDENE MALONONITRILE
CAS: 2698-41-1

$ClC_6H_4CHC(CN)_2$

Synonyms: CS; OCBM; chlorobenzylidene malononitrile; 2-chlorobenzylidene malononitrile

Physical Form. White, crystalline solid

Uses. Riot-control agent

Exposure. Inhalation; skin absorption

Toxicology. *o*-Chlorobenzylidene malononitrile (CS) aerosol is a potent lacrimator and upper respiratory irritant.

In human experiments, concentrations ranging from 4.3 to 6.7 mg/m³ were barely tolerated when reached gradually over a period of 30 minutes.[1] Following cessation of exposure, a burning sensation and deep pain in the eyes persisted for 2 to 5 minutes. Severe conjunctivitis lasted for 25 to 30 minutes, and erythema of the eyelids, with some blepharospasm, was present for 1 hour. There was a burning sensation in the throat, with cough, followed by a constricting sensation in the chest; no therapy other than removal from exposure was necessary.

At a concentration of 1.5 mg/m³, three of four men developed headache during a 90-minute exposure; one subject developed slight eye and nose irritation.[1] On the skin, the powder caused a burning sensation, which was greatly aggravated by moisture; erythema and vesiculation resembling second-degree burns were produced.

In animals, the manifestation of lethal toxicity is different following intravenous, intraperitoneal, oral, and inhalation routes. After intravenous administration, there is rapid onset of signs characteristic of effects on the nervous system as a result of the alkylating properties of CS.[2] High doses of intraperitoneal CS result in expression of the cyanogenic potential of the malononitrile radical. By the oral route, local inflammation in the gastrointestinal tract contributes to toxicity. Lethal toxicity from inhalation is from lung damage leading to asphyxia or, in the case of delayed deaths, bronchopneumonia secondary to respiratory tract damage. Rats survived a 10-minute exposure at 1800 mg/m³, but 20 of 20 succumbed following 60 minutes at 2700 mg/m³.

o-Chlorobenzylidene malononitrile did not cause a mutagenic response when tested in a variety of assays that examined point mutations, germinal gene mutations, chromosomal breaks, and mitotic chromosome misdistribution.[3] Although limited, a study of the repeated inhalation toxicity of CS in mice, rats, and guinea pigs did not find a relationship between tumors in a particular site and total dose of CS.[4] F344N rats exposed at 0.075, 0.25, or 0.75 mg/m³ and B6C3F$_1$ mice exposed at 0.75 or 1.5 mg/m³, 6 hr/day, 5 days/week, for 2 years had no compound-related incidences of neoplasm.[5] Nonneoplastic lesions occurred primarily in the nasal passages and included hyperplasia and squamous metaplasia of the respiratory epithelium.

The 1995 ACGIH ceiling–threshold limit value (C-TLV) for *o*-chlorobenzylidene malononitrile is 0.05 ppm (0.39 mg/m³) with a notation for skin absorption.

REFERENCES

1. Punte CL, Owens EJ, Gutentag PJ: Exposure to ortho-chlorobenzylidene malononitrile. *Arch Environ Health* 6:366–374, 1963
2. Ballantyne B, Swanston DW: The comparative acute mammalian toxicity of 1-chloroacetophenone (CN) and 2-chlorobenzylidene malononitrile (CS). *Arch Toxicol* 40:75–95, 1978
3. Wild D, Eckhardt K, Harnasch D, et al: Genotoxicity study of CS (ortho-chlorobenzylidene malonitrile) in salmonella, drosophila and mice. *Arch Toxicol* 54:167–170, 1983
4. Marrs TC, Colgrave HF, Cross NL, et al: A repeated dose study of the toxicity of inhaled 2-chlorobenzylidene malononitrile (CS) aerosol in three species of laboratory animal. *Arch Toxicol* 52:183–198, 1983
5. National Toxicology Program: *Toxicology and*

Carcinogenesis Studies of CS₂ (94% o-Chlorobenzylidene Malononitrile) in F344/N Rats and B6C3F₁ Mice. Technical Report No 377. Research Triangle Park, NC, National Institutes of Health, 1990

CHLOROBROMOMETHANE
CAS: 74-97-5

CH₂BrCl

Synonyms: Monochloromonobromomethane; bromochloromethane; methylene chlorobromide; monobromochloromethane; chloromethyl bromide

Physical Form.   Colorless liquid

Uses.   Firefighting agent

Exposure.   Inhalation

Toxicology.   Chlorobromomethane is a mild irritant of the eyes and mucous membranes; at high concentrations, it causes central nervous system depression.

Exposure of three firefighters to unknown but very high vapor concentrations was characterized by disorientation, headache, nausea, and irritation of the eyes and throat. Two of the three became comatose; of the two, one had convulsive seizures, and the other had respiratory arrest, from which he was resuscitated.[1] Recovery was slow but complete. Some effects may have been due to the inhalation of thermal decomposition products.

Prolonged skin contact may cause dermatitis.[1] The liquid in the eye causes an immediate burning sensation, followed by corneal epithelial injury and conjunctival edema.[2]

Concentrations near 30,000 ppm were lethal to rats within 15 minutes; toxic signs included loss of coordination and narcosis. This level of exposure produced pulmonary edema and, in cases of delayed deaths, there was interstitial pneumonitis.[3] Concentrations as low as

3000 ppm for 15 minutes produced light narcosis in rats. No toxic effects were observed in rats, rabbits, and dogs exposed 7 hr/day, 5 days/week, for 14 weeks to 1000 ppm.[4]

Metabolic studies of inhaled chlorobromomethane in rats have shown production of carbon monoxide, halide ions, and other reactive intermediates.[5]

Chlorobromomethane has a distinctive odor at 400 ppm; however, the odor is not disagreeable and does not provide sufficient warning properties.

The 1995 ACGIH threshold limit value–time-weighted average (TLV-TWA) for chlorobromomethane is 200 ppm (1060 mg/m³).

REFERENCES

1. Rutstein HR: Acute chlorobromomethane toxicity. *Arch Environ Health* 7:40–444, 1963
2. Grant WM: *Toxicology of the Eye*, 3rd ed, pp 210–211. Springfield, IL, Charles C Thomas, 1986
3. Comstock CC, Fogleman RW, Oberst FW: Acute narcotic effects of monochlorobromomethane vapor in rats. *AMA Arch Ind Hyg Occup Med* 7:526–528, 1953
4. Svirbely JL, Highman B, Alford WC, et al: The toxicity and narcotic action of monochloromonobromomethane with special reference to inorganic and volatile bromide in the blood, urine and brain. *J Ind Hyg Toxicol* 29:382–389, 1947
5. Gargas ML et al: Metabolism of inhaled dihalomethanes in vivo: differentiation of kinetic constants for two independent pathways. *Toxicol Appl Pharmacol* 82:211–223, 1986

---

p-CHLORO-*m*-CRESOL
CAS: 59-50-7

CH₃C₆H₃ClOH

Synonyms: PCMC; 4-chloro-*m*-cresol; 3-methyl-4-chlorophenol; Candaseptic

Physical Form.   Crystals, usually with a phenolic odor

Uses.   Antiseptic

Exposure.   Inhalation

Toxicology.   *p*-Chloro-*m*-cresol   (PCMC) causes kidney damage in male rats after chronic exposure.

PCMC was evaluated for chronic toxicity and carcinogenicity in Wistar rats (50/sex/dose level) when administered in the feed for 24 months at doses of 0, 400, 2000, and 10,000 ppm.[1] Over 104 weeks, rats ingested: 21, 103.1, and 558.9 mg (males) and 27.7, 134.3, and 743.5 mg (females). All the effects observed were associated with the highest exposure level. These included decreased body weight gain for both sexes, increased incidence of female rats found in poor condition of health, increased relative kidney weights for both sexes, and gross and micropathological evidence of kidney damage (i.e., papillary necrosis, cortical dilation, and fibrosis) in males only. There was no indication of carcinogenic potential of the substance up to and including the 10,000-ppm exposure level.

In a developmental toxicity study, PCMC was administered to pregnant Wistar rats (number and sex of rats/group was not reported) by oral gavage at doses of 0, 30, 100, and 300 mg/kg bw/day on days 6 through 15 of gestation.[2] The 100 mg/kg bw/day dose was reported as maternally toxic on account of reduction in body weight gain and food consumption in dams. At 300 mg/kg bw/day, a 25% mortality of the dams was reported. Clinical signs of toxicity were observed on several days during the administration period, but these signs were not reported. A statistically significant increase of early resorptions and a decrease in mean fetal weights were observed only in the 300 mg/kg bw/day dose group.

A threshold limit value has not been assigned to PCMC.

REFERENCES

1. Bayer AG: Initial submission: chronic toxicity and carcinogenicity study in Wistar rats (administration in feed for 104 weeks) (Interim Report) with cover letter dated 3/27/92,

TSCATS/422992, Doc 88-920001578, Environmental Protection Agency/Office of Toxic Substances 1992

2. Miles Inc: Initial submission: embryotoxicity report on *p*-chloro-*m*-cresol, TSCATS/421561, Doc 88-920000850, Environmental Protection Agency/Office of Toxic Substances 1992

CHLORODIBROMOMETHANE
CAS: 124-48-1

CHBr₂Cl

Synonyms: Dibromochloromethane; monochlorodibromomethane

Physical Form.   Colorless liquid

Uses.   One of four common trihalomethanes formed after chlorination of water supplies; in the past, used to make fire extinguisher fluids, spray can propellants, refrigerator fluids, and pesticides; only small amounts currently produced for laboratory use

Exposure.   Inhalation; skin absorption; ingestion

Toxicology.   Chlorodibromomethane is a central nervous system depressant at extremely high concentrations; it is toxic to the liver and kidneys of rodents and induces hepatocellular tumors in mice after long-term exposure.

In animal studies, the oral LD$_{50}$ typically ranges between 800 and 1200 mg/kg.[1,2] Acute signs of intoxication include sedation, flaccid muscle tone, ataxia, and prostration; death is due to CNS depression. In cases in which death does not occur until several days after acute exposure, hepatic and renal injury may be the cause of death.

It should be noted that, in humans, opportunities for exposure to acutely lethal doses of chlorodibromomethane are remote. In animals, no direct effects of oral exposure to chlorodi-bromomethane have been noted for the respiratory, cardiovascular, hematologic, or musculoskeletal systems or on the skin or eyes. One study indicated that short-term oral exposure of mice to doses of 125 mg/kg/day could produce significant changes in both the humoral and cell-mediated immune systems.[3]

In the drinking water of rats, 137 and 165 mg/kg/day of chlorodibromomethane for 90 days produced mild toxicity in the liver; the observed vacuolar changes due to fatty infiltration were reversible after a 90-day recovery period.[4]

Administered by gavage to rats and mice for 13 weeks, 250 mg/kg/day of chlorodibromomethane caused hepatic and renal toxicity in male and female rats and in male mice.[5] Results of 2-year gavage studies showed fatty metamorphosis and cytoplasmic changes in the livers of rats receiving up to 80 mg/kg/day; in mice receiving up to 100 mg/kg/day of the chemical, hepatic lesions included necrosis and hepatocytomegaly in males, and calcification and fatty change in females.[5,6] Evidence of nephrosis was seen in male mice and female rats.

In the same 2-year gavage study, chlorodibromomethane significantly increased the incidence of hepatocellular adenomas, as well as the combined incidence of hepatocellular adenomas or carcinomas, in the high-dose female mice. The incidence of hepatocellular carcinomas was significantly increased in the high-dose male mice although the combined incidence of hepatocellular adenomas or carcinomas was only marginally significant.

Under the conditions of the gavage studies, there was no evidence of carcinogenicity in rats receiving doses of 40 or 80 mg/kg/day for 2 years; there was equivocal evidence of carcinogenicity in male mice receiving 100 mg/kg/day for 2 years, and some evidence of carcinogenicity in female mice receiving 50 or 100 mg/kg/day for 2 years.

There is no clear epidemiologic evidence for the carcinogenicity of chlorodibromomethane in humans. However, a number of studies suggest an association between chronic ingestion of trihalomethanes in chlorinated drinking water and increased risk of bladder or colon cancer.[6] These studies cannot provide information on whether any observed effects are due to chlorodibromomethane or to one or more of the hundreds of other by-products that also are present in chlorinated drinking water.

The IARC has determined that there is limited evidence for the carcinogenicity of chlorodibromomethane in experimental animals and inadequate evidence in humans.[7]

No teratogenic effects were observed in rats given 200 mg/kg/day during gestation.[8] At doses of 685 mg/kg/day, which caused marked maternal toxicity, there were significant decreases in litter size, gestational survival, and postnatal body weight and survival.[9]

A 1995 TLV has not been established.

REFERENCES

1. Bowman FJ, Borzelleca JF, Munson AE: Toxicity of some halomethanes in mice. *Toxicol Appl Pharmacol* 44:213–225, 1978
2. Chu I, Secours V, Marino I, et al: Acute toxicity of four trihalomethanes in male and female rats. *Toxicol Appl Pharmacol* 52:351–353, 1980
3. Munson AE, Sain LE, Sanders, VM, et al: Toxicology of organic drinking water contaminants: Trichloromethane, bromodichloromethane, dibromochloromethane, and tribromomethane. *Environ Health Perspect* 46:117–126, 1982
4. Chu I, Villeneuve DC, Secours VE, et al: Trihalomethanes. II. Reversibility of toxicological changes produced by chloroform, bromodichloromethane, chlorodibromomethane and bromoform in rats. *J Environ Sci Health* B17:225–240, 1982
5. Dunnick JK, Haseman JK, Lilja HS, et al: Toxicity and carcinogenicity of chlorodibromomethane in Fischer 344/N Rats and B6C3F1 Mice. *Fund Appl Toxicol* 5:1128–1136, 1985
6. National Toxicology Program: *Toxicology and Carcinogenesis Studies of Chlorodibromomethane (CAS No. 124-48-1) in F344/N Rats and B6C3F₁ Mice (Gavage Studies)*, Technical Report Series No 282, NIH Pub No 88-2538. Research Triangle Park, NC, U S Department of Health and Human Services, Public Health Service, National Institutes of Health, National Toxicology Program, 1985
7. *IARC Monographs on the Evaluation of Carcinogenic Risks to Humans*, Vol 52, *chlorinated drinking-water; chlorination by-products; some other halogenated compounds; cobalt and cobalt compounds*, pp 243–263. Lyon, International Agency for Research on Cancer, 1991
8. Ruddick JA, Villeneuve DC, Chu I, et al: A teratological assessment of four trihalomethanes in the rat. *J Environ Sci Health* 18:333–349, 1983
9. Borzelleca JF, Carchman RA: *Effects of Selected Organic Drinking Water Contaminants on Mice Reproduction*. EPA 600/1-82-009. NTIS No PB 82-259847. Research Triangle Park, NC, US Environmental Protection Agency, Office of Research and Development, 1982

CHLORODIFLUOROMETHANE
CAS: 75-45-6

$CHClF_2$

Synonyms: Freon 22; monochlorodifluoromethane; difluoromonochloromethane

Physical Form. Colorless, nearly odorless, nonflammable gas

Uses. Aerosol propellant; refrigerant; low-temperature solvent

Exposure. Inhalation

Toxicology. Chlorodifluoromethane gas causes central nervous system effects at very high concentrations.

There have been few reports of adverse health effects in workers despite nearly 50 years of commercial use of chlorodifluoromethane.[1] Fatalities have been reported, however, in connection with intentional inhalation, with death due to acute respiratory arrest.[2]

The incidence of cardiac palpitations was compared in two employee groups.[3] One group of 118 hospital laboratory employees was exposed to an average concentration of 300 ppm chlorodifluoromethane during its use as a tissue preservative. The control group of 85 employees came from a different department and had no chemical exposure. The number of employees exhibiting palpitations was significantly higher in the exposed group than in the control group. An epidemiologic study involving workers exposed to chlorofluorocarbons, including chlorodifluoromethane, showed no increased mortality due to heart, circulatory, or malignant disorders.[1]

Animal studies found an LC_{50} of 277,000 ppm for a 30-minute exposure in mice, and a threshold concentration of 300,000 ppm for death in rabbits.[4] Chlorodifluoromethane was thought to have an irritative effect on the respiratory system or a stimulative effect on the parasympathetic system, which caused a great amount of mucous fluid, rattling in the chest, and high cyanosis.[4] The cause of death was thought to be respiratory insufficiency from aspiration of mucous fluid into the lungs. Exposure of rats and guinea pigs for 2 hours to levels of 75,000 to 100,000 ppm caused excitation and/or dysfunction in equilibrium.[5] Narcosis occurred at 200,000 ppm, and animals died at 300,000 to 400,000 ppm.

Studies in the dog and other species show that high concentrations (above 50,000 ppm) in association with injected epinephrine are required to produce cardiac arrythmias.[1] This is a relatively low order of potency by comparison to other chlorofluorocarbons.[1]

Pregnant rats exposed to 50,000 ppm, 6 hr/day, on days 6 to 15 of gestation had decreased body weight gain, and their offspring had an increased incidence of anophthalmia (absent eyes).[1] At this dose, chlorodifluoromethane did not affect the pregnant rabbit or her offspring, nor was there any effect on male fertility in the rat or the mouse.

Evaluation of tumor data from lifetime studies showed an increased incidence of fibrosarcomas, some involving the salivary gland, in male rats chronically exposed to 50,000 ppm.[1] Negative results were obtained for females. Other studies in mice or in rats receiving oral doses were negative or inconclusive.[6] The IARC has determined that there is limited evidence for carcinogenicity to animals and inadequate evidence for carcinogenicity to humans.[6]

The 1995 ACGIH threshold limit value–time-weighted average (TLV-TWA) for chlorodifluoromethane is 1000 ppm (3540 mg/m³).

REFERENCES

1. Litchfield MH, Longstaff E: Summaries of toxicological data: the toxicological evaluation of chlorofluorocarbon 22 (CFC22). *Food Chem Toxic* 22:465–475, 1984
2. Fitzgerald RL, Fishel CE, Bush LLE: Fatality due to recreational use of chlorodifluoromethane and chloropentafluoroethane. *J Foren Sci* 38:476–482, 1993
3. Speizer FE, Wegman DH, Ramirez A: Palpitation rate associated with fluorocarbon exposure in a hospital setting. *New Eng J Med* 292:624, 1975
4. Sakata M, Kazama H, Miki A, et al: Acute toxicity of fluorocarbon-22: toxic symptoms, lethal concentration and its fate in rabbit and mouse. *Toxicol Appl Pharmacol* 59:64–70, 1981
5. Weigand W: Examinations of the inhalation toxicology of the fluoroderivatives of methane, ethane and cyclobutane. Zentr Zbl Arbeitsmed 2:149, 1971
6. *IARC Monographs on the Evaluation of the Carcinogenic Risk of Chemicals to Humans*, Vol 41, *Some Halogenated Hydrocarbons and Pesticide Exposures*, pp 237–252. Lyon, International Agency for Research on Cancer, 1986

CHLORODIPHENYL, 42% CHLORINE
CAS: 53469-21-9

$C_{12}H_7Cl_3$

Synonyms: Arochlor 1242; polychlorinated biphenyl; PCB

Physical Form. Straw-colored liquid

Uses. Dielectric in capacitors and transformers; investment casting processes; heat-exchange fluid; hydraulic fluid

Exposure. Skin absorption; ingestion; inhalation

Toxicology. Chlorodiphenyl, 42% chlorine (a polychlorinated biphenyl, or PCB) is an irritant of the eyes and mucous membranes, is toxic to the liver, and causes an acneform dermatitis (chloracne). It is a liver carcinogen in animals.

In humans, systemic effects are anorexia, nausea, edema of the face and hands, and abdominal pain.[1] In a survey of 34 workers exposed to concentrations of up to 2.2 mg/m³, complaints were a burning sensation of the face and hands, nausea, and a persistent (uncharacterized) body odor.[1] One worker had chloracne, and five had an eczematous rash on the legs and hands.[1] Although hepatic function tests were normal, the mean blood level of chlorodiphenyl in the exposed group was approximately 400 ppb, whereas none was detected in the control group.[1]

Cases of mild to moderate skin irritation and chloracne have been reported in workers exposed to 0.1 mg/m³ for several months. Levels of 10 mg/m³ were unbearably irritating, presumably to mucous membranes and skin.[2] Chloracne does not appear to occur at concentrations below 0.1 mg/m³.

Chloracne usually is persistent and affects the face, ears, neck, shoulders, arms, chest, and abdomen (especially around the umbilicus and on the scrotum). The most sensitive areas are below and to the outer side of the eye (malar crescent) and behind the ear. The skin frequently is dry with noninflammatory comedones and pale yellow cysts containing sebaceous matter and keratin. Some evidence of liver disease is often seen in association with PCB-induced chloracne.[3]

Some studies of occupationally exposed groups have revealed evidence of liver injury by serum enzyme studies or other liver-function tests. Adverse effect and dose-effect relationships have not been consistent within and between studies, raising the possibility that other factors (e.g., alcohol intake, other exposures) could be responsible.[2] Review of these studies indicates that some liver effects may have occurred with repeated exposures at concentrations below 0.1 mg/m³, assuming that PCBs were responsible. Several deaths due to toxic hepatitis have been reported among workers exposed to mixtures of PCBs with chlorinated naphthalenes; such effects have not been observed with PCB exposure alone.[2]

A cross-sectional survey of 205 capacitor manufacturing workers with a geometric mean serum PCB level of 18.2 ppb, standard deviation (SD) 2.88, found no statistically significant correlations between PCB levels and clinical chemistry results, including SGOT, GGTP, and LDH levels.[4] The primary dielectric used in the plant was Aroclor 1242. However, another cross-sectional survey of 120 railroad transformer workers with mean plasma PCB levels of 33.4 ppb did reveal statistically significant correlations of PCB level with serum triglyceride and SGOT (but not SGPT or GGTP) levels.[5] There was a significant correlation between self-reported direct dermal contact with PCBs and the plasma PCB level. In a survey of 80 heavily exposed capacitor or transformer manufacturing workers in Italy with mean blood PCB levels of about 340 ppb, there was a correlation between blood PCB levels and abnormal liver findings (including hepatomegaly and increased GGTP, SGOT, and SGPT levels).[6] Even in this latter group, except for a few cases of chloracne, no other symptoms or findings referable to PCB exposure were present. The biological significance of these generally mild elevations in serum enzymes is also unclear.

Industrial hygiene studies support the notion that the dermal and dermal/oral, rather than the respiratory, route of exposure are the predominant contributors to body burden among workers occupationally exposed to PCBs.[7]

The toxic effects of PCBs in humans are further illustrated by a 1968 outbreak of poisoning ("Yusho" disease) in Japan that involved more than 1000 people who ingested PCB-contaminated rice bran oil for a period of several months.[8,9] The contamination of the oil (estimated 1500–2000 ppm) occurred when heat-transfer pipes immersed in the oil during processing developed pin-sized holes. The clinical aspects of the poisoning included chloracne, brown pigmentation of the skin and nails, distinctive hair follicles, increased eye discharge, swelling of eyelids, transient visual disturbance, and systemic gastrointestinal symptoms with jaundice. In some patients, symptoms persisted 3 years after PCB exposure was discontinued. Infants born to poisoned mothers had decreased birth weights and showed skin discoloration. Chemical analysis of the contaminated rice oil revealed significant amounts of polychlorinated dibenzofurans (PCDFs) as well as PCBs.[10] High concentrations of PCDFs were found in blood and adipose tissue of Yusho victims. In contrast, in a group of workers occupationally exposed to PCBs, PCB levels were higher than in the Yusho victims, but PCDFs were not generally detected.[10,11] Animal experiments have reproduced some findings seen in Yusho victims with administration of PCDFs but not PCBs. Thus, it appears that PCDFs were the main causative agents in the induction of Yusho disease.[10,11]

PCBs are poorly metabolized and tend to accumulate in animal tissues, including human tissues.[8] The accumulation, particularly in tissues and organs rich in lipids, appears to be higher in the case of penta- and more highly chlorinated biphenyls. Studies have revealed PCBs in human fat tissue and blood plasma. PCBs in amounts greater than 2 ppm were reported in 198 of 637 (31%) samples of human fat tissue taken from the general population of 18 states and the District of Columbia. PCB residues ranging up to 29 ppb were also found in 43% of 616 plasma samples collected from volunteers in a southeastern US county. A recent study of nonoccupationally exposed adults in Southern California found mean plasma PCB levels of 5 ± 4 ppb ($\pm$SD), with a range of 1 to 37 ppb.[12] Other studies of the general US population have revealed mean plasma or serum levels of PCBs between 2.1 and 24.4 ppb.

A cohort study of 544 male and 1557 female workers employed between 1946 and 1978 in a capacitor manufacturing plant using PCBs in Italy found statistically significant excesses of total cancer deaths in males (14 observed vs. 7.6 expected) and females (12 observed vs. 5.3 expected), cancer of the GI tract in males (6 observed vs. 2.2 expected), and hematologic neoplasms in females (4 observed vs. 1.1 expected).[13] Of the six GI tract malignancies in males, the primary sites were stomach (2), pancreas (2), liver (1), and biliary tract (1). There was an excess of hematologic neoplasms in males (3 observed vs. 1.1 expected), but this excess was not statistically significant. The authors qualified their conclusions regarding ex-

cess malignancies because of the small number of deaths in the cohort, the occurrence of some tumors in workers with minimal exposure or short latency intervals, and the disparate sites and types of tumors.

An update of a retrospective cohort mortality study of 2588 workers exposed to PCBs in two capacitor manufacturing plants revealed no excess of all cancers (SMR 78), stomach cancer (SMR 36), pancreatic cancer (SMR 54), or lymphatic and hematopoietic cancer (SMR 68). The only statistically significant excess was for cancer of the liver and biliary passages (5 observed vs. 1.9 expected, SMR 263). Both Aroclor 1254 (54% chlorine) and 1242 (42% chlorine) had been used at different times in both plants. Although the workers studied had positions involving greater exposure to PCBs than other workers in the plants, historical levels of exposure were unknown. Four of the five cases of liver and biliary tract cancer occurred in women in plant 2. Although all five workers were first employed in the 1940s and early 1950s, when exposures were presumed to be the highest, analysis did not reveal that risk was associated with time since first employment or length of employment in "PCB-exposed" jobs. Attempts to confirm the site of origin of the cancer as liver by review of records were unsuccessful in two of the cases; in one, no records were available and, in the other, records indicated that the primary site was unknown. The author tempered his conclusions because of these limitations and the small number of cases identified.[14]

Guinea pigs died at intervals up to 21 days after the first of 11 daily applications of 34.5 mg to the skin; at necropsy, the liver showed fatty degeneration and central atrophy; rats, however, survived 25 daily applications, and only slight changes in the liver were observed.[15]

After application of radiolabeled PCB, 42% chlorine, to the skin of guinea pigs and rhesus monkeys, 33% and 15% to 34%, respectively, of the applied doses were absorbed.[16]

All PCB mixtures adequately tested in mice and rats have shown carcinogenic activity.[2] For example, of 20 rats fed Aroclor 1242 at 100 ppm in the diet for 24 months, 11 developed liver tumors, of which three were hepatomas.

Evidence from bioassays suggests that the less highly chlorinated PCBs (e.g., Aroclor 1242) have less carcinogenic potential than the more highly chlorinated mixtures (e.g., Aroclor 1254).[2]

The IARC concluded in 1987 that there was limited evidence for carcinogenicity of PCBs to humans and that there was suggestive evidence for carcinogenicity to animals, based on a number of studies in mice and rats in which oral administration resulted in benign and malignant liver neoplasms.[17]

Adverse reproductive effects, including fetotoxicity and teratogenicity, have been observed in animals fed PCBs in the diet. PCBs have been observed in human cord blood and in tissues of newborn humans and animals.[2]

The reproductive histories of 200 women exposed to PCBs during the production of capacitors were compared with those of 200 controls. There was only a slight relationship between esimated PCB levels in serum and decreased birth weight.[18]

The 1995 ACGIH threshold limit value-time weighted average (TLV-TWA) for chlorodiphenyl (42% chlorine) is 1 mg/m^3 with a notation for skin absorption.

REFERENCES

1. Ouw HK, Simpson GR, Siyali DS: Use and health effects of Arochlor 1242, a polychlorinated biphenyl, in an electrical industry. *Arch Environ Health* 31:189–194, 1976
2. National Institute for Occupational Safety and Health: *Criteria for a Recommended Standard . . . Occupational Exposure to Polychlorinated Biphenyls.* DHEW (NIOSH) Pub No 77-225. Washington, DC, US Government Printing Office, 1977
3. von Oettingen WF: *The Halogenated Aliphatic, Olefinic, Cyclic, Aromatic, and Aliphatic-Aromatic Hydrocarbons Including the Halogenated Insecticides, Their Toxicity and Potential Dangers.* US Public Health Service Pub No 414, pp 311–313. Washington, DC, U S Government Printing Office, 1955
4. Acquevella JF et al: Assessment of clinical, metabolic, dietary, and occupational correlations with serum polychlorinated biphenyl levels among employees at an electric capacitor manufacturing plant. *J Occup Med* 28:1177–1180, 1986
5. Chase KH et al: Clinical and metabolic abnormalities associated with occupational exposure to polychlorinated biphenyls (PCBs). *J Occup Med* 24:109–114, 1982
6. Maroni M et al: Occupational exposure to polychlorinated biphenyls in electrical workers. II. Health effects. *Br J Ind Med* 38:55–60, 1981
7. Lees PSJ, Corn M, Breysse P: Evidence for dermal absorption as the major route of body entry during exposure of transformer maintenance and repairmen to PCBs. *Am Ind Hyg Assoc J* 48:257–264, 1987
8. Lloyd JW, Moore RM Jr, Woolf BS, Stein HP: Polychlorinated biphenyls. *J Occup Med* 18:109–113, 1976
9. Kuratsune M, Yoshimura T, Matsuzaka J, Yamasuchi A: Yusho, a poisoning caused by rice oil contaminated with polychlorinated biphenyls. *HSMHA Health Repts* 36:1083–1091, 1971
10. Masuda Y, Yoshimura H: Chemical analysis and toxicity of polychlorinated biphenyls and dibenzofurans in relation to Yusho. *J Toxicol Sci* 161–175, 1982
11. Kunita N et al: Causal agents of Yusho. *Am J Ind Med* 5:45–58, 1984
12. Sahl JD et al: Polychlorinated biphenyl concentrations in the blood plasma of a selected sample of non-occupationally exposed Southern California working adults. *Scien Total Environ* 46:9–18, 1985
13. Bertazzi PA et al: Cancer mortality of capacitor manufacturing workers. *Am J Ind Med* 11:165–176, 1987
14. Brown DP: *Mortality of Workers Exposed to Polychlorinated Biphenyls—An Update.* Cincinnati, OH, National Institute for Occupational Safety and Health (NIOSH), US Department of Health and Human Services, 1986
15. Miller JW: Pathological changes in animals exposed to a commercial chlorinated diphenyl. *Pub Health Repts* 59:1085–1093, 1944
16. Werter RC et al: Polychlorinated biphenyls (PCB): dermal absorption, systemic elimination, and dermal wash efficiency. *J Toxicol Environ Health* 12:511–519, 1983
17. *IARC Monographs on the Evaluation of the Carcinogenic Risk of Chemicals to Humans, Chemicals, Industrial Processes, and Industries Associated with Cancer in Humans,* Suppl 7. Lyon, Inter-

national Agency for Research on Cancer, 1987

18. Taylor PR, Stelman JM, Laurence CE: The relationship of polychlorinated biphenyls to birth weight and gestational age in the offspring of occupationally exposed mothers. *Am J Epidem* 129:395–406, 1989

CHLORODIPHENYL, 54% CHLORINE
CAS: 11097-69-1

$C_6H_2Cl_3C_6H_3Cl_2$

Synonyms: Aroclor 1254; polychlorinated biphenyl; PCB

Physical Form.   Viscous liquid

Uses.   Dielectric in capacitors and transformers; investment casting processes; heat-exchange fluid; hydraulic fluid

Exposure.   Skin absorption; ingestion; inhalation

Toxicology.   Chlorodiphenyl, 54% chlorine (a polychlorinated biphenyl, or PCB) is an irritant of the eyes and mucous membranes. It is toxic to the liver of animals, and severe exposure may produce a similar effect in humans. It also causes an acne form dermatitis (chloracne). It is a liver carcinogen in animals.

Note.   For a full description of the toxicology of this compound, see "Chlorodiphenyl, 42% Chlorine," which immediately precedes this discussion. Special characteristics of the 54% chlorine compound are given below.

Rats exposed to 5.4 mg/m^3 of the 54% chlorine compound for 7 hours daily for 4 months showed increased liver weight and injury to liver cells; 1.5 mg/m^3 for 7 months also produced histopathological evidence of liver damage, which was considered to be of a reversible character.[1] The minimal lethal dose when the liquid was applied to the skin of rabbits was 1.5 g/kg.[2] The vapor and the liquid are moderately

irritating to the eye; contact of the chemical with skin leads to removal of natural fats and oils, with subsequent drying and cracking of the skin.[2]

After application of radiolabeled PCB, 54% chlorine, to the skin of guinea pigs, 56% of the applied dose was absorbed.[3]

Administration of a PCB mixture (mean chlorine content 54%) twice a week for 6 weeks via a stomach tube to rats at relatively low dose levels led to histopathological changes in the liver, increases in cholesterol and triglyceride levels, and serum enzyme increases. At the 2 mg/kg dose, centrilobular hepatic necrosis and elevated cholesterol levels were observed. Increases in bilirubin and triglyceride levels occurred only at 50 mg/kg, and increases in SGOT (AST) and SGPT (ALT) only at doses above 50 mg/kg.[4]

All PCB mixtures adequately tested in mice and rats have shown carcinogenic activity.[1] For example, hepatomas developed in 9 of 22 BALB/cj male mice fed Arochlor 1254 at 300 ppm for 11 months. Of 27 rats fed Aroclor 1254 at 100 ppm in the diet for 24 months, 19 developed liver tumors, 6 of which were hepatomas, compared with one neoplastic nodule in 23 controls. Evidence from bioassays suggests that the more highly chlorinated PCBs (e.g., Aroclor 1254) have more carcinogenic potential than the less highly chlorinated mixtures (e.g., Aroclor 1242).[5]

IARC concluded in 1982 that there was inadequate evidence for carcinogenicity of PCBs to humans but that there was sufficient evidence for carcinogenicity to animals, based on a number of studies in mice and rats in which oral administration resulted in benign and malignant liver neoplasms.[6]

Adverse reproductive effects have been observed in animals fed PCB in the diet.[1] Fetal resorptions were common, and dose-related incidences of terata were found in pups and piglets when females were fed Aroclor 1254 at 1 mg/kg/day or more. PCBs have been observed in human cord blood and in tissues of newborn humans and animals.[1]

The 1995 ACGIH threshold limit value-time weighted average (TLV-TWA) for chlorodiphenyl (54% chlorine) is 0.5 mg/m³ with a notation for skin.

REFERENCES

1. Treon JF, Cleveland FP, Cappel JW, Atchley RW: The toxicity of the vapors of Arochlor 1242 and Arochlor 1254. *Am Ind Hyg Assoc Q* 17:204–213, 1956
2. Hygienic guide series: chlorodiphenyls. *Am Ind Hyg Assoc J* 26:92–94, 1965
3. Werter RC et al: Polychlorinated biphenyls (PCBs): dermal absorption, systemic elimination, and dermal wash efficiency. *J Toxicol Environ Health* 12:511–519, 1983
4. Baumann M et al: Effects of polychlorinated biphenyls at low dose levels in rats. *Arch Environ Contam Toxicol* 12:509–515, 1983
5. National Institute for Occupational Safety and Health: *Criteria for a Recommended Standard . . . Occupational Exposure to Polychlorinated Biphenyls*. DHEW (NIOSH) Pub No 77-225. Washington, DC, US Government Printing Office, 1977
6. *IARC Monographs on the Evaluation of the Carcinogenic Risk of Chemicals to Humans, Chemicals, Industrial Processes, and Industries Associated with Cancer in Humans*, Suppl. 4, pp 217–219. Lyon, International Agency for Research on Cancer, 1982

CHLOROFORM
CAS: 67-66-3

CHCl₃

Synonyms: Trichloromethane; methenyl chloride; methane trichloride

Physical Form. Colorless liquid

Uses. In manufacture of fluorocarbons for refrigerants, aerosol propellants, plastics; purifying antibiotics; solvent; photographic processing; dry cleaning

Exposure. Inhalation

Toxicology. Chloroform is a central nervous system depressant and hepatotoxin; renal and cardiac damage may also occur. It is carcinogenic in experimental animals.

Chloroform was abandoned as an anesthetic agent because of the frequency of cardiac arrest during surgery and of delayed death due to hepatic injury.[1] Concentrations used for the induction of anesthesia were in the range of 20,000 to 40,000 ppm, followed by lower maintenance levels.[2] Continued exposure to 20,000 ppm results in respiratory failure, cardiac arrhythmia, and death.[1] Effects of damage to the liver typically are not observed for 24 to 48 hours post-exposure.[1] Symptoms include progressive weakness, prolonged vomiting, delirium, coma, and death. Increased serum bilirubin, ketosis, and lowered blood prothrombin and fibrinogen are reported. Death usually occurs on the fourth or fifth day, and autopsy shows massive hepatic necrosis.

In experimental human exposures, 14,000 to 16,000 ppm caused rapid loss of consciousness; 4100 ppm or less caused serious disorientation, whereas single exposures of 1000 ppm caused dizziness, nausea, and aftereffects of fatigue and headache.[2] Prolonged exposure to concentrations ranging from 77 to 237 ppm caused lassitude; digestive disturbances; frequent, burning urination; and mental dullness; whereas 20 to 70 ppm produced milder symptoms.[3] Of 68 chemical workers exposed regularly to concentrations of 10 to 200 ppm for 1 to 4 years, nearly 25% had hepatomegaly.[1] However, another group exposed repeatedly to about 50 ppm experienced no signs or symptoms.[1]

High concentrations of vapor cause conjunctival irritation and blepharospasm.[4] Liquid chloroform splashed in the eye causes immediate burning pain and conjunctival irritation; the corneal epithelium may be injured, but regeneration is prompt, and the eye returns to normal in 1 to 3 days.[4] Applied to the skin, chloroform causes burning pain, erythema, and vesiculation.[1]

In acute animal studies, target organs identical to those observed in humans (central nervous system, liver, and kidney) have been identified.[5] Recent studies in mice and rats have also shown that exposure to concentrations ranging up to 300 ppm for 6 hr/day for 7 days can produce concentration-dependent lesions in the nasal passages.[6]

Evidence for the carcinogenicity of chloroform in experimental animals following chronic oral administration includes statistically significant increases in renal epithelial tumors in male rats, hepatocellular carcinomas in mice, and renal tumors in male mice.[7-9] Chloroform also has been shown to promote growth and metastasis of murine tumors.[10] In these studies, the carcinogenicity of chloroform is organ-specific primarily to the liver and kidneys; these organs also are the target of acute chloroform toxicity and covalent binding by reactive intermediates (phosgene, carbene, chlorine ion) of chloroform metabolism.[1] Typically, doses of chloroform that do not produce necrosis are not carcinogenic. This suggests that the increased proliferation of liver and kidney cells during regeneration after necrosis may be involved in the development of tumors.[5] Furthermore, most studies suggest that chloroform is not genotoxic, which supports the case for an epigenetic mechanism of carcinogenicity. Chloroform also possesses antitumorigenic properties when administered in the drinking water of animals previously treated with the hepatocarcinogens ethylnitrosourea and ethylnitrosamine, or 1,2-dimethylhydrazine, a gastrointestinal tract carcinogen.[11]

Small increases in rectal, bladder, and colon cancer have been observed in several studies of human populations with chlorinated drinking water. Because other possible carcinogens were present along with chloroform, it is impossible to identify chloroform as the sole carcinogenic agent. Based on sufficient animal evidence and limited epidemiological evidence, the IARC regards chloroform as a probable human carcinogen.[12]

In animals, chloroform causes some fetal loss and delays in fetal development when administered during gestation at levels of 100 ppm or more.[13] Teratogenic effects such as cleft palate were observed in the mouse only at doses associated with maternal toxicity.[14]

Several substances alter the toxicity of chloroform in animals—most probably by modifying the metabolism to a reactive intermediate.[1] Factors that potentiate chloroform's toxic effects include ethanol, polybrominated biphenyls, steroids, and ketones. Disulfiram, its metabolites, and a high-carbohydrate diet appear to protect somewhat against chloroform toxicity.[1]

The 1995 ACGIH threshold limit value–time-weighted average (TLV-TWA) for chloroform is 10 ppm (49 mg/m^3) with an A2 suspected human carcinogen designation.

REFERENCES

1. US Environmental Protection Agency: Health assessment document for chloroform. Final report. Washington, DC, Office of Health and Environmental Assessment, September 1985
2. National Institute for Occupational Safety and Health: *Criteria for a Recommended Standard . . . Occupational Exposure to Chloroform.* DHEW (NIOSH) 75-114. Washington DC, US Government Printing Office, 1974
3. Challen PJ, Hickish DE, Bedford J: Chronic chloroform intoxication. *Br J Ind Med* 15: 243–249, 1958
4. Winslow SG, Gerstner HB: Health aspects of chloroform—a review. *Drug Chem Toxicol* 1:259–275, 1978
5. Agency for Toxic Substances and Disease Registry (ATSDR): *Toxicological Profile for Chloroform.* US Department of Health and Human Services, Public Health Service, TP-92/07, p 141, 1993
6. Mery S, Larson JL, Butterworth BE, et al: Nasal toxicity of chloroform in male F-344 rats and female B6C3F1 mice following a 1-week inhalation exposure. *Toxicol Appl Pharmacol* 125:214-227, 1994
7. National Cancer Institute (NCI). *Report on Carcinogenesis Bioassay of Chloroform.* PB-264018. Springfield, VA, National Technical Information Service, 1976
8. Jorgenson TA, Meierhenry EF, Rushbrook CJ, et al: Carcinogenicity of chloroform in drinking water to male Osborne Mendel rats and female B6C3F$_1$ mice. *Fund Appl Toxicol* 5:760–769, 1985
9. Roe FJC, Palmer AK, Worden AN: Safety evaluation of toothpaste containing chloro-

form. I. Long-term studies in mice. *J Environ Pathol Toxicol* 2:799–819, 1979

10. Capel ID, Dorrell HM, Jenner M, et al: The effect of chloroform ingestion on the growth of some murine tumors. *Eur J Cancer* 15:1485–1490, 1979

11. Daniel FB, DeAngelo AB, Stober JA, et al: Chloroform inhibition of 1,2-dimethylhydrazine-induced gastrointestinal tract tumors in the Fischer 344 Rat. *Fund Appl Toxicol* 13:40–45, 1989

12. *IARC Monographs on the Evaluation of the Carcinogenic Risk of Chemicals to Man*, Vol 20, *Some halogenated hydrocarbons*, pp 401–417. Lyon, International Agency for Research on Cancer, 1979

13. Murray FA, Schwetz BA, McBride JB, et al: Toxicity of inhaled chloroform in pregnant mice and their offspring. *Toxicol Appl Pharmacol* 50:515–522, 1979

14. Schwetz BA, Leong BKJ, Gehring PJ: Embryo- and fetotoxicity of inhaled chloroform in rats. *Toxicol Appl Pharmacol* 25:442–451, 1974

bis(CHLOROMETHYL) ETHER
CAS: 542-88-1

ClCH₂OCH₂Cl

$ClCH_2OCH_2Cl$

Synonyms: BCME; chloromethyl ether; chloro-(chloromethoxy)methane; dichloromethyl ether; symmetrical-dichloro-dimethyl ether; dimethyl-1-1'-dichloroether

Physical Form. Colorless liquid

Uses. Chemical intermediate

Exposure. Inhalation; skin absorption

Toxicology. Bis(chloromethyl) ether (BCME) is a mucous membrane and respiratory irritant; it is a recognized human carcinogen.

In humans, concentrations of 3 ppm are reported to be distinctly irritating. A fatal case of accidental acute poisoning of a research chemist by BCME has been reported.[1] Increased frequency of chronic cough and low-end expiratory flow rates have been described occurring in a dose-related fashion with exposure to BCME and CMME.[2]

A retrospective study of 136 BCME workers employed at least 5 years revealed 5 cases of lung cancer, which represented a ninefold increase in lung cancer risks; 0.54 cases would have been expected to occur in the plant population.[3] The predominant histologic type of carcinoma was small-cell–undifferentiated; exposure ranged from 7.5 to 14 years, and the mean induction period was 15 years. In addition, abnormal sputum cytology was observed in 34% of 115 current workers with exposure to BCME for 5 or more years, contrasted with 11% in a control group.

In another study, 6 cases of lung cancer occurred among 18 experimental technical department workers, a group known to experience very high BCME exposure; other cases of lung cancer were reported among 50 production workers.[4] Oat-cell carcinomas occurred in 5 of 8 cases.

BCME is also found as an impurity (1% to 7%) in the related chloromethyl methyl ether (CMME); 14 cases of lung cancer, mainly of oat-cell type, were reported in a chemical plant where exposure to CMME occurred.[5] In the reported epidemiologic studies, insufficient evidence is available to separate the carcinogenic effects of the two compounds.[6]

A follow-up of CMME (BCME) workers found no increased risk of respiratory cancer among those exposed less than 1 year, with up to a twelvefold increase among those exposed 10 years or more.[7] Latency did not appear to be inversely related to dose but, instead, peaked at approximately 20 years from initial exposure.

Features implicating BCME as the primary causative agent include: (1) early age at death, (2) development of lung cancer among nonsmokers as well as cigarette smokers, and (3) unusual histological-type small-cell or oat-cell carcinoma rather than the squamous-cell carcinoma common among male smokers.

Exposure to 1 ppm for 6 hr/day, 5 days/week, for 82 days caused lung tumors in 26 of 47 animals, with an average of 5.2 tumors per tumor-bearing animal; 20 of 49 controls developed lung tumors, with 2.2 tumors per tumor-bearing animal.[8] In 19 rats exposed to 0.1 ppm BCME 6 hr/day for 101 exposures, five squamous-cell carcinomas of the lung and five esthesioneuroepithelliomas arising from the olfactory epithelium were observed.[9] Cutaneous application of 2 mg BCME applied to mice 3 times a week for 325 days caused papillomas in 13 of 20 animals; 12 of these papillomas progressed to squamous-cell carcinomas.[10]

The IARC has concluded that there is sufficient evidence of carcinogenicity of BCME to both humans and animals.[11]

The 1995 ACGIH threshold limit value–time-weighted average (TLV-TWA) for bis(chloromethyl)ether is 0.001 ppm (0.0047 mg/m^3) with an A1 confirmed human carcinogen designation.

REFERENCES

1. Environmental Protection Agency: Ambient water quality criteria for chloroalkyl ethers, pp C-25, C-27. Springfield, VA, NTIS (USEPA), 1980
2. Weiss W: Chloromethyl ethers, cigarettes, cough and cancer. *J Occup Med* 18:194–199, 1976
3. Lemen RA et al: Cytologic observations and cancer incidence following exposure to BCME. *Ann NY Acad Sci* 271:71–80, 1976
4. Theiss AM, Hay W, Zeller H: Zur Toxikologie von Dichlorodimethylather—Verdacht, auf kanzerogene Wirking auch beim Menschen. (Toxicology of bis(chloromethyl)ether—suspicion of carcinogenicity in man.) Zbl Arbeitsmed 23:97–102, 1973
5. Figueroa WG, Raszkowski R, Weiss W: Lung cancer in chloromethyl methyl ether workers. *New Eng J Med* 288:1096–1097, 1973
6. *IARC Monographs on the Evaluation of the Carcinogenic Risk of Chemicals to Man*, Vol 4, Some aromatic amines, hydrazine and related substances, N-nitroso compounds and miscellaneous alkylating agents, pp 231–238. Lyon, International Agency for Research on Cancer, 1974
7. Maher KV, DeFonso LR: Respiratory cancer among chloromethyl ether workers. *J Natl Cancer Inst* 78:839–843, 1987
8. Leong BKJ, Macfarland HN, Reese WH Jr: Induction of lung adenomas by chronic inhalation of bis(chloromethyl)ether. *Arch Environ Health* 22:663–666, 1971
9. Laskin S et al: Tumors of the respiratory tract induced by inhalation of bis(chloromethyl)-ether. *Arch Environ Health* 23:135–136, 1971
10. Van Duuren BL et al: Alpha-haloethers: a new type of alkylating carcinogen. *Arch Environ Health* 16:472–476, 1968
11. *IARC Monographs on the Evaluation of Carcinogenic Risks to Humans, Overall Evaluations of Carcinogenicity: An Updating of IARC Monographs Vols 1–42*, Suppl 7, pp 131–132. Lyon, International Agency for Research on Cancer, 1987

CHLOROMETHYL METHYL ETHER
CAS: 107-30-2

C_2H_5ClO

Synonyms: CMME; dimethylchloroether; methyl chloromethyl ether

Physical Form.   Colorless liquid

Uses.   Chemical intermediate; preparation of ion-exchange resins

Exposure.   Inhalation

Toxicology.   Chloromethyl methyl ether (CMME) exposure has been associated with an increased incidence of human lung cancer.

Among 111 CMME workers observed during a 5-year period, there were four cases of lung cancer; this was eight times the incidence of a control group of plant workers with similar smoking histories.[1] Evidence of a lung cancer risk was further supported by the retrospective identification of a total of 14 cases among chemical operators in a plant engaged in synthesis of CMME. Except for one case of doubtful exposure, the duration of exposure was 3 to 14 years, and the age at diagnosis ranged from 33 to 55 years. During the synthetic process, fumes were often visible. The employees considered it a good day if the entire building had to be evacuated only three or four times per 8-hour shift because of noxious fumes. Three of the men diagnosed with lung cancer had never smoked, and one had smoked a pipe only; the other 10 had smoked one or more packs of cigarettes per day. Oat-cell carcinoma was histologically confirmed in 12 cases, whereas the doubtful exposure case was squamous-cell carcinoma; the cell type in one case was not determined.[1]

In another study of 669 workers exposed during 1948 to 1972, 19 died of lung cancer although only 5.6 cases were expected.[2] There were higher relative risks for workers exposed to intermediate to high levels of CMME for 1 or more years.[2]

In a study of 276 men exposed to CMME and followed through 1980 at a plant in the United Kingdom in operation since 1948, there were 10 deaths from lung cancer, with a relative risk of 10.97 compared to an unexposed group.[3] The occurrence of lung cancer appeared to be related to both the estimated exposure level and the duration of exposure. Among a subgroup of 51 workers who began work after the process was enclosed in 1972, no deaths from lung cancer had been observed through 1980. In another factory where 394 men had been exposed to CMME at lower estimated exposure levels, no excess of lung cancer was observed.[3]

Lung cancer occurred at a higher rate among potentially exposed CMME workers at a factory in France (rate ratio 5.0 compared to nonexposed workers and 7.6 compared to an external referent population).[4] The average age at diagnosis was 10.5 years lower than nonexposed cases, and the predominately small-cell cancers of the exposed were mostly oat-cell.

It should be noted that commercial CMME contains 1% to 7% of highly carcinogenic bis(chloromethyl) ether (BCME); in the reported epidemiological studies, insufficient evidence is available to differentiate the carcinogenic effects of the two compounds.[5] Further, when CMME is hydrolyzed, HCl and formaldehyde are produced, which may recombine to form BCME. Therefore, although findings may reflect the carcinogenicity of BCME, commercial grade CMME also must be considered to be a carcinogen, although perhaps of a lower potency than that of BCME.

CMME is a mucous membrane and respiratory irritant in both humans and animals.[6,7] Human exposure to CMME has been reported to cause breathing difficulties, sore throat, fever, and chills.[6] Acute exposure of rats and hamsters resulted in pulmonary edema and hemorrhage and necrotizing bronchitis.[7] An increased frequency of chronic cough and of low-end expiratory flow rates has been observed in a dose-related fashion with exposure to CMME and BCME.[8]

Technical-grade CMME (contaminated with BCME), on subcutaneous injection in mice, has produced local sarcomas.[5] Dermal application of mice, followed by a phorbol ester promoter, resulted in an apparent excess of skin papillomas and carcinomas. Inhalation studies in mice showed an equivocally increased occurrence of lung tumors compared to unexposed controls.[5]

The IARC has concluded that there is sufficient evidence for carcinogenicity of technical-grade CMME to both humans and animals.[9]

ACGIH has designated chloromethyl methyl ether as an A2 suspected human carcinogen without an assigned numerical threshold limit value (TLV).

REFERENCES

1. Figueroa WG, Raszkowski R, Weiss W: Lung cancer in chloromethyl methyl ether workers. *New Eng J Med* 288:1096–1097, 1973
2. DeFonso LR, Kelton SC Jr: Lung cancer following exposure to chloromethyl methyl ether. *Arch Environ Health* 31:125–130, 1976
3. McCallum RI, Woolley V, Petrie A: Lung cancer associated with chloromethyl methyl ether manufacture: an investigation at two factories in the United Kingdom. *Br J Ind Med* 40:384–389, 1983
4. Gowers DS, DeFonso LR, Schaffer P, et al: Incidence of respiratory cancer among workers exposed to chloromethyl-ethers. *Am J Epidemiol* 137:31–42, 1993
5. *IARC Monographs on the Evaluation of the Carcinogenic Risk of Chemicals to Man*, Vol 4, Some

aromatic amines, hydrazine and related substances, N-nitroso compounds and miscellaneous alkylating agents, pp 239–244. Lyon, International Agency for Research on Cancer, 1974
6. Van Duuren BL et al: Alpha-haloethers: a new type of alkylating carcinogen. *Arch Environ Health* 16:472–476, 1968
7. *Ambient Water Quality Criteria for Chloroalkyl Ethers*, pp C-25, C-27. Springfield, VA, National Technical Information Service, US Environmental Protection Agency, 1980
8. Weiss W: Chloromethyl ethers, cigarettes, cough and cancer. *J Occup Med* 18:194–199, 1976
9. *IARC Monographs on the Evaluation of the Carcinogenic Risk of Chemicals to Humans*, Suppl 4, pp 64–66. Lyon, International Agency for Research on Cancer, 1982

nary edema and cellular necrosis of the heart, liver, and kidneys.[2-3]

The 1995 ACGIH threshold limit value–time weighted average (TLV-TWA) for 1-chloro-1-nitropropane is 2 ppm (10 mg/m^3).

REFERENCES

1. Stokinger HE: Aliphatic nitro compounds, nitrates, nitrites. In Clayton GD, Clayton FE (eds): *Patty's Industrial Hygiene and Toxicology*, 3rd ed, rev, Vol 2C, *Toxicology*, pp 4162–4164. New York, Wiley-Interscience, 1982
2. Machle W et al: The physiological response of animals to certain mononitroparaffins. *J Ind Hyg Toxicol* 27:95–102, 1945
3. Browning E: *Toxicity and Metabolism of Industrial Solvents*, pp 292–293. Amsterdam, Elsevier, 1965

1-CHLORO-1-NITROPROPANE
CAS: 600-25-9

$CH_3CH_2CHClNO_2$

Synonyms: None

Physical Form. Colorless liquid

Uses. Fungicide

Exposure. Inhalation

Toxicology. 1-Chloro-1-nitropropane is an irritant of the eyes and mucous membranes. It is a pulmonary irritant in animals, and severe exposure is expected to cause the same effect in humans.

Systemic effects in humans have not been reported.

The lethal oral dose in rabbits is 0.05 to 0.10 g/kg, which is approximately 5 times more toxic than the unchlorinated mononitroparaffin.[1] Rabbits exposed to 2600 ppm for 2 hours died, but 2200 ppm for 1 hour was nonlethal. Effects included irritation of the eyes and mucous membranes, and autopsy revealed pulmo-

CHLOROPENTAFLUOROETHANE
CAS: 76-15-3

C_2ClF_5

Synonyms: F-115; FC 115; Refrigerant 115; Propellant 115; Freon 115

Physical Form. Colorless gas

Uses. Refrigerant; aerosol propellant

Exposure. Inhalation

Toxicology. Chloropentafluoroethane has low inhalation toxicity and little potential for cardiac sensitization.

Inhalation studies with chloropentafluoroethane in anesthetized dogs, rats, and monkeys showed that exposure to 100,000 to 250,000 ppm, under certain conditions, caused an increase in blood pressure, accelerated heart rate, and depression of myocardial contractility and sensitized the heart to epinephrine.[1-3] Compared to other chlorofluorocarbons, it is ranked among the least potent for cardiac sensitization.[4]

In a NIOSH Health Hazard Evaluation of refrigeration workers exposed far below the TLVs for chloropentafluoroethane and chlorodifluoromethane, 27 workers were medically evaluated.[5,6] Seventy-one percent complained of dizziness and light-headedness compared to 21% of controls. Palpitations were reported in 36% of the exposed workers and none of the nonexposed. No clinical neurological or electroneurophysiological abnormalities were detected in eight of the refrigeration repair workers followed up for 3 years during continuous employment.[6]

Death from acute respiratory arrest has occurred following intentional inhalation of an azeotrophic mixture of chlorodifluoromethane and chloropentafluoroethane.[7]

REFERENCES

1. Belej MA, Aviado DM: Cardiopulmonary toxicity of propellants for aerosols. *J Clin Pharmacol* 15:105–115,1975
2. Aviado DM, Belej MA: Toxicity of aerosol propellants in the respiratory and circulatory systems. V. Ventricular function in the dog. *Toxicology* 3:79–86, 1975
3. Friedman SA, Cammarato M, Aviado DM: Toxicity of aerosol propellants in the respiratory and circulatory systems. II. Respiratory and bronchopulmonary effects in the rat. *Toxicology* 1:345–355, 1973
4. Aviado DM: Toxicity of aerosol propellants in the respiratory and circulatory system. 10. Proposed classification. *Toxicology* 3:321–332, 1975
5. Health Hazard Evaluation Report No HETA-81-043-1207, Refrigeration workers. Salt Lake City, UT, DHHS, NIOSH, Cincinnati, OH, 1981
6. Campbell DD, Lockey JE, Petajan JH, et al: Health effects among refrigeration repair workers exposed to fluorocarbons. *Br J Ind Med* 43:107–111, 1986
7. Fitzgerald RL, Fishel CE, Bush LLE: Fatality due to recreational use of chlorodifluoromethane and chloropentafluoroethane. *J Forens Sci* 38:476–482, 1993

CHLOROPICRIN
CAS: 76-06-2

CCl_3NO_2

Synonyms: Trichloronitromethane; nitrochloroform

Physical Form. Colorless, slightly oily liquid

Uses. Fumigant for cereals and grains; soil insecticide; war gas

Exposure. Inhalation

Toxicology. Chloropicrin is a severe irritant of the eyes, mucous membranes, skin, and lungs.

A lethal exposure for humans is stated to be 119 ppm for 30 minutes, with death resulting from pulmonary edema. Particular injury occurs in the medium and small bronchi.[1] In addition to pulmonary irritation, human exposure results in lacrimation, cough, nausea, vomiting, and skin irritation; persons injured by inhalation of chloropicrin vapor are said to be more susceptible to subsequent exposures.[1]

A concentration of 15 ppm could not be tolerated longer than 1 minute, even by persons acclimated to chloropicrin; exposure to 4 ppm for a few seconds is temporarily disabling because of irritant effects. Concentrations of 0.3 to 0.37 ppm have resulted in painful eye irritation in 3 to 30 seconds.[1]

A man accidentally exposed to residual spray of undetermined concentration had dry cough, and his nasal and pharyngeal mucosa were red and edematous.[2]

In mice, exposure to 9 ppm caused a 50% decrease in respiratory rate. Lesions included ulceration and necrosis of the respiratory epithelium and moderate damage to lung tissue.[3] Rats administered, via oral gavage, 10, 20, 40, or 80 mg/kg for 10 consecutive days or 32 mg/kg for 90 consecutive days had inflammation, necrosis, acantholysis, hyperkeratosis, and epithelial hyperplasia of the forestomach.[4]

The 1995 ACGIH threshold limit value–

time-weighted average (TLV-TWA) for chloropicrin is 0.1 ppm (0.67 mg/m³).

REFERENCES

1. Stokinger HE: Aliphatic nitro compounds, nitrates, nitrites. In Clayton GD, Clayton FE (eds): *Patty's Industrial Hygiene and Toxicology*, 3rd ed, Vol 2C, *Toxicology*, pp 4164–4166. New York, Wiley-Interscience,1982
2. TeSlaa G, Kaiser M, Biederman L, et al: Chloropicrin toxicity involving animal and human exposure. *Vet Hum Toxicol* 28:323–324, 1986
3. Buckley LA et al: Respiratory tract lesions induced by sensory irritants at the LD₅₀ concentrations. *Toxicol Appl Pharmacol* 74:417–429, 1984
4. Condie LW, Daniel FB, Olson GR, et al: Ten and ninety day toxicity studies of chloropicrin in Sprague-Dawley rats. *Drug Chem Toxicol* 17:125–137, 1994

β-CHLOROPRENE
CAS: 126-99-8

C_4H_5Cl

Synonyms: Chlorobutadiene; 2-chloro-1,3-butadiene; chloroprene

Physical Form.   Colorless liquid

Uses.   In the manufacture of synthetic rubber

Exposure.   Inhalation; skin absorption

Toxicology.   Chloroprene causes central nervous system abnormalities as well as skin and eye irritation. Reproductive, mutagenic, embryotoxic, and carcinogenic effects have been reported.

Exposure of workers to high concentrations for short periods led to temporary unconsciousness; one fatality occurred after a 3- to 4-minute exposure inside an unventilated polymerization vessel containing chloroprene vapor.[1] Experimental exposure of humans to 973 ppm led to nausea and giddiness in resting subjects in 15 minutes and in subjects performing light work in 5 to 10 minutes.[1] Extreme fatigue and unbearable chest pain occurred after approximately 1 month of exposure to levels ranging from 56 ppm to greater than 334 ppm. Irritability, personality changes, and reversible hair loss also were reported

A significant rise in the number of chromosome aberrations was observed in a number of studies of workers exposed to chloroprene at 5 ppm and below.[2] Functional disturbances in spermatogenesis and morphological abnormalities of sperm were observed among workers occupationally exposed to 0.28 and 1.94 ppm chloroprene.[3,4] A threefold excess of miscarriages in the wives of these workers also was reported.

Two Russian studies suggested an increased incidence of lung and skin cancers in chloroprene-exposed workers compared to a variety of control groups.[5] A US study of cancer mortality among two cohorts of males engaged in the production and/or polymerization of chloroprene concluded that there was no significant excess of lung cancer deaths.[6] However, there was a disproportionately high incidence of lung cancer cases in maintenance workers who had potentially high exposure to chloroprene. A recent study of chloroprene workers confirmed a significant increased risk for liver, lung, and lymphatic cancers among maintenance mechanics who have the highest occupational chloroprene exposure.[7] A dose-response relationship appeared to exist in a cohort of 1213 workers, with low-exposure groups having low SMRs and high-exposure groups having the highest risk of cancer.[2] Since most reported effects have involved mixed exposures to multiple substances and to short-chain polymers of chloroprene, the reported symptoms cannot all be assigned to the monomer alone.[5]

Contact with skin may cause chemical burns. Conjunctivitis and focal necrosis of the cornea have been reported from eye exposure.[5]

In acute animal studies, the concentrations that killed at least 70% of animals with an 8-hour exposure were 170 ppm for mice, 700 ppm for cats, 2000 ppm for rabbits, and 4000 to 6000 ppm for rats.[8] Symptoms included inflammation of the mucous membranes of the eyes and nose, followed by central nervous system depression and death from respiratory failure. Repeated exposure of rats 6 hr/day, 5 days/week, for 4 weeks caused skin and eye irritation and growth depression at 40 ppm; at 160 and 625 ppm, it resulted in loss of hair, morphologic liver damage, and increased mortality.[9]

Exposure of male rats at concentrations of 120 to 6277 ppm and of male mice at concentrations of 12 to 152 ppm for 8 hours resulted in sterility or impotence in 13 of 19 rats and in 8 of 14 mice versus a mean of 0.5 in the two control groups.[7] Degenerative changes in the testes were observed in some of the exposed animals. A significant increase in embryonic mortality was observed in female rats fertilized by males exposed to 1 ppm 4 hr/day for 48 days.[2,5] Hydrocephalus and cerebral herniation occurred in all fetuses from rat dams given oral doses of 0.5 mg/kg during 14 days of pregnancy. Inhalation of 1.11 ppm for 2 days of pregnancy also caused increases in these anomalies.[2,5] In a nother study, neither embryotoxic nor convincing teratological effects were found after exposing rats at 1, 10, or 25 ppm.[10]

Chloroprene did not cause a significant increase in chromosomal aberrations or sister chromatid exchanges in mice treated 6 hr/day for 12 days with up to 80 ppm; however, the mitotic index in the bone marrow of exposed mice was significantly increased.[11,12]

No carcinogenic effects were found in rats following oral, subcutaneous, or intratracheal administration, or in mice by skin application.[2] The IARC working group considered the studies inadequate to evaluate the carcinogenic risk in animals.[2]

The 1995 ACGIH threshold limit value–time-weighted average (TLV-TWA) for β-chloroprene is 10 ppm (36 mg/m³) with a notation for skin absorption.

REFERENCES

1. Nystrom AE: Health hazards in the chloroprene industry and their prevention. *Acta Med Scand* Suppl 132:5–125, 1948
2. *IARC Monographs on the Evaluation of the Carcinogenic Risk of Chemicals to Humans.* Vol 19, Some monomers, plastics and synthetic elastomers, and acrolein, pp 131–156. Lyon, International Agency for Research on Cancer, 1979
3. Sanotskii IV: Aspects of the toxicology of chloroprene: immediate and long-term effects. *Environ Health Perspect* 17:85–93, 1976
4. Baranski B: Effects of the workplace on fertility and related reproductive outcomes. *Environ Health Perspect* (Suppl) 101:81–90, 1993
5. National Institute for Occupational Safety and Health: *Criteria for a Recommended Standard . . . Occupational Exposure to Chloroprene.* DHEW (NIOSH) Pub No 77-210, pp 1–176. Washington DC, US Government Printing Office, 1977
6. Pell S: Mortality of workers exposed to chloroprene. *J Occup Med* 20:21–29, 1978
7. Li SQ, Dong QN, Liu YQ, et al: Epidemiologic study of cancer mortality among chloroprene workers. *Biomed Environ Sci* 2:141–149, 1989
8. von Oettingen WF et al: 2-Chloro-butadiene (chloroprene): its toxicity and pathology and the mechanism of its action. *J Ind Hyg Toxicol* 18:240–270, 1936
9. Clary JJ et al: Toxicity of β-Chloroprene (2-chlorobutadiene-1,3): acute and subacute toxicity. *Toxicol Appl Pharmacol* 46:375–384, 1978
10. Culik R et al: β-Chloroprene (2-chlorobutadiene-1,3) embryotoxic and teratogenic studies in rats (abstract 194). *Toxicol Appl Pharmacol* 37:172, 1976
11. Tice RR, Boucher R, Luke CA, et al: Chloroprene and isoprene: cytogenic studies in mice. *Mutagenesis* 3:141–146, 1988
12. Shelby MD: Results of NTP-sponsored mouse cytogenic studies on 1,3-butadiene, isoprene, and chloroprene. *Environ Health Perspect* 86:71–73, 1990

o-CHLOROSTYRENE
CAS: 2039-87-4

C_8H_7Cl

Synonyms: 2-Chlorostyrene

Physical Form. Liquid

Uses. Organic synthesis; preparation of specialty polymers

Exposure. Inhalation

Toxicology. By analogy to styrene, *o*-chlorostyrene is expected to cause central nervous system depression and possibly irritation of the eyes, nose, and mucous membranes.

In an inhalation study, groups of rats, rabbits, guinea pigs and one dog were exposed to 101 ppm *o*-chlorostyrene 7 hr/day, 5 days/week, for 6 months.[1] There were no adverse effects on any species.

The 1995 threshold limit value–time-weighted average (TLV-TWA) is 50 ppm or 283 mg/m^3 with a short-term excursion level of 100 ppm or 425 mg/m^3.

REFERENCES

1. Torkelson, TR: Summary report of toxicity data to TLV Committee. The Dow Chemical Company, Biochemical Research Laboratories, Midland, MI, February 15, 1967

CHLOROTHALONIL
CAS: 1897-45-6

$C_6Cl_4CN_2$

Synonyms: Chloroalonil; 1,3-dicyanotetrachlorobenzene; tetrachlorisophthalonitrile

Physical Form. Colorless, odorless crystals

Uses. Fungicide

Exposure. Inhalation; ingestion; skin absorption

Toxicology. Chlorothalonil is an irritant to the skin and eyes and has been reported to produce allergic contact dermatitis in exposed workers.

Patch testing demonstrated that between 10% and 28% of 88 Japanese farmers were sensitive to chlorothalonil and other pesticides. Thirty-five of these farmers had acute dermatitis. Photosensitization was involved in some cases. Reactions were also observed in greenhouse workers, vegetable farmers, and others

with pesticide-induced dermatitis. Four cases of severe recurrent contact dermatitis have been reported in workers exposed to chlorothalonil-containing wood preservatives.[1]

Whether chlorothalonil is a true dermal sensitizer in humans or strictly a skin irritant remains controversial.[2] Some investigators suggest that repeated exposure results in an enhanced irritant response, while others suggest that it is a potent contact allergen.[2] It is noted that relatively few cases of allergy to chlorothalonil have been reported despite widespread use for over 20 years. Furthermore, at a plant that produces the chemical, cases of work-related contact dermatitis have not been reported for years following adoption of good hygienic practices.[2]

The oral LD$_{50}$ of chlorothalonil in female mice is 6000 mg/kg whereas in rats, it is greater than 10,000 mg/kg.[1]

Technical-grade chlorothalonil was tested for possible carcinogenicity in rats and mice. In rats, chlorothalonil was administered in the diet at average doses of 5063 and 10,126 ppm for 80 weeks, followed by observation for 30 to 31 weeks. Adenomas and carcinomas of the renal tubular epithelium were observed.[3] Mice were administered time-weighted average doses of 2688 or 5375 ppm (males) or 3000 or 6000 ppm (females). No evidence of carcinogenicity was found in mice.[3]

The International Agency for Research on Cancer has stated that there is limited evidence that chlorothalonil is carcinogenic to experimental animals, based on the above study. No data are available to evaluate chlorothalonil for possible carcinogenicity to humans, but the Environmental Protection Agency has classified it as a probable human carcinogen.[1,4]

Chlorothalonil was not mutagenic in two fungal systems. It did not induce reversion mutations in *Salmonella typhimurium*.[1]

An ACGIH threshold limit value has not been adopted for chlorothalonil.

REFERENCES

1. *IARC Monographs on the Evaluation of Carcinogenic Risks to Humans*, Vol. 30, Miscellaneous

pesticides, pp 319–326. Lyon, International Agency for Research on Cancer, 1983
2. Eilrich GL, Chelsky M: Letters to the editor. *Contact Dermatitis* 25:141–144, 1991
3. National Cancer Institute: *Bioassay of Chlorothalonil for Possible Carcinogenicity*, pp vii–viii. Bethesda, MD, National Institutes of Health, DHEW (NIH) Publ No 78-841, 1978
4. Environmental Protection Agency (EPA): Updated list of classified carcinogenic pesticides. *Pesticide Toxic Chem News* 18:33–36, 1990

---

o-CHLOROTOLUENE
CAS: 95-49-8

C_7H_7Cl

Synonyms: 2-Chloro-1-methylbenzene; 2-chlorotoluene; *o*-tolyl chloride

Physical Form. Colorless liquid

Uses. Solvent; synthesis of dyes, pharmaceuticals, and synthetic rubber compounds

Exposure. Inhalation

Toxicology. *o*-Chlorotoluene causes central nervous system depression in animals and is expected to cause similar effects in humans.

Rats exposed for 6 hours to 4000 ppm became uncoordinated in 1.5 hours, followed in another half-hour by prostration and tremor.[1] Rats exposed to 14,000 ppm exhibited incoordination, vasodilation, labored respiration, and narcosis, all of which were reversible.

In another study, mice, rats, and guinea pigs were exposed to 4400 ppm.[2] Mice developed gasping, ataxia, and convulsions after 30 minutes of exposure. Rats and guinea pigs showed gasping, hyperpnea, ataxia, and convulsions after 45 minutes of exposure. All animals were comatose in 60 minutes. All mice and rats died, as did 7 of 10 guinea pigs.

Moderate skin irritation was noted on rabbit skin after application for 24 hours. A single instillation into the eyes of rabbits produced moderate conjunctival irritation, which was reversible by the fifth day of observation.

The 1995 threshold limit value–time-weighted average (TLV-TWA) is 50 ppm (259 mg/m³).

REFERENCES

1. Ely, T: Communication to TLV Committee from Laboratory of Industrial Medicine, Eastman Kodak Co, Rochester, NY, November 3, 1971
2. Hazleton Laboratories, Inc.: Acute inhalation exposure—rats, mice, and guinea pigs; primary skin irritation—rabbits; acute eye irritation—rabbits. Orthochlorotoluene. Final Report. Projects No 157-147 and 157-148. Hazleton Laboratories, Inc, Falls Church, VA, June 1, 1966

CHLORPYRIFOS
CAS: 2921-88-2

$C_9H_{11}C_{13}NO_3PS$

Synonyms: O,O-Diethyl-O-(3,5,6-trichloro-2-pryidinyl) phosphorothioate; Dursban; Dowco 179; ENT 27311; Eradex; Lorsban; NA 2783; OMS-0971; Pyrinex

Physical Form. White crystalline solid

Uses. Insecticide

Exposure. Inhalation; skin absorption; ingestion

Toxicology. Chlorpyrifos is an anticholinesterase agent but has only moderate capacity to reduce red blood cell cholinesterase and little capacity to cause systemic injury.

Signs and symptoms of overexposure are due to the inactivation of the enzyme cholinesterase, which in turn results in the accumulation of acetylcholine at synapses in the nervous system, skeletal and smooth muscles, and secretory glands. The sequence of the development of systemic effects varies with the route of entry.

The onset of signs and symptoms usually is prompt but may be delayed up to 12 hours.[1-5]

Chlorpyrifos does not have enough vapor pressure to present a vapor hazard; however, if it is dispersed as a mist, particulate inhalation is possible. Five of seven spray workers exposed to 0.5% chlorpyrifos emulsion showed more than 50% reduction in cholinesterase within 2 weeks.[5] Symptoms were not reported.

Human subjects who ingested chlorpyrifos once daily for 4 weeks showed depression of plasma cholinesterase but were symptomless at a dose of 0.1 mg/kg.[5] When four repeated doses were applied to the skin of human volunteers for 12 hours each, doses of 25 mg/kg depressed plasma cholinesterase but caused no symptoms. Chlorpyrifos and its principal metabolite, 3,5,6-trichloro-2-pyridinol, are rapidly eliminated, predominantly in the urine.[6]

An 8.5-year morbidity survey of employees engaged in the manufacture of chlorpyrifos did not show any statistically significant differences in illness or prevalence of symptoms between exposed and unexposed groups.[7] Potentially exposed employees did report symptoms of dizziness, malaise, and fatigue relatively more often than did subjects from the comparison group; however, there was no relationship of their symptoms to exposure levels.

In animal studies, repeated inhalation of chlorpyrifos at 287 $\mu g/m^3$ (near the theoretical maximum vapor concentration) for 13 weeks caused no treatment-related changes in urinalysis, hematology, clinical chemistry, terminal body and organ weights, or pathology.[8] Induction of delayed polyneuropathy in animals occurs only at doses that exceed the LD_{50}.[9]

Chlorpyrifos was not carcinogenic in rats fed up to 3.0 mg/kg/day for 2 years.[10]

There was no evidence of teratological or reproductive effects in rats fed 1.0 mg/kg/day during a three-generation study; it was not teratogenic in mice at gavage doses of up to 25 mg/kg/day.[11] Repeated intraperitoneal injection of Dursban (active ingredient chlorpyrifos) to pregnant rats at doses of 0.03, 0.1, or 0.3 mg/kg caused increased incidences of embryolethality, physical abnormalities, and early postnatal neurotoxicity.[12] It was noted that the method of exposure (ip) and solvents present

in Dursban may have contributed to the adverse effects.

The persistent strong odor is probably due to the sulfur content of the pesticide.

The 1995 ACGIH threshold limit value–time-weighted average (TLV-TWA) for chlorpyrifos is 0.2 mg/m^3 with a notation for skin absorption.

Note: For a description of diagnostic signs, differential diagnosis, and medical control, including clinical laboratory tests as well as specific treatment of overexposure to anticholinesterase insecticides, see "Parathion."

REFERENCES

1. Koelle GB (ed): Cholinesterases and anticholinesterase agents. *Handbuch der Experimentellen Pharmakologie*, Vol 15, pp. 989–1027. Berlin, Springer-Verlag, 1963
2. Koelle GB: Anticholinesterase agents. In Goodman LS, Gilman A (eds): *The Pharmacological Basis of Therapeutics*, 5th ed, pp 456–466. New York, Macmillan Co, 1975
3. Hayes WJ Jr: *Clinical Handbook on Economic Poisons. Emergency Information for Treating Poisoning.* US Public Health Service Pub No 476, pp 12–23, 35–37. Washington, DC, US Government Printing Office, 1963
4. Namba T, Nolte CT, Jackrel J, Grob D: Poisoning due to organophosphate insecticides. *Am J Med* 50:475, 1971
5. Chlorpyrifos. *Documentation of TLVs and BEIs*, 5th ed, p 138. Cincinnati, OH, American Conference of Governmental Industrial Hygienists (ACGIH), 1986
6. Nolan RJ et al: Chlorpyrifos: pharmacokinetics in human volunteers. *Toxicol Appl Pharmacol* 73:8–15, 1984
7. Brenner FE, Bond GG, McLaren EA, et al: Morbidity among employees engaged in the manufacture or formulation of chlorpyrifos. *Br J Ind Med* 46:133–137, 1989
8. Calhoun RA, Dittenber DA, Lomax LG, et al: Chlorpyrifos: a 13-week nose-only vapor inhalation study in Fischer 344 rats. *Fund Appl Toxicol* 13:616–618, 1989
9. Capodicasa E, Scapellato ML, Moretto A, et al: Chlorpyrifos-induced delayed polyneuropathy. *Arch Toxicol* 65:150–155, 1991
10. McCollister SB, Kociba RJ, Humiston CG, et al: Studies of the acute and long-term oral

toxicity of chlorpyrifos (0,0-diethyl-0-(3,5,6-trichloro-2-pyridyl)phosphorothioate). *Food Cosmet Toxicol* 12:45–61, 1974

11. Deacon MM et al: Embryotoxicity and fetotoxicity of orally administered chlorpyrifos in mice. *Toxicol Appl Pharmacol* 54:31–40, 1980

12. Muto MA, Lobelle F Jr, Bidanset JH, et al: Embryotoxicity and Neurotoxicity in rats associated with prenatal exposure to Dursban. *Vet Hum Toxicol* 34:498–501, 1992

CHROMIUM (METAL AND INORGANIC COMPOUNDS, AS CR)
CAS: 7440-47-3 (Metal)

Cr

Compounds: Chromium can have a valence of 2, 3, or 6. Chromium compounds vary greatly in their toxic and carcinogenic effects. For this reason, ACGIH divides chromium and its inorganic compounds into a number of groupings:

1. *Chromium metals and alloys:* including chromium metal, stainless steels, and other chromium-containing alloys.
2. *Divalent chromium compounds* (Cr^{2+}) (chromous compounds): including chromous chloride ($CrCl_2$) and chromous sulfate ($CrSO_4$).
3. *Trivalent chromium compounds* (Cr^{3+}) (chromic compounds): including chromic oxide (Cr_2O_3), chromic sulfate ($Cr_2[SO_4]_3$), chromic chloride ($CrCl_3$), chromic potassium sulfate ($KCr[SO_4]_2$), and chromite ore ($FeO\ CdCr_2O_3$).
4. *Hexavalent chromium compounds* (Cr6+): including chromium trioxide (CrO_3)—the anhydride of chromic acid chromates (e.g., Na_2CrO_4), dichromates, (e.g., $Na_2Cr_2O_7$), and polychromates.

   Certain hexavalent chromium compounds have been demonstrated to be carcinogenic on the basis of epidemiologic investigations of workers and experimental studies with animals. In general, these compounds tend to be of low solubility in water and, thus, are subdivided into two subgroups:
   a. Water-soluble hexavalent chromium compounds: including chromic acid and its anhydride, and the monochromates and dichromates of sodium, potassium, ammonium, lithium, cesium, and rubidium.
   b. Water-insoluble hexavalent chromium compounds: including zinc chromate, calcium chromate, lead chromate, barium chromate, strontium chromate, and sintered chromium trioxide.[1]

Physical Form. Most chromium compounds are solids at room temperature

Uses. In stainless and alloy steels, refractory products, tanning agents, pigments, electroplating, catalysts, and corrosion-resistant products

CHROMIUM METAL AND DIVALENT AND TRIVALENT COMPOUNDS

Exposure. Inhalation

Toxicology. Chromium metal is relatively nontoxic. There is little evidence of significant toxicity from chromic or chromous salts, probably because of poor penetration of skin and mucous membranes. Dermatitis from some chromic salts has been reported.

Four workers engaged in the production of ferrochrome alloys developed a nodular type of pulmonary disease with impairment of pulmonary function; air concentrations of chromium averaged 0.26 mg/m³, although other fumes and dusts also were present.[2] Chest roentgenograms are said to have revealed only "exaggerated pulmonic markings" in workers exposed to chromite dust.[1] The lungs of other workers exposed to chromite dust have been shown to be the seat of pneumoconiotic changes consisting of slight thickening of interstitial tissue and interalveolar septa, with histological fibrosis and hyalinization.[3] A refractory plant using chromite ore to make chromite brick had no excess of lung cancer deaths over a 14-year period, and it was concluded that chromite alone probably is not carcinogenic.[4] Exposure to chromium metal does not give rise to pulmonary fibrosis.[1]

Chromite ore roast mixed with sheep fat implanted intrapleurally in rats produced sarcomas coexisting with squamous-cell carcinomas of the lungs; the same material implanted in the thigh of rats produced fibrosarcomas.[5] However, the IARC concluded that these studies were inadequate to evaluate fully the carcinogenicity of this compound.[6] Other animal studies have found no increase in the incidence of tumors with chromium metal and chromite ore.[6] The IARC has determined that there is inadequate evidence in humans and animals for the carcinogenicity of metallic chromium and chromium[III] compounds.

Unlike nickel, chromium metal does not produce allergic contact dermatitis.[7] Some patients exhibit positive patch tests to divalent chromium compounds, but these compounds are considerably less potent as sensitizers than hexavalent chromium compounds. A case of chromium (chromic) sulfate-induced asthma in a plating worker, confirmed by specific challenge testing and the presence of IgE antibodies, has been reported.[8]

These compounds do not appear to cause other effects associated with the hexavalent chromium compounds, such as chrome ulcers, irritative dermatitis, or nasal septal perforation.[7]

HEXAVALENT CHROMIUM

Exposure.   Inhalation

Toxicology. The water-soluble hexavalent chromium compounds, such as chromic acid mist and certain chromate dusts, are severe irritants of the nasopharynx, larynx, lungs, and skin; exposure to certain, mainly water-insoluble, hexavalent chromium compounds, appears to be related to an increased risk of lung cancer.

Hexavalent chromium compounds have been implicated as responsible for such effects as ulcerated nasal mucosae, perforated nasal septa, rhinitis, nosebleed, perforated eardrums, pulmonary edema, asthma, kidney damage, erosion and discoloration of the teeth, primary irritant dermatitis, sensitization dermatitis, and skin ulceration.[9]

Chromic Acid. Workers exposed to chromic acid or chromates in concentrations of 0.11 to 0.15 mg/m^3 developed ulcers of the nasal septum and irritation of the conjunctiva, pharynx, and larynx, as well as asthmatic bronchitis.[10] A worker exposed to unmeasured but massive amounts of chromic acid mist for 4 days developed severe frontal headache, wheezing, dyspnea, cough, and chest pain on inspiration; after 6 months, the worker still experienced chest pain on inspiration and cough.[10]

In an industrial plant where the airborne chromic acid concentrations measured from 0.18 to 1.4 mg/m^3, moderate irritation of the nasal septum and turbinates was observed after 2 weeks of exposure, ulceration of the septum was present after 4 weeks, and there was perforation of the septum after 8 weeks.[10] A worker exposed to an unmeasured concentration of chromic acid mist for 5 years developed jaundice and was found to be excreting significant amounts of chromium; liver function was mildly to moderately impaired in four other workers with high urinary chromium excretion.[10]

Erosion and discoloration of the teeth have been attributed to chromic acid exposure. Papillomas of the oral cavity and larynx were found in 15 of 77 chrome platers exposed for an average of 6.6 years to chromic acid mist at air concentrations of chromium of 0.4 mg/m3.[4]

A concentrated solution of chromic acid in the eye causes severe corneal injury; chronic exposure to the mist causes conjunctivitis. Prolonged exposure to chromic acid mist causes dermatitis, which varies from a dry erythematous eruption to a weeping eczematous condition.

Chromates. Epidemiological studies from around the world have consistently shown excess risks for lung cancer in workers involved in chromate and chromate pigment production.[6] The epidemiological studies do not clearly implicate specific compounds but do implicate chromium[VI] compounds.[11] In one report, the relative risk of dying from respiratory cancer among chromate workers was over 20 times the rate for a control population; the latent period was relatively short.[9] In most studies, a positive correlation between duration of exposure and lung cancer death was found.[12]

Some less soluble hexavalent chromium compounds (lead chromate and zinc chromate pigments, calcium chromate) are carcinogenic in rats, producing tumors at the sites of administration by several routes. Lead chromate also produces renal carcinomas following intramuscular administration in rats.[9]

The IARC has concluded that there is sufficient evidence in humans for the carcinogenicity of chromium[VI] compounds as encountered in the chromate production, chromate pigment production, and chromate-plating industries. In experimental animals, there is sufficient evidence for the carcinogenicity of calcium chromate, zinc chromates, strontium chromate, and lead chromate.[6]

Chromium[VI] compounds have been consistently genotoxic, inducing a wide variety of effects, including DNA damage, gene mutation, sister chromatid exchange, chromosomal aberrations, cell transformation, and dominant lethal mutations.[6] Chromium[VI] compounds have caused developmental effects in rodents in the absence of maternal toxicity following oral administration.[11]

Chrome ulcer, a penetrating lesion of the skin, occurs chiefly on the hands and forearms where there has been a break in the epidermis; it is believed to be due to a direct necrotizing effect of the chromate ion. The ulcer is relatively painless, heals slowly, and produces a characteristic depressed scar. Sensitization dermatitis with varying degrees of eczema has been reported numerous times and is the single most common manifestation of chromium toxicity, affecting not only industrial workers but also the general population.[9]

The 1995 ACGIH threshold limit value–time-weighted average (TLV-TWA) for chromium metal and chromium[III] compounds is 0.5 mg/m³ with an A4 not classifiable as a human carcinogen designation; for water-soluble chromium[VI] compounds, the TLV-TWA is 0.05 mg/m³ with an A1 confirmed human carcinogen designation and, for insoluble chromium[VI] compounds, it is 0.01 mg/m³, also with an A1 designation.

REFERENCES

1. Chromium. *Documentation of the TLVs and BEIs*, 5th ed, p 139. Cincinnati, OH, American Conference of Governmental Industrial Hygienists (ACGIH), 1986
2. Princi F, Miller LH, Davis A, Cholak J: Pulmonary disease of ferroalloy workers. *J Occup Med* 4:301-310, 1962
3. Mancuso TF, Hueper WC: Occupational cancer and other health hazards in a chromate plant: a medical appraisal. 1. Lung cancers in chromate workers. *Ind Med Surg* 20:358–363, 1951
4. Committee on Biological Effects of Atmospheric Pollutants, Division of Medical Sciences, National Research Council: *Chromium*, pp 42–73, 125–145. Washington, DC, National Academy of Sciences, 1974
5. Hueper, WC: Experimental studies in metal cancerigenesis. X. Cancerigenic effects of chromite ore roast deposited in muscle tissue and pleural cavity of rats. *AMA Arch Ind Health* 18:284–291, 1958
6. *IARC Monographs on the Evaluation of Carcinogenic Risks to Humans*, Vol 49, Chromium, nickel and welding, pp 49–256. Lyon, International Agency for Research on Cancer, 1990
7. Burrows D: The dichromate problem. *Int J Dermatol* 21:215–220, 1984
8. Novey H, Habib M, Wells ID: Asthma and IgE antibodies induced by chromium and nickel salts. *J Allergy Clin Immunol* 72:407–412, 1983
9. Enterline PE: Respiratory cancer among chromate workers. *J Occup Med* 16:523–526, 1974
10. Franchini I et al: Mortality experience among chromeplating workers. *Scand J Work Environ Health* 9:247–252, 1983
11. Agency for Toxic Substances and Disease Registry (ATSDR): *Toxicological Profile for Chromium*. US Department of Health and Human Services, Public Health Service, p 227, TP-92/08, 1993
12. Cohen MD, Kargacin B, Klein CB, et al:

Mechanisms of chromium carcinogenicity and toxicity. *Crit Rev Toxicol* 23:255–281, 1993

CHROMYL CHLORIDE
CAS: 14977-61-8

CrO_2Cl_2

Synonyms: Chromium dioxychloride; chromium dioxide dichloride; chromium oxychloride; chromium chloride oxide

Physical Form. Dark red liquid with an unpleasant odor

Uses. In organic oxidations and chlorinations; as a solvent for chromium oxide; in making chromium complexes and dyes

Exposure. Inhalation

Toxicology. Chromyl chloride is a severe irritant.

Although information is not available on exposure levels, the vapor is expected to cause eye, nose, and throat irritation; there may be difficulty breathing and lung injury.[1] On brief contact with the skin, the liquid will produce second- and third-degree burns; it is very injurious to the eyes. Ingestion causes burning of the mouth and stomach. The toxicity of chromyl chloride is mediated by products formed during hydrolysis; in water, it reacts violently to form hydrochloric acid, chromic acid, and chlorine gases.

NIOSH has classified chromyl chloride as an inferred carcinogen.[2]

The 1995 ACGIH threshold limit value–time-weighted average (TLV-TWA) for chromyl chloride is 0.025 ppm (0.16 mg/m³).

REFERENCES

1. Weiss G (ed): *Hazardous Chemicals Data Book, Environmental Health Review No. 4*, p 257. Noyes Data Co, Park Ridge, NJ, 1980

2. National Institute for Occupational Safety and Health: *Criteria for a Recommended Standard . . . Occupational Exposure to Chromium (VI)*. DHEW (NIOSH) Pub No 76-129, Washington, DC, US Government Printing Office, 1975

CHRYSENE
CAS: 218-01-9

$C_{18}H_{12}$

Synonyms: 1,2-Benzo[a]phenanthrene; 1,2-benzphenanthrene; benzo[a]phenanthrene; 1,2,5,6-dibenzonaphthalene

Physical Form. Colorless rhombic plates

Uses. Laboratory reagent; formed during the pyrolysis of organic matter

Exposure. Inhalation; ingestion; skin absorption

Toxicology. There is limited evidence that chrysene is an animal carcinogen.

There is no information regarding the toxicity of chrysene to humans.[1] The LD_{50} for chrysene given intraperitoneally is greater than 320 mg/kg bw in the mouse.[2]

The International Agency for Research on Cancer (IARC) has concluded that there is limited evidence that chrysene is carcinogenic to experimental animals.[2] Chrysene produced skin tumors following skin application to mice and has been shown to be active as a tumor initiator. Local tumors were observed following its subcutaneous injection in mice. Perinatal administration of chrysene to mice by subcutaneous or intraperitoneal injection increased the incidence of liver tumors.[2]

There are no reports directly correlating human chrysene exposure and tumor development, in part perhaps because chrysene does not occur in isolation but, rather, occurs as only one component of a mixture of polycyclic aromatic hydrocarbons. There are, however, a number of reports associating human cancer and exposure to mixtures of polycyclic aromatic hydrocarbons that include chrysene as a component, for example, among coke oven workers.[3]

Chrysene was mutagenic to *Salmonella typhimurium* in the presence of an exogenous metabolic system. It induced sister chromatid exchanges in one mouse study and chromosomal aberrations in one hamster study.[2] Chrysene is metabolically activated to a 1,2-diol-3,4-epoxide, which is mutagenic and carcinogenic in experimental animals and forms covalent adducts with DNA.[2,4]

ACGIH has classified chrysene as an A2 suspected human carcinogen with no assigned threshold limit value.

REFERENCES

1. Agency for Toxic Substances and Disease Registry (ATSDR): *Toxicological Profile for Chrysene*. US Department of Health and Human Services, Public health Service, pp 73, 1990
2. *IARC Monographs on the Evaluation of Carcinogenic Risks to Humans*, Vol. 32, *Polynuclear aromatic compounds, Part 1, Chemical, environmental and experimental data*, pp 247–261. Lyon, International Agency for Research on Cancer, 1983
3. *IARC Monographs on the Evaluation of Carcinogenic Risks to Humans*, Vol. 34, *Polynuclear aromatic compounds, Part 3, Industrial exposures in aluminium production, coal gasification, coke production, and iron and steel founding*, pp 124–125. Lyon, International Agency for Research on Cancer, 1984
4. Perera FP et al: Detection of polycyclic aromatic hydrocarbon-DNA adducts in white blood cells of foundry workers. *Cancer Res* 48:2288–2291, 1988

COAL DUST

Synonyms: None

Physical Form. Solid

Uses. Fuel; in the production of coke, coal gas, and coal tar compounds

Exposure. Inhalation

Toxicology. The inhalation of coal dust causes coal workers' pneumoconiosis (CWP).

Simple coal workers' pneumoconiosis has no clinically distinguishing symptoms because many miners have a slight cough and blackish sputum, which are of no help in establishing whether or not the disease is present.[1] Simple CWP is diagnosed according to the number of small opacities present in the chest film; the small opacities may be linear (irregular) or rounded (regular); however, the latter are more commonly seen in CWP and are most frequently located in the upper lung zones.[1] The primary lesion of simple CWP is the coal macule, a focal collection of coal dust particles with a little reticulin and collagen accumulation, measuring up to 5 mm in diameter. Focal emphysema may be associated.

Simple CWP often occurs concomitantly with chronic bronchitis and emphysema.[1] Although CWP is associated with several respiratory impairments, it is not associated with shortened life span; the importance of this benign condition is the fact that it is a precursor of progressive massive fibrosis (PMF).[1] However, simple CWP does not progress in the absence of further exposure.

Any opacity greater than 1 cm in a coal miner is classified as complicated pneumoconiosis or PMF unless there is evidence to suggest another disease such as tuberculosis.[1]

Complicated pneumoconiosis (PMF) is associated with a reduction in ventilatory capacity, low diffusing capacity, abnormalities of gas exchange, low arterial oxygen tension, pulmonary hypertension, and premature death; it may appear several years after exposure has ceased and may progress in the absence of further dust exposure.[1] Macroscopically, the lesions consist of a mass of black tissue that is often adherent to the chest wall. The lesions are of a rubbery consistency and are relatively well defined. Unlike conglomerate silicosis, which consists of matted aggregates of whorled silicotic nodules, the massive lesion of PMF is amorphous, irregular, and relatively homogeneous. In some instances, its center may contain a cavity filled with a black liquid. Cavitation is a consequence of ischemic necrosis or secondary infection by tuberculosis. In PMF, the vascular bed of the affected region is destroyed. Obstructive airway disease is common and is probably a consequence of the distortion and narrowing of the bronchi and bronchioles produced by the conglomerate mass.

The percentage of miners showing definite radiographic evidence of either simple or complicated pneumoconiosis has varied considerably in different geographical areas; factors responsible for this difference include respirable dust levels, the number of years of exposure, and the physical and chemical composition of the coal.[2] Prevalence of CWP has been associated with coal "rank," with higher-ranking coals consisting of an older coal with a higher percentage of carbon and a higher prevalence of CWP.[3]

A study of 9076 US miners from 1969 to 1971 showed an overall prevalence of CWP of 30%, with PMF occurring in 2.5% of the sample.[2] A more recent study of US coal workers through 1988 has found a reduction in the incidence of pneumoconiosis, coinciding with a reduction in workplace exposures to 2 mg/m³ after 1969.[3] In Great Britain, the prevalence of all categories of CWP in working miners has fallen from 13.4% in 1959–1960 to 5.2% in 1978; for PMF, the rate in 1978 was 0.4%.[4]

Various studies have examined the incidence of gastric cancer with coal dust exposure because various carcinogenic substances have been identified in coal and because coal dust may reach the GI tract through the pulmonary clearance system.[5] Although early reports have found an increased standardized mortality ratio for gastric cancer, a recent matched case-control study found no evidence of a dose-response relationship between coal mining and gastric cancer.

The 1995 ACGIH threshold limit value–time-weighted average (TLV-TWA) for coal dust is 2 mg/m³, for dust containing less than 5% crystalline silica.

REFERENCES

1. Morgan WKC: Coal workers' pneumoconiosis. In Morgan WKC, Seaton A (eds): *Occupational Lung Diseases*, 2nd ed, pp 377–488. Philadelphia, PA, WB Saunders, 1984
2. Morgan WKC, Burgess DB, Jacobson G, et al: The prevalence of coal workers' pneumoconiosis in US coal miners. *Arch Environ Health*, 27:221–226, 1973
3. Attfield MD, Castellan RM: Epidemiological data on US coal miners' pneumoconiosis, 1960–1988. *Am J Pub Health* 82:964–970, 1992
4. Parkes WR: *Occupational Lung Disorders*, 2nd ed, p 178. London, Butterworths, 1982
5. Swaen GMH, Aerdts CWHM, Slangen JJM: Gastric cancer in coal miners: Final report. *Br J Ind Med* 44:777–779, 1987

COAL TAR PITCH VOLATILES
CAS: 65996-93-2

Synonyms: CTPV; particulate polycyclic organic matter (PPOM); particulate polycyclic aromatic hydrocarbons (PPAH); polynuclear aromatics (PNAs)

Physical Form.   As stated by ACGIH:[1]

The pitch of coal tar is the black or dark brown amorphous residue that remains after the redistillation process. The volatiles contain a large quantity of lower-molecular-weight polycyclic hydrocarbons. As these hydrocarbons (naphthalene, fluorene, anthracene, acridine, phenanthrene) sublime into the air, there is an increase of benzo(a)-pyrene (BaP or 3,4-benzpyrene) and other higher-weight polycyclic hydrocarbons in the tar and in the fumes. Polycyclic hydrocarbons, known to be carcinogenic, are of this large molecular type.

Sources.   Emissions from coke ovens, from coking of coal tar pitch, and from Soderberg aluminum reduction electrolytic cells

Uses. For base coatings and paints; for roofing and paving; as a binder for carbon electrodes

Exposure. Inhalation

Toxicology. Epidemiologic evidence suggests that workers intimately exposed to the products of combustion or distillation of bituminous coal are at increased risk of cancer at many sites, including lungs, kidney, and skin.[2]

The chemical composition and particle size distribution of coal tar pitch volatiles (CTPV) from different sources are significant variables in determining toxicity.[3,4]

In a study of 22,010 US male aluminum reduction workers with over 5 years' employment in the industry, there was a slight positive association with lung cancer (standardized mortality ratio or SMR = 121), which was somewhat stronger in Soderberg workers (SMR = 162).[4] There was a slight but not statistically significant excess of leukemia (SMR = 170) and lymphoma (SMR = 125) in pot room workers.

In a more detailed analyses of the mortality experience of this cohort up to year-end 1977, the results of other studies relative to an excess of lung cancer were not confirmed, but there were indications of a higher than expected mortality in pancreatic cancer, lymphohematopoietic cancers, genitourinary cancer, nonmalignant respiratory disease, and benign and unspecified neoplasms.[6]

A recent case-cohort study of aluminum production plant workers showed a clear excess of lung cancer risk in men who had worked in Soderberg potrooms in jobs with high exposure to coal tar pitch volatiles and indicated that the risk was not due to confounding by smoking.[7] The rate ratio for lung cancer rose with cumulative exposure to coal tar pitch volatiles measured as benzene-soluble material to 2.25 at 10 to 19 mg/m[3]–years benzene-soluble matter, but did not rise further at higher exposures.

A study in Canada of 5891 men in two aluminum reduction plants found the mortality from lung cancer related to "tar-years" of exposure; the SMR for persons exposed for more than 21 years to the higher levels of tars was 2.3 times that of persons not exposed to tars.[8]

A follow-up study of this cohort through 1977 showed excess deaths from respiratory disease; pneumonia and bronchitis; malignant neoplasms of the stomach and esophagus, bladder, and lung; malignant neoplasms (all sites); Hodgkin's disease; and hypertensive disease. Mortality from malignant neoplasms of the bladder and lung was related to the number of tar-years and to years of exposure.[9]

Exposure to coke oven emissions is a cause of lung and kidney cancer. A major study of US coke oven workers showed that mortality from lung cancer for full topside workers is 9 times the expected rate; for partial topside workers, it is almost 2.5 times the expected rate; and, for side oven workers, it is 1.7 times the expected rate.[10] All these rates are based on 5 or more years of exposure in the job category. As the length of employment increases, so does the mortality experience. For example, for employees with 20 or more years of employment topside, the lung cancer rate is 20 times the expected rate. In addition to the risk of lung cancer, the relative risk of mortality from kidney cancer for all coke oven workers is 7.5.[11]

Certain industrial populations exposed to coal tar products have a demonstrated risk of skin cancer. Substances containing polycyclic hydrocarbons or polynuclear aromatics (PNAs), such as coal tar pitch and cutting oils, which may produce skin cancer, also produce contact dermatitis.[4] Although allergic dermatitis is readily induced by PNAs in guinea pigs, it only rarely is reported in humans from occupational contact with PNAs. Incidences in humans have resulted largely from the therapeutic use of coal tar preparations.[6]

Components of pitch and coal tar produce cutaneous photosensitization; skin eruptions usually are limited to areas exposed to ultraviolet light.[4,12,13] Most of the phototoxic agents will induce hypermelanosis of the skin; if chronic photodermatitis is severe and prolonged, leukoderma may occur.[12] Some oils containing PNAs have been associated with follicular and sebaceous gland changes, which commonly take the form of acne.[4]

Coal tar fumes were mutagenic in a modified Ames test.[14] Fumes generated at 316°C contained significantly higher concentrations of polycyclic aromatic hydrocarbons (PAHs) than those generated at 232°C, and the mutagenic activity generally paralleled the PAH content.

Biological monitoring of 1-hydroxypyrene (a PAH metabolite) in urine has been a useful indicator of PAH exposure in coke oven workers.[15]

The 1995 ACGIH threshold limit value–time-weighted average (TLV-TWA) for coal tar pitch volatiles is 0.2 mg/m³ as benzene solubles, with an A1 confirmed human carcinogen designation.

REFERENCES

1. Coal tar pitch volatiles. *Documentation of TLVs and BEIs*, 6th ed, pp 327–329. Cincinnati, OH, American Conference of Governmental Industrial Hygienists (ACGIH), 1991
2. National Institute for Occupational Safety and Health, US Department of Health, Education, and Welfare: *Criteria for a Recommended Standard . . . Occupational Exposure to Coke Oven Emissions.* (HSM) Pub No 73-11016, pp III-1–III-14, V-1–V-9. Washington, DC, US Government Printing Office, 1973
3. Hittle DC, Stukel JJ: Particle size distribution and chemical composition of coal tar fumes. *Am Ind Hyg Assoc J* 37:199–204, 1976
4. Scala RA: Toxicology of PPOM. *J Occup Med* 17:784–788, 1975
5. Cooper C, Gaffey W: *A Mortality Study of Aluminum Workers.* New York, The Aluminum Association, Inc, 1977
6. Rockette HE, Arena VC: Mortality studies of aluminum reduction plant workers: pot room and carbon department. *J Occup Med* 25:549–557, 1983
7. Armstrong B, Tremblay C, Baris D, et al: Lung cancer mortality and polynuclear aromatic hydrocarbons: a case cohort study of aluminum production workers in Arvida, Quebec, Canada. *Am J Epidemiol* 139:250–262, 1994
8. Gibbs GW, Horowitz I: Lung cancer mortality in aluminum plant workers. *J Occup Med* 21:347–353, 1979
9. Gibbs GW: Mortality of aluminum reduction plant workers, 1950 through 1977. *J Occup Med* 27:761–770, 1985
10. Lloyd JW: Long-term mortality study of steelworkers. V. Respiratory cancer in coke plant workers. *J Occup Med* 13:53–68, 1971
11. Redmond CK, Ciocco A, Lloyd JW, Rush HW: Long-term mortality study of steelworkers. VI. Mortality from malignant neoplasms among coke oven workers. *J Occup Med* 14:621–629, 1972
12. Committee on Biological Effects of Atmospheric Pollutants, Division of Medical Sciences, National Research Council: *Particulate Polycyclic Organic Matter*, pp 166–246, 307–354. Washington, DC, National Academy of Sciences, 1972
13. National Institute for Occupational Safety and Health: *Criteria for a Recommended Standard . . . Occupational Exposure to Coal Tar Products.* DHEW (NIOSH) Pub No 78-107. Washington, DC, US Government Printing Office, 1977
14. Machado ML, Beatty PW, Fetzer JC, et al: Evaluation of the relationship between PAH content and mutagenic activity of fumes from roofing and paving asphalts and coal tar pitch. *Fund Appl Toxicol* 21:492–499, 1993
15. Jongeneelen FJ: Biological exposure limit for occupational exposure to coal tar pitch volatiles at coke ovens. *Int Arch Occup Environ Health* 63:511–516, 1992

COBALT
CAS: 7440-48-4

Co

Synonyms: None

Physical Form. Gray solid

Uses. Alloys; carbides; high-speed steels; paints; electroplating

Exposure. Inhalation

Toxicology. Cobalt inhalation primarily affects the lungs; it is a skin irritant and sensitizer and is carcinogenic in experimental animals.

Three types of lung disease have been reported in the cemented tungsten carbide industry: (1) an interstitial fibrotic process, (2) an interstitial pneumonitis, which often disappears when exposure ceases, and (3) an obstructive airways syndrome. The latter may result from simple irritation but, in addition, a distinct form of occupational asthma occurs.[1] Cobalt, which is used as a binder for the tungsten carbide crystals, has been implicated as the etiologic agent.[2,3]

Among 12 workers engaged in the manufacturing of, or grinding with, tungsten carbide tools who developed interstitial lung disease, there were 8 fatalities; serial chest roentgenograms over a period of 3 to 12 years revealed gradually progressive densities of a linear and nodular nature, which gradually involved major portions of both lungs. Cough, production of scanty mucoid sputum, dyspnea on exertion, and reduced pulmonary function occurred early in the course of the disease.[3] Disease is seldom seen without at least 10 years of exposure, but shorter periods have been reported.[2]

The obstructive airways syndrome appears to be an allergic response and is characterized by wheezing, cough, and shortness of breath while at work.[1,3] There is no evidence that this type of disease progresses to interstitial fibrosis. In a report of nine cases, the syndrome did not develop until after 6 to 18 months of exposure.[4]

On screening 1039 tungsten carbide workers, interstitial lung disease was observed in 0.7%, and work-related wheezing occurred in 10.9%.[5]

Occupational exposure to cobalt dust has been associated with cardiomyopathy characterized by functional effects on the ventricles and enlargement of the heart.[6]

Cobalt and its compounds produce an allergic dermatitis of an erythematous papular type that usually occurs in skin areas subjected to friction, such as the ankle, elbow flexures, and sides of the neck.[7] Ocular effects have included congestion of the conjunctiva.[8]

Animal studies have reported developmental effects (stunted fetuses and decreased pup weight and viability) following oral expo-

sure to cobalt at doses that also produced maternal toxicity.[9] Testicular atrophy was reported in rats exposed to 19 mg cobalt/m^3 (as cobalt sulfate) for 16 days.[10] Male mice exposed for 13 weeks at 1.14 mg cobalt/m^3 had a decrease in sperm motility and, at 11.4 mg cobalt/m^3, there was testicular atrophy; at the high dose, female mice had a significant increase in length of the estrous cycle.[10]

Rhabdomyosarcomas developed in rats injected intramuscularly with the powder of either pure cobalt metal or cobalt oxide.[11] In other studies, implantation of cobalt caused local fibrosarcomas in rabbits, but inhalation studies in hamsters did not reveal any increase in tumors from cobalt oxide.[8]

Epidemiological studies to determine the carcinogenicity of cobalt in humans have been confounded by concurrent exposure to other known carcinogens, such as nickel and arsenic, and small study populations.[12] The IARC has classified cobalt and compounds as group 2B possible human carcinogens.[13]

In mammalian cells, in vitro cobalt compounds have caused DNA strand breaks, sister chromatid exchanges, and aneuploidy but not chromosomal aberrations.[14] Cobalt salts are generally nonmutagenic in procaryotic assays.[14]

The 1995 ACGIH threshold limit value–time-weighted average (TLV-TWA) for elemental cobalt and inorganic compounds is 0.02 mg/m^3 with an A3 animal carcinogen designation.

REFERENCES

1. Morton WKC, Seaton A: *Occupational Lung Diseases*, 2nd ed, pp 486–489. Philadelphia, WB Saunders, 1975
2. Miller CW, Davis MW, Goldman A, Wyatt JP: Pneumoconiosis in the tungsten-carbide tool industry. *AMA Arch Ind Hyg Occup Med* 8:453, 1953
3. Coates EO Jr, Watson JHL: Diffuse interstitial lung disease in tungsten-carbide workers. *Ann Intern Med* 75:709, 1971
4. Coates EO Jr: Hypersensitivity bronchitis in tungsten-carbide workers. *Chest* 64:390, 1973
5. Cugell DW, Morgan WKC, et al: The respi-

ratory effects of cobalt. *Arch Intern Med* 150:177–183, 1990

6. Horowitz SF, Fischbein A, Matza D: Evaluation of right and left ventricular function in hard metal workers. *Br J Ind Med* 45:742–746, 1988
7. Browning E: *Toxicity of Industrial Metals*, 2nd ed, pp 132–142. London, Butterworth, 1969
8. Agency for Toxic Substances and Disease Registry (ATSDR): *Toxicological Profile for Cobalt*. US Department of Health and Human Services, Public Health Service, TP-91/10, p 140, 1991
9. Domingo JL, Paternain JL, Llobet JM, et al: Effects of cobalt on postnatal development and late gestation in rats upon oral administration. *Rev Esp Fisiol* 41:293–298
10. National Toxicology Program: *Cobalt Sulfate Heptahydrate (CAS No. 10026-24-1) in F344/N Rats and B6C3F₁ Mice (Inhalation Studies)*. Technical Report Series No 5, NIH Pub No 91-3124, National Institutes of Health. Triangle Park, NC, US Department of Health and Human Services, Public Health Service, 1991
11. Heath JC: The histogenesis of malignant tumors induced by cobalt in the rat. *Br J Cancer* 14:478, 1960
12. Jensen AA, Tuchsen F: Cobalt exposure and cancer risk. *CRC Rev Toxicol* 20:427–437, 1990
13. *IARC Monographs on the Evaluation of Carcinogenic Risks to Humans*. Vol 52, Chlorinated drinking water; chlorination by-products; some other halogenated compounds; cobalt and cobalt compounds. International Agency for Research on Cancer, Lyon, 1991
14. Beyersmann D, Hartwig A: The genetic toxicology of cobalt. *Toxicol Appl Pharmacol* 115:137–145, 1992

COBALT HYDROCARBONYL
CAS: 16842-03-8

C_4HCoO_4

Synonyms: Cobalt carbonyl hydride; tetracarbonylhydrocobalt

Physical Form.   Flammable gas with an offensive odor

Uses.   Catalyst in organic reactions

Exposure.   Inhalation

Toxicology.   Cobalt hydrocarbonyl is expected to be a pulmonary irritant.

The 30-minute LC_{50} in rats was determined to be 165 mg/m³.[1] Definitive toxicity data for cobalt hydrocarbonyl do not exist because of the rapid decomposition in air of the chemical to a solid particulate. In most cases, exposures are primarily to inorganic cobalt compounds.

By analogy to nickel carbonyl, acute effects from animal exposures are expected to be pulmonary edema, congestion, and hemorrhage. In humans, nickel carbonyl causes an acute flu-like syndrome, which subsides and is followed, after 12 to 36 hours, by an acute respiratory syndrome. Exposure to cobalt hydrocarbonyl may be expected to produce similar effects.

The 1995 threshold limit value–time-weighted average (TLV-TWA) is 0.1 mg/m³ as Co.

REFERENCES

1. Palmes ED, Nelson N, Laskin S, Kuschner M: Inhalation toxicity of cobalt hydrocarbonyl. *Am Ind Hyg Assoc J* 20:453–468, 1959

COPPER (DUST AND FUME)
CAS: 7440-50-8

CU, CuO

Synonyms: None

Physical Form.   Reddish solid

Sources.   Copper and brass manufacture; welding of copper-containing metals

Exposure.   Inhalation

Toxicology.   Copper fume causes irritation of

the upper respiratory tract and metal fume fever, an influenzalike illness.

Respiratory, gastrointestinal, and dermal effects have been observed in workers exposed to copper dust and fumes.[1] Exposure to concentrations of 1 to 3 mg/m^3 for short periods resulted in altered taste response but no nausea; levels from 0.02 to 0.4 mg/m^3 produced no complaints.

Typical metal fume fever, a 24- to 48-hour illness characterized by chills, fever, aching muscles, dryness in the mouth and throat, and headache has been reported in several workers exposed to copper fume.[2,3] With metal fume fever, leukocytosis is usually present, with counts of 12,000 to 16,000/mm^3; recovery is usually rapid, and there are no sequelae.[4] Most workers develop an immunity to these attacks, but it is quickly lost, and attacks tend to be more severe on the first day of the workweek.[4]

Gastrointestinal effects, including anorexia, nausea, and occasional diarrhea, have been attributed to swallowing of the dust.[1]

Copper dust may cause respiratory irritation.[1] Workers have experienced mucosal irritation of the mouth, eyes, and nose. Lung damage after chronic exposure to fumes in industry has not been described.[5] The higher incidence of respiratory cancer reported in copper smelters is due to the presence of arsenic in the ore.[5]

Transient irritation of the eyes has followed exposure to a fine dust of oxidation products of copper produced in an electric arc.[6]

The 1995 ACGIH threshold limit value–time-weighted average (TLV-TWA) for copper is 0.2 mg/m^3 for the fume and 1 mg/m^3 for dusts and mists as Cu.

REFERENCES

1. Agency for Toxic Substances and Disease Registry (ATSDR): *Toxicological Profile for Copper.* US Department of Health and Human Services, Public Health Service, pp 1–143, 1990
2. Committee on Medical and Biologic Effects of Environmental Pollutants: *Copper*, pp 55–58. Washington, DC, National Academy of Sciences, 1977
3. Cohen SR: A review of the health hazards from copper exposure. *J Occup Med* 16:621–624, 1974
4. McCord CP: Metal fume fever as an immunological disease. *Ind Med Surg* 29:101–106, 1960
5. Triebig G, Schaller KH: Copper. In Alessio L et al (eds): *Biological Indicators for the Assessment of Human Exposure to Industrial Chemicals*, pp 57–66. Luxembourg, Office for Official Publications of the European Communities, 1984
6. Grant WM: *Toxicology of the Eye*, 3rd ed, pp 260–269. Springfield, IL, Charles C Thomas, 1986

COTTON DUST, RAW

Synonyms: None

Physical Form. Fibers

Source. Cotton processing

Exposure. Inhalation

Toxicology. Raw cotton dust causes a respiratory syndrome termed byssinosis.

The initial symptoms are tightness in the chest, cough, wheezing, and dyspnea in varying degrees of severity on the first day of the workweek (grade 1 byssinosis).[1-3] Symptoms usually disappear an hour or so after the individual leaves work but may recur on the first day of each workweek. With continued exposure, the symptoms also appear on subsequent days of the week (grade 2 byssinosis). There usually is a decrease in the FEV$_1$ and in vital capacity on the first day of the workweek after 3 to 4 hours of exposure; the changes in airway resistance and decreased flow rates have been attributed to narrowing of small airways due to bronchoconstriction. Eventually, obstructive airway disease, which is irreversible, occurs (grade 3 byssinosis).

Although a loss in lung function has been documented in cotton textile workers, no clear evidence of increased mortality has been reported.[4] A review of 2895 consecutive autopsies showed no significant differences in the prevalence of emphysema, interstitial fibrosis, or cor pulmonale between 283 employees of a cotton textile mill and the nontextile population.[5] In another postmortem study of 49 cotton workers, the incidence of emphysema was associated with cigarette smoking, with 16 of 36 smokers showing centrilobular emphysema versus 1 of 13 nonsmokers.[6] Another study of women with advanced byssinosis confirmed the association between emphysema and cigarette smoking rather than cotton dust exposure.[7]

Two more recent studies of cotton workers also found no excess mortality from respiratory disease but differed in other findings. In the first report, of 3458 British cotton industry workers, mortality from respiratory disease was reduced overall but, for subjects who initially reported byssinotic symptoms, the mortality from respiratory disease was slightly raised, and it increased with length of service.[8] The mortality from lung cancer was lower than expected, and it decreased with length of service. A mortality study of 1065 women employed in Finnish cotton mills did not confirm low mortality from respiratory cancer.[9] Instead, an increase in lung (3 vs. 1.9 expected) and gastrointestinal (13 vs. 6.6 expected) cancers was reported. Cotton dust exposure also appeared to increase the morbidity of renal disease and rheumatoid arthritis.

A syndrome known as "mill fever," which may or may not be related to the development of byssinosis, has been described in some persons unaccustomed to breathing cotton dust.[2] Shortly after exposure, there is development of malaise, cough, fever, chills, and upper respiratory symptoms; these may recur daily for days to months until acclimatization takes place and symptoms disappear. Tolerance may be lost temporarily after a period of absence from exposure or of exposure to a greater concentration of dust. The exact prevalence of "mill fever" among new employees is unknown, but estimates range from 10% to 80%.[1]

Epidemiological studies have indicated that prevalence of byssinosis among cotton workers can be correlated with the average concentration of lint-free dust of particle diameter under 15 μ and with the number of years of exposure.[2] Specifically, in a follow-up study of 66 cotton

textile workers with an additional 10 years' exposure, the prevalence of byssinosis increased from 23% to 43% in the female workers and from 23% to 52% in males.[10]

There is little evidence of a threshold below which zero prevalence is found.[2] The slopes of the prevalence-dustiness curves obtained by different investigators vary considerably.[2]

Little is known about the identity of the byssinogenic agent or the underlying mechanisms; the active agent is thought to be water-extractable, filterable through a 0.22-μ filter, nonvolatile at 40°C, nondialyzable, and present in the harvested cotton rather than the lint.[11]

The 1995 ACGIH threshold limit value–time-weighted average (TLV-TWA) for cotton dust is 0.2 mg/m^3.

REFERENCES

1. Harris TR, Merchant JA, Kilburn KH, Hamilton JD: Byssinosis and respiratory diseases of cotton mill workers. *J Occup Med* 14:199–206, 1972

2. National Institute for Occupational Safety and Health: *Criteria for a Recommended Standard . . . Occupational Exposure to Cotton Dust.* DHEW (NIOSH) Pub No 75-118, pp 12–60. Washington, DC, US Government Printing Office, 1974

3. Department of Labor: Standard for exposure to cotton dust. *Federal Register* 41:56498–56525, 1976

4. Elwood PC et al: Respiratory disability in ex-cotton workers. *Br J Ind Med* 43:580–586, 1986

5. Moran TJ: Emphysema and other chronic lung disease in textile workers: an 18-year autopsy study. *Arch Environ Health* 38:267–276, 1983

6. Pratt PC et al: Epidemiology of pulmonary lesions in nontextile and cotton textile workers—a retrospective autopsy analysis. *Arch Environ Health* 35:133–138, 1980

7. Honeybourne D, Pickering CAC: Physiological evidence that emphysema is not a feature of byssinosis. *Thorax* 41:6–11, 1986

8. Hodgson JT, Jones RD: Mortality of workers in the British cotton industry in 1968–1984. *Scand J Work Environ Health* 16:113–120, 1990

9. Koskela RS, Klockars M, Jarvinen E: Mortal

ity and disability among cotton mill workers. *Br J Ind Med* 47:384–391, 1990

10. Zuskin E, Ivankovic D, Schachter EN, et al: A ten-year follow-up study of cotton textile workers. *Am Rev Respir Dis* 143:301–305, 1991
11. Brown DF, McCall ER, Piccolo B, Tripp VW: Survey of effects of variety and growing location of cotton on cardroom dust composition. *Am Ind Hyg Assoc J* 38:107–115, 1977

CRESOL (ALL ISOMERS)
CAS: 1319-77-3

C_7H_8O

Synonyms: Cresylic acid; tricresol; methylphenol; *o*-cresol; *m*-cresol; *p*-cresol; hydroxytoluene

Physical Form. Ortho and para isomers are solids; the meta isomer and isomer mixtures are yellowish liquids

Uses. Antiseptics; disinfectants; solvent; insecticides; resins; flame-retardant; plasticizers

Exposure. Skin absorption; inhalation

Toxicology. All isomers of cresol cause skin and eye burns; exposure may also cause impairment of kidney and liver function, as well as central nervous system and cardiovascular disturbances.

Skin and eye contact are the major concerns of occupational exposure.[1] Signs and symptoms related to skin contact are a burning sensation, erythema, skin peeling, localized anesthesia and, occasionally, ochronosis, a darkening of the skin.[1] Hypersensitivity also has been reported.[2]

Cresols are rapidly absorbed through the skin, producing systemic effects.[1] About 20 ml of a 90% cresol solution accidentally poured over an infant's head caused chemical burns, cyanosis, unconsciousness, and, within 4 hours,

death.[3] Histopathological examination showed hepatic necrosis, cerebral edema, acute tubular necrosis of the kidneys, and hemorrhagic effusions from the peritoneum, pleura, and pericardium. The blood contained 12 mg cresol/100 dl.

Inhalation of appreciable amounts of cresol vapor is unlikely under normal conditions because of the low vapor pressure; however, hazardous concentrations may be generated at elevated temperatures.[1] Seven workers exposed to cresol vapor at unspecified concentrations for 1.5 to 3 years had headaches, which were frequently accompanied by nausea and vomiting.[1] Four of the workers also had elevated blood pressure, signs of impaired kidney function, blood calcium imbalance, and marked tremors. Of 10 subjects exposed to 1.4 ppm *o*-cresol vapor, 8 experienced upper respiratory tract irritation.[1]

Several cases of ingestion have shown cresol to be corrosive to body tissues and to cause toxic effects on the vascular system, liver, kidneys, and pancreas.[1]

Rats survived 8 hours of inhaling air saturated with the vapor.[4] Irritation of the nose and eyes and some deaths were observed in mice exposed to saturated concentrations 1 hr/day for 10 days.[5] Animal experiments have produced varying results with regard to concentrations necessary to produce death.[1] In general, the ortho and para isomers are considered equal in toxicity, with the meta isomer regarded as the least toxic.[1]

At doses that were maternally toxic, *o*- and *p*-isomers induced slightly elevated incidences of minor variations in the offspring of exposed rats and rabbits.[6]

Cresol isomers promoted dimethylbenzanthracene-induced papillomas in mice when applied as 20% solutions in benzene twice weekly for 11 weeks; no carcinomas were produced.[7]

Cresol mixtures and the *o*- and *p*-isomers have been found to be weakly genotoxic in some in vitro assays, inducing sister chromatid exchange and chromosomal aberrations in Chinese hamster ovary cells.[6] Results were negative with the *m*-isomer, as were all in vivo assays.

In the eyes of rabbits, undiluted cresols

caused permanent opacification and vascularization; a drop of a 33% solution applied to rabbit eyes and removed with irrigation within 60 seconds caused only moderate injury, which was reversible.[8]

In rat liver tissue, *p*-cresol was five- to tenfold more toxic than the *o*- or *m*-isomers, as determined by the degree of cell killing.[9] Furthermore, the toxicity of *p*-cresol was dependent on the formation of a reactive intermediate, and it was suggested that the mechanism of toxicity for *p*-cresol may differ from that of the *o*- and *m*-isomers.

The odor of cresol is recognized at concentrations as low as 5 ppm.[2] The 1995 ACGIH threshold limit value–time-weighted average (TLV-TWA) for all isomers of creosol is 5 ppm (22 mg/m³) with a notation for skin absorption.

REFERENCES

1. National Institute for Occupational Safety and Health: *Criteria for a Recommended Standard . . . Occupational Exposure to Cresol.* DHEW (NIOSH) Pub No 78-133, p 117. Washington, DC, US Government Printing Office, 1978
2. Deichmann WB, Keplinger ML: Phenols and phenolic compounds. In Clayton GD, Clayton FE (eds): *Patty's Industrial Hygiene and Toxicology*, 3rd ed, rev, Vol 2A, *Toxicology*, pp 2597–2601. New York, Wiley-Interscience, 1981
3. Green MA: A household remedy misused—fatal cresol poisoning following cutaneous absorption—a case report. *Med Sci Law* 15:65–66, 1975
4. Smyth HF Jr: Improved communication—hygienic standards for daily inhalation. *Am Ind Hyg Assoc Q* 17:129–185, 1956
5. Campbell J: Petroleum cresylic acids—a study of their toxicity and the toxicity of cresylic disinfectants. *Soap Sanit Chem* 17:103–111, 1941
6. Agency or Toxic Substances and Disease Registry (ATSDR): *Toxicological Profile for Cresols: o-Cresol, p-Cresol, m-Cresol.* US Department of Health and Human Services, Public Health Service, TP-91/11, p 148, 1992
7. Boutwell RK, Bosch DK: The tumor-promoting action of phenol and related compounds for mouse skin. *Cancer Res* 19:413–424, 1959
8. Grant WM: *Toxicology of the Eye*, 3rd ed, pp 283–284. Springfield, IL, Charles C Thomas, 19869.
9. Thompson DC, Perera K, Fisher R, et al: Cresol isomers: comparison of toxic potency in rat liver slices. *Toxicol Appl Pharmacol* 125:51–58, 1994

CROTONALDEHYDE
CAS: 4170-30-3

CH₃CHCHCHO

Synonyms: β-Methyl acrolein; 2-butenal; crotonic aldehyde

Physical Form.   Colorless liquid

Uses.   In manufacture of *n*-butyl alcohol; solvent in purification of oils

Exposure.   Inhalation

Toxicology.   Crotonaldehyde is an irritant of the eyes, mucous membranes, and skin.

Exposure of humans to 4 ppm for 10 minutes caused lacrimation and upper respiratory irritation; at 45 ppm, there was conjunctival irritation after a few seconds.[1,2]

In eight cases of corneal injury reported from industrial exposure to crotonaldehyde, healing was complete in 48 hours; the severity of exposure was not specified.[3]

Rats did not survive exposure to 1650 ppm for 10 minutes; effects included respiratory distress, an excitatory stage, and terminal convulsions; autopsy revealed bronchiolar damage.[2] Pulmonary edema has also been observed in rats after fatal exposure to 1500 ppm for 30 minutes or 100 ppm for 4 hours. Administered in the drinking water for 113 weeks, 0.6 mm crotonaldehyde induced neoplastic lesions in rats; 2 of 27 animals had hepatocellular carcinomas, and 9 of 27 had neoplastic lesions.[4] Altered liver-cell foci occurred in 23 of the 27 animals. Crotonaldehyde is classified as a group B carcinogen—justifiably suspected of having carcinogenic potential.

The 1995 ACGIH threshold limit value–time-weighted average (TLV-TWA) for crotonaldehyde is 2 ppm (5.7 mg/m³).

REFERENCES

1. Pattle RE, Cullumbine H: Toxicity of some atmospheric pollutants. *Br Med J* 2:913–916, 1956
2. Rinehart WE: The effect of single exposures to crotonaldehyde vapor. *Am Ind Hyg Assoc J* 28:561–566, 1967
3. Grant WM: *Toxicology of the Eye*, 3rd ed, pp 284–285. Springfield, IL, Charles C Thomas, 1986
4. Chung FL, Tanaka T, Hecht SS: Induction of liver tumors in F344 Rats by crotonaldehyde. *Cancer Res* 46:1285–1289, 1986

CUMENE
CAS: 98-82-8

$C_6H_5C_3H_7$

Synonyms: Cumol; 2-phenylpropane; isopropylbenzene

Physical Form. Colorless liquid

Uses. As thinner for paints and lacquers; as component of high-octane aviation fuel; in production of styrene; in organic synthesis

Exposure. Inhalation; skin absorption

Toxicology. Cumene is an eye and mucous membrane irritant. At high concentrations, it causes narcosis in animals; it is expected that severe exposure will produce the same effect in humans.

Concentrations lethal to humans are not expected to be encountered at room temperature because of the low volatility of cumene.[1] If inhalation of high concentrations of the vapor did occur, dizziness, incoordination, and unconsciousness could be expected.[2] In animals, cumene narcosis is characterized by slow induction and long duration, suggesting a cumulative action.[3] There are no reports of systemic effects in humans.

The LC_{50} for rats was 8000 ppm for a 4-hour exposure.[4] The LC_{50} was 2040 ppm for a 7-hour exposure to mice; the effect was central nervous system depression.[3] A 20-minute exposure to concentrations ranging from 2000 to 8000 ppm caused neurobehavioral effects in mice, including gait disturbances, impaired psychomotor coordination, decreased arousal and rearing, and changes in posture.[5]

Repeated inhalation by rabbits and rats of 2000 ppm caused ataxia and lethargy.[6] No significant changes were noted in rats exposed 8 hr/day to 500 ppm for 150 days.[2] In rats, guinea pigs, monkeys, and dogs exposed at either 224 ppm 8 hr/day, 5 days/week, for 6 months or 30 ppm continuously for 90 days, there were no adverse effects on symptoms, body weight, or histology.[7]

The LD_{50} for penetration of rabbit skin was 12.3 ml/kg after 14 days.[4] Contact of the liquid with the skin causes erythema and irritation.[8] Eye contamination may produce conjunctival irritation.[1]

It is generally agreed that cumene has no damaging effect on the hematopoietic system in spite of its chemical similarity to benzene.[5]

The 1995 TLV-TWA is 50 ppm (246 mg/m³) with a notation for skin absorption.

REFERENCES

1. AIHA hygenic guide series: cumene, Vol 1. Akron, OH, American Industrial Hygiene Association, 1978
2. Sandmeyer EE: Aromatic hydrocarbons. In Clayton GD, Clayton FE (eds): *Patty's Industrial Hygiene and Toxicology*, 3rd ed, Vol 2B, *Toxicology*, pp 3309–3310. New York, Wiley-Interscience, 1981
3. Werner HW, Dunn RC, von Oettingen WF: The acute effects of cumene vapors in mice. *J Ind Hyg Toxicol* 26:264–268, 1944
4. Smyth HF Jr, Carpenter CP, Weil CS: Range-

finding toxicity data: List IV. *AMA Arch Ind Hyg Occup Med* 4:119–122, 1951

5. Tegeris JS, Balster RL: A comparison of the acute behavioral effects of alkylbenzenes using a functional observational battery in mice. *Fund Appl Toxicol* 22:240–250, 1994
6. Lee EW: Cumene. In Snyder R (ed): *Ethel Browning's Toxicity and Metabolism of Industrial Solvents*, 2nd ed, Vol I, *Hydrocarbons*, pp 96–104. New York, Elsevier, 1987
7. Jenkins, LJ Jr, Jones RA, Siegel J: Long-term inhalation screening studies of benzene, toluene, *o*-xylene and cumene on experimental animals. *Toxicol Appl Pharmacol* 16:818–823, 1970
8. Gerarde HW: Toxicological studies on hydrocarbons. *AMA Arch Ind Health* 19:403–418, 1959

CYANAMIDE
CAS: 420-04-2

CH_2N_2

Synonyms: Carbodiimide; cyanoamine; hydrogen cyanamide; cyanogen nitride; carbamonitrile

Physical Form. Crystalline solid

Uses. Fumigants; metal cleaners; in production of synthetic rubber; chemical synthesis

Exposure. Ingestion; inhalation

Toxicology. Cyanamide is an irritant of the eyes, mucous membranes, and skin; it is an inhibitor of aldehyde dehydrogenase and can cause an "Antabuse" effect with ethanol ingestion.

Cyanamide is severely irritating and caustic to the eyes, skin, and respiratory tract.[1]

Concurrent exposure to cyanamide and ethanol produces tachycardia, decreased diastolic blood pressure, hypertension, increased respiration rate, and symptoms of alcohol intoxication because of a buildup of acetaldehyde.[2]

In a six-month study of rats, oral doses of 2.7 or 25 mg/kg/day caused no hepatic changes.[3] A two-generation study of reproduction and fertility in rats used oral doses of 2, 7, or 25 mg/kg/day.[4] Maternal toxicity was observed. Decreases in dam weight, number of corpora lutea, number of implantations, and number of neonates was attributable to the toxic effects in the dams. There were no findings in the F_1 generation.

The 1995 ACGIH threshold limit value–time-weighted average (TLV-TWA) for cyanamide is 2 mg/m³.

REFERENCES

1. Grant WM: *Toxicology of the Eye*, 3rd ed, p 286. Springfield, IL, Charles C Thomas, 1986
2. Hills BW, Venable HL: The interaction of ethyl alcohol and industrial chemicals. *Am J Indust Med* 3:321–333, 1982
3. Obach R et al: Lack of hepatotoxicity after long-term administration of cyanamide in rats: a histological and biochemical study. *Acta Pharmacol Toxicol* 57:279–284, 1985
4. Vallies J et al: A two-generation reproduction-fertility study of cyanamide in the rat. *Pharmacol Toxicol* 61:20–25, 1987.

CYANIDES

NaCN: 143-33-9
KCN: 151-50-8
Ca(CN)₂: 592-01-8

Synonyms/compounds: Sodium cyanide; potassium cyanide; calcium cyanide ("black cyanide")

Physical Form. Powders, granules, or flakes

Uses. Extraction of gold and silver; electroplating; hardening of metals; coppering; zincing; bronzing; manufacture of mirrors; reclamation of silver from photographic film

Exposure. Inhalation; skin absorption; ingestion

Toxicology. The alkali salts of cyanide can cause rapid death from metabolic asphyxiation.

The cyanide ion exerts an inhibitory action on certain metabolic enzyme systems, most notably cytochrome oxidase, the enzyme involved in the ultimate transfer of electrons to molecular oxygen.[1] Because cytochrome oxidase is present in practically all cells that function under aerobic conditions, and because the cyanide ion diffuses easily to all parts of the body, cyanide quickly halts practically all cellular respiration. The venous blood of a patient dying of cyanide is bright red and resembles arterial blood because the tissues have not been able to utilize the oxygen brought to them.[2] Cyanide intoxication produces lactic acidosis, the result of an increased rate of glycolysis and production of lactic acid.[3]

In the presence of even weak acids, hydrogen cyanide gas is liberated from cyanide salts. A concentration of 270 ppm hydrogen cyanide has long been said to be immediately fatal to humans. A more recent study, however, states that the estimated LC_{50} to humans for a 1-minute exposure is 3404 ppm.[1] Other studies indicate that 270 ppm is fatal after 6 to 8 minutes, 181 ppm after 10 minutes, and 135 ppm after 30 minutes.[1]

If large amounts of cyanide have been absorbed, collapse usually is instantaneous, the patient falling unconscious, often with convulsions, and dying almost immediately. Symptoms of intoxication from less severe exposure include weakness, headache, confusion and, occasionally, nausea and vomiting.[1,2] The respiratory rate and depth usually are increased initially, becoming, at later stages, slow and gasping. Coma and convulsions occur in some cases. If cyanosis is present, it usually indicates that respiration either has ceased or has been very inadequate for a few minutes. In one case of nonfatal ingestion of 600 mg of potassium cyanide, the clinical course was marked by acute pulmonary edema and lactic acidosis.[3]

Most reported cases of occupational cyanide poisoning have involved workers with a mixture of repeated acute or subacute exposures and chronic or prolonged low-level exposures, making it unclear whether symptoms simply resulted from multiple acute exposures with acute intoxication. Some symptoms persisted after cessation of such exposures, perhaps because of the effect of anoxia from inhibition of cytochrome oxidase. Symptoms from claimed "chronic" exposure are similar to those reported after acute exposures, such as weakness, nausea, headache, and vertigo.[1] A study of 36 former workers in a silver-reclaiming facility who were repeatedly exposed to cyanide demonstrated some residual symptoms 7 or more months after cessation of exposure; frequent headache, eye irritation, easy fatigue, loss of appetite, and epistaxis occurred in at least 30% of these workers.[4]

Cyanide solutions or cyanide aerosols generated in humid atmospheres have been reported to cause irritation of the upper respiratory tract (primarily nasal irritation) and skin.[1] Skin contact with solutions of cyanide salts can cause itching, discoloration, or corrosion, which are most likely due to the alkalinity of the solutions. Skin irritation and mild systemic symptoms (e.g., headache, dizziness) have been caused by solutions as dilute as 0.5% potassium cyanide.[1]

Skin contact with aqueous cyanide solutions for long periods has caused caustic burns; these cases may be fatal because of significant skin absorption.[1]

No studies are available to evaluate the carcinogenic risk of cyanide exposure in humans or animals.[5] The cyanide salts are not mutagenic in a variety of genotoxic assays.[5]

At high levels, cyanide acts so rapidly that its odor has no value as a warning.[2] At lower levels, the odor may provide some warning although many individuals are unable to recognize the scent of "bitter almonds."[3]

The 1995 ACGIH ceiling threshold limit for hydrogen cyanide is 4.7 ppm (5 mg/m^3) and, for the cyanide salts, the ceiling limit is 5 mg/m^3 as CN, with a notation for skin absorption.

Diagnosis. *Signs and Symptoms:* Asphyxia and death can occur from high exposure levels, whereas weakness, headache, confusion, nausea, and vomiting result from lesser exposures. Other signs and symptoms include: increased rate and depth of respiration; slow and gasping respiration; pulmonary edema; eye, nose, and skin irritation; and lactic acidosis.

Differential Diagnosis. The diagnosis usually is self-evident from the patient's history; but, even if it is not completely established and cyanide poisoning is suspected, the recommended therapy should be considered.

Special Tests. A blood level of cyanide in excess of $0.2 \mu g/ml$ suggests a toxic reaction but, because of tight binding of cyanide to cytochrome oxidase, serious poisoning may occur with only modest blood levels, especially several hours after poisoning. In view of the likelihood of lactic acidosis, there should be measurement of blood pH, plasma bicarbonate, and blood lactic acid.[3]

Treatment. Cyanide is one of the few toxic substances for which a specific antidote exists, which functions as follows.[6,7] First, amyl nitrite (by inhalation) and then sodium nitrite (intravenously) are administered to form methemoglobin, which binds firmly with free cyanide ions, thus trapping any circulating cyanide ions. The formation of 10% to 20% methemoglobin usually does not involve appreciable risk, and yet it provides a large amount of cyanide-binding substance. Second, sodium thiosulfate is administered intravenously to increase the rate of conversion of cyanide to the less toxic thiocyanate. Although early literature suggests the use of methylene blue, it must not be administered because it is a poor methemoglobin former; moreover, it promotes the conversion of methemoglobin back to hemoglobin.[7]

Note: The therapeutic regimen (amyl nitrite/sodium nitrite plus sodium thiosulfate) is the "classical" one, which has long been recommended. However, there is growing support for a newer regimen of 100% oxygen, together with intravenous hydroxocobalamin and/or sodium thiosulfate.[3,8] Hydroxocobalamin reverses cyanide toxicity by combining with cyanide to form cyanocobalamin (vitamin B_{12}); one advantage is that hydroxocobalamin apparently is of low toxicity. The only side effects have been occasional urticaria and a brown-red discoloration of the urine.[4] Nonspecific supportive therapy, including the correction of acidosis, maintenance of blood pressure, and assisted ventilation as needed, is an essential part of the treatment; in two cases, patients have recovered from mas-sive cyanide poisoning with supportive therapy only.[3]

Caution: This section should not be used as a guide for treatment of cyanide when specific exposures have occurred. Treatment protocols are constantly changing, and medical toxicologists and poison control centers should be consulted for medical advice.

REFERENCES

1. National Institute for Occupational Safety and Health: *Criteria for a Recommended Standard . . . Occupational Exposure to Hydrogen Cyanide and Cyanide Salts (NaCN, KCN, and Ca(CN)₂)*, DHEW (NIOSH) Pub No 77-108, pp 37–95, 106–114, 160–173, 178. Washington, DC, US Government Printing Office, 1976
2. Gosselin RE, Smith RP, Hodge HC: *Clinical Toxicology of Commercial Products, Section III*, 5th ed, pp 123–130. Baltimore, MD, Williams and Wilkins, 1984
3. Graham DL, Laman D, Theodore J, Robin ED: Acute cyanide poisoning complicated by lactic acidosis and pulmonary edema. *Arch Intern Med* 137:1051–1055, 1977
4. Blanc P, Hogan M, Mallin K, et al: Cyanide intoxication among silver-reclaiming workers. *JAMA* 253:367–371, 1985
5. Agency for Toxic Substances and Disease Registry (ATSDR): *Toxicological Profile for Cyanide*. US Department of Health and Human Services, Public Health Service, p 123, 1993
6. Chen KK, Rose CL: Nitrite and thiosulfate therapy in cyanide poisoning. *JAMA* 149:113–119, 1952
7. Wolfsie JH: Treatment of cyanide poisoning in industry. *AMA Arch Ind Hyg Occup Med* 4:417–425, 1951
8. Berlin C: Cyanide poisoning—a challenge. *Arch Intern Med* 137:993–994, 1977

CYANOGEN

CAS: 460-19-5

$(CN)_2$

Synonyms: Carbon nitride; dicyanogen; nitriloacetonitrile; oxalonitrile; prussite

Physical Form. Gas

Uses. Organic synthesis; fuel gas for welding and cutting heat-resistant metals; rocket and missile propellant; fumigant

Exposure. Inhalation

Toxicology. Cyanogen reacts with water, acids, and acid salts to produce hydrogen cyanide, which causes death from metabolic asphyxiation.

Exposure of humans to 16 ppm caused eye and nose irritation.[1] A concentration of 270 ppm hydrogen cyanide has long been said to be immediately fatal to humans. A more recent study, however, states that the estimated LC_{50} to humans for a 1-minute exposure is 3404 ppm.[2] Others state that 270 ppm is fatal after 6 to 8 minutes, 181 ppm after 10 minutes, and 135 ppm after 30 minutes.[2]

If large amounts of cyanide have been absorbed, collapse usually is instantaneous, the patient falling unconscious, often with convulsions, and dying almost immediately.[2,3] Symptoms of intoxication from less severe exposure include weakness, headache, confusion, vertigo, fatigue, anxiety, dyspnea and, occasionally, nausea and vomiting.[2-4] The respiratory rate and depth usually are increased initially and, at later stages, become slow and gasping. Coma and convulsions occur in some cases. If cyanosis is present, it usually indicates that respiration either has ceased or has been very inadequate for a few minutes.

Cyanide ion exerts an inhibitory action on certain metabolic enzyme systems, most notably cytochrome oxidase, the enzyme involved in the ultimate transfer of electrons to molecular oxygen.[2] Because cytochrome oxidase is present in practically all cells that function under aerobic conditions and because the cyanide ion diffuses easily to all parts of the body, cyanide quickly halts practically all cellular respiration. The venous blood of a patient dying of cyanide is bright red and resembles arterial blood because the tissues have not been able to utilize the oxygen brought to them.[3] Cyanide intoxication produces lactic acidosis, probably the result of an increased rate of glycolysis and the production of lactic acid.[4]

Studies in rats suggested that cyanogen is tenfold less acutely toxic than hydrogen cyanide.[1] In rats and monkeys exposed to 11 or 25 ppm cyanogen for 6 hr/day, 5 days/week, for 6 months, there were no effects on hematologic or clinical chemistry values.[5] Mean body weight was reduced in rats at the higher level.

The 1995 ACGIH threshold limit value–time-weighted average TLV-TWA is 10 ppm (21 mg/m^3).

Note: For a description of diagnostic signs, differential diagnosis, clinical laboratory tests, and specific treatment of overexposure to cyanogen, see "Cyanides."

REFERENCES

1. McNerney JM, Schrenk HH: The acute toxicity of cyanogen. *Am Ind Hyg Assoc J* 21:121–124, 1960
2. National Institute for Occupational Safety and Health: *Criteria for a Recommended Standard... Occupational Exposure to Hydrogen Cyanide Salts (NaCN, KCN, and Ca[CN]2),* DHEW (NIOSH) Pub No 77-108, pp 37–95, 106–114, 170–173, 178. Washington, DC, US Government Printing Office, 1976
3. Gosselin RE, Smith RP, Hodge HC: *Clinical Toxicology of Commercial Products, Section III,* 5th ed, pp 123–130. Baltimore, MD, Williams and Wilkins, 1984
4. Graham DL, Laman D, Theodore J, Robin ED: Acute cyanide poisoning complicated by lactic acidosis and pulmonary edema. *Arch Intern Med* 137:1051–1055, 1977
5. Lewis TR, Anger WK, Te Vault RK: Toxicity evaluation of subchronic exposures to cyanogen in monkeys and rats. *J Environ Pathol Toxicol Oncol* 5:151–163, 1984

CYANOGEN CHLORIDE
CAS:506-77-4

ClCN

Synonyms: Chlorine cyanide; chlorocyanogen; chlorocyanide

Physical Form. Colorless liquid or gas

Uses. Organic synthesis; poison gas used by the military

Exposure. Inhalation

Toxicology. Cyanogen chloride is a severe irritant of the eyes and respiratory tract and can also cause death from metabolic asphyxiation.

In man, concentrations of 159 and 48 ppm have been reported as fatal after 10 and 30 minutes, respectively.[1] A concentration of 20 ppm was considered intolerable after 1 minute, and 1 ppm for 10 minutes was irritating. Symptoms of exposure include severe irritation of the eyes and respiratory tract, with hemorrhagic exudate of the bronchi and trachea, and pulmonary edema.

In addition to the irritant effects, cyanogen chloride may also cause interference with cellular metabolism via the cyanide radical. The cyanide ion exerts an inhibitory action on certain metabolic enzyme systems, most notably cytochrome oxidase, the enzyme involved in the ultimate transfer of electrons to molecular oxygen.[2] Since cytochrome oxidase is present in practically all cells that function under aerobic conditions and, because the cyanide ion diffuses easily to all parts of the body, cyanide quickly halts practically all cellular respiration. The venous blood of a patient dying of cyanide is bright red and resembles arterial blood because the tissues have not been able to utilize the oxygen brought to them.[3] Cyanide intoxication produces lactic acidosis, probably the result of increased rate of glycolysis and production of lactic acid.[4]

In animal studies, dogs were exposed to 2.3 to 3.6 mg/l cyanogen chloride for 1 to 2 minutes.[1] Following exposure, half of the animals received artificial respiration and amyl nitrite therapy; 8% of the untreated animals survived versus 77% of the amyl nitrite-treated dogs. It was concluded that the combined therapy of oxygen plus amyl nitrite and artificial respiration is indicated in cases with both pulmonary and nitrile effects.

The 1995 threshold limit value–ceiling is 0.3 ppm (0.75 mg/m^3).

Note: For a description of diagnostic signs, differential diagnosis, clinical laboratory tests and specific treatment of overexposure to cyanogen chloride, see "Cyanides."

REFERENCES

1. Hartung R: Cyanides and nitriles. In Clayton GD, Clayton FE (eds): *Patty's Industrial Hygiene and Toxicology*, 3rd ed, Vol 2C, *Toxicology*, pp 4859–4861. New York, Wiley-Interscience, 1982
2. National Institute for Occupational Safety and Health: *Criteria for a Recommended Standard . . . Occupational Exposure to Hydrogen Cyanide Salts (NaCN, KCN, and CA[CN]₂)*, DHEW (NIOSH) Pub No 77-108, pp 37–95, 106–114, 170–173, 178. Washington, DC, US Government Printing Office, 1976
3. Gosselin RE, Smith RP, Hodge HC: *Clinical Toxicology of Commercial Products, Section III*, 5th ed, pp 123–130. Baltimore, MD, Williams and Wilkins, 1984
4. Graham DL, Laman D, Theodore J, Robin ED: Acute cyanide poisoning complicated by lactic acidosis and pulmonary edema. *Arch Intern Med* 137:1051–1055, 1977

CYCLOHEXANE
CAS: 110-82-7

C_6H_{12}

Synonyms: Hexahydrobenzene; benzene hexahydride; hexamethylene

Physical Form. Colorless liquid

Uses. Chemical intermediate; solvent for fats, oils, waxes, resins, and rubber

Exposure. Inhalation

Toxicology. Cyclohexane is irritating to the eyes and mucous membranes; at high concentrations, it causes narcosis in animals, and it is expected that severe exposure will produce the same effect in humans.

A concentration of 300 ppm is detectable by odor and is somewhat irritating to the eyes and mucous membranes.[1] At higher concentrations, the vapor may cause dizziness, nausea, and unconsciousness.[2] Unlike benzene, cyclohexane is not associated with hematologic changes.

Rabbits exposed to 786 ppm cyclohexane for 50 periods of 6 hours each showed minor microscopic changes in the liver and kidneys; lethargy, light narcosis, increased respiration, diarrhea, and some deaths were observed during a total of 60 hours of exposure to 7444 ppm; 1-hour exposure to 26,752 ppm caused rapid narcosis, tremor and, rarely, opisthotonos and was lethal to all exposed rabbits.[3] In mice, exposure to 18,000 ppm produced tremors within 5 minutes, disturbed equilibrium within 15 minutes, and recumbency at 25 minutes.[2] Lethal concentrations administered by inhalation or orally to animals caused generalized vascular damage with severe degenerative changes in the heart, lung, liver, kidney, and brain.[2]

Repeated intraperitoneal administration of 1.5 g/kg caused evidence of renal tubular injury in rats; effects were attributed to cyclohexanol, the main metabolite of cyclohexane.[4] Concentrations of cyclohexanol in urine and cyclohexane in whole blood and serum have shown significant correlations with occupational exposure levels.[5]

Cyclohexane defats the skin on repeated contact.[2]

The 1995 ACGIH threshold limit value–time-weighted average (TLV-TWA) for cyclohexane is 300 ppm (1030 mg/m³).

REFERENCES

1. Gerarde HW: The alicyclic hydrocarbons. In Fassett DW, Irish DD (eds): *Industrial Hygiene and Toxicology*, 2nd ed, Vol 2, *Toxicology*, pp 938–940. New York, Interscience, 1963
2. Sandmeyer EE: Alicyclic hydrocarbons. In Clayton GD, Clayton FE (eds): *Patty's Industrial Hygiene and Toxicology*, 3rd ed, Vol 2B, *Toxicology*, pp 3227–3228. New York, Wiley-Interscience, 1981
3. Treon JF, Crutchfield WE Jr, Kitzmiller KV: The physiological response of animals to cyclohexane, methylcyclohexane, and certain derivatives of these compounds. *J Ind Hyg Toxicol* 25:323–347, 1943
4. Bernard AM, deRussis R, Normand JC, et al: Evaluation of the subacute nephrotoxicity of cyclohexane and other industrial solvents in the Sprague-Dawley rat. *Toxicol Letts* 45:271–280, 1989
5. Yasugi T, Kawai T, Mizunuma K, et al: Exposure monitoring and health effect studies of workers occupationally exposed to cyclohexane vapor. *Int Arch Occup Environ Health* 65:343–350, 1994

CYCLOHEXANOL
CAS: 108-93-0

$C_6H_{11}OH$

Synonyms: Hexahydrophenol; cyclohexyl alcohol

Physical Form. Colorless, viscous liquid

Uses. Solvent for oils, resins, and ethyl cellulose; in manufacture of soap and plastics

Exposure. Inhalation; skin absorption

Toxicology. Cyclohexanol causes irritation of the eyes, nose, and throat; at high concentrations, it causes narcosis in animals, and it is expected that severe exposure will produce the same effect in humans.

Human volunteers exposed to a vapor concentration of 100 ppm for 3 to 5 minutes experienced eye, nose, and throat irritation.[1] Headache and conjunctival irritation have resulted from prolonged exposure to "excessive" but undefined concentrations.[2]

Rabbits exposed 6 hr/day to 272 ppm over a 10-week period showed slight eye irritation;

at 997 ppm, additional effects were salivation, lethargy, narcosis, mild convulsive movements, and some deaths.[3] Lethal doses of cyclohexanol produced slight necrosis of the myocardium and damage to the lungs, liver, and kidneys.[3] The application of 10 ml of cyclohexanol to the skin of a rabbit for 1 hr/day for 10 days induced narcosis, hypothermia, tremors, and athetoid movements; necrosis, exudative ulceration, and thickening of the skin occurred in the area of contact.[2]

Mice fed diets containing 1% cyclohexanol during gestation produced offspring with an increased mortality during the first 3 weeks of life.[2]

The 1995 threshold limit value–time-weighted average (TLV-TWA) is 50 ppm (206 mg/m^3) with a notation for skin absorption.

REFERENCES

1. Nelson KW: Sensory response to certain industrial solvent vapors. *J Ind Hyg Toxicol* 25:282–285, 1943
2. Rowe VK, McCollister SB: Alcohols. In Clayton GD, Clayton FE (eds): *Patty's Industrial Hygiene and Toxicology*, 3rd ed, Vol 2C, *Toxicology*, pp 4643–4649. New York, Wiley-Interscience, 1982
3. Treon JF, Crutchfield WE Jr, Kitzmiller KV: The physiological response of animals to cyclohexane, methylcyclohexane, and certain derivatives of these compounds. *J Ind Hyg Toxicol* 25:323–347, 1943

CYCLOHEXANONE
CAS: 108-94-1

$C_6H_{10}O$

Synonyms: Pimelic ketone; hexanon; sextone

Physical Form. Colorless liquid

Uses. Industrial solvent for cellulose acetate resins, vinyl resins, rubber, and waxes; sovent-

sealer for polyvinyl chloride; in printing industry; coating solvent in audio- and videotape production

Exposure. Inhalation; skin absorption

Toxicology. Cyclohexanone causes eye, nose, and throat irritation; at high concentrations, it produces lethargy and narcosis in animals, and it is expected that severe exposure will cause the same effect in humans.

In most human subjects, exposure to 25 ppm of the vapor for 5 minutes did not cause effects, but 50 ppm was irritating, especially to the throat; exposure to 75 ppm resulted in more noticeable eye, nose, and throat irritation.[1] One recent case history suggested that exposure to cyclohexanone at high levels for many years may be associated with epileptic seizures.[2]

Rabbits exposed to 190 ppm for 6 hr/day for 50 days showed slight liver and kidney injury. At 309 ppm, there was slight conjunctival irritation and, at 1414 ppm, lethargy was observed. At 3082 ppm, effects were incoordination, salivation, labored breathing, narcosis, and some deaths.[3] Five of six rats survived exposure to 2000 ppm for 4 hours, but 4000 ppm caused coma and death of all six. Narcosis, hypothermia, and decreased respiration were observed in guinea pigs exposed to 4000 ppm for 6 hours.[4] Recovery from narcosis was slow, and three of ten animals died within 4 days of exposure.

Rats and mice administered cyclohexanone in their drinking water for 2 years showed marginal evidence of carcinogenic activity.[5] Male rats receiving 3300 ppm had a 13% incidence of adrenal cortex adenomas versus 2% in controls; the incidence of this neoplasm did not increase in the higher-dose males or in any of the female rats. Mice had a statistically significant increase in incidence of lymphomas/leukemias among the females given 6500 ppm but not among the group given the higher doses. Thus, a dose-related trend in increased neoplasms was not observed among any of the groups.

The IARC has determined that there is inadequate evidence for the carcinogenicity of cyclohexanone in experimental animals and that no data are available on the carcinogenicity to humans.[6]

Rats exposed on gestation days 5 through 20 to concentrations of up to 500 ppm for 7 hr/day showed no significant fetotoxic effects.[7]

Eye contact with liquid cyclohexanone may cause corneal injury.[8] The liquid is a defatting agent, and prolonged or repeated skin contact may produce irritation or dermatitis.[8] Allergic contact dermatitis has been reported in a patient with repeated, direct contact with 100% cyclohexanone solution.[9]

The main metabolite of cyclohexanone is cyclohexanol, which is excreted in the urine.[10] A good correlation has been shown between postshift urinary cyclohexanol levels (corrected for creatinine) and occupational exposure to cyclohexanone.[10]

Cyclohexanone has an odor similar to peppermint, and harmful concentrations are not likely to be voluntarily tolerated.

The 1995 ACGIH threshold limit value–time-weighted average (TLV-TWA) for cyclohexanone is 25 ppm (100 mg/m^3) with a notation for skin absorption.

References

1. Nelson KW et al: Sensory response to certain industrial solvent vapors. *J Ind Hyg Toxicol* 25:282, 1943
2. Jacobsen M, Baelum J, Bonde JP: Temporal epileptic seizures and occupational exposure to solvents. *Occup Environ Med* 51:429–430, 1994
3. Treon JF, Crutchfield WE Jr, Kitzmiller KV: The physiological response of animals to cyclohexanone, methylcyclohexane and certain derivatives of these compounds. II. Inhalation. *J Ind Hyg Toxicol* 25:323, 1943
4. Specht H, Miller JW, Valaer PJ, Sayers RR: *Acute Response of Guinea Pigs to the Inhalation of Ketone Vapors.* National Institute of Public Health Bulletin No 176. Washington, DC, US Government Printing Office, 1940
5. Lijinsky W, Kovatch RM: Chronic toxicity study of cyclohexanone in rats and mice. *J Natl Cancer Inst* 77:941–949, 1986
6. *IARC Monographs on the Evaluation of Carcinogenic Risks to Humans*, Vol 47, Some organic solvents, resin monomers and related compounds, pigments and occupational exposures in paint manufacture and painting, pp 157–169, Lyon, International Agency for Research on Cancer, 1989
7. Samimi BS, Harris SB, DePeyster A: Fetal effects of inhalation exposure to cyclohexanone vapor in pregnant rats. *Toxicol Ind Health* 5:1035–1043, 1989
8. Hygienic guide series: cyclohexanone. *Am Ind Hyg Assoc J* 26:630, 1965
9. Sanmartin O, de la Cuadra J: Occupational contact dermatitis from cyclohexanone as a PVC adhesive. *Contact Dermatitis* 27:189–190, 1992
10. Ong CN, Chia SE, Phoon WH, et al: Monitoring of exposure to cyclohexanone through the analysis of breath and urine. *Scand J Work Environ Health* 17:430–435, 1991

CYCLOHEXENE
CAS: 110-83-8

C_6H_{10}

Synonyms: 1,2,3-Tetrahydrobenzene

Physical Form.   Colorless liquid

Uses.   In manufacture of adipic acid, maleic acid, hexahydrobenzoic acid, and aldehyde; stabilizer for high-octane gasoline

Exposure.   Inhalation

Toxicology.   Cyclohexene is regarded as a mild respiratory irritant and central nervous system depressant by analogy to the observed effects of chemically similar substances.

No acute or chronic effects have been reported in humans.

Mice lost their righting reflex at approximately 9000 ppm, and 15,000 ppm was lethal.[1]

Dogs inhaling cyclohexene vapor (concentration not stated) exhibited symptoms characterized by muscular quivering and incoordination.[2] A 6-month inhalation study of various species repeatedly exposed at 75, 150, 300, or 600 ppm showed a lower weight gain for rats exposed at the highest level.[3] Increased alkaline phosphatase was found with exposures, but no other biochemical or hematological abnormalities were observed.

The liquid defats the skin on direct contact.

The 1995 ACGIH threshold limit value–time-weighted average (TLV-TWA) for cyclohexene is 300 ppm (1010 mg/m³).

REFERENCES

1. Sandemeyer EE: Alicyclic hydrocarbons. In Clayton GD, Clayton FE (eds): *Patty's Industrial Hygiene and Toxicology*, 3rd ed, Vol 2B, *Toxicology*, pp 3233–3266. New York, Wiley-Interscience, 1981
2. Fairhall LT: *Industrial Toxicology*, pp 278–279. Baltimore, MD, Williams and Wilkins, 1949
3. Laham S: Inhalation toxicity of cyclohexene (Abstract 152). *Toxicol Appl Pharmacol* 37:15–156, 1976

CYCLOHEXIMIDE
CAS: 66-81-9

$C_{15}H_{23}NO_4$

Synonyms: Actidione; Acti-aid; isocycloheximide; Naramycin

Physical Form.   Colorless crystals

Uses.   Fungicide; growth regulator

Exposure.   Ingestion; skin contact

Toxicology.   Cycloheximide is a teratogen and reproductive toxin in experimental animals.

There is no information concerning toxic effects in humans although the probable lethal oral dose for humans is 5 to 50 mg/kg.

The lowest LD_{50} reported for cycloheximide is 2 mg/kg following oral administration in the rat.[1] In animal experiments, cycloheximide is irritating to the skin and eyes.

Several studies in rats, mice, and rabbits demonstrate that cycloheximide is em-bryotoxic, fetotoxic, and teratogenic.[2] Intraperitoneal administration of doses as low as 250 μg/kg on day 10 of gestation produced central nervous system, craniofacial, and cardiovascular system abnormalities in rats. Musculoskeletal abnormalities were produced in mice following intraperitoneal administration of doses as low as 30 mg/kg on day 9 of gestation. Subcutaneous administration of 5 mg/kg cycloheximide on day 11 caused postimplantation mortality in the mouse. Effects on fertility were observed in pregnant rabbits administered as little as 5 μg/kg on day 1 of gestation.

Cycloheximide is genotoxic in *Escherichia coli* with metabolic activation and in the mouse sperm-morphology assay. Carcinogenicity bioassays in the mouse and rat are inconclusive.[2]

The ACGIH has not established a threshold limit value for cycloheximide.

REFERENCES

1. Sax, N.I.: Cycloheximide. *Dangerous Properties of Industrial Materials Report* 9:55–64, 1989
2. *Registry of Toxic Effects of Chemical Substances*, pp 13059–13061. Washington, DC, National Institute of Occupational Safety and Health, 1990

CYCLOHEXYLAMINE
CAS: 108-91-8

$C_6H_{13}N$

Synonyms: Aminocyclohexane; aminohexahydrobenzene; CHA; hexahydroaniline; hexahydrobenzenamine

Physical Form.   Colorless to slightly yellow liquid with a strong, fishy odor

Uses.   In production of rubber-processing chemicals; corrosion inhibitor in boiler feed water; in production of insecticides, plasticizers, and dry-cleaning soaps; metabolite of the sweetener cyclamate

Exposure.   Inhalation; eye and skin contact

Toxicology.   Cyclohexylamine is an irritant of the mucous membranes, eyes, and skin.

In three cases of acute human exposure from industrial accidents, symptoms included light-headedness, drowsiness, anxiety and apprehension, and nausea; slurred speech and vomiting also occurred in one case.[1] In human patch tests, a 23% solution caused severe irritation and possible sensitization. However, guinea pig sensitization tests did not confirm a potential for sensitization.[2]

In a multispecies study, rabbits, guinea pigs, and rats were exposed 7 hr/day, 5 days/week, to levels of 150, 800, and 1200 ppm.[1] At 1200 ppm, all animals except one rat died after a single exposure. At 800 ppm, fractional mortality occurred after repeated exposures. At 150 ppm, four of five rats and two guinea pigs survived 70 hours of exposure, but one rabbit died after only 7 hours. Effects were irritation of the respiratory tract and eyes, with the development of corneal opacities.

When the undiluted liquid was applied to the skin of guinea pigs and kept in contact under an occluding cuff, the LD_{50} was between 1 and 5 ml/kg; edema, necrosis, and persistent eschars were observed.[2] One drop of a 50% aqueous solution in the eye of a rabbit caused complete destruction of the eye.

Cyclohexylamine has long been known to be pharmacologically active and has sympathomimetic effects on the heart and blood pressure.[3] However, it is not particularly potent.[4]

Cyclohexylamine is a metabolite of the artificial sweetener sodium cyclamate, with the amount of conversion varying considerably from person to person.[5]

Cyclohexylamine has been studied for carcinogenicity in two studies in mice, one of which was a multigeneration study, and in four studies in rats.[6] There were no differences in tumor incidence between treated and control animals. Several studies have shown no evidence of mutagenicity or teratogenicity.

Chromosome damage was induced in bone marrow cells of rats by intraperitoneal injection of 10 to 50 mg/kg per day for 5 days, and in peripheral blood cells of fetal lambs treated in utero with 50 to 250 mg/kg.[7,8]

The 1995 ACGIH threshold limit value–time-weighted average (TLV-TWA) for cyclohexylamine is 10 ppm (41 mg/m³).

REFERENCES

1. Watrous RM, Schulz HN: Cyclohexylamine, p-chloronitrobenzene, 2-aminopyridine: toxic effects in industrial use. *Ind Med Surg* 19:317, 1950
2. Sutton WL: Aliphatic and alicyclic amines. In Fassett DW, Irish DD (eds): *Industrial Hygiene and Toxicology*, 2nd ed, Vol. 2, *Toxicology*, pp 2037–2059. New York, Interscience, 1963
3. Eichelbaum M, Hengstmann JH, Rost HD, et al: Pharmacokinetics, cardiovascular and metabolic actions of cyclohexylamine in man. *Arch Toxicol* 31:243, 1974
4. Buss NE, Renwick AG: Blood pressure changes and sympathetic function in rats given cyclohexylamine by intravenous infusion. *Toxicol Appl Pharmacol* 115:211–215, 1992
5. Kojima S, Ichibagase H: Studies on synthetic sweetening agents: Vol III. Cyclohexylamine, a metabolite of sodium cyclamate. *Chem Pharm Bull (Tokyo)* 14:971, 1966
6. *IARC Monographs on the Evaluation of the Carcinogenic Risk of Chemicals to Man*, Vol 22, Some nonnutritive sweetening agents, p 93. Lyon, International Agency for Research on Cancer, 1980
7. Legator MS et al: Cytogenetic studies in rats of cyclohexylamine, a metabolite of cyclamate. *Science* 165:1139, 1969
8. Turner JH, Hutchinson DL: Cyclohexylamine mutagenicity: an in vivo evaluation utilizing fetal lambs. *Mutat Res* 26:407, 1974

CYCLONITE
CAS: 121-82-4

$C_3H_6N_6O_6$

Synonyms: Hexahydro-1,3,5-trinitro-1,3,5-triazine; RDX; Hexogen

Physical Form.   Colorless crystals

Uses. High explosive; rodenticide

Exposure. Inhalation

Toxicology. Cyclonite is a convulsant.

Workers exposed to cyclonite in an explosives plant complained of nausea and exhibited vomiting, epileptiform seizures, and unconsciousness, which lasted a few minutes to 24 hours, with periods of stupor, nausea, vomiting, and weakness.[1] Recovery was complete, with no sequelae.

In an epidemiological study at a munitions plant where workers were exposed to 0.28 mg/m³ TWA, there were no abnormalities of the hematologic, hepatic, or renal systems.[2]

In male rats dosed by gavage at doses up to 60 mg/kg, spontaneous seizures occurred at 12.5 mg/kg, the lowest dose used.[3]

The 1995 threshold limit value–time-weighted average (TLV-TWA) is 1.5 mg/m³ with a notation for skin.

REFERENCES

1. Kaplan AS, Berghout CF, Peczenik A: Human intoxication from RDX. *Arch Environ Health* 10:877–883, 1965
2. Hathaway JA, Buck CR: Absence of health hazards associated with RDX manufacture and use. *J Occup Med* 19:269–272, 1977
3. Burdette LJ, Cook LL, Dyer RS: Convulsant properties of cyclotrimethylenetrinitramine (RDX): spontaneous, audiogenic, and amygdaloid kindled seizure activity. *Toxicol Appl Pharmacol* 92:436–444, 1988

Uses. In manufacture of resins; in organic synthesis

Exposure. Inhalation; minor skin absorption

Toxicology. Cyclopentadiene is an irritant; repeated exposures have caused mild liver and kidney injury in experimental animals.

In human volunteers, the vapor was irritating at both 250 and 500 ppm.[1]

The oral LD_{50} in rats was 0.82 g/kg.[2] Rats exposed 7 hr/day to 500 ppm for 35 days (over a period of 53 days) developed centrilobular, cloudy swelling of liver cells and cloudy vacuolization of renal tubular epithelium.[1] Dogs exposed 39 times to 400 ppm for 6 hours, followed by 16 exposures at 800 ppm, had no ill effects, as determined by observation, clinical tests, or histological examination.[1] Repeated daily exposure in four species at 250 ppm for 6 months also caused no symptoms. Applied to the skin of rabbits, the liquid caused marked irritation, exudates in the pleural and peritoneal cavities, and hyperemia of the kidneys.[1] The dermal LD_{50} was 6.72 ml/kg for the rabbit.[2]

The 1995 threshold limit value–time-weighted average (TLV-TWA) is 75 ppm (203 mg/m³).

REFERENCES

1. Cyclopentadiene. *Documentation of the TLVs and BEIs*, 6th ed, pp 369–370, Cincinnati, OH, American Conference of Governmental Industrial Hygienists (ACGIH), 1991
2. Smyth HF Jr, Carpenter CP, Weil CS et al: Range-finding toxicity data: List V. *Arch Ind Hyg Occup Med* 10:61–68, 1954

CYCLOPENTADIENE
CAS: 542-92-7

C_5H_6

Synonyms: 1,3-Cyclopentadiene; *p*-pentine; pentole; pyropentylene

Physical Form. Colorless liquid

CYCLOPENTANE
CAS: 287-92-3

C_5H_{10}

Synonym: Pentamethylene

Physical Form. Liquid

Uses/Sources. As a laboratory reagent; in the manufacture of pharmaceuticals; found in solvents and in petroleum ether

Exposure. Inhalation

Toxicology. Cyclopentane is a central nervous system depressant and irritant.

Only limited information is available on human exposures. Symptoms of acute exposure to high concentrations are expected to be excitement, loss of equilibrium, stupor, coma, and respiratory failure.[1] Because cyclopentane is not sufficiently stable to occur naturally in large quantities, most exposures involve a mixture of substances.[1] In the Italian shoe industry, exposure to glue solvents containing up to 18% cyclopentane has been associated with polyneuropathy.[2] However, it is assumed that *n*-hexane is present in these solvents and accounts for the polyneuropathy.

In experiments with mice, the minimal narcotic concentration, loss of reflexes, and lethal dose all occurred at 110 mg/l.[1]

Repeated exposures of rats to 8000 ppm for 6 hr/day for 12 weeks resulted in decreased body weight gains in females; no effects were found in males or females exposed to up to 1100 ppm, 6 hr/day, for 3 weeks.[1] Applied to the skin of guinea pigs, cyclopentane produced slight erythema and dryness.[1]

By analogy with other cyclohydrocarbons, cyclopentane might also be expected to cause upper respiratory tract and eye irritation.

REFERENCES

1. Sandmeyer EE: Alicyclic hydrocarbons. In Clayton GD and Clayton FE (eds): *Patty's In-dustrial Hygiene and Toxicology*, 3rd ed, Vol 2B, pp 3223–3226, New York, Wiley-Interscience, 1981
2. Abbritti G, Siracusa A, Cianchetti C, et al: Shoe-maker's polyneuropathy in Italy: the aetiological problem. *Br J Ind Med* 33:92–99, 1976

CYMENE
CAS: o-Cymene—527-84-4
  m-Cymene—535-77-3
  p-Cymene—99-87-6

$C_{10}H_{14}$

Synonyms: Isopropyltoluene; isopropylmethylbenzene

Physical Form. Colorless liquid with a sweet, aromatic odor

Uses/Sources. Diluent for lacquers, varnishes, and dyes; in the production of resins; component of fragrances; by-product in the manufacture of sulfite paper pulp

Exposure. Inhalation

Toxicology. Cymene, which may occur in ortho, meta, or para forms, is an irritant of the skin and mucous membranes and may cause central nervous system effects; in animals, subcutaneous injection has produced hematological changes.

In a very early report, *p*-cymene produced headache and nausea in volunteers who ingested 3 to 4 g/day for 2 to 3 days (equivalent to 43–57 mg/kg/day).[1] A severe case of blood dyscrasia was found in a man who had been exposed to 340 ppm *p*-cymene for 20 years while working in a sulfite pulp mill.[2] There was also exposure to a variety of other substances during this time, including acetone, sulfur dioxide, acetaldehyde, methyl alcohol, formic acid, and terpenes. No similar cases of hematologic effects following human exposure to *p*-cymene have been reported.

The lowest lethal concentration for rats was 5000 ppm when they were exposed for 45 minutes.[3] At this concentration, signs included dyspnea, twitching of the whiskers, and ataxia, which were followed by hyperreactivity to auditory stimuli. Other signs included rigid tails, carpopedal spasm, generalized quivering, profuse salivation, and hypothermia. In mice, the LD_{50} was 4370 ppm; effects were characteristic of central nervous system excitation, such as tremor and convulsions, which lasted 2 to 3 hours.[1] Inhibitory effects of the central nervous system followed over the next 48 hours and included lethargy, shallow breathing, and coma.

Subcutaneous injection of rabbits with 2 ml of *p*-cymene for 2 days caused an increased number of immature hematopoietic cells in the peripheral blood.

On the skin, *p*-cymene may cause erythema, dryness, and defatting. However, 4% *p*-cymene in petroleum did not produce irritation in 25 humans after a 48-hour closed patch test or after 10 daily applications to the same spot on the backs of subjects.[4] Undiluted *p*-cymene applied to rabbit skin for 24 hours under occlusion was moderately irritating. The LD_{50} by skin absorption is greater than 5 g/kg in rabbits.

A threshold limit value has not been established for *o*-, *m*-, or *p*-cymene.

REFERENCES

1. Lee EW: *p*-Cymene. In Snyder R (ed): *Ethel Browning's Toxicity and Metabolism of Industrial Solvents*, 2nd ed, Vol I, *Hydrocarbons*, pp 105–111. New York, Elsevier, 1987
2. Carlson GW: Aplastic anemia following exposure to products of the sulfite pulp industry: a report of one case. *Ann Intern Med* 24:277–284, 1946
3. Furnas DW, Hine CH: Neurotoxicity of some selected hydrocarbons. *AMA Arch Ind Health* 18:9–15, 1958
4. Opdyke DLJ: *p*-cymene. *Food Cosmet Toxicol* 2:327–343, 1974

DDT
CAS: 50-29-3

$C_{14}H_9Cl_5$

Synonyms: 1,1,1-Trichloro-2,2-bis(*p*-chlorophenyl)ethane; dichlorodiphenyltrichloroethane

Physical Form. White crystalline solid

Uses. Insecticide; use banned in many temperate countries but still widely used in the tropics

Exposure. Inhalation; skin absorption; ingestion

Toxicology. DDT affects the nervous system at high doses and causes paresthesias, tremor, and convulsions.

Ingestion by humans of 10 mg/kg is sufficient to cause effects in some, and convulsions have frequently occurred after ingestion of 16 mg/kg; 285 mg/kg has been taken without fatal results.[1] The onset of effects, usually occurring 2 or 3 hours after ingestion, is characterized by paresthesias of the tongue, lips, and face; the subject soon develops tremor, a sense of apprehension, dizziness, confusion, malaise, headache, and fatigue; in severe intoxication, convulsions may occur and there may be paresis of the hands.[1] Ingestion of very large doses induces vomiting.[1] Recovery is well advanced or complete in 24 hours except in the most serious cases; three persons who each ingested an estimated 20 g of DDT showed a residual weakness of the hands after 5 weeks. Heavy exposure to the dust may cause skin and eye irritation.[1]

Although chronic poisoning in humans has not been described, continued absorption of DDT by humans results in storage of DDT and metabolites, including DDE [2,2-bis(*p*-chlorophenyl)-1,1-dichloroethylene] in fat.[2-4] In a study of 20 workers exposed to DDT for 11 to 19 years and with a calculated daily intake of 18 mg/person (calculated from DDA content in fat and DDA excretion in urine), the sum of isomers and metabolites of DDT in the fat was 38 to 647 ppm (compared to an average of 8 ppm for the general population); while DDE was the major excretory product in the general population, DDA [2,2-bis(p-chlorophenyl)acetic acid] was the major excretory product in DDT-exposed workers.[2]

Large oral doses of DDT in rats caused focal and centrolobular necrosis of the liver.[1] However, in clinical evaluation and laboratory studies of 31 workers exposed to equivalent oral intakes of 3.6 to 18 mg daily for an average of 21 years, there was no evidence of hepatotoxicity; an observed increase in activity of hepatic microsomal enzymes was not accompanied by clinical evidence of detriment to general health.[5]

The hepatocarcinogenicity of DDT by the oral route has been demonstrated and confirmed in several strains of mice. Liver-cell tumors have been produced in both sexes, and, in CF mice, were found to have metastasized to the lungs.[6] However, the tumorigenic potential of DDT was negligible in monkeys after dosing for 15 to 22 years. Of 35 monkeys administered 20 mg DDT/kg, 5 days/week, for 130 months, only one developed hepatocellular carcinoma following a latency period of 20 years.[7]

Embryotoxicity and fetotoxicity have been reported in animal studies in the absence of maternal toxicity.[8] In mice, exposure to DDT during gestation and in the neonatal stage has also caused developmental neurotoxicity in the form of behavioral deficits in the learning process, which persisted into adulthood.

DDT is genotoxic, inducing chromosomal aberrations in human and nonhuman systems.[8]

The 1995 ACGIH threshold limit value–time-weighted average for DDT is 1.0 mg/m³.

REFERENCES

1. Hayes WJ Jr: *Pesticides Studied in Man*, pp 180–205. Baltimore, MD, Williams and Wilkins, 1982
2. Laws ER Jr, Curley A, Biros FJ: Men with intensive occupational exposure to DDT—a clinical and chemical study. *Arch Environ Health* 15:766–755, 1967
3. Hayes WJ Jr, Dale WE, Pirkle CI: Evidence of safety of long-term, high, oral doses of DDT for man. *Arch Environ Health* 22:119–135, 1971
4. Wolfe HR, Armstrong JF: Exposure of formulating plant workers to DDT. *Arch Environ Health* 23:169–176, 1971
5. Laws ER Jr, Maddrey WC, Curley A, Burse VW: Long-term occupational exposure to DDT. *Arch Environ Health* 27:318–321, 1973
6. *IARC Monographs on the Evaluation of the Carcinogenic Risk of Chemicals to Man*, Vol. 5, Some organochlorine pesticides, pp 90–124. Lyon, International Agency for Research on Cancer, 1974
7. Thorgeirsson UP, Dalgard DW, Reeves J, et al: Tumor incidence in a chemical carcinogenesis study of nonhuman primates. *Reg Toxicol Pharmacol* 19:130–151, 1994
8. Agency for Toxic Substances and Disease Registry (ATSDR): *Toxicological Profile for 4,4'-DDT, 4,4'-DDE, 4,4'-DDD*. US Department of Health and Human Services, Public Health Service, p 131, 1992

DECABORANE
CAS: 17702-41-9

$B_{10}H_{14}$

Synonyms: Decaboron tetradecahydride

Physical Form. White crystalline solid

Uses. In rocket propellants; in polymer synthesis; corrosion inhibitor; fuel addditive; mothproofing agent

Exposure. Inhalation; skin absorption

Toxicology. Decaborane affects the nervous system and causes signs of both hyperexcitability and narcosis.

In humans, the onset of symptoms is frequently delayed for 24 to 48 hours after exposure; dizziness, headache, and nausea are common; other symptoms of mild intoxication include light-headedness, drowsiness, incoordination, and fatigue; more severe intoxication results in tremor, localized muscle spasms, and convulsive seizures.[1-3] Muscle spasm usually subsides after 24 hours, whereas light-headedness and fatigue may remain for up to 3 days.[3]

The 4-hour inhalation LC50 for mice was 26 ppm; signs included restlessness, depressed breathing, generalized weakness, and corneal opacities.[4] Rats exhibited normal activity during 4-hour exposures to concentrations ranging up to 95 ppm.

Exposure of rabbits to 56 ppm for 6 hours was fatal; effects included dyspnea, coarse movements of the head, weakness, rigid hindquarters, absence of eye reflexes, and convulsive seizures.[5] By percutaneous application, the rabbit LD50 was 113 mg/kg.[6] The hazard from skin absorption is considered to be high.[7]

Cumulative toxic effects occurred in various animal species receiving repeated small doses of decaborane by oral, intraperitoneal, or cutaneous routes.[7] The rate of recovery was markedly delayed in some animal species surviving repeated doses compared to those that had received a single large dose. In dogs repeatedly given oral doses of 3 mg/kg, the effects on the central nervous system were not pronounced, but there was damage to the liver and kidneys.

Intravenous administration of 4 to 10 mg/kg produced bradycardia and an initial transient hypertensive effect in the anesthetized dog.[8]

Toxicity is thought to occur from the decomposition of decaborane to a stable intermediate which, in turn, inhibits intracellular pyridoxal phosphate–requiring enzymes.[9]

Rapid olfactory fatigue excludes odor as a satisfactory early warning device.[2]

The 1995 ACGIH threshold limit value–time-weighted average for decaborane is 0.05 ppm (0.25 mg/m^3) with a short-term excursion limit of 0.15 pm (0.75 mg/m^3) and a notation for skin absorption.

REFERENCES

1. Lowe HJ, Freeman G: Boron hydride (borane) intoxication in man. *AMA Arch Ind Health* 16:523–533, 1957

2. Chemical Safety Data Sheet SD-84, *Boron Hydrides,* pp 5–18. Washington, DC, Manufacturing Chemists Association, 1961
3. Roush G Jr: The toxicology of the boranes. *J Occup Med* 1:46–52, 1959
4. Svirbely JL: Acute toxicity studies of decaborane and pentaborane by inhalation. *Arch Ind Health* 10:298–304, 1954
5. Krackow EH: Toxicity and health hazards of boron hydrides. *AMA Arch Ind Hyg Occup Med* 8:335–339, 1953
6. Svirbely JL: Toxicity tests of decaborane for laboratory animals. I. Acute toxicity studies. *Arch Ind Health* 11:132–137,1955
7. Svirbely JL: Toxicity tests of decaborane for laboratory animals. II. Effect of repeated doses. *Arch Ind Health* 11:138–141, 1955
8. Tadepalli AS, Buckley JP: Cardiac and peripheral vascular effects of decaborane. *Toxicol Appl Pharmacol* 29:210–222, 1974
9. Naeger LL, Leibman KC: Mechanisms of decaborane toxicity. *Toxicol Appl Pharmacol* 22:517–527, 1972

DECALIN
CAS: 91-17-8

$C_{10}H_{18}$

Synonyms: Decahydronaphthalene; bicyclo-(4.4.0)decane; naphthane

Physical Form   Colorless liquid

Uses.   Solvent for naphthalene, fats, resins, oils; alternative to turpentine in lacquers, shoe polishes, and waxes; component in motor fuels and lubricants

Toxicology.   Decalin is an irritant of the eyes and mucous membranes; in animals, it causes species- and sex-specific kidney damage.

No serious poisonings with decalin in humans have been reported.[1] Inhalation of 100 ppm produces upper respiratory irritation.[2] Contact with the liquid produced vesicular eczema, accompanied by intense itching, in a worker exposed to decalin and some detergents.[3]

The LC_{50} in rats was estimated to be 710 ppm for a 4-hour exposure and, in mice, the LC_{50} was 1085 ppm, also for a 4-hour exposure.[2,4] A 4-hour inhalation exposure of 8 rats at 1000 ppm caused tremors, convulsions, and death in 3 of the animals.[5]

Subchronic exposure of rats, mice, and guinea pigs to 50 or 250 ppm, 6 hr/day, for 1 month produced different effects in the different species. Rats exhibited increased cytoplasmic hyaline droplet formation in the renal tubule epithelium, whereas mice exposed to the higher concentration developed hepatocellular cytoplasmic vacuolization. Guinea pigs had signs of alveolar irritation that was not dose-related.[5]

An additional, longer-term study of dogs, rats, and female mice exposed to 5 or 50 ppm for 90 days was also conducted.[2] No distinct exposure-related effects were noted in dogs or in female rats; mild, reversible liver damage was noted in the mice. In male rats, decalin exposure produced nephropathy.

Further studies of decalin exposure in rats have characterized the specific sequence of renal alterations: first, the variable occurrence of light-microscopically evident proximal convoluted-tubule epithelial-cell necrosis, presumably a reflection of cellular injury associated with excessive protein accumulation (hyaline droplets); then, the occurrence of granular casts at the junction of the inner and outer bands of the outer zone of the medulla; and, finally, chronic nephrosis occurring secondarily to tubular obstruction by granular casts.[6] It is not known how this excessive protein accumulation (specifically, $\alpha_{2\mu}$-globulin, a low-molecular-weight glycoprotein) results in renal tubular cell death, but it does not appear to be caused through an autolytic process induced by lysosomal enzymeleakage. Recent studies have shown that, although decalin exposure induced enlarged lysosomes in renal tubular cells of treated male rats, the lysosomes remained intact.[7]

Other reports have confirmed the species and sex specificity for kidney toxicity by decalin.

Relevance to human exposure has not been established.[8]

Dropped into rabbit eyes, decalin caused no injury; administered systemically to rabbits and guinea pigs, it produced cataracts.[3]

Decalin has a mild terpenelike odor that may not provide adequate warning of exposure.

A threshold limit value has not been established for decalin.

REFERENCES

1. Longacre SL: Decalin. In Snyder R (ed): *Ethel Browning's Toxicity and Metabolism of Industrial Solvents*, 2nd ed, Vol I, *Hydrocarbons*, pp 242–249, New York, Elsevier, 1987
2. Gaworski CL, Haun CC, MacEwen JD, et al: A 90-day vapor inhalation toxicity study of decalin. *Fund Appl Toxicol* 5:785–793, 1985
3. Browning E: Cyclic hydrocarbons. 14. Decalin. In *Toxicity and Metabolism of Industrial Solvents*, pp 138–140, New York, Elsevier, 1965
4. Gaworski CL, Leahey HF: Subchronic inhalation toxicity of decalin. *Proceedings, 10th Conference on Environmental Toxicology*, University of California, Irvine, pp 226–237, November, 1979
5. Gage JC: The subacute inhalation toxicity of 109 industrial chemicals. *Br J Ind Med* 27:1–18, 1970
6. Kanerva RL, McCracken MS, Alden CL, et al: Morphogenesis of decalin-induced renal alterations in the male rat. *Food Chem Toxicol* 25:53–61, 1987
7. Eurell TE, Eurell JC, Schaeffer DJ, et al: Lysosomal changes in renal proximal tubular epithelial cells of male Sprague-Dawley Rats following decalin exposure. *Toxicol Path* 18:637–642, 1990
8. Stone LC, McCracken MS, Kanerva RL, et al: Development of a short-term model of decalin inhalation nephrotoxicity in the male rat. *Food Chem Toxicol* 25:3541–47,1987

DEMETON

CAS: 8065-48-3

$C_8H_{19}O_3PS_2$

Synonyms: Mixture of *O,O*-diethyl *S*-(and *O*)-2-[(ethylthio)ethyl] phosphorothioates; Mercaptofos; Demox; Systox

Physical Form. Pale yellow to light brown oily liquid

Uses. Acaricide; insecticide

Exposure. Inhalation; skin absorption; ingestion

Toxicology. Demeton is an anticholinesterase agent.

At least four fatal, several severe nonfatal, and a number of mild cases of demeton intoxication have been reported. Both animal experiments and human exposures suggest that the toxicity and potency of demeton are similar to those of parathion.[1] Signs and symptoms of overexposure are caused by the inactivation of the enzyme cholinesterase, which results in the accumulation of acetylcholine at synapses in the nervous system, skeletal and smooth muscle, and secretory glands.[1-3] The sequence of the development of systemic effects varies with the route of entry. The onset of signs and symptoms is usually prompt but may be delayed up to 12 hours. After inhalation, respiratory and ocular effects are the first to appear, often within a few minutes of exposure. Respiratory effects include tightness in the chest and wheezing both of which are caused by bronchoconstriction and excessive bronchial secretion; laryngeal spasms and excessive salivation may add to the respiratory distress; cyanosis may also occur. Ocular effects include miosis, blurring of distant vision, tearing, rhinorrhea, and frontal headache.

After ingestion, gastrointestinal effects, such as anorexia, nausea, vomiting, abdominal cramps, and diarrhea appear within 15 minutes to 2 hours. After skin absorption, localized sweating and muscular fasciculations in the immediate area occur, usually within 15 minutes to 4 hours; skin absorption is somewhat greater at higher ambient temperatures and is increased by the presence of dermatitis.[1-3]

With severe intoxication by all routes, an excess of acetylcholine at the neuromuscular junctions of skeletal muscle causes weakness aggravated by exertion, involuntary twitchings, fasciculations and, eventually, paralysis. The most serious consequence is paralysis of the respiratory muscles. Effects on the central nervous system include giddiness, confusion, ataxia, slurred speech, Cheyne–Stokes respiration, convulsions, coma, and loss of reflexes. The blood pressure may fall to low levels, and cardiac irregularity, including complete heart block, may occur. Complete symptomatic recovery usually occurs within 1 week; increased susceptibility to the effects of anticholinesterase agents persists for up to several weeks. Daily exposure to concentrations that are insufficient to produce symptoms following a single exposure may result in the onset of symptoms. Continued daily exposure may be followed by increasingly severe effects.

Note: For a description of diagnostic signs, differential diagnosis, and medical control, including clinical laboratory tests, as well as specific treatment of overexposure to anticholinesterase insecticides, see "Parathion."

The 1995 ACGIH threshold limit value-time-weighted average (TLV-TWA) for demeton is 0.01 ppm (0.11 mg/m³) with a notation for skin absorption.

REFERENCES

1. Hayes WJ Jr: Organic phosphorus pesticides. In Hayes, WJ Jr: *Pesticides Studied in Man*, pp 284–435. Baltimore, MD, Williams and Wilkins, 1982
2. Koelle GB (ed): Cholinesterases and anticholinesterase agents. *Handbuch der Experimentellen Pharmakologie*, Vol 15, pp 989–1027. Berlin: Springer-Verlag, 1963
3. Taylor P: Anticholinesterase agents. In Gilman AG et al (eds): *Goodman and Gilman's Pharmacological Basis of Therapeutics*, 7th ed, pp 110–129. New York, MacMillan, 1985

DIACETONE ALCOHOL
CAS:123-42-2

(CH₃)₂C(OH)CH₂COCH₃

$(CH_3)_2C(OH)CH_2COCH_3$

Synonyms: 4-Hydroxyl-4-methyl-2-penta-none; diacetonyl alcohol; diacetone; dimethyl acetonyl carbinol

Physical Form. Colorless liquid

Uses. Solvent for pigments, cellulose, resins, oils, fats, and hydrocarbons; hydraulic brake fluid; antifreeze

Exposure. Inhalation; minor skin absorption

Toxicology. Diacetone alcohol causes irritation of the eyes and respiratory tract; at high concentrations, it causes narcosis in animals, and it is expected that severe exposure will cause the same effect in humans.

Most human subjects exposed to 100 ppm for 15 minutes complained of eye, nose, and throat irritation; exposure to 400 ppm also caused chest discomfort.[1,2]

Animals exposed to 2100 ppm for 1 to 3 hours exhibited restlessness, mucous membrane irritation, and drowsiness.[3] Rats exposed to 1500 ppm for 8 hours survived.[4] Injection of 3 ml/kg or intragastric administration of 5 ml/kg diacetone alcohol in rabbits caused respiratory depression, narcosis, and death.[5] A temporary decrease in the number of erythrocytes in the blood of rats was observed for 1 to 4 days following intragastric administration of 2 ml/kg of diacetone alcohol; hepatic lesions characterized by vacuolization and granulation of the parenchymal cells were noted, but recovery was complete in 7 days.[6]

The liquid defats the skin and may produce dermatitis with prolonged or repeat contact; in the eyes, it causes moderate to marked irritation and transient corneal damage.[3]

The 1995 ACGIH threshold limit value–time-weighted average (TLV-TWA) for diacetone alcohol is 50 ppm (238 mg/m³).

REFERENCES

1. Silverman L, Schulte HF, First MW: Further studies on sensory response to certain industrial solvent vapors. *J Ind Hyg Toxicol* 28:262–266, 1946
2. Shell Chemical Corp.: Diacetone alcohol, SC 57–84. *Ind Hyg Bull Toxicity Data Sheet,* 1957
3. Krasavage WJ et al: Ketones. In Clayton GD, Clayton FE (eds): *Patty's Industrial Hygiene and Toxicology,* 3rd ed, Vol 2C, *Toxicology,* pp 4754–4756. New York, Wiley-Interscience, 1982
4. Smyth HF: Improved communication—hygienic standards for daily inhalation. *Am Ind Hyg Assoc Q* 17:129–185, 1956
5. Walton DC, Kehr EF, Lovenhart AS: A comparison of the pharmacological action of diacetone alcohol and acetone. *J Pharmacol* 33:175–183, 1928
6. Keith HM: Effect of diacetone alcohol on the liver of the rat. *Arch Pathol* 13:707–712, 1932

2,4-DIAMINOTOLUENE
CAS: 95-80-7

C₇H₁₀N₂

$C_7H_{10}N_2$

Synonyms: TDA; 2,4-TDA; Toluene-2,4-diamine; 3-amino-*p*-toluidine; 1,3-diamino-4-methylbenzene

Physical Form. Crystalline solid

Uses. Intermediate in the production of toluene diisocyanate, which is used to produce polyurethane; in the production of dyes; developer for direct dyes; in dyeing furs

Exposure. Inhalation; skin absorption

Toxicology. Diaminotoluene (TDA) causes hepatocellular carcinomas in both sexes of rats and in female mice.

TDA was tested for carcinogenicity in the diet in both sexes of F344 rats at time-weighted average doses of 79, 176 (males), and 171 (females) ppm TDA for 103 weeks.[1] Rats of both sexes had hepatocellular carcinomas or neoplastic nodules. The significance of these tumors in both sexes was supported by a high incidence of associated nonneoplastic lesions of the liver. Female rats also had carcinomas or adenomas of the mammary gland in a dose-related manner and at a higher incidence than in controls.

Both sexes of B6C3F$_1$ mice were fed dietary levels of 100 or 200 ppm TDA for 101 weeks.[1] The female mice had excess hepatocellular carcinomas. No tumors occurred in excess in the male mice.

Amine-induced asthma was studied in polyurethane foam workers.[2] A total of 12 subjects occupationally exposed to polyurethane foams who complained of wheezing and breathlessness at work were tested. Subjects were given methacholine, toluene diisocyanate, or TDA, and forced expiratory volume in 1 second (FEV$_1$) was measured. High sensitivity to methacholine was seen in 10 subjects. All subjects reacted to TDI, but no subject responded to TDA. The authors concluded that TDA does not produce asthma in workers exposed to polyurethane foams.

Dinitrotoluene (DNT) and TDA are intermediates in the production of toluene diisocyanate and polyurethane plastics.[3] At a chemical complex, 84 workers exposed to DNT/TDA (classified by intensity and recency of exposure) and 119 nonexposed workers were studied for possible reproductive disorders. Each worker was the subject of: a physician's urogenital examination, a reproductive and fertility questionnaire, an estimation of testicular volume, an assessment of serum follicle–stimulating hormone, and an analysis of semen for sperm count and morphology. No differences were found between the exposed and control groups among any of these variables.

A threshold limit value for TDA has not been assigned.

REFERENCES

1. National Toxicology Program: *Bioassay of 2,4-Diaminotoluene for Possible Carcinogenicity (CAS No 95-80-7)*. Technical Report Series 162. National Technical Information Service, US Department of Commerce, Springfield, VA, 1979

2. Candura F, Moscato G: Do amines induce occupational asthma in workers manufacturing polyurethane foams? *Br J Indust Med* 41:552–553, 1984

3. Hamill PV, Steinberger E, Levine RJ, et al: The epidemiologic assessment of male reproductive hazard from occupational exposure to TDA and DNT. *J Occup Med* 24:985–993, 1982

DIAZOMETHANE
CAS: 334-88-3

CH$_2$N$_2$

Synonyms: Azimethylene; diazirine

Physical Form.   Yellow gas

Uses.   Powerful methylating agent for acidic compounds such as carboxylic acids, phenols, enols; not manufactured for sale and distribution because of toxicity and explosivity

Exposure.   Inhalation

Toxicology.   Diazomethane is a severe pulmonary irritant.

Exposure to the gas is extremely dangerous, causing irritation of the eyes, chest pain, cough, fever, and severe asthmatic attacks. A chemist briefly exposed to an unknown concentration in a laboratory developed a violent cough and shortness of breath, leading to severe pulmonary edema; symptoms completely subsided within 2 weeks.[1] In a fatal incident, another chemist exposed to an unknown concentration of diazomethane, as well as other irritant gases, experienced immediate respiratory distress leading to pneumonitis and death on the fourth day after exposure.[1]

A physician exposed to diazomethane from a laboratory spill noted only a faint odor but immediately experienced severe headache, cough, mild anterior chest pain, generalized aching of muscles, and a sensation of overwhelming tiredness.[2] Within 5 minutes, he was stuporous and, on admission to a hospital, was markedly flushed and feverish; he recovered in approximately 48 hours. Subsequent exposure to trace amounts of the gas produced wheezing, cough, and malaise, leading to the suspicion that this substance may also have a sensitizing effect on the respiratory system. Skin exposure has produced irritation and denudation.[3]

Exposure of cats to 175 ppm for 10 minutes resulted in pulmonary edema and hemorrhage, with death occurring in 3 days.[1] Limited animal studies indicate that diazomethane is carcinogenic in mice (increased incident of lung tumors following skin application) and rats (exposure to the gas caused lung tumors). The carcinogenic risk to humans has not been determined.[4]

The warning properties of diazomethane are poor.[3]

The 1995 ACGIH threshold limit value–time-weighted average (TLV-TWA) for diazomethane is 0.2 ppm (0.34 mg/m³).

REFERENCES

1. Reinhardt CF, Brittelli MR: Heterocyclic and miscellaneous nitrogen compounds. In Clayton GD, Clayton FE (eds): *Patty's Industrial Hygiene and Toxicology*, 3rd ed, rev, Vol 2A, *Toxicology*, pp 2784–2786. New York, Wiley-Interscience, 1981
2. Lewis CE: Diazomethane poisoning: report of a case suggesting sensitization reaction. *J Occup Med* 6:91–93, 1964
3. Sunderman WF: Diazomethane. In International Labour Office, *Encyclopaedia of Occupational Health and Safety*, Vol I, A–K, pp 383–384. New York, McGraw-Hill, 1971
4. *IARC Monographs on the Evaluation of the Carcinogenic Risk of Chemicals to Man*, Vol 7, Some anti-thyroid and related substances, nitrofurans and industrial chemicals, pp 223–230. Lyon, International Agency for Research on Cancer, 1974

DIBENZ[a,h]ANTHRACENE
CAS: 53-70-3

$C_{22}H_{14}$

Synonyms: DBA; dibenzo[a,h]anthracene; 1,2,5,6-dibenzanthracene

Physical Form. Colorless solid

Sources. Major component of polynuclear aromatic hydrocarbons, also known as polycyclic aromatic hydrocarbons; usually bound to small particulate matter present in urban air, industrial and natural combustion emissions, and cigarette smoke.

Exposure. Inhalation

Toxicology. Dibenz[a,h]anthracene (DBA) is carcinogenic to experimental animals.

The IARC considers that there is sufficient evidence that DBA is carcinogenic to experimental animals.[1-2] It has been shown to cause skin papillomas and carcinomas in mice when applied dermally 3 times/week for a lifetime.[3-5]

The genetic toxicity of DBA has been evaluated in a variety of short-term genetic toxicology assays and was positive in most systems.[5] DBA undergoes metabolism to form several reactive intermediates. The 3,4-dihydrodiol metabolite of DBA is thought to be further metabolized to a 3,4-diol-1,2-epoxide, the ultimately mutagenic metabolite. Thus, the genotoxicity of DBA is dependent on metabolic activation, either exogenously supplied or endogenously present, and the ratio of enzymatic activation and detoxification pathways.

Most human exposure to DBA in the environment or workplace occurs when it is particle-bound and a component of complex mixtures of polycyclic aromatic hydrocarbons. Thus, it has not been possible to study the effects of human exposure to DBA alone.

No threshold limit value has been assigned for DBA.

REFERENCES

1. *IARC Monographs on the Evaluation of the Carcinogenic Risk of Chemicals to Humans*, Vol 32, Polynuclear aromatic compounds, Part 1, Chemical, environmental and experimental data, pp 299–308. Lyon, International Agency for Research on Cancer, December 1983
2. *IARC Monographs on the Evaluation of the Carcinogenic Risk of Chemicals to Humans*, Vol 3, Certain polycyclic aromatic hydrocarbons and heterocyclic compounds, pp 178–196. International Agency for Research on Cancer, 1973.
3. Wynder EL and Hoffman D: A study of tobacco carcinogenesis. VII. The role of higher polycyclic hydrocarbons. *Cancer* 12:1079–1086, 1959
4. Van Duuren BL, et al: Carcinogenicity of epoxides, lactones, and peroxy compounds. VI. Structure and carcinogenic activity. *J Natl Cancer Inst* 39:1217–1228, 1967
5. Agency for Toxic Substances and Disease Registry: *Toxicological Profile for Dibenz[a,h]anthracene*, pp 23–40, ATSDR/TP-88/13. Public Health Service, Centers for Disease Control, Atlanta, GA, 1990

DIBORANE
CAS: 19287-45-7

B_2H_6

Synonyms: Boroethane; boron hydride

Physical Form.   Gas

Uses.   High-energy fuel; reducing agent; initiator of polymerization of ethylene, vinyl, and styrene

Exposure.   Inhalation

Toxicology.   Diborane is a pulmonary irritant.

In humans, overexposure results in a sensation of tightness in the chest, leading to precordial pain, shortness of breath, nonproductive cough and, sometimes, nausea.[1-4] Prolonged exposure to low concentrations causes headache, light-headedness, vertigo, chills and, less frequently, fever. Fatigue or weakness occurs and may persist for several hours; tremor or muscular fasciculations occur infrequently and are usually localized and of short duration. Diborane gas has not been found to have significant effects on contact with skin or mucous membranes, although high concentrations may cause eye irritation.[5]

The LC_{50} for rats was 50 ppm for 4 hours; in other animal experiments, acute exposure caused pulmonary edema and hemorrhage, as well as temporary damage to the liver and kidneys.[6] Repeated exposure of dogs at about 5 ppm for 6 hr/day resulted in death after 10 to 25 exposures; 1 of 2 animals survived repeated exposure at 1 to 2 ppm for 6 months.[7] Repeated respiratory insult was thought to be the underlying cause of death.

The threshold of odor detection is approximately 3.3 ppm; the repulsive odor is described as rotten eggs, sickly sweet, musty, or foul.[6]

The 1995 ACGIH threshold limit value–time-weighted average (TLV-TWA) for diborane is 0.1 ppm (0.11 mg/m^3).

REFERENCES

1. Lowe HJ, Freeman G: Boron hydride (boron) intoxication in man. *AMA Arch Ind Health* 16:523–533, 1957
2. Cordasco EM, Cooper RW, Murphy JV, Anderson C: Pulmonary aspects of some toxic experimental space fuels. *Dis Chest* 41:68–74, 1962
3. Roush G Jr: The toxicology of the boranes. *J Occup Med* 1:46–52, 1959
4. Rozendaal HM: Clinical observations on the toxicology of boron hydrides. *AMA Arch Ind Hyg Occup Med* 4: 257–260, 1951
5. Chemical Safety Data Sheet, SD-84, Boron hydrides, pp 5–7. Washington, DC, MCA, Inc, 1961
6. Holzmann RT (ed): *Production of the Boranes and Related Research*, pp 289–294, 329–331, 433–489. New York, Academic Press, 1967
7. Comstock CC et al: Research Report No 258.

Washington, DC, US Army Chemical Corps, Medical Laboratories, March 1954

1,2-DIBROMO-3-CHLOROPROPANE
CAS: 96-12-8

$C_3H_5Br_2Cl$

Synonyms: DBCP; dibromochloropropane

Physical Form. Colorless to yellow liquid

Uses. Formerly as an agricultural nematocide (use banned in the United States in 1977)

Exposure. Inhalation; skin absorption

Toxicology. 1,2-Dibromo-3-chloropropane (DBCP) causes sterility in male workers resulting from a selective effect on seminiferous tubules. It causes cancer in mice and rats, and it is an irritant of the eyes and mucous membranes, as well as a mild depressant of the central nervous system in animals.

DBCP has caused oligospermia and aspermia in male workers.[1] Initial documentation of these effects occurred in workers engaged in the production of DBCP at an agricultural chemical plant in Lathrop, California. Of 41 exposed workers, 3 were women and 11 were men with previous vasectomies. Of the 27 remaining men, 11 had abnormally low sperm counts of less than 1 million/ml; all had been exposed for at least 3 years. None with sperm counts above 40 million had been exposed for more than 3 months.[2]

Subsequent studies in this and three other DBCP plants showed a total of more than 100 cases of oligospermia or aspermia. Exposures in one plant were estimated at 100 to 600 ppb.[1]

A larger clinical-epidemiologic study of these men was undertaken to determine the exposure-effect relationships involved. Of 142 nonvasectomized men providing semen samples, 107 had been exposed to DBCP, and 35 had not been exposed. There was a clear-cut difference in both the distribution of sperm counts and the median counts between the exposed and nonexposed men. Of the exposed, 13.1% were azoospermic, 16.8% were severely oligospermic, and 15.8% were mildly oligospermic.[3] Sperm concentration and serum follicle-stimulating hormone (FSH) levels in 44 of these workers were reassessed 5 to 8 years after exposure was terminated in 1977. Of the 8 originally azoospermic workers, 2 produced sperm during the follow-up, although only 1 had normal sperm counts. No increase in sperm production could be detected in men who had low sperm counts in 1977, and elevated serum FSH levels did not drop in oligospermic or azoospermic men. These results suggested that permanent destruction of germinal epithelium occurs in most DBCP-sterile persons.[4]

However, two other follow-up studies of workers exposed to DBCP indicated that spermatogenesis recovery does occur and may be dose-dependent.[5,6] A subsequent study reported recovery among 30 azoospermic and oligospermic workers who had a minimum of 18 months of exposure during 1976 to 1977.[7] A maximum of 11 years of follow-up data were examined. Of the 26 azoospermic subjects who voluntarily participated in follow-up, 19 (73.0%) showed evidence of spermatogenesis recovery. Thirteen azoospermic subjects recovered to normospermic levels; however, their mean most recent sperm count (44.4 million/ml) was significantly lower than the mean (88.8 million/ml) of the 17 oligospermic subjects who recovered to normospermic levels. The lack of spermatogenesis recovery was definitively shown to be job- and, possibly, age-related. The follicle-stimulating hormone level in 1977 was significantly associated with azoospermia, as well as the likelihood of return to normospermia among the azoospermic subjects.

Male exposure to DCBP has also been associated with an increased frequency of spontaneous abortions in wives of exposed workers, but congenital abnormalities have not been observed among children of workers who received sufficient DBCP exposure to induce oligospermia.[8,9]

Although the testes appear to be the main target of DBCP toxicity, other effects reported by exposed workers include headache, nausea, light-headedness, and weakness.[9]

In animal studies, effects of exposure include increased mortality, gonadal atrophy, and carcinomas. The LC_{50} for rats was 368 ppm for 1 hour and 103 ppm for 8 hours.[10] Irritation of the eyes and respiratory tract was observed at levels of 60 ppm and higher. Moderate depression of the central nervous system was manifested as sluggishness and ataxia.

In rats of both sexes given 50 to 66 exposures to 12 ppm over 70 to 90 days, 40% to 50% of the animals died.[10] In most cases, death was attributed to lung infection. The most striking observation at autopsy was severe atrophy and degeneration of the testes. There were also degenerative changes of the seminiferous tubules, reduction in sperm count, and abnormal development of sperm cells. Other effects were mild damage to the liver and kidneys.

The liquid applied undiluted in the eye of a rabbit caused transient irritation.[10] An LD_{50} of 1.4 g/kg was obtained when the material was applied undiluted for 24 hours to the rabbit skin. Repeated application (20 times) to the skin of a rabbit caused slight crustiness. However, the dermis and subcutaneous tissue showed extensive necrosis.

In a study of carcinogenesis, DBCP was orally administered to rats and mice 5 times/week at maximally tolerated doses and at half those doses.[11-13] As early as 10 weeks after initiation of treatment, there was a high incidence of squamous-cell carcinomas of the stomach in both species. In female rats, there were also mammary adenocarcinomas. Chronic inhalation resulted in carcinomas of the respiratory tract in mice and multiple-site tumors in rats.[14]

DBCP is genotoxic in microbial and mammalian assays.[9] The mechanism for DBCP-induced testicular toxicity may be related to direct DNA damage. Binding of DBCP metabolites to testicular-cell DNA has been demonstrated. Alternatively, inhibition of sperm carbohydrate metabolism could also account for DBCP toxicity to epididymal sperm.

The odor of DBCP was detected at 1.7 ppm, the only level tested.[8]

The ACGIH has not established a threshold limit value for 1,2-dibromo-3-chloropropane.

REFERENCES

1. Department of Labor: Emergency temporary standard for occupational exposure to 1,2 dibromo-3-chloropropane (DBCP). *Federal Register* 42:45535, 1977
2. Whorton D, Krauss RM, Marshall S, et al: Infertility in male pesticide workers. *Lancet* 2:1259–1261, 1977
3. Whorton D, Milby TH, Krauss RM, et al: Testicular function in DBCP exposed pesticide workers. *J Occup Med* 21:161–166, 1979
4. Eaton M, Schenker M, Whorton M, et al: Seven year follow-up of workers exposed to 1,2-dibromo-3-chloropropane. *J Occup Med* 28:1145–1150, 1986
5. Lantz GD, Cunningham GR, Huckins C, et al: Recovery from severe oligospermia after exposure to dibromochloropropane. *Fertil Steril* 35:46–53, 1981
6. Potashnik G: A four-year reassessment of workers with dibromochloropropane-induced testicular function. *Andrologia* 15:164–170, 1983
7. Olsen GW, Laham JM, Bodner KM, et al: Determinants of spermatogenesis recovery among workers exposed to 1,2-dibromo-3-chloropropane. *J Occup Med* 32:979–984, 1990
8. Kharrazi M, Potashnik G, Goldsmith JR: Reproductive effects of dibromochloropropane. *Isr J Med Sci* 10:403–406, 1980
9. Agency for Toxic Substances and Disease Registry (ATSDR): *Toxicological Profile for 1,2-Dibromo-3-chloropropane.* US Department of Health and Human Services, Public Health Service, p 140, TP-91/12, 1992
10. Torkelson TR, Sadek SE, Rowe VK: Toxicologic investigations of 1,2-dibromo-3-chloropropane. *Toxicol Appl Pharmacol* 3:545–559, 1961
11. Olson WA, Habermann R, Weisburger E, et al: Brief communication: Induction of stomach cancer in rats and mice by halogenated aliphatic fumigants. *J Natl Cancer Inst* 51:1993–1995, 1973
12. Ward JM, Habermann RT: Pathology of stomach cancer in rats and mice induced with the agricultural chemicals ethylene dibromide

and dibromochloropropane. *Bull Soc Pharmacol Environ Pathol* 74 (Series 2, Issue 2):10–11, 1974

13. Powers MB, Voelker R, Page N, et al: Carcinogenicity of ethylene dibromide (EDB) and 1,2-dibromo-3-chloropropane and oral administration in rats and mice (Abstract). *Toxicol Appl Pharmacol* 33:171–172, 1975

14. National Toxicology Program: *Carcinogenesis Bioassay of 1,2-Dibromo-3-chloropropane (CAS No. 96-12-8) in F344 Rats and B6C3F₁ Mice (Inhalation Study)*. Technical Report Series No. 206 NTP-81-21, Research Triangle Park, NC, US Department of Health and Human Services, Public Health Service, National Institutes of Health 1982

2-N-DIBUTYLAMINOETHANOL
CAS: 102-81-8

$(C_4H_9)_2NCH_2CH_2OH$

Synonyms: DBAE; β-*n*-dibutylaminoethyl alcohol; *N,N-* dibutylethanolamine

Physical Form. Colorless liquid

Uses. Organic syntheses

Exposure. Inhalation; skin absorption

Toxicology. 2-N-dibutylaminoethanol (DBAE) is a skin and eye irritant; in animals, it causes kidney, liver, and hematological effects.

Effects in humans have not been reported.

Rats exposed at 70 ppm had tremors, convulsive seizures, and eye and nasal irritation within 4 hours of exposure. On subsequent days of 6-hour exposures, mild tremors were evident, and one death in five animals occurred on the fourth day. Effects at the end of 5 days included a 57% average body weight loss, a twofold increase in liver and kidney to body weight ratios, a tenfold increase in total serum bilirubin, an elevated hematocrit, and a slight increase in clotting times.[1] Exposure at 33 ppm for 5 days resulted in growth failure but no mortality; animals at this exposure level appeared essentially normal except for some occasional nose rubbing suggestive of mild irritation. Rats exposed for 6 months to 22 ppm were comparable to controls throughout the exposure period.

At high oral dose levels (4–8 g/kg), rats exhibited periods of depression, followed by tremors, incoordination, clonicotonic convulsions, and death. At lower dose levels (0.5–1.0 g/kg), animals appeared depressed during the first day. On the day after dosing, surviving rats appeared normal except for mild diarrhea. The acute oral LD₅₀ for the neutralized DBAE was 1.78 g/kg. No histopathological changes in the heart, liver, kidneys, adrenals, spleen, brain, or testes were observed in rats sacrificed 24 hours after a dose of 1.2 g/kg.

The LD₅₀ by percutaneous absorption was 1.68 g/kg for the rabbit.[2] Applied to the skin of rabbits, the liquid caused necrosis within 24 hours and, instilled in rabbit eyes, it produced corneal necrosis.

The nauseating odor of DBAE may provide adequate warning of overexposure because it is unlikely that individuals would stay in badly contaminated areas for any length of time.[1]

The 1995 ACGIH threshold limit value–time-weighted average (TLV-TWA) for dibutylaminoethanol is 0.5 ppm (3.5 mg/m³) with a notation for skin absorption.

REFERENCES

1. Cornish HH, Dambrauskas T, Beatty LD: Oral and inhalation toxicity of 2-N-dibutylaminoethanol. *Am Ind Hyg Assoc J* 30:46–51, 1969
2. Smyth HF, Carpenter CP, Weil CS, et al: Range-finding toxicity data. List V. *Arch Ind Hyg Occup Med* 10:61–68, 1954

2,6-DI-*tert*-BUTYL-*p*-CRESOL
CAS:128-37-0

$C_{15}H_{24}O$

Synonyms: BHT; butylated hydroxytoluene; DBPC; 2,6-bis(1,1-dimethylethyl)-4-methyl-phenol

Physical Form.   White, crystalline solid

Uses.   Antioxidant used to preserve fat-containing foods and stabilize rubber, plastics, petroleum

Exposure.   Ingestion

Toxicology.   2,6-Di-*tert*-butyl-*p*-cresol, or BHT, is of relatively low acute toxicity in animals, and there is no evidence of either acute or chronic effects among exposed workers.

The IARC stated that there is limited evidence for the carcinogenicity of BHT in experimental animals.[1] Several feeding studies have been conducted in mice and rats, with these results: either no difference in tumor incidence or increased pulmonary tumors (mice), or pituitary adenomas (rats) at the low dose but not the high dose.[2-4] When tested in combination with other chemicals (usually known mutagens or carcinogens), BHT enhanced, reduced, or had no effect on the DNA-damaging, mutagenic, and clastogenic activities.[1] In most studies, it reduced the activity of indirectly acting mutagens or carcinogens. In mice, a single intraperitoneal injection or feeding of BHT caused pulmonary alveolar-cell necrosis and proliferation.[5,6] The metabolism of BHT to a lung-toxic metabolite appears to be mediated by a cytochrome-P450 system.[5] It also induced hepatomegaly in rats in a feeding study.[6]

The 1995 ACGIH threshold limit value–time-weighted average (TLV-TWA) for 2,6-di-*tert*-butyl-*p*-cresol is 10 mg/m³.

REFERENCES

1. *IARC Monographs on the Evaluation of the Carcinogenic Risk of Chemicals to Humans*, Vol 40, Some naturally occurring and synthetic food components, furocoumarins and ultraviolet radiation, pp 161–206. Lyon, International Agency for Research on Cancer, 1986
2. Clapp NK et al: Selective sex-related modification of diethylnitrosamine-induced carcinogenesis in BALB/c mice by concomitant administration of butylated hydroxytoluene. *J Natl Cancer Inst* 61:177–182, 1978
3. National Cancer Institute: *Bioassay of Butylated Hydroxytoluene (BHT) for Possible Carcinogenicity (CAS No 128-37-0)*, Technical Report Series 150, Bethesda, MD, 1979
4. Hirose M et al: Chronic toxicity of butylated hydroxytoluene in Wistar rats. *Food Cosmet Toxicol* 19:147–151, 1981
5. Kehrer JP, Kacew S: Systemically applied chemicals that damage lung tissue. *Toxicology* 35:251–293, 1985
6. Deichmann WB, Clemmer JJ, Rakoczy R, Bianchine J: Toxicity of ditertiarybutylmethylphenol. *Arch Ind Health* 11:93–101, 1955

DIBUTYL PHENYL PHOSPHATE
CAS: 2528-36-1

$C_{14}H_{23}PO_4$

Synonyms: DBPP; Phosphoric acid, dibutyl phenyl ester

Physical Form.   Slightly yellow liquid

Uses.   Component in hydraulic fluids

Exposure.   Inhalation; skin absorption

Toxicology.   Dibutyl phenyl phosphate (DBPP) has caused skin irritation in humans after repeated or prolonged contact.

Skydrol 500B-4 fire-resistant hydraulic fluid, a proprietary phosphate ester mixture composed principally of DBPP and tributyl phosphate, was evaluated in an inhalation study.[1] Rats were exposed to respirable levels of 5, 100, and 3 00 mg/m³ for 6 hr/day, 5 days/week. After 6 weeks of exposure, 10 rats/sex/

group were euthanized and assessed for indications of toxicity. Another 15 rats/sex/group were studied after a total of 13 weeks of exposure. The only clinical sign of toxicity was a reddish nasal discharge with accompanying oral salivation in mid- and high-exposure animals of both sexes, indicative of an irritant response. Reduced body weights, increased liver weights, and decreased erythrocyte counts, hemoglobin levels, and hematocrit values were observed in high-exposure female rats after 13 weeks of Skydrol exposure. High-exposure male rats also had increased liver weights and decreased hematocrit values after 13 weeks.

DBPP was administered to male and female rats in their diets in separate subchronic (91-day) and two-generation reproduction studies.[2] Dose levels were 5, 50, and 250 mg/kg/day in both studies. In the reproduction study, cross-fostering was performed between some high-exposure and control litter offspring and dams following a second mating of F_0 animals. Compared to control animals, body weights were consistently lower in high-exposure adult animals in both studies. High-exposure rats in the subchronic study had decreased erythrocyte counts and hematocrit and hemoglobin levels. They also had increased liver weights. In the reproduction study, mating and fertility indices were comparable among the parental animals in both generations, but survivability among high-exposure pups reared by control dams appeared to be decreased. Urinary bladder histopathologic changes consisting of mononuclear cell infiltration and transitional epithelial hyperplasia were noted in mid- and high-exposure rats from both studies. The no-observable-adverse-effect level in both these studies was 5 mg/kg/day.

DBPP was tested for its potential to cause organophosphorus compound–induced delayed neurotoxicity (OPIDN) in the adult hen.[3] The acute oral LD_{50} of DBPP was estimated to be 1500 mg/kg and was used as a test dose. For the acute delayed neurotoxicity test, hens were given two test doses of DBPP 21 days apart and killed 21 days after the second dose. None of the hens given DBPP exhibited nerve damage or clinical signs that were different from untreated control animals. A single dose of TOCP resulted in paralysis and a histopathological profile typical of a distal neuropathy. These results indicate that DBPP is unlikely to cause OPIDN with any single sublethal dose.

DBPP is not considered to be a primary irritant or a sensitizing agent, based on patch testing of 50 human volunteers.[4] Repeated or prolonged contact with the skin has caused drying and cracking of exposed skin.

The 1995 ACGIH threshold limit value–time-weighted average (TLV-TWA) is 0.3 ppm (3.5 mg/m³) with a notation for skin absorption.

REFERENCES

1. Healy CE, Nair RS, Ribelin WE, Bechtel CL: Subchronic rat inhalation study with Skydrol 500B-4 fire resistant hydraulic fluid. *Am Ind Hyg Assoc J* 53:175–180, 1992
2 Healy CE, Nair RS, Lemen JK, Johannsen FR: Subchronic and reproduction studies with dibutyl phenyl phosphate in Sprague-Dawley rats. *Fund Appl Toxicol* 16:117–127, 1991
3. Carrington CD, Lapadule DM, Othman M, et al: Assessment of the delayed neurotoxicity of tributyl phosphate, tributoxyethyl phosphate, and dibutylphenyl phosphate. *Toxicol Ind Health* 6:415–424, 1990.
4. Monsanto Chemical Company: *Material Safety Data Bulletin for Dibutyl Phenyl Phosphate.* Monsanto Chemical Co, St Louis, MO, October 28, 1985

DIBUTYL PHOSPHATE
CAS: 107-66-4

$(n\text{-}C_4H_9O)_2(OH)PO$

Synonyms: Dibutyl hydrogen phosphate; di-*n*-butyl phosphate

Physical Form. Pale amber liquid

Uses. Organic catalyst; antifoaming agent

Exposure. Inhalation

Toxicology. Dibutyl phosphate is an irritant of the eyes and mucous membranes.

Data on effects in humans are sparse; workers exposed to unspecified concentrations of vapor complained of respiratory irritation and headache.[1] It is a moderately strong acid and could be expected to be irritating on contact.

In rats, the oral LD_{50} is 3.2 g/kg.[2]

The 1995 ACGIH threshold limit value–time-weighted average (TLV-TWA) for dibutyl phosphate is 1 ppm (8.6 mg/m³) with a short-term excursion limit of 2 ppm 17 mg/m³.

REFERENCES

1. Dibutyl phosphate. *Documentation of the TLVs and BEIs*, 6th ed, p 399. Cincinnati, OH, American Conference of Governmental Industrial Hygienists (ACGIH), 1991
2. Sweet DV (ed): *Registry of Toxic Effects of Chemical Substances*, 1985–1986 ed. US Department of Health and Human Services, p 1947. Washington, DC, US Government Printing Office, 1987

DIBUTYL PHTHALATE
CAS: 84-74-2

$C_{16}H_{22}O_4$

Synonyms: DBP; butyl phthalate; 1,2-benzenedicarboxylic acid dibutyl ester; phthalic acid dibutyl ester

Physical Form. Colorless or slightly colored oily liquid

Uses. Plasticizer in the production of polymer products; in cosmetics; manometer fluid; insect repellant

Exposure. Inhalation; ingestion

Toxicology. Dibutyl phthalate (DBP) is of low-order acute toxicity; reproductive and developmental effects have been reported in animal studies.

A chemical worker who accidently swallowed 10 g (about 140 mg/kg) developed nausea, dizziness, headache, pain, and irritation in the eyes, conjunctivitis, and toxic nephritis. He recovered completely after 2 weeks.[1]

There were no positive reactions to 5% DBP among 53 subjects given a 48-hour closed patch test.[2] Cosmetic formulations containing up to 9% DBP ranged from nonirritating to slightly irritating in various patch-test procedures. Sensitization and photosensitization did not occur.[3]

The single-dose oral LD_{50} has been estimated to be between 20,000 and 25,000 mg/kg for the rat, with some deaths occurring at 10,000 mg/kg.[4] Rats exposed to concentrations as low as 0.5 mg/m³ of DBP mist for 6 hours per day for 6 months had smaller weight gains and greater brain and lung weights than controls.[5] At 50 mg/m³, the effects were more pronounced. In some rodent studies, DBP exposure has been found to have hepatic effects; specifically, DBP may interfere with mitochondrial respiration.[6]

Undiluted DBP instilled in rabbit eyes caused no observable irritation up to 48 hours post-instillation.[7]

Reduced testes weights and histological evidence of testicular injury were found in rats and guinea pigs but not hamsters or mice fed 2 g/kg/day DBP for 10 days, indicating a species-specific response.[8] The basis of this species variation may be related to species differences in the ability to conjugate monobutyl phthalate, the primary metabolite of DBP, with glucuronic acid.[6]

In a continuous breeding study, mice given 1.0% DBP in their diets for 7 days prior to, and during, a 98-day cohabitation period had significant reproductive effects, including a reduction in the number of litters and a reduction in the proportion of pups born alive.[9] These effects were not observed at lower dose levels. Oral or intraperitoneal administration of DBP in pregnant animals at high doses relative to the LD_{50} produced an increased number of resorptions, increased fetal deaths, neural tube defects, cleft palate, skeletal abnormalities, and maternal toxicity.[10-13]

Carcinogenesis was not observed in 18-month or longer feeding studies in rats.[3] DBP has tested negative or marginally genotoxic in gene mutation and chromosomal aberration studies.[6]

The 1995 ACGIH threshold limit value–time-weighted average (TLV-TWA) is 5 mg/m^3.

REFERENCES

1. Krauskopf LG: Studies on the toxicity of phthalates via ingestion. *Environ Health Perspect* 3:61–72, 1973
2. Kaaber S et al: Skin sensitivity to denture base materials in the burning mouth syndrome. *Contact Dermatitis* 5:90–96, 1979
3. Final report on the safety assessment of dibutyl phthalate, dimethyl phthalate, and diethyl phthalate. *J Am Coll Toxic* 4:267–303, 1985
4. White RD, Earnest DL, Carter DE: The effect of intestinal esterase inhibition on the in vivo absorption and toxicity of di-*n*-butyl phthalate. *Food Chem Toxicol* 21:99–101, 1983
5. Kawano M: Toxicological studies on the phthalate esters. 1. Inhalation effects of dibutyl phthalate on rats. *Japan J Hyg* 35:684–692, 1980
6. Agency for Toxic Substances and Disease Registry (ATSDR): *Toxicology Profile for Di-n-butylphthalate.* US Department of Health and Human Services, Public Health Service, TP-90–10, p 103, 1990
7. Lawrence WH, Malik M, Turner JE, et al: Toxicological investigation of some acute, short-term, and chronic effects of administering di-2-ethylhexyl phthalate (DEHP) and other phthalate esters. *Environ Res* 9:1–11, 1975
8. Gangolli SD: Testicular effects of phthalate esters. *Environ Health Perspect* 45:77–84, 1982
9. Lamb JC IV, Chapin RE, Teague J, et al: Reproductive effects of four phthalic acid esters in the mouse. *Tox Appl Pharm* 88:255–269, 1987
10. Peters JW, Cook RM: Effect of phthalate esters on reproduction in rats. *Environ Health Perspect* 3:91–94, 1973
11. Singh AR, Lawrence WH, Autian J: Teratogenicity of phthalate esters in rats. *J Pharm Sci* 61:51–55, 1972
12. Shiota K, Nishimura H: Teratogenicity of di(2-ethylhexyl) phthalate (DEHP) and di-*n*-butyl phthalate (DBP) in mice. *Environ Health Perspect* 45:65–70, 1982
13. Ema M, Amano H, Itami T, et al: Teratogenic evaluation of di-*n*-butyl phthalate in rats. *Tox Letts* 69:197–203, 1993

DICHLOROACETYLENE
CAS: 7572-29-4

Cl_2C_2

Synonyms: Acetylene, dichloro; dichloroethyne

Physical Form. Liquid

Uses/Sources. By-product in synthesis of vinylidene chloride; decomposition product of trichloroethylene under alkaline conditions

Exposure. Inhalation

Toxicology. Dichloroacetylene is a neurotoxin; it is carcinogenic in experimental animals.

Exposure of humans to dichloroacetylene in a variety of settings has caused headache, dizziness, nausea, vomiting, eye irritation, mucous membrane irritation, and neurological disorders, manifested as paresis and neuralgia in several cranial and cervical nerves.[1-5] In some cases, the cranial nerve involvement persisted for several days to years. Extreme nausea occurred among individuals exposed to levels as low as 0.5 to 1.0 ppm.[4]

After a single exposure of rabbits to 17 ppm for 6 hours, the sensory trigeminal nucleus was severely affected.[6] Other effects included tubular and focal necrosis in the collecting tubules of the kidney and fatty degeneration of the liver.[6,7]

In a carcinogenicity inhalation study, rats and mice were exposed to 9 ppm, 6 hr/day, 1 day/week, for 12 months; 2 ppm, 6 hr/day, 1 day/week, for 18 months; or 2 ppm, 6 hr/day, 2 days/week, for 18 months.[8] There was a significant increase in cystic kidney tumors in all exposed animals. Male mice were the most susceptible, with kidney tumors in 90% of exposed animals. Female rats showed an excess of malignant lymphomas.

The selective renal carcinogenicity of dichloroactylene may be due to a bioactivation mechanism that involves glutathione S-conjugate formation, translocation to the kidneys and subsequent renal metabolism to yield reactive electrophiles presumably responsible for carcinogenicity.[9]

The IARC stated that there is limited evidence for the carcinogenicity of dichloroacetylene to experimental animals.[10]

The 1995 proposed ACGIH ceiling-threshold limit value (C-TLV) for dichloroacetylene is 0.1 ppm (0.39 mg/m3) with an A3-animal carcinogen designation.

REFERENCES

1. Humphrey JH, McClelland M: Cranial nerve palsies with herpes following general anaesthesia. a report from the central Middlesex County Hospital. *Br Med J* 1:315–318, 1946
2. Defalque RJ: Pharmacology and toxicology of trichloroethylene: critical review of the world literature. *Clin Pharmacol Ther* 2:665–688, 1961
3. Greim H, Wolff T, Hofler M, Lahaniatis E: Formation of dichloroacetylene from trichloroethylene in the presence of alkaline material—possible cause of intoxication after abundant use of chloroethylene-containing solvents. *Arch Toxicol* 256:74–77, 1984
4. Saunders RA: A new hazard in closed environment atmospheres. *Arch Environ Health* 14:380–384, 1967
5. Henschler D, Broser F, Hopf HC: "Polyneuritis cranialis" following poisoning with chlorinated acetylenes while handling vinylidene chloride copolymers (Ger) *Arch Toxicol* 26:62–75, 1970
6. Reichert D, Liebaldt G, Henschler D: Neurotoxic effects of dichloroacetylene. *Arch Toxicol* 37:23–28, 1976
7. Reichert D, Henschler D, Bannasch P: Nephrotoxic and hepatotoxic effects of dichloroa-

cetylene. *Food Cosmet Toxicol* 16:227–235, 1978
8. Reichert D, Spengler U, Romen W, Henschler D: Carcinogenicity of dichloroacetylene: an inhalation study. *Carcinogenesis* 5:1411–1420, 1984
9. Dekant W, Vamvakas S, Koob M, et al: A mechanism of haloalkene-induced renal carcinogenesis. *Environ Health Perspect* 88:107–110, 1990
10. *IARC Monographs on the Evaluation of the Carcinogenic Risk of Chemicals to Humans*, Vol 39, Some Chemicals used in plastics and elastomers, pp 369–378. Lyon, International Agency for Research on Cancer, 1985

---

o-DICHLOROBENZENE
CAS: 95-50-1

$C_6H_4Cl_2$

Synonyms: 1,2-Dichlorobenzene; dichlorobenzol; dichloricide

Physical Form. Clear liquid

Uses. Organic synthesis (primarily 3,4-dichloroaniline); solvent; in dye manufacture

Exposure. Inhalation

Toxicology. o-Dichlorobenzene is a skin and eye irritant. At high doses, it causes central nervous system depression and liver and kidney damage in animals. Heavy exposure is expected to produce the same effects in humans.

In humans, eye irritation is not usually evident below 20 ppm but becomes noticeable at 25 to 30 ppm and painful to some at 60 to 100 ppm if exposures last longer than a few minutes.[1] Some acclimation may occur but not to a great extent. Workers exposed to concentrations ranging from 1 to 44 ppm and averaging 15 ppm showed no indication of injury or of untoward hematological effect.[2] Accidental exposure of 26 subjects to unspecified levels 8 hr/day for 4 days caused eye, nose, and throat irritation.[3] Of the 26 subjects, 10 reported dizziness, severe headache, fatigue, and nausea. Chromosome studies showed significant alterations in the leukocytes of exposed workers, which appeared reversible 6 months later.

The liquid left on the skin may produce blistering and, later, the area may become pigmented.[2] Sensitization dermatitis has been reported.[2]

Rats died from exposure to 977 ppm for 7 hours but survived when exposed for only 2 hours; animals survived exposure to 539 ppm for 3 hours but, at necropsy, showed marked centrilobular necrosis of the liver, as well as cloudy swelling of the tubular epithelium of the kidney.[2] During exposure, rats exhibited drowsiness, unsteadiness, eye irritation, difficulty in breathing, and anesthesia. Several species of animals exposed for periods of 6 or 7 months to 93 ppm for 7 hours daily showed no adverse effects.[2]

Recent studies with male F344 rats have shown that *o*-dichlorobenzene was more toxic to the liver and kidneys than the meta and para isomers following a single administration.[4] In addition to isomer specificity, strain-specific differential toxicity has also been demonstrated, with Sprague-Dawley rats being relatively resistant to the acute hepatic toxicity of *o*-dichlorobenzene.[5]

Repeated dermal application to rats was fatal.[6] The liquid instilled in the rabbit eye produced apparent distress and slight conjunctival irritation.[2]

There was no evidence of carcinogenicity in rats or mice receiving 60 or 120 mg/kg by gavage 5 times per week for 2 years.[7]

The odor of *o*-dichlorobenzene is perceptible to most people at 2 to 4 ppm.[1]

The 1995 ACGIH threshold limit value–time-weighted average (TLV-TWA) for *o*-dichlorobenzene is 25 ppm (150 mg/m³) with a short-term excursion limit of 50 ppm (301 mg/m³).

REFERENCES

1. Hygienic guide series: *o*-dichlorobenzene. *Am Ind Hyg Assoc J* 25:320–323, 1964
2. Hollingsworth RL, Rowe VK, Oyen R, et al: Toxicity of *o*-dichlorobenzene. *AMA Arch Ind Health* 17:180–187, 1958
3. Zapata-Gayon C et al: Clastogenic chromosomal aberrations in 26 individuals accidentally exposed to ortho dichlorobenzene vapors in the National Medical Center in Mexico City. *Arch Environ Health* 37:231–235, 1982
4. Valentovic MA, Ball JG, Anestis D, et al: Acute hepatic and renal toxicity of dichlorobenzene isomers in Fischer 344 rats. *J Appl Toxicol* 13:1–7, 1993
5. Stine ER, Gunawardhana L, Sipes IG: The acute hepatotoxicity of the isomers of dichlorobenzene in Fischer-344 and Sprague-Dawley rats: isomer-specific and strain-specific differential toxicity. *Toxicol Appl Pharmacol* 109:472–481, 1991
6. US Environmental Protection Agency: *Health Assessment Document for Chlorinated Benzenes, Final Report.* Washington, DC, Office of Health and Environmental Assessment, January 1985
7. National Toxicology Program: *Toxicology and Carcinogenesis Studies of 1,2-Dichlorobenzene (o-Dichlorobenzene) (CAS No 95-50-1) in F344/N Rats and B6C3F₁ Mice (Gavage Studies)*, TRS-255. DHHS (NIH) Pub No 86–2511. Research Triangle Park, NC, US Department of Health and Human Services, October 1985

ρ-DICHLOROBENZENE
CAS: 106-46-7

$C_6H_4Cl_2$

Synonyms: 1,4-dichlorobenzene; ρ-chlorophenyl chloride; paracide

Physical Form. Colorless or white crystals

Uses. Disinfectant and deodorant; chemical intermediate; moth control

Exposure. Inhalation

Toxicology. ρ-Dichlorobenzene vapor is an irritant of the eyes and upper respiratory tract and is toxic to the liver. It is carcinogenic in experimental animals.

In five cases of intoxication by inhalation from household or occupational exposure to ρ-dichlorobenzene used as a mothproofing agent, one person with only moderate exposure suffered severe headache, periorbital swelling, and profuse rhinitis, which subsided 24 hours after cessation of exposure.[1] The other four persons, who had heavier and more prolonged exposure, developed anorexia, nausea, vomiting, weight loss, and hepatic necrosis with jaundice; two died, and another developed cirrhosis. Although these five cases were temporarily associated with known exposure to ρ-dichlorobenzene in four different settings, it is unclear how thoroughly other potential causes for these findings were excluded.

In 58 workers exposed for an average of 4.8 years (range 8 months to 25 years) to ρ-dichlorobenzene at levels of 10 to 725 ppm, there was no evidence of hematologic effects in spite of the structural similarity to benzene, a potent bone marrow depressant. Painful irritation of the eyes and nose was noted at levels between 50 and 80 ppm, and pain was severe at 160 ppm.

Solid particles of ρ-dichlorobenzene in the human eye cause pain.[2] The solid material produces a burning sensation when held in contact with the skin, but the resulting irritation is slight; warm fumes or strong solutions of ρ-dichlorobenzene may irritate the intact skin slightly on prolonged or repeated contact.[2,3] A case of allergic purpura induced by ρ-dichlorobenzene has been reported.[4]

In a study of workers engaged in synthesizing or otherwise handling ρ-dichlorobenzene, it was concluded that urinary excretion of 2,5-dichlorophenol (a metabolite of ρ-dichlorobenzene) can serve as an index of exposure.[5]

Administration of ρ-dichlorobenzene to rats for 13 weeks caused renal tubular cell degeneration in males receiving 300 mg/kg or more; in mice, hepatocellular degeneration was observed in both sexes at doses above 600 mg/kg, but renal damage did not occur at doses up to 1800 mg/kg for 13 weeks.[6]

In male rats given ρ-dichlorobenzene by gavage at 150 and 300 mg/kg for 2 years, there was a significant dose-related increased incidence of tubular cell adenocarcinomas of the kidney; no excess was observed in female rats or in either sex of mice.[6] It has been proposed that ρ-dichlo-robenzene causes an increase in protein droplet formation in the kidney of male rats, leading to cell death and subsequent cell proliferation, which may play a critical role in the carcinogenesis process.[7] The presence of α_{2u}-globulin is essential for the development of this syndrome, and rats that do not synthesize this protein, such as the NCI–Black–Reiter, do not develop renal disease following exposure.[8] ρ-Dichlorobenzene also increased the incidence of hepatocellular adenomas and carcinomas, as well as nonneoplastic liver lesions in male and female mice dosed at 600 mg/kg for 2 years. The National Toxicology Program study concluded that there was clear evidence of carcinogenicity for male rats and for both male and female mice.[6]

In a long-term inhalation study in male and female rats and female mice, there was no evidence of carcinogenicity following exposure at 75 or 500 ppm for 5 hr/day, 5 days/week, for 76 weeks (rats) or 57 weeks (mice).[9] Although there has been a report of 5 cases of blood dyscrasias, including leukemia, among individuals exposed to o- or ρ-dichlorobenzene, the IARC has concluded that the human data are inadequate to evaluate the carcinogenicity of dichlorobenzenes.[10]

Exposure of rats to ρ-dichlorobenzene vapor concentrations up to 538 ppm for two generations resulted in F_0 and F_1 adult toxicity, including reduced body weights in both sexes and kidney effects (hyaline droplet neuropathy and renal tubular cell hyperplasia) in males, but with no effects on reproduction. Postnatal toxicity in F_1 and F_2 litters was observed at the high dose.[11]

ρ-Dichlorobenzene is not genotoxic in both in vivo and in vitro assay systems.[12]

The 1995 ACGIH threshold limit value–time-weighted average (TLV-TWA) for ρ-dichlorobenzene is 10 ppm (60 mg/m^3) with an A3 animal carcinogen designation.

REFERENCES

1. Cotter LH: Paradichlorobenzene poisoning from insecticides. *NY State J Med* 53:1690–1692, 1953

2. Hollingsworth RL, Rowe VK, Oyen F, et al:

Toxicity of paradichlorobenzene: determinations on experimental animals and human subjects. *AMA Arch Ind Health* 14:138–147, 1956

3. Hygienic guide series: ρ-dichlorobenzene. *Am Ind Hyg Assoc J* 25:323–325, 1964
4. Nalbandian RM, Pearce JF: Allergic purpura induced by exposure to ρ-dichlorobenzene. *JAMA* 194:238–239, 1965
5. Pagnotto LD, Walkley JE: Urinary dichlorophenol as an index of para-dichlorobenze exposure. *Am Ind Hyg Assoc J* 26:137–142, 1965
6. National Toxicology Program: *Toxicology and Carcinogenesis Studies of 1,4-Dichlorobenzene (CAS No 106-46-7) in F344/N Rats and B6C3F₁ Mice (Gavage Studies).* DHHS (NIH) Pub No 87-2575. Research Triangle Park, NC, US Department of Health and Human Services, 1987
7. Charbonneau M, Strasser J Jr, Lock EA, et al: Involvement of reversible binding to α_{2u}-globulin in 1,4-dichlorobenzene-induced nephrotoxicity. *Tox Appl Pharm* 99:122–132, 1989
8. Dietrich DR, Swenberg JA: NCI–Black–Reiter (NBR) male rats fail to develop renal disease following exposure to agents that induce α-2u-globulin(α_{2u})nephropathy. *Fund Appl Toxicol* 16:749–762, 1991
9. Loeser F, Litchfield MH: Review of recent toxicology studies on ρ-dichlorobenzene. *Food Chem Toxicol* 21:825–832,1983
10. *IARC Monographs on the Evaluation of Carcinogenic Risks to Humans: Overall Evaluations of Carcinogenicity: An Updating of IARC Monographs Vols 1 to 42.* Suppl 7, pp 192–193. Lyon, International Agency for Research on Cancer, 1987
11. Neeper-Bradley TL, Tyl RW, Fisher LC, et al: Reproductive toxicity study of inhaled paradichlorobenzene (PDCB) vapor in CD rats. *Teratology* 39:470–471, 1989. Abstract
12. Agency for Toxic Substances and Disease Registry (ATSDR): *Toxicological Profile for 1,4-Dichlorobenzene.* US Department of Health and Human Services, Public Health Service, p 111, TP-92/10, 1993

3,3'-DICHLOROBENZIDINE
CAS: 91-94-1

$C_{12}H_{10}C_{12}N_2$

Synonyms: DCB; 4,4'-diamino-3,3'-dichlorobiphenyl; o,o'-dichlorobenzidine

Physical Form. Colorless crystals

Uses. In the production of yellow and red pigments for the printing ink, textile, paper, paint, rubber, plastic, and related industries.

Exposure. Inhalation; skin absorption

Toxicology. 3,3'-Dichlorobenzidine (DCB) is carcinogenic in several animal species.

The acute LD_{50} of DCB in rats has been estimated to be 7100 mg/kg for the free base and 3800 mg/kg for the dihydrochloride salt.[1] Considering these high LD_{50}s, acute lethality in man following oral exposure is not expected to be very likely.[2]

Dermatitis was cited as the only verified health problem encountered by workers in contact with DCB at a DCB manufacturing plant.[1] Applied to the skin of rabbits, DCB dihydrochloride caused no discernible reaction; instilled in the rabbit eye, 20 mg (as 0.1 ml of 20% corn oil suspension) produced erythema, pus, and corneal opacity. No effects were reported when 100 mg of the free base was placed in rabbit eyes.

Existing animal data show that DCB induces tumors at a variety of sites in several animal species.[2]

Of 111 rats given 20 mg DCB by injection or gastric intubation 6 days/week for 10 to 20 months, 17 had tumors of the zymbal gland (a specialized sebaceous gland adjacent to the external ear canal), 13 had mammary tumors, 8 had skin tumors, 5 had malignant lymphomas, 3 had urinary bladder tumors, 3 had salivary gland tumors, and 2 had intestinal tumors; no tumors were found in 130 control rats.[3]

Of 44 male rats fed 1000 ppm for 12 months, 9 developed granulocytic leukemia, and 8 developed zymbal gland tumors; mammary gland tumors were found in rats of both sexes.[4]

In hamsters, 0.3% DCB in the diet produced transitional cell carcinomas of the bladder and some liver-cell tumors.[5] Liver tumors were also found in mice exposed to DCB.[3] Female dogs fed 8 mg/kg/day for a period of 6 to 7 years had hepatocellular carcinomas and papillary transitional cell carcinomas of the urinary bladder; tumors were absent in untreated controls.[6]

There are no reports in which DCB exposure has been conclusively linked to cancer in humans.[1] However, DCB exposure may have been a factor in some cases of bladder cancer attributed to benzidine, because these substances are often produced together and DCB bears a close structural similarity to benzidine.[7] A British plant handling 3,3'-dichlorobenzidine had a site incidence of bladder cancer 2 to 3 times that predicted for males employed between 1972 and 1987; the cause of this apparent excess could not be identified because of potential exposure to many other chemicals.[8] Since that time, the incidence of bladder cancer appears to have fallen to background levels and has been attributed to an alteration in hygiene standards.[8]

Studies in several test systems have shown DCB to be genotoxic in vitro and in vivo and suggest that this effect most likely mediates the carcinogenicity of the chemical.[2] In vitro, DCB has induced sister chromatid exchanges, unscheduled DNA synthesis, and positive responses in bacterial *Salmonella typhimurium* assays; in vivo DCB induced micronuclei in polychromatic erythrocytes in male mice and fetuses.[9-12]

Because of demonstrated potent carcinogenicity in animals, DCB may be regarded as a probable human carcinogen, and exposure by any route should be avoided.[2]

3,3'-Dichlorobenzidene has no TLV exposure limit and is classified as an A2 suspected human carcinogen.

REFERENCES

1. Gerarde HW, Gerarde DF: Industrial experience with 3,3'-dichlorobenzidine: an epidemiological study of a chemical manufacturing plant. *J Occup Med* 16:322–344, 1974
2. Agency for Toxic Substances and Disease Registry (ASTDR): *Toxicological Profile for 3,3'-Dichlorobenzidene.* US Public Health Service, p 87, December 1989
3. Pliss GB: Dichlorobenzidine as a blastomogenic agent. *Vop Onkol* 5:524–533, 1959
4. Stula EF, Sherman H, Zapp JA Jr, Clayton JW Jr: Experimental neoplasia in rats from oral administration of 3,3'-dichlorobenzidine, 4,4'-methylene-bis(2-chloroaniline), and 4,4'-methylene-bis(2-methylaniline). *Toxicol Appl Pharmacol* 31:159–176, 1975
5. Sellakumar AR, Montesano R, Saffiotti U: Aromatic amines carcinogenicity in hamsters. *Proc Am Assoc Cancer Res* 10:78, 1969.
6. Stula EF, Barnes JR, Sherman H et al: Liver and urinary bladder tumors in dogs from 3,3'-dichlorobenzidine. *J Environ Pathol Toxicol* 1:475–490, 1978
7. *IARC Monographs on the Evaluation of the Carcinogenic Risk of Chemicals to Man*, Vol 4, Some aromatic amines, hydrazine and related substances, N-nitroso compounds and miscellaneous alkylating agents, pp 49–55. Lyon, International Agency for Research on Cancer, 1974
8. Leeser JE, Cowan JB: Epidemiology update (Letters to the Editor): *JOM* 35:892, 1993
9. Shiraishi Y: Hypersensitive character of Bloom syndrome B-lymphoblastoid cell lines usable for sensitive carcinogen detection. *Mut Res* 175:179–187, 1986
10. Ashby J, Mohammed R: UDS activity in the rat liver of the human carcinogens benzidene and 4-aminobiphenyl and the rodent carcinogens 3,3'-dichlorobenzidine and direct black 38. *Mutagenesis* 3:69–71, 1988
11. Iba MM, Thomas PE: Activation of 3,3'-dichlorobenzidine in rat liver microsomes to mutagens: involvement of cytochrome P-450. *Carcinogenesis* 9:717–723, 1988.
12. Cihak R, Vontorvoka M: Benzidine and 3,3'-dichlorobenzidine (DCB) induce micronuclei in the bone marrow and the fetal liver of mice after gavage. *Mutagenesis* 2:267–270, 1987

DICHLORODIFLUOROMETHANE
CAS:75-71-8

CCl_2F_2

Synonyms: Freon 12; Refrigerant 12; Isotron; Halon; Genetron 12; Frigen 12

Physical Form. Colorless gas

Uses. Refrigerant; aerosol propellant; in plastics; blowing agent

Exposure. Inhalation

Toxicology. Dichlorodifluoromethane causes central nervous system depression at very high concentrations.

Volunteers exposed to 200,000 ppm for a short time experienced significant eye irritation as well as central nervous system effects.[1] The effects disappeared within minutes after return to fresh air. Exposure at 110,000 ppm for 11 minutes caused a marked decrease in consciousness, amnesia, and cardiac arrhythmias; at 40,000 ppm for 80 minutes, there was generalized paresthesia, tinnitus, apprehension, and slurred speech.[1] Two volunteers exposed to 10,000 ppm for 2.5 hours showed slight psychomotor impairment.[2]

Chronic exposure of volunteers to 1000 ppm, 8 hr/day, for 17 days caused no subjective symptoms, no cardiac abnormalities, and no pulmonary function abnormalities.[3]

Sniffing aerosols of fluorochlorinated hydrocarbons has caused sudden death from cardiac arrest probably as a result of cardiac arrhythmias from sensitization of the myocardium to epinephrine.[4]

Refrigerator repairmen exposed to dichlorodifluoromethane and chlorodifluoromethane (peak exposures 1300–10,000 ppm) showed no clear connection between exposure and cardiac arrhythmia as determined by ambulatory electrocardiograms.[5] The investigators suggest that subjects with compromised cardiac function may be more susceptible to the arrhythmogenic potential of fluorocarbons but that, in general, a higher-exposure concentration on the order

of 100,000 to 200,000 ppm may be necessary to provoke cardiac arrhythmias.

In rats, when dichlorodifluoromethane was administered at various concentrations with 20% oxygen for 30 minutes, the following effects were observed: 200,000 ppm, no observable effects; 300,000 ppm, muscular twitching and tremor; 800,000 ppm, coma, corneal reflexes absent; 800,000 ppm for 4 and 6 hours was not lethal, and the animals suffered no permanent effects.[6]

Chronic exposure of rats 6 hr/day for 90 days at 10,000 ppm and of dogs at 5000 ppm caused no adverse effects as determined by observation, clinical tests, or histologic examination.[7]

A variety of reproductive and carcinogenic studies have found no significant effects.[1]

The 1995 ACGIH threshold limit value–time-weighted average (TLV-TWA) for dichlorodifluoromethane is 1000 ppm (4950 mg/m³).

REFERENCES

1. Haskell Laboratory, Dupont: Toxicity Review, Freon 12, pp 1–26. November 1982
2. Azar A et al: Experimental human exposures to fluorocarbon 12 (dichlorodifluoromethane). *Am Ind Hyg Assoc J* 33:207–216, 1972
3. Stewart RD et al: Physiological response to aerosol propellants. *Environ Health Perspect* 26:275–285, 1978
4. Reinhardt CF et al: Cardiac arrhythmias and aerosol "sniffing." *Arch Environ Health* 22:265–279, 1971
5. Poika-Antii M, Heikkila J, Saarinen L: Cardiac arrhythmias during occupational exposure to fluorinated hydrocarbons. *Br J Ind Med* 47:138–140, 1990
6. Lester D, Greenberg LA: Acute and chronic toxicity of some halogenated derivatives of methane and ethane. *Arch Ind Hyg Occup Med* 2:335–344, 1950
7. Leuschner F et al: Report of subacute toxicological studies with several fluorocarbons in rats and dogs by inhalation. *Drug Res* 33:1475–1476, 1983

1,3-DICHLORO-5,5-DIMETHYLHYDANTOIN
CAS: 118-52-5

$C_5H_6Cl_2O_2N_2$

Synonyms: Dactin; Halane; DCDMH

Physical Form. White powder

Uses. Chlorinating agent; disinfectant; laundry bleach; in water treatment; intermediate for drugs; insecticides; polymerization catalyst

Exposure. Inhalation

Toxicology. 1,3-Dichloro-5,5-dimethylhydantoin powder in contact with water yields hypochlorous acid, which is an irritant of the eyes and mucous membranes.

There is a single report of a workman exposed to concentrations exceeding 0.2 mg/m³ who experienced cough and chest discomfort.[1]

The LD_{50} for rats when administered orally as a 10% aqueous suspension was 542 mg/kg; at necropsy, gastrointestinal hemorrhages were found.

The 1995 ACGIH threshold limit value–time-weighted average (TLV-TWA) for 1,3-dichloro-5,5-dimethylhydantoin is 0.2 mg/m³ with a short-term excursion limit of 0.4 mg/m³.

REFERENCES

1. 1,3-Dichloro-5,5-Dimethylhydantoin. *Documentation of the TLVs and BEIs*, 6th ed, pp 423–424. Cincinnati, OH, American Conference of Governmental Hygienists, 1991

1,1-DICHLOROETHANE
CAS: 75-34-3

CH_3CHCl_2

Synonyms: Ethylidene dichloride

Physical Form. Colorless liquid

Uses. Cleansing agent; degreaser; solvent for plastics, oils, and fats; grain fumigant; chemical intermediate; formerly used as an anesthetic

Exposure. Inhalation

Toxicology. At high concentrations, 1,1-dichloroethane causes central nervous system depression.

There have been no reported cases of human overexposure by inhalation. In the past, 1,1-dichloroethane was used as an anesthetic at levels of approximately 25,000 ppm.[1] Use was discontinued when it was discovered that cardiac arrhythmias might be induced. Cardiovascular toxicity has not been reported in animals following exposure.

Rats exposed to 32,000 ppm for 30 minutes survived but died after 2.5 hours of exposure.[2] The most consistent findings in animals exposed to concentrations of above 8000 ppm for up to 7 hours were pathologic changes in the kidney and liver and, at much higher concentrations, near 64,000 ppm, damage to the lungs as well. No adverse clinical effects were noted in rats, rabbits, or guinea pigs exposed to 1000 ppm for 13 weeks, which followed a prior 13-week exposure to 500 ppm.[3] Under the same conditions, renal injury was apparent in cats, as evidenced by increased serum urea and creatinine levels.

No histopathological alterations were noted in the liver, kidneys or lungs of male mice that ingested up to 2500 mg/l 1,1-dichloroethane in drinking water for 52 weeks.[4]

A significant increase in endometrial stromal polyps, a benign neoplasm, occurred in female mice administered up to 3.3 g/kg/day 1,1-dichloroethane by gavage for 78 weeks.[5] There was also a dose-related trend for the incidence of hemangiosarcomas and mammary adenocarcinomas in female rats and hepatocellular carcinoma in male mice. High mortality in all animal groups obscured results. The National Cancer Institute determined that there was no conclusive evidence for carcinogenicity, but 1,1-dichloroethane should be treated with caution by analogy to other chloroethanes shown to be carcinogenic in laboratory animals.[5,6]

The liquid applied to the intact or abraded skin of rabbits produced slight edema and very slight necrosis after the sixth of 10 daily applications. When instilled in the eyes of rabbits, there was immediate, moderate conjunctival irritation and swelling, which subsided within a week.[2]

Although the liquid may be absorbed through the skin, it is apparently not absorbed in amounts sufficient to produce systemic injury.

Exposure of rats to 6000 ppm 7 hr/day on days 6 through 15 of gestation was associated with an increased incidence of delayed ossification of sternebrae.[7] Maternal toxicity was limited to decreased weight gain

Odor cannot be relied on to provide warning of overexposure.

The 1995 ACGIH threshold limit value–time-weighted average (TLV-TWA) for 1,1-dichloroethane is 100 ppm (405 mg/m³).

REFERENCES

1. Browning E: *Toxicity and Metabolism of Industrial Solvents*, pp 247–252. New York, Elsevier, 1965
2. Hygienic guide series: 1,1-Dichloroethane (ethylidene chloride). *Am Ind Hyg Assoc J* 32:67–71, 1971
3. Hofmann HT, Birnstiel H, Jobst P: Inhalation toxicity of 1,1 and 1,2 dichloroethane. (German) *Arch Toxikol* 27:248–265, 1971
4. Klaunig JE, Ruch RJ, Pereira MA: Carcinogenicity of chlorinated methane and ethane compounds administered in drinking water to mice. *Environ Health Perspect* 69:89–95,1986
5. National Cancer Institute: *Bioassay of 1,1-Dichloroethane for Possible Carcinogenicity*. Carcinogenesis Technical Report Series No. 66, NCI-CG-TR-66, p 82. DHEW (NIH) Pub No 78–1316, 1978
6. NIOSH: *Current Intelligence Bulletin 27, Chloroethanes: Review of Toxicity*, p 22. DHEW (NIOSH) Pub No 78–181, 1978
7. Schwetz BA et al: Embryo and fetoxicity of inhaled carbon tetrachloride, 1,1-dichloroethane and methyl ethyl ketone in rats. *Toxicol Appl Pharmacol* 28:452–464, 1974

1,2-DICHLOROETHYLENE
CAS: 540-59-0

cis-**1,2-DICHLOROETHYLENE**
CAS: 156-59-2

trans-**1,2-DICHLOROETHYLENE**
CAS: 156-60-5

$C_2H_2Cl_2$

Synonyms: Acetylene dichloride; 1,2-dichloroethene

Physical Form. Colorless liquid

Uses. 1,2-Dichloroethylene used as a solvent for organic materials and as an intermediate in the synthesis of other chlorinated compounds

Source. May be produced by the chlorination of acetylene; often a by-product in the manufacture of other chlorinated compounds

Exposure. Inhalation; ingestion; skin

Toxicology. 1,2-Dichloroethylene causes central nervous system depression at high concentrations; liver, lung, and heart damage have been reported in animal studies.

1,2-Dichloroethylene is a mixture of two geometric isomers, *cis* and *trans*; the proportion of the *cis* to *trans* isomer varies from mixture to mixture, depending on the manufacturer's specifications. The properties of the mixtures are expected to reflect those of the individual isomers.

There has been only one report of industrial poisoning, a fatality caused by very high vapor inhalation in a small exclosure.[1] The isomeric concentration of the vapor was not reported, nor were the level and duration of the exposure or symptoms of toxicity. In another early report, exposure to the *trans* isomer at 2200 ppm caused nausea, drowsiness, fatigue, vertigo, and increased intracranial pressure in two human subjects.[1]

In mice, the LC_{50} for a single 6-hour inhalation exposure was 22,000 ppm for the *trans* isomer.[2]

A very limited rat study reported the following after inhalation of the *trans* isomer: exposure for 8 hours at 3000 ppm was associated with pathological changes in the heart, described as fibrous swelling of the myocardium and hyperemia; at 1000 ppm for one day, pathological changes in the lungs included pulmonary capillary hyperemia, alveolar septal distention, and pulmonary edema; hematological effects at this level included a reduction in the number of circulating leukocytes and erythrocytes; pathological changes in the liver consisted of lipid accumulation and fatty degeneration following an 8-hour exposure at 200 ppm.[3]

In a more recent report, the acute oral LD_{50} for *trans*-1,2-dichloroethylene administered by gavage was 8000 mg/kg for male rats and 9900 mg/kg for the females.[4] Signs associated with lethal doses included those of pulmonary hyperemia and central nervous system depression, including ataxia, loss of righting reflex, and depressed respiration.

Rats receiving approximate daily doses of 500, 1500 or 3000 mg *trans*-1,2-dichloroethylene in their drinking water for 90 days had no significant adverse effects, as determined by hematological, serological, or urinary parameters.[4] There were no compound-related gross or histological effects although there were dose-dependent increases in kidney weights and ratios in treated females. The authors concluded that toxicity from exposure to *trans*-1,2-dichloroethylene in drinking water at 1 µg/liter is low and probably does not constitute a serious health hazard. It should be noted, however, that adequate information is not available on possible chronic effects.

Administered by inhalation to rats 6 hr/day on days 7 through 16 of gestation, 12,000 ppm *trans*-1,2-dichloroethylene caused fetal toxicity in the form of reduced fetal weights; overt maternal toxicity was also observed at this dose and was expressed as a significant reduction in weight gain and in feed consumption.[5] Increased incidences of alopecia, lethargy, salivation, and ocular iritation were also observed in the treated dams. *trans*-1,2-Dichloroethylene was not considered to be uniquely toxic to the rat conceptus.

Mild burning of the eyes after acut e exposure to either *trans*-1,2-dichloroethylene vapor or aerosol was reported by two subjects in a 1936 self-experimentation study. However, dichloroethylene has been employed in combination with ether as a general anesthetic in at least 2000 cases, with no evidence of ocular toxicity.[6]

The 1995 ACGIH threshold limit value–time-weighted average (TLV-TWA) for 1,2-dichloroethylene is 200 ppm (793 mg/m³).

REFERENCES

1. von Oettingen WF: *The Halogenated Aliphatic, Olefinic, Cyclic, Aromatic Hydrocar bons, Including the Halogenated Insecticides, Their Toxicity and Potential Dangers*, US Public Health Service Pub No 414, pp 198–202. Washington, DC, US Government Printing Office, 1955
2. Gradiski D, Bonnet P, Raoult G, et al: Comparative acute inhalation toxicity of the principal chlorinated aliphatic solvents. *Arch Mal Prof Med Trav Secur Soc* 39:249–257, 1978
3. Freundt KJ, Liebaldt GP, Lieberwirth E: Toxicity studies on *trans*-1,2-dichloroethylene. *Toxicology* 7:141–153, 1977
4. Hayes JR, Condie LW Jr, Egle JL Jr, Borzelleca JF: The acute and subchronic toxicity in rats of *trans*-1,2-dichloroethylene in drinking water. *J Am Coll Toxicol* 6:471–478, 1987
5. Hurtt ME, Valentine R, Alvarez L: Developmental toxicity of inhaled *trans*-1,2,dichloroethylene in the rat. *Fund Appl Toxicol* 20:225–230, 1993
6. Grant WM: *Toxicology of the Eye*, 3rd ed, p 326. Springfield, IL, Charles C Thomas, 1986

DICHLOROETHYL ETHER
CAS: 111-44-4

$C_4H_8C_{12}O$

Synonyms: bis(2-Chloroethyl) ether; chlorex; 1-chloro-2-(β-chloroethoxy)ethane

Physical Form. Colorless liquid

Uses. Solvent for resins, wax, oils, turpentine; insecticide

Exposure. Inhalation; skin absorption

Toxicology. Dichloroethyl ether is a severe respiratory and eye irritant; high levels cause narcosis in animals, and severe exposure is expected to cause the same effects in humans. The inhalation hazard is limited by its relatively low volatility; skin absorption is more hazardous.

In experimental human exposure, 500 ppm caused intolerable irritation to the eyes and nose, with cough, lacrimation, and nausea; at 100 ppm, there was some irritation, whereas at 35 ppm, there were no effects.[1]

In guinea pigs, concentrations of 500 to 1000 ppm were fatal after 5 to 8 hours of exposure; effects were immediate lacrimation and nasal irritation, followed by unsteadiness and coma; autopsy findings were pulmonary edema, pulmonary hemorrhage, and occasional complete consolidation.[1]

Repeated oral administration of 300 mg/ kg daily to both sexes of two strains of mice for 80 weeks induced a significant elevated incidence of tumors, mostly hepatomas.[2] The IARC has determined that there is limited evidence of carcinogenicity in animals and that no adequate data are available to assess human exposures.[3]

Both the pure liquid and a 10% solution dropped in the eye of a rabbit caused moderate discomfort, conjunctival irritation, and corneal injury, which generally healed within 24 hours.[3] On the skin of rabbits, the liquid had no local effect, but a sufficient amount penetrated the skin to cause death within a day; the 24-hour LD_{50} for skin absorption was 90 mg/kg.[4]

The 1995 ACGIH threshold limit value–time-weighted average (TLV-TWA) for dichloroethyl ether is 5 ppm (29 mg/m³) with a short-term excursion limit of 10 ppm (58 mg/m³) and a notation for skin absorption.

REFERENCES

1. Schrenk HH, Patty FA, Yant WP: Acute response of guinea pigs to vapors of some new commercial organic compounds. VII. Dichloroethyl ether. *Pub Health Rept* 48:1389–1398, 1933
2. Innes JRM et al: Bioassay of pesticides and industrial chemicals for tumorigenicity in mice: a preliminary note. *J Natl Cancer Inst* 42:1101–1114, 1969
3. *IARC Monographs on the Evaluation of Carcinogenic Risks to Humans, Overall Evaluations of Carcinogenicity: An Updating of IARC Monographs Vols 1 to 42*, Suppl 7, p 58. Lyon, International Agency for Research on Cancer, 1987
4. Kirwin C, Sandmeyer E.: Ethers. In Clayton GD, Clayton FE (eds): *Pattys Industrial Hygiene and Toxicology*, 3rd ed, Vol 2, *Toxicology*, pp 2517–1519. New York, Wiley-Interscience, 1981

DICHLOROFLUOROMETHANE
CAS: 75-43-4

$CHFCl_2$

Synonyms: Dichloromonofluoromethane; fluorodichloromethane; Freon 21; Refrigerant 21; FC-21

Physical Form. Colorless gas

Uses. Refrigerant; propellant

Exposure. Inhalation

Toxicology. Dichlorofluoromethane at high concentrations causes asphyxia in animals; repeated or prolonged exposure to lower concentrations results in liver damage.

Acute or chronic effects from human exposure have not been reported. In liquid form, this substance may cause frostbite.

Exposure of guinea pigs to 400,000 ppm with 18% oxygen was fatal, and death was preceded by dyspnea, tremor, and convulsive movements, but not narcosis.[1] Animals died at 102,000 ppm with congested lungs, kidneys, and liver, but survived 52,000 ppm, showing tremor, incoordination, and irregular breathing.[1]

In rats, 90-day exposures to 1000 and 5000 ppm caused bilateral hair loss, extensive liver damage, and excessive mortality.[2] The chronic toxicity of dichlorofluoromethane appears to be quite different from difluorinated methanes and more similar to the hepatotoxin chloroform.[3] In mice, 100,000 ppm induced arrhythmias and sensitized the heart to epinephrine.

After exposure at 10,000 ppm on days 6 through 15 of gestation, 15 of 25 pregnant female rats had no viable fetuses or implantation sites on the uterine wall.[3]

The 1995 ACGIH threshold limit value–time-weighted average (TLV-TWA) for dichlorofluoromethane is 10 ppm (42 mg/m³).

REFERENCES

1. von Oettingen WF: *The Halogenated Aliphatic, Olefinic, Cyclic, Aromatic, and Aliphatic-Aromatic Hydrocarbons Including the Halogenated Insecticides, Their Toxicity and Potential Dangers.* US Public Health Service, Pub No 414, pp 73–75. Washington, DC, US Government Printing Office, 1955
2. Trochimowicz HJ et al: Ninety-day inhalation toxicity studies on two fluorocarbons. *Toxicol Appl Pharmacol* 41:299 (Abstract), 1977
3. Dichlorofluoromethane. Documentation of TLVs and BEIs, 6th ed, pp 434–435. Cincinnati, OH, American Conference of Governmental Industrial Hygienists, 1991

1,1-DICHLORO-1-NITROETHANE
CAS: 594-72-9

$CH_3CCl_2NO_2$

Synonyms: Ethide

Physical Form.   Colorless liquid

Uses.   Fumigant insecticide

Exposure.   Inhalation

Toxicology.   1,1-Dichloro-1-nitroethane is a pulmonary, skin, and eye irritant in animals; it is expected that severe exposure will cause the same effects in humans.

No effects in humans have been reported.

Exposure of rabbits to 2500 ppm for 40 minutes was fatal, but exposure to 170 ppm for 30 minutes was nonlethal; autopsy revealed pulmonary edema and hemorrhage, with damage to the heart, liver, and kidneys. At high concentrations, effects included lacrimation, increased nasal secretion, sneezing, cough, pulmonary rales, and weakness. Application of the liquid to the skin of rabbits caused irritation and edema.[1,2] This compound is considerably more irritating to skin and mucous membranes of animals than is 1-chloro-1-nitropropane and exhibits greater toxicity by inhalation.[3]

The 1995 ACGIH threshold limit value–time-weighted average (TLV-TWA) for 1,1-dichloro-1-nitroethane is 2 ppm (12 mg/m³).

REFERENCES

1. Machle W, Scott EW, Treon JF, et al: The physiological response of animals to certain chlorinated mononitroparaffins. *J Ind Hyg Toxicol* 27:95–102, 1945
2. Negherbon WO (ed): *Handbook of Toxicology,* Vol. III, *Insecticides,* pp 212–213. Philadelphia, WB Saunders, 1959
3. Stokinger HE: Aliphatic nitro compounds, nitrates, nitrites. In Clayton GD, Clayton FE (eds): *Patty's Industrial Hygiene and Toxicology,*

3rd ed, rev, Vol 2C, *Toxicology*, pp 4162–4164. New York, Wiley-Interscience, 1982

2,4-DICHLOROPHENOL
CAS:120-83-2

$C_6H_3OHCl_2$

Synonyms: DCP; 2,4-DCP; 1,3-dichloro-4-hydroxybenzene

Physical Form. White solid

Uses. Intermediate in production of herbicidal chlorophenoxy acids such as 2,4-dichlorophenoxyacetic acid

Exposure. Inhalation

Toxicology. 2,4-Dichlorophenol (2,4-DCP) has caused liver, kidney, and bone marrow damage in animals.

In a recent report, an accidental death has been attributed to absorption of 2,4-DCP through the skin.[1] A 33-year-old male, while disposing of industrial waste, splattered portions of his right thigh and right arm with a pure solution of 2,4-DCP. He washed himself without undressing and, shortly thereafter (within 20 min), he experienced a seizure and collapsed. Resuscitation efforts failed. It was determined that less than 10% of his body surface was contaminated, which resulted in blood concentrations of 24.3 mg/l. Other drugs, including ethanol, were not detected in a toxicological screen. The authors suggest that the blood level of 2,4-DCP in this case is in accordance with reported lethal blood concentrations of phenol.

The oral LD_{50} in rats was 2830 mg/kg.[2] Typical effects associated with acute lethal oral doses include restlessness and increased respiratory rate, which appear quickly, followed shortly by tremors, convulsions, dyspnea, coma, and death.[3]

In an NTP report, exposure of rats to concentrations as high as 2000 mg/kg/day in the diet for up to 13 weeks did not cause mortality; 2600 mg/kg/day did not affect survival of mice at this duration, but all mice died when exposed to 5200 mg/kg for 4 weeks.[4] Renal tubular necrosis was found in mice at the highest dose level, but no effect was seen in mice fed 2600 mg/kg/day or in rats fed 2000 mg/kg for 13 weeks.

Both erythoid and myelocitic elements of bone marrow were depleted in rats fed 500 mg/kg/day for 13 weeks.[4] Mice fed 325 mg/kg/day or more for 13 weeks had dose-related increases in hepatic necrosis.[5]

Feeding tests with rats and mice for periods of up to 103 weeks, at doses as high as 440 mg/kg/day for rats and 1300 mg/kg/day for mice, showed no evidence for carcinogenic activity due to 2,4-DCP.[4]

Topical application to mice of 0.3% dimethylbenzanthracene in benzene as an initiator, followed by twice-weekly applications of 20% 2,4-DCP in benzene, produced papillomas in 75% and carcinomas in 6% at 24 weeks; 62% had carcinomas after 39 weeks.[6] There is no evidence that 2,4-DCP acting alone induces papillomas or carcinomas.[3]

Female rats were given 3, 30, or 300 ppm in drinking water from 3 weeks of age through breeding and parturition (Group 1) or for 24 months (Group 2).[3,4,7] Animals from Group 1 were bred to untreated males at 90 days of age; litter sizes at 300 ppm were significantly smaller than controls. The percent of stillborn pups increased at all doses. In Group 2, liver weights were significantly increased in the 300 ppm group. Spleen weights were higher and thymuses were smaller than in the control group. Delayed-type hypersensitivity responses in treated animals were significantly suppressed when compared with controls. Tumor incidence, latency, or type was not different from controls.

Oral exposure of pregnant rats to 750 mg/kg/day for 10 gestational days induced slightly decreased fetal weight, delayed ossification of sternal and vertebral arches, and some early embryonic deaths.[8] Maternal deaths also occurred at this dose, indicating that 2,4-DCP was not selectively toxic to embryos or fetuses.

No effects were noted in dams or offspring exposed at 375 mg/kg/day.

A 1995 ACGIH threshold limit value has not been established for 2,4-dichlorophenol.

REFERENCES

1. Kintz P, Tracqui A, Mangin P: Accidental death caused by the absorption of 2,4-dichlorophenol through the skin. *Arch Toxicol* 66:298–299, 1992
2. Vernot EH et al: Acute toxicity and skin corrosion data for some organic and inorganic compounds and aqueous solutions. *Toxicol Appl Pharm* 42:417–423, 1977
3. Agency for Toxic Substances and Disease Registry (ASTDR): *Toxicological Profile for 2,4-Dichlorophenol*, TP-91/14. US Department of Health and Human Services, Public Health Service, July 1992
4. National Toxicology Program (NTP): *Toxicology and Carcinogenesis Studies of 2,4-Dichlorophenol in F344/N Rats and B6C3F1 Mice (Feed Studies)*, Technical Report Series No 353. Research Triangle Park, NC, US Department of Health and Human Services, Public Health Service, National Institutes of Health, 1989
5. Borzelleca JF, Hayes JR, Condie LW, et al: Acute and subchronic toxicity of 2,4-dichlorophenol in CD-1 mice. *Fund Appl Toxicol* 5:478–486, 1985
6. Boutwell RK, Bosch DK: The tumor-promoting action of phenol and related compounds for mouse skin. *Cancer Res* 19:413–424, 1959
7. Exon JH, Henningsen GM, Osborne CA, et al. Toxicologic, pathologic, and immunotoxic effects of 2,4-dichlorophenol in rats. *J Toxicol Environ Health* 14:723–730, 1984
8. Rodwell DE, Wilson RD, Nemec MD, et al: Teratogenic assessment of 2,4-dichlorophenol in Fischer 344 rats. *Fund Appl Pharmacol* 13:635–640, 1989

2,4-DICHLOROPHENOXYACETIC ACID
CAS: 94-75-7

$C_8H_6C_{12}O_3$

Synonyms: 2,4-D; Hedonal; component of Agent Orange

Physical Form. Crystalline solid

Uses. Herbicide

Exposure. Inhalation; ingestion; skin absorption

Toxicology. 2,4-Dichlorophenoxyacetic acid (2,4-D) causes signs of both hypo- and hyperexcitation of the central nervous system.

One fatal case of poisoning involved a suicidal person who ingested not less than 6500 mg and experienced violent convulsions; there were no significant findings at autopsy.[1] In another fatality from suicidal ingestion of a mixture of 2,4-D and two other related herbicides, progressive hypotension, coma, tachypnea, and abdominal distention preceded death. An autopsy revealed nonspecific findings. Concentrations of 2,4-D measured in blood and urine were 520 and 670 mg/l, respectively.[2] A single dose of 3.6 g of 2,4-D administered intravenously to a patient for treatment of disseminated coccidioidomycosis caused stupor, hyporeflexia, fibrillary twitching of some muscles, and urinary incontinence; 24 hours after the dose, the patient still complained of profound muscular weakness, which subsided after an additional 24 hours.[3,4]

Contact of the material with the skin may cause dermatitis.[3,4] Dermal absorption and ingestion of aerosol droplets trapped in the nose appear to be the primary routes of entry in spraying operations.

Peripheral neuropathy has been reported to occur occasionally following exposure to 2,4-D but more frequently following exposure to another phenoxyherbicide, 2,4,5-T (2,4,5-trichlorophenoxyacetic acid) or its contaminants, including 2,3,7,8-TCDD (2,3,7,8-tetrachlorodibenzo-*p*-dioxin).[5] A study of workers employed in the manufacture of 2,4-D and 2,4,5-T found a statistically significant increased frequency of mild slowing of nerve conduction velocity in the sural sensory and median motor nerves; there were no associated symptoms.[6]

Several case-control studies of soft-tissue sarcoma and lymphoma have suggested an increased risk among workers exposed to phenoxyacetic acid herbicides, including 2,4-D.[7,8] In one study involving primarily 2,4-D exposure, there was an increased risk for malignant lymphoma of the non-Hodgkin's type but not for soft-tissue sarcomas.[9] The IARC has deemed the evidence implicating 2,4-D as inadequate.[7] Concomitant exposure to other known carcinogenic substances and insufficient accumulation of person-years of observation are two primary limiting factors in establishing the risks associated with 2,4-D exposures.[10,11] Large cohort studies of agricultural and forestry workers exposed to these herbicides have not subsequently confirmed any increased incidence of malignancy.[7,12] For 878 chemical workers potentially exposed to 2,4-D at any time between 1945 and 1983, an analysis by production area, duration of exposure, and cumulative dose showed no patterns suggestive of a causal association between 2,4-D exposure and any particular cause of death.[13] Particular attention was given to deaths from brain neoplasms in this cohort because a recent unpublished study reported an increased incidence of astrocytomas in male rats fed 45 mg/kg/day in the diet for 2 years. No brain neoplasms were observed.[13] Four additional years of mortality follow-up on this cohort through 1986 have not revealed any patterns suggestive of a causal association between 2,4-D exposure and any particular cause of death, including cancer.[14] A case-control study of Vietnam era veterans with soft-tissue sarcoma did not find an association with potential exposure to Agent Orange (a 1:1 mixture of 2,4-D and 2,4,5-T).[15] A subsequent mortality study of these veterans did find an elevated standardized proportionate mortality ratio for soft-tissue sarcoma, but the study was not based on adequate numbers of deaths or adequate exposure data.[16]

A recent review of epidemiological studies of chlorophenoxy herbicides found no consistent or conclusive evidence linking 2,4-D to human carcinogenesis. It was further stated that, in general, animal studies, conducted under current test guidelines, have also shown no evidence of carcinogenicity supporting the results of epidemiological studies.[17]

A two- to threefold increased risk of birth defects among children of Vietnam veterans exposed to Agent Orange has been suggested by several epidemiologic studies, but these studies have been criticized on a number of grounds, including exposure assessment, outcome verification, and potential for recall bias.[18] Animal studies have not demonstrated clear-cut adverse effects of phenoxyherbicide exposure on reproductive outcomes.[18,19]

Because 2,4-D is readily absorbed through the skin, measurements of ambient air concentrations do not necessarily reflect the total absorbed dose.[20] Immunochemical determination of 2,4-D in urine has provided effective measurement of human exposure levels.

The 1995 ACGIH threshold limit value–time-weighted average (TLV-TWA) for 2,4-D is 10 mg/m^3.

REFERENCES

1. Nielsen K, Kaempe B, Jensen-Holm J: Fatal poisoning in man by 2,4-diphenoxyacetic acid (2,4-D): Determination of the agent in forensic materials. *Acta Pharmacol Toxicol* 22:224–234, 1965

2. Fraser AD, Isner AF, Perry RA: Toxicologic studies in a fatal overdose of 2,4-D, Mecoprop, and Dicamba. *J Forens Sci* 29:1237–1241, 1984

3. Seabury JH: Toxicity of 2,4-dichlorophenoxyacetic acid for man and dog. *Arch Environ Health* 7:202–209, 1963

4. Hayes WJ Jr: *Clinical Handbook on Economic Poisons, Emergency Information for Treating Poisoning*, US Public Health Service Pub No 476, pp 106–109. Washington, DC, US Government Printing Office, 1963

5. Kolmodin-Hedman B, Hoglund S, Akerblom M: Studies on phenoxy acid herbicides. I. Field study: occupational exposure to phenoxy acid herbicides (MCPA, Dichloroprop,

Meco-prop, and 2,4-D) in agriculture. *Arch Toxicol* 54:257–265, 1983

6. Singer R et al: Nerve conduction velocity studies of workers employed in the manufacture of phenoxy herbicides. *Environ Res* 29:297–311, 1982

7. *IARC Monographs on the Evaluation of the Carcinogenic Risk of Chemicals to Humans:* Chemicals, industrial processes and industries associated with cancer in humans, Suppl 4, pp 101–103, 211–212. Lyon, International Agency for Research on Cancer, 1982

8. Hardell L, Eriksson M: The association between soft-tissue sarcomas and exposure to phenoxyacetic acids. A new case-referent study.*Cancer* 62:652–656, 1988

9. Hoar SK, Blair A, Holmes FF, et al: Agricultural herbicide use and risk of lymphoma and soft-tissue sarcoma. *JAMA* 256:1141–1147, 1986

10. Johnson ES: Review. Association between soft-tissue sarcomas, malignant lymphomas, and phenoxy herbicides/chlorophenols: evidence from occupational cohort studies. *Fund Appl Toxicol* 14:219–234, 1990

11. Bond GG, Bodner KM, Cook RR: Phenoxy herbicides and cancer: Insufficient epidemiological evidence for a causal relationship. *Fund Appl Pharmacol* 12:172–188, 1989

12. Wiklund K, Holme L: Soft-tissue sarcoma risk in Swedish agricultural and forestry workers. *J Natl Cancer Inst* 76:229–234, 1986

13. Bond GG, Wetterstroem NH, Roush GJ, et al: Cause specific mortality among employees engaged in the manufacture, formulation, or packaging of 2,4-dichlorophenoxyacetic acid and related salts. *Br J Ind Med* 45:98–105, 1988

14. Bloemen LJ, Mandel JS, Bond GG, et al: An update of mortality among chemical workers potentially exposed to the herbicide 2,4-dichlorophenoxyacetic acid and its derivatives. *J Occup Med* 35:1208–1212, 1993

15. Kang HK et al: Soft-tissue sarcomas and military service in Vietnam: A case comparison group analysis of hospital patients. *J Occup Med* 28:1215–1218, 1986

16. Kogan MD, Clapp RW: Soft-tissue sarcoma mortality among Vietnam veterans in Massachusetts, 1972–1983. *Int J Epidem* 17:39–43, 1988

17. Bond GG, Rossbacher R: A review of potential human carcinogenicity of the chlorophen-

oxy herbicides MCPA, MCPP, and 2,4-DP. *Br J Ind Med* 50:340–348, 1993

18. Hatch MC, Stein ZA: Agent Orange and risks to reproduction: the limits of epidemiology. *Teratogen, Carcinogenesis Mutagen* 6:185–202, 1986

19. Pesticide Fact Sheet: 2,4-D. Washington, DC, Environmental Protection Agency, 1986

20. Knopp D: Assessment exposure to 2,4-dichlorophenoxyacetic acid in the chemical industry: results of a five year biological monitoring study. *Occup Environ Med* 51:152–159, 1994

1,3-DICHLOROPROPENE
CAS: 542-75-6

$C_3H_4Cl_2$

Synonyms: 1,3-DCP; α-chloroallyl chloride; 1,3-dichloropropylene; Telone; Telone II; DD fumigants

Physical Form. Clear to amber-colored liquid

Uses. Widely used as a preplanting soil fumigant for the control of nematodes

Exposure. Inhalation; skin absorption

Toxicology. 1,3-Dichloropropene (1,3-DCP) is an irritant of the eyes, mucous membranes, and skin; exposures in animals have been associated with contact hypersensitivity and damage to the nasal tissues, lungs, liver, kidneys, and urinary bladder. It is considered to be carcinogenic to experimental animals.

A truck spill in 1985 resulted in exposure of an estimated 80 people.[1,2] Signs and symptoms were headache in 6 persons, mucous membrane irritation in 5, dizziness in 5, and chest discomfort in 4. Eleven of 41 persons tested had slightly elevated SGOT and/or SGPT values. In 28 persons interviewed 12 weeks post-exposure, complaints were headache (12), abdominal discomfort (6), chest discomfort (5), and malaise

(5). In one case, the diagnosis was pneumonia, based on persistent dyspnea and cough.

In a report of two firefighters who were simultaneously exposed at a chemical spill, lymphomas appeared simultaneously in the two men six years later, and they died within several months of each other.[3] The IARC noted that, because firefighters are exposed to a large number of chemicals, the role of 1,3-dichloropropene could not be evaluated.[4,5]

In a recently reported case, accidental ingestion of 1,3-DCP by a 27-year-old worker resulted in gastrointestinal distress, adult respiratory distress syndrome, hematological and hepatorenal impairment, and death within 40 hours from multiorgan failure.[6] Initial symptoms, hospital admittance, included acute gastrointestinal distress, sweating, tachypnea, and tachycardia. The chemical was toxicologically identified by gas chromatography, and initial blood levels were 1.13 μmol/l in blood and 0.20 μM/l in urine.

The oral LD_{50} in male rats was 713 mg/kg and, in females, it was 470 mg/kg.[7] (LD_{50} values as low as 100 mg/kg have been reported for rats; this range in values is attributed to different rat strains and to differences in the 1,3-DCP formulations used.[8])

Acute dermal application of dilute or full-strength DCP rapidly produced erythema and edema in rats, rabbits, and guinea pigs.[8] Delayed-type hypersensitivity reactions and contact sensitization have also been reported in guinea pigs and humans.[8]

A National Toxicology Program carcinogenicity study in rats and mice used Telone II, a technical-grade dichloropropene (89% 1,3-dichloropropene, 2.5% 1,2-dichloropropane, 1.5% trichloropropene isomer, and 1% epichlorohydrin) which was administered by oral gavage 3 times per week for 104 weeks.[9] This exposure produced tumors of the urinary bladder, lung, and forestomach in mice and of the liver and forestomach in rats. The IARC stated that there is sufficient evidence for animal carcinogenicity of 1,3-dichloropropene technical-grade but inadequate evidence for carcinogenicity to humans.[5]

In a subsequently published report, an inhalation carcinogenicity study with technical-grade material, rats and mice were exposed to 0, 5, 20, or 60 ppm 6 hr/day, 5 days/week, for up to 2 years.[10] There were morphologic changes in the nasal tissue of rats exposed to 60 ppm and mice exposed to 20 and 60 ppm. Mice exposed to 20 or 60 ppm had hyperplasia of the epithelial lining of the urinary bladder. Rats showed no increased tumor incidence. Male mice showed an increased incidence of bronchoalveolar adenomas in the 60-ppm group.

Rats of both sexes were exposed in an inhalation reproduction study to technical-grade at 0, 10, 30, or 90 ppm for 6 hr/day, 5 days/week, for two generations.[11] There were no adverse effects on the reproductive and neonatal parameters. Parental effects were focal hyperplasia and/or focal degenerative changes in the olfactory epithelium at 90 ppm.

A number of genotoxic effects have been reported for 1,3-DCP, including increased DNA strand breaks, sister chromatid exchanges, and mitotic aberrations in Chinese hamster cells.[8,12]

The 1995 ACGIH threshold limit value–time-weighted average (TLV-TWA) for 1,3-dichloropropene is 1 ppm (4.5 mg/m^3) with a notation for skin absorption.

REFERENCES

1. Hayes WJ Jr: *Pesticides Studied in Man.* pp 162–163. Baltimore, MD Williams and Wilkins, 1982
2. Flessel P, Goldsmith JR, Kahn E, Wesolowski JJ: Acute and possible longterm effects of 1,3-dichloropropene. *Morbid Mortal Weekly Rept* 27:50, 1978
3. Markovitz A, Crosby WH: A soil fumigant, 1,3-dichloropropene, as possible cause of hematologic malignancies. *Arch Int Med* 144:1409–1411, 1984
4. *IARC Monographs on the Evaluation of the Carcinogenic Risks to Humans, Vol 41*, Some halogenated hydrocarbons and pesticide exposures, pp 113–130. Lyon, International Agency for Research on Cancer, 1986
5. *IARC Monographs on the Evaluation of the Carcinogenic Risks to Humans*, Suppl 7, *Overall Evaluations of Carcinogenicity: An Updating of IARC Monographs Vols 1–142*, pp 195–196.

Lyon, International Agency for Research on Cancer, 1987

6. Hernandez AF, Martin-Rubi JC, Ballesteros JL, et al: Clinical and pathological findings in fatal 1,3-dichloropropene intoxication. *Hum and Exp Toxicol* 13:303–306, 1994

7. Torkelson TR, Oyen F: The toxicity of 1,3-dichloropropene as determined by repeated exposure of laboratory animals. *Am Ind Hyg Assn J* 38:217–223, 1977

8. Agency for Toxic Substances and Disease Registry (ASTDR): *Toxicological Profile for 1,3-Dichloropropene.* US Department of Health and Human Services, Public Health Service, p 123, September 1992

9. National Toxicology Program: *Toxicology and Carcinogenesis Studies of Telone II (Technical Grade 1,3-Dichloropropene [CAS No 542-75-6] Containing 1.0% Epichlorohydrin as a stabilizer) in F344/N Rats and B6C3F₁ Mice (Gavage Studies).* NTP Technical Report Series 269; pp 9–109, 1985

10. Lomax LG, Stott W, Johnson K, et al: The chronic toxicity and oncogenicity of inhaled technical-grade 1,3-dichloropropene in rats and mice. *Fund Appl Toxicol* 12:418–431, 1989

11. Breslin WJ, Kirk H, Streeter C, et al: 1,3-Dichloropropene: two-generation inhalation reproduction study in Fischer 344 rats. *Fund Appl Toxicol* 12:129–143, 1989

12. Martelli A, Allavena A, Ghia M, et al: Cytotoxic and genotoxic activity of 1,3-dichloropropene in cultured mammalian cells. *Tox Appl Pharmacol* 120:114–119, 1993

2,2-DICHLOROPROPIONIC ACID
CAS: 75-99-0

CH_3CCl_2COOH

Synonyms: Dalapon; Dalzpon; Radapon

Physical Form. Liquid

Uses. Herbicide marketed as sodium salt or a mixture of the sodium and magnesium salts used to control grasses in a wide variety of crops and in a number of noncrop applications, such as along drainage ditches and railroads and in industrial areas.

Exposure. Inhalation

Toxicology. 2,2-Dichloropropionic acid is expected to be an irritant of the eyes and skin.

In human subjects, approximately 50% of five consecutive daily doses of 0.5 mg/kg was excreted in the urine within an 18-day period.[1] Acute toxicity data indicate that it has a low order of toxicity in mammals, with a range of oral LD_{50} values of 4 to 9 g/kg. The dry powder or a concentrated solution can be irritating to the eyes or skin if not removed by washing.

Short-term multiple-dose studies suggest that the toxicity of the compound is not cumulative. Cattle that received a 1-g/kg daily oral dose for 10 days showed some signs of toxicity but rapidly recovered when dosing ceased. Slight cloudy swelling of the convoluted tubules and hypertrophy or swelling of the glomerular cells of the kidney were the only findings in a bull calf receiving 1 g/kg/day. Dogs were dosed by gavage for an 81-day period, initially with 50 mg/kg/day, with dosages adjusted upward until the animals were receiving 1000 mg/kg/day. Vomiting ensued at this high dose level, and the study was terminated at 81 days. Except for vomiting, no other signs of toxicity were evident. Extensive hematological and biochemical parameters were all normal, as were the organ-to-body weight ratios.

In a 97-day rat study, there were no effects in male rats fed dalapon in the diet at levels up to 115 mg/kg/day. In female rats, there were slight, statistically significant increases in average kidney weights at the 34.6 mg/kg/day level. At 346 or 1150 mg/kg/day, both male and female rats showed growth retardation, increased liver and kidney weights, and slight histopathological changes in the liver and kidneys. This study established a no-observed-adverse-effect level (NOAEL) of 11.5 mg/kg/day for oral intake in the rat.

A 1-year study conducted with dogs showed significant increases in the average kidney weight in animals receiving 100 mg/kg/day but not in those receiving 50 mg/kg/day. All other parameters were comparable to controls.

In a 2-year rat study, significant increases were noted in the kidney weight in rats receiving 50 mg/kg/day but not in those receiving 15 or 5 mg/kg/day. In this chronic study, a dose of 15 mg/kg/day was identified as the NOAEL. In a 2-year mice study, increased liver weights were noted at 200 mg/kg/day in the diet. No associated lesions were noted on histologic examination of the livers. There were also increased incidences of benign lung adenomas and cystadenomas of the harderian gland in male mice fed dalapon for 2 years. No tumors were found in rats fed dalapon for 2 years. In a three-generation reproduction study in rats, no reproductive effects were found in rats administered dalapon in the diet at levels up to 3000 ppm (150 mg/kg/day).

In a rabbit teratogenicity study, decreased body weights were noted in pups from dams receiving oral doses of 300 mg/kg/day. However, no fetal effects were noted at 30 or 100 mg/kg/day. Similarly, the mean of the pup weights was depressed when pregnant female rats received 1000 or 1500 mg/kg/day in the diet from days 6 through 15 of gestation but not when they received 500 mg/kg/day. No other effects on the fetuses were observed.

Dichloropropionic acid was not mutagenic in a variety of organisms, including *Salmonella typhimurium, Escherichia coli*, T4 bacteriophage, *Streptomyces coelicolor, Saccharomyces cerevisiae*, and *Aspergillus nidulans*.

No reports of adverse effects in individuals who manufacture or utilize the compound were found. In addition, no human toxicity data or epidemiological studies were found for this compound.

The 1995 threshold limit value–time-weighted average (TLV-TWA) is 1 ppm (5.8 mg/m³).

REFERENCES

1. Drinking water criteria document for dalapon; final draft. DART/T/91000934, NTIS Technical Report (NTIS/PB90-215427) April 1990

DICHLOROTETRAFLUOROETHANE
CAS: 76-14-2

$C_2Cl_2F_4$

Synonyms: Refrigerant 114; CFC-114; Freon 114; 1,2-dichloro-1,1,2,2-tetrafluoroethane

Physical Form. Colorless gas

Uses. Refrigerant; aerosol propellant; solvent; fire extinguisher

Exposure. Inhalation

Toxicology. Dichlorotetrafluoroethane causes asphyxia at extremely high concentrations.

Although dichlorotetrafluoroethane has not been directly implicated, sniffing aerosols of other fluorochlorinated hydrocarbons has caused sudden death due to cardiac arrest, probably a result of sensitization of the myocardium to epinephrine.[1] The liquid spilled on the skin may cause frostbite.

Exposure to 200,000 ppm for 16 hours was fatal to dogs; single 8-hour exposures produced tremor and convulsions but no fatalities; repeated exposures at 140,000 to 160,000 ppm for 8 hours caused incoordination, tremor and, occasionally, convulsions, but all dogs survived.[2] At 47,000 ppm for 2 hours, guinea pigs developed irregular respiration.[2] At 25,000 ppm, 1 of 12 dogs developed serious arrhythmia following intravenous epinephrine.[1]

Chronic administration of 10,000 ppm to rats and 5000 ppm to dogs, 6 hr/day for 90 days caused no effects, as determined by clinical, biochemical, and histological examinations.[3]

A 40% solution applied to rabbit skin was without effect. Repeated spraying caused irritation of the mucous membrane of rabbit eyes.[2]

The 1995 ACGIH threshold limit value–time-weighted average (TLV-TWA) for dichlorotetrafluoroethane is 1000 ppm (6990 mg/m³).

REFERENCES

1. Reinhardt CF, et al: Cardiac arrhythmias and aerosol "sniffing." *Arch Environ Health* 22:265–279, 1971
2. Dichlorotetrafluoroethane. *Documentation of the TLVs and BEIs*, 6th ed, pp 443–445. Cincinnati, OH, American Conference of Governmental Industrial Hygienists, 1991
3. Leuschner F et al: Report on subacute toxicological studies with several fluorocarbons in rats and dogs by inhalation. *Drug Res* 33:1475–1476, 1983

DICHLORVOS
CAS: 62-73-7

$C_4H_7Cl_2O_4P$

Synonyms: 2,2-Dichlorovinyl dimethyl phosphate; DDVP

Physical Form. Oily liquid

Uses. Insecticide; commodity or space fumigant; "pest strips"

Exposure. Inhalation; skin absorption; ingestion

Toxicology. Dichlorvos (DDVP) is an anticholinesterase agent.

Signs and symptoms of overexposure are caused by the inactivation of the enzyme cholinesterase, which results in the accumulation of acetylcholine at synapses in the nervous system, skeletal and smooth muscle, and secretory glands.[1,2] The sequence of the development of systemic effects varies with the route of entry. The onset of signs and symptoms is usually prompt but may be delayed up to 12 hours. After inhalation, respiratory and ocular effects are the first to appear, often within a few minutes of exposure. Respiratory effects include tightness in the chest and wheezing owing to bronchoconstriction and excessive bronchial secretion; laryngeal spasm and excessive saliva-

tion may add to the respiratory distress; cyanosis may also occur.[3] Ocular effects include blurring of distant vision, tearing, rhinorrhea, and frontal headache. After ingestion, gastrointestinal effects such as anorexia, nausea, vomiting, abdominal cramps, and diarrhea appear within 15 minutes to 2 hours. Following skin absorption, localized sweating and muscular fasciculations in the immediate area occur, usually within 15 minutes to 4 hours; skin absorption is some what greater at higher ambient temperatures and is increased by the presence of dermatitis.[1,2]

With severe intoxication by all routes, an excess of acetylcholine at the neuromuscular junctions of skeletal muscle causes weakness aggravated by exertion, involuntary twitchings, fasciculations and, eventually, paralysis. The most serious consequence is paralysis of the respiratory muscles. Effects on the central nervous system include giddiness, confusion, ataxia, slurred speech, Cheyne–Stokes respiration, convulsions, coma, and loss of reflexes. The blood pressure may fall to low levels, and cardiac irregularities, including complete heart block, may occur.

Complete symptomatic recovery usually occurs within a week; increased susceptibility to the effects of anticholinesterase agents persists for up to several weeks after exposure. Daily exposure to concentrations that are insufficient to produce symptoms following a single exposure may result in the onset of symptoms. Continued daily exposure may be followed by increasingly severe effects.

In a study of 13 workers exposed for 12 months to an average concentration of DDVP of 0.7 mg/m³, the erythrocyte cholinesterase activity was reduced by approximately 35%, whereas the serum cholinesterase activity was reduced by 60%. The results of other tests and of thorough medical examination conducted at regular intervals were entirely normal.[4]

DDVP has been shown to cause a persistent irritant contact dermatitis in one worker with negative patch tests and appears to be capable of inducing an allergic contact dermatitis.[5-6]

DDVP is an alkylating agent, causing methylation of DNA in vitro. However, there is no evidence of mutagenicity in man or other mammals, presumably because of DDVP's rapid degradation. Several animal bioassays for carcinogenicity have not demonstrated any statistically significant excesses of tumors. However, because of limitations in these studies, the IARC has concluded that the data are inadequate to evaluate the carcinogenicity of this agent.[7]

The 1995 ACGIH threshold limit value–time-weighted average for dichlorvos is 0.1 ppm (0.90 mg/m^3) with a notation for skin absorption.

Note: For a description of diagnostic signs, differential diagnosis, and medical control, including clinical laboratory tests, as well as specific treatment of overexposure to anticholinesterase insecticides, see "parathion."

REFERENCES

1. Koelle GB (ed): Cholinesterases and anticholinesterase agents. *Handbuch der Experimentellen Pharmakologie*, Vol 15, pp 989–1027. Berlin, Springer-Verlag, 1963
2. Taylor P: Anticholinesterase agents. In Gilman AG et al (eds): *Goodman and Gilman's Pharmacological Basis of Therapeutics*, 7th ed, pp 110–129. New York, Macmillan, 1985
3. Hayes WJ Jr: *Toxicology of Pesticides*, pp 379–428. Baltimore, MD, Williams and Wilkins, 1975
4. Menz M, Luetkemeir H, Sachsse K: Long-term exposure of factory workers to dichlorvos (DDVP) insecticide. *Arch Environ Health* 28:72–76, 1971
5. Mathias CGT: Persistent contact dermatitis from the insecticide dichlorvos. *Contact Dermatitis* 9:217–218, 1983
6. Matsushita T et al: Allergic contact dermatitis from organophosphorus insecticides. *Industrial Health* 23:145–153, 1985
7. *IARC Monographs on the Evaluation of the Carcinogenic Risk of Chemicals to Humans*, Vol 20, Some halogenated hydrocarbons, pp 97–123. Lyon, International Agency for Research on Cancer, 1979

DICYCLOPENTADIENE
CAS: 77-73-6

$C_{10}H_{12}$

Synonyms: Bicyclopentadiene; 1,3-cyclopentadiene dimer; 3a,4,7,7a-tetrahydro-4,7-methanoindene; DCPD

Physical Form. Colorless crystals when pure; liquid

Uses. Chemical intermediate in the manufacture of pesticides; in the production of resin coatings

Exposure. Inhalation

Toxicology. Dicyclopentadiene is an irritant to the eyes, skin, and upper respiratory tract; at high concentrations, it is a central nervous system depressant.

Voluntary human exposure to 1 ppm for 7 minutes produced mild eye and throat irritation; olfactory fatigue occurred in one subject within 24 minutes, but no fatigue occurred during a 30-minute exposure at 5.5 ppm.[1] Workers accidentally exposed to vapors for 5 months experienced headaches during the first 2 months of exposure but lacked symptoms during the next 3 months, indicating a certain degree of adaptation.

Animal studies show considerable difference in sensitivity for the various species. The 4-hour LC_{50} for guinea pigs and rabbits was approximately 770 ppm, whereas rats were slightly more sensitive, with a 4-hour LC_{50} of 660 ppm, and mice were the most sensitive species, with a 4-hour LC_{50} of 145 ppm. All species followed a general pattern of eye irritation, loss of coordination and, if death ensued, convulsions. For example, dogs exposed at 773 ppm had irritation of the eyes, nose, and extremities within 60 minutes; at 458 ppm, tremors occurred within 15 minutes, and signs of irritation, including lacrimation, were apparent within 50 minutes; 272 ppm produced tremors within 3 hours.

Rats repeatedly exposed to 332 ppm for 6 hr/day for 10 days succumbed. At autopsy, there was hemorrhage of the lungs and blood in the intestines and, in females, there was also hemorrhage of the thymus. Rats exposed at the 2 lower concentrations (146 and 72 ppm) exhibited no adverse clinical signs, and no gross lesions were apparent at necropsy. Subchronic exposure of rats for 7 hr/day for 3 months caused some kidney damage and lung involvement in the form of chronic pneumonia and bronchiectasis at both 74 and 35 ppm.

No consistent adverse effects were found in dogs exposed to 32, 23, or 9 ppm, 7 hr/day, for 89 days. Minimal changes in biochemical test values were reported at the highest exposure level, but no dose-related pathological changes were noted among any of the groups.

Applied to the skin of rabbits, undiluted dicyclopentadiene caused minor irritation and, when instilled in the eye, only trace injury occurred.

Dicyclopentadiene has a camphorlike odor with a 100% recognition threshold of 0.02 ppm, however, there may not be noticeable irritation below 10 ppm.[2]

The 1995 ACGIH threshold limit value–time-weighted average (TLV-TWA) for dicyclopentadiene is 5 ppm (27 mg/m^3).

REFERENCES

1. Kinkead ER, Pozzani UC, Geary DL et al: The mammalian toxicity of dicyclopentadiene. *Tox Appl Pharm* 20:552–561, 1971
2. Dicyclopentadiene. Materials Safety Data Sheet No 340, rev B. Corporate Research and Development, Schenectady, NY, February 1984

DICYCLOPENTADIENYL IRON
CAS: 102-54-5

$C_{10}H_{10}Fe$

Synonyms: Ferrocene; bicyclopentadienyl iron

Physical Form. Orange crystalline solid or orange needles

Uses. In ultraviolet stabilizers and smoke depressants for polymers; to increase the burn rate of rocket propellants; to prevent erosion of space capsule shields; to improve the viscosity of lubricants; to catalyze polymerization reactions; to catalyze combustion; some derivatives used as hematinic agents

Exposure. Inhalation; ingestion

Toxicology. Dicyclopentadienyl iron causes hepatic cirrhosis and changes in blood parameters.

The toxicological properties of dicyclopentadienyl iron have not been extensively investigated. However, it has been used as a preventive and therapeutic iron-deficiency drug, and its utilization is listed as tolerable.

In rats, the inhalation LD_{50} is greater than 150 mg/m^3; various oral LD_{50}s ranging from 1000 to 2000 mg/kg have been cited.[1] For mice, LD_{50}s of 600 to 1550 mg/kg have been reported. There were no fatalities in rats administered 10 treatments of 200 mg/kg over a 2-week period.

Repeated inhalation exposure of F344/N rats and B6C3F$_1$ mice at 0, 2.5, 5.0, 10, 20, or 40 mg of vapor for 6 hr/day for 2 weeks caused exposure-related lesions in the nasal turbinates of both species.[2] The lesions were centered primarily in the olfactory epithelium and were morphologically diagnosed as subacute, necrotizing inflammation. Exposure-related effects on organ weights were also seen; rats had decreased liver weights, whereas, in mice, liver, spleen, kidney, and brain weights were decreased, and thymus weights increased. The investigators suggested that these alterations were secondary effects brought on by the nasal lesions. Exposures for 13 weeks in rats and mice at 0, 3.0, 10, or 30 mg/m³ caused histopathological lesions in the larnyx, trachea, lungs, liver (kidneys only in mice) and, most notably, the nasal epithelium.[3] Lesions in the nasal olfactory epithelium consisted of pigment accumulation, necrotizing inflammation, metaplasia, and epithelial regeneration. Although increases in lung burdens of iron were found, there were no exposure-related changes in respiratory function, lung biochemistry, bronchoalveolar lavage cytology, total lung collagen, clinical chemistry, or hematology parameters.

Dogs receiving daily oral doses of 300 mg/kg had a reversible drop in hemoglobin, packed cell volume, and erythrocyte count during the first 4 weeks of treatment.[4] The 300 mg/kg/day eventually (time not specified) resulted in hepatic cirrhosis. The dicyclopentadienyl iron–dosed animals had high liver-iron levels (up to 30–40 times those of controls), which remained elevated after the agent was discontinued. Twenty-six months after the end of dosing, the treated dogs had liver-iron levels roughly 30 times higher than those of controls. Testicular hypoplasia was evident in males treated for 6 months.

Of 20 mice given 28 weekly subcutaneous injections of 5 mg dicyclopentadienyl iron, 17 survived for 9 months, and there were no tumors.[5]

The 1995 ACGIH threshold limit value–time-weighted average for dicyclopentadienyl iron is 10 mg/m³.

REFERENCES

1. Ferrocene: National Toxicology Program Executive Summaries, US Department of Health, Education and Welfare, pp 1–13, February 28, 1984
2. Sun JD, Dahl AR, Gillett NA, et al: Two-week, repeated inhalation exposure of F344/N rats and B6C3F₁ mice to ferrocene. *Fund Appl Toxicol* 17:150–158, 1991
3. Nikula KJ, Sun JD, Barr EB, et al: Thirteen-week, repeated inhalation exposure of F344/N rats and B6C3F₁ mice to ferrocene. *Fund Appl Toxicol* 21:127–139, 1993
4. Yeary RA: Chronic toxicity of dicyclopentadie-
nyl iron (ferrocene) in dogs. *Toxicol Appl Pharmacol* 15:666–676, 1969
5. Haddow A, Horning ES: On the carcinogenicity of an iron-dextran complex. *J Natl Cancer Inst* 24:109–147, 1960

DIELDRIN
CAS: 60-57-1

$C_{12}H_8Cl_6O$

Synonyms: Compound 497; Octalox; HEOD

Physical Form.   Light tan to brown powder

Uses.   Insecticide

Exposure.   Skin absorption; inhalation; ingestion

Toxicology.   Dieldrin is a convulsant; it causes liver cancer in mice.

A number of poisonings, including a few fatalities, have occurred among workers involved in spraying or manufacturing of dieldrin. Early symptoms of intoxication may include headache, dizziness, nausea, vomiting, malaise, sweating, and myoclonic jerks of the limbs; clonic and tonic convulsions, and, sometimes coma follow and may occur without the premonitory symptoms.[1-3] In some patients, convulsions are followed by agitation, hyperactivity, and temporary personality changes, including weeping, mania, and inappropriate behavior. Recovery is generally prompt over several weeks and complete, although a few patients have been described with persistent symptoms for several months, and recurrent convulsions have occurred rarely.[1] The half-life of dieldrin in man is reportedly as long as 0.73 years. Dieldrin is well absorbed dermally, which may be the primary route of occupational exposure.[1]

Electroencephalogram changes, including bilateral spikes, spike and wave complexes, and slow theta waves, occur in sufficiently exposed workers. Electroencephalograms have been used successfully in monitoring workers and justifying removal from exposure, but this test has been supplanted by measurement of blood levels. In a study of five aldrin/dieldrin workers who had suffered one or more convulsive seizures and/or myoclonic limb movements, the probable concentration of dieldrin in the blood during intoxication ranged from 16 to 62 μg/100 g of blood; in healthy workers the concentrations of dieldrin ranged up to 22 μg/100 g of blood.[4]

The hepatocarcinogenicity of dieldrin in mice has been confirmed in several experiments and, in some cases, the liver-cell tumors metastasized.[5] No excess of tumors has been observed in a number of bioassays in rats and one bioassay in Syrian golden hamsters.[5,6]

A study of 870 men employed in the manufacture of aldrin, dieldrin, and endrin found no increase in mortality from all cancers; there were apparent increases in mortality from cancers of the esophagus, rectum, and liver, based on very small numbers.[7] In another study, follow-up of 232 workers with similar exposures revealed 9 cancer deaths with 12 expected.[8] The IARC has determined that there is inadequate evidence for carcinogenicity to humans and limited evidence for carcinogenicity to animals.[9]

Developmental effects have been noted in animals following a single very large dose in midgestation. A dose of 30 mg/kg administered orally to pregnant golden hamsters during the period of fetal organogenesis caused a high incidence of fetal deaths, congenital anomalies, and growth retardation.[10] No information was available on the health of the maternal animals, but it should be noted that this dosage approaches reported LD_{50}'s.[11] No developmental effects were observed in rats exposed to concentrations as high as 6 mg/kg/day on days 7 through 15 of gestation or in mice exposed up to 4 mg/kg/day on days 6 through 14 of gestation.[11] Decreased fertility has been reported in some reproductive studies.

Dieldrin is genotoxic in some assays, but it is thought not to act directly on DNA.[11]

The 1995 ACGIH threshold limit value–time-weighted average for dieldrin is 0.25 mg/m³ with a notation for skin absorption.

REFERENCES

1. Hayes WJ Jr, Laws ER, Jr: *Handbook of Pesticide Toxicology*, Vol 2, *Classes of Pesticides*, pp 828–839, New York, Academic Press, 1991
2. Committee on Toxicology: Occupational dieldrin poisoning. *JAMA* 172:2077–2080, 1960
3. Hoogendam I, Versteeg JPJ, de Vlieger M: Nine years' toxicity control in insecticide plants. *Arch Environ Health* 10:441–448, 1965
4. Brown VKH, Hunter CG, Richardson A: A blood test diagnostic of exposure to aldrin and dieldrin. *Br J Ind Med* 21:283–286, 1964
5. *IARC Monographs on the Evaluation of the Carcinogenic Risk of Chemicals to Man*, Vol 5, Some organochlorine pesticides, pp 125–156. Lyon, International Agency for Research on Cancer, 1974
6. Ashwood-Smith MJ: The genetic toxicology of aldrin and dieldrin. *Mut Res* 86:137–154, 1981
7. Ditraglia D, Brown DP, Namekata T, et al: Mortality study of workers employed at organochlorine pesticide manufacturing plants. *Scand J Work Environ Health* 7 (Suppl 4):140–146, 1981
8. Ribbens PH: Mortality study of industrial workers exposed to aldrin, dieldrin and endrin. *Int Arch Occup Environ Health* 56:75–79, 1985
9. *IARC Monographs on the Evaluation of Carcinogenic Risks to Humans. Overall Evaluations of Carcinogenicity. an updating of IARC Monographs, Vols 1–42*, Suppl 7, pp 196–197. Lyon, International Agency for Research on Cancer, 1987
10. Ottolenghi AD, Haseman JK, Suggs F: Teratogenic effects of aldrin, dieldrin, and endrin in hamsters and mice. *Teratology* 9:11–16, 1974
11. Agency for Toxic Substances and Disease Registry (ATSDR): *Toxicological Profile for Aldrin/Dieldrin*. US Department of Health and Human Services, Public Health Service, TP-92/01 p 184, 1993

DIEPOXYBUTANE
CAS: 1464-53-5

$C_4H_6O_2$

Synonyms: 2,2'-bioxirane; 1,1'-bi(ethylene oxide); butadiene diepoxide; butadiene dioxide; 2,4-diepoxybutane; dioxybutadiene

Physical Form. Colorless liquid

Uses. Curing of polymers; cross-linking of textile fibers; prevention of microbial spoilage

Exposure. Inhalation, skin absorption

Toxicology. Diepoxybutane is a mucous membrane irritant. Certain stereoisomers of diepoxybutane have been shown to cause cancer in laboratory animals.

In man, minor, accidental exposure to diepoxybutane caused eyelid swelling, painful eye irritation, and upper respiratory tract irritation within 6 hours.[1] On contact with the skin, it is expected to produce burns and blisters.

The oral LD_{50} of diepoxybutane in rats is 78 mg/kg bw, whereas the inhalation LC_{50} is 371 mg/m³ for a 4-hour exposure. Rats exposed via inhalation experienced lacrimation, clouding of the cornea, labored breathing, and lung congestion.[1] Leukopenia and lymphopenia were produced in rabbits following 6 intramuscular injections of 25 mg/kg.[1]

Both D,L- and meso-1,2:3,4-diepoxybutane induced skin papillomas and carcinomas in mice following dermal application.[1] Lung tumors were produced in mice following intraperitoneal administration of 27, 108, or 192 mg/kg (total dose). The tumor incidences at these doses were 55%, 64%, and 78%, respectively, compared to 31% in control mice.[1]

The International Agency for Research on Cancer has indicated that there is sufficient evidence for carcinogenicity of diepoxybutane to experimental animals. There are no data regarding possible carcinogenicity to humans.[2] Diepoxybutane causes mutations in bacteria and has been found to alkylate DNA.[1]

The ACGIH has not established a threshold limit value for diepoxybutane.

REFERENCES

1. *IARC Monographs on the Evaluation of the Carcinogenic Risk of Chemicals to Man*, Vol. 11, Cadmium, nickel, some epoxides, miscellaneous industrial chemicals and general considerations on volatile anesthetics, pp 115–123. Lyon, International Agency for Research on Cancer, 1976
2. *IARC Monographs on the Evaluation of Carcinogenic Risks to Humans, Overall Evaluations of Carcinogenicity: An Updating of IARC Monographs Vols 1 to 42*, Suppl 7, p 62. Lyon, International Agency for Research on Cancer, 1987

DIETHANOLAMINE
CAS: 111-42-2

$HO(CH_2)_2NH(CH_2)_2OH$

Synonyms: DEA; 2,2-iminodiethanol; dihydroxydiethylamine

Physical Form. Either a colorless liquid or crystals at ambient temperatures.

Uses. Reacts with long-chain fatty acids to form ethanolamine soaps, which are used extensively as emulsifiers, thickeners, wetting agents, and detergents in cosmetic formulations; dispersing agent in agricultural chemicals; chemical intermediate; corrosion inhibitor

Exposure. Inhalation

Toxicology. In animal studies, target organs of diethanolamine (DEA) toxicity have included bone marrow, kidney, testis, skin, and central nervous system.

Limited reports of DEA toxicity are available in humans. Clinical skin testing of cosmetic products containing DEA showed mild skin irritation in concentrations above 5%.[1]

The oral LD_{50} in rats has ranged from 0.71 ml/kg to 2.83 g/kg.[2,3] The effects of ip administration to rats of doses of 100 or 500 mg/kg were assessed at 4 and 24 hours after dosing.[4] In the liver and kidneys, there was cytoplasmic vacuolization. The high doses caused renal tubular degeneration. In rats fed 0.17 g/kg for 90 days, effects included cloudy swelling and degeneration of kidney tubules, as well as fatty degeneration of the liver.[3,5]

Rats exposed to doses of DEA ranging from 160 to 5000 ppm in drinking water for 13 weeks exhibited dose-dependent hematological changes, tubular necrosis of the kidney with decreased renal function, demyelination of the brain and spinal cord, and degeneration of the seminiferous tubules.[6] Hematological changes consisted of a moderate, poorly regenerative anemia that did not appear to involve hemolysis but rather decreased hematopoiesis. Renal tubular epithelial necrosis was more pronounced in female rats than males. Demyelination in the brain was not associated with apparent neurological signs, but long-term effects of this lesion are unknown. Degeneration of the seminiferous tubule epithelium was associated with dose-related decreases in testis and epididymis weights, with reduced sperm motility and sperm count in the cauda epididymis.

In the same study, topical application of DEA doses ranging from 32 to 500 mg/kg for 13 weeks caused skin lesions characterized by ulceration, inflammation, hyperkeratosis, and acanthosis. Other target organs in the skin application study were similar to, but less severe than, those observed in the drinking water study. Differences in dose-response relationships between the two studies were attributed to the limited dermal absorption of DEA in rats.

In a follow-up study in mice, exposure to DEA via the drinking water or by topical application caused dose-dependent toxic effects in the liver (hepatocellular cytological alterations and necrosis), kidney (nephropathy and tubular epithelial necrosis in males), heart (cardiac myocyte degeneration), and skin (site of application: ulceration, inflammation, hyperkeratosis and acanthosis).[7] Doses ranged from 630 to 10,000 ppm in the drinking water and from 80 to 1250 mg/kg in the topical application study.

The mechanism of DEA toxicity is unknown but may be related to its high tissue accumulation and effects on phospholipid metabolism, resulting in alterations in membrane structure and function.[7]

The liquid applied to the skin of rabbits under semiocclusion for 24 hours on 10 consecutive days caused only minor irritation.[1] With long contact time, DEA is irritating to rabbit eyes at concentrations of 50% and above.[1] Toxicity resulting from direct contact may be in part due to irritation associated with the alkalinity of this chemical.[7]

In the presence of *N*-nitrosating agents, DEA may give rise to *N*-nitrosodiethanolamine, a known animal carcinogen.[1]

The 1995 ACGIH threshold limit value–time-weighted average (TLV-TWA) for DEA is 0.46 ppm (2 mg/m^3) with a notation for skin absorption.

REFERENCES

1. Beyer KH Jr et al: Final report on the safety assessment of triethanolamine, diethanolamine and monoethanolamine. *J Am Coll Toxicol* 2:193–235, 1983
2. Smyth HF Jr, Weil CS, West JS, Carpenter, CP: An exploration of joint toxic action. II. Equitoxic versus equivolume mixtures. *Toxicol Appl Pharmacol* 17:498–503, 1970
3. Mellon Institute. Submission of Data by FDA. Mellon Institute of Industrial Research, University of Pittsburgh, Special Report on the Acute and Subacute Toxicity of Mono-, Di-, and Triethanolamine, Carbide and Carbon Chem Div, UCC Industrial Fellowship No 274-13 (Report 13–67), 1950
4. Grice HC et al: Corrrelation between serum enzymes, isozyme patterns, and histologically detectable organ damage. *Food Cosmet Toxicol* 9:847, 1971
5. Smyth HF Jr et al: Range-finding toxicity data. List IV. *AMA Arch Ind Hyg Occup Med* 4:119–122, 1951
6. Melnick RL, Mahler J, Bucher JR, et al: Toxicity of diethanolamine. 1. Drinking water and topical application exposures in F344 rats. *J Appl Toxicol* 14:1–9, 1994

7. Melnick RL, Mahler J, Bucher JR, et al: Toxicity of diethanolamine. 2. Drinking water and topical application exposures in B6C3F₁ mice. *J Appl Toxicol* 14:11–19, 1994

DIETHYLAMINE
CAS: 109-89-7

(C₂H₅)₂NH

Synonyms: Diethamine; *N*-ethyl-ethanamine

Physical Form.   Colorless liquid

Uses.   In the rubber and petroleum industry; in flotation agents; in resins, dyes, and pharmaceuticals

Exposure.   Inhalation

Toxicology.   Diethylamine is an irritant of eyes, mucous membranes, and skin.

Volunteers exposed to concentrations increasing from 0 to 12 ppm for 60 minutes (average concentration 10 ppm) reported distinct nasal and eye irritation and moderate to strong odor detection.[1] Acute nasal reactions, as determined by acoustic rhinometry and rhinomanometry, were not recorded with exposure to 25 ppm for 15 minutes.

Exposure to high vapor concentrations may cause severe cough and chest pain; heavy, repeated, or prolonged exposure could result in pulmonary edema.[2,3] Contact of the liquid with eyes causes corneal damage. In one reported case, the liquid splashed into the eye caused intense pain.[4] In spite of emergency irrigation and treatment, the cornea became swollen and cloudy; some permanent visual impairment resulted. Prolonged or repeated contact of the eyes with the vapor at concentrations slightly below the irritant level often results in corneal edema, with consequent foggy vision and the appearance of halos around lights.[1] Dermal contact with the liquid causes vesiculation and necrosis of the skin.[3]

In rats, exposure to the saturated vapor is lethal in 5 minutes, and the 4-hour inhalation LC_{50} is 4000 ppm.[5]

Rabbits repeatedly exposed to 50 ppm for 7 hr/day, 5 days/week, for 6 weeks showed corneal damage, pulmonary irritation, moderate peribronchitis, and slight thickening of the vascular walls; at 100 ppm, for the same exposure period, there was striking parenchymatous degeneration of the heart muscle in all exposed animals.[6]

Sneezing, tearing, reddened noses, and lesions of the nasal mucosa were observed in rats exposed at 200 ppm for 6.5 hr/day, 5 days/week, for 24 weeks.[7] Histopathologic examinations showed squamous metaplasia, suppurative rhinitis, and lymphoid hyperplasia of the respiratory epithelium.

Diethylamine has an ammonialike odor that is detectable at 0.13 ppm.[8]

The 1995 ACGIH threshold limit value–time-weighted average (TLV-TWA) for diethylamine is 5 ppm (15 mg/m³) with a short-term excursion limit of 15 ppm (45 mg/m³) and an A4, not classifiable as a human carcinogen, designation; there is a notation for skin absorption.

REFERENCES

1. Lundqvist GR, Yamagiwa M, Pedersen OF, et al: Inhalation of diethylamine—acute nasal effects and subjective response. *Am Ind Hyg Assoc J* 53:181–185, 1992
2. Chemical Safety Data Sheet SD-97, Diethylamine, pp 15–16. Washington, DC, MCA, Inc, 1971
3. Hygienic guide series: diethylamine. *Am Ind Hyg Assoc J* 21:266–267, 1960
4. Grant WM: *Toxicology of the Eye*, 3rd ed, p 333. Springfield, IL, Charles C Thomas, 1986
5. Smyth HF Jr et al: Range-finding toxicity data, List IV. *AMA Arch Ind Hyg Occup Med* 4:109–122, 1951
6. Brieger H, Hodes WA: Toxic effects of exposure to vapors of aliphatic amines. *Arch Ind Hyg Occup Med* 3:287–291, 1951
7. Lynch DW, Moorman WJ, Stober P, et al: Subchronic inhalation of diethylamine vapor in Fischer-344 Rats: organ system toxicity. *Fund Appl Toxicol* 6:559–565, 1986
8. Amoore JE, Hautala E: Odor as an aid to chem-

ical safety: odor thresholds compared with threshold limit values and volatilities for 214 industrial chemicals in air and water dilution. *J Appl Toxicol* 3:272–290. 1983

2-DIETHYLAMINOETHANOL
CAS: 100-37-8

$(C_2H_5)_2NCH_2CH_2OH$

Synonyms. Diethyl ethanolamine; DEAE; *n,n*-diethylethanolamine

Physical Form. Colorless liquid

Uses. Anticorrosive agent; chemical intermediate for the production of emulsifiers, detergents, solubilizers, cosmetics, drugs, and textile finishing agents

Exposure. Inhalation; skin absorption

Toxicology. Diethylaminoethanol (DEAE) is an irritant of the eyes, mucous membranes, and skin in animals.

A laboratory worker attempting to remove animals from an inhalation chamber containing approximately 100 ppm became nauseated and vomited within 5 minutes after a fleeting exposure; no irritation of the eyes or throat was noted duirng this brief exposure.[1] Other persons in the same room also complained of a nauseating odor but showed no ill effects.

After DEAE was released through a leak in the steam heating system into the air of a large office building, most of the 2500 employees experienced irritative symptoms of the respiratory tract.[2] In addition, 14 workers developed asthma for the first time within 3 months of exposure. Bronchial hyperreactivity may have resulted from significant airway irritation by the amine.

Rats exposed to 500 ppm, 6 hours daily, for 5 days exhibited marked eye and nasal irritation, and a number of animals had corneal opacity by the end of the third day; the mortality rate

was 20% and, at autopsy, findings were acute purulent bronchiolitis and bronchopneumonia.[1] Exposure to 25 ppm for 14 days caused respiratory tract epithelial hyperplasia, squamous metaplasia, and clinical rales.[3]

The liquid is a severe skin irritant; in the guinea pig, it is a skin sensitizer.[4] It is also a severe eye irritant and may produce permanent eye injury.

The 1995 ACGIH threshold limit value–time-weighted average (TLV-TWA) for 2-diethylaminoethanol is 2 ppm (9.6 mg/m^3) with a notation for skin absorption.

REFERENCES

1. Cornish HH: Oral and inhalation toxicity of 2-diethylaminoethanol. *Am Ind Hyg Assoc J* 26:479–484, 1965
2. Gadon ME, Melius JM, McDonald GJ, et al: New-onset asthma after exposure to the steam system additive 2-diethylaminoethanol. *JOM* 36:623–626, 1994
3. Hinz JP, Thomas JA, Ben-Dyke R: Evaluation of the inhalation toxicity of diethylethanolamine (DEEA) in rats. *Fund Appl Toxicol* 18:418–424, 1992
4. Miller FA, Scherberger RF, Tischer KS, Webber AM: Determination of microgram quantities of diethanolamine, 2-methylaminoethanol, and 2-diethylaminoethanol in air. *Am Ind Hyg Assoc J* 28:330–334, 1967

DIETHYLENE TRIAMINE
CAS: 111-40-0

$(NH_2CH_2CH_2)_2NH$

Synonyms: 2,2–Diaminodiethylamine; DETA

Physical Form. Yellow, viscous liquid

Uses. Hardener and stabilizer for epoxy resins; solvent for dyes, acid gases, and sulfur

Exposure. Inhalation; skin contact

Toxicology. Diethylene triamine (DETA) is a skin, eye, and respiratory irritant; it also causes skin and pulmonary sensitization.

On the skin, DETA is a potent primary irritant, causing edema and, sometimes, necrosis.[1] Repeated contact with the liquid may lead to skin sensitization, and an asthmatic type of response may result from repeated inhalation of the vapors.[2,3]

In rats, oral LD_{50}s of 1 and 2 g/kg have been reported; there were no deaths following an 8-hour exposure to saturated vapors at room temperature.[2] The dermal LD_{50} in rabbits was 1.09 ml/kg. A 10% solution applied to the skin of mice produced dermal ulceration.

In a lifetime study, 25-μl aliquots of a 5% solution applied 3 times a week to the skin of male mice caused a low incidence of dermatitis, hyperkeratosis, and necrosis.[4] There were no treatment-related skin tumors, nor was there any evidence of an increased incidence of internal tumors. Systemic effects were also absent, indicating limited absorption of the compound.

In the eye, solutions of 15% or more caused lasting corneal damage, but a 5% solution caused only minor injury.[2] The injury caused by single applications appears to be attributable to the highly alkaline character of DETA rather than to some other innate toxicity.[1]

DETA has a strong ammonialike odor, but it does not provide adequate warning of hazardous concentrations.[3]

The 1995 ACGIH threshold limit value–time-weighted average (TLV-TWA) for diethylene triamine is 1 ppm (4.2 mg/m³) with a notation for skin absorption.

REFERENCES

1. Grant WM: *Toxicology of the Eye*, 3rd ed, p. 336. Springfield, IL, Charles C Thomas, 1986
2. Beard RR, Noe JT: Aliphatic and alicyclic amines. In Clayton GD and Clayton FE (eds.): *Patty's Industrial Hygiene and Toxicology*, 3rd ed, rev, Vol 2B, *Toxiciology*, pp 3146–3164. New York, Wiley-Interscience, 1981
3. Hygienic guide series: Diethylene triamine. *Am Ind Hyg Assoc J* 21:268–269, 1960
4. DePass LR, Fowler EH, Weil CS: Dermal oncogenic studies on various ethyleneamines in

male C3H mice. *Fund Appl Toxicol* 9:807–811,1987

DIETHYLHEXYL ADIPATE
CAS: 103-23-1

$C_{22}H_{42}O_4$

Synonyms: DEHA; bis(2-ethylhexyl) adipate; octyl adipate

Physical Form. Colorless or very pale amber liquid

Uses. Plasticizer in polyvinyl chloride films, sheeting, extrusions, and plastisols; solvent and emollient in cosmetics

Exposure. Inhalation; skin contact

Toxicology. Diethylhexyl adipate (DEHA) is of low acute toxicity but is carcinogenic in mice by the oral route.

There are no reports of effects in humans from specific exposure to DEHA, but fumes generated at high temperatures may cause throat and eye irritation.[1]

DEHA has a low acute toxicity, as indicated by the relatively high oral LD_{50} in rats of 9.1 g/kg.[2] Skin absorption is expected to be low because the dermal LD_{50} in rabbits is 15 g/kg.

In a National Toxicology Program carcinogenicity study, rats and mice of both sexes were fed DEHA at 12,000 and 25,000 ppm in the diet for 103 weeks.[3] DEHA was noncarcinogenic in rats but caused hepatocellular carcinomas in female mice in both dose groups and hepatocellular adenomas in males at the higher dose. The IARC has determined that there is limited evidence of carcinogenicity to animals and that no adequate data are available to assess human carcinogenicity.[4]

The liquid in contact with the skin of rabbits under occlusion for 24 hours produced mild skin irritation.[5] In the eye, examination after 24 hours revealed no irritation from DEHA.

The ACGIH has not established a threshold limit value for diethylhexyl adipate.

REFERENCES

1. Smith TF et al: Evaluation of Emissions from Simulated Commercial Meat Wrapping Operation Using PVC Wrap. *Am Ind Hyg Assoc J* 44:176, 1983
2. Smyth HF Jr, Carpenter CP, Weil CS: Range-finding toxicity data: List IV. *Arch Ind Hyg Occup Med* 4:119, 1951
3. National Toxicology Program: NTP Technical Report, *Carcinogenesis Bioassay of Di(2-ethylhexl) Adipate in F344 Rats and B6C3F₁ Mice.* NTP-80-29, DHHS (NIH) Pub No 81-1768. Washington, DC, US Government Printing Office, 1982
4. *IARC Monographs on the Evaluation of Carcinogenic Risks to Humans.* Suppl 7, *Overall Evaluations of Carcinogenicity: An Updating of IARC Monographs, Vols 1–42,* p 62. Lyon, International Agency for Research on Cancer, 1987
5. Wickhen Products Inc: FYI-OTS-0684-0286, Suppl Seq H. Di(2-ethylhexyl) Adipate: Animal Toxicology Studies. Report from Kolmar Research Center, May–August 1967. Washington, DC, Office of Toxic Substances, US Environmental Protection Agency, 1984

DI(2-ETHYLHEXYL) PHTHALATE
CAS: 117-81-7

$C_{24}H_{38}O_4$

Synonyms: DEHP; bis(2-ethylhexyl)phthalate; diethylhexyl phthalate; di-*sec*-octyl phthalate

Physical Form. Clear to slightly colored oily liquid

Uses. Commonly used plasticizing agent for PVCs and, as such, a component of blood bank bags, surgical tubing, and other products, including food wrappers and children's toys

Exposure. Inhalation

Toxicology. The acute toxicity of di(2-ethylhexyl) phthalate (DEHP) is low; chronic exposure has been associated with liver damage, testicular injury, and teratogenic and carcinogenic effects in experimental animals.

Two male volunteers developed mild gastric disturbance after swallowing 5 or 10 g.[1] Dermally applied, it was judged to be moderately irritating and, at most, only slightly sensitizing to human skin.

The oral LD_{50} for rats is 26 g/kg and, for rabbits, it is 34 g/kg.[1] A single oral dose of 2 g/kg of DEHP to dogs caused no toxicity. The lethal effects appear to be cumulative because the chronic LD_{50} value for intraperitoneal administration to mice 5 times per week for 10 weeks was 1.36 g/kg, compared to a single-dose value of 37.8 g/kg.[2] DEHP is poorly absorbed through the skin, but two of six rabbits died several days after dermal exposure to 19.7 g/kg.[3]

Acute, intermediate, and chronic oral exposure to DEHP has been found to have significant effects on rodent liver.[4] Effects may include hyperplasia, within 24 hours of exposure, accompanied by an increase in relative liver weight; proliferation of hepatic peroxisomes; altered enzyme functions; changes in the morphology of the bile ducts; and, eventually, the appearance of precancerous altered cell foci and tumors.[4] The amount of damage in the rodent liver is influenced by dose and duration of exposure, diet, age, and exposure to other chemicals.[4] There are also distinct species differences in the toxicity of DEHP to the liver. For example, monkeys exposed to DEHP showed either no increase or a nonsignificant increase in liver weight with doses of 10 to 2000 mg/kg/day for 14 to 25 days.

In 2-year chronic studies, DEHP caused a significant increase in hepatocellular carcinomas in female rats fed diets containing 12,000 ppm, in male mice ingesting 6000 ppm, and in female mice ingesting 3000 or 6000 ppm.[5]

Investigators have suggested that the production of hepatic tumors with DEHP in rodents may be associated with the ability of this substance to induce proliferation of hepatic peroxisomes because no significant genotoxicity of DEHP has been found.[1,6,7] Differences between rodent and human metabolism and susceptibility to peroxisome proliferation suggest that humans may be less susceptible to the carcinogenic effects of DEHP. (An alternate theory holds that the increased cell division, or hyperplasia, induced by DEHP leads to the tumorigenic state.[4]) The IARC determined that there is sufficient evidence for the carcinogenicity of DEHP in mice and rats and that no adequate data were available to evaluate the risk to humans.[3]

Embryolethal and teratogenic effects have been reported in animal studies. DEHP administration in the diet of mice throughout gestation resulted in an increased incidence of exophthalmia, exencephaly, tail defects, major vessel malformations, and skeletal defects at doses (0.10% and 0.15%) that produced maternal toxicity, and at a dose (0.05%) that did not produce significant maternal toxicity.[8] There were also increased resorptions and late fetal deaths, a decreased number of live fetuses, and reduced fetal weights at the two higher dose levels. In contrast, DEHP was not teratogenic in rats at the doses tested but did produce maternal and some embryofetal toxicity at 1.0%, 1.5%, and 2.0% of the diet. Inhalation exposure of up to 0.3 mg/l, 6 hr/day, during gestation failed to produce developmental toxicity in rats.[9]

Mice given diets containing 0.1% and 0.3% DEHP for 7 days prior to, and during, a 98-day cohabitation period had dose-dependent decreases in male and female fertility and in the number of pups born alive.[10]

DEHP-induced testicular injury has been reported in a number of studies.[4,11] Administration of 20,000 mg/kg in the diet of rats produced seminiferous tubular degeneration and testicular atrophy within 7 days, 12,500 mg/kg produced similar effects within 90 days, and 6000 ppm was effective by the end of 2 years of exposure. Testicular damage has been found to be more severe in young rats than in older rats, and damage appears to be reversible if DEHP is withdrawn from the diet before sexual maturity is reached.

DEHP was not mutagenic in a wide variety of microbial and mammalian genotoxic assays.[4]

The 1995 ACGIH threshold limit value–time-weighted average for DEHP is 5 mg/m^3, with a short-term excursion limit of 10 mg/m^3.

REFERENCES

1. Thomas JA, Thomas MJ: Biological effects of di-(2-ethylhexyl) phthalate and other phthalic acid esters. *Crit Rev Toxicol* 13:283–317, 1984
2. Lawrence WH, Malik M, Turner JE, et al: A toxicological investigation of some acute, short-term and chronic effects of administering di-2-ethylhexyl phthalate (DEHP) and other phthalate esters. *Environ Res* 9:1–11, 1975
3. *IARC Monographs on the Evaluation of the Carcinogenic Risk of Chemicals to Humans*, Vol 29, Some industrial chemicals and dyestuffs, pp 269–294. Lyon, International Agency for Research on Cancer, May 1982
4. Agency for Toxic Substances and Disease Registry (ATSDR): *Toxicological Profile for Di(2-ethylhexyl) Phthalate.* US Department of Health and Human Services, Public Health Service, p 147, TP-92/05, 1992
5. National Toxicology Program: *Carcinogenesis bioassay of di-(2-ethylhexyl) Phthalate (CAS No 117-81-7) in Fischer 344 Rats and B6C3F$_1$ Mice (Feed Study).* DHHS (NIH) Pub No 82-1773, NTP-80-37, pp 1–127. Washington, DC, US Government Printing Office, March 1983
6. Turnbull D, Rodricks JV: Assessment of possible carcinogenic risk to humans resulting from exposure to di-(2-ethylhexyl) phthalate (DEHP). *J Am College Toxicol* 4:111–145, 1985
7. Melnick RL, Morrissey RE, Tomaszewski KE: Studies by the National Toxicology Program on di-(2-ethylhexyl) phthalate. *Toxicol Ind Health* 3:99–163, 1987
8. Tyl RW, Price CJ, Marr MC, et al: Developmental toxicity evaluation of dietary di-(2-ethylhexyl) phthalate in Fischer 344 rats and CD-1 mice. *Fund Appl Toxicol* 10:395–412, 1988
9. Merkle J, Klimisch H, Jack R: Developmental toxicity in rats after inhalation exposure of di-(2-ethylhexyl) phthalate (DEHP). *Toxicol Letts* 42:215–223, 1988
10. Lamb JC, Chapin RE, Teague J, et al: Reproductive effects of four phthalic acid esters in the mouse. *Toxicol Appl Pharmacol* 88:255–269, 1987
11. Dostal LA, Chapin RE, Stefanski SA, et al:

Testicular toxicity and reduced Sertoli cell numbers in neonatal rats by di-(2-ethylhexyl) phthalate and the recovery of fertility as adults. *Toxicol Appl Pharmacol* 95:104–121, 1988

2. Smyth HF Jr, Carpenter CP, Weil CS, et al: Range-finding toxicity data. List V. *Arch Ind Hyg Occup Med* 10:61–68, 1954
3. Amoore JE, Hautala E: Odor as an aid to chemical safety: odor thresholds compared with threshold limit values and volatilities for 214 industrial chemicals in air and water dilution. *J Appl Toxicol* 3:272–290, 1983

DIETHYL KETONE
CAS: 96-22-0

$C_2H_5COC_2H_5$

DIETHYL PHTHALATE
CAS: 84-66-2

$C_6H_4(COOC_2H_5)_2$

Synonyms: 3-Pentanone; DEK; dimethylacetone; methacetone, propione

Physical Form. Colorless liquid

Uses. In organic synthesis

Exposure. Inhalation

Toxicology. Diethyl ketone is expected to be an irritant and central nervous system depressant at high concentrations.

Limited toxicological information is available for diethyl ketone. Based on analogy with other methyl ketones, high vapor concentrations are expected to irritate the conjunctiva of the eyes and mucous membranes of the nose and throat.[1] Excessive inhalation may produce dizziness, headache, nausea, vomiting, and ataxia.

In rats, a 4-hour exposure at 8000 ppm was fatal to 4 of 6 animals.[2] The liquid applied to rabbit skin caused mild irritation, and 50 mg instilled in the eye produced moderate irritation. Repeated skin contact would be expected to cause dermatitis by defatting.[1]

A distinct acetonelike odor is detectable at 2 ppm.[3]

The 1995 ACGIH threshold limit value–time-weighted average (TLV-TWA) for diethyl ketone is 200 ppm (705 mg/m³).

REFERENCES

1. Diethyl Ketone. Material Safety Data Sheet No 478, Corporate Research and Development, Schenectady, NY, March 1982

Synonyms: 1,2-benzenedicarboxylic acid diethyl ester; DEP

Physical Form. Colorless liquid

Uses. Plasticizer for cellulose ester plastic films and sheets; in molded plastics; in manufacture of varnishes; cosmetics

Exposure. Inhalation

Toxicology. Diethyl phthlate (DEP) is of low toxicity.

Exposure to the heated vapor may produce some transient irritation of the nose and throat.[1] Although skin sensitization to DEP is extremely rare, it has been reported.[2] No systemic effects are known pertaining to its occupational use.

The lowest lethal doses in rabbits and guinea pigs were determined to be 4.0 and 5.0 g/kg, respectively, when administered by gavage.[3] There were no adverse effects in rats fed 1.25 g/kg/day or in dogs fed 2.5 g/kg/day for 6 weeks or more.[4]

Diethyl phthalate is absorbed following dermal application. In rat studies, approximately 50% of the applied dose of 30 to 40 mg/kg was excreted unchanged within 1 week, and the unexcreted dose remained in the area of application.[5]

Diethyl phthalate administered in the diet to rats during major organogenesis increased the incidence of fetal lumbar ribs only at 3200 mg/kg/day, a maternally toxic dose.[6] In another report, there also was an increased incidence of supernumerary ribs, but no other embryo/fetal effects, in the offspring of rats fed 5% DEP on gestational days 6 through 15; maternal toxicity was evident as reduced body weight gain.[7]

No effects on the male reproductive system have been found in rats in a number of investigations.[3] Both positive and negative results have been obtained in a variety of genotoxic assays.[3]

The 1995 ACGIH threshold limit value–time-weighted average for diethyl phthalate is 5 mg/m³.

REFERENCES

1. Sandmeyer EE, Kirwin CJ Jr: Esters. In Clayton GD, Clayton FE (eds): *Patty's Industrial Hygiene and Toxicology*, 3rd ed, Vol. 2A, *Toxicology*, p 2344. New York, Wiley-Interscience, 1981
2. Oliwiecki S, Beck MH, Chalmers RJG: Contact dermatitis from spectacle frames and hearing aids containing diethyl phthalate. *Contact Derm* 25:264–265, 1991
3. Agency for Toxic Substances and Disease Registry (ATSDR): *Toxicological Profile for Diethyl Phthalate*. US Department of Health and Human Services, Public Health Service, p 85, 1993
4. Shibko SI, Blumenthal H: Toxicology of phthalic acid esters used in food-packaging material. *Environ Health Perspect* 3:131, 1973
5. Elsisi AE, Carter DE and Sipes IG: Dermal absorption of phthalate diesters in rats. *Fund Appl Toxicol* 12:70N77, 1989
6. Price CJ, Sleet RB, George JD, et al: Developmental toxicity evaluation of diethyl phthalate (DEP) in CD rats. *Teratology* 39:473–474, 1989. (Abst)
7. Field EA, Price CJ, Sleet RB, et al: Developmental toxicity evaluation of diethyl and dimethyl phthalate in rats. *Teratology* 48:33–44, 1993

DIETHYL SULFATE
CAS: 64-67-5

$C_4H_{10}O_4S$

Synonyms: Diethyl monosulphate; ethyl sulfate; sulfuric acid diethyl ester; diethyl tetraoxosulfate

Physical Form. Colorless, oily liquid

Uses. Ethylating agent; Accelerator in the sulfation of ethylene; intermediate in the production of ethyl alcohol from ethylene and sulfuric acid

Exposure. Ingestion; inhalation; skin absorption

Toxicology. Diethyl sulfate is highly toxic by inhalation, ingestion, or skin or eye contact and is carcinogenic in experimental animals.

There is no information on acute toxicity in humans. However, diethyl sulfate is less volatile and is considered less acutely toxic than dimethyl sulfate, which has been shown to produce severe irritation to mucous membranes and the respiratory tract.[1,2]

Animal experiments demonstrated an oral LD_{50} of 350 mg/kg in the rat and 647 mg/kg in the mouse. The lowest dose by inhalation that resulted in death in the rat was 250 ppm for a 4-hour exposure.[2]

An historical cohort study of 743 workers at a plant manufacturing isopropyl alcohol and ethanol showed excess mortality (standardized mortality ratio of 504) from upper respiratory (laryngeal) cancers, based on four cases. These persons had spent most of their time working in the strong acid–ethanol plant, which produced high concentrations of diethyl sulfate.[3] A subsequent case-control study nested in an expanded cohort at this plant indicated that the increased risk was related to exposure to sulfuric acid.[3] An association between estimated exposure to diethyl sulfate and risk for brain tumor (gliomas) was suggested in a study of workers at a petrochemical plant. The International Agency for Research on Cancer (IARC) has noted that there has been no measurement of exposure to diethyl sulfate in these studies and that concomitant exposure to mists and vapors from strong inorganic acids, primarily sulfuric acid, may play a role in increasing these risks.[3]

Local tumors were produced in rats following subcutaneous administration for 49 weeks. A small group of rats receiving 25 or 50 mg/kg diethyl sulfate by gavage had a low incidence of squamous-cell carcinomas of the forestomach, whereas 6 of 24 animals had benign papillomas of the forestomach. In another experiment, a single subcutaneous dose of diethyl sulfate (85 mg/kg) was administered to three pregnant rats on day 15 of gestation. Malignant tumors of the nervous system were observed in 2 of 30 offspring on days 285 and 541 offspring-life. No tumors of this type had been observed in controls.

The IARC has concluded that there is limited evidence for carcinogenicity to humans from diethyl sulfate and sufficient evidence for carcinogenicity to animals.[3]

Diethyl sulfate is an alkylating agent. It induced unscheduled DNA synthesis in human cells in vitro. It induced chromatid breaks in mouse embryos treated transplacentally and dominant lethal mutations in mice, as well as a variety of mutagenic and clastogenic effects in rodent cells in vitro.[3]

A threshold limit value has not been established for diethyl sulfate.

REFERENCES

1. Sandodonata, J: *Monographs on Human Exposure to Chemicals in the Workplace: Diethyl sulfate.* Syracuse, NY, Syracuse Research Corporation, 1985
2. Sandmeyer, EE: In Clayton GD, Clayton FE (eds): *Patty's Industrial Hygiene and Toxicology,* 3rd ed, rev, pp 2094–2096. New York, Wiley-Interscience, 1981
3. *IARC Monographs on the Evaluation of the Carcinogenic Risk to Humans,* Vol 54, Occupational exposures to mists and vapours from strong inorganic acids; and other industrial chemicals, pp 213–228, Lyon, International Agency for Research on Cancer, 1992

DIFLUORODIBROMOMETHANE
CAS: 75-61-6

CF_2Br_2

Synonyms: Halon 1202; Freon 12-B2

Physical Form. Colorless liquid or gas

Uses. Fire-extinguishing agent

Exposure. Inhalation

Toxicology. Difluorodibromomethane causes respiratory irritation and narcosis in animals, and severe exposure is expected to produce the same effects in humans.

No effects have been reported from industrial exposures.

Rats exposed to 4000 ppm for 15 minutes showed pulmonary edema, while 2300 ppm daily for 6 weeks resulted in the death of more than half the animals.[1] At 2300 ppm, dogs showed rapid and progressive signs of intoxication after a few days of exposure, with weakness and loss of balance followed by convulsions. At autopsy, these dogs had pulmonary congestion, centrilobular necrosis of the liver, and evidence of central nervous system damage. Other dogs tolerated daily exposures of 350 ppm for 7 months without signs of intoxication.

The 1995 ACGIH threshold limit value–time-weighted average (TLV-TWA) for difluorodibromomethane is 100 ppm (858 mg/m³).

REFERENCES

1. Difluorodibromomethane. *Documentation of the TLVs and BEIs,* 6th ed, pp 469–70. Cincinnati, OH, American Conference of Governmental Industrial Hygienists, 1991

DIGLYCIDYL ETHER
CAS: 2238-07-5

$C_6H_{10}O_3$

Synonyms: Bis(2,3-epoxy propyl)ether; DGE; di(2,3-epoxy propyl) ether

Physical Form. Colorless liquid

Uses. Diluent for epoxy resins; stabilizer of chlorinated organic compounds

Exposure. Inhalation; skin absorption

Toxicology. Diglycidyl ether, DGE, causes severe irritation of the eyes, respiratory tract, and skin; hematopoietic effects have been observed in animals.

Because of its toxicity, DGE generally is not used outside experimental laboratories.[1] No systemic effects have been reported in humans.

The LC_{50} for mice was 30 ppm for 4 hours, but exposure at 200 ppm for 8 hours was not lethal to rats.[2] Rabbits exposed to 24 ppm for 24 hours had leukocytosis at autopsy. There were acute changes in the lungs and kidneys, as well as atrophied testes.[3] At 12 ppm, there was thrombocytopenia and, at 6 ppm, basophilia.

In rats, three or four exposures at 20 ppm for 4 hours produced intense cytoplasmic basophilia, grossly distorted lymphocytic nuclei with indistinct cellular membranes, and lowered leukocyte and marrow cell counts.[3] Repeated exposure of rats to 3 ppm caused increased mortality, decreased body weight, and leukopenia. Exposures to 0.3 ppm did not cause significant changes. Cutaneous applications greater than 100 mg/kg also caused leukopenia, weight loss, and death.

The oral LD_{50} was 0.17 g/kg in mice and 0.45 g/kg in rats; following intragastric administration, effects were incoordination, ataxia, depressed motor activity, and coma.[2]

Diglycidyl ether is extremely damaging to skin, producing ecchymoses and necrosis. In one long-term study, skin painting 3 times per week for 1 year caused hyperkeratosis, epithelial hyperplasia, and skin papillomas.[4]

Instilled in rabbit eyes, DGE is a severe irritant. Exposure to vapor at 3 ppm for 24 hours produced erythema and edema of the conjunctiva in rabbits and, at 24 ppm, corneal opacity was produced.[3]

The 1995 ACGIH threshold limit value–time-weighted average (TLV-TWA) for diglycidyl ether is 0.1 ppm (0.53 mg/m³).

REFERENCES

1. National Institute for Occupational Safety and Health: *Criteria for a Recommended Standard . . . Occupational Exposure to Glycidyl Ethers,* DHEW (NIOSH) Pub No 78–166. Washington, DC, US Government Printing Office, 1978
2. Hine CH et al: The toxicology of glycidol and some glycidyl ethers. *Arch Ind Health* 14:250–264, 1956
3. Hine CH et al: Effects of diglycidyl ether on blood of animals. *Arch Environ Health* 2:31–44, 1961
4. McCammon CJ, Kotkin P, Falk HL: The cancerogenic potency of certain diepoxides. *Proc Assoc Cancer Res* 2:229–230, 1957

DIISOBUTYL KETONE
CAS: 108-83-8

$[(CH_3)_2CHCH_2]_2CO$

Synonyms: Diisopropyl acetone; isovalerone; 2,6-dimethyl-4-heptanone; DBK

Physical Form. Colorless liquid

Uses. Solvent; dispersant for resins; intermediate in the synthesis of pharmaceuticals and insecticides

Exposure. Inhalation

Toxicology. Diisobutyl ketone vapor is an irritant of the eyes and mucous membranes; at high concentrations, it causes narcosis in animals, and it is expected that severe exposure will produce the same effect in humans.[1]

Human subjects exposed to 100 ppm for 3 hours noted slight lacrimation and throat irritation, as well as slight headache and dizziness on returning to fresh air.[2] In another study, the majority of subjects experienced eye irritation above 25 ppm, and nose and throat irritation above 50 ppm within 15 minutes.[3]

The liquid is a defatting agent, and prolonged or repeated skin contact may cause dermatitis.

Although diisobutyl ketone may be more toxic and irritative than lower-molecular-weight ketones, at equivalent concentrations, it poses less of an inhalation hazard because of its relatively low volatility.[1]

Exposure of female rats to 2000 ppm for 8 hours caused narcosis, and 7 of 12 rats died; however, male rats survived the same treatment, as did both sexes of one other strain of rats.[2] Damage to the lungs, liver, and kidneys was observed at autopsy. Repeated exposures to rats over 30 days resulted in increased liver and kidney weights at 920 and 530 ppm, but there were no effects at 125 ppm.[2]

The 1995 ACGIH threshold limit value–time-weighted average (TLV-TWA) for diisobutyl ketone is 25 ppm (145 mg/m³).

REFERENCES

1. National Institute for Occupational Safety and Health: *Criteria for a Recommended Standard . . . Occupational Exposure to Ketones*, NIOSH, Pub No 78–174, pp 79–80, 134, 187. Washington, DC, US Government Printing Office, June 1978
2. Carpenter CP, Pozzani UC, Weil CS: Toxicity and hazard of diisobutyl ketone vapors. *AMA Arch Ind Hyg Occup Med* 8:377–381, 1953
3. Silverman L, Schulte HF, First MW: Further studies on sensory response to certain industrial solvent vapors. *Ind Hyg Toxicol* 28:262–266, 1946

DIISOPROPYLAMINE
CAS: 108-18-9

$(CH_3)_2CHNHCH(CH_3)_2$

Synonyms: N-(1-Methylethyl)-2-propanamine

Physical Form. Colorless liquid

Uses. Chemical intermediate

Exposure. Inhalation; skin absorption

Toxicology. Diisopropylamine is an eye irritant in humans; it is a pulmonary irritant in animals, and severe exposure is expected to produce the same effect in humans.

Workers exposed to concentrations between 25 and 50 ppm complained of disturbances of vision described as "haziness."[1] In two instances, there were also complaints of nausea and headache. Prolonged skin contact with an irritant of this nature is likely to cause dermatitis.

Exposure of several species of animals to 2207 ppm for 3 hours was fatal; effects were lacrimation, corneal clouding, and severe irritation of the upper respiratory tract; at autopsy, findings included pulmonary edema and hemorrhage. Repeated exposure to 600 ppm, 7 hr/day for 40 days, caused deaths in all rabbits and some guinea pigs; cats and rats survived but had cloudiness of the cornea, with loss of vision. It was determined that the ocular effects of diisopropylamine are due to direct contact with the vapor because no corneal effects occurred from subcutaneous injection in guinea pigs.

Diisopropylamine has a strong odor of ammonia.

The 1995 ACGIH threshold limit value–time-weighted average (TLV-TWA) for diisopropylamine is 5 ppm (21 mg/m³) with a notation for skin absorption.

REFERENCES

1. Treon JF, Sigmon H, Kitzmiller KV, Heyroth FF: The physiological response of animals to

respiratory exposure to the vapors of diisopropylamine. *J Ind Hyg Toxicol* 31:142–145, 1949

DIMETHOXYETHYL PHTHALATE
CAS: 117-82-8

$C_6H_4[COOCH_2CH_2OCH_3]_2$

Synonyms: DMEP: 1,2-benzenedicarboxylic acid bis(2-methyoxyethyl)ester; bis(methoxyethyl)phthalate; dimethyl cellosolve phthalate

Physical Form. Light-colored, clear liquid

Uses. Plasticizer; solvent

Exposure. Inhalation

Toxicology. Dimethoxyethyl phthalate (DMEP) causes teratogenic, reproductive, and fetotoxic effects in animals.

Acute lethality data indicate that DMEP exhibits slight to moderate toxicity. The oral LD_{50} in rats ranged from 3.2 to 6.4 g/kg.[1] Exposu re of rats to 1595 ppm for 6 hours caused deaths of all animals, whereas 770 ppm for 6 hours was not lethal.[2] The dermal LD_{50} in guinea pigs was greater than 10 ml/kg, suggesting very little absorption.

Fetotoxic and teratogenic effects were observed in rats administered 0.374 ml/kg of DMEP via intraperitoneal injection on days 5, 10, and 15 of gestation.[3] Resorptions occurred at an incidence of 27.6%, and teratogenic effects included skeletal and gross abnormalities. Pregnant rats given a single intraperitoneal injection of 0.6 ml/kg on day 10, 11, 12, 13, or 14 of gestation had offspring with skeletal malformations.[4] A single intraperitoneal injection of 2.38 ml/kg in mice resulted in a marked reduction in the incidence of pregnancies and litter size per pregnancy.[5] In another study, DMEP administered intraperitoneally to rats on day 12 of gestation produced hydronephrosis, short limbs and tails, and rare heart defects, including dilated ductus arteriosus and aortic arch and ventral polydactyly.[6]

It has been hypothesized that DMEP acts by in vivo hydrolysis to 2-methoxyethanol, also a known teratogen which, in turn, is metabolized to methoxyacetic acid, the proximate teratogen.[6,7]

Oral administration of 1000 mg/kg by gavage to male rats for a total of 12 treatments over 16 days caused reduced testes weight and increases in abnormal sperm heads.[8]

DMEP did not cause dermal sensitization in guinea pigs.[2] In the eyes of rabbits, it was not irritating.

A threshold limit value for dimethoxyethyl phthalate has not been established by the ACGIH.

REFERENCES

1. Eastman Kodak Co: Material Data Sheet from Eastman Kodak Co to T O'Bryan, FYI-OTS-0884-0329 Supp, Seq C. Washington, DC, Office of Toxic Substances, US Environmental Protection Agency, 1984
2. Sandmeyer EE, Kirwin CJ: Esters. In Clayton GD, Clayton FE (eds): *Patty's Industrial Hygiene and Toxicology*, 3rd ed, Vol 2A, *Toxicology*, pp 2346–2351. New York, Wiley-Interscience, 1981
3. Singh AR, Lawrence WH, Autian, J: Teratogenicity of phthalate esters in rats. *J Pharmacol Sci* 61(1):51–55,1972
4. Parkhie MR, Webb M, Norcross MA: Dimethoxyethyl phthalate. Embryopathy, teratogenicity, fetal metabolism and the role of zinc in the rat. *Environ Health Perspect* 45:89,1982
5. Singh AR, Lawrence WH, Autian J: Mutagenic and anti-fertility sensitivities of mice to di-2-ethylhexyl phthalate (DEHP) and dimethoxyethyl phthalate (DMEP). *Toxicol Appl Pharmacol* 29:35–46, 1974
6. Ritter EJ, Scott WJ Jr, Randall JL, et al: Teratogenicity of dimethoxyethyl phthalate and its metabolites methoxyethanol and methoxyacetic acid in the rat. *Teratology* 32:25–31, 1985
7. Campbell J, Holt D and Webb M: Dimethoxyethylphthalate metabolism: Teratogenicity of the diester and its metabolites in the pregnant rat. *J Appl Toxicol* 4:35–41, 1984
8. Eastman Kodak Co: Basic toxicity of bis(methoxyethyl) phthalate from Eastman Kodak Co to

Document Control Officer. FYI-OTS-0385-0329, Seq D. Washington, DC, Office of Toxic Substances, US Environmental Protection Agency, 1985

DIMETHYL ACETAMIDE
CAS: 127-19-5

$CH_3CON(CH_3)_2$

Synonyms: Acetyldimethylamine; *N,N*-dimethyl acetamide; DMAC

Physical Form.   Colorless liquid

Uses.   Commercial solvent especially for textile fibers

Exposure.   Inhalation; skin absorption

Toxicology.   Dimethylacetamide (DMAC) causes liver damage.

Workers repeatedly exposed to 20 to 25 ppm developed jaundice; appreciable skin absorption was thought to have occurred.[1] Nine patients with neoplastic disease were given daily doses of 400 mg/kg by an unspecified route for 3 or more days as a therapeutic trial; they experienced depression, lethargy, confusion, and disorientation; on the last (fourth or fifth) day of therapy or within 24 hours thereafter, the patients had visual and auditory hallucinations, perceptual distortions and, at times, delusions; after 24 hours, these effects gradually subsided.[2]

Rats exposed at 288 ppm, 6 hr/day for 2 weeks, showed nasal irritation, transient increase in blood cholesterol, and liver hypertrophy; testicular atrophy was evident 2 weeks post-exposure.[3]

Repeated dermal application of the liquid to dogs at a dosage level of 4.0 mg/kg for 6 weeks caused severe fatty infiltration of the liver.[3] Repeated exposure of rats to a concentration of 195 ppm for 6 months resulted in focal necrosis of the liver; exposure to 40 ppm for the same period of time caused no adverse effects.[4]

The approximate lethal dose for skin absorption in pregnant rats and rabbits was 7.5 and 5.0 g/kg, respectively.[5] Cutaneous application of DMAC resulted in a marked incidence of embryo mortality at doses that did not affect maternal body weight or produce any signs of maternal toxicity. Teratogenic effects (three fetuses from one dam with encephalocele; one of eight with diffuse subcutaneous edema) were found in rats only when DMAC was applied on gestation days 10 and 11 at a total dose of 2400 mg/kg.[5] In another study, DMAC administered by gavage to rats caused treatment-related malformations of the fetal heart, major vessels, and oral cavity but only at levels (400 mg/kg/day) that also produced significant maternal toxicity and other indicators of fetal toxicity, including increased postimplantation loss and skeletal variation.[6]

Dimethylacetamide has a significant antitumor effect in animals.[2]

In practice, the dermal absorption factor is considered to be so significant that no air concentration, however low, will provide protection if skin contact with the liquid is permitted.

The 1995 ACGIH threshold limit value–time-weighted average (TLV-TWA) for dimethyl acetamide is 10 ppm (36 mg/m³) with a notation for skin absorption.

REFERENCES

1. Dimethyl acetamide. *Documentation of TLVs and BEIs*, 6th ed, pp 477–478. Cincinnati, OH, American Conference of Governmental Industrial Hygienists, 1991
2. Weiss AJ et al: Dimethylacetamide: A hitherto unrecognized hallucinogenic agent. *Science* 136:151–152, 1962
3. Kelly DP et al: Subchronic inhalation toxicity of dimethylacetamide in rats. *The Toxicologist* 4:65 (Abstract), 1984
4. Horn HJ: Toxicology of dimethylacetamide. *Toxicol App Pharmacol* 3:12–24, 1961
5. Stula EF, Krauss WC: Embryotoxicity in rats and rabbits from cutaneous application of amide-type solvents and substituted ureas. *Toxicol Appl Pharmacol* 41:35–55, 1977
6. Johannsen FR, Levinskas GJ, Schardein JL: Teratogenic response of dimethylacetamide in rats. *Fund Appl Toxicol* 9:550–556, 1987

DIMETHYLAMINE
CAS: 124-40-3

(CH₃)₂NH

$(CH_3)_2NH$

Synonyms: N-Methylmethanamine; DMA

Physical Form. Gas, liquefying at 7°C

Uses. In manufacture of pharmaceuticals; stabilizer in gasoline; in production of insecticides and fungicides; in manufacture of soaps and surfactants

Exposure. Inhalation; skin contact

Toxicology. Dimethylamine is an irritant of the skin and respiratory tract.

Dermatitis and conjunctivitis are occasionally observed in chemical workers after prolonged exposure.[1] No systemic effects from industrial exposure have been reported.

The LC_{50} for rats exposed 6 hours to dimethylamine and observed for 48 hours postexposure was 4540 ppm.[2] Clinical observations were characterized by signs of eye irritation immediately after onset of exposure, followed by gasping, secretion of bloody mucus from the nose, salivation, and lacrimation within 1 hour of exposure. Corneal opacity was generally observed after 3 hours of exposure. Death was often preceded by convulsions. Rats exposed to nonlethal concentrations (600–2500 ppm for 6 hours) showed signs of eye irritation, moderate gasping, and slight bloody nasal discharge. At autopsy, findings included severe congestion, ulcerative rhinitis, and necrosis of the nasal turbinates. At concentrations above 2500 ppm, peripheral emphysema, bronchopneumonia, hepatic necrosis, and corneal ulceration were noted.

At lower concentrations, 175 to 500 ppm, less damage to the lower respiratory tract occurred, but inflammation, ulcerative rhinitis, and early squamous metaplasia were observed in the respiratory nasal mucosa.[3] Various species survived 5 ppm of continuous exposure for 90 days without signs of toxicity but, at autopsy,

some showed mild inflammatory changes in the lungs.[4]

Chronic exposure 6 hr/day, 5 days/week for 1 year to concentrations ranging from 10 to 175 ppm caused concentration-related lesions in the nasal passages in rats and mice.[5] The respiratory epithelium in the anterior nasal passages and the olfactory epithelium were primarily affected. Hyperplasia of basophilic cells adjacent to the basement membrane was present in the high-dose rats only. At 10 ppm, there was minimal loss of olfactory sensory cells in a few mice and rats. In a subsequent report, male rats were exposed to 175 ppm DMA 6 hr/day for 1,2,4, or 9 days or 2 years.[6] Severe tissue destruction occurred in the anterior nose following a single 6-hour exposure; however, there was little evidence of progression of the lesions, even after 2 years. The findings indicated a possible regional susceptibility to DMA toxicity or a degree of adaptation by the rat to continued exposure. Despite damage to the nasal epithelium, the mucociliary apparatus continued to function in the exposed rats, and this clearance system responded to alterations of nasal structure by modification of mucus flow patterns.[6]

Intraperitoneal injection of 2.5 or 5 mM/kg/day into pregnant CD-1 mice on days 1 through 17 of gestation did not cause any obvious maternal or fetal effects.[7] Added to mouse embryo cultures, dimethylamine inhibited embryo development.[7]

Skin contact with the liquid causes necrosis, and a drop in the eye may result in severe corneal injury or permanent corneal opacity.

A "fishy" odor is detectable at 0.5 ppm, which may provide warning of overexposure.

The 1995 ACGIH threshold limit value–time-weighted average (TLV-TWA) for dimethylamine is 5 ppm (9.2 mg/m³) with a short-term excursion limit (STEL) of 15 ppm (27.6 mg/m³).

REFERENCES

1. Chemical Safety Data Sheet SD-57, Methylamines, pp 17–19. Washington, DC, MCA, Inc, 1955
2. Steinhagen WH et al: Acute inhalation toxicity

and sensory irritation of dimethylamine. *Am Ind Hyg Assoc J* 3:411–417, 1982
3. McNulty MJ: Biochemical toxicology of inhaled dimethylamine. *CIIT Activities.* Chemical Industry Institute of Toxicology, pp 1–4. August 1983
4. Coon RA, Jones RA, Jenkins LJ Jr, et al: Animal inhalation studies on ammonia, ethylene glycol, formaldehyde, dimethylamine, and ethanol. *Toxicol Appl Pharmacol* 16:646–655, 1970
5. Buckley LA, Morgan KT, Swenberg JA, et al: The toxicity of dimethylamine in F-344 rats and B6C3F$_1$ mice following a 1-year inhalation exposure. *Fund Appl Toxicol* 5:341–352, 1985
6. Gross EA, Patterson DL, Morgan KT: Effects of acute and chronic dimethylamine exposure on the nasal mucociliary apparatus of F-344 rats. *Toxicol Appl Pharm* 90:359–376, 1987
7. Guest I, Varma DR: Developmental toxicity of methylamines in mice. *J Toxicol Environ Health* 32:319–330, 1991

4-DIMETHYLAMINOAZOBENZENE
CAS: 60-11-7

$C_6H_5N_2C_6H_4N(CH_3)_2$

Synonyms: p-Dimethylaminoazobenzene; butter yellow; DAB

Physical Form.   Yellow solid

Uses.   For coloring polishes and wax products

Exposure.   Inhalation

Toxicology.   4-Dimethylaminoazobenzene (DAB) is a potent carcinogen in animals.

In humans, contact dermatitis was observed in 90% of factory workers handling DAB.[1] There have been no reports of an increased cancer incidence among exposed persons.[2]

Two of 10 dogs survived ingestion of 20 mg/kg/day for 38 months of continuous treatment, followed by 48 months of intermittent treatment; both developed bladder papillomas.[3] Oral administration of 1, 3, 10, 20, or 30 mg/ day produced liver tumors in rats; the induction time was inversely proportional to the daily dose, ranging between 34 days for the 30 mg/ day dose and 700 days for the 1 mg/day dose.[4] In rats fed 5 mg DAB/day for 40 to 200 days and then kept for their life span on a normal diet, there was a 20% to 81% incidence of liver carcinoma.[5]

Cutaneous application of 1 ml of a 2% solution of DAB in acetone two times/week for 90 weeks caused skin tumors in all 6 male rats treated. Squamous-cell, basal-cell, and anaplastic carcinomas were observed; there were no tumors in controls given acetone alone.[6]

Because of its demonstrated carcinogenicity in animals, human exposure to DAB by any route should be avoided. In recent years, however, this compound has been used only in laboratories as a model of tumorigenic activity in animals.[7] It is of little occupational health importance.

The ACGIH has not established a threshold limit value for 4-dimethylaminoazobenzene.

REFERENCES

1. National Research Council: *Food Colors*, p 7. Washington, DC, National Academy of Sciences, 1971
2. *IARC Monographs on the Evaluation of Carcinogenic Risk to Man*, Vol 8, Some aromatic azo compounds, pp 125–146, Lyon, International Agency for Research on Cancer, 1975
3. Nelson SA, Woodward G: Tumors of the urinary bladder, gall bladder and liver in dogs fed o-aminoazotoluene or p-dimethylaminoazobenzene. *J Natl Cancer Inst* 13:1497–1509, 1953
4. Druckrey H, Kupfmuller, K: Quantitative analyse der krebsentstehung. *Z Naturforsch* 3b:254–266, 1948
5. Druckrey H: Quantitative aspects in chemical carcinogenesis. In Trichaut R (ed): *Potential Carcinogenic Hazards from Drugs.* UICC Monograph Series 7:60–78, 1967
6. Fare G: Rat skin carcinogenesis by topical applications of some azo dyes. *Cancer Res* 26:2405–2408, 1966
7. Stokinger HE: Occupational Carcinogenesis. In Clayton GD, Clayton FE (eds): *Patty's Industrial Hygiene and Toxicology*, 3rd ed, rev, Vol.

IIB, *Toxicology*, p 2893. New York, Wiley-Interscience, 1981

DIMETHYLANILINE
CAS: 121-69-7

$C_8H_{11}N$

Synonyms: Dimethylphenylamine; aminodimethylbenzene; *N,N*-dimethylaniline

Physical Form.   Yellow to brown, oily liquid

Uses.   Intermediate in the manufacture of dyes; solvent; rubber vulcanizing agent; stabilizer

Exposure.   Inhalation; skin absorption

Toxicology.   Dimethylaniline absorption causes anoxia as a result of the formation of methemoglobin.

Few reports of industrial experience are available from which to form an accurate appraisal of the health hazards of dimethylaniline; it is said to be less potent than aniline as a cause of methemoglobin but more of a central nervous system depressant. The effects of methemoglobinemia are cyanosis (especially of the lips, nose, and earlobes), weakness, dizziness, and severe headache.[1]

In dogs, the repeated subcutaneous injection of 1.5 g caused vomiting, weakness, cyanosis, methemoglobinemia and hyperglobulinemia.[2] Rats survived an 8-hour exposure to concentrated vapor.[3] The single-dose oral LD_{50} for rats was 1.41 ml/kg, and the single-dose dermal LD_{50} for rabbits was 1.77 ml/kg.

Continuous exposure of rats by inhalation to 0.0055 and 0.3 mg/m³ for 100 days resulted in methemoglobinemia, lowered erythrocyte hemoglobin, leukopenia, and reticulocytosis, as well as reduced muscle chronaxy.[4] Doses of up to 500 mg/kg administered by gavage to rats and mice for 13 weeks caused cyanosis and decreased motor activity, as well as hemosiderosis

in the spleen, liver, kidney, and testes.[5] Bone marrow hyperplasia was observed in rats, and mice had increased hematopoiesis in the liver. In general, all toxic effects could be attributed to chronic methemoglobinemia, erythrocyte destruction, and erythrophagocytosis.

Mice and rats administered up to 30 mg/kg by gavage, 5 days/week for 103 weeks, had an increased incidence of forestomach papillomas (female mice) and an increase in splenic sarcomas (male rats), which exceeded normal historical controls.[6]

The IARC has determined that there is inadequate evidence in humans for the carcinogenicity of dimethylaniline and limited evidence in animals.[4] Overall dimethylaniline is not classifiable as to its carcinogenicity in humans.

Dimethylaniline induced gene mutation, sister chromatid exchange, and chromosomal aberrations in cultured mammalian cells. It was not mutagenic in *Salmonella typhimurium*.[4]

The liquid was slightly irritating to the clipped skin of rabbits within 24 hours of a 0.01-ml application, and 0.005 ml caused severe burns when instilled in rabbit eyes.[3]

The 1995 ACGIH threshold limit value–time-weighted average (TLV-TWA) for dimethylaniline is 5 ppm (25 mg/m³) with a short-term excursion level of 10 ppm (50 mg/m³) and a notation for skin absorption.

Note: For a description of diagnostic signs, differential diagnosis, and medical control, including clinical laboratory treatment of overexposure to methemoglobin-forming agents, see "Aniline."

REFERENCES

1. Beard RR, Noe JT: Aromatic nitro amino compounds. In Clayton GD, Clayton FE (eds): *Patty's Industrial Hygiene and Toxicology*, 3rd ed, Vol 2A, *Toxicology*, pp 2413–2489. New York, Wiley-Interscience, 1981
2. von Oettingen WF: *The Aromatic Amino and Nitro Compounds, Their Toxicity and Potential Dangers*. US Public Health Service Publication No 271, pp 15–16. Washington, DC, US Government Printing Office, 1941

3. Smyth HF Jr et al: Range-finding toxicity data. List VI. *Am Ind Hyg Assoc J* 23:95–107, 1962

4. *IARC Monographs on the Evaluation of the Carcinogenic Risks to Humans*, Vol 57, Occupational exposures of hairdressers and barbers and personal use of hair colourants; some hair dyes, cosmetic colourants, industrial dyestuffs and aromatic amines, pp 337–350. Lyon, International Agency for Research on Cancer, 1993

5. Abdo KM, Jokinen MP, Hiles R: Subchronic (13-week) toxicity studies of *N,N*-dimethylaniline administered to Fischer 344 rats and B6C3F₁ mice. *J Toxicol Environ Health* 29:77–88, 1990

6. US National Toxicology Program: *Toxicology and Carcinogenesis Studies of* N,N-*Dimethylaniline (CAS No. 121-69-7) in F344/N rats and B6C3F₁ mice (gavage studies).* NTP TR 360, NIH Pub No. 90-2815, US Department of Health and Human Services, 1989

DIMETHYL CARBAMOYL CHLORIDE
CAS: 79-44-7

$(CH_3)_2NCOCl$

Synonyms: Dimethylcarbamic chloride; (dimethylamino)carbonyl chloride; dimethyl carbamyl chloride; DMCC

Physical Form.   Liquid

Uses.   Chemical intermediate in the manufacture of carbamate drugs and pesticides

Exposure.   Inhalation

Toxicology.   Dimethyl carbamoyl chloride is a skin, eye, and respiratory irritant; it is carcinogenic in experimental animals by skin application and by subcutaneous or intraperitoneal injection.

One case of eye irritation and one of impaired liver function have been observed in workers exposed to dimethyl carbamoyl chloride.[1]

When rats were exposed to saturated vapors, 5 of 6 or 6 of 6 died following 1 or 2 hours, respectively. Dimethyl carbamoyl chloride damaged the mucous membranes of the nose, throat, and lungs and caused difficulty in breathing, sometimes several days after exposure.[1] Rats tolerated an 8-minute exposure to the saturated vapors and survived 14 days postexposure. Fifty-one of 100 rats exposed 6 hr/day for 15 days at 10 ppm succumbed.

Applied to the skin of rats and rabbits, the undiluted liquid produced skin irritation, with subsequent degeneration of the epidermis; skin sensitization tests in guinea pigs were negative.

In long-term animal studies, dimethyl carbamoyl chloride produced local tumors by each of three routes of administration.[2] Two milligrams applied to the skin of mice 3 times/week for 492 days caused skin papillomas in 40 out of 50 animals; of these, 30 progressed to skin carcinomas. Following weekly subcutaneous injections of 5 mg for 26 weeks, 36 of 50 female mice developed local sarcomas, and 3 had local squamousp-cell carcinomas. Weekly intraperitoneal injections of 1 mg of the chemical for up to 450 days resulted in papillary tumors of the lung in 14 of 30 treated mice, compared with 10 of 30 mice given the vehicle alone. Eight treated mice and 1 control developed local sarcomas, and 1 treated mouse developed a squamous-cell carcinoma of the skin.

In another study, exposure of rats and male hamsters by inhalation induced a high incidence of nasal tract carcinomas.[3]

No cancer deaths or X-ray indications of lung cancer were found in an investigation of 39 dimethyl carbamoyl chloride production workers, 26 processing workers, and 42 former workers, aged 17 to 65, exposed for periods ranging from 6 months to 12 years.[1]

The IARC has determined that there is sufficient evidence for cacinogenicity to animals and inadequate evidence for humans.[4] In the absence of adequate human data, dimethyl carbamoyl chloride should be treated as if it presented a carcinogenic risk to humans.

ACGIH has classified dimethyl carbamoyl as A2, a suspected human carcinogen with no threshold limit value.

REFERENCES

1. *IARC Monographs on the Evaluation of Carcinogenic Risk to Man*, Vol 12, Some carbamates, thiocarbamates and carbazides, pp 77–84. Lyon, International Agency for Research on Cancer, 1976
2. Van Duuren BL, Goldschmidt BM, Katz C, et al: Carcinogenic activity of alkylating agents. *J Natl Cancer Inst* 53:695–700, 1974
3. Sellakumar AR, Laskin S, Kuschner M, et al: Inhalation carcinogenesis by dimethylcarbamoyl chloride in Syrian golden hamsters. *J Environ Pathol Toxicol* 4:107–115, 1980
4. *IARC Monographs on the Evaluation of Carcinogenic Risk to Humans, Overall Evaluations of Carcinogenicity: An Updating of IARC Monographs Vols 1 to 42*, Suppl 7, pp 199–200. Lyon, International Agency for Research on Cancer, 1987

DIMETHYLFORMAMIDE
CAS: 68-12-2

(CH₃)₂NCHO

$(CH_3)_2NCHO$

Synonyms: DMF; DMFA; *N,N*-dimethylformamide

Physical Form. Colorless liquid

Uses. Solvent

Exposure. Inhalation; skin absorption

Toxicology. Dimethylformamide (DMF) is toxic to the liver.

Subjective complaints of exposed workers have included nausea, vomiting, and anorexia.[1] Air concentration measurements may not define the total exposure experience because DMF is readily absorbed through the skin as well as the lungs. A worker who was splashed with the liquid over 20% of the body surface initially suffered only dermal irritation and hyperemia. Abdominal pain began 62 hours after the exposure and became progressively more severe, with vomiting; blood pressure was elevated to 190/100. The effects gradually subsided and abated entirely by the seventh day after the exposure.[2] Some workers have noted flushing of the face after inhalation of the vapor, especially with coincident ingestion of alcohol.[2]

Hepatomegaly, jaundice, and altered liver-function tests have been reported in accidental poisonings with DMF. An outbreak of toxic liver disease was associated with DMF exposure at a fabric-coating factory.[3] Of 58 workers, 36 had elevations of either aspartate aminotransferase or alanine aminotransferase. Serologic tests excluded known infectious causes of hepatitis in all but 2 cases. After modification of work practices and removal of the most severely affected from exposure, improvement in liver enzyme abnormalities and symptoms occurred in most patients. Medical surveillance of the working population for 14 months revealed no further cases of toxic liver disease, indicating that DMF was the causative agent of the outbreak.[4]

Epidemiological studies indicated that DMF exposure may be associated with an increased risk of developing testicular cancer. Three workers (ranging in age from 25 to 36 years) at a small leather tannery developed testicular tumors with common histological features after latency periods of 8 to 14 years.[5] The men performed similar job tasks and had virtually continuous exposure via skin contact to dye containing DMF.

In another report, clusters of embryonal-cell carcinomas of the testes were observed in two groups of aircraft mechanics who had worked extensively with DMF.[6] At the first rework shop, 3 of 153 men had testicular germ-cell tumors, and evaluation of an occupationally identical shop revealed 4 additional cases.

A cohort study of 2530 employees potentially exposed to DMF at a fiber-producing plant showed a significant excess in incidences of buccal cavity and pharynx cancer and malignant melanoma.[7] A significant excess of prostate cancer was observed among workers exposed to DMF and acrylonitrile.

A case-control study of four plants where DMF was produced or used showed no statistically significant association between ever having been exposed to DMF and subsequent development of cancers of the buccal cavity and pharynx, liver, prostate, and testes as well as malignant melanoma.[8] Although prostate cancer was significantly elevated at one plant when examined by plant site, it did not appear to be related to exposure level or duration.

The IARC has determined that there is limited evidence for the carcinogenicity of dimethylformamide in humans.[9]

In both mice and rats exposed 6 hr/day, 5 days/week, for 12 weeks, the no-effect dose was below 150 ppm, and the maximum tolerated dose was below 600 ppm. At doses of up to 1200 ppm, there were few signs of overt toxicity and, at necropsy, the only treatment-related lesions occurred in the liver.[10]

Metabolic studies of DMF show quantitative differences in human and rodent pathways, suggesting that rodent studies may not be indicative of human results.[11]

Subchronic studies in monkeys showed no exposure-related adverse health effects or reproductive effects following exposure 6 hr/day, 5 days/week, for 13 weeks to concentrations of up to 500 ppm.[12]

Inhalation and epicutaneous exposures of DMF by rats have produced no teratogenic effects and only slight evidence of embryotoxicity at levels producing some maternal toxicity.[13,14]

In a number of short-term assays, DMF was not mutagenic or genotoxic.[15]

The 1995 ACGIH threshold limit value–time-weighted average for dimethylformamide is 10 ppm (30 mg/m³) with a notation for skin absorption.

REFERENCES

1. Hazard Data Bank: Sheet No 77. Dimethyl formamide. *The Safety Practitioner*, pp 48–49. May, 1986
2. Potter HP: Dimethylformamide-induced abdominal pain and liver injury. *Arch Environ Health* 27:340–341, 1973
3. Redlich CA, Beckett WS, Sparer J, et al: Liver disease associated with occupational exposure to the solvent dimethylformamide. *Ann Int Med* 108:680–686, 1988
4. Fleming LE, Shalat SL, Redlich CA: Liver injury in workers exposed to dimethylformamide. *Scand J Work Environ Health* 16:289–292, 1990
5. Levine SM, Baker DB, Langrigan PJ, et al: Testicular cancer in leather tanners exposed to dimethylformamide. *Lancet* (Letter) 2:1153, 1987
6. Ducatman AM, Conwell DE, Crawl J: Germ cell tumors of the testicle among aircraft repairmen. *J Urol* 136:834–836, 1986
7. Chen JL, Fayerweather WE, and Pell S: Cancer incidence of workers exposed to dimethylformamide and/or acrylonitrile. *J Occ Med* 30:813–818, 1988
8. Walrath J, Fayerweather WE, Gilby PG, et al: A case-control study of cancer among DuPont employees with potential for exposure to dimethylformamide. *J Occ Med* 31:432–438, 1989
9. IARC Monographs on the Evaluation of Carcinogenic Risks to Humans Vol 47, Some organic solvents, resin monomers and related compounds, pigments and occupational exposures in paint manufacture and painting, pp 171–197. Lyon, International Agency for Research on Cancer, World Health Organization, 1989
10. Craig DK, Wier RJ, Wagner W, et al: Subchronic inhalation toxicity of dimethylformamide in rats and mice. *Drug Chem Toxicol* 7:551–571, 1984
11. Mraz J, Cross H, Gescher A, et al: Differences between rodents and humans in the metabolic toxification of *N,N*-dimethylformamide. *Toxicol Appl Pharmacol* 98:507–516, 1989
12. Hurtt ME, Placke ME, Killinger JM, et al: 13-Week inhalation toxicity study of dimethylformamide (DMF) in Cynomolgus monkeys. *Fund Appl Toxicol* 18:596–601, 1992
13. Kennedy GL Jr: Biological effects of acetamide, formamide and their monomethyl and dimethyl derivatives. *CRC Crit Rev Tox* 17:129–182, 1986
14. Hansen E, Meyer O: Embryotoxicity and teratogenicity study in rats dosed epicutaneously with dimethylformamide (DMF). *J Appl Toxicol* 10:333–338, 1990
15. Antoine JL, Arany J, Leonard A, et al: Lack of mutagenic activity of dimethylformamide. *Toxicology* 26:207–212, 1983

1,1-DIMETHYLHYDRAZINE
CAS: 57-14-7

$C_2H_8N_2$

Synonyms: *asym*-Dimethylhydrazine; unsymmetrical dimethylhydrazine; dimazine; UDMH

Physical Form.   Colorless liquid

Uses.   Base in rocket fuel formulations; intermediate in organic synthesis

Exposure.   Inhalation; skin absorption

Toxicology.   1-1-Dimethylhydrazine is a respiratory irritant and convulsant; it is carcinogenic in experimental animals.

Accidental human exposures have resulted in eye irritation, a choking sensation, chest pain, dyspnea, lethargy, nausea, and skin irritation.[1] Based on the results of exposure of dogs, the effects expected in humans from exposure for 60 minutes are: at 100 ppm, irritation of eyes and mucous membranes; at 200 ppm, marked central nervous system stimulation and perhaps death; at 900 ppm, convulsions and death.[2] Impairment of liver function (elevated SGPT levels) has been reported in 47 of 1193 workers exposed to 1,1-dimethylhydrazine under variable conditions; in a few of these cases, fatty infiltration of the liver was also demonstrated by liver biopsy, although alcohol intake may have been a factor in some cases.[3]

Exposure of dogs to 111 ppm for 3 hours caused vomiting, convulsions, and death; at autopsy, pulmonary edema and hemorrhage were present but were believed to be a secondary manifestation of the convulsive seizures rather than a primary effect of the agent.[4] Dogs repeatedly exposed to 25 ppm developed vomiting, diarrhea, ataxia, convulsions, and hemolytic anemia.[4] At 5 ppm for 26 weeks, dogs had slightly decreased body weight, hemolytic anemia, and hemosiderosis of the spleen.[5]

Applied to the shaved skin of dogs, the liquid was mildly irritating and rapidly absorbed; the dermal LD_{50} was between 1.2 and 1.7 g/kg.[6] In the eye, it caused mild conjunctivitis.[7]

Administration of 0.1% in the drinking water of 50 male and 50 female Swiss mice resulted in a high incidence of angiosarcomas in various organs (79%); tumors of the lungs (71%), kidneys (10%), and liver (7%) were also observed.[8] In another study, mice given daily gavage doses of 0.5 mg, 5 days/week, for 40 weeks showed inconclusive evidence of lung tumor induction.[7] Chronic inhalation of 1,1-dimethylhydrazine by rats produced benign tumors of the lung and pituitary.[9] A broad distribution of tumors occurred in mice following inhalation with the respiratory system and liver most severely affected. Lesions of the respiratory system included inflammation and other indications of chronic insult, including a variety of rare, but benign, tumors of the upper respiratory system and the more common lung adenoma; liver lesions included a variety of benign and malignant tumors. These lesions were seen sporadically at 0.05 ppm.

The carcinogenic risk to humans has not been determined, but 1,1-dimethylhydrazine is classified as a suspected human carcinogen based on animal results. The National Institute for Occupational Safety and Health (NIOSH) has also noted that the role of nitrosodimethylamine, a contaminant of 1,1-dimethylhydrazine, must be considered in evaluating the tumorigenicity of 1,1-dimethylhldrazine.[7] In one follow-up report, investigators, using pure 1,1-dimethylhydrazine, were able to demonstrate that previous oncogenic findings and findings of their study could not be explained by the contaminant.[9]

In a report on embryotoxicity, intraperitoneal administration of 10, 30, or 60 mg/kg/day in rats on days 6 to 15 of pregnancy caused a dose-dependent reduction in maternal weight gain and slight embryotoxicity in the form of reduced 20-day fetal weights in the high-dose group.[10]

1,1 Dimethylhydrazine was genotoxic in a wide variety of assays.[11]

The ammoniacal or fishy odor has been variously been reported as detectable between 6 and 14 ppm and below 0.3 ppm.[7]

The proposed 1995 ACGIH threshold limit value–time-weighted average for 1,1-dimethylhydrazine is 0.01 ppm (0.025 mg/m³)

with an A3 animal carcinogen designation and a notation for skin absorption.

REFERENCES

1. Shook BS, Cowart DH: Health hazards associated with unsymmetrical dimethylhydrazine. *Ind Med Surg* 26:333–336, 1957
2. 1,1-Dimethylhydrazine—emergency exposure limits. *Am Ind Hyg Assoc J* 25:582–584, 1964
3. Petersen P, Bredahl E, Lauritsen O, Laursen T: Examination of the liver in personnel working with liquid rocket propellant. *Br J Ind Med* 27:141–146, 1970
4. Jacobson KH, Clem JH, Wheelwright HJ, et al: The acute toxicity of the vapors of some methylated hydrazine derivatives. *AMA Arch Ind Health* 12:609–616, 1955
5. Reinhart WE, Donati E, Green EA: The subacute and chronic toxicity of 1,1-dimethylhydrazine vapor. *Am Ind Hyg Asoc J* 21:207–210, 1960
6. Smith EB, Clark DA: Absorption of unsymmetrical dimethylhydrazine (UDMH) through canine skin. *Toxicol Appl Pharmacol* 8:649–659, 1971
7. National Institute for Occupational Safety and Health: *Criteria for a Recommended Standard . . . Occupational Exposure to Hydrazines.* DHEW (NIOSH) Pub No 78–212, p 269. Washington, DC, US Government Printing Office, 1974
8. *IARC Monographs on the Evaluation of the Carcinogenic Risk of Chemicals to Man*, Vol 4, Some aromatic amines, hydrazine and related substances, N-nitroso compounds and miscellaneous alkylating agents, pp 137–143. Lyon, International Agency for Research on Cancer, 1974
9. Keller WC: Toxicity assessment of hydrazine fuels. *Aviat Space Environ Med* 59 11 (Suppl): A100-6, 1988
10. Keller WC, Olson CT, Back KC, et al: Teratogenic assessment of three methylated hydrazine derivatives in the rat. *J Toxicol Environ Health* 13:125–131, 1984
11. Matheson D, Brusick D, Jagannath D: Genetic activity of 1,1-dimethylhydrazine and methylhydrazine in a battery of in vitro and in vivo assays. *Mutat Res* 53:93–96, 1978

DIMETHYL HYDROGEN PHOSPHITE
CAS: 868-85-9

(CH$_3$O)$_2$POH

Synonyms: DMHP; dimethoxyphosphine oxide; dimethyl phosphite; methyl phosphonate

Physical Form. Colorless liquid with a mild odor

Uses. Flame retardant on Nylon 6 fibers; intermediate in the production of pesticides and herbicides; stabilizer in oil and plaster; additive to lubricants

Exposure. Inhalation

Toxiclogy. Dimethyl hydrogen phosphite (DMHP) is an irritant of the eyes, mucous membranes, and skin; it causes neurological impairment and reversible cataracts in animals; it is carcinogenic in rats and causes testicular atrophy in mice.

No human cases of intoxication have been reported.[1]

The acute oral LD$_{50}$s for dimethyl hydrogen phosphite were 3300 and 3000 mg/kg bw for male and female Fischer 344/N rats, respectively, and 2800 mg/kg bw for male B6C3F$_1$ mice.[1]

Rats exposed to airborne levels of 934 ppm, 6 hr/day for 3 days, died.[2] Effects observed included irritation of the skin, eyes, and mucous membranes, neuromuscular impairment, and lung congestion. Rats exposed to 431 ppm, 6 hr/day for 5 days, survived but exhibited the same irritant effects as seen at 934 ppm.

In a month-long study, rats were exposed to 12, 35, 119, or 198 ppm for 6 hr/day, 5 days/week.[3,4] In the high-dose group, 27 of 40 animals were dead by day 27; in the 119-ppm group, 2 animals died on days 14 and 23. There was neurological impairment at 198 and 119 ppm, which usually reversed by the following morning. Necrosis and purulent inflammation of the skin were thought to be the only lesions that may have caused death. Although there was treatment-related irritation of the eyes and nares, there was no treatment-related irritation of the trachea or lungs. Lenticular opacities occurred at 35 ppm and above, which progressed to cataracts in the 119- and 198-ppm groups. In rats killed 2 weeks post-treatment, the process of cataract formation had stopped; at 4 weeks, the formation of normal lens fibers was evident.

Rats treated with 200 mg/kg/day by gavage for 4, 5, or 6 weeks showed early treatment-related changes in the lungs (significant increases in serum angiotensin) and possible preneoplastic changes in the forestomach, characterized by epithelial hyperplasia, hyperkeratosis, subepithelial and submucosal inflammation, and edema.[5]

In a carcinogenic study, male and female rats were given DMHP by gavage 5 days/week for 103 weeks.[6] At 200 mg/kg, there were increases in alveolar/bronchiolar carcinomas, squamous cell carcinomas of the lung, and carcinomas of the stomach in male rats. The IARC determined that there is limited evidence for the carcinogenicity of dimethyl hydrogen phosphite in experimental animals and that it is not classifiable as to its carcinogenicity to humans.[1]

Limited data indicate that DMHP may have testicular effects; calcification and atrophy of the testes were observed in mice in the course of chronic and subchronic oral studies at 200 mg/kg for 103 weeks, and 375 and 750 mg/kg for 13 weeks, respectively.[6]

Dimethyl hydrogen phosphite was not mutagenic to several strains of *Salmonella typhimurium*, but it did cause sister chromatid exchanges and chromosomal aberrations in the Chinese hamster ovary (CHO) cell line.[1]

A threshold limit value (TLV) has not been established for dimethyl hydrogen phosphite.

REFERENCES

1. *IARC Monographs on the Evaluation of Carcinogenic Risks to Humans.* Some flame retardants and textile chemicals, and exposures in the textile manufacturing industry. Vol 48, pp 85–93. Lyon, International Agency for Research on Cancer, 1990
2. Mobil Research and Development Corporation. TSCA sec. 8(e) Submission 8EHQ-0381-0366 follow-up. A five day inhalation toxicity study of MCTR-174-79 in the rat. Performed by Bio/dynamics Inc, Washington, DC, Office of Toxic Substances, US Environmental Protection Agency, 1981
3. Mobil Oil Corporation. TSCA sec. 8(e) Submission 8EHQ-0381-0366 follow-up. A four week inhalation toxicity study in the rat. Prepared by Bio/dynamics, Inc, Washington, DC, Office of Toxic Substances, US Environmental Protection Agency, 1981
4. Mobil Oil Corporation. TSCA sec. 8(e) Submission 8EHQ-0381-0366 follow-up. Histopathologic observations on a four week inhalation toxicity study of MCTR-242-79 in the rat. Prepared by Toxicity Research Laboratories, Ltd, Washington, DC, Office of Toxic Substances, US Environmental Protection Agency, 1981
5. Nomeir AA, Uraih LC: Pathological and biochemical effects of dimethylhydrogen phosphite in Fischer 344 rats. *Fund Appl Toxicol* 10:114-124, 1988
6. National Toxicology Program: NTP Technical report on the toxicology and carcinogenesis studies of dimethyl hydrogen phosphite (CAS No 868-85-9) in F344 Rats and B7CSF Mice (gavage studies). NTP TR 287. Research Triangle Park, NC, National Toxicology Program, 1984

DIMETHYL METHYLPHOSPHONATE
CAS: 756-79-6

$C_3H_9O_3P$

Synonyms: Phosphonic acid, methyl, dimethyl ester; DMMP

Physical Form. Solid

Uses. Flame retardant; preignition additive for gasoline; antifoam agent; plasticizer and stabilizer; textile conditioner; antistatic agent; used experimentally to mimic the physical and spectroscopic (but not biologic) properties of anticholinesterase agents

Exposure.   Inhalation

Toxicology.   Dimethyl methylphosphonate (DMMP) administered to male rats is a reproductive toxicant and carcinogen. Effects in humans are unknown.

The acute oral LD_{50} is estimated to be greater than 3000 mg/kg for rats and greater than 6000 mg/kg for mice.[1] Clinical signs reported in rats and mice following doses of up to 6810 mg/kg included inactivity, unsteady gait, and prostration. In 15-day studies, compound-related deaths occurred in rats at 5000 mg/kg/day and above and, in mice at 10,000 mg/kg/day; the only compound-related lesions observed were stomach lesions in the mice.

DMMP administered to male Fischer rats by gavage 5 days/week for 90 days at dosages of 250, 500, 1000, and 2000 mg/kg caused a dose-related decrease in sperm count, sperm motility, and the male fertility index.[2] When the rats were mated, treated males sired fewer litters with fewer pups per litter, and untreated dams had more resorptions. The percentage of resorptions was 6.1% in the control group; percentages increased to 14.9%, 37.8%, and 79.1% in the 250, 500, and 1000 mg/kg groups, respectively. Histological abnormalities of the testis were observed only in the high-dose group and were characterized by lack of spermatogenesis or by degeneration, vacuolization, and necrosis of cells in the spermatogenic tubules. Microscopic changes in the prostate were also observed in some of the high-dose animals, and abnormalities of the kidney (tubular cell regeneration, hyalin droplet degeneration, and cellular infiltrate) were seen in some animals from each of the dosed groups.

Further study of the reproductive lesions in male rats showed elongating spermatids and morphologic alterations in Sertoli cells, as well as functional defects in spermatozoa following administration of 1750 mg/kg for up to 12 weeks.[3]

In chronic studies, DMMP was administered by gavage in corn oil for up to two years at doses of 500 or 1000 mg/kg/day to rats and at doses of 1000 or 2000 mg/kg/day to mice.[1,4] Survival in dosed male rats was reduced, in part because of renal toxicity. Lesions of the kidney included increased severity of spontaneous age-related nephropathy including calcification, hyperplasia of the tubular and transitional epithelium, tubular cell adenocarcinomas, and transitional cell papillomas and carcinomas. Similar lesions were not seen in female rats or in mice of either sex, although reduced survival in male mice prevented adequate analysis. The authors noted that the spectrum of lesions observed in male rats after DMMP treatment is similar to that seen after chronic administration of a variety of other chemicals, including unleaded gasoline, hydrocarbon solvents, and 1,4-dichlorobenzene, which suggests that a common mechanism may be responsible for the lesions. It cannot be determined if the kidney tumors seen in male rats after DMMP administration are predictive of kidney tumors in humans.

Under the conditions of the 2-year gavage studies, it was determined that there was some evidence of carcinogenicity for male rats, no evidence of carcinogenicity for female rats and female mice, and inadequate evidence for male mice.[1]

A threshold limit value has not been established for dimethyl methylphosphonate.

REFERENCES

1. National Toxicology Program: *Toxicology and Carcinogenesis Studies of Dimethyl Methylphosphonate in F344/N Rats and B6C3F₁ Mice*, pp 1–64. NTP TR 323, NIH Pub 87-2579. Research Triangle Park, North Carolina, National Toxicology Program, 1987
2. Dunnick JK, Gupta BN, Harris MW, et al: Reproductive toxicity of dimethyl methylphosphonate (DMMP) in the male Fischer 344 rat. *Toxicol Appl Pharmacol* 72:379–387, 1984
3. Chapin RE, Dutton SL, Ross MD, et al: Development of reproductive tract lesions in male F344 rats after treatment with dimethyl methylphosphonate. *Exper Molecular Path* 41:126–140, 1984
4. Dunnick JK, Eustis SL, Haseman JK: Development of kidney tumors in the male F344/N rat after treatment with dimethyl methylphosphonate. *Fund Appl Toxicol* 11:91–99, 1988

2,4-DIMETHYLPHENOL
CAS:105-67-9

$C_6H_3(CH_3)_2OH$

Synonyms: *m*-xylenol; 1-hydroxy-2,4-dimethylbenzene; 2,4-xylenol; Lysol Brand Disinfectant; soluble concentrate (1.5% solution)

Physical Form. Solid

Uses. Disinfectant; manufacture of pharmaceuticals, plastics, insecticides, fungicides, rubber chemicals, wetting agents, and dyestuffs

Exposure. Inhalation; skin absorption

Toxicology. 2,4-Dimethylphenol is expected to be an irritant of the eyes, mucous membranes, and skin, by analogy to other phenols.

The oral LD_{50} for rats was 3.2 g/kg; the dermal LD_{50} in mice was 1.04 g/kg.[1]

Gavage administration of 1200 mg/kg for 10 days was lethal to male and female rats; at 600 mg/kg for the same time period, there was a significant increase in relative liver weight in females and alterations in hematologic and clinical chemistries in males.[2] Hyperkeratosis and epithelial hyperplasia of the forestomach were observed in rats following administration of 180 or 540 mg/kg for 90 days.[2] The higher dose also caused reduced body weights and some deaths in treated animals.

In a dermal carcinogenicity study in mice, twice weekly application of 20% 2,4-dimethyl phenol in benzene (after a single pretreatment with 0.3% dimethylbenzanthracene in benzene as an initiator) produced papillomas in 50% and carcinomas in 11% of animals at 15 weeks; by 23 weeks, 18% had developed carcinomas.[3] When 10% 2,4-dimethylphenol in benzene was applied twice weekly in the absence of an initiator, 31% had papillomas at 20 weeks and no carcinomas were observed. By 24 weeks, 12% had carcinomas.

2,4-Dimethylphenol was tested for mutagenicity in the *Salmonella* microsome preincubation assay using the standard protocol of the National Toxicology Program and five strains of *Salmonella;* results were negative.[4]

The ACGIH has not established a threshold limit value for 2,4-dimethylphenol.

REFERENCES

1. Uzhdovini ER, et al: Acute toxicity of lower phenols. *Gig Tr Prof Zabol* 2:58–59 (in Russian CA 81:418q)
2. Daniel FB, Robinson M, Olson GR, et al: Ten and ninety-day toxicity studies of 2,4-dimethylphenol in Sprague-Dawley rats. *Drug Chem Toxicol* 16:351–368, 1993
3. Boutwell RK, Bosch DK: The tumor-promoting action of phenol and related compounds for mouse skin. *Cancer Res* 19:413–424, 1959
4. Mortelmans K et al. Salmonella mutagenicity tests: II. Results from testing of 270 chemicals. *Environ Mutagen* 8:(Suppl 7)1–119, 1986

DIMETHYL PHTHALATE
CAS:131-11-3

$C_{10}H_{10}O_4$

Synonynms: 1,2-benzenedicarboxylic acid dimethyl ester; phthalic acid dimethyl ester; methyl phthalate

Physical Form. Oily liquid

Uses. Plasticizer; insect repellent for application to the skin

Exposure. Inhalation (of spray or mist); skin absorption

Toxicology. Dimethyl phthalate (DMP) is of low-order acute toxicity.[1]

A solution, including 2% DMP in petrolatum, was nonirritating to humans following 48-hour patch tests.[2]

Rats fed 4.0% and 8.0% in the diet showed slight but significant changes in growth;

chronic nephritis was seen at the higher dose, but mortality rates were the same as controls.[3] Applied to 10% of the body surface of rabbits for 90 days, 4.0 ml/kg caused some deaths with pulmonary edema and slight renal damage; no skin irritation was observed.[4] The undiluted liquid instilled into rabbit eyes produced no grossly observable irritation for up to 48 hours.[5]

Intraperitoneal injection of pregnant rats with 10%, 33%, or 50% of the LD_{50} (3.4 ml/kg) on days 5, 10, and 15 of gestation resulted in litters with a higher number of skeletal abnormalities.[6] These results were not confirmed in a more recent study in which mice were administered up to 5% DMP in the drinking water during gestation.[7] Although treatment with 5% DMP resulted in increased relative maternal liver weight and reduced body weight gain, there was no effect on any parameter of embryo/fetal development.

In genotoxic assays, DMP was determined to be a weak bacterial mutagen.[8]

The 1995 ACGIH threshold limit value–time-weighted average (TLV-TWA) for dimethyl phthalate is 5 mg/m³.

REFERENCES

1. Final report on the safety assessment of dibutyl phthalate, dimethyl phthalate, and diethyl phthalate. *J Am Coll Toxicol* 4:267–303, 1985
2. Schulsinger C, Mollgaard K: Polyvinyl chloride dermatitis not caused by phthalates. *Contact Dermatitis* 6:477–480, 1980
3. Lehman AJ: Insect repellents. *Assoc Food Drug Office US Q Bull* 19:87–99, 1955
4. Draize J et al: Toxicological investigations of compounds proposed for use as insect repellents. J Pharmacol Exp Ther 93:26–39, 1948
5. Lawrence WH et al: Toxicological investigation of some acute, short-term, and chronic effects of administering di-2-ethylhexyl phthalate (DEHP) and other phthalate esters. *Environ Res* 9:1–11, 1975
6. Singh AR, Lawrence WH, Autian J: Teratogenicity of phthalate esters in rats. *J Pharmacol Sci* 61:51–55, 1972
7. Field EA, Price CJ, Sleet RB, et al: Developmental toxicity evaluation of diethyl and dimethyl phthalate in rats. *Teratology* 48:33–44, 1993
8. Kozumbo WJ, Rubin RJ: Mutagenicity and metabolism of dimethyl phthalate and its binding to epidermal and hepatic macromolecules. *J Toxicol Environ Health* 33:29–46, 1991

DIMETHYL SULFATE
CAS:77-78-1

(CH₃)₂SO₄

$(CH_3)_2SO_4$

Synonyms: Sulfuric acid dimethyl ester; DMS

Physical Form.   Colorless, oily liquid

Uses.   Methylating agent in the manufacture of many organic chemicals

Exposure.   Inhalation; skin absorption

Toxicology.   Dimethyl sulfate (DMS) is highly toxic; it is a severe irritant of the eyes, mucous membranes, and skin; it is carcinogenic in experimental animals.

When DMS comes into contact with a moist mucosal surface, it is slowly hydrolyzed into sulfuric acid, methanol, and methyl hydrogen sulfate.[1] The methanol can be absorbed into the circulation, leading to neurotoxic effects, while the sulfuric acid and methyl hydrogen sulfate induce severe irritative and erosive actions to the mucosa.

Several human deaths have occurred from occupational exposure, and it has been estimated that inhalation of 100 ppm for 10 minutes would be fatal.[2,3] A major cause of mortality in DMS intoxication is respiratory failure, a consequence of mucosal inflammation and edema of major airways and of noncardiogenic pulmonary edema. Often, exposure of humans produces no immediate effects other than occasional slight eye and nose irritation; after a latent period of up to 10 hours or more, there is onset of headache and giddiness, with intense conjunctival irritation, photophobia, and angioneurotic edema, followed by inflammation of the pharyngolaryngeal mucosa, dysphonia, aphonia, dysphagia, productive cough, oppression in the chest, dyspnea, and cyanosis.[2,4] Vomiting and diarrhea may intervene.[2,4] Dysuria may occur for 3 to 4 days; there may be persistence of laryngeal edema for up to 2 weeks and of photophobia for several months. Other effects include delirium, fever, convulsions, and coma, as well as damage to heart, liver, and kidneys.[2,5] The long-term sequelae of acute DMS poisonings have been examined in 62 patients followed for 2 to 12 years.[1] Hoarseness remained in 33% of the moderately to severely intoxicated patients. Mild ventilatory disturbances were demonstrated in 5 cases, and mild to moderate pulmonary function abnormalities were observed in 3 patients. No abnormalities were found in ECG and routine blood examinations. No evidence of pulmonary neoplasms was found on follow-up chest X rays. In another recent case, there was persistent cough and mucopurulent sputum 10 months after exposure, with repeated infective episodes, probably secondary to mucosal damage by DMS.[6] It was not known if more extensive fibrosis would develop with time.

Contact of the liquid with the eyes or skin causes very severe burns as a result of its powerful vesicant action.[2] In an incident of moderate skin contact with the liquid, generalized intoxication occurred even though there was prompt treatment of the skin; vapor inhalation lasted for a few minutes, at the most.[5]

In mice and rats, inhalation at 0.1 to 4.0 ppm throughout pregnancy caused preimplantation losses and embryotoxic effects including anomalies of the cardiovascular system.[3]

Dimethyl sulfate is carcinogenic to animals following its inhalation or subcutaneous injection, producing mainly local tumors and, after prenatal exposure, producing tumors of the nervous system.[7] Of 15 rats surviving exposure to 10 ppm, 1 hr/day for 19 weeks, 3 developed squamous-cell carcinoma of the nasal cavity, 1 developed a glioma of the cerebellum, and another developed lymphosarcoma of the thorax with metastases in the lungs. Several early deaths from inflammation of the nasal cavity and pneumonia were also reported.[7] A statistically significant increase in lung adenomas was observed in a group of 90 mice exposed at 4 ppm for 4 hr/day, 5 days per week. A single intravenous dose of 20 mg/kg given to 8 pregnant rats on day 15 of gestation induced malignant tumors, including 3 tumors of the nervous system, in 7 of 59 offspring that were observed for over 1 year.[8]

Despite anecdotal case reports of cancer in exposed individuals, no significant increase in mortality or in deaths from lung cancer was found in a group of workers exposed for various periods between 1932 and 1972.[3,7]

The IARC has determined that there is sufficient evidence of carcinogenicity to animals; it should be assumed to be a potential human carcinogen.[3,7]

The 1995 ACGIH threshold limit value–time-weighted average (TLV-TWA) for dimethyl sulfate is 0.1 ppm (0.5 mg/m^3) with a notation for skin absorption and an A2 suspected human carcinogen designation.

REFERENCES

1. Ying W, Jing X, Qin-wai W: Clinical report on 62 cases of acute dimethyl sulfate intoxication. *Am J Ind Med* 13:455–462, 1988
2. Browning E: *Toxicity and Metabolism on Industrial Solvents*, pp 713-721. Amsterdam, Elsevier, 1965
3. World Health Organization: *Environmental Health Criteria 48*, Dimethyl sulfate, p 55. Geneva, World Health Organization, 1985
4. Dimethyl Sulfate. *Documentation of the TLVs and BEIs*, 6th ed, pp 497-499. Cincinnati, OH, American Conference of Governmental Industrial Hygienists (ACGIH), 1991
5. Fassett DW: Esters. In Patty FA (ed): *Industrial Hygiene and Toxicology*, 2nd ed, Vol 2, *Toxicology*, pp 1927–1930. New York, Interscience, 1963
6. Ip M, Wong KL, Wong KF: Lung injury in dimethyl sulfate poisoning. *J Occup Med* 31:141–143, 1989
7. *IARC Monographs on the Evaluation of the Carcinogenic Risks of Chemicals to Humans*, Suppl 4, pp 119–120. Lyon, International Agency for Research on Cancer, 1982
8. *IARC Monographs on the Evaluation of the Carcinogenic Risk of Chemicals to Man*, Vol 4, Some aromatic amines, hydrazine and related substances, *N*-nitroso compounds and miscellane-

ous alkylating agents, pp 271–276. Lyon, International Agency for Research on Cancer, 1974

DINITRO-*o*-CRESOL
CAS: 534-52-1

$CH_3C_6H_2OH(NO_2)_2$

Synonyms: DNOC; 2-methyl-4,6-dinitrophenol; dinitrol

Physical Form. Yellow crystalline solid

Uses. Herbicide; insecticide; intermediate in the synthesis of fungicides; polymerization inhibitor for vinyl aromatic compounds

Exposure. Inhalation; skin absorption; ingestion

Toxicology. Dinitro-*o*-cresol (DNOC) causes an increase in metabolic rate that results in hyperpyrexia. Severe exposure may cause coma and death. Exposure also causes a yellow pigmentation of the skin, hair, sclera, and conjunctivas.

In a report of eight fatalities among agricultural sprayers, symptoms of intoxication included fatigue, profuse sweating, excessive thirst, and weight loss, which were incorrectly attributed to heat strain.[1] Decline was rapid, with hyperpnea, tachycardia, and fever; death occurred within 48 hours of exposure. A number of other fatalities from hyperthermia have been reported.[2]

The risk of serious intoxication increases during hot weather.[1] A nonfatal case of intoxication resulting from exposure to 4.7 mg/m³ resulted in fever, tachycardia, hyperpnea, profuse sweating, cough, shortness of breath, and a marked increase in basal metabolic rate.[2] The clinical picture resembled that of thyroid crisis.

Lethal doses may be absorbed through the skin; local irritation is usually slight. Skin application of 50 g of a 25% dinitro-*o*-cresol ointment to a 4-year-old boy caused vomiting,

headache, yellow-stained skin and sclera, elevated pulse and respiratory rate, unconsciousness, and death within 3.5 hours.[2] Autopsy showed diffuse petechial hemorrhages in the intestinal mucosa and brain, as well as pulmonary edema.

In two cases, ingestion of 50 and 140 g was lethal.[2]

In human volunteers given 75 mg/day orally for 5 days, the earliest symptom was an exaggerated sense of well-being at blood levels of dinitro-*o*-cresol of approximately 20 μg/g.[3] At a level near 40 μg/g of blood, symptoms were headache, lassitude, and malaise; yellow coloration of the sclera appeared on the fourth day of exposure and persisted for 5 days; urinary excretion of unchanged dinitro-*o*-cresol was so slow that blood levels of 1 to 1.5 μg/g were still detectable 40 days after the last dose was administered.[4]

Blood levels appear to correlate with the severity of intoxication.[2] Individuals with concentrations of 40 μg/g of whole blood or greater are most likely to develop toxic effects. Those with ranges between 20 and 40 μg/g may or may not show adverse effects, and most with blood levels below 20 μg/g are not affected.[2]

The development of bilateral cataracts has been reported in chronic intoxication as a result of repeated ingestion of dinitro-*o*-cresol for ill-advised therapeutic purposes; cataracts have not been observed following industrial or agricultural exposure.[2] Contact with the eyes or absorption of DNOC by any route can cause a characteristic yellow staining of the conjunctiva and sclera of the eye.[5] DNOC stains human skin yellow on contact. Although the yellow staining of the skin and sclera may be unsightly, such cosmetic effects are not regarded as adverse.[5]

In reproductive studies, DNOC did not affect either sperm counts or testicular weights in mice given single doses in the range of 3 to 12 mg/kg/day for 5 days.[6] Intermediate-duration feeding and gavage studies have suggested that the ovaries and uterus may be target organs of DNOC.[5] In addition, male rats fed DNOC for 90 days had aspermatogenesis. However, these results have not been confirmed in other studies using similar dosing protocols.[5]

No developmental effects were observed in mice given 8 mg/kg/day of DNOC from days 11 to 14 of gestation.[7] The numbers for corpora lutea, implantations, live embryos, resorbed embryos, pre- and postimplantation loss, weight of embryos, and malformations did not differ significantly from those for controls. Treatment of female mice with 5 mg/kg/day for four days during the second trimester of pregnancy resulted in chromosomal aberrations in embryonic livers.[7] The same treatment during the first trimester did not significantly increase the frequency of chromosomal aberrations.

Investigators have concluded that dinitro-o-cresol affects metabolism by uncoupling the oxidative phosphorylation process, resulting in increased cellular respiration (increased oxygen consumption) and decreased formation of adenosine triphosphate (ATP), which contains high-energy phosphate bonds.[2] Therefore, energy generated in the body cannot be converted to its usual form (ATP) and is released as heat instead. Toxicity is cumulative.[2]

The 1995 ACGIH threshold limit value–time-weighted average (TLV-TWA) for dinitro-o-cresol is 0.2 mg/m^3 with a notation for skin absorption.

REFERENCES

1. Bidstrup PL, Payne DJH: Poisoning by dinitro-ortho-cresol, report of eight fatal cases occurring in Great Britain. *Br Med J* 2:16–19, 1951
2. National Institute for Occupational Safety and Health: *Criteria for a Recommended Standard . . . Occupational Exposure to Dinitro-ortho-cresol.* DHEW (NIOSH) Pub No 78-131, p 147. Washington, DC, US Government Printing Office, 1978
3. Bidstrup PL, Bonnell JAL, Harvey DG: Prevention of acute dinitro-ortho-cresol (DNOC) poisoning. *Lancet,* 262:794–795, 1952
4. Harvey DG, Bidstrup PL, Bonnell JAL: Poisoning by dinitro-ortho-cresol: some observations on the effects of dinitro-ortho-cresol administered by mouth to human volunteers. *Br Med J* 2:13–16. 1951.
5. Agency for Toxic Substances and Disease Registry: *Toxicological Profile for Dinitrocresols.* US Department of Health and Human Services, Public Health Service, Atlanta, GA, pp 1–117, 1994
6. Quinto I, De Marinis E, Mallardo M, et al: Effect of DNOC, Ferbam and Imidan exposure on mouse sperm morphology. *Mutat Res* 224:405–408, 1989
7. Nehez M, Paldy A, Selypes A, et al: The teratogenic and mutagenic effects of dinitro-o-cresol-containing herbicide on the laboratory mouse. *Ecotoxicol Environ Safety* 5:38–44, 1981

DINITROBENZENE (ALL ISOMERS)

CAS: 528-29-0; 99-65-0; 100-25-4

$C_6H_4(NO_2)_2$

Synonyms: Dinitrobenzol; DNB

Physical Form. Colorless or yellowish needles or plates

Uses. Synthesis of dyestuffs; explosives; in celluloid production

Exposure. Inhalation; skin absorption

Toxicology. All isomers of dinitrobenzene (DNB) cause anoxia due to the formation of methemoglobin; moderate exposure causes respiratory tract irritation, and chronic exposure results in anemia. Testicular toxicity has been reported in laboratory animals following ingestion of *m*-DNB.

Exposed workers have complained of a burning sensation in the mouth, dry throat, and thirst; somnolence, staggering gait, and coma have been observed with more intense exposures.[1] Most signs and symptoms of overexposure are due to the loss of oxygen carrying-capacity of the blood.

The onset of symptoms of methemoglobinemia is insidious and may be delayed for up to 4 hours; headache is commonly the first symptom and may become quite intense as the severity of methemoglobinemia progresses.[2]

Cyanosis occurs when the methemoglobin concentration is 15% or more; blueness in the lips, the nose, and the earlobes is usually recognized by fellow workers.[2]

The subject usually feels well, has no complaints, and is insistent that nothing is wrong until the methemoglobin concentration approaches approximately 40%.[2] At methemoglobin concentrations of over 40%, there is usually weakness and dizziness; at up to 70% concentration, there may be ataxia, dyspnea on mild exertion, tachycardia, nausea, vomiting, and drowsiness.[2] Coma may ensue with methemoglobin levels at about 70%, and the lethal level is estimated to be at 85% to 90%.[3]

Five workers at an Ohio rubber plant became ill with symptoms, including yellow discoloration of the hands, blue discoloration of the lips and nail beds, headache, nausea, chest pain, dizziness, confusion, and difficulty in concentrating; one worker suffered a seizure.[4] Medical examinations showed that blood methemoglobin levels ranged from 3.8% to 41.2%. Effects were attributed to dermal exposure to an adhesive containing 1 wt% by weight p-DNB. Following replacement of the adhesive, symptoms disappeared, and methemoglobin levels were within normal limits.

In another report of acute intoxication by m-DNB dust, six workers developed cyanosis, followed by slight to moderate anemia.[5] Prolonged recovery (of approximately 1 month) from the anemia was characteristic of the cases, but no adverse health effects were attributable to the exposure in a 10-year follow-up.

The ingestion of alcohol aggravates the toxic effects of dinitrobenzene. In general, higher ambient temperatures increase susceptibility to cyanosis from exposure to methemoglobin-forming agents. Chronic exposure of workers causes anemia; there are scattered reports of liver injury. Visual impairment has occurred in the form of reduced visual acuity and central scotomas, particularly for red and green colors; yellow discoloration of the conjunctiva and sclera is common.[1,6] Yellow-brown discoloration of the hair and exposed skin of workers has also been observed.[1]

A number of studies have shown that m-DNB is a potent testicular toxicant in laboratory animals. Subchronic ingestion of 20 mg/l in drinking water caused testicular atrophy in rats.[7] At higher levels of 200 mg/l, more than 50% of the seminiferous tubules were collapsed, with neither germinal cells nor Sertoli cells present. Male rats gavaged 5 day/week with 3.0 mg/kg/day did not sire litters when bred with females during treatment week 10.[8] Diminished sperm production, decreased cauda epididymal sperm reserves, nonmotile spermatozoa, atypical sperm motility, decreased weights of the testes and epididymis, and seminiferous tubular atrophy were also observed. Sperm production was also decreased in males dosed at 1.5 mg/kg/day. Single acute exposure of rats to 48 mg/kg caused alterations in testis weight and sperm motility; histological changes included maturation depletion of mid- and late spermatids and immature germ cells in the epididymis.[9] Fertilizing ability was lost by 5 to 6 weeks post-treatment, and some animals failed to recover in 5 months. Susceptibility to the reproductive effects of m-DNB varied with the age of the animals in this study. Increases in plasma lactate dehydrogenase isozyme C4 (LDH-C4) were found to precede noticeable histological findings of testicular damage in rats.[10] LDH-C4 may be used as a biochemical marker of acute testicular damage.

Marked differences in species susceptibility to m-DNB have also been observed.[11] Hamsters showed no testicular lesions at dose levels up to 50 mg/kg, whereas damage to rat testicular tubules is readily apparent at 25 mg/kg. Similarly, m-DNB induced substantially less methemoglobin in the hamster than in the rat (15% vs. 80% at 25 mg/kg dose).

Follow-up studies have demonstrated that m-DNB exerts a direct effect on the germinal epithelium and not through alterations in hypothalamic and pituitary control of gonadal function.[12] No reproductive effects have been reported in humans.

In vitro studies show that m-DNB is mutagenic in *Salmonella typhimurium*.[13]

The 1995 ACGIH threshold limit value–time-weighted average (TLV-TWA) for all isomers of dinitrobenzene is 0.15 ppm (1.0 mg/m³) with a notation for skin absorption.

Note: For a description of diagnostic signs, differential diagnosis, and medical control, including clinical laboratory treatment of overexposure to methemoglobin-forming agents, see "Aniline."

REFERENCES

1. von Oettingen WE: *The Aromatic Amino and Nitro Compounds, Their Toxicity and Potential Dangers,* US Public Health Service Bulletin No 271, pp 94–103. Washington, DC, US Government Printing Office, 1941
2. Hamblin DP: Aromatic nitro and amino compounds. In Patty FA (ed): *Industrial Hygiene and Toxicology,* 2nd ed, Vol 2, *Toxicology,* pp 2105–2131, 2138–2140. New York, Interscience, 1963
3. Chemical Safety Data Sheet SD-21, Nitrobenzene, pp 5–6, 12–14. Washington, DC, MCA, Inc, 1967
4. Hazards Evaluations and Technical Assistance Branch, NIOSHCDC: Methemoglobin due to occupational exposure to dinitrobenzene—Ohio, 1986. *MMWR* 37:353–354, June 10, 1988
5. Okubo T, Shigeta S: Anemia cases after acute *m*-dinitrobenzene intoxication due to an occupational exposure. *Ind Health* 20:297–304, 1982
6. Grant WM: *Toxicology of the Eye,* 2nd ed, p 409. Springfield, Charles C Thomas, 1974
7. Cody TE, Witherup S, Hastings L et al: 1,3-Dinitrobenzene: toxic effects in vivo and in vitro. *J Toxicol Environ Health* 7:829–847, 1981
8. Linder RE, Hess RA, Strader LF: Testicular toxicity and infertility in male rats treated with 1,3-dinitrobenzene. *J Toxicol Environ Health* 19:477–489, 1986
9. Linder RE, Strader LF, Barbee RR, et al: Reproductive toxicity of a single dose of 1,3-dinitrobenzene in two ages of young male rats. *Fund Appl Toxicol* 14:284–298, 1990
10. Reader SJC, Shingles C, Stonard MD: Acute testicular toxicity of 1,3-dinitrobenzene and ethylene glycol monomethyl ether in the rat: evaluation of biochemical effect markers and hormonal responses. *Fund Appl Toxicol* 16:61–70, 1991
11. Obasaju MF, Katz DF, Miller MG: Species differences in susceptibility to 1,3-dinitrobenzene-induced testicular toxicity and methemoglobinemia. *Fund Appl Toxicol* 16:257–266, 1991
12. Rehnberg GL, Linder RE, Goldman JM, et al: Changes in testicular and serum hormone concentrations in the male rat following treatment with *M*-dinitrobenzene. *Toxicol Appl Pharmacol* 95:255–264, 1988
13. Agency for Toxic Substances and Disease Registry (ATSDR): *Toxicological Profile for 1,3-Dinitrobenzene and 1,3,5-Trinitrobenzene.* US Department of Health and Human Services, Public Health Service, pp 1–93, 1993

2,4-DINITROPHENOL
CAS: 51-28-5

$C_6H_4N_2O_5$

Synonyms: 2,4-DNP; Aldifen; Chemox PE; Dinofan; Fenoxyl Carbon N; Maroxol 50; Caswell No. 392; Sulfo Black B; Nitro Kleenup

Physical Form.   Yellow solid

Uses.   In manufacture of dyes, other organic chemicals, wood preservatives, photographic developer, and explosives

Exposure.   Inhalation

Toxicology.   2,4-Dinitrophenol (2,4-DNP) uncouples oxidative phosphorylation from electron transport, resulting in diminished production of ATP, with the energy dissipated as heat, which can lead to fatal hyperthermia.[1]

Fatal cases of 2,4-DNP poisoning were reported among workmen in the munitions industry in France.[2] Workers were exposed to airborne vapor and dust of 2,4-DNP and had direct dermal contact with the solid material, although duration and levels of exposure were not reported. Deaths were preceded by sudden onset of extreme fatigue, elevation of body temperature to 40°C or higher, profuse sweating, thirst and labored respiration. No characteristic lesions were found at autopsy.

Two workers exposed for a few months to mists and dust of 2,4-DNP in a US chemical plant developed fever, profuse sweating, and restlessness.[3] Following treatment and rest, they returned to work, collapsed and died. Workroom air levels, measured after the deaths, were found to be at least 40 mg/m^3, and significant dermal exposure may also have occurred.

During the 1930s, 2,4-DNP was used extensively as a weight loss agent.[4] Cataracts developed in a small percentage of patients who took the agent, and there are at least 164 cases in the published literature.[5] Representative case reports that provided doses indicate that cataracts developed in the patients at doses ranging from 1.86 to 3.6 mg/kg/day, but no correlation with duration of exposure could be established. Individual susceptibility to 2,4-DNP-induced cataractogenesis appears to vary widely. Development of agranulocytosis, peripheral neuritis, and dermal effects such as rash, pruritus, urticaria, and maculopapular skin lesions were also observed.[6-7]

2,4-DNP and its metabolite 2-amino-4-nitrophenol may be found in the urine of exposed humans, but a correlation of levels to the amount of exposure to 2,4 DNP has not been made.[5]

A threshold limit value has not been established for 2,4-DNP.

REFERENCES

1. Loomis WF, Lipmann F: Reversible inhibition of the coupling between phosphorylation and oxidation. *J Biol Chem* 173:807–808, 1948
2. Perkins RG: A study of the munitions intoxications in France. *Public Health Rep* 34:2335–2374, 1919
3. Gisclard JB, Woodward MM: 2,4-Dinitrophenol poisoning: A case report. *J Ind Hyg Toxicol* 28:47–51, 1946
4. Parascandola J: Dinitrophenol and bioenergetics: An historical perspective. *Mol Cell Biochem* 5:69–77, 1974
5. Agency for Toxic Substances and Disease Registry: *Toxicological Profile for Dinitrophenols*, Draft for public comment, pp 7–53. ASTDR, 1600 Clifton Road NE, E-29, Atlanta, GA 30333, 1993
6. Horner WD: Dinitrophenol and its relation to formation of cataracts. *Arch Ophthalmol (Paris)* 27:1097–1121, 1942
7. Tainter ML, Stockton AB, Cutting WC: Dinitrophenol in the treatment of obesity: final report. *JAMA* 105:332–337, 1935

DINITROTOLUENE (all isomers)
CAS: 25321-14-6

$CH_3C_6H_3(NO_2)_2$

Synonyms: DNT; dinitrotoluol

Physical Form.   Yellow crystals

Uses.   In the production of toluene diisocyanate which, in turn, is used to produce polyurethane foams; explosives; dyes

Exposure.   Inhalation; skin absorbtion

Toxicology.   Dinitrotoluene (DNT) exposure causes anoxia owing to the formation of methemoglobin; in animal studies, chronic exposure to the 2,6-DNT isomer has been associated with hepatocellular carcinomas.

There are six isomers of DNT with technical or commercial DNT composed primarily of 2,4-DNT (80%) and 2,6-DNT (20%). The lethal doses of the various DNT isomers range from 309 mg/kg for 3,5-DNT to 1102 mg/kg for 2,3-DNT in male rats; in male mice, the 3,5-isomer has an LD$_{50}$ of 611 mg/kg, whereas the 2,4-isomer has an LD$_{50}$ of 1924 mg/kg.[1] The individual isomers were generally less toxic in mice than in rats, and the lethal dose for cats was much lower than for rodents (27 mg/kg for the 2,4-isomer).

In humans, very early reports found pallor, cyanosis, and anemia as common symptoms in workers exposed to presumably high concentrations of technical DNT.

Hematological effects have been observed in a variety of animal studies. The most common findings are methemoglobinemia, anemia, reticulocytosis, and an increase in Heinz bodies.

Cyanosis was observed in rats administered 60 mg/kg/day of 2,4-DNT for 5 days.[2] Severe anemia occurred in dogs administered 25 mg/kg/day and rats administered 206 mg/kg/day for 13 weeks; mild anemia was seen in mice given 441 mg/kg/day for the same duration.[3] In chronic studies, hematological effects have been observed, but the animals have often exhibited "compensated anemia," an adaptive response to the DNT exposure.

Neurologic signs, consisting of tremors followed by extensor rigidity, were noted in one dog receiving 10 mg/kg/day of 2,4-DNT for 8 weeks; minimal signs in other animals consisted of incoordination and stiffness, particularly in the hind legs.[1]

Chronic studies in rats have shown isomer-specific hepatocarcinogenesis in F344 rats. Administration of 7 or 14 mg/kg/day of 2,6-DNT for 1 year produced hepatocellular carcinomas in 85% and 100% of the animals, respectively.[4] The majority of the tumors had a trabecular pattern, and pulmonary metastases were present. In contrast, a diet of 27 mg/kg/day of 2,4-DNT for 1 year caused no tumors. Treatment with 35 mg/kg/day of technical-grade-DNT, containing 76% 2,4-DNT and 18% 2,6-DNT, resulted in a 47% incidence of hepatocellular tumors. The results demonstrated that the 2,6-isomer is a potent and complete hepatocarcinogen, under the test conditions, whereas the 2,4-isomer is nonhepatocarcinogenic. The results also explain the inconsistent results that had been reported in previous bioassays: in an initial study by the National Cancer Institute, 2,4-DNT was found to be nonhepatogenic, whereas a CIIT study produced a 100% incidence in the same strain from a technical-grade DNT.[5,6] In the NCI bioassay, 2,6-DNT constituted less than 5% of the DNT whereas, in the CIIT study, it was over 18% of the mixture. Chronic studies are not available on the other isomers.

In an attempt to determine if the carcinogenicity observed in animal studies was predictive for humans, the mortality experience of munitions workers with opportunity for substantial DNT exposure was examined. No evidence of carcinogenic effect was found, but an unsuspected excess from ischemic heart disease was noted. Additional analyses showed evidence of a 15-year latency period and suggested a relationship with duration and intensity of exposure.[7]

A more recent study of nearly 5000 DNT-exposed workers found an excess of liver and biliary cancer among those employed at least 5 months at the study facility between 1949 and 1980 and with at least one day on a job with probable DNT exposure.[8] The six observed cases were statistically significant based on comparison with an internal referent group of unexposed workers. The authors noted these limitations: relatively few workers with long duration of exposure and the lack of quantitative information on exposure levels to DNT and other chemicals. A retrospective cohort study of this same population did not find increased mortality from ischemic heart disease.[9]

Animal studies have shown that oral exposure to DNT can result in adverse effects on reproduction. Observed effects have included decreased sperm production, testicular atrophy, changes in Sertoli-cell morphology, degenerated seminiferous tubules, and decreased fertility.[1] It has been suggested that DNT acts on Sertoli cells, resulting in both inhibition of spermatogenesis and changes in testicular-pituitary endocrine activity.[10] A study of 30 workers exposed to DNT and other chemicals found a decrease in sperm counts relative to controls that were abnormally high, a slight change in one category of abnormal sperm, and a slight increase in spontaneous abortions for wives.[1] Other studies reported no detectable differences in sperm levels or fertility rates as a result of occupational exposure.[11]

Dinitrotoluene was not found to be teratogenic following oral administration to rats; embryo/fetal toxicity was observed only at a dose that also produced 46.2% maternal mortality.[12]

The DNTs appear to cause mutations in *Salmonella typhimurium* assays following metabolic activation.[12]

All 6 isomers have been found to be nonirritating in the eye of rabbits. Applied to the skin of rabbits, 2,4-, 2,6-, and 3,5-DNT were nonirritating, whereas 2,3-, 3,4-, and 2,5-were mildly to moderately irritating.[1]

The 1995 ACGIH threshold limit value–time-weighted average (TLV-TWA) for dinitrotoluene is 0.15 mg/m^3 with an A2 suspected human carcinogen designation and a notation for skin absorption.

REFERENCES

1. Rickert DE, Butterworth BE, Popp JA: Dinitrotoluene: acute toxicity, oncogenicity, genotoxicity, and metabolism. *CRC Crit Rev Toxic* 13:217–234, 1983
2. Lane RW, Simon GS, Dougherty RW, et al: Reproductive toxicity and lack of dominant lethal effects of 2,4-dinitrotoluene in the male rat. *Drug Chem Toxicol* 8:265–280, 1985
3. Lee CC, Hong CB, Ellis HV, et al: Subchronic and chronic toxicity studies of 2,4-dinitrotoluene. Part II. CD rats. *J Am Coll Toxicol* 4:243–256, 1985
4. Leonard TB, Graichen ME, Popp JA: Dinitrotoluene isomer-specific hepatocarcinogenesis in F344 rats. *JNCI* 79: 1313–1319, 1987
5. National Cancer Institute: *Bioassay of 2,4-Dinitrotoluene for Possible Carcinogenicity. CAS No 121-14-2, NCI-CG-TR-54.* US Department of Health, Education and Welfare, Public Health Service, National Institutes of Health, *1978*
6. Chemical Industry Institute of Toxicology: CIIT Chemical Safety Studies Dinitrotoluene Final Report, Docket No 12362, CIIT, Research Triangle Park, NC, 1979
7. Levine RJ, Andjelkovich DA, Kersteter SL, et al: Heart disease in workers exposed to dinitrotoluene. *JOM* 28:811–816, 1986
8. Stayner LT, Dannenberg AL, Bloom T, et al: Excess hepatobiliary cancer mortality among munitions workers exposed to dinitrotoluene. *JOM* 35:291–296, 1993
9. Stayner LT, Danneberg AL, Thun M, et al: Cardiovascular mortality among munitions workers exposed to nitroglycerin and dinitrotoluene. *Scan J Work Environ Health* 18:34–43, 1992
10. Bloch E, Gondos B, Gatz M, et al: Reproductive toxicity of 2,4-dinitrotoluene in the rat. *Toxicol Appl Pharmacol* 94:466–472, 1988
11. Agency for Toxic Substances and Disease Registry (ATSDR): *Toxicological Profile for 2,4-Dinitrotoluene and 2,6-Dinitrotoluene.* US Public Health Service, pp 1–111, 1989
12. Price CJ, Tyl RW, Marks TA, et al: Teratologic evaluation of dinitrotoluene in the Fischer 344 rat. *Fund Appl Toxicol* 5:948–961, 1985

DIOXANE
CAS:123-91-1

$C_4H_8O_2$

Synonyms: 1,4-Diethylene dioxide; diethylene ether; 1,4-dioxacyclohexane; 1,4-dioxane; *p*-dioxane; dioxyethylene ether

Physical Form. Colorless liquid

Uses. Solvent; stabilizer in chlorinated solvents

Exposure. Inhalation; skin absorption

Toxicology. Dioxane is an irritant of the eyes and mucous membranes; on prolonged exposure, it is toxic to the liver and kidneys. It is carcinogenic in experimental animals.

Human volunteers exposed to 50 ppm for 6 hours reported eye irritation throughout the exposure.[1] At 300 ppm for 15 minutes, there was transient eye, nose, and throat irritation.[2] Exposure to 1600 ppm for 10 minutes caused immediate burning of the eyes with lacrimation and, at 5500 ppm for 1 minute, slight vertigo was also noted.[3]

Five deaths due to heavy exposure for 5 weeks have been reported.[4] Signs and symptoms of poisoning included epigastric pain, anorexia, and vomiting, followed by oliguria, anuria, coma, and death. At autopsy, there was liver necrosis, kidney damage and edema of the lungs and brain. Another fatal case involved a one week exposure to levels ranging from 208 to 605 ppm and possibly higher with concurrent skin exposure.[5] Epigastric pain, increased blood pressure, convulsions, and unconsciousness preceded death. Studies of workers exposed at levels of up to 24 ppm for periods of up to 50 years found no increase in chronic disease, no excess total deaths, no excess cancer deaths, or common cause of death.[6]

Applied to human skin, dioxane causes dryness without other signs of irritation; hypersensitivity has been reported.[6]

In animal experiments, guinea pigs exposed

to 30,000 ppm for 3 hours exhibited narcosis after 87 minutes and died within 2 days.[7] The LC_{50} for rats was 14,000 ppm for 4 hours.[8] Repeated exposure of several animal species to 1000 ppm produced damage to kidneys and liver, and repeated inhalation of 800 ppm over 30 days resulted in fatal kidney injury in some of the exposed rabbits.[7,9]

The liquid applied to rabbit and guinea pig skin was rapidly absorbed and produced signs of incoordination and narcosis. Repeated applications caused liver and kidney damage.[6] Instilled in a rabbit's eye, dioxane produced hyperemia and purulent conjunctivitis.[10]

High doses of dioxane by oral administration produced malignant tumors of the nasal cavity and liver in rats, and tumors of the liver and gallbladder in guinea pigs.[11] Rats administered either 0.5% or 1.0% (v/v) in the drinking water had squamous-cell carcinomas of the nasal turbinates; hepatocellular adenomas were seen in the dosed females.[12] Subsequent re-examination of this study suggested that inhalation of dioxane-containing drinking water may have been responsible for the tumors of the nasal passage.[13] However, in another study, inhalation of 111 ppm, 7 hr/day, 5 days/week for 2 years, did not result in any increased tumor incidence in rats.[14]

A mortality study of 165 workers who had been exposed to low concentrations of 1,4 dioxane (since 1954) did not show any increased cancer risk.[15]

The IARC has determined that there is sufficient evidence of carcinogenicity to animals and inadequate evidence to humans.[11]

Administered to rats by gavage on days 6 through 15 of gestation, 1.0 ml/kg/day caused slight embryo and maternal toxicity in the form of reduced weights. There were no teratogenic effects.[16]

The warning properties are inadequate to prevent overexposure. Although dioxane has a low odor threshold (3–6 ppm), it is not unpleasant, and individuals acclimatize within a few minutes.[6]

The 1995 ACGIH threshold limit value–time-weighted average for dioxane is 25 ppm (90 mg/m³) with a notation for skin absorption.

REFERENCES

1. Young JD, Braum WH, Rampy LW: Pharmacokinetics of 1,4-dioxane in humans. *J Toxicol Environ Health* 4:507–520, 1977
2. Silverman L, Schulte HF, First MW: Further studies on sensory response to certain industrial solvent vapors. *J Ind Hyg Toxicol* 28:262–266, 1946
3. Yant WP: Acute response of guinea pigs to vapors of some new commercial organic compounds—VI. Dioxan. *Publ Health Rept* 45:2023–2032, 1930
4. Barber H: Haemorrhagic nephritis and necrosis of the liver from dioxane poisoning. *Guys Hosp Rept* 84:267–280, 1934
5. Johnstone RT: Death due to dioxane? *Arch Ind Health* 20:445–447, 1959
6. National Institute for Occupational Safety and Health: *Criteria for a Recommended Standard . . . Occupational Exposure to Dioxane. DHEW (NIOSH) Pub No 77–226. Washington, DC, US* Government Printing Office, 1977
7. Fairley A et al: The toxicity to animals of 1:4 dioxan. *J Hyg* 34:486–501, 1934
8. Pozzani UC et al: The toxicologic basis of threshold limit values—5. The experimental inhalation of vapor mixtures by rats, with notes upon the relationship between single dose inhalation and single dose oral data. *Am Ind Hyg Assoc J* 20:364–369, 1959
9. Smyth HF Jr: Improved communication—hygienic standards for daily inhalation. *Am Ind Hyg Assoc J* 17:129–185, 1956
10. von Oettingen WF, Jirouch EA: The pharmacology of ethylene glycol and some of its derivatives in relation to their chemical constitution and physical chemical properties. *J Pharmacol Exp Ther* 42:355–372, 1931
11. *IARC Monographs on the Evaluation of Carcinogenic Risk to Humans. Overall Evaluations of Carcinogenicity: An Updating of IARC Monographs Vols 1–42*, Suppl 7, p 201. Lyon, International Agency for Research on Cancer, 1987
12. National Cancer Institute: *Bioassay of Dioxane for Possible Carcinogenicity*, Carcinogenesis Technical Report Series No 80, p 107, National Institutes of Health, Bethesda, MD, 1978
13. Goldsworthy TL, Monticello TM, Morgan KT, et al: Examination of potential mechanisms of carcinogenicity of 1,4-dioxane in rat

nasal epithelial cells and hepatocytes. *Arch Toxicol* 65:1–9, 1991

14. Torkelson TR, Leong BKJ, Kociba JR, et al: 1,4-Dioxane—II. Results of a 2-year inhalation study in rats. *Toxicol Appl Pharmacol* 30:287–298, 1974
15. Buffler PA, Wood SM, Suarez L, et al: Mortality follow-up of workers exposed to 1,4-Dioxane. *J Occup Med* 20:255–259, 1978
16. Giavini E, Vismara C, Broccia ML: Teratogenesis study of Dioxane in rats. *Toxicol Lett* 26:85–88, 1985

DIPHENYLAMINE
CAS:122-39-4

$(C_6H_5)_2NH$

Synonyms: N-phenylbenzeneamine; N-phenylaniline; N,N-diphenylamine; N-diphenylaniline; DPA

Physical Form.   Colorless solid

Uses.   Rubber antioxidant and accelerator; fungicide; in veterinary medicine; stabilizer for nitrocellulose explosives and celluloids; manufacture of dyes

Exposure.   Inhalation

Toxicology.   Diphenylamine causes kidney and liver damage in animals.

In a two-year feeding study of beagle dogs of both sexes, 0.01%, 0.1%, or 1% diphenylamine was administered in the diet.[1] Decreased weight gain and anemia were noted at the two higher levels. Increases in liver and kidney weights were observed at the highest level. Rats fed diets ranging from 0.5% to 2.5% for 1 to 2 years had cystic dilation of renal tubules and a reversible anemia.[2] Diphenylamine treatment did not cause an increase in neoplasms in either species.

Cystic lesions of the proximal nephron occurred in newborn offspring of pregnant rats treated with commercial diphenylamine during gestation. No significant cystic tubule changes were identified in pups whose dams were administered chromatographically pure diphenylamine. An impurity present in diphenylamine, N,N,N'-triphenyl-p-phenylenediamine, has been identified as inducing the polycystic kidney disease.[3]

Diphenylamine was found to be not mutagenic in *Salmonella typhimurium* in the Ames test.[4]

The 1995 ACGIH threshold limit value–time-weighted average (TLV-TWA) for diphenylamine is 10 mg/m^3.

REFERENCES

1. Thomas JO, Ribelin WE, Woodward JR, Deeds F: The chronic toxicity of diphenylamine for dogs. *Toxicol Appl Pharmacol* 11:184–194, 1967
2. Thomas, JO et al: Chronic toxicity of diphenylamine to albino rats. *Toxicol Appl Pharmacol* 10:362–374, 1967
3. Clegg, S et al: Identification of a toxic impurity in commercial diphenylamine. *J Environ Sci Bull* B16:125–130,1981
4. Florin I et al: Screening of tobacco smoke constituents for mutagenicity using the Ames test. *Toxicology* 15:219–232, 1980

1,2-DIPHENYLHYDRAZINE
CAS: 122-66-7

$C_{12}H_{12} N_2$

Synonyms: Hydrazobenzene; N,N'-diphenylhydrazine, *sym*-diphenylhydrazine

Physical Form.   White crystalline solid

Uses.   Formerly used as a starting material in the production of benzidine for dyes; in production of certain drugs.

Exposure.   Ingestion; inhalation; skin absorption

Toxicology. 1,2-Diphenylhydrazine is a liver toxin in rodents, and it appears to be carcinogenic in experimental animals.

No information is available on the toxicity of 1,2-diphenylhydrazine in humans.

In rats and mice given 1,2-diphenylhydrazine in the diet for four weeks, the lethal ranges were 54 mg/kg/day and above for rats and 390 mg/kg/day and above for mice.[1] Gross pathological examinations showed intestinal hemorrhages in mice that died.

Chronic oral administration (78 weeks of treatment, followed by observation) of 4 or 15 mg/kg/day in male rats and 2 or 5 mg/kg/day in females caused significantly increased mortality in the high-dose females. In mice, mortality was increased for males and females at 52 mg/kg/ day. Causes of the mortality in the rats or mice were not indicated. Statistically increased incidences of interstitial inflammation of the lungs were observed in treated male rats and in low-dose females but not in mice. Treatment also produced degenerative alterations in the liver of rats (fatty metamorphosis) and female mice (coagulative necrosis) and, in treated male rats, there was stomach hyperkeratosis and acanthosis. In these same animal studies, 1,2-diphenylhydrazine caused increased incidences of hepatocellular carcinoma and Zymbal's gland carcinomas in male rats; neoplastic nodules of the liver and mammary adenocarcinomas were observed in female rats; and, in female mice, there was an increased incidence of hepatocellular carcinomas.

Animals did not show histological alterations in reproductive organs in chronic studies, but reproductive function was not evaluated.[1,2]

1,2-Diphenylhydrazine is a solid with a low vapor pressure at ambient temperature, which makes inhalation exposure of this substance in the vapor state unlikely. Exposure to dusts of 1,2-diphenylhydrazine is conceivable.[2]

Limited information is available on the metabolism of 1,2-diphenylhydrazine.[2] Two of the known metabolites, aniline and benzidine, may contribute to the toxicity and/or carcinogenicity of the substance.

A threshold limit value has not been established for diphenylhydrazine.

REFERENCES

1. National Cancer Institute: *Bioassay of Hydrazobenzene for Possible Carcinogenicity*, Technical Report Series No 92. DHEW Pub No (NIH) 78-1342, Bethesda, MD, National Institutes of Health, 1978
2. Agency for Toxic Substances and Disease Registry (ASTDR): *Toxicological Profile for 1,2-Diphenylhydrazine*. US Department of Health and Human Services, Public Health Service, Atlanta, GA, pp 1–71, 1990

DIPROPYLENE GLYCOL METHYL ETHER
CAS: 34590-94-8

$CH_3OC_3H_6OC_3H_6OH$

Synonyms: Dipropylene glycol monomethyl ether; DPGME; DPM

Physical Form. Colorless liquid

Uses. Solvent for nitrocellulose and synthetic resins

Exposure. Inhalation

Toxicology. Dipropylene glycol methyl ether (DPGME) at very high concentrations causes narcosis in animals, and it is expected that severe exposure will produce the same effect in humans. Because the propylene glycol ethers are metabolized differently from the ethylene glycol ethers, they are not associated with potent teratogenic, spermatotoxic, or hematopoietic effects.[1]

Concentrations expected to be hazardous to humans are disagreeable and not tolerated; in addition, concentrations above 200 ppm (40 % saturated atmosphere) are difficult to attain, suggesting that these levels would not normally be encountered in the work environment.[2] Vapor concentrations reported as 300 ppm caused eye and nasal irritation in humans.[3] No evidence of skin irritation or sensitization was observed when the undiluted liquid was applied to the skin of 250 subjects for prolonged periods or after repeated applications.[3]

A single 7-hour exposure of rats to 500 ppm resulted in mild narcosis with rapid recovery.[3] Repeated daily inhalation exposures to 300 to 400 ppm for over 100 days produced minor histopathologic liver changes in rabbits, monkeys, and guinea pigs; rats initially experienced slight narcosis but developed tolerance to this effect after a few weeks.[3] Daily exposure of rats and rabbits to 200 ppm for 13 weeks caused no effects.[2] Topical administration of 10 mg/kg, 5 times/week for 13 weeks, to shaved rabbit skin caused six deaths among seven animals.[3]

The LD_{50} for rats was 5.4 ml/kg; the low oral toxicity indicates that there is practically no likelihood that toxic amounts of these materials would be swallowed in ordinary handling and use.[3]

Direct contact of the eyes with the liquid or with high vapor concentrations may cause transient irritation.[2]

The 1995 ACGIH threshold limit value–time-weighted average for dipropylene glycol methyl ether is 100 ppm (606 mg/m³) with a short term excursion limit of 150 ppm (909 mg/m³) and a notation for skin absorption.

REFERENCES

1. Miller RR, Hermann EA, Calhoun LL, et al: Metabolism and disposition of dipropylene glycol monomethyl ether (DPGME) in male rats. *Fund Appl Toxicol* 5:721–726, 1985
2. Landry TD, Yano BL: Dipropylene glycol monomethyl ether: a 13-week inhalation toxicity study in rats and rabbits. *Fund Appl Toxicol* 4:612–617, 1984
3. Rowe VK et al: Toxicology of mono-, di-, and tri-propylene glycol methyl ethers. *AMA Arch Ind Hyg Occup Med* 9:509–525, 1954

DIPROPYL KETONE
CAS:123-19-3

$(CH_3CH_2CH_2)_2CO$

Synonyms: 4-Heptanone; DPK; propyl ketone; butyrone; heptan-4-one

Physical Form.   Colorless liquid

Uses.   Solvent for nitrocellulose, oils, resins, and polymers, and in flavorings

Exposure.   Inhalation

Toxicology.   Dipropyl ketone causes narcosis in animals, and it is expected that severe exposure in humans will produce the same effect.

The oral LD_{50} in rats was 3.73 g/kg.[1] The LC_{50} for 6 hours' exposure was 2690 ppm in the rat; 6 hours at 1600 ppm caused narcosis.[2] Repeated exposure for 6 hr/day to 1200 ppm for 5 days/week for 2 weeks caused marginal liver enlargement.

The liquid on the skin of guinea pigs under occlusive wrap caused slight irritation.[2] In the eye of the rabbit, there was slight irritation.

There are no reports of adverse effects in humans.

The 1995 ACGIH threshold limit value–time-weighted average (TLV-TWA) for dipropyl ketone is 50 ppm (233 mg/m³).

REFERENCES

1. Carpenter CP, Weil CS, Smyth HF Jr: Range-finding toxicity data. List VIII. *Toxicol Appl Pharmacol* 28:313–319, 1974
2. Krasavage WJ, O'Donoghue JL, Divincenzo GD: Ketones. In Clayton GD, Clayton FE (eds): *Patty's Industrial Hygiene and Toxicology*, 3rd ed, Vol 2C, *Toxicology*, pp 4763–4764. New York, Wiley-Interscience, 1982

DIQUAT
CAS: 2764-72-9

$C_{12}H_{12}N_2$

DIQUATE DIBROMIDE
CAS: 85-00-7

$C_{12}H_{12}Br_2N_2$

Synonym: 1,1-ethylene-2,2-dipyridylium di-bromide

Physical Form. Yellow crystals; available commercially as aqueous solutions (15% to 25% w/v) and as water-soluble granules (2.5%)

Uses. Contact herbicide

Exposure. Inhalation; skin absorption

Toxicology. Diquat causes gastrointestinal damage and, in animals, produces cataracts; no cataracts have been reported in exposed workers.

The estimated lethal dose for humans is 6 to 12 g of diquat.[1] In one report of a human fatality from ingestion, initial effects were gastrointestinal symptoms, ulceration of mucous membranes, acute renal failure, and liver damage.[2] Pulmonary signs included interalveolar exudation, but there was no evidence of proliferative or fibroplastic changes characteristic of paraquat intoxication.

Skin contact with concentrated solutions may lead to a color change and softening of the fingernails.[1] Exposure to the dust or mist can cause nosebleeds.

The primary toxic effect of diquat in animals is gastrointestinal damage resulting in diarrhea with consequent dehydration.[3]

The oral LD_{50} in rats ranged from 230 to 440 mg/kg.[4] Effects included dilated pupils, lethargy, and labored respiration. Lung changes characteristic of paraquat were not observed. When applied daily to the skin of rabbits at 40 mg/kg, 4 of 6 rabbits died after 8 to 20 applications. Prior to death, there was weight loss, incoordination, and muscular weakness.

Prolonged exposure to diquat is necessary to produce cataracts, and a clear dose-response relationship has been established in chronic feeding studies in animals. At a level of 1000 ppm, complete opacities occur in rats within 6 months; at 50 ppm for 12 months, only some of the animals exhibit slight opacities.[1] Lens opacities developed within 11 months in dogs fed 15 mg/kg/day and within 17 months at 5 mg/kg/day.[4] Dogs tolerated 1.7 mg/kg/day for four years without developing cataracts.

Rats exposed to 1.9 mg/m^3 for 4 hr/day, 6 days/week for 5 months, showed inflammatory changes in the peribronchial and perivascular connective tissues.[5] Long-term studies have shown no carcinogenic potential.[4,5] Most mutagenicity data suggest that diquat is not mutagenic.[1]

In a multigeneration study of reproductive effects, levels of 500 ppm or 125 ppm did not affect fertility, litter production, or litter size and did not cause congenital abnormalities.[6] Lens opacities were found in the parents and F_1 and F_2 generation receiving 500 ppm, but not at the 125 ppm level.

The 1995 ACGIH threshold limit value–time-weighted average for diquat is 0.5 mg/m^3 for total dust and 0.1 mg/m^3 for the respirable fraction of dust; there is a notation for skin absorption.

REFERENCES

1. Hayes WJ Jr, Laws ER Jr: *Handbook of Pesticide Toxicology,* Vol 3, *Classes of Pesticides,* pp 1376–1380. New York, Academic Press, 1991
2. Schonborn H, Schuster HP, Kossling FK: Klinik und morphologie der akuten peroralen diquat intoxikation. *Arch Toxikol* 27:204, 1971
3. Crabtree HC, Lock EA, Rose MS: Effects of diquat on the gastrointestinal tract of rats. *Toxicol Appl Pharmacol* 41:585–595, 1977
4. Clark DG, Hurst, EW: The toxicity of diquat. *Br J Indust Med* 27:51, 1970
5. Bainova A, Zlateva M, Vulcheva VI: Chronic inhalation toxicity of dipyridilium herbicides. *Khig Zdraveopazvane* 15:25, 1972
6. World Health Organization: *Pesticide Residues Series,* No 2, pp 1–243, Geneva, WHO 1973

DISULFIRAM
CAS: 97-77-8

$C_{10}H_{20}N_2S_4$

Synonyms: Antabuse; bis(diethylthiocarbamoyl) disulfide; TETD; tetraethylthiuram disulfide; Thiuram E

Physical Form.   White crystalline solid

Uses.   Rubber accelerator and vulcanizer; activator of thiazole accelerators; plasticizer in neoprene; pharmaceutical grade used in treatment of alcoholism

Exposure.   Inhalation

Toxicology.   Disulfiram causes an "Antabuse-alcohol" syndrome, hepatotoxicity, and allergic contact dermatitis.

Most of the human experience with disulfiram has come from its use as an avoidance therapy for alcoholism. Metabolites of disulfiram inhibit aldehyde dehydrogenase, resulting in elevated levels of acetaldehyde following ethanol ingestion. Side effects include flushing of the face, tachycardia, severe headache, apprehension, hyperpnea, hypotension, dizziness, nausea, vomiting, and fainting.[1]

Disulfiram metabolites include diethyldithiocarbamate and its metabolites, the moieties that irreversibly inhibit aldehyde dehydrogenase, and carbon disulfide, thought to be responsible for the occasional polyneuritis and psychotic episodes.[2-3] Several episodes of hepatotoxicity have also been reported.[4-5] Type IV allergic contact dermatitis has been reported in a few individuals.[6]

In a lifetime carcinogenicity bioassay, disulfiram was not carcinogenic in either rats or mice when fed in the diet.[7] The highest doses were 600 ppm in rats and 2000 ppm in mice.

The 1995 threshold limit value–time-weighted average (TLV-TWA) is 2 mg/m³.

REFERENCES

1. Petersen EN: The pharmacology and toxicology of disulfiram and its metabolites. *Acta Psychiatr Scand Suppl* 369:7–13, 1992
2. Johansson B: A review of the pharmacokinetics and pharmacodynamics of disulfiram and its metabolites. *Acta Psychiatr Scand* (Suppl 369):15–26, 1992
3. Kane, FJ Jr: Carbon disulfide intoxication from overdosage of disulfiram. *Am J Psych* 127:690–694, 1970
4. Keefe EB, Smith FW: Disulfiram hypersensitivity hepatitis. *JAMA* 230:435–436, 1974
5. Eisen HJ, Ginsberg AL: Disulfiram hepatotoxicity. *Ann Int Med* 83:673–675, 1975
6. von Hintzenstern J, Heese A, Koch HU, et al: Frequency, spectrum and occupational relevance of type IV allergies to rubber chemicals. *Contact Dermatitis* 24:244–252, 1991
7. National Toxicology Program: *Bioassay of Tetraethylthiuram Disulfide for Possible Carcinogenicity (CAS No 97-77-9)*, Technical Report Series No 166. National Technical Information Service, US Department of Commerce, Springfield, VA, 1979

DISULFOTON
CAS: 298-04-4

$C_8H_{19}O_2PS_3$

Synonyms: O,O-Diethyl-S-ethylmercaptoethyldithiophosphate; Disyston; Frumin AL; Solvirex; Dithiosystox; Thiodementon

Physical Form.   Pure material is a colorless liquid; technical grade is a dark yellow liquid

Uses.   Systemic insecticide and acaracide

Exposure.   Inhalation; skin absorption

Toxicology.   Disulfoton is an anticholinesterase agent.

Exposure to disulfoton can result in inhibition of cholinesterase activity in blood and at nerve synapses of muscles, secretory organs, and nervous tissue in the brain and spinal cord.[1] Central nervous system signs and symptoms include anxiety, restlessness, depression of respiratory and circulatory centers, ataxia, convulsions, and coma.

Nicotinic signs of intoxication include muscle weakness, tremor and fasciculations , and involuntary twitching. Muscle weakness that affects the respiratory muscles may contribute to dyspnea and cyanosis. Tachycardia may result from stimulation of sympathetic ganglia in cardiac tissue and may mask the bradycardia because of the muscarinic action of disulfoton on the heart. Nicotinic action at the sympathetic ganglion may also result in pallor, high blood pressure, and hyperglycemia.

Muscarinic signs include miosis, increased salivation, sweating, urination and defecation, nausea and vomiting, and increased bronchial secretions.

Severe signs and symptoms of disulfoton intoxication (miosis, salivation, monoplegia) were observed in a man 2 to 3 hours after he consumed 3 to 4 tablespoons of disulfoton.[2]

There was no evidence of a carcinogenic response in mice fed 2.08 mg/kg/day for 23 months or in beagle dogs fed up to 0.098 mg/kg/day for 2 years.[3-4]

Note: For a description of diagnostic signs, differential diagnosis, and medical control, including clinical laboratory tests, as well as specific treatment of intoxication by anticholinesterase insecticides, see "Parathion."

The 1995 threshold limit value–time-weighted average (TLV-TWA) for disulfoton is 0.1 mg/m³ with a notation for skin absorption.

REFERENCES

1. Taylor P: Anticholinesterase agents. In Gilman AG et al (eds): *Goodman and Gilman's The Pharmacological Basis of Therapeutics*, 8th ed, pp 131-149. New York: Macmillan, 1990
2. Yashiki M, Kojima T, Ohtani M, et al: Determination of disulfoton and its metabolites in the body fluids of a Di-Syston intoxication case. Forens *Sci Int* 48:145–154, 1990
3. Hayes RH: Oncogenicity study of disulfoton technical on mice. Study No 80-271-04. Mobay Chemical Corporation, Corporate Toxicology Department, Stilwell, KS, 1983
4. Hoffman K, Weischer CH, Luckhaus G, et al: S 276 (disulfoton) chronic toxicity study on dogs (two-year feeding experiment). Report Nos. 5618 and 45287. AG Bayer, West Germany, 1975

DIVINYL BENZENE
CAS: 1321-74-0

$C_{10}H_{10}$

Synonyms: Diethenylbenzene; DVB; 1,4-divinyl benzene; vinylstyrene

Physical Form. Straw-colored liquid

Uses. Comonomer for preparation of cross-linked polymers in production of ion-exchange beads and gel permeation chromatography polystyrene beads; polymerization monomer for synthetic rubber, drying oils, and casting resins

Exposure. Inhalation

Toxicology. Divinyl benzene is an irritant of eyes, nose, and mucous membranes.

Mild respiratory irritation occurred in workers exposed to 0.4- to 4-ppm divinyl benzene.[1] Mild irritation was also reported from skin and eye contact.

A single, 2-hour exposure of five rats to fume generated from polymerizing divinyl benzene at 120°C yielded a level of 27,317 ppm and produced peripheral vasodilation, lethargy, salivation, bilateral corneal opacity and dyspnea.[2] When the temperature of the polymerizing divinyl benzene was kept at 80°C, yielding a concentration in the chamber of 3312 ppm, effects were ataxia, tachypnea, ocular irritation, and rhinitis. Exposure of rats for 7 hours to 645 ppm resulted in no observable effects.

Instillation of 0.1 ml into the eyes of rabbits for 30 seconds caused irritation and conjunctivitis, the latter of which was still present 8 days later.[3] On the abdominal skin of rabbits, a mixture of divinyl benzene and ethyl vinyl benzene repeatedly applied and occluded for 2 weeks caused slight erythema, edema, and moderated exfoliation at the application site.

The 1995 threshold limit value–time-weighted average (TLV-TWA) for divinyl benzene is 10 ppm (53 mg/m^3).

REFERENCES

1. DOW Chemical Company: Communication to the TLV Committee, in ACGIH: *Documentation of the TLVs and BEIs.* 6th ed, pp 540–542. Cincinnati, OH, American Conference of Governmental Industrial Hygienists, 1991
2. Leong BKJ, Rampy LW, Kociba RJ: Preliminary studies on the toxicological properties of divinyl benzene. Unpublished Report. Chemical Biology Research, DOW Chemical USA, Midland, MI, January 28, 1986
3. Henck JW: Divinyl benzene—HP (70–85%): acute toxicological properties and industrial handling hazards. Unpublished Report. Toxicology Research Laboratory, Dow Chemical Company, Midland, MI, June 18, 1980

ENDOSULFAN
CAS: 115-29-7

$C_9H_6Cl_6O_3S$

Synonyms: Thiodan®

Physical Form. Technical endosulfan is a semiwaxy solid containing 90% to 95% of a 70%:30% mixture of the alpha- and beta-stereoisomers

Uses. Insecticide

Exposure. Inhalation

Toxicology. Endosulfan is a convulsant.

Convulsions were reported in nine workers exposed to the endosulfan-containing insecticide Thiodan® during bagging of the product.[1] Other effects noted prior to the convulsions were malaise, nausea, vomiting, dizziness, confusion, and weakness. The level and duration of exposure were not indicated.

Accidental or intentional ingestion of endosulfan has resulted in death in humans. In two cases of suicide, the dose was up to 100 ml of Thiodan® and, in three other poisonings, the doses were not specified.[2] Initial signs of poisoning included gagging, vomiting, diarrhea, agitation, writhing, cyanosis, dyspnea, and coma.

Signs of acute endosulfan intoxication in animals are similar to those seen in humans and include hyperexcitability, dyspnea, decreased respiration, fine tremor, and tonic-clonic convulsions. Oral LD_{50} values range from 7.4 mg/kg in mice to 40 to 125 mg/kg for male rats.[3–5] Female rats were 4 to 5 times more sensitive to the acute effects than the male rats.[4]

A 2-year carcinogenicity study was performed with endosulfan in the diet up to 952 ppm for male rats and 445 ppm for females; concentrations for male mice ranged up to 6.9 ppm and for female mice up to 3.9 ppm.[6] Animals were fed endosulfan for the first 80% of their life span and then observed for the remaining 20%. At both doses in rats, there was a high incidence of toxic nephropathy in both sexes and testicular atrophy in males. In both species, high early mortality was observed in the male groups, so that no conclusions could be drawn regarding carcinogenicity. There was no evidence of carcinogenicity in the female mice or rats.

The 1995 threshold limit value–time-weighted average (TLV-TWA) for endosulfan is 0.1 mg/m^3 with a notation for skin absorption.

REFERENCES

1. Ely TS, Macfarlane JW, Galen WP, et al: Convulsions in Thiodan workers. *J Occup Med* 9:35–37, 1967
2. Terziev Z, Dimitrova N, Rusev F: Forensic

medical and forensic chemical study of acute lethal poisonings with Thiodan. *Folia Med* 16:325–329, 1974

3. Gupta PK, Murthy RC, Chandra SV: Toxicity of endosulfan and manganese chloride: cumulative toxicity rating. *Toxicol Lett* 7:221–228, 1981

4. Hoechst: Summary and evaluation of the toxicity data for endosulfan–substance technical. Hoechst AG, Frankfurt, Germany. Rept No 90.0848, 1990

5. Agency for Toxic Substances and Disease Registry: *Toxicological Profile for Endosulfan.* ATSDR/TP-91/16, Public Health Service, Centers for Disease Control, Atlanta, GA, pp 7–65, 1993

6. National Toxicology Program: *Bioassay of endosulfan for possible carcinogenicity (CAS No 115-29-7)*, Technical Report Series No 62. National Technical Information Service, US Department of Commerce, Springfield, VA, 1978

ENDRIN

CAS:72-20-8

$C_{12}H_8Cl_6O$

Synonyms: Compound 269; Experimental Insecticide 269; OMS-197

Physical Form. White crystalline solid

Uses. All uses of endrin in the United States were canceled by the manufacturer in 1986; previously used as an insecticide, avicide, and rodenticide

Exposure. Inhalation; skin absorption; ingestion

Toxicology. Endrin primarily affects the central nervous system; at high concentrations, it is a convulsant.

In humans, the first effect of endrin intoxication is frequently a sudden epileptiform convulsion, which may occur from 30 minutes to 10 hours after overexposure; it lasts for several minutes and is usually followed by a stuporous state for 15 minutes to 1 hour.[1,2] Severe poisoning results in repeated violent convulsions and in some cases, status epilepticus.[3] The electroencephalogram (EEG) may show dysrhythmic changes, which frequently precede convulsions; withdrawal from exposure usually results in a normal EEG within 1 to 6 months.[2] In most cases, recovery is rapid, but headache, dizziness, lethargy, weakness, and anorexia may persist for 2 to 4 weeks.[2] In less severe cases of endrin intoxication, the complaints are headache, dizziness, leg weakness, abdominal discomfort, nausea, vomiting, insomnia, agitation, and, occasionally, slight mental confusion.[1,3]

Poisonings resulting in convulsions have occurred in manufacturing workers. Recovery following occupational exposures is usually complete within 24 hours. Unlike dieldrin, which persists in the body, endrin is rapidly eliminated and apparently does not accumulate, even in fatty tissue.[3,4] However, endrin is the most acutely toxic of the cyclodiene compounds, which also include chlordane, heptachlor, dieldrin, and aldrin.[4]

Ingestion of endrin has resulted in numerous fatalities.[2,4] In one nonfatal incident, ingestion of bread made with endrin-contaminated flour caused sudden convulsions in three people; in one of them, the serum endrin level was 0.053 ppm 30 minutes after the convulsion and 0.038 ppm after 20 hours; in the other two cases, no endrin was detected in the blood at 8.5 or 19 hours, respectively, after convulsions.

Single doses of 2.5 mg/kg of endrin administered orally to pregnant golden hamsters during the period of fetal organogenesis caused a high incidence of fetal death, congenital anomalies, growth retardation, and maternal toxicity.[5]

Rats fed a diet of 50 or 100 ppm endrin for 2 years developed degenerative changes in the liver.[1] The IARC has concluded that animal bioassays in mice and rats have been inadequate to evaluate the carcinogenicity of endrin.[6] Limited studies of endrin-exposed workers have not detected increased mortality due to cancer.[7] Endrin is not mutagenic in in vitro microbial assays.[7]

The 1995 ACGIH threshold limit value–time-weighted average (TLV-TWA) for endrin is 0.1 mg/m^3 with a notation for skin absorption.

REFERENCES

1. Jager KW: *Aldrin, Dieldrin, Endrin and Telodrin—An Epidemiological and Toxicological Study of Long-Term Occupational Exposure*, pp 78–87, 217–218, 225–234. Amsterdam, Elsevier, 1970
2. Coble Y, Hildebrandt P, Davis J, et al: Acute endrin poisoning. *JAMA*, 202:489–493, 1967
3. Hayes WJ Jr: *Pesticides Studied in Man*, pp 247–251. Baltimore, MD, Williams and Wilkins, 1982
4. Acute convulsions associated with endrin poisoning—Pakistan. *MMWR* 33:687–688, 693, 1984
5. Ottolenghi AD, Haseman JK, Suggs F: Teratogenic effects of aldrin, dieldrin, and endrin in hamsters and mice. *Teratology* 9:11–16, 1974
6. *IARC Monographs on the Evaluation of the Carcinogenic Risk of Chemicals to Man*, Vol 5, Some organochlorine pesticides, pp 157–171. Lyon, International Agency for Research on Cancer, 1974
7. Agency for Toxic Substances and Disease Registry (ATSDR): *Toxicological Profile for Endrin and Endrin Aldehyde*. US Department of Health and Human Services, Public Health Service, p 1–97, TP-90-14, 1990

ENFLURANE
CAS: 13838-16-9

CHF$_2$OCF$_2$CHClF

Synonyms: Ethrane®; 2-chloro-1,1,2-trifluoroethyl difluoromethyl ether

Physical Form.   Liquid

Uses.   Anesthetic in clinical anesthesia

Exposure.   Inhalation

Toxicology.   Enflurane is a general anesthetic used for inducing clinical anesthesia.

Exposure of humans at 15,000 to 20,000 ppm causes anesthesia.[1] At levels of 4200 to 5300 ppm for 30 minutes, cognitive tests indicated a performance decrement for remembering word pairs.[2]

No information on the effects of repeated or prolonged occupational exposure to enflurane is available thus far.

The 1995 ACGIH threshold limit value–time-weighted average (TLV-TWA) for enflurane is 75 ppm (566 mg/m^3).

REFERENCES

1. Gion H, Saidman LJ: The minimum alveolar concentration in man. *Anesthesiology* 35:361–364, 1971
2. Cook TL, Smith M, Winter PM, et al: Effect of subanesthetic concentrations of enflurane and halothane on human behavior. *Anesth Analg* 57:434–440, 1978

EPICHLOROHYDRIN
CAS: 106-89-8

C$_3$H$_5$OCl

Synonyms: 1-Chloro-2,3-epoxypropane; 3-chloro-1,2-epoxypropane; (chloromethyl)ethylene oxide; chloromethyloxirane; 3-chloro-1,2-propylene oxide; alpha-epichlorohydrin

Physical Form.   Colorless liquid

Uses.   In the manufacture of epoxy and phenoxy resins

Exposure.   Inhalation; skin absorption

Toxicology.   Epichlorohydrin is a severe irritant of skin, eye, and respiratory tract. Repeated or prolonged exposure can cause lung, liver, and kidney damage. It causes chromosomal aberrations in lymphocytes and is carcinogenic in experimental animals.

According to one industrial report, exposure at 20 ppm for 1 hour caused temporary burning of the eye and nasal passages.[1] At 40 ppm, irritation was more persistent, lasting 48 hours.[1] Pulmonary edema and renal lesions may result from exposure to concentrations greater than 100 ppm. In one worker acutely exposed to unspecified but probably very high concentrations, immediate effects were nausea, vomiting, headache, and dyspnea, with conjunctival and upper respiratory irritation. During the two years following the incident, bronchitis, liver damage, and hypertension were observed.[1]

Exposed workers had a marked increase in percentage of lymphocytes with chromatid breaks, chromosome breaks, severely damaged cells, and abnormal cells.[2]

Skin contact causes itching, erythema, and severe burns, which appear after a latent period ranging from several minutes to days, depending on the intensity of exposure. One worker who failed to remove contaminated shoes for 6 hours developed severe skin damage, with painful enlarged lymph nodes in the groin.[1] Skin sensitization has also been reported.[3]

Mice showed signs of irritation, along with gradual development of cyanosis and muscular relaxation of the extremities, and finally died from depression of the respiratory system following repeated 1-hour exposures to 2370 ppm.[4]

Rats repeatedly exposed to 120 ppm 6 hr/day experienced labored breathing, profuse nasal discharge, weight loss, leukocytosis, and increased urinary protein excretion. At autopsy, there was lung, liver, and kidney damage.[5] Respiratory distress was observed at 56 ppm during multiple exposures, while 17 ppm for 19 days produced no effects. Function of the liver and kidney was altered in rats receiving 5.2 or 1.8 ppm for 4 hours.[1]

Male rats administered 5 oral doses of 20 mg/kg had a temporary fertility loss, whereas a single 100 mg/kg dose caused spermatocele formation and probable permanent sterility.[6] Fifty inhalation exposures at 50 ppm for 6 hours each caused transient infertility in male rats; no changes were observed in reproductive parameters of female rats; rabbits remained fertile.[7] There was no evidence of teratogenicity in rat

fetuses at doses that caused death in some of the treated dams.[8]

No detrimental effect on fertility has been found in occupationally exposed workers where exposure levels are estimated to be less than 1 ppm.[9]

A number of studies indicate that epichlorohydrin induces tumors of localization dependent on the mode of application. A high incidence (100% for females and 81% for males vs. none in controls) of squamous-cell carcinomas of the forestomach occurred in rats administered 10 mg/kg 5 times/week for up to 2 years by gastric intubation.[10] Administered in the drinking water, epichlorohydrin also caused squamous-cell carcinomas of the forestomach in rats.[11] Exposure to 100 ppm, 6 hr/day for 30 days, produced a high incidence of malignant tumors of the nasal cavity in rats.[12] An increase in local sarcomas occurred in mice given weekly subcutaneous injections.[13]

A variety of epidemiological studies have not found increased cancer mortality among exposed workers.[14–16] Initial reports associating epichlorohydrin exposure with lung cancer and also heart disease mortality have not been confirmed.[17] In the most recent mortality study of 1064 male employees with exposure to epichlorohydrin, there were no significantly elevated SMRs for all malignant neoplasms, lung cancer, circulatory disease, or arteriosclerotic heart disease compared to controls.[18]

The carcinogenic risk to humans cannot be fully assessed however, because of insufficient latency periods, limited number of deaths, and indeterminate levels and duration of exposure.

The IARC has determined that there is sufficient evidence of carcinogenicity in animals and inadequate evidence to humans.[19]

The proposed 1995 ACGIH threshold limit value–time-weighted average (TLV-TWA) for epichlorohydrin is 0.1 ppm (0.38 mg/m³) with an A2 suspected human carcinogen designation and a notation for skin absorption.

REFERENCES

1. *Criteria for a Recommended Standard . . . Occupational Exposure to Epichlorohydrin.* DHEW (NIOSH) Pub No 76-206, pp 1–152. Na-

tional Institute for Occupational Safety and Health, US Department of Health, Education and Welfare. Washington, DC, US Government Printing Office, 1976

2. Picciano D: Cytogenic investigation of occupational exposure to epichlorohydrin. *Mut Res* 66:169–173, 1979

3. Beck MH, King CM: Allergic contact dermatitis to epichlorohydrin in a solvent cement. *Contact Derm* 9:315, 1983

4. Freuder E, Leake CD: The toxicity of epichlorohydrin. *Univ Calif Berk Publ Pharmacol* 2:69–77, 1941

5. Gage JC: The toxicity of epichlorohydrin vapour. *Br J Ind Med* 16:11–14, 1959

6. Cooper ERA et al: Effects of alpha-chlorohydrin and related compounds on the reproductive organs and fertility of the male rat. *J Reprod Fertil* 38:379–386, 1974

7. John JA et al: Inhalation toxicity of epichlorohydrin: effects on fertility in rats and rabbits. *Toxicol Appl Pharmacol* 68:415–423, 1983

8. Marks TA et al: Teratogenic evaluation of epichlorohydrin in the mouse and rat and glycidol in the mouse. *J Toxicol Environ Health* 9:87–96, 1982

9. Venable JR et al: A fertility study of male employees engaged in the manufacture of glycerine. *J Occup Med* 22:87–91, 1980

10. Wester PW et al: Carcinogenicity study with epichlorohydrin (CEP) by gavage in rats. *Toxicology* 36:325–339, 1985

11. Konishi Y, Kawabata A, Denda A, et al: Forestomach tumors induced by orally administered epichlorohydrin in male Wistar rats. *Gann* 71:922–923, 1980

12. Laskin S, Sellakumar AR, Kuschner M, et al: Inhalation carcinogenicity of epichlorohydrin in noninbred Spague–Dawley rats. *J Natl Cancer Inst* 65:751–757, 1980

13. Van Duuren BL et al: Carcinogenic activity of alkylating agents. *J Natl Cancer Inst* 53:695–700, 1974

14. Tassignon JP, Bos GD, Craigen AA, et al: Mortality in European cohort occupationally exposed to epichlorohydrin (ECH). *Int Arch Occup Environ Health* 51:325–336, 1983

15. Bond GG, Flores GH, Shellenberger RJ, et al: Mortality among a large cohort of chemical manufacturing employees. *J Natl Cancer Inst* 75:859–869, 1985

16. Tsai SP, Cowles SR, Lynne-Tackett D, et al: Morbidity prevalance study of workers with potential exposure to epichlorohydrin. *Br J Ind Med* 47:392–399, 1990

17. Enterline PE, Henderson V, Marsh G: Mortality of workers potentially exposed to epichlorohydrin. *Br J Ind Med* 47:69–76, 1990

18. Olsen GW, Lacy SE, Chamberlin SR, et al: Retrospective cohort mortality study of workers with potential exposure to epichlorohydrin and allyl chloride. *Am J Ind Med* 25:205–218, 1994

19. *IARC Monographs on the Evaluation of the Carcinogenic Risk of Chemicals to Humans.* Suppl 4, pp 122–124. Lyon, International Agency for Research on Cancer, 1982

EPN
CAS: 2104-64-5

$C_{14}H_{14}NO_4PS$

Synonyms: O-Ethyl O-p-nitrophenyl phenylphosphonothioate; EPN-300

Physical Form. Light yellow to brown solid

Uses. Acaricide; insecticide

Exposure. Inhalation; skin absorption; ingestion

Toxicology. EPN is an anticholinesterase agent.

A few deaths have been reported following poisoning by EPN, most resulting from suicidal ingestion, but at least one death has been associated with EPN spraying. It is moderately to highly toxic in animals but less potent than parathion.[1]

Signs and symptoms of overexposure are caused by the inactivation of the enzyme cholinesterase; this inactivation results in the accumulation of acetylcholine at synapses in the nervous system, skeletal and smooth muscle, and secretory glands.[1-3] The sequence of the development of systemic effects varies with the route of entry. The onset of signs and symptoms is usually prompt but may be delayed up to 12 hours. After inhalation, respiratory and ocular effects are the first to appear, often within a few minutes of exposure. Respiratory effects include tightness in the chest and wheezing owing to bronchoconstriction and excessive bronchial secretion; laryngeal spasm and excessive salivation may add to the respiratory distress; cyanosis may also occur. Ocular effects include miosis, blurring of distant vision, tearing, rhinorrhea, and frontal headache.

After ingestion, gastrointestinal effects, such as anorexia, nausea, vomiting, abdominal cramps, and diarrhea, appear within 15 minutes to 2 hours. After skin absorption, localized sweating and muscular fasciculations in the immediate area usually occur within 15 minutes to 4 hours; skin absorption is somewhat greater at higher ambient temperatures and is increased by the presence of dermatitis.[2,3]

With severe intoxication by all routes, an excess of acetylcholine at the neuromuscular junctions of skeletal muscle causes weakness aggravated by exertion, involuntary twitchings, fasciculations and, eventually, paralysis. The most serious consequence is paralysis of the respiratory muscles. Effects on the central nervous system include giddiness, confusion, ataxia, slurred speech, Cheyne–Stokes respiration, convulsions, coma, and loss of reflexes. The blood pressure may fall to low levels, and cardiac irregularities, including complete heart block, may occur.[1–3]

Complete symptomatic recovery usually occurs within a week; increased susceptibility to the effects of anticholinesterase agents persists for up to several weeks after exposure. Daily exposure to concentrations that are insufficient to produce symptoms following a single exposure may result in the onset of symptoms. Continued daily exposure may be followed by increasingly severe effects.

No significant effects on plasma or red blood cell cholinesterase activity occurred in volunteers given 6 mg of EPN for up to 47 days; 9 mg appears to be the threshold for toxicity.[4]

Delayed neuropathy characterized by distal axonal degeneration is a systemic health effect caused by some organophosphate pesticides and is not due to anticholinesterase inhibition. EPN is neurotoxic to atropine-protected hens, producing polyneuropathy pr ogressing to paralysis and some deaths following ingestion of 5 to 10

mg/kg/day. There are no reports, however, of neurotoxicity from EPN in man.[1]

EPN was not teratogenic or fetotoxic to mice at maternally nontoxic doses.[5]

The 1995 ACGIH threshold limit value–time-weighted average (TLV-TWA) for EPN is 0.1 mg/m^3 with a notation for skin absorption.

Note. For a description of diagnostic signs, differential diagnosis, and medical control, including clinical laboratory tests, as well as specific treatment of overexposure to anticholinesterase insecticides, see "Parathion."

REFERENCES

1. Hayes WJ Jr: Organic phosphorus pesticides. In *Pesticides Studied in Man*, pp 284–435. Baltimore, MD, Williams and Wilkins, 1982
2. Koelle GB (ed): Cholinesterases and anticholinesterase agents. In *Handbuch der Experimentellen Pharmakologie*, Vol 15, pp 989–1027. Berlin, Springer-Verlag, 1963
3. Taylor P: Anticholinesterase agents. In Gilman AG et al (eds): *Goodman and Gilman's The Pharmacological Basis of Therapeutics*, 7th ed, pp 110–129. New York, Macmillan, 1985
4. Moeller HC, Rider JA: Plasma and red blood cell cholines terase activity as indications of the threshold of incipient toxicity of ethyl-*p*-nitrophenyl thionobenzenephosphonate (EPN) and malathion in human beings. *Toxicol Appl Pharmacol* 4:123–130, 1962
5. Courtney KD, Andrews JE, Springer J, et al: Teratogenic evaluation of the pesticides Baygon, Carbofuran, Dimethoate, and EPN. *J Environ Sci Health* B20 4:373–406, 1985

1,2-EPOXYBUTANE
CAS: 106-88-7

C$_4$H$_8$O

Synonyms: 1,2-Butene oxide; butylene oxide; 1,2-butylene oxide; ethyl ethylene oxide; ethyl oxirane

Physical Form. Colorless liquid with pungent odor

Uses. Primarily a stabilizer for chlorinated hydrocarbon solvents; also a chemical intermediate in the production of butylene glycols

Exposure. Inhalation

Toxicology. 1,2-Epoxybutane exposure causes body weight effects and nasal lesions in experimental animals; chronic exposure is carcinogenic to rats but not to mice.

No adverse effects from 1,2-epoxybutane exposure have been reported in humans.

All rats exposed to 6550 ppm died during the 4-hour exposure period; at 2050 ppm, ocular discharge and dyspnea were observed, and eye irritation occurred at a 1400 ppm level.[1] In mice, 2050 ppm was lethal to all, and 1420 ppm was lethal to 4 out of 5 mice of each sex. In 14-day studies, mortality was seen at 3200 ppm in male rats and, at 1600 ppm, in female rats and mice of both sexes. Compound-related lesions included pulmonary hemorrhage and rhinitis in rats at 1600 ppm, and nephrosis in mice at 800 ppm; final body weights of surviving animals were significantly reduced compared to the controls in these exposure groups.

No deaths were observed in rats at concentrations of up to 800 ppm or in mice at concentrations of up to 400 ppm in an NTP study lasting 13 weeks (6 hr/day, 5 days/week). Nasal cavity lesions and reduced body weight were seen in rats exposed at 800 ppm. In mice, renal tubular necrosis was found at 800 ppm, a dose that was lethal. Inflammation of the nasal turbinates was observed in female mice at 100 ppm and above and, in male mice, at 200 ppm and above. In an earlier study, slight growth retardation was observed in rats and mice exposed at 600 ppm for 13 weeks; inflammatory and degenerative changes in the nasal mucosa were observed in both species. Myeloid hyperplasia in bone marrow occurred in male rats only.[2] No effects were noted at 75 or 150 ppm.

Rats exposed for 2 years to 400 ppm had increased incidence of papillary adenomas of the nasal cavity; the incidences of alveolar/bronchiolar adenomas or carcinomas (combined) were also increased in the male rats but not in the females.[1] Nonneoplastic lesions of the nasal cavity included inflammation, epithelial hyperplasia, and squamous metaplasia of the nasal epithelium, as well as atrophy of the olfactory sensory epithelium. Mice exposed at 50 or 100 ppm for 2 years had no significant increases in the incidence of neoplastic lesions of the nasal cavity. Treatment-related, nonneoplastic nasal changes were similar to those seen in rats.

In a combined exposure experiment, oral administration of trichloroethylene containing 1,2-epoxybutane induced squamous-cell carcinomas of the forestomach in mice, whereas administration of the trichloroethylene alone did not.[3]

A 10% solution applied to the shaved skin of mice 3 times per week for 77 weeks caused no visible skin reaction and no tumors.[4]

The IARC has determined that there is limited evidence for the carcinogenicity of 1,2-epoxybutane in experimental animals.[5]

Exposure to 1000 ppm prior to and during gestation did not cause any teratogenic effects in rats; fetal growth and viability were not affected in spite of depressed maternal body weight gain.[6] Rabbits exposed at 250 or 1000 ppm for 7 hours per day during gestational days 0 to 24 had maternal deaths at both exposure concentrations. No teratogenic effects were observed although the pregnancy rate was reduced in the high-dose group.

Instilled in the eyes of rabbits, 1,2-epoxybutane caused corneal injury.[7]

A threshold limit value has not been established for 1,2-epoxybutane although US manufacturers have recommended a voluntary time-weighted average–threshold limit value of 40 ppm.

REFERENCES

1. National Toxicology Program: Toxicology and carcinogenesis studies of 1,2-epoxybutane (CAS No. 106-88-7) in F344/N Rats and B6C3F₁ mice. Technical Report No 329; NIH Publication No 88-2585. Research Triangle Park, NC, US Department of Health and Human Services, 1988
2. Miller RR, Quast JF, Ayers JA, et al: Inhalation toxicity of butylene oxide. *Fund Appl Toxicol* 1:319–324, 1981

3. Henschler D, Elsasser H, Romen W, et al: Carcinogenicity study of trichloroethylene, with and without epoxy stabilizers, in mice. *J Cancer Res Clin Oncol* 107:149–156, 1984
4. Van Duuren BL, Langseth L, Goldschmidt BM, et al: Carcinogenicity of epoxides, lactones, and peroxy compounds. VI. Structure and carcinogenic action. *J Natl Cancer Inst* 39:1217–1228, 1967
5. *IARC Monographs on the Evaluation of Carcinogenic Risk to Humans*, Vol 47, Some organic solvents, resins, monomers and related compounds, pigments and occupational exposure in paint manufacture and painting, pp 217–228. Lyon, International Agency for Research on Cancer, 1989
6. Hardin BD, Niemeier RW, Sikov MR, et al: Reproductive-toxicologic assessment of the epoxides ethylene oxide, propylene oxide, butylene oxide, and styrene oxide. *Scand J Work Environ Health* 9:94–102, 1983
7. Weil CS, Condra N, Haun C, et al: Experimental carcinogenicity and acute toxicity of representative epoxides. *Am Ind Hyg Assoc J* 24:305–325, 1963

EPOXY RESINS

Synonyms: Epoxies; Epon resins

Physical Form. Uncured resins are long-chained prepolymers that are viscous liquids or solids; the cured resins are strong, solid polymers

Uses. Molding compounds; surface coatings; adhesives; laminating or reinforcing plastics

Exposure. Inhalation; skin contact

Introduction. Epoxy resins are polymers containing more than one epoxide group (a three-membered ring containing two carbon atoms and one oxygen atom).[1] An epoxy resin system is composed of two primary components: (1) the uncured resin, and (2) the curing agent (also referred to as the hardener, catalyst, accelerator, activator, or cross-linking agent).

Uncured resins are oligomers of relatively low molecular weight that may be a liquid or a solid. Before epoxy resins can become useful products, they must be cured with the addition of a curing agent. Curing involves the cross-linkage by polymerization of the reactive epoxy groups into a three-dimensional matrix.

In addition to the two primary components, there are several other components that may be included: diluents/solvents, fillers, and pigments. Diluents, which may represent 10% to 15% of resin volume, are added primarily to reduce viscosity. There are two types of diluents: reactive and nonreactive. Reactive diluents, primarily the glycidyl ethers, contain epoxy groups, which will take part in the curing process. Nonreactive diluents include a variety of organic solvents. Some uncured resins (liquids) are primary skin irritants or sensitizers or both. Toxicity generally decreases with increase in molecular weight and epoxy number. The resins with the greatest potential for sensitization are those with molecular weights under 500.[2] None of the uncured resins possess significant volatility; thus, inhalation poses little hazard.[1] The vast majority of epoxy resins are manufactured by the reaction between epichlorohydrin and bisphenol A, producing DGEBA (diglycidyl ether of bisphenol A) resins. After the initial manufacture of uncured resin, epichlorohydrin is probably not present during the subsequent mixing and polymerization steps.

Toxicology. The toxicity of epoxy resin systems results from the toxicity of the various components, each of which must be considered.

Curing agents account for much of the potential hazard associated with the use of epoxy resins.[1,2] There are several major types of curing agents: aliphatic amines, aromatic amines, cycloaliphatic amines, acid anhydrides, polyamides, and catalytic curing agents. The latter two types are true catalysts in that they do not participate in the curing process.

The aliphatic amines, including triethylene tetramine (TETA) and diethylene triamine (DETA), are highly alkaline (pH 13 to 14), caustic, and volatile and may cause severe burns.[3] They can cause skin irritation and sensitization as well as respiratory tract irritation. Eye irritation with conjunctivitis and corneal edema (resulting in "halos" around lights) may occur. Asthmatic symptoms suggesting respiratory tract sensitization have been described.[1] In ski manufacturing workers using epoxy resins, 3-(dimethylamino) propylamine has been shown to cause declines in FEV_1 and flow rates and to produce work-related respiratory symptoms (e.g., cough, tightness in the chest).[4]

Aromatic amines are generally solids and less irritating than aliphatic amines. 4,4'-Methylene dianiline (MDA) has caused outbreaks of reversible toxic hepatitis, apparently after skin absorption. Severe symptoms, including elevated AST, alkaline phosphatase, bilirubin, and liver enlargement, have been observed in some workers using MDA as a curing agent with epoxy resins.[5] m-Phenylenediamine is a strong irritant and allergic sensitizer; like MDA, it stains the skin and nails yellow.[2] 4,4'-Diaminodiphenyl sulfone (DDS) is tumorigenic in experiments.[1]

Acid anhydrides can cause severe eye and skin irritation and burns, dependent on the concentration and duration of contact.[1] Inhalation of high concentrations can cause significant respiratory tract irritation. Phthalic anhydride (PA), tetrachlorophthalic anhydride (TCPA), and trimellitic anhydride can induce asthma in epoxy resin workers; frequently, a dual (immediate and late) asthmatic response has been documented. Specific IgE antibodies on RAST testing have been demonstrated in patients with TCPA asthma.[6] One worker developed asthma on grinding epoxy resin cured with phthalic anhydride, presumably as a result of the release of some unreacted residual phthalic anhydride during grinding of a cured molding.[7]

Polyamides, reaction products of aliphatic amines and fatty acids, are considerably less toxic than the aliphatic amines but are moderately irritating to the skin and extremely irritating to the eyes.[1,2]

Isophorone diamine, a cycloaliphatic amine, has been reported to cause skin sensitization.[7]

Glycidyl ethers, reactive diluents in epoxy resin systems, are characterized by the presence of the 2,3-epoxypropyl group and an ether linkage to another organic group. Virtually all these substances are liquids with low vapor pressures at room temperature. Dermal contact is the major route of exposure. Vapor pressures become more appreciable at higher temperatures, which may occur during the curing process. Some glycidyl ethers commonly used in epoxy resin systems are allyl glycidyl ether (AGE), n-butyl glycidyl ether (BGE), o-cresyl glycidyl ether (CGE), isopropyl glycidyl ether (IGE), phenyl glycidyl ether (PGE), resorcinol diglycidyl ether, and 1,4-butanediol diglycidyl ether.[1,8] In humans exposed to glycidyl ethers, adverse effects have generally been limited to irritation and sensitization.[8] PGE and BGE have produced severe skin irritation in humans, causing burns and blistering. AGE has produced skin and eye irritation in humans. Skin sensitivity to AGE, BGE, and PGE has been documented in some humans occupationally exposed to epoxy resins.[1,8] In animals, glycidyl ethers have produced CNS effects, including muscular incoordination, reduced motor activity, agitation and excitement, deep depression, narcotic sleep, and coma. PGE has produced CNS depression with dermal administration; BGE and AGE have produced depression after inhalation exposure.[8] Experimental inhalation of glycidyl ethers has resulted in pulmonary irritation and inflammation, including pneumonitis and peribronchiolitis. For example, rats exposed to PGE at 10 ppm for 7 hr/day, 5 days/week, for 10 weeks had peribronchial and perivascular inflammatory infiltrates. Exposure to some glycidyl ethers, usually by injection, has been demonstrated to produce testicular abnormalities, alteration of leukocyte counts, atrophy of lymphoid tissue, and bone marrow cytotoxicity.[8]

Solvents used as nonreactive diluents include acetone, cellosolve, methyl ethyl ketone, methyl isobutyl ketone, methylene chloride, 1,1,1-trichloroethane, toluene, and xylene. Skin and eye irritation and, in higher concentrations, CNS depression and respiratory irritation, may result from exposure to these solvents as diluents for epoxy resin systems. These solvents may dehydrate and defat the skin, which may render the skin more vulnerable to the irritating and sensitizing components of epoxy resin formulations.[1,2]

Fillers used in epoxy resins are normally

inert, finely divided powders. Common fillers include calcium carbonate, clay (bentonite), talc, silica, diatomaceous earth, and asbestos. Workers exposed to excessive amounts of some of these dusts may experience lung damage.[1]

The curing process renders the resin essentially inert and nontoxic. At room temperature, full curing may take several days; incompletely cured resins may cause skin irritation and sensitization.[1] Respiratory symptoms may result from inhalation of cured epoxy dusts during grinding, presumably on account of the release of residual curing agent.[1,7] Skin irritation and sensitization have been associated with epoxy resin exposure.

Dermatitis from epoxy resin components usually develops first on the hands, particularly between the fingers, in the finger webs, on the dorsum of the hands, and on the wrists. It may vary in severity from erythema to a marked bullous eruption.[2] When sensitization occurs, the eruption is typically pruritic, with small vesicles on the fingers and hands resembling dyshidrotic eczema. The eruption may spread to other areas of the body that accidentally contact resin components such as the face and neck. In highly sensitized individuals, vapors from the curing agent or reactive diluents may cause recurrence of itching and redness in the absence of direct skin contact.[2]

Prevention of epoxy dermatitis requires meticulous attention to avoiding skin contact during mixing and application, use of protective clothing such as PVC gloves, good housekeeping, regular hand washing before eating and breaks, and prohibition of eating and smoking in the work area. In some cases, sensitized workers may need to be completely removed from the work area and further exposure.[2]

There are no reports of carcinogenic, mutagenic, teratogenic, or reproductive effects to humans of uncured resins, curing agents, or glycidyl ethers, but there are some positive animal studies.[1,8] Animal experiments using DGEBA resins have generally indicated no carcinogenic activity, but the results are inconclusive.[1] Diglycidyl resorcinol ether administered by gavage to rats and mice for two years caused an increased incidence of forestomach tumors.[9] Mutagenicity tests using various liquid and solid epoxy resins have yielded some positive and some negative results.[1]

Of the aromatic amine curing agents, diaminodiphenyl sulfone is tumorigenic in animal experiments, whereas 4,4'-methylenedianiline is a suspect animal carcinogen.[1] Many of the glycidyl ethers produce a mutagenic response in the Ames assay and in some other short-term tests.[8] Glycidyl ethers are rapidly metabolized to less cytotoxic substances and rapidly conjugate with skin proteins on dermal contact. Their low volatility decreases the possibility of significant systemic absorption via inhalation. Together, these factors reduce the likelihood of conjugation with nuclear macromolecules in somatic or germ cells, which otherwise might result in carcinogenic or teratogenic effects.[8]

A threshold limit value is not established for epoxy resins.

REFERENCES

1. Acres Consulting Service: Occupational Health Survey—Worker Exposure to Epoxy Resins and Associated Substances. Niagara Falls, Ontario, Canada, Occupational Health and Safety Division, Ontario Ministry of Labor, 1984

2. Adams RM: *Occupational Skin Disease*, pp 241–250. Philadelphia, JB Lippincott, 1978

3. Birmingham DJ: Clinical observations on the cutaneous effects associated with curing epoxy resins. AMA *Arch Ind Health* 19:365–367, 1959

4. Brubaker RE et al: Evaluation and control of a respiratory exposure to 3-(dimethylamino)-propylamine. *J Occup Med* 21:688–690, 1979

5. Bastion PG: Occupational hepatitis caused by methylene-dianiline. *Med J Australia* 141:533–535, 1984

6. Howe W et al: Tetrachlorophthalic anhydride asthma: evidence for specific IgE antibody. *J Allerg Clin Immun* 71:5–11, 1983

7. Ward MJ, Davies D: Asthma due to grinding epoxy resin cured with phthallic anhydride. *Clin Allerg* 12:165–168, 1982

8. National Institute for Occupational Safety and Health: *Criteria for a Recommended Standard . . . Occupational Exposure to Glycidyl Ethers*, DHEW (NIOSH) Pub 78-166. Washington, DC, US Government Printing Office, 1978

9. Murthy ASK, McConnell EE, Huff JE, et al: Forestomach neoplasms in Fischer F344/N rats and B6C3F$_1$ mice exposed to diglycidyl

resorcinol ether—an epoxy resin. *Food Chem Toxicol* 28:723–729, 1990

ference of Governmental Industrial Hygienists (ACGIH), 1991

2. Sax NI: *Dangerous Properties of Industrial Materials*, 6th ed, p 313. Van Nostrand Reinhold, 1984

ETHANE
CAS: 74-84-0

CH_3CH_3

Synonyms: Bimethyl; dimethyl; methylmethane; ethyl hydride

Physical Form. Colorless gas

Uses. In the production of ethylene, vinyl chloride, and chlorinated hydrocarbons; component of bottled fuel gas

Exposure. Inhalation

Toxicology. Ethane is considered to be toxicologically inert and is classified as a simple asphyxiant gas.

At extremely high concentrations, ethane displaces oxygen from the air and blood.[1] The first symptoms of oxygen deprivation are rapid respirations and air hunger, followed by diminished mental alertness and impaired muscular coordination.[2] Emotional instability may ensue, and fatigue occurs rapidly. In severe cases, there may be nausea and vomiting, prostration, loss of consciousness, and finally, convulsions, coma, and death.[2] Early symptoms may develop when the asphyxiant exceeds a concentration of 33% in the mixture of air and gas; marked symptoms occur when concentrations approach 50%, and 75% is fatal in a matter of minutes.

Atmospheres deficient in oxygen do not provide adequate warning of hazardous concentrations, and ethane itself is odorless.[1] Particular concern exists over the explosive hazard of ethane and, for this reason, exposure limits of 1000 ppm have been recommended.

References

1. Ethane. *Documentation of the TLVs and BEIs,* 6th ed, p 559. Cincinnati, OH, American Con-

ETHANOLAMINE
CAS: 141-43-5

$NH_2CH_2CH_2OH$

Synonyms: 2-Aminoethanol; 2-hydroxyethylamine; ethylolamine; colamine

Physical Form. Liquid

Uses. To remove carbon dioxide and hydrogen sulfide from natural gas and other gases; synthesis of surface active agents; softening of hides

Exposure. Inhalation

Toxicology. Ethanolamine is an eye and respiratory tract irritant.

No systemic effects from industrial exposure have been reported. The liquid applied to the human skin for 1.5 hours caused marked erythema.[1]

Dogs and cats exposed to 990 ppm for 4 days survived, but 4 of 6 guinea pigs died from exposure to 233 ppm for 1 hour; pathologic changes were chiefly those of pulmonary irritation, with some nonspecific changes in the liver and kidneys.[1] In animals exposed repeatedly to 66 to 100 ppm, there was some mortality during the 24 to 30 days of exposure, and all animals were lethargic.[2] No mortality or pathology resulted from 90-day continuous exposure of dogs to 26 ppm, of rats to 12 ppm, or of guinea pigs to 15 ppm.[2] The liquid produced moderate irritation on the skin of rabbits and severe irritation in the eyes of rabbits.[1]

Ethanolamine administered by gavage to pregnant rats on days 6 to 15 of gestation at levels of 50, 300, or 500 mg/kg/day caused dose-dependent increases in intrauterine deaths, malformations, and intrauterine growth retardation.[3] Sex of the pups and intrauterine position with respect to contiguous rat siblings were important factors in the degree of development exhibited.

The odor is described as ammonialike or musty at 25 ppm but is detected by means of a sensation at 3 ppm.[2]

The 1995 ACGIH threshold limit value–time-weighted average for ethanolamine is 3 ppm (7.5 mg/m³) with a short-term excursion level (STEL) of 6 ppm (15 mg/m³).

REFERENCES

1. Beard RR, Noe JT: Aliphatic and alicyclic amines. In Clayton GD, Clayton FE (eds): *Patty's Industrial Hygiene and Toxicology*, 3rd ed, rev, Vol 2B, *Toxicology*, p 3168. New York, Wiley-Interscience, 1981
2. Weeks MH et al: The effects of continuous exposure of animals to ethanolamine vapor. *Am Ind Hyg Assoc J* 21:374–381, 1960
3. Mankes RF: Studies on the embryopathic effects of ethanolamine in Long-Evans rats: preferential embryopathy in pups contiguous with male siblings in utero. *Terat Carcin Mutagen* 6:403–417, 1986

2-ETHOXYETHANOL
CAS: 110-80-5

$C_2H_5OCH_2CH_2OH$

Synonyms: Ethylene glycol monoethyl ether; Cellosolve; EGEE; 2-EE

Physical Form. Colorless liquid

Uses. Photoresist in the semiconductor industry; solvent for nitrocellulose lacquers and alkyd resins; in dyeing textiles and leather; in cleaners and varnish removers

Exposure. Inhalation; skin absorption

Toxicology. 2-Ethoxyethanol (EE) is of low acute toxicity, but repeated or chronic exposures have caused hematotoxic, fetotoxic, teratogenic, and testicular effects in experimental animals.

In mice, the LC_{50} for 7 hours was 1820 ppm; death was attributed to pulmonary edema and kidney injury.[1] Dogs repeatedly exposed to 840 ppm for 12 weeks developed a slight decrease in red cells and hemoglobin and an increase in immature white cells. In female rats exposed to 125 ppm for 4 hours, there was an increase in erythrocyte osmotic fragility, an effect that has been noted from other glycol ethers and in other species as well.[2]

Teratology studies in rats and rabbits have demonstrated both embryo/fetotoxicity and congenital malformations following exposure by oral, inhalation, or dermal routes. Exposures of pregnant rabbits at 160 ppm resulted in significant increases in cardiovascular, renal, and ventral body wall defects, minor skeletal changes, and fetal resorptions, with minimal maternal toxicity. Similarly, exposure of pregnant rats at 200 ppm resulted in fetal growth suppression and an increase in cardiovascular defects and wavy ribs, in the absence of significant maternal toxicity.[3] Dermal exposure of pregnant rats led to increased fetal resorptions, cardiovascular malformations, and skeletal variations.[3] A no-effect level of 10 to 50 ppm for reproductive effects in animals has been observed.[3]

In the drinking water of mice, continuously housed as breeding pairs, 0.5% had no effect on fertility, but 1% significantly reduced the numbers of litters produced.[4] Cross-breeding studies showed that the fertility of each sex was severely reduced at 2%, and reduced substantially at 1%.

In rats, a single 3-hour exposure to 4500 ppm caused testicular atrophy, as did 500 ppm for 11 days, whereas 250 ppm had no testicular effect.[5,6] Oral doses of 300 mg/kg/day for 6 weeks reduced testicular weight and spermatid counts, and some effects were detected at doses of 150 mg/kg/day in mated rats.[7]

Reproductive and testicular effects have also been reported in humans with exposure to EE, but the significance of these studies cannot be evaluated because of concomitant exposures, population bias, and uncertainty of exposure levels. An evaluation of 73 painters exposed to 9.9 mg/m^3 EE (range 0–80.5 mg/m^3) found an increased prevalence of oligospermia and azoospermia, and an increased odds ratio for lower sperm count.[8] In another report of workers exposed to EE (0–24 ppm) in a metal castings process, no effect was found on semen volume, sperm viability, motility, velocity, or morphology; some differences in the proportion of abnormal sperm shapes were observed.[9] In a recent case-control study, ethoxyacetic acid, the primary metabolite of EE and its acetate, was detected in 39 out of 1019 infertile men versus 6 out of 475 normal fertile controls (odds ratio 3.11).[10] The presence of ethoxyacetic acid in the urine was strongly associated with exposure to solvents, especially paint products.

The liquid instilled in the eyes of animals caused immediate discomfort, some conjunctival irritation, and a slight transitory irritation of the cornea, which was readily reversible.[11] Repeated and prolonged contact of the liquid with the skin of rabbits caused only a mild irritation, but toxic amounts were readily absorbed through the skin.

Because EE is well absorbed through the skin, ambient monitoring of environmental exposure level is not considered an accurate method of determining absorbed dose. Biological monitoring of the ethoxyethanol metabolite 2-ethoxyacetic acid in urine has been shown to be an effective indicator of absorbed dose in workers.[12]

The 1995 ACGIH threshold limit value–time-weighted average (TLV-TWA) for 2-ethoxyethanol is 5 ppm (18 mg/m^3) with a notation for skin absorption.

REFERENCES

1. Browning E: *Toxicity and Metabolism of Industrial Solvents*, pp 601–605. Amsterdam, Elsevier, 1965
2. Carpenter CP et al: The toxicity of butyl Cellosolve solvent. *AMA Arch Ind Health* 14:114–131, 1956
3. Hardin BD: Reproductive toxicity of the glycol ethers. *Toxicology* 27:91–102, 1983
4. Lamb JC, Gulati DK, Russell VS, et al: Reproductive toxicity of ethylene glycol monoethyl ether tested by continuous breeding of CD-1 mice. *Environ Health Perspect* 57:85–90, 1984
5. Doe JE: Further studies on the toxicology of the glycol ethers with emphasis on rapid screening and hazard assessment. *Environ Health Perspect* 57:199–206, 1984
6. Foster PMD, Creasy DM, Foster JR, et al: Testicular toxicity of ethylene glycol monomethyl and monoethyl ethers in the rat. *Toxicol Appl Pharmacol* 69:385–399, 1983
7. Hurtt ME, Zenick H: Decreasing epididymal sperm reserves enchances the detection of ethoxyethanol-induced spermatotoxicity. *Fund Appl Toxicol* 7:348–353, 1986
8. Welch LS, Schrader SM, Turner TW, et al: Effects of exposure to ethylene glycol ethers on shipyard painters; II. Male reproduction. *Am J Ind Med* 14:509–526, 1988
9. Ratcliffe JM, Schrader SM, Clapp DE, et al: Semen quality in workers exposed to 2-ethoxyethanol. *Br J Ind Med* 46:399–406, 1989
10. Veulemans H, Steeno O, Masschelein R, et al: Exposure to ethylene glycol ethers and spermatogenic disorders in man: a case control study. *Br J Ind Med* 50:71–78, 1993
11. Rowe VK: Derivatives of glycols. In Patty FA (ed): *Industrial Hygiene and Toxicology*, 2nd ed, Vol 2, *Toxicology*, pp 1547–1550. New York, Interscience, 1963
12. Sohnlein B, Letzel S, Weltle D, et al: Occupational chronic exposure to organic solvents. XIV. Examinations concerning the evaluation limit value for 2-ethoxyethanol and 2-ethoxyethyl acetate and the genotoxic effects of these glycol ethers. *Int Arch Occup Environ Health* 64:479–484, 1993

2-ETHOXYETHYL ACETATE
CAS: 111-15-9

$C_2H_5OCH_2CH_2OOCCH_3$

Synonyms: Cellosolve acetate; ethylene glycol monoethyl ether acetate; EGEEA; 2-ethoxy ethanol acetate; 2-EEA; EEA; EEAc; ethyl glycol acetate

Physical Form. Colorless liquid

Uses. In the coatings industry, especially in the semiconductor industry; solvent for nitrocellulose and some resins

Exposure. Inhalation; skin absorption

Toxicology. 2-Ethyoxyethyl acetate (EEA) is irritating to the eyes, nose, and throat; in animal studies, it is myelotoxic, spermatoxic, and teratogenic. At high concentrations, EEA causes central nervous system depression.

Guinea pigs survived exposure to saturated vapor concentrations (4000 ppm), but two such exposures of cats for 4 to 6 hours caused narcosis, kidney damage, and death.[1] Exposure for 8 hours to 1500 ppm was fatal to 2 of 6 rats. Mice, guinea pigs, and a rabbit survived twelve 8-hour exposures to 450 ppm, but another rabbit and two cats died before the end of the exposure period; kidney damage was observed at autopsy.[2] Dogs survived 120 daily exposures to 600 ppm with slight eye and nose irritation but without apparent systemic injury as determined by histopathology and hematological tests.[3]

A number of developmental toxicity studies have been conducted on EEA.[4-7] In rabbits, inhalation exposure to 100 to 300 ppm resulted in maternal toxicity, including clinical signs and alterations in hematology (reduced hemoglobin).[4] Developmental toxicity was seen as an increased incidence of totally resorbed litters above 200 ppm and an increase in nonviable fetuses at 300 ppm; fetal ossification was observed above 100 ppm; and the incidence of total malformations was 100% at 300 ppm. Similar effects were observed in rats, with maternal developmental toxicity at 100 to 300 ppm and teratogenic effects at 200 to 300 ppm.

In another experiment, exposure of rats to 600 ppm on days 7 to 15 of gestation caused 100% fetal resorptions, 390 ppm caused skeletal and cardiovascular defects, and one cardiac malformation occurred at 130 ppm.[5]

Mice given oral doses of 500 mg/kg/day for 5 weeks had testicular atrophy; both red and white blood cell formations were also affected at this level.[6]

Because EEA is well absorbed through the skin, ambient monitoring of environmental exposure levels is not considered an accurate method of determining absorbed dose. Biological monitoring of the EEA metabolite 2-ethoxyacetic acid in urine has been shown to be an effective indicator of absorbed dose in workers.[8] Cytogenic examinaton of persons exposed to EEA did not show an increase in sister chromatid exchanges or in micronulei.[8]

The 1995 ACGIH threshold limit value–time-weighted average (TLV-TWA) for 2-ethoxyethyl acetate is 5 ppm (27 mg/m³) with a notation for skin absorption.

REFERENCES

1. Rowe VK, Wolf MA: Derivatives of glycols. In Clayton GD, Clayton GE (eds): *Patty's Industrial Hygiene and Toxicology*, 3rd ed, Vol 2C, *Toxicology*, pp 4024–4026. New York, Wiley-Interscience, 1982
2. Lehmann KB, Flury F: *Toxicology and Hygiene of Industrial Solvents*, p 289. Baltimore, MD, Williams and Wilkins, 1943
3. Smyth HF Jr: Improved communication—hygienic standards for daily inhalation. *Am Ind Hyg Assoc Q* 17:129–184, 1956
4. Tyl RW, Pritts IM, France KA, et al: Developmental toxicity evaluation of inhaled 2-ethoxyethanol acetate in Fischer 344 rats and New Zealand White rabbits. *Fund Appl Toxicol* 10:20–39, 1988
5. Nelson BK, Setzer JV, Brightwell WS, et al: Comparative inhalation teratogenicity of four glycol ether solvents and an amino derivative in rats. *Environ Health Perspect* 57:261–271, 1984
6. Nagano K, Nakayama E, Oobayashi H, et al: Experimental studies on toxicity of ethylene

glycol ethers in Japan. *Environ Health Perspect* 57:75–84, 1984

7. Doe JE: Ethylene glycol monoethyl ether and ethylene glycol monoethyl ether acetate teratology studies. *Environ Health Perspect* 57:33–41, 1984

8. Sohnlein B, Letzel S, Welte D, et al: Occupational chronic exposure to organic solvents. *Int Arch Occup Environ Health* 64:479–484, 1993

ETHYL ACETATE
CAS 141-78-6

CH₃COOC₂H₅

$CH_3COOC_2H_5$

Synonyms: Acetic ether; ethyl ester; ethyl ethanoate

Physical Form. Colorless liquid

Uses. Lacquer solvent; artificial fruit essences

Exposure. Inhalation

Toxicology. Ethyl acetate causes respiratory tract irritation; at very high concentrations, it produces narcosis in animals, and it is expected that severe exposure will cause the same effect in humans.

Unacclimated human subjects exposed to 400 ppm for 3 to 5 minutes experienced nose and throat irritation.[1] However, no adverse symptoms were observed in workmen exposed at 375 to 1500 ppm for several months.[2] In rare instances, exposure may cause sensitization resulting in inflammation of the mucous membranes and in eczematous eruptions.[3]

Cats exposed to 9000 ppm for 8 hours suffered irritation and labored breathing; 20,000 ppm for 45 minutes caused deep narcosis, whereas 43,000 ppm for 14 to 16 minutes was fatal; at autopsy, findings were pulmonary edema with hemorrhage and hyperemia of the respiratory tract.[3] Repeated exposure of rabbits to 4450 ppm resulted in secondary anemia with leucocytosis, hyperemia, and damage to the liver.[3]

Ethyl acetate has a fruity odor detectable at 10 ppm.[4]

The 1995 ACGIH threshold limit value–time-weighted average for ethyl acetate is 400 pm (1440 mg/m³).

REFERENCES

1. Nelson KW et al: Sensory response to certain industrial solvent vapors. *J Ind Hyg Toxicol* 25:282–285, 1943

2. Patty FA: Potential exposures in industry. In Patty FA (ed): *Industrial Hygiene and Toxicology*, 2nd ed, Vol 2, *Toxicology*, p 2278, New York, Interscience, 1963

3. von Oettingen WF: The aliphatic acids and their esters: toxicity and potential dangers. *AMA Arch Ind Health* 21:28–65, 1960

4. Hygienic Guide Series: Ethyl acetate. *Am Ind Hyg Assoc J* 26:201–203, 1964

ETHYL ACRYLATE
CAS: 140-88-5

C₅H₈O₂

$C_5H_8O_2$

Synonyms: Ethyl 2-propenoate;2-propenoic acid ethyl ester; acrylic acid ethyl ester

Physical Form. Colorless liquid

Uses. A monomer widely used in the production of polymers and copolymers for manufacturing textiles, latex paints, paper coatings, dirt release agents, and specialty plastics

Exposure. Inhalation

Toxicology. Ethyl acrylate is an irritant of the skin, eyes, respiratory tract, and mucous membranes of the gastrointestinal tract; it has a history of dermal sensitization. It is carcinogenic in experimental animals.

The vapor is moderately irritating at 4

ppm, and it was believed that workers would not tolerate 25 ppm for any length of time.[1] However, in another report, prolonged exposure to 50 to 75 ppm is said to have produced drowsiness, headache, and nausea.[2] Skin sensitization has occurred from industrial exposure; a 4% concentration in petrolatum produced sensitization reactions in 10 of 24 volunteers.[3]

In rats, 2000 ppm for 4 hours was fatal, with death attributed to severe pulmonary irritation; 1000 ppm for 4 hours was not fatal but caused irritation of the skin.[4] Repeated exposure to 500 ppm was fatal to rats, and 275 ppm was lethal to rabbits and guinea pigs.[5] Irritation of the eyes, nose, and mouth, as well as lethargy, dyspnea, and convulsive movements, preceded death. At autopsy, there were pulmonary edema and degenerative changes in liver, kidneys, and heart muscle. The epidermis and dermis are the primary target tissues when the liquid is applied to the skin.

Gavage administration of a single dose causes profound gastric toxicity, which includes concentration and time-dependent mucosal and submucosal edema and vacuolization of the forestomach.[6] These studies suggest that ethyl acrylate is an acutely irritating chemical, causing lesions in tissues directly exposed to it.[7]

Chronic exposure of mice and rats to 25, 75, or 225 ppm caused concentration-dependent lesions within the nasal cavity.[8] There was no indication of an oncogenic response in any organ or tissue.[8]

Ethyl acrylate applied to the skin of mice 3 times per week for life caused dermatitis, dermal fibrosis, epidermal necrosis, and hyperkeratosis; neoplastic changes were not observed.[9]

Ethyl acrylate was carcinogenic in rats and mice when administered by gavage in corn oil, producing squamous-cell carcinomas of the forestomach.[10] There were also dose-related increases in the incidences of nonneoplastic lesions, including hyperkeratosis, hyperplasia, and inflammation. In a follow-up to this study, rats administered 200 mg/kg/day by gavage for 6 months had an increase in forestomach epithelial hyperplasia, which was reversible.[11] In contrast, animals treated for 12 months on the same dosing regime developed forestomach squamous-cell carcinomas and papillomas. The

authors conclude that ethyl acrylate carcinogenesis is a consequence of promotion of spontaneously initiated cells.

The IARC has determined that there is sufficient evidence for carcinogenicity in experimental animals based on the gavage studies.[2]

In a recent cohort study, an excess mortality from cancer of the colon and rectum was observed in a group of men employed extensively in the early 1940s in jobs entailing the highest exposures to vapor-phase ethyl acrylate and methyl methacrylate.[13] The excess mortality appeared in those with the equivalent of three years' exposure and after a latency period of 20 years. Two cohorts with later dates of hire did not show excess mortality. The authors acknowledge the possibility of confounding exposures in the first cohort and further suggest that the role of ethyl acrylate in inducing tumors is not supported by any known biological mechanisms. Specifically, there is no evidence that ethyl acrylate can cause carcinogenesis at distant sites.

Ethyl acrylate was negative in most genotoxic assays.[11]

Exposure of pregnant rats to 150 ppm, 6 hr/day, during days 6 through 15 of gestation caused some maternal toxicity and a slight, but not statistically significant, increase in malformed fetuses; at 50 ppm, there was neither maternal toxicity nor an adverse effect on the fetus.[13]

One drop of the liquid instilled in the eye of the rabbit caused corneal necrosis within 24 hours.[4]

The odor is detectable below 1 ppm and should serve as a good warning property.[1,4]

The 1995 ACGIH threshold limit value–time-weighted average for ethyl acrylate is 5 ppm (20 mg/m^3) with a short-term excursion limit of 15 ppm (61 mg/m^3) and an A2 suspected human carcinogen designation.

REFERENCES

1. Hygienic Guide Series: Ethyl acrylate. *Am Ind Hyg Assoc J* 27:571–574, 1966.
2. *IARC Monographs on the Evaluation of the Carcinogenic Risk of Chemicals to Humans*, Vol 39,

Some chemicals used in plastics and elastomers, pp 81–98. Lyon, International Agency for Research on Cancer, 1986

3. Opdyke DLJ: Monographs on fragrance raw materials, ethyl acrylate. *Food Cosmet Toxicol* (Suppl) 13:801–802, 1975

4. Pozzani UC, Weil CS, Carpenter CP: Subacute vapor toxicity and range-finding data for ethyl acrylate. *J Ind Hyg Toxicol* 31:311–316, 1949

5. Treon JR et al: The toxicity of methyl and ethyl acrylate. *J Ind Hyg* 31:317–326, 1949

6. Ghanayem BI, Maronpot RR, and Matthews HB: Ethyl acrylate-induced gastric toxicity: I. Effect of single and repetitive dosing. *Toxicol Appl Pharmacol* 80:323–335, 1985

7. Ghanayem BI, Burka LT, and Matthews HB: Ethyl acrylate distribution, macromolecular binding, excretion and metabolism in male Fischer 344 rats. *Fund Appl Toxicol* 9: 389–397, 1987

8. Miller RR, Young JT, Kociba RJ, et al: Chronic toxicity and oncogenicity bioassay of inhaled ethyl acrylate in Fischer 344 rats and B6C3F$_1$ mice. *Drug Chem Toxicol* 8:1–42, 1985

9. DePass LR, Fowler EH, Meckley DR, et al: Dermal oncogenicity bioassays of acrylic acid, ethyl acrylate and butyl acrylate. *J Toxicol Environ Health* 14:115–120, 1984

10. National Toxicology Program: *Carcinogenesis Studies of Ethyl Acrylate (CAS No 140-88-5) in F344/N Rats and B6C3F$_1$ Mice (Gavage Studies)*. Technical Report Series 259, DHHS (NIH) Pub No 87-2515, pp 1–224. Research Triangle Park, NC, National Institutes of Health, 1986

11. Ghanayem BI, Sanchez IM, Maronpot RR, et al: Relationship between the time of sustained ethyl acrylate forestomach hyperplasia and carcinogenicity. *Environ Health Perspect* 101(S5):277–280, 1993

12. Walker AM, Cohen AJ, Loughlin JE, et al: Mortality from cancer of the colon or rectum among workers exposed to ethyl acrylate and methyl methacrylate. *Scand J Work Environ Health* 17:7–19, 1991

13. Murray JS et al: Teratological evaluation of inhaled ethyl acrylate in rats. *Toxicol Appl Pharmacol* 60:106–111, 1981

ETHYL ALCOHOL
CAS: 64-17-5

C_2H_5OH

Synonyms: Ethanol; anhydrol; ethyl hydrate; ethyl hydroxide; grain alcohol

Physical Form. Clear, colorless, mobile, flammable liquid

Uses. Solvent

Exposure. Inhalation; ingestion

Toxicology. Ethyl alcohol is an irritant of the eyes and mucous membranes and causes central nervous system depression; chronic, excessive ingestion is associated with developmental effects and various cancers.

Few adverse effects have been reported in humans from dermal or inhalation exposures in industrial settings.[1] Exposure of humans at 5000 to 10,000 ppm has caused transient irritation of the eyes and nose, and cough.[1,2] At 15,000 ppm, effects were continuous lacrimation and cough. A level of 20,000 ppm was judged as just tolerable; above this level, the atmosphere was described as intolerable and suffocating on even brief exposure.[2]

Chronic exposure to the vapor may result in irritation of mucous membranes, headache, and symptoms of central nervous system depression, such as lack of concentration and somnolence.[3] However, in current industrial practice, the vapor is considered to be practically devoid of systemic hazard from inhalation.

Ethanol is not appreciably irritating to skin, even with repeated or prolonged exposure.[1] Splashed in the eye, the liquid causes immediate burning and stinging sensations, with reflex closure of the lids and tearing.[4]

Intoxication from ingestion is related to the blood alcohol levels: at 0.05% to 0.15%, there are slight impairment of visual acuity, muscular incoordination, and changes in reaction time, mood, and personality; at 0.15% to 0.30%, there are slurred speech, slowed reaction time, and increasing muscular incoordination; at blood levels approaching 0.50%, there is severe intoxication, with blurred or double vision, stupor, nausea, coma, and respiratory depression; death can occur from respiratory or circulatory failure.[1] Ethyl alcohol consumption can also increase the metabolism and, sometimes, the toxicity, of other chemicals. Chronic ingestion can cause damage to the liver, including fatty infiltration, necrosis, fibrosis, and cirrhosis.

Ethyl alcohol is a developmental toxin in humans. Excessive consumption is associated with fetal alcohol syndrome, which is characterized by joint, limb, and cardiac anomalies and behavioral and cognitive impairment.[1,5]

According to the IARC, sufficient evidence of carcinogenicity for alcoholic beverages has been established in humans.[5] Epidemiological studies clearly indicate that consumption of alcoholic beverages is causally related to cancers of the oral cavity, pharynx, larynx, esophagus, and liver. Both positive and negative results have been reported for the genotoxicity of ethyl alcohol.[5] Increased frequencies of chromosomal aberrations, sister chromatid exchanges, and aneuploides were observed in the peripheral lymphocytes of alcoholics.

The 1995 ACGIH threshold limit value–time-weighted average for ethyl alcohol is 1000 ppm (1880 mg/m³).

REFERENCES

1. Lington AW, Bevan C: Alcohols. In Clayton GD, Clayton FE (eds): *Patty's Industrial Hygiene and Toxicology*, 4th ed, Vol 2D, *Toxicology*, pp 2616–2622. New York, Wiley-Interscience, 1994
2. Lester D, Greenberg LA: The inhalation of ethyl alcohol by man. *Q J Stud Alcohol* 12:167–168, 1951
3. Hygienic Guide Series: Ethyl alcohol (ethanol). *Am Ind Hyg Assoc Q* 17:94–95, 1956
4. Grant WM: *Toxicology of the Eye*, 3rd ed, pp 53–59. Springfield, IL, Charles C Thomas, 1986
5. *IARC Monographs on the Evaluation of Carcinogenic Risks to Humans*, Vol 44. Alcohol drinking. Lyon, International Agency for Research on Cancer, World Health Organization, pp 416, 1988

ETHYLAMINE
CAS: 75-04-7

$C_2H_5NH_2$

Synonyms: Monoethylamine; aminoethane; ethanamine

Physical Form. Liquid

Uses. In resin chemistry; stabilizer for rubber latex; intermediate for dyestuffs and pharmaceuticals; in oil refining

Exposure. Inhalation

Toxicology. Ethylamine is an irritant of the eyes, mucous membranes, and skin.

Eye irritation and corneal edema in humans have been reported from industrial exposure.[1]

Exposure of rats to 8000 ppm for 4 hours was fatal to 2 of 6 animals within 14 days.[2] Rabbits survived exposures to 50 ppm daily for 6 weeks but showed pulmonary irritation and some myocardial degeneration; corneal damage was observed 2 weeks after exposure.[3] In the rabbit eye, 1 drop of a 70% solution of ethylamine caused immediate, severe irritation. A 70% solution dropped on the skin of guinea pigs caused prompt skin burns, leading to necrosis; when held in contact with guinea pig skin for 24 hours, there was severe skin irritation with extensive necrosis and deep scarring.[1]

The odor is like that of ammonia. The 1995 ACGIH threshold limit value–time-weighted average (TLV-TWA) for ethylamine is 5 ppm (9.2 mg/m³) with a short-term excursion limit of 15 ppm (27.6 mg/m³) and a notation for skin absorption.

REFERENCES

1. Ethylamine. *Documentation of the TLVs and BEIs*, 6th ed, p 577. Cincinnati, OH, American

Conference of Governmental Industrial Hygienists (ACGIH), 1991

2. Smyth HF Jr et al: Range-finding toxicity data: List V. *AMA Arch Ind Hyg Occup Med* 10:61–68, 1954

3. Brieger H, Hodes WA: Toxic effects of exposure to vapors of aliphatic amines. *AMA Arch Ind Hyg Occup Med* 3:287–291, 1951

ETHYL AMYL KETONE
CAS: 541-85-5

$C_8H_{16}O$

Synonyms: 5-Methyl-3-heptanone; ethyl *sec*-amyl ketone; EAK

Physical Form.   Colorless liquid

Uses.   Solvent for resins; organic intermediate

Exposure.   Inhalation

Toxicology.   Ethyl amyl ketone is an irritant of the eyes and mucous membranes; at very high concentrations, it produces central nervous system depression in animals, and it is expected that severe exposure will cause the same effect in humans.

Humans exposed to 25 ppm experienced irritation of the eyes and respiratory tract and detected a strong odor; at 100 ppm, irritation of mucous membranes, headache, and nausea were too severe to tolerate for more than a few minutes.[1] Eye contact with the liquid causes transient corneal injury. Prolonged or repeated cutaneous contact may lead to drying and cracking of the skin.

Three of six mice and no rats died after a 4-hour exposure to 3000 ppm, whereas exposure to 6000 ppm for 8 hours caused death in all exposed mice and in four of six rats; all animals developed signs of eye and respiratory tract irritation; varying degrees of ataxia, prostration, respiratory distress, and narcosis were ob-

served.[1] Surviving animals recovered with no apparent adverse effects.

The 1995 threshold limit value–time-weighted average (TLV-TWA) for ethyl amyl ketone is 25 ppm (131 mg/m³).

REFERENCES

1. Krasavage WJ et al: Ketones. In Clayton GD, Clayton FE (eds): *Patty's Industrial Hygiene and Toxicology*, 3rd ed, Vol 2C, *Toxicology*, pp 4767–4768. New York, Wiley-Interscience, 1982

ETHYL BENZENE
CAS: 100-41-4

C_8H_{10}

Synonyms: Ethylbenzol; phenylethane

Physical Form.   Colorless liquid

Uses.   Primarily in the production of styrene; industrial solvent; constituent of asphalt and naptha; in fuels

Exposure.   Inhalation; skin absorption

Toxicology.   Ethyl benzene is an irritant of the skin, eyes, and mucous membranes; at high concentrations, it causes neurological and respiratory depression.

Humans exposed briefly to 1000 ppm experienced eye irritation, but tolerance developed rapidly; 2000 ppm caused lacrimation, nasal irritation, and vertigo; 5000 ppm produced intolerable irritation of the eyes and nose.[1]

When chronic exposures exceed 100 ppm, complaints include fatigue, sleepiness, headache, and mild irritation of the eyes and respiratory tract.[2]

The rate of absorption of ethyl benzene through the skin of the hand and forearm in human subjects was 22 to 33 mg/cm²/hr, indicating that skin absorption could be a major route of uptake of liquid ethyl benzene.[3,4]

In guinea pigs, exposure to 10,000 ppm caused immediate, intense eye and nose irritation, ataxia, narcosis, and death in 2 to 3 hours; 5000 ppm was lethal during or after 8 hours of exposure; 2000 ppm produced ataxia in 8 hours, and 1000 ppm caused eye irritation.[1]

Inhalation of ethyl benzene at 600 ppm for 186 days by rats and guinea pigs resulted in slight changes in liver and kidney weights and slight testicular histopathology in rabbits and monkeys.[5] Exposure of rabbits to 230 ppm, 4 hr/day for 7 months, resulted in changes in blood cholinesterase activity, leukocytosis, reticulocytosis, and dystrophic changes in the liver and kidneys.[6]

Exposure to 782 ppm for 4 weeks caused an increase in platelet counts in male rats and an increase in total leukocyte count in female rats; hematological parameters did not change for mice or rabbits exposed to the same or higher concentrations.[7] In spite of its chemical similarity, ethyl benzene does not appear to cause the same damage to the hematopoietic system as benzene does.[8]

In a preliminary report, rats exposed to 100, 250, 500, 750, or 1000 ppm for 90 days developed lung lesions and hyperplastic bronchial/mediastinal lymph nodes.[9] These effects were not seen in mice similarly exposed.

In an extremely limited study of rats administered 500 mg/kg/day by gavage, 5 days/week for 2 years, there was an increased incidence in the number of total tumors (31/77 vs. 23/94 for controls).[10,11] There was no increase in any specific type of tumor, and only one dose level was used.[11]

Pregnant rats exposed at 100 or 1000 ppm, 6 hr/day on days 1 to 19 of gestation, had offspring with significant increase in extra rib formation; at the higher dose, maternal toxicity was indicated by increased liver, kidney, and spleen weights.[12] Rabbits similarly exposed on days 1 to 24 of gestation had significantly fewer live pups per litter at both exposure levels.[12]

In another study, rats exposed at 150, 300, or 600 ppm for 24 hours per day, from days 7 to 15 of pregnancy, had mild maternal toxicity at all exposure levels, increased incidences of

skeletal retardation, and decreased fetal weights at the highest concentration.[11] The incidence of extra ribs, and of litters with internal malformations, increased at the highest concentration.

Ethyl benzene is not mutagenic in most test systems, but it has caused a mutagenic effect in mouse lymphoma cells and has induced a marginal yet significant increase in sister chromatid exchanges in human lymphocytes at toxic doses.[4]

Two drops of the liquid in the eyes of a rabbit caused slight conjunctival irritation but no corneal injury.[5] The liquid in contact with the skin of a rabbit caused erythema, exfoliation, and vesiculation.[5]

The 1995 threshold limit value–time-weighted average (TLV-TWA) for ethyl benzene is 100 ppm (434 mg/m³) with a short-term excursion limit of 125 ppm (543 mg/m³).

REFERENCES

1. Yant WP, Schrenk HH, Waite CP, Patty FA: Acute response of guinea pigs to vapors of some new commercial organic compounds. *Public Health Rept* 45:1241–1250, 1930
2. Bardodej Z, Bardodejova E: Biotransformation of ethylbenzene, styrene and alpha-methylstyrene in man. *Am Ind Hyg Assoc J* 31:206–209, 1970
3. Dutkiewicz T, Tyras H: A study of the skin absorption of ethylbenzene in man. *Br J Ind Med* 24:330–332, 1967
4. Agency for Toxic Substances and Disease Registry (ASTDR): *Toxicological Profile for Ethylbenzene.* US Department of Health and Human Services. TP-90-15, December 1990
5. Wolf MA et al: Toxicological studies of certain alkylated benzenes and benzene. *AMA Arch Ind Health* 14:387–398, 1956
6. Haley TJ: A review of the literature on ethylbenzene. *Dangerous Properties Industrial Materials Report*, pp 2–4. July/August 1981
7. Cragg ST, Clarke EA, Daly IW, et al: Subchronic inhalation toxicity of ethylbenzene in mice, rats, and rabbits. *Fund Appl Toxicol* 13:399–408, 1989
8. Gerarde HW: Toxicological studies on hydrocarbons. *AMA Arch Ind Health* 13:468–474, 1956
9. National Toxicology Program (NTP): Sub-

chronic and Chronic Toxicity Study of Ethyl-benzene; 90-Day Subchronic Study Report on Inhalation Exposure of F344/N Rats and B6C3F$_1$ Mice. Draft, National Institute of Health, October 1988

10. Maltoni C, Conti B, Cotti G, et al: Experimental studies on benzene carcinogenicity at the Bologna Institute of Oncology: current results and ongoing research. *Am J Ind Med* 7:415, 1985

11. Ethylbenzene: European Chemical Industry. Joint Assessment of Commodity Chemicals No 7. Brussels, Belgium, August 1, 1986

12. Hardin BD et al: Testing of selected workplace chemicals for teratogenic potential. *Scand J Work Environ Health* 7(Suppl 4):66–75, 1981

ETHYL BROMIDE

CAS: 74-96-4

C_2H_5Br

Synonyms: Bromoethane; hydrobromic ether; bromic ether

Physical Form.   Colorless liquid

Uses.   Ethylating agent in synthesis of pharmaceuticals; refrigerant

Exposure.   Inhalation

Toxicology.   Ethyl bromide is a respiratory irritant and causes hepatic and renal toxicity; at high concentrations, it causes narcosis.

The former use of ethyl bromide as a human anesthetic (at concentrations approaching 100,000 ppm) produced respiratory irritation and caused some fatalities, either immediately, from respiratory or cardiac arrest, or delayed, from effects on the liver, kidneys, or heart.[1] At autopsy, findings were pulmonary edema and marked fatty degeneration of the liver, kidneys, and heart. Relatively little experience with this substance in industry has been reported, but exposure of volunteers to 6500 ppm for 5 min-utes produced vertigo, slight headache, and mild eye irritation.[1]

Guinea pigs exposed to 50,000 ppm for 98 minutes died within an hour following exposure.[2] Exposure to 24,000 ppm for 30 minutes was fatal within 3 days; at autopsy, findings were pulmonary edema and centrilobular necrosis of the liver; exposure to 3200 ppm for 9 hours produced lung irritation, and death occurred after 1 to 5 days. The 1-hour LC$_{50}$ for male rats was 27,000 ppm and, for mice, it was 16,200 ppm.[3]

In inhalation studies conducted by the National Toxicology Program, acute, subchronic, and chronic effects of ethyl bromide were examined in mice and rats.[4] All mice and 3 of 5 female rats died before the end of a 4-hour exposure to 5000 ppm; rats and mice exposed to 2000 ppm, 6 hr/day, died before the end of 14-day studies. In 14-week studies, 1600 ppm was lethal to some animals and caused compound-related lesions, including muscle atrophy and atrophy of the testis and uterus, thought to be secondary to body weight loss; rats also had minimal to moderate multifocal mineralization in the cerebellum and minimal to severe hemosiderosis of the spleen.

A variety of effects (dependent on species and sex) were seen in the 2-year studies with exposures of 100, 200, or 400 ppm, 6 hr/day, 5 days/week.[4] There was some evidence of carcinogenic activity of ethyl bromide for male F344/N rats, as indicated by increased incidences of pheochromocytomas of the adrenal gland (control, 8/40; 100 ppm, 23/45; 200 ppm, 18/46; 400 ppm, 21/46); neoplasms of the brain and lung may also have been related to exposure to ethyl bromide. For female F344/N rats, there was equivocal evidence of carcinogenic activity, as indicated by marginally increased incidences of neoplasms of the brain and lung. In the high-dose rats, alveolar epithelial hyperplasia was increased, as were the incidences of epithelial hyperplasia and squamous metaplasia of the nasal cavity. For male B6C3F$_1$ mice, there was equivocal evidence of carcinogenic activity, based on marginally increased incidences of neoplasms of the lung. There was clear evidence of carcinogenic activity for female B6C3F$_1$ mice, as indicated by neoplasms of the uterus.

Ethyl bromide was mutagenic in *Salmonella* assays with and without microsomal activation when tested in an enclosed system; it also induced sister chromatid exchange in Chinese hamster ovary cells.[4]

Applied to the skin of mice, the liquid produced local necrosis.[5] Prolonged or repeated contact of ethyl bromide to the skin may lead to significant absorption of the compound. Instilled in rabbit eyes, it was an irritant.

The etherlike odor of ethyl bromide is detectable only at concentrations well above 200 ppm and, therefore, will not give warning of hazardous concentrations.[5]

The 1995 threshold limit value–time-weighted average (TLV-TWA) for ethyl bromide is 5 ppm (22 mg/m³) with a skin notation and an A2 suspected human carcinogen designation.

REFERENCES

1. von Oettingen WF: *The Halogenated Aliphatic, Olefinic, Cyclic, Aromatic, and Aliphatic-Aromatic Hydrocarbons Including the Halogenated Insecticides, Their Toxicity and Potential Dangers.* US Public Health Service Pub No 414, pp 134–138. Washington, DC, US Government Printing Office, 1955
2. Sayers RR, Yant WP, Thomas BGH, Berger LB: Physiological response attending exposure to vapors of methyl bromide, methyl chloride, ethyl bromide and ethyl chloride. *Pub Health Bull* 185:1–56, 1929
3. Vernot EH et al. Acute toxicity and skin corrosion data for some organic and inorganic compounds and aqueous solutions. *Toxicol Appl Pharmacol* 42:412–417, 1977
4. National Toxicology Program: *Toxicology and Carcinogenesis Studies of Bromoethane (Ethyl Bromide) (CAS No. 74-96-4) in F344/N Rats and B6C3F Mice (Inhalation Studies).* NTP-TR 363, NIH Pub No 90-2818. US Department of Health and Human Services, pp 1–186, 1989
5. Hygienic Guide Series: Ethyl bromide. *Am Ind Hyg Assoc* 26:192–195, 1978

ETHYL BUTYL KETONE
CAS: 106-35-4

$C_7H_{14}O$

Synonyms: 3-Heptanone; EBK

Physical Form. Colorless liquid

Uses. Solvent and intermediate for organic materials

Exposure. Inhalation

Toxicology. Ethyl butyl ketone (EBK) is mildly irritating to the skin and eyes of animals and causes narcosis at high concentrations.

No adverse effects have been reported in humans.

Rats survived a 4-hour exposure to 2000 ppm, but 4000 ppm for 4 hours was fatal.[1] The oral LD_{50} in rats was 2.76 g/kg, and the LD_{50} for penetration of rabbit skin was greater than 20 ml/kg.[1]

Rats given 1.0% EBK in drinking water for 120 days showed no signs of neurotoxicity.[2] Exposure of rats at 700 ppm, 72 hr/week for 24 weeks, was also without neurotoxic effect.[3] Extremely large gavage doses, 2 g/kg/day, 5 days/week for 14 weeks, were required to produce signs of neurotoxicity; 2 of 2 rats had hind-limb weakness and tail drag.[4] Neuropathology showed central-peripheral-distal axonopathy, characterized by giant axonal swelling and neurofilamentous hyperplasia.[4]

Dropped into rabbit eyes or applied to skin, the liquid has caused mild irritation.[1]

The 1995 ACGIH threshold limit value–time-weighted average (TLV-TWA) for ethyl butyl ketone is 50 ppm (234 mg/m³).

REFERENCES

1. Smyth HF Jr, Carpenter CP, Weil CS: Range-finding toxicity data. List III. *J Ind Hyg Toxicol* 31:60–62, 1949
2. Homan ER, Maronpot RR: Neurotoxic evalua-

tion of some aliphatic ketones. *Toxicol Appl Pharmacol* 45:312 (Abstract), 1978

3. Katz GV et al: Comparative neurotoxicity and metabolism of ethyl *n*-butyl ketone and methyl *n*-butyl ketone in rats. *Toxicol Appl Pharmacol* 52:153–158, 1980
4. O'Donoghue JL et al: Further studies on ketone neurotoxicity and interactions. *Toxicol Appl Pharmacol* 72:201–209, 1984

ETHYL CHLORIDE
CAS 75-00-3

CH₃CH₂Cl

CH_3CH_2Cl

Synonyms: Chloroethane; monochloroethane; hydrochloric ether

Physical Form. Colorless gas

Uses. Blowing agent in foamed plastics; in the production of tetraethyl lead; formerly an inhalation anesthetic agent

Exposure. Inhalation; skin absorption

Toxicology. Ethyl chloride at high concentrations causes central nervous system depression.

In the past, concentrations of 40,000 ppm were used clinically to produce anesthesia.[1] Sudden and unforeseen fatalities from ethyl chloride anesthesia have been reported. Concentrations of 20,000 ppm or above have reportedly caused increased respiratory rate, cardiac depression, dizziness, eye irritation, and abdominal cramps.[1] Exposure to 19,000 ppm resulted in mild analgesia after 12 minutes, and 13,000 ppm caused slight symptoms of inebriation.[2]

Chronic effects from industrial exposure have not been reported although skin absorption is said to occur. In liquid form, this substance may cause frostbite.

Guinea pigs exposed to 40,000 ppm appeared uncoordinated in 3 minutes, had eye irritation, and were unable to stand after 40 minutes; some animals died from exposure for 9 hours, but exposure for 4.5 hours was nonfatal; histopathological changes in the lungs, liver, and kidneys were observed in sacrificed animals of the latter group.[3]

Two-week repeated exposure of rats and dogs to 4000 or 10,000 ppm caused no treatment-related effects except for slight increases in liver to body weight ratios in male rats.[4] Similarly, the only observed effect in mice exposed for 11 days, 23 hr/day, to up to 5000 ppm was an increase in relative liver weight and a slight increase in hepatocellular vacuolation.[5] Neurobehavioral observation, clinical chemistry, hematology studies, and necropsy failed to show other effects, indicating that ethyl chloride was well tolerated despite the unusually long exposure periods.

Histopathological examination of reproductive organs showed no evidence of toxicity in rats and dogs exposed at 10,000 ppm for 2 weeks or rats and mice exposed at 19,000 ppm for 13 weeks.[4,6]

Pregnant mice exposed to 5000 ppm, 6 hr/day, on days 6 through 15 of gestation had no overt maternal toxicity; there was slight delayed ossification of skull bones in the offspring.[1]

In a chronic inhalation study, 86% (43/50) of female mice exposed at 15,000 ppm, 6 hr/day, for 102 weeks had highly malignant uterine carcinomas versus none (0/49) in the controls.[6] The incidence of hepatocellular carcinomas was also increased. Male mice had an increase in alveolar and bronchial adenomas, but results were confounded by poor survival. Both male and female rats had marginally significant increases in epithelial tumors and brain astrocytomas, respectively. More recent studies have suggested that the mechanism of uterine tumor induction in mice is species-specific, a high-dose phenomenon, and may be related to glutathione conjugation rather than to other metabolic pathways.[7]

The genotoxic potency of ethyl chloride appears to be low. It was negative in in vivo micronucleus tests, but it has produced both positive and negative results in bacterial gene mutation assays.[1]

The proposed 1995 ACGIH threshold

limit value–time-weighted average for ethyl chloride is 100 ppm (264 mg/m³) with an A3 animal carcinogen designation and a notation for skin absorption.

REFERENCES

1. Agency for Toxic Substances and Disease Registry (ATSDR): *Toxicological Profile for Chloroethane*. US Public Health Service, US Environmental Protection Agency, p 89, 1989
2. von Oettingen WF: *The Halogenated Aliphatic, Olefinic, Cyclic, Aromatic, Aliphatic-Aromatic Hydrocarbons Including the Halogenated Insecticides, Their Toxicity and Potential Dangers*. US Public Service Pub No 414, pp 128–134. Washington, DC, US Government Printing Office, 1955
3. Sayers RR, Yant WP, Thomas BH, Burger LB: Physiological response attending exposure to vapors of methyl bromide, methyl chloride, ethyl bromide and ethyl chloride. *Pub Health Bull* 185:1–56, 1929
4. Landry TD, Ayres JA, Johnson KA, et al: Ethyl chloride: a two-week inhalation toxicity study and effects on liver non-protein sulfhydryl concentrations. *Fund Appl Toxicol* 2:230–234, 1982
5. Landry TD, Johnson KA, Phillips JE, et al: Ethyl chloride: 11-day continuous exposure inhalation toxicity study in B6C3F₁ Mice. *Fund Appl Toxicol* 13:516–522, 1989
6. National Toxicology Program (NTP): *Technical Report on the Toxicology and Carcinogenesis Studies of Chloroethane in F344/N Rats and B6C3F₁ Mice*. NIH Pub No 89-2801, 1989
7. Fedtke N, Certa H, Ebert R, et al: Species differences in biotransformation of ethyl chloride. II. GSH-dependent metabolism. *Arch Toxicol* 68:217–223, 1994

ETHYLENE
CAS: 74-85-1

C_2H_4

Synonyms: Ethene; acetene; bicarburetted hydrogen; olefiant gas

Physical Form. Colorless gas

Uses. Chemical intermediate in the manufacture of polyethylene, ethylene oxide, ethylene dichloride, and ethyl benzene; fruit- and vegetable-ripening agent

Exposure. Inhalation

Toxicology. Ethylene is of low toxicity and has traditionally been regarded as a simple asphyxiant.

Concentrations of less than 2.5% are physiologically inert; at very high concentrations, there may be narcosis, unconsciousness, and asphyxia due to oxygen displacement.[1] Humans exposed to as much as 50% ethylene in air in which the oxygen is decreased to 10% may experience unconsciousness, and death may occur at 8% oxygen. Exposure to 37% for 15 minutes may result in memory disturbances.[1]

Ethylene inhaled at 11.5 g/m³ (10,000 ppm) for 4 hours was hepatotoxic in rats pretreated with the polychlorinated biphenyl Arochlor 1254, given orally at a dose of 300 μm/kg daily for 3 days to induce liver enzymes. It is not toxic without such treatment.[2,3]

Rats exposed to 300, 1000, or 3000 ppm, 6 hr/day 5 days/week, for up to 2 years showed no statistically significant evidence of chronic toxicity or oncogenic effects.[4] (A subsequent review of the same data by other investigators found that the incidence of mononuclear cell leukemia was somewhat increased in both sexes at the highest dose level.[5]) Metabolic studies in rats and mice indicate that ethylene may be metabolized to ethylene oxide, an agent with genotoxic and carcinogenic potential.[2,5] Ethylene was not genotoxic in *Salmonella typhimurium* nor did it induce micronuclei in the bone marrow of rats or mice.[2]

The IARC has determined that there is inadequate evidence of carcinogenicity of ethylene in humans and experimental animals.[2]

Ethylene is not irritating to the skin and eyes.[1] The gas has a faintly sweet odor that probably does not provide adequate warning of hazardous concentrations. Because of the highly flammable and explosive characteristics of ethylene, it should be handled cautiously.[1]

The ACGIH regards ethylene as a simple asphyxiant and does not recommend threshold limit values because the limiting factor is available oxygen. The minimal oxygen content should be 18% by volume under normal atmospheric pressure.

REFERENCES

1. Sandmeyer EE: Aliphatic hydrocarbons. In Clayton GD, Clayton FE (eds): *Patty's Industrial Hygiene and Toxicology*, 3rd ed, Vol 2B, *Toxicology*, pp 3198–3199, Wiley-Interscience, 1981
2. *IARC Monographs on the Evaluation of Carcinogenic Risks to Humans*, Vol 60, pp 45–71, Some industrial chemicals, Lyon, International Agency for Research on Cancer, 1994.
3. Connolly RB, Jaeger RJ: Acute hepatotoxicity of ethylene and halogenated ethylenes after PCB pretreatment. *Environ Health Perspect* 21:131, 1977
4. Hamm TE Jr, Guest D, Dent JG: Chronic toxicity and oncogenicity bioassay of inhaled ethylene in Fischer-344 rats. *Fund Appl Toxicol* 4:473–478, 1984
5. Rostron C: Ethylene metabolism and carcinogenicity. *Food Chem Toxic* 24:70, 1985

ETHYLENE CHLOROHYDRIN
CAS: 107-07-3

ClCH$_2$CH$_2$OH

Synonyms: β-Chloroethyl alcohol; glycol chlorohydrin; 2-chloroethanol

Physical Form.   Colorless liquid

Uses.   Production of ethylene glycol and ethylene oxide; solvent for cellulose acetate, cellulose ethers, and various resins

Exposure.   Inhalation; skin absorption

Toxicology.   Ethylene chlorohydrin is an irritant of the skin and eyes and is toxic to the liver, kidneys, cardiovascular system, and central nervous system.

Several human fatalities have resulted from inhalation, dermal contact, or ingestion of ethylene chlorohydrin. Typically, neurotoxic symptoms were described, and death was attributed to cardiac and respiratory collapse.[1] One fatality was caused by exposure to an estimated 300 ppm for 2.25 hours.[2] In another fatal case, autopsy showed pulmonary edema and damage to the liver, kidneys, and brain.[2]

Exposure to the vapor has caused irritation of the eyes, nose, and throat; visual disturbances; vertigo, incoordination, and paresthesias; and nausea and vomiting.[2,3] More severe exposure has also caused headache, severe thirst, delirium, low blood pressure, cyanosis, collapse, shock, and coma. In some cases, there have been albumin, casts, and red blood cells in the urine.

Ethylene chlorohydrin is highly irritating to mucous membranes but produces little reaction on contact with rabbit skin.[1] Toxic amounts can be absorbed through the skin without causing dermal irritation; the dermal LD$_{50}$ for rabbits is 68 mg/kg.[4] This value extrapolated to humans suggests that a volume slightly more than a teaspoon could be lethal with prolonged contact.[4]

The liquid instilled in rabbit eyes caused moderately severe injury, but human eyes have recovered from corneal burns within 48 hours.[5]

Inhalation exposures of 15 minutes a day at concentrations of approximately 1000 ppm were fatal to rats within a few days.[6]

Significant levels of fetotoxicity and maternal toxicity but no teratogenicity were found in rabbits administered 36 mg/kg/day intravenously.[7]

In 2-year dermal studies, there was no evidence of carcinogenicity in rats given 50 or 100 mg/kg/day or mice given 7.5 or 15 mg per animal per day.[1] Increased risks for pancreatic cancer and lymphopoietic cancers associated with a chlorohydrin plant that primarily produced ethylene chlorohydrin have been attributed to by-products of the process, including ethylene dichloride; ethylene chlorohydrin itself has not been associated with the occurrence of tumors.[8]

Skin contact is particularly hazardous because the absence of signs of immediate irritation prevents any warning when the skin is wetted by the substance.[3]

The 1995 ACGIH ceiling–threshold limit value (C-TLV) for ethylene chlorohydrin is 1 ppm (3.3 mg/m^3) with a notation for skin absorption.

REFERENCES

1. National Toxicology Program: *Toxicology and Carcinogenesis Studies of 2-Chloroethanol (Ethylene Chlorohydrin) (CAS No 107-07-3) in F344/ N Rats and Swiss CD-1 Mice (Dermal Studies).* DHHS (NTP) TR-275, p 194. Washington, DC, US Government Printing Office, November, 1985
2. Bush AF, Abrams HK, Brown HV: Fatality and illness caused by ethylene chlorohydrin in an agricultural occupation. *J Ind Hyg Toxicol* 26:352–358, 1949
3. Hygienic Guide Series: Ethylene chlorohydrin. *Am Ind Hyg Assoc J* 22:513–515, 1961
4. Lawrence W et al: Toxicity of ethylene chlorohydrin. I. Acute toxicity studies. *J Pharm Sci* 60:568–571, 1971
5. Grant WM: *Toxicology of the Eye*, 2nd ed, pp 266–267. Springfield, IL, Charles C Thomas, 1974
6. Goldblatt M, Chiesman W: Toxic effects of ethylene chlorohydrin. Part I. Clinical. *Br J Ind Med* 1:207–223, 1944
7. Research Triangle Institute: Teratologic evaluation of ethylene chlorohydrin (CAS No 107-07-3) in New Zealand white rabbits. Final report. Washington, DC, National Institute of Environmental Health Sciences, 1983
8. Benson LO, Teta MJ: Mortality due to pancreatic and lymphopoietic cancers in chlorohydrin production workers. *Br J Ind Med* 50:710–716, 1993

ETHYLENEDIAMINE
CAS: 107-15-3

NH$_2$CH$_2$CH$_2$NH$_2$

Synonyms: 1,2-Diaminomethane; EDA

Physical Form. Colorless liquid

Uses. Catalytic agent in epoxy resins; dyes; solvent stabilizer; neutralizer in rubber products

Exposure. Inhalation

Toxicology. Ethylenediamine is a potent skin sensitizer and is an irritant of the eyes and mucous membranes.

In human subjects, inhalation of 400 ppm for 5 to 10 seconds caused intolerable nasal irritation; 200 ppm caused tingling of the face and slight nasal irritation; 100 ppm was inoffensive.[1]

Most of the information regarding the skin sensitization potential of ethylenediamine has come from its use as a stabilizer in pharmaceutical preparations, especially in Mycolog cream, where it has reportedly caused many cases of sensitization.[2,3] Results of skin-patch tests, conducted between 1972 and 1974, showed that 6% of the 3216 patients tested exhibited sensitivity to a 1% ethylenediamine-HCL solution.[4] Although ethylenediamine is a potent sensitizer, industrial exposure rarely leads to sensitization and dermatitis because exposure is not prolonged or intimate and because normal skin is usually involved.[2] In clinical practice, the ethylenediamine in Mycolog cream is often applied to damaged skin, which is more readily sensitized than the relatively normal skin of most industrial workers.[2] A follow-up study of 16 patients who had exhibited a strong contact allergy to ethylenediamine in 1974 or 1975 showed that, in 25% of the cases, the sensitivity had disappeared after a period of 10 years in which the allergen had been avoided.[5]

In a case of asthma resulting from ethylenediamine exposure in a 30-year-old male, initial symptoms of sneezing, nasal discharge, and productive cough began 2.5 years after employment and progressed during the following 5 months. An inhalation provocation test with ethylenediamine produced tightness in the chest, cough, wheezing, and a 26% reduction in FEV$_1$ 4 hours post-exposure. The reaction was reproducible on a different day and was specific; a similar reaction was not demonstrated with other chemicals to which the subject was exposed.[6]

A recent study of EDA-sensitized workers

(as determined by EDA-associated rhinitis, coughing, and wheezing) in an industrial population suggested that smoking may decrease the latency between first exposure to EDA and onset of respiratory symptoms.[7]

Exposure of rats to 4000 ppm for 8 hours was uniformly fatal, whereas 2000 ppm was not lethal.[8] Rats exposed daily for 30 days to 484 ppm did not survive; injury to lungs, liver, and kidneys was observed; at 132 ppm, there was no mortality.[1]

No reproductive toxicity was found in rats exposed to 0.50 g/kg/day for two generations.[9] A reduction in body weight gain and changes in liver and kidney weights were observed in the F_0 and F_1 parent rats. A microscopic liver lesion occurred in the F_1 rats, with greater prevalence and severity in the females. Ethylenediamine was not genotoxic in a variety of in vivo and in vitro tests, nor was it carcinogenic in lifetime skin-painting studies in mice.[10]

In the eye of a rabbit, the liquid caused extreme irritation and corneal damage; partial corneal opacity was produced by a 5% solution.[8] The undiluted liquid applied to the shaved skin of rabbits and left uncovered produced severe irritation and necrosis.[8]

The 1995 ACGIH threshold limit value–time-weighted average (TLV-TWA) for ethylenediamine is 10 ppm (25 mg/m³) with a notation for skin absorption.

REFERENCES

1. Pozzani UC, Carpenter CP: Response of rats to repeated inhalation of ethylenediamine vapors. *AMA Arch Ind Hyg Occup Med* 9:223–226, 1954
2. Fisher AA: *Contact Dermatitis*, 2nd ed, pp 40–41. Philadelphia, Lea & Febiger, 1973
3. Baer R, Ramsey DL, Biondi E: The most common contact allergens. *Arch Dermatol* 108:74–78, 1973
4. North American Contact Dermatitis Group: The frequency of contact sensitivity in North America, 1972–74. *Contact Derm* 1:277–280, 1975
5. Nielsen M, Jorgensen J: Persistence of contact sensitivity to ethylenediamine. *Contact Derm* 16:275–276, 1987
6. Lam S, Chan-Yeung M: Ethylenediamine-induced asthma. *Am Rev Resp Dis* 121:151–155, 1980
7. Aldrich FD, Strange AW, Geesaman RE: Smoking and ethylenediamine sensitization in an industrial population. *J Occ Med* 29:311–314, 1987
8. Smyth HF Jr et al: Range-finding toxicity data: List IV. *AMA Arch Ind Hyg Occup Med* 4:119–122, 1951
9. Yang RSH et al: Two-generation reproduction study of ethylenediamine in Fischer 344 rats. *Fund Appl Toxicol* 4:539–546, 1984
10. Slesinski RS et al: Assessment of genotoxic potential of ethylenediamine: in vitro and in vivo studies. *Mut Res* 124:299–314, 1983

ETHYLENE DIBROMIDE
CAS: 106-93-4

$C_2H_4Br_2$

Synonyms: 1,2 Dibromoethane; EDB

Physical Form.   Clear liquid

Uses.   Fumigant (now banned for soil and grain use); in gasoline as a lead scavenger; chemical intermediate in the industrial synthesis of other brominated compounds

Exposure.   Inhalation; skin absorption

Toxicology.   Ethylene dibromide (EDB) is a severe mucous membrane, eye, and skin irritant. It is a testicular toxicant and causes liver and kidney damage; it is carcinogenic in experimental animals.

In an early report, accidental use of ethylene dibromide as a human anesthetic produced general weakness, vomiting, diarrhea, chest pains, coughing, shortness of breath, cardiac insufficiency, and uterine hemorrhaging.[1] Death occurred 44 hours after inhalation. Postmortem examination showed upper respiratory tract irritation, swelling of the pulmonary lymph glands, advanced states of parenchymatous degeneration of the heart, liver, and kidneys, and hemorrhages in the respiratory tract.

Two workers collapsed while inside a tank that was later found to contain a 0.1% to 0.3% EDB solution.[2] Removed after 20 to 45 minutes in the tank, one man was intermittently comatose, while the other was delirious and combative. Both experienced vomiting, diarrhea, abdominal pain, and burning of the eyes and throat. Metabolic acidosis and acute renal and hepatic failure ensued. Death occurred 12 and 64 hours later, respectively, despite supportive measures.

Skin contact produces intensive burning pain preceding hyperemia, which develops into blisters.[1] Skin sensitization has been reported.[1]

Acute exposure of experimental animals resulted in adverse effects similar to those described for humans. Rats did not survive when exposed to the vapor for longer than 6 minutes at 3000 ppm; minimum lethal concentration for an 8-hour exposure was 200 ppm; these exposures caused hepatic necrosis, pulmonary edema, and cloudy swelling of renal tubules.[3] Depression of the CNS was observed in rats exposed at higher concentrations, and deaths occurred within 24 hours from respiratory or cardiac failure. At lower concentrations, deaths due to pneumonia occurred as a result of injury to the lungs and were delayed for up to 12 days post-exposure.

Four species of animals tolerated daily inhalation of 25 ppm for 6 months without adverse effects.[3]

Application of a 10% solution or the undiluted liquid to rabbit skin caused marked CNS depression and death within 24 hours.[3] A dermal LD_{50} of 400 mg/kg was estimated.[1]

An increased incidence of skin carcinomas and lung tumors has been found in mice receiving repeated skin applications.[4] Rats and mice chronically exposed to 10 or 40 ppm had increased incidence of tumors at multiple sites.[5] Animal studies have shown increased toxic and carcinogenic effects when EDB is administered with Disulfiram, a widely used drug in alcoholism-control programs.[6]

Human epidemiologic studies to observe carcinogenic effects are inconclusive because of small cohort size, incomplete exposure data, and insufficient latencies.[7]

EDB is toxic to the male reproductive system in several species. Testicular atrophy was seen in rats and mice with chronic gavage administration of 41 mg/kg/day or 107 mg/kg/day, respectively.[8] Abnormal spermatozoa and decreased spermatozoic concentration occurred in bulls fed EDB.[9]

Intraperitoneal injection of 10 mg/kg for 5 days to male rats caused a decrease in average litter size in females mated after 3 weeks of exposure and no litters at 4 weeks.[10] Continuous exposure to 32 ppm during gestation caused minor skeletal anomalies in rats and mice.[11]

Adverse reproductive effects have also been reported in humans. Fumigators chronically exposed to EDB showed statistically significant decreases in sperm count and percentages of viable and motile sperm and increases in sperm with specific abnormalities as compared to controls.[12] Decreases in sperm velocity and semen volume have been reported in another group of fumigators who were exposed to EDB seasonally.[13] No adverse effects were found on sperm counts of 50 workers exposed to less than 5.0 ppm.[14]

Ethylene dibromide is a potent mutagen, producing a broad spectrum of mutations in a variety of in vivo and in vitro assays.[15]

The ACGIH has designated ethylene dibromide as an A2 suspected human carcinogen, with no assigned threshold limit value.

REFERENCES

1. National Institute for Occupational Safety and Health. *Criteria for a Recommended Standard . . . Occupational Exposure to Ethylene Dibromide.* DHEW (NIOSH) Pub No 77-221. ab Washington, DC, US Government Printing Office, 1977
2. Letz GA, Pond SM, Osterloh JD, et al: Two fatalities after acute occupational exposure to ethylene dibromide. *JAMA* 252:2428–2431, 1984
3. Rowe VK, Spencer HC, McCollister DD, et al: Toxicity of ethylene dibromide determined on experimental animals. *AMA Arch Ind Hyg Occup Med* 6:158–173, 1952

4. Van Duuren BL, Goldschmidt BM, Loewengart G, et al: Carcinogenicity of halogenated olefinic and aliphatic hydrocarbons in mice. *J Natl Cancer Inst* 63:1433–1439, 1979

5. National Cancer Institute: *Carcinogenesis Bioassay of 1,2-Dibromoethane (Inhalation Study)*, TR-210 (CAS No 106-93-4), Carcinogenesis Testing Program. DHHS (NIH) Pub No 81-1766. Washington, DC, US Government Printing Office, 1981

6. Wong LCK, Winston JM, Hong CB, et al: Carcinogenicity and toxicity of 1,2-dibromoethane in the rat. *Toxic Appl Pharmacol* 63:155–165, 1982

7. Ott MG, Scharnweber HC, Langner RR: Mortality experience of 161 employees exposed to ethylene dibromide in two production units. *Br J Ind Med* 37:163–168, 1980

8. National Cancer Institute: *Bioassay of 1,2-Dibromomethane for Possible Carcinogenicity*. Bethesda, MD, NTIS No PB 288428, 1978

9. Amir D, Colcani R: Effect of dietary ethylene dibromide on bull semen. *Nature* 206:99–100, 1965

10. Edwards K, Jackson H, Jones A: Studies with alkylating esters—II. A chemical interpretation through metabolic studies of the infertility effects of ethylene dimethanesulphonate and ethylene dibromide. *Biochem Pharmacol* 19:1783–1789, 1970

11. Short RD Jr, Minor JL, Ferguson B, et al: *Toxicity Studies of Selected Chemicals, Task I—The Developmental Toxicity of Ethylene Dibromide Inhaled by Rats and Mice During Organogenesis*. Report No EPA-560/6-76-018. Washington, DC, US Environmental Protection Agency, Office of Toxic Substances, 1976

12. Ratcliffe JM, Schrader SM, Steenland K, et al: Semen quality in papaya workers with long term exposure to ethylene dibromide. *Br J Ind Med* 44:317–326, 1987

13. Schrader SM, Turner TW, Ratcliffe JM: The effects of ethylene bromide on semen quality: a comparison of short-term and chronic exposure. *Reproduct Toxicol* 2:191–198, 1988

14. Ter Haar G: An investigation of possible sterility and health effects from exposure to ethylene dibromide. In *Banbury Report 5—Ethylene Dichloride: A Potential Health Risk?* pp 167–188. Cold Spring Harbor, NY, Cold Spring Harbor Laboratory, 1980

15. Agency for Toxic Substances and Disease Registry (ATSDR): *Toxicological Profile for 1,2-Dibromomethane*. US Department of Health and Human Services, Public Health Service, p 148, TP-91/13, 1992

ETHYLENE DICHLORIDE
CAS: 107-06-2

$C_2H_4Cl_2$

Synonyms: 1,2-Dichloroethane; dichloroethane; ethylene chloride

Physical Form. Colorless liquid

Uses. Manufacture of vinyl chloride; antiknock agent; fumigant, insecticide; degreaser compounds; rubber cements

Exposure. Inhalation; ingestion

Toxicology. Ethylene dichloride is a central nervous system depressant and causes injury to the liver and kidneys; in chronic gavage studies, it is carcinogenic to experimental animals.

In one fatality, exposure to concentrated vapor in a tank for 30 minutes caused drowsiness, nausea, and respiratory distress; coma developed 20 hours after initial exposure.[1] Serum levels of lactate and ammonia were increased, followed by elevation of glutamic transaminases, lactic dehydrogenase, and creatine phosphokinase. Ornithine carbamyl transferase and glutamic oxaloacetic transaminase of mitochondrial origin were remarkably high. Multiple organ failure developed, and the patient died in cardiac arrhythmia on the fifth day. At autopsy, the lungs were severely congested and edematous; diffuse degenerative changes of the myocardium, extensive centrilobular necrosis of the liver, and acute tubular necrosis of the kidneys were noted.

Workers exposed to 10 to 200 ppm complained of lacrimation, dizziness, insomnia, vomiting, constipation, and anorexia; liver tenderness on palpation, epigastric pain, and elevated urobilinogen were observed.[2] Impairment of the central nervous system and increased morbidity, especially diseases of the liver and bile ducts, were found in workers chronically exposed to ethylene dichloride at concentrations below 40 ppm and averaging 10 to 15 ppm.[2]

Ingestion of quantities estimated between 8 and 200 ml have been reported to be lethal, with a toxic response similar to that of cases of inhalation.[3,4]

Interactions between ethylene dichloride and other substances have been reported in animal studies. Specifically, a combination treatment with disulfiram caused testicular atrophy (not seen with either agent alone) and lowered the ethylene dichloride dose at which liver effects occurred.[5] Increased hepatotoxicity has also been observed in some animals given phenobarbital along with ethylene dichloride.

Eye contact with either the liquid or high concentrations of vapor causes immediate discomfort, with conjunctival hyperemia and slight corneal injury; corneal burns from splashes recover quickly with no scarring. Prolonged skin exposure, as from contact with soaked clothing, produces severe irritation, moderate edema, and necrosis; systemic effects may ensue because the liquid is readily absorbed through the skin.[2]

Animal studies indicate that ethylene dichloride has little ability to adversely affect the reproductive or developmental processes except at maternally toxic levels.[4]

For intermediate duration studies, the lethal oral dose depended on the method of administration.[6,7] Administered in the drinking water for 13 weeks, 8000 ppm was relatively nontoxic to two strains of rats, causing elevated liver weights and minimal histological evidence of kidney damage in F344/N female rats. Gavage administration of 240 mg/kg in male rats and 300 mg/kg in females for 13 weeks was lethal; necrosis of the cerebellum occurred in one-third of the treated animals.

Chronic administration by gavage of 95 or 47 mg/kg/day for 78 weeks caused a significant increase in hemangiosarcomas of the circulatory system in rats.[8] Squamous-cell carcinomas of the forestomach were significantly increased in male rats, and high-dose females had in-creased incidences of mammary gland adenocarcinomas and fibroadenomas. A variety of tumors have been similarly induced in mice.[8] Intraperitoneal and inhalation studies in animals have not shown a significant carcinogenic response.[9,10]

Pronounced increases were seen for total cancer, lymphatic and hematopoietic cancers, and leukemia in a mortality study of chlorohydrin production workers.[11] The investigators attributed the excesses to ethylene dichloride exposure based on probable exposures of the workers; however, concomitant exposure to other chemicals precludes identifying the etiologic agent(s).

Ethylene dichloride has been shown to alkylate DNA, and it is genotoxic in a variety of in vivo and in vitro assays.[10]

The IARC has determined that there is sufficient evidence of carcinogenicity in mice and rats and that, in the absence of adequate human data, ethylene dichloride should be regarded as possibly carcinogenic to humans.[12]

Most subjects could detect ethylene dichloride at a concentration of 6 ppm.[1]

The 1995 ACGIH threshold limit value–time-weighted average (TLV-TWA) for ethylene dichloride is 10 ppm (40 mg/m^3).

REFERENCES

1. Nouchi T, Miura H, Kanayama M, et al: Fatal intoxication by 1,2-dichloroethane—a case report. *Int Arch Occup Environ Health* 54:111–113, 1984
2. National Institute for Occupational Safety and Health: *Criteria for a Recommended Standard . . . Occupational Exposure to Ethylene Dichloride (1,2-Dichloroethane).* DHEW (NIOSH) Pub No 76-139. Washington, DC, US Government Printing Office, 1976
3. Yodaiken RE, Babcock JR: 1,2-Dichloroethane poisoning. *Arch Environ Health* 26:281–284, 1973
4. *Health Assessment Document for 1,2-Dichloroethane (Ethylene Dichloride).* Final Report, EPA/600/8-84/006F. Washington, DC, US Environmental Protection Agency, Office of Toxic Substances, September 1985
5. Igwe OJ, Que Hee SS, Wagne WD: Interac-

tion between 1,2-dichloroethane and disul-firam. *Fund Appl Toxicol* 6:733–746, 1986

6. Morgan DL, Bucher JR, Elwell MR: Comparative toxicity of ethylene dichloride in F344/N, Sprague–Dawley and Osborne–Mendel rats. *Food Chem Toxicol* 28:839–845, 1990
7. National Toxicology Program: *NTP Report on the Toxicity Studies of 1,2-Dichloroethane in F344/N Rats, Sprague–Dawley Rats and Osborne–Mendel Rats, and B6C3F₁ Mice (Drinking Water and Gavage Studies)*. US Department of Health and Human Services, Public Health Service, National Institute of Health, Research Triangle Park, NC, 1991
8. National Cancer Institute: *Bioassay of 1,2-Dichloroethane for Possible Carcinogenicity*, TR-55. DHEW (NIH) Pub No 78-1361. Washington, DC, US Government Printing Office, 1978
9. *IARC Monographs on the Evaluation of the Carcinogenic Risk of Chemicals to Humans*, Vol 20, pp 422–448. Lyon, International Agency for Cancer Research, 1979
10. Agency for Toxic Substances and Disease and Registry (ATSDR): *Toxicological Profile for Dichloroethane*. US Department of Health and Human Services, Public Health Service, p 131, 1992
11. Benson LO, Teta MJ: Mortality due to pancreatic and lymphopoietic cancers in chlorohydrin production workers. *Br J Ind Med* 50:710–716, 1993
12. *IARC Monographs on the Evaluation of Carcinogenic Risks to Humans. Overall Evaluations of Carcinogenicity: An Updating of IARC Monographs, Vols 1–42*, p 60. Lyon, International Agency for Research on Cancer, 1987

ETHYLENE GLYCOL
CAS: 107-21-1

CH₂OHCH₂OH

Synonyms: 1,2 Dihydroxyethane; 1,2-ethanediol; ethylene alcohol; ethylene dihydrate

Physical Form. Clear, colorless liquid

Uses. Antifreeze and coolant mixtures for motor vehicles; in hydraulic fluids and heat exchangers; solvent

Exposure. Inhalation; ingestion

Toxicology. Ethylene glycol aerosol causes irritation of the upper respiratory tract; ingestion can cause central nervous system depression, severe metabolic acidosis, liver and kidney damage, and pulmonary edema.

Inhalation is not usually a hazard because the low vapor pressure precludes excessive vapor exposure. Exposure to the vapor from the liquid heated to 100°C has been reported to cause nystagmus and coma of 5- to 10-minute duration.[1] Human volunteers exposed to an aerosol of 12 ppm for 20 to 22 hr/day for 4 weeks complained of throat irritation and headache.[2] At 56 ppm, there was more pronounced irritation of the upper respiratory tract and, at 80 ppm of aerosol, the irritation and cough were intolerable.

The chief hazard from ethylene glycol is associated with ingestion of large quantities in a single dose. Several metabolites are responsible for the clinical syndrome, which can be divided into three stages.[3] During the first 12 hours, central nervous system manifestations predominate. If the intoxication is mild, the patient appears to be drunk but without the breath odor of alcohol. In more severe cases, there will be convulsions and coma. Other signs may include nystagmus, ophthalmoplegia, papilledema, depressed reflexes, and tetanic convulsions. The central nervous system manifestations are related to the aldehyde metabolites of ethylene glycol, which reach their maximum concentrations 6 to 12 hours after ingestion.

In the second stage, cardiopulmonary symptoms become prominent, consisting of mild hypertension, tachypnea, and tachycardia. Widespread capillary damage is assumed to be the primary lesion. If the patient survives the first two stages, renal complications may be expected at 24 to 72 hours post-ingestion. Albuminuria and hematuria are common findings, and oxalate crystals are excreted in the urine. Glycoaldehyde, glycolic acid, and glyoxylate are the putative agents for kidney damage.[3]

The most significant laboratory findings in ethylene glycol intoxication are severe metabolic acidosis from the accumulation of glycolate and the presence of high anion gap.[3] Low arterial pH and bicarbonate levels are often observed. Nonspecific findings are leukocytosis and increased amounts of protein in the cerebrospinal fluid. Chelation of calcium oxalate may cause hypocalcemia which, when severe enough, can lead to tetany and cardiac dysfunction.[3] The minimum lethal dose is on the order of 100 ml in adults, although much higher doses have reportedly been survived.[4]

The effects of the liquid in the eyes of rabbits are immediate signs of moderate discomfort, with mild conjunctivitis but no significant corneal damage.[5] In one human incident of a splash in the eye, there was reversible conjunctival inflammation.[6] The liquid produces no significant irritant action on the skin.

In developmental studies, maternal deaths from kidney failure occurred at 2000 mg/kg/day in rabbits and at 3000 mg/kg/day in mice.[7] Rat dams survived 5000 mg/kg/day.[8] Developmental toxicity, including teratogenicity, occurred in mice and rats at doses of 500 mg/kg/day and 1250 mg/kg/day, respectively. Maternal toxicity was not evident at these levels.[7] Rabbit fetuses did not exhibit developmental toxicity, even at doses that were maternally lethal.[8] No teratogenic effects were observed in rats and mice following inhalation or dermal exposure, suggesting that route of exposure is critical to producing fetal effects.[8]

No evidence of a carcinogenic effect was found in mice or rats administered up to 1000 mg/kg/day for 2 years or in female mice fed up to 50,000 ppm or in males fed up to 25,000 ppm ethylene glycol in the diet for 2 years.[9,10]

Ethylene glycol was found to be nonmutagenic in the *Salmonella typhimurium* assays; it did not induce sister chromatid exchanges or chromosomal aberrations in Chinese hamster ovary cells.[10]

The proposed 1995 ACGIH ceiling–threshold limit value (C-TLV) for ethylene glycol as an aerosol is 39.4 ppm (100 mg/m³) with an A4, not classifiable as a human carcinogen, designation.

Treatment. Ethyl alcohol has been used successfully to treat ethylene glycol poisoning.[11] Ethyl alcohol acts by competing for alcohol dehydrogenase, which is necessary to convert ethylene glycol to its toxic intermediates.

REFERENCES

1. Troisis FM: Chronic intoxication by ethylene glycol vapour. *Br J Ind Med* 7:65, 1950
2. Wills JH, Coulston F, Harris ES, et al: Inhalation of aerosolized ethylene glycol by man. *Clin Toxicol* 7:463, 1974
3. Linnanvuo-Laitinen M, Huttunen K: Ethylene glycol intoxication. *Clin Toxicol* 24:167–174, 1986
4. Parry MF, Wallach R: Ethylene glycol poisoning. *Am J Med* 57:143, 1974
5. McDonald TO, Roberts MD, Borgman, AR: Ocular toxicity of ethylene chlorohydrin and ethylene glycol in rabbits' eyes. *Toxicol Appl Pharmacol* 21:143, 1972
6. Sykowsky P: Ethylene glycol toxicity. *Am J Ophthal* 34:1599, 1951
7. Price CJ, Kimmell CA, Tyl RW, et al: The developmental toxicity of ethylene glycol in rats and mice. *Toxicol Appl Pharmacol* 81:113–127, 1985
8. Tyl RW, Price CJ, Marr MC, et al: Developmental toxicity evaluation of ethylene glycol by gavage in New Zealand white rabbits. *Fund Appl Toxicol* 20:402–412, 1993
9. DePass LR, Garman RH, Woodside MD, et al: Chronic toxicity and oncogenicity studies of ethylene glycol in rats and mice. *Fund Appl Toxicol* 7:566–572, 1986
10. National Toxicology Program: *Toxicology and Carcinogenesis Studies of Ethylene Glycol in B6C3F₁ Mice (Feed Studies)*. U.S. Department of Health and Human Services, Public Health Service, National Institutes of Health, NTP TR-413, NIH Pub No 91-3144, 1993
11. Curtin L, Kramer J, Wine H, et al: Complete recovery after massive ethylene glycol ingestion. *Arch Intern Med* 152:1311–1313, 1992

ETHYLENE GLYCOL DINITRATE
CAS: 628-96-6

$CH_2NO_3CH_2NO_3$

Synonyms: EGDN; nitroglycol; 1,2-dinitroethane

Physical Form. Oily liquid

Uses. As an explosive, usually mixed with nitroglycerin (NG) in the manufacture of dynamite

Exposure. Inhalation; skin absorption. Data on toxic effects are reported chiefly from industrial exposures to EGDN-NG mixed vapors

Toxicology. Ethylene glycol dinitrate (EGDN) causes vasodilatation and cardiac effects.

Intoxication results in a characteristic intense throbbing headache, presumably from cerebral vasodilatation, often associated with dizziness and nausea and, occasionally, with vomiting and abdominal pain.[1,2] More severe exposure also causes hypotension, flushing, palpitation, low levels of methemoglobinemia, delirium, and depression of the central nervous system. Aggravation of these symptoms after alcohol ingestion has been observed. On repeated exposure, a tolerance to headache develops but is usually lost after a few days without exposure. At times, persistent tachycardia, diastolic hypertension, and reduced pulse pressure have been observed. On rare occasions, a worker may have an attack of angina pectoris a few days after cessation of repeated exposures, a manifestation of cardiac ischemia. Sudden death due to unheralded cardiac arrest has also been reported under these circumstances.[3]

Volunteers exposed to the vapor of a mixture of EGDN and nitroglycerin (NG), at a combined concentration of 2 mg/m³, experienced headache and fall in blood pressure within 3 minutes of exposure; a mean concentration of 0.7 mg/m³ for 25 minutes also produced lowered blood pressure and slight headache.[4]

A mortality study of 4061 workers employed in a Scottish explosives factory and followed from 1965 to 1980 revealed an excess of deaths from acute myocardial infarction in the younger group of workers exposed to both NG and EGDN. This excess was not observed in workers considered to have been exposed to NG only.[5]

EGDN is readily absorbed through the intact skin.

The 1995 ACGIH threshold limit value–time-weighted average (TLV-TWA) for ethylene glycol dinitrate is 0.05 ppm (0.3 mg/m³) with a notation for skin absorption.

REFERENCES

1. Trainor DC, Jones RC: Headaches in explosive magazine workers. *Arch Environ Health* 12:231–234, 1966
2. Einert CE et al: Exposure to mixtures of nitroglycerin and ethylene glycol dinitrate. *Am Ind Hyg Assoc J* 24:435–447, 1963.
3. Carmichael P, Lieben J: Sudden death in explosives workers. *Arch Environ Health* 7:424–439, 1963
4. Lund RP, Haggendal J, Johnsson G: Withdrawal symptoms in workers exposed to nitroglycerin. *Br J Ind Med* 25:136–138, 1968
5. Craig R, Gillis CR, Hole DJ, et al: Sixteen year follow-up of workers in an explosives factory. *J Soc Occup Med* 35:107–110, 1985

ETHYLENE GLYCOL MONOBUTYL ETHER
CAS: 111-76-2

$C_4H_9OCH_2CH_2OH$

Synonyms: Butyl Cellosolve; 2-butoxyethanol; EGBE; EGMBE

Physical Form. Colorless liquid

Uses. Widely used solvent and cleaning agent

Exposure.   Inhalation; skin absorption

Toxicology. Ethylene glycol monobutyl ether (EGBE) is an irritant of the eyes and mucous membranes and, in animals, it is a hemolytic agent.

Exposure of humans to high concentrations (300–600 ppm) of the vapor for several hours would be expected to cause respiratory and eye irritation, narcosis, and damage to the kidney and liver.[1]

Human subjects exposed to 195 ppm for 8 hours had discomfort of the eyes, nose, and throat, although there were no objective signs of injury and no increase in erythrocytic fragility.[2] Similar symptoms occurred at 113 ppm for 4 hours.[2] No clinical signs of adverse effects or subjective complaints occurred among 7 male volunteers exposed to 20 ppm for 2 hours.[3]

The 4-hour LC_{50} values were 486 ppm for male rats and 450 ppm for female rats; toxic effects included narcosis, respiratory difficulty, and kidney damage.[4] Acute or prompt deaths are likely to be due to the narcotic effects of the substance, whereas delayed deaths are usually attributable to congested lungs and severely damaged kidneys. In a 9-day study, rats exposed to 245 ppm, 6 hr/day, had significant depression of red blood cell counts and hemoglobin with increases in nucleated erythrocytes, reticulocytes, and lymphocytes.[4] Decreased body weight gains and increased liver weights were also found. Toxic effects showed substantial reversal 14 days post-exposure. In a 90-day study, only mild hematologic alterations were observed in rats exposed to 77 ppm, 30 hr/week. Repeated gavage administration to rats of 222, 443, or 885 mg/kg/day for 90 days caused a significant dose-dependent decrease in hemoglobin concentration and red blood cell counts.[5] Secondary effects included increased spleen weights, splenic congestion, and increased hemosiderin deposition in the liver and kidneys. EGBE had no adverse effects on testes, bone marrow, thymus, or white blood cells. The mechanism for EGBE-induced red blood cell depression in rats is unknown, but acid metabolites may be involved.[3,4] There appear to be strikingly different hematologic effects among species; differences in metabolism are probably responsible.[4] It has been suggested that the hematologic effects are of lesser consequence in humans in contrast to rodents since acute exposures of 200 ppm produced no alterations in erythrocyte fragility.[2]

EGBE appears to be less hazardous than other monoalkyl ethers of ethylene glycol with regard to reproductive effects.[4] Mice treated orally with 1000 mg/kg for 5 weeks had no change in absolute or relative testis weights.[6] Exposure of pregnant rats at 100 ppm or rabbits at 200 ppm during organogenesis resulted in maternal toxicity and embryotoxicity, including decreased number of viable implantations per litter.[7] Slight fetotoxicity in the form of poorly ossified or unossified skeletal elements was also observed in rats. Teratogenic effects were not observed in either species.[7]

In continuous breeding studies in mice, EGBE affected reproductive parameters, including the number of litters per pair, the number of live pups per litter, the proportion of pups born alive, and adjusted live pup weights only at levels (1% and 2% in drinking water) that resulted in significant mortality of the dams.[8,9] In males, testes and epididymis weights were normal, as were sperm number and motility, even at generally toxic doses.[9] At the 0.5% dose level, EGBE did not significantly affect the fertility or reproductive performance of either first- or second-generation mice.[8]

Daily skin application to rabbits of 150 mg/kg as a 43.8% aqueous solution for 13 weeks caused no adverse effects.[7] The LD_{50} for rabbits was 0.45 ml/kg (0.40 g/kg) when confined to the skin for 24 hours.[1] The liquid is not significantly irritating to the skin; instilled directly into the eye, it produces pain, conjunctival irritation, and transient corneal injury.[1]

The 1995 ACGIH threshold limit value–time-weighted average (TLV-TWA) for ethylene glycol monobutyl ether is 25 ppm (121 mg/m^3) with a notation for skin absorption.

REFERENCES

1. Rowe VK, Wolf MA: Derivatives of glycols. In Clayton GD, Clayton FE (eds): *Patty's Industrial Hygiene and Toxicology*, 3rd ed, rev, Vol 2C,

Toxicology, pp 3931–3030. New York, Wiley-Interscience, 1982

2. Carpenter CP et al: The toxicity of Butyl Cellosolve solvent. *AMA Arch Ind Health* 14:114–131, 1956

3. Johanson G et al: Toxicokinetics of inhaled 2-butoxyethanol (ethylene glycol monobutyl ether) in man. *Scand J Work Environ Health* 12:594–602. 1986

4. Dodd DE et al: Ethylene glycol monobutyl ether: acute 9-day, and 90-day vapor inhalation studies in Fischer 344 rats. *Toxicol Appl Pharmacol* 68:405–414, 1983

5. Krasavage WJ: Subchronic oral toxicity of ethylene glycol monobutyl ether in male rats. *Fund Appl Toxicol* 6:349–355, 1986

6. Nagano K et al: Testicular atrophy of mice induced by ethylene glycol mono alkyl ether. *Japan J Ind Health* 21:29–35, 1979

7. Glycol Ethers Program Panel Research Status Report, p 5. Washington, DC, Chemical Manufacturers Association, April 22, 1985

8. National Toxicology Program: *Ethylene Glycol Monobutyl Ether: Reproduction and Fertility Assessment in CD-1 Mice When Administered in Drinking Water*. National Institute of Environmental Health Sciences, NTP-85-155, Final Report, pp 1–240. May 1985

9. Heindel JJ, Gulati DK, Russell VS, et al: Assessment of ethylene glycol monobutyl and monophenyl ether reproductive toxicity using a continuous breeding protocol in Swiss CD-1 mice. *Fund Appl Toxicol* 15:683–696, 1990

ETHYLENE OXIDE
CAS: 75-21-8

C_2H_4O

Synonyms: 1,2-Epoxyethane; oxirane; dimethylene oxide

Physical Form. Colorless gas

Uses. Sterilizing agent; fumigant; insecticide; reagent in organic chemical synthesis

Exposure. Inhalation

Toxicology. Ethylene oxide is an irritant of the eyes, respiratory tract, and skin; at high concentrations, it causes central nervous system depression; it is carcinogenic in female mice.

In humans, the early symptoms are irritation of the eyes, nose, and throat and a peculiar taste; effects that may be delayed are headache, nausea, vomiting, dyspnea, cyanosis, pulmonary edema, drowsiness, weakness, and incoordination.[1]

Contact of solutions of ethylene oxide with the skin of human volunteers caused characteristic burns; after a latent period of 1 to 5 hours, effects were edema and erythema, and progression to vesiculation, with a tendency to coalescence into blebs, and desquamation. Complete healing without treatment usually occurred within 21 days with, in some cases, residual brown pigmentation. Application of the liquid to the skin caused frostbite; three of the eight volunteers were said to have become sensitized to ethylene oxide solutions.[2] The undiluted liquid or solutions may cause severe eye irritation or damage.

Exposure of several species of animals to concentrations calculated to be greater than 1000 ppm for 2 hours caused lacrimation and nasal discharge, followed by gasping and labored breathing; corneal opacity was observed in guinea pigs. Delayed effects occurred after several days and included vomiting, diarrhea, dyspnea, pulmonary edema, paralysis of hind quarters, convulsions, and death; at autopsy, findings were degenerative changes of the lungs, liver, and kidneys.[3] The LC_{50}s for mice, dogs, and rats exposed 4 hours were 835, 960, and 1460 ppm, respectively.[4]

A number of cases of subacute sensory motor polyneuropathy have been described among sterilizing workers exposed to ethylene oxide.[5,6] Findings have included weakness with bilateral foot drop, sensory loss, loss of reflexes, and neuropathologic changes on EMG in the lower extremities. In some cases, sural nerve biopsy showed axonal degeneration. Removal from exposure resulted in resolution of symptoms in 1 to 7 months. The abnormalities have been consistent with a distal "dying-back" axonopathy with secondary demyelination similar to that seen with other peripheral neurotoxins, such as *n*-hexane.[5] An animal model of distal axonal degeneration with pathologic confirmation has been described in rats exposed to 500 ppm ethylene oxide 3 times/week for 13 weeks.[7]

Two epidemiological studies of workers exposed to ethylene oxide revealed increased rates of leukemia.[8,9] In one study, 2 cases of leukemia (0.14 expected) and 3 stomach cancers (0.4 expected) were observed. The other study found 3 cases of leukemia (0.2 expected). Because these workers had exposures to other potential carcinogens, the findings cannot be linked with certainty to ethylene oxide. The small cohort size, the small number of deaths, and uncertainties about exposure levels have also been noted.[10] A number of other studies have not found an increased rate of cancer mortality from ethylene oxide exposure. A mortality study of over 18,000 ethylene oxide workers from 14 plants producing medical supplies and foodstuffs did not find an excess of leukemia or of brain, stomach, or pancreatic cancers.[11] There was, however, an increase in non-Hodgkin's lymphoma in male workers. A recent follow-up of 1896 ethylene oxide production workers did not find an increase in mortality from leukemia, non-Hodgkin's lymphoma, or brain, pancreatic, or stomach cancers.[12]

In a chronic inhalation bioassay in rats exposed for 6 hr/day, 5 days/week, for 2 years to 100, 33, or 10 ppm ethylene oxide, there was a dose-related increased occurrence of mononuclear cell leukemia in both sexes at all concentrations. There was also an increased occurrence of primary brain tumors at 100 and 33 ppm in both sexes and, in male rats, peritoneal mesotheliomas arising from the testicular serosa at 100 and 33 ppm.[13]

The IARC has determined that there is limited evidence in humans for the carcinogenicity of ethylene oxide and sufficient evidence in experimental animals.[14]

Hospital staff exposed to ethylene oxide in sterilizing operations during pregnancy were found to have a higher frequency of spontaneous abortions (16.7%) compared, by a questionnaire study, with a control group (5.6%). Analysis of a hospital discharge register confirmed the findings. The association persisted after analysis for potential confounding factors, such as age and smoking status.[15] There is animal evidence of adverse reproductive effects, including decreased fertility and reduced sperm count and motility in males and increased fetal losses and malformed fetuses in females.[16]

Ethylene oxide causes dose-related increases in the frequency of chromosomal aberrations and sister chromatid exchange in peripheral lymphocytes and micronuclei in bone marrow of exposed workers.[14] It is a potent genotoxic agent in a wide variety of procaryotic and eucaryotic assays and induces dose-related increases in the formation of adducts with DNA and hemoglobin.[10]

The 1995 ACGIH threshold limit value–time-weighted average for ethylene oxide is 1 ppm (1.8 mg/m^3) with an A2 suspected human carcinogen designation.

REFERENCES

1. Glaser ZR: *Special Occupational Hazard Review with Central Recommendations for the Use of Ethylene Oxide as a Sterilant in Medical Facilities.* DHEW (NIOSH) Pub No 77-200. Washington, DC, US Government Printing Office, 1977
2. Sexton RJ, Henson EV: Experimental ethylene oxide human skin injuries. *AMA Arch Ind Hyg Occup Med* 2:549–564, 1950
3. Hollingsworth RL et al: Toxicity of ethylene oxide determined on experimental animals. *AMA Arch Ind Health* 13:217–227, 1956
4. Jacobson KH, Hackley EB, Feinsilver L: The toxicity of inhaled ethylene oxide and propylene oxide vapors. *AMA Arch Ind Health* 13:237–244, 1956
5. Finelli PF, Morgan TF, Yaar I, et al: Ethylene oxide–induced polyneuropathy: a clinical and electrophysiologic study. *Arch Neurology* 40:419–421, 1983
6. Kuzuhara S: Ethylene oxide polyneuropathy: report of 2 cases with biopsy studies of nerve and muscle. *Clin Neurology* 22:707–713, 1982
7. Ohnishi A et al: Ethylene oxide induces central peripheral distal axonal degeneration of the lumbar primary neurones in rats. *Br J Ind Med* 42:373–379, 1985
8. Hogstedt C, Malmqvist N, Wadman B: Leukemia in workers exposed to ethylene oxide. *JAMA* 241:1132–1133, 1979
9. Hogstedt C, Aringer L, Gustavsson A: Epide-

miologic support for ethylene oxide as a cancer-causing agent. *JAMA* 255:1575–1578, 1986

10. Agency for Toxic Substances and Disease Registry (ATSDR): *Toxicological Profile for Ethylene Oxide.* US Department of Health and Human Services, Public Health Service, p 109, TP-90-16, 1990
11. Steenland K, Stayner L, Halperin W, et al: Mortality among workers exposed to ethylene oxide. *N Engl J Med* 324:1402–1407, 1991
12. Teta MJ, Benson LO, Vitale JN: Mortality study of ethylene oxide workers in chemical manufacturing: a 10 year update. *Br J Ind Med* 50:704–709, 1993
13. Snellings W, Weil C, Maronpot R: A two-year inhalation study of the carcinogenic potential of ethylene oxide in Fischer 344 rats. *Toxicol Appl Pharmacol* 75:105–117, 1984
14. *IARC Monographs on the Evaluation of the Carcinogenic Risk of Chemicals to Humans,* Vol 60, Some industrial chemicals, pp 73–145. Lyon, International Agency for Research on Cancer, 1994.
15. Hemminki K, Mutinen P, Saloniemi I, et al: Spontaneous abortion in hospital staff engaged in sterilizing instruments with chemical agents. *Br Med J* 285:1461–1463, 1982
16. Landrigan PJ, Meinhardt TJ, Gordon J, et al: Ethylene oxide: an overview of toxicologic and epidemiologic research. *Am J Ind Med* 6:103–115, 1984

ETHYLENE THIOUREA
CAS: 96-45-7

$C_3H_6N_2S$

Synonyms: ETU; imidazolidinethione; 2-imidazoline-2-thiol; 2-mercaptomidazoline

Physical Form. White crystalline solid

Uses. Accelerator in the curing of polychloroprene (Neoprene) and polyacrylate rubber; intermediate in the manufacture of antioxidants, insecticides, fungicides, dyes, pharmaceuticals, and synthetic resins

Exposure. Inhalation

Toxicology. Ethylene thiourea (ETU) is an antithyroid substance and animal carcinogen.

Clinical examination and thyroid-function tests carried out over a period of 3 years on 13 exposed workers showed one subgroup, the mixers, to have significantly lower levels of total thyroxine than other workers; one person had an appreciably raised level of thyroid-stimulating hormone and was considered to be hypothyroid.[1] There was no evidence of any clinical effect in any of the workers. Background air concentrations at the plants generally ranged up to 240 μg/m³, but levels up to 330 μg/m³ were registered on one individual's personal sampler.

Two previous studies found only slight differences in total thyroxine and triiodothyroxine in exposed workers (concentrations unspecified).

In groups of rats fed 125 or 625 ppm for up to 90 days, marked increases in serum thyroid-stimulating hormone were found.[2] The high-dose group also exhibited decreases in iodide uptake by the thyroid and in serum triiodothyronine and thyroxine levels. The majority of rats at both these exposure levels had enlarged red thyroids. Clinical signs of poisoning included excessive salivation, hair loss, and scaly skin texture by day 8 in the 625 ppm group. The no-effect level for dietary ETU in rats was considered to be 25 ppm.

In rats, ETU produced a high incidence of follicular carcinoma of the thyroid after oral administration in three studies.[3–5] Doses in one of these studies were 5, 25, 125, 250, or 500 ppm.[2] At the two highest dose levels, animals of both sexes had thyroid carcinomas, although male rats had a higher incidence. The lower dose levels produced thyroid follicular hyperplasia. ETU is believed to induce thyroid tumors through the suppression of thyroxin synthesis which, in turn, triggers hypersecretion of thyroid-stimulating hormone by the pituitary; prolonged TSH secretion may result in follicular-cell hypertrophy, hyperplasia, and adenomas and carcinomas of the thyroid.[1] In mice, repeated oral administration of the maximal tole rated dose of 215 mg/kg ETU produced liver tumors. The mechanisms by which ETU and other in vitro nongenotoxic chemicals cause liver neoplasms are not known. The IARC has determined that there is sufficient evidence for carcinogenicity to animals and inadequate evidence for humans.[6]

A recent study confirmed that ethylene thiourea was carcinogenic in male and female rats, as shown by increased incidences of thyroid follicular-cell neoplasms following treatment of up to 250 ppm in the diet for 2 years.[7] In mice, concentrations ranging from 100 to 1000 ppm for 2 years caused liver and pituitary tumors in addition to thyroid tumors. Perinatal exposure up to 8 weeks, followed by a control diet for 2 years, was not carcinogenic in rats or mice. Combined perinatal/adult ETU exposures produced the same carcinogenic effects as adult-only exposures.

ETU was a potent teratogen in rats at doses that produced no maternal toxicity or fetal deaths.[8] At doses greater than 10 mg/kg, there were neural tube closure defects, hydrocephalus, and other brain malformations, and limb defects.[8] Other anomalies observed in rats included the urogenital and ocular systems. Treatment only on gestation day 11 with ETU primarily caused hydronephrosis.[9] Doses of 80 mg/kg decreased the brain weight of rabbits. ETU was also teratogenic after oral dosing in cats and mice and after skin and inhalation exposures in rats.[9]

At present, there is no threshold limit value for ETU.[10] Suggested precautions include exclusion of women of childbearing potential from exposure, encapsulation of the powder into master batch rubber, personal respiratory protection at a level of 30 ppm and above, and local exhaust ventilation and general hygiene.[10]

REFERENCES

1. Smith DA: Ethylene thiourea: thyroid function in two groups of exposed workers. *Br J Ind Med* 41:362–366, 1984
2. Freudenthal RI et al: Dietary subacute toxicity of ethylene thiourea in the laboratory rat. *J Environ Pathol Toxicol* 1:147–161, 1977
3. Graham SL et al: Effects of one year administration of ethylene thiourea upon the thyroid of the rat. *J Agric Feed Chem* 21:324, 1973
4. Graham SL et al: Effects of prolonged ethylene thiourea ingestion on the thyroid of the rat. *Food Cosmet Toxicol* 13:493, 1975
5. Weisburger EK et al: Carcinogenicity tests of certain environmental and industrial chemicals. *J Natl Cancer Inst* 67:75, 1981
6. *IARC Monographs on the Evaluation of Carcinogenic Risks to Humans. Overall Evaluations of Carcinogenicity: An Updating of IARC Monographs Vols 1–42*, Suppl 7, pp 207–208. Lyon, International Agency for Research on Cancer, 1987
7. Chhabra RS, Eustis S, Haseman JK, et al: Comparative carcinogenicity of ethylene thiourea with or without perinatal exposure in rats and mice. *Fund Appl Toxicol* 18:405–417, 1992
8. Khera KS: Ethylene thiourea: teratogenicity study in rats and rabbits. *Teratology* 7:243, 1973
9. Daston GP, Rehnberg BF, Carver B, et al: Functional teratogens of the rat kidney. *Fund Appl Toxicol* 11:401–415, 1988.
10. Z-Imidazolidinethione. In Sax, NI: *Dangerous Properties of Industrial Materials Report*, pp 106–111, Van Nostrand Reinhold, May/June 1987

ETHYLENIMINE
CAS: 151-56-4

$(CH_2)_2NH$

Synonyms: Aziridine; ethyleneimine; azirane; azacyclopropane; dihydroazirine

Physical Form. Colorless liquid

Uses. Organic syntheses; production of polyethyleneimines used in the paper industry and as flocculation aids in the clarification of effluents

Exposure. Inhalation; skin absorption

Toxicology. Ethylenimine is a severe irritant

of the eyes, mucous membranes, and skin and causes pulmonary edema; in experimental animals, it is carcinogenic.

More than 100 cases of significant acute effects following exposure have been reported in the past 30 years, including fatalities from inhalation and skin contact.[1] The effects of overexposure are usually delayed for one-half hour to 3 hours and include nausea, vomiting, headache, dizziness, irritation of eyes and nose, laryngeal edema, bronchitis, dyspnea, pulmonary edema, and secondary bronchial pneumonia.[1,2] In experimental human studies, eye and nose irritation occurred at concentrations of 100 ppm and above.[3]

Severe corneal damage and death resulted from placing 0.005 ml of the liquid in the eyes of rabbits; severe eye burns in humans have resulted from direct contact. On the skin, the liquid is a potent irritant and vesicant, which may produce sensitization.[4]

The LC_{50} in mice was 2236 ppm for 10 minutes; signs of exposure were irritation of eyes and nose, delayed onset of pulmonary edema, and renal tubular damage with proteinuria, hematuria, and elevated blood urea nitrogen.[5] In other exposed animals, a decrease in the white blood cell count and a depression of all blood elements have also been observed.[1]

Animal studies have confirmed the carcinogenic potential of ethylenimine. In one study, rats given subcutaneous injections twice weekly for 33 weeks developed sarcomas at the injection site.[6] Ethylenimine administered to mice by gavage for 3 weeks, followed by dietary administration for 77 weeks, caused a significant increase in hepatomas and pulmonary tumors.[6]

Although animal studies have found ethylenimine to be carcinogenic, an epidemiologic study of 144 workers with up to 40 years' experience showed no evidence of carcinogenicity.[7] It should, however, be handled as a suspected human carcinogen.[2]

The odor and irritant thresholds do not provide sufficient warning of overexposure.[4] The 1995 ACGIH threshold limit value–time-weighted average (TLV-TWA) for ethylenimine is 0.5 ppm (0.88 mg/m³) with a notation for skin absorption.

REFERENCES

1. Reinhardt CF, Brittelli MR: Heterocyclic and miscellaneous nitrogen compounds. In Clayton GD, Clayton FE (eds): *Patty's Industrial Hygiene and Toxicology*, 3rd ed, rev, Vol 2A, *Toxicology*, pp 2672–2676. New York, Wiley-Interscience, 1981
2. Theiss AM et al: Aziridines. In International Labour Office: *Encyclopaedia of Occupational Health and Safety*, Vol I, pp 228–230. New York, McGraw-Hill, 1983
3. Carpenter CP, Smyth HF Jr, Shaffer CB: The acute toxicity of ethylene imine to small animals. *J Ind Hyg Toxicol* 30:2–6, 1948
4. Hygienic Guide Series: Ethyleneimine. *Am Ind Hyg Assoc J* 26:86–88, 1965
5. Silver SD, McGrath FP: A comparison of acute toxicities of ethylene imine and ammonia to mice. *J Ind Hyg Toxicol* 30:7–9, 1948
6. *IARC Monographs on the Evaluation of the Carcinogenic Risk of Chemicals to Man*, Vol 9, Some aziridines, *N*-, *S*- and *O*- mustards and selenium. pp 37–46, Lyon, International Agency for Research on Cancer, 1975
7. Ethylenimine. *Documentation of the TLVs and BEIs*, 6th ed, pp 628–630. Cincinnati, OH, American Conference of Governmental Industrial Hygienists, 1986

ETHYL ETHER
CAS: 60-29-7

$C_2H_5OC_2H_5$

Synonyms: Diethyl ether; ethoxyethane; ethyl oxide; ether; anesthesia ether; sulfuric ether

Physical Form. Colorless liquid

Uses. Solvent in the manufacture of dyes, plastics, and cellulose acetate rayon; anesthetic agent

Exposure. Inhalation

Toxicology. Ethyl ether causes eye and respiratory irritation and, at high concentrations, it produces central nervous system depression and narcosis.

Concentrations of ethyl ether ranging from 100,000 to 150,000 are required for induction of human anesthesia; however, exposure at this concentration may also produce fatalities from respiratory arrest.[1,2] Maintenance of surgical anesthesia is achieved at 50,000 ppm, and the lowest anesthetic limit is 19,000 ppm.[1] Continued inhalation of 2000 ppm in human subjects may produce dizziness; however, concentrations up to 7000 ppm have been tolerated by some workers for variable periods of time without untoward effects.[3] Initial symptoms of acute overexposure include vomiting, respiratory tract irritation, headache, and either depression or excitation. In some persons, chronic exposure results in anorexia, exhaustion, headache, drowsiness, dizziness, excitation, and psychic disturbances.[2] Albuminuria has been reported.[1] Tolerance may be acquired through repeated exposures.[2]

Ethyl ether is a mild skin irritant; repeated exposure causes drying and cracking.[2] The vapor is irritating to the eyes, and the undiluted liquid in the eyes causes painful inflammation of a transitory nature.[3] Human subjects found 200 ppm irritating to the nose but not to the eyes or throat.[4]

There is a large margin of safety between the concentration that causes nasal irritation and the concentrations that cause anesthesia, permanent damage, and death.

The 1995 ACGIH threshold limit value–time-weighted average (TLV-TWA) for ethyl ether is 400 ppm (1210 mg/m³) with a short-term excursion limit (STEL) of 500 ppm (1520 mg/m³)

REFERENCES

1. Sandmeyer EE, Kirwin CJ Jr: Ethers. In Clayton GD, Clayton FE (eds): *Patty's Industrial Hygiene and Toxicology*, 3rd ed, rev, Vol 2A. *Toxicology*, pp 2507–2511. New York, Wiley-Interscience, 1981
2. Hygienic Guide Series: Ethyl ether. *Am Ind Hyg Assoc J* 27:85–87, 1966
3. Chemical Safety Data Sheet SD-29, Ethyl Ether, pp 17–18. Washington, DC, MCA, Inc, 1965
4. Nelson KW et al: Sensory response to certain industrial solvent vapors. *J Ind Hyg Toxicol* 25:282–285, 1943

ETHYL FORMATE
CAS: 109-94-4

HCOOC₂H₅

$HCOOC_2H_5$

Synonyms: Formic ether; ethyl methanoate; ethyl formic ester

Physical Form. Clear liquid

Uses. Food flavoring; in organic syntheses; fumigant

Exposure. Inhalation

Toxicology. Ethyl formate causes irritation of the eyes and nose; at very high concentrations, it causes narcosis in animals, and it is expected that severe exposure will produce the same effect in humans.

In humans, a concentration of 330 ppm caused slight irritation of the eyes and rapidly increasing nasal irritation.[1] No chronic systemic effects have been reported in humans.[2]

Rats survived 4 hours of inhalation at 4000 ppm, but 8000 ppm was fatal to five of six animals.[3] Cats exposed to 5000 ppm for 20 minutes showed eye irritation; 10,000 ppm for 80 minutes caused narcosis, followed by death.[1] Pulmonary edema and death were observed in dogs exposed to 10,000 ppm for 4 hours.[4]

When applied to the skin of mice, ethyl formate showed no evidence of tumorigenic activity in 10 weeks.[4]

The liquid is only slightly irritating to the skin but, dropped into the eye, it causes moderate injury to the cornea.[3]

The 1995 ACGIH threshold limit value–time-weighted average (TLV-TWA) for ethyl formate is 100 ppm (303 mg/m³).

REFERENCES

1. Ethyl formate. *Documentation of the TLVs and BEIs*, 6th ed, pp 633–634. Cincinnati, OH, American Conference of Governmental Industrial Hygienists (ACGIH), 1991
2. Smyth HF Jr: Improved communication— hygienic standards for daily inhalation. *Am Ind Hyg Assoc Q* 17:129–185, 1956
3. Smyth HF Jr et al: Range-Finding Toxicity Data: List V. *AMA Arch Ind Hyg Occup Med* 10:61–68, 1954
4. Sandmeyer EE, Kirwin CJ: Esters. In *Patty's Industrial Hygiene and Toxicology*, 3rd ed, Vol 2A, *Toxicology*, pp 2263–2267. Wiley-Interscience, 1981

2-ETHYLHEXYL ACRYLATE
CAS: 103-11-7

$C_{11}H_{20}O_2$

Synonyms: EHA; 2-propenoic acid 2-ethylhexyl ester

Physical Form. Colorless liquid; commercial form contains hydroquinone (1000 ppm) or hydroquinone methyl ether (15 or 200 ppm) to prevent polymerization

Uses. As a plasticizing comonomer for the production of resins used in adhesives, latex paints, textile and leather finishes, and coatings for paper

Exposure. Inhalation; skin contact

Toxicology. By analogy to effects caused by other acrylates, 2-ethylhexyl acrylate (EHA) is expected to be an irritant of the eyes, nose, and skin.

Dermal sensitization to EHA has been documented from exposure to its presence in adhesive tape.[1] This potential has been confirmed in the guinea pig.[2]

In a lifetime dermal oncogenesis study in mice, 20 mg EHA in acetone was applied 3 times weekly for the life span of the mice.[3] There were 40 mice in the group at the start of the study. Two animals developed squamous-cell carcinomas, and 4 other animals had squamous-cell papillomas. The first tumor was observed after 11 months of treatment. None of the acetone-treated controls developed tumors. There was an apparent increase in the frequency of chronic nephritis in the EHA-treated mice—68% compared to the controls (15%). Treatment with EHA may have exacerbated the onset and development of this condition, which is normally seen in aged mice.

In another study, skin tumors, including papillomas, squamous-cell carcinomas, and malignant melanomas, were seen in mice receiving skin applications of 21% or 86.5% EHA in 25-μl acetone 3 times per week for 2 years.[4]

No skin tumors were observed in another strain of mice receiving up to 85% EHA in acetone for up to 2 years. Hyperkeratosis and hyperplasia occurred in all treated groups.[5]

The IARC has determined that there is limited evidence of carcinogenicity in experimental animals and inadequate evidence in humans for the carcinogenicity of ethylhexyl acrylate.[6]

The ACGIH has not established a threshold limit value for 2-ethylhexyl acrylate.

REFERENCES

1. Jordan WP: Cross-sensitization patterns in acrylate allergies. *Contact Derm* 1:13, 1975
2. Waegemaekers TH, Van Der Walle HB: The sensitizing potential of 20-ethylhexyl acrylate in the guinea pig. *Contact Derm* 9:372, 1983
3. DePass LR, Maronpot RR, Weil CS: Dermal oncogenicity bioassays of monofunctional and multifunctional acrylates and acrylate-based oligomers. *J Toxicol Environ Health* 16:55, 1985
4. Wenzel-Hartung RP, Brune H, Klimisch HJ: Dermal oncogenicity study of 2-ethylhexyl acrylate by epicutaneous application in male C3H/HeJ mice. *J Cancer Res Clin Oncol* 115:543–549, 1989
5. Mellert W, Kuhborth B, Gembardt C, et al: Two year carcinogenicity study in the male NMRI mouse with 2-ethylhexyl acrylate by epicutaneous administration. *Food Chem Toxicol* 32:233–237, 1994

6. *IARC Monographs on the Evaluation of Carcinogenic Risks to Humans.* Vol 60, pp 475–486. Some industrial chemicals. Lyon, International Agency for Research on Cancer, 1994

ETHYLIDENE NORBORNENE
CAS: 16219-25-3

C_9H_{12}

Synonyms: Ethylidenebicyclo(2,2,1)hept-2-ene; ENB; 2-norbornene, 5-ethylidene

Physical Form. Colorless liquid (stabilized with 100 ppm of *tert*-butyl catechol because of its reactivity with oxygen)

Uses. As the third monomer in EPDM (ethylene-propylenediene monomer)

Exposure. Inhalation

Toxicology. Ethylidene norbornene is an irritant of the eyes and mucous membranes and causes testicular atrophy in animals.

Humans exposed to 11 ppm for 30 minutes noted some irritation of the eyes and nose, and transient eye irritation at 6 ppm.[1]

The 4-hour LC_{50} values varied in different species from 730 ppm for mice to 3100 ppm for rabbits.[1]

Beagle dogs were exposed at 93 ppm for 7 hr/day, 5 days/week, for a total of 89 exposures and showed testicular atrophy, hepatic lesions, and slight blood changes.[1] Less marked effects were seen at 61 ppm, and no effects were seen at 22 ppm. Rats were exposed similarly to 237 ppm and exhibited testicular atrophy, hepatic lesions, and hydrothorax. The rat oral LD_{50} was 2527 mg/kg, and the rabbit dermal LD_{50} was 8189 mg/kg.[2]

The 1995 ACGIH ceiling–threshold limit value (C-TLV) for ethylidene norbornene is 5 ppm (25 mg/m^3).

REFERENCES

1. Kinkead ER, Pozzani UC, Geary DL, Carpenter CP: The mammalian toxicity of ethylidene norbornene. *Toxicol Appl Pharmacol* 20:250–259, 1971
2. Smyth HF Jr, Carpenter CP, Weil CS, et al: Range-finding toxicity data: List VII. *Am Ind Hyg Assoc J* 30:470–476, 1967

ETHYL MERCAPTAN
CAS: 75-08-1

C_2H_5SH

Synonyms: Ethanethiol; ethyl hydrosulfide; ethyl sulfhydrate; ethyl thioalcohol; thioethanol; thioethyl alcohol

Physical Form. Flammable gas liquefying at 34.7°C

Uses. Stenching agent for liquefied petroleum gases; adhesive stabilizer; manufacture of plastics, insecticides, and antioxidants

Exposure. Inhalation

Toxicology. Ethyl mercaptan causes irritation of mucous membranes; at high concentrations, it causes narcosis in animals, and it is expected that severe exposure will cause the same effects in humans.

At concentrations below those necessary to produce toxic effects, ethyl mercaptan is extremely malodorous, and voluntary exposure to high concentrations is unlikely to occur. Observations on humans are limited to a single brief report of exposure of workers to 4 ppm for 3 hours daily over 5 to 10 days; the workers experienced headache, nausea, fatigue, and irritation of mucous membranes.[1]

In animals, ethyl mercaptan vapor causes mucous membrane irritation and narcosis and, at near-lethal levels, by analogy to other mercaptans, it may produce pulmonary edema. It appears to be several times less acutely toxic than hydrogen sulfide or methyl mercaptan.

In rats, the LD_{50} for 4 hours was 4420 ppm; effects included irritation of mucous membranes, increased respiration, incoordination, staggering gait, weakness, partial skeletal muscle paralysis, light to severe cyanosis, and mild to heavy sedation.[2] Animals that survived single near-lethal doses by the intraperitoneal and oral routes were frequently found to have had liver and kidney damage at autopsy up to 20 days post-treatment.[2]

The liquid dropped in the eyes of rabbits caused slight to moderate irritation. Chronic inhalation exposures in rats, mice, and rabbits over 5 months showed no significant effects at 40 ppm.[1]

The odor threshold of ethyl mercaptan is less than 1 ppb.[3] The 1995 ACGIH threshold limit value–time-weighted average (TLV-TWA) for ethyl mercaptan is 0.5 ppm (1 mg/m^3).

REFERENCES

1. Ethyl mercaptan. *Documentation of the TLVs and BEIs*, 6th ed, pp 636–637. Cincinnati, OH, American Conference of Governmental Industrial Hygienists (ACGIH), 1991
2. Fairchild EJ, Stokinger HE: Toxicologic studies on organic sulfur compounds—1. Acute toxicity of some aliphatic and aromatic thiols (mercaptans). *Am Ind Hyg Assoc J* 19:171–189, 1958
3. *The Merck Index*, 11th ed, p 588. Budavari S, et al eds: Rahway, NJ, Merck and Co, Inc, 1989

N-ETHYLMORPHOLINE
CAS: 100-74-3

$C_6H_{13}NO$

Synonyms: 4-Ethylmorpholine

Physical Form. Colorless liquid

Uses. Catalyst in polyurethane foam production

Exposure. Inhalation; skin absorption

Toxicology. N-Ethylmorpholine is an irritant of the eyes and mucous membranes.

In an experimental study, humans exposed to 100 ppm for 2.5 minutes experienced irritation of eyes, nose, and throat, while 50 ppm produced less irritation.[1] Workers exposed to low vapor concentrations for several hours reported temporary fogged vision with rings around lights; corneal edema was observed.[2] Corneal edema is thought to occur when air concentrations of substituted morpholines exceed 40 ppm. The symptoms usually appear at the end of the workday and clear within 3 to 4 hours after cessation of exposure.[3]

The liquid instilled in the eye of a rabbit caused corneal haziness, with sloughing and irregularities of the surface characteristic of severe desiccation.[3] On the skin of a rabbit, the undiluted liquid produced no reaction, an effect surprisingly unlike that of unsubstituted morpholine, which is a severe skin irritant.[4]

The 1995 ACGIH threshold limit value–time-weighted average (TLV-TWA) for n-ethylmorpholine is 5 ppm (24 mg/m^3) with a notation for skin absorption.

REFERENCES

1. N-Ethylmorpholine. *Documentation of the TLVs and BEIs*, 6th ed, pp 638–639. Cincinnati, OH, American Conference of Governmental Industrial Hygienists (ACGIH), 1991
2. Dernehl CU: Health hazards associated with polyurethane foams. *J Occup Med* 8:59–62, 1966
3. Mellerio J, Weale RA: Miscellanea: hazy vision in amine plant operatives. *Br J Ind Med* 23:153–154, 1966
4. Smyth HF Jr et al: Range-finding toxicity data: List V. *AMA Arch Ind Hyg Occup Med* 10:61–68, 1954

ETHYL SILICATE
CAS: 78-10-4

$Si(OC_2H_5)_4$

Synonyms: Tetraethyl silicate; ethyl orthosilicate; silicic acid, tetramethyl ester

Physical Form. Colorless liquid

Uses. For arresting decay and disintegration of stone; for manufacture of weatherproof and acidproof mortars and cement

Exposure. Inhalation

Toxicology. Ethyl silicate is an irritant of the eyes and mucous membranes.

In humans, the eyes and nose are affected by brief exposures as follows: 3000 ppm is extremely irritating and not tolerable; 1200 ppm causes lacrimation and stinging; 700 ppm produces mild stinging; and, at 250 ppm, there is slight tingling.[1] Repeated or prolonged skin contact with the liquid may cause dermatitis because of its solvent effect.[2]

Exposure of guinea pigs to 2530 ppm for 4 hours was lethal to more than half of the animals; death was usually delayed and a result of pulmonary edema; effects were irritation of the eyes and the nose, lacrimation, tremor, dyspnea, and narcosis; some surviving animals developed a delayed but profound anemia.[1] Exposure to 1000 ppm for up to three 7-hour periods was fatal to 4 of 10 rats; autopsy findings were marked tubular degeneration and necrosis of the kidneys, mild liver damage, and slight pulmonary edema and hemorrhage.[2] In rats exposed to 125 ppm, 7/hr/day for 15 to 20 days, slight to moderate kidney damage was observed, but no pathological changes were detected in the liver or the lungs. Instillation of the liquid into the rabbit eye caused immediate, marked irritation that was reversible.[2]

The 1995 threshold limit value–time-weighted average (TLV-TWA) is 10 ppm (85 mg/m³).

REFERENCES

1. Smyth HF Jr, Seaton J: Acute response of guinea pigs and rats to inhalation of the vapors of tetraethyl orthosilicate (ethyl silicate). *J Ind Hyg Toxicol* 23:288–296, 1940
2. Rowe VK, Spencer HC, Bass SL: Toxicological studies on certain commercial silicones and hydrolyzable silane intermediates. *J Ind Hyg Toxicol* 30:332–353, 1948

FENTHION
CAS:55-38-9

$C_{10}H_{15}O_3PS$

Synonyms: Baycid; Baytex; Entex; Lebaycid; Mercaptophos; Tiguvon; phenthion

Physical Form. Yellow to tan oily liquid

Uses. Organothiophosphate insecticide

Exposure. Inhalation; skin absorption

Toxicology. Fenthion is an anticholinesterase agent and may also cause delayed neurotoxicity and ocular damage.

Signs and symptoms of overexposure are caused by the inactivation of the enzyme cholinesterase, which results in the accumulation of acetylcholine at synapses in the nervous system, skeletal and smooth muscle, and secretory glands. The sequence of the development of systemic effects varies with the route of entry. The onset of signs and symptoms is usually prompt but may be delayed up to 12 hours.[1-4]

After inhalation, respiratory and ocular effects are the first to appear, often within a few minutes after exposure. Respiratory effects include tightness in the chest and wheezing due to bronchoconstriction and excessive bronchial secretion; laryngeal spasms and excessive salivation may add to the respiratory distress, and cyanosis may also occur. Ocular effects include miosis, blurring of distant vision, tearing, rhinorrhea, and frontal headache.

After ingestion, gastrointestinal effects, such as anorexia, nausea, vomiting, abdominal cramps, and diarrhea appear within 15 minutes to 2 hours. After skin absorption, localized sweating and muscular fasciculations in the immediate area usually occur within 15 minutes to 4 hours; skin absorption is somewhat greater at higher ambient temperatures and is increased by the presence of dermatitis.[3]

With severe intoxication by all routes, an excess of acetylcholine at the neuromuscular junctions of skeletal muscle causes weakness aggravated by exertion, involuntary twitchings, fasciculation and, eventually, paralysis. The most serious consequence is paralysis of the respiratory muscles. Effects on the central nervous system include giddiness, confusion, ataxia, slurred speech, Cheyne–Stokes respiration, convulsions, coma, and loss of reflexes. The blood pressure may fall to low levels, and cardiac irregularities, including complete heart block, may occur.[2]

An intermediate syndrome of neurotoxic effects in humans has been described for fenthion, in which effects developed 24 to 96 hours after poisoning.[5] Patients developed paralysis of proximal limb muscles, neck flexors, motor cranial nerves, and respiratory muscles.

Fenthion has also exhibited delayed neurotoxicity, wherein the initial cholinergic crisis was delayed 5 days and recurred 24 days after ingestion.[6,7] Psychosis was a persistent manifestation. Because of the high lipid solubility of fenthion, toxin analysis of repeated fat biopsies was an essential component of patient management.[6]

In a study designed to determine the potential retinal changes in 79 subjects exposed to fenthion, 15 of the 79 workers examined had macular changes, characterized by perifoveal irregularity of pigmentation and hypopigmentation of 1/8 to 1/3 disk diameter.[8] Symptoms reported were diminution of vision, bright light aversion, flashes of light, black dots in front of the eyes, and visual blurring. Other causes of macular involvement in these workers were excluded. Mean exposure duration of subjects with macular involvement was 7.9 years.

A series of studies originating from Japan reported a more advanced visual disease syndrome, Saku disease, which correlated with increasing organophosphate exposure.[9] Ocular effects are dose-dependent and range in severity from lenticular changes to the more serious histopathological changes in the ciliary body and retina. Although the association between Saku disease and organophosphate exposure remains controversial (because of the lack of similar reports from around the world, the poor quality of some of the Japanese studies, and the similarity of Saku symptoms with common ocular diseases), animal studies have also shown the occurrence of ocular toxicity from organophosphate exposure.[10] Acute exposure of fenthion in rats has been associated with long-term changes in electroretinograms, whereas chronic exposure has produced permanent ocular degeneration. A single 100 mg/kg dose of fenthion administered subcutaneously to rats caused a long-lasting pertur bation in muscarinic receptor function.[11]

Two-year feeding studies in rats (3–75 ppm) and mice showed no indication of carcinogenic effects.[12,13] Fenthion was not teratogenic in tests on mice and rats.[12,14]

The 1995 threshold limit value–time-weighted average (TLV-TWA) is 0.2 mg/m³, with a notation for potential skin absorption.

Note. For a description of diagnostic signs, differential diagnosis, clinical laboratory tests, and specific treatment of overexposure to anticholinesterase insecticides, see "Parathion."

REFERENCES

1. Hayes WJ Jr: Organic phosphorus pesticides. In *Pesticides Studied in Man.* pp 284–435. Baltimore, MD, Williams and Wilkins, 1982
2. Taylor P: Anticholinesterase agents. In Gilman AG et al (eds): *Goodman and Gilman's The Pharmacological Basis of Therapeutics*, 7th ed, pp 110–129. New York, Macmillan, 1985
3. Koelle GB (ed): Cholinesterases and anticholinesterase agents. *Handbuch der Experimentellen Pharmakologie*, Vol 15, pp 989–1027. Berlin, Springer-Verlag, 1963
4. Namba T, Nolte CT, Jackrel J, Grob D: Poi-

soning due to organophosphate insecticides. *Am J Med* 50:475–492, 1971

5. Senanayake N, Karalliedde L: Neurotoxic effects of organosphosphorus insecticides: intermediate syndrome. *New Engl J Med* 316:761–763, 1987
6. Merrill DG, Mihm FG: Prolonged toxicity of organophosphate poisoning. *Crit Care Med* 10:550–551, 1982
7. Cherniak MG: Toxicological screening for organophosphorus-induced delayed neurotoxicity: complications in toxicity testing. *Neurotoxicology* 249–271, 1988
8. Misra UK, Nag D, Misra NK, et al: Some observations on the macula of pesticide workers. *Hum Toxicol* 4:135–145, 1985
9. Dementi B: Ocular effects of organophosphates: a historical perspective of Saku Disease. *J Appl Toxicol* 14:119–129,1994
10. Boyes WK, Tandon P, Barone S Jr, et al: Effects of organophosphates on the visual system of rats. *J Appl Toxicol* 14:135–143, 1994
11. Tandon P, Padilla S, Barone S Jr, et al: Fenthion produces a persistent decrease in muscarinic receptor function in the adult rat retina. *Toxicol Appl Pharmacol* 125:271–280, 1994
12. Food and Agriculture Organization: *Pesticide Residues in Food: 1978 Evaluations.* Rome, 1979
13. National Cancer Institute: Bioassay of Fenthion for possible carcinogenicity. CAS No. 55-38-9. Carcinogenesis Technical Report Series, No 103, pp 1–104. US Department of Health, Education and Welfare, Public Health Service, National Institutes of Health, Bethesda, MD. 1979
14. World Health Organization: *1975 Evaluation of Some Pesticide Residues in Food.* Geneva, WHO, 1976

Uses.   Fungicide

Exposure.   Inhalation

Toxicology.   Ferbam is an irritant of the eyes and respiratory tract; in animals, it causes central nervous system depression, and it is expected that severe exposure will cause the same effect in humans.

In humans, the dust is irritating to the eyes and the respiratory tract; it causes dermatitis in some individuals.[1] Large oral doses cause gastrointestinal disturbances.[1]

In guinea pigs given ferbam by stomach tube, the lethal range was 450 to 2000 mg/kg; the animals became stuporous and died in coma.[2] Ten of 20 rats died from a diet containing 0.5% ferbam for 30 days; there was a slight and ill-defined tendency toward anemia. At autopsy, there was no evidence of a regularly appearing tissue injury; minor abnormalities of the lung, liver, kidney, and bone marrow were observed in a few animals.[2] Animal experiments revealing an increased acetaldehyde level after ingestion of alcohol suggest that ferbam, like other dithiocarbamates, may be capable of causing an Antabuse-like reaction.[3]

Following oral administration of ferbam to mice and rats, no carcinogenic effects were seen, but the IARC has concluded that insufficient data are available to evaluate fully the carcinogenicity of this compound.[3] Administration of ferbam to pregnant rats in high but sublethal doses yielded evidence of embryo/fetotoxicity.[3]

The 1995 ACGIH threshold limit value–time-weighted average (TLV-TWA) for ferbam is 10 mg/m^3.

FERBAM
CAS: 14484-64-1

$[(CH_3)_2NCS_2]_3Fe$

Synonyms: Ferric dimethyldithiocarbamate; Cormate; Fermacide

Physical Form.   Black solid

REFERENCES

1. AMA Council on Pharmacy and Chemistry: Outlines of information on pesticides. Part 1. Agricultural fungicides. *JAMA* 157:237–241, 1955
2. Hodge HC et al: Acute and short-term oral toxicity tests of ferric dimethyldithiocarbamate (ferbam) and zinc dimethyldithiocarbamate (ziram). *J Am Pharm Assoc* 41:662–665,1952
3. Ferbam. *IARC Monographs on the Evaluation of*

Carcinogenic Risk of Chemicals to Man, Vol 12, Some carbamates, thiocarbamates and carbazides, pp 121–129. Lyon, International Agency for Research on Cancer, 1982

American Conference of Industrial Hygienists, 1991

FERROVANADIUM DUST
CAS: 12604-58-9

FeV

Synonyms: None

Physical Form.   Gray to black dust

Uses.   Added to steel to produce fineness of grain, toughness, and resistance to high temperature and torsion

Exposure.   Inhalation

Toxicology.   Ferrovanadium dust is a mild irritant of the eyes and respiratory tract.

Workers exposed to unspecified concentrations developed slight irritation of the eyes and respiratory tract.[1] Systemic effects have not been reported from industrial exposure.

Animals exposed for 1 hour on alternate days for 2 months to very high concentrations (1000–2000 mg/m³) developed chronic bronchitis and pneumonitis.[2] No active intoxication occurred in animals exposed at concentrations as high as 10,000 mg/m³.

The 1995 ACGIH threshold limit value–time-weighted average (TLV-TWA) for ferrovanadium dust is 1 mg/m³, with a short-term excursion limit (STEL) of 3 mg/m³.

REFERENCES

1. Roberts WC: The ferroalloy industry—hazards of the alloys and semimetallics: Part II. *J Occup Med* 7:71–77, 1965
2. Ferrovanadium. *Documentation of TLVs and BEIs*, 6th ed, pp 651–652, Cincinnati, OH,

FLUORANTHENE
CAS: 206-44-0

$C_{16}H_{10}$

Synonyms: FA; 1,2-Benzacenaphthene; benzo(jk)fluorene

Physical Form.   Yellow solid

Uses/Sources.   Component of polynuclear aromatic hydrocarbons, also known as polycyclic aromatic hydrocarbons; usually bound to small particulate matter present in urban air, industrial and natural combustion emissions, and cigarette smoke

Exposure.   Inhalation

Toxicology.   Fluoranthene (FA) is a lung carcinogen in mice.

A 24-week lung adenoma bioassay using newborn mice was employed to determine the tumorigenicity of fluoranthene (FA).[1] A 6.5-fold elevation of lung tumor incidence (58%) and a 12-fold increase in numbers (1.08 tumors/mouse) was observed in animals treated with the highest dose (3.5 mg/mouse), but no increase in tumor incidence was induced by 700 µg/mouse. Male mice surviving for 24 weeks in the FA treatment group that developed lung tumors had 2 to 3 times more tumors than comparably treated females.

Fluoranthene induced lung and liver tumors 6 to 9 months after intraperitoneal injection of 0.7, 1.75, and 3.5 mg FA into preweanling CD-1 mice.[2] There was a dose-dependent increase in lung tumors, with a maximum tumor incidence of nearly 45%.

FA and FA-DNA adducts in various tissues were isolated in various tissues of male rats 24 hours after a single ip injection of [³H]FA.[3] Formation and distribution of DNA adducts after chronic administration of FA in the diet were also studied. FA-derived radioactivity was widely distributed throughout the animal after a single dose, and excreta contained the greatest amounts of radioactivity at all dose levels.

The in vitro metabolism of FA was assessed by incubating 3-[3H]FA with rat hepatic microsomal enzymes.[4] The major metabolite produced was FA 2,3-diol, accounting for 29% to 43% of the total extractable metabolites. This study indicated that a major metabolic activation pathway of FA involved the formation of the FA 2,3-diol and the subsequent oxidation of this diol to an FA 2,3-diol-1,10b-epoxide, resulting in the production of mutagenic species.

No threshold limit value has been established for FA.

REFERENCES

1. Busby WF Jr, Goldman ME, Newberne PM, Wogan GN: Tumorigenicity of fluoranthene in a newborn mouse lung adenoma bioassay. *Carcinogenesis* 5:1311–1316, 1984
2. Wang JS, Busby WF Jr: Induction of lung and liver tumors by fluoranthene in a preweanling CD-1 mouse bioassay. *Carcinogenesis* 14:1871–1874, 1993
3. Gorelick NJ, Hutchins DA, Tannenbaum SR, Wogan GN: Formation of DNA and hemoglobin adducts of fluoranthene after single and multiple exposures. *Carcinogenesis* 10:1579–1587, 1989
4. Babson JR, Russo-Rodriguez SE, Wattley RV, et al: Microsomal activation of fluoranthene to mutagenic metabolites. *Toxicol Appl Pharmacol* 85:355–366, 1986

FLUORIDES

F

Compounds: Sodium fluoride; calcium fluoride; fluorspar; cryolite

Physical Form. Various

Sources/Uses. Grinding, drying, and calcining of fluoride-containing minerals; metallurgical processes such as aluminum reduction and steelmaking; kiln firing of brick and ceramic materials; melting of raw material in glassmaking; preservative; rodenticide; insecticide; additive to water to prevent dental caries

Exposure. Inhalation; ingestion

Toxicology. Fluoride causes irritation of the eyes and respiratory tract; absorption of excessive amounts of fluoride over a long period of time results in increased radiographic density of bone.[1]

Workers exposed to an airborne fluoride concentration of 5 mg/m[3] complained of eye and respiratory tract irritation and nausea.[2] The lethal oral dose of sodium fluoride for humans has been estimated to be 32 to 65 mg F/kg of body weight.[3] Effects from ingestion are diffuse abdominal pain, diarrhea, and vomiting; excessive salivation, thirst, and perspiration; painful spasms of the limbs; and sometimes albuminuria.[4,5]

Repeated exposure to excessive concentrations of fluoride over a period of years results in increased radiographic density of bone and eventually may cause crippling fluorosis (osteosclerosis due to deposition of fluoride), now an exceedingly rare phenomenon.[1] The gross changes in the skeleton are quite distinctive and characteristic; as the amount of fluoride in the bone increases, exostoses may develop, especially on the long bones; the sacrotuberous and sacrosciatic ligaments begin to calcify, vertebrae occasionally fuse together, and typical stiffness of the spinal column develops.[1] The absorption of 20 to 80 mg of fluoride daily may be expected to lead to crippling fluorosis in 10 to 20 years; this condition has not been reported in the United States from industrial exposure.[6]

Mottled appearance and altered form of teeth are produced only when excessive amounts of fluoride are ingested during the period of formation and calcification of teeth, which occurs during the first 8 years of life in humans; after calcification has been completed, fluoride does not have an adverse effect on the teeth.[5]

The morbidity experience of a small cohort of 431 Danish cryolite ($AlNa_3F_6$) workers employed for at least 6 months between 1924 and 1961, and followed until 1981, showed an apparent excess number of respiratory cancers.[6] Cancer morbidity showed no apparent correlation with length of employment or time from first exposure. Since detailed information on predictors for respiratory cancer, such as smoking habits, was not available, a possible contributing effect of fluoride was not excluded by the authors. In a recent follow-up of this cohort, the increase in lung cancer remained (standard incidence ratio 1.35), and an increase in bladder tumors (SIR 1.84) was observed. The cancer incidence was not related to length of exposure, and other confounding effects, such as smoking, alcohol, and multiple chemical exposures in addition to fluoride, may have contributed to the observed increases.[7]

In 1987, the IARC concluded that evidence for carcinogenicity of fluoride to animals was inadequate, based on review of three experiments in three different strains of mice by oral administration of sodium fluoride (NaF), and that evidence for carcinogenicity to humans was also inadequate, based on review of several studies on water fluoridation and cancer.[8] A subsequent NTP study reported equivocal evidence of carcinogenicity based on osteosarcomas in 4 of 80 male rats administered 175 ppm sodium fluoride in the drinking water for 2 years.[9] No evidence of carcinogenicity was found in another strain of rats that ingested up to 13.7 mg/day for 2 years.[10]

Sodium fluoride did not produce genotoxic damage in a variety of assays.[11]

Fluoride concentration in the urine has been used as a biologic indicator of fluoride.[12,13] Most absorbed fluoride is excreted rapidly in the urine. A portion is stored in bone, but a nearly equal amount is mobilized from bone and excreted. Some storage of fluoride occurs from the ingestion of as little as 3 mg/day. Evidence from several sources indicates that urinary fluoride concentrations not exceeding 5 mg/l in preshift samples taken after 2 days off work are not associated with detectable osteosclerosis and that such changes are unlikely at urinary levels of 5 to 8 mg/l.[12] Preshift urinary

fluoride concentration is considered to be a measure of the worker's body (skeletal) burden of fluoride, whereas the postshift sample is taken to be representative of exposure conditions during that workshift.[1] NIOSH recommends that urinary postshift fluoride analysis should be made available at an interval not exceeding every 3 months to at least one-fourth of all workers subject to occupational exposure to fluoride.[14] Participating workers should be rotated to provide all exposed workers the opportunity for urinalysis every year. Spot urine samples are collected at the conclusion of the workshift after 4 or more consecutive days of exposure.

The 1995 ACGIH threshold limit value-time-weighted average (TLV-TWA) for fluorides is 2.5 mg/m³, as F.

REFERENCES

1. Hodge HC, Smith FA: Occupational fluoride exposure. *J Occup Med* 19:12–39, 1977
2. Elkins HB: *Chemistry of Industrial Toxicology*, pp 72–73, New York, John Wiley and Sons, 1943
3. Perry WG, Smith FA, Kent MB: The halogens. In Clayton GD, Clayton FE (eds): *Patty's Industrial Hygiene and Toxicology*, 4th ed, Vol II, Part F, *Toxicology*, pp 4471–4476. New York, John Wiley and Sons, 1994
4. World Health Organization, *Fluorides and Human Health*, pp 225–271. Geneva, WHO, 1970
5. Committee on Biologic Effects of Atmospheric Pollutants, Division of Medical Sciences, National Research Council: *Fluorides*, pp 163–221. Washington, DC, National Academy of Sciences, 1971
6. Grandjean P, Juel J, Jensen OM: Mortality and cancer morbidity after heavy occupational fluoride exposure. *Am J Epidemiol* 121:57–64, 1985
7. Grandjean P, Olsen J, Jensen OM, et al: Cancer Incidence and mortality in workers exposed to fluoride. *J Natl Cancer Inst* 84:1903–1909, 1992
8. *IARC Monographs on the Evaluation of Carcinogenic Risks to Humans. Overall Evaluations of Carcinogenicity: An Updating of IARC Monographs Vols 1–42*, Suppl 7, pp 208–210. Lyon,

International Agency for Research on Cancer, 1987

9. National Toxicology Program: *Technical Report on the Toxicology and Carcinogenesis Studies of Sodium Fluoride in F344/N Rats and B6C3F₁ Mice (Drinking Water Studies)*. NTP TR-393. DHEW Pub No NIH 90-2848, 1990

10. Maurer JK, Cheng MC, Boysen BG: Two-year carcinogenicity study of sodium fluoride in rats. *J Natl Cancer Inst* 82:1118–1126, 1990

11. Tong CC, McQueen CA, Brat SV, et al: The lack of genotoxicity of sodium fluoride in a battery of cellular tests. *Cell Biol Toxicol* 4:173–186, 1988

12. Biological monitoring guides: fluorides. *Am Ind Hyg Assoc J* 32:274–279, 1971

13. Largent EJ: Rates of elimination of fluoride stored in the tissues of man. *AMA Arch Ind Hyg Occup Med* 6:37–42, 1952

14. National Institute for Occupational Safety and Health: *Criteria for a Recommended Standard ... Occupational Exposure to Inorganic Fluorides*. DHEW (NIOSH) Pub No 76-103, pp 19–100. Washington, DC, US Government Printing Office, 1975

FLUORINE
CAS: 7782-41-4

F_2

Synonyms: None

Physical Form. Yellow gas

Uses. Conversion of uranium tetrafluoride to uranium hexafluoride; oxidizer in rocket-fuel systems; in manufacture of various fluorides and fluorocarbons

Exposure. Inhalation

Toxicology. Fluorine is a severe irritant of the eyes, mucous membranes, skin, and lungs.

Because fluorine is the most reactive of the elements, free fluorine is rarely found in nature. Fluorine reacts with water to produce ozone and hydrofluoric acid.[1] In humans, the inhalation of high concentrations causes laryngeal spasm and bronchospasm, followed by the delayed onset of pulmonary edema.[2,3] At sublethal levels, severe local irritation and laryngeal spasm will preclude voluntary exposure to a high concentration unless the individual is trapped or incapacitated.[2] Two human subjects found momentary exposure to 50 ppm intolerable; 25 ppm was tolerated briefly, but both subjects developed sore throat and chest pain, which persisted for 6 hours.[4] Short-term exposures to concentrations of up to 10 ppm were tolerated without discomfort.[3]

The LC_{50} in mice for 60 minutes was 150 ppm; effects were irritation of the eyes and nose and the delayed onset of labored breathing and lethargy; autopsy findings included marked pulmonary congestion and hemorrhage.[5] Mice exposed to sublethal concentrations had pulmonary irritation and delayed development of focal necrosis in the liver and kidneys.[5]

A blast of fluorine gas on the shaved skin of a rabbit caused a second-degree burn; lower concentrations cause severe burns of insidious onset that result in ulceration and are similar to burns produced by hydrogen fluoride.[1,4]

The 1995 ACGIH threshold limit value–time-weighted average (TLV-TWA) for fluorine is 1 ppm (1.6 mg/m³) with a short-term excursion limit of 2 ppm (3.1 mg/m³).

REFERENCES

1. Largent EJ: Fluorine and compounds. In International Labour Office: *Encyclopaedia of Occupational Health and Safety*, Vol 1, pp 557–559. New York, McGraw-Hill, 1971

2. Ricca PM: Exposure criteria for fluorine rocket propellants. *Arch Environ Health* 12:339–407, 1966

3. Ricca PM: A survey of the acute toxicity of elemental fluorine. *Am Ind Hyg Assoc J* 31:22–29, 1970

4. Hygienic guide series: fluorine. *Am Ind Hyg Assoc J* 26:624–627, 1965

5. Keplinger ML, Suissa LW: Toxicity of fluorine short-term inhalation. *Am Ind Hyg Assoc J* 29:10–18, 1968

FORMALDEHYDE
CAS: 50-00-0

HCHO

Synonyms: Methanal; formic aldehyde; oxomethane; oxymethylene; methylene oxide; methyl aldehyde

Physical Form. Gas (*Note:* Formalin is a 37% to 50% solution by weight of formaldehyde gas.)

Uses. Manufacture of formaldehyde resins used as adhesives in particleboard, plywood, and insulating materials; countertops and wall paneling; coating for fabrics to impart permanent press characteristics; in manufacture of rubber, photographic film, leather, cosmetics, embalming fluids, and insulation; disinfectants; fumigants

Exposure. Inhalation

Toxicology. Formaldehyde is an irritant of the eyes and respiratory tract; it causes both primary irritation and sensitization dermatitis; at high levels, it is carcinogenic in experimental animals and, although results are equivocal in humans, it is considered a suspected human carcinogen.

Mild eye irritation with lacrimation and other transient symptoms of mucous membrane irritation have been observed in some persons at concentrations of 0.1 to 0.3 ppm. For most people, however, a tingling sensation in the eyes, nose, and posterior pharynx is not experienced until 2 to 3 ppm.[1,2] Some tolerance occurs, so that repeated 8-hour exposures at this level are possible.[3] At 4 to 5 ppm, irritation of mucous membranes increases and lacrimation becomes evident. This level may be tolerated by some for short periods but, after 30 minutes, discomfort becomes quite pronounced.

Concentrations of 10 ppm can be withstood for only a few minutes; profuse lacrimation occurs in all subjects, even those acclimated to lower levels. Between 10 and 20 ppm, it becomes difficult to take a normal breath; burning of the nose and throat becomes more severe and extends to the trachea, producing cough. On cessation of exposure, the lacrimation subsides promptly, but the nasal and respiratory irritation may persist for about an hour. It is not known at which levels serious inflammation of the bronchi and lower respiratory tract would occur in humans; it is expected that 5- or 10-minute exposures to levels of 50 to 100 ppm would cause very serious injury. Acute irritation of the human respiratory tract from inhalation of high levels of formaldehyde have caused pulmonary edema, pneumonitis, and death.[1]

Solutions of 25% to 44% splashed in the eyes have caused severe injury and corneal damage.

Formaldehyde is one of the most common causes of occupational skin disease; the major effects of formaldehyde on the skin are irritant dermatitis and allergic contact dermatitis.[4] Irritant dermatitis results from direct injury to the skin and is characterized by redness and thickening of the affected areas. In more severe cases, there may be blistering, scaling, and the formation of fissures.

Dermal sensitization to formaldehyde is an often reported phenomenon. Following skin contact, a symptom-free induction period typically ensues for 7 to 10 days.[4] With subsequent contact, there is itching, redness, swelling, multiple small blisters, and scaling in sensitized individuals. Repeated contact tends to cause more severe reactions, and sensitization usually persists for life.

A number of studies have suggested that formaldehyde causes asthma and/or exacerbates pre-existing respiratory conditions. Small, transient declines in lung-function parameters over the course of a workshift have been the most consistent findings.[4] In a recent report, statistically significant postshift declines in FEV_1/$FVC_\%$, $FEF_{25\%-75\%}$, $FEF_{50\%}$, and $FEF_{75\%}$ were observed in workers exposed to less than 3.0 ppm.[5] In this same group of workers, there was no evidence of permanent respiratory impairment after a mean exposure time of 10 years. Although other studies have also failed to implicate formaldehyde as a cause of permanent respiratory impairment, a small group of resin workers exposed for more than 5 years had lower $FEV_1/FVC_\%$ and $FEF_{50\%}/FVC$ ratios over the workweek, suggesting the possibility of chronic lung-function shifts.[6]

A few case reports have suggested that formaldehyde may cause asthmatic symptoms by acting as an immunologic sensitizer.[4] Other investigators feel that there is insufficient evidence that formaldehyde causes immunologic disease since IgG and IgE antibodies to formaldehyde have not been demonstrable and antibodies to formaldehyde-albumin complexes, while present in people exposed to formaldehyde, do not correlate well with clinical symptomatology.[7] In a few anecdotal cases, prolonged exposure has been associated with impaired central nervous system function, including abnormal balance, constricted visual fields, delayed blink latency, and deficits in cognitive-function tests.[8]

In a number of reproductive tests in rodents and dogs, formaldehyde did not cause adverse outcomes in the offspring, except at maternally toxic doses.[9] In one report, there was a significant increase in the incidence of abnormal sperm in rats following 200 mg/kg administered as a single gavage dose; in other studies, there was no effect on sperm in mice after administration of 100 mg/kg/day for 5 days, nor were there changes in reproductive function in rats treated with 0.1 ppm in the drinking water for 6 months.

Formaldehyde has been shown to be carcinogenic in two strains of rats, resulting in squamous-cell cancers of the nasal cavity following repeated inhalation of about 14 ppm. In one study, 51 of 117 male and 42 of 115 female Fischer 344 rats developed this tumor, but no nasal tumors were seen at 0 or 2 ppm. No other neoplasm was increased significantly. In a similar study of mice, this nasal tumor occurred in two male mice at 14.3 ppm. None of the excesses were statistically significant except for the high-exposure data in rats.[10]

A large number of epidemiological studies have now been completed on persons with potential exposure to formaldehyde. Although a variety of excess cancers, including brain, bladder, colon, skin, kidney, and leukemias, have been reported, the evidence for a possible involvement of formaldehyde is strongest for respiratory cancers.[11] A case-control study of men with histologically confirmed primary epithelial cancer of the nasal cavities or accessory sinuses found the relative risk approximately doubled for nasal cancer, particularly squamous-cell carcinoma, in formaldehyde-exposed workers.[12] Another case-control study of 759 histologically verified cancers of the nasal cavity showed an association between squamous-cell carcinoma and formaldehyde exposure.[13] Among industrial workers exposed to formaldehyde-containing particulates, the risk of death from cancer of the nasopharynx increased with cumulative exposure to formaldehyde from an SMR of 192 for <0.5 ppm-years to 403 for 0.5 to <5.5 ppm-years, and 746 for >5.5 ppm-years.[14] The 5 workers with formaldehyde exposure who died from nasal pharyngeal cancer had held jobs where exposures had excursions to levels exceeding 4 ppm. An increased risk of nasopharyngeal cancer was found among individuals who resided in mobile homes where exposure to formaldehyde averaged as much as 0.5 ppm and easily reached or exceeded 1 ppm.[4,15] The relative risk went from 2.1 for those who resided in mobile homes for 1 to 9 years, to 5.5 for those who resided there for over 10 years. A relative risk of 6.7 for nasopharyngeal cancer was found for people with both occupational and residential exposure.[15,16]

Slight excesses in the occurrence of lung cancer have been noted in several studies. In a large cohort mortality analysis of 26,000 workers, there were significant excesses of lung cancer in those with more than 20 years since initial formaldehyde exposure.[17] Reanalysis of this same data has also shown a significant trend of increasing lung cancer risk with increasing formaldeyde exposure.[18] In the other large study of industrial workers, there was a significant excess of mortality from lung cancer at one plant, where 73% of workers were exposed to formaldehyde at levels estimated to be over 2 ppm.[19] An additional 8-year follow-up of this cohort found no cases of nasopharyngeal cancer (vs. 1.3 expected), one nasal cancer death (vs. 1.7 expected), and slight excesses of lung cancer, respiratory disease, and stomach cancer, which did not correlate with estimated cumulative dose or time since first exposure.[20] An enlarged and updated cohort study of formaldehyde-exposed workers from a US chemical plant found statistically significant elevated standardized mortality ratios for total mortality, ischemic heart disease, nonmalignant respiratory disease, and cancers of the lung, skin, and central nervous system; among long-term workers, there was no clear evidence of an association between lung cancer mortality and formaldehyde exposure.[21] The authors suggested that unmeasured occupational or nonoccupational factors may have played a role in the significant excesses found in short-term workers.

Although the results of certain of the published cancer studies have been challenged and are subject to different interpretations, the IARC has determined that the body of evidence suggests sufficient evidence for carcinogenicity to animals but limited evidence for carcinogenicity to humans.[11,22] It has also been suggested that formaldehyde is not carcinogenic at low levels, based on the evidence that formaldehyde is formed naturally in food, that it is a normal human metabolite, and that a threshold for carcinogenicity exists in animal experiments.[9]

Formaldehyde is genotoxic in a variety of systems, inducing gene locus mutations, cell transformations, deletions, and chromosomal aberrations.[23]

The odor is perceptible to previously unexposed persons at or below 1 ppm.

The 1995 ACGIH threshold limit value–ceiling (TLV-C) is 0.3 ppm (0.37 mg/m^3) with an A2 suspected human carcinogen designation.

REFERENCES

1. National Institute for Occupational Safety and Health: *Criteria for a Recommended Standard . . . Occupational Exposure to Formaldehyde.* DHEW (NIOSH) Pub No 77-126, pp 21–81. Washington, DC, US Government Printing Office, 1976
2. Casteel SW et al: Formaldehyde: Toxicology and hazards. *Vet Hum Toxicol* 29:31–33, 1987
3. Nordman H, Keskmen H, Tuppurainen M: Formaldehyde asthma—rare or overlooked? *J Allergy Clin Immunol* 75(1):91–99, 1985
4. Department of Labor Occupational Safety and Health Administration: Occupational exposure to formaldehyde: final rule. *Federal Register*, Vol 52, No 233:46168–46312, December 4, 1987
5. Horvath EP et al: Effects of formaldehyde on the mucous membranes and lungs. *JAMA* 259:701–707, 1988
6. Schoenberg JB, Mitchell CA: Airway disease caused by phenolic (phenolformaldehyde) resin exposure. *Arch Environ Health* 30:574–577, 1975
7. Chang CC, Gershwin ME: Perspectives on formaldehyde toxicity: separating fact from fantasy. *Reg Toxicol Pharmacol* 16:150–160, 1992
8. Kilburn K: Neurobehavioral impairment and seizures from formaldehyde. *Arch Environ Health* 49:37–44, 1994
9. Restani P, Galli CL: Oral toxicity of formaldehyde and its derivatives. *Crit Rev Toxicol* 21:315–328, 1991
10. Kers W et al: Carcinogenicity of formaldehyde in rats and mice after long-term inhalation exposure. *Cancer Res* 43:4382–4392, 1983
11. *IARC Monographs on the Evaluation of Carcinogenic Risks to Humans.* Suppl 7:211–216. Lyon, International Agency for Research on Cancer, 1987
12. Hayes RB et al: Cancer of the nasal cavity and paranasal sinuses and formaldehyde exposure. *Int J Cancer* 37:487–492, 1986
13. Olsen JH, Asnaes S: Formaldehyde and the risk of squamous cell carcinomas of the sinonasal cavities. *Br J Ind Med* 43:769–774, 1986
14. Blair A et al: Cancers of the nasopharynx and oropharynx and formaldehyde exposure. *J Natl Cancer Inst* 78:191–192, 1987
15. Vaughan TL et al: Formaldehyde and cancers of the pharynx, sinus, and nasal cavity. II. Residential exposures. *Int J Cancer* 38:685–688, 1986
16. Vaughan TL et al: Formaldehyde and cancers of the pharynx, sinus, and nasal cavity. I. Occupational exposures. *Int J Cancer* 38:677–685, 1986
17. Blair A, Stewart P, O'Berg M, et al: Mortality among industrial workers exposed to formaldehyde. *J Natl Cancer Inst* 76:1071–1084, 1986
18. Sterling TD, Weinkam JJ: Mortality from respiratory cancers (including lung cancer) among workers employed in formaldehyde industries. *Am J Ind Med* 25:593–602, 1994
19. Acheson ED, Barnes HR, Gardner MJ, et al:

Formaldehyde in the British chemical industry: an occupational cohort study. *Lancet* 1:611–616, 1984

20. Gardner MJ, Pannett B, Winter PD, et al: A cohort study of workers exposed to formaldehyde in the British chemical industry: an update. *Br J Ind Med* 50:827–834, 1993
21. Marsh GM, Stone RA, Esmen NA, et al: Brief communications: mortality patterns among chemical plant workers exposed to formaldehyde and other substances. *JNCI* 86:384–386, 1994
22. Ad Hoc Panel: Epidemiology of chronic occupational exposure to formaldehyde. Report of the Ad Hoc Panel on Health Aspects of Formaldehyde. *Toxicol Ind Health* 4:77–90, 1988
23. Heck H d'A, Casanova M, Starr TB: Formaldehyde toxicity—new understanding. *Crit Rev Toxicol* 20:397–426, 1990

FORMIC ACID
CAS: 64-18-6

HCOOH

Synonyms: Methanoic acid; formylic acid, hydrogen carboxylic acid

Physical Form. Colorless liquid

Uses. Preservative of silage; reducer in dyeing wool; lime descaler; chemical intermediate

Exposure. Inhalation

Toxicology. Formic acid vapor is a severe irritant of the eyes, mucous membranes, and skin.

Exposure causes eye irritation with lacrimation, nasal discharge, throat irritation, and cough.[1] A worker splashed in the face with hot formic acid developed marked dyspnea with dysphagia and died within 6 hours.[2] Workers exposed to a mixture of formic and acetic acids at an average concentration of 15 ppm of each complained of nausea.[3] Twelve farmers exposed to 7.3 mg formic acid/m³ for 8 hours had increased renal ammoniagenesis and urinary calcium at 30 hours post-exposure.[4]

The liquid on the skin causes burns with vesiculation; keloid formation at the site of the burn often results.[2] Although ingestion of the liquid is unlikely in ordinary industrial use, the highly corrosive nature of the substance can produce serious burns of the mouth and esophagus.[3] Other clinical features include gastrointestinal irritation, vomiting, hematemesis, and abdominal pain.[5] Cicatricial stenosis may appear after recovery. The major complications are acute renal failure and disseminated intravascular coagulation, while pulmonary aspiration with a secondary pneumonia may occur. Occasionally, a direct toxic pneumonitis may occur.

Formic acid is an inhibitor of cytochrome oxidase at the terminal end of the respiratory chain in mitochondria and causes histotoxic hypoxia at the cellular level.[6,7] Therefore, persons with cardiovascular disease may be considered at special risk to the effects of formic acid.[6]

The 1995 ACGIH threshold limit value–time-weighted average (TLV-TWA) for formic acid is 5 ppm (9.4 mg/m³) with a short-term excursion limit of 10 ppm (19 mg/m³).

REFERENCES

1. Henson EV: Toxicology of the fatty acids. *J Occup Med* 1:339–345, 1959
2. von Oettingen WF: The aliphatic acids and their esters—toxicity and potential dangers. *AMA Arch Ind Health* 20:517–522, 530–531, 1959
3. Guest D et al: Aliphatic carboxylic acids. In Clayton GD, Clayton FE (eds): *Patty's Industrial Hygiene and Toxicology*, 3rd ed, rev, Vol 2C, *Toxicology*, pp 4903–4909. New York, Wiley-Interscience, 1982
4. Liesivuori J, Laitinen J, Savolainen H: Kinetics and renal effects of formic acid in occupationally exposed farmers. *Arch Toxicol* 66:522–524, 1992
5. Moore DF, Bentley AM, Dawling S, et al: Folinic acid and enhanced renal elimination in formic acid intoxication. *Clin Toxicol* 32:199–204, 1994
6. Liesivuori J, Kettunen A: Farmers' exposure

to formic acid vapour in silage making. *Ann Occup Hyg* 27:327–329, 1983

7. Liesivuori J, Savolainen H: Urinary formic acid indicator of occupational exposure to formic acid and methanol. *Am Ind Hyg Assoc J* 48:32–34, 1987

FURFURAL
CAS: 98-01-1

$C_4H_4O_2$

Synonyms: 2 Furaldehyde; pyromucic aldehyde

Physical Form. Colorless to reddish-brown, oily liquid

Uses. Solvent refining of lubricating oils, resins, and other organic materials; insecticide, fungicide, germicide; intermediate for tetrahydrofuran, furfural alcohol, phenolic, and furan polymers

Exposure. Inhalation; skin absorption

Toxicology. Furfural is an irritant of the eyes, mucous membranes, and skin and is a central nervous system depressant.

Although the vapor is an irritant, the liquid has a relatively low volatility, so that inhalation by workers of significant quantities is unlikely.[1] Exposure of workers to levels of 1.9 to 14 ppm caused complaints of eye and throat irritation and headache.[1] The liquid or vapor is irritating to the skin and may cause dermatitis, allergic sensitization, and photosensitization.[2] Dermal absorption has been found to be significant in humans. A 15-minute whole-hand immersion in the liquid resulted in absorption of an amount equivalent to an 8-hour inhalation exposure of 10 to 20 mg/m³ (3–5 ppm) vapor.[3]

Exposure of cats to 2800 ppm for 30 minutes resulted in fatal pulmonary edema.[1] Inhalation of 260 ppm for 6 hours was fatal to rats but produced no deaths in mice or rabbits.[2]

Slight liver changes were seen in dogs exposed daily to 130 ppm for 4 weeks.[2] Symptoms following oral administration of 50 to 100 mg/kg in rats were weakness, ataxia, coma, and death.[1]

Rats exposed at 40 ppm, 1 hr/day, for periods of 7, 15, or 30 days had pulmonary irritation, parenchymal injury, and regenerative proliferation of pneumocytes, the severity of which depended on duration of exposure.

Survival was reduced in groups of rats receiving 90 mg/kg/day for 13 weeks by gavage, and cytoplasmic vacuolization of hepatocytes was increased in exposed males. In mice, centrilobular coagulative necrosis and/or multifocal subchronic inflammation of the liver occurred at doses of up to 1200 mg/kg.[5]

In 2-year gavage studies, there was some evidence of carcinogenic activity in male rats based on increased incidences of cholangiocarcinomas and bile duct dysplasia and fibrosis. There was also some evidence of carcinogenicity in female mice, based on increased incidences of hepatocellular adenomas. Male mice showed clear evidence of carcinogenicity, based on increased incidences of hepatocellular adenomas and carcinomas.[5]

In another experiment with hamsters, exposure to furfural vapor 7 hr/day, 5 days/week, for 1 year caused irritation of the nasal mucosa and growth retardation but no evidence of carcinogenic effects.[6]

Drops of a 10% aqueous solution in the eyes of animals caused immediate discomfort; the lids and conjunctivas became red and swollen, but these effects disappeared after 24 hours.[7]

The 1995 ACGIH threshold limit value–time-weighted average (TLV-TWA) for furfural is 2 ppm (7.9 mg/m³) with a notation for skin absorption.

REFERENCES

1. Brabec MJ: Aldehydes and acetals. In Clayton GD, Clayton FE (eds): *Patty's Industrial Hygiene and Toxicology*, 3rd ed, Vol 2A, *Toxicology*, pp 2665–2666. New York, Wiley-Interscience, 1981

2. Hygienic guide series: furfural. *Am Ind Hyg Assoc J* 26:196, 1965

3. Flek J, Sedisec V: The absorption, metabolism and excretion of furfural in man. *Int Arch Occup Environ Health* 41:159–168, 1978
4. Gupta GD, Misra A, Agarwal DK: Inhalation toxicity of furfural vapours: an assessment of biochemical response in rat lungs. *J Appl Toxicol* 11:343–347, 1991
5. National Toxicology Program: *Toxicology and Carcinogenesis Studies of Furfural in F344/N Rats and B6C3F Mice (Gavage Studies)*, pp 1–76. US Department of Health and Human Services, NTP TR-382, 1990
6. Feron VJ, Kruysse A: Effects of exposure to furfural vapor in hamster simultaneously treated with benzo(a)pyrene or diethylnitrosamine. *Toxicology* 11:127–144, 1978.
7. Grant WM: *Toxicology of the Eye*, 3rd ed, p 449. Springfield, IL, Charles C Thomas, 1986

FURFURYL ALCOHOL
CAS: 98-00-0

$C_6H_6O_2$

Synonyms: 2-Furyl carbinol; 2-furanmethanol; furfural alcohol

Physical Form. Colorless liquid, which turns dark in air

Uses. Solvent for cellulose esters, resins, and dyes; liquid propellant; binder in foundry cores; manufacture of resins, including furfuryl alcohol resin (furan resin) and furfuryl alcohol-formaldehye resins

Exposure. Inhalation; skin absorption

Toxicology. Furfural alcohol is an eye, nose, and throat irritant; exposure to high concentrations causes central nervous system depression.

Workers exposed to 8.6 and 10.8 ppm in a foundry core-making operation experienced no discomfort, but two persons exposed to 15.8 ppm (and 0.33 ppm formaldehyde) experienced lacrimation and a desire to leave the area. In another foundry core-making operation, no ill effects were seen after exposures of about 6 to 16 ppm.[1]

Large doses injected subcutaneously in dogs caused depressed respiration, lowered body temperature, salivation, diarrhea, diuresis, and signs of narcosis.[2]

In rats exposed to 700 ppm, effects included excitement followed by eye irritation and drowsiness.[3] The rat LC_{50} for 4 hours was 233 ppm.[4] Repeated daily exposure of rats to an average of 19 ppm caused moderate respiratory irritation.[3] Intravenous injection into rabbits and cats caused depression of the central nervous system; death occurred at doses of 800 to 1400 mg/kg.[5]

Eye contact in rabbits resulted in reversible inflammation and corneal injury with opacity.[1] Animal experiments indicated that the liquid is well absorbed through the skin, with a dose-related mortality; mild skin irritation may also result from contact.[1] Prolonged inhalation exposures of rats to 25, 50, and 100 ppm resulted in decreased weight gain and, at the two highest doses, biochemical changes in the brain suggestive of mitochondrial damage, glial-cell degeneration, and early demyelization.[6]

The odor is detectable at 8 ppm.[4] Mixing with acids results in polymerization, a highly exothermic reaction that may result in explosions.[1]

The 1995 ACGIH threshold limit value–time-weighted average (TLV-TWA) for furfural alcohol is 10 ppm (40 mg/m^3) with a short-term excursion limit (STEL) of 15 ppm (60 mg/m^3) and a notation for skin absorption.

REFERENCES

1. National Institute for Occupational Safety and Health: *Criteria for a Recommended Standard . . . Occupational Exposure to Furfuryl Alcohol.* DHEW (NIOSH) Pub No 79-133. Washington, DC, US Government Printing Office, 1979
2. Erdmann E: Uber das Kaffeeol und die Physiologische Wirkung des darin Enthaltenen Furfuralkols. *Arch Exp Pathol Pharmacol* 48:233–261, 1902
3. Comstock CC, Oberst FW: Inhalation Toxicity of Aniline, Furfuryl Alcohol and Their Mix-

tures in Rats and Mice. Chemical Corps Medical Laboratories Research Report No 139, October 1952
4. Jacobson KH et al: The toxicology of an aniline-furfuryl alcohol-hydrazine vapor mixture. *Am Ind Hyg Assoc J* 19:91–100, 1958
5. Fine EA, Wills JH: Pharmacologic studies of furfuryl alcohol. *Arch Ind Hyg Occup Med* 1:625–632, 1950
6. Savelainen H, Pfaffli P: Neurotoxicity of furfuryl alcohol vapor in prolonged inhalation exposure. *Environ Res* 31:420–427, 1983

GASOLINE
CAS:8006-61-9

Synonyms: Motor fuel; petrol

Physical Form. Liquid gasoline is a complex mixture of at least 150 hydrocarbons, with about 60% to 70% alkanes, 25% to 30% aromatics, and 6% to 9% alkenes. The small-chain, low-carbon-numbered components are more volatile and, thus, in higher percentages in the vapor phase than the larger and heavier molecules.[1] The concentrations of aromatics, the more toxic of the components, are depleted to about 2% in the vapor phase. The light alkanes, the less toxic, are enriched to about 90%. Benzene is also present and represents a component of major concern

Uses. Fuel for spark-ignited internal combustion engines

Exposure. Inhalation

Toxicology. Gasoline is an irritant of the eyes and mucous membranes and is a central nervous system depressant.

Exposure of humans to 900 ppm for 1 hour caused slight dizziness and irritation of the eyes, nose, and throat.[2] At 2000 ppm for 1 hour, there was dizziness, mucous membrane irritation, and anesthesia; 10,000 ppm caused nose and throat irritation in 2 minutes, dizziness in 4 minutes, and signs of intoxication in 4 to 10 minutes.[2] At high concentrations, coma and death may result in a few minutes without any accompanying respiratory struggle or postmortem signs of anoxia.[3]

On skin contact, gasoline vaporizes rapidly and has little if any irritant effect.[4] If occluded, or if the liquid remains in continued contact with the skin, a severe chemical burn can occur. Repeated exposures may cause defatting of the skin.

On ingestion, gasoline produced local irritation, central nervous system depression, and congestion and capillary hemorrhage in visceral organs.[5] Aspiration of the liquid into the lungs produced chemical pneumonitis. Intentional use of leaded gasoline as an intoxicant has resulted in encephalopathy from the tetraethyl lead.[6]

Octane-improving additives to gasoline, such as methylcyclopentadienyl manganese tricarbonyl (MMT), do not appear to influence toxicity, based on acute animal tests.[5,7]

In a 2-year inhalation study, rats and mice were exposed to 0, 67, 292, or 2056 ppm, 6 hr/day, 5 days per week.[8] The major finding was a time- and dose-related increase in the incidence of kidney lesions in the male rats. These lesions consisted of cortical multifocal tubular basophilia (indicative of areas of cell regeneration), protein casts, and interstitial inflammation. There was epithelial cell shedding, and the casts were found within dilated renal tubules, commonly at the corticomedullary junction.

The pattern of renal tubule degeneration, regeneration, dilation, and hyaline deposition (termed *light hydrocarbon nephrophathy*) is produced in male rats of three strains (Sprague-Dawley, Fischer-344, and Harlan–Wistar) but not in female rats of those strains nor in male or female mice, cats, dogs, or monkeys.[5] In three instances, male rats that showed light hydrocarbon nephropathy at 3 months developed tumors after 2 years. The hydrocarbons most likely to be associated with light hydrocarbon nephropathy were branched-chain aliphatic compounds containing at least 6, and probably not more than 8 to 10, carbon atoms.[9] Aromatic hydrocarbons were without activity.

Additional mechanistic studies suggest that rat renal tumors involve a rat-specific protein α_{2u}-globulin. This protein binds with branched aliphatics, which then accumulate in renal tubule cells, resulting in cell death and, in turn, a proliferative sequence that increases renal tubule tumors.[5,10] The α_{2u}-globulin protein is species-specific to rats and gender-specific to males.

It does not appear that the nephrotoxicity attributable to α_{2u}-globulin syndrome is relevant to humans. Most epidemiological studies have not shown an association between gasoline exposure and renal cancer risk.[11] However, a recent case-control study from Finland reported a significant association between renal cell cancer and gasoline that was dose-dependent.[12]

In general, gasoline is not considered to be genotoxic.[13]

The IARC concluded that limited evidence exists for the carcinogenicity of unleaded gasoline in animals.[14] The epidemiological studies were inadequate in demonstrating increased carcinogenic risk in humans.[14] The IARC Working Group did note that some components of gasoline, especially benzene, are carcinogenic in humans, and concluded that gasoline is possibly carcinogenic in humans.

Although anecdotal reports have suggested a link between gasoline exposure during pregnancy and developmental effects in humans, animal studies have not confirmed the toxicity of gasoline in the fetus.[13,15] Rats exposed to 1600 ppm during days 6 through 15 of gestation had no evidence of maternal toxicity or adverse effects on the fetuses.

The 1995 threshold limit value–time-weighted average (TLV-TWA) for gasoline is 300 ppm (890 mg/m^3) with a short-term excursion limit of 500 ppm (1480 mg/m^3).

REFERENCES

1. Page NP, Mehlman M: Health effects of gasoline refueling vapors and measured exposures at service stations. *Toxicol Ind Health* 5:869–890, 1989
2. Gerarde HW: Aromatic hydrocarbons. In Patty F (ed): *Industrial Hygiene and Toxicology*, 2nd ed. New York, Interscience, 1963
3. Reese E, Kimbrough RD: Acute toxicity of gasoline and some additives. *Environ Health Perspect* 101 (Suppl 6):115–131, 1993
4. Weaver NK: Gasoline toxicology, implications for human health. *Ann NY Acad Sci* 534:441–451, 1988
5. Scala RA: Motor gasoline toxicity. *Fund Appl Toxicol* 10:553–562, 1988
6. Fortenberry JD: Gasoline sniffing. *Am J Med* 79:740–744, 1985
7. Abbott PJ: Methylcyclopentadienyl manganese tricarbonyl (MMT) in petrol: the toxicological issues. *Sci Total Environ* 67:247–255, 1987
8. MacFarland HN, Ulrich CE, Holdsworth CE, et al: A chronic inhalation study with unleaded gasoline vapor. *J Am Coll Toxicol* 3:231–248, 1984
9. Scala RA: Comments on structure-activity relationships, summary and concluding remarks, pp 1–4. Unpublished addendum to Workshop on the Kidney Effects of Hydrocarbons. Boston, 1984
10. Raabe GK: Review of the carcinogenic potential of gasoline. *Environ Health Perspect* 101 (Suppl 6):35–38, 1993
11. Mclaughlin JK: Renal cell cancer and exposure to gasoline: a review. *Environ Health Perspect* 101 (Suppl 6):111–114, 1993
12. Partanen T, Heikkila P, Hernberg S, et al: Renal cell cancer and occupational exposure to chemical agents. *Scand J Work Environ Health* 17:231–239, 1991
13. Agency for Toxic Substances and Disease Registry (ATSDR): *Toxicological Profile for Automotive Gasoline*. US Department of Health and Human Services, Public Health Service, p 131, 1993
14. *IARC Monographs on the Evaluation of Carcinogenic Risks to Humans*, Vol 45, Occupational exposures in petroleum refining; crude oil and major petroleum fuels, pp 159–201. Lyon, International Agency for Research on Cancer, 1989
15. Skalko RG: Reproductive and developmental toxicity of the components of gasoline. *Environ Health Perspect* 101 (Suppl 6):143–149, 1993

GERMANIUM TETRAHYDRIDE
CAS: 7782-65-2

GeH₄

Synonyms: Germane; germanium hydride

Physical Form.   Colorless gas

Uses.   Doping agent for solid-state electronic components

Exposure.   Inhalation

Toxicology.   Germanium tetrahydride is apparently a hemolytic agent.

There is little information on the toxicity of this compound to humans. It is reported that inhalation by humans of germanium tetrahyride may cause lung problems, but no details were given.[1]

Germanium tetrahydride was lethal to mice following inhalation of 610 mg/m³ for 4 hours.[1] Degenerative changes were observed in the liver and kidneys of rodents exposed to high, one-time concentrations of 0.26 to 1.4 g/m³.[1,2] Nonspecific changes in the blood were also observed.[2] Nervous system effects, including excitation, impairment of locomotor activity, listlessness, hypothermia, and convulsions, were observed in mice prior to death following inhalation exposure to 2 g/m³.[2]

The 1995 ACGIH threshold limit value–time-weighted average (TLV-TWA) for germanium tetrahydride is 0.2 ppm (0.63 mg/m³).

REFERENCES

1. Furst A: Biological testing of germanium. *Toxicol Ind Health* 3:167–181, 1987
2. Vouk VB: Germanium. In Friberg L, Nordberg GF, Vouk VB (eds): *Handbook on the Toxicology of Metals*, 2nd ed., pp 255–263. Amsterdam, Elsevier, 1986

GLUTARALDEHYDE
CAS: 111-30-8

OCH(CH₂)₃CHO

Synonyms: Cidex (2% alkaline glutaraldehyde aqueous solution); 1,5-pentanedial; 1,5-pentanedione; glutaric dialdehyde; glutaral

Physical Form.   Colorless crystalline solid, soluble in water and organic solvents

Uses.   Broad-spectrum antimicrobial cold sterilant/disinfectant for hospital equipment; tanning agent for leather; tissue fixative; crosslinking agent for proteins; preservative in cosmetics; therapeutic agent for warts, hyperhidrosis, and dermal mycotic infections; in X-ray processing solutions and film emulsion

Toxicology.   Glutaraldehyde is an irritant of the upper respiratory tract and may be capable of inducing asthma in some individuals; it is a skin irritant and can cause an allergic contact dermatitis.

Glutaraldehyde has caused an allergic contact dermatitis in hospital workers using it as a cold sterilant or in handling recently processed X-ray film. It appears to be a strong sensitizer.[1–3] In general, reactions present as a vesicular dermatitis of the hands and forearms. Rubber gloves do not appear to afford complete protection. In unsensitized individuals, it acts as a mild skin irritant.

Glutaraldehyde also can produce eye and skin irritation when its solutions are aerosolized.[4] It has a low vapor pressure at room temperature, which reduces the potential for inhalation exposures.

Four nurses who were sterilizing endoscopes with glutaraldehyde developed symptoms of asthma and rhinitis temporally related to exposures to glutaraldehyde. Three of the four nurses, however, had a prior history of mild seasonal asthma.[4] On specific provocation testing, one patient had an increase in nasal airway resistance, with a dual immediate and late response pattern. Another patient had a delayed 22% decline in FEV₁ 80 minutes after the final exposure to glutaraldehyde. The occurrence of late reactions suggested that the underlying mechanism involved sensitization rather than an irritant effect.[4]

Swedish hospital workers exposed to low glutaraldehyde concentrations (below 0.2 ppm) had an increased frequency of reported nose and throat symptoms; skin symptoms, such as exzema and rash; and general symptoms, such as headache and nausea, compared to unexposed controls.[5]

Animal experiments demonstrate that solutions containing 25% or more glutaraldehyde cause a significant degree of skin irritation and eye injury; dilute solutions (5% of less) have low acute toxicity.[6]

In mice, oronasal exposure to 2.6 ppm led to a 50% decrease in respiratory rate.[7] Mice exposed at 0.3, 1.0, and 2.6 ppm, 6 hr/day for 4, 9, and 14 days, had lesions of the respiratory epithelium, including squamous metaplasia, focal necrosis, and keratin exudate, which were dose-dependent at the lower exposure levels. Lesions persisted 2 weeks after exposure but were decreasing 4 weeks after the end of exposure. No exposure-related lesions were observed in the lungs of exposed mice.

Results of short-term tests have been variable; glutaraldehyde was weakly positive in the Ames test in *Salmonella* in one study, and was negative in several other studies. The chemical tested negative in several genotoxicity tests in mammalian cells, including sister chromatid exchange and unscheduled DNA synthesis tests. The authors concluded that glutaraldehyde's in vitro reactivity with proteins, along with its nonreactivity with nucleic acids, made genotoxic effects unlikely.[6]

The 1995 ACGIH STEL/ceiling limit for glutaraldehyde is 0.2 ppm (0.82 mg/m³).

REFERENCES

1. Goncalo S et al: Occupational contact dermatitis to glutaraldehyde. *Contact Derm* 10:183–184, 1984
2. Hansen KS: Glutaraldehyde occupational dermatitis. *Contact Derm* 9:81–82, 1983
3. Fisher AA: Reactions to glutaraldehyde with particular reference to radiologists and X-ray technicians. *Cutis* 28:113–122, 1981
4. Corrado OJ et al: Asthma and rhinitis after exposure to glutaraldehyde in endoscopy units. *Hum Toxicol* 5:325–327, 1986
5. Norback D: Skin and respiratory symptoms from exposure to alkaline glutaraldehyde in medical services. *Scand J Work Environ Health* 14:366–371, 1988
6. Slesinski RS et al: Mutagenicity evaluation of glutaraldehyde in a battery of in vitro bacterial and mammalian test systems. *Food Chem Toxicol* 21:621–629, 1983
7. Zissu D, Gagnaire F, Bonnet P: Nasal and pulmonary toxicity of glutaraldehyde in mice. *Tox Letts* 71:53–62, 1994

GLYCIDOL
CAS: 556-52-5

$C_3H_6O_2$

Synonym: 2,3-Epoxy-1-propanol

Physical Form.   Colorless liquid

Uses.   Stabilizer in the manufacture of vinyl polymers; in preparation of glycerol, glycidyl ethers, esters, and amines; in pharmaceuticals; in sanitary chemicals

Exposure.   Inhalation

Toxicology.   Glycidol is an irritant of the eyes, upper respiratory tract, and skin; at high concentrations, it causes narcosis in animals, and it is expected that severe exposure will have the same effect in humans. It is carcinogenic and mutagenic in experimental animals.

The acute hazard to humans from vapor exposure appears to be relatively slight, as ample warning in the form of eye, nose, and throat irritation occurs at low concentrations; no chronic effects have been reported in humans.[1]

The LC_{50} in mice was 450 ppm for a 4-hour exposure; in rats, the LC_{50} for 8 hours was 580 ppm; labored breathing, lacrimation, salivation, and nasal discharge were seen, and pneumonitis was observed at autopsy.[1] Rats repeatedly exposed to 400 ppm showed only slight eye irritation and mild respiratory distress, with no evidence of systemic toxicity.

The oral LD_{50} was 0.45 g/kg for mice and 0.85 g/kg for rats; symptoms included central nervous system depression characterized by incoordination, ataxia, coma, and death.[1] Animals surviving exposure showed reversible excitation and tremor; lacrimation and labored breathing also were observed. In 16-day studies, focal demyelination of the brain occurred in mice given 300 mg/kg/day by gavage.[2] In the same study, male rats receiving 300 mg/kg/day had edema and degeneration of the epididymal stroma and atrophy of the testes. Longer exposures of 13 weeks resulted in cerebellar necrosis and demyelination of the medulla, renal tubular cell degeneration, and thymic lymphoid necrosis in rats and demyelinaton of the medulla and thalamus and renal tubular cell degeneration in mice. A reduction in sperm count and sperm motility and testicular atrophy occurred in males of both species at doses of up to 300 or 400 mg/kg/day for mice and rats, respectively.[2] Glycidol was not teratogenic to mice receiving up to 200 mg/kg/day by gavage on days 6 through 15 of gestation.[3]

In experimental animals, glycidol was a broadly acting, multipotent carcinogen. Rats administered 75 mg/kg or 37.5 mg/kg, 5 days/ week, by gavage for up to 2 years developed mesotheliomas (80%), mammary adenocarcinomas (33%), forestomach tumors (22%), tumors of the oral mucosa (14%), Zymbal gland tumors (12%), brain gliomas (12%), follicular cell tumors of the thyroid (12%), and intestinal tumors (8%).[2] Mice received a slightly lower dose and developed a smaller spectrum of tumors, including tumors of the mammary gland, Harderian gland, and forestomach. In hamsters (20 male and 20 female) administered 12 mg twice weekly by gavage for 60 weeks, there were more tumors in treated animals than in controls.[4] However, the spleen was the only notable target organ, and the number of hamsters with spleen hemangiosarcomas was small. (*Note:* The study used only one dose and a small number of animals.)

Application of the liquid to animal skin caused moderate irritation. Chronic topical administration of a 5% solution to mice did not cause skin tumors or any visible skin reaction.[5] In the eyes, glycidol produced severe irritation; despite the severity of primary injury, no blindness or permanent defects in the cornea, lens, or iris resulted from the applications.

In *Salmonella typhimurium*, glycidol is a potent direct-acting mutagen.[2] In vitro, it caused an increased number of sister chromatid exchanges and chromosomal aberrations in Chinese hamster ovary cells and human lymphocytes.

The 1995 ACGIH threshold limit value–time-weighted average (TLV-TWA) for glycidol is 25 ppm (76 mg/m³).

REFERENCES

1. Hine CH et al: The toxicology of glycidol and some glycidyl ethers. *AMA Arch Ind Health* 14:250–264, 1956
2. National Toxicology Program (NTP): *Toxicology and Carcinogenesis Studies of Glycidol in F344/N Rats and B6C3F₁ Mice.* NTP Technical Report Series No 374, 1990
3. Marks TA, Gerling FS, Staples RE: Teratogenic evaluation of epichlorohydrin in the mouse and rat and glycidol in the mouse. *J Toxicol Environ Health* 9:87–96, 1982
4. Lijinsky W, Kovatch RM: A study of the carcinogenicity of glycidol in Syrian hamsters. *Toxicol Ind Health* 8:267–271, 1992
5. Van Duuren BL et al: Carcinogenicity of epoxides, lactones and peroxy compounds, VI. Structure and carcinogenic activity. *J Natl Cancer Inst* 39:1217–1228, 1967

GRAPHITE (NATURAL)
CAS: 7782-42-5

C (with traces of Fe, SiO₂, etc)

Synonyms: Plumbago; black lead; mineral carbon

Physical Form. Usually soft, black scales; crystals rare

Uses. For pencils, refractory crucibles, pigment, lubricant, polishing compounds, electroplating

Exposure. Inhalation

Toxicology. Natural graphite dust causes graphite pneumoconiosis.

The earliest roentgenologic changes may be the disappearance of normal vascular markings with the later appearance of pinpoint and nodular densities in all lung fields.[1,2] Massive lesions, when present, are caused by large cysts filled with black fluid. The pleura is often involved; hydrothorax, pneumothorax, and pleural thickening may occur.

At autopsy, the lungs are gray-black to black; histologically, there are widely scattered particles, spicules, and plates of graphite, often within intra-alveolar phagocytes amid diffuse interstitial fibrosis and, occasionally, pneumonitis. There are also interwoven bands of collagen, similar to those found in silicosis, which frequently are the most prominent feature of the fibrotic lesions occupying the lung and the bronchial lymph nodes. Symptoms include expectoration of black sputum, dyspnea, and cough.

Of 344 workers in a graphite mine in Ceylon, 78 had radiographic abnormalities, including small, rounded, and irregular opacities, large opacities, and enlargement of hilar shadows. Some affected workers had cough, dyspnea, and/or digital clubbing.[3]

It has generally been believed that the capacity of inhaled natural graphite dust to cause a disease is largely the result of its crystalline silica component.[4]

The 1995 ACGIH threshold limit value–time-weighted average (TLV-TWA) for graphite (except graphite fiber) is 2 mg/m^3 as respirable dust.

REFERENCES

1. Pendergrass EP et al: Observations on workers in the graphite industry—Part One. *Med Radiogr Photogr* 43:70–99, 1967
2. Pendergrass EP et al: Observations on workers in the graphite industry—Part Two. *Med Radiogr Photogr* 44:2–17, 1968
3. Ranasinka KW, Uragoda CG: Graphite pneumoconiosis. *Br J Ind Med* 29:178–183, 1972
4. Hanoa R: Graphite pneumonoconiosis: a review of etiologic and epidemiologic aspects. *Scand J Work Environ Health* 9:303–314, 1983

GRAPHITE (SYNTHETIC)
CAS: 7782-42-5

C

Synonyms: None

Physical Form. A crystalline form of carbon made from high-temperature treatment of coal or petroleum products; same properties as natural graphite; chemically inert

Uses. Similar to those of natural graphite in refractories and electrical products

Exposure. Inhalation

Toxicology. Pure synthetic graphite acts as an inert or nuisance dust.

In contrast to the several reports of pneumoconiosis in workers exposed to natural graphite (qv), there was, until recently, only the rare anecdotal report of significant pulmonary findings due to exposure to synthetic graphite.[1,2] One man who had spent 17 years turning and grinding synthetic graphite bars developed simple pneumoconiosis, with cough, dyspnea, reduced pulmonary function, and X-ray changes.[1] At autopsy, there was emphysema, with scattered fine black nodules (microscopic to 5 mm diam) with some strands of fibrous tissue. Many of the nodules consisted of almost acellular collagen. There were traces of iron in the nodules and the hilar nodes. Ashed material from the lung showed little or no birifringent particles, indicating the absence of siliceous material. The lung contained 8.8% to 9.5% carbon by dry weight.

Despite a general belief that pneumoconiosis in the United States ceased to be a problem after World War II, 5 workers involved in the manufacture of carbon electrodes have now been reported to have developed this condition following exposures after 1940.[3] The variability in clinical findings that characterizes these cases suggests a mixed dust exposure.

Synthetic graphite injected peritoneally in mice produces a reaction characteristic of an inert material. On the basis of experimental evidence, and the rarity of reports of adverse effects of exposure in man, it is concluded that pure synthetic graphite acts only as an inert dust.

The 1995 ACGIH threshold limit value–time-weighted average (TLV-TWA) for synthetic graphite is 2 mg/m³.

REFERENCES

1. Lister WB, Wimborne D: Carbon pneumoconiosis in a synthetic graphite worker. *Br J Ind Med* 29:108–110, 1972
2. Hanoa R: Graphite pneumoconiosis. A review of etiologic and epidemiologic aspects. *Scand J Work Environ Health* 9:303–314, 1983
3. Petsonk EL: Pneumoconiosis in carbon electrode workers. *J Occup Med* 30:887–891, 1988

HAFNIUM (AND COMPOUNDS)
CAS: 7440-58-6

Hf

Compounds: Hafnium chloride ($HfCl_4$); hafnyl chloride ($HfOCl_4$)

Physical Form. Hard, shiny, ductile stainless steel–colored metal or dull-gray powder

Uses/Sources. Used in control rods in nuclear reactors and in manufacture of light bulb filaments; found in all zirconium-containing minerals

Exposure. Inhalation

Toxicology. Hafnium dust is very low in toxicity. No health hazards have been recognized from the industrial handling of hafnium powder other than those arising from fire or explosion.[1]

Hafnium salts are mild irritants of the eye and the skin and have produced liver damage in animals.[2] In mice, the LD_{50} of hafnyl chloride by intraperitoneal injection was 112 mg/kg.[2] In cats, intravenous administration of hafnyl chloride at 10 mg/kg was fatal. Rats fed a diet containing 1% for 12 weeks showed slight changes in the liver, consisting of perinuclear vacuolization of the parenchymal cells and coarse granularity of the cytoplasm.[1] The application of 1 mg of hafnium chloride to the eyes of rabbits produced transient irritation. Topical application of hafnium chloride crystals to unabraded rabbit skin produced transient edema and erythema; application to abraded skin caused ulceration.[2]

The 1995 ACGIH threshold limit value–time-weighted average (TLV-TWA) for hafnium is 0.5 mg/m³.

REFERENCES

1. Chemical Safety Data Sheet, SD-92, Zirconium and hafnium powder, pp 5–6, 10. Washington, DC, MCA, Inc, 1966
2. Haley TJ, Raymond K, Komesu N, Upham HC: The toxicologic and pharmacologic effects of hafnium salts. *Toxicol Appl Pharmacol* 4:238–246, 1962

HALOTHANE
CAS: 151-67-7

CF₃CHClBr

Synonyms: 2-Bromo-2-chloro-1,1,1-trifluoroethane; bromochlorotrifluoroethane; Fluothane

Physical Form. Colorless liquid

Use. Clinical anesthetic

Exposure. Inhalation

Toxicology. Halothane causes central nervous system depression, affects the cardiovascular system and, occasionally, causes hepatitis.

Halothane is used as a clinical anesthetic, and all levels of central nervous system depression can be expected, including amnesia, analgesia, anesthesia, and respiratory depression. Levels ranging from 5000 to 30,000 ppm can induce anesthesia, whereas 5000 to 15,000 can maintain it.[1] A 30-minute exposure to 4000 ppm caused amnesia and impairment of manual dexterity, whereas similar exposure to 1000 ppm did not alter the outcome on various psychomotor tests.[2]

Levels of halothane associated with anesthesia may also reduce cardiac output by 20% to 50%.[3] Tachyarrhythmias may occur in the presence of halothane.[4]

Hepatitis occasionally occurs in patients after clinical anesthesia. Typically, 2 to 5 days after anesthesia, a fever develops, accompanied by anorexia, nausea, and vomiting.[5] There may be a progression to hepatic failure, and death occurs in about 50% of these patients. The incidence of the syndrome is 1 in 10,000 anesthetic administrations and is seen most often after repeated administration of halothane over a short period of time.

Epidemiological studies of occupationally exposed populations have examined possible carcinogenic and teratogenic effects of chronic exposure to the operating-room environment. In one study, a high rate of miscarriages (18/31) was observed among pregnant anesthetists.[6] Pregnancies among anesthetists and nurses in anesthetic departments ended in spontaneous abortion or premature delivery approximately twice as often (20% vs. 10%) as among unexposed women.[7] In a third study, female anesthetists were found to have a higher frequency of involuntary infertility (12% vs. 6%) and spontaneous abortion (18.3% vs. 14.7%) than unexposed women.[8] A national study reported that women chronically exposed to the operating-room environment had increased risks of cancer, diseases of the liver and the kidney, spontaneous abortion, and congenital anomalies in their children.[9,10] It has been noted that all the epidemiological studies to date have involved either mixed exposures, or exposure to unmeasured concentrations of halothane.[11]

In animal studies, macroscopically visible injuries to fetuses have occurred following exposure to 1600 ppm, 6 hr/day for multiple days during gestation.[11] Retardations occurred in mice similarly exposed at 1000 ppm, and 3000 ppm caused minor skeletal variations. No visible damage to the offspring was apparent in rabbits treated 1 hr/day to 22,000 ppm during organogenesis.

No carcinogenicity in animals has been reported.[12] The IARC has stated that there is inadequate evidence for carcinogenicity to humans.[13]

The 1995 ACGIH threshold limit value–time-weighted average (TLV-TWA) for halothane is 50 ppm (404 mg/m³).

REFERENCES

1. Halothane. *Documentation of the TLVs and BEIs*, 6th ed, pp 721–723. Cincinnati, OH, American Conference of Governmental Industrial Hygienists (ACGIH), 1991
2. Cook TL, Smith M, Winter PM, et al: Effect of subanesthetic concentrations of enflurane and halothane on human behavior. *Anesth Analg* 57:434–440, 1978
3. Marshall BE et al: Pulmonary venous admixture before, during and after halothane: oxygen anesthesia in man. *J Appl Physiol* 27:653–657, 1969
4. Marshall BE, Wollman H: General anesthetics. In Gilman AG, Goodman LS, Gilman A (eds): *Goodman and Gilman's The Pharmacological Basis of Therapeutics*, 6th ed, pp 277–283. New York, Macmillan, 1980
5. Summary of the National Halothane Study. *JAMA* 197:775–788, 1966
6. Vaisman AI: Working conditions in surgery and their effect on the health of anesthesiologists. *Eksp Khir Anesteziol* 3:44–49, 1967
7. Askrog V, Harvald B: Teratogen effeck af inhalations—anaestetika. *Nord Med* 83:498–500, 1970
8. Knill-Jones RP, Moir DD, Rodrigues LV,

Spence AA: Anaesthetic practice and pregnancy. *Lancet* 2:1326–1328, 1972

9. Cohen EN, Brown BW, Bruce DL, et al: Occupational disease among operating room personnel: a national survey. Report of an ad hoc committee on the effect of trace anesthetics on the health of operating room personnel, American Society of Anesthesiologists. *Anesthesiology* 41:321–340, 1974

10. National Institute for Occupational Safety and Health: *Criteria for a Recommended Standard...Occupational Exposure to Waste Anesthetic Gases and Vapors*, p 255. Washington, DC, US Government Printing Office, 1977

11. Baeder C, Albrecht M: Embryotoxic/teratogenic potential of halothane. *Int Arch Environ Health* 62:263–271, 1990

12. Eger EI II et al: A test of the carcinogenicity of enflurane, isoflurane, halothane, methoxyflurane, and nitrous oxide in mice. *Anesth Analg* 57:678–694, 1978

13. *IARC Monographs on the Evaluation of Carcinogenic Risk to Humans, Overall Evaluations of Carcinogenicity: An Updating of IARC Monographs Vols 1–42*, Suppl 7, pp 93–95. Lyon, International Agency for Cancer, 1987

HELIUM
CAS:7440-59-7

He

Synonyms: None

Physical Form. Gas

Uses. Inert gas shield in arc welding; in air ships; in mixtures with neon and argon for electronic tubes and "neon" signs

Exposure. Inhalation

Toxicology. Helium is a simple asphyxiant.

Helium is among a number of gases that have no significant physiological action and act primarily as simple asphyxiants by displacing oxygen from the environment.[1] Atmospheres

deficient in oxygen do not provide adequate warning.

No threshold limit value has been established for helium. The limiting factor is available oxygen, which should be 18% by volume under normal atmospheric pressure.

REFERENCES

1. ACGIH: *Documentation of the TLVs and BEIs.* 6th ed, p 725. Cincinnati, OH, American Conference of Governmental Industrial Hygienists, 1991

HEPTACHLOR
CAS: 76-44-8

$C_{10}H_5Cl_7$

Synonyms: 2-Chlorochlordene; Drinox; E-3314; ENT 15,152; Vesicol 104

Physical Form. White to tan crystalline solid

Uses. Insecticide for control of boll weevil and termite

Exposure. Inhalation; skin absorption; ingestion

Toxicology. Heptachlor is a convulsant in animals and causes liver damage.

There have been occasional anecdotal reports of blood dyscrasias following exposure to heptachlor, but exposure levels are not available.[1]

In rats, the oral LD_{50} was 90 mg/kg; within 30 to 60 minutes after administration, effects were tremor and convulsions; liver necrosis was noted.[2] Multiple applications of a solution to the skin of rats of 20 mg/kg were toxic, indicating a marked cumulative action.[2] Reversible histological changes in the rat liver have occurred following dosages of 0.35 mg/kg for 50 weeks.[1] Rats given heptachlor in the diet at 6 mg/kg body weight developed cataracts after 4.5 to 9.5

months of feeding.[3] This observation has not been replicated in a number of subsequent animal studies.[1] In animals, heptachlor is more potent than chlordane, to which it is closely related chemically.[4]

Chronic oral exposure to heptachlor increased the incidence of liver carcinomas in 3 species of mice and 1 species of rats.[5] A study of two cohorts of workers exposed to chlordane and heptachlor at two different production facilities failed to demonstrate any overall excess of cancer. There was 1 death from liver cancer, with 0.59 expected. There was a slight excess of lung cancer (12 observed, 9 expected), but this was not statistically significant.[6]

The IARC has determined that there is sufficient evidence in experimental animals for the carcinogenicity of heptachlor but inadequate evidence of carcinogenicity in humans.[7] Heptachlor is not considered genotoxic in mammalian systems.[5]

The 1995 ACGIH threshold limit value–time-weighted average (TLV-TWA) for heptachlor is 0.05 mg/m^3, with an A3 animal carcinogen designation and a notation for skin absorption.

REFERENCES

1. Hayes WJ Jr: *Pesticides Studied in Man*, pp 233–234. Baltimore, MD, Williams and Wilkins, 1982
2. von Oettingen WF: *The Halogenated Aliphatic, Olefinic, Cyclic, Aromatic, and Aliphatic-Aromatic Hydrocarbons Including the Halogenated Insecticides, Their Toxicity and Potential Dangers*. US Public Health Service Pub No 414, pp 326–327. Washington, DC, US Government Printing Office, 1955
3. Mestitzova M: On reproduction studies and the occurrence of cataracts in rats after long-term feeding of the insecticide heptachlor. *Experientia* 23:42–43, 1967
4. Council on Pharmacy and Chemistry: The present status of chlordane. *JAMA* 158:1364–1367, 1955
5. Agency for Toxic Substances and Disease Registry (ATSDR): *Toxicological Profile for Heptachlor/Heptachlor Epoxide*. US Department of Health and Human Services, Public Health Service, pp 1–131, TP-92/11, 1993
6. Wang HH, MacMahon B: Mortality of workers employed in the manufacture of chlordane and heptachlor. *J Occup Med* 21:745–748, 1979
7. *IARC Monographs on the Evaluation of the Carcinogenic Risk of Chemicals to Humans*, Vol 53, Occupational exposures in insecticide application, and some pesticides, pp 115–175. Lyon, International Agency for Research on Cancer, 1991

HEPTACHLOR EPOXIDE
CAS: 1024-57-3

$C_{10}H_5Cl_7$

Synonyms: Epoxyheptachlor; 1,4,5,6,7,8,8a-heptachloro-2,3-epoxy-3a,4,7,7a-tetra-hydro-4,7-methanoindene

Physical Form.   White crystalline solid

Sources.   Not commercially produced; formed as a metabolite of heptachlor in mammals

Exposure.   Consequent to exposure to heptachlor

Toxicology.   Heptachlor epoxide is a liver carcinogen in rodents.

Heptachlor epoxide is a metabolic product of heptachlor.[1] Trace amounts of heptachlor epoxide have been found in fat in the general population in most countries.[2] Heptachlor epoxide has been found in tissues of stillborn infants, indicating transplacental transfer.[3]

Heptachlor epoxide is more toxic than heptachlor.[3] The acute oral LD$_{50}$ for heptachlor epoxide in rodents and rabbits ranged from 39 to 144 mg/kg.[3] Following dietary exposure of rats, heptachlor epoxide caused hepatic-cell vacuolization at all dose levels (0.5–10 ppm for up to 108 weeks). Degeneration, hepatomegaly, and regeneration were also reported. Like heptachlor, the ability of heptachlor epoxide to induce lethality following acute exposure may involve its ability to interfere with nerve action or release of neurotransmitters and to inhibit the function of the receptor for γ-aminobutyric acid.[3]

In 2 three-generation studies with rats administered either heptachlor, heptachlor epoxide, or a mixture of the two in the diet, the number of resorbed fetuses increased and the fertility decreased with succeeding generations. No adverse effects on reproductive capacity were reported in male mice receiving single oral doses of 7.5 or 15 mg/kg heptachlor:heptachlor epoxide (25%:75%) in a dominant lethal assay.[3]

Mice fed heptachlor epoxide in the diet, at 10 mg/kg for 24 months, showed a significant excess of liver carcinomas. In another study, an excess of liver carcinomas was observed in female rats given 5 and 10 mg/kg in the diet.[2]

The International Agency for Research on Cancer has concluded that there is sufficient evidence that heptachlor epoxide is carcinogenic in experimental animals. There are no data to evaluate the potential carcinogenicity of heptachlor epoxide in humans.[2] The majority of genotoxic assays suggest that heptachlor epoxide is not genotoxic.[3]

The 1995 ACGIH threshold limit value–time-weighted average (TLV-TWA) for heptachlor epoxide is 0.05 mg/m[3], with an A3 animal carcinogen designation and a notation for skin absorption.

REFERENCES

1. Hayes WJ Jr: *Pesticides Studied in Man*, pp. 233–234. Baltimore, MD, Williams and Wilkins, 1982
2. *IARC Monographs on the Evaluation of Carcinogenic Risks to Humans*, Vol. 53, Occupational exposures in insecticide application, and some pesticides, pp 115–175. Lyon, International Agency for Research on Cancer, 1991
3. Agency for Toxic Substances and Disease Registry (ATSDR): *Toxicological Profile for Heptachlor/Heptachlor Epoxide*, US Department of Health and Human Services, Public Health Service, p 131, TP-92/11, 1993

n-HEPTANE
CAS: 142-82-5

$CH_3(CH_2)_5CH_3$

Synonyms: Dipropyl methane; heptyl hydride; heptane

Physical Form. Volatile, flammable liquid

Uses. As standard in testing knock of gasoline engines; solvent

Exposure. Inhalation

Toxicology. *n*-Heptane causes central nervous system depression.

Human subjects exposed to 1000 ppm for 6 minutes, or to 2000 ppm for 4 minutes, reported slight vertigo.[1] At 5000 ppm for 4 minutes, effects included marked vertigo, inability to walk a straight line, hilarity, and incoordination, but there were no complaints of eye, upper respiratory tract, or mucous membrane irritation. In some subjects, a 15-minute exposure at 5000 ppm produced a state of stupor lasting for 30 minutes after exposure. These subjects also reported loss of appetite, slight nausea, and a taste resembling gasoline for several hours after exposure.

Dermal application resulted in immediate irritation, characterized by erythema and hyperemia. The subjects complained of a painful burning sensation and, after 5 hours, blisters formed on the exposed areas.[2]

n-Heptane induced anesthesia in mice at 8000 ppm; at 32,000 ppm for 5 minutes, mice developed irregular respiratory patterns, followed by deep narcosis; at 48,000 ppm, 3 of 4 mice had respiratory arrest within 4 minutes.[3] Chronic inhalation studies in rats exposed to 400 or 3000 ppm 6hr/day, 5day/week for 26 weeks, found no evidence of neurological disturbances or organ toxicity.[4] Except for increased serum alkaline phosphatase levels in females at 3000 ppm, blood chemistry showed no hematological, renal, or liver abnormalities.

Although *n*-heptane exposure produces narcotic effects, it has not been shown to cause the type of peripheral neuropathy associated with *n*-hexane at the same exposure levels.[5] A metabolic study of heptane in rats and humans showed that only a very small amount of 2,5-heptanedione, the purported neurotoxic metabolite responsible for peripheral neuropathy, is produced.[6] Clinical damage to the peripheral nervous system following *n*-heptane exposure, therefore, seems unlikely.[6]

The 1995 ACGIH threshold limit value–time-weighted average (TLV-TWA) for *n*-heptane is 400 ppm (1640 mg/m³) with a ceiling limit of 500 ppm (2050 mg/m³).

REFERENCES

1. Patty FA, Yant WP: Report of investigations—odor intensity and symptoms produced by commercial propane, butane pentane, hexane and heptane vapor, No 2979. US Department of Commerce, Bureau of Mines, 1929
2. National Institute for Occupational Safety and Health: *Criteria for a Recommended Standard ... Occupational Exposure to Alkanes* (C5-C8). DHEW (NIOSH) Pub No 77-151. Washington, DC, US Government Printing Office, 1977
3. Swann HE Jr, Kwon BK, Hogan GK, Snellings WM: Acute inhalation toxicology of volatile hydrocarbons. *Am Ind Hyg Assoc J* 35:511–518, 1974
4. American Petroleum Institute: A 26-week inhalation toxicity study of heptane in the rat. API Contract No PS-29; pp 1–32, 1980
5. Takeuchi Y, Ono Y, Hisanaga N, et al: A comparative study of the neurotoxicity of *n*-pentane *n*-hexane, and *n*-heptane in the rat. *Br J Ind Med* 37:241–247, 1980
6. Perbellini L, Brugnone F, Cocheo V, et al: Identification of the *n*-heptane metabolites in rat and human urine. *Arch Toxicol* 58:229–234, 1986

HEXACHLOROBENZENE
CAS: 118-74-1

C_6Cl_6

Synonyms: Hexachlorobenzol; perchlorobenzene; HCB; pentachlorophenyl chloride

Physical Form. White crystalline solid

Uses. No current commercial use of HCB in the United States; was used as a pesticide until 1985; minor amounts have been used in the production of synthetic rubber

Exposure. Ingestion; inhalation

Toxicology. Hexachlorobenzene (HCB) has caused porphyria cutanea tarda, enlarged liver, painless arthritis, enlarged thyroid, and neurological symptoms following ingestion of contaminated seed grain.

Evidence of the human health effects of HCB exposure comes primarily from Turkey where, between 1955 and 1959, 4000 people consumed grain treated with HCB. The consumption level was estimated to be 0.05 to 0.2 g/day for several years. The majority of the affected patients were children.[1] There was a high rate of mortality in infants of lactating mothers known to have ingested the bread, and children born to porphyric mothers did not survive.[2] Other manifestations included the development of a condition resembling porphyria cutanea tarda, with abnormal porphyrin metabolism and skin lesions, hyperpigmentation, liver enlargement, hirsutism, short stature (in affected children), thyroid enlargement, painless arthritis, and neurological findings, including weakness, paresthesias, cogwheeling, and myotonia.[3]

A study of 32 of the affected individuals demonstrated that abnormal porphyrin metabolism and symptoms persisted 20 years after HCB ingestion.[1]

In rodents, the liver is a primary target organ for HCB effects. Exposure to 2000 ppm in the diet caused increased porphyrin levels, microscopic lesions in the liver, elevation of serum enzyme levels, and induction of liver microsomal enzymes.[4] Male rats exposed at 40 ppm in the diet for 130 weeks developed chronic nephrosis, and renal tubular damage was noted in rats exposed to 10 mg/kg/day for 15 weeks.[5] Nephropathy is dependent on the presence of α_{2u}-globulin and is specific to male rats.[6] Exposure of animals to 100, 200, or 500 ppm in the diet caused a 2.5 to 3-fold increase in thyroid size.[7] In another study of rats, HCB has been demonstrated to cause hyperparathyroidism and osteosclerosis.[5] Hexachlorobenzene is immunotoxic in a number of animal studies, interfering with humoral and cellular immune functions in dogs, rats, mice, and monkeys.[2]

There is no information on in utero developmental effects in humans exposed to HCB, but oral exposure of young children has caused small or atrophied hands, short stature, pinched facies, osteoporosis of the carpal, metacarpal, and phalangeal bones, and painless arthritic changes.[2] HCB has been demonstrated to cross the placenta in humans and in rodents.[1] HCB residues have been detected in human milk and adipose tissue and in the blood of the umbilical cord of newborn infants and their mothers. Teratogenic effects were not demonstrated in rats following exposure to up to 120 mg/kg/day during organogenesis. Cleft palate and renal agenesis were observed in mice at 100 mg/kg/day.[1] Parameters such as fertility index and gestational indices have not been affected in rats at HCB doses of up to 40 ppm.[8]

No excess of cancer was reported in two follow-up studies of affected individuals in Turkey about 20 to 30 years after consumption of contaminated grain had ceased.[9,10] In mice, liver tumors were observed following exposure to HCB at 12 to 24 mg/kg/day in the diet but not at 6 mg/kg/day.[1] Hepatomas, hepatocellular carcinomas, bile duct adenomas, and renal cell adenomas were observed in rats following dietary administration.[11] Liver tumors were also observed in 100% of surviving females and 16% of males following dietary administration to rats for 90 weeks. In another study, increased incidence of parathyroid adenomas and adrenal pheochromocytomas was observed in male and female rats; liver neoplastic nodules were reported in females of the F_1 generation in a two-generation feeding study.

The International Agency for Research on Cancer has stated that there is inadequate evidence to link HCB with human cancer. However, there is sufficient evidence for carcinogenicity of HCB in experimental animals.[11]

In an in vivo experiment in rats, HCB did not induce dominant lethal mutations. Chromosomal aberrations were not induced in cultured Chinese hamster ovary cells, nor were mutations induced in bacteria.[2,11] Hexachlorobenzene does not appear to be genotoxic.

The 1995 ACGIH threshold limit value–time-weighted average for hexachlorobenzene is 0.025 mg/m^3 with an A3 animal carcinogen designation and a notation for skin absorption.

REFERENCES

1. *IARC Monographs on the Evaluation of Carcinogenic Risk to Humans*, Vol. 20, Some halogenated hydrocarbons, pp 155–178. Lyon, International Agency for Research on Cancer, 1979

2. Agency for Toxic Substances and Disease Registry (ATSDR): *Toxicological Profile for Hexachlorobenzene.* US Department of Health and Human Services, Public Health Service, 1994

3. Hexachlorobenzene. *Dangerous Properties of Industrial Materials Report* 9:99–101, 1989

4. Wada O et al: Behavior of hepatic microsomal cytochromes after treatment of mice with drugs known to disturb porphyrin metabolism in liver. *Biochem Pharmacol* 17:595–603, 1968

5. Andrews JE et al: Hexachlorobenzene-induced hyperparathyroidism and osteosclerosis in rats. *Fund Appl Toxicol* 12:242–251, 1989

6. Bouthillier L, Greselin E, Brodeur J, et al: Male rat specific nephrotoxicity resulting from subchronic administration of hexachlorobenzene. *Toxicol Appl Pharmacol* 110:315–326, 1991

7. Smith AG et al: Goitre and wasting induced in hamsters by hexachlorobenzene. *Arch Toxicol* 60:343–349, 1987

8. Arnold DL et al: Long-term toxicity of hexachlorobenzene in the rat and the effect of

dietary vitamin A. *Food Chem Toxicol* 23:779–793, 1985

9. Peters HA et al: Epidemiology of hexachloro-benzene-induced porphyria in Turkey. *Arch Neurol* 39:744–749, 1982
10. Cripps DJ et al: Porphyria turcica due to hexa-chlorobenzene: 20 to 30 year follow-up study on 204 patients. *Br J Dermatol* 111:413–422, 1984
11. *IARC Monographs on the Evaluation of Carcin ogenic Risks to Humans, Overall Evaluations of Carcinogenicity: An Updating of IARC Mono-graphs Vols 1–42*, Suppl 7, pp 219–220. Lyon, International Agency for Research on Cancer, 1987.

HEXACHLOROBUTADIENE

CAS: 87-68-3

C_4Cl_6

Synonyms: HCBD; hexachloro-1,3-butadiene; perchlorobutadiene

Physical Form. Colorless liquid

Sources/Uses. An unwanted by-product in the production of tetrachloroethylene, trichloroethylene, carbon tetrachloride, and chlorine; formerly used as a pesticide outside the United States

Exposure. Inhalation; skin absorption; ingestion

Toxicology. Hexachlorobutadiene (HCBD) causes kidney damage, including renal cancer in experimental animals; it also produces central nervous system effects and causes hepatic disorders at very high concentrations.

In the only report of human exposures cited by the IARC, vineyard workers exposed to hexachlorobutadiene (0.8–30 mg/m³) and polychlorobutane (0.12–6.7 mg/m³) showed multiple toxicological effects contributing to the development of hypotension, cardiac disease,

chronic bronchitis, disturbances of nervous system function, and chronic hepatitis.[1]

In rats, 4- to 7-hour exposures to concentrations ranging from 133 to 500 ppm caused death.[2] Guinea pigs and cats died after exposures to 160 ppm for 53 minutes or from 7.5-hour exposures to 35 ppm.

Repeated exposures to 250 ppm (twice for 4 hours) or 100 ppm (twelve 6-hour exposures) caused eye and nose irritation and respiratory difficulty in rats; at autopsy, there was degeneration of renal tubules, along with injury to the adrenals.[3] Fifteen exposures to 5 or 10 ppm resulted in no observed toxic effect, except for retarded weight gain at the higher dose.

Small groups of rats, rabbits, and guinea pigs exposed to 3 ppm, 7 hr/day for approximately 5 months, had liver and kidney damage, whereas those exposed to 1 ppm were not adversely affected.[4]

A 30-day dietary study at 30, 65, and 100 mg/kg/day in rats caused renal toxicity in the form of increased kidney to body weight ratio and renal tubular degeneration, necrosis, and regeneration. Other adverse effects included reduced body weight gain at 10 mg/kg/day and minimal hepatocellular swelling at 100 mg/kg/day.[2]

In a chronic dietary study, ingestion by rats of 20 mg/kg/day for up to 2 years resulted in a statistically significant increase in renal tubular adenomas and adenocarcinomas, some of which metastasized to the lungs.[5] Other toxicological effects included decreased body weight gain, increased mortality, increased excretion of urinary coproporphyrin, increased terminal weights of the kidneys, and increased renal tubular epithelial hyperplasia. At the intermediate dose level of 2.0 mg/kg/day, effects were limited to an increased urinary excretion of urinary coproporphyrin and increased hyperplasia of the renal tubular epithelium. Lifetime ingestion of the lowest dose level of 0.2 mg/kg/day caused no treatment-related effects.

In 1979, the IARC determined, on the basis of a lifetime feeding study, that there was limited evidence for carcinogenicity in rats, but no data are available to determine carcinogenicity in humans.[1] In 1987, the IARC still regarded this substance as not classifiable as to carcinogenicity in humans.[6]

Studies of the mutagenicity of hexachlorobutadiene and its metabolites concluded that hexachlorobutadiene exerts genotoxic effects after metabolic activation.[7] This hypothesis may be important for the evaluation of the carcinogenic potential, as it is generally accepted that a minimum risk threshold for genotoxic substances cannot be assigned. Alternatively, it has also been reported that HCBD induces little genotoxicity in vivo and that renal tumors have been observed only at doses that induce severe chronic nephrosis. Accordingly, chronic cytotoxicity to the renal proximal tubular cells by HCBD intermediates formed in the kidney may account for HCBD-induced renal carcinogenesis.[8]

Reproductive indices, including pregnancy rate, gestational survival, neonatal survival, and morphological alterations in neonates, were not affected when male and female rats were fed up to 20 mg/kg/day for 90 days prior to mating and during gestation and lactation.[9]

Hexachlorobutadiene is classified as an A2 suspected human carcinogen with a 1995 TLV-TWA of 0.02 ppm (0.21 mg/m³) and a notation for skin.

REFERENCES

1. *IARC Monographs on the Evaluation of the Carcinogenic Risk of Chemicals to Man*, Vol 20, Some halogenated hydrocarbons pp 179–193. Lyon, International Agency for Research on Cancer, 1979
2. Kociba RJ, Schwetz BA, Keyes DG, et al: Chronic toxicity and reproductive studies of hexachlorobutadiene in rats. *Environ Health Perspectives* 21:49–53, 1977
3. Gage JC: The subacute inhalation toxicity of 109 industrial chemicals. *Br J Ind Med* 27:1–15, 1970
4. The Dow Chemical Co, Midland, MI, unpublished data, cited by Torkelson TR, and Rose VK. In Clayton G, Clayton FE (eds): Patty's *Industrial Hygiene and Toxicology*, 3rd ed, rev, Vol 2B, *Toxicology*, p 3582. New York, Wiley-Interscience, 1981
5. Kociba RJ, Keyes DG, Jersey GC, et al: Results of a two year chronic toxicity study with hexachlorobutadiene in rats. *Am Ind Hyg Assoc J* 38:589–602, 1977
6. *IARC Monographs on the Evaluation of the Carcinogenic Risk to Humans, Overall Evaluations of Carcinogenicity: An Updating of IARC Monographs Vols 1–42*, Supp 7, p 64. Lyon, International Agency for Research on Cancer, 1987
7. Reichert D, Neudecker T, Schutz R: Mutagenicity of hexachlorobutadiene, perchlorobutenoic acid and perchlorobutenoic acid chloride. *Mut Res* 137:89–93, 1984
8. Dekant W, Vamvakas S, Anders MW: Bioactivation of hexachlorobutadiene by glutathione conjugation. *Food Chem Toxic* 28:285–293, 1990
9. Schweitz BA, Norris JM, Kociba RJ, et al: Results of a reproduction study in rats fed diets containing hexachlorobutadiene. *Toxicol Appl Pharmacol* 42:387–398, 1977

HEXACHLOROCYCLOPENTADIENE
CAS: 77-47-4

C_5Cl_6

Synonyms: HCCPD; HCCP; HEX; perchlorocyclopentadiene

Physical Form. Yellow to amber-colored liquid

Uses. Intermediate in the manufacture of chlorinated pesticides; intermediate in the manufacture of flame retardants

Exposure. Inhalation

Toxicology. Hexachlorocyclopentadiene is a lacrimator and severe irritant of the mucous membranes, respiratory tract, and skin.

A large amount of hexachlorocyclopentadiene was dumped into a municipal sewage system and caused exposure to 145 sewage treatment workers.[1,2] Exposures were estimated to range from less than 0.05 ppm to 20 ppm for several seconds to 15 minutes. The major complaints were eye irritation, headache, and throat irritation. Medical examination of 41 workers 3 days after the exposure showed proteinuria and elevation of serum lactic dehydrogenase levels. These findings had resolved 3 weeks later.

In a recent study of male operators employed in a chemical plant, it was concluded that long-term exposure to a mixture of chlorinated hydrocarbons, including hexachlorocyclopentadiene, below or near the current threshold limit values did not lead to clinically significant effects on the liver or kidney, as determined by biochemical function tests.[3]

Hexachlorocyclopentadiene appears to be more toxic when inhaled than when ingested.[4] The reported oral LD_{50} for rats is 425 mg/kg and, for mice, it is 680 mg/kg. The 4-hour LC_{50} values range from 1.6 to 3.5 ppm for rats and mice. Rats, rabbits, and guinea pigs exposed for 7 hr/day, 5 days/week, to 0.15 ppm for 30 weeks survived.[5] Exposure at 0.34 ppm caused death in the mice and rats after 20 exposures. Effects observed were lacrimation, salivation, gasping respiration, and tremor. Severe pulmonary edema and acute necrotizing bronchitis and bronchiolitis were evident, as were degenerative changes in the brain, heart, liver, adrenal glands, and kidneys. The liquid on the skin of monkeys caused severe irritation.[5]

Exposure of rats to 0.5 ppm 6 hr/day 5 days/week for 2 weeks, caused lesions in the olfactory and bronchiolar epithelium, along with inflammatory exudate in the lumens of the respiratory tract.[6]

In a 13-week oral gavage study in rats and mice at doses up to 150 mg/kg/day, there was irritation of the forestomach in both sexes of both species, as well as a high incidence of toxic nephrosis in the females only of both species.[7]

There was no evidence of carcinogenicity in rats or mice exposed to 0.01, 0.05, or 0.2 ppm for 6 hr/day for 2 years.[8] Pigmentation of the respiratory epithelium occurred in both species, and squamous metaplasia of the laryngeal epithelium occurred in female rats. Genotoxic assays have been uniformly negative.

No evidence of teratogenicity was found following oral exposure in three species.[4]

The 1995 ACGIH threshold limit value–

time-weighted average (TLV-TWA) is 0.01 ppm (0.11 mg/m³).

REFERENCES

1. Morse DL, Kominsky JR, Wisseman CL, Landrigan PJ: Occupational exposure to hexachlorocyclopentadiene. How safe is sewage? *JAMA* 241:2177–2179, 1979
2. Kominsky JR, Wisseman CL, Morse DL: Hexachlorocyclopentadiene contamination of a municipal waste-water treatment plant. *Am Ind Hyg Assoc J* 41:552–556, 1980
3. Boogaard PJ, Rocchi, PSJ, van Sittert NJ: Effects of exposure to low concentrations of chlorinated hydrocarbons on the kidney and liver of industrial workers. *Br J Ind Med* 50:331–339, 1993
4. World Health Organization: Hexachlorocyclopentadiene. Environmental Health Criteria 120. Geneva, WHO, 1991
5. Treon JF, Cleveland FP, Cappel, J: *J AMA Arch Environ Health* 11:459, 1955
6. Rand GM et al: Effects of inhalation exposure to hexachlorocyclopentadiene on rats and monkeys. *J Toxicol Environ Health* 9:743–760, 1982
7. Abdo KM et al: Toxicity of hexachlorocyclopentadiene: Subchronic (13-week) administration by gavage to F344 rats and B6C3F₁ mice. *J Appl Toxicol* 4:75–81, 1984
8. National Toxicology Program: *Toxicology and Carcinogenesis Studies of Hexachlorocyclopentadiene (CAS No 77-47-4) in F344 Rats and B6C3F₁ Mice (Inhalation Studies)*, NTP TR 437, NIH Pub No 93-3168. US Department of Health and Human Services, Public Health Service, 1994

HEXACHLOROETHANE
CAS: 67-72-1

CCl₃CCl₃

Synonyms: Carbon hexachloride; perchloroethane

Physical Form. Colorless crystals

Uses. Chemical intermediate in the manufacture of pyrotechnics, insecticides, and other chlorinated materials

Exposure. Inhalation; skin absorption

Toxicology. Hexachloroethane is an eye irritant and causes kidney and central nervous system effects in animals. At high doses, it is carcinogenic to mice.

Exposure of workmen to fumes from hot hexachloroethane resulted in blepharospasm, photophobia, lacrimation, and reddening of the conjunctiva but no corneal injury or permanent damage.[1] No chronic effects have been reported from industrial exposure although significant skin absorption is said to occur.[2]

Rats exposed to 5900 ppm for 8 hours showed ataxia, tremor, and convulsions, and 2 of 6 died.[1] At 260 ppm for 8 hours, there were no toxic signs, but repeated exposure to this concentration 6 hr/day, 5 days/week, caused tremor, red exudate around the eyes and, after 4 weeks, some deaths. Dogs exposed at 260 ppm developed tremor, ataxia, hypersalivation, and facial muscular fasciculations and held their eyelids closed during the exposure; 3 of 4 survived 6 weeks of repeated exposures. No treatment-related effects were found in a number of species repeatedly exposed at 48 ppm.[1]

Rats fed 62 mg/kg/day for 16 weeks exhibited no overt toxicity.[2] Kidney effects characterized by increased kidney weights and microscopic changes (tubular atrophy, degeneration, hypertrophy, and/or dilation) were observed in males at 15 and 62 mg/kg/day; in females, tubular atrophy and degeneration of the kidneys were observed only at the highest dose. Both sexes also had increased liver weights at 62 mg/kg/day.[2]

The dermal LD_{50} for male rabbits was greater than 32 g/kg.[2] Applied to rabbit skin for 24 hours, the dry material caused no skin irritation, whereas a water paste caused slight redness.[1] In the eyes of 5 of 6 rabbits, 1 gram of the crystal overnight caused moderate corneal opacity, iritis, severe swelling, and discharge.

Gavage administration of 590 and 1179 mg/kg/day to mice for 78 weeks caused a significant increase in the incidence of hepatocellular

carcinomas, whereas no increase in these tumors was observed in rats given 212 or 423 mg/kg/day. A nonsignificant increase in renal tumors was seen in rats, and tubular nephropathy occurred in both species.[3] Based on these studies, the IARC determined that there is limited evidence for carcinogenicity in experimental animals by hexachloroethane.[4] Additional carcinogenicity studies were subsequently completed for rats.[5] In 2-year gavage studies, there was clear evidence of carcinogenicity in male rats administered 20 mg/kg, 5 days/week, based on increased incidences of renal neoplasms. Marginally increased incidences of pheochromocytomas of the adrenal gland may also have been related to hexachloroethane administration in males. There was no evidence of carcinogenicity for female rats administered 80 or 160 mg/kg for the 2-year duration although the severity of nephropathy was increased in dosed females as well as in males.[5]

Hexachloroethane has a camphorlike odor, readily sublimes, and, when heated to decomposition, emits phosgene.[1] Sublimation of hexachloroethane may contribute to exposure control problems. Sedimented HCE dust may accumulate on fluorescent tube illuminators and other warm surfaces and act as an exposure reservoir adding to exposure levels.[6]

Hexachloroethane exposure can be determined from blood plasma. In one group of workers, plasma levels increased nearly 100-fold despite the use of personal protective equipment.[6]

The 1995 ACGIH threshold limit value–time-weighted average (TLV-TWA) is 1 ppm (9.7 mg/m^3) with an A2 suspected human carcinogen designation and a notation for skin absorption.

REFERENCES

1. Weeks MH, Angerhofer RA, Bishop R, et al: The toxicity of hexachloroethane in laboratory animals. *Am Ind Hyg Assoc J* 40:187–199, 1979
2. Gorzinski SJ et al: Subchronic oral toxicity, tissue distribution and clearance of hexachloroethane in the rat. *Drug Chem Toxicol* 8:155–169, 1985
3. National Cancer Institute: *Bioassay of Hexachlo-*

roethane for Possible Carcinogenicity, TR-68. DHEW Pub No (NIH) 78-1318. Washington, DC, US Government Printing Office, 1978

4. *IARC Monographs on the Evaluation of the Carcinogenic Risks of Chemicals to Humans,* Vol 20, Some halogenated hydrocarbons, Vol 20, pp 467–473. Lyon, International Agency for Research on Cancer, 1979

5. National Toxicology Program: *Toxicology and Carcinogenesis Studies of Hexachloroethane (CAS No 67-72-1) in F344/N Rats (Gavage Studies).* NTP TR-361 NIH Pub No 89-2816, pp 1–120, US Department of Health and Human Services, 1989

6. Selden A, Nygren M, Kvarnlof A, et al: Biological monitoring of hexachloroethane. *Int Arch Occup Environ Health* 65: S111–S114, 1993

HEXACHLORONAPHTHALENE
CAS: 1335-87-1

$C_{10}H_2Cl_6$

Synonyms: Halowax 1014

Physical Form. Waxy, yellow-white solid

Uses. In synthetic wax; in electric wire insulation; in lubricants

Exposure. Inhalation; skin absorption

Toxicology. Hexachloronaphthalene is toxic to the liver and causes chloracne.

Human fatalities due to acute yellow atrophy of the liver have occurred with repeated exposure to penta- and hexachloronaphthalene.[1,2] Air measurements showed concentrations averaging 1 to 2 mg/m³. Other workers experienced jaundice, nausea, indigestion, and weight loss.

The most common problem, a severe acneiform dermatitis termed chloracne, typically occurs from long-term contact with the fume or dust or from shorter contact with the hot vapor.[3] The reaction is usually slow to appear, and the skin may take months to return to normal.

Repeated exposure of rats to an average concentration of 8.9 mg/m³ of a mixture of penta- and hexachloronaphthalene produced jaundice and was fatal; the liver showed a marked fatty degeneration and centrilobular necrosis.[3] At 1.16 mg/m³, minor liver injury still occurred.

The 1995 ACGIH threshold limit value–time-weighted average (TLV-TWA) for hexachloronaphthalene is 0.2 mg/m³ with a notation for skin absorption.

REFERENCES

1. Hygienic Guide Series: Chloronaphthalenes. *Am Ind Hyg Assoc J* 27:89–92, 1966
2. Elkins HB: *The Chemistry of Industrial Toxicology,* 2nd ed, pp 151–152. New York, John Wiley & Sons, 1959
3. Deichmann WB: Halogenated cyclic hydrocarbons. In Clayton GD, Clayton FE (eds): *Patty's Industrial Hygiene and Toxicology,* 3rd ed, Vol 2B, *Toxicology,* pp 3669–3675. New York, Wiley-Interscience, 1981

HEXAFLUOROACETONE
CAS: 684-16-2

C_3F_6

Synonyms: HFA; acetone, hexafluoro; perfluoro-2-propanone

Physical Form. Colorless gas, which reacts vigorously with water to form hydrates

Uses. In the synthesis of polymer, pharmaceutical, and agricultural chemicals; solvent for polyamides, polyesters, and polyacetals; in the synthesis of hexafluoroisopropanol

Exposure. Inhalation; skin absorption

Toxicology. Hexafluoroacetone affects the lungs, liver, and kidneys and causes testicular damage and teratogenesis in rats.

Upper respiratory tract irritation has been reported in humans following exposure to 4 ppm hexafluoroacetone dihydrate.[1]

In dogs, 5000 ppm was lethal to 1 of 2 animals following a 45-minute exposure.[2] Deaths occurred within 3 days post-exposure; lung hemorrhage and edema were observed, whereas the trachea, spleen, liver, kidney, and urinary bladder appeared normal.[2]

The oral LD_{50} for the trihydrate in rats is 190 mg/kg; moderate signs of central nervous system depression were observed, which abated in the survivors after 2 days.[2]

LC_{50}s of 900, 570, 275, and 200 ppm have been reported in rats for exposure times of 0.5, 1, 3, and 4 hours, respectively.[1] Exposure of rats to 200 ppm or above for 4 hours caused injury to the liver, kidneys, and thymus.[3] Pulmonary edema and congestion were seen in the lungs, and surviving males had testicles that were small and that, on gross examination and microscopically, showed aspermatogenesis, destruction of the stem cells, and effects on the interstitial tissue.

Exposure of rats and beagle dogs to 12 ppm for 6 hr/day, 5 days/week for 13 weeks, produced severe testicular damage and slight hypoplasia of the spleen, thymus, lymph nodes, and bone marrow.[4] Similar exposure at 0.1 ppm caused no effects. In another report, rats exposed to 25 ppm, 6 hr/day for 3 months, had increased mortality, but no testicular effects were detected.[5]

Hexafluoroacetone sesquihydrate was applied dermally to male rats at doses of 13, 39, or 130 mg/kg/day for 14 days.[6] At the highest dose, all rats developed severe testicular atrophy, while 50% of the animals at the medium dose had the same effects. No effects were observed at the low dose.

In a teratology study, hexafluoroacetone trihydrate was applied to the skin of pregnant rats from day 6 to 16 of gestation.[7] Teratogenic effects were seen at 5 and 25 mg/kg/day, and consisting of gross external, internal soft-tissue, and skeletal abnormalities.

The liquid is a severe skin irritant; one drop of the dihydrate produced marked erythema and blanching in guinea pig skin, but no irritation was seen when the dihydrate was diluted to 10%.[1,3] Instilled in rabbit eyes, hexafluoroacetone sesquihydrate produced severe, extensive injury, including corneal opacity, scar tissue, and chronic conjunctivitis.

The 1995 ACGIH threshold limit value–time-weighted average (TLV-TWA) for hexafluoroacetone is 0.1 ppm (0.68 mg/m³) with a notation for skin absorption.

REFERENCES

1. Kennedy Jr, GL: Toxicology of fluorine-containing monomers. *Crit Rev Toxicol* 21:149–170, 1990
2. Borzelleca JF, Lester D: Acute toxicity of some perhalogenated acetones. *Toxicol Appl Pharmacol* 7:592–597, 1965
3. EI Du Pont de Nemours & Co, Inc: Inhalation toxicity of hexafluoroacetone compounds. Haskell Laboratory Report No 46-62. Wilmington, DE, January 25, 1962.
4. Ibid., Thirteen week inhalation exposures of rats and dogs to hexafluoroacetone (HFA). No 4-71. January 7, 1971
5. Douglas GR, Nestmann ER, McKague AB, et al: Determination of potential hazard from pulp and paper mills: mutagenicity and chemical analysis. *Carcinogen Mutagen Environ* 5:151, 1985
6. Gillies PJ, Lee KP: Effects of hexafluoroacetone on testicular morphology and lipid metabolism in the rat. *Toxicol Appl Pharmacol* 68:188–197, 1983
7. Brittelli MR, Culik R, Dashiell OL, Fayerweather WE: Skin absorption of hexafluoroacetone: Teratogenic and lethal effects in the rat fetus. *Toxicol Appl Pharmacol* 47:35–40, 1979

HEXAMETHYLENE DIISOCYANATE
CAS: 822-06-0

$C_8H_{12}N_2O_2$

Synonyms: HDI; HMDI; 1,6-diisocyanato-hexane; 1,6-hexamethylene diisocyanate

Physical Form. Pale yellow liquid

Uses. Cross-linking agent (hardener) in the production of polyurethane materials such as car paints, dental materials, and contact lenses

Exposure. Inhalation

Toxicology. Hexamethylene diisocyanate (HDI) is an irritant of the eyes, mucous membranes, and skin and a sensitizer of the respiratory tract.

Severe eye injury, including conjunctivitis, glaucoma, keratitis, and corneal damage, can occur with exposure to HDI.[1] By analogy with toluene diisocyanate threshold levels for irritation would be expected to be in the range of 50 ppb.[1] If the breathing zone concentration of diisocyanates reaches 0.5 ppm, the possibility of respiratory response is imminent.[2] Depending on the length of exposure and level of concentrations above 0.5 ppm, respiratory symptoms may develop, with a latent period of 4 to 8 hours. Symptoms include increased secretions, cough, pain of respiration and, if severe enough, some restriction of air movement owing to a combination of secretions, edema, and pain. On removal from exposure, the symptoms may persist for 3 to 7 days.

A second type of response to isocyanates is allergic sensitization of the respiratory tract.[3] It usually develops after some months of exposure. The onset of symptoms may be insidious, becoming progressively more pronounced with continued exposure. Initial symptoms are often nocturnal dyspnea and/or nocturnal cough with progression to asthmatic bronchitis.

Productive cough and shortness of breath developed in a spray painter 12 to 18 months after introduction of a spray paint that con-tained HDI.[4] When exposed to a diagnostic spray mist containing 5% HDI for 5 minutes, an 18% drop in respiratory function was noted in 10 minutes and a 41% drop was seen in 3 hours. The worker also had an enhanced nonspecific reactivity to inhaled histamine that persisted for 18 months after the worker ceased exposure to HDI.

Cross-reactivity of diisocyanates was investigated in 24 exposed workers with respiratory symptoms.[5] All workers had been exposed to toluene diisocyanate (TDI). In inhalation challenge tests, 16 gave asthmatic reactions to TDI at levels ranging from 0.0001 to 0.02 ppm. Five gave nonimmediate (late) reactions only, and 11 gave combined (dual) reactions. Eight of these 16 also reacted to methylene diisocyanate (MDI). Of the 8 TDI and MDI reactors, 4 had histories of exposure only to TDI, and 2 of those 4 also reacted to HDI. Of 9 subjects tested with HDI, 3 gave asthmatic reactions, and all 3 also reacted to TDI and MDI. Reactions to MDI and HDI were elicited only in TDI reactors. Among the possible explanations for these findings are cross-reactivity between the different isocyanates, an irritant or pharmacological effect in subjects with hyperreactive airways, or both.

Isocyanates mediate their toxicity through a high degree of chemical reactivity.[1] These reactions can result in cross-linkages of biological macromolecules, which lead to the denaturation of proteins, loss of enzyme function, and the formation of immunological reactivities.

HDI was not mutagenic against a variety of *Salmonella* assays with or without metabolic activation.

The 1995 ACGIH threshold limit value–time-weighted average (TLV-TWA) for hexamethylene diisocyanate is 0.005 ppm (0.034 mg/m³).

REFERENCES

1. Von Burg R: Toxicology update: hexamethylene diisocyanate. *J Appl Toxicol* 13:435–439, 1993
2. Rye WA: Human response to isocyanate exposure. *J Occup Med* 15:306–307, 1973

3. National Institute for Occupational Safety and Health: *Criteria for a Recommended Standard... Occupational Exposure to Di-isocyanates.* DHEW (NIOSH) Pub No 78-215. Washington DC, US Government Printing Office, 1978

4. Cockcroft DW, Mink JT: Isocyanate-induced asthma in an automobile spray painter. *J Can Med Assoc* 121:602–604, 1979

5. O'Brien IM, Harries MG, Burge PS, Pepys J: Toluene diisocyanate-induced asthma: I. Reactions to TDI, MDI, HDI, and histamine. *Clin Allergy* 9:1–6, 1979

HEXAMETHYL PHOSPHORAMIDE
CAS: 680-31-9

$C_{16}H_{18}N_3OP$

Synonyms: HMPA; HMPT; HPT; hexamethyl phosphoric triamide

Physical Form.   Colorless liquid

Uses.   Solvent for polymers; polymerization catalyst; stabilizer against thermal degradation in polystyrene; UV stabilizer in polyvinyl and polyolefin resins

Exposure.   Inhalation; skin absorption

Toxicology.   Hexamethyl phosphoramide (HMPA) is carcinogenic in experimental animals.

Effects from human exposures have not been reported.

Nasal tumors were induced in rats by inhalation exposure to hexamethyl phosphoramide for 6 to 24 months at levels of 50, 100, 400, and 4000 ppb, 6 hr/day 5 days week, but not in rats exposed to 10 ppb for 24 months.[1,2] Most nasal tumors were epidermoid carcinomas and developed from the respiratory epithelium or subepithelial nasal glands, both of which revealed squamous metaplasia or dysplasia in the anterior nasal cavity. Other effects were: keratinized squamous metaplasia of the trachea (4000 ppb); dose-related increases in tracheitis and desquamation of the tracheal epithelium, and bronchitis, desquamation, and regeneration of the bronchial epithelium (100, 400, and 4000 ppb); bone marrow erythropoietic hyperplasia (males, 4000 ppb); testicular atrophy (males, 4000 ppb); and degenerative changes in the convoluted tubules of the kidneys.[2,3]

Dogs also showed squamous metaplasia of the nasal cavity following inhalation exposure to HMPA for 5 months at 400 and 4000 ppb.

In a teratology study, rats were given daily oral doses of 200 mg/kg/day from day 7 to day 20 of gestation; no abnormalities were seen in offspring.[4]

The IARC has stated that hexamethyl phosphoramide is carcinogenic to rats following administration by inhalation.[5]

Hexamethyl phosphoramide is classified as an A2 suspected human carcinogen because of its demonstrated carcinogenesis in experimental animals, with no exposure levels permissible and a notation for skin.

REFERENCES

1. Lee KP, Trochimowicz HJ: Morphogenesis of nasal tumors in rats exposed to hexamethyl-phosphoramide by inhalation. *Environ Res* 33:106–118, 1983

2. EI Du Pont Haskell Laboratory: Pathology Report No 37-80, Hexamethylphosphoramide (HMPA). H-8419-MR-1785—Textile Fibers Dept 2-year inhalation test CD Rats. EPA/OTS doc FYI-OTS-0382-0040. Wilmington, DE, 1980

3. Zapp JA Jr: Inhalation toxicity of hexamethyl-phosphoramide. *Am Ind Hyg Assoc J* 36:915–919, 1975

4. Kimbrough RD, Gaines TB: Toxicity of hexamethylphosphoramide in rats. *Nature (London)* 211:146–147, 1966

5. *IARC Monographs on the Evaluation of the Carcinogenic Risk of Chemicals to Man*, Vol 15, Some fumigants, the herbicides 2,4-D and 2,4,5-T, chlorinated dibenzodioxins and miscellaneous industrial chemicals, p 217. Lyon, International Agency for Research on Cancer, 1977

n-HEXANE
CAS: 110-54-3

$CH_3(CH_2)_4CH_3$

Synonym: Hexane

Physical Form. Colorless, very volatile liquid solvent and thinner

Uses. Solvent

Exposure. Inhalation; skin absorption

Toxicology. *n*-Hexane is an upper respiratory irritant and central nervous system depressant; chronic exposure causes peripheral neuropathy.

In human subjects, 2000 ppm for 10 minutes produced no effects, but 5000 ppm resulted in dizziness and confusion.[1] Other investigators reported slight nausea, headache, and irritation of the eyes and throat at 1500 ppm.[2] In industrial practice, mild symptoms of narcosis such as dizziness have been observed when concentrations of solvents containing various isomers of hexane exceeded 1000 ppm but symptoms were not observed for exposures below 500 ppm.[3]

Dermal exposure to hexane caused immediate irritation, characterized by erythema and hyperemia.[4] Subjects complained of painful burning sensations with itching and, after 5 hours, blisters formed on the exposed areas.[4]

Polyneuropathy has been reported following chronic occupational exposure to vapors containing *n*-hexane at concentrations typically in the 400 to 600 ppm range, with some ceiling exposures up to 2500.[4-6] One person developed polyneuropathy after 1 year of exposure at 54 to 200 ppm.[5] Initial symptoms may include sensation disturbances, muscle weakness, and distal symmetric pain in the legs after 2 to 6 months of exposure.[5] Clinical changes are muscle atrophy, hypotonic decreased muscle strength, foot drop, and paresthesias in the arms and legs. Characteristic electroneurophysiological findings include a noticeable fall in nerve conduction velocities, profound amplitude reduction of compound muscle action potentials and sensory action potentials, and prolongation of distal latencies.[6] Evoked potential studies show prolongation of conduction times in the visual, auditory, and somatosensory pathways of the central nervous system. Changes in color vision, in retinal pigmentation, and in perifoveal capillaries were found in workers exposed to 420 to 1280 ppm for more than 5 years.[5] Peripheral nerve biopsies show significant swelling of the nerve with thinning of the myelin sheath. Functional disturbances commonly progress for 2 to 3 months after cessation of exposure. Recovery may be expected within a year but, in some cases, clinical polyneuropathy has remained after 2 years.[5] A follow-up of 11 patients with moderate to severe *n*-hexane-induced polyneuropathy found that sensory functions were regained earlier than motor functions and that abnormal color vision and muscle atrophy persisted for up to 4 years.[7]

A recent anecdotal report has suggested that prolonged exposure (30 years) to low-grade levels of *n*-hexane (10–100 mg/m³) may also cause polyneuropathy.[8]

Experimental animals continuously exposed to pure *n*-hexane developed the same clinical, electrophysiologic, and histopathologic changes found in humans exposed to mixed vapors containing *n*-hexane.[9] Continuous inhalation by rats of 400 ppm caused axonapathy.[10] Interestingly, intermittent exposure of rats to 10,000 ppm 6 hr/day, 5 days/week for 13 weeks, caused only slight paranodal axonal swelling.[11] It is postulated that 2,5-hexanedione, a metabolite of *n*-hexane and purported neurotoxic agent, must build to an effective concentration. With continuous exposure, there is no recovery during each day or week.

In regard to reproductive effects, the only difference found in rats exposed to 1000 ppm during gestation was in their offspring, which weighed less than expected at ages 1 week through 6 weeks.[12]

In genotoxic assays, commercial hexane, consisting of *n*-hexane and other six-carbon isomers, did not produce chromosomal mutations either in vitro or in vivo.[13] *n*-Hexane was not mutagenic in bacteria.

Urinary concentration of 2,5-hexanedione has been used in the biological monitoring of workers exposed to *n*-hexane and is considered to be a reliable indicator of alveolar and percutaneous absorption.[14] Variability between environmental concentrations of *n*-hexane and 2,5-hexanedione levels has been attributed to variable use of protective clothing.

The neurotoxic properties of *n*-hexane are potentiated by exposure to methyl ethyl ketone (qv). Because other compounds may also have this effect, human exposure to mixed solvents containing any neurotoxic hexacarbon compound should be minimized.[9]

The 1995 ACGIH threshold limit value–time-weighted average (TLV-TWA) for *n*-hexane is 50 ppm (176 mg/m³).

REFERENCES

1. Patty FA, Yant WP: Report of investigations—odor intensity and symptoms produced by commercial propane, butane, pentane, hexane, and heptane vapor, No 2979. US Dept of Commerce, Bureau of Mines, 1929
2. Drinker P, Yaglou CP, Warren MF: The threshold toxicity of gasoline vapor. *J Ind Hyg Toxicol* 25:225–232, 1943
3. Elkins HB: *Chemistry of Industrial Toxicology*, p 101. New York, John Wiley & Sons, 1959
4. National Institute for Occupational Safety and Health: *Criteria for a Recommended Standard . . . Occupational Exposure to Alkanes*, (C5–C8) DHEW (NIOSH) Pub No 77-151. Washington, DC US Government Printing Office, 1977
5. Jorgensen NK, Cohr KH: *n*-Hexane and its toxicological effects. *Scand J Work Environ Health*, 7:157–168, 1981
6. Chang YC: An electrophysiological follow up of patients with *n*-hexane polyneuropathy. *Br J Ind Med* 48:12–17, 1991
7. Chang YC: Patients with *n*-hexane induced polyneuropathy: a clinical follow up. *Br J Ind Med* 47:485–489, 1990
8. Barregard L, Sallsten G, Nordborg C, et al: Polyneuropathy possibly caused by 30 years of low exposure to *n*-hexane. *Scand J Work Environ Health* 17:205–207,1991
9. Spencer PS, Schaumburg HH, Sabri MI, et al: The enlarging view of hexacarbon and neu-
rotoxicity. *CRC Crit Rev Toxicol* 7:279–356, 1980
10. Schaumburg HH, Spencer PS: Degeneration in central and peripheral nervous systems produced by pure *n*-hexane: an experimental study. *Brain* 99:183–192, 1976
11. Cavender FL et al: A 13-week vapor inhalation study of *n*-hexane in rats with emphasis on neurotoxic effects. *Fund Appl Toxicol* 4:191–201, 1984
12. Bus JS et al: Perinatal toxicity and metabolism of *n*-hexane in Fischer-344 rats after inhalation exposure during gestation. *Toxicol Appl Pharmacol* 511:295–302, 1979
13. Daughtrey WC, Putman DL, Duffy J, et al: Cytogenetic studies on commercial hexane solvent. *J Appl Toxicol* 14:161–165, 1994
14. Cardona A, Marhuenda D, Marti J, et al: Biological monitoring of occupational exposure to *n*-hexane by measurement of urinary 2,5-hexanedione. *Int Arch Occup Environ Health* 65:71–74, 1993

---

sec-**HEXYL ACETATE**
CAS: 108-84-9

$C_8H_{16}O_2$

Synonyms: Methyl amylacetate; 4-methyl pentyl 2-acetate; 1,3-dimethylbutyl acetate; methyl isoamyl acetate

Physical Form.   Clear liquid

Uses.   In lacquer industry; in fragrances

Exposure.   Inhalation

Toxicology.   *sec*-Hexyl acetate causes irritation of the eyes and upper respiratory tract; at concentrations approaching saturation, it causes narcosis in animals, and it is expected that similar exposure will cause the same effect in humans.

Human volunteers exposed to 100 ppm for 15 minutes experienced eye irritation and objected to the odor and taste; nose and throat irritation occurred at levels greater than 100 ppm.[1] No chronic or systemic effects in humans have been reported.

Four of six rats survived exposure to 4000 ppm for 4 hours, but 8000 ppm was lethal to all the animals.[2]

The liquid was poorly absorbed through rabbit skin but did cause moderate irritation.[2,3] Little corneal injury resulted from eye instillation.[3]

The 1995 ACGIH threshold limit value–time-weighted average (TLV-TWA) for *sec*-hexyl acetate is 50 ppm (295 mg/m^3).

REFERENCES

1. Silverman L, Schulte HF, First MW: Further studies on sensory response to certain industrial solvent vapors. *J Ind Hyg Toxicol* 28:262–266, 1946
2. Smyth HF Jr, Carpenter CP, Weil CS, Pozzani UC: Range-finding toxicity data. *Arch Ind Hyg Occup Med* 10:61–68, 1954
3. Carpenter CP et al: Range-finding toxicity data: List VIII. *Toxicol Appl Pharmacol* 28:313–319, 1974

HEXYLENE GLYCOL
CAS: 107-41-5

$C_6H_{14}O_2$

Synonyms: 2-Methyl-2,4-pentanediol; 2,4-dihydroxy-2-methyl pentane; Isol; Pinakon

Physical Form. Liquid with a mild odor

Uses. Fuel and lubricant additive; solvent in cosmetics; solvent in petroleum refining; coupling agent in hydraulic brake fluid and printing inks; gasoline anti-icer additive

Exposure. Inhalation

Toxicology. Hexylene glycol is an irritant of the eyes and mucous membranes, and causes narcosis at high levels.

Sensory response evaluations in humans indicated that exposure to 50 ppm for 15 minutes produced slight odor and slight eye irritation.[1] At 100 ppm for 5 minutes, the odor was plainly detectable, and slight nasal and respiratory discomfort was noticed by unacclimated subjects. At 1000 ppm for 5 minutes, various degrees of eye irritation and throat and respiratory discomfort were noted.

The oral LD$_{50}$ in rats was 4.79 g/kg, with death being preceded by narcosis.[1] No adverse effects were detected in rats given 590 mg/kg/day for 8 months.

The liquid in the rabbit eye caused appreciable irritation and corneal injury, which was slow to heal.[2] Mild to moderate irritation occurred from the liquid applied to the skin of rabbits. Skin absorption is minimal; the dermal LD50 for rabbits was 12.3 g/kg.[1]

The irritant and sensitizing properties of hexylene glycol compared to propylene glycol were investigated in 823 eczema patients by routine patch testing.[3] Edema and erythema reactions occurred in 2.8% of the patients exposed to hexylene glycol compared to 3.8% reacting to propylene glycol.

The 1995 ACGIH STEL/ceiling limit for hexylene glycol is 25 ppm (121 mg/m^3).

REFERENCES

1. Rowe VK, Wolf MA: Glycols. In Clayton GD, Clayton FE (eds): *Patty's Industrial Hygiene and Toxicology*, 3rd ed, rev, Vol 2C, *Toxicology*, pp 3881–3884. New York, Wiley-Interscience, 1982
2. Smyth HF Jr, Carpenter CP: Further experience with the range-finding test in the industrial toxicology laboratory. *J Ind Hyg Toxicol* 30:63–68, 1948
3. Kinnunen T, Hannuksela M: Skin reactions to hexylene glycol. *Contact Derm* 21:154–158, 1989

HYDRAZINE
CAS: 302-01-2

NH₂NH₂

Synonyms: Hydrazine anhydrous; diamide; diamine; nitrogen hydride

Physical Form.   Colorless oily liquid; fuming in air

Uses.   Reducing agent; in the production of plastic blowing agents, herbicides, and rocket propellants

Exposure.   Inhalation; skin absorption

Toxicology.   Hydrazine is a severe skin and mucous membrane irritant, a convulsant, a hepatotoxin, and a moderate hemolytic agent; it is carcinogenic in experimental animals and is considered a possible human carcinogen.

In humans, the vapor is immediately irritating to the nose and throat and causes dizziness and nausea; itching, burning, and swelling of the eyes develop over a period of several hours.[1] Severe exposure of the eyes to the vapor causes temporary blindness, lasting for about 24 hours.[2] The liquid in the eyes or on the skin causes severe burns.[1] Hydrazine and its salts will also produce skin irritation and allergic reactions in humans.

Hydrazine is absorbed through the skin. In one case attributed to hydrazine hydrate exposure, systemic effects included weakness, vomiting, excited behavior, and tremors; the chief histological findings were severe tracheitis and bronchitis, fatty degeneration of the liver, and nephritis.[3]

The LC₅₀ values for rats and mice were 570 and 252 ppm, respectively.[4] The exposed rodents were restless and had breathing difficulties and convulsions. Exposure of mice, rats, dogs, and monkeys to 1.0 and 5.0 ppm, 6 hr/day 5 days/week, or at levels of 0.2 and 1.0 ppm continuously, had a variety of effects. Increased mortality occurred in mice and was attributed to liver damage; rats showed a dose-related

growth depression; dogs also had increased mortality and developed depressed erythrocyte counts, hematocrit values, and hemoglobin concentrations at higher doses; there were no effects in monkeys.[1] Lipid deposition in the kidneys of monkeys has been reported following intraperitoneal administration of hydrazine.[1]

Recent studies in rats have shown that acute doses of hydrazine cause hepatic steatosis accompanied by depletion of ATP and reduced glutathione (GSH), as well as hepatic accumulation of triglycerides. Biochemical effects from repeated exposures, however, included depletion of triglycerides and induction of nitrophenol hydroxylase activity in addition to changes in other microsomal enzymes.[5]

Hydrazine or hydrazine salts are carcinogenic in mice after oral administration (pulmonary adenocarcinoma, hepatocarcinoma) or intraperitoneal injection (pulmonary carcinoma), and in rats after oral administration (pulmonary adenocarcinoma).[6] Hydrazine induced a significantly greater incidence of nasal tumors, primarily benign, in rats and in hamsters after 1 year of intermittent inhalation exposure at levels of up to 5.0 ppm.[7]

A group of 427 men with varying degrees of occupational exposure to hydrazine showed no increase in overall mortality (49 deaths vs. 61.47 expected) or deaths from lung cancer (5 vs. 6.65 expected) although two of these deaths occurred within the highest exposure category (relative risk 1.2).[8] In other case reports, choroidal melanoma was observed in one man who had been exposed to hydrazine for 6 years, and chronic myeloid leukemia was reported in two patients with long-lasting exposure to hydrazine.[9,10]

The IARC has determined that there is sufficient evidence of carcinogenicity in animals and inadequate evidence for humans.[6]

The proposed 1995 ACGIH threshold limit value–time-weighted average (TLV-TWA) for hydrazine is 0.01 ppm (0.013 mg/m³) with a notation for skin absorption. Hydrazine is classified A3, a carcinogen in experimental animals at relatively high doses by routes of administration that are not considered relevant to worker exposure. According to the ACGIH, available evidence suggests that the agent is not likely to cause cancer in humans except under uncommon or unlikely routes or levels of exposure.

REFERENCES

1. National Institute for Occupational Safety and Health: *Criteria for a Recommended Standard ... Occupational Exposure to Hydrazines.* DHEW (NIOSH) 78-172. Washington, DC, US Government Printing Office, June 1978
2. Comstock CC, Lawson LH, Greene EA, Oberst FW: Inhalation toxicity of hydrazine vapor. *AMA Arch Ind Hyg Occup Med* 10:476–490, 1954
3. Sotanieme E et al: Hydrazine toxicity in the human—report of a fatal case. *Ann Clin Res* 3:30–33, 1971
4. Jacobson KG et al: The acute toxicity of the vapors of some methylated hydrazine derivatives. *Arch Ind Health* 12:609–616, 1955
5. Jenner AM, Timbrell JA: Effect of acute and repeated exposure to low doses of hydrazine on hepatic microsomal enzymes and biochemical parameters in vivo. *Arch Toxicol* 68:240–245, 1994
6. *IARC Monographs on the Evaluation of the Carcinogenic Risk of Chemicals to Man*, Chemicals, industrial processes and industries associated with cancer in humans, Suppl 4, pp 136–138. Lyon, International Agency for Research on Cancer, 1982
7. Vernot EH et al: Long-term inhalation toxicity of hydrazine. *Fund Appl Toxicol* 5:1050–1064, 1985
8. Wald N et al: Occupational exposure to hydrazine and subsequent risk of cancer. *Br J Ind Med* 41:31–34, 1984
9. Albert DM, Puliafito CA: Choroidal melanoma: Possible exposure to industrial toxins. *N Eng J Med* 296:634–635, 1977
10. Freund M, Eisert J, Anagnou J, et al: Zwei Falle von chronisch myeloisher Leukamie mit Hydrazin-Exposition. *Zbl Arbeitsmed* 35:375–377, 1985

HYDROGENATED TERPHENYLS
CAS: 61788-32-7

Synonym: Terphenyls, hydrogenated

Physical Form: Liquid; the hydrogenated terphenyls are complex mixtures of ortho-, meta-, and paraterphenyls in various stages of hydrogenation; five such stages exist for each of these three isomers

Uses. Heat-transfer media and plasticizers; as coolants, they are 40% hydrogenated (HB-40)

Exposure. Inhalation

Toxicology. Hydrogenated terphenyls have caused lung, kidney, and liver changes in animals.

The oral LD_{50} in rats for 40% hydrogenated terphenyls (reactor coolant) was 17.5 g/kg; for irradiated reactor coolant, it was 6 g/kg.[1] Ingestion by mice for 16 weeks of the irradiated mixture at 1200 mg/kg was lethal, whereas the nonirradiated mixture was not lethal, but did cause irreversible interstitial nephritis. At 250 mg/kg, no lesions were observed for the 16-week period of exposure.

Mice exposed for 8 weeks to an irradiated mixture of hydrogenated terphenyl at 2000 mg/m³ showed transient changes in Type II cells of the alveolar epithelium and some proliferation of the smooth endoplastic reticulum in the liver.[2]

The 1995 ACGIH threshold limit value–time-weighted average (TLV-TWA) for hydrogenated terphenyls is 0.5 ppm (4.9 mg/m³).

REFERENCES

1. Adamson IYR, Weeks JL: LD_{50} and chronic toxicity of reactor terphenyls. *Arch Environ Health* 27:69-73, 1973
2. Adamson IYR, Bowden DH, Wyatt JP: Acute toxicity of reactor polyphenyls on the lung. *Arch Environ Health* 19:499–504, 1969

HYDROGEN BROMIDE
CAS: 10035-10-6

HBr

Synonyms: Hydrobromic acid; anhydrous hydrobromic acid

Physical Form. Colorless, nonflammable gas

Uses/Sources. In manufacture of organic and inorganic bromides; reducing agent; catalyst in oxidations; alkylation of aromatic compounds; can be generated during the pyrolysis of a variety of materials

Exposure. Inhalation

Toxicology. Hydrogen bromide gas is an irritant of the eyes, mucous membranes, and skin.

There are no systemic effects reported from industrial exposure. Experimental exposure of humans to 5 ppm for several minutes caused nose and throat irritation in most persons, and a few were affected at concentrations of 3 to 4 ppm.[1] Solutions in contact with the eyes, skin, or mucous membranes may cause burns.[2]

The 1-hour inhalation LC_{50} was 2860 ppm for rats and 815 ppm for mice.[3] Rats exposed to 1300 ppm for 30 minutes and euthanized 24 hours after exposure showed tissue injury confined to the nasal region, including epithelial and submucosal necrosis, accumulations of inflammatory cells, exudates, and the extravasation of erythrocytes.[4] Intratracheal administration of the same dose produced some mortality and major tissue disruption in the trachea, including epithelial, submucosal, glandular, and cartilage necrosis, and accumulations of inflammatory cells and exudates.

The 1995 ACGIH ceiling–threshold limit value (C-TLV) for hydrogen bromide is 3 ppm (9.9 mg/m³).

REFERENCES

1. Hydrogen bromide. *Documentation of the TLVs and BEIs*, 6th ed, pp 771–772. Cincinnati, OH, American Conference of Governmental Industrial Hygienists (ACGIH), 1991
2. Alexandrov DD: Bromine and compounds. In International Labour Office: *Encyclopaedia of Occupational Health and Safety*, 3rd ed, rev, Vol I, A-K. p 327. Geneva, 1983
3. Registry of Toxic Effects of Chemical Substances (RTECS). *DHHS*. Public Health Service, Vol 3, p 2717. Washington, DC, US Government Printing Office, April 1987
4. Stavert DM, Archuleta DC, Behr MJ, et al: Relative acute toxicities of hydrogen fluoride, hydrogen chloride, and hydrogen bromide in nose- and pseudo-mouth-breathing rats. *Fund Appl Toxicol* 16:636–655, 1991

HYDROGEN CHLORIDE
CAS: 7647-01-0

HCl

Synonyms: HCl; hydrochloric acid, aqueous; muriatic acid

Physical Form. Colorless gas (aqueous solution is hydrochloric acid)

Uses/sources. Production of chlorinated organic chemicals; production of dyes and dye intermediates; steel pickling; oil well acidizing operations to dissolve subsurface dolomite or limestone; formed during thermal decomposition of PVC

Exposure Inhalation

Toxicology. Hydrogen chloride is a strong irritant of the eyes, mucous membranes, and skin.

The major effects of acute exposure are usually limited to the upper respiratory tract and are sufficiently severe to encourage prompt withdrawal from a contaminated atmosphere.[1] Exposure to the gas immediately causes cough, burning of the throat, and a choking sensation. Effects are usually limited to inflammation and occasionally ulceration of the nose, throat, and larynx.[2] Acute exposures causing significant trauma are usually limited to people who are prevented from escaping; in such cases, laryngeal spasm or pulmonary edema may occur.

In workers, exposure to 50 to 100 ppm for 1 hour was barely tolerable; short exposure to 35 ppm caused irritation of the throat, and 10 ppm was considered the maximal concentration allowable for prolonged exposure.[3] In one study, workers chronically exposed to hydrogen chloride did not exhibit the pulmonary-function changes observed in naive subjects exposed to similar concentrations; this observation suggests acclimatization of the workers to hydrogen chloride.[4]

Ten young adult asthmatics showed no adverse respiratory health effects following multiple inhalation challenge with 0.8 ppm and 1.8 ppm HCL.[5]

Exposure of the skin to a high concentration of the gas or to a concentrated solution of the liquid (hydrochloric acid) causes burns; repeated or prolonged exposure to dilute solutions may cause dermatitis.[2] Erosion of exposed teeth may also occur from repeated or prolonged exposure. Although ingestion is unlikely, hydrochloric acid causes severe burns of the mouth, esophagus, and stomach with consequent pain, nausea, and vomiting.[6]

Exposure of mice to 1300 ppm for 30 minutes caused tissue injury to the nasal region, including epithelial and submucosal necrosis, accumulations of inflammatory cells, and exudates.[7] Mice administered the same concentration by tracheal tubes (to simulate mouth breathing) had major tissue damage in the trachea, including epithelial, submucosal, glandular and cartilage necrosis; peripheral lung damage was manifested by histopathological changes in the larger conducting airways.

Rodent studies may be of limited value in determining human effects because of the increased sensitivity of rats compared to that of primates. A comparison of the lethality data indicates that the mouse (LLD = 3200 ppm, 5 min) is more sensitive than the rat (LLD = 15,250–32,250 ppm, 5 min) or the baboon (LLD = 16,570–30,000+ ppm, 5 min).[8] For longer exposure periods, the LLDs for the rat and baboon also diverge; for a 30-minute

exposure, the LLD is less than 3000 ppm for rats and greater than 5000 ppm for baboons.

No evidence of an association between HCL exposure and lung cancer was found in a nested case-control study of chemical workers.[9] This result was consistent with a rodent bioassay in which chronic exposure to 10 ppm, 6 hr/day for life, did not cause any neoplastic lesions.[10]

Warning properties are good, and most people can detect 5 ppm.[1]

The 1995 ACGIH STEL/ceiling for hydrogen chloride is 5 ppm (7.5 mg/m^3).

REFERENCES

1. Committee on Medical and Biologic Effects of Environmental Pollutants: *Chlorine and Hydrogen Chloride*, pp 138–144. Washington, DC, National Academy of Sciences, 1976
2. Chemical Safety Data Sheet SD-39, Hydrochloric acid, pp 5–6, 24–26. Washington, DC, MCA, Inc, 1970
3. Henderson Y, Haggard HW: *Noxious Gases*, p 126. New York, Reinhold, 1943
4. Toyama T, Kondo T, Nakamura K: Environments in acid aerosol producing workplaces and maximum flow rate of workers. *Jap J Ind Health* 4:15–22, 1962
5. Stevens B, Koenig JQ, Rebolledo V, et al: Respiratory effects from the inhalation of hydrogen chloride in young adult asthmatics. *JOM* 34:923–926, 1992
6. Poteshman NL: Corrosive gastritis due to hydrochloric acid ingestion. *Am J Roentgenol Radium Ther Nucl Med* 99:182–185, 1967
7. Stavert DM, Archuleta DC, Behr MJ, et al: Relative acute toxicities of hydrogen fluoride, hydrogen chloride, and hydrogen bromide in nose- and pseudo-mouth-breathing rats. *Fund Appl Toxicol* 16:636–655, 1991
8. Hinderer RK, Kaplan HL: Assessment of the inhalation toxicity of hydrogen chloride gas to man, pp 2–4. *Dangerous Properties of Industrial Materials Report*, Van Nostrand Reinhold, March/April 1986
9. Bond GG, Flores GH, Stafford BA, et al: Lung cancer and hydrogen chloride exposure: results from a nested case control study of chemical workers. *JOM* 33:958–961, 1991
10. Sellakumar AR et al: Carcinogenicity of formaldehyde and hydrogen chloride in rats. *Toxic Appl Pharmacol* 81:401–406, 1985

HYDROGEN CYANIDE
CAS: 74-90-8

HCN

Synonyms: Hydrocyanic acid; aero liquid HCN; prussic acid; formonitrile

Physical Form. Colorless gas liquefying at 26°C (may be found in the workplace both as a liquid and a gas)

Uses. Rodenticide and insecticide; fumigant; chemical intermediate for the manufacture of synthetic fibers, plastics, and nitrites

Exposure. Inhalation; skin absorption; ingestion

Toxicology. Hydrogen cyanide can cause rapid death due to metabolic asphyxiation.

The cyanide ion exerts an inhibitory action on certain metabolic enzyme systems, most notably cytochrome oxidase, the enzyme involved in the ultimate transfer of electrons to molecular oxygen.[1] Because cytochrome oxidase is present in practically all cells that function under aerobic conditions, and because the cyanide ion diffuses easily to all parts of the body, cyanide quickly halts practically all cellular respiration. The venous blood of a patient dying of cyanide poisoning is bright red and resembles arterial blood because the tissues have not been able to utilize the oxygen brought to them.[2] Cyanide intoxication produces lactic acidosis, the result of increased rate of glycolysis and production of lactic acid.[3]

A concentration of 270 ppm hydrogen cyanide has long been quoted as being immediately fatal to humans. A more recent study, however, states that the estimated LC_{50} to humans for a 1-minute exposure is 3404 ppm.[1] Others state that 270 ppm is fatal after 6 to 8 minutes, 181 ppm after 10 minutes, and 135 ppm after 30 minutes.[1]

If large amounts of cyanide have been absorbed, collapse is usually instantaneous, the patient falling unconscious, often with convulsions, and dying almost immediately.[1,2]

Symptoms of intoxication from less severe exposure include weakness, headache, confusion, vertigo, fatigue, anxiety, dyspnea and occasionally nausea and vomiting. Respiratory rate and depth are usually increased initially and, at later stages, become slow and gasping. Coma and convulsions occur in some cases. If cyanosis is present, it usually indicates that respiration has either ceased or has been very inadequate for a few minutes.

Hydrogen cyanide has recently been recognized in significant concentrations in some fires, as a combustion product of wool, silk, and many synthetic polymers; it may play a role in toxicity and deaths from smoke inhalation.[4]

Most reported cases of chronic cyanide poisoning involve workers with a mixture of repeated acute or subacute exposures, making it unclear whether symptoms resulted simply from multiple acute exposures with acute intoxication or from prolonged, chronic exposure. Some symptoms persisted after cessation of such exposures, perhaps because of the effect of anoxia from inhibition of cytochrome oxidase. Symptoms from chronic exposure are similar to those reported after acute exposures, such as weakness, nausea, headache, and vertigo.[1] A study of 36 former workers in a silver-reclaiming facility who were chronically exposed to cyanide demonstrated some residual symptoms 7 or more months after cessation of exposure; frequent headache, eye irritation, easy fatigue, loss of appetite, and epistaxis occurred in at least 30% of these workers.[5]

Liquid hydrogen cyanide, hydrogen cyanide in aqueous solution (hydrocyanic acid), and concentrated vapor are absorbed rapidly through the intact skin and may cause poisoning with little or no irritant effect on the skin itself.[1] The liquid in the eye may cause some local irritation; the attendant absorption may be hazardous.[6]

Cyanide is one of the few toxic substances for which a specific antidote exists, and it functions as follows.[7,8] First, amyl nitrite (by inhalation) and then sodium nitrite (intravenously) are administered to form methemoglobin, which binds firmly with free cyanide ions. This traps any circulating cyanide ions. The formation of 10% to 20% methemoglobin usually does not involve appreciable risk yet provides a large amount of cyanide-binding substance. Second, sodium thiosulfate is administered intravenously to increase the rate of conversion of cyanide to the less toxic thiocyanate. Although early literature suggests the use of methylene blue, it must not be administered because it is a poor methemoglobin former and, moreover, promotes the conversion of methemoglobin back to hemoglobin.[7]

At high levels, cyanides act so rapidly that odor has no value as forewarning.[1] At lower levels, the odor may provide some forewarning, although many individuals are unable to recognize the bitter almond scent.[3]

Note: For a description of diagnostic signs, differential diagnosis, clinical laboratory tests, and specific treatment of overexposure to hydrogen cyanide and other cyanogens, see "Cyanides."

The 1995 ACGIH ceiling threshold limit value (C-TLV) for hydrogen cyanide is 4.7 ppm (5 mg/m³) with a notation for skin absorption.

REFERENCES

1. National Institute for Occupational Safety and Health: *Criteria for a Recommended Standard . . . Occupational Exposure to Hydrogen Cyanide and Cyanide Salts (NaCN, KCN, and Ca(CN)2)*. DHEW (NIOSH) Pub No 77-108, pp 37–95, 106–114, 170–173, 178. Washington, DC, US Government Printing Office, 1976
2. Gosselin RE, Smith RP, Hodge HC: *Clinical Toxicology of Commercial Products*, Section III, 5th ed, pp 123–130. Baltimore, MD, Williams and Wilkins, 1984
3. Graham DL, Laman D, Theodore J, Robin ED: Acute cyanide poisoning complicated by lactic acidosis and pulmonary edema. *Arch Intern Med* 137:1051–1055, 1977
4. Becker CE: The role of cyanide in fires. *Vet and Hum Toxicol* 27:487–490, 1985
5. Blanc P et al: Cyanide intoxication among silver-reclaiming workers. *JAMA* 253:367–371, 1985
6. Hygienic Guide Series: Hydrogen cyanide. *Am Ind Hyg Assoc J* 31:116–119, 1970
7. Chen KK, Rose CL: Nitrite and thiosulfate therapy in cyanide poisoning. *JAMA* 149:113–119, 1952

HYDROGEN FLUORIDE
CAS: 7664-39-3

HF

Synonyms: None

Physical Form. Gas, liquefying at 19.5°C; aqueous solution is hydrofluoric acid

Uses. Catalyst for production of high-octane gasoline; aqueous solution for frosting, etching, and polishing glass, for removing sand from metal casings, and for etching silicon wafers in semiconductor manufacture

Exposure. Inhalation; skin contact

Toxicology. Hydrogen fluoride (HF), as a gas, is a severe respiratory irritant and, in solution, causes severe and painful burns of the skin and eyes.

From accidental, occupational, and volunteer exposures, it is estimated that the lowest lethal concentration for a 5-minute human exposure to hydrogen fluoride is in the range of 50 to 250 ppm.[1] The LC_{50} for 5, 15, and 60 minutes are considered to be 500 to 800 ppm, 450 to 1000 ppm, and 30 to 600 ppm, respectively.[1] Inhalation of HF produces transient choking and coughing. After an asymptomatic period of several hours up to 1 to 2 days, fever, cough, dyspnea, cyanosis, and pulmonary edema may develop.

Death from pulmonary edema occurred within 2 hours in three of six workers splashed with 70% solution, despite prompt showering with water. The HF concentration in the breathing zone was estimated to be above 10,000 ppm.[2] A chemist exposed to HF splashes on the face and upper extremities developed pulmonary edema 3 hours after exposure and died 10 hours later.[3] Persistent respiratory symptoms, including hoarseness, coughing fits, and nose bleeds but with normal pulmonary function, were observed in one subject who survived a massive exposure. Acute renal failure of uncertain cause has also been documented after an ultimately fatal inhalation exposure.[4] Significant systemic absorption by dermal or inhalation exposure may result in hypocalcemia and hypomagnesemia; cardiac arrhythmias may result as a consequence.[5,6]

In human subjects, exposure to 120 ppm for 1 minute caused conjunctival and respiratory irritation with stinging of skin.[7] At 30 ppm for several minutes, there was mild irritation of the eyes, nose, and respiratory tract; 2.6 to 4.8 ppm, 6 hr/day for periods up to 50 days, caused slight irritation of nose, eyes, and skin but no signs or symptoms of pulmonary irritation.[7,8]

Repeated exposure to excessive concentrations of fluoride over a period of years may result in an increased radiographic density of bone and eventually may cause crippling fluorosis (osteosclerosis due to deposition of fluoride in bone).[7] The early signs of increased bone density from fluoride deposition are most apparent in the lumbar spine and pelvis and can be detected by X ray.

Biological monitoring of urinary fluoride concentration provides an indication of total fluoride intake. Data indicate that a postshift urinary fluoride level of less than 8 mg/l, averaged over an extended period of time, will not lead to osteosclerosis, although a minimal or questionable increase in bone density might develop after many years of occupational exposure.[7]

HF solutions (hydrofluoric acid) in contact with skin result in marked tissue destruction; undissociated HF readily penetrates skin and deep tissue, where the corrosive fluoride ion can cause necrosis of soft tissues and decalcification of bone; the destruction produced is excruciatingly painful.[6,9–11] The fluoride ion also attacks enzymes (eg, of glycolysis) and cell membranes. The process of tissue destruction and neutralization of the hydrofluoric acid is prolonged for days, which is not the case with other, rapidly neutralized acids.[9–11] Because of the insidious manner of penetration, a relatively mild or minor exposure can cause a serious

burn. When skin contacts solutions of less than 20%, the burn manifests itself by pain and erythema, with a latent period of up to 24 hours; with 20% to 50% solutions, the burn becomes apparent 1 to 8 hours following exposure; solutions above 50% cause immediate pain, and tissue destruction is rapidly apparent.[9] Delayed recognition of contact with dilute solutions, with consequently delayed irrigation, often results in more severe burns.[6] Depending on the severity of the burn, it may demonstrate erythema alone, central blanching with peripheral erythema, swelling, vesiculation, serous crusting and, with more serious burns, ulceration, blue-gray discoloration, and necrosis may be noted.[5,6]

Severe eye injuries from splashes may occur. In one case of eye burns from a fine spray of hydrofluoric acid in the face, considerable loss of epithelium occurred despite immediate and copious flushing with water and irrigation for 3 hours with a 0.5 % solution of benzethonium chloride; within 19 days, there was recovery of normal vision.[12]

The 1995 ACGIH threshold limit value ceiling is 3 ppm (2.6 mg/m^3) as F.

REFERENCES

1. Halton DM et al: *Toxicity Levels to Humans During Acute Exposure to Hydrogen Fluoride*, p 40. Ottawa, Canada, Atomic Energy Control Board, November 28, 1984
2. Mayer L, Geulich J: Hydrogen fluoride (HF) inhalation and burns. *Arch Environ Health* 7:445–447, 1963
3. Kleinfeld M: Acute pulmonary edema of chemical origin. *Arch Environ Health* 10:942–946, 1965
4. Braun J et al: Intoxication following the inhalation of hydrogen fluoride. *Arch Toxicol* 56:50–54, 1984
5. White JW: Hydrofluoric acid burns. *Cutis*, 34:241–244, 1984
6. Edelman P: Hydrofluoric acid burns. In La-Dou J (ed): *State of the Art Reviews: Occupational Medicine—The Microelectronics Industry.* pp 89–103, Philadelphia, Hanley & Belfus, 1986
7. National Institute for Occupational Safety and Health: *Criteria for a Recommended Stan-*

dard . . . *Occupational Exposure to Hydrogen Fluoride*. DHEW (NIOSH) Pub No 76-143, pp 106–115. Washington, DC, US Government Printing Office, 1976

8. Largent EJ: *Fluorosis—The Health Aspects of Fluorine Compounds*, pp 34–39, 43–48. Columbus, Ohio State University Press, 1961
9. Dibbell DG et al: Hydrofluoric acid burns of the hand. *J Bone Joint Surg* 52A:931–936, 1970
10. Reinhardt CF, Hume WG, Linch AL, Wetherhold JM: Hydrofluoric acid burn treatment. *Am Ind Hyg Assoc J* 27:166–171, 1966
11. Wetherhold JM, Shepherd FP: Treatment of hydrofluoric acid burns. *J Occup Med* 7:193–195, 1965
12. Grant WM: *Toxicology of the Eye*, 3rd ed, pp 490–492. Springfield, IL, Charles C Thomas, 1986

HYDROGEN PEROXIDE (90%)

CAS: 7722-84-1

90% H_2O_2

Synonyms: None

Physical Form. Liquid

Uses. In synthesis of compounds; bleaching agent, especially for textiles and paper; disinfectant; rocket fuel

Exposure. Inhalation

Toxicology. Hydrogen peroxide is an irritant of the eyes, mucous membranes, and skin.

In humans, inhalation of high concentrations of vapor or mist may cause extreme irritation and inflammation of the nose and throat.[1,2] Severe systemic poisoning may also cause headache, dizziness, vomiting, diarrhea, tremors, numbness, convulsions, pulmonary edema, unconsciousness, and shock.[3]

Exposure for a short period of time to mist or diffused spray may cause stinging of the eyes and lacrimation.[1,2] Splashes of the liquid in the eyes may cause severe damage, including ulceration of the cornea; there may be a delayed appearance of damage to the eyes, and corneal ulceration has, on rare occasions, appeared even a week or more after exposure.[1]

Skin contact with the liquid for a short time will cause a temporary whitening or bleaching of the skin; if splashes on the skin are not removed, erythema and the formation of vesicles may occur.[1] Although ingestion is unlikely to occur in industrial use, it may cause irritation of the upper gastrointestinal tract; decomposition of the hydrogen per oxide will result in the rapid liberation of oxygen, which may distend the esophagus or stomach and cause severe damage.

Repeated exposure of dogs to 7 ppm for 6 months caused sneezing, lacrimation, and bleaching of hair; at autopsy, there was local atelectasis.[4]

A number of investigators have shown that hydrogen peroxide in vitro leads to genetic damage through the formation of free radicals.[5] It is not known whether such damage presents a danger to the mammalian organism or whether various enzymes protect against damage.

Additional hazards are the possibility of explosion when higher-strength hydrogen peroxide is mixed with organic compounds, and violent decomposition of the liquid if contaminated by metallic ions or salts.[3] Since hydrogen peroxide is such a strong oxidizer, it can set fire to combustible materials when spilled on them.[3]

The 1995 ACGIH threshold limit value–time-weighted average (TLV-TWA) for hydrogen peroxide is 1 ppm (1.4 mg/m³).

REFERENCES

1. Chemical Safety Data Sheet SD-53, Hydrogen peroxide, pp 5, 30–31. Washington, DC, MCA, Inc, 1969
2. Hygienic Guide Series: Hydrogen peroxide (90%). *Am Ind Hyg Assoc* 18:275–276, 1957
3. Woodbury CM: Hydrogen peroxide. In International Labour Office, *Encyclopaedia of Occupational Health and Safety*, 3rd ed, rev, Vol 1, pp 1088–1090. Geneva, 1983
4. Oberst FW, Comstock CC, Hackley EB: Inhalation toxicity of ninety per cent hydrogen per-

oxide vapor—acute, subacute, and chronic exposures of laboratory animals. *AMA Arch Ind Hyg Occup Med* 10:319–327, 1954

5. Speit G et al: Characterization of sister chromatid exchange induction by hydrogen peroxide. *Environ Mut* 4:135–142, 1982

HYDROGEN SELENIDE
CAS: 7783-07-5

H₂Se

Synonyms: Selenium hydride

Physical Form.   Colorless gas

Sources.   Produced by reaction of acids or water with metal selenides

Exposure.   Inhalation

Toxicology.   Hydrogen selenide gas is an irritant of the eyes and mucous membranes; it also causes pulmonary irritation and liver damage in animals.

In humans, a concentration of 1.5 ppm is said to produce intolerable irritation of the eyes and nose.[1] Five workers exposed to hydrogen selenide (and possibly other selenium compounds as well) at concentrations of less than 0.2 ppm for 1 month developed nausea, vomiting, diarrhea, metallic taste, garlic odor of the breath, dizziness, lassitude, and fatigue; following cessation of exposure, there was a gradual regression of symptoms during the succeeding months.[2] Urinary selenium levels of the workers ranged from 0- to 13.1-µg selenium/100-ml urine; there was no correlation between symptoms and urinary levels of selenium.[2]

Guinea pigs exposed to 10 ppm for 2 hours exhibited immediate irritation of the eyes and nose; a high percentage of the animals died, apparently from pneumonitis.[3] In guinea pigs, the LC_{50} for 8 hours was 1 mg/m³ (0.3 ppm); pulmonary irritation and liver damage were observed.

The 1995 ACGIH threshold limit value–time-weighted average (TLV-TWA) for hydrogen selenide is 0.05 ppm (0.16 mg/m³).

REFERENCES

1. Grant WM: *Toxicology of the Eye*, 3rd ed, pp 806–809. Springfield, IL, Charles C Thomas, 1986
2. Buchan RF: Industrial selenosis. *J Occup Med* 3:439–456, 1947
3. Hygienic Guide Series: Hydrogen selenide. *Am Ind Hyg Assoc J* 20:514–515, 1959

HYDROGEN SULFIDE
CAS: 7783-06-4

H₂S

Synonyms: Sulfureted hydrogen; hydrosulfuric acid

Physical Form.   Gas

Sources.   By-product of many industrial processes; around oil wells and in areas where petroleum products are processed, stored, or used; decay of organic matter; naturally occurring in coal, natural gas, oil, volcanic gases and sulfur springs.

Exposure.   Inhalation

Toxicology.   Hydrogen sulfide is an irritant of the eyes and respiratory tract at low concentrations; at higher levels, it causes respiratory paralysis, with consequent asphyxia, and is rapidly fatal.

Hydrogen sulfide intoxication in man has generally been categorized as acute, subacute, or chronic, depending on the nature of the predominant clinical signs and symptoms.[1] Acute intoxication refers to the effects of a single exposure to massive concentrations that rapidly produce signs of respiratory distress. Inhalation of 1000 ppm or more can cause coma after a single breath and can be rapidly fatal as a result of respiratory paralysis.[1-5] At slightly lower levels, the gas may be rapidly absorbed through the lung into the blood, which initially induces hyperpnea followed by apnea.

The sequelae of acute poisoning appear to be quite variable and depend on duration, as well as level, of exposure. Patients who have been unconscious in high levels of hydrogen sulfide atmosphere for longer than 5 minutes may have persistent neurological and neuropsychological impairment years after exposure as a result of hydrogen sulfide–induced hypoxia.[6]

Subacute intoxication refers to the effects caused by continuous exposure for up to several hours to concentrations ranging from 100 to 1000 ppm.[1-5] Pulmonary edema is a potentially fatal complication of intoxication and is common after exposure to 250 ppm for prolonged periods of time. Symptoms of gastrointestinal disturbances, including nausea, abdominal cramps, vomiting, and severe diarrhea, have been reported and frequently occur in subacute intoxication.

Exposure to levels above 50 ppm for 1 hour can produce acute conjunctivitis, with pain, lacrimation, and photophobia; in severe form, this can progress to keratoconjunctivitis and vesiculation of the corneal epithelium. Prolonged exposure to 50 ppm also causes rhinitis, pharyngitis, bronchitis, and pneumonitis.

Reports of adverse effects of hydrogen sulfide on humans as a result of chronic intoxication are less well established. It has been postulated that exposures below 50 ppm over long periods may cause certain neurasthenic symptoms such as fatigue, headache, dizziness, and irritability. Others suggest that the signs and symptoms referred to as chronic poisoning are actually the results of recurring acute exposures or the sequelae of acute poisoning.

A number of toxicological mechanisms have been proposed for hydrogen sulfide : at extremely high concentrations, it may exert a direct paralyzing effect on respiratory centers; hydrogen sulfide is also known to inhibit cytochromec oxidase, resulting in altered oxidative metabolism; it can also disrupt critical disulfide bonds in essential cellular proteins.[5]

Skin absorption appears to be minimal in humans.

In the only available epidemiological study, no significant increase in cancer incidence was found for individuals residing downwind from two natural gas refineries, which emit primarily sulfur compounds, including hydrogen sulfide.[7]

Rats exposed to 100 ppm of hydrogen sulfide, 6 hr/day during days 6 through 20 of gestation, showed no signs of maternal toxicity or adverse effects on the developing fetus.[8] In another report, rat dams and pups were exposed 7 hr/day, to 20, 50 or 75 ppm from day 1 of gestation until day 21 postpartum. Blood glucose was significantly elevated in dams on day 21 postpartum at all exposure levels, but the toxicological significance of this effect has not been established.[9]

The odor is offensive and characterized as "rotten eggs," with a threshold ranging from 0.0005 to 0.13 ppm; it is unreliable as a warning signal because the gas exerts a paralyzing effect on the olfactory apparatus above 150 ppm; at these concentrations, the odor has been characterized as sickeningly sweet.[1]

The 1995 threshold limit value–time-weighted average (TLV-TWA) is 10 ppm (14 mg/m^3) with a STEL/ceiling of 15 ppm (21 mg/m^3).

REFERENCES

1. World Health Organization: *Environmental Health Criteria 19: Hydrogen Sulfide.* Geneva, WHO, 1981
2. National Institute for Occupational Safety and Health: *Criteria for a Recommended Standard . . . Occupational Exposure to Hydrogen Sulfide.* DHEW (NIOSH), pp 22–64. Washington, DC, US Government Printing Office, 1977
3. Stine RJ, Slosberg B, Beacham BE: Hydrogen sulfide intoxication—a case report and discussion of treatment. *Ann Intern Med* 85:756, 1976
4. Milby TH: Hydrogen sulfide intoxication—review of the literature and report of unusual accident resulting in two cases of nonfatal poisoning. *J Occup Med* 4:431, 1962
5. Beauchamp RO Jr, Bus JS, Popp JA, et al: A critical review of the literature on hydrogen sulfide toxicity. *CRC Crit Rev Toxicol* 13:25–97, 1984

6. Tvedt B, Skyberg K, Aaserud O, et al: Brain damage caused by hydrogen sulfide: a follow-up study of six patients. *Am J Ind Med* 20:91–101, 1991
7. Schechter MT, Spitzer WO, Hutcheon ME: Cancer downwind from sour gas refineries: the perception and the reality of an epidemic. *Environ Health Perspect* 79:283–290, 1989
8. Saillenfait AM, Bonnet P, deCeaurriz J: Effects of inhalation exposure to carbon disulfide and its combination with hydrogen sulfide on embryonal and fetal development in rats. *Toxicol Lett* 48:57–66, 1989
9. Hayden LJ, Goeden H, Roth SH: Exposure to low levels of hydrogen sulfide elevates circulating glucose in maternal rats. *J Toxicol Environ Health* 31:45–52, 1990

HYDROQUINONE
CAS:123-31-9

$C_6H_4(OH)_2$

Synonyms: 1,4-Benzenediol; 1,4-dihydroxybenzene; *p*-hydroxybenzene; hydroquinol; quinol; Tecquinol

Physical Form.   White crystalline solid

Uses.   Photographic reducer and developer; antioxidant; stabilizing agent for some polymers; intermediate in the manufacture of some dyes and pigments; in cosmetic formulations

Exposure.   Inhalation

Toxicology.   Hydroquinone is moderately toxic and primarily affects the eyes.

Acute exposure to quinone vapor and hydroquinone dust causes conjunctival irritation, whereas chronic exposure produces changes characterized as: (1) brownish discoloration of the conjunctiva and cornea confined to the interpalpebral tissue; (2) small opacities of the cornea; and (3) structural changes in the cornea that result in loss of visual acuity.[1,2] The pigmentation changes are reversible, but the more slowly developing structural changes in the cornea may progress. Pigmentation may appear with less than 5 years of exposure, but this is uncommon and usually is not associated with serious injury to the eye.[2]

Ingestion of 5 to 12 g of hydroquinone has been reported to be fatal.[3-5] In one nonfatal case of hydroquinone ingestion of approximately 1 g, tinnitus, dyspnea, cyanosis, and extreme sleepiness were observed.[3] Although acute, high-dose oral ingestion produces noticeable CNS effects in humans, no effects have been observed in workers exposed to lower concentrations in actual industrial situations.[3] No signs of toxicity were found in subjects who ingested 300 to 500 mg of hydroquinone daily for 3 to 5 months.[6]

Repeated skin contact with hydroquinone creams (generally 5% or more hydroquinone) produced skin irritation, allergic sensitization, dermatitis, and depigmentation.[3] Excessive use of skin-lightening preparations containing hydroquinone have produced severe and irreversible cutaneous damage.[5] Deleterious effects start with darkening and coarsening of the skin, followed by a hyperpigmented papular condition. Histologically, there is increased basophilia of the collagen, followed by the formation of yellow fibers that swell and break down to form an amorphous eosinphilic material.

Oral LD_{50}s of 70, 200, and 550 mg/kg have been reported for cats, dogs, and guinea pigs, respectively.[5] In 14-day and 13-week studies, mice administered up to 500 mg/kg by gavage and rats administered up to 1000 mg/kg had lethargy, tremors, and convulsions.[7] The central nervous system, forestomach, and liver were identified as target organs in both species, and renal toxicity was identified in rats.

In two-year studies, rats were given 0, 25, or 50 mg/kg hydroquinone by gavage 5 days/week, while doses for mice were 0, 50, or 100 mg/kg on the same schedule.[7] There was evidence of carcinogenicity in male rats as indicated by increased incidences of tubular cell adenomas of the kidney, in female rats as shown by increases in mononuclear cell leukemia, and in female mice based on increases in hepatocellular neoplasms, mainly adenomas. There was no evidence of carcinogenicity in male mice.

Pellets of cholesterol containing 2 mg of hydroquinone implanted in mice bladders caused an excessive number of bladder carcinomas.[4] In other studies, rats fed up to 1% hydroquinone in their diets for 2 years did not develop tumors, nor did hydroquinone initiate significant numbers of tumors in skin-painting studies of mice.[3,4]

Pregnant rats given up to 300 mg/kg hydroquinone by gavage on the 6th through 15th day of gestation had a slight but significant reduction in body weight gain and feed consumption. This effect was associated with a slightly reduced mean fetal body weight but no other significant effects were noted in the rat conceptus.[8] In rabbits hydroquinone at 150 mg/kg on gestation days 6 to 18 produced minimal developmental alterations in the presence of maternal toxicity.[9] Based on these studies, it was concluded that hydroquinone is not selectively toxic to the developing conceptuses and does not appear to be a developmental toxicant.[8]

Hydroquinone induces alterations of the DNA in eucaryotic cells (micronuclei, chromosomic aberrations, and disintegrations) but is nonmutagenic in *Salmonella* tester strains with or without metabolic activation.[5]

The 1995 ACGIH threshold limit value–time-weighted average (TLV-TWA) for hydroquinone is 2 mg/m³.

REFERENCES

1. Anderson B, Oglesby F: Corneal changes from quinone-hydroquinone exposure. *AMA Arch Ophthalmol* 59:495–501, 1958
2. Sterner JH, Oglesby FL, Anderson B: Quinone vapors and their harmful effects. I. Corneal and conjunctival injury. *J Ind Hyg Toxicol* 29:60–73, 1947
3. National Institute for Occupational Safety and Health: *Criteria for a Recommended Standard . . . Occupational Exposure to Hydroquinone.* DHEW (NIOSH) Pub No 78-155, p 182. Washington, DC, US Government Printing Office, 1978
4. *IARC Monographs on the Evaluation of the Carcinogenic Risk of Chemicals to Man*, Vol 15, Some fumigants, the herbicides, 2,4-D and 2,4,5-T chlorinated dibenzodioxins and miscellaneous industrial chemicals, pp 155–175. Lyon, International Agency for Research on Cancer, 1977
5. Devillers J, Boule P, Vasseur P, et al: Environmental and health risks of hydroquinone. *Ectoxic Environ Saf* 19:327–354, 1990
6. Carlson AJ, Brewer NR: Toxicity studies on hydroquinone. *Proc Soc Exp Biol Med* 84:684–688, 1953
7. Kari FW, Bucher J, Eustis SL, et al: Toxicity and carcinogenicity of hydroquinone in F344/N rats and B6C3F₁ mice. *Food Chem Toxic* 9:737–747, 1992
8. Krasavage WJ, Blacker AM, English C, et al: Hydroquinone: a developmental toxicity study in rats. *Fund Appl Toxicol* 18:370–375, 1992
9. Murphy SJ, Schroeder RE, Blacker AM, et al: A study of developmental toxicity of hydroquinone in the rabbit. *Fund Appl Toxicol* 19:214–221, 1992

HYDROXYLAMINE (AND SALTS)
CAS: 7803-49-8

NH_2OH

Synonyms: Oxammonium; hydroxyl ammonium

Physical Form. Colorless flakes or crystals.

Uses. Reducing agent used in photographic processing, leather tanning, manufacturing of nylon and other polymers; as a stabilizer for natural rubber; to prevent the development of objectionable tastes and odors during the refining of fatty materials.

Exposure. Inhalation

Toxicology. Hydroxylamine and its salts are irritants of eyes, mucous membranes, and skin; higher levels cause methemoglobinemia.

Workers exposed to hydroxylamine sulfate for 1 day at unspecified air levels showed blood methemoglobinemia concentrations of 25%.[1] Dusts and mists of hydroxylamine sulfate are irritants of the mucous membranes and eyes. Although details are lacking, repeated exposure to the sulfate is reported to have caused respiratory sensitization with asthmalike symptoms.

Hydroxylamine hydrochloride is highly irritating to the skin, eyes, and mucous membranes and has caused contact dermatitis in workers exposed for 2 to 60 days.[2] Hydroxylamine itself is only moderately irritating to the skin. Hydroxylamine sulfate on the skin of rabbits was irritating at levels as low as a 10 mg dose.[3] It is considered to be a potential skin sensitizer.

Carcinogenicity of hydroxylamine and its salts has not been demonstrated. Several studies have shown a decreased incidence of spontaneous mammary tumors in mice exposed to the sulfate and hydrochloride.[3-6] There was some indication of an increase in the incidence of spontaneous mammary tumors when the sulfate was administered to older animals whose mammary glands were already well developed.

Embryotoxic effects have occurred in rabbits exposed to hydroxylamine hydrochloride by intracoelomic injection.[7] Subcutaneous or intravenous injection of pregnant rabbits with 50 to 650 mg hydroxylamine hydrochloride on gestational day 12 caused death or sacrifice of all rabbits within 30 hours.[8] All maternally injected rabbits exhibited severe cyanosis, presumably due to methemoglobinemia. At 8 hours, all embryos were dead from cardiovascular effects, which are considered to be secondary to the severe maternal toxicity. Hydroxylamine is a direct-acting developmental toxicant only under conditions of direct embryonic exposure; intracoelomic injection into the chorionic cavity of developing embryos of 100 μg of hydroxylamine caused deaths in 31 of 32 embryos. In general, exposure to hydroxylamine would kill the mother before levels within the embryo became sufficiently high to cause direct developmental toxicity.

A threshold limit value (TLV-TWA) has not been established for this substance.

REFERENCES

1. Hydroxylamine sulfate: Product Safety Data Sheet. Morristown, NJ, Allied Corporation, 1983

2. Folesky Von H, Nickel H, Rothe A, Zschunke E: Allergisches Ekzem durch Salze des Hydroxylamins (Oxammonium). *Z Gesamte Hyg Grenzgeb* 17:353–356, 1971
3. Yamamoto RS, Weisburger EK, Korzis J: Chronic administration of hydroxylamine and derivatives in mice. *Proc Soc Exp Biol Med* 124:1217–1220, 1967
4. Harman D: Prolongation of the normal lifespan and inhibition of spontaneous cancer by antioxidants. *J Gerontol* 16:247–257, 1961
5. Evarts RP, Brown CA: Morphology of mammary gland, ovaries, and pituitary gland of hydroxylamine-fed C3H/HeN mice. *Lab Invest* 37:53–63, 1977
6. Evarts RP, Brown CA, Atta GJ: The effects of hydroxylamine on the morphology of the rat mammary gland and on the induction of mammary tumors by 7,12-dimethylbenz(a)anthracene. *Exp Mol Pathol* 30:337–348, 1979
7. DeSesso JM: Demonstration of the embryotoxic effects of hydroxylamine in the New Zealand white rabbit. *Anat Rec* 196:45A–46A, 1980
8. DeSesso JM, Goeringer GC: Developmental toxicity of hydroxylamine: an example of a maternally mediated effect. *Tox Ind Health* 6:109–121, 1990

2-HYDROXYPROPYL ACRYLATE
CAS: 999-61-1

CH₂CHCOOCH₂CHOHCH₃

$CH_2CHCOOCH_2CHOHCH_3$

Synonyms: HPA; 1,2-propanediol-1-acrylate; propylene glycol monoacrylate

Physical Form. Clear to light yellow liquid

Uses. Monomer used in manufacture of thermosetting resins for surface coatings.

Exposure. Inhalation; skin absorption

Toxicology. 2-Hydroxypropyl acrylate is an irritant of the eyes, nose, respiratory tract, and skin.

Inhalation exposure of rats, mice, dogs, and

rabbits to 5 ppm for 6 hr/day, 5 days/week over 30 days, caused nasal and respiratory tract irritation.[1]

The dermal LD_{50} in rabbits was approximately 0.17 g/kg.[2] Animals that survived developed severe irritation, moderate edema, and moderate to severe necrosis. Direct contact with the eye caused severe eye burns.

The 1995 threshold limit value–time-weighted average (TLV-TWA) for 2-hydroxypropyl acrylate is 0.5 ppm (2.8 mg/m³) with a notation for skin absorption.

REFERENCES

1. DOW Chemical Company: Communication to TLV Committee, Midland, MI, 1977. In *Documentation of the TLVs and BEIs*, 6th ed, pp 793–794. Cincinnati, OH, American Conference of Governmental Industrial Hygienists, 1991
2. Smyth HF, Carpenter CP, Weil CS, et al: Range-finding toxicity data: List VII. *Am Ind Hyg Assoc J* 30:470–476, 1969

INDENE
CAS: 95-13-6

C_9H_8

Synonym: Indonaphthene

Physical Form. Colorless liquid

Uses. Preparation of coumarone-indene resins

Exposure. Inhalation

Toxicology. Indene is expected to be an irritant of the mucous membranes.

Oral doses of 2.5 ml of a 1:1 v/v mixture in olive oil were fatal to rats.[1] An historical study indicates that exposure of rats to 800 to 900 ppm for 7 hr/day for 6 exposures caused hemorrhagic

liver necrosis in some of the rats, as well as focal necrosis of the kidneys.[2] No deaths occurred from these exposures.

Indene vapor inhalation exposure of human subjects has not been reported. By analogy to related hydrocarbons, inhalation of indene can be expected to cause irritation of the mucous membranes.

The 1995 threshold limit value–time-weighted average (TLV-TWA) for indene is 10 ppm (48 mg/m³).

REFERENCES

1. Gerarde HW: *Toxicology and Biochemistry of Aromatic Hydrocarbons*, pp 202, 209–216. New York: Elsevier, 1960
2. Cameron GR, Doniger CR: The toxicity of indene. *J Pathol Bacteriol* 49:529–533, 1939

INDENO(1,2,3-CD)PYRENE
CAS: 193-39-5

$C_{22}H_{12}$

Synonyms: IP; 2,3-Phenylenepyrene; *o*-phenylenepyrene; indeneopyrene

Physical Form. Yellow solid

Uses/Sources. A component of polynuclear aromatic hydrocarbons, also known as polycyclic aromatic hydrocarbons; usually bound to small particulate matter present in urban air, industrial and natural combustion emissions, and cigarette smoke.

Exposure. Inhalation

Toxicology. Indeno(1,2,3-cd)pyrene (IP) is a complete carcinogen and an initiator for skin carcinogenesis in the mouse.[1]

Groups of female mice were painted with IP, either in dioxane or in acetone, 3 times weekly for 12 months.[2] A concentration of 0.1%

produced a total of 6 papillomas and 3 carcinomas, the first tumors appearing at 9 months. A concentration of 0.5% produced a total of 7 papillomas and 5 carcinomas, the first tumors appearing at 3 months.

The same study demonstrated that 10 paintings at 2-day intervals, for a total dose of 250 µg IP, initiated skin carcinogenesis. In 30 mice subsequently treated with croton oil in acetone, a total of 10 papillomas in 5 animals was produced.

A threshold limit value has not been established.

REFERENCES

1. *IARC Monographs on the Evaluation of the Carcinogenic Risk of Chemicals to Humans*, Vol 3, Certain polycyclic aromatic hydrocarbons and heterocyclic compounds, pp 229–237. International Agency for Research on Cancer, 1973
2. Hoffman D, Wynder EL: Beitrag zur carcinogenen Wirkung von Dibenzopyrenen. *Z Krebsforsch* 68:137, 1966

INDIUM AND COMPOUNDS
CAS: 7440-74-6

In

Synonyms: Indium sesquioxide; indium trichloride; indium nitrate; indium antimonide; indium arsenide

Physical Form.   Solid

Uses.   In the manufacture of semiconductors; in the manufacture of glass, graphite, and cathode oscillographs; in metal alloys to prevent corrosion and metal fatigue

Exposure: Inhalation

Toxicology.   Indium (In) and compounds cause injury to the lungs, liver, and kidneys in animals.

There are no reports of toxicity in humans. There was no evidence of irritation when indium was applied to the skin.

A range of oral LD_{50}s has been reported in animals, depending on the route of administration and the type of compound.[1] Administered parenterally to rats, rabbits, and dogs, indium trichloride ($InCl_3$) had an acute lethal dose range from 0.33 to 3.6 mg of In/kg.[2] Indium sesquioxide (In_2O_3) was less toxic, with intraperitoneal doses of 955 mg/kg fatal to all rats within 9 days. Gross signs of In poisoning from intraperitoneal or intravenous administration have included reduced food and water consumption, with accompanying weight loss, and degenerative changes in the liver and kidneys.[1–3]

A single intratracheal dose of 1.3 mg In/kg as $InCl_3$ given to female Fischer 344 rats caused severe upper and lower pulmonary damage that was present 8 weeks after dosing.[4] In addition, damage to the alveolar and bronchial/bronchiolar epithelial cells initiated inflammatory and repair processes that led to the rapid development of fibrosis.

In another report, rats exposed to the sesquioxide (In_2O_3) dust by inhalation at levels ranging from 24 to 97 mg/m^3 for a total of 224 hours had widespread alveolar edema and alteration of the alveolar walls resembling alveolar proteinosis in which alveolar clearance was reduced.[5] The lesion exhibited no change during exposure or after a 12-week postexposure period, including no evidence of wound healing or fibrosis. Lack of a fibrotic response in this study may be due to the relative insolubility of In_2O_3 (compared to $InCl_3$) and less reactivity with biomembranes.[4]

When ingested by rats, In_2O_3 was practically nontoxic; incorporated in the diet, 8% for 3 months caused no effects on growth mortality or tissue morphology.[1] Indium trichloride caused marked growth depression at 4% in the diet over the same period.

In cellular studies, indium exposure has been associated with a general suppression of protein synthesis and the induction of heme oxygenase which, in turn, is associated with the reduction of enzyme activities dependent on cytochrome P-450.[6,7] The significance of these alterations in the synthesis and maintenance of various enzyme systems in relation to a possible carcinogenic response has not been determined.[8]

The 1995 ACGIH threshold limit value–time-weighted average (TLV-TWA) for indium and compounds as In is 0.1 mg/m³.

REFERENCES

1. Stokinger HE: The metals. In Clayton GD, Clayton FE (eds): *Patty's Industrial Hygiene and Toxicology*, 3rd ed, rev, Vol 2A *Toxicology*, pp 1654–61. New York, Wiley-Interscience, 1981
2. Downs WK, Scott JK, Steadman LT, Maynard EA, University of Rochester Atomic Energy Report UR-588. Rochester, NY, November 1959
3. McCord CP, Meek SF, Harrold GC, Huessner CE: Physiologic properties of indium and its compounds. *Ind Hyg Toxicol* 24:243–254, 1942
4. Blazka ME, Dixon D, Haskins E, et al: Pulmonary toxicity to intratracheally administered indium trichloride in Fischer 344 rats. *Fund Appl Toxicol* 22:231–239, 1994
5. Leach LJ, Scott JK, Armstrong RD, et al: Atomic Energy Commission Research and Development Report UR-590. University of Rochester, Rochester, NY 1961
6. Fowler BA, Yamauchi H, Akkerman M: Comparison of the induction of stress proteins after administration of gallium arsenide and indium arsenide in hamster proximal tubule epithelial cells. *Toxicologist*
7. Woods JS, Carver GT, Fowler BA: Altered regulations of hepatic heme metabolism by indium chloride. *Toxicol Appl Pharmacol* 49:455–461, 1979
8. Fowler BA, Yamauchi H, Conner EA, et al: Cancer risks for humans from exposure to the semiconductor metals. *Scand J Work Environ Health* 19 (Suppl 1):101–103, 1993

IODINE
CAS: 7553-56-2

I_2

Synonyms: None

Physical Form. Crystalline solid; blue-black scales or plates

Uses. In synthesis of organic chemicals; photographic film; disinfectant in drinking water

Exposure. Inhalation; ingestion

Toxicology. Iodine is an irritant of the eyes, mucous membranes, and skin; it is a pulmonary irritant in animals, and it is expected that severe exposure will cause the same effect in humans.

Exposed workers (concentration and time unspecified) experienced a burning sensation in the eyes, lacrimation, blepharitis, rhinitis, stomatitis, and chronic pharyngitis; after brief accidental exposure in a laboratory, technicians reported headache and a feeling of tightness in the chest.[1,2]

Iodine is an essential nutritional element and is required by the thyroid.[3] However, ingestion of as little as 2 to 3 g may be fatal. Chronic absorption of iodine causes iodism, a syndrome characterized by insomnia, conjunctivitis, rhinitis, bronchitis, tremor, tachycardia, parotitis, diarrhea, and weight loss.[4,5] Iodine absorbed by the lungs is changed to iodide and eliminated mainly in the urine.

In an experimental investigation, four human subjects tolerated 0.57 ppm iodine vapor for 5 minutes without eye irritation, but all experienced eye irritation in 2 minutes at 1.63 ppm.[3] In patients exposed to air saturated with iodine vapor for 3 to 4 minutes for therapeutic purposes, there was brown staining of the corneal epithelium and subsequent spontaneous loss of the layer of tissue; recovery occurred within 2 to 3 days.[6] Iodine in crystalline form or in strong solutions is a severe skin irritant; it is not easily removed from the skin, and the lesions resemble thermal burns with brown staining.[5]

Hypersensitivity to iodine characterized by a skin rash has been reported.[4,5] Iodine is absorbed through the skin in small amounts from a tincture or from vapor applied to the skin.[3]

Intratracheal administration to dogs of the vapor at 36 mg iodine/kg body weight was fatal after about 3 hours; the animals developed cough, difficulty in breathing, and rales; autopsy findings were pulmonary edema, subpleural hemorrhage, and an increased iodine content of the thyroid and urine.[1]

Administered in the drinking water of rats for 100 days, 1, 3, 10, or 100 mg/l of iodine caused no signs of overt toxicity, but some modifications in thyroid function occurred.[7] Specifically, there was a dose-related trend in increased plasma thyroxine levels and a statistically significant increase in the thyroxine to triiodothyronine ratio.

The 1995 ACGIH ceiling–threshold limit value (C-TLV) for iodine is 0.1 ppm (1.0 mg/m³).

REFERENCES

1. Luckhardt AB, Koch FC, Schroeder WF, Weiland AH: The physiological action of the fumes of iodine. *J Pharmacol Exp Ther* 15:1–21, 1920
2. Heyroth F: Halogens. In Patty FA (ed): *Industrial Hygiene and Toxicology*, 2nd ed, Vol 2, *Toxicology*, pp 854–856. New York, Interscience, 1963
3. Hygienic Guide Series: Iodine. *Am Ind Hyg Assoc J* 26:423–426, 1965
4. Seymour WB Jr: Poisoning from cutaneous application of iodine, a rare aspect of its toxicologic properties. *Arch Int Med* 59:952–966, 1937
5. Peterson JE: Iodine. In International Labour Office: *Encyclopaedia of Occupational Health and Safety*, 3rd ed, rev, Vol I, pp 1153–1154. New York, McGraw-Hill, 1983
6. Grant WM: *Toxicology of the Eye*, 3rd ed, pp 519–520. Springfield, IL, Charles C Thomas, 1986
7. Sherer TT, Thrall KD, Bull RJ: Comparison of toxicity induced by iodine and iodide in male and female rats. *J Toxicol Environ Health* 32:89–101, 1991

IODOFORM
CAS: 75-47-8

CHI_3

Synonym: Triiodomethane

Physical Form.   Yellow or green-yellow solid

Uses.   In veterinary medicine as an antiseptic on superficial lesions; formerly used in medicine as a germicide

Exposure.   Inhalation

Toxicology.   Iodoform causes central nervous system depression and damage to the kidneys, liver, and heart.

The 7-hour LC_{50} for iodoform in rats was 165 ppm, and death of the animals was attributed to cardiopulmonary collapse.[1] Exposure of rats to 14 ppm for 7 hr/day for 7 consecutive days showed only mineralized deposits in the medullary renal tubules.

When used as a topical anesthetic in medical applications, iodoform produced central nervous system depression with vomiting, coma, and damage to the kidneys, liver, and heart.[2]

A 78-week bioassay for possible carcinogenicity of technical-grade iodoform was conducted using rats and mice.[3] Iodoform in corn oil was administered by gavage to groups of 50 male and 50 female animals of each species. Administration was 5 days a week for a period of 78 weeks, followed by an observation period of 34 weeks for rats and 13 or 14 weeks for mice. The high time-weighted average dosages of iodoform were, respectively, 142 and 55 mg/kg/day for male and female rats and 93 and 47 mg/kg/day for male and female mice. There was no evidence of carcinogenicity.

The 1995 threshold limit value–time-weighted average (TLV-TWA) for iodoform is 0.6 ppm or 10 mg/m³.

REFERENCES

1. Tansy MF, Werley M, Landin W: Subacute inhalation toxicity testing with iodoform vapor. *Toxicol Environ Health* 8:59–70, 1981
2. Sell DA, Reynolds ES: Liver parenchymal cell injury. VIII. Lesions of membranous cellular components following iodoform. *J Cell Biol* 41:736–752, 1969
3. National Cancer Institute: *Bioassay of Iodoform for Possible Carcinogenicity (CAS 75-47-8)*. Technical Report Series No 110, 1978

IRON OXIDE FUME
CAS: 1309-37-1

Fe_2O_3

Synonyms: Ferric oxide fume

Physical Form. Fume

Source. Result of welding and silver finishing

Exposure. Inhalation

Toxicology. Inhalation of iron oxide fume or dust causes a benign pneumoconiosis (siderosis).

Iron oxide alone does not cause fibrosis in the lungs of animals, and it is probable that the same applies to humans.[1] Exposures of 6 to 10 years are usually required before changes recognizable by X ray occur; the retained dust produces X ray shadows that may be indistinguishable from fibrotic pneumoconiosis.[2,3] Of 25 welders exposed chiefly to iron oxide for an average of 18.7 (range 3–32) years, 8 had reticulonodular shadows on chest X rays consistent with siderosis, but there was no reduct ion in pulmonary function; exposure levels ranged from 0.65 to 47 mg/m^3.[4]

In another study, the X rays of 16 welders with an average exposure of 17.1 (range 7–30) years also suggested siderosis; their spirograms were normal. However, the static and functional compliance of the lungs was reduced; some of the welders were smokers.[5] The welders with the lowest compliance complained of dyspnea.

Welders are typically exposed to a complicated mixture of dust and fume of metallic oxides, as well as irritant gases, and are subject to mixed-dust pneumoconiosis with possible loss of pulmonary function; this should not be confused with benign pneumoconiosis caused by iron oxide.[1] Although an increased incidence of lung cancer has been observed among hematite miners exposed to iron oxide, presumably as a result of concomitant radon gas exposure, there is no evidence that iron oxide alone is carcinogenic to man or animals.[6]

The 1995 ACGIH threshold limit value–time-weighted average (TLV-TWA) for iron oxide fume is 5 mg/m^3 as total particulate as Fe.

REFERENCES

1. Jones JG, Warner CG: Chronic exposure to iron oxide, chromium oxide, and nickel oxide fumes of metal dressers in a steelworks. *Br J Ind Med* 29:169–177, 1972
2. Sentz FC Jr, Rakow AB: Exposure to iron oxide fume at arc air and power-burning operations. *Am Ind Hyg Assoc J* 30:143–146, 1969
3. Harding HE, McLaughlin AIG, Doig AT: Clinical, radiographic, and pathological studies of the lungs of electric arc and oxyacetylene welders. *Lancet* 2:394–398, 1958
4. Kleinfeld M, Messite J, Kooyman O, Shapiro J: Welders' siderosis. *Arch Environ Health* 19:70–73, 1969
5. Stanescu DC et al: Aspects of pulmonary mechanics in arc welders' siderosis. *Br J Ind Med* 24:143–147, 1967
6. Stokinger HE: A review of world literature finds iron oxides noncarcinogenic. *Am Ind Hyg Assoc J* 45(2):127–133, 1984

IRON PENTACARBONYL
CAS: 13463-40-6

FE(CO)₅

$FE(CO)_5$

Synonyms: Pentacarbonyl iron; iron carbonyl

Physical Form. Colorless to yellow liquid

Uses. Strong reducing agent; in manufacture of high-frequency coils used in radios and television sets; antiknock agent in motor fuels

Exposure. Inhalation

Toxicology. Iron pentacarbonyl is a pulmonary irritant similar to nickel carbonyl.

Iron pentacarbonyl is approximately one-third as potent as nickel carbonyl when inhaled by rats for 30 minutes.[1] Effects from inhalation of high concentrations of the chemical are expected to be similar to those of nickel carbonyl, which include frontal headache, vertigo, nausea, vomiting and, sometimes, substernal and epigastric pain.[2,3] Generally, these early effects disappear when the subject is removed to fresh air.

There may be an asymptomatic interval between recovery from initial symptoms and onset of delayed symptoms, which tend to develop 12 to 36 hours following exposure. Constrictive pain in the chest is characteristic of the delayed onset of pulmonary effects, followed by cough, hyperpnea, and cyanosis, leading to profound weakness. Except for the pronounced weakness and hyperpnea, the physical findings and symptoms resemble those of a viral or an influenzal pneumonia.

Iron pentacarbonyl is relatively benign when administered orally. In a study of iron deficiency anemia, single doses of 10 g were tolerated by 20 nonanemic volunteers with no evidence of toxicity and only minor gastrointestinal side effects.[4] Daily doses of up to 3 g/day for 8 to 28 days resulted in no evidence of toxicity other than gastrointestinal irritation.

The 1995 ACGIH threshold limit value–time-weighted average (TLV-TWA) is 0.1 ppm (0.23 mg/m^3) as Fe with a STEL/ceiling of 0.2 ppm (0.45 mg/m^3) as Fe.

REFERENCES

1. Sunderman F, West B, Kincaid J: Toxicity study of Fe pentacarbonyl. *Arch Ind Health* 19:11–13, 1959
2. Committee on Medical and Biologic Effects of Environmental Pollutants, Division of Medical Sciences, National Research Council: *Nickel*, pp 113–128, 231–268. Washington, DC, National Academy of Sciences, 1975
3. Jones CC: Nickel carbonyl poisoning. *Arch Environ Health* 26:245–248, 1973
4. Gordeuk VR et al: Carbonyl iron therapy for iron deficiency anemia. *Blood* 67:745–752, 1986

ISOAMYL ACETATE
CAS: 123-92-2

CH₃COOCH₂CH(CH₃)C₂H₅

$CH_3COOCH_2CH(CH_3)C_2H_5$

Synonyms: Amyl acetate; banana oil; pear oil; amylacetic ester; 3-methyl butyl acetate; 3-methyl-1-butanol acetate

Physical Form. Colorless liquid

Uses. Solvent; flavoring in water and syrups

Exposure. Inhalation

Toxicology. Isoamyl acetate is an irritant of the eyes and mucous membranes; at high concentrations, it causes narcosis in animals, and it is expected that severe exposure will cause the same effect in humans.

Several grades of technical amyl acetate are known; isoamyl acetate is the major component of some grades, whereas other isomers predominate in other grades.[1]

Men exposed to 950 ppm isoamyl acetate for 30 minutes had irritation of the nose and throat, headache, and weakness.[1]

Cats exposed to 1900 ppm for six 8-hour exposures showed irritation of the eyes, salivation, weakness, and loss of weight; lung irritation was noted at necropsy. A 24-hour exposure to 7200 ppm caused light narcosis and delayed death due to pneumonia.[2] Dogs exposed to 5000 ppm for 1 hour had nasal irritation and drowsiness.[2] Isoamyl acetate may cause skin irritation.

Amyl acetate has a banana- or pearlike odor detectable at 7 ppm.[1]

The 1995 ACGIH threshold limit value–time-weighted average (TLV-TWA) for isoamyl acetate is 100 ppm (532 mg/m^3).

REFERENCES

1. Hygienic Guide Series: Amyl acetate. *Am Ind Hyg Assoc J* 26:199–202, 1965
2. Sandmeyer EE, Kirwin CJ: Esters. In Clayton GD, Clayton FE (eds): *Patty's Industrial Hygiene and Toxicology*, 3rd ed, rev, Vol 2, *Toxicology*, p 2274. Wiley-Interscience, 1981

ISOAMYL ALCOHOL
CAS: 123-51-3

$(C_2H_5)_2CHOH$

Synonyms: 3-Methylbutanol-1; isobutyl carbinol; isopentyl alcohol

Physical Form. Colorless liquid

Uses. Solvent; chemical synthesis; manufacture of smokeless powders, artificial silk, and lacquers

Exposure. Inhalation

Toxicology. Isoamyl alcohol is an irritant of the eyes and mucous membranes; at high concentrations, it causes narcosis in animals, and it is expected that severe exposure will produce the same effect in humans.

Human volunteers exposed to 100 ppm for 3 to 5 minutes experienced throat irritation and, at 150 ppm, also eye and nose irritation.[1,2] No chronic systemic effects have been reported in humans.

Rats survived 8-hour exposure to 2000 ppm. Oral administration of 0.7 g/kg produced stupor and loss of voluntary movement in half the treated rabbits; the LD$_{50}$ was 3.4 g/kg.[3]

Instilled in rabbit eyes, isoamyl alcohol caused severe burns with moderately severe corneal necrosis.[4] Topical application produced minimal skin irritation.[4]

A total of 10 malignant tumors were found in 24 rats injected subcutaneously with 0.04 ml/kg isoamyl alcohol for 95 weeks; control animals had no malignancies.[4]

Isoamyl alcohol has a disagreeable, pungent odor.

The 1995 ACGIH threshold limit value–time-weighted average (TLV-TWA) for isoamyl alcohol is 100 ppm (361 mg/m^3) with a short-term excursion limit (STEL) of 125 ppm (452 mg/m^3).

REFERENCES

1. Nelson KW et al: Sensory response to certain industrial solvent vapors. *J Ind Hyg Toxicol* 25:282–285, 1943
2. Smyth HF Jr: Improved communication—hygienic standards for daily inhalation. *Am Ind Hyg Assoc Q* 17:129–185, 1956
3. Munch JC: Aliphatic alcohols and alkyl esters: Narcotic and lethal potencies to tadpoles and to rabbits. *J Ind Med* 41:31–33, 1972
4. Rowe VK, McCollister SB: Alcohols. In Clayton, GD, Clayton FE (eds): *Patty's Industrial Hygiene and Toxicology*, 3rd ed, rev, Vol 2C, *Toxicology*, pp 4594–4599. New York, Wiley-Interscience, 1982

ISOBUTANE
CAS: 75-28-5

C_4H_{10}

Synonyms: 2-Methylpropane; trimethylmethane

Physical Form. Colorless gas

Uses. In the production of propylene glycols and oxides, and polyurethane foams and resins; component of motor fuels and aerosal propellants; industrial gas carrier and general fuel source

Exposure. Inhalation

Toxicology. Isobutane is of generally low toxicity; at extremely high concentrations, it may produce cardiac effects and narcosis.

Humans exposed to isobutane at concentrations of 250, 500, or 1000 ppm for periods of 1 minute to 8 hours did not exhibit any untoward physiological responses as determined by continuous ECG telemetry, spirometric measures, blood count, urinalysis, and a battery of cognitive tests.[1] Repetitive exposures at 500 ppm for up to 8 hr/day for 10 days also were without any measurable untoward effect.

In mice, exposure to 520,000 ppm was lethal to 100% of the animals within an average of 28 minutes.[2] Near the LC$_{50}$ dose, mice exhibited central nervous system depression, rapid and shallow respiration, and apnea.[3] At concentrations of 350,000 ppm, loss of posture occurred after 25 minutes; exposure to 150,000 ppm for 60 minutes or 230,000 ppm for 26 minutes produced light anesthesia.

In dogs, 450,000 ppm for 10 minutes caused anesthesia; exposure to 200,000 ppm for 10 minutes produced respiratory depression, bronchospasm, and decreased pulmonary compliance.[2,4]

In various animal studies, isobutane has been found to sensitize the myocardium to epinephrine. Concentrations of 50,000 ppm predisposed the dog heart to cardiac arrythmias induced by catecholamines.[5] Monkeys administered 50,000 to 100,000 ppm for 5 minutes via tracheal cannulation had tachycardia, arrhythmias, and myocardial depression.[6]

Repeated exposure of monkeys to 4000 ppm for up to 90 days caused no signs of toxicity.[7]

The vapor exerts no effect on the skin or the eyes.

A threshold limit value has not been established for isobutane.

REFERENCES

1. Stewart RD, Herrmann AA, Baretta ED, et al: Acute and repetitive human exposure to isobutane. *Scand J Work Environ Health* 3:234–243, 1977
2. Stoughton RW, Lamson PD: The relative anesthetic activity of the butanes and the pentanes. *J Pharmacol Exp Ther* 58:74077, 1936
3. Aviado DM, Zakheri S, Watanabe T: *Nonfluorinated Propellants and Solvents for Aerosols*, pp 49–81. Cleveland, OH, CRC Press, 1977
4. Aviado DM: Toxicology of aerosol propellants in the respiratory and circulatory systems. X. Proposed classifications. *Toxicology* 3:321–332, 1975
5. Reinhardt CF, Azar A, Maxfield ME, et al: Cardiac arrhythmia and aerosol sniffing. *Arch Environ Health* 22:265–279, 1971
6. Belej MA, Smith DG, Aviado DM: Toxicity of aerosol propellants in the monkey. *Toxicology* 2:381–395, 1974
7. Moore AF: Final report on the safety assessment of isobutane isopentane, *n*-butane, and propane. *J Am College Toxicol* 1:127–142, 1982

ISOBUTYL ACETATE
CAS 110-19-0

$CH_3COOCH_2CH(CH_3)_2$

Synonyms: Acetic acid, isobutyl ester; 2-methylpropyl acetate

Physical Form. Colorless liquid

Uses. Solvent; flavoring

Exposure. Inhalation

Toxicology. At high concentrations, isobutyl acetate causes narcosis in animals, and it is expected that severe exposure will cause the same effect in humans; it is considered to be a respiratory tract and eye irritant by analogy with *n*-butyl acetate.

Rats survived exposure to 4000 ppm, but 8000 ppm for 4 hours was fatal to 4 of 6 rats.[1]

Exposure of rats to 21,000 ppm for 150 minutes was fatal to all animals exposed; no symptoms were observed at 3000 ppm for 6 hours.[2]

Isobutyl acetate has a fruity odor.

The 1995 ACGIH threshold limit value–time-weighted average (TLV-TWA) for isobutyl acetate is 150 ppm (713 mg/m^3).

REFERENCES

1. Smyth HF Jr et al: Range-finding toxicity data: List VI. *Am Ind Hyg Assoc J* 23:95–107, 1962
2. Sandmeyer EE, Kirwin CJ: Esters. In Clayton GD, Clayton FE (eds): *Patty's Industrial Hygiene and Toxicology*, 3rd ed, Vol 2A, *Toxicology*, p 2273. Wiley-Interscience, 1981

ISOBUTYL ALCOHOL
CAS: 78-83-1

$(CH_3)_2CHCH_2OH$

Synonyms: 2-Methylpropanol-1; 2-methyl-1-propanol; isopropylcarbinol

Physical Form. Colorless liquid

Uses. Lacquers, paint removers, cleaners, and hydraulic fluids; in manufacture of isobutyl esters

Exposure. Inhalation

Toxicology. At high concentrations, isobutyl alcohol causes narcosis in animals, and it is expected that severe exposure in humans would produce the same effect.

The liquid on the skin of a human subject was a mild irritant and caused slight erythema and hyperemia.[1] No evidence of eye irritation was noted in humans with repeated 8-hour exposures to 100 ppm.[1] No chronic systemic effects have been reported in humans.

Intermittent exposure of mice to 6400 ppm for 136 hours produced narcosis; exposure to

10,600 ppm for 300 minutes or 15,950 ppm for 250 minutes was fatal.[1]

Rats survived a 2-hour exposure to the saturated vapor (about 16,000 ppm), but 2 of 6 died following a 4-hour exposure to 8000 ppm.[2]

One drop of isobutyl alcohol in a rabbit eye caused moderate to severe irritation without permanent corneal injury.[1]

A variety of malignant tumors developed in rats dosed twice weekly for life by oral intubation or subcutaneous injection.[1] Control animals had no malignancies. The carcinogenic risk to humans has not been determined.

The 1995 ACGIH threshold limit value–time-weighted average (TLV-TWA) for isobutyl alcohol is 50 ppm (152 mg/m^3).

REFERENCES

1. Rowe VK, McCollister SB: Alcohols. In Clayton GD, Clayton FE (eds): *Patty's Industrial Hygiene and Toxicology*, 3rd ed, Vol 2C, *Toxicology*, pp 4578–4582. New York, Wiley-Interscience, 1982
2. Smyth HF Jr, Carpenter CP, Weil CS, Pozzani UC: Range-finding toxicity data. List V. *Arch Ind Hyg Occup Med* 10:61–68, 1954

ISOOCTYL ALCOHOL
CAS: 26952-21-6

$C_8H_{17}OH$

Synonyms: Isooctanol; 2-ethylhexanol; 2-ethylhexyl alcohol

Physical Form. Liquid; a mixture of closely related isomeric, primary alcohols with branched chains

Uses. Intermediate in the manufacture of 2-ethylhexyl acetate, a lacquer solvent; solvent for nitrocellulose, urea, resins, enamels, alkyd varnishes, and lacquers; used in ceramics, paper coatings, textiles, and latex rubbers

Exposure.   Inhalation; skin absorption

Toxicology.   Isooctyl alcohol is a mucous membrane irritant and central nervous system depressant in animals.

Exposure of mice, rats, and guinea pigs to 227 ppm for 6 hours produced no mortality.[1] Central nervous system depression was observed, as was labored respiration and local irritation of the mucous membranes of the eyes and nose.

Male and female rats fed diets containing 0.01%, 0.05%, 0.25%, or 1.25% for 90 days showed histological evidence of liver and kidney effects at the highest level.[2]

Dermal application of up to 2.6 g/kg resulted in no deaths and no signs of percutaneous toxicity; moderate irritation of the skin was observed. Instillation of the liquid into the eye of a rabbit produced erythema and edema of the conjunctiva, tearing, and mucous secretion but no corneal injury.

The 1995 ACGIH threshold limit value–time-weighted average (TLV-TWA) for isooctyl alcohol is 50 ppm (266 mg/m^3) with a notation for skin absorption.

REFERENCES

1. Scala RA, Burtis EG: Acute toxicity of a homologous series of branch-chain primary alcohols. *Am Ind Hyg Assoc J* 34:493–499, 1973
2. Union Carbide Corporation, unpublished data. In Rowe VK, McCollister SB: Alcohols. In Clayton GD, Clayton FE (eds): *Patty's Industrial Hygiene and Toxicology*, 3rd ed, Vol 2C, *Toxicology*, pp 4620–4623. New York: Wiley Interscience, 1982

ISOPHORONE
CAS: 78-59-1

C$_9$H$_{14}$O

Synonyms: Isoacetophorone; isoforon; trimethyl cyclohexenone

Physical Form.   Water-white liquid

Uses.   Solvent for lacquers, resins, and plastics

Exposure.   Inhalation; skin absorption

Toxicology.   Isophorone is an irritant of the eyes and mucous membranes.

Human subjects exposed briefly to 25 ppm experienced irritation of the eyes, nose, and throat.[1] Workers exposed to 5 to 8 ppm for 1 month complained of fatigue and malaise, which disappeared when air levels were reduced to 1 to 4 ppm.[2] Repeated or prolonged skin contact with the liquid may cause dermatitis because of its defatting action.[2] Although isophorone may be more toxic and irritative than lower-molecular-weight ketones at equivalent concentrations, it poses less of an inhalation hazard because of its relatively low volatility.[2]

Repeated exposures of animals at concentrations of 50 ppm or more resulted in evidence of damage to kidney and lung and, to a lesser extent, to liver. No effects, however, were seen at 25 ppm. More recent feeding studies with pure compound in rats, mice, and beagle dogs have not demonstrated specific toxicity.[3]

A 2-year gavage study at 250 and 500 mg/kg demonstrated a dose-related statistically significant excess of tubular cell adenomas and adenocarcinomas of the kidney in male rats, a number of preputial gland tumors in dosed male rats, and a probable increased incidence of hepatocellular neoplasms in high-dose male mice.[3]

A recent study has indicated that renal effects may be specific to certain strains of male rats that synthesize α_{2u}-globulin.[4] Monkeys, guinea pigs, dogs, mice, female rats, and male NBR rats that do not synthesize the hepatic form of α_{2u}-globulin do not develop renal disease in response to isophorone.

The proposed 1995 ACGIH ceiling–threshold limit value (C-TLV) for isophorone is 5 ppm (28 mg/m^3) with an A3 animal carcinogen designation.

REFERENCES

1. Silverman L, Schulte HF, First MW: Further studies on sensory response to certain industrial solvent vapors. *J Ind Hyg Toxicol* 28:262–266, 1946

2. National Institute for Occupational Safety and Health: *Criteria for a Recommended Standard . . . Occupational Exposure to Ketones*. DHEW (NIOSH) Pub No 78-173, pp 44, 86–87, 126–134, 176, 189–190, 242. Washington, DC, US Government Printing Office, June 1978
3. Bucher JR, Huff J, Kluwe WM: Toxicology and carcinogenesis studies of isophorone in F344 rats and B6C3F$_1$ mice. *Toxicology* 39:208–219, 1986
4. Dietrich DR, Swenberg JA: NCI–Black Reiter (NBR) male rats fail to develop renal disease following exposure to agents that induce α-2u-globulin α$_{2u}$) nephropathy. *Fund Appl Toxicol* 16:749–762, 1991

ISOPHORONE DIISOCYANATE
CAS:4098-71-9

$C_{12}H_{18}N_2O_2$

Synonyms: IPDI; 3-isocyanatomethyl-3,5,5-trimethyl cyclohexylisocyanate

Physical Form. Liquid

Uses. Polyurethane paints and varnishes; as an elastomer in casting compounds, flexible textile coatings

Exposure. Inhalation; skin absorption

Toxicology. Isophorone diisocyanate (IPDI) is a sensitizer of the respiratory tract and the skin.

By analogy to toluene diisocyanate, exposure of humans to sufficient concentrations is expected to cause irritation of the eyes, nose, and throat; a choking sensation; and a productive cough of paroxysmal type with retrosternal soreness and chest pain.[1,2]

Higher concentrations would be expected to produce a sensation of oppression or constriction of the chest. There may be bronchitis and severe bronchospasm; pulmonary edema may also occur. Upon removal from exposure, the symptoms may persist for 3 to 7 days.[3]

Although the acute effects may be severe, their importance is overshadowed by respiratory sensitization in susceptible persons. The onset of symptoms of sensitization may be insidious, becoming progressively more pronounced with continued exposure over a period of days to months. Initial symptoms are often nocturnal dyspnea and/or nocturnal cough, with progression to asthmatic bronchitis.[1]

A 50-year-old spray painter developed severe asthma soon after the introduction of a new paint containing IPDI.[3] A bronchial challenge test with the paint gave a positive response.

In another case, a spray painter developed tightness of the chest and dyspnea shortly after using a paint containing IPDI.[4] The symptoms disappeared after a few days off work but recurred shortly after resumption of work.

IPDI has been shown to provoke allergic dermatitis in exposed workers.[5]

Skin and eye irritation in rabbits is considered moderate.[1]

The 1995 ACGIH threshold limit value–time-weighted average (TLV-TWA) is 0.005 ppm (0.045mg/m^3).

REFERENCES

1. National Institute for Occupational Safety and Health: *Criteria for a Recommended Standard . . . Occupational Exposure to Toluene Diisocyanate*. DHEW (NIOSH) Pub No (HSM) 73-11022. Washington, DC, US Government Printing Office, 1973
2. Elkins HB, McCarl GW, Brugsch HG, Fahy JP: Massachusetts experience with toluene diisocyanate. *Am Ind Hyg Assoc J* 23:265–272, 1962
3. Clarke CW, Aldons PM: Isophorone diisocyanate induced respiratory disease. *Aust N Z J Med* 11:290–292, 1981
4. Tyrer FH: Hazards of spraying with two-pack paints containing isocyanates. *J Soc Occup Med* 29:22–24, 1979
5. Lachapelle JM, Lachapelle-Ketelaer MJ: Cross-sensitivity between isophorone diamine and isophorone diisocyanate (IPDI). *Contact Derm* 5:55, 1979

2-ISOPROPOXYETHANOL
CAS: 109-59-1

$(CH_3)_2CHOCH_2CH_2OH$

Synonyms: IPE; ethylene glycol monoisopropyl ether; Isopropyl Cellosolve®; isopropyl glycol

Physical Form. Liquid

Uses. Solvent in latex paints, lacquers, and other coatings, resins, coalescing aids, and coupling solvents

Exposure. Inhalation

Toxicology. 2-Isopropoxyethanol (IPE) causes hemolytic anemia in experimental animals.

In a subacute inhalation toxicity study, rats of both sexes were exposed for 6 hr/day, 5 days/week, for 4 weeks.[1] Recovery groups were kept for an observation period of 14 days without treatment. At 891, 441, or 142 ppm, hemolytic anemia was observed. Mild hemolytic anemia was found in female rats exposed to 100 ppm, but it had disappeared after the 14-day recovery period.

Higher plasma bilirubin values were observed in groups exposed to 891 ppm, and decreased urinary pH values occurred in groups exposed to 891 or 441 ppm. A concentration-related increase in absolute and relative spleen weight in the 441 and 891 ppm groups was accompanied by extramedullary hematopoiesis and brown pigment accumulation in the spleen. The no-observed-adverse-effect level was 30 ppm.

The 1995 threshold limit value–time-weighted average (TLV-TWA) is 25 ppm (106 mg/m³) with a notation for skin absorption.

REFERENCES

1. Arts JHE, Reuzel PGJ, Woutersen RA, et al: Repeated-dose (28-day) inhalation toxicity of isopropyl ethylene glycol ether in rats. *Inhal Toxicol* 4:43–55, 1992

ISOPROPYL ACETATE
CAS: 108-21-4

$CH_3COOCH(CH_3)_2$

Synonyms: 2-Propyl acetate; acetic acid, isopropyl ester

Physical Form. Colorless liquid

Uses. Solvent

Exposure. Inhalation

Toxicology. Isopropyl acetate is an irritant of the eyes; at extremely high concentrations, it causes narcosis in animals, and it is expected that severe exposure will produce the same effect in humans.

Human subjects exposed to 200 ppm for 15 minutes experienced some degree of eye irritation; there was little objection to the odor.[1] No systemic effects have been reported in humans.

Exposure of rats to 32,000 ppm was fatal to 5 of 6 animals after 4 hours; 16,000 ppm for 4 hours was fatal to 1 of 6 rats.[2] The oral LD_{50} for rats was 6.75 g/kg.[2]

The 1995 ACGIH threshold limit value–time-weighted average (TLV-TWA) is 250 ppm (1040 mg/m³) with a STEL/ceiling of 310 ppm (1290 mg/m³).

REFERENCES

1. Silverman L, Schulte HF, First MW: Further studies on sensory response to certain industrial solvent vapors. *J Ind Hyg Toxicol* 28:262–266, 1946

2. Smyth HF Jr, Carpenter CP, Weil CS, Pozzani UC: Range-finding toxicity data. List V. *Arch Ind Hyg Occup Med* 10:61–68, 1954

ISOPROPYL ALCOHOL

CAS: 67-63-0

CH$_3$CHOHCH$_3$

Synonyms: Isopropanol; 2-propanol; dimethyl carbinol

Physical Form. Colorless liquid

Uses. In manufacture of acetone; solvent; in skin lotions, cosmetics and pharmaceuticals; most commonly available commercially as rubbing alcohol (70% isopropanol)

Exposure. Inhalation; ingestion

Toxicology. Isopropyl alcohol is an irritant of the eyes and mucous membranes; at very high doses, it causes central nervous system depression.

Human subjects exposed to 400 ppm for 3 to 5 minutes experienced mild irritation of the eyes, nose, and throat; at 800 ppm, the irritation was not severe, but the majority of subjects considered the atmosphere uncomfortable.[1]

Occupational poisoning by isopropyl alcohol has not been reported. Toxicity in man is based largely on accidental ingestion. An oral dose of 25 ml in 100 ml of water produced hypotension, facial flushing, bradycardia, and dizziness. Other symptoms following ingestion have included vomiting, depression, headache, coma, and shock.[2] Renal insufficiency, including anuria followed by oliguria, nitrogen retention, and edema, may be a complication of isopropyl alcohol poisoning. Estimates of fatal doses are between 160 and 240 ml. Death following ingestion often occurs in 24 to 36 hours from respiratory paralysis.[2] In a recent report, a newborn was exposed for 2 hours to 70% isopropyl, which had been accidentally placed in the humidifier of the infant's ventilator.[3] Despite supportive care, he became cyanotic and bradycardic; 12.5 hours after exposure, he became asystolic and died.

Studies indicate that isopropyl alcohol may be substantially better absorbed by the dermal route than had previously been believed, although significant toxicity by this route would require prolonged exposure.[4] Delayed dermal absorption rather than inhalation may account for a number of pediatric poisonings that have occurred following repeated or prolonged sponged bathing with isopropyl alcohol to reduce fever. In several cases symptoms have included respiratory distress, stupor, and coma.[2] Recovery was complete within 36 hours. Hypersensitivity characterized by delayed eczematous reactions has occasionally been observed following dermal contact with isopropyl alcohol.[2]

Rats exposed to 12,000 ppm for 4 hours survived, but exposure for 8 hours was lethal to half the animals.[5] Mice exposed to 3250 ppm for 460 minutes developed ataxia, prostration and, finally, narcosis. Guinea pigs exposed to 400 ppm for 24 successive hours had slight changes in the mucosa of the nose and trachea, whereas exposure to 5500 for the same amount of time caused severe pathological degeneration of the respiratory mucosa.[6] In the eye of a rabbit, 70% isopropyl alcohol caused conjunctivitis, iritis, and corneal opacity.[5]

Early epidemiologic studies suggested an association between the manufacture of isopropyl alcohol and paranasal sinus cancer.[7,8] The risk for laryngeal cancer may also have been elevated in these workers.[8] The increased cancer incidence, however, appears to be associated with some aspect of the strong-acid manufacturing process rather than the isopropyl alcohol itself. It is unclear whether the cancer risk is due to the presence of diisopropyl sulfate, which is an intermediate in the process, to isopropyl oils, which are formed as by-products, or to other factors, such as sulfuric acid.[8] Isopropyl alcohol has not been tested adequately in animals to assess carcinogenicity.[8]

No evidence of teratogenicity was observed in rats treated with doses of up to 1200 mg/kg/day on gestation days 6 through 15 or in rabbits administered up to 480 mg/kg/day on gestation days 6 through 18.[9] No evidence of developmental toxicity, as determined by pathological findings, organ weights, or behavioral tests, was observed in rats administered up to 1200 mg/kg/day on gestation day 6 through postnatal day 21.[10]

When absorbed, isopropyl alcohol is oxidized in the liver at the hydroxyl moiety and converted to acetone.[11] Occupational exposure to isopropyl alcohol can be biomonitored by means of urinalysis for acetone following exposures as low as 70 ppm.[11] The acetone metabolite may also be responsible for the enhanced toxicity of carbon tetrachloride following pretreatment of animals with isopropyl alcohol.[2] Extra caution is in order when isopropyl alcohol is used concurrently with carbon tetrachloride in an industrial setting.

The odor threshold is 40 to 200 ppm. The 1995 ACGIH threshold limit value–time-weighted average (TLV-TWA) for isopropyl alcohol is 400 ppm (983 mg/m^3) with a short-term excursion limit of 500 ppm (1230 mg/m^3).

REFERENCES

1. Nelson KW et al: Sensory response to certain industrial solvent vapors. *J Ind Hyg Toxicol* 25:282–285, 1943
2. Zakhari S et al: *Isopropanol and Ketones in the Environment*, pp 3–54. Cleveland, OH, CRC Press, Inc, 1977
3. Vicas IM, Beck R: Fatal inhalation isopropyl alcohol poisoning in a neonate. *J Toxicol Clin Tox* 31:473–481, 1993
4. Martinez TT, Jaeger RW, deCastro FJ, et al: A comparision of the absorption and metabolism of isopropyl alcohol by oral, dermal, and inhalation routes. *Vet Hum Toxicol* 28:233–236, 1986
5. Rowe VK, McCollister SB: Alcohols. In Clayton GD, Clayton FE (eds): *Patty's Industrial Hygiene and Toxicology*, 3rd ed, rev, Vol 2C, *Toxicology*, pp 4561–4571. New York, Wiley-Interscience, 1982
6. Ohashi Y, Nakai Y, Ikeoka, H, et al: An experimental study on the respiratory toxicity of isopropyl alcohol. *J Appl Toxicol* 8:67–71, 1987
7. Weil CS, Smyth HF Jr, Nale TW: Quest for a suspected industrial carcinogen. *J Ind Hyg Occup Med* 5:535–547, 1952
8. *IARC Monographs on the Evaluation of Carcinogenic Risks to Humans. Overall Evaluations of Carcinogenicity: An Updating of IARC Monographs Vols 1 to 42*, Suppl 7, pp 229–230. Lyon, International Agency for Research on Cancer, 1987
9. Tyl RW, Masten LW, Marr MC, et al: Developmental toxicity evaluation of isopropanol by gavage in rats and rabbits. *Fund Appl Toxicol* 22:139–151, 1994
10. Bates HK, McKee RH, Bieler GS, et al: Developmental neurotoxicity evaluation of orally administered isopropanol in rats. *Fund Appl Toxicol* 22:152–158, 1994
11. Kawai T, Yasugi T, Horiguchi S, et al: Biological monitoring of occupational exposure to isopropyl alcohol vapor by urinalysis for acetone. *Int Arch Occup Health* 62:409–413, 1990

ISOPROPYLAMINE
CAS: 75-31-0

$(CH_3)_2CHNH_2$

Synonyms: 2-Aminopropane

Physical Form.   Liquid

Uses.   Chemical synthesis of dyes, pharmaceuticals

Exposure.   Inhalation

Toxicology.   Isopropylamine is an irritant of the eyes, mucous membranes, and skin.

Human subjects experienced irritation of the nose and throat after brief exposure to 10 to 20 ppm.[1] Workers complained of transient visual disturbances (halos around lights) after exposure to the vapor for 8 hours, probably as a result of mild corneal edema, which usually cleared within 3 to 4 hours.[2] The liquid is also capable of causing severe eye burns, which may result in permanent visual impairment.[2] Isopropylamine in both liquid and vapor forms is irritating to the skin and may cause skin burns; repeated lesser exposures may result in dermatitis.[2]

All rats exposed to 8000 ppm for 4 hours died within 14 days, but 6 of 6 survived a 4-hour exposure at 4000 ppm.[3]

The odor is like ammonia and becomes definite at 5 to 10 ppm.[1]

The 1995 ACGIH threshold limit value–

time-weighted average (TLV-TWA) for isopropylamine is 5 ppm (12 mg/m³) with a short-term excursion limit of 10 ppm (24 mg/m³).

REFERENCES

1. Beard RR, Noe JT: Aliphatic and alicyclic amines. In Clayton GD, Clayton FE (eds): *Patty's Industrial Hygiene and Toxicology*, 3rd ed, rev, Vol 2B, *Toxicology*, pp 3154–3155. New York, Wiley-Interscience, 1981
2. Chemical Safety Data Sheet SD-72, Isopropylamine, pp 13–15. Washington, DC, MCA, Inc, 1959
3. Smyth HJ Jr et al: Range-finding toxicity data: List IV. *AMA Arch Ind Hyg Occup Med* 4:119–122, 1951

N-ISOPROPYLANILINE
CAS: 768-52-5

$C_9H_{13}N$

Synonym: N-IPA

Physical Form. Liquid

Uses. In dyeing acrylic fibers; chemical intermediate

Exposure. Inhalation; skin absorption

Toxicology. N-Isopropylaniline absorption causes methemoglobinemia in animals, and the same effect is expected in humans.

Rats exposed to levels of n-isopropylaniline at 5, 20, or 100 mg/m³ for 14 weeks showed no mortality or gross toxicity.[1] Elevated methemoglobin levels were observed in all exposure groups. Slight signs of toxicity in the high-dose group consisted of decreased body weight gain, increased spleen and kidney weights, and increased hemosiderin in the spleen.

The oral LD_{50} was 560 mg/kg, and the dermal LD_{50} was 3550 mg/kg.[2] Slight eye and skin irritation were noted in acute toxicity studies with rabbits.

By analogy to methemoglobinemia caused by aniline in humans, the formation of methemoglobinemia often is insidious.[3] Following skin absorption, onset of symptoms may be delayed for up to 4 hours. Headache commonly is the first symptom and may become intense as the severity of methemoglobinemia progresses. Cyanosis occurs when the methemoglobin concentration is 15% or more. Blueness develops first in the lips, nose, and earlobes, and is usually recognized by fellow workers. The individual usually feels well, has no complaints, and insists that nothing is wrong until the methemoglobin level approaches approximately 40%. At higher levels, there are weakness and dizziness and, at levels near 70%, there may be ataxia, dyspnea on mild exertion, and tachycardia. Lethal levels are estimated to be 85% to 90%.

For a discussion of differential diagnosis, special tests and treatment of methemoglobinemia, see "Aniline."

The 1995 ACGIH threshold limit value–time-weighted average (TLV-TWA) is 2 ppm (11 mg/m³) with a notation for skin absorption.

REFERENCES

1. Monsanto Company: Three-month rat inhalation study with N-isopropylaniline. Project No ML-86-278, Study No 86100. Monsanto Company, Environmental Health Laboratory, St Louis, MO, 1988
2. Monsanto Company: Material Safety Data Sheet—N-Isopropylaniline. Monsanto Company, St Louis, MO, 1985
3. Hamblin DO: Aromatic nitro and amino compounds. In Fassett DW, Irish DD (eds): *Industrial Hygiene and Toxicology*, 2nd ed, pp 2105–2133, 2242. New York, Wiley-Interscience, 1963

ISOPROPYL ETHER
CAS: 108-20-3

$(CH_3)_2CHOCH(CH_3)_2$

Synonyms: Diisopropyl ether; 2-isopropoxy-propane

Physical Form. Colorless liquid

Uses. Solvent; chemical intermediate

Exposure. Inhalation

Toxicology. Isopropyl ether is a mild irritant of the eyes and mucous membranes; at high concentrations, it causes narcosis in animals, and it is expected that severe exposure will produce the same effect in humans.

Human subjects exposed to 800 ppm for 5 minutes reported irritation of the eyes and nose, and the most sensitive reported respiratory discomfort.[1] Of the volunteers exposed to 300 ppm for 15 minutes, 35% objected to the odor rather than the irritation.[2]

Animals (monkey, rabbit, and guinea pig) survived a 1-hour exposure to 30,000 ppm with signs of anesthesia; 60,000 ppm for 1 hour was lethal.[1] The lethal concentration for rats was 16,000 ppm for a 4-hour exposure. In rabbits, repeated skin application of the liquid for 10 days caused dermatitis.[1] The liquid dropped in the eye of a rabbit caused minor injury.

The 1995 ACGIH threshold limit value–time-weighted average (TLV-TWA) for isopropyl ether is 250 ppm (1040 mg/m³) with a short-term excursion limit of 310 ppm (1300 mg/m³).

REFERENCES

1. Kirwin C, Sandmeyer E: Ethers. In Clayton GD, Clayton FE (eds): *Patty's Industrial Hygiene and Toxicology*, 3rd ed, Vol 2, *Toxicology*, pp 2511–2512. New York, Wiley-Interscience, 1981
2. Silverman L, Schulte HF, First MW: Further studies on sensory response to certain industrial solvent vapors. *J Ind Hyg Toxicol* 28:262–266, 1946

ISOPROPYL GLYCIDYL ETHER
CAS: 4016-14-2

$C_6H_{12}O_2$

Synonyms: IGE

Physical Form. Colorless liquid

Uses. Reactive diluent for epoxy resins; stabilizer for organic compounds; chemical intermediate for synthesis of ethers and esters

Exposure. Inhalation

Toxicology. Isopropyl glycidyl ether (IGE) causes both primary irritation and sensitization dermatitis; in animals, it causes irritation of the eyes and mucous membranes, and it is expected that severe exposure will cause the same effects in humans.

Systemic effects have not been demonstrated in workers exposed to IGE.[1]

A technician who handled both IGE and phenyl glycidyl ether developed localized dermatitis on the back of the hands; patch testing showed sensitization to both substances.[2] Dermatitis has occurred in workers with repeated skin contact.[3]

In mice, the LC_{50} was 1500 ppm for 4 hours.[2] Rats repeatedly exposed to levels of 400 ppm exhibited slight eye and respiratory irritation. Large oral doses produced central nervous system depression, but this effect was not seen from inhalation exposure.

Moderate irritation resulted from instillation of the liquid in the eyes of rabbits and from application to the skin of rabbits.[2]

Isopropyl glycidyl ether was mutagenic in the *Drosophila* sex-linked recessive lethal (SLRL) assay and induced reciprocal translocations.[4]

The 1995 TLV-TWA is 50 ppm (238 mg/m³) with a STEL/ceiling of 75 ppm (356 mg/m³).

REFERENCES

1. National Institute for Occupational Safety and Health: *Criteria for a Recommended Standard . . . Occupational Exposure to Glycidyl Ethers.* DHEW (NIOSH) Pub No 78-166, p 197. Washington, DC, US Government Printing Office, 1978
2. Hine CH et al: The toxicology of glycidol and some glycidyl ethers. *AMA Arch Ind Health* 14:250–264, 1956
3. Hine CH, Rowe VK: Epoxy Compounds. In Patty FA (ed): *Industrial Hygiene and Toxicology.* 2nd ed, Vol 2, *Toxicology,* pp 1637–1638. New York, Interscience, 1963
4. Foureman P, Mason JM, Valencia R, et al: Chemical mutagenesis testing in *Drosophila.* IX. Results of 50 coded compounds tested for the National Toxicology Program. *Environ Molec Mut* 23:51–63, 1994

KETENE
CAS: 463-51-4

CH₂CO

Synonyms: Ethenone; carbomethane; keten

Physical Form. Gas

Uses. In organic chemical syntheses; in conversion of higher acids into their anhydrides; for acetylation in the manufacture of cellulose acetate and aspirin

Exposure. Inhalation

Toxicology. Ketene is a severe pulmonary irritant in animals and is expected to produce the same effect in humans.

For mice, monkeys, cats, and rabbits, the least concentrations that caused death after a 10-minute exposure were 50, 200, 750, and 1000 ppm, respectively.[1] Few signs appeared during the exposure period but, after a latent period of variable duration, there were dyspnea, cyanosis, and signs of severe pulmonary damage; death was often preceded by convulsions. Significant pathological changes were confined to the lungs and consisted of generalized alveolar edema and congestion. Several species tolerated exposure to 1 ppm for 6 hr/day for 6 months without apparent chronic injury.[1] Exposure of mice to concentrations in excess of 5 ppm for 10 minutes protected mice 3 to 14 days later against otherwise lethal exposures to pulmonary edema–producing agents.[2] A high degree of tolerance to the acute effects of ketene itself has also been reported.[3] By analogy to effects on the skin caused by other severe irritants, repeated or prolonged exposure is expected to cause dermatitis.

The 1995 ACGIH threshold limit value–time-weighted average (TLV-TWA) for ketene is 0.5 ppm (0.86 mg/m³) with a short-term excursion limit of 1.5 ppm (2.6 mg/m³).

REFERENCES

1. Treon JF et al: Physiologic response of animals exposed to airborne ketene. *J Ind Hyg Toxicol* 31:209–218, 1949
2. Stokinger HE: Toxicologic interactions of mixtures of air pollutants. *Intl J Air Poll* 2:313–326, 1960
3. Mendenhall RM, Stokinger HE: Tolerance and cross-tolerance development to atmospheric pollutants ketene and ozone. *J Appl Physiol* 14:923–926, 1959

LEAD (INORGANIC COMPOUNDS)
CAS: 7439-92-1

Pb

Synonyms/Compounds: Metallic lead; lead oxide; lead salts, inorganic

Physical Form. Solid

Uses. Storage batteries; paint; ink; ceramics; automobile radiator repair; ammunition

Exposure. Inhalation; ingestion

Toxicology. Prolonged absorption of lead or its inorganic compounds results in severe gastrointestinal disturbances and anemia; with more serious intoxication, there is neuromuscular dysfunction, whereas the most severe lead exposure may result in encephalopathy.

The onset of symptoms of lead poisoning, or plumbism, is often abrupt; presenting complaints may include weakness, weight loss, lassitude, insomnia, and hypotension.[1-4] Associated with these is a disturbance of the gastrointestinal tract, such as constipation, anorexia, and abdominal discomfort, or actual colic, which may be excruciating. Physical signs are usually facial pallor, malnutrition, abdominal tenderness, and pallor of the eye grounds. The anemia often associated with lead poisoning is of the hypochromic, normocytic type, with reduction in mean corpuscular hemoglobin; stippling of erythrocytes and reticulocytosis are evident. On gingival tissues, a line or band of punctate blue or blue-black pigmentation (lead line) may appear but only in the presence of poor dental hygiene; this is not pathognomonic of lead poisoning.[3]

Occasionally, the alimentary symptoms are relatively slight and are overshadowed by neuromuscular dysfunction, accompanied by signs of motor weakness, which may progress to paralysis of the extensor muscles of the wrist ("wrist drop") and less often of the ankles ("foot drop").[2,3] Encephalopathy, the most serious result of lead poisoning, occurs frequently in children who have ingested inorganic lead compounds but rarely in adults, except for exposure to organic lead.[1-4]

Subtle, often subclinical, neurological effects have been demonstrated in workers with relatively low blood lead levels, below 40 to 60 $\mu g/100$ ml blood. Performance of lead workers on various neuropsychological tests was mildly reduced, relative to a control group, at mean levels of 49 $\mu g/100$ ml blood and, in a prospective follow-up study, at levels between 30 and 45 $\mu g/100$ ml blood.[5-7] In some of these studies,

the lead-exposed workers reported significantly more complaints of nonspecific subjective symptoms, such as anxiety, depressed mood, poor concentration, and forgetfulness.[5] However, a recent evaluation of 21 studies found inadequate evidence of decreased neurobehavioral test performance in adults with cumulative low-level exposure to lead.[8]

Mild neurophysiological changes, including reductions in motor and sensory nerve conduction velocities (sometimes still within the normal range), have been documented in lead-exposed workers compared with control groups, with blood lead levels less than 40 $\mu g/100$ ml blood.[9] A prospective follow-up study of workers with blood lead levels of 30 to 50 $\mu g/100$ ml blood demonstrated mild slowing of conduction velocities.[10]

Nephropathy has been associated with chronic lead poisoning.[2,3,11] A study of two large cohorts of heavily exposed lead workers followed through 1980 demonstrated a nearly threefold excess of deaths attributed to chronic nephritis or "other hypertensive disease," primarily kidney disease.[12] Most of the excess deaths occurred before 1970, among men who began work before 1946, suggesting that current lower levels of exposure may reduce the risk. Experimental animal studies suggest that there may be a threshold for lead nephrotoxicity and, in workers, nephropathy occurred only in those with blood levels over 62 $\mu g/dl$ for up to 12 years.[13]

The role of chronic low-level lead exposure in the pathogenesis of hypertension remains controversial. Although results have been mixed, overall, the studies may suggest a small positive association between blood lead and blood pressure.[14]

Following absorption, inorganic lead is distributed in the soft tissues, with the highest concentrations being in the kidneys and the liver.[4] In the blood, nearly all circulating inorganic lead is associated with the erythrocytes.[4] Over a period of time, the lead is redistributed, being deposited mostly in bone and also in teeth and hair.[3,4] Lead absorption is cumulative; elimination of lead from the body is slow, requiring a considerably longer period than that needed for the storage of toxic amounts.[1,4] Asymptomatic lead workers, when subjected to a sudden increase in exposure to, and absorption of, lead, often respond with an episode of typical lead poisoning.[1] Removal of the worker from exposure to abnormal quantities of lead often leads to a seemingly sudden and apparent complete recovery; this has occurred even when the individual has a considerable quantity of residual lead in the body.[1]

Epidemiologic studies have not shown a clear relation between lead exposure and the incidence of cancer.[15,16] A study of 4347 lead-exposed workers in a copper smelter failed to demonstrate any significant excess of neoplasms.[17] A study of two large cohorts of lead workers (3519 battery plant workers and 2300 lead production workers) followed through 1980 demonstrated statistically significant elevation in the standardized mortality ratio (SMR) for gastric (SMR = 168) and lung cancer (SMR = 125) in the battery plant workers only. Citing the absence of prior evidence from other studies for these associations, and their inability to assess and correct for possible confounding factors (such as diet, alcohol, and smoking), the authors considered these findings quite tentative. There were no excess deaths from malignancies of the kidney or other sites in either cohort.[12]

There are several reports that certain lead compounds, including lead acetate and lead phosphate, administered to animals in high doses are carcinogenic, primarily producing renal tumors.[18,19] The International Agency for Research on Cancer has concluded that the evidence for carcinogenicity of lead to humans is inadequate, although there is sufficient evidence of carcinogenicity of some lead salts to animals.[15] (Note: Those salts demonstrating carcinogenicity in animals are soluble, whereas human beings are primarily exposed to insoluble metallic lead and lead oxide.)

Reproductive effects from lead exposure have been documented in animals and human beings of both sexes. High occupational exposure levels in pregnant women have been associated with increased incidences of spontaneous abortions, miscarriages, and still births.[16] Some studies also seem to indicate that prenatal exposure to lower levels of lead may increase the risk

of preterm delivery and reduced birth weight.[20] Lead penetrates the placental barrier and has caused congenital abnormalities in animals.[3,21] There is no conclusive evidence, however, that low-level lead exposure leads to an increased incidence of malformations in humans.[22] Excessive exposure to lead during pregnancy has resulted in neurologic disorders in infants; low levels of exposure may be related to neurobehavioral deficits or delays.[16]

In battery workmen with a mean occupational exposure to lead of 8.5 (1 to 23) years, and with blood lead concentrations of 53 to 75 μg/100 ml of blood, there was an increased frequency of abnormalities of sperm, including hypospermia, compared with a control group.[23]

The proposed 1995 ACGIH threshold limit value–time-weighted average (TLV-TWA) for lead, including elemental and inorganic compounds as Pb, is 0.05 mg/m^3, with an A3 animal carcinogen designation.

Diagnosis. *Signs and Symptoms:* Among the effects of lead poisoning are: weakness, lassitude, insomnia; facial pallor and pallor of the eye grounds; anorexia, weight loss, malnutrition; constipation, abdominal discomfort and tenderness, colic; anemia; lead line on gingival tissues; signs of motor weakness, including paralysis of the extensor muscles of the wrists and, less often, of the ankles; encephalopathy; nephropathy. A detailed neurologic examination with electromyography and nerve conduction velocity may be useful when peripheral nerve damage is suspected.

Special Tests. Measurement of blood lead concentration is the most widely used biomarker of lead exposure.[16] Blood lead determination is an exacting laboratory procedure, requiring constant attention to quality control. A blood lead level greater than 10 μg/dl indicates that excessive lead exposure may be occurring. The current biological exposure index for lead in blood of exposed workers is 50 μg/dl, which represents the threshold for effects in some adults.

Other indicators of lead exposure relate to the inhibition by lead of the synthesis of heme. The inhibition of delta-aminolevulinic acid dehydrase (ALA-D), an enzyme involved in porphyrin synthesis, leads to an increase in levels of delta-aminolevulinic acid (ALA) in blood and urine.[3] The blood and urine levels of coproporphyrin III and free erythrocyte protoporphyrins (FEP) are also usually elevated.[3] FEP combines with zinc in the blood to form zinc protoporphyrin (ZPP), which is the moiety assayed.[24]

The ZPP test is now widely used in biological monitoring for lead absorption, in conjunction with blood lead levels. One advantage of the ZPP test is its ability to "average" the effects of lead absorption over a time period of several months, reflecting the 120-day average life span of the red blood cell. Though there is a delay in the increase in ZPP after initial lead exposure, the ZPP will remain elevated longer than the blood lead level following cessation of exposure. The blood lead level is a better indicator of acute lead intoxication, whereas the ZPP is a better indicator of chronic intoxication.[4] The ZPP is not completely specific for lead effects in that iron deficiency will also result in an increase in the ZPP level. The normal ZPP level is generally below 40 to 50 $\mu g/100$ ml blood in nonoccupationally exposed populations.[24]

ALA-D, ALA, coproporphyrin III assays, and blood examinations for hemoglobin, reticulocytes, and stippled red cells are useful in the assessment of worker health, but no one of these measurements alone is an accepted specific index of lead absorption.[24]

The edetate calcium disodium (EDTA) mobilization test for lead is used for estimating both current and previous absorption of increased amounts of lead.[25–27] In this test, a single dose of 1 g EDTA in 250 ml of 5% dextrose is infused intravenously over a period of 1 hour. Urine is then collected quantitatively for 24 hours (4 days in subjects with renal insufficiency). The upper limit of normal in healthy adult subjects is 500 to 600 μg of lead excreted in the urine. The use of X-ray fluorescence of bone to determine lead burden has been proposed as an alternative to the lead mobilization test.[28]

Treatment. The primary therapy for lead intoxication is cessation of exposure. Prophylactic chelation therapy to prevent the rise of blood lead levels is illegal in the United States. Prior to diagnostic or therapeutic chelation treatment, workers must be notified in writing why they are receiving this treatment.

In adults, the use of chelation therapy should be reserved for those with significant symptoms or signs of toxic reactions. Occasionally, adult patients may have no clinically evident symptoms or signs of toxic reactions to lead. These individuals should be removed from exposure and followed up carefully, but chelation therapy is not appropriate. Elevated blood lead levels associated with significant symptoms warrant chelation. Chelation therapy reduces lead in tissues to which it is not tightly bound, such as blood, kidney, liver, and trabecular bone. Unfortunately, with long-term exposures, a significant fraction of the total body burden of lead will be tightly bound to compact bone and brain.

In acute lead poisoning or acute exacerbations of chronic lead poisoning with severe neurologic or gastrointestinal symptoms, chelating agents such as edetate calcium disodium, dimercaprol, penicillamine, and dimercaptosuccinic acid may be administered.[28] The patient should be hospitalized and the treatment overseen by a physician who has had experience with chelation. In acute encephalopathy, both dimercaprol and edetate are used until blood lead levels are less than 40 $\mu g/dL$. Therapy should be used for 5 days and, if further chelation is required, a minimum interval of 48 to 72 hours should intervene.

Precautions. EDTA has caused proteinuria, microscopic hematuria, large epithelial cells in the urinary sediment, renal failure from proximal tubule damage, hypercalcemia, and fever. It should not be used during periods of anuria. Safe administration of EDTA requires the following determinations on the first, third, and fifth day of each course of therapy: serum electrolytes; urea nitrogen, creatinine, calcium, phosphorus, and alkaline phosphatase measurements in blood; and routine urinalysis. The patient should also be monitored for irregularities of cardiac rhythm.

REFERENCES

1. Kehoe RA: Occupational lead poisoning. Clinical types. *J Occup Med* 14:298–300, 1972

2. National Institute for Occupational Safety and Health: *Criteria for a Recommended Standard ... Occupational Exposure to Inorganic Lead.* DHEW (HSM) Pub No 73-22020. Washington, DC, US Government Printing Office, 1972

3. Committee on Biologic Effects of Atmospheric Pollutants, Division of Medical Sciences, National Research Council: *Lead— Airborne Lead in Perspective.* Washington, DC, National Academy of Sciences, 1972

4. Klaassen CD: Heavy metals and heavy-metal antagonists. In Goodman LS, Bilman AG (eds): *Goodman and Gilman's The Pharmacological Basis of Therapeutics,* 6th ed, pp 1616–1622. New York, Macmillan, 1980

5. Jeyaratnam J et al: Neuropsychological studies on lead workers in Singapore. *Br J Ind Med* 43:626–629, 1986

6. Williamson AM, Teo RKC: Neurobehavioral effects of occupational exposure to lead. *Br J Ind Med* 43:374–380, 1986

7. Mantere P et al: A prospective follow-up study on psychological effects in workers exposed to low levels of lead. *Scand J Work Environ Health* 10:43–50, 1984

8. Balbus-Kornfeld JM, Stewart W, Bolla KI, et al: Cumulative exposure to inorganic lead and neurobehavioral test performance in adults: an epidemiological review. *Occup Environ Med* 52:2–12, 1995

9. Zi-giang Chen et al: Peripheral nerve conduction velocity in workers occupationally exposed to lead. *Scand J Work Environ Health* 11 (Suppl 4):26–28, 1985

10. Seppalainem AM, Hernberg S, Vesanto R, et al: Early neurotoxic effects of occupational lead exposure: a prospective study. *Neurotoxicology* 4:181–192, 1983

11. Vitale LF, Joselow MM, Wedeen RP, Pawlow M: Blood lead—an inadequate measure of occupational exposure. *J Occup Med* 17:155–156, 1975

12. Cooper WC, Wong O, Kheifets L: Mortality among employees of lead battery plants and lead-producing plants. 1947–1980. *Scand J Work Environ Health* 11:331–345, 1985

13. Beck BD: Symposium overview: an update on exposure and effects of lead. *Fund Appl Toxicol* 18:1–16, 1992

14. Hertz-Picciotto I, Croft J: Review of the relation between blood lead and blood pressure. *Epi Rev* 15:352–373, 1993

15. *IARC Monographs on the Evaluation of the Carcinogenic Risk of Chemicals to Humans: Overall Evaluations of Carcinogenicity: An Updating of IARC Monographs, Vols 1–42.* Lyon, International Agency for Research on Cancer, 1987

16. Agency for Toxic Substances and Disease Registry (ATSDR): *Toxicological Profile for Lead.* US Department of Health and Human Services, Public Health Service, pp 307, TP-92/12, 1993

17. Gerhardsson L, Lundstrom NG, Nordberg G, et al: Mortality and lead exposure: a retrospective cohort study of Swedish smelter workers. *Br J Ind Med* 43:707–712, 1986

18. Mao P, Molnar JJ: The fine structure and histochemistry of lead-induced renal tumors in rats. *Am J Pathol* 50:571–581, 1967

19. Boyland E, Dukes CE, Grover PL, Mitchley BCV: The induction of renal tumors by feeding lead acetate to rats. *Br J Cancer* 16:283–288, 1962

20. Andrews KW, Savitz DA, Hertz-Picciotta I: Prenatal lead exposure in relation to gestational age and birth weight: a review of epidemiologic studies. *Am J Ind Med* 26:13–32, 1994

21. Ferm VH, Carpenter SJ: Developmental malformations resulting from the administration of lead salts. *Exp Mol Pathol* 7:208–213, 1967

22. Ernhart CB: A critical review of low-level prenatal lead exposure in the human: I. Effects on the fetus and newborn. *Repro Toxicol* 6:9–19, 1992

23. Lancranjan I, Popescu HI, Gavanescu O, et al: Reproductive ability of workmen occupationally exposed to lead. *Arch Environ Health* 30:396–401, 1975

24. Lauwerys R: *Industrial Chemical Exposure: Guidelines for Biological Monitoring,* pp 27–34. Davis, CA, Biomedical Publications, 1982

25. Selander S: Treatment of lead poisoning: a comparison between the effects of sodium calcium edate and penicillamine administered orally and intravenously. *Br J Ind Med* 24:272–283, 1967

26. Emmerson BT: Chronic lead nephropathy: the diagnostic use of calcium EDTA and the association with gout. *Aust Ann Med* 12:310–324, 1963

27. Lilis R, Fischbein A: Chelation therapy in workers exposed to lead. *JAMA* 235:2823–2824, 1976

28. Isselbacher KJ, Braunwald E, Wilson JD, et al (eds): *Harrison's Principles of Internal Medicine,* 13th ed, Vol 2, pp 2463–2464, New York, McGraw-Hill, 1994

LEAD ARSENATE
CAS: 10102-48-4

$Pb_3(AsO_4)_2$

Synonyms: Arsinette; Ortho L10 Dust; Gypsine; Soprabel; Talbot

Physical Form. White powder (required to be colored pink in most of United States.)

Uses. Insecticide; control of tapeworms in cattle, goats, and sheep

Exposure. Inhalation; ingestion

Toxicology. Lead arsenate may cause lead and/or arsenic intoxication; arsenic symptoms are likely to predominate in acute intoxication, whereas prolonged inhalation of lead arsenate may induce the symptoms of lead intoxication.[1]

Some of the effects of acute arsenic intoxication are nausea, vomiting, diarrhea, and irritation; inflammation and ulceration of the mucous membranes and skin; and kidney damage.[2] Among the effects of chronic arsenic poisoning are increased pigmentation and keratinization of the skin, dermatitis, and epidermoid carcinoma. Other effects that are seen after ingestion but that are not common from industrial exposure are muscular paralysis, visual disturbances, and liver and kidney damage.[2]

Effects of lead intoxication include damage to the central and peripheral nervous systems, to the kidneys, and to the blood-forming mechanism, which may lead to anemia.[3] Symptoms include colic, loss of appetite, and constipation; excessive tiredness and weakness; and nervous irritability. In peripheral neuropathy, the distinguishing clinical feature of lead intoxication is a predominance of motor impairment, with minimal or no sensory abnormalities. There is a tendency for the extensor muscles of the hands and feet to be affected. Lead intoxication has also resulted in kidney damage with few, if any, symptoms appearing until permanent damage has occurred.

In a follow-up mortality study in 1973 of a cohort of 1231 individuals (primarily orchardists) who had participated in a 1938 mortality study, it was concluded that excess mortality did not occur consistently from exposure to lead arsenate spray.[4,5] In contrast, two other independent studies reported a significant excess of lung cancer among other cohorts of this same population.[2] In a study of workers engaged in the formulation and packaging of lead arsenate and calcium arsenate, there was an excess of lung cancer, which was dose-related.[6] In vineyard workers chronically exposed to lead, calcium, and copper arsenate dust in Germany and France, there are numerous reports of skin cancer—including basal-cell and squamous-cell carcinomas—Bowen's disease, and lung cancer.[7]

The IARC has concluded that there is sufficient evidence that inorganic arsenic compounds, including lead arsenate, are skin and lung carcinogens in humans.[7]

The 1995 ACGIH threshold limit value–time-weighted average (TLV-TWA) for lead arsenate as $Pb_3 (AsO_4)_2$ is 0.15 mg/m³.

REFERENCES

1. Clarkson TW: Inorganic and organometal pesticides. In Hayes WJ Jr, Laws ER Jr (eds): *Handbook of Pesticide Toxicology*, Vol 2, pp 531–537, Academic Press, New York 1991
2. Department of Labor: Standard for exposure to inorganic arsenic. *Federal Register* 40:3392–3404, 1975
3. Department of Labor: Occupational exposure to lead. *Federal Register* 40:45934–45948, 1975
4. Nelson WC, Lykins MH, Mackey J, et al: Mortality among orchard workers exposed to lead arsenate spray: a cohort study. *J Chron Dis* 26:105–118, 1973
5. Neal PA et al: *A Study of the Effect of Lead Arsenate Exposure on Orchardists and Consumers of Sprayed Fruit.* US Public Health Service Bull No 267, pp 47–165, 171–181. Washington, DC, US Government Printing Office, 1941
6. Ott MG, Holder BB, Gordon HL: Respiratory cancer and occupational exposure to arsenicals. *Arch Environ Health* 29:250–255, 1974
7. *IARC Monographs on the Evaluation of the Carcinogenic Risk of Chemicals to Humans:* Some

metals and metallic compounds, Vol 23, pp 39–41. Lyon, International Agency for Research on Cancer, 1980

LEAD CHROMATE
CAS:7758-97-6

PbCrO$_4$

Synonyms: Chrome yellow; CI pigment yellow 34; CI 77600

Physical Form. Yellow crystals or powder, insoluble in water

Uses. Pigment

Exposure. Inhalation

Toxicology. Lead chromate is a suspected human carcinogen of the lung and can cause chronic lead poisoning.

Lead chromate could potentially pose a double hazard and cause signs and symptoms of chronic lead intoxication (severe gastrointestinal disturbances, anemia, neuromuscular dysfunction, nephritis, and encephalopathy), and chromium VI toxicity (sensitization dermatitis, primary irritant dermatitis, ulcerated nasal mucosa and skin, and nephropathy), although the latter has not been specifically observed from lead chromate.

Lead poisoning from lead chromate in the chromate pigment industry has been documented.[1] Evidence of lung cancer attributable solely to lead chromate in the industry has not been consistent.[2,3] Long-term mortality was studied in a group of 57 chromate pigment workers who suffered clinical lead poisoning, mostly between 1930 and 1945.[1] One death was attributed to lead poisoning, and there were significant excesses of deaths from nephritis and cerebrovascular disease. The deaths from nephritis followed service exceeding 10 years, whereas the risk of cerebrovascular disease was unrelated to duration of exposure, and even affected men employed for under 1 year. Other contemporary workers at the factories showed no excess mortality from cerebrovascular disease.

Lung cancer mortality among 1152 men working at three English chromate pigment factories was studied from the 1930s to the 1940s and some until 1981.[2] Workers exposed only to lead chromate at one factory experienced no increased risk in cause-specific mortality. Workers at two other factories were exposed to both lead and zinc chromate, and lung cancer mortality was significantly raised among those with high or medium exposure for at least 1 year before 1955. After that time, working conditions were improved and workers starting after that date did not have excess lung cancer deaths. The results provided no indication that lead chromate induced lung cancer, even under conditions conducive to lead poisoning.

In contrast, another study of 548 men at three lead chromate facilities showed that workers exposed at two of the facilities had a threefold excess of lung cancer.[3,4] Workers at the third facility, who had zinc chromate exposure as well as lead chromate exposure, had a significant excess of lung cancer and stomach cancer. An industrial hygiene survey indicated that nearly half of the samples at the three facilities reached or exceeded the OSHA standards for lead and chromium.

Chronic animal studies have also yielded varying results. Intratracheal implantation of lead chromates in rats failed to significantly increase the carcinogenic response after two years.[5] Intrapleural administration caused a 9% incidence of lung tumors in rats within 19 to 21 months.[6] Intramuscular injection resulted in lymphomas, renal tumors, fibrosarcomas, and rhabdomyosarcomas at the site of injection in rats.[7]

The IARC has concluded that there is sufficient evidence in experimental animals and in humans for the carcinogenicity of lead chromate.[8]

The 1995 ACGIH threshold limit value–time-weighted average (TLV-TWA) for lead chromate is 0.05 mg/m^3, as Pb and 0.012 mg/m^3 as Cr, with an A2 suspected human carcinogen designation.

REFERENCES

1. Davies JM: Long-term mortality study of chromate pigment workers who suffered lead poisoning. *Br J Ind Med* 41:158–169, 1984
2. Davies JM: Lung cancer mortality among workers making lead chromate and zinc chromate pigments at three English factories. *Br J Ind Med* 41:158–169, 1984
3. Equitable Environmental Health, Inc: An epidemiologic study of workers in lead chromate plants. Final report submitted to the Dry Color Manufacturers Assoc, June 25, 1976
4. Current Intelligence Bulletin 4, *Chrome Pigment*. NIOSH, US Department of Health and Human Services, Cincinnati OH, 6 pp, 1976
5. Levy LS, Martin PA, Bidstrup PL: Investigation of the potential carcinogenicity of a range of chromium containing materials on rat lung. *Br J Ind Med* 43:243–256, 1986
6. Heuper WC: Environmental carcinogenesis and cancers. *Cancer Res* 21:842–857, 1961
7. Furst A, Schlauder M, Sasmore DP: Tumorigenic activity of lead chromate. *Cancer Res* 36:1779–1783, 1976
8. *IARC Monographs on the Evaluation of the Carcinogenic Risk of Chemicals to Humans*, Vol 49, Chromium. nickel and welding. Lyon, International Agency for Research on Cancer, 1990

LINDANE
CAS: 58-89-9

$C_6H_6Cl_6$

Synonyms: 1,2,3,4,5,6-Hexachlorocyclohexane, gamma isomer; gamma HCH; gamma benzene hexachloride; gamma BHC; Kwell

Physical Form. Crystalline solid

Uses. Insecticide

Exposure. Inhalation; skin absorption; ingestion

Toxicology. Lindane causes central nervous system effects.

Exposure to the vapor causes irritation of the eyes, nose, and throat, severe headache, and nausea.[1] Lindane levels in the blood do not appear to increase with increased duration of exposure but primarily reflect recent lindane absorption.[2] Production workers exposed to air levels of 31 to 1800 $\mu g/m^3$ had blood levels of 1.9 to 8.3 ppb.[1]

Lindane has been suspected as a cause of aplastic or hypoplastic anemia in a number of cases reported from various countries.[3] Although one report tabulated 46 case reports of bone marrow injury temporally associated with environmental exposure to lindane, the authors questioned the association on several grounds.[3] In 17 cases, there was exposure to other toxic agents, including benzene and chloramphenicol. In 8 cases, investigation of the bone marrow did not reveal aplasia or hypoplasia. In some cases, documentation of exposure was limited. Moreover, no cases have been reported following the therapeutic use of lindane (Kwell) as a scabicide in children or adults, even though lindane is well absorbed dermally. Cross-sectional studies of workers chronically exposed to lindane during manufacture have failed to reveal any hematologic conditions or significant differences in hemoglobin or total leukocyte count relative to a control population.[4] Although some statistically significant differences were found in some hematologic parameters, such as increases in polymorphonuclear leukocyte counts and reticulocyte counts compared with the control group, the results were still largely within the reference range and of questionable biological significance. No significant differences were observed for transaminases (AST, ALT) or other liver function studies.[4]

Accidental ingestion has caused fatalities; effects were repeated, violent clonic convulsions, sometimes superimposed on a continuous tonic spasm. Respiratory difficulty and cyanosis, secondary to the convulsions, were common.[5] Following nonfatal accidental ingestions, symptoms have included malaise, dizziness, nausea, and vomiting. Agitation, collapse, convulsions, loss of consciousness, muscle tremor, fever, and cyanosis have commonly been observed. Most patients who survive recover completely over 1 to 3 days; protracted illness is rare.[2]

Minor liver lesions have been reported in rats at dosages as low as 2.6 to 5.0 mg/kg/day. After repeated high doses, degenerative changes have been reported in the kidney, pancreas, and testes of rodent species.[2] Feeding of 1500 ppm in the diet to rats for 90 days, a maximally tolerated dose, resulted in testicular atrophy, with spermatogenic arrest and apparent inhibition of androgen synthesis by Leydig cells.[6]

Administered to mice for 80 weeks, lindane caused a significant increase in hepatocellular tumors in low-dose males but not in other groups of mice or in rats. The IARC has concluded that there is limited evidence of carcinogenicity to animals.[7] Lindane was not genotoxic in a variety of assays.[7]

The 1995 threshold limit value–time-weighted average (TLV-TWA) for lindane is 0.5 mg/m³ with a notation for skin absorption.

REFERENCES

1. Hygienic Guide Series: Hexachlorocyclohexane, gamma isomer—lindane. *Am Ind Hyg Assoc J* 33:36–59, 1972
2. Hayes WJ Jr, Laws ER Jr: *Handbook of Pesticide Toxicology*, Vol 2, *Classes of Pesticides*, pp 791–816, New York, Academic Press, 1991
3. Morgan DP, Stockdale EM, Roberts RJ, Walter AW: Anemia associated with exposure to lindane. *Arch Environ Health* 35:307–310, 1980
4. Brassow HL, Baumann K, Hehnert G: Occupational exposure to hexachlorocyclohexane. II. Health conditions of chronically exposed workers. *Int Arch Occup Environ Health* 48:81–87, 1981
5. Hayes WJ Jr: *Clinical Handbook on Economic Poisons, Emergency Information for Treating Poisoning*, US Public Health Service Pub No 476, pp 50–55. Washington, DC, US Government Printing Office, 1963
6. Shivanandappa T, Krishnakumari MK: Hexachlorocyclohexane-induced testicular dysfunction in rats. *Acta Pharmacol Toxicol* 52:12–17, 1983
7. *IARC Monographs on the Evaluation of the Carcinogenic Risks to Humans*, Suppl 7, *Overall Evaluations of Carcinogenicity: An Updating of Monographs Vols 1–42*, pp 220–222. Lyon, International Agency for Research on Cancer, 1987

LIQUIFIED PETROLEUM GAS
CAS: 68476-85-7

Mixture: C_3H_6, C_3H_8, C_4H_8 and C_4H_{10}

Synonyms: LPG; bottled gas; liquified hydrocarbon gas

Physical Form. Gas or liquid

Uses. Fuel; in production of chemicals

Exposure. Inhalation

Toxicology. Liquified petroleum gas is practically nontoxic below the explosive limits but may cause asphyxia by oxygen displacement at extremely high concentrations.[1]

No chronic systemic effects have been reported from occupational exposure. The vapor is not irritating to the eyes, nose, or throat.[2] Direct contact with the liquid may cause burns or frostbite to the eyes and skin.[3] Olefinic impurities may lend a narcotic effect. At extremely high concentrations, the limiting toxicological factor is available oxygen. Minimal oxygen content should be 18% by volume under normal atmospheric pressure. Generally, flammability and explosive hazards outweigh the biological effects.[1]

The 1995 TLV-TWA is 1000 ppm (1800 mg/m³).

REFERENCES

1. Deichmann WB, Gerarde HW: *Toxicology of Drugs and Chemicals*, p 345. New York, Academic Press, 1969
2. Weiss G: *Hazardous Chemical Data Book*, p 568. Park Ridge, NJ, Noyes Data Corporation, 1980
3. Sandmeyer EE: Aliphatic hydrocarbons. In Clayton GD, Clayton FE (eds): *Patty's Industrial Hygiene and Toxicology*, 3rd ed, rev, Vol 2B, *Toxicology*, pp 3175–3220. New York, Wiley-Interscience, 1981

LITHIUM HYDRIDE
CAS: 7580-67-8

LiH

Synonyms: None

Physical Form. White crystals that darken on exposure to light

Uses. Reducing agent; condensing agent with ketones and acid esters; desiccant; in hydrogen generators

Exposure. Inhalation; ingestion

Toxicology. Lithium hydride is a severe irritant of the eyes, mucous membranes, and skin.

The toxicity of lithium hydride differs markedly from that of the soluble salts of lithium because of its vigorous chemical reactivity with water, which produces acute irritation and corrosion of biological tissues.[1]

The explosion of a cylinder of lithium hydride led to eye contact and swallowing of a small amount of the dust by a technician.[2] The resulting burns caused scarring of both corneas and strictures of the larynx, trachea, bronchi, and esophagus; death occurred 10 months later.

Exposure of humans in the range of 0.025 to 0.1 mg/m^3 caused some nasal irritation; tolerance was acquired with continuous exposure.[3] At 0.5 to 1.0 mg/m^3, severe nasal irritation, cough, and some eye irritation were noted; in the range of 1.0 to 5.0 mg/m^3, all effects were severe, and some skin irritation was also experienced.

Exposure of animals to concentrations above 5 mg/m^3 caused sneezing and cough with secondary pulmonary emphysema; levels of 10 mg/m^3 corroded the body fur and skin of the legs, and there was occasional inflammation of the eyes and nasal septum.[1] The lesions of the nose and legs were attributed to the alkalinity of lithium hydroxide, the hydrolysis product of lithium hydride.

Powdered lithium hydride may ignite spontaneously in humid air or on contact with most mucous surfaces; resulting tissue effects may have features of both thermal and alkali burns.[4]

The 1995 TLV-TWA is 0.025 mg/m^3.

REFERENCES

1. Spiegl CJ et al: Acute inhalation toxicity of lithium hydride. *AMA Arch Ind Health* 14:468–470, 1956
2. Cracovaner AJ: Stenosis after explosion of lithium hydride. *Arch Otolaryngol* 80:87–92, 1964.
3. Stokinger HE: The metals. In Clayton GD, Clayton FE (eds): *Patty's Industrial Hygiene and Toxicology*, 3rd ed, rev, Vol 2A, pp 1728–1740. New York: Wiley-Interscience, 1981
4. Gosselin RE et al: *Clinical Toxicology of Commercial Products*, 5th ed. Baltimore, MD, Williams and Wilkins, 1984

MAGNESITE
CAS: 546-93-0

MgCO$_3$

Synonyms: Magnesium carbonate

Physical Form. Solid

Uses. Chemical intermediate for magnesium salts; component of pharmaceuticals, cosmetics, dentifrices, and free-running table salt; agent in heat insulation and refractory applications

Exposure. Inhalation

Toxicology. Magnesite is considered to be a nuisance dust.

Among 619 workers in a magnesite plant with 6 to 20 years of employment, 13 cases of pneumoconiosis were observed, mainly among workers exposed to calcined magnesite.[1] The workers were exposed to dust from crude or calcined magnesite, which also contained 1% to 3% silicon dioxide.

In several reports, the severity of the pneumoconioses caused by the action of magnesite ore dusts was found to be a function of the crystalline silica content.[2]

Adverse health effects have not been reported for workers exposed to magnesite containing no asbestos and <1% crystalline silica.[3] No cases of human systemic magnesium intoxication from inhalation of magnesite have been reported.

The 1995 threshold limit value–time-weighted average (TLV-TWA) for magnesite is 10 mg/m³, total dust containing no asbestos and <1% crystalline silica.

REFERENCES

1. Zeleneva NI: Hygienic, clinical, and experimental data on magnesite pneumoconiosis. *Gig Truda Prof Zabol* 14:21–24, 1970
2. Tokmurzina PU, Dzangosina DM: The biological aggressiveness of some types of dust from Kazakhstan iron and manganese ores. *Gig Truda Prof Zabol* 14:51–54, 1970
3. Magnesite. *Documentation of the TLVs and BEIs.* 6th ed, pp 867–868. Cincinnati, OH, American Conference of Governmental Industrial Hygienists, 1991

Examination of 95 workers exposed to an unspecified concentration of magnesium oxide dust revealed slight irritation of the eyes and nose; the magnesium level in the serum of 60% of those examined was above the normal upper limit of 3.5 mg/dl.[1]

Experimental subjects exposed to fresh magnesium oxide fume developed metal fume fever, an illness similar to influenza; effects were fever, cough, oppression in the chest, and leukocytosis.[2] There are no reports of metal fume fever resulting from industrial exposure to magnesium oxide fume.[2,3]

The 1995 threshold limit value–time-weighted average (TLV-TWA) for magnesium oxide fume is 10 mg/m³.

REFERENCES

1. Stokinger HE: The metals. In Clayton GD, Clayton FE (eds): *Patty's Industrial Hygiene and Toxicology*, 3rd ed, Vol 2, *Toxicology*, pp 1740–1748. New York, Wiley–Interscience, 1981
2. Drinker KR, Thomson RM, Finn JL: Metal fume fever: The effects of inhaled magnesium oxide fume. *J Ind Hyg* 9:187–192, 1927
3. Hygienic Guide Series: Magnesium. *Am Ind Hyg Assoc J* 21:97–98, 1960

MAGNESIUM OXIDE FUME
CAS: 1309-48-4

MgO

Synonyms: None

Physical Form. Fume

Sources. From manufacture of refractory crucibles, fire bricks, magnesia cements, and boiler scale compounds

Exposure. Inhalation

Toxicology. Magnesium oxide fume is a mild irritant of the eyes and nose.

MALATHION
CAS: 121-75-5

$C_{10}H_{19}O_6PS_2$

Synonyms: Diethyl mercaptosuccinate, S-ester with *O,O*-dimethyl phosphorodithioate; Malathon; carbophos; Cythion 4049

Physical Form. Colorless to light-amber liquid

Uses. Insecticide

Exposure. Inhalation; skin absorption; ingestion

Toxicology. Malathion is an anticholinesterase agent, but it is of a relatively low order of toxicity compared to other organophosphates.

Signs and symptoms of intoxication by anticholinesterase agents are caused by the inactivation of the enzyme cholinesterase, which results in the accumulation of acetylcholine at synapses in the nervous system, skeletal and smooth muscle, and secretory glands.[1-4] After inhalation of extremely high concentrations of malathion, ocular and respiratory effects may appear simultaneously. Ocular effects include miosis, blurring of distant vision, and tearing; rhinorrhea, and frontal headache were also noted. Respiratory effects include tightness in the chest, wheezing, laryngeal spasms, and excessive salivation. Peripheral effects include excessive sweating, muscular fasciculations, and weakness. Effects on the central nervous system include giddiness, confusion, ataxia, slurred speech, and convulsions. After ingestion, anorexia, nausea, vomiting, abdominal cramps, and diarrhea also appear.

Malathion itself has only a slight direct inhibitory action on cholinesterase, but one of its metabolites, malaoxon, is an active inhibitor.[4] Both malathion and malaoxon are rapidly detoxified by esterases in the liver and other organs. This rapid metabolism is the apparent reason for the lower toxicity of malathion compared to other organophosphates. Malaoxon inactivates cholinesterase by phosphorylation of the active site of the enzyme to form the "dimethylphosphoryl enzyme." Over the following 24 to 48 hours, there is a process, called aging, of conversion to the "monomethylphosphoryl enzyme." Aging is of clinical interest in the treatment of poisoning because cholinesterase reactivators, such as pralidoxime (2-PAM, Protopam) chloride, are ineffective after aging has occurred.

The relative safety of malathion to humans has been demonstrated repeatedly. In a group of workers with an average exposure of 3.3 mg/m³ for 5 hours (maximum of 56 mg/m³), the cholinesterase levels in the blood were not significantly lowered, and no one exhibited signs of cholinesterase inhibition.[5] In a human experiment in which four men were exposed 1 hour daily for 42 days to 84.8 mg/m³, there was moderate irritation of the nose and the conjunctiva, but there were no cholinergic signs or symptoms.[6]

Almost all reports of fatalities from malathion have involved ingestion.[4] The acute oral lethal dose is estimated to be somewhat below 1.0 g/kg. Nonlethal intoxication has occurred in agricultural workers but usually has been the result of gross exposures with concomitant skin absorption.[4]

Malathion has caused skin sensitization, and dermatitis may occur under conditions of heavy field use.[7]

In rats, malathion was not teratogenic when administered by gastric intubation on days 6 through 15 of gestation at doses as high as 300 mg/kg.[8]

National Cancer Institute studies showed that administration of 4700 or 8150 mg/kg for 80 weeks or 2000 or 4000 mg/kg for 103 weeks in the diets of rats was not carcinogenic.[9-11] Subsequent data re-evaluation by NTP confirmed these conclusions.[12] Mice fed diets containing 8000 or 16,000 mg/kg for 80 weeks also had no significant increase in tumor incidence.[9] The IARC determined that there is no available evidence to suggest that malathion is likely to present a carcinogenic risk to humans.[13]

The 1995 ACGIH threshold limit value–time-weighted average (TLV-TWA) for malathion is 10 mg/m³.

Note: For a description of diagnostic signs, differential diagnosis, and medical control, including clinical laboratory tests, as well as specific treatment of overexposure to anticholinesterase insecticides, see "Parathion."

REFERENCES

1. Grob D: Anticholinesterase intoxication in man and its treatment. In Koelle GB (ed): *Handbuch der Experimentellen Pharmakologie*, Vol 15, *Cholinesterases and Anticholinesterase Agents*, (in English) pp 989–1027. Berlin, Springer-Verlag, 1963
2. Taylor P: Anticholinesterase agents. In Gilman AG et al (eds): *Goodman and Gilman's The Pharmacological Basis of Therapeutics*, 7th ed, pp 110–129. New York, Macmillan, 1985
3. Hayes WJ Jr: *Clinical Handbook on Economic*

Poisons, Emergency Information for Treating Poisoning, US Public Health Service Pub No 476, pp 12–23. Washington, DC, Government Printing Office, 1963

4. National Institute for Occupational Safety and Health, US Department of Health, Education and Welfare: *Criteria for a Recommended Standard . . . Occupational Exposure to Malathion.* (NIOSH) 76-205, 183 pp. Washington, DC, US Government Printing Office, 1976

5. Culver D, Caplan P, and Batchelor GS: Studies of human exposure during aerosol application of malathion and chlorthion. *AMA Arch Ind Health* 1:516–523, 13:37–50, 1956

6. Golz HH: Controlled human exposures to malathion aerosols *AMA Arch Ind Health* 19:516–523, 1959

7. Milby TH, Epstein, WL: Allergic contact sensitivity to malathion. *Arch Environ Health* 9:434–437, 1964

8. Khera KC et al: Teratogenicity studies on linuron, malathion, and methoxychlor in rats. *Toxicol Appl Pharmacol* 45:435–444, 1978

9. National Cancer Institute: Bioassay of malathion for possible carcinogenicity, TR-24. DHEW (NIH) Pub No 78-824. Washington, DC, US Department of Health, Education and Welfare, 1978

10. National Cancer Institute: *Bioassay of Malathion for Possible Carcinogenicity*, TR-192. DHEW (NIH) Pub No 78-1748. Washington, DC, US Department of Health, Education and Welfare, 1979

11. National Cancer Institute: *Bioassay of Malathion for Possible Carcinogenicity*, TR-135. DHEW (NIH) Pub No (NIH) 79-1390. Washington, DC, US Department of Health, Education and Welfare, 1979

12. Huff JE et al: Malathion and malaoxon: histopathological reexamination of the National Cancer Institute's carcinogenesis studies. *Environ Res* 37:154–173, 1985

13. *IARC Monographs on the Evaluation of the Carcinogenic Risk of Chemicals to Humans*, Vol 30, Miscellaneous pesticides, pp 103–129. Lyon, International Agency for Research on Cancer, 1983

MALEIC ANHYDRIDE
CAS: 108-31-6

$C_4H_2O_3$

Synonyms: 2,5 Furandione; *cis*-butenedioic anhydride; toxilic anhydride

Physical Form. White crystalline solid

Uses. In the manufacture of polyester resins, fumaric acid, agricultural pesticides, and alkyl resins

Exposure. Inhalation

Toxicology. Maleic anhydride is a severe irritant of the eyes; it is an irritant and sensitizer of both the skin and respiratory tract and may produce asthma on repeated exposures.

Workers exposed to vapors from heated maleic anhydride developed an intense burning sensation in the eyes and throat, with cough and vomiting; exposure to high fume concentrations caused photophobia, double vision, and a visual phenomenon of seeing rings around lights.[1,2] Exposure of humans to a concentration of 1.5 to 2 ppm resulted in nasal irritation within 1 minute and eye irritation after 15 to 20 minutes.[3] Among workers repeatedly exposed to 1.25 to 2.5 ppm, effects were ulceration of nasal mucous membranes, chronic bronchitis and, in some cases, asthma.[3] In one case, a worker exposed to dust concentrations below 1 mg/m^3 developed cough, rhinitis, breathlessness, and wheezing about 1 month after initial exposure.[4] Symptoms developed within minutes of exposure to the dust, which occurred during the loading of chemicals into a reactor. Within 3 months, his symptoms worsened, and he was admitted to the hospital for an acute asthmatic attack. The patient had a positive challenge test to maleic anhydride but was negative to phthalic anhydride to which he was concomitantly exposed.

The dust on dry skin may result in a delayed burning sensation but, on moist skin, the sensation is almost immediate, producing erythema, which may progress to vesiculation.[3] Prolonged or repeated exposure also may cause dermatitis.

In rats, maleic anhydride has an oral LD_{50} of 1050 mg/kg. It is corrosive to the skin and eyes of rabbits, with a dermal LD_{50} of 2620 mg/kg.[5] An inhalation study of rats, hamsters, and monkeys exposed to 1.1, 3.3, or 9.8 mg/m³, respectively, 6 hr/day, 5 days/week for 6 months, revealed dose-related signs of nasal and ocular irritations, including discharge, sneezing, gasping, and coughing for all species.[5] No treatment-related effects were observed in hematology, clinical chemistry, urinalysis, and pulmonary-function tests. Although microscopic evaluation showed evidence of nasal irritation, there was no evidence of systemic toxicity directly attributable to maleic anhydride.

In a study in which rats were injected subcutaneously with 1 mg maleic anhydride in oil twice weekly for 61 weeks, two of three surviving animals developed fibrosarcomas, which appeared 80 weeks after the start of the experiment.[6] Administered in the diet of rats for 2 years, it was not carcinogenic.[7]

Pregnant rats treated orally with up to 140 mg/kg/day from day 6 to 15 of gestation had no treatment-related effects on fetal development.[8] In a multigenerational study, no adverse effects on fertility or pups were observed at doses up to 55 mg/kg/day over two generations; at 150 mg/kg/day, maleic anhydride was toxic to parental animals, causing renal cortical necrosis in both females and males.[8]

The 1995 ALGIH threshold limit value–time-weighted average (TLV-TWA) for maleic anhydride is 0.25 ppm (1.0 mg/m³).

REFERENCES

1. Grant WM: *Toxicology of the Eye*, 3rd ed, pp 574–575. Springfield, IL, Charles C Thomas, 1986
2. Chemical Safety Data Sheet SD-88, Maleic anhydride, pp 5–6, 11–13. Washington, DC, MCA Inc, 1962
3. Hygienic Guide Series: Maleic anhydride. *Am Ind Hyg Assoc J* 31:391–394, 1970
4. Lee HS, Wang YT, Cheong TH, et al: Occupational asthma due to maleic anhydride. *Br J Ind Med* 48:283–285, 1991
5. Short RD, Johannsen FR, Ulrich CE: A 6-month multispecies inhalation study with maleic anhydride. *Fund Appl Toxicol* 10:517–524, 1988
6. Dickens F, Jones HEH: Further studies on the carcinogenic and growth-inhibitory activity of lactones and related substances. *Br J Cancer* 17:100–108, 1963
7. CIIT (Chemical Industry Institute of Toxicology): Chronic dietary administration of maleic anhydride—final report, CIIT Docket No. 114N3. Research Triangle Park, NC, 1983
8. Short RD, Johannsen FR, Levinskas GJ, et al: Teratology and multigeneration reproduction studies with maleic anhydride in rats. *Fund Appl Toxicol* 7:359–366, 1986

MANGANESE (AND COMPOUNDS)
CAS: 7439-96-5

Mn

Compounds: Manganese dioxide; manganese tetroxide; manganous chloride; manganous sulfate

Physical Form. Elemental manganese is a silver solid

Uses/Sources. In manufacture of alloys, dry-cell batteries, glass, inks, ceramics, paints, welding rods, rubber and wood preservatives, and fungicides; mining and processing of manganese ores

Exposure. Inhalation

Toxicology. The major concern of humans exposed to manganese is its effects on the central nervous system following chronic exposure.

The neurologic disorder known as chronic manganese poisoning, or manganism, occurs after variable periods of heavy exposure which have ranged from 6 months to 3 years.[1,2] The disease begins insidiously with headache, asthenia, irritability and, occasionally, psychotic behavior.[1] The latter occurs most frequently in miners rather than in industrial workers. Manganese psychosis consists of transitory psychological disturbances, such as hallucinations, compulsive behavior, and emotional instability.[3] Severe somnolence, followed by insomnia, is often found early in the disease. As manganese exposure continues, symptoms include generalized muscle weakness, speech impairment, incoordination, and impotence; tremor, paresthesia, and muscle cramps have been noted.[1,3,4] In the advanced stage, the subject exhibits excessive salivation, inappropriate emotional reactions, and Parkinson-like symptoms, such as masklike facies, severe muscle rigidity, and gait disorders.[1] Manganism is reversible if it is limited to psychological disturbances and the subject is removed from exposure. Established neurologic signs and symptoms tend to persist or even progress in the absence of additional exposure.[5]

Exposure levels associated with advanced manganism typically have been very high; 150 cases were found in three mines where levels reached 450 mg/m³.[2] More recent studies report cases showing neurologic symptoms and a few signs at lower concentrations. Of 36 workers exposed to magnesium dioxide dust ranging from 6.8 mg/m³ to 42.2 mg/m³, 8 exhibited symptoms of manganism.[6] Neurologic screening of 117 workers with exposures greater than 5 mg/m³ revealed 7 cases with definite signs and symptoms.[7] Comparison of 369 workers exposed to 0.3 to 20 mg/m³ suggested that slight neurologic disturbances may occur at exposures of less than 5 mg/m³, but the disturbances seem to be more prevalent at higher exposures.[8] Low-level exposure to manganese ranging from 0.19 to 1.39 mg/m³ for 1 to 45 years has reportedly caused alterations in neurophysiological and psychological parameters, which were interpreted as preclinical signs of manganism.[9]

One of the striking aspects of manganism is its similarity to Parkinson's disease.[10] In both conditions, neuropathological changes occur in the basal ganglia, with selective destruction of dopaminergic neurons.

An association between manganese exposure and pulmonary effects, including pneumonia, chronic bronchitis, and airway disability, has been observed. Extrapolation from animal studies suggests that it is unlikely that manganese could be the sole etiologic agent responsible for serious pathological changes in the lungs. Instead, it is possible that susceptibility to infection is increased.[1]

Acute poisoning by manganese is rare but may occur following ingestion of large amounts of manganese compounds or from inhalation. Freshly formed manganese oxide fumes at high concentrations may cause metal fume fever. This influenzalike illness is characterized by chills, fever, sweating, nausea, and cough. The syndrome begins 4 to 12 hours after exposure and lasts for 24 hours without causing permanent damage.[11]

Anecdotal reports have suggested that exposure to high levels of manganese dusts results in decreased libido and impotence.[10] In animal studies, growth and maturation of the testes was delayed following oral exposure.[12] Intratracheal administration of a single dose of 160 mg manganese/kg in rabbits resulted in degeneration of the seminiferous tubules, with loss of spermatogenesis and complete infertility within 8 months.[13]

Repeated subcutaneous or intraperitoneal injection of manganese dichloride caused increased incidences of lymphosarcomas in mice.[14] Chronic oral exposure of rats to manganese sulfate led to a slight increase in pancreatic tumors that was not dose-responsive.[10] There is no information relating manganese exposure to cancer occurrence in humans.[10] Manganese compounds, in general, are not genotoxic in in vivo assays.

The proposed 1995 ACGIH threshold limit value for manganese and inorganic compounds as Mn is 0.2 mg/m³.

REFERENCES

1. *Health Assessment Document for Manganese.* Final report—PB84-229954, pp 1–353. Washington, DC, US Environmental Protection Agency, 1984
2. Rodier J: Manganese poisoning in Moroccan miners. *Br J Ind Med* 12:21–35, 1955
3. Cook DG, Fahn S, Brait KA: Chronic manganese intoxication. *Arch Neurol* 30:59–64, 1974

4. Hine CH, Pasi A: Manganese intoxication. *West J Med* 123:101–107, 1975

5. Barbeau A et al: Role of manganese in dystonia. *Adv Neurol* 14:339–352, 1976

6. Emara AM, El-Ghawabi SH, Madkour OI, et al: Chronic manganese poisoning in the dry battery industry. *Br J Ind Med* 28:78–82, 1971

7. Eriksson H, Magiste K, Plantin LO, et al: Effects of manganese oxide on monkeys as revealed by a combined neurochemical, histological and neurophysiological evaluation. *Arch Toxicol* 61:46–52, 1987

8. Saric M, Lucic-Palaic S: Possible synergism of exposure to airborne manganese and smoking habit in occurrence of respiratory symptoms. In Walton WH (ed): *Inhaled Particles*, IV, pp 773–779. New York, Pergamon Press, 1977

9. Wennberg A, Iregren A, Struwe G, et al: Manganese exposure in steel smelters a health hazard to the nervous system. *Scand J Work Environ Health* 17:255–262, 1991

10. Agency for Toxic Substances and Disease Registry (ATSDR): *Toxicological Profile for Manganese.* US Department of Health and Human Services, Public Health Service, pp 136, TP-91/19, 1992

11. Piscator M: Health hazards from inhalation of metal fumes. *Environ Res* 11:268–270, 1976

12. Laskey JW, Rehnberg GL, Hein JF: Effects of chronic manganese (Mn_3O_4) exposure on selected reproductive parameters in rats. *J Toxicol Environ Health* 9:677–687, 1982

13. Chandra SV, Ara R, Nagar N, et al: Sterility in experimental manganese toxicity. *Acta Biol Med* 30:857–862, 1973

14. DiPaolo JA: The potentiation of lymphosarcomas in mice by manganese chloride. *Fed Proc* 23:393, 1964 (Abstract)

MANGANESE CYCLOPENTADIENYL TRICARBONYL

CAS: 12079-65-1

$C_5H_5—Mn(CO)_3$

Synonyms: MCT; cymantrene; cyclopentadienyl manganese carbonyl

Physical Form. Liquid

Uses. Octane enhancer for gasoline

Exposure. Inhalation; skin absorption

Toxicology. Manganese cyclopentadienyl tricarbonyl (MCT) causes convulsions and pulmonary edema in laboratory animals.

The oral LD_{50} in rats was 22 mg/kg, and the expected dose for convulsions was 32 mg/kg.[1] Phenobarbital pretreatment prevented the convulsions and pulmonary damage ordinarily caused by a 50 mg/kg intraperitoneal dose of MCT.

The pneumotoxicity of MCT in rats was compared to that of manganese methylcyclopentadienyl carbonyl by subcutaneous administration of 0.5, 1.0, or 2.5 mg/kg of both compounds.[2] MCT was twice as potent in causing large increases in pulmonary lavage albumin and protein content.

Russian literature indicates that rats exposed 4 hr/day to 1 mg/m³ for 11 months showed no outward indications of toxicity, but there was decreased diuresis, as well as some protein excretion in the urine.[3] MCT penetrated the tails of rats and caused death.

The 1995 ACGIH threshold limit value–time-weighted average (TLV-TWA) for manganese cyclopentadienyl tricarbonyl is 0.1 mg/m³ as Mn, with a notation for skin.

REFERENCES

1. Penney DA, Hogberg K, Traiger GJ, Hanzlik RP: The acute toxicity of cyclopentadienyl manganese tricarbonyl in the rat. *Toxicology* 34:341–347, 1985

2. Clay RJ, Morris JB: Comparative pneumotoxicity of cyclopentadienyl manganese tricarbonyl and methylcyclopentadienyl manganese tricarbonyl. *Toxicol Appl Pharmacol* 98:434–443, 1989

3. Arkipova OG et al: Toxicity within a factory of the vapor of new antiknock compound, manganese cyclopentadienyl-tricarbonyl. *Gigiena Sanitoriya* 20:40–44, 1965

MANGANESE TETROXIDE
CAS: 1317-35-7

Mn₃O₄

Mn_3O_4

Synonyms: Manganese oxide; trimanganese tetroxide

Physical Form.   Powder

Source.   Fume generated whenever manganese oxides are heated in air; ferromanganese fume, generated in the pouring and casting of molten ferromanganese, is largely manganese tetroxide.

Exposure.   Inhalation

Toxicology.   Manganese tetroxide affects the central nervous system, and toxicity occurs mostly in chronic form (manganism).

The neurologic disorder known as chronic manganese poisoning occurs after variable periods of heavy exposure which have ranged from 6 months to 3 years.[1,2] The disease begins insidiously with headache, asthenia, and irritability. As exposure continues, symptoms include generalized muscle weakness, speech impairment, incoordination, and impotence. Tremor, paresthesia, and muscle cramps have been noted.[3] In the advanced stage, the subject exhibits excessive salivation, inappropriate emotional reactions, and Parkinson-like symptoms, such as masklike facies, severe muscle rigidity, and gait disorders.

In a report of 5 cases of manganism in a steel plant, 3 resulted from exposure to ferromanganese fume and 2 from exposure to ferromanganese dust.[4] As indicated above, ferromanganese fume is primarily manganese tetroxide. Two of the workers exposed to the fume worked in a pig-casting operation where the exposure was estimated to have been 13.3 mg/m³ for 5 years.

It is generally held that manganese fume is more hazardous than equivalent concentrations of manganese-containing dust.

The 1995 ACGIH threshold limit value–time-weighted average (TLV-TWA) for manganese fume is 1 mg/m³, with a STEL/ceiling of 3 mg/m³.

REFERENCES

1. Health Assessment Document for Manganese. Final report—PB84-229954, pp 1–353. Washington, DC, US Environmental Protection Agency, 1984
2. Rodier J: Manganese poisoning in Moroccan miners. *Br J Ind Med* 12:21–35, 1955
3. Cook DG, Fahn S, Brain KA: Chronic manganese intoxication. *Arch Neurol* 30:59–64, 1974
4. Whitlock CM, Amuso SJ, Bittenbender JB: Chronic neurological disease in manganese steel workers. *Am Ind Hyg Assoc J* 27:454–459, 1966

MERCURY
CAS: 7439-97-6

Hg

Synonyms: Quicksilver; mercury vapor; mercury liquid; mercury salts

Physical Form.   Silver-white, heavy, liquid metal

Uses.   In electrical apparatus; in measurement and control systems, such as thermometers and sphygmomanometers; agricultural and industrial poisons; catalyst; in antifouling paint; in dental practice; in gold mining

Exposure.   Inhalation; skin absorption; ingestion

Toxicology.   Acute exposure to high concentrations of mercury vapor causes severe respiratory damage, whereas chronic exposure to lower levels is primarily associated with central nervous system damage and renal effects.

Inhalation of mercury vapor may produce a metal-fume-feverlike syndrome, including chills, nausea, general malaise, tightness in the chest, and respiratory symptoms.[1] High concentrations cause corrosive bronchitis and interstitial pneumonitis.[2] In the most severe cases, the patient will succumb because of respiratory insufficiency.[2] In one episode involving four workers, it was estimated that mercurial pneumonitis resulted from exposure for several hours to concentrations ranging between 1 and 3 mg/m^3.[3]

With chronic exposure to mercury vapor, early signs are nonspecific and include weakness, fatigue, anorexia, loss of weight, and disturbances of gastrointestinal function.[2] This syndrome has been termed *asthenic-vegetative syndrome*, or *micromercurialism*. At higher exposure levels, a characteristic mercurial tremor appears, which begins with intentional tremors of fingers, eyelids, and lips and may progress to generalized trembling of the entire body and violent chronic spasms of the extremities.[2,4] Parallel to the development of tremor, mercurial erethism develops. This is characterized by behavioral and personality changes, increased excitability, loss of memory, insomnia, and depression. In severe cases, delirium and hallucination may occur. Another characteristic feature of mercury intoxication is severe salivation and gingivitis. Chronic changes in the cornea and lens have also been described.[5]

It has been estimated that the probability of manifesting typical mercurialism with tremor and behavioral changes will increase with exposures to concentrations of 0.1 mg/m^3 or higher.[2]

Renal damage has been reported following both acute and chronic exposure.[6,7] Mercury is known to accumulate in the kidneys, and case studies have described increased creatinine excretion, proteinuria, hematuria, and degeneration of the convoluted tubules in exposed individuals. Increased levels of the urinary enzyme NAG (N-acetyl-β-glucosaminidase), compared with controls, have been observed in chronically exposed workers.[8,9]

Ingestion of mercuric salts causes corrosive ulceration, bleeding, and necrosis of the gastrointestinal tract, usually accompanied by shock and circulatory collapse.[2,4] If the patient survives the gastrointestinal damage, renal failure occurs within 24 hours as a result of necrosis of the proximal tubular epithelium, followed by oligu-ria, anuria, and uremia. Chronic low-dose exposure to mercury salts, or probably even elemental mercury vapor, may also induce an immunologic glomerular disease.[2,4]

Applied locally, mercury may cause sensitization dermatitis.[1,2]

In several epidemiological studies, no increased risk of congenital abnormalities, stillbirths, or spontaneous abortions was observed with occupational exposure to mercury.[7] Exposure of pregnant rats on gestational days 10 through 15 at 0.5 mg/m^3 resulted in an increased incidence of resorptions; gross cranial defects occurred at this dose when it was administered throughout the entire gestational period.[10]

Intraperitoneal injection of metallic mercury in rats has produced sarcomas.[1] The sarcomas develop without exception at those sites in direct contact with the metal, suggesting a foreign body reaction rather than chemical carcinogenesis. Mercuric chloride was tested for carcinogenicity in two-year gavage studies in mice and rats.[11] Three of 49 high-dose male mice had renal tubule tumors and, in male rats, there was an increase in squamous-cell papillomas of the forestomach.

There is no conclusive evidence from epidemiological studies that mercury increases cancer risk in humans.[12] In the few studies in which increases have been reported, concomitant exposure to other known carcinogens has confounded the results. The IARC has determined that there is inadequate evidence in humans for the carcinogenicity of mercury and mercury compounds.[12] In animals, there is inadequate evidence for carcinogenicity of metallic mercury and limited evidence for the carcinogenicity of mercuric chloride.

Genotoxic assays have given both positive and negative results.[7]

Blood and urine mercury concentrations are commonly used as biomarkers of mercury exposure.[7]

The 1995 threshold limit value–time-weighted average (TLV-TWA) for mercury (inorganic forms, including metallic mercury) is 0.025 mg/m^3, with an A4, not classifiable as a human carcinogen, designation.

Diagnosis. *Signs and Symptoms:* Paresthesias of the lips, hands, and feet; ataxia; dysarthria; impairment of vision and hearing; emotional disturbances; spasticity, with jerking movements of the limbs, head, or shoulders; dizziness; hypersalivation and lacrimation; nausea, vomiting, diarrhea, and constipation; kidney damage; dermatitis; skin burns.

Differential Diagnosis: Effects seen with alkyl mercury may mimic organic lead syndromes, Parkinsonism, hereditary ataxias, cerebrocerebellar degeneration, neurosyphilis, and metastatic lesions of the central nervous system.

Special Tests: An electromyograph may be useful in determining the extent of nerve dysfunction. The mercury content of the blood may indicate the extent of absorption.

Treatment. The aim of treatment is to remove mercury compound from the body. In severe cases, hemodialysis and infusion of chelating agents, such as cysteine, *N*-acetylcysteine, or *N*-acetylpenicillamine may be considered; Ca-EDTA should not be used because of nephrotoxicity. Dimercaprol (BAL) enhances mercury excretion through the bile, as well as the urine and is the agent of choice if there is renal impairment from mercury. A newer chelator, 2,3-dimercaptosuccinic acid, hold promise of less toxicity and more specific therapy.[13]

REFERENCES

1. National Institute for Occupational Safety and Health: *Criteria for a Recommended Standard . . . Occupational Exposure to Inorganic Mercury.* DHEW (NIOSH) Pub No 73-11024. Washington, DC, US Government Printing Office, 1973
2. Berlin M: Mercury. In Friberg L et al (eds): *Handbook on the Toxicology of Metals,* 2nd ed, Vol II, *Specific Metals,* pp 387–445. Amsterdam, Elsevier, 1986
3. Milne J, Christophers A, De Silva P: Acute mercurial pneumonitis. *Br J Ind Med* 27:334–338, 1970
4. Goyer RA: Toxic effects of metals. In Klaasen CD et al (eds): *Casarett and Doull's Toxicology. The Basic Science of Poisons,* 3rd ed, pp 605–609. New York, Macmillan, 1986
5. Rosenman KD et al: Sensitive indicators of inorganic mercury toxicity. *Arch Environ Health* 41:208–215, 1986
6. Zalups RK, Lash LH: Advances in understanding the renal transport and toxicity of mercury. *J Toxicol Environ Health* 42:1–44, 1994
7. Agency for Toxic Substances and Disease Registry (ASTDR): *Toxicological Profile for Mercury.* US Department of Health and Human Services, Public Health Service, pp 217, 1993
8. Barregard L, Hultberg B, Schutz A, et al: Enzymuria in workers exposed to inorganic mercury. *Intl Arch Occup Health* 61:65–69, 1988
9. Piikivi L, Ruokonen A: Renal function and long-term low mercury vapor exposure. *Arch Environ Health* 44:146–149, 1989
10. Steffek AJ, Clayton R, Siew C, et al: Effects of elemental mercury vapor exposure on pregnant Sprague–Dawley rats. *Teratology* 35:59 (Abstract)
11. National Toxicology Program: NTP *Technical Report on the Toxicology and Carcinogenesis of Mercuric Chloride (CAS No. 7487-94-7) in F344/N Rats and B6C3F₁ Mice (Gavage Studies).* TR-408, US Department of Health and Human Services, Public Health Service, National Institutes of Health, Research Triangle Park, NC, 1991
12. *IARC Monographs on the Evaluation of Carcinogenic Risks to Humans,* Vol 58, Beryllium, cadmium, mercury, and exposures in the glass manufacturing industry, pp 289–324, Lyon, International Agency for Research on Cancer, 1993
13. Mofenson HC, Caraccio TR, Greensher J: Acute poisonings. In Rakel RE (ed): *Conn's Current Therapy,* p 1030. Philadelphia, WB Saunders Co, 1988

MERCURY (ALKYL COMPOUNDS)
CAS: Varies with Compound

RHgX

Compounds: Methyl mercury; ethyl mercury chloride, dimethyl mercury

Physical Form. Colorless liquids

Uses. Fungicides in seed dressings and folial

sprays; preservative solutions for wood, paper pulp, textiles, and leather

Exposure. Inhalation; skin absorption; ingestion

Toxicology. Organo (alkyl) mercury compounds cause dysfunction of the central nervous system and kidneys and are irritants of the eyes, mucous membranes, and skin; methyl mercury causes developmental effects in humans.

Methyl and ethyl mercury compounds have similar toxicological properties, and there is no sharp demarcation between acute and chronic poisoning.[1] Once a toxic dose has been absorbed and retained for a period of time, functional disturbances and damage occur. The latency period for a single toxic dose may vary from one to several weeks; longer latency periods on the order of years have been reported for chronic exposures.[1,2]

Symptoms of poisoning include numbness and tingling of the lips, hands, and feet (paresthesia); ataxia; dysarthria; concentric constriction of the visual fields; impairment of hearing; and emotional disturbances.[3]

With severe intoxication, clonic seizures may occur, and the symptoms are usually irreversible.[1,3] Severe intoxication also results in incontinence; periods of spasticity and jerking movements of the limbs, head, or shoulders; and bouts of groaning, moaning, shouting, or crying. Less frequent symptoms are dizziness, hypersalivation, lacrimation, nausea, vomiting, and diarrhea or constipation.[4] The pathological changes in the CNS are characterized by general neuron degeneration in the cerebral cortex, especially the visual areas of the occipital cortex, and by gliosis.[1]

An epidemic of intoxication from ingestion of fish contaminated with methyl mercury occurred in the Minamata district in Japan and, as a result, methyl mercury intoxication is often referred to as Minamata disease.[4] Infants born to mothers with exposure to large amounts of methyl mercury had microencephaly, mental retardation, and cerebral palsy with convulsions. In an incidence in Iraq, ingestion by pregnant women of wheat products contaminated with methyl mercury fungicide caused similar symptoms of neurological damage and mental retardation.[2] The fetus is particularly sensitive to the effects of methyl mercury, which interferes with organ development. Toxic concentrations inhibit the normal migration of nerve cells from the central parts of the neurotube toward the peripheral parts of the brain cortex and thus inhibit the normal development of the fetal brain.[1] Differences between fetal and adult hematocrits may result in differing mercury concentrations in the two; studies suggest that the difference in sensitivity between the fetus and adult organism is close to a factor of 2.[1] It has been suggested that women of childbearing age should have no occupational exposure to alkyl mercury.[2]

The biological half-life in humans for methyl mercury is about 70 days; since elimination is slow, irregular, and individualized, there is a considerable risk of an accumulation of mercury to toxic levels.[3] A precise relationship between atmospheric levels of alkyl mercury and concentrations of mercury in blood or urine has not been shown.[3] Clinical observations indicate that concentrations of 50 to 100 μg mercury/100 ml of whole blood may be associated with symptoms of intoxication; concentrations of around 10 to 20 μg mercury/100 ml are not associated with symptoms.[3] In a study of 20 workers engaged in the manufacture of organic mercurials and exposed for 6 years to mercury concentrations in air between 0.01 and 0.1 mg/m^3, there was no evidence of physical impairment or clinical laboratory abnormalities.[5] Low levels of methyl mercury in the blood do not seem to affect the results of behavioral performance tests.[6]

Methyl mercury concentrations in hair can be used as an indicator of mercury concentration in blood, with a ratio of blood to hair of 1 to 250.[2] Under occupational conditions, the possibility of external contamination of hair should be kept in mind.

The alkyl mercury halides are irritating to the eyes, mucous membranes, and skin and may cause severe dermatitis and burns; skin sensitization has occasionally occurred.[7,8]

The emission of mercury-containing vapor from certain latex paints has been demonstrated. Symptoms of acrodynia in a four-year-old boy (leg cramps; a generalized rash; pruritis; sweating; tachycardia; intermittent low-grade fever; marked personality change; erythema and desquamation of the hands, feet, and nose; weakness of the pelvic and pectoral girdles; and lower extremity nerve dysfunction) developed sequentially beginning 10 days after the inside of his home was painted with 64 liters of latex paint containing phenylmercuric acetate.[9] This individual and others of a group from recently painted interiors had elevated levels of mercury in urine. The Environmental Protection Agency announced that compounds containing mercury could no longer be lawfully added to interior latex paint after August 20, 1990.[9]

Epidemiologic studies of methyl mercury–exposed populations have not shown any evidence of a carcinogenic effect.[9] In chronic animal studies, methyl mercury chloride in the diet caused an increase in renal adenomas and adenocarcinomas in male mice.[10] The IARC has determined that there is inadequate evidence in humans for the carcinogenicity of mercury compounds but that there is sufficient evidence for the carcinogenicity of methyl mercury chloride in experimental animals.[11] Organomercury compounds exert a direct effect on chromosomes by inhibiting the spindle mechanism, which results in clastogenic effects.[11]

Methyl mercury vapor is detectable by smell at concentrations well below that which could prove hazardous on intermittent exposure.[12]

The 1995 threshold limit value–time-weighted average (TLV-TWA) is 0.01 mg/m³, with a short-term exposure limit of 0.03 mg/m³.

REFERENCES

1. Berlin M: Mercury. In Friberg L et al: *Handbook on the Toxicology of Metals*, 2nd ed, Vol II, *Specific Metals*, pp 418–445. Amsterdam, Elsevier, 1986
2. Inskip MJ, Piotrowski JK: Review of the health effects of methyl mercury. *J Appl Toxicol* 5:113–133, 1985
3. Report of an International Committee: Maximum allowable concentrations of mercury compounds. *Arch Environ Health* 19:891–905, 1969
4. Rustam H, Von Burg R, Amin-Zaki L, El Hassani S: Evidence for a neuromuscular disorder in methyl mercury poisoning—clinical and electrophysiological findings in moderate to severe cases. *Arch Environ Health* 30:190–195, 1975
5. Dinman BD, Evans EE, Linch AL: Organic mercury—environmental exposure, excretion, and prevention of intoxication in its manufacture. *AMA Arch Ind Health* 18:248–260, 1958
6. Valcinkas J et al: Neurobehavioral assessment of Mohawk Indians for subclinical indications of methyl mercury neurotoxicity. *Arch Environ Health* 41:269–272, 1986
7. Dales LG: The neurotoxicity of alkyl mercury compounds. *Am J Med* 53:219–232, 1972
8. American National Standards Institute, Inc.: *American: National Standard Acceptable Concentrations of Organo (Alkyl) Mercury, ANSI Z37.30-1969*. New York, American National Standards Institute, Inc, 1970
9. Agocs MM, Etzel RA, Parrish RG, et al: Mercury exposure from interior latex paint. *N Eng J Med* 323:1096–1101, 1990
10. Agency for Toxic Substances and Disease Registry (ATSDR): *Toxicological Profile for Mercury*. US Department of Health and Human Services, Public Health Service, 1992
11. *IARC Monographs on the Evaluation of Carcinogenic Risks to Humans*, Vol 58, Beryllium, cadmium, mercury, and exposures in the glass manufacturing industry, pp 289–330. Lyon, International Agency for Research on Cancer, 1993
12. Junghans RP: A review of the toxicity of methyl mercury compounds with application to occupational exposures associated with laboratory uses. *Environ Res* 31:1–31, 1983

MESITYL OXIDE
CAS: 141-79-7

(CH₃)₂CCHCOCH₃

$(CH_3)_2CCHCOCH_3$

Synonyms: Methyl isobutenyl ketone; isopropylideneacetone; 4-methyl-3-pentene-2-one

Physical Form. Oily, colorless liquid

Uses. Solvent; chemical intermediate

Exposure. Inhalation

Toxicology. Mesityl oxide is an irritant of the eyes and mucous membranes; at high concentrations, it causes narcosis in animals, and it is expected that severe exposure will produce the same effect in humans.

Human subjects exposed to 25 ppm for 15 minutes experienced eye irritation; at 50 ppm, there was also nasal irritation and a persistent unpleasant taste that remained with many subjects 3 to 6 hours after the exposure.[1] Liquid mesityl oxide produces dermatitis with sustained skin contact.[2]

Rats and guinea pigs exposed 8 hr/day to 500 ppm for 10 days had nose and eye irritation and developed slight kidney injury; slight liver and lung injury were observed in a few animals; 13 of 20 animals died from 30 exposures of 8 hours each at 500 ppm, whereas all animals tested at 250 ppm survived.[3,4] Guinea pigs exposed to 2000 ppm for up to 422 minutes died during or following exposure. Signs of eye and respiratory tract irritation with gradual loss of corneal and auditory reflexes preceded coma and death.

Exposure of rats to irritant levels of mesityl oxide (above 137 ppm) caused leukopenia.[5] This hematological effect was regarded as an associative response to the sensory irritation that can act as a stressor to laboratory animals.

The strong peppermint or honeylike odor is detectable at 12 ppm; severe overexposure is unlikely because of local irritation and odor; however, olfactory fatigue may occur.[3,6]

The irritation and systemic effects resulting from mesityl oxide exposure appear to be more serious than those produced by the lower ketones.[7]

The 1995 ACGIH threshold limit value–time-weighted average (TLV-TWA) is 15 ppm (60 mg/m³) with a STEL/ceiling of 25 ppm (100 mg/m³).

REFERENCES

1. Silverman L, Schulte HF, First MW: Further studies on sensory response to certain industrial solvent vapors. *J Ind Hyg Toxicol* 28:262–266, 1946
2. Shell Chemical Corporation: Safety Data Sheet SC: 57-105. Mesityl oxide, pp 1–3. *Industrial Hygiene Bulletin*, 1957
3. Smyth HF Jr, Seaton J, Fischer L: Response of guinea pigs and rats to repeated inhalation of vapors of mesityl oxide and isophorone. *J Ind Hyg Toxicol* 24:46–50, 1942
4. Specht H, Miller JW, Valaer PJ, Sayers RR: Acute response of guinea pigs to the inhalation of ketone vapors. *National Institutes of Health Bulletin* No 176, 1940
5. Brondeau MT, Bonnet P, Guenier JP, et al: Adrenal-dependent leucopenia after short-term exposure to various airborne irritants in rats. *J Appl Toxicol* 10:83–86, 1990
6. Shell Chemical Corporation: Toxicity Data Sheet SC: 57-106. Mesityl Oxide. *Industrial Hygiene Bulletin*, 1957
7. National Institute for Occupational Safety and Health: *Criteria for Recommended Standard . . . Occupational Exposure to Ketones.* DHEW (NIOSH) Pub No 78-173. Washington, DC, US Government Printing Office, 1978

METHACRYLIC ACID
CAS: 79-41-4

C₄H₆O₂

Synonyms: 2-Methyl-2-propenoic acid; 2-methylenepropionic acid; alpha-methacrylic acid

Physical Form. Colorless liquid

Uses. Manufacture of methacrylic resins and plastics

Exposure. Inhalation

Toxicology. Methacrylic acid is an irritant of the eyes, nose, throat, and skin and is corrosive on contact.

Rats exposed to 1300 ppm for 5 hr/day for 5 days showed nose and eye irritation but no adverse findings in blood and urine tests.[1] Exposure of rats to 300 ppm for 6 hr/day for 20 days resulted in no clinical signs, but histopathological findings showed slight renal congestion.

Applied to the depilated guinea pig abdomen for 24 hours under an occlusive wrap, the liquid produced severe irritation.[2] The liquid also produced severe irritation when instilled in rabbit eyes.

In vitro, methacrylic acid has been found to be toxic to rat embryos, producing growth retardation and abnormal development, primarily abnormal neurulation.[3]

The 1995 threshold limit value–time-weighted average (TLV-TWA) is 20 ppm (70 mg/m³).

REFERENCES

1. Gage JC: The subacute inhalation toxicity of 109 industrial chemicals. *Br J Ind Med* 27:1, 1970
2. Guest D, Katz GV, Astill BD: Aliphatic carboxylic acids. In Clayton GD, Clayton FE (eds): *Patty's Industrial Hygiene and Toxicology*, 3rd ed, Vol 2C, *Toxicology*, pp 4952–4958. New York, Wiley-Interscience, 1982
3. Rogers JG, Greenaway JC, Mirkes PE, et al: Methacrylic acid as a teratogen in rat embryo culture. *Teratology* 33:113–117, 1986

METHANE
CAS: 74-82-8

CH₄

CH_4

Synonyms: Marsh gas; methyl hydride

Physical Form. Colorless gas

Uses/Sources. Constituent in cooking and illuminating gas; in the production of ammonia, methanol, and chlorohydrocarbons; occurs in natural gas and is produced by the decomposition of organic matter

Exposure. Inhalation

Toxicology. Methane acts as a simple asphyxiant by causing oxygen deprivation at very high concentrations.

Methane is practically inert and has no demonstrated physiological or toxicological effects.[1,2] A concentration of 87% has been shown to cause asphyxiation in mice, whereas 90% produces respiratory arrest. These effects are thought to be due entirely to dilution of atmospheric oxygen.

Healthy people at sea level can tolerate atmospheres with oxygen concentrations as low as 15% without symptoms; methane can cause asphyxiation in healthy individuals only when it is present in very high concentrations or when atmospheric oxygen has been otherwise reduced.

Methane exposure can occur in coal miners when methane is trapped within coal seams. Because methane is lighter than air, it accumulates first at the top of an enclosed space; thus, collapsing from loss of consciousness can be lifesaving.

On the skin, liquefied methane can cause frostbite.[1,2]

At concentrations below those required to produce any severe oxygen deprivation, methane presents an explosion hazard, and a time–weighted average of 1000 ppm has been recommended as a threshold limit for exposure.[1]

REFERENCES

1. Sandmeyer EE: Aliphatic hydrocarbons. In Clayton GD, Clayton FE (eds): *Patty's Industrial Hygiene and Toxicology*, 3rd ed, Vol 2B, *Toxicology*, p 3180. New York, Wiley-Interscience, 1981
2. Low LK, Meeks JR, Mackerer CR: Methane. In Snyder R (ed): *Ethel Browning's Toxicity and Metabolism of Industrial Solvents*, 2nd ed, Vol 1, *Hydrocarbons*, pp 255–257. New York, Elsevier, 1987

METHOMYL
CAS: 16752-77-5

C₅H₁₀N₂O₂S

$C_5H_{10}N_2O_2S$

Synonyms: Lannate; DuPont 1179; S-methyl-N[(methylcarbamoyl)oxy]thioacetimidate

Physical Form. Crystalline solid

Uses. Carbamate insecticide for broad-spectrum control of pests

Exposure.   Inhalation

Toxicology.   Methomyl is a short-acting car-bamate anticholinesterase agent that is rapidly metabolized and demonstrates little evidence of cumulative toxicity.

Exposure to methomyl can result in inhibition of cholinesterase activity in blood and at nerve synapses of muscles, secretory organs, and nervous tissue in the brain and spinal cord.[1] Central nervous system signs and symptoms include anxiety, restlessness, depression of respiratory and circulatory centers, ataxia, convulsions, and coma.

Nicotinic signs of intoxication include muscle weakness, tremor and fasciculations, and involuntary twitching. Muscle weakness that affects the respiratory muscles may contribute to dyspnea and cyanosis. Tachycardia may result from stimulation of sympathetic ganglia in cardiac tissue and may mask the bradycardia because of the muscarinic action on the heart. Nicotinic action at the sympathetic ganglion may also result in pallor, high blood pressure, and hyperglycemia.

Muscarinic signs include miosis, increased salivation, sweating, urination and defecation, vomiting and nausea, and increased bronchial secretions.

In a survey of occupationally acquired disease in workers at a pesticide plant, 11% of 102 workers were hospitalized from exposure to methomyl and 3,4-dichloroaniline.[2] On clinical evaluation, 5 (46%) of 11 packaging workers, the group with the highest exposure to methomyl, had experienced blurred vision or pupillary constriction.

Long-term carcinogenicity studies in mice and rats administered methomyl up to 1000 ppm in the diet showed no evidence of carcinogenic effects.[3]

The 1995 threshold limit value–time-weighted average (TLV-TWA) for methomyl is 2.5 mg/m³.

REFERENCES

1. Taylor P: Anticholinesterase agents. In Gilman AG et al (eds): *Goodman and Gilman's The Phar-macological Basis of Therapeutics*, 8th ed pp 131–149. New York: Macmillan, 1990
2. Morse DL, Baker, Jr EK, Kimbrough RD, et al: Propanil chloracne and methomyl toxicity in workers of a pesticide manufacturing plant. *Clin Toxicol* 15:13–21, 1979
3. Kaplan AM, Sherman H: Toxicity studies with methyl *N*[[(methylamino)carbonyl]oxy]ethanimidothioate. *Toxicol Appl Pharmacol* 40:1–17, 1970

METHOXYCHLOR
CAS: 72-43-5

$C_{16}H_{15}Cl_3O_2$

Synonyms: 1,1,1-Trichloro-2,2-bis(para-methoxyphenyl)ethane; methoxy-DDT; Marlate; Prentox; Methoxcide

Physical Form.   Crystalline solid

Uses.   Insecticide

Exposure.   Inhalation; ingestion

Toxicology.   Methoxychlor is a convulsant of low toxicity; in animals, it is a reproductive toxicant.

No adverse effects on health or clinical laboratory data were found in groups of volunteers given 2 mg/kg/day for 8 weeks.[1]

The oral LD_{50} in rats ranged from 5.0 to 7.0 g/kg.[2] Dogs fed a daily diet containing 4 g/kg body weight developed signs of chlorinated hydrocarbon intoxication, including fasciculations, tremor, hyperesthesia, tonic seizures, and tetanic convulsions after 5 to 8 weeks. Most of the dogs died within 3 weeks after onset of effects.[3] Rabbits given oral daily doses of 200 mg/kg died after 4 to 15 doses; autopsy findings included mild liver damage and nephrosis.[4] In mice given 5 mg orally over 3 days and in rats given 20 mg, there was a uterotropic effect manifested as a marked increase in the weight of the uterus.[5]

Methoxychlor is a weakly estrogenic compound that has been shown to alter fertility in male and female rats and to cause development effects. Administration of 1000 mg/kg in the diet of pregnant rats caused vaginal defects in their offspring.[6] Reduced fertility in both sexes was also noted when the offspring reached maturity.[6] Subchronic administration of methoxychlor in the diet of rats at 25, 50, 100, or 200 mg/kg/day from weaning through adulthood produced a variety of effects: in females, methoxychlor accelerated the age at vaginal opening (a morphological indicator of puberty), cycles were in constant estrus, ovarian luteal function was inhibited, and implantation was blocked; in males, treatment reduced growth, seminal vesicle weight, caudal epididymal weight, caudal sperm content, and pituitary weight.[7] Puberty was delayed in the two highest-dosage groups, but the fertility of treated males was not reduced when they were mated with untreated females.[7]

Intragastric administration of 200 mg/kg/day to rats on days 6 to 15 of pregnancy caused a decrease in the number and weight of the fetuses and delayed bone ossification.[8] Recent studies have shown that female mice exposed only during their first pregnancy, then allowed to mate again, delivered second litters (F_{1b}) with reproductive alterations in the form of significant advancement in vaginal opening time, even though the F_{1b} litters had not been directly exposed to methoxychlor.[9]

Female mice fed up to 2000 mg/kg and males given 3500 mg/kg in the diet for 78 weeks showed no statistically significant increase in the incidence of benign and malignant tumors that could be attributed to methoxychlor.[10] Chronic feeding studies in rats, at 850 and 1400 mg/kg for males and females, respectively, also showed no significant carcinogenic responses, although high tumor rates in controls may have masked detection.[10] Based on NCI results and several earlier animal studies, the IARC has determined that no evidence has been provided that methoxychlor is carcinogenic in experimental animals.[11]

Studies on the genotoxicity of methoxychlor have generally yielded negative results in prokaryotic assays, mixed results in in vitro eukaryotic systems, and negative results in in vivo studies.[2]

The 1995 ACGIH threshold limit value–time-weighted average (TLV-TWA) for methoxychlor is 10 mg/m^3.

REFERENCES

1. Stein AA et al: Safety evaluation of methoxychlor in human volunteers. *Toxicol Appl Pharmacol* 7:499 (Abstract), 1965
2. Agency for Toxic Substances and Disease Registry (ATSDR): *Toxicological Profile for Methoxychlor.* US Department of Health and Human Services, Public Health Service, pp 107, 1992
3. Tegeris AS, Earl FL, Smalley HE Jr, Curtis JM: Methoxychlor toxicity. *Arch Environ Health* 13:776–787, 1966
4. Negherbon WO: *Handbook of Toxicology*, Vol 3, pp 467–469. Philadelphia, WB Saunders Co, 1957
5. Tullner WW: Uterotrophic action of the insecticide methoxychlor. *Science* 133:647–648, 1961
6. Harris SJ, Cecil HC, Bitman J, et al: Effect of several dietary levels of technical methoxychlor on reproduction in rats. *J Agric Food Chem* 22:969–973, 1974
7. Gray LE Jr, Ostby J, Ferrell J, et al: A dose-response analysis of methoxychlor-induced alterations of reproductive development and function in the rat. *Fund Appl Toxicol* 12:92–108, 1989
8. Khera KS, Whalen C, Trivett G: Teratogenicity studies on linuron, malathion, and methoxychlor in rats. *Toxicol Appl Pharmacol* 45:435–444, 1978
9. Swartz WJ, Corkern M: Effects of methoxychlor treatment on pregnant mice on female offspring of the treated and subsequent pregnancies. *Reprod Toxicol* 6:431–437, 1992
10. National Cancer Institute: *Bioassay of Methoxychlor for Possible Carcinogenicity*, TRS-35. DHEW (NIH) Pub No 78-835, p 91. Washington, DC, 1978
11. *IARC Monographs on the Evaluation of the Carcinogenic Risks of Chemicals to Humans.* Vol 20, Some halogenated hydrocarbons, pp 259–281. Lyon, International Agency for Research on Cancer, 1979

2-METHOXYETHANOL
CAS: 109-86-4

CH₃OCH₂CH₂OH

Synonyms: Ethylene glycol monomethyl ether; EGME; Methyl Cellosolve; Dowanol EM; 2ME

Physical Form. Liquid

Uses. Solvent; jet fuel anti-icing additive; in the semiconductor industry in manufacture of printed circuit boards

Exposure. Inhalation; skin absorption

Toxicology. 2-Methoxyethanol (2ME) affects the central nervous system and depresses the hematopoietic system; in animals, it causes adverse reproductive effects, including teratogenesis, testicular atrophy, and infertility.

Cases of toxic encephalopathy and macrocytic anemia have been reported from industrial exposures that may have been as low as 60 ppm.[1] Symptoms were headache, drowsiness, lethargy, and weakness. Manifestations of central nervous system instability included ataxia, dysarthria, tremor, and somnolence. These effects were usually reversible. In acute exposures, the central nervous system effects were the more pronounced, whereas prolonged exposure to lower concentrations primarily produced evidence of depression of erythrocyte formation. When exposure was reduced to 20 ppm, no further cases occurred.

Two workers exposed primarily through skin contact showed signs of encephalopathy; one had bone marrow depression; the other had pancytopenia.[2]

The LC$_{50}$ for a 7-hour exposure of rats was 1480 ppm; death was due to lung and kidney injury.[3] Rabbits exposed to 800 ppm and 1600 ppm for 4 to 10 days showed irritation of the upper respiratory tract and lungs, severe glomerulonephritis, hematuria, and albuminuria.[3] Oral doses of 100 mg/kg/day for 4 days produced hemorrhagic bone marrow, thymic atro-

phy, lymphocytopenia, and neutropenia in rats.[4] Instilled in rabbit eyes, 2-ME caused immediate pain, conjunctival irritation , and slight corneal cloudiness, which cleared in 24 hours.[3]

Adverse reproductive effects have been reported in a number of species.[5] Testicular atrophy was observed in rats and mice exposed at 1000 ppm for 9 days and in rabbits exposed for 13 weeks at 300 ppm.[5,6] Slight-to-severe microscopic testicular changes occurred at 30 to 100 ppm in rabbits. At 500 ppm for 5 days, there was temporary infertility in male rats and abnormal spermhead morphology in mice.[5]

Exposure of pregnant rabbits to 50 ppm, 6 hr/day, on gestational days 6 through 18, induced significant increases in the incidence of malformations, especially of the skeletal and cardiovascular systems, and in the number of resorptions.[7] At this exposure level, decreases in maternal body weight gain, as well as decreased fetal weight, occurred.[7] Only slight fetotoxicity was observed in mice and rats similarly exposed. In another study, fetal cardiovascular and skeletal defects occurred in rats exposed at the 50 ppm level on days 7 to 15 of gestation.[5] By gavage, 250 mg/kg on days 7 to 14 caused increased embryonic deaths and gross fetal defects in mice.[8] Further studies on developmental phase-specific effects in mice showed exencephaly to be related to exposure between gestation days 7 to 10, whereas paw anomalies were maximal after administration on gestational day 11.[9] In rabbits, the most sensitive species tested to date, the minimally toxic fetal dose was 10 ppm, and the no-observed-effect level was 3 ppm.[5] Recent studies with pregnant cynomolgus monkeys showed that 12 mg/kg given by daily gavage throughout organogenesis (days 20 to 45) induced embryonic death; at 36 mg/kg, all 8 pregnancies ended in death of the embryo, and one of the dead embryos was missing a digit on each forelimb.[10] A single dermal dose of 500 mg/kg or greater, administered to pregnant rats, also produced significant increases in external, visceral, and skeletal malformations in the offspring.[11]

In one epidemiologic study, a small group of occupationally exposed workers showed no clinically significant differences in fertility or hematologic indices.[12] A survey of 73 painters who worked in a large shipyard found an increased prevalence of oligospermia and azoospermia, and an increased odds ratio for a lower sperm count per ejaculate. The authors attributed these effects to exposure to 2-ME and 2-ethoxyethanol, although it should be noted that shipyard painters may be exposed to a variety of other agents, including cadmium, zinc, iron, and lead, that may affect sperm quality.

Recent studies have focused on the immunotoxic effects of glycol ethers; some investigators have suggested that the immune system may be more sensitive than the reproductive system to the toxic effects of 2-ME.[13] Rats receiving 50 to 20 mg/kg/day for 10 days had decreases in thymus weights in the absence of decreased body weights, and lymphoproliferative responses to concanavallin A and phytohemagglutinin were also reduced. In another report, dose-related increases in natural killer cell cytotoxic activities and decreases in specific antibody production were observed following 2-ME exposure in the drinking water.

NIOSH recommended exposure limits of 0.1 ppm as a time-weighted average for up to 10-hour days during a 40-hour workweek; it is also recommended that dermal contact be prohibited.[15]

The 1995 ACGIH threshold limit value–time-weighted average (TLV-TWA) for 2-methoxyethanol is 5 ppm (16 mg/m³) with a notation for skin absorption.

REFERENCES

1. Zavon MR: Methyl cellosolve intoxication. *Am Ind Hyg Assoc J* 24:36–41, 1963
2. Ohi G, Wegman DH: Transcutaneous ethylene glycol monomethyl ether poisoning in the work setting. *J Occup Med* 20:675–676, 1978
3. Rowe VK, Wolf MA: Derivatives of glycols. In Clayton GD, Clayton FE (eds): *Patty's Industrial Hygiene and Toxicology*, 3rd ed, rev, Vol 2C, *Toxicology*, pp 3911–3919. New York, Wiley-Interscience, 1982
4. Grant D, Sulsh S, Jones HB, et al: Acute toxicity and recovery in the hemopoietic system of rats after treatment with ethylene glycol monomethyl and monobutyl ethers. *Toxicol Appl Pharmacol* 77:187–200, 1985
5. National Institute for Occupational Safety and Health: *NIOSH Current Intelligence Bulletin 39, Glycol Ethers*. DHHS (NIOSH) Pub No 83-112, p 22. Washington, DC, US Government Printing Office, May 2, 1982
6. Miller RR et al: Comparative short-term inhalation toxicity of ethylene glycol monomethyl ether and propylene glycol monomethyl ether in rats and mice. *Toxicol Appl Pharmacol* 61:368–377, 1981
7. Hanley TR Jr et al: Comparison of the teratogenic potential of inhaled ethylene glycol monomethyl ether in rats, mice and rabbits. *Toxicol Appl Pharmacol* 75:409–422, 1984
8. Nagano K et al: Embryotoxic effects of ethylene glycol monomethyl ether in mice. *Toxicology* 20:335–343, 1981
9. Horton VL, Sleet RB, John-Greene JA, et al: Developmental phase-specific and dose-related teratogenic effects of ethylene glycol monomethyl ether in CD-1 mice. *Toxicol Appl Pharmacol* 80:108–118, 1985
10. Scott WJ, Fradkin R, Wittfoht W, et al: Teratologic potential of 2-methoxyethanol and transplacental distribution of its metabolite, 2-methoxyacetic acid, in non-human primates. *Teratology* 39:363–373, 1989
11. Feuston MH, Kerstetter SL, Wilson PD: Teratogenicity of 2-methoxyethanol applied as a single dermal dose to rats. *Fund Appl Toxicol* 15:448–456, 1990
12. Cook RR et al: A cross-sectional study of ethylene glycol monomethyl ether process employees. *Arch Environ Health* 37:346–351, 1982
13. Smialowicz RJ, Riddle MM, Luebke RW, et al: Immunotoxicity of 2-methoxyethanol following oral administration in Fischer 344 rats. *Toxicol Appl Pharmacol* 109:494–506, 1991
14. Exon JH, Mather GG, Bussiere JL, et al: Effects of subchronic exposure of rats to 2-methoxyethanol or 2-butoxyethanol: thymic atrophy and immunotoxicity. *Fund Appl Toxicol* 16:830–840, 1991
15. National Institute for Occupational Safety and Health: *Criteria for a Recommended Standard . . . Occupational Exposure to Ethylene Glycol Monomethyl Ether, Ethylene Glycol Monoethyl Ether, and Their Acetates*. US Department of Health and Human Services, Public Health Service, Pub No 91-119, Cincinnati, OH, 1991

2-METHOXYETHYL ACETATE
CAS: 110-49-6

$CH_3COOCH_2CH_2OCH_3$

Synonyms: Ethylene glycol monomethyl ether acetate; EGMEA; 2-MEA; methyl cellosolve acetate; methyl glycol acetate

Physical Form. Colorless liquid

Uses. In the lacquer industry; in textile printing; in the manufacture of photographic film, coatings, and adhesives

Exposure. Inhalation; skin absorption

Toxicology. 2-Methoxyethyl acetate affects the central nervous system, the hematopoietic system, and the reproductive system in animals. Similar effects are to be expected in humans.

Mice and rabbits tolerated 1-hour exposure to 4500 ppm with only irritation of mucous membranes; guinea pigs survived the 1-hour exposure but succumbed days later.[1] Repeated exposure to 500 ppm for 8 hr/day caused narcosis and death in cats, and 1000 ppm for 8 hr/day was lethal to rabbits; all animals showed kidney injury.[1] Anemia was observed in cats repeatedly exposed to 200 ppm for 4 to 6 hours.

Dose-related increases in testicular atrophy and leukopenia have been reported in mice following administration of 63 to 2000 mg/kg, 5 days/week, for 5 weeks.[2]

2-Methoxyethyl acetate is hydrolyzed in vivo to form 2-methoxyethanol, which is subsequently metabolized to 2-methoxyacetic acid, a purported teratogenic substance.[3] Consequently, the acetate is expected to show similar profiles of developmental and reproductive toxicity as 2-methoxyethanol (qv). In a recent case report, a woman who was extensively exposed to 2-methoxyethyl acetate, both dermally and probably by inhalation during pregnancy, gave birth to two sons with hypospadias.[4] Since family history and medical examination showed no overt risks other than the significant exposure of the mother and, since 2-methoxyethyl acetate can cause teratogenic effects in animals, the malformations were attributed to the exposure.

The liquid is mildly irritating to the eyes of rabbits but not to the skin; prolonged contact can result in significant absorption.[1]

The 1995 ACGIH threshold limit value–time-weighted average for 2-methoxyethyl acetate is 5 ppm (24 mg/m³) with a notation for skin absorption. NIOSH has recommended exposure limits of 0.1 ppm as a time-weighted average for a 10-hour day during a 40-hour workweek and also recommended that dermal contact be prohibited.[5]

REFERENCES

1. Rowe VK, Wolf MA: Derivatives of glycols. In Clayton GD, Clayton FE (eds): *Patty's Industrial Hygiene and Toxicology*, 3rd ed, Vol 2C, *Toxicology*, pp 4022–4024. New York, Wiley-Interscience, 1982
2. Nagano K, Nakayama E, Oobayashi H, et al: Experimental studies on toxicity of ethylene glycol alkyl ethers in Japan. *Environ Health Perspect* 57:75–84, 1984
3. Hardin BD: Reproductive toxicity of the glycol ethers. *Toxicology* 27:91–102, 1983
4. Bolt HM: Maternal exposure to ethylene glycol monomethyl ether acetate and hypospadia in offspring: a case report. *Br J Ind Med* 47:352–353, 1990
5. Wess JA: Reproductive toxicity of ethylene glycol monomethyl ether, ethylene glycol monoethyl ether and their acetates. *Scand J Work Environ Health* 18 (Suppl 2):43–45, 1992

4-METHOXYPHENOL
CAS: 150-76-5

$C_7H_8O_2$

Synonyms: Hydroquinone monomethyl ether; 4-hydroxyanisole

Physical Form. Solid

Uses. Inhibitor for acrylic monomers; stabilizer for chlorinated hydrocarbons and ethyl cellulose; UV inhibitor

Exposure. Inhalation

Toxicology. 4-Methoxyphenol is expected to cause liver and renal toxicity with narcosis but only at high levels of exposure.

4-Methoxyphenol is moderately potent acutely as evidenced by an oral LD_{50} of 1600 mg/kg for the rat.[1] The gross signs of acute intoxication included paralysis and anoxia at lower doses and narcosis at higher doses.

During the industrial handling of 4-methoxyphenol, 2 of 8 process workers developed skin depigmentation.[2]

In medicine, 4-methoxyphenol is known as 4-hydroxyanisole. It is a depigmenting agent that has been shown to have activity against malignant melanoma when given intra-arterially in humans.[3] An intravenous-dose escalation study was carried out with the aim of obtaining maximum plasma concentrations in a 5-day schedule. Eight patients entered this study, which was stopped because of drug toxicity after three patients had been treated at the third-dose escalation of 15 g/m². Two patients had WHO grade 4 liver toxicity;[3] one also had grade 4 renal toxicity, and another had grade 4 hemoglobin toxicity. Extrapolated plateau plasma levels between 112 and 860 µM/l were obtained, which in vitro studies suggested would be cytotoxic.

The 1995 threshold limit value–time-weighted average (TLV-TWA) for methoxyphenol is 5 mg/m³.

REFERENCES

1. Hodge HC, Sterner JH, Maynard EA, Thomas J: Short-term toxicity tests on the mono and dimethyl ethers of hydroquinone. *J Ind Hyg Toxicol* 31:79–92, 1949
2. Chivers CP: Two cases of occupational leucoderma following contact with hydroquinone monomethyl ether. *Br J Ind Med* 29:105–107, 1972
3. Rustin GJ, Stratford MR, Lamont A, et al: Phase I study of intravenous 4-hydroxyanisole. *Eur J Cancer* 28:1362–1364, 1992

METHYL ACETATE
CAS: 79-20-9

$CH_3C(O)OCH_3$

Synonyms: Acetic acid, methyl ester

Physical Form. Colorless, highly volatile liquid

Uses. Solvent for lacquers, oils, and resins

Exposure. Inhalation

Toxicology. Methyl acetate is irritating to the eyes and mucous membranes; at high concentrations, it causes narcosis in animals, and it is expected that severe exposure will produce the same effect in humans.

Human exposure to 10,000 ppm for a short period of time resulted in eye, nose, and throat irritation, which persisted after cessation of exposure.[1] In a man exposed to unmeasured concentrations, effects were general CNS depression, headaches, and dizziness, followed by blindness of both eyes caused by atrophy of the optic nerve.[2] The toxic action on the optic nerve is possibly related to the presence of methanol after hydrolysis of methyl acetate.[3]

Cats exposed to 5000 ppm showed eye irritation and salivation; at 18,500, there were dyspnea, convulsions, and narcosis; 54,000 ppm was lethal within minutes.[1] Repeated exposure at 6600 ppm resulted in weight loss and weakness.

Prolonged contact with the liquid may cause dryness, cracking, and irritation of the skin.

The 1995 TLV-TWA is 200 ppm (606 mg/m³) with a STEL/ceiling of 250 ppm (757 mg/m³).

REFERENCES

1. Sandmeyer EE, Kirwin CJ: Esters. In Clayton GD, Clayton FE (eds): *Patty's Industrial Hygiene and Toxicology*, 3rd ed, p 2272. New York, Wiley-Interscience, 1981
2. Lund A: Toxic amblyopia after inhalation of methyl acetate. *J Ind Hyg Toxicol* 28:35 (Abstract), 1946
3. Hygienic Guide Series: Methyl acetate. *Am Ind Hyg Assoc J* 25:317–319, 1964

METHYL ACETYLENE
CAS: 74-99-7

CH₃CCH

CH_3CCH

Synonyms: Allylene; propyne; propine

Physical Form. Colorless gas

Uses. Propellant; in welding

Exposure. Inhalation

Toxicology. At high concentrations, methyl acetylene causes narcosis in animals, and it is expected that severe exposure will produce the same effect in humans.

Rats exposed to 42,000 ppm became hyperactive within the first 7 minutes and, at the end of 7 minutes, they appeared lethargic and ataxic.[1] After 95 minutes, the animals were completely anesthetized. There was no mortality when the exposure was terminated at the end of 5 hours, and most of the animals recovered completely within 40 minutes. Edema and alveolar hemorrhage were present in animals killed at termination of the single exposure, whereas bronchiolitis and pneumonitis were observed in rats sacrificed 9 days post-exposure.

Two dogs and 20 rats were exposed to 28,700 ppm, 6 hr/day, 5 days/week for 6 months; after 7 minutes of exposure, ataxia was noted in the rats and, after 13 minutes, ataxia and mydriasis were observed in the dogs.

Within 15 minutes, the dogs also exhibited staggering, marked salivation, and muscular fasciculations. There was a 40% mortality rate among exposed rats versus a 10% mortality rate in the control animals.

Methyl acetylene has a "sweet" odor similar to that of acetylene.

The 1995 ACGIH threshold limit value–time-weighted average (TLV-TWA) is 1000 ppm (1640 mg/m³).

REFERENCES

1. Horn HJ, Weir RJ Jr, Reese WH: Inhalation toxicology of methylacetylene. *AMA Arch Ind Health* 15:20–25, 1957

METHYL ACRYLATE
CAS: 96-33-3

CH₂CHCOOCH₃

CH_2CHCOOCH_3

Synonyms: 2-Propenoic acid methyl ester; acrylic acid methyl ester; methyl propenoate

Physical Form. Colorless liquid

Uses. As a monomer, polymer, and copolymer in the manufacture of acrylic fibers

Exposure. Inhalation; skin absorption

Toxicology. Methyl acrylate is a lacrimating agent and an irritant of the mucous membranes.

The lowest dose reported to have any irritant effect in humans is 75 ppm.[1]

The liquid is readily absorbed by mucous membranes and through the skin. The dermal LD_{50} in rabbits was 1.3 g/kg. It was moderately to severely irritating to the rabbit skin. The liquid tested in the eye caused mild, reversible injury.

In rats, the LD_{50} for 4 hours was 1350 ppm.[2] Behavior of the animals suggested irritation of the eyes, nose, and respiratory tract, with labored breathing. At necropsy, there were no discernible gross abnormalities of the major organs. In the same study, rats were exposed to methyl acrylate at 110 ppm, 4 hr/day, 5 days/week, for 32 days. There were no overt signs of central nervous system or respiratory effects although the animals huddled with their eyes closed, possibly indicating some eye discomfort.

No exposure-related clinical signs or lesions of systemic toxicity and no oncogenic responses were observed in rats exposed by inhalation at concentrations of 0, 15, 45, or 135 ppm, 6 hr/day, 5 days/week, for 24 consecutive months.[3] Dose-related changes occurred in the anterior portion of the olfactory epithelium and consisted of atrophy of the neurogenic epithelial cells, followed by progressive hyperplasia of the reserve cells and, ultimately, loss of the upper epithelial cell layer. Opacity and neovascularization of the cornea were also observed in methyl acrylate–exposed animals.

The IARC has determined that there is inadequate evidence for the carcinogenicity of methyl acrylate to experimental animals and that no data are available to assess the risk to humans.[4]

The proposed 1995 ACGIH threshold limit value–time-weighted average (TLV-TWA) is 2 ppm (7 mg/m³) with a notation for skin absorption and an A4 designation—not classifiable as a human carcinogen: there are inadequate data on which to classify the agent in terms of its carcinogenicity in humans and/or animals.

Methyl acrylate was not found to be mutagenic in the *Salmonella*/microsome assay, but it increased chromosomal aberrations in vitro and tested positive in the micronucleus assay in mice.[3]

REFERENCES

1. Sandmeyer EE, Kirwin CJ: Esters. In Clayton GD, Clayton FE (eds): *Patty's Industrial Hygiene and Toxicology*, 3rd ed, Vol 2B, *Toxicology*, pp 2293–2296. New York, Wiley-Interscience, 1981
2. Oberly R, Tansy MF: LC₅₀ values for rats acutely exposed to vapors of acrylic and metha-crylic acid esters. *J Toxicol Environ Health* 16:811, 1985.
3. Reininghaus W, Koestner A, Klimisch HJ: Chronic toxicity and oncogenicity of inhaled methyl acrylate and *n*-butyl acrylate in Sprague–Dawley Rats. *Food Chem Toxicol* 29:329–339, 1991
4. *IARC Monographs on the Evaluation of the Carcinogenic Risk of Chemicals to Humans*, Vol 39, Some chemicals used in plastics and elastomers, pp 99–112. Lyon, International Agency for Research on Cancer, 1986

METHYLACRYLONITRILE
CAS: 126-98-7

$CH_2C(CH_3)CN$

Synonyms: 2-Methyl-2-propenenitrile; 2-cyanopropene-1; isopropene cyanide; isopropenylnitrile; methacrylonitrile

Physical Form. Colorless liquid

Uses. Widely used monomer in the production of plastic elastomers and coatings

Exposure. Inhalation; skin absorption

Toxicology. Methylacrylonitrile is a potent neurotoxin.

The approximate LC_{50} for mice exposed to airborne concentrations for 1 hour was 630 ppm and, for a 4-hour exposure, it was 400 ppm.[1] Exposure to 75 ppm for 8 hours caused no deaths, but respiratory difficulties and convulsions were observed. In a study with rats exposed at concentrations of 3180 to 5700 ppm, the clinical symptoms, rapid unconsciousness with convulsions and lethality, suggested that the acute toxicity is caused predominantly by metabolically formed cyanide.[2] Cyanide reacts readily with cytochrome oxidase in mitochondria and inhibits cellular respiration. Cyanide antidotes were also effective against methylacrylonitrile toxicity. Metabolic studies have suggested that methylacrylonitrile may also exert toxic effects by interacting directly with the cytoplasmic (hemoglobin) and membrane proteins of red blood cells.[3]

The oral LD_{50} in mice was 20 to 25 mg/kg and, in rats, 25 to 50 mg/kg.[4] Symptoms included weakness, tremors, cyanosis, and convulsions. When beagle dogs were exposed 7 hr/day, 5 days/week, to 13.5 ppm over a period of 90 days, 2 of 3 animals exhibited convulsions and loss of motor control in the hind limbs about halfway through the exposure period.[5] No effects occurred at 3.2 ppm.

The liquid was rapidly absorbed through the skin of a rabbit, and caused death after 3 hours at a dose of 2.0 ml/kg.[1] Skin irritation at the site of application was negligible. One drop in the eye of a rabbit caused transient irritation.

The 1995 ACGIH threshold limit value–time-weighted average (TLV-TWA) for methylacrylonitrile is 1 ppm (2.7 mg/m^3) with a notation for skin absorption.

Treatment. Treatment must be rapid to be effective.[6] Administration of nitrite, which oxidizes hemoglobin to methemoglobin which, in turn, competes with cytochrome oxidase for the cyanide ion, can be effective. Amyl nitrite may be administered by inhalation, while sodium nitrite is prepared for intravenous administration (10 ml of a 3% solution). Alternatively, 4-dimethylaminophenol, which also oxidizes hemoglobin to methemoglobin, can be used in a dose of 3 mg/kg intravenously or intramuscularly to accelerate detoxification; following nitrite therapy, thiosulfate is administered intravenously (50 ml of a 25% aqueous solution), and the thiocyanate formed is readily excreted in the urine.

REFERENCES

1. McOmie WA: Comparative toxicities of methacrylonitrile and acrylonitrile. *J Ind Hyg Toxicol* 31:113, 1949
2. Peter H, Bolt HM: Effect on antidotes on the acute toxicity of methacrylonitrile. *Int Arch Occup Environ Health* 55:175–177, 1985
3. Cavazos R Jr, Farooqui MYH, Day WW, et al: Disposition of methacrylonitrile in rats and distribution in blood components. *J Appl Toxicol* 9:53–57, 1989
4. Hartung R: Cyanides and Nitriles. In Clayton GD, Clayton FE (eds): *Patty's Industrial Hygiene and Toxicology*, 3rd ed, rev, Vol 2C, *Toxicology*, pp 4867–4868. New York, Wiley-Interscience, 1972
5. Pozzani UC, Kinhead ER, King JJ: The mammalian toxicity of methacrylonitrile. *Am Ind Hyg Assoc J* 29:202, 1968
6. Klaasen CD: Nonmetallic environmental toxicants: air pollutants, solvents and vapors, and pesticides. In Gilman AG et al (eds): *Goodman and Gilman's The Pharmacological Basis of Therapeutics*, 7th ed, pp 1642–1643. New York, Macmillan, 1985

METHYLAL
CAS: 109-87-5

$CH_2(OCH_3)_2$

Synonyms: Dimethoxymethane; formal; methylene dimethyl ether

Physical Form. Colorless liquid

Uses. Solvent; fuel; perfume

Exposure. Inhalation

Toxicology. Methylal is an irritant of the eyes and mucous membranes and, at high concentrations, it causes central nervous system depression.

In humans, methylal has been used as an anesthetic in a number of surgical operations; however, anesthesia was produced more slowly than with ether, and the effect of methylal was more transitory.[1]

In guinea pigs exposed to a concentration near 154,000 ppm, effects included vomiting, lacrimation, sneezing, cough, and nasal discharge; coma occurred in 20 minutes and death in 2.5 hours.[1]

The LC_{50} for a 7-hour exposure of mice

was 18,354 ppm.[1] Exposure 7 hr/day to 11,300 ppm for 1 week caused mild eye and nose irritation, incoordination, and, after 4 hours light narcosis; the exposure was fatal to 6 of 50 mice. Animals exposed to toxic concentrations often developed marked fatty changes in the liver, kidney, and heart and inflammatory changes in the lungs.[1] Rats were unaffected by eight 6-hour exposures to 4000 ppm.[2]

Methylal can cause superficial irritation of the eyes.[1] Frequent or prolonged skin contact with the liquid may cause dermatitis due to a defatting action.

The liquid has a chloroformlike odor and pungent taste.

The 1995 ACGIH threshold limit value–time-weighted average (TLV-TWA) is 1000 ppm (3110 mg/m^3).

REFERENCES

1. Weaver FL Jr, Hough AR, Highman B, Fairhall LT: The toxicity of methylal. Br J Ind Med 8:279–283, 1951
2. Gage JC: The subacute inhalation toxicity of 109 industrial chemicals. Br J Ind Med 27:1–18, 1970

METHYL ALCOHOL
CAS 67-56-1

CH$_3$OH

Synonyms: Methanol; wood spirits; carbinol; wood alcohol; wood naphtha; methylol; Columbian spirits; colonial spirits

Physical Form. Colorless liquid

Uses. In production of formaldehyde; in paints, varnishes, cements, inks, and dyes

Exposure. Inhalation; skin absorption

Toxicology. Methyl alcohol causes optic neuropathy, metabolic acidosis, and respiratory depression.

Although methyl alcohol poisoning has occurred primarily from the ingestion of adulterated alcoholic beverages, symptoms also can occur from inhalation or absorption through the skin.[1,2] Impairment of vision and death from absorption by the latter routes were reported in the early literature.[2] Typically, within 18 to 48 hours after ingestion, patients develop nausea, abdominal pain, headache, and abnormally slow, deep breathing. This is accompanied by visual symptoms ranging from blurred or double vision and changes in color perception to constricted visual fields and complete blindness.[1,3] The most severely poisoned patients become comatose and may die; those who recover from coma may be found blind.[3] One of the most striking features of methyl alcohol poisoning is acidosis; the degree of acidosis has been found to parallel closely the severity of poisoning.[1] Accumulated evidence suggests that chronic exposure to 1200 to 8300 ppm can lead to impaired vision.[1] Exposure to vapor concentrations ranging from 365 to 3080 ppm may result in blurred vision, headache, dizziness, and nausea.[4]

In the eyes, the liquid has caused superficial lesions of the cornea that were of a nonserious nature.[1] Prolonged or repeated skin contact will cause dermatitis, erythema, and scaling.[1]

The presence of an asymptomatic latent period following ingestion suggests that methyl alcohol must be metabolized before toxicity is fully manifest.[1] This concept also explains the discrepancy between plasma concentrations of methyl alcohol and clinical signs of toxicity.[5] Furthermore, methyl alcohol poisoning is ameliorated by ethanol, a substance with greater affinity than methyl alcohol for alcohol dehydrogenase, which is responsible for the initial step in metabolism.[5] The metabolite formate appears to be the mediator of ocular injury and acidosis.[6] The individual variations in activity of the alcohol dehydrogenase systems, which are responsible for the oxidative metabolism of methyl (and ethyl) alcohol, may well account for the wide variation in the individual responses observed with methyl alcohol poisoning.[1]

In developmental animal studies, methyl

alcohol produced malformations in mice and rats following inhalation of 15,000 or 20,000 ppm, respectively, for 6 to 7 hr/day during gestation; slight maternal toxicity was also observed.[7] In very limited chronic studies, methyl alcohol was found not to be carcinogenic. It was not genotoxic in a variety of in vivo and in vitro assays.[7]

The 1995 ACGIH threshold limit value–time-weighted average (TLV-TWA) for methyl alcohol is 200 ppm (262 mg/m³) with a short-term excursion limit of 250 ppm (328 mg/m³) and a notation for skin absorption.

Diagnosis. *Signs and Symptoms:* Headache; blurred vision, constricted visual fields; shortness of breath; dizziness, vertigo.

Differential Diagnosis: The combination of visual disturbances and metabolic acidosis, together with a history of exposure and the presence of formic acid in the urine, is confirmation of methyl alcohol intoxication.

Special Tests: Formic acid in the urine may be assayed, and blood pH and plasma bicarbonate may be measured.

Treatment. Treatment includes: (1) emesis or gastric lavage, rehydration, correction of acidosis, and folate to enhance formate oxidation; (2) intravenous ethanol when plasma methanol concentrations are higher than 20 mg/dl, when ingested doses are greater than 30 ml, or when there is evidence of acidosis or visual abnormalities; (3) hemodialysis when plasma methanol concentrations are greater than 40 mg/dl or when metabolic acidosis is unresponsive to bicarbonate given intravenously.[5] Serum formate concentration may be a more direct indicator of toxicity; levels exceeding 20 mg/dl may be expected to produce acidosis or ocular injury.[6]

REFERENCES

1. National Institute for Occupational Safety and Health, US Department of Health, Education and Welfare: *Criteria for a Recommended Standard . . . Occupational Exposure to Methyl Alcohol.* DHEW (NIOSH) 76-148, pp 68–75. Washington, DC, US Government Printing Office, 1976

2. Henson EV: The toxicology of some aliphatic alcohols—Part II. *J Occup Med* 1:497–502, 1960

3. Grant WM: *Toxicology of the Eye*, 3rd ed, pp 591–596. Springfield, IL, Charles C Thomas, 1986

4. Frederick LJ et al: Investigation and control of occupational hazards associated with the use of spirit duplicators. *Am Ind Hyg Assoc J* 45:51–55, 1984

5. Ekins BR et al: Standardized treatment of severe methanol poisoning with ethanol and hemodialysis. *West J Med* 142:337–340, 1985

6. Osterloh JD et al: Serum formate concentrations in methanol intoxication as a criterion for hemodialysis. *Ann Intern Med* 104:200–203, 1986

7. Lington AW, Bevan C: Alcohols. In Clayton GD, Clayton FE (eds): *Patty's Industrial Hygiene and Toxicology*, 4th ed, rev, Vol II. Part D, *Toxicology*, pp 2600–2609. New York, Wiley-Interscience, 1994

METHYLAMINE
CAS: 74-89-5

CH_3NH_2

Synonyms: Monomethylamine; aminomethane

Physical Form. Gas

Uses. In the tanning and dyeing industries; fuel additive; chemical intermediate in the production of pharmaceuticals, insecticides, and surfactants

Exposure. Inhalation

Toxicology. Methylamine is a severe irritant of the eyes and skin; in animals, repeated inhalation causes upper respiratory tract irritation.

In humans, brief exposure at 20 to 100 ppm is said to produce transient irritation of the eyes, nose, and throat.[1] No symptoms of irritation are produced from longer exposures at less than 10 ppm. On the basis of the irritant properties of methylamine, it is possible that severe exposure may cause pulmonary edema.

In rats, when administered orally as the base in a 40% aqueous solution, the LD_{50} was 0.1 to 0.2 g/kg.[1] Repeated exposures of rats at 750 ppm, 6 hr/day, 5 days/week, for 2 weeks caused severe body weight loss, liver damage, hematopoietic abnormalities, and some deaths.[2] Histopathological effects, which were also observed with similar dosing at 250 ppm, included necrosis and ulceration of the respiratory mucosa of the nasal turbinates and atrophy with regeneration of the olfactory mucosa. Repeated exposures at 75 ppm produced marginal changes in the olfactory mucosa.

In the eyes of rabbits, 1 drop of a 5% aqueous solution caused conjunctival hemorrhage, superficial corneal opacities, and edema.[3] On the skin of animals, a 40% solution caused necrosis.[1]

The ammonialike odor is detectable at less than 5 ppm.

The 1995 ACGIH threshold limit value–time-weighted average (TLV-TWA) is 5 ppm (6.4 mg/m³) with a STEL/ceiling of 15 ppm (19 mg/m³).

REFERENCES

1. Beard RR, Noe JT: Aliphatic and alicyclic amines. In Clayton GD, Clayton FE (eds): *Patty's Industrial Hygiene and Toxicology*, 3rd ed, rev, Vol 2B, *Toxicology*, pp 3135–3173. New York, Wiley-Interscience, 1981
2. Kinney LA, Valentine R, Chen HC, et al: Inhalation toxicology of methylamine. *Inhal Toxicol* 2:29–35, 1990
3. Grant WM: *Toxicology of the Eye*, 3rd ed, pp 606–607. Springfield, IL, Charles C Thomas, 1986

METHYL *n*-AMYL KETONE
CAS: 110-43-0

$CH_3COC_5H_{11}$

Synonym: 2-Heptanone

Physical Form.   Liquid

Uses.   Organic solvent

Exposure.   Inhalation

Toxicology.   Methyl *n*-amyl ketone is irritating to the eyes and mucous membranes; at high concentrations, it causes narcosis in animals, and it is expected that severe exposure will produce the same effect in humans.

There have been no reports of effects in humans, and the concentration at which irritation may be produced is not known.[1] However, both sensory and pulmonary irritation can be expected with sufficient exposure. Sensory irritation is characterized by immediate eye and nose irritation, which may increase to sensations of burning and pain and is due to interaction between the substance and receptors in the trigeminal nerve.[2] Vapors reaching the lower respiratory tract as well as the lungs may interact with the nerves in these regions, causing dyspnea and breathlessness, or pulmonary irritation.

In guinea pigs, exposure to 4800 ppm caused narcosis and death in 4 to 8 hours; 2000 ppm was strongly narcotic, and 1500 ppm was irritating to the mucous membranes.[3]

Rats and monkeys exposed to 1025 ppm methyl *n*-amyl ketone for 6 hr/day, 5 days/week, for 9 months showed no evidence of neuropathy or clinical signs of illness. Microscopic examination revealed no tissue damage.[4]

The liquid has a marked fruity odor and a pearlike flavor.[1]

The 1995 ACGIH threshold limit value–time-weighted average (TLV-TWA) is 50 ppm (233 mg/m³).

REFERENCES

1. National Institute for Occupational Safety and Health: *Criteria for a Recommended Standard . . . Occupational Exposure to Ketones.* DHEW (NIOSH) Pub No 78-172. Washington, DC, US Government Printing Office, 1978
2. Hansen LF, Nielsen GD: Sensory irritation and pulmonary irritation of *n*-methyl ketones: receptor activation mechanisms and relationships with threshold limit values. *Arch Toxicol* 68:193–202, 1994
3. Specht H, Miller JW, Valaer RJ, Sayers RR: Acute response of guinea pigs to the inhalation of ketone vapors. *National Institutes of Health Bulletin No 176*, 1940
4. Johnson BL et al: Neurobehavioral effects of methyl *n*-amyl butyl ketone and methyl *n*-amyl ketone in rats and monkeys: summary of NIOSH investigations. *J Environ Pathol Toxicol* 2:113–133, 1979

N-METHYL ANILINE
CAS: 100-61-8

$C_6H_5NHCH_3$

Synonyms: Monomethylaniline; *n*-methylaminobenzene

Physical Form. Colorless or slightly yellow liquid that becomes brown on exposure to air

Uses. Chemical syntheses

Exposure. Inhalation; skin absorption

Toxicology. N-methyl aniline causes anoxia in animals as a result of the formation of methemoglobin.

There are no reports of human intoxication from exposure to *n*-methyl aniline. Overexposure would be expected to produce the effects of methemoglobinemia, including cyanosis (especially in the lips, nose, and earlobes), weakness, dizziness, and severe headache.

Animal fatalities occurred from daily exposure to 7.6 ppm; signs of intoxication included prostration, labored breathing, and cyanosis. Methemoglobinemia developed promptly in rabbits and cats; the rabbits also exhibited mild anemia and bone marrow hyperplasia.[1] Animals that died had pulmonary involvement ranging from edema to interstitial pneumonia, as well as occasional centrilobular hepatic necrosis, and moderate kidney damage.

Applied to the skin of rabbits, 3 g/kg of body weight caused death.[2] The minimum lethal dose in rabbits was 280 mg/kg when administered orally; signs of intoxication included weight loss, dyspnea, prostration, cyanosis, and occasional terminal convulsions.[2]

N-methyl aniline (1.95 g/kg of food) given together with sodium nitrite (1.0 g/l of drinking water) to Swiss mice resulted in a 17% incidence of lung adenomas and a 14% incidence of malignant lymphomas; there were no carcinogenic effects in animals treated with *n*-methyl aniline alone, suggesting that in vivo nitrosation is necessary for forming carcinogenic nitrosamines.[3]

Note: For a description of diagnostic signs, differential diagnosis, and medical control, including clinical laboratory treatment of overexposure to methemoglobin-forming agents, see "Aniline."

The 1995 ACGIH threshold limit value–time-weighted average (TLV-TWA) is 0.5 ppm (2.2 mg/m³) with a notation for skin absorption.

REFERENCES

1. Treon JF et al: The toxic properties of xylidine and monomethylaniline. II. The comparative toxicity of xylidine ($C_6H_3[CH_3]2HH_2$) and monomethylaniline ($C_6H_5NH[CH_3]$) inhaled as vapor in air by animals. *AMA Arch Ind Hyg Occup Med* 1:506–524, 1950
2. Treon JF, Deichman WB, Sigmon HE, et al: The toxic properties of xylidine and monomethylaniline. I. The comparative toxicity of xylidine and monomethylaniline when administered orally or intravenously to animals or applied on their skin. *J Ind Hyg Toxicol* 31:1–20, 1949
3. Greenblatt M, Mirvish S, So BT: Nitrosamine studies: induction of lung adenomas by concur-

rent administration of sodium nitrite and secondary amines in Swiss mice. *J Natl Cancer Inst* 46:1029–1034, 1971

METHYL BROMIDE
CAS: 74-83-9

CH₃Br

Synonyms: Bromomethane; monobromomethane; isobrome

Physical Form. Colorless gas

Uses. Fumigant of soil and stored foods for the control of insects, fungi, and rodents; methylating agent; previously used as a refrigerant and fire extinguishing agent

Exposure. Inhalation; skin absorption

Toxicology. Methyl bromide is a neurotoxin and causes convulsions; very high concentrations cause pulmonary edema; chronic exposure causes peripheral neuropathy.

There are numerous reports of human intoxication from accidental exposure associated with its use in fire extinguishers and as a fumigant.[1] Estimates of concentrations that have caused human fatalities range from 8000 ppm for a few hours to 60,000 ppm for a brief exposure. The onset of toxic symptoms is usually delayed, and the latent period may be from 30 minutes to several hours. Early symptoms include headache, visual disturbances, nausea, vomiting, and malaise.[2] In some instances, there is eye irritation, vertigo, and intention tremor of the hands; the tremor may progress to twitchings and, finally, to convulsions of the Jacksonian type, which are first restricted to one extremity but gradually spread to the entire body.[1,3] Severe exposure may lead to pulmonary edema.[4] Tubular damage in the kidneys has been observed in fatal cases.[2] Some of those who have recovered from severe intoxication have had persistent central nervous system ef-

fects, including vertigo, depression, hallucinations, anxiety, and inability to concentrate.[2]

Eight of 14 workers repeatedly exposed to the vapor (concentration unmeasured) for 3 months developed peripheral neuropathy; all recovered within 6 months.[5]

It is unlikely that the bromide ion resulting from metabolic conversion of methyl bromide plays a significant role in the toxicity of methyl bromide.[6] Blood bromide levels following methyl bromide poisoning are much lower than those associated with intoxication by inorganic bromide salts. Concentrations of 100 mg/l have been associated with death following methyl bromide exposure, whereas blood bromide levels of 1000 mg/l or greater have been observed following therapeutic administration of inorganic bromides in the absence of signs of intoxication.[6] A recent report of six methyl bromide poisonings showed serum bromide concentrations at the time of hospital admission to be a poor predictor of survival.[7] One fatal case had an antemortem bromide level of 108 mg/l, whereas a survivor measured 321 mg/l. In another instance, 9 workers exposed to 200 ppm or more on 2 consecutive days had varying symptoms ranging from headache to severe reactive myoclonus and convulsions.[8] A direct association between serum bromide concentrations and neurological symptoms was absent.

Contact with the eye by the gas or liquid results in transient irritation and conjunctivitis.[9] Minor skin exposure to the liquid produced erythema and edema.[10] Prolonged or repeated contact resulted in deeper burns with delayed vesiculation.[10] It is doubtful that significant cutaneous absorption occurs. Although victims of skin exposure may show symptoms of neurotoxicity, inhalation is considered the likely cause.[10]

Toxicological studies in animals indicate a steep concentration-response curve for methyl bromide and clear species and sex differences in sensitivity.[11] Inhalation exposure of up to 120 ppm, 6 hr/day, for 13 weeks resulted in 17% mortality in male mice but no mortality in female mice or in rats of both sexes. No methyl-bromide–induced histological lesions were observed in either species, including mice killed in a moribund state. At 160 ppm for 6 hr/day, there was high mortality in rats and mice. Primary target organs were the brain, kidney, nasal cavity, heart, adrenal gland, liver, and testis. Nephrosis was probably a major cause of moribundity and death of mice, whereas neuronal necrosis may have been the principal lesion contributing to the early death of some rats. At 66 ppm, rats and guinea pigs showed no response for up to 6 months of exposure, but rabbits and monkeys developed paralysis within 3 months; the paralysis was particularly severe in rabbits, which also had pulmonary lesions. No toxic response was observed at 17 ppm.

Repeated exposure at 70 ppm of female rats before and during pregnancy did not cause maternal or embryo toxicity, but severe neurotoxicity and mortality were produced in the rabbit.[12] In a two-generation reproduction study of rats fed diets containing 500 ppm total bromine, food consumption was lower in the F_1 parental females, and F_2 pups had lowered body weights.[13] No other treatment-related changes were found for clinical signs, estrus cycle, sperm count and morphology, mating, fertility, gestation, litter size, pup viability, and gross or histopathological examination.

In a 90-day study, 50 mg/kg administered by gavage 5 days per week caused squamouscell carcinomas of the forestomach in 13 of 20 rats; a dose-related incidence of hyperplasia was observed at the 2 and 10 mg/kg level.[14] A second study using the same experimental design found that the early hyperplastic lesions of the forestomach regressed after discontinuation of treatment and should not be considered neoplasms.[15] Rats fed diets containing 500 ppm total bromine following fumigation with methyl bromide for 2 years showed no evidence of a carcinogenic response.[16] Inhalation of up to 90 ppm, 6/hr/day, 5 days/week, for 29 months caused degenerative and hyperplastic changes of the nasal olfactory epithelium, an increased incidence of lesions in the heart, hyperkeratosis in the esophagus and forestomach, but no increase in tumor incidence in rats.[17]

Methyl bromide is genotoxic in a number of in vivo and in vitro assays and does not require metabolic activation.[18] This is consistent with the fact that it is a direct-acting alkylating agent that can methylate DNA.

In certain circumstances, the concentration of bromide ion in the blood can be used as a biomarker of bromomethane exposure.[18] In general, unexposed individuals have levels less than 15 ppm, levels of 80 ppm may occur without any clinical signs, and levels of 150 to 400 ppm are observed in people with moderate to severe symptoms. The bromide ion is cleared from the blood with a half-life ranging from 3 to 15 days. Consequently, the correlation between severity of effects and bromide ion levels is most apparent within the first 2 days of exposure, and there may be little correlation later.

Methyl bromide itself has poor warning properties, but warning agents such as chloropicrin are frequently added.

The 1995 ACGIH threshold limit value–time-weighted average (TLV-TWA) for methyl bromide is 5 ppm (19 mg/m³) with a notation for skin absorption.

REFERENCES

1. von Oettingen WF: *The Halogenated Aliphatic, Olefinic, Cyclic, Aromatic, and Aliphatic-Aromatic Hydrocarbons Including the Halogenated Insecticides, Their Toxicity and Potential Dangers.* US Public Health Service Pub No 414, pp 15–30. Washington, DC, US Government Printing Office, 1955

2. Hine CH: Methyl bromide poisoning. *J Occup Med* 11:1–10, 1969

3. Greenberg JO: The neurological effects of methyl bromide poisoning. *Ind Med* 40:27–29, 1971

4. Rathus EM, Landy PJ: Methyl bromide poisoning. *Br J Ind Med* 18:53–57, 1961

5. Kantarjian AD, Shaheen AS: Methyl bromide poisoning with nervous system manifestations resembling polyneuropathy. *Neurology* 13:1054–1058, 1963

6. Hayes WJ Jr: *Pesticides Studied in Man*, pp 140–142. Baltimore, MD, Williams and Wilkins, 1982

7. Marraccini JV et al: Death and injury caused by methyl bromide, an insecticide fumigant. *J Forens Sci* 28:601–607, 1983

8. Hustinx WNM, van de Lar RTH, van Huffelen AC: Systemic effects of inhalational methyl bromide poisoning: a study of nine cases occupationally exposed due to inadvertent spread during fumigation. *Br J Ind Med* 50:155–159, 1993

9. Grant WM: *Toxicology of the Eye*, 2nd ed,

pp 680–685. Springfield, IL, Charles C Thomas, 1974

10. Jarowenko DG, Mancusi-Ungaro HR: The care of burns from methyl bromide (case report). *J Burn Care Rehab* 6:119–123, 1985

11. Eustis SL, Haber SB, Drew RT, et al: Toxicology and pathology of methyl bromide in F344 rats and B6C3F₁ mice following repeated inhalation exposure. *Fund Appl Toxicol* 11:594–610 1988

12. Sikov MR, Cannon WC, Carr DB: *Teratologic Assessment of Butylene Oxide, Styrene Oxide and Methyl Bromide*. US Department of Health and Human Services. DHHS (NIOSH) Pub No 81-124, July 1981

13. Kaneda M, Hatakenaka N, Teramoto S, et al: A two-generation reproduction study in rats with methyl bromide-fumigated diets. *Food Chem Toxicol* 31: 533–542, 1993

14. Danse LHJC, van Velsen FL, Van Der Heijden CA: Methylbromide: carcinogenic effects in the rat forestomach. *Toxicol Appl Pharmacol* 72:262–271, 1984

15. Boorman GA, Hong HL, Jameson CW, et al: Regression of methyl-bromide–induced forestomach lesions in the rat. *Toxicol Appl Pharmacol* 86:131–139, 1986

16. Mitsumori K, Maita K, Kosaka T, et al: Two-year oral chronic toxicity and carcinogenicity study in rats of diets fumigated with methyl bromide. *Food Chem Toxicol* 28:109–119, 1990

17. Reuzel PGJ, Dreef-van der Meulen HC, Hollanders VMH, et al: Chronic inhalation toxicity and carcinogenicity study of methyl bromide in Wistar rats. *Food Chem Toxic* 29:31–39, 1991

18. Agency for Toxic Substances and Disease Registry (ATSDR): *Toxicological Profile for Bromomethane*. US Department of Health and Human Services, Public Health Service, p 104, TP-91/06, 1992

METHYL BUTYL KETONE
CAS: 591-78-6

$C_6H_{12}O$

Synonyms: 2-Hexanone; *n*-butyl methyl ketone; MBK; MNBK; propylacetone

Physical Form. Colorless liquid

Uses. Industrial solvents for adhesives, lacquers, paint removers, and acrylic coatings

Exposure. Inhalation; skin absorption

Toxicology. At high concentrations, methyl butyl ketone (MBK) may produce ocular and respiratory irritation, followed by central nervous system depression and narcosis. Chronic inhalation causes peripheral neuropathy.

Human volunteers exposed to a vapor concentration of 1000 ppm for several minutes developed moderate eye and nasal irritation.[1]

Although MBK is considered to be only a mild sensory irritant with acute exposure, an outbreak of neuropathy among workers in a coated-fabrics plant in 1973 revealed the more serious consequences of chronic exposure.[2] Workers exposed for 6 to 12 months with extensive skin exposure to the mixed vapor of methyl butyl ketone (averaging 9.2 ppm in front of printing machines and 36 ppm behind) developed peripheral neuropathy.[3–6] The neurologic pattern was one of a distal motor and sensory disorder, with minimal loss of tendon reflexes.[5] In workers with prominent motor involvement, initial symptoms included slowly developing weakness of the hands, with difficulty in pincer movement in the grasping of heavy objects, or weakness of the ankle extensors, resulting in a slapping gait. In other cases, the initial symptoms were intermittent tingling and paresthesias in the hands or feet. Nerves affected could be sensory nerves or motor nerves, or both. Nerve biopsies usually showed enlarged axons, diminished numbers of myelinated nerve fibers, an increased neurofilament accumulation, and increased Wallerian degeneration. In some cases, the condition progressed slowly for several months after cessation of exposure; in moderate to severe cases, improvement occurred over a period of up to 8 months although workers did not always recover fully.[2–6] Body weight reduction was the only other toxicological effect noted.

The 4-hour LC_{50} for the rat was 8000 ppm.[2] In guinea pigs, exposure to 10,000 to 20,000 ppm was potentially lethal in 30 to 60 minutes; concentrations greater than 20,000 ppm killed the animals within a few minutes; at 6000 ppm, there were signs of narcosis after 30 minutes, deep anesthesia after 1 hour, and death after approximately 6.5 hours.[1] A maximum of 3000 ppm for 1 hour did not cause serious disturbances.

Animals continually exposed to concentrations of 100 to 600 ppm developed signs of peripheral neuropathy after 4 to 8 weeks; in cats, the conduction velocity of the ulnar nerve was less than one-half of normal after exposure for 7 to 9 weeks.[4] In these animals, histologic examination revealed focal denudation of myelin from nerve fibers with or without axonal swelling. In rats and monkeys, adverse effects on neurophysiologic indicators of nervous system integrity were found with 9-month exposures to 100 ppm, 6 hr/day, 5 days/week.[7] Methyl butyl ketone neuropathies, however, occurred after only 4 months' exposure at 1000 ppm. Four months of intermittent respiratory exposure of rats to 1300 ppm caused severe symmetric weakness in the hind limbs.[8]

Damage due to hexacarbons such as MBK has also been found in the optic tract and hypothalamus of the cat. These findings are significant because it is possible that such central nervous system damage is permanent, whereas the peripheral nervous system shows regeneration.[9]

Testicular atrophy of the germinal epithelium was seen in male rats administered 660 mg/kg by gavage for 90 days.[10] A reduction in total circulating white blood cells has also been reported following MBK exposure.[1]

Pregnant rats exposed to 1000 or 2000 ppm MBK during 21 days of gestation had reduced weight gain; a significant decrease in the number and weight of live offspring was also observed in the high-dose group. Behavioral alterations, including deficits in avoidance conditioning and increased activity, occurred in the offspring of both groups.[11]

2,5-Hexanedione was found to be a major metabolite of methyl butyl ketone in several animal species; peripheral neuropathy occurred in rats after daily subcutaneous injection of 2,5-hexanedione at a dose of 340 mg/kg, 5 days/week, for 19 weeks.[12-14] Nonneurotoxic aliphatic monoketones, such as methyl ethyl ketone, enhance the neurotoxicity of MBK. In one rat study, the longer the carbon chain length of the nonneurotoxic monoketone, the greater the

potentiating effect on MBK. It is expected that exposure to a subneurotoxic dose of MBK, plus high doses of some aliphatic monoketones, would also produce neurotoxicity. In addition, MBK itself potentiates the toxicity of other chemicals.[2]

MBK can cause mild eye irritation and minor transient corneal injury. Repeated skin contact may be irritating because of the ability of MBK to defat the skin, resulting in dermatitis.[1]

MBK has an acetonelike odor detectable at 0.076 ppm.[15] The 1995 ACGIH threshold limit value–time-weighted average (TLV-TWA) is 5 ppm (20 mg/m³) with a notation for skin absorption.

REFERENCES

1. Krasavage WJ et al: Ketones. In Clayton GD, Clayton FE (ed): *Patty's Industrial Hygiene and Toxicology*, 3rd ed, Vol 2C, *Toxicology*, pp 4741–4747. New York, Wiley-Interscience, 1982
2. Bos PMJ, de Mik G, Bragt PC: Critical review of the toxicity of methyl *n*-butyl ketone: risk from occupational exposure. *Am J Ind Med* 20:175–194, 1991
3. Billmaier R et al: Peripheral neuropathy in a coated fabrics plant. *J Occup Med* 16:665–671, 1974
4. Mendell JR et al: Toxic polyneuropathy produced by methyl *n*-butyl ketone. *Science* 185:787–789, 1974
5. Allen N et al: Toxic polyneuropathy due to methyl *n*-butyl ketone. *Arch Neurol* 32:209–218, 1975
6. Gilchrist MA et al: Toxic peripheral polyneuropathy. *Morb Mort Weekly Rept* 23:9–10, 1974
7. Johnson BL et al: Neurobehavioral effects of methyl *n*-butyl ketone and methyl *n*-amyl ketone in rats and monkeys: a summary of NIOSH investigations. *J Environ Path Toxicol* 2:113–133, 1979
8. Spencer PS et al: Nervous system degeneration produced by the industrial solvent methyl *n*-butyl ketone. *Arch Neurol* 32:219–222, 1975
9. Schaumberg HH, Spencer PS: Environmental hydrocarbons produce degeneration in cat hypothalamus and optic tract. *Science* 199:199–200, 1978
10. Krasavage WJ et al: The relative neurotoxic-

ity of methyl-*n*-butyl ketone and *n*-hexane and their metabolites. *Toxicol Appl Pharmacol* 52:433–441, 1980

11. Peters MA, Hudson PM, Dixon RL: The effect totigestational exposure to methyl-*n*-butyl ketone has on postnatal development and behavior. *Ecotoxicol Environ Safety* 5:291–306, 1981
12. Scala RA: Hydrocarbon neuropathy. *Ann Occup Hyg* 19:293–299, 1976
13. Raleigh RL, Spencer PS, Schaumberg HH: Methyl-*n*-butyl ketone. *J Occup Med* 17: 286, 1975
14. Granvil CP, Sharkawi M, Plaa GL: Metabolic fate of methyl *n*-butyl ketone, methyl isobutyl ketone and their metabolites in mice. *Tox Lett* 70:263–267, 1994
15. ASTDR (Agency for Toxic Substances and Disease Registry): *Toxicological Profile for 2-Hexanone*. US Department of Health and Human Services, Public Health Service, TP-91/18, p 92, 1991

METHYL CHLORIDE
CAS: 74-87-3

CH₃Cl

Synonyms: Chloromethane; monochloromethane

Physical Form. Colorless gas

Uses. Chemical intermediate, especially in industrial methylating reactions; blowing agent for plastic foams; rarely, as a refrigerant

Exposure. Inhalation

Toxicology. Methyl chloride causes kidney and liver damage and is a central nervous system depressant. In experimental animals, it is a reproductive toxin and a teratogen.

Human fatalities have occurred from a single severe exposure or prolonged exposures to lower concentrations.[1] Acute poisoning in humans is characterized by a latent period of several hours, followed by dizziness, drowsiness, staggering gait, and slurred speech; nausea, vomiting, and diarrhea; double vision; weakness, paralysis, convulsions, cyanosis, and coma.[1-3] Renal or hepatic damage and anemia also occur. Recovery from an acute exposure usually occurs within 5 to 6 hours but may take as long as 30 days or more in massive exposures.[1] In one study, however, 10 survivors of methyl chloride poisoning experienced mild neurological or psychiatric sequelae 13 years after the incident.[4] Recurrence of symptoms after apparent recovery without further exposure has been observed in the immediate postexposure period.

Six workers chronically exposed to 200 to 400 ppm for 2 to 3 weeks developed symptoms of intoxication, including confusion, blurring of vision, slurred speech, and staggering gait; symptoms disappeared over a period of 1 to 3 months after removal from exposure.[1]

Concentrations ranging from 150,000 to 300,000 ppm are expected to kill most animals in a short time; levels of 20,000 to 40,000 are considered dangerous within 60 minutes.

Mice exposed continuously to 100 ppm or intermittently to 400 ppm for 11 days had histopathological evidence of brain lesions characterized by degeneration and atrophy of the granular layer of the cerebellum.[5] Daily exposure of mice to 1000 ppm for 2 years induced a functional limb muscle impairment and atrophy of the spleen.[6] At 2400 ppm, administered daily, there were renal and hematopoietic effects, and the mice were moribund by day 9.[5] For rats exposed to 3500 ppm, 6 hr/day for up to 12 days, clinical signs included severe diarrhea, incoordination of the forelimbs and, in a few animals, hind limb paralysis and convulsions.[7]

Daily exposure of male rats to 1500 ppm for 10 weeks caused severe testicular degeneration; no males sired litters during a subsequent 2-week breeding period.[8]

An increase in fetal heart defects was observed in mice following 12 days of repeated exposure in utero to 500 ppm.[9]

Rats (F344) and mice (B6C3F₁) were exposed at 0, 50, 225, or 1000 ppm 6 hr/day, 5 days/week, for 2 years. An excess of tumors was found only in male mice of the highest-exposure group; cystadenomas and adenomas of the renal cortex and papillary cystadenomas were reported.[10,11] Subsequent mechanistic studies have shown that methyl chloride does not exhibit direct methylation of DNA in vivo.[11] It has been suggested that methyl chloride, at high doses, is metabolically transformed to formaldehyde, which causes DNA-protein cross-links and DNA single-strand breaks.[12] It has also been noted that such a mechanism is not likely to be operative in humans at low exposure concentrations.[11]

NIOSH recommends that methyl chloride be considered a potential occupational teratogen and carcinogen.[10]

The IARC states that there is inadequate evidence for the carcinogenicity of methyl chloride to experimental animals and humans.[13]

The 1995 ACGIH threshold limit value–time-weighted average (TLV-TWA) for methyl chloride is 50 ppm (103 mg/m³) with a STEL/ceiling of 100 ppm (207 mg/m³) and a notation for skin absorption.

REFERENCES

1. Scharnweber HC, Spears GN, Cowles SR: Chronic methyl chloride intoxication in six industrial workers. *J Occup Med* 16:112–113, 1974
2. Spevak L, Nadj V, Felle D: Methyl chloride poisoning in four members of a family. *Br J Ind Med* 33:272–274, 1976
3. Hansen H, Weaver NK, Venable FS: Methyl chloride intoxication. *AMA Arch Ind Hyg Occup Med* 8:328–334, 1953
4. Gudmundsson G: Letter. Methyl chloride poisoning 13 years later. *Arch Environ Health* 32:236–237, 1977
5. Landry TD et al: Neurotoxicity of methyl chloride in continuously versus intermittently exposed female C57BL/6 mice. *Fund Appl Toxicol* 5:87–98, 1985
6. Pavokv KL et al: Major findings in a twenty-four month inhalation toxicity study of methyl chloride in mice and rats. *Toxicologist* 2:161, 1982
7. Morgan KT et al: Histopathology of acute toxic response in rats and mice exposed to methyl chloride by inhalation. *Fund Appl Toxicol* 2:293–299, 1982
8. Hamm TE Jr: Reproduction in Fischer-344 rats exposed to methyl chloride by inhalation for two generations. *Fund Appl Toxicol* 5:568–577, 1985
9. Wolkowski-Tyl R et al: Evaluation of heart malformations in B6C3F₁ mouse fetuses induced by in utero exposure to methyl chloride. *Teratol* 27:197–206, 1983
10. NIOSH: *Current Intelligence Bulletin 43. Monohalomethanes. Methyl Chloride CH₃Cl Methyl Bromide CH₃Br Methyl Iodide CH₃I.* DHHS (NIOSH) Pub No 84-117, p 22. Cincinnati, OH, Sept 27, 1984
11. Bolt HM, Gansewendt B: Mechanisms of carcinogenicity of methyl halides. *CRC Rev Toxicol* 23:237–253, 1993
12. Ristau C, Bolt HM, Vangala RR: Formation and repair of DNA lesions in kidneys of male mice after acute exposure to methyl chloride. *Arch Toxicol* 64:254–256, 1990
13. *IARC Monographs on the Evaluation of the Carcinogenic Risk of Chemicals to Humans*, Vol 41, Some halogenated hydrocarbons and pesticide exposures, p 176. Lyon, International Agency for Research on Cancer, 1986

METHYL 2-CYANOACRYLATE
CAS:137-05-3

$C_5H_5NO_2$

Synonyms: Mecrylate; methyl cyanoacrylate

Physical Form. Colorless liquid

Uses. In high-bond-strength, fast-acting glues (eg, Krazy Glue); surgical tissue adhesive

Exposure. Inhalation

Toxicology. Methyl 2-cyanoacrylate is an irritant of the eyes and nose and can induce occupational asthma.

Nose and eye irritation occur at levels of 2 to 5 ppm; at 20 ppm, there is lacrimation and rhinorrhea, and concentrations greater than 50 ppm produce painful irritation.[1]

A 52-year-old man exposed to undetermined concentrations of methyl cyanoacrylate in an adhesive developed respiratory symptoms after 1 month on the job.[2] Eleven weeks after stopping work, the patient was challenged by working with the adhesive for 25 minutes. This provoked a 42% fall in FEV_1 15 hours after the challenge, along with symptoms of rhinitis during most of the day.

The LC_{50} in rats was 101 ppm for 6 hours.[3] Repeated exposure of rats to 31.3 ppm, 6 hr/day, 5 days/week, for 12 exposures caused no signs of mucous membrane irritation. The acute dermal toxicity is low, with the dermal LD_{50} in guinea pigs greater than 10 ml/kg.

Applied as an adhesive to rabbit or human eyes, some reports described corneal haze and inflammation; other reports with highly purified material indicated less toxicity.[4] Mistaken use in the eyes as eyedrops has caused immediate brief smarting and firm gluing of the eyelids together.[5] Acetone on a swab can be used to unglue the lids and remove the glue from the cornea with minimal, if any, injury to the corneal epithelium.[4]

Methyl 2-cyanoacrylate was positive in the Ames test with and without activation by metabolic enzymes.[6]

The 1995 ACGIH threshold limit value–time-weighted average (TLV-TWA) for methyl 2-cyanoacrylate is 2 ppm (9.1 mg/m³) with a short-term excursion limit of 4 ppm (18 mg/m³).

REFERENCES

1. McGee WA, Oglesby FI, Raleigh RI, Fassett DW: The determination of a sensory response to alkyl 2-cyanoacrylate vapor in air. *Am Ind Hyg Assoc J* 29:558–561, 1968
2. Lozewicz S, Davison AG, Hopkirk A, et al: Occupational asthma due to methyl methacrylate and cyanoacrylates. *Thorax* 40:836–839, 1985
3. Methyl 2-cyanoacrylate. *Documentation of the TLVs for Substances in Workroom Air*, 6th ed, pp 965–966. Cincinnati, OH, American Conference of Governmental Hygienists (ACGIH), 1991
4. Grant WM: *Toxicology of the Eye*, 3rd ed, p 291. Springfield, IL, Charles C Thomas, 1986
5. Morgan SJ, Astbury NJ: Inadvertent self administration of superglue: a consumer hazard. *Br Med J* 289:226–227, 1984
6. Rietveld ED, Garnaat MA, Seutter-Berlage F: Bacterial mutagenicity of some methyl 2-cyanoacrylates and methyl 2-cyano-3-phenylacrylates. *Mutat Res* 188:97–104, 1987

METHYLCYCLOHEXANE
CAS: 108-87-2

C_7H_{14}

Synonyms: Cyclohexylmethane; hexahydrotoluene

Physical Form. Colorless liquid

Uses. Solvent; in organic synthesis

Exposure. Inhalation

Toxicology. At high concentrations, methylcyclohexane causes narcosis in animals, and it is expected that severe exposure will produce the same effect in humans.

No effects have been reported in humans.

Rabbits did not survive exposure for 70 minutes to 15,227 ppm; conjunctival congestion, dyspnea, rapid narcosis, and severe convulsions preceded death.[1] Exposure to 10,000 ppm, 6 hr/day, for a total of 10 days resulted in convulsions, narcosis, and death.[1] There were no signs of intoxication in rabbits exposed to 2880 ppm for a total of 90 hours, but slight cellular injury was observed in the liver and kidneys.[1] The liquid on the skin of a rabbit caused local irritation, thickening, and ulceration.[2]

The 1995 ACGIH threshold limit value–time-weighted average (TLV-TWA) is 400 ppm (1610 mg/m³).

REFERENCES

1. Treon JF, Crutchfield WE Jr, Kitzmiller KV: The physiological response of animals to cyclo-

hexane, methylcyclohexane, and certain derivatives of these compounds. II. Inhalation *J Ind Hyg Toxicol* 25:323–347, 1943
2. Treon JF, Crutchfield WE Jr, Kitzmiller KV: The physiological response of rabbits to cyclohexane, methylcyclohexane, and certain derivatives of these compounds. I. Oral administration and cutaneous application. *J Ind Hyg Toxicol* 25:199–214, 1943

METHYLCYCLOHEXANOL
CAS: 25639-42-3

CH₃C₆H₁₀OH

$CH_3C_6H_{10}OH$

Synonyms: Hexahydrocresol; hexahydromethylphenol; methylhexalin

Physical Form. Colorless, viscous liquid

Uses. Solvent for lacquers; blending agent in textile soaps; antioxidant in lubricants

Exposure. Inhalation; skin absorption

Toxicology. In animals, methylcyclohexanol is a mild irritant of the eyes and mucous membranes and, at high concentrations, it causes signs of narcosis. It is expected that severe exposure will produce the same effects in humans.

Headache and irritation of the ocular and upper respiratory membranes may result from prolonged exposure to excessive concentrations of the vapor.[1]

Rabbits exposed 6 hr/day to 503 ppm for 10 weeks had conjunctival irritation and slight lethargy.[2] There were no signs of intoxication at 232 ppm for a total exposure of 300 hours.

The minimal lethal dose for rabbits by oral administration was 2 g/kg; rapid narcosis and convulsive movements preceded death.[3] Sublethal doses caused narcosis with spasmodic head jerking; salivation and lacrimation were also observed; hepatocellular degeneration was apparent at autopsy.

Repeated cutaneous applications to rabbits

of large doses of methylcyclohexanol caused skin irritation and thickening, weakness, tremor, narcosis, and death.[3]

Methylcyclohexanol can be detected by its odor at 500 ppm, a concentration capable of causing upper respiratory irritation.[1]

The 1995 ACGIH threshold limit value–time-weighted aveage (TLV-TWA) is 50 ppm (234 mg/m³).

REFERENCES

1. Rowe VK, McCollister SB: Alcohols. In Clayton GD, Clayton FE (eds): *Patty's Industrial Hygiene and Toxicology*, 3rd ed, rev, Vol 2C, *Toxicology*, pp 4649–4652. New York, Wiley-Interscience, 1982
2. Treon JF, Crutchfield WE Jr, Kitzmiller KV: The physiological response of rabbits to cyclohexane, methylcyclohexane, and certain derivatives of these compounds. II. Inhalation. *J Ind Hyg Toxicol* 25:323–347, 1943
3. Treon JF, Crutchfield WE Jr, Kitzmiller KV: The physiological response of rabbits to cyclohexane, methylcyclohexane, and certain derivatives of these compounds. I. Oral administration and cutaneous application. *J Ind Hyg Toxicol* 25:199–214, 1943

---

o-METHYLCYCLOHEXANONE
CAS: 583-60-8

CH₃C₅H₉CO

$CH_3C_5H_9CO$

Synonyms: 2-Methylcyclohexanone

Physical Form. Clear to pale yellow liquid

Uses. Solvent; rust remover

Exposure. Inhalation; skin absorption

Toxicology. In animals *o*-methylcyclohexanone is an irritant of the eyes and mucous membranes and, at high concentrations, it causes narcosis; it is expected that severe exposure would produce the same effects in humans.

Several species of animals exposed to 3500 ppm suffered marked irritation of the mucous membranes, became incoordinated after 15 minutes of exposure, and were prostrate after 30 minutes.[1] Conjunctival irritation, lacrimation, salivation, and lethargy were observed in rabbits exposed to 1822 ppm for 6 hr/day for 3 weeks.[2] Exposure of mice to 450 ppm for an unspecified time period resulted in severe irritation of the eyes and respiratory tract.[3]

Repeated cutaneous application to rabbits of large doses of the liquid caused irritation of the skin, tremor, narcosis, and death; the minimum lethal dose was between 4.9 and 7.2 g/kg.[3]

There are no reports of chronic or systemic effects in humans, probably because of the chemical's irritant properties and warning acetonelike odor at levels below those causing serious effects. Furthermore, lethal concentrations of vapors are not expected at temperatures commonly encountered in the workplace.[1]

The 1995 ACGIH threshold limit value–time-weighted average (TLV-TWA) is 50 ppm (229 mg/m³) with a STEL/ceiling level of 75 ppm (344 mg/m³) with a notation for skin absorption.

REFERENCES

1. Krasavage WJ et al: Ketones. In Clayton GD, Clayton FE (eds): *Patty's Industrial Hygiene and Toxicology*, 3rd ed, rev, Vol 2C, *Toxicology*, pp 4782–4784. New York: Wiley-Interscience, 1982
2. Treon JF, Crutchfield WE Jr, Kitzmiller KV: The physiological response of rabbits to cyclohexane, methylcyclohexane, and certain derivatives of these compounds. II. Inhalation. *J Ind Hyg Toxicol* 25:323–347, 1943
3. Treon JF, Crutchfield WE Jr, Kitzmiller KV: The physiological response of rabbits to cyclohexane, methylcyclohexane, and certain derivatives of these compounds. I. Oral administration and cutaneous application. *J Ind Hyg Toxicol* 25:199–214, 1943

2-METHYLCYCLOPENTADIENYL MANGANESE TRICARBONYL
CAS:12108-13-3

$C_9H_7MNO_3$

Synonyms: MMT; Combustion Improver-2; CI-2; Antiknock-33

Physical Form. Liquid

Uses. Octane enhancer in gasoline; reduces smoke emissions from home, commercial, industrial, and marine burners

Exposure. Inhalation; skin absorption

Toxicology. 2-Methylcyclopentadienyl manganese tricarbonyl (MMT) causes central nervous system effects and liver, kidney, and pulmonary damage in animals.

Accidental exposure to workers has caused metallic taste in the mouth, headache, nausea, gastrointestinal upset, dyspnea, tightness in the chest, and paresthesia.[1] A quantity of 5 to 15 ml that was spilled on one hand and wrist of a worker caused nausea and headache within 3 to 5 minutes.[2]

Toxic symptoms in various animal species were similar and consisted of excitement and hyperactivity; tremor; spasms; slow, labored respiration; clonic convulsions, and terminal coma.[1] On histological examination, there was degeneration and necrosis of liver cells and renal tubules, perivascular edema and swelling of the lungs, and degeneration of the cells of the cerebral cortex.

The oral LD_{50} of MMT in mice, rats, rabbits, and guinea pigs ranged from 58 to 905 mg/kg.[3] The dermal LD_{50} in rabbits was 1.35 g/kg. In rats, the 4-hour inhalation LC_{50} was 76 mg/m³.[4] No deaths occurred in cats, rabbits, guinea pigs, mice, or rats after 150 exposures to 6.4 mg/m³ for 7 hours.[5]

Chronic oral administration of 0.5 g/kg in the diet for 12 months caused suppressed weight gain in mice.[6]

Oxidative metabolism in rats appeared to

be an important detoxifying mechanism; the intraperitoneal LD_{50} was 12.1 mg/kg for MMT, but two major metabolites, hydroxymethyl-cyclopent adienyl manganese tricarbonyl and carboxycyclopentadienyl manganese tricarbonyl, caused no significant toxicity, even at doses of 250 mg/kg.[7]

The 1995 ACGIH threshold limit value–time-weighted average (TLV-TWA) for 2-methylcyclopentadienyl manganese tricarbonyl is 0.2 mg/m³ as Mn with a notation for skin absorption.

REFERENCES

1. Toxicology of methylcyclopentadienyl manganese tricarbonyl (MMT) and related manganese compounds emitted from mobile and stationary sources. Environmental Toxicology Research Laboratory, NERC-EPA, Cincinnati, OH, Report No AMRL-TR-73-125, Paper No 20, pp 251–270, 1974
2. US Navy Smoke Abatement Additive (Combustion Improver No 2 (CI-2)). COMNA-VAIRPACNOTE 470, NAVAIRPAC 742, Sec 1, p 2
3. World Health Organization: *Environmental Health Criteria 17: Manganese*, WHO, Geneva, 1981
4. Hinderer RK: Toxicity studies of methylcyclopentadienyl manganese tricarbonyl (MMT). *Am Ind Hyg Assoc J* 40:164–167, 1979
5. Ethyl Corporation: Supplement to toxicological data sheet, MMT, 1968
6. Komura J, Sakamoto M: Disposition, behavior, and toxicity of methylcyclopentadienyl manganese tricarbonyl in the mouse. *Arch Environ Contam Toxicol* 23:473–475, 1992
7. Cox DN, Traiger GJ, Jacober SP, Hanzlik RP: Comparison of the toxicity of methylcyclopentadienyl manganese tricarbonyl with that of its two major metabolites. *Environ Lett* 39:1–5, 1987

4,4'-METHYLENE BIS(2-CHLOROANILINE)
CAS: 101-14-4

$CH_2(C_6H_4ClNH_2)_2$

Synonyms: Methylene bis(ortho-chloroaniline); DACPM; MBOCA; MOCA

Physical Form. Colorless crystals

Uses. Curing agent for polyurethanes and epoxy resins

Exposure. Inhalation; skin absorption

Toxicology. 4,4' Methylene bis(2-chloroaniline), or MOCA, is carcinogenic in experimental animals.

Acute effects have not been reported in humans. Sprayed on the skin, the liquid caused a burning sensation of the eyes and skin, as well as nausea.[1]

Rats fed 1000 ppm MOCA in a standard diet for 2 years developed lung tumors; there were 25 adenomatoses and 48 adenocarcinomas in 88 rats.[2] Accompanying liver changes included hepatocytomegaly, necrosis, bile duct proliferation, and fibrosis.[1] In 88 control animals, there were 2 lung adenomatoses. MOCA in a low-protein diet caused lung tumors in rats of both sexes, liver tumors in males, and malignant mammary tumors in females.

Repeated subcutaneous injection of MOCA in 34 rats (total dose, 25 g/kg for 620 days) resulted in 9 liver-cell carcinomas and 7 lung carcinomas; 13 of 50 control animals developed tumors, but no malignant tumors of the liver or the lungs were observed.[3]

MOCA was fed to male and female mice for 18 months at a dose of either 1 or 2 g/kg; in female mice, but not in males, a statistically significant incidence of hepatoma was observed.[4] In addition, a higher incidence of hemangiosarcomas and hemangiomas was observed in treated animals compared to controls.[4] Urinary bladder tumors (primarily papillary transitional cell carcinomas) occurred in dogs given 100 mg MOCA by capsule for up to 9.0 years.[5]

There was no evidence that MOCA was tumorigenic in a study of 31 active workers exposed from 6 months to 16 years.[6] Quantitative analysis of the workers' urine confirmed exposure to the chemical. In addition, the records were reviewed for 178 employees who, at one time, had worked with MOCA but who thereafter had had no further exposure for at least 10 years. The general health of exposed workers with respect to illness, absenteeism, and medical history was similar to that of the total plant population. Two deaths in this group due to malignancy had been diagnosed prior to any work with, or exposure to, MOCA.[6] For the plant population in general, there were 115 cancer deaths/100,000 over a 15-year period compared with the national death rate for cancer of 139/100,000 population.

Three noninvasive papillary tumors of the bladder were identified in a screening of 540 MOCA workers; two tumors occurred in men with completely normal urine screening who were under 30, had never smoked, and had no previous occupational exposure to known bladder carcinogens.[7]

Exposure to MOCA was believed to be the cause of urinary frequency and mild hematuria in 2 of 6 exposed workers; however, a variety of other materials, including toluene diisocyanate, polyester resins, polyether resins, and isocyanate-containing resins, also were present.[8]

The IARC has determined that there is inadequate evidence for carcinogenicity to humans and sufficient evidence for carcinogenicity to animals. On the basis of animal experiments, however, it was concluded that MOCA probably is carcinogenic to humans and that exposure by all routes should be monitored carefully.[1]

MOCA is genotoxic in a wide variety of assays. It also forms adducts with DNA both in vivo and in vitro.[1]

Because MOCA can be absorbed in significant amounts through the skin, determination of levels in breathing zones is probably not an accurate indication of absorbed dose. Biological monitoring of MOCA in urine has been used, but this indicates only recent exposure, and results depend critically on the timing of the sampling. Determination of hemoglobin adducts has been proposed as a biomonitoring method for MOCA exposure.[9]

The 1995 threshold limit value–time-weighted average (TLV-TWA) for 4,4' methylene bis(2-chloroaniline) is 0.01 ppm (0.11 mg/m^3) with an A2 suspected human carcinogen designation and a notation for skin absorption.

REFERENCES

1. *IARC Monographs on the Evaluation of Carcinogenic Risks to Humans*, Vol 57, Occupational exposures of hairdressers and barbers and personal use of hair colourants; some hair dyes, cosmetic colourants, industrial dyestuffs and aromatic amines, pp 271–301, Lyon, International Agency for Research on Cancer, 1993
2. Stula EF, Sherman H, Zapp JA Jr, Wesley-Clayton J Jr: Experimental neoplasia in rats from oral administration of 3,3'-dichlorobenzidine, 4,4'-methylene bis(2-chloroaniline), and 4,4'-methylene-bis(2-methylaniline). *Toxicol Appl Pharmacol* 31:159–176, 1975
3. Steinhoff D, Grundmann E: Zur Cancerogen Wirkung von 3,3' Dichlor-4,4'-diaminodiphenylmethan bein Ratten. *Naturwiss* 58:578, 1971
4. Russfield AB, Homburger F, Boger E, et al: The carcinogenic effect of 4,4' methylene-bis(2-chloroaniline) in mice and rats. *Toxicol Appl Pharmacol* 31:47–54, 1975
5. Stula EF, Barnes JR, Sherman H: Urinary bladder tumors in dogs from 4,4'-methylene-bis(2-chloroaniline) (MOCA). *J Environ Pathol Toxicol* 1:31–50, 1977
6. Linch AL, O'Conner GB, Barnes JR, et al: Methylene-bis-ortho-chloroaniline (MOCA): Evaluation of hazards and exposure control. *Am Ind Hyg Assoc J* 32:802–819, 1971
7. Ward E, Halperin W, Thun M, et al: Screening workers exposed to 4,4'methylene bis(2-chloroaniline) for bladder cancer by cytoscopy. *J Ocup Med* 32:865–868, 1990
8. Mastromatteo E: Recent occupational health experiences in Ontario. *J Occup Med* 7:502–511, 1965
9. Bailey E, Brooks AG, Farmer PB, et al: Monitoring exposure to 4,4'-methylene-bis(2-chloroaniline) through the gas chromatography-mass spectrometry measurements of adducts to hemoglobin. *Environ Health Perspect* 99:175–177, 1993

METHYLENE BISPHENYL ISOCYANATE
CAS: 101-68-8

$CH_2(C_6H_4NCO)_2$

Synonyms: Methylene diisocyanate; MDI; diphenylmethane diisocyanate

Physical Form. Liquid; aerosol

Uses. In the production of polyurethane foams and plastics

Exposure. Inhalation

Toxicology. Methylene bisphenyl isocyanate (MDI) is an irritant of the eyes and mucous membranes and a sensitizer of the respiratory tract.

If the breathing zone concentration reaches 0.5 ppm, the possibility of respiratory response is imminent.[1] Depending on the length of exposure and level of concentration above 0.5 ppm, respiratory symptoms may develop with a latent period of 4 to 8 hours. Symptoms include increased secretion, cough, pain on respiration and, if severe enough, some restriction of air movement due to a combination of secretions, edema, and pain. On removal from exposure, the symptoms may persist for 3 to 7 days.[1]

A second type of response to isocyanates is allergic sensitization of the respiratory tract. It usually develops after some months of exposure.[1-4] The onset of symptoms may be insidious, becoming progressively more pronounced with continued exposure. Initial symptoms are often nocturnal dyspnea and/or nocturnal cough, with progression to asthmatic bronchitis.[3] Asthma, characterized by bronchial hyperreactivity, cough, wheeze, tightness in the chest, and dyspnea, was observed in 12 of 78 foundry workers exposed to MDI concentrations greater than 0.02 ppm.[5] Inhalation provocation tests on 6 out of 9 of the asthmatics resulted in specific asthmatic reaction to MDI.[5] Persons who are sensitized must not be exposed to any concentration of MDI and must be removed from any work involving potential exposure to MDI.

MDI is not a significant eye or skin irritant, but it may produce skin sensitization.[3]

The 1995 ACGIH threshold limit value–time-weighted average (TLV-TWA) is 0.005 ppm (0.051 mg/m³).

REFERENCES

1. Rye WA: Human response to isocyanate exposure. *J Occup Med* 15:306–307, 1973
2. Woolrich PF, Rye WA: Urethanes. *J Occup Med* 11:184–190, 1969
3. National Institute for Occupational Safety and Health: *Criteria for a Recommended Standard . . . Occupational Exposure to Diisocyanates.* DHEW (NIOSH) Pub No 78-215. Washington, DC, US Government Printing Office, 1978
4. Tanser AR, Bourke MP, Blandford AG: Isocyanate asthma: respiratory symptoms caused by diphenyl-methane-diisocyanate. *Thorax* 28: 596–600, 1973
5. Johnson A et al: Respiratory abnormalities among workers in an iron and steel foundry. *Br J Ind Med* 42:94–100, 1985

METHYLENE CHLORIDE
CAS: 75-09-2

CH_2Cl_2

Synonyms: Dichloromethane; methylene dichloride; methylene bichloride

Physical Form. Colorless liquid

Uses. Multipurpose solvent; paint remover; in the manufacture of photographic film; in aerosol propellants; in urethane foam

Exposure. Inhalation; skin absorption

Toxicology. Methylene chloride is a mild central nervous system depressant and an eye, skin, and respiratory tract irritant; it is carcinogenic in experimental animals and is considered a suspected human carcinogen.

Concentrations in excess of 50,000 ppm are thought to be immediately life-threatening.[1] Four workers exposed to unmeasured but high levels of methylene chloride for 1 to 3 hours had eye and respiratory tract irritation and reduced hemoglobin and red blood cell counts; all became comatose, and one died.[2]

A chemist repeatedly exposed to concentrations ranging from 500 to 3600 ppm developed signs of toxic encephalopathy.[3] A healthy young worker engaged in degreasing metal parts had a brief exposure to an undetermined but very high concentration of vapor; he complained of excessive fatigue, weakness, sleepiness, light-headedness, chills, and nausea; pulmonary edema developed after several hours, but all signs and symptoms had cleared within 18 hours of terminating the exposure.[4] In human experiments, inhalation of 500 to 1000 ppm for 1 or 2 hours resulted in light-headedness.[5]

Volunteers exposed at 300 to 800 ppm for at least 40 minutes had altered responses to various sensory and psychomotor tests.[6] No effects were seen in volunteers exposed to 250 ppm for up to 7.5 hours.[6] Although an excess in self-reported neurologic symptoms was found in workers repeatedly exposed at 75 to 100 ppm, no significant deleterious effects were observed on clinical examination, which included measurement of motor conduction velocity, electrocardiogram, and psychological tests.[7]

Limited epidemiologic studies initially found no specific cause for excess deaths in workers chronically exposed to methylene chloride.[6] There is no clear evidence of liver or kidney damage in humans in spite of many reports of fatty degeneration in the liver and tubular degeneration in the kidneys of exposed animals.[8] A recent evaluation of workers exposed to high levels of methylene chloride averaging 475 ppm for 10 years found no adverse health effects, as determined by selected liver, cardiac, and neurological tests.[9] In another report, no firm evidence of central nervous system effects was found in retired mechanics who had had long-term exposure to methylene chloride.[10]

Methylene chloride is metabolized by two pathways.[11] One pathway produces carbon monoxide via mixed-function oxidase enzymes, which results in the subsequent formation of carboxyhemoglobin (COHb). Carbon dioxide is produced from the pathway involving glutathione transferase. The metabolism to COHb is saturable, with disproportionately less carboxyhemoglobin formed and more unchanged methylene chloride expired as exposure increases. CNS effects are thought to be due to methylene chloride itself or methylene chloride in combination with other sources of COHb but not to the metabolism of methylene chloride to COHb alone. Serious poisonings from methylene chloride have been reported in the absence of significant elevation of COHb levels. Elevated COHb levels may persist for several hours following removal from exposure, as fat and other tissues continue to release accumulated amounts of methylene chloride.[3] Although the elevated COHb levels associated with moderate methylene chloride exposure are not expected to cause adverse effects in healthy individuals, those with a compromised cardiovascular system may not be able to tolerate the added cardiovascular stress.[6]

Contact with the liquid is irritating to the skin, and prolonged contact may cause severe burns.[12] In a thumb immersion experiment, an intense burning sensation was noted within 2 minutes, and mild erythema and exfoliation were observed after 30 minutes of immersion; the erythema and paresthesia subsided within an hour post-exposure.[13] Marked irritative conjunctivitis and lacrimation were noted at concentrations sufficient to produce unconsciousness.[2] Splashed in the eye, it is painfully irritating but is not likely to cause serious injury.[1]

Limited animal studies have suggested that methylene chloride is slightly fetotoxic at doses that also produce maternal toxicity; in rats and mice exposed at 1250 ppm on days 6 through 15 of gestation, delayed ossification of sternebrae and increased incidence of extra sternebrae were noted, respectively.[14]

A number of long-term animal studies have explored the carcinogenic potential of methylene chloride. A 1986 NTP study with B6C3F$_1$ mice exposed at 2000 or 4000 ppm, 6 hr/day, 5 days/week, for 2 years showed "clear evidence of carcinogenicity," as indicated by increased incidences of alveolar-bronchiolar and hepato-cellular neoplasms.[15] There was also a significant increase in benign mammary gland neoplasms in similarly exposed rats.

Epidemiological studies of occupationally exposed workers have not found statistically significant excesses in deaths from lung or liver cancer. However, a cohort mortality study of 1013 hourly male Eastman Kodak workers chronically exposed to methylene chloride in the manufacture of photographic products found an increased incidence of pancreatic deaths (8 vs. 3.1 expected).[16] A continuing study of the same Eastman Kodak cohort resulted in an updated report of mortality experience for an additional 4 years through 1988.[17] Mean exposure was 26 ppm (8-hour time-weighted average) for 23 years; median follow-up from first exposure was 33 years. A comparison with death rates in both the general population and industrial referents showed nonsignificant deficits in observed-expected ratios for such hypothesized causes as lung and liver cancer and ischemic heart disease. Overall, mortality from 1964 to 1988 ($n = 238$) was significantly decreased for both referent groups. No additional pancreatic cancer deaths occurred since the 1984 study; 8 have been observed versus 4.2 expected. An analysis of dose response for selected causes of death demonstrated no statistically significant trend according to either career methyelene chloride exposure or latency.

Assessment of the carcinogenic risk to humans from review of animal data is complicated by the results of pharmacokinetic studies that have associated methylene chloride carcinogenicity with a specific metabolic pathway.[18,19] This glutathione S-mediated pathway appears to proceed slowly in humans compared to mice and only at high-exposure doses. Therefore, extrapolation from high dose to low dose and between species may not provide accurate risk assessment of human exposure.

The IARC has determined that there is sufficient evidence for the carcinogenicity of methylene chloride in animals and inadequate evidence in humans.[20]

Methylene chloride has given positive and negative results in a wide variety of genotoxic assays. It may be a weak mutagen in mammalian systems.[11]

Although a number of methods have been proposed for the biological monitoring of occupational methylene chloride exposure, measurement of urinary methylene chloride levels may be the most suitable. The measurement of urinary methylene chloride is noninvasive, is not influenced by smoking, as are COHb or carbon monoxide levels in alveolar air, and may reflect cumulative exposures more accurately.[21]

The 1995 ACGIH threshold limit value–time-weighted average (TLV-TWA) for methylene chloride is 50 ppm (174 mg/m^3) with an A2 suspected human carcinogen designation.

REFERENCES

1. Hygienic Guide Series: Dichloromethane. *Am Ind Hyg Assoc J* 26:633–636, 1965
2. Moskowitz S, Shapiro H: Fatal exposure to methylene chloride vapor. *AMA Arch Ind Hyg Occup Med* 6:116–123, 1952
3. Methylene chloride. *Documentation of the TLVs and BEIs*, 6th ed, pp 981–987. Cincinnati, OH, American Conference of Governmental Industrial Hygienists (ACGIH), 1991
4. Hughes JP: Hazardous exposure to some so-called safe solvents. *JAMA* 156:234–237, 1954
5. Stewart RD, Fischer TN, Hosko MJ, et al: Experimental human exposure to methylene chloride. *Arch Environ Health* 25:342–348, 1972
6. Illing HPA, Shillaker RO: Toxicity review 12. Dichloromethane (methylene chloride). *Health and Safety Executive*, p 87. London, Her Majesty's Stationery Office, 1985
7. Cherry N, Venables H, Waldron HA, et al: Some observations on workers exposed to methylene chloride. *Br J Ind Med* 38:351–355, 1981
8. World Health Organization: *Environmental Health Criteria 32:* Methylene chloride, p 55. WHO, Geneva, 1984
9. Soden KJ: An evaluation of chronic methylene chloride exposure. *J Occup Med* 35:282–286, 1993
10. Lash AA, Becker CE, So Y, et al: Neurotoxic effects of methylene chloride: Are they long lasting in humans? *Br J Ind Med* 48:418–426, 1991

11. Agency for Toxic Substances and Disease Registry (ATSDR): *Toxicological Profile for Methylene Chloride.* US Department of Health and Human Services, Public Health Service, p 111, TP-92/13, 1993

12. National Institute for Occupational Safety and Health: *Criteria for a Recommended Standard . . . Occupational Exposure to Methylene Chloride.* DHEW (NIOSH) Pub No 76-138. Washington, DC, US Government Printing Office, 1976

13. Stewart RD, Dodd HC: Absorption of carbon tetrachloride, trichloroethylene, tetrachloroethylene, methylene chloride, and 1,1,1-trichloroethane through the human skin. *Am Ind Hyg Assoc J* 25:439–446, 1964

14. Schwetz BA, Leong BJ, Gehring PJ: The effect of maternally inhaled trichloroethylene, perchloroethylene, methyl chloroform and methylene chloride on embryonal and fetal development in mice and rats. *Toxicol Appl Pharmacol* 32:84–96, 1975

15. National Toxicology Program: *Toxicology and Carcinogenesis Studies of Dichloromethane (Methylene Chloride) (CAS No 75-09-2) in F344/N Rats and B6C3F₁ Mice (Inhalation Studies).* TR-306, DHHS (NIH) Pub No 86-2562. Washington, DC, US Government Printing Office, 1986

16. Hearne FT, Grose F, Pifer JW, et al: Methylene chloride mortality study: dose-response characterization and animal model comparison. *J Occup Med* 29:217–228, 1987

17. Hearne FT, Pifer JW, Grose F: Absence of adverse mortality effects in workers exposed to methylene chloride: an update. *J Occup Med* 32:234–240, 1990

18. Andersen ME, Clewell HJ III, Gargas ML, et al: Physiologically based pharmacokinetics and the risk assessment process for methylene chloride. *Toxicol Appl Pharmacol* 87:185–205, 1987

19. Reitz RH, Mendrala AL, Guengerich FP: In vitro metabolism of methylene chloride in human and animal tissues: use in physiologically based pharmacokinetic models. *Toxicol Appl Pharmacol* 97:230–246, 1989

20. *IARC Monographs on the Evaluation of the Carcinogenic Risk of Chemicals to Humans,* Vol 41, Some halogenated hydrocarbons and pesticide exposures. Lyon, International Agency for Research on Cancer, World Health Organization, 1986

21. Ghittori S, Marraccini P, Franco G, et al: Methylene chloride exposure in industrial workers. *Am Ind Hyg Assoc J* 54:27–31, 1993

4,4'-METHYLENE DIANILINE
CAS: 101-77-9

$C_{13}H_{14}N_2$

Synonyms: 4,4'-Diaminodiphenylmethane; DDM; MDA; 4-(4-aminobenzl)aniline

Physical Form. Light brown crystalline solid

Uses. In the production of methylene diphenyl diisocyanate (MDI), which is used to produce polyurethanes; hardening agent for epoxy resins, anticorrosive materials, printed-circuit parts, dyestuff intermediates, and filament-wound pipe and wire coatings

Exposure. Skin absorption; inhalation; ingestion

Toxicology. 4,4'-Methylene dianiline (MDA) is a human hepatotoxin; it is carcinogenic in experimental animals and is considered a suspected human carcinogen.

Occupational exposure of 12 male workers whose hands were in contact with MDA several hours per day caused toxic hepatitis.[1] The clinical pattern of the cases included right-upper-quadrant pain, high fever, and chills, with subsequent jaundice. A skin rash was seen in 5 of the cases. Percutaneous absorption was considered to be the major route of exposure because workers in the same occupational setting who did not have direct skin contact with MDA were not affected. All patients recovered within 7 weeks, and follow-up more than 5 years later showed no biochemical or clinical evidence of chronic hepatic disease.

Over a 9-year period (1967–1976), 11 cases of jaundice were reported from a company that mixed preground MDA with silicon dioxide.[2] In one instance, transient signs of myocardial damage in addition to transient signs of hepatic damage were observed following MDA exposure from a defective filter system.[3]

Ingestion of bread prepared with MDA-contaminated flour led to an outbreak of 84 cases of jaundice in Epping in the United Kingdom.[4] Liver biopsies from 7 of the patients showed partial inflammation, eosinophil infiltration, cholangitis, cholestasis, and varying degrees of hepatocellular damage. All patients made a good clinical recovery, with no evidence of progressive liver damage.[5]

In another case, ingestion of MDA in potassium carbonate and γ-butyrolactone resulted in severe systemic toxicity and visual dysfunction.[6] Transient effects included ECG abnormalities, bradycardia, and hypotension, which suggested myocardial involvement, and glycosuria with normoglycemia, which indicated renal tubular dysfunction. Liver effects included slight hepatomegaly 6 weeks post-ingestion, which quickly resolved, and disappearance of jaundice 2 to 3 weeks later. Liver biopsy 1 year after the poisoning showed normal hepatocytes and a preservation of hepatic architecture, but disturbed liver-function tests were still evident 18 months after the incident. Most significant, however, was the development of toxic optic neuritis, with severe visual dysfunction. Investigation of the retina revealed gross malfunction of the retinal pigment epithelium, reflected clinically as impaired visual acuity with severe loss of central visual field, color discrimination, and dark adaptation. Eighteen months later, there was little improvement, and all visual indices remained subnormal, with little likelihood of further recovery.

Support for the role of MDA in causing visual disturbances is found in animal studies.[7] Oral doses of 25 to 50 mg/kg in cats caused retinal damage. The changes observed in the affected eyes consisted of severe granular degeneration of the rods and cones and proliferation of the pigmented epithelial cells of the retina. The neuronal structures located beyond the pigmented layer remained intact. No visual disturbances were induced by MDA in the rabbit, guinea pig, and rat. In another study, degeneration of the inner and outer segments of the photoreceptor cells and the pigmented epithe-lial-cell layer of the retinas of guinea pigs resulted from a total inhaled dose of 24 mg/kg.

Chronic oral exposure of rats and mice to MDA and its dihydrochloride is carcinogenic.[8] Treatment-related increases in the incidences of thyroid follicular-cell adenomas and hepatocellular neoplasms were observed in mice following chronic ingestion of MDA in drinking water.[9] In rats, increases in the incidences of thyroid follicular-cell carcinoma and hepatic nodules were observed in males, and thyroid follicular-cell adenomas occurred in females. Although not statistically significant, certain uncommon tumors, such as bile-duct adenomas, papillomas of the urinary bladder, and granulosa-cell tumors of the ovary, also were reported. These tumors are of low incidence in historical controls. In another report, MDA acted as a promoter of thyroid tumors in rats.[10]

An epidemiologic study of workers potentially exposed to MDA (and numerous other agents) in the helicopter parts manufacturing industry showed limited evidence of an association between MDA and bladder cancer, colon cancer, lymphosarcoma, and reticulosarcoma.[11] A recent follow-up of 10 workers who had significant exposure to MDA between 1967 and 1976 revealed one case of a pathologically confirmed bladder cancer.[12] Although not statistically significant, these cases are of interest because of findings of bladder tumors in animals and the structural similarity of MDA to known human bladder carcinogens such as benzene.

The IARC has determined that there is sufficient evidence for the carcinogenicity of 4,4'-methylene dianiline and its dihydrochloride to experimental animals and that it is possibly carcinogenic to humans.[8]

Reports of allergic sensitivity to MDA are confounded by mixed exposures to chemicals such as epoxy resins and isocyanates, which makes it difficult to relate specific cause with effect. MDA does appear to cause an intense yellow staining reaction involving the skin (especially fingers and palms), nails, and occasionally hair in exposed workers.[13] The staining should serve as a marker for potential systemic exposure.

MDA has a faint amine odor, but the odor

is not offensive enough to be useful as a warning property.[7]

Monitoring atmospheric levels of MDA may not be useful, as skin absorption may be a more significant route of exposure. Concentrations of *N*-acetyl MDA, a major metabolite of MDA, in the urine may be used to reflect overall exposure.[14]

The 1995 ACGIH threshold limit value–time-weighted average (TLV-TWA) for MDA is 0.1 ppm (0.81 mg/m³) with an A2 suspected human carcinogen designation and a notation for skin absorption.

REFERENCES

1. McGill DB, Motto JD: An industrial outbreak of toxic hepatitis due to methylenedianiline. *N Eng J Med* 291:278–282, 1974
2. Dunn GW, Guirguis SS: *Proceedings of the 19th International Congress on Occupational Health*, VI, ISS chemical hazards, pp 639–644, Geneva, International Commission on Occupational Health, 1980
3. Brooks LJ et al: Acute myocardiopathy following tripathway exposure to methylenedianiline. *JAMA* 242:1527–1528, 1979
4. Kopelman H et al: The liver lesion of the Epping jaundice. *Quart J Med* 35:553–564, 1966
5. Kopelman H: The Epping jaundice after two years. *Postgrad Med J* 44:78–81, 1968
6. Roy CW et al: Methylene dianiline: A new toxic cause of visual failure and hepatitis. *Human Toxicol* 4:61–66, 1985
7. Leong BKJ et al: Retinopathy from inhaling 4,4′-methylenedianiline aerosols. *Fund Appl Toxicol* 9:645–658, 1987
8. *IARC Monographs on the Evaluation of the Carcinogenic Risk of Chemicals to Humans*, Vol 39, Some chemicals used in plastics and elastomers, pp 349–365. Lyon, International Agency for Research on Cancer, 1985
9. National Toxicology Program: *Carcinogenesis Studies of 4,4′-Methylenedianiline Dihydrochloride (CAS No 13552-44-8) in F344/N Rats and B6C3F Mice (Drinking Water Studies)*, Technical Report No 248. Research Triangle Park, NC, 1983
10. Hiasa Y et al: 4,4′-Diaminodiphenylmethane: Promoting effect on the development of thyroid tumors in rats treated with N-bis(2-hydroxypropyl)nitrosamine. *J Natl Cancer Inst* 72:471–476, 1984
11. National Institute for Occupational Safety and Health: *Current Intelligence Bulletin 47, 4,4′-Methylenedianiline (MDA)*, rev, pp 1–19. US Department of Health and Human Services, Public Health Service, Centers for Disease Control, July 25, 1986
12. Liss GM, Guirguis SS: Follow-up of a group of workers intoxicated with 4,4′methylenedianiline. *Am J Ind Med* 26:117–124, 1994
13. Cohen SR: Yellow staining caused by a 4,4′-methylenedianiline exposure. *Arch Derm* 121:1022–1027, 1985
14. Cocker J et al: Assessment of occupational exposure to 4,4′-diaminodiphenylmethane (methylene dianiline) by gas chromatography–mass spectrometry analysis of urine. *Br J Ind Med* 43:620–625, 1986

METHYLENE BIS-(4-HEXYLISOCYANATE)
CAS: 5124-30-1

$C_{15}H_{22}N_2O_2$

Synonyms: HMDI; hydrogenated MDI; dicyclohexylmethane-4,4′-diisocyanate; bis(4-isocyanalocyclohexyl)methane

Physical Form. Liquid

Uses. In the manufacture of polymers

Exposure. Inhalation

Toxicology. Methylene bis-(4-hexylisocyanate) (HMDI) is an irritant of the eyes, nose, and upper respiratory tract and causes dermal sensitization but apparently not respiratory sensitization.

Eleven of 15 workers who were exposed to HMDI showed allergic and nonallergic skin reactions.[1] Six suffered from vertigo with or without headaches, and 4 showed obstructive ventilatory disorders, tachycardia, and hypotension (ECG normal). All were treated with oral antihistamines and local steroid application. The signs of the intoxication disappeared after 10 to 14 days of treatment. There was no difference in the clinical syndrome between the atopic and the nonatopic workers.

In another case study, a small polyurethane molding plant employing poor hygienic techniques in which a number of employees developed contact dermatitis was described.[2] Three employees were examined. Patch testing in 2 of these revealed positive reactions suggesting allergic sensitization to an HMDI and to the catalyst methylenedianiline.

A recent study examined immune responses in mice following topical exposure to three allergenic diisocyanates: diphenylmethane-4,4′-diisocyanate (MDI), dicyclohexylmethane-4,4′-diisocyanate (HMDI), and isophorone diisocyanate (IPDI).[3] Contact and respiratory sensitizers induce differential immune responses in mice characteristic of TH1 and TH2 T-helper cell activation, respectively. All three chemicals are contact allergens. MDI is, in addition, a known human respiratory allergen. HMDI and IPDI did not produce an immunologic response in the mouse similar to that of MDI. These findings suggest that HMDI has much less potential to cause respiratory sensitization in humans than MDI does.

HMDI is a strong eye and skin irritant in animals.[4]

The 1995 ACGIH threshold limit value–time-weighted average (TLV-TWA) is 0.005 ppm (0.054 mg/m³).

REFERENCES

1. Israeli R, Smirnov V, Sculsky M: Symptoms of intoxication due to dicyclohexyl-methane-4,4′-diisocyanate exposure. *Int Arch Occup Environ Health* 48:179–184, 1981
2. Emmett EA: Allergic contact dermatitis in polyurethane plastic moulders. *J Occup Med* 18:802–804, 1976
3. Dearman RJ, Spence LM, Kimber I: Characterization of murine immune responses to allergenic diisocyanates. *Toxicol Appl Pharmacol* 112:190–197, 1992
4. US Environmental Protection Agency: Generic health hazard assessment of the chemical class diisocyanates. Final report. EPA Contract No 68-02-3990. US EPA, Washington, DC, May 5, 1987

METHYL ETHYL KETONE
CAS: 78-93-3

$CH_3COCH_2CH_3$

Synonyms: 2-Butanone; MEK

Physical Form.   Colorless liquid

Uses.   Solvent

Exposure.   Inhalation

Toxicology.   Methyl ethyl ketone (MEK) is an irritant of the eyes, mucous membranes, and skin; at high concentrations, it causes narcosis in animals, and it is expected that severe exposure in humans will produce the same effect.[1]

In humans, short-term exposure to 300 ppm was "objectionable," causing headache and throat irritation; 200 ppm caused mild irritation of eyes; 100 ppm caused slight nose and throat irritation.[2] No significant neurobehavioral effects (as determined by a series of psychomotor tests) were found in volunteers from 4-hour exposures to methyl ethyl ketone at 200 ppm; significant odor and irritant effects were reported.[3]

Several workers exposed to both the liquid and the vapor at 300 to 600 ppm for an unspecified time period complained of numbness of the fingers and arms; one worker complained of numbness in the legs and a tendency for them to "give way under him."[4] Many workers in this plant developed dermatitis from contact with the liquid; two workers developed dermatitis of the face from vapor exposure alone.

Three cases of polyneuropathy occurred in shoe factory workers exposed to combined methyl ethyl ketone and acetone vapors, as well as methyl ethyl ketone and toluene vapors at concentrations below 200 ppm.[5] Skin absorption also occurred. Although not highly neurotoxic itself, MEK may potentiate substances known to cause neuropathy.

In animal studies, death of rats and mice occurred within a few hours at concentrations of 90,000 ppm and above.[6] Guinea pigs exposed to 10,000 ppm had signs of eye and nose irritation, which developed rapidly, and narcosis occurred after 5 hours.[7] Exposure of rats to 6000 ppm, 8 hr/day, 7 days/week, did not result in any obvious motor impairment; all animals died from bronchopneumonia during the seventh week.[8]

Animal studies have shown MEK to enhance the development of, or increase the severity of, neurotoxic effects due to methyl *n*-butyl ketone, ethyl butyl ketone, *n*-hexane, and 2,5-hexanedione.[9-12] Methyl ethyl ketone exposure did not, however, potentiate the neurobehavioral test decrements produced by acetone.[13] Exposure to 200 ppm MEK or 100 ppm MEK plus 125 ppm acetone for 4 hours did not produce any significant effects in a variety of behavioral performance tests, whereas exposure to 250 ppm acetone caused some mild decrements.

Rats exposed to 3000 ppm during days 6 through 15 of gestation produced litters with an increased incidence of a minor skeletal variation and delay in ossification of fetal bones.[14] Similar effects, including reduced fetal weight, a low incidence of cleft palate, fused ribs, missing vertebrae, and syndactyly, were reported in the offspring of mice exposed at 3000 ppm on days 6 through 15 of gestation.[15] Slight maternal toxicity in the form of increased relative liver weight was also noted.

Methyl ethyl ketone was not genotoxic in a variety of in vivo and in vitro assays.[6]

Methyl ethyl ketone can be recognized at 25 ppm by its odor, which is similar to that of acetone but more irritating; its warning properties should prevent inadvertent exposure to toxic levels.[7] In determining worker exposure to MEK, end of shift urine levels appear to be the most reliable biological indicator of occupational exposure.[16]

The 1995 ACGIH threshold limit value–time-weighted average (TLV-TWA) for methyl ethyl ketone is 200 ppm (590 mg/m³) with a short-term excursion limit of 300 ppm (885 mg/m³).

REFERENCES

1. National Institute for Occupational Safety and Health, US Department of Health, Education and Welfare: *Criteria for a Recommended Standard . . . Occupational Exposure to Ketones.* DHEW (NIOSH) 78-173. Washington, DC, US Government Printing Office, 1978
2. Nelson KW, Ege JF Jr, Ross M, et al: Sensory response to certain industrial solvent vapors. *J Ind Hyg Toxicol* 25:282–285, 1943
3. Dick RB, Krieg EF Jr, Setzer J, et al: Neurobehavioral effects from acute exposures to methyl isobutyl ketone and methyl ethyl ketone. *Fund Appl Toxicol* 19:453–473, 1992
4. Smith AR, Mayers MR: Poisoning and fire hazards of butanone and acetone. *Ind Hyg Bull* 23:175–176, 1944
5. Dyro FM: Methyl ethyl ketone polyneuropathy in shoe factory workers. *Clin Toxicol* 13:371–376, 1978
6. Agency for Toxic Substances and Disease Registry (ATSDR): *Toxicological Profile for 2-Butanone.* US Department of Health and Human Services, Public Health Service, p 118, TP-91/08, 1992
7. Krasavage WJ, O'Donoghue JL, DiVincenzo GD: Ketones. In Clayton GD, Clayton FE (eds): *Patty's Industrial Hygiene and Toxicology*, 3rd ed, Vol 2C, *Toxicology*, pp 4728–4733. New York, John Wiley and Sons, Inc, 1982
8. Altenkirch H, Stoltenburg G, Wagner H: Experimental studies on hydrocarbon neuropathies induced by methyl-ethyl-ketone (MEK). *J Neurol* 219:159–170, 1978
9. Saida K, Mendell J, Weiss H: Peripheral nerve changes induced by methyl *n*-butyl ketone and potentiation by methyl ethyl ketone. *J Neuropathol Exp Neurol* 35:207–225, 1976
10. O'Donoghue J, Krasavage W, DiVincenzo G, et al: Further studies on ketone neurotoxicity and interactions. *Toxicol Appl Pharmacol* 72:201–209, 1984
11. Altenkirch H, Wagner H, Stoltenburg-Didinger G, et al: Potentiation of hexacarbon neurotoxicity by methyl ethyl ketone. *Neurobehav Toxicol Teratol* 4:623–627, 1982
12. Ralston WH et al: Potentiation of 2,5-hexanedione neurotoxicity by methyl ethyl ketone. *Toxicol Appl Pharmacol* 81:319–327, 1985
13. Dick RB, Setzer JV, Taylor BJ, et al: Neurobehavioral effects of short duration exposures

to acetone and methyl ethyl ketone. *Br J Ind Med* 46:111–121, 1989

14. Deacon MM et al: Embryo and fetotoxicity of inhaled methyl ethyl ketone in rats. *Toxicol Appl Pharmacol* 59:620–622, 1981

15. Schwetz BA, Mast TJ, Weigel RJ et al: Developmental toxicity of inhaled methyl ethyl ketone in Swiss mice. *Fund Appl Toxicol* 16:742–748, 1991

16. Ong CN, Sia GL, Ong HY, et al: Biological monitoring of occupational exposure to methyl ethyl ketone. *Int Arch Occup Environ Health* 63:319–324, 1991

METHYL ETHYL KETONE PEROXIDE
CAS:1338-23-4

$C_{18}H_{16}O_4$ or
$C_{18}H_{18}O_4$

Synonyms: 2-Butanone peroxide; MEKP; MEK peroxide

Physical Form.   Liquid

Uses.   Reactive free-radical–generating chemical used as a curing agent for unsaturated polyester resins; hardening agent for fiberglass reinforced plastics

Exposure.   Inhalation

Toxicology.   Methyl ethyl ketone peroxide (MEKP) is a skin and eye irritant.

MEKP has caused irritant dermatitis with direct contact; only rarely has it caused allergic contact dermatitis from occupational exposure.[1]

Exposure of the eyes has resulted in mild to severe injury.[2] The severity of ocular injury was dependent on the length of time from exposure to adequate lavage. Delayed keratitis has also been reported.

In a case of accidental ingestion, massive peripheral zonal hepatic necrosis developed in a 47-year-old man.[3] The clinical course was characterized by temporary cardiac arrest, abdominal burns, severe metabolic acidosis, rapid hepatic failure, rhabdomyolysis, and respiratory insufficiency. The patient died 4 days later from hepatic coma associated with blood coagulation disorders. Microscopic examination showed massive periportal hepatic necrosis. The pathogenic mechanism may involve lipid peroxidation caused by free oxygen radicals derived from the MEKP.[3]

The 4-hour LC_{50} for rats was 200 ppm; the oral LD_{50} was 484 mg/kg.[4] Rats dosed by oral gavage with approximately 96 mg/kg, 3 times a week for 7 weeks, died; histopathologic study revealed liver damage. Two drops of 40% MEKP in dimethyl phthalate in rabbit eyes caused severe damage; at 3%, a moderate reaction occurred, lasting for 2 days, followed by rapid improvement. The maximum nonirritating strength on rabbit skin was 1.5%.

In a tumor-promoting study, using ultraviolet radiation in the UVB region as a tumor initiator, MEKP showed weak promoting activity.[5]

The 1995 ACGIH ceiling–threshold limit value (C-TLV) for methyl ethyl ketone peroxide is 0.2 ppm (1.5 mg/m³).

REFERENCES

1. Stewart B, Beck MH: Contact sensitivity to methyl ethyl ketone in a paint sprayer. *Cont Derm* 26:52–53, 1992

2. Fraunfelder FT, Coster DJ, Drew R, et al: Ocular injury induced by methyl ethyl ketone peroxide. *Am J Ophthalmol* 110:635–640, 1990

3. Karhunen PJ, Ojanpera I, Lalu K, et al: Peripheral zonal hepatic necrosis caused by accidental ingestion of methyl ethyl ketone peroxide. *Hum Exp Toxicol* 9:197–200, 1990

4. Floyd EP, Stokinger HE: Toxicity studies of certain organic peroxides and hydroperoxides. *Am Ind Hyg Assoc J* 19:205–212, 1958

5. Logani MK, Sambuco CP, Forbes PD, Davies RE: Skin-tumour promoting activity of methyl ethyl ketone peroxide—a potent lipid-peroxidizing agent. *Food Chem Toxicol* 22:879–882, 1984

METHYL FORMATE
CAS: 107-31-3

HCOOCH₃

$HCOOCH_3$

Synonyms: Methyl methanoate; formic acid, methyl ester

Physical Form. Colorless liquid

Uses. Solvent; chemical intermediate; insecticide, fumigant; refrigerant

Exposure. Inhalation; skin absorption

Toxicology. Methyl formate is an irritant of the eyes and respiratory tract; at high concentrations, it causes narcosis in animals, and it is expected that severe exposure will produce the same effect in humans.

Workers exposed to the vapor of a solvent containing 30% methyl formate, in addition to ethyl formate, methyl acetate, and ethyl acetate, complained of irritation of mucous membranes, oppression in the chest, dyspnea, symptoms of central nervous system depression, and temporary visual disturbances; air concentrations were not determined.[1] No effects were noted from experimental human exposures to 1500 ppm for 1 minute.[2]

Exposure of guinea pigs to 10,000 ppm for 3 hours was fatal; effects were eye and nose irritation, incoordination, and narcosis; autopsy revealed pulmonary edema.[2]

Methyl formate has a distinct and pleasant odor, but an odor threshold has not been reported.[2]

The 1995 ACGIH threshold limit value–time-weighted average (TLV-TWA) is 100 ppm (246 mg/m³) with a STEL/ceiling of 150 ppm (368 mg/m³).

REFERENCES

1. von Oettingen WF: The aliphatic acids and their esters—toxicity and potential dangers. *AMA Arch Ind Health* 20:517–531, 1959

2. Schrenk HH, Yant WP, Chornyak J, Patty FA: Acute response of guinea pigs to vapors of some new commercial organic compounds. XIII. Methyl formate. *Pub Health Rept* 51:1329–1337, 1936

METHYL HYDRAZINE
CAS: 60-34-4

CH_3NH-NH_2

Synonyms: Monomethylhydrazine; MMH

Physical Form. Liquid

Uses. Rocket fuel; solvent; chemical intermediate

Exposure. Inhalation; skin absorption

Toxicology. Methyl hydrazine causes respiratory irritation, methemoglobinemia, and convulsions; it is carcinogenic in experimental animals.

Volunteers exposed to 90 ppm for 10 minutes had slight redness in the eyes and experienced a tickling sensation of the nose. The only clinical abnormality found during the 60-day follow-up period was the presence of Heinz bodies in 3% to 5% of the erythrocytes by the seventh day.

As a reducing agent, methyl hydrazine causes characteristic oxidative damage to human erythrocytes in vitro. Effects include formation of Heinz bodies and production of methemoglobin.[1]

Exposure of dogs to 21 ppm for 4 hours resulted in convulsions and some deaths; postmortem examination revealed no lesions attributable primarily to methyl hydrazine although secondary manifestations, probably due to convulsions, included pulmonary hemorrhage and edema; convulsions but not death occurred at 15 ppm.[2] In the dogs that survived exposure, there was evidence of moderately severe intravascular hemolysis. The hemolytic effect was most pronounced 4 to 8 days after exposure, and blood values returned to normal within 3 weeks. In another study, additional signs, including eye irritation, tremor, ataxia, diarrhea, and cyanosis, were noted in dogs.[3] Dogs exposed at 5 ppm for 6 hr/day for 6 months had at least a twofold increase in methemoglobin and reductions in erythyrocytes, hemoglobin concentrations, and hematocrit values; the effect was reversible and was not observed at the 1 ppm level.[4]

Applied to the shaved skin of dogs, the liquid was rapidly absorbed, producing toxic signs; at the site of application, the skin became red and edematous.[5]

Administered intraperitoneally to rats on days 6 through 15 of pregnancy, 10 mg/kg/day caused slight maternal toxicity in the form of reduced weight gains but was not selectively embryotoxic or teratogenic.[6]

Results from various long-term animal cancer studies are equivocal. Mice administered 0.001% methyl hydrazine sulfate in drinking water for life developed an increase in lung tumors, whereas 0.01% methyl hydrazine enhanced the development of lung tumors by shortening latent periods; control incidences were not clearly defined in this study.[7] In two other mice studies that may not have allowed for a sufficient latency period, no evidence of carcinogenicity was found.[8,9]

Chronic inhalation exposure by mice (up to 2 ppm, 6 hr/day, 5 days/week, for 1 year) or hamsters (up to 5 ppm, 6 hr/day, 5 days/week, for 1 year) caused a significant increase in rare tumors of the upper respiratory system, including papillomas, adenomas, and osteomas.[10] These benign tumors were thought to be the result of chronic insult to the system. An increase in liver tumors (hemangioma, hemangiosarcoma, adenoma, and carcinoma) also occurred in mice.

The odor threshold is 1 to 3 ppm; it is described as ammonialike or fishy.[1]

The proposed 1995 ACGIH threshold limit value–time-weighted average for methyl hydrazine is 0.01 ppm (0.019 mg/m³) with a notation for skin absorption and an A3 animal carcinogen designation.

Note: For a description of diagnostic signs, differential diagnosis, and medical control, including clinical laboratory treatment of overexposure to methemoglobin-forming agents, see "Aniline."

REFERENCES

1. National Institute for Occupational Safety and Health: *Criteria for a Recommended Standard . . . Occupational Exposure to Hydrazines.* DHEW (NIOSH) Pub No 78-172, p 269. Washington, DC, US Government Printing Office, 1978
2. Jacobson KH et al: The acute toxicity of the vapors of some methylated hydrazine derivatives. *AMA Arch Ind Health* 12:609–616, 1955
3. Haun CC, MacEwen JD, Vernot EH, Eagan GF: Acute inhalation toxicity of monomethylhydrazine vapor. *Am Ind Hyg Assoc J* 31:667–677, 1970
4. MacEwen JD, Haun CC: *Chronic Exposure Studies with Monomethylhydrazine,* p 15. NTIS AD 751 440. Springfield, VA, National Technical Information Service, US Department of Congress, 1971
5. Smith EB, Clark DA: The absorption of monomethylhydrazine through canine skin. *Proc Soc Exp Biol Med* 131:226–232, 1969
6. Keller WC et al: Teratogenic assessment of three methylated hydrazine derivatives in the rat. *J Toxicol Environ Health* 13:125–131, 1984
7. Toth B: Hydrazine, methylhydrazine and methylhydrazine sulfate carcinogenesis in Swiss mice—failure of ammonium hydroxide to interfere in the development of tumors. *Int J Cancer* 9:109–118, 1972
8. Kelly MG et al: Comparative carcinogenicity of *n*-isopropyl-a-(2-methylhydrazino)-*p*-toluamide HCL (procarbazine hydrochloride), its degradation products, other hydrazines, and isonicotinic acid hydrazide. *J Natl Cancer Inst* 42:337–344, 1969
9. Roe FJC et al: Carcinogenicity of hydrazine and 1,1-dimethylhydrazine for mouse lung. *Nature* 216:375–376, 1967
10. Keller WC: Toxity assessment of hydrazine fuels. *Avia Space Environ Med* 59 (11, Suppl): A100–A106, 1988

METHYL IODIDE
CAS: 74-88-4

CH₃I

Synonyms: Iodomethane; monoiodomethane

Physical Form. Colorless liquid

Uses. Chemical intermediate; in microscopy because of its high refractive index

Exposure. Inhalation; skin absorption

Toxicology. Methyl iodide is a neurotoxin and convulsant and has caused pulmonary edema. It is carcinogenic in experimental animals.

The latency period between exposure and onset of symptoms ranges from hours to days.[1] Initial symptoms are lethargy, somnolence, slurred speech, ataxia, dysmetria, and visual disturbances. Neurologic dysfunction may progress to convulsions, coma, and death. If recovery occurs, neurologic findings recede over several weeks and are followed by psychiatric disturbances such as paranoia, delusions, and hallucinations.

A chemical worker accidentally exposed to an unknown concentration of the vapor developed giddiness, diarrhea, sleepiness, and irritability, with recovery in a week; when re-exposed 3 months later, he experienced drowsiness, vomiting, pallor, incoordination, slurred speech, muscular twitching, oliguria, coma, and death.[2] At autopsy, there were bronchopneumonia and pulmonary hemorrhages, with accumulation of combined iodine in the brain.

Experimental application of the liquid to human skin produced a stinging sensation and slight reddening in 10 minutes; after 6 hours of contact, there was spreading erythema, followed by formation of vesicles.[3] Absorption through the skin is said to occur.[4] Splashed in the eye, the liquid causes conjunctivitis.[4]

In rats, reported LC_{50} values are 1750, 900, and 232 ppm for 0.5-, 1-, and 4-hour exposures, respectively.[3-5]

Local sarcomas occurred in rats following subcutaneous injection with 10 mg/kg weekly for 1 year or with a single 50 mg/kg dose.[6] Tumors occurred between 500 and 700 days after the first injection and, in most cases, pulmonary metastases were observed.[6] Repeated intraperitoneal injection of 44 mg/kg in mice reduced survival and caused an increased incidence of lung tumors.[7]

Methyl iodide is considered a potent methylating agent; it methylates hemoglobin in experimental animals and man.[8] It is mutagenic in short-term genotoxic assays and does not require activation. In DNA binding studies in rats, adducts were found in all organs examined, with the highest levels in the stomach and forestomach, following both oral and inhalation administration.[8]

NIOSH has determined that there is sufficient evidence of carcinogenicity in animals to indicate a potential for human carcinogenicity.[4]

The IARC states that there is limited evidence for the carcinogenicity of methyl iodide to experimental animals, but no evaluation could be made of the carcinogenicity of methyl iodide to humans.[9]

The 1995 ACGIH threshold limit value–time-weighted average for methyl iodide (TLV-TWA) is 2 ppm (12 mg/m³) with an A2 suspected human carcinogen designation and a notation for skin absorption.

REFERENCES

1. Appel GB, Galen R, O'Brien J, Schoenfeldt R: Methyl iodide intoxication—a case report. *Ann Intern Med* 82:534–536, 1975
2. von Oettingen WF: *The Halogenated Aliphatic, Olefinic, Cyclic, Aromatic, and Halogenated Insecticides, Their Toxicity and Potential Dangers.* US Public Health Service Pub No 414, pp 30–32. Washington, DC, US Government Printing Office, 1955
3. Buckell M: The toxicity of methyl iodide. I. Preliminary survey. *Brit J Ind Med* 7:122–124, 1950
4. National Institute for Occupational Safety and Health: *Current Intelligence Bulletin 43, Monohalomethanes.* DHHS (NIOSH) Pub No 84-117, 1984

5. Deichmann WB, Gerarde HW: *Toxicology of Drugs and Chemicals*, p 756. New York, Academic Press, 1969
6. Preussman R: Direct alkylating agents as carcinogens. *Food Cosmet Toxicol* 6:576–577, 1968
7. Poirier LA, Stoner GD, Shimkin MB: Bioassay of alkyl halides and nucleotide base analogs by pulmonary tumor response in Strain A mice. *Can Res* 35:1411–1415, 1975
8. Bolt HM, Gansewendt B: Mechanisms of carcinogenicity of methyl halides. *CRC Rev Toxicol* 23:237–253, 1993
9. *IARC Monographs on the Evaluation of the Carcinogenic Risk of Chemicals to Humans*, Vol 41, Some halogenated hydrocarbons and pesticide exposures, p 223. Lyon, International Agency for Research on Cancer, 1986

METHYL ISOAMYL KETONE
CAS:110-12-3

$CH_3COCH(C_2H_5)_2$

Synonyms: MIAK; 5-methyl-2-hexanone; 2-methyl-5-hexanone

Physical Form. Liquid

Uses. Solvent for nitrocellulose, cellulose acetate, butyrate, acrylics, and vinyl polymers

Exposure. Inhalation

Toxicology. Methyl isoamyl ketone (MIAK) is an irritant of the eyes and, at high concentrations, causes narcosis.

Rats exposed to 2000 ppm, 6 hr/day, 5 days/week, for 2 weeks exhibited lethargy and decreased response to noise.[1] When exposed over a period of 90 days to 1000 ppm, there were nose and eye irritation, gel-like casts in seminal fluid of males, and increases in liver and kidney weight. Microscopic examination revealed hepatocyte hypertrophy and renal hyaline droplet formation in males. The toxicity of MIAK following inhalation exposure was not as extensive or severe as that resulting from an earlier study

in which male rats were dosed orally with 2000 mg/kg/day for 13 weeks.

A single 6-hour exposure of rats to 3207 ppm caused eye irritation, decreased respiratory rate, narcosis, and the death of 1 of 4 rats.[2] MIAK produced slight eye irritation in the eyes of rabbits. Repeated daily applications to the backs of guinea pigs resulted in irritation. Slight skin sensitization was observed in 1 of 5 guinea pigs injected with MIAK and Freund's complete adjuvant; this is not compelling evidence of a sensitization potential.

The 1995 ACGIH threshold limit value–time-weighted average (TLV-TWA) is 50 ppm (234 mg/m^3).

REFERENCES

1. Katz GV, Renner ER Jr, Terhaar CJ: Subchronic inhalation toxicity of methyl isoamyl ketone in rats. *Fund Appl Toxicol* 6:498–505, 1986
2. Krasavage WJ, O'Donoghue JL, Divicenzo GD: Ketones. In Clayton GD, Clayton FE (eds): *Patty's Industrial Hygiene and Toxicology*, 3rd ed, Vol 2C, *Toxicology*, pp 4759–4760. New York, Wiley-Interscience, 1982

METHYL ISOBUTYL CARBINOL
CAS 108-11-2

$C_6H_{14}O$

Synonyms: Methyl amyl alcohol; 4-methyl-2-pentanol

Physical Form. Colorless liquid

Uses. Solvent; inorganic syntheses; in brake fluids

Exposure. Inhalation; skin absorption

Toxicology. Methyl isobutyl carbinol is an eye irritant; at high concentrations, it causes narcosis in animals, and it is expected that severe exposure in humans will produce the same effect.

Human subjects exposed to 50 ppm for 15 minutes had eye irritation.[1] No acute, chronic, or systemic effects have been reported in humans.

Five of six rats died following exposure to 2000 ppm for 8 hours; there were no deaths after exposure for 2 hours to the saturated vapor.[2] The single-dose oral toxicity for rats was 2.6 g/kg; the dermal LD_{50} in rabbits was 3.6 ml/kg.[2]

The 1995 ACGIH threshold limit value–time-weighted average (TLV-TWA) for methyl isobutyl carbinol is 25 ppm (104 mg/m^3) with a short-term excursion limit of 40 ppm (167 mg/m^3) and a notation for skin absorption.

REFERENCES

1. Silverman L, Schulte HF, First MW: Further studies on sensory response to certain industrial solvent vapors. *J Ind Hyg Toxicol* 28:262–266, 1946
2. Smyth HF Jr, Carpenter CP, Weil CS: Range-finding toxicity data: List IV. *AMA Arch Ind Hyg Occup Med* 4:119–122, 1951

METHYL ISOBUTYL KETONE
CAS: 108-10-1

$C_6H_{12}O$

Synonyms: Hexone; MIBK; 4-methyl-2-pentanone

Physical Form. Colorless liquid

Uses. In paints, glues, and cleaning agents; used in the plastic and petrol industries

Exposure. Inhalation

Toxicology. Methyl isobutyl ketone (MIBK) is an irritant of the eyes, mucous membranes, and skin; high concentrations cause narcosis in animals, and it is expected that severe exposure will cause the same effects in humans.

Exposures to 80 to 500 ppm produced weakness, loss of appetite, headache, eye irritation, sore throat, and nausea.[1] At 200 ppm, the eyes of most persons were irritated, and 100 ppm was the highest concentration most volunteers estimated to be acceptable for an 8-hour exposure.[2] Volunteers exposed to 50 ppm for 2 hours showed no significant effects on the performance of reaction time tasks or tests of mental arithmetic; irritation in the nose and throat was reported by 3 of 8 subjects at this level.[3]

Exposure of rats to 4000 ppm for 4 hours caused death; 2000 ppm for 4 hours was not fatal.[4] A 2-week exposure of rats to 200 ppm produced toxic nephrosis of the proximal tubules and increased liver weights.[5] A 90-day continuous exposure at 100 ppm produced no significant changes.[5] In a more recent report of rats and mice exposed 6 hr/day for 2 weeks to 100, 500, or 2000 ppm, the only observed histological changes were increases in regenerative tubular epithelia and hyalin droplets in kidneys of male rats exposed at the two highest levels.[6] Exposure of both species to methyl isobutyl ketone at levels of up to 1000 ppm for 14 weeks was without significant toxicological effect except for an increase in the incidence and extent of hyalin droplets in the kidneys of male rats. The relevance of kidney tubular effects to humans is not known.

Studies in mice have shown that methyl isobutyl ketone can enhance the ethanol-induced loss of righting reflex by reducing the elimination rate of ethanol.[7] Human response to ethanol may be affected by MIBK, and simultaneous exposure to alcoholic beverages and MIBK should be avoided.

The liquid splashed in the eyes may cause pain and irritation. Repeated or prolonged skin contact may cause defatting of the skin, with primary irritation and desquamation.[8]

Results of a number of genotoxic assays show that MIBK exhibits very little, if any, mutagenic activity.[9] Existing studies also demonstrate that MIBK is not teratogenic and exhibits low reproductive toxicity.

Methyl isobutyl ketone has a characteristic camphorlike odor detectable at 100 ppm.[1]

The 1995 ACGIH threshold limit value–time-weighted average (TLV-TWA) for methyl isobutyl ketone is 50 ppm (205 mg/m³) with a short-term excursion limit of 75 ppm (307 mg/m³).

REFERENCES

1. National Institute for Occupational Safety and Health: *Criteria for a Recommended Standard . . . Occupational Exposure to Ketones.* DHEW (NIOSH) Pub No 78-173. Washington, DC, US Government Printing Office, 1978
2. Silverman L, Schulte HF, First MW: Further studies on sensory response to certain industrial solvent vapors. *J Ind Hyg Toxicol* 28:262–266, 1946
3. Hjelm EW, Hagberg M, Iregren A, et al: Exposure to methyl isobutyl ketone: toxico-kinetics and occurrence of irritative and CNS symptoms in man. *Int Arch Occup Environ Health* 62:19–26, 1990
4. Smyth HF Jr, Carpenter CP, Weil CS: Range-finding toxicity data: List IV. *AMA Arch Ind Hyg Occup Med* 4:119–122, 1951
5. MacEwen JD et al: *Effect of 90-Day Continuous Exposure to Methyl Isobutyl Ketone on Dogs, Monkeys and Rats*, p 23. NTIS AD 730 291. Springfield, VA, National Technical Information Service, US Department of Commerce, 1971
6. Phillips RD, Moran EJ, Dodd DE, et al: A 14-week vapor inhalation toxicity study of methyl isobutyl ketone. *Fund Appl Toxicol* 9:380–388, 1987
7. Cunningham J, Sharkawi M, Plaa GL: Pharmacological and metabolic interactions between ethanol and methyl *n*-butyl ketone, methyl isobutyl ketone, methyl ethyl ketone or acetone in mice. *Fund Appl Toxicol* 13:102–109, 1989
8. Hygienic Guide Series: Methyl isobutyl ketone. *Am Ind Hyg Assoc J* 27:209–211, 1966
9. Strickland GD: Methyl ethyl ketone and methyl isobutyl ketone not carcinogenic. *Environ Health Perspect* 101:566, 1993

METHYL ISOCYANATE
CAS: 624-83-9

CH₃CNO

Synonyms: Isocyanic acid-methyl ester; MIC

Physical Form. Liquid; aerosol

Uses. In production of polyurethane foams and plastics; chemical intermediate

Exposure. Inhalation; skin absorption

Toxicology. Methyl isocyanate (MIC) is an irritant of the eyes, mucous membranes, and skin; at high doses, it is extremely toxic and can cause death from pulmonary edema.

Isocyanates cause pulmonary sensitization in susceptible individuals; if this occurs, further exposure should be avoided, because extremely low levels of exposure may trigger an asthmatic episode; cross-sensitization to unrelated materials probably does not occur.[1]

Experimental exposure of 4 human subjects to MIC for 1 to 5 minutes caused the following effects: 0.04 ppm, no effects; 2 ppm, lacrimation and irritation of the nose and throat; 4 ppm, symptoms of irritation more marked; 21 ppm, unbearable irritation of eyes, nose, and throat.[2]

The accidental release of several tons of methyl isocyanate in 1984 at Bhopal, India, resulted in a very heavy death toll and, in survivors, significant impairment of health.[3,4] Immediate symptoms were difficulty breathing, skin and eye irritation, vomiting, and unconsciousness. Only a few deaths were recorded in the first few hours, with the maximum fatalities occurring between 24 and 72 hours after exposure. The predominant cause of death was cardiac arrest following severe pulmonary edema. Lung-function abnormalities have persisted years after exposure. Ophthalmic effects included lacrimation, lid edema, photophobia, and ulceration of the corneal epithelium. A follow-up study 3 years after exposure showed excess irritation, eyelid infection, cataracts, and decrease in visual acuity, but corneal erosion was resolved.[5]

Reproductive effects at the time of the incident included a 44% loss of fetuses in 865 pregnant women (15% expected) and the neonatal death rate increased from 3% to 15%.[3] Reproductive toxicity of methyl isocyanate has been confirmed in animal studies; exposure has caused increased resorptions, reduced pup weight, and reduced neonatal survival.

In genotoxic assays, both positive and negative results have been reported.[3]

The 1995 threshold limit value–time-weighted average (TLV-TWA) for methyl isocyanate is 0.02 ppm (0.047 mg/m³) with a notation for skin absorption.

REFERENCES

1. Rye WA: Human responses to isocyanate exposure. *J Occup Med* 15:306–307, 1973
2. Methyl isocyanate. *Documentation of TLVs and BEIs*, 6th ed, Vol II, pp 1022–1024. Cincinnati, OH, American Conference of Governmental Industrial Hygienists, 1991
3. Varma DR, Guest I: The Bhopal accident and methyl isocyanate toxicity. *J Toxicol Environ Health* 40:513–529, 1993
4. Mehta PS, Mehta AS, Mehta SJ, et al: Bhopal tragedy's health effects—review of methyl isocyanate toxicity. *JAMA* 264:1781–2787, 1990
5. Andersson N, Ajwani MK, Mahashabde S, et al: Delayed eye and other consequences from exposure to methyl isocyanate: 93% follow-up of exposed and unexposed cohorts in Bhopal. *Br J Ind Med* 47:553–558, 1990

METHYL ISOPROPYL KETONE
CAS: 563-80-4

(CH₃)₂CHCOCH₃

Synonyms: MIPK; 3-methyl-2-butanone

Physical Form. Liquid

Uses. Solvent for nitrocellulose lacquers

Exposure. Inhalation

Toxicology. Methyl isopropyl ketone (MIPK), by analogy to other aliphatic ketones, is expected to be an irritant of the eyes, mucous membranes, and skin; at high concentrations, it causes narcosis in animals, and it is expected that severe exposure in humans will produce the same effect.

In a range-finding study, exposure for 4 hours to 5700 ppm was fatal to rats.[1] The oral LD_{50} for male rats and mice was 3200 mg/kg. Signs of intoxication included weakness, prostration, and ataxia.

The 1995 threshold limit value–time-weighted average (TLV-TWA) for methyl isopropyl ketone is 200 ppm (705 mg/m³).

REFERENCES

1. Carpenter CP, Weil CS, Smyth HF Jr: Range-finding toxicity data: List VIII. *Toxicol Appl Pharmacol* 28:313–319, 1974

METHYL MERCAPTAN
CAS: 74-93-1

CH₃SH

Synonyms: Methanethiol; mercaptomethane; thiomethyl alcohol; methyl sulfhydrate

Physical Form. Flammable gas liquefying at 6°C; odor of rotten cabbage

Uses. Intermediate in manufacturing of jet fuels, pesticides, fungicides, plastics; synthesis of methionine; emission from paper pulp mills; odoriferous additive to natural gas

Exposure. Inhalation

Toxicology. Methyl mercaptan causes coma at high levels; hematological effects have also been reported.

In a fatal human exposure, a worker engaged in emptying metal gas cylinders of methyl mercaptan was found comatose at the work site; he developed expiratory wheezes, elevated blood pressure, tachycardia, and marked rigidity of extremities.[1] Methemoglobinemia and severe hemolytic anemia developed with hematuria and proteinuria but were brief in duration; deep coma persisted until death due to pulmonary embolus 28 days after exposure. It was determined that the individual was deficient in erythrocyte glucose-6-phosphate dehydrogenase, which was the likely cause of the hemolysis and formation of methemoglobin.

In a nonfatal incident, a worker in a refinery inhaled methyl mercaptan and was comatose for 9 hours. Although not dyspneic, the individual was cyanotic and experienced convulsions; recovery occurred by the fourth day. Ten days later, the worker was treated successfully for a lung abcess.

Although details are lacking, one report states that effects in animals exposed to methyl mercaptan were restlessness and muscular weakness, progressing to paralysis, convulsions, respiratory depression, and cyanosis.[1] Rats exposed via inhalation to methyl mercaptan at 1400 ppm, but not 1200 ppm, for 15 minutes became lethargic and comatose.[2] Exposure of rats to various concentrations for 4 hours allowed a determination of an LC_{50} of 675 ppm, thus making it slightly less acutely toxic than hydrogen sulfide (LC_{50} 444 ppm). A subchronic toxicity study in young male rats exposed at 2, 17, and 57 ppm for 3 months showed a dose-related decreased weight gain (about 15% at 57 ppm) but no clear pathologic or biochemical test alterations. There were some minor microscopic hepatic alterations of questionable significance in the exposed animals.

There are no reports of developmental, reproductive, or genotoxic effects.[4]

The toxic potential of methyl mercaptan is due to its reversible inhibition of cytochrome c oxidase at the end of the respiratory chain of mitochondria.[5]

The 1995 ACGIH threshold limit value–time-weighted average (TLV-TWA) for methyl mercaptan is 0.5 ppm (0.98 mg/m³).

REFERENCES

1. Shults WT, Fountain EN, Lynch EC: Methanethiol poisoning. *JAMA* 211:2153–2154, 1970
2. Zieve L, Doizaki WM, Zieve FJ: Synergism between mercaptans and ammonia or fatty acids in the production of coma: a possible role for mercaptans in the pathogenesis of hepatic coma. *J Lab Clin Med* 83:16–28, 1974
3. Tansy M et al: Acute and subchronic toxicity studies of rats exposed to vapors of methyl mercaptan and other reduced-sulfur compounds. *J Toxicol Environ Health* 8:71–88, 1981
4. Agency for Toxic Substances and Disease Registry (ATSDR): *Toxicological Profile for Methyl Mercaptan.* US Department of Health and Human Services, Public Health Service. TP-91/20, 1992
5. Wilms J, Lub J, Wever R: Reactions of mercaptans with cytochrome c oxidase and cytochrome c. *Biochim Biophys Acta* 589:324, 1980

METHYL METHACRYLATE
CAS: 80-62-6

$C_5H_8O_2$

Synonyms: Methacrylic acid, methyl ester; methyl 2-methylpropenoic acid; methyl α-methyl acrylate; methyl methylacrylate; 2-(methyoxycarbonyl)-1-propene

Physical Form. Colorless liquid; commercial form contains a small amount of hydroquinone or hydroquinone monomethyl ether to inhibit spontaneous polymerization

Uses. In production of polymethyl methacrylate polymers for use in acrylic sheets and moldings, extrusion powder, acrylic surface coatings, printing inks, and adhesives used in surgery and dentistry

Exposure. Inhalation

Toxicology. Methyl methacrylate is an irritant of eyes, skin, and mucous membranes.

The toxic effects are due to the monomer; the polymer appears inert. The severity of effects is believed to be inversely proportional to the degree of polymerization.

Workers exposed to either 11 to 33 mg/m[3] or 100 to 200 mg/m[3] had dose-dependent increases in incidences of neurasthenia, laryngitis, and hypotension.[1] In another study of 91 exposed and 43 nonexposed workers at five plants producing polymethyl methacrylate sheets, exposures ranged from 4 to 49 ppm, and there were no detectable clinical signs or symptoms.[2] In a survey of 152 workers exposed to concentrations ranging from 0.5 to 50 ppm, 78% reported a high incidence of headache, 30% pain in the extremities, 10% irritability, 20% loss of memory, and 21% excessive fatigue and sleep disturbances.[3]

Handlers of methyl methacrylate cement have developed paresthesia of the fingers.[4] Dental technicians who use bare hands to mold methyl methacrylate putty had significantly slower distal sensory conduction velocities from the digits, implicating mild axonal degeneration in the area of contact with methyl methacrylate.[5] The toxic effect on the nervous tissues may be due to diffusion into the nerve cells causing lysis of the membrane lipids and destruction of the myelin sheath.

Humans have developed strong skin reactions when rechallenged with the liquid.[6] Allergic contact dermatitis has been reported in workers handling methacrylate sealants, including methyl methacrylate.[7] In another report, 5 subjects were shown by bronchial provocation tests to have occupational asthma due to methyl methacrylate or cyanoacrylates.[8]

Acute inhalation exposure of dogs to 11,000 ppm led to central nervous system depression, a drop in blood pressure, liver and kidney damage, and death due to respiratory arrest.[9] Mice exposed to 1520 ppm for 2 hours twice daily for 10 days showed no significant histological changes in heart, liver, kidney, or lungs. In male rats exposed to methyl methacrylate vapor at 116 ppm, 7 hr/day 5 days/week, for 5 months, the tracheal mucosa was denuded of cilia, and the number of microvilli on the epithelium was reduced.[10] Exposure of pregnant rats to 27,500 ppm (nose-only inhalation) on days 6 through 15 of gestation for 54 minutes resulted in maternal toxicity and significant increases in fetal deaths, hematomas, and skeletal anomalies.[11] At 2028 ppm, for 6 hr/day during days 6 through 15 of gestation, there was decreased maternal food consumption and body weight gain in pregnant rats exposed by vapor inhalation but no embryo or fetal toxicity or malformations.[12]

In a 2-year inhalation study, there was no evidence of carcinogenicity of methyl methacrylate for male rats exposed at 500 or 1000 ppm, for female rats exposed at 250 or 500 ppm, or for male and female mice exposed at 500 or 1000 ppm.[13] There was inflammation of the nasal cavity and degeneration of the olfactory sensory epithelium in rats and mice; epithelial hyperplasia of the nasal cavity was also observed in exposed mice.

A mortality study of three cohorts engaged in the manufacturing and polymerizaton of acrylate monomers revealed an excess colon cancer rate among men employed extensively during the 1940s in jobs entailing the highest exposures to vapor-phase ethyl acrylate and methyl acrylate monomer.[14] The excess mortality appeared only after the equivalent of 3 years' employment followed by a latency of 20 years. The two cohorts with later dates of hire showed no excess mortality.[14] A mortality study of 2671 men exposed to methyl methacrylate alone found a nonsignificantly increased mortality from all cancers but no significant risk at any particular site with increasing dose.[15]

The IARC has determined that there is inadequate evidence in humans for the carcinogenicity of methyl methacrylate and that there is evidence suggesting the lack of carcinogenicity in experimental animals.[1]

Methyl methacrylate is genotoxic in vitro, inducing gene mutation, sister chromatid exchange, micronuclei, and chromosomal aberrations in a variety of assays.[1]

Methanol concentrations in blood and urine have been found to correlate with methyl methacrylate exposure.[16] The lack of specificity of methanol to methyl methacrylate exposure limits its usefulness as a biological indicator.

The 1995 ACGIH threshold limit value–time-weighted average (TLV-TWA) is 100 ppm (410 mg/m[3]).

REFERENCES

1. *IARC Monographs on the Evaluation of Carcinogenic Risks to Humans*, Vol 60, Some industrial chemicals, pp 445–474. Lyon, International Agency for Research on Cancer, 1994
2. Cromer J, Kronoveter K: A study of methyl methacrylate exposures and employee health. US Department of Health, Education, and Welfare, National Institute for Occupational Safety and Health, Cincinnati, OH. DHEW (NIOSH) Pub No 77-119. Washington, DC, US Government Printing Office, 1976
3. Blagodatin VM, Golova IA, Bladokatkina NK, et al: Establishing the maximum permissible concentration of the methyl ester of methacrylic acid in the air of a work area. *Gig Tr Prot Zabol* 6:5–8, 1976
4. Kassis V, Vedel P, Darre E: Contact dermatitis to methyl methacrylate. *Contact Derm* 11:26–28, 1984
5. Seppalainen A, Rajaniemi R: Local neurotoxicity of methyl methacrylate among dental technicians. *Am J Ind Med* 5:471–477, 1984.
6. Speakman CR et al: Monomeric methyl methacrylate: studies on toxicity. *J Ind Med* 14:292, 1945
7. Conde-Salazar L et al: Occupational allergic contact dermatitis from anaerobic acrylic sealants. *Contact Derm* 18:129–132, 1988
8. Lozewicz S et al: Occupational asthma due to methyl methacrylate and cyanoacrylates. *Thorax* 40:836–839, 1985
9. McLaughlin et al: Pulmonary toxicity of methyl methacrylate vapors: an environmental study. *Arch Environ Health* 34:336–338, 1979
10. Tansy M et al: Chronic biological effects of methyl methacrylate vapor. III. Histopathology, blood chemistry, and hepatic and ciliary function in the rat. *Environ Res* 21:117–125, 1980
11. Nicholas CA, Lawrence WH, Autian J: Embryotoxicity and fetotoxicity from maternal inhalation of methyl methacrylate monomer in rats. *Toxicol Appl Pharmacol* 50:451–458, 1979
12. Solomon HM, McLaughlin JE, Swenson RE, et al: Methyl methacrylate: inhalation developmental toxicity study in rats. *Teratology* 48:115–125, 1993
13. National Toxicology Program: *NTP Technical Report on the Toxicology and Carcinogenesis Studies of Methyl Methacrylate (CAS No 80-62-6) in F344/N Rats and B6C3F₁ Mice (Inhalation Studies)*. DHHS (NTP) TR-314, pp 1–202. Research Triangle Park, NC, US Department of Health and Human Services, October 1986
14. Walker AM, Cohen AJ, Loughlin JE, et al: Mortality from cancer of the colon or rectum among workers exposed to ethyl acrylate and methyl acrylate. *Scand J Work Environ Health* 17:7–19, 1991
15. Collins JJ, Page LC, Caporossi JC, et al: Mortality patterns among men exposed to methyl methacrylate. *J Occup Med* 31:41–46, 1989
16. Mizunuma K, Kawai T, Yasugi T, et al: Biological monitoring and possible health effects in workers occupationally exposed to methyl methacrylate. *Int Arch Occup Environ Health* 65:227–232, 1993

METHYL PARATHION
CAS: 298-00-0

$(CH_3O)_2P(S)OC_6H_4NO_2$

Synonyms: O,O-Dimethyl O-p-nitrophenyl phosphorothioate; Metron; Nitrox; parathion-methyl; Metacide, metaphos

Physical Form. White solid (pure); tan to brown solid (technical grade)

Uses. Insecticide

Exposure. Inhalation; skin absorption; ingestion

Toxicology. Methyl parathion is an anticholinesterase agent.

Signs and symptoms of overexposure are caused by the inactivation of the enzyme cholinesterase; inactivation results in the accumulation of acetylcholine at synapses in the nervous system, skeletal and smooth muscle, and secretory glands. The sequence of the development of systemic effects varies with the route of entry. The onset of signs and symptoms is usually prompt but may be delayed up to 12 hours.[1-5] After inhalation, respiratory and ocular effects are the first to appear, often within a few minutes of exposure. Respiratory effects include tightness in the chest and wheezing due to bronchoconstriction and excessive bronchial secretion; laryngeal spasms and excessive salivation may add to the respiratory distress; cyanosis may also occur. Ocular effects include miosis, blurring of distant vision, tearing, rhinorrhea, and frontal headache.

After ingestion, gastrointestinal effects, such as anorexia, nausea, vomiting, abdominal cramps, and diarrhea appear within 15 minutes to 2 hours. After skin absorption, localized sweating and muscular fasciculations in the immediate area usually occur within 15 minutes to 4 hours; skin absorption is somewhat greater at higher ambient temperatures and is increased by the presence of dermatitis.[1–3]

With severe intoxication by all routes, an excess of acetylcholine at the neuromuscular junctions of skeletal muscle causes weakness aggravated by exertion, involuntary twitchings, fasciculations and, eventually, paralysis. The most serious consequence is paralysis of the respiratory muscles. Effects on the central nervous system include giddiness, confusion, ataxia, slurred speech, Cheyne–Stokes respiration, convulsions, coma, and loss of reflexes. The blood pressure may fall to low levels, and cardiac irregularities, including complete heart block, may occur.[2]

Complete symptomatic recovery usually occurs within a week; increased susceptibility to the effects of anticholinesterase agents persists for up to several weeks after exposure.[4] Daily exposure to concentrations that are insufficient to produce symptoms following a single exposure may result in the onset of symptoms. Continued daily exposure may be followed by increasingly severe effects.

Deaths from occupational exposure have been reported, usually following massive accidental exposures.[1] Data from human poisonings by methyl parathion are not sufficiently detailed to identify the range between the doses producing first symptoms and those producing severe or fatal intoxication.[4] The probable oral lethal dose is 5 to 50 mg/kg. Most animal data and limited human data indicate that methyl parathion is somewhat less acutely toxic than parathion.[4]

Methyl parathion itself is not a strong cho-linesterase inhibitor, but one of its metabolites, methyl paraoxon, is an active inhibitor. Methyl paraoxon inactivates cholinesterase by phosphorylation of the active site of the enzyme to form the "dimethylphosphoryl enzyme." Over the following 24 to 48 hours, there is a process, called aging, of conversion to the "monomethylphosphoryl enzyme." Aging is of clinical interest in the treatment of poisoning because cholinesterase reactivators such as pralidoxime (2-PAM, Protopam) chloride are ineffective after aging has occurred. Measurement of metabolites of methyl parathion, namely, para-nitrophenol and dimethylphosphate, in the urine has been used to monitor exposure of workers.[6]

Methyl parathion, administered intraperitoneally at maternally lethal doses, was teratogenic to mice, producing cleft palate and rib abnormalities. High-dose administration to rats, sometimes producing maternal toxicity, resulted in evidence of embryo-fetotoxicity with increased resorptions and growth retardation.[6]

A two-year bioassay of methyl parathion in mice and rats did not demonstrate any increased incidence of tumors in dosed animals.[7] The IARC has concluded that there is no evidence that methyl parathion is carcinogenic to experimental animals.[6]

Methyl parathion is not strongly genotoxic; it has produced both positive and negative results in both eukaryotic and prokaryotic assays.[8]

There is no evidence that methyl parathion can induce delayed peripheral neuropathy in man or experimental animals.[6]

The 1995 ACGIH threshold limit value–time-weighted average (TLV-TWA) for methyl parathion is 0.2 mg/m³ with a notation for skin absorption.

Note: For a description of diagnostic signs, differential diagnosis, and medical control, including clinical laboratory tests, as well as specific treatment of overexposure to anticholinesterase insecticides, see "Parathion."

REFERENCES

1. Hayes WJ Jr: *Pesticides Studied in Man*, pp 284–435. Baltimore, MD, Williams and Wilkins, 1982

2. Taylor P: Anticholinesterase agents. In Gilman AG et al (eds): *Goodman and Gilman's The Pharmacological Basis of Therapeutics*, 7th ed, pp 116–129. New York, Macmillan, 1985
3. Koelle GB (ed): Cholinesterases and anticholinesterase agents. *Handbuch der Experimentellen Pharmakologie*, Vol 15, pp 989–1027. Berlin, Springer-Verlag, 1963
4. National Institute for Occupational Safety and Health: *Criteria for a Recommended Standard . . . Occupational Exposure to Methyl Parathion*. DHEW (NIOSH) Pub No 77-106, pp 31–68. Washington, DC, US Government Printing Office, 1976
5. Namba T, Nolte CT, Jackrel J, Grob D: Poisoning due to organophosphate insecticides. *Am J Med* 50:475–492, 1971
6. *IARC Monographs on the Evaluation of the Carcinogenic Risk of Chemicals to Humans*, Vol 30, Miscellaneous pesticides, pp 131–152. Lyon, International Agency for Research on Cancer, 1983
7. National Cancer Institute: *Bioassay of Methyl Parathion for Possible Carcinogenicity*, TR-157. DHEW (NIH) Pub No 79-1713. Washington, DC, US Government Printing Office, 1979
8. Agency for Toxic Substances and Disease Registry (ATSDR): *Toxicological Profile for Methyl Parathion*. US Department of Health and Human Services, Public Health Service, p 148, TP-91/21, 1992

METHYL PROPYL KETONE
CAS: 107-87-9

$CH_3COC_3H_7$

Synonyms: 2-Pentanone; ethyl acetone

Physical Form. Colorless liquid

Uses. Solvent

Exposure. Inhalation

Toxicology. Methyl propyl ketone is an irritant of the eyes and mucous membranes; at high concentrations, it causes narcosis in animals, and it is expected that severe exposure will produce the same effect in humans.

At 400 ppm, 6 of 10 subjects reported eye irritation; the 4 nonresponders experienced irritation when the concentration was raised to 600 ppm. In other reports, brief exposures of humans to 2000 to 4000 ppm were very irritating; 1500 ppm had a strong odor and caused irritation of the eyes and nose.[2] There have been no reports of chronic or systemic effects in humans.

In guinea pigs, exposure to 50,000 ppm for 50 minutes or 13,000 ppm for 300 minutes was fatal.[3] Animals survived 810 minutes at 5000 ppm, but narcosis occurred in 460 to 710 minutes.[2] Although methyl propyl ketone has not been reported to be nephro- or hepatotoxic in rats, it has been shown to potentiate kidney and liver injury produced by chloroform.[4] Applied to the skin of rabbits, the undiluted liquid was only slightly irritating within 24 hours.[5]

The 1995 ACGIH threshold limit value–time-weighted average (TLV-TWA) for methyl propyl ketone is 200 pm (705 mg/m³) with a short-term excursion limit of 250 ppm (881 mg/m³).

REFERENCES

1. Hansen LF, Nielsen GD: Sensory irritation and pulmonary irritation of *n*-methyl ketones: receptor activation mechanisms and relationships with threshold limit values. *Arch Toxicol* 68:193–202, 1994
2. Henson EV: Toxicology of some aliphatic ketones. *J Occup Med* 1:607–613, 1959
3. Yant WP et al: Acute response of guinea pigs to vapors of some new commercial organic compounds. *Pub Health Rep* 51:392–399, 1936
4. Hewitt WR, Brown EM: Nephrotoxic interactions between ketonic solvents and halogenated aliphatic chemicals. *Fund Appl Toxicol* 4:902–908, 1984
5. National Institute for Occupational Safety and Health: *Criteria for a Recommended Standard . . . Occupational Exposure to Ketones*. DHEW (NIOSH) Pub No 78-173, p 244. Washington, DC, US Government Printing Office, 1978

N-METHYL-2-PYRROLIDONE
CAS: 872-50-4

C₅H₉NO

C_5H_9NO

Synonyms: NMP; M-Pyrol; methylpyrrolidone

Physical Form. Almost colorless liquid with a mild aminelike odor

Uses. Solvent for high-temperature resins; in petrochemical processing, surface coatings, dyes and pigments, and industrial and domestic cleaning compounds; in agricultural and pharmaceutical formulations

Toxicology. N-Methyl-2-pyrrolidone (NMP) is of low systemic toxicity but produces skin irritation with prolonged contact and severe eye irritation.

NMP produced no skin irritation on patch testing for 24 hours in 50 volunteers.[1] A few mild transient reactions were noted after repeated applications. There was no evidence of contact sensitization.

Of 12 workers, 10 experienced acute irritant contact dermatitis of the hands after 2 days of direct contact.[2] In the most severe case, a woman with no previous skin problems who wore latex gloves intermittently had painful swelling of the fingers of both hands with redness and vesicles on the palms. The affected skin later became thickened and showed a brownish discoloration. Another worker noticed small vesicles on the forehead, probably from scratching with contaminated fingers. All cutaneous reactions cleared within 3 weeks of termination of exposure. Gas chromatograph analysis of the NMP used at the factory did not reveal any contaminating compounds.

Although NMP is of low volatility, another study suggests that concentrations as low as 0.7 ppm in air for short periods of time may cause severe eye irritation and headaches.[3]

In rats, the oral LD₅₀ was approximately 4.2 ml/kg. In rabbits, the dermal LD₅₀ was between 4 and 8 mg/kg.[1] Repeated skin application in lower doses, 0.4 and 0.8 mg/kg/day, resulted in mild skin irritation in rabbits. Although NMP is a severe eye irritant in rabbits, producing conjunctivitis and corneal opacity after instillation, it did not appear to produce permanent eye damage. Rats exposed to vapor from NMP heated for 6 hours or to saturated room temperature air for 6 hr/day for 10 days showed no evidence of toxic effects.[1]

In subchronic inhalation studies, rats exposed to 1.0 mg/l, 6 hr/day, 5 days/week, for 4 weeks, exhibited lethargy, respiratory difficulty, and excessive mortality.[4] Rats had focal pneumonia, bone marrow hypoplasia, and atrophy of lymphoid tissue in the spleen and thymus. The lesions were reversible in surviving animals following 2 weeks of recovery. No carcinogenic effects were observed in rats exposed to 0.04 or 0.4 mg/l for 2 years; male mice had slightly reduced mean body weight at the higher dose.

Subacute 90-day feeding studies in rats, mice, and beagle dogs demonstrated no apparent clinically significant toxic effects in the treated animals; several minor, statistically significant changes in laboratory parameters, such as GGPT and platelet counts, were noted at high doses in some treated groups but not consistently in all three studies.[1]

Dermal application studies in female rats showed no evidence of teratogenic effects although lower weight gains in the maternal animals and skeletal variations in the offspring were observed at the highest dose (750 mg/kg/day); the latter effect was thought to be due to maternal toxicity.[1] Pregnant rats exposed to 0.1 or 0.36 mg/l, 6 hr/day, on days 6 to 15 of gestation, had sporadic lethargy and irregular respiration in the first 3 days but no other clinical signs or pathological lesions.[5] No abnormal development was detected in the offspring.

A 1995 TLV has not been established.

REFERENCES

1. N-*Methylpyrrolidone—Summary of Toxicity Information.* Wayne, NJ, GAF Corporation, 1983
2. Leira HL, Tiltnes A, Svendsen K, et al: Irritant

cutaneous reactions to *N*-methyl-2-pyrroli-done (NMP). *Contact Derm* 27:148–150, 1992
3. Beaulieu HJ, Schmerber KR: M-pyrol (NMP) use in the microelectronic industry. *Appl Occup Environ Hyg* 6:874–880, 1991
4. Lee KP, Chromey NC, Culik R, et al: Toxicity of *n*-methyl-2-pyrrolidone (NMP): Terato-genic, subchronic and two-year inhalation studies. *Fund Appl Toxicol* 9:222–235, 1987
5. Becci PJ, Knickerbocker MJ, Reagan EL: Ter-atogenicity study of *n*-methylpyrrolidone after dermal application to Sprague–Dawley rats. *Fund Appl Toxicol* 2:73–76, 1982

METHYL SILICATE
CAS: 681-84-5

(CHO₃)₄Si

(CHO$_3$)$_4$Si

Synonyms: Tetramethoxy silane; tetramethyl orthosilicate; methyl orthosilicate; tetra-methyl silicate

Physical Form. Liquid

Uses. For coating screens of television pic-ture tubes; mold binders; corrosion-resistant coatings; in catalyst preparation; silicone inter-mediate

Exposure. Inhalation

Toxicology. Methyl silicate is a severe eye irritant and an irritant of the nose and throat.

Application of undiluted methyl silicate to the eyes of rabbits caused marked edema and necrosis of the eyelid.[1] Exposure of rats to 250 ppm for 4 hours caused death in all 6 animals, whereas none died after exposure to 125 ppm for 4 hours.

Exposure of rats for 6 hr/day, 5 days/week, for 4 weeks to 15 ppm caused corneal lesions in some of the rats, along with reductions in total serum protein, lactate dehydrogenase, and serum albumin.[2] Rats exposed to 30 ppm exhib-ited irritation of the upper respiratory tract and bronchiolar inflammation. No adverse effects were noted at 10 ppm.

Human experience indicates that methyl silicate exposure has a delayed action on the eyes, causing slight or no immediate effect, which is followed, after a latent period of several hours, by potentially serious injury to the eyes.[3]

The 1995 ACGIH threshold limit value–time-weighted average (TLV-TWA) for methyl silicate is 1 ppm (6 mg/m³).

REFERENCES

1. Smyth HF Jr, Carpenter CP, Weil CS: Range-finding toxicity data: List IV. *Arch Ind Hyg Occup Med* 4:119–122, 1951
2. Kolesar GB, Siddiqui WH, Geil RG, et al: Subchronic inhalation toxicity of tetrameth-oxysilane in rats. *Fund Appl Toxicol* 13:285–295, 1989
3. Grant WM: *Toxicology of the Eye*, 3rd ed, p 627. Springfield, IL, Charles C Thomas, 1986

α-METHYL STYRENE
CAS: 98-83-9

C₉H₁₀

C$_9$H$_{10}$

Synonyms: 1-Methyl-1-phenylethylene; iso-propenylbenzene; β-phenylpropylene

Physical Form. Colorless liquid

Uses. In the formulation of polymers and resins

Exposure. Inhalation

Toxicology. α-Methyl styrene is an irritant of the eyes and mucous membranes; severe ex-posure may result in central nervous system de-pression.

Humans briefly exposed to 600 ppm expe-rienced strong eye and nasal irritation; at 200 ppm, the odor was objectionable whereas, at 100 ppm, the odor was strong but was tolerated without excessive discomfort.[1]

Guinea pigs and rats exposed 7 hr/day to 3000 ppm for 3 to 4 days died; at 800 ppm for 27 days, there were slight changes in liver and kidney weight and some reduction in growth.[1] Exposure 7 hr/day to 200 ppm for 139 days caused no adverse effects in several species.

The liquid dropped in the eyes of rabbits caused slight conjunctival irritation; applied to rabbit skin, it produced erythema.

The odor of α-methyl styrene is detectable at 50 ppm; the odor and irritant properties provide good warning of toxic levels.

The 1995 ACGIH threshold limit value–time-weighted average (TLV-TWA) is 50 ppm (242 mg/m³) with a short-term excursion level of 100 ppm (483 mg/m³).

REFERENCES

1. Wolf MA, Rowe VK, McCollister DD, et al: Toxicological studies of certain alkylated benzenes and benzene. *Arch Ind Health* 14:387–398, 1956

MEVINPHOS
CAS: 7786-34-7

$C_7H_{13}O_6P$

Synonym: 2-Methoxycarbonyl-1-methylvinyl dimethyl phosphate

Physical Form. Light yellow to orange liquid

Uses. Insecticide

Exposure. Skin absorption; inhalation; ingestion

Toxicology. Mevinphos is an anticholinesterase agent.

Signs and symptoms of overexposure are caused by the inactivation of the enzyme cholinesterase, which results in the accumulation of acetylcholine at synapses in the nervous system, skeletal and smooth muscle, and secretory glands. The sequence of the development of systemic effects varies with the route of entry. The onset of signs and symptoms is usually prompt but may be delayed up to 12 hours.[1–3] After inhalation, respiratory and ocular effects are the first to appear, often within a few minutes after exposure. Respiratory effects include tightness in the chest and wheezing due to bronchoconstriction and excessive bronchial secretion. Laryngeal spasms and excessive salivation may add to the respiratory distress; cyanosis may also occur. Ocular effects include miosis, blurring of distant vision, tearing, rhinorrhea, and frontal headache.

After ingestion, gastrointestinal effects, such as anorexia, nausea, vomiting, abdominal cramps, and diarrhea appear within 15 minutes to 2 hours. After skin absorption, localized sweating and muscular fasciculations in the immediate area occur, usually within 15 minutes to 4 hours; skin absorption is somewhat greater at higher ambient temperatures and is increased by the presence of dermatitis.[1–3]

With severe intoxication by all routes, an excess of acetylcholine at the neuromuscular junctions of skeletal muscle causes weakness, aggravated by exertion, involuntary twitchings, fasciculations and, eventually, paralysis. The most serious consequence is paralysis of the respiratory muscles. Effects on the central nervous system include giddiness, confusion, ataxia, slurred speech, Cheyne–Stokes respiration, convulsions, coma, and loss of reflexes. The blood pressure may fall to low levels, and cardiac irregularities, including complete heart block, may occur. Complete symptomatic recovery usually occurs within 1 week; increased susceptibility to the effects of antic holinesterase agents persists for up to several weeks after exposure. Daily exposure to concentrations that are insufficient to produce symptoms following a single exposure may result in the onset of symptoms. Continued daily exposure may be followed by increasingly severe effects.

A group of 31 farm workers who entered a field only 2 hours after it was sprayed with mevinphos developed a variety of initial symptoms, including eye irritation, headache, visual disturbances, dizziness, nausea, vomiting, chest pain, shortness of breath, pruritis, eyelid and arm fasciculations, excessive sweating, and diarrhea.[4] Following cessation of exposure, headache, dizziness, visual disturbances, and nausea persisted for 5 to 8 weeks or more in a significant number of the field workers. Despite symptoms suggesting moderate organophosphate intoxication, mean plasma and red blood cell cholinesterase depression was only 16% and 6%, respectively, when measured against a presumed baseline obtained in these workers long after the exposure.[4] Another study of 16 cauliflower workers poisoned by residues of mevinphos and phosphamidon (a less potent organophospate) demonstrated persistent headaches, blurred vision, and weakness in a number of workers 5 to 9 weeks or more following the exposure.[5]

In many of these occupational cases, the dermal route of exposure may predominate.[6] In a recent report, it was found that, when greenhouse workers wore long-sleeved shirts, approximately 6% of the pesticide reached the skin, compared to 38% for workers wearing short sleeves.[6]

In two cases of moderate intoxication from mevinphos, urinary excretion of dimethyl phosphate (a metabolite of mevinphos) was almost complete 50 hours after exposure.[7] Although a number of other organophosphorus pesticides also yield dimethyl phosphate, the presence of significant amounts of this metabolite in the urine may be useful in estimating the absorption of mevinphos.

Mevinphos inactivates cholinesterase by phosphorylation of the active site of the enzyme to form the "dimethylphosphoryl enzyme." Over the following 24 to 48 hours, there is a process, called aging, of conversion to the "monomethylphosphoryl enzyme." Aging is of clinical interest in the treatment of poisoning because cholinesterase reactivators such as pralidoxime (2-PAM, Protopam) chloride are ineffective after aging has occurred.

The 1995 ACGIH threshold limit value–time-weighted average (TLV-TWA) for mevinphos is 0.01 ppm (0.092 mg/m^3) with a short-term excursion limit of 0.03 ppm (0.27 mg/m^3) and a notation for skin absorption.

Note: For a description of diagnostic signs, differential diagnosis, and medical control, including clinical laboratory tests, as well as specific treatment of overexposure to anticholinesterase insecticides, see "Parathion."

REFERENCES

1. Koelle GB (ed): Cholinesterases and anticholinesterase agents. *Handbuch der Experimentellen Pharmakologie*, Vol 15, pp 989–1027. Berlin, Springer-Verlag, 1963
2. Taylor P: Anticholinesterase agents. In Gilman AG et al (eds): *Goodman and Gilman's The Pharmacological Basis of Therapeutics*, 7th ed, pp 110–129. New York, MacMillan, 1985
3. Hayes WJ Jr: Organic phosphorus pesticides. *Pesticides Studied in Man*, pp 284–435. Baltimore, MD, Williams and Wilkins, 1982
4. Coye MJ et al: Clinical confirmation of organophosphate poisoning of agricultural workers. *Am J Ind Med* 10:399–409, 1986
5. Midtling JE, Barnett PG, Coye MJ: Clinical management of field worker organophosphate poisoning. *West J Med* 142:514–518, 1985
6. Kangas J, Laitinen S, Jauhiainen A, et al: Exposure of sprayers and plant handlers to mevinphos in Finnish greenhouses. *AMA Ind Hyg Assoc J* 54:150–157, 1993
7. Holmes JHG, Starr HG Jr, Hanisch RC, von Kaulla KN: Short-term toxicity of mevinphos in man. *Arch Environ Health* 29:84–89, 1974

MICA
CAS: 12001-26-2

$K_2Al_4(Al_2Si_6O_20)(OH)_4$-*Muscovite*
Containing <1% quartz

Synonyms: Mica is a nonfibrous silicate occurring in plate form and including 9 different species; muscovite is a hydrated aluminum potassium silicate, also called white mica; phlogopite is an aluminum potassium magnesium silicate, also called amber mica; other forms are biotite, lepidolite, zinnwaldite, and roscoelite

Physical Form. Light gray to dark-colored flakes or particles

Uses. Insulation in electrical equipment; in

manufacture of roofing shingles and wallpaper; in oil refining; in rubber manufacture

Exposure. Inhalation

Toxicology. Mica dust causes pneumoconiosis.

In a study of 57 workers exposed to mica dust, 5 of 6 workers exposed to concentrations in excess of 25 mppcf for more than 10 years had pneumoconiosis.[1] The most characteristic finding by chest X ray was fine granulation of uneven density; there was a tendency, in some cases, to a coalescence of shadows. The symptoms most frequently reported were chronic cough and dyspnea; complaints of weakness and weight loss were less frequent.[1] Only 1 of 6 workers exposed more than 10 years at concentrations in excess of 25 mppcf failed to show evidence of pneumoconiosis.

A group of mica miners were said to show a higher incidence of pneumoconiosis than miners of other minerals, but some quartz was present in the dust to which they were all exposed.[2]

In a recent case report, a 63-year-old male with a long history of extensive exposure to mica presented (30 years after initial exposure) with complaints of progressive shortness of breath and a chronic nonproductive cough. Pulmonary-function tests revealed restrictive lung function and a mild reduction in the total lung capacity. Chest radiographs and lung biopsy showed extensive interstitial fibrosis with heavy mica deposition. The presence of mica was confirmed spectroscopically, whereas asbestos and other silicates were not identified, suggesting that mica was the fibrogenic agent in this case. The authors note that the long latency and chronic exposure associated with the disease indicate that mica is not as fibrogenic as other pneumoconiotic agents.

The 1995 ACGIH threshold limit value–time-weighted average (TLV-TWA) is 3 mg/m[3], respirable dust.

REFERENCES

1. Dreessen WC et al: Pneumoconiosis among mica and pegmatite workers. *Pub Health Bull* 250:1–74, 1940
2. Vestal TF, Winstead JA, Joliet PV: Pneumoconiosis among mica and pegmatite workers. *Ind Med* 12:11–14, 1943
3. Landas SK, Schwartz DA: Mica-associated pulmonary interstitial fibrosis. *Am Rev Respir Dis* 144:718–721, 1991

MOLYBDENUM AND COMPOUNDS
CAS: 7439-98-7

Mo

Synonyms: Soluble compounds include molybdenum trioxide, ammonium molybdate, and sodium molybdate; insoluble compounds include molybdenum disulfide and molybdenum dioxide

Physical Form. Silverish metal or dark powder

Uses. In manufacture of special-purpose steel; in ceramic glazes, enamels, and pigments; lubricant; corrosion inhibitor; additive to fertilizer

Toxicology. Molybdenum and compounds are considered to be of relatively low toxicity.

Workers at a molybdenum-roasting plant with TWA exposures of approximately 9.5 mg Mo/m[3] to soluble dusts had increased plasma and urine levels of molybdenum; the only adverse biochemical findings were large elevations in serum ceruloplasmin levels and some increase in serum uric acid levels.[1]

No evidence of systemic disease or dermatitis attributable to molybdenum (especially molybdenum disulfide) was seen by a plant physician reporting on 50 years of operation.[1]

An increased incidence of nonspecific symptoms, including weakness, fatigue, anorexia, headaches, and joint and muscle pains, was reported among Soviet mining and metallurgy workers exposed to 60 to 600 mg/m[3] molybdenum.[2] Signs of gout and elevated uric acid concentrations have been observed among inhabitants of areas of Armenia where the soil is rich in molybdenum. This effect apparently results from the induction of the enzyme xanthine oxidase, for which molybdenum is a cofactor.

Insoluble molybdenite, MoS_2, was practically nontoxic in animal studies; guinea pigs with exposure to 230 mg Mo/m^3 for 25 days showed only increases in respiration rate; rats ingesting as much as 500 mg/day for 44 days showed no toxic signs.

More soluble and more active molybdenum compounds, including calcium molybdate and molybdenum trioxide, were fatal at oral daily doses of over 100 mg/day.[1]

Guinea pigs exposed to molybdenum trioxide dust at a concentration of 200 mg molybdenum/m^3 for 1 hour daily for 5 days developed nasal irritation, diarrhea, weight loss, and incoordination.[3]

In livestock, chronic molybdenum poisoning, known as teart disease, is caused by a diet high in molybdenum and low in copper. Symptoms include anemia, gastrointestinal disturbances, bone disorders, and growth retardation.[4]

The metabolism of molybdenum is closely associated with that of copper; molybdenum toxicity in animals can be alleviated by the administration of copper. High intake of molybdenum in rats resulted in a substantial reduction in the activity of sulfide oxidase in the liver.[5] The reduced activity of this enzyme leads to accumulation of sulfide in the tissues and subsequent formation of highly undissociated copper sulfide, thus removing copper from the metabolic process. This is a possible explanation for the induction of copper deficiency by molybdate.

The TLV-TWA is 5 mg/m^3 for the more active molybdenum compounds, such as molybdenum trioxide and the soluble molybdates; the TLV-TWA for molybdenum metal and insoluble compounds is 10 mg/m^3.

REFERENCES

1. Stokinger HE: The metals. In Clayton GD, Clayton FE (eds): *Patty's Industrial Hygiene and Toxicology*, 3rd ed, Vol 2, *Toxicology*, pp 807–1819. New York, Wiley-Interscience, 1981
2. Lener J, Bibr B: Effects of molybdenum on the organism (a review). *J Hyg Epidemiol Microbiol Immunol* 29:405–419, 1984
3. Browning E: *Toxicity of Industrial Materials*, 2nd ed, pp 243–248. London, Butterworths, 1969
4. Friberg L, Lener J: Molybdenum. In Friberg L et al (eds): *Handbook on the Toxicology of Metals*, Vol. II. *Specific Metals*, pp 445–461. Amsterdam, Elsevier, 1986
5. Halverson AW, Phifer JH, Monty KJ: A mechanism for the copper-molybdenum interrelationship. *J Nutr* 71:79–100, 1960

MORPHOLINE
CAS:110-91-8

C_4H_9NO

Synonyms: Diethylenimide oxide; diethylene imidoxide; tetrahydro-2H-1,4-oxazine

Physical Form. Clear liquid with ammonialike odor

Uses. Solvent for resins, waxes, casein, and dyes; in corrosion inhibitors; intermediate for rubber-processing chemicals; brightener for detergents

Exposure. Inhalation; skin absorption

Toxicology. Morpholine vapor is an irritant of the eyes, nose, and throat.

In industrial use, some instances of skin and respiratory tract irritation have been observed, but no chronic effects have been reported.[1] A human exposure to 12,000 ppm for 1.5 minutes in a laboratory produced nose irritation and cough; mouth pipetting of the liquid caused a severe sore throat and reddened mucous membranes.[2] Workers exposed for several hours to low vapor concentrations complained of foggy vision with rings around lights, the results of corneal edema, which cleared within 3 to 4 hours after cessation of exposure.[1]

Repeated daily exposure of rats to 18,000 ppm for 8 hours was lethal to some animals; those dying had damage to lungs, liver, and kidneys.[2] Rats and guinea pigs survived an 8-hour exposure at 12,000 ppm. Sublethal signs from inhalation include lacrimation, rhinitis, and inactivity.[1] Rats exposed by inhalation to 250 ppm for 6 hr/day, 5 days/week, showed signs of irritation after 1 week; animals examined after 7 to 13 weeks of exposure had focal erosions and squamous metaplasia in the maxilloturbinates.[3]

Oral doses of undiluted unneutralized morpholine caused irritation of the intestinal tract with hemorrhage. Applied to the skin of rabbits, it caused skin burns and systemic injury, including necrosis of the liver and kidney; the dermal LD_{50} was 0.5 ml/kg. The liquid dropped in the eye of a rabbit caused moderate injury, with ulceration of the conjunctiva and corneal clouding.[4]

Rats given 10 g morpholine/kg in the diet, plus 0.2% sodium nitrite in the drinking water, had a significantly increased incidence of liver tumors compared to controls.[5] The carcinogenic response is attributed to the in vitro production of N-nitrosomorpholine. In another study, morpholine alone produced a low number of tumors of the liver, lung, and brain, and it was suggested that an unknown nitrate source reacted with the morpholine to form the carcinogenic N-nitrosomorpholine.[6]

Morpholine has also been tested for carcinogenicity by inhalation exposure in rats. Exposure to 10, 50, or 150 ppm, 6 hr/day, 5 days/week, for up to 104 weeks, was associated with dose-related increases in inflammation of the cornea, inflammation and squamous metaplasia of the turbinate epithelium, and necrosis of the turbinate bones in the nasal cavity; at these dosages, no significant increase in the incidence of tumors was reported.[7]

The IARC has determined that there is inadequate evidence for the carcinogenicity of morpholine in experimental animals and that morpholine is not classifiable as to its carcinogenicity in humans.[8]

The 1995 ACGIH threshold limit value–time-weighted average (TLV-TWA) for morpholine is 20 ppm (71 mg/m³) with a notation for skin absorption.

REFERENCES

1. Reinhardt CF, Brittelli MR: Heterocyclic and Miscellaneous Nitrogen Compounds. In Clayton GD, Clayton FE (eds): *Patty's Industrial Hygiene and Toxicology*, 3rd ed, rev, Vol 2A, *Toxicology*, pp 2693–2696. New York, Wiley-Interscience, 1981
2. Shea TE: The acute and subacute toxicity of morpholine. *J Ind Hyg Toxicol* 21:236–245, 1939
3. Conaway CC, Coate WB, Voelker RW: Subchronic inhalation toxicity of morpholine in rats. *Fund Appl Toxicol* 4:465–472, 1984
4. Grant WM: *Toxicology of the Eye*, 2nd ed, pp 722–723. Springfield, IL, Charles C Thomas, 1974
5. Mirvish SS et al: Liver and forestomach tumors and other forestomach lesions in rats treated with morpholine and sodium nitrite, with and without sodium ascorbate. *J Natl Cancer Inst* 71:81–84, 1983
6. Shank RC, Newberne PM: Dose-response study of the carcinogenicity of dietary sodium nitrite and morpholine in rats and hamsters. *Food Cosmet Toxicol* 14:1–8, 1976
7. Harbison RD, Marino DJ, Conaway CC, et al: Chronic morpholine exposure of rats. *Fund Appl Toxicol* 12:491–507, 1989
8. *IARC Monographs on the Evaluation of Carcinogenic Risks to Humans*, Vol 47, Some organic solvents, resin monomers and related compounds, pigments and occupational exposures in paint manufacture and painting, pp 199–213. Lyon, International Agency for Research on Cancer, 1989

MUSTARD GAS
CAS: 505-60-2

$C_4H_8Cl_2S$

Synonyms: Sulfur mustard; bis-2-chloroethyl sulfide; di-2-chloroethyl sulfide; 1,1-thiobis (2-chloroethane); HD (chemical agent symbol)

Physical Form. Colorless, odorless, oily liquid

Uses. As a vesicant in chemical warfare. US stockpiles were thought to exist through the early 1970s; several other countries currently maintain large stockpiles that present an imminent danger from accidental or intentional exposure.[1] Also used in small quantities as a model compound in biological studies on alkylating agents

Exposure. Inhalation; skin contact

Toxicology. Mustard gas causes skin and eye injury; following inhalation, pulmonary damage may occur. Chronic exposure has been associated with an increased risk of respiratory cancer in humans.

Mustard gas is primarily a vesicant, with blisters being formed by either liquid or vapor contact.[2] It attacks the eyes and lungs and is a systemic poison, so that protection must be provided for the entire body. Insidious in its action, there is no pain at the time of exposure, and the first symptoms typically appear in 4 to 6 hours. The higher the concentration, the shorter the interval between exposure to the agent and the first symptoms. After several hours, the gross biological evidences of injury begin to appear as edema, hyperemia, and irritation. In the eye, the corneal epithelium becomes edematous, the lids and conjunctivas become red and swollen, and the patient experiences burning discomfort and photophobia, including tearing and blepharospasm.[3] Areas of contaminated skin become inflamed and blistered. Burns caused by mustard gas are severe and require long healing periods. Following inhala-

tion of the agent, pulmonary edema and long-term dyspnea may occur.[4]

The toxic effects of mustard gas are primarily related to its alkylating ability.[5] In an aqueous environment, mustard gas rearranges and loses one or two molecules of hydrogen chloride; then, mustard gas, minus its chlorides, becomes firmly attached through one or both of its b-carbon atoms to tissue components, altering their functional and physiochemical properties. Cytotoxic effects have specifically been related to a double alkylation reaction in which the two reactive ends of the mustard gas molecule attach to strands of DNA, forming cross-links that prevent cell replication.

Exposure to mustard gas, in light of its strong alkylating ability, is considered to be a possible cause of cancer in man. Two types of exposures have been studied in particular: acute exposure resulting from the use of the gas in war and chronic exposure in the course of its manufacture.

The mortality of British and American veterans of the 1914–1918 war who were acutely exposed to mustard gas has been investigated. British soldiers who received a pension for mustard gas poisoning were found to have a high mortality from chronic bronchitis (217 observed vs. 21 expected) and increased mortality from cancer of the lung and pleura (29 observed vs. 14 expected).[6] However, most of the exposed men also had chronic bronchitis, and a similar excess of lung and pleural cancers was found in pensioners with bronchitis who had not been exposed to the gas. United States veterans with mustard gas injury had significantly increased mortality from pneumonia and tuberculosis. There was some increased risk of respiratory cancer in the exposed group, but the extent of the increase was not large.[7] A further study involving an additional 10 years of follow-up produced similar results.[8]

Studies of the effects of occupational exposure to mustard gas have provided a stronger association between exposure and respiratory cancer. Among 495 Japanese workers engaged in the manufacture of mustard gas between 1929 and 1945, 33 died from cancers of the respiratory tract compared to 0.9 expected.[9] In an earlier report of this same cohort, it was stated that the working environment attained mustard gas concentrations of 0.55 to 0.07 mg/l and that protective measures were neither fully effective nor generally applied.[10]

A study of 3354 British workers employed in the manufacture of mustard gas during the 1939–1945 war and traced for mortality to the end of 1984, found large and highly significant excesses, compared to national death rates, from cancer of the larynx (11 observed vs. 4 expected), pharynx (15 observed vs. 2.73 expected), and all other buccal cavity and upper respiratory sites combined (lip, tongue, salivary gland, mouth, and nose: 12 observed vs. 4.29 expected).[11] For lung cancer deaths, there were 200 observed cases compared to 138 expected. Significant excesses were also observed for deaths from acute and chronic nonmalignant respiratory disease. The relative risk of both lung cancer and nonmalignant respiratory disease was substantially reduced. However, if comparison rates for the nearest urban area was used rather than national rates, the risk for cancer of the pharynx and lung was significantly related to duration of employment. Furthermore, the risk of respiratory cancer was not localized to individuals employed in process areas where exposures occurred to high levels of short duration, suggesting that the risk of cancer was due more to lower-level ambient exposure of longer duration. Significant excess mortality was also observed for cancers of the esophagus and stomach, but there was no consistent relation with time since first exposure or with duration of exposure. The authors conclude that the results provide strong evidence that exposure to mustard gas can cause cancers of the upper respiratory tract and some evidence that it can cause lung cancer and nonmalignant respiratory disease.

Mustard gas has been tested for carcinogenicity in mice, producing lung tumors after inhalation or intravenous injection and local sarcomas after subcutaneous injection.[12]

The IARC has determined that there is sufficient evidence for carcinogenicity in humans and limited evidence in animals.[12]

Mustard gas is highly genotoxic.[13] In vitro assays in both prokaryotic and eukaryotic systems support a mechanism of DNA alkylation leading to cross-link formation, inhibition of DNA synthesis and repair, point mutation, and chromosome and chromatid aberration formation.

ACGIH has not established a threshold limit value for mustard gas.

REFERENCES

1. Somani SM, Sabu SR: Toxicodynamics of sulfur mustard. *Int J Clin Pharm Ther Toxicol* 27:419–438, 1989
2. Kirk-Othmer: *Encyclopedia of Chemical Technology*, 3rd ed, Vol 5, pp 395–397. New York, John Wiley and Sons, 1979
3. Grant WM: *Toxicology of the Eye*, 3rd ed, pp 643–645. Springfield, IL, Charles C Thomas, 1986
4. Aasted A, Darre E, Wulf HC: Mustard gas: clinical, toxicological, and mutagenic aspects based on modern experience. *Ann Plastic Surg* 19:330–333, 1987
5. Gilman A, Phillips FS: The biological actions and therapeutic applications of the b-chloroethyl amines and sulfides. *Science* 103:409–415, 1946
6. Case RAM, Lea AJ: Mustard gas poisoning, chronic bronchitis and lung cancer: an investigation into the possibility that poisoning by mustard gas in the 1914–18 war might be a factor in the production of neoplasia. *Brit J Prev Soc Med* 9:62–72, 1955
7. Beebe GW: Lung cancer mortality in World War I veterans: possible relation to mustard gas injury and 1918 influenza epidemic. *J Natl Cancer Inst* 25:1231–1252, 1960
8. Norman JE: Lung cancer mortality in World War I veterans with mustard gas injury: 1919–1965. *JNCI* 54:311–317, 1975
9. Wada S, Nishimoto Y, Miyanishi M, et al: Mustard gas as a cause of respiratory neoplasia in man. *Lancet* 1:1611–1613, 1968
10. Yamada A, Hirose F, Nagai M, et al: Five cases of cancer of the larynx found in persons who suffered from occupational mustard gas poisoning. *Gann* 48:366–368, 1957
11. Easton DF, Peto J, Doll R: Cancer of the respiratory tract in mustard gas workers. *Br J Ind Med* 45:652–659, 1988
12. *IARC Monographs on the Evaluation of the Carcinogenic Risk of Chemicals to Humans. Overall Evaluations of Carcinogenicity: An Updating of IARC Monographs Vols 1–42*, Suppl 7, pp 259–

260. Lyon, International Agency for Research on Cancer, 1987

13. Agency for Toxic Substances and Disease Registry (ATSDR): *Toxicological Profile for Mustard "Gas."* US Department of Health and Human Services, Public Health Service, TP-91/22, p 65, 1992

NALED
CAS: 300-76-5

$C_4H_7Br_2Cl_2O_4P$

Synonyms: 1,2-Dibromo-2, 2-dichloroethyl dimethyl phosphate; Dibrom

Physical Form. Light straw-colored liquid with slightly pungent odor

Uses. Acaricide; insecticide

Exposure. Inhalation; skin absorption; ingestion

Toxicology. Naled is an anticholinesterase agent.

Signs and symptoms of overexposure are caused by the inactivation of the enzyme cholinesterase, which results in the accumulation of acetylcholine in the nervous system, skeletal and smooth muscle, and secretory glands.[1-3] The sequence of the development of systemic effects varies with the route of entry. The onset of signs and symptoms is usually prompt but may be delayed up to 12 hours. After inhalation of the vapor, respiratory and ocular effects are the first to appear, often within a few minutes of exposure. Respiratory effects include tightness in the chest and wheezing due to bronchoconstriction and excessive bronchial secretion; laryngeal spasm and excessive salivation may add to the respiratory distress; cyanosis may also occur. Ocular effects include miosis, blurring of distant vision (due to spasm of accommodation), tearing, rhinorrhea, and frontal headache.

After ingestion of the liquid, gastrointestinal effects such as anorexia, nausea, vomiting, abdominal cramps, and diarrhea appear within 15 minutes to 2 hours. Following skin absorption, localized sweating and muscular fasciculations usually occur in the immediate area within 15 minutes to 4 hours; skin absorption is somewhat greater at higher ambient temperatures and is enhanced by the presence of dermatitis.

With severe intoxication, an excess of acetylcholine at the neuromuscular junctions of skeletal muscle causes weakness aggravated by exertion, involuntary twitchings, fasciculations and, eventually, paralysis. The most serious consequence is paralysis of the respiratory muscles. Effects on the central nervous system include giddiness, confusion, ataxia, slurred speech, Cheyne–Stokes respiration, convulsions, coma, and loss of reflexes. The blood pressure may fall to low levels, and cardiac irregularities, including complete heart block, may occur. Complete symptomatic recovery usually occurs within a week; increased susceptibility to the effects of anticholinesterase agents persists for up to several weeks after exposure. Daily exposure to concentrations that are insufficient to produce symptoms following a single exposure may result in the onset of symptoms. Continued daily exposure may be followed by increasingly severe effects.

Dermatitis occurred on the arms, face, neck, and abdomen of 9 of 12 persons working in a field of flowers that had been freshly sprayed with a solution of naled; 3 of 4 workers patch-tested were positive to a 60% solution of naled in xylene and negative to xylene alone.[4]

In the eye, the liquid may be expected to cause injury.

Naled inactivates cholinesterase by phosphorylation of the active site of the enzyme to form the "dimethylphosphoryl enzyme." Over the following 24 to 48 hours, there is a process, called aging, of conversion to the "monomethylphosphoryl enzyme." Aging is of clinical interest in the treatment of poisoning because cholinesterase reactivators such as pralidoxime (2-PAM, Protopam) chloride are ineffective after aging has occurred.

The 1995 ACGIH threshold limit value–time-weighted average (TLV-TWA) is 3 mg/m^3 with a notation for skin absorption.

Note: For a description of diagnostic signs, differential diagnosis, and medical control, including clinical laboratory tests, as well as specific treatment of overexposure to anticholinesterase insecticides, see "Parathion."

REFERENCES

1. Koelle GB (ed): Cholinesterases and anticholinesterase agents. *Handbuch der Experimentellen Pharmakologie*, Vol 15, pp 989–1027. Berlin, Springer-Verlag, 1963
2. Taylor P: Anticholinesterase agents. In Gilman AG et al (eds): *Goodman and Gilman's The Pharmacological Basis of Therapeutics*, 7th ed, pp 110–129. New York, Macmillan, 1985
3. Hayes WJ Jr: *Pesticides Studied in Man*, pp 312–313. Baltimore, MD, Williams and Wilkins, 1982
4. Edmundson WF, Davies JE: Occupational dermatitis from naled—a clinical report. *Arch Environ Health* 15:89–91, 1967

NAPHTHA, COAL TAR
CAS: 64742-95-6

Synonyms: Naphtha solvent, high flash naphtha, refined naphtha, and heavy naphtha describe various fractions and grades

Physical Form. Light yellow liquid with boiling ranges between 110 and 190°C

Uses. Solvent

Exposure. Inhalation

Toxicology. Coal tar naphtha is a central nervous system depressant.

Coal tar naphtha is primarily a mixture of toluene, xylene, cumene, benzene, and other aromatic hydrocarbons; it is distinguished from petroleum naphtha, which is constituted mainly of aliphatic hydrocarbons.[1]

There are no well-documented reports of industrial injury resulting from the inhalation of coal tar naphtha.[1] However, severe exposure is expected to cause light-headedness, drowsiness, and possibly irritation of the eyes, nose, and throat.

Nephrotoxicity of naphtha, as evidenced by an increased prevalence of albuminuria, erythrocyturia, and leukocyturia, was suggested in one study of newspaper pressroom workers with low levels of exposure.[2] In another report, no evidence of naphtha-associated renal effects was found in 248 workers with exposures ranging from 6 to 790 mg/m³ and lengths of exposure ranging from 0.8 to 7.3 years.[3] Differences in formulations of naphthas may account for some of the inconsistencies observed between studies. In animal experiments, variations in the proportion of alkanes to alkenes and aromatics and of highly branched and straight-chain paraffins have produced 100-fold changes in the dose of naphtha necessary to cause toxicity.[3]

No signs of neurotoxicity were observed in rats exposed for 90 days to concentrations of up to 1500 ppm. Histopathological examination of peripheral nervous tissue of exposed animals revealed no degenerative changes.[4]

Exposure of pregnant mice to near-lethal levels of 1500 ppm was maternally toxic and caused increased fetal mortality, reduced weight, delayed ossification, and an increased incidence of cleft palate.[5] At 500 ppm, there was reduced maternal and fetal weight gain. No developmental or maternal toxicity was seen at 100 ppm. Similar studies in rats reported developmental delays only at doses that were maternally toxic. No significant adverse effects on reproductive parameters were found in rats exposed for three generations at doses that produced severe toxicity.[5]

Naphtha was not genotoxic in a number of in vivo and in vitro assays.[4]

Skin contact with the liquid may result in drying and cracking due to defatting action.

Coal tar naphtha, a mixture of hydrocarbons, has been deleted from the ACGIH listing of TLVs in favor of reference to its chemical components.

REFERENCES

1. Browning E: *Toxicity and Metabolism of Industrial Solvents*, pp 141–144. New York: Elsevier, 1965

2. Hashimoto DM, Kelsey KT, Seitz T, et al: The presence of urinary cellular sediment and albuminuria in newspaper press workers exposed to solvents. *J Occup Med* 33:516–526, 1991
3. Rocskay AZ, Robins TG, Schork MA, et al: Renal effects of naphtha exposure among automotive workers. *J Occup Med* 35:617–622, 1993
4. Douglas JF, McKee RH, Cagen SZ, et al: A neurotoxicity assessment of high flash aromatic naphtha. *Toxicol Ind Health* 9:1047–58, 1993
5. McKee RH, Wong ZA, Schmitt S, et al: The reproductive and developmental toxicity of high flash aromatic naphtha. *Toxicol Ind Health* 6:441460, 1990

NAPHTHALENE
CAS: 91-20-3

$C_{10}H_8$

Synonyms: Naphthalin; tar camphor; white tar

Physical Form.   White crystalline solid with a characteristic "mothball" odor

Uses.   Insect repellant; feedstock for synthesis of a variety of compounds, especially phthalic anhydride

Exposure.   Inhalation; ingestion

Toxicology.   Naphthalene is a hemolytic agent and an irritant of the eyes; it may cause cataracts.

Severe intoxication from ingestion results in characteristic manifestations of marked intravascular hemolysis and its consequences, including potentially fatal hyperkalemia.[1,2] Initial symptoms include eye irritation, headache, confusion, excitement, malaise, profuse sweating, nausea, vomiting, abdominal pain, and irritation of the bladder; there may be progression to jaundice, hematuria, hemoglobinuria, renal tubular blockade, and acute renal shutdown.[1,2] Hematologic features include red cell fragmentation, icterus, severe anemia with nucleated red cells, leukocytosis, and dramatic decreases in hemoglobin, hematocrit, and red cell count; sometimes there is formation of Heinz bodies and methemoglobin.[3] Naphthalene itself is nonhemolytic; several metabolites, including α-naphthol, are hemolytic.[3] Individuals with a hereditary deficiency of the enzyme glucose-6-phosphate dehydrogenase in red blood cells (and, consequently, decreased concentrations of reduced glutathione) are particularly susceptible to the hemolytic properties of naphthalene.[3]

The vapor causes eye irritation at 15 ppm; eye contact with the solid may result in conjunctivitis, superficial injury to the cornea, chorioretinitis, scotoma, and diminished visual acuity. Cataracts and ocular irritation have been produced experimentally in animals and have been described in humans.[4] Of 21 workers exposed to high concentrations of fume or vapor for 5 years, 8 had peripheral lens opacities. In other studies, no abnormalities of the eyes have been detected in workers exposed to naphthalene for several years.[4]

Reportedly, headache, nausea, and confusion may occur after inhalation of vapor. Occupational poisoning from vapor exposure is rare.[3] Naphthalene on the skin may cause hypersensitivity dermatitis.[1]

In acute and subchronic experiments in CD-mice, naphthalene failed to induce either hemolytic anemia or cataract formation, even at doses that produced mortality.[5] In chronic studies, female B6C3F$_1$ mice had a significantly increased incidence of pulmonary alvelor/bronchiolar adenomas following 2 years' exposure at 30 ppm.[6] The increased incidence did not occur in males or in low-dose females.

Napthalene was not mutagenic in a variety of bacterial assays, but it did cause sister chromatid exchanges and chromosomal aberrations in Chinese hamster ovary cells.[7]

Naphthalene was not teratogenic in a number of developmental studies although a trend toward dose-related malformations was seen in rats administered up to 450 mg/kg/day on gestation days 6 through 15.[7,8]

The 1995 ACGIH threshold limit value–time-weighted average (TLV-TWA) for naphthalene is 10 ppm (52 mg/m^3) with a short-term excursion limit of 15 ppm (79 mg/m^3).

REFERENCES

1. Hygienic Guide Series: Naphthalene. Am Ind Hyg Assoc J 28:493–496, 1967
2. Gidron E, Leurer J: Naphthalene poisoning. Lancet 1:228–230, 1956
3. Gosselin RE, Smith RP, Hodge HC: *Clinical Toxicology of Commercial Products, Section III,* 5th ed, pp 307–311. Baltimore, MD, Williams and Wilkins, 1984
4. Grant WM: *Toxicology of the Eye,* 3rd ed, pp 650–653. Springfield, IL, Charles C Thomas, 1986
5. Shopp GM, White KL Jr, Holsapple MP, et al: Naphthalene toxicity in CD-1 mice: general toxicology and immunotoxicology. *Fund Appl Toxicol* 4:406–419, 1984
6. National Toxicology Program: *Toxicology and Carcinogenesis Studies of Naphthalene (CAS No 91-20-3) in B6C3F$_1$ Mice (Inhalation Studies).* US Department of Health and Human Services, Public Health Service, National Institutes of Health, NTP-TR-410, NIH Pub No 92-3141, 1992
7. Agency for Toxic Substances and Disease Registry (ATSDR): *Toxicological Profile for Naphthalene, 1-Methylnaphthalene, and 2-Methylnaphthalene.* US Department of Health and Human Services, Public Health Service, p 139, 1993
8. National Toxicology Program: *Developmental Toxicity of Naphthalene (CAS: No 91-20-3) Administered by Gavage to Sprague–Dawley (CD) Rats on Gestational Days 6 Through 15.* US Department of Health and Human Services, Public Health Service, National Institutes of Health, TER-91006, 1991

β-NAPHTHYLAMINE
CAS: 91-59-8

$C_{10}H_9N$

Synonyms: 2-Aminonaphthalene; BNA; 2-naphthylamine

Physical Form. Colorless crystals that darken on oxidation

Uses. Formerly used in the manufacture of dyes and antioxidants; rarely used for industrial research purposes today

Toxicology. β-Naphthylamine (BNA) is a potent bladder carcinogen.

A cohort study of a factory population revealed 7 cases of bladder or kidney cancer in 735 person-years of exposure, an attack rate of 952 per 100,000 among BNA workers.[1] Of 48 BNA workers employed in a coal tar dye plant, 12 developed bladder tumors.[2] The time elapsed from first exposure to first abnormal signs or symptoms (dysuria, frequency, hematuria) ranged from 1 to 35 years, with a mean of 18 years. The time elapsed from first exposure to diagnosis of bladder malignancy ranged from 2 to 42 years, with a mean of 23 years.[2] Workers employed at the last facility in the United States that manufactured BNA had a remarkable and significantly increased incidence of bladder cancer (13 observed vs. 3.3 expected).[3] The mortality incidence from bladder cancer in this cohort was not as profound, with two deaths observed versus 0.7 such deaths expected.[4] The authors suggest that an inadequate latency period and/or the high survival rate for bladder cancer could account for the small number of deaths.

Dyestuff workers exposed to β-naphthylamine and benzidine prior to 1972 showed alterations in some T-lymphocyte subpopulations some 20 years later.[5] Specifically, there was a decreased number of circulating CD4+ T-lymphocytes in exposed workers. Measurement of this T-lymphocyte subpopulation may provide a useful biological marker of past exposure to aromatic amines.

Bladder tumors were induced in 24 of 34 dogs that were fed 6.25 to 50 mg/kg/day for 6 to 26 months; carcinomas were present in 9 of 11 dogs that received 100 to 200 g BNA, whereas 6 of 22 carcinomas occurred in dogs receiving total doses of less than 100 g.[6] All dogs treated with the carcinogen had multiple tumors.

In monkeys, intragastric administration of 37 to 2400 mg/kg/week for up to 250 weeks caused 9 transitional cell carcinomas of the bladder and three papillary adenomas.[7]

The IARC has determined that there is sufficient evidence of carcinogenicity from

BNA exposure, either alone or as an impurity in other humans and animals.[8] Because of demonstrated high carcinogenicity in humans and animals, exposure by any route should be avoided.

ACGIH classifies β-naphthylamine as A1, a confirmed human carcinogen, and, thus, there is no TLV.

REFERENCES

1. Mancuso TF, El-Attar AA: Cohort study of workers exposed to beta-napthylamine and benzidine. *J Occup Med* 9:277–285, 1967
2. Goldwater LJ, Rossa AJ, Kleinfeld M: Bladder tumors in a coal tar dye plant. *Arch Environ Health* 11:814–817, 1965
3. Schulte PA, Ringen K, Hemstreet GP, et al: Risk assessment of a cohort exposed to aromatic amines: initial results. *JOM* 27:115–121, 1985
4. Stern FB, Murthy LI, Beaumont JJ, et al: Notification and risk assessment for bladder cancer of a cohort exposed to aromatic amines. III. Mortality among workers exposed to aromatic amines in the last beta-naphthylamine manufacturing facility in the United States. *JOM* 27:495–500, 1985
5. Araki S, Tanigawa T, Ishizu S, et al: Decrease of CD4-positive T-lymphocytes in workers exposed to benzidine and beta-naphthylamine. *Arch Environ Health* 48:205–208, 1993
6. Conzelman GM Jr, Moulton JE: Dose-response relationships of the bladder tumorigen 2-naphthylamine: a study in beagle dogs. *J Natl Cancer Inst* 49:193–205, 1972
7. Conzelman GM Jr et al: Induction of transitional cell carcinomas of the urinary bladder in monkeys fed 2-naphthylamine. *J Natl Cancer Inst* 42:825–836, 1969
8. *IARC Monographs on the Evaluation of Carcinogenic Risks to Humans. Overall Evaluations of Carcinogenicity: An Updating of IARC Monographs Vol 1–42*, Suppl 7, pp 261–262. Lyon, International Agency for Research on Cancer, 1987

NICKEL (AND INORGANIC COMPOUNDS)
CAS: 7440-02-0

Ni

Compounds: Nickel carbonate; nickel oxide; nickel subsulfide; nickel sulfate

Physical Form. Elemental nickel is a silver-white metal; salts are crystals

Uses/Sources. Corrosion-resistant alloys; electroplating; in production of catalysts; nickel-cadmium batteries; nickel subsulfide (Ni_3S_2) is encountered in the smelting and refining of certain nickel ores and may be formed in petroleum refining from the use of nickel catalysts

Toxicology. Metallic nickel and certain nickel compounds cause sensitization dermatitis. Nickel refining has been associated with an increased risk of nasal and lung cancer.

"Nickel itch" is a dermatitis resulting from sensitization to nickel; the first symptom is usually pruritis, which occurs up to 7 days before skin eruption appears.[1,2] The primary skin eruption is erythematous or follicular; it may be followed by superficial discrete ulcers, which discharge and become crusted, or by eczema. The eruptions may spread to areas related to the activity of the primary site, such as the elbow flexure, eyelids, or sides of the neck and face.[2] In the chronic stages, pigmented or depigmented plaques may be formed. Nickel sensitivity, once acquired, is apparently not lost; of 100 patients with positive patch tests to nickel, all reacted to the metal when retested 10 years later.[3]

A worker who had developed cutaneous sensitization also developed apparent asthma from inhalation of nickel sulfate; immunologic studies showed circulating antibodies to the salt, and controlled exposure to a solution of nickel sulfate resulted in decreased pulmonary function and progressive dyspnea; the possibility of hypersensitivity pneumonitis could not be excluded.[4]

Pneumoconiosis has been reported among

workers exposed to nickel dust, but exposure to known fibrogenic substances could not be excluded.[5] Nasal irritation, damage to the nasal mucosa, perforation of the nasal septum, and loss of smell have only occasionally been reported in workers exposed to nickel aerosols and other contaminants.[6]

The severe acute systemic effects found with nickel carbonyl exposure are not associated with inorganic nickel.[5]

Epidemiologic studies have shown an increased incidence of cancers among nickel refinery workers.[6-8]

A mortality update of a cohort of 967 Clydach, Wales, refinery workers, employed for at least 5 years and followed to 1971, showed significant risks of both lung and nasal cancers among those hired before 1930.[9] The SMR for lung cancer was 623 (observed = 137/21.98), and for nasal cancer, 28718 (expected = 567/0.195). Latency for nickel-induced lung cancer was approximately 14 years and, for nasal cancer, 15 to 24 years. No case of nasal cancer occurred among those entering employment after 1930, and lung cancer rates dropped steeply after this date. The reduction was attributed to industrial hygiene improvements and process changes made in the 1920s. The respiratory cancers were related primarily to exposure to soluble nickel compounds at >1 mg nickel/m^3 and to exposure to less soluble compounds at >10 mg nickel/m^3.[10]

An excess of sinus cancers occurred in a cohort of 1852 West Virginia nickel alloy workers employed prior to 1948, when calcining of nickel sulfide matte was done at the plant.[11]

In one of the largest studies, an excess of lung and nasal cancers was found in a cohort of 54,724 Canadian workers.[6-8] The respiratory cancer risk was confined to the sintering, calcining, and leaching occupational group. There was no excess among miners, concentrators, smelters or other groups.

Other cancers, including prostatic and laryngeal, have been significantly elevated in certain studies but are less convincingly associated with nickel refinery work.[8]

In a recent evaluation of epidemiological studies to date, it was concluded that most of the respiratory cancer seen among the nickel refinery workers could be attributed to exposure to a mixture of oxidic and sulfidic nickel at very high concentrations.[10] Exposure to large concentrations of oxidic nickel in the absence of sulfidic nickel was also associated with increased lung and nasal cancer risks. There was also evidence that soluble nickel exposure increased the risk of these cancers and that it may enhance risks associated with exposure to less soluble forms of nickel. There was no evidence that metallic nickel was associated with increased lung and nasal cancer risks. The interaction between smoking and nickel exposure appears to be additive rather than multiplicative.[12]

Animal studies suggest, in general, an inverse relationship between solubility and carcinogenic potential; nickel metal, nickel oxide, and nickel subsulfide may exert variable degrees of carcinogenic potential in vivo and in vitro, whereas most nickel salts are noncarcinogenic.[7] The differences in activity may result from the ability of different compounds to enter the cell and be converted to nickel ion, the purported carcinogenic species.[7]

The IARC has determined that there is sufficient evidence for carcinogenicity to humans for nickel and nickel compounds.[13] In vitro and in vivo studies indicate that nickel is genotoxic. It has been reported to interact with DNA, resulting in cross-links and strand breaks.[12]

In experimental animals, a range of reproductive effects can be induced by nickel; in male rats, exposure to nickel salts results in degenerative changes in the testes and epididymis and in effects on spermatogenesis.[7] Exposure of pregnant animals has been associated with delayed embryonic development, increased resorptions, and an increase in structural malformations.[14] It has been noted, however, that doses used are high and may not relate at all to human exposure.[14] There are no reports indicating that exposure to nickel has caused malformations in humans. Reproductive effects do not seem likely to occur as a result of occupational exposures if other toxic effects are prevented.

Biological monitoring of nickel exposure has been accomplished primarily by determination of nickel in serum and urine.[12]

The proposed 1995 ACGIH threshold

limit value–time-weighted average for nickel, elemental, insoluble, and soluble compounds as Ni is 0.05 mg/m^3 with an A1 confirmed human carcinogen designation.

REFERENCES

1. Browning E: *Toxicity of Industrial Metals*, 2nd ed, pp 249–260. London, Butterworths, 1969
2. Fisher AA: *Contact Dermatitis*, 2nd ed, pp 96–102. Philadelphia, Lea & Febiger, 1973
3. Veien NK: Nickel sensitivity and occupational skin disease. *Occ Med: State of the Art Reviews* 9:81–95, 1994
4. McConnell LH, Fink JN, Schlueter DP, et al: Asthma caused by nickel sensitivity. *Ann Intern Med* 78:888–890, 1973
5. Norseth T: Nickel. In Friberg L et al (eds): *Handbook on the Toxicology of Metals*, 2nd ed, Vol II, *Specific Metals*, pp 462–481. Amsterdam, Elsevier, 1986
6. Mastromatteo E: Nickel. *Am Ind Hyg Assoc J* 10:589–601, 1986
7. US Environmental Protection Agency: *Health Assessment Document for Nickel and Nickel Compounds. Final Report.* Washington, DC, Office of Health and Environmental Assessment, September 1986
8. Wong O, et al: *Critical Evaluation of Epidemiologic Studies of Nickel-Exposed Workers—Final Report*, pp 1–99. Berkeley, CA, Environmental Health Associates, Inc, 1983
9. Doll R, Mathews JD, Morgan LG, et al: Cancers of the lung and nasal sinuses in nickel workers: a reassessment of the period of risk. *Br J Ind Med* 34:102–105, 1977
10. Doll R: Report on the international committee on nickel carcinogenesis in man. *Scan J Work Environ Health* 16:1–82, 1990
11. Enterline PE, Marsh GM: Mortality among workers in a nickel refinery and alloy manufacturing plant in West Virginia. *J Natl Cancer Inst* 68:925–933, 1982
12. Agency for Toxic Substances and Disease Registry (ATSDR): *Toxicological Profile for Nickel*, p 147, TP-92/14, 1993
13. *IARC Monographs on the Evaluation of Carcinogenic Risks to Humans.* Vol 49, Chromium, nickel and welding, pp 257–445. Lyon, International Agency for Research on Cancer, 1990
14. *Health Effects Document on Nickel*, pp 1–204. Department of Environmental Medicine, Odense University, Odense, Denmark. Submitted to Ontario Ministry of Labour, 1986

NICKEL CARBONYL
CAS: 13463-39-3

Ni(CO)$_4$

Synonyms: Nickel tetracarbonyl

Physical Form. Colorless liquid

Uses. Purification intermediate in refining nickel; catalyst in the petroleum, plastic, and rubber industries

Exposure. Inhalation

Toxicology. Nickel carbonyl is a severe pulmonary irritant.

Initial symptoms usually include frontal headache, vertigo, nausea, vomiting, and sometimes substernal and epigastric pain; generally, these early effects disappear when the individual is removed to fresh air.[1,2] It is estimated that exposure to 30 ppm nickel carbonyl for 30 minutes may be lethal to humans.[2]

There may be an asymptomatic interval between recovery from initial symptoms and the onset of delayed symptoms, which tend to develop 12 to 36 hours following exposure. Constrictive pain in the chest is characteristic of the delayed onset of pulmonary effects, followed by cough, hyperpnea, and cyanosis, leading to profound weakness; gastrointestinal symptoms may also occur. The temperature seldom rises above 101°F, and leukocytosis above 12,000/cm^3 is infrequent. Physical signs are compatible with pneumonitis or bronchopneumonia. Except for the pronounced weakness and hyperpnea, the physical findings and symptoms resemble those of a viral pneumonia.[2,3]

Terminally, delirium and convulsions frequently occur; death has occurred from 3 to 13 days after exposure to nickel carbonyl. In subjects who recover from nickel carbonyl intoxication, convalescence is usually protracted (2 to 3 months) and is characterized by excessive fatigue on slight exertion.

A close correlation exists between the clinical severity of acute nickel carbonyl intoxication and the urinary concentration of nickel during the first 3 days after exposure; hospitalization should be considered in all cases in which the urinary nickel content exceeds 0.5 mg/l of urine.[2]

Controversy as to whether nickel carbonyl causes cancer arose from observation of increased incidence of cancer of the paranasal sinuses and lungs among workers in nickel refineries. Suspicion of carcinogenicity focused primarily on nickel carbonyl vapor although there were concurrent exposures to respirable particles of nickel, nickel subsulfide, and nickel oxide.[1] Subsequent studies have shown an increased risk of lung and sinus cancer in nickel refineries where nickel carbonyl was not used in the process.[4] Furthermore, the incidence of respiratory cancer decreased greatly by 1930 despite continued exposure of workers to the same levels of nickel carbonyl through 1957.

Administration of nickel carbonyl to rats by repeated intravenous injection was associated with an increased incidence of various malignant tumors.[5] Inhalation exposure of rats was associated with a few pulmonary malignancies not reaching statistical significance.

The IARC has determined that there is limited evidence for the carcinogenicity of nickel carbonyl in experimental animals and that, overall, nickel compounds are carcinogenic to humans.

The 1995 proposed ACGIH threshold limit value–time-weighted average (TLV-TWA) for nickel, including nickel carbonyl, is 0.05 mg/m³ as Ni, with an A1 confirmed human carcinogen designation.

Treatment: The use of sodium diethyldithiocarbamate (Dithriocarb) in the hospital treatment of acute nickel carbonyl poisoning has been described.[6]

REFERENCES

1. Committee on Medical and Biologic Effects of Environmental Pollutants, Division of Medical Sciences, National Research Council: *Nickel,* pp 113–128, 164–171, 231–268. Washington, DC, National Academy of Sciences, 1975
2. Hygienic Guide Series: Nickel carbonyl. *Am Ind Hyg Assoc J* 29:304–307, 1968
3. Jones CC: Nickel carbonyl poisoning. *Arch Environ Health* 26:245–248, 1973
4. Nickel carbonyl. *Documentation of TLVs and BEIs,* 6th ed, pp 1076–1078. Cincinnati, OH, American Conference of Governmental Industrial Hygienists (ACGIH), 1991
5. *IARC Monographs on the Evaluation of the Carcinogenic Risks to Humans,* Vol 49, pp 257–445. Lyon, International Agency for Research on Cancer, 1990
6. Sunderman FW: The treatment of acute nickel carbonyl poisoning with sodium diethyldithiocarbamate. *Ann Clin Res* 3:182–185, 1971

NICOTINE
CAS: *54-11-5*

$C_{10}H_{14}N_2$

Synonyms: 1-Methyl-2-(3-pyridyl)pyrrolidine; Black Leaf

Physical Form. Colorless to pale yellow, oily liquid; turns brown on exposure to air or light

Uses/Sources. Insecticide; in tanning; present in tobacco

Exposure. Inhalation; skin absorption; ingestion

Toxicology. Nicotine is a potent and fast-acting poison; it is rapidly absorbed from all routes of entry, including the skin.

Nicotine acts on the central nervous system, autonomic ganglia, adrenal medulla, and neuromuscular junctions; initial stimulation is followed by a depressant phase of action.[1,2] The resulting physiologic effects are often complex and unpredictable. Small doses of nicotine cause nausea, vomiting, diarrhea, headache, dizziness, and neurologic stimulation resulting in tachycardia, hypertension, hyperpnea, tachypnea, sweating, and salivation.[1,2] With severe intoxication, there are convulsions and cardiac arrhythmias. In fatal cases, death nearly always occurs within 1 hour and may occur within a few minutes.[3] Autopsy from fatal nicotine poisoning has shown marked dilation of the right side of the heart, mild pulmonary edema, hemorrhagic gastritis, brain edema, and renal hyperemia.[4]

Many of the acute physiological effects of smoking, chewing or inhaling tobacco are attributed to nicotine, but the chronic effects of smoking, such as lung cancer, emphysema, and heart disease, are thought to be due to the nitrosamines, polycyclic aromatic hydrocarbons, and carbon monoxide that are also present.[4]

Nicotine, absorbed dermally, is probably the cause of "green-tobacco sickness," a self-limited illness consisting of pallor, vomiting, and prostration that is seen in men handling tobacco leaves in the field.[3]

Nicotine is teratogenic in mice; skeletal system malformations occurred in the offspring of pregnant mice injected subcutaneously with nicotine between days 9 to 11 of pregnancy.[5]

The 1995 ACGIH threshold limit value–time-weighted average (TLV-TWA) is 0.5 mg/m^3, with a notation for skin absorption.

REFERENCES

1. Friedman PA: Poisoning and its management. In Petersdorf RG et al (eds): *Harrison's Principles of Internal Medicine*, 10th ed, p 1271. New York, McGraw-Hill, 1983
2. Taylor P: Ganglionic stimulating and blocking agents. In Gilman AG et al (eds): *Goodman and Gilman's The Pharmacological Basis of Therapeutics*, 7th ed, pp 217–218. New York, Macmillan, 1985
3. Gosselin RE, Smith RP, Hodge HC: *Clinical Toxicology of Commercial Products, Section III*, 5th ed, pp 311–314. Baltimore, MD, Williams and Wilkins, 1984
4. Nicotine. *Documentation of TLVs and BEIs*, 6th ed, pp 1083–1085. Cincinnati, OH, American Conference of Governmental Hygienists (ACGIH), 1991
5. Nishimura H, Nakai K: Developmental anomalies in offspring of pregnant mice treated with nicotine. *Science* 127:877–878, 1958

NITRIC ACID
CAS: 7697-37-2

HNO$_3$

Synonyms: Aquafortis; azotic acid; hydrogen nitrate

Physical Form. Colorless or yellowish liquid with a suffocating odor

Uses. Production of fertilizers in the form of ammonium nitrate; photoengraving; steel etching; dye intermediates; explosives

Exposure. Inhalation

Toxicology. From topical contact, nitric acid causes corrosion of the skin and other tissues and, from inhalation, acute pulmonary edema or chronic obstructive pulmonary disease.

When nitric acid is exposed to air or comes in contact with organic matter, it decomposes to yield a mixture of oxides of nitrogen, including nitric oxide and nitrogen dioxide, the latter being more hazardous than nitric acid.[1] Exposure to high concentrations of nitric acid vapor and nitrogen oxides causes pneumonitis and pulmonary edema, which may be fatal; onset of symptoms, such as dryness of the throat and nose, cough, chest pain, and dyspnea, may or may not be delayed.[2]

In a recent report, three pulp mill workers died after inhaling fumes for approximately 10 to 15 minutes from a nitric acid tank explosion (concentrations not available).[3] No significant respiratory complaints were apparent during the initial examination. However, 4 to 6 hours later, the workers became cyanotic, with frothy fluid escaping from the nose and mouth. All died in less than 24 hours. Necropsy showed bronchiolar epithelial necrosis, marked capillary engorgement, and slight interstitial edema of the alveoli; the lungs were 5 times heavier than normal and released abundant frothy fluid from all lobes. The delayed manifestations of lung injury were consistent with the formation of nitrogen dioxide and other nitrous oxides from the nitric acid and subsequent cellular damage from the formation of chemical free-radicals and acids from hydration of nitogen dioxide. Pulmonary edema was a consequence of increased microvascular permeability, initiated by the nitrogen dioxide–mediated capillary injury. Additional findings in this study also implicated neutrophils and serum-derived mediators in the pathogenesis of the pulmonary edema.

Healthy volunteers exposed to 500 μg/m for 4 hours showed no evidence of proximal airway or distal lung injury.[4] However, other reports have suggested that prolonged exposure to low concentrations of the vapor may lead to chronic bronchitis and/or loss of appetite.[5]

The vapor and mist have been reported to erode exposed teeth.[1] However, in cases of dental erosion attributed to nitric acid, concomitant exposure to sulfuric acid, a potent cause of dental erosion was known to occur. Ingestion of the liquid will cause immediate pain and burns of the gastrointestinal tract.

In contact with the eyes, the liquid produces severe burns, which may result in permanent damage and visual impairment.[2] On the skin, the liquid or concentrated vapor produces immediate, severe, and penetrating burns; concentrated solutions cause deep ulcers and stain the skin a bright yellow or yellowish-brown color.[1,2] Dilute solutions of nitric acid produce mild irritation of the skin and tend to harden the epithelium without destroying it.

The 1995 TLV-TWA is 2 ppm (5.2 mg/m³) with a STEL/ceiling of 4 ppm (10 mg/m³).

REFERENCES

1. National Institute for Occupational Safety and Health: *Criteria for a Recommended Standard . . . Occupational Exposure to Nitric Acid.* DHEW (NIOSH) Pub No 76-141, pp 35–36. Washington, DC, US Government Printing Office, 1976
2. Hygienic Guide Series: Nitric acid. *Am Ind Hyg Assoc J* 25:426–428, 1964
3. Hajela R, Janigan DT, Landrigan PL, et al: Fatal pulmonary edema due to nitric acid fume inhalation in three pulp-mill workers. *Chest* 97:487–489, 1990
4. Aris R, Christian D, Tager I, et al: Effects of nitric acid gas alone or in combination with ozone on healthy volunteers. *Am Rev Respir Dis* 148:965–73, 1993
5. Nitric acid. Hazard Data Bank, Sheet No 82. *The Safety Practitioner*, pp 46–47, October 1986

NITRIC OXIDE
CAS: 10102-43-9

NO

Synonyms: Nitrogen monoxide; mononitrogen monoxide

Physical Form. Colorless gas

Uses. In manufacture of nitric acid; for bleaching rayon; stabilizer

Exposure. Inhalation

Toxicology. Nitric oxide is a vasodilator and, at higher concentrations, causes formation of methemoglobin.

In human volunteers, significant lung vasodilator effects have been observed at 10 to 40 ppm.[1] Studies indicate that nitric oxide stimulates guanylate cyclase, which leads to smooth muscle relaxation and vasodilation. Because nitric oxide is rapidly inactivated in hemoglobin, internal organs other than the lungs are unlikely to be affected by vasodilation.[1]

In animals, methemoglobin formation is seen at concentrations above 10 ppm. Exposure of mice to 5000 ppm for 6 to 8 minutes was lethal, as was 2500 ppm for 12 minutes; cyanosis occurred after a few minutes, the red eye grounds became gray-blue, and breathlessness appeared, with paralysis and convulsions; spectroscopy of the blood showed methemoglobin.[2]

Some recent studies in mice have suggested that concentrations of 2 to 10 ppm may reduce host resistance to infection.[1]

Nitric oxide is converted spontaneously in air to nitrogen dioxide; hence, some of the latter gas is invariably present whenever nitric oxide is found in the air.[2] At concentrations below 50 ppm, however, this reaction is slow, and substantial concentrations of nitric oxide may occur with negligible quantities of nitrogen dioxide.[2] It is likely that the effects of concomitant exposure to nitrogen dioxide will become manifest before the methemoglobin effects due to nitric oxide can occur. Nitrogen dioxide may cause irritation of the eyes, nose, and throat, as well as delayed pulmonary edema.[2]

The 1995 ACGIH threshold limit value–time-weighted average (TLV-TWA) for nitric oxide is 25 ppm (31 mg/m³).

REFERENCES

1. Gustafsson LE: Experimental studies on nitric oxide. *Scand J Work Environ Health* 19(Suppl 2):44–48, 1993
2. National Institute for Occupational Safety and Health: *Criteria for a Recommended Standard . . . Occupational Exposure to Oxides of Nitrogen (Nitrogen Dioxide and Nitric Oxide).* DHEW (NIOSH) Pub No 76-149, pp 46–50, 75–76. Washington, DC, US Government Printing Office, 1976

---

p-NITROANILINE
CAS: 100-01-6

$NH_2C_6H_4NO_2$

Synonyms: PNA; 1-amino-4-nitrobenzene

Physical Form.   Yellow crystals

Uses.   Chemical intermediate in the manufacture of antioxidants, antiozonants, dye colors, and pigments

Exposure.   Inhalation; skin absorption

Toxicology. *p*-Nitroaniline (PNA) absorption, whether from inhalation of the vapor or absorption of the solid through skin, causes anoxia as the result of the formation of methemoglobin; jaundice and anemia have been reported from chronic exposure.

Signs and symptoms of overexposure are due to loss of oxygen-carrying capacity of the blood. The onset of symptoms of methemoglobinemia is often insidious and may be delayed for up to 4 hours; headache is commonly the first symptom and may become quite intense as the severity of methemoglobinemia progresses.[1] Cyanosis develops early in the course of intoxication; blueness in the lips, the nose, and the earlobes is usually recognized by fellow workers. Cyanosis occurs when the methemoglobin concentration is 15% or more. The individual usually feels well, has no complaints, and insists that nothing is wrong until the methemoglobin concentration approaches approximately 40%. At methemoglobin concentrations of over 40%, there usually is weakness and dizziness; methemoglobin levels above 50% are rarely observed with *p*-nitroaniline exposure; however, concentrations of up to 70% would be expected to cause ataxia, dyspnea on mild exertion, tachycardia, nausea, vomiting, and drowsiness; methemoglobin of about 75% usually results in collapse, coma, and even death.[1,2] There are no reports of chronic effects from single exposures, but prolonged or excessive exposures may cause liver damage.[1,2]

p-Nitroaniline is mildly irritating to the eyes and may cause some corneal damage.[2]

Ingestion of alcohol aggravates the toxic effects of *p*-nitroaniline.[2]

In general, higher ambient temperatures increase susceptibility to cyanosis from exposure to methemoglobin-forming agents.[3]

Exposure of rats to aerosol/vapor of PNA at 30 mg/m³ for 4 weeks produced a significant increase in methemoglobin levels.[4] In subchronic studies, administration of PNA at 3, 10, or 30 mg/kg/day for 90 days produced a dose-related increase in methemoglobin; decreases in hematocrit, hemoglobin, and/or red blood cell count were indicative of anemia; histopathological changes in the spleen included congestion, hemosiderosis, and excessive extramedullary hematopoiesis.[5] Chronic studies in male rats administered 0, 0.25, 1.5, or 9.0 mg/kg/day by gavage for a period of 2 years yielded similar results: blood methemoglobin levels were elevated in the mid- and high-dosage group, and anemia and increased spleen weights were observed in the high-dose groups.[6] No treatment-related increase in tumor incidence occurred at these PNA levels. In a reproductive study, the same doses were administered to male and female rats, prior to mating and during mating, gestation, and lactation to the F_0 and F_1 generations.[6] No significant effects were seen in mating, pregnancy, or fertility indices.

The 1995 ACGIH threshold limit value–time-weighted average (TLV-TWA) is 3 mg/m³, with a notation for skin absorption.

Note: For a description of diagnostic signs, differential diagnosis, and medical control, including clinical laboratory treatment of overexposure to methemoglobin-forming agents, see "Aniline."

REFERENCES

1. Beard RR, Noe JT: Aromatic nitro and amino compounds. In Clayton GD, Clayton FE (eds): *Patty's Industrial Hygiene and Toxicology*, 3rd ed, Vol 2A, *Toxicology*, pp 2413–2489. New York, Wiley-Interscience, 1981
2. Chemical Safety Data Sheet SD-94, para-Nitroaniline, pp 5–6, 11–13. Washington, DC, MCA, Inc, 1966
3. Linch AL: Biological monitoring for industrial exposure to cyanogenic aromatic nitro and amino compounds. *Am Ind Hyg Assoc J* 35:426–432, 1974
4. Nair RS, Johannsen FR, Levinskas GJ, et al: Subchronic inhalation toxicity of *p*-nitroaniline and *p*-nitrochlorobenzene in rats. *Fund Appl Toxicol* 6:618–627, 1986
5. Houser RM, Stout LD, Ribelin WE: The subchronic toxicity of *p*-nitroaniline administered to male and female Sprague–Dawley rats for 90 days. *Toxicologist* 3:128, 1983
6. Nair RS, Auletta CS, Schroeder RE, et al: Chronic toxicity, oncogenic potential, and reproductive toxicity of *p*-nitroaniline in rats. *Fund Appl Toxicol* 15:607–621, 1990

NITROBENZENE
CAS: 98-95-3

$C_6H_5NO_2$

Synonyms: Nitrobenzol; oil of mirbane

Physical Form. Almost water-white, oily liquid, turning yellow with exposure to air

Uses. Chemical intermediate for the production of aniline and other products

Exposure. Inhalation; skin absorption

Toxicology. Nitrobenzene causes anoxia as a result of the formation of methemoglobin; in experimental animals, chronic exposure has been associated with lesions of the liver, spleen, and kidney and with testicular atrophy; it is carcinogenic to mice and rats.

Exposure of workers to 40 ppm for 6 months resulted in some cases of intoxication and anemia; concentrations ranging from 3 to 6 ppm caused headache and vertigo in 2 of 39 workers; increased methemoglobin and sulfhemoglobin levels and Heinz bodies were observed in the blood.[1]

Signs and symptoms of overexposure are due to the loss of oxygen-carrying capacity of the blood. The onset of symptoms of methemoglobinemia is often insidious and may be delayed up to 4 hours; headache is commonly the first symptom and may become quite intense as the severity of methemoglobinemia progresses.[2] Cyanosis develops early in the course of intoxication, characterized by blueness of the lips, nose, and earlobes, usually recognized first by fellow workers, and occurring when the methemoglobin level is 15% or more. The individual usually feels well, has no complaints, and insists that nothing is wrong until the methemoglobin concentration approaches 40%. At methemoglobin concentrations ranging from 40% to 70%, there is headache, weakness, dizziness, ataxia, dyspnea on mild exertion, tachycardia, nausea, vomiting and drowsiness.[2,3] Coma may ensue with methemoglobin levels above

70%; and the lethal level is estimated to be 85% to 90%.[3]

Hepatotoxicity, manifested by alterations in liver function, including hyperbilirubinemia, and by decreased prothrombin activity, is associated with exposure in both animals and humans.[4]

Inhalation exposure of rats and mice (10–25 ppm over 2 weeks or 5–50 ppm over 13 weeks) caused methemoglobinemia, encephalopathy, and lesions in the liver (hepatocyte necrosis and hepatomegaly), kidney (hyaline nephrosis), and spleen (extramedullary hematopoiesis and proliferative capsular lesions).[5,6]

In a 2-year inhalation study, nitrobenzene was carcinogenic in mice and rats with differing target organs based on species, sex, and strain.[7] Male B6C3F$_1$ mice exposed at concentrations of up to 50 ppm had increased incidences of pulmonary alveolar/bronchiolar and thyroid follicular-cell neoplasms, whereas females had mammary gland neoplasms. In rats, exposures of up to 25 ppm resulted in hepatocellular and renal neoplasms (male F344 rats), endometrial stromal neoplasms (female F344), and hepatocellular neoplasms (male CD rats).

Nonneoplastic effects from chronic exposure included methemoglobinia, anemia, lesions of the olfactory epithelium and, in the CD males, an increased incidence of testicular atrophy.[7] Degenerative testicular lesions have also occurred in rats exposed to single oral doses of 50 to 450 mg/kg.[8] Male rats repeatedly administered up to 100 mg/kg body weight by gavage daily showed atrophy of the seminiferous tubules of the testis, but male fertility was not affected.[9]

No evidence of teratogenesis or adverse fetal effects was apparent in the offspring of rats exposed at concentrations of 40 ppm for 6 hr/day from days 6 through 15 of pregnancy.[8]

Ingestion of alcohol aggravates the toxic effects of nitrobenzene.[3] In general, higher ambient temperatures increase susceptibility to cyanosis from exposure to methemoglobin-forming agents.[10] p-Nitrophenol and p-aminophenol are metabolites of nitrobenzene, and their presence in the urine is an indication of exposure.[11] Nitrobenzene is mildly irritating to the eyes; it may produce dermatitis as a result of primary irritation or sensitization.[3]

The 1995 ACGIH threshold limit value–time-weighted average (TLV-TWA) is 1 ppm (5 mg/m^3) with a notation for skin absorption.

Note: For a description of diagnostic signs, differential diagnosis, and medical control, including clinical laboratory treatment of over-exposure to methemoglobin-forming agents, see "Aniline."

REFERENCES

1. Pacseri I, Magos L, Batskor IA: Threshold and toxic limits of some amino and nitro compounds. *AMA Arch Ind Health* 18:1–8, 1958
2. Hamblin DO: Aromatic nitro and amino compounds. In Patty FA (ed): *Industrial Hygiene and Toxicology*, 2nd ed, Vol 2, *Toxicology*, pp 2105–2147. New York, Interscience, 1963
3. Chemical Safety Data Sheet SD-21, Nitrobenzene, pp 5–6, 12–14. Washington, DC, MCA, Inc, 1967
4. Beauchamp RO et al: A critical review of the literature on nitrobenzene toxicity. *CRC Crit Rev Toxicol* 11:33–84, 1983
5. Hamm TE Jr: Ninety-day inhalation toxicity study of nitrobenzene in F344 rats, CD rats, and B6C3F$_1$ mice: final study report. CIIT, Research Triangle Park, NC, 1984
6. Medinsky MA, Irons RD: Sex, strain and species differences in the response of rodents to nitrobenzene vapors. In Rickert DE (ed): *The Toxicity of Nitroaromatic Compounds*, pp 35–51. New York, Hemisphere Publishing, 1985
7. Cattley RC, Everitt JI, Gross EA, et al: Carcinogenicity and toxicity of inhaled nitrobenzene in B6C3F$_1$ mice and F344 and CD rats. *Fund Appl Toxicol* 22:328–340, 1994
8. Dodd DE et al: Reproductive and fertility evaluations in CD rats following nitrobenzene inhalation. *Fund Appl Toxicol* 8:493–505, 1987
9. Mitsumori K, Kodama Y, Uchida O et al: Confirmation study, using nitrobenzene, of the combined repeat dose and reproductive/developmental toxicity test protocol proposed by the Organization for Economic Cooperation and Development (OECD). *J Toxicol Sci* 19:141–149, 1994
10. Linch AL: Biological monitoring for indus-

trial exposure to cyanogenic aromatic nitro and amino compounds. *Am Ind Hyg Assoc J* 35:426–432, 1974

11. Ikeda M, Kita A: Excretion of *p*-nitrophenol and *p*-aminophenol in the urine of a patient exposed to nitrobenzene. *Br J Ind Med* 21:210–213, 1964

o-NITROCHLOROBENZENE
CAS: 88-73-3

$NO_2C_6H_4Cl$

Synonyms: 2-chloronitrobenzene; 1-chloro-2-nitrobenzene; 2 CNB; ONCB

Physical Form. Yellow solid

Uses. Chemical intermediate in manufacture of dyes, picric acid, lumber preservatives, and diaminophenol hydrochloride (a photographic developer)

Exposure. Inhalation; skin absorption

Toxicology. *o*-Nitrochlorobenzene (ONCB) absorption causes anoxia through formation of methemoglobin.

Numerous cases of cyanosis in workers exposed to ONCB and related compounds occurred in the period 1935–1965.[1]

Signs and symptoms of overexposure are caused by the loss of oxygen-carrying capacity of the blood. The onset of symptoms of methemoglobinemia is often insidious and may be delayed up to 4 hours. Headache is commonly the first symptom and may become quite intense as the severity of methemoglobinemia progresses.[1] Cyanosis develops early in the course of intoxication; it is characterized by blueness of the lips, nose, and earlobes, usually recognized first by fellow workers, and occurs when the methemoglobin concentration approaches 40%. At methemoglobin concentrations ranging from 15% to 70%, there is head-ache, weakness, dizziness, ataxia, dyspnea on mild exertion, tachycardia, nausea, vomiting, and drowsiness. Coma may ensue with methemoglobin levels of about 70%; the lethal level is estimated to be 85% to 90%.

In general, higher ambient temperatures increase susceptibility to cyanosis from exposure to methemoglobin-forming agents.[1]

The acute oral LD_{50} of ONCB in rats is 560 mg/kg, whereas the dermal LD_{50} in rabbits is 400 mg/kg.[2] In subchronic inhalation studies, rats were exposed to 10, 30, or 60 mg/m³, 6 hr/day, 5 days/week, for 4 weeks. Animals exposed to the mid- and high concentrations showed a significant increase in blood methemoglobin and a significant decrease in hemoglobin, hematocrit, and red blood cell counts. Spleen and liver weights were also significantly increased for these two groups; microscopic changes, observed only in the spleen, included an increased degree of extramedullary hematopoiesis and hemosiderosis.

In carcinogenicity bioassays, it was found that ONCB produced an increase in the incidence of multiple tumors in male rats at the low dose (1000 mg/kg diet for 6 months, followed by 500 mg/kg diet for 12 months and control diets for an additional 6 months) but not at the high dose (2000 mg/kg diet for 6 months, followed by 100 mg/kg diet for 12 months and control diets for 6 months).[3] *o*-Nitrochlorobenzene produced an increase in hepatocellular carcinomas in female mice at high (6000 mg/kg diet) and low dose (3000 mg/kg diet) levels, and in male mice at low dose levels but not at high dose levels. Because of the inconsistency of the dose-response effects, the high doses used, and the long latent periods before tumor development, ONCB was not regarded as a very potent carcinogen under the conditions of the test.

There is no TLV established for the *o*-isomer of nitrochlorobenzene.

Note: For a description of diagnostic signs, differential diagnosis, and medical control, including clinical laboratory treatment of overexposure to methemoglobin-forming agents, see "Aniline"

REFERENCES

1. Linch AL: Biological monitoring for industrial exposure to cyanogenic aromatic nitro and amino compounds. *Am Ind Hyg Assoc J* 35:426, 1974
2. Nair RS, Johannsen FR, Levinskas GJ, et al: Assessment of toxicity of *o*-nitrochlorobenzene in rats following a 4 week inhalation exposure. *Fund Appl Toxicol* 7:609–614, 1986
3. Weisburger EK et al: Testing of 21 environmental aromatic amines or derivatives for long-term toxicity or carcinogenicity. *J Environ Pathol Toxicol* 2:325, 1978

p-NITROCHLOROBENZENE
CAS: 100-00-5

$NO_2C_6H_4Cl$

Synonyms: PCNB; *p*-Chloronitrobenzene; 4-CNB

Physical Form.   Yellowish crystals

Uses.   In manufacture of dyes, rubber, and agricultural chemicals

Exposure.   Inhalation; skin absorption

Toxicology.   Absorption of *p*-nitrochlorobenzene causes anoxia from the formation of methemoglobin.

Signs and symptoms of overexposure are due to the loss of oxygen-carrying capacity of the blood. The onset of symptoms of methemoglobinemia is often insidious and may be delayed for up to 4 hours. Headache is commonly the first symptom and may become quite intense as the severity of methemoglobinemia progresses.[1] Cyanosis develops early in the course of intoxication; blueness appears first in the lips, the nose, and the earlobes and is usually recognized by fellow workers. Cyanosis occurs when the methemoglobin concentration is 15% or more. The subject usually feels well, has no complaints, and is insistent that nothing is wrong until the methemoglobin concentration approaches approximately 40%. At methemoglobin concentrations over 40%, there is weakness and dizziness; at closer to 70% concentration, there may be ataxia, dyspnea on mild exertion, tachycardia, nausea, vomiting, and drowsiness. The ingestion of alcohol aggravates the toxic effects of *p*-nitrochlorobenzene. In general, higher ambient temperatures increase susceptibility to cyanosis from exposure to methemoglobin-forming agents.

Four workers exposed to an unmeasured concentration of the vapor for a period of 2 to 4 days developed methemoglobinemia; in these cases, there was an initial collapse, a slate-gray appearance, dyspnea, and mild anemia 1 week after exposure.[2]

The acute oral LD_{50} in rats was 530 mg/kg.[3] In a 4-week inhalation study, exposure to 0.82, 2.5, or 7.5 ppm (5, 15, or 45 mg/m^3) 6 hr/day, 5 days/week, caused a dose-related increase in methemoglobin levels and decreases in hemoglobin, hematocrit, and red blood cell counts.[4] Microscopic changes in the spleen included congestion, increased extramedullary hematopoiesis, and hemosiderosis.

No increase in tumor incidence was seen in rats fed up to 1000 ppm in the diet for 2 years; in mice, results were equivocal, with high-dose animals showing an increase in vascular tumors and low-dose males showing an increase in liver tumors.[5]

Administered to pregnant rabbits or rats, fetal effects were observed only at doses that produced severe maternal toxicity.[6]

Applied to the skin or eyes of rabbits, PNCB did not cause irritation; it was absorbed, producing methemoglobinemia, Heinz bodies in erythrocytes, anemia, hematuria, and hemoglobinuria.[3] The acute dermal LD_{50} for rabbits was 3400 mg/kg.

p-Nitrochlorobenzene has a pleasant, aromatic odor.

The 1995 ACGIH threshold limit value–time-weighted average (TLV-TWA) is 0.1 ppm (0.64 mg/m^3) with a notation for skin absorption.

Note: For a description of diagnostic signs, differential diagnosis, and medical control, including clinical laboratory treatment of overexposure to methemoglobin-forming agents, see "Aniline."

REFERENCES

1. Hamblin DO: Aromatic nitro and aminocompounds. In Patty FA (ed): *Industrial Hygiene and Toxicology*, 2nd ed, Vol 2, *Toxicology*, pp 2105–2119, 2130–2131. New York, Wiley-Interscience, 1963
2. Renshaw A, Ashcroft GV: Four cases of poisoning by mononitrochlorobenzene and one by acetanicide occurring in a chemical works: with an explanation of the toxic symptoms produced. *J Ind Hyg* 8:67–73, 1926
3. p-Nitrochlorobenzene. *Documentation of the TLVs and BEIs*, 6th ed, pp 1100–1102, Cincinnati, OH, American Conference of Governmental Industrial Hygienists (ACGIH), 1992
4. Nair RS, Johannsen FR, Levinskas GJ, et al: Subchronic inhalation toxicity of p-nitroaniline and p-nitrochlorobenzene in rats. *Fund Appl Toxicol* 6:618–627, 1986
5. Weisburger EK, Russfield AB, Homburger F, et al: Testing of twenty-one environmental aromatic amines or derivatives for long-term toxicity or carcinogenicity. *J Environ Toxicol* 2:325–356, 1978
6. Nair RS, Johannsen FR, Schroeder RE: Evaluation of teratogenic potential of p-nitroaniline and p-nitrochlorobenzene in rats and rabbits. In Rickert DE (ed): Chemical Industry Institute of Toxicology Series. *Toxicity of Nitroaromatic Compounds*, pp 61–86. Washington, DC, Hemisphere Publishing, 1985

4-NITRODIPHENYL
CAS:92-93-3

$C_{12}H_9NO_2$

Synonyms: 4-Nitrobiphenyl; p-nitrobiphenyl; PNB

Physical Form. White crystals

Uses. Plasticizer; fungicide; wood preservative

Exposure. Inhalation; skin absorption

Toxicology. 4-Nitrodiphenyl, or p-nitrobiphenyl (PNB), is a urinary bladder carcinogen in dogs.

There are no reports on carcinogenicity of PNB in man.[1] However, PNB was used as an intermediate in the preparation of 4-aminobiphenyl, a recognized human bladder carcinogen, and bladder tumors found in men exposed to 4-aminobiphenyl may have been partially due to PNB.[2]

Three of four dogs fed 0.3 g of PNB (in capsules) 3 times/week for up to 33 months developed bladder tumors.[2] The total dose administered ranged from 7 to 10 g/kg in the affected dogs; the animal that did not develop bladder tumors was the largest and therefore had received less of the compound per kilogram of body weight (5.5 g/kg). The tumors produced by PNB were identical histologically to those produced by 4-aminobiphenyl.[2]

The case for the carcinogenicity of PNB is supported by: (1) the induction of urinary bladder cancer in dogs after administration of PNB; (2) the evidence that PNB is metabolized, in vivo, to 4-aminobiphenyl (a potent carcinogen); and (3) the possibility that the cases of human urinary bladder cancer attributed to 4-aminobiphenyl may also have been induced by exposure to PNB.[1]

The 1995 ACGIH threshold limit value (TLV) for PNB is 0. It is classified as a confirmed human carcinogen, and exposure by any route—respiratory, oral, or skin—should be avoided.

REFERENCES

1. *IARC Monographs on the Evaluation of the Carcinogenic Risk of Chemicals to Man*, Vol 4, Some aromatic amines, hydrazine and related substances, N-nitroso compounds and miscellaneous alkylating agents, pp 113–117. Lyon, International Agency for Research on Cancer, 1974
2. Deichmann WB et al: Para-nitrobiphenyl—a new bladder carcinogen in the dog. *Ind Med Surg* 27:634–637, 1958

NITROETHANE
CAS: 79-24-3

$C_2H_5NO_2$

Synonyms: None

Physical Form. Colorless, oily liquid

Uses. Common industrial solvent; more recently, a commercial artificial nail remover

Exposure. Inhalation

Toxicology. In animals, nitroethane is a respiratory irritant and, at high concentrations, it causes narcosis and liver damage; methemoglobin has been reported following ingestion by humans.

Accidental ingestion of less than 1 ounce of an artificial fingernail remover containing 100% nitroethane resulted in life-threatening methemoglobinemia in a 20-month-old child.[1] The child was initially asymptomatic but, on hospital admission, 10 hours later, was short of breath and visibly cyanotic. Methemoglobin concentrations reached 40.1%. The patient received 15 mg methylene blue intravenously, with resolution of cyanosis. One hour later, the child's methemoglobin concentration dropped to 5.7%, and other laboratory findings and vital signs were within normal limits. He was discharged 1 day later with a methemoglobin concentration of 1.5%.

The delayed onset of symptoms (10 hours in this case) suggests a possible metabolism of nitroethane to a more toxic nitrite compound, which may, in turn, be responsible for the induction of methemoglobin.[1]

Rabbits died from exposure to 5000 ppm for 3 hours but survived 3 hours at 2500 ppm.[2,3] Exposure to the higher concentrations caused irritation of mucous membranes, lacrimation, dyspnea, pulmonary rales and, in a few animals, pulmonary edema; convulsions were rare and of brief duration.[2] Autopsy of animals exposed to lethal concentrations showed mild to severe liver damage and nonspecific changes in the kidneys. Nitroethane was not hepatotoxic after administration of 9 mM/kg to mice.[4] It also was not genotoxic in *Salmonella typhimurium* tester strains.

The liquid is a mild skin irritant because of its solvent action.

The odor of nitroethane is detectable at 163 ppm; the odor and irritant properties do not provide sufficient warning of toxic concentrations.[2,3]

The 1995 ACGIH threshold limit value–time-weighted average (TLV-TWA) for nitroethane is 100 ppm (307 mg/m³).

REFERENCES

1. Hornfeldt CS, Rabe WH: Nitroethane poisoning from an artificial fingernail remover. *Clin Toxicol* 32:321–324, 1994
2. Machle W, Scott EW, Treon J: The physiological response of animals to some simple mononitroparaffins and to certain derivatives of these compounds. *J Ind Hyg Toxicol* 22:315–332, 1940
3. American Industrial Hygiene Association: Hygienic Guide Series. Nitroethane. 3 pp. 1978
4. Dayal R, Gescher A, Harpur ES, et al: Comparison of the hepatotoxicity in mice and the mutagenicity of three nitroalkanes. *Fund Appl Toxicol* 13:341–348, 1989

NITROGEN DIOXIDE
CAS: 10102-44-0

NO_2

Synonyms: None

Physical Form. Gas

Uses/Sources. Intermediate in nitric and sulfuric acid production; in nitration of organic compounds and explosives; found in vehicle emissions and fossil fuel combustion

Exposure. Inhalation

Toxicology. Nitrogen dioxide is a respiratory irritant; at high concentrations, it causes pulmonary edema and, rarely among survivors, bronchiolitis obliterans.

Brief exposure of humans to concentrations of about 250 ppm causes cough, production of mucoid or frothy sputum, and increasing dyspnea.[1,2] Within 1 to 2 hours, the person may develop pulmonary edema with tachypnea, cyanosis, fine crackles and wheezes through the lungs, and tachycardia. Alternatively, there may be only increasing dyspnea and cough over several hours, with symptoms then gradually subsiding over a 2- to 3-week period. The condition may then enter a second stage of abruptly increasing severity; fever and chills precede a relapse, with increasing dyspnea, cyanosis and recurring pulmonary edema. Death may occur either in the initial or second stage of the disease; a severe second stage may follow a relatively mild initial stage. The subject who survives the second stage usually recovers over 2 to 3 weeks; however, some cases do not return to normal but experience varying degrees of impaired pulmonary function.

The radiographic features in the acute initial stage vary from normal to those of typical pulmonary edema; most reports mention a pattern of nodular shadows on the chest film at the outset.[1,2] The roentgenogram may then clear, only to show miliary mottling as the second stage commences, progressing to the development of a confluent pattern. Results of pulmonary-function tests in the acute stage show reduction in lung volume and diffusing capacity; similar findings are recorded in the second stage.

Pathologic examination of the acute lesion shows extensive mucosal edema and inflammatory-cell exudation. The delayed lesion shows the histologic appearance of bronchiolitis obliterans; small bronchi and bronchioles contain an inflammatory exudate that tends to undergo fibrinous organization, eventually obliterating the lumen.

Humans exposed to nitrogen dioxide for 60 minutes can expect the following effects: at 100 ppm, pulmonary edema and death; at 50 ppm, pulmonary edema, with possible subacute or chronic lesions in the lungs; and, at 25 ppm,

respiratory irritation and chest pain.[3] A concentration of 50 ppm is moderately irritating to the eyes and nose; 25 ppm is irritating to some people.[1] Exposure of healthy and asthmatic volunteers at 4 ppm for 75 minutes caused a small but significant decrease in systolic blood pressure; there were no significant effects on airway resistance, symptoms, heart rate, skin conductance, or self-reported emotional state.[4] However, in an earlier study, human volunteers exposed to 5 ppm for 15 minutes and 2.5 ppm for 2 hours showed increased airway resistance.[5]

Most reported cases of severe illness from nitrogen dioxide have been accidental exposures to explosion or combustion of nitroexplosives, nitric acid, the intermittent process of arc or gas welding (especially in a confined space), or entry into an agricultural silo that was not vented.[1,6]

Less severe respiratory complaints have been reported in 116 individuals exposed to nitrogen dioxide during two hockey games in an indoor ice arena.[7] The gas was emitted from the malfunctioning engine of an ice resurfacer. Air concentrations were not recorded, although air sampling under simulated conditions detected 4 ppm nitrogen dioxide; levels were probably higher during the games. Of interest was the occurrence of cough, dyspnea, chest pain, and mild hemoptysis, principally among the hockey players and cheerleaders, who were actively exercising and, therefore, had a higher minute ventilation rate and greater lung tissue exposure than spectators.

In experimental animals, nitrogen dioxide induces several types of pulmonary toxicity.[8] Decreased pulmonary function occurs in mice after chronic exposure to 0.2 ppm with daily excursions to 0.8 ppm. Effects on lung morphology were seen in rats exposed to 10 ppm for 36 hours and included cilia loss and hypertrophy of the bronchiolar epithelium. In guinea pigs, acute exposure to 4 ppm caused increased airway hyperresponsiveness toward histamine.

Animal experimentation has also indicated that, in addition to irritation and pathological changes, nitrogen dioxide exposure may decrease host resistance to infection.[8,9] An increased mortality in mice infected with pneumonia-causing organisms was found subsequent to exposure at 0.5 ppm for 7 days and longer.[10] Nitrogen dioxide can adversely affect lung defense mechanisms by reducing the efficacy of mucociliary clearance, the alveolar macrophage, and the immune system.[11]

Individuals with a history of reactive airway disease (asthma) are known to be especially susceptible to symptoms arising out of exposure to low levels of nitrogen dioxide.[12]

Nitrogen dioxide does not appear to be teratogenic, mutagenic or directly carcinogenic.[8]

The odor threshold is of the order of 0.12 ppm.

The 1995 ACGIH threshold limit value–time-weighted average (TLV-TWA) for nitrogen dioxide is 3 ppm (5.6 mg/m^3) with a short-term exposure limit of 5 ppm (9.4 mg/m^3).

REFERENCES

1. National Institute for Occupational Safety and Health: *Criteria for a Recommended Standard . . . Occupational Exposure to Oxides of Nitrogen (Nitrogen Dioxide and Nitric Oxide).* DHEW (NIOSH) Pub No 76-149, pp 76–85. Washington, DC, US Government Printing Office, 1976
2. Morgan WKC, Seaton A: Occupational Lung Diseases, pp 330–335, 344–345. Philadelphia, WB Saunders Co, 1975
3. Emergency Exposure Limits: Nitrogen Dioxide. *Am Ind Hyg Assoc J* 25:580–582, 1964
4. Linn WS, Solomon JC, Trim SC, et al: Effects of exposure to 4 ppm nitrogen dioxide in healthy and asthmatic volunteers. *Arch Environ Health* 40:234–239, 1985
5. Kerr HD et al: Effects of nitrogen dioxide on pulmonary function in human subjects: an environmental chamber study. *Environ Res* 19:392–404, 1979
6. Scott EG, Hunt WB Jr: Silo filler's disease. *Chest* 63:701–706, 1973
7. Hedberg K et al: An outbreak of nitrogen dioxide–induced respiratory illness among ice hockey players. *JAMA* 262:3014–3017, 1989
8. Moldeus P: Toxicity induced by nitrogen dioxide in experimental animals and isolated cell systems. *Scan J Work Environ Health* 19(S2):28–34, 1993
9. State of California Air Resources Board: Short-term ambient air quality standard for nitrogen dioxide. PO Box 2815, Sacramento, CA, 1985
10. Gardner DE et al: Influence of exposure node on the toxicity of NO_2. *Environ Health Persp* 30:23–29, 1979
11. Samet JM, Utell MJ: The risk of nitrogen dioxide: What have we learned from epidemiological and clinical studies? *Toxicol Ind Health* 6:247–262, 1990
12. Bauer MA, Utell MJ, Morrow PE, et al: Inhalation of 0.30 ppm nitrogen dioxide potentiates exercise-induced bronchospasm in asthmatics. *Am Rev Respir Dis* 134:1203–1208, 1986

NITROGEN TRIFLUORIDE
CAS: 7783-54-2

NF_3

Synonyms: Nitrogen fluoride

Physical Form. Colorless gas

Uses. Oxidizing agent in fuel combustion

Exposure. Inhalation

Toxicology. Nitrogen trifluoride causes anoxia in animals through the formation of methemoglobin.

Although there are no reports of human intoxication from nitrogen trifluoride, the initial effects of methemoglobinemia include cyanosis (especially in the lips, nose, and earlobes), weakness, dizziness, and severe headache.[1] At higher methemoglobin concentrations of up to 70%, there may be ataxia, dyspnea on mild exertion, and tachycardia. Coma may ensue with methemoglobin levels of about 70%, and the lethal level in humans is estimated to be 85% to 90%.

Rats died from exposure to 10,000 ppm for

60 to 70 minutes; the methemoglobin concentrations at the time of death were equivalent to 60% to 70% of available hemoglobin.[2] Animals exposed to nearly lethal concentrations suffered severe respiratory distress and cyanosis as a result of methemoglobinemia; severely affected animals showed incoordination, collapse, and convulsions. Rats repeatedly exposed to 100 ppm for 4.5 months appeared normal, but autopsy findings indicated injury to the liver and kidneys.[3] Dogs surviving exposure to 9600 ppm for 60 minutes exhibited Heinz body anemia, decreased hematocrit levels, decreased hemoglobin levels, reduced red blood cell count, and clinical signs consistent with anoxia from methemoglobin formation; some eye irritation was observed during exposure.[4]

Nitrogen trifluoride provides no odor-warning properties at potentially dangerous levels.

The 1995 ACGIH threshold limit value-time weighted average (TLV-TWA) is 10 ppm (29 mg/m^3).

Note: For a description of diagnostic signs, differential diagnosis, and medical control, including clinical laboratory treatment of overexposure to methemoglobin-forming agents, see "Aniline."

REFERENCES

1. Hamblin DO: Aromatic Nitro and Amino Compounds. In Patty FA (ed): *Industrial Hygiene and Toxicology*, 2nd ed, Vol 2, *Toxicology*, pp 2105–2119. New York, Interscience, 1963
2. Dost FN, Reed DJ, Wang CH: Toxicology of nitrogen trifluoride. *Toxicol Appl Pharmacol* 17:585–595, 1970
3. Torkelson TR, Oyen F, Sadek SE, Rowe VK: Preliminary toxicologic studies on nitrogen trifluoride. *Toxicol Appl Pharmacol* 4:770–781, 1962
4. Vernot EH, Haun CC, MacEwen JD, Egan GF: Acute inhalation toxicology and proposed emergency exposure limits of nitrogen trifluoride. *Toxicol Appl Pharmacol* 26:1–13, 1973

NITROGLYCERIN
CAS: 55-63-0

$CH_2NO_3CHNO_3CH_2NO_3$

Synonyms: 1,2,3-Propanetriol trinitrate; glycerol trinitrate; nitroglycerol; NG; trinitroglycerol; NTG; trinitrin

Physical Form. Oily liquid at room temperature; colorless in pure form and pale yellow or brown in commercial form

Uses. In manufacture of dynamite, gun powder, and rocket propellants; as a therapeutic agent, primarily to alleviate angina pectoris. *Note:* Workers engaged in the product ion or use of dynamite are potentially exposed to mixed vapors of nitroglycerin (NG) and ethylene glycol dinitrate (EGDN).

Exposure. Inhalation; skin absorption

Toxicology. Nitroglycerin (NG) is a vasodilator and has been associated with acute episodes of angina pectoris, myocardial infarction, and sudden death.

Initial exposure to NG (or NG:EGDN mixtures) characteristically results in an intense throbbing headache, which begins in the forehead and moves to the occipital region.[1] Volunteers developed mild headaches when exposed to NG:EGDN vapor at concentrations of 0.5 mg/m^3 for 25 minutes.[2] It has been suggested that at least some workers may develop headaches at concentrations as low as 0.1 mg/m^3.[1]

Other signs and symptoms associated with initial exposure include dizziness, nausea, palpitations, and decreases in systolic, diastolic, and pulse pressures.[1] These initial signs and symptoms, including headache, are indicative of a shift in blood volume from the central to the peripheral circulatory system, initiated by dilation of the blood vessels.

After 2 to 4 days of repeated NG exposure, tolerance to the vasodilatory activity occurs, probably as a result of compensatory vasoconstriction. Tolerance may be lost during periods without NG exposure, such as weekends and holidays.[3]

Chronic, repeated exposures to NG and NG mixtures have also been associated with more serious cardiovascular effects, including angina pectoris and sudden death.

Signs and symptoms of ischemic heart disease were observed in 9 munitions workers involved in handling a nitroglycerin-cellulose mixture.[4] Within 1 to 4 years of initial exposure, these workers developed nonexertional chest pain, which was relieved either by therapeutic nitroglycerin or by returning to work after the weekend. Coronary angiography performed in 5 of the patients showed no obstructive lesions. In 1 patient, observed while in withdrawal state, coronary artery spasm was demonstrated and readily reversed by sublingual nitroglycerin.

Like the attacks of angina pectoris, sudden deaths occurred most frequently during brief periods away from work and, in particular, on Sunday nights or Monday mornings.[5] In most cases, there were no premonitory signs or symptoms, although some subjects had anginal episodes during brief periods away from work. At autopsy, atherosclerotic plaques, with or without thrombosis, have been found in the coronary arteries of workers, but their coronary arteries were generally not occluded to the same extent as those of unexposed workers who had died suddenly.[1]

The pathogenesis of the sudden death syndrome has been postulated to be due to withdrawal of coronary vasodilators (eg, NG), resulting in vasoconstriction with acute hypertension, or with myocardial ischemia in workers adapted to, and dependent on, NG to maintain a minimum level of coronary flow.[3] A second contributing mechanism for coronary artery toxicity due to NG may relate to the so-called aging of the vessels due to repeated dilation.[6] Other theories suggest that sudden deaths may be related to peripheral vasodilation consequent to re-exposure of NG.[5]

Estimates of exposure levels associated with sudden death have not been made because workers typically absorb considerable amounts of NG through the skin in addition to inhalation.[1] Skin contact may also cause an irritant dermatitis resembling poison ivy and, occasion-ally, allergic contact dermatitis has been reported.[7]

Epidemiological studies have suggested that the effects of long-term workplace exposure to NG may not be completely reversed after exposure is terminated. Former workers may be at increased risk for cardiovascular mortality for months to years after exposure has ceased.

A cohort study of 5668 NG-exposed workers found an increased standardized mortality ratio for deaths from ischemic heart disease.[8] The increase was more pronounced for those with 10 or more years of exposure and was statistically significant for the 40-to 49-year age group, whereas a deficit of cardiovascular mortality had been anticipated as a result of pre-placement and annual medical examinations designed to exclude persons with cardiovascular abnormalities. These results were confirmed in a recent retrospective cohort mortality study which found a significant excess of ischemic heart disease mortality among workers actively exposed to nitroglycerin and under the age of 45.[9] (*Note:* This study failed to detect a chronic cardiovascular effect because excess risk was associated only with workers actively exposed to nitroglycerin.)

An excess of deaths from acute myocardial infarction was also confirmed in a younger group of workers exposed to NG and EGDN in a Scottish explosives factory and followed for 16 years.[10]

In a case-control study in Sweden, a relative risk of 2.5 for cardiocerebrovascular disease was found in explosives workers with over 20 years' experience; most of the deaths occurred months or years after exposure had ceased.[11]

It is generally recognized that workers exposed to either NG or EGDN have reduced tolerance for alcohol.[1] Animal studies suggest that NG may decrease the activity of alcohol dehydrogenase, thereby decreasing the rate of alcohol metabolism.[1]

The NIOSH-recommended exposure limit for NG, EGDN, or a mixture of the two was set at a level to prevent significant changes in the diameter of cerebral blood vessels during initial exposure, as indicated by the occurrence of headache or by decrease in blood pressure, thereby preventing the development of compensatory vasoconstrictive mechanisms that may eventually result in more serious effects.[1]

Individuals with pre-existing ischemic heart disease should not be assigned to work where significant exposure to NG may occur.[12]

The 1995 ACGIH time-weighted average–threshold limit value (TLV-TWA) for nitroglycerin is 0.05 ppm (0.46 mg/m^3) with a notation for skin absorption.

REFERENCES

1. National Institute for Occupational Safety and Health, US Department of Health, Education and Welfare: *Criteria for a Recommended Standard . . . Occupational Exposure to Nitroglycerin and Ethylene Glycol Dinitrate.* DHEW (NIOSH) Pub No 78-167, p 215. Washington DC, US Government Printing Office, 1978
2. Trainor DC, Jones RC: Headaches in explosive magazine workers. *Arch Environ Health* 12:231–234, 1966
3. Sivertsen E: Glyceryltrinitrate as a problem in industry. *J Clin Lab Invest* 44(Suppl)173: 81–84, 1984
4. Lange RL et al: Nonatheromatous ischemic heart disease following withdrawal from chronic industrial nitroglycerin exposure. *Circulation* 46:666–678, 1972
5. Carmichael P, Lieben J: Sudden death in explosives workers. *Arch Environ Health* 7:424–439, 1963
6. Klaassen CD et al (eds): *Casarett and Doull's Toxicology: The Basic Science of Poisons*, 3rd ed, p 399. New York, Macmillan, Pub Co, 1986
7. Kanerva L, Laine R, Jolanki R, et al: Occupational allergic contact dermatitis caused by nitroglycerin. *Cont Derm* 24:356–362, 1991
8. Reeve G et al: Cardiovascular disease among nitroglycerin-exposed workers. *Am J Epidemiol* 118:418, 1983
9. Stayner LT, Dannenberg AL, Thun M, et al: Cardiovascular mortality among munitions workers exposed to nitroglycerin and dinitrotoluene. *Scand J Work Environ Health* 18:34–43, 1992
10. Craig R et al: Sixteen year follow-up of workers in an explosives factory. *J Soc Occup Med* 35:107–110, 1985
11. Hogstedt C, Axelson O: Nitroglycerin-nitroglycol exposure and mortality in cardiocerebrovascular diseases among dynamite workers. *J Occup Med* 19:675, 1977
12. Rosenman KD: Cardiovascular disease and environmental exposure. *Br J Ind Med* 36:85–97, 1979

NITROMETHANE
CAS: 75-52-5

CH_3NO_2

Synonyms: Nitrocarbol

Physical Form. Colorless, oily liquid

Uses. Solvent in chemical synthesis; fuel for professional and model racing cars; in explosive mixtures

Exposure. Inhalation

Toxicology. In animals, nitromethane affects the central nervous system, causing convulsions and narcosis at high doses; it is also a mild pulmonary irritant and may cause liver damage.

The human oral lethal dose is estimated to be between 0.5 and 5.0 g/kg; no systemic effects have been reported.[1]

Rabbits died from exposure to 10,000 ppm for 6 hours; initial effects were weakness, ataxia, and muscular incoordination, followed by convulsions.[2,3] The same concentration for 3 hours was not fatal. Autopsy of animals exposed to lethal concentrations showed focal necrosis in the liver and moderate kidney damage. Lower concentrations produced slight irritation of the respiratory tract, followed by mild narcosis, weakness, and salivation, but no evidence of eye irritation.

In a subchronic inhalation study, rabbits exposed to 98 ppm, 7 hr/day, 5 days/week, for 6 months showed hemoglobin depression with some methemoglobin, elevated serum carbamyl transferase, and thyroxin depression.[4] For rats similarly exposed at 745 ppm, there were altered hematocrit, hemoglobin, and erythrocyte counts, altered prothrombin time, and increased thyroid weight.

Of 10 rats given 0.1% nitromethane in drinking water for 15 weeks, 4 died. Survivors did not gain weight as expected and had liver abnormalities.[1]

The 1995 ACGIH threshold limit value–time-weighted average (TLV-TWA) is 20 ppm (50 mg/m³).

REFERENCES

1. Nitromethane. National Toxicology Program Executive Summaries, pp 1–12, 1983
2. Stokinger HE: Aliphatic nitro compounds, nitrates, nitrites. In Clayton DG, Clayton FE (eds): *Patty's Industrial Hygiene and Toxicology*, 3rd ed, rev, Vol 2C, *Toxicology*, pp 4153–4155. New York, Wiley-Interscience, 1982
3. Machle W, Scott EW, Treon J: Physiological response of animals to some simple mononitroparaffins and to certain derivatives of these compounds. *J Ind Hyg Toxicol* 22:315–332, 1940
4. Lewis TR, Ulrich CE, Busey WM: Subchronic inhalation toxicity of nitromethane and 2-nitropropane. *J Environ Pathol Toxicol* 2:233–249, 1979

1-NITROPROPANE
CAS: 108-03-2

$CH_3CH_2CH_2NO_2$

Synonym: 1-NP

Physical Form. Liquid

Uses. Solvent for organic materials; propellant fuel; gasoline additive

Exposure. Inhalation

Toxicology. 1-Nitropropane vapor is an irritant of the eyes; in animals, it also causes liver

damage and mild respiratory tract irritation.[1] There are no reports of systemic effects from industrial exposures.

Rabbits died from exposure to 5000 ppm for 3 hours, but 10,000 ppm for 1 hour was not lethal.[2] Effects were conjunctival irritation, lacrimation, slow respiration with some rales, incoordination, ataxia, and weakness.[2] Autopsy of animals exposed to lethal concentrations revealed severe fatty infiltration of the liver and moderate kidney damage.[2]

Rats exposed 7 hr/day, 5 days/week, at 100 ppm for up to 21 months showed no effects on appearance, behavior, serum chemistry, or hematology; body and organ weights were unchanged.[3] There were no histopathologic effects on the liver and, in particular, no induction of hepatocarcinomas. This contrasts with similar exposures to 2-nitropropane, which produce severe hepatotoxicity and hepatocellular carcinomas at this level. Further studies have suggested that the lack of a carcinogenic effect of 1-nitropropane may be associated with the fact that it does not induce cell proliferation in the liver, whereas the carcinogenic isomer 2-nitropropane induces marked and rapid induction of cell proliferation in this organ.[4]

1-Nitropropane is mutagenic in V79 cells and can induce unscheduled DNA synthesis in rat hepatocytes, but it was not mutagenic in *Salmonella typhimurium* assays nor did it produce sister chromatid exchanges or chromosomal aberrations in vitro.

The 1995 ACGIH threshold limit value–time-weighted average (TLV-TWA) for 1-nitropropane is 25 ppm (91 mg/m³).

REFERENCES

1. Silverman L, Schulte HF, First MW: Further studies on sensory response to certain industrial solvent vapors. *J Ind Hyg Toxicol* 28:262, 1946
2. Machle W, Scott EW, Treon JF: The physiological response of animals to some simple mononitroparaffins and to certain derivatives of these compounds. *J Ind Hyg Toxicol* 22:315, 1940
3. Griffin TB, Stein AA, Coulston F: Inhalation exposure of rats to vapors of 1-nitropropane

at 100 ppm. *Ecotoxicol Environ Safety* 6:268–282, 1982

4. Cunningham ML, Matthews HB: Relationship of hepatocarcinogenicity and hepatocellular proliferation induced by mutagenic noncarcinogens vs carcinogens. *Toxicol Appl Pharmacol* 110:505–513, 1991

2-NITROPROPANE
CAS: 79-46-9

CH₃CHNO₂CH₃

CH$_3$CHNO$_2$CH$_3$

Synonyms: Isonitropropane; nitroisopropane; dimethylnitromethane; 2-NP; NiPar S-20; NiPar S-30

Physical Form. Liquid

Uses. Industrial solvent; chemical intermediate; component in inks and paints

Exposure. Inhalation

Toxicology. 2-Nitropropane is a pulmonary irritant and hepatotoxin. Inhalation of vapor produces hepatocellular carcinomas in rats, and it is a suspected human carcinogen.

Workers exposed to hot vapor containing an unspecified concentration of xylene and 20 to 45 ppm of 2-nitropropane developed occipital headache, anorexia, nausea, vomiting and, in some cases, diarrhea.[1] Substitution of methyl ethyl ketone for 2-nitropropane eliminated the problem. Workers exposed to 30 to 300 ppm of 2-nitropropane complained of irritation of the respiratory tract.[2] A number of fatalities have been reported in association with 2-nitropropane.[3,4] The deaths all involved application of paint coatings in poorly ventilated areas. In all cases, liver failure was the primary cause of death, and postmortem findings were of massive hepatocellular destruction. Descriptions of prodromal symptoms have included typical central nervous system effects of solvent exposure, including headache, nausea, and vomiting. In the most recently reported cases, two construction workers became ill after applying an epoxy resin coating containing 2-nitropropane in an enclosed area.[4] One man died 10 days later from fulminant hepatic failure; the second man survived but has had persistently elevated serum aminotransferase activity. The serum concentration of 2-nitropropane on admission of the man who died was 13 mg/liter versus 8.5 mg/l for his co-worker. Extrapolating from animal pharmacokinetic studies, the serum concentrations would be consistent with 6 hours of inhalation in the 600 ppm range.

Chronic health effects in humans from exposure to 2-nitropropane have not been adequately determined, although a retrospective mortality study of 1481 employees and former employees of a 2-nitropropane production facility with up to 27 years of exposure found no increase in cancer of the liver or other organs and no unusual disease mortality pattern.[5]

Rabbits died from exposure to a concentration near 2400 ppm for 4.5 hours, but 1400 ppm was not lethal.[6] High concentrations caused lethargy, weakness, difficult breathing, cyanosis, prostration, and occasional convulsions; low levels of methemoglobin and the formation of Heinz bodies in erythrocytes were observed. Autopsy of animals exposed to lethal concentrations revealed pulmonary edema and hemorrhage, as well as liver damage.[6] The 6-hour LC_{50} in the male rat is 400 ppm.[7]

Rats exposed to 207 ppm daily for 6 months developed hepatic neoplasms; hepatocellular hyperplasia and necrosis occurred after 3 months' exposure at this concentration.[7] In another series of inhalation experiments on rats, 200 ppm produced hepatocellular carcinomas in both sexes; 100 ppm resulted in liver tumors in males after 12 months' exposure and, in females, after 18 months. At 25 ppm for up to 22 months' exposure, no tumors or other hepatic lesions were produced.[8] The authors further suggested that damage to the liver parenchymal cells is an essential precursor to the induction of hepatocarcinoma in the rat. Hepatocellular carcinomas occur only when the degree of exposure is sufficient to cause severe hepatotoxicity, followed by hyperregeneration, with some of the newly regenerated cells becoming autonomous, leading to neoplasia. More recent studies have shown that 10-day gavage treatment of rats with up to 2 mM/kg 2-nitropropane caused an increased incidence of cell proliferation; similar treatment with the noncarcinogenic isomer 1-nitropropane did not cause an increase in cell proliferation.[9]

The IARC has determined that there is sufficient evidence for carcinogenicity in rats.[10] There is suspected carcinogenic potential for humans.

2-Nitropropane is genotoxic in a variety of assays, including the Ames/*Salmonella* assay, in vitro sister chromatid exchange and chromosome aberrations, and unscheduled DNA synthesis assay.[9]

Although the early literature stated that the odor threshold was above 80 ppm and, therefore, not capable of providing adequate warning of exposure, a more recent study has determined that a lower threshold of approximately 5 ppm exists that should provide some warning of exposure, especially if workers are familiarized with the odor.[11]

The 1995 ACGIH threshold limit value–time-weighted average (TLV-TWA) for 2-nitropropane is 10 ppm (36 mg/m³) with an A2 suspected human carcinogen designation.

REFERENCES

1. Skinner JB: The toxicity of 2-nitropropane. *Ind Med* 16:441–443, 1947
2. American Industrial Hygiene Association: Hygienic Guide Series. Nitropropane. 3 pp. 1978.
3. Hine CH, Pasi A, Stephens BG: Fatalities following exposure to 2-nitropropane. *JOM* 20:333–337, 1978
4. Harrison R, Letz G, Pasternak G, et al: Fulminant hepatic failure after occupational exposure to 2-nitropropane. *Ann Int Med* 107:466–468
5. Bolender FL: 2-NP Mortality Epidemiology Study of the Sterlington, LA, Employees: An Update Report to the International Minerals & Chemical Corp, Northbrook, IL, 1983
6. Treon JF, Dutra FR: Physiological response of experimental animals to the vapor of 2-nitropropane. *AMA Arch Ind Hyg Occup Med* 5:52–61, 1952
7. Lewis TR, Ulrich CE, Busey WM: Subchronic inhalation toxicity of nitromethane and 2-nitropropane. *J Environ Path Toxicol* 2:233–249, 1979
8. Griffin TB, Stein AA, Coulston F: Inhalation exposure of rats to vapors of 1-nitropropane at 100 ppm. *Ecotoxicol Environ Safety* 6:268–282, 1982
9. Cunningham ML, Matthews HB: Relationship of hepatocarcinogenicity and hepatocellular proliferation induced by mutagenic noncarcinogens vs carcinogens. *Toxicol Appl Pharmacol* 110:505–513, 1991
10. *IARC Monographs on the Evaluation of the Carcinogenic Risk of Chemicals to Humans*, Vol 29, 2-Nitropropane, pp 331–343. Lyon, International Agency for Research on Cancer, 1982
11. Crawford GN, Garrison RP, McFee DR: Odor threshold determination for 2-nitropropane. *Am Ind Hyg Assoc J* 45:B-7–B-8, 1984

N-NITROSODIMETHYLAMINE
CAS: 62-75-9

(CH₃)N₂O

Synonyms: Dimethylnitrosamine; DMNA; DMN; NDMA

Physical Form. Yellow liquid

Uses/Sources. Industrial solvent; formerly used in synthesis of rocket fuel; contaminant in food and beverages, and a pollutant in ambient air, especially in the rubber and tire industries

Exposure. Inhalation; skin absorption

Toxicology. *N*-Nitrosodimethylamine (DMN) is a liver toxin and is carcinogenic in many species of test animals.

Two men accidentally exposed to DMN developed toxic hepatitis.[1] There are no reports of chronic effects from human exposure.[2]

The LC₅₀ for rats exposed to DMN vapor for 4 hours (and observed for 14 days) was 78 ppm; for similarly exposed mice, the LC_{50} was 57 ppm.[3] Dogs exposed for 4 hours to 16 to 144 ppm developed vomiting, polydipsia, and anorexia; most exposed dogs died, but one survivor showed residual liver damage 7 months after exposure.[3]

Swiss mice fed a diet containing 0.005% DMN for 1 week developed tumors of the kidney and lung.[4] Hamsters fed a diet containing 0.0025% for 11 weeks developed liver tumors.[5] A consistent observation following oral administration of DMN in rats has been that long-term treatment with doses compatible with a favorable survival rate leads to liver tumors, whereas short-term treatment with high doses produces renal tumors.[2]

Hamsters receiving weekly subcutaneous injections of DMN for life developed tumors; 3 of 10 females receiving weekly injections of 4.3 mg/kg developed liver tumors; at 21.5 mg/kg/week, there were 8 liver tumors and 5 kidney tumors; in 10 male animals receiving 2.8 mg/kg/week, there were 5 liver tumors and 1 kidney tumor.[6]

Intraperitoneal injection of 6 mg/kg once weekly for 10 weeks in mice resulted in a statistically significant increase in vascular tumors, mainly in the retroperitoneum in females. There was a low incidence of hepatic vascular tumors in both sexes.[7] Pregnant mice treated with the maximum nonfetotoxic dose of DMN on gestation day 16 or 19 had significant transplacental carcinogenic effects, causing an increase in hepatocellular carcinomas and sarcomas.[8] One intracranial schwannoma was attributed to DMN, because such tumors are extremely rare in mice.[8]

Chronic exposure to hepatotoxic doses of DMN has also been found to suppress humoral and cellular immunity in mice.[9] DMN is genotoxic in a wide variety of assays, inducing DNA synthesis, chromosomal aberrations, sister chromatid exchange, and bacterial mutations.[10]

The IARC has determined that there is sufficient evidence of carcinogenicity in animals and that, although no data are available for humans, the agent is probably carcinogenic in humans.

The ACGIH has classified N-nitrosodimethylamine as an A2 suspected human carcinogen with no assigned threshold limit value.

REFERENCES

1. Freund HA: Clinical manifestation and studies in parenchymatous hepatitis. *Ann Int Med* 10:1144–1155, 1937
2. *IARC Monographs on the Evaluation of the Carcinogenic Risk of Chemicals to Man.* Vol 17, Some N-nitroso compounds, pp 125–175. Lyon, International Agency for Research on Cancer, 1978
3. Jacobson KH, Wheelwright HJ Jr, Clem JH, Shannon RN: Studies on the toxicology of N-nitrosodimethylamine vapor. *AMA Arch Ind Health* 12:617–622, 1955
4. Terracini B, Palestro G, Gigliardi RM, Montesano R: Carcinogenicity of dimethylnitrosamine in Swiss mice. *Br J Cancer* 20:871–876, 1966
5. Tomatis L, Magee PN, Shubik P: Induction of liver tumors in the Syrian golden hamster by feeding dimethylnitrosamine. *J Natl Cancer Inst* 33:341–345, 1964
6. Mohr U, Haas H, Hilfrich J: The carcinogenic effects of dimethylnitrosamine and nitrosomethylurea in European hamsters (*Cricetus cricetus L*). *Br J Cancer* 29:359–364, 1974
7. Cardesa A, Pour P, Althoff J, Mohr U: Vascular tumors in female Swiss mice after intraperitoneal injection of dimethylnitrosamine. *J Natl Cancer Inst* 51:201–205, 1973
8. Anderson LM, Hagiwara A, Kovatch RM et al: Transplacental initiation of liver, lung, neurogenic and connective tissue tumors by N-nitroso compounds in mice. *Fund Appl Toxicol* 12:604–620, 1989.
9. Desjardins R, Fournier M, Denizeau F, et al: Immunosuppression by chronic exposure to N-nitrosodimethylamine (NDMA) in mice. *J Toxicol Environ Health* 37:351–361, 1992
10. Agency for Toxic Substances and Disease Registry (ATSDR): *Toxicological Profile for N-Nitrosdimethylamine*, US Public Health Service, p 119, 1989.

N-NITROSODIPHENYLAMINE
CAS: 86-30-6

$(C_6H_5)_2N_2O$

Synonyms: NDPhA; diphenyl nitrosamine

Physical Form. Yellow to brown or orange powder or flakes

Uses. Vulcanization retarder in the rubber industry

Exposure. Inhalation

Toxicology. *N*-Nitrosodiphenylamine (ND-PhA) is an animal carcinogen and causes bladder tumors in male and female rats.

Early carcinogenic studies in rats and mice in which NDPhA was administered orally or by intraperitoneal injection showed no evidence of carcinogenicity.[1-5] However, a more recent study demonstrated carcinogenesis in rats.[6,7] NDPhA was administered in the diet to rats and mice at the maximum tolerated dose for each species and at one-half that amount. A significant incidence of bladder tumors occurred in male (40%) and female (90%) rats at 240 mg/kg and 320 mg/kg, respectively. Few bladder tumors were seen in the mice.

IARC has determined that there is limited evidence for carcinogenicity in experimental animals and that no evaluation of the carcinogenicity to humans can be made.[8]

ACGIH has not established a threshold limit value for *N*-nitrosodiphenylamine.

REFERENCES

1. Argus MF, Hoch-Ligeti C: Comparative study of the carcinogenic activity of nitrosamines. *J Natl Cancer Inst* 27:695–709, 1961
2. Boyland E, Carter RL, Gorrod JW, Roe FJC: Carcinogenic properties of certain rubber additives. *Europ J Cancer* 4:233–239, 1968
3. National Cancer Institute. *Evaluation of Carcinogenic, Teratogenic, and Mutagenic Activities of Selected Pesticides and Industrial Chemicals*, Vol I, *Carcinogenic Study*. Washington, DC, US Government Printing Office, 1968
4. Innes JRM et al: Bioassay of pesticides and industrial chemicals for tumorigenicity in mice: a preliminary note. *J Natl Cancer Inst* 42(6):1101–1106, 1969
5. Druckrey H, Preussmann R, Ivankovic S, Schmahl D: Organotrope carcinogene Wirkungen bei 65 verschiedenen *N*-Nitroso- Verbindungen an BD-Ratten. *Z Krebsforsch* 69:103–201, 1967
6. National Cancer Institute: *Bioassay of* N-*Nitrosodiphenyl amine for Possible Carcinogenicity*. DHEW (NIH) Pub No 79-1720. Washington, DC, US Government Printing Office, 1979
7. Cardy RH, Lijinsky W, Hilderbrandt PW: Neoplastic and non-plastic urinary bladder lesions induced in Fischer 344 rats and B6C3F$_1$ hybrid mice by *N*-nitrosodiphenylamine. *Ectotoxicol Environ Safety* 3:29–35, 1979
8. *IARC Monographs on the Evaluation of the Carcinogenic Risk of Chemicals to Humans*, Vol 27, pp 213–225. Lyon, International Agency for Research on Cancer, 1982

N-NITROSODI-*n*-PROPYLAMINE
CAS: 621-64-7

$(NO)N(CH_2CH_2CH_3)_2$

Synonyms: NDPA; Di-*n*-propylnitrosamine; DPNA; dipropylnitrosamine

Physical Form. Liquid

Uses/Sources. Research chemical; impurity in herbicides isopropalin and trifluralin; contaminant in wastewater from chemical factories, cheese, brandy, and other liquors. *N*-nitrosamines are frequently produced during rubber processing and may be airborne in the workplace.

Exposure. Inhalation

Toxicology. *N*-nitrosodi-*n*-propylamine (NDPA) is an animal carcinogen.

The IARC considers that there is "sufficient evidence" that NDPA is carcinogenic to experimental animals.[1] NDPA administered to rats in the drinking water at 2.6 mg/kg/day for 5 days/week for 30 weeks caused liver carcinomas, nasal cavity carcinomas, tongue carcinomas, and esophageal papillomas and carcinomas.[2]

NDPA has exhibited genotoxicity in bacteria (*S. typhimurium, E. coli*) and mammalian cells (mouse lymphoma, Chinese hamster) and caused DNA effects (fragmentation, unscheduled synthesis, repair) in rat hepatocytes and chromosome aberrations in Chinese hamster cells.[3]

Rats that received a single lethal dose of NDPA showed centrilobular necrosis and fatty degeneration of the liver.[4] Specific doses that caused this effect were not listed, but the oral LD_{50} was determined to be 480 mg/kg.

A threshold limit value–time-weighted average (TLV-TWA) for NDPA has not been assigned.

REFERENCES

1. *IARC Monographs on the Evaluation of the Carcinogenic Risk of Chemicals to Humans, Overall Evaluations of Carcinogenicity: An Updating of IARC Monographs Vols 1–42.* Suppl 7, p 68. Lyon, International Agency for Research on Cancer, 1987
2. Lijinsky W, Reuber MD: Comparative carcinogenesis by some aliphatic nitrosamines in Fischer rats. *Cancer Lett* 14:297–302, 1981
3. Agency for Toxic Substances and Disease Registry: *Toxicological Profile for N-Nitrosodi-n-propylamine.* Public Health Service, Centers for Disease Control, Atlanta, GA, pp 21–29, December 1989
4. Druckrey H, Preussman R, Ivankovic S, et al: Organotropic carcinogenic effects of 65 different *N*-nitroso compounds on BD rats. *Z Krebsforsch* 69:103–121, 1967

N-NITROSOMORPHOLINE
CAS: 59-89-2

$C_4H_8N_2O_2$

Synonyms: NMOR; NNM; 4-nitrosomorpholine

Physical Form. Yellow crystals

Uses. Solvent for polyacrylonitrile; present during rubber manufacturing

Exposure. Inhalation; skin absorption

Toxicology. *N*-nitrosomorpholine (NMOR) is carcinogenic in animals.

There is no information available concerning toxic effects in humans.

The LD_{50} of NMOR in rats by oral and intraperitoneal routes was 320 mg/kg.

N-nitrosomorpholine causes centrilobular hepatic necrosis in rats.[1] Hepatocellular carcinomas were observed in 14 of 16 rats administered NMOR in the drinking water at doses of 8 mg/kg bw/day for life. Continuous oral exposure of Sprague–Dawley rats to 6, 12, or 24 mg/kg bw resulted in a dose-dependent increase in the total number of preneoplastic foci of altered hepatocytes and in the incidence of hepatocellular adenomas and carcinomas.[2] The induction of liver tumors by NMOR has been confirmed in several strains of rats.[1] Epithelial kidney tumors were observed in 47 of 69 rats to which NMOR had been administered in the drinking water at 120 or 500 mg/l for 3 to 14 weeks. NMOR is also carcinogenic in hamsters following subcutaneous injection, producing tumors of the respiratory system (mainly nasal cavity and trachea). The International Agency for Research on Cancer has stated that there is sufficient evidence for carcinogenicity of *N*-nitrosomorpholine to experimental animals. There are no data to evaluate its possible carcinogenicity in humans.[1]

N-nitrosomorpholine is mutagenic in bacterial assays in the presence of activated liver microsomal fractions.

The ACGIH has not established a threshold limit value for *N*-nitrosomorpholine.

REFERENCES

1. *IARC Monographs on the Evaluation of the Carcinogenic Risk of Chemicals to Humans*, Vol. 17, Some *N*-nitroso compounds, pp 263–275. Lyon, International Agency for Research on Cancer, 1978
2. Weber E, Bannasch P: Dose and time dependence of the cellular phenotype in rat hepatic preneoplasia and neoplasia induced by continuous oral exposure to *N*-nitrosomorpholine. *Carcinogen* 15:1235–42, 1994

NITROTOLUENE

CAS: 88-72-2; 99-08-1; 99-99-0 (*o, m, p*)

CH₃C₆H₄NO₂

Synonyms: Methylnitrobenzene; nitrotoluol, nitrophenylmethane

Physical Form. Ortho and meta isomers are yellowish liquid; para is a yellow solid

Uses. All isomers are used in the synthesis of dyestuffs, explosives, and agricultural chemicals

Exposure. Inhalation; skin absorption

Toxicology. Nitrotoluene has a low potency for producing methemoglobin and subsequent anoxia. Chronic exposure to other aromatic nitro compounds has caused anemia, and it is expected that nitrotoluene may cause the same effect.

Signs and symptoms of overexposure are due to the loss of oxygen-carrying capacity of the blood. The onset of symptoms of methemoglobinemia is often insidious and may be delayed up to 4 hours; headache, which is commonly the first symptom, may become quite intense as the severity of methemoglobinemia progresses.[1] Cyanosis develops when the methemoglobin concentration is 15% or more; blueness develops first in the lips, nose, and earlobes, and is usually recognized by fellow workers. Until the methemoglobin concentration approaches approximately 40%, the individual feels well, has no complaints, and typically insists that nothing is wrong. At methemoglobin concentrations over 40%, there usually is weakness and dizziness; up to 70% concentration, there may be ataxia, dyspnea on mild exertion, tachycardia, nausea, vomiting, and drowsiness.[1]

In general, higher ambient temperatures increase susceptibility to cyanosis from exposure to methemoglobin-forming agents.[2]

In subchronic animal studies, *o*-, *m*-, or *p*-nitrotoluene was administered in the feed to rats and mice at doses ranging from 625 to 10,000 ppm for 13 weeks.[3] Decreased body weights occurred in rats and mice receiving the higher dose levels and were most pronounced in rats receiving the ortho isomer. In mice, the only treatment-related lesion was degeneration and metaplasia of the olfactory epithelium in animals receiving *o*-nitrotoluene. In male rats, all isomers produced kidney toxicity consisting of hyaline droplet nephropathy and an associated increase in the renal concentration of α_{2u}-globulin. Treatment-related hepatic lesions, which occurred only in male rats receiving *o*-nitrotoluene, and consisted of cytoplasmic vacuolization and oval-cell hyperplasia. Elevations in liver weights were observed at the higher dose levels in rats and mice treated with any of the three isomers. Spleens of male and female rats had a mild increase in hematopoiesis, hemosiderin deposition, and/or congestion. All isomers impaired testicular function in the rat, as shown by testicular degeneration and reduction in the density, motility, and number of sperm cells. Mesotheliomas of the epididymis occurred in the *o*-nitrotoluene male rats at 5000 ppm, and mesothelial-cell hyperplasia occurred at 10,000 ppm.

Administered by oral gavage to rats for 6 months, all three isomers produced splenic lesions.[4] The meta and para isomers produced testicular atrophy, whereas *o*-nitrotoluene caused renal lesions.[4]

Metabolism and genetic toxicity have been reported to differ with the isomer of nitrotoluene.

Only the ortho isomer induces DNA excision repair in the in vivo–in vitro hepatocyte unscheduled DNA synthesis assay.[5] Furthermore, o-nitrotoluene binds to hepatic DNA to a much greater extent than m- or p-nitrotoluene, and investigators suggest that it may act similarly to the rodent hepatocarcinogen 2,6-dinitrotoluene.[6]

The 1995 threshold limit value–time-weighted average (TLV-TWA) for nitrotoluene is 2 ppm (11 mg/m³) with a notation for skin absorption.

Note: For a description of diagnostic signs, differential diagnosis, and medical control, including clinical laboratory treatment of overexposure to methemoglobin-forming agents, see "Aniline."

References

1. Hamblin DO: Aromatic nitro and amino compounds. In Patty FA (ed): *Industrial Hygiene and Toxicology*, 2nd ed, pp 2105–2119, 2148–2149. New York, Wiley-Interscience, 1963
2. Linch AL: Biological monitoring for industrial exposure to cyanogenic aromatic nitro and amino compounds. *Am Ind Hyg Assoc J* 35:426–432, 1974
3. Dunnick JK, Elwell MR, Bucher JR: Comparative toxicities of o-, m-, and p-nitrotoluene in 13-week feed studies in F344 rats and B6C3F₁ mice. *Fund Appl Toxicol* 22:411–421, 1994
4. Ciss M et al: Toxicological study of nitrotoluenes: Long-term toxicity. *Dakar Medical* 25:293, 1980
5. Doolittle DJ et al: The influence of intestinal bacteria, sex of the animal, and position of the nitro group on the hepatic genotoxicity of nitrotoluene isomers in vivo. *Cancer Res* 43:2836, 1983
6. Rickert DE et al: Hepatic macromolecular covalent binding of mononitrotoluenes in Fischer-344 rats. *Chem Biol Interact* 52:131–139, 1984

NITROUS OXIDE
CAS: 10024-97-2

N_2O

Synonyms: Laughing gas; nitrogen oxide; dinitrogen monoxide

Physical Form. Colorless gas with a slightly sweet odor

Uses. Anesthetic agent; foaming agent for whipped cream; as oxidant for organic compounds; in rocket fuels

Exposure. Inhalation

Toxicology. Nitrous oxide is an asphyxiant at high concentrations; prolonged exposure has been associated with damage to the hematopoietic system, the central nervous system, and the reproductive system.

Until recent times, the only toxicological hazards attributable to nitrous oxide were those common to asphyxiants, with death or permanent brain injury occurring only under conditions of hypoxia.[1] A number of untoward and toxic effects have now been associated with exposure. One of the earliest findings was that patients given 50% nitrous oxide and 50% oxygen for prolonged periods to induce continuous sedation developed bone marrow depression and granulocytopenia. The bone marrow usually returned to normal within a matter of days once the nitrous oxide was removed, but several deaths from aplastic anemia have been recorded.[1,2]

Central nervous system toxicity from either social abuse of nitrous oxide or extremely heavy occupational exposure has been characterized by symptoms of numbness, paresthesias, impaired equilibrium, and difficulty concentrating.[2] In severe cases, the patient became incontinent, impotent, and unable to walk. Neurologic signs included ataxic gait, muscle weakness, impaired sensation, and diminished reflexes.

Acute exposure to levels of 200,000 ppm

and above causes deterioration of performance on tests of reaction time; it has been suggested that the threshold at which nitrous oxide starts to affect performance lies between 80,000 and 120,000 ppm.[3,4] Other studies have examined the effects of trace levels of nitrous oxide on performance tests, with conflicting results. In one unconfirmed study, volunteers exposed to 50 ppm for up to 4 hours showed decrements in audiovisual performance tests.[5] In another report, similar exposures did not produce any changes in a battery of psychomotor tests, including an audiovisual task, but there was a nonsignificant trend for mood factors such as tiredness to occur.[6]

A number of epidemiologic studies have shown a correlation between occupational exposure to nitrous oxide and adverse reproductive effects. In a survey of female dental assistants, there was a 100% increase in spontaneous abortions among those exposed to nitrous oxide compared to those not exposed; a 52% increase in spontaneous abortions also was observed among wives of dentists.[7] Despite various limitations to the studies, including participant bias, inadequate reporting of exposure levels, and possible confounding factors, ACGIH has determined that there is sufficient evidence that nitrous oxide poses a reproductive hazard to women.[1]

Although a number of animal studies demonstrate that nitrous oxide exposure can cause congenital anomalies, equivocal evidence exists for such effects in humans. In one report, the offspring of chairside female dental assistants exposed to nitrous oxide had a 50% higher incidence of congenital anomalies than the offspring of unexposed assistants.[7] However, the incidence was not related to the extent of nitrous oxide exposure, and the incidence was not greater than that occurring in the wives of dentists (both exposed and unexposed). An increased incidence of reproductive problems also has been reported in the wives of men exposed to nitrous oxide, that is, women who were not directly exposed themselves. A survey of 49,585 anesthesia personnel found a 25% increase in the incidence of congenital abnormalities in the children of male anesthesiologists compared to a control group composed of the children of male physicians who worked outside the operating room.[8]

In general, numerous animal studies suggest that the production of teratogenic effects requires prolonged exposure to high concentrations during particular times of pregnancy.[9] For example, in rats, the exposure threshold for teratogenic effects appears to lie between 350,000 ppm and 500,000 ppm, the former producing no adverse effects and the latter producing cervical rib defects as well as other defects.[10] Groups of rats exposed at 600,000 ppm for 24 hours on each of days 6 through 12 of gestation exhibited an increased incidence of cervical rib defects and an increased incidence of right-sided aortic arch and left-sided umbilical artery (abnormalities indicative of altered laterality), but only following exposure on day 8 of gestation. Increases in skeletal malformations and hydrocephalus occurred following exposure on day 9 of gestation. An increase in fetal deaths occurred from exposure on days 8 and 11.[9]

There were no significant changes in sperm count or in sperm morphology in a group of male anesthesiologists exposed to nitrous oxide compared to an unexposed group.[11] Concentrations were estimated to range from 5 to 300 ppm, which is substantially lower than the concentrations that have been shown to have a deleterious effect on sperm in experimental animals.[10]

Nitrous oxide exerts a variety of its adverse effects by oxidizing vitamin B_{12} and rendering it inactive as a coenzyme in many essential metabolic processes.[12] One vitamin B_{12}-dependent enzyme in particular, methionine synthetase, is involved in cell division and is necessary for DNA production. Adverse reproductive and hematologic effects caused by nitrous oxide are thought to be due to inactivation or dysfunction of methionine synthetase, resulting in impairment of cell division.

The possible carcinogenicity of nitrous oxide has been studied in dentists and chairside assistants with occupational exposures. No effect was observed in male dentists, but a 2.4-fold increase in cancer of the cervix in heavily exposed female assistants was reported.[7] Other epidemiologic reports of workers exposed to waste anesthetic gases have been negative.[1] Carcinogenic bioassays in animals have yielded negative results. Nitrous oxide was not genotoxic in a variety of assays.[1]

The 1995 ACGIH threshold limit value–time-weighted average (TLV-TWA) for nitrous oxide is 50 ppm (90 mg/m³).

REFERENCES

1. Nitrous oxide. *Documentation of the TLVs and BEIs*, 6th ed, pp 1134–1138. Cincinnati, OH, American Conference of Governmental Industrial Hygienists, 1991
2. Eger EI, Gaskey NJ: A review of the present status of nitrous oxide. *J Assoc Nurs Anesth* 54:9–36, 1986
3. Cook TL, Smith M, Starkweather JA, et al: Behavioural effects of trace and subanesthetic halothane and nitrous oxide in man. *Anesthesiology* 49:419–424, 1978
4. Allinson RH, Shirley AW, Smith G: Threshold concentration of nitrous oxide affecting psychomotor performance. *Br J Anesth* 51:177–180, 1979
5. Bruce DL, Bach MJ: Effects of trace anesthetic gases on behavioural performance of volunteers. *Br J Anesth* 48:871–876, 1976
6. Venables H, Cherry N, Waldron HA, et al: Effects of trace levels of nitrous oxide on psychomotor performance. *Scand J Work Environ Health* 9:391–396, 1983
7. Cohen EN, Brown BW, Wu ML, et al: Occupational disease in dentistry and chronic exposure to trace anesthetic gases. *J Am Dent Assoc* 101:21–31, 1980
8. American Society of Anesthesiologists: Occupational disease among operating room personnel: a national study. Report of an ad hoc committee on the effect of trace anesthetics on operating room personnel. *Anesthesiology* 41:321–340, 1980
9. Fujinaga M, Baden JM, Mazze RI: Susceptible periods of nitrous oxide teratogenicity in Sprague–Dawley rats. *Teratology* 40:439–444, 1989
10. Mazze RI, Wilson AI, Rice SA, et al: Reproduction and fetal development in rats exposed to nitrous oxide. *Teratology* 30:259–265, 1984
11. Wyrobek AJ, Brodsky J, Gordon L, et al: Sperm studies in anesthesiologists. *Anesthesiology* 55:527–532, 1981
12. Nunn JF, Chanarin I: Nitrous oxide inactivates methionine synthetase. In Eger EI (ed):

Nitrous Oxide, pp 211–233, New York, Elsevier, 1985

NONANE
CAS: 111-84-2

$CH_3(CH_2)_7CH_3$

Synonyms: *n*-Nonane

Physical Form. Liquid

Uses. Solvent; organic synthesis; distillation chaser; major ingredient of such petroleum fractions as VM&P Naphtha, Stoddard solvent, and gasoline

Exposure. Inhalation

Toxicology. Nonane is an irritant of the eyes and skin.

The 4-hour LC_{50} in rats was 3200 ppm.[1] Rats exposed for 6 hr/day for 7 days to 1500 ppm had mild tremor, slight incoordination, and slight irritation of eyes and extremities. A no-adverse-effect level for 65 days, 6 hr/day, 5 days/week, was 590 ppm in rats.

Nonane could be expected to dry and defat skin, resulting in irritation and dermatitis, by analogy to other liquid paraffin hydrocarbons. Aspiration into the lung could be expected to cause chemical pneumonitis.

The 1995 ACGIH threshold limit value-time weighted average (TLV-TWA) for nonane is 200 ppm (1050 mg/m³).

REFERENCES

1. Carpenter CP et al: Petroleum hydrocarbon toxicity studies XVII: Animal response to nonane. *Toxicol Appl Pharmacol* 44:53, 1978

NONYLPHENOL

CAS: *mixed isomers—25154-52-3*
    2-nonylphenol—136-83-4
    4-nonylphenol—104-40-5

$C_9H_{19}(C_6H_4)OH$

Synonyms: 2-Nonylphenol; *o*-nonylphenol; 4-nonylphenol; *p*-nonylphenol

Physical Form. Clear, straw-colored liquid; technical grade is a mixture of isomers, predominantly para-substituted

Uses. Intermediate in the production of nonionic ethoxylated surfactants; intermediate in the manufacture of phosphite antioxidants for the plastics and rubber industries

Exposure. Inhalation

Toxicology. Nonylphenol is a severe irritant of the eyes and skin.

Reports of the oral LD_{50} in rats for the mixed isomers have ranged from 580 to 1537 mg/kg; the dermal LD_{50} in rabbits was between 2000 and 3160 mg/kg.[1-4] Nonylphenol is considered to be a corrosive agent that may cause burns and blistering of the skin.[3] When the liquid was applied to the shaved skin of a rabbit and left in place for 4 hours, there was skin necrosis 48 hours after the application.[5] No skin sensitization occurred in tests with guinea pigs.[6] When tested on black guinea pigs and black mice, irritation was observed, but nonylphenol did not induce depigmentation.[7]

The liquid in the eye of the rabbit as a 1% solution caused severe corneal damage.[1,2]

Leukoderma was reported in two women engaged in degreasing metal parts with synthetic detergents containing polyoxyethylene, nonyl- or octylphenylether. Analysis revealed contamination with free alkylphenol, possibly octylphenol, or nonylphenol. Although a relationship between the cases of leukoderma and octyl- and nonylphenol exposure was suggested, it could not be confirmed.[8]

Nonylphenol has recently been shown to have estrogenic properties triggering mitotic activity in rat endometrium and cell proliferation in estrogen-sensitive tumor cells.[9] Potential toxicity to humans who may be exposed to nonylphenol through leaching of plastics has not been determined.

A threshold limit value–time-weighted average has not been established for nonylphenol.

REFERENCES

1. Smyth HF Jr et al: Range-finding toxicity data: List VI. *Am Ind Hyg Assoc J* 23:95–107, 1962
2. Smyth HF Jr et al: Range-finding toxicity data: List VII. *Am Ind Hyg Assoc J* 30:470–476, 1969
3. Texaco Chemical Company. FYI-OTS-06845-0402 FLWP Seq I. Material Safety Data Sheet. Washington, DC, US Environmental Protection Agency, Office of Toxic Substances, 1985
4. Monsanto Industrial Chemicals Co. FYI-OTS-0685-0402 FLWP Seq G. Material Safety Data Sheet. Washington, DC, US Environmental Protection Agency, Office of Toxic Substances, 1985
5. Texaco Chemical Company. FYI-OTS-0685-0402 FLWP Seq I. DOT Corrosivity Study in Rabbits. Washington, DC, US Environmental Protection Agency, Office of Toxic Substances, 1985
6. Texaco Chemical Company. FYI-OTS-0685-0402 FLWP Seq. I. Dermal Sensitization Study. Washington, DC, US Environmental Protection Agency, Office of Toxic Substances, 1985
7. Gellin GA, Maibach HI, Misiaszek MH, Ring M: Detection of environmental depigmenting substances. *Cont Derm* 5:201–213, 1979
8. Ikeda M, Ohtsuji H, Miyahara S: Two cases of leukoderma, presumably due to nonyl- or octylphenol in synthetic detergents. *Ind Health* 8:192–196, 1970
9. Soto AM, Justicia H, Wray JW, et al: *p*-Nonylphenol: an estrogenic xenobiotic released from "modified" polystyrene. *Environ Health Perspec* 92:167–173, 1991

NUISANCE PARTICULATES
Containing no asbestos and <1% crystalline silica

Synonyms: Particulates not otherwise classified; PNOC

Physical Form. Total dust as described here includes air-suspended particles of greater than respirable diameter

Source. Ubiquitous

Exposure. Inhalation

Toxicology. As stated by the ACGIH:

In contrast to fibrogenic dusts which cause scar tissue to be formed in lungs when inhaled in excessive amounts, so-called nuisance dusts have a long history of little adverse effect on lungs and do not produce significant organic disease or toxic effect when exposures are kept under reasonable control. The nuisance dusts have also been called biologically inert dusts, but the latter term is inappropriate to the extent that there is no dust which does not evoke some cellular response in the lung when inhaled in sufficient amount. However, the lung-tissue reaction caused by inhalation of nuisance dusts has the following characteristics: the architecture of the air spaces remains intact; collagen (scar tissue) is not formed to a significant extent; the tissue reaction is potentially reversible.

Excessive concentrations of nuisance dusts in the workroom air may seriously reduce visibility, may cause unpleasant deposits in the eyes, ears and nasal passages, or cause injury to the skin or mucous membranes by chemical or mechanical action per se or by the rigorous skin cleansing procedures necessary for their removal.[1]

Recent animal studies have found that sub-chronic exposure to nuisance dusts at levels equal to the threshold limit value have induced mild inflammatory response in the lungs and sufficient accumulation of particles to slow lung clearance.[2] The investigators suggest that exposure to nuisance dust at a level that will impair pulmonary clearance should be avoided to prevent excessive accumulation of dust in the lungs.

The 1995 ACGIH threshold limit value–time-weighted average (TLV-TWA) is 10 mg/m^3, total dust, containing no asbestos and <1% crystalline silica.

REFERENCES

1. Particulates not otherwise classified (PNOC). *Documentation of TLVs and BEIs*, 6th ed, pp 1166–1167. Cincinnati, OH, American Conference of Governmental Industrial Hygienists, 1991
2. Henderson RF, Barr EB, Cheng YS, et al: The effect of exposure pattern on the accumulation of particles and the response of the lung to inhaled particles. *Fund Appl Toxicol* 19:367-374, 1992

OCTACHLORONAPHTHALENE
CAS: 2234-13-1

$C_{10}Cl_8$

Synonym: Halowax

Physical Form. Waxy solid

Uses. In electric cable insulation; additive to lubricants.

Exposure. Inhalation; skin absorption

Toxicology. The higher chlorinated naphthalenes may cause severe injury to the liver.

Exposure of workers by inhalation or skin absorption to lower chlorinated naphthalenes (penta- and hexachloro-) causes a severe acneiform dermatitis chloracne.[1] Surprisingly, on human volunteers, octachloronaphthalene was entirely nonacnegenic.[2] There is no informa-tion on systemic effects in humans. In animals, systemic toxicity from chlorinated naphtha-lenes appears to be limited to liver injury char-acterized as acute yellow atrophy.[1,2]

The 1995 threshold limit value–time-weighted average (TLV-TWA) for octachloro-naphthalene is 0.1 mg/m^3, with a short-term excursion limit of 0.3 mg/m^3 and a notation for skin absorption.

REFERENCES

1. Deichmann WB: Halogenated cyclic hydro-carbons. In Clayton GD, Clayton FE (eds): *Patty's Industrial Hygiene and Toxicology*, 3rd ed, Vol 2B, *Toxicology*, 3669–3675. New York, Wi-ley-Interscience, 1981
2. Shelley WB, Kligman AM: The experimental production of acne by penta- and hexachloro-naphthalenes. *Arch Dermatol* 75:689-695, 1957

OCTANE
CAS: 111-65-9

$CH_3(CH_2)_6CH_3$

Synonyms: None

Physical Form. Colorless liquid; *n*-octane has 17 isomers with similar properties

Uses. Constituent in motor and aviation fuels; industrial solvent; in organic synthesis

Toxicology. In animals, octane is a mucous membrane irritant and, at high concentrations, it causes narcosis; it is expected that severe ex-posure in humans will produce the same effects.

The health effects of octane (both the iso- and *n*- forms) are thought to be similar to those of *n*-heptane except that octane is approxi-mately 1.2 to 2 times more toxic.[1] Octane has not been shown to cause the type of peripheral neuropathy associated with *n*-hexane.

In the only report involving human expo-sure, the liquid applied to the forearm for 1 hour and to the thigh for 5 hours caused erythema, hyperemia, inflammation, and pigmentation. The volunteers experienced burning and itch-ing at the site of application, and some blister formation occurred with the 5-hour exposures.

There was no narcosis in mice exposed to isooctane at 8000 ppm for 5 minutes; at 16,000 ppm, however, there was sensory irritation throughout a 5 minute exposure, and 1 of 4 mice died; at 32,000 ppm, effects were irritation and irregular respiration, and all 4 mice died within 4 minutes of exposure.[2]

The 1995 ACGIH threshold limit value-time-weighted average (TLV-TWA) is 300 ppm (1400 mg/m^3) with a STEL/ceiling of 375 ppm (1750 mg/m^3).

REFERENCES

1. Low LK, Meeks JR, Mackerer CR: *n*-Octane. In Snyder R (ed): *Ethyl Browning's Toxicity and Metabolism of Industrial Solvents*, 2nd ed, Vol I, *Hydrocarbons*, pp 307–311. New York, Else-vier, 1987
2. Swann HE Jr, Kwon BK, Hogan GK, Snellings WM: Acute inhalation toxicology of volatile hydrocarbons. *Am Ind Hyg Assoc J* 35:511, 1974

OIL MIST (MINERAL)
CAS: 8012-95-1

Varying chemical composition

Synonyms: Petrolatum liquid; mineral oil; par-affin oil

Physical Form. Colorless, oily, odorless, and tasteless liquid

Uses. Mineral oil is a lubricant and is used as a solvent for inks in the printing industry

Exposure. Inhalation

Toxicology. Highly refined mineral oil mist is of low toxicity.

A single case of lipoid pneumonitis suspected to result from repeated exposure to very high concentrations of oil mist was reported in 1950; this occurred in a cash register serviceman whose heavy exposure took place over 17 years of employment.

A review of exposures to mineral oil mist averaging below 15 mg/m³ (but higher in some jobs) in several industries disclosed a striking lack of reported cases of illness related to these exposures.[2] A study of oil mist exposures in machine shops, at mean concentrations of 3.7 mg/m³ and a maximum concentration of 110 mg/m³, showed no increase in respiratory symptoms or decrement in respiratory performance attributable to oil mist inhalation among men employed for many years.[3] Similar results were found in a 5-year study of 460 printing pressmen exposed to a respirable concentration of less than 5 mg/m³.[4,5] An increased prevalence of slight basal lung fibrosis was found in oil cable workers in Norway; it is noted that cable oils are only mildly refined and short-term excursion exposures may have approached 4000 mg/m³.[6,7]

Early epidemiologic studies linked cancers of the skin and scrotum with exposure to mineral oils.[8] These effects have been attributed to contaminants such as polycyclic aromatic hydrocarbons (PAH) and/or additives with carcinogenic properties present in the oil. Solvent refining and, to some extent, hydroprocessing selectively extract PAH and reduce carcinogenicity.[9] Later studies, which have also reported excess numbers of scrotal cancer, have failed to characterize the composition of the mineral oil and the exposure levels.[10] The IARC has determined that there is inadequate evidence that the fully solvent refined oils are carcinogenic to experimental animals in feeding or skin-painting studies.[11] The IARC's determination that there is sufficient evidenc e for carcinogenicity in humans is based on epidemiological studies of uncharacterized mineral oils containing additives and impurities; there is inadequate evidence for carcinogenicity to humans for highly refined oils.[11] Most mineral oils in use today present no hazard because of refining techniques; however, because individual oils may vary in composition, an assessment must be made on each product.[12]

The proposed 1995 ACGIH threshold limit value–time-weighted average (TLV-TWA) for mineral oil mist is 5 mg/m³ (or the severely refined product; the TLV-TWA) is 0.2 mg/m³ with an A1 confirmed human carcinogen designation for the mildly refined product (as cyclohexane soluable particulate containing polynuclear aromatic hydrocarbons).

REFERENCES

1. Proudfit JP, Van Ordstrand HS, Miller CW: Chronic lipid pneumonia following occupational exposure. *AMA Arch Ind Hyg Occup Med* 1:105–111, 1950
2. Hendricks NV et al: A review of exposures to oil mist. *Arch Environ Health* 4:139–145, 1962
3. Ely TS, Pedley SF, Hearne FT, Stille WT: A study of mortality, symptoms, and respiratory function in humans occupationally exposed to oil mist. *J Occup Med* 12:253–261, 1970
4. Lippman M, Goldstein DH: Oil mist studies, environmental evaluation and control. *Arch Environ Health* 21:591–599, 1970
5. Goldstein DH, Benoit JN, Tyroler HA: An epidemiologic study of an oil mist exposure. *Arch Environ Health* 21:600–603, 1970
6. Skyberg K, Ronneberg A, Kamoy JI, et al: Pulmonary fibrosis in cable plant workers exposed to mist and vapor of petroleum distillates. *Environ Res* 40:261–273, 1986
7. Skyberg K, Skaug V, Gylseth B, et al: Subacute inhalation toxicity of mineral oils, C15-C20 alkylbenzenes, and polybutene in male rats. *Environ Res* 53:48–61, 1990
8. *IARC Monographs on the Evaluation of the Carcinogenic Risk of Chemicals to Humans*, Suppl 4, pp 227–228. Lyon, ab International Agency for Research on Cancer, October 1982
9. Bingham E: Carcinogenic potential of petroleum hydrocarbons: a critical review of the literature. *J Environ Pathol Toxicol* 3:483–563, 1980
10. Jarvolm B et al: Cancer morbidity among men exposed to oil mist in the metal industry. *J Occup Med* 23:333–337, 1981
11. *IARC Monographs on the Evaluation of the Carcinogenic Risk of Chemicals to Humans, Overall*

Evaluations of Carcinogenicity: An Updating of IARC Monographs, Vols 1–42, Suppl 7, pp 252–254. Lyon, International Agency for Research on Cancer, 1987

12. Kane ML et al: Toxicological characteristics of refinery streams used to manufacture lubricating oils. *J Ind Med* 5:183–200, 1984

OSMIUM TETROXIDE
CAS: 20816-12-0

OsO₄

Synonym: Osmic acid

Physical Form. Colorless crystals or yellow crystalline mass with acrid chlorinelike odor

Uses. Oxidizing agent

Exposure. Inhalation

Toxicology. Osmium tetroxide is an irritant of the eyes, mucous membranes, and skin.

A laboratory investigator briefly exposed to a high concentration of vapor experienced a sensation of chest constriction and difficulty in breathing.[1] Irritation of the eyes is usually the first symptom of exposure to low concentrations of the vapor; lacrimation, a gritty feeling in the eyes, and the appearance of rings around lights are frequently reported. In most cases, recovery occurs within a few days.[2] Workers exposed to fume concentrations of up to 0.6 mg/m³ developed lacrimation, visual disturbances and, in some cases, frontal headache, conjunctivitis, and cough.[1]

Rabbits exposed for 30 minutes to vapor at estimated concentrations of 130 mg/m³ developed irritation of mucous membranes and labored breathing; at autopsy, there was bronchopneumonia, as well as slight kidney damage.[1] A 4-hour exposure at 400 mg/m³ was lethal to rats.[3]

Application of a drop of 1% solution of osmium tetroxide to a rabbit eye caused severe corneal damage, permanent opacity, and superficial vascularization.[4] Osmium compounds have a caustic action on the skin, resulting in eczema and dermatitis.[2]

The 1995 ACGIH threshold limit value–time-weighted average (TLV-TWA) is 0.0002 ppm (0.0016 mg/m³) as Os, with a STEL/ceiling of 0.0006 ppm (0.0047 mg/m³) as Os.

REFERENCES

1. McLaughlin AIG, Milton R, Perry KMA: Toxic manifestations of osmium tetroxide. *Br J Ind Med* 3:183–186, 1946
2. Hygienic Guide Series: Osmium and its compounds. *Am Ind Hyg Assoc J* 29621–29623, 1968.
3. Registry of Toxic Effects of Chemical Substances. RTECS DHHS (NIOSH) Pub No 86-103, p 1315. Washington, DC, US Department of Health and Human Services, 1985
4. Grant WM: *Toxicology of the Eye*, 3rd ed, p 682. Springfield, IL., Charles C Thomas, 1976

OXALIC ACID
CAS: 144-62-7

C₂O₄H₂

Synonym: Ethanedioic acid

Physical Form. Crystalline solid

Uses. Chemical synthesis; bleaches; metal polish; rust remover

Exposure. Inhalation; ingestion

Toxicology. Oxalic acid is an irritant of the eyes, mucous membranes, and skin; severe intoxication results in convulsions.

There is little reported information on industrial exposure although chronic inflammation of the upper respiratory tract has been described in a worker exposed to hot vapor arising from oxalic acid.[1] Ingestion of as little as 5 g has caused fatalities; there is rapid onset of shock, collapse, and convulsions. The convulsions are thought to be the result of hypocalcemia due to the calcium-complexing action of oxalic acid, which depresses the level of ionized calcium in body fluids. Marked renal damage from deposition of calcium oxalate may occur.[1] A study of railroad-car cleaners with heavy exposure to oxalic acid solutions found an increased incidence of urinary stones. There was a 53% incidence of urolithiasis in exposed workers compared to a rate of 12% in unexposed workers from the same company.[2]

Gross contact of the hands with solutions of oxalic acid (5.3% and 11.5% in two reported cases) used as cleaning solutions caused tingling, burning, soreness, and cyanosis of the fingers.[3]

Splashes of solutions in the eyes have produced epithelial damage from which recovery has been prompt.[4]

The single oral LD_{50} for a 5 wt% oxalic acid solution was 9.5 ml/kg for male rats and 7.5 ml/kg for females.[5] Applied to rabbit skin, a single exposure of 20 g/kg of the solution was not lethal.

Reproductive toxicity has been noted in animal studies.[6] Male mice administered 8400 mg/kg for 7 days prior to mating had decreased fertility. Female mice given the same dose 7 days prior to mating and throughout 21 days of gestation showed embryotoxicity and fetotoxicity.[6] Female rats had disrupted estrous cycles when maintained on diets containing 2.5% and 5% oxalic acid.[7]

The 1995 ACGIH threshold limit value–time-weighted average (TLV-TWA) for oxalic acid is 1 mg/m³, with a short-term exposure limit of 2 mg/m³.

REFERENCES

1. Fassett DW: Oxalic acid and derivatives. In International Labour Office: *Encyclopaedia of Occupational Health and Safety*, Vol II, p 984. New York, McGraw-Hill, 1972
2. Laerum E, Arseth S: Urolithiasis in railroad shopmen in relation to oxalic acid exposure at work. *Scand J Work Environ Health* 11:97–100. 1985
3. Klauder JV, Shelanski L, Gabriel K: Industrial uses of compounds of fluorine and oxalic acid. *AMA Arch Ind Health* 12:412–419, 1955
4. Grant WM: *Toxicology of the Eye*, 3rd ed, pp 685–686. Springfield, IL, Charles C Thomas, 1986
5. Vernot EH et al: Acute toxicity and skin corrosion data for some organic and inorganic compounds and aqueous solutions. *Toxicol Appl Pharmacol* 42:417–423, 1977
6. Von Burg R: Toxicology update: oxalic acid and sodium oxalate. *J Appl Toxicol* 14:233–237, 1994
7. Goldman M, Doering GJ, Neilson RG: Effect of dietary ingestion of oxalic acid on growth and reproduction in male and female Long–Evans rats. *Res Commun Chem Pathol Pharmacol* 18:369–372, 1977

OXYGEN DIFLUORIDE
CAS: 7783-41-7

OF₂

Synonyms: Fluorine monoxide; oxygen fluoride; fluorine oxide

Physical Form.   Colorless gas

Uses.   Oxidant in missile propellant systems

Exposure.   Inhalation

Toxicology.   Oxygen difluoride is a severe pulmonary irritant in animals; exposure is expected to cause the same effect in humans.

In humans, inhalation of the gas at fractions of a part per million produced intractable headache.[1] Although there are no reports of effects on the eyes or skin of humans, it would be expected that the gas under pressure impinging on the eyes or skin would produce serious burns.[2]

In monkeys and dogs, the LC_{50} was 26 ppm for 1 hour; signs of toxicity were lacrimation, dyspnea, muscular weakness, and vomiting; at autopsy, massive pulmonary edema and hemorrhage were observed.[3] In mice, exposure to a low concentration (1 ppm for 60 minutes) produced tolerance to subsequent exposures 8 days later at levels that would otherwise have been fatal (4.25 ppm for 60 minutes).

The 1995 ACGIH ceiling threshold limit value (C-TLV) for oxygen difluoride is 0.05 ppm (0.11 mg/m³).

REFERENCES

1. Oxygen difluoride. *Documentation of the TLVS and BEIs*, 6th ed, pp 1153–1154. Cincinnati, OH, American Conference of Governmental Industrial Hygienists, 1991
2. Hygienic Guide Series: Oxygen difluoride. *Am Ind Hyg Assoc J* 28:194–196, 1967
3. Davis UV: Acute toxicity of oxygen difluoride. In: *Proceedings of the First Annual Conference on Environmental Toxicology*, AMRL-TR-70-102, pp 329–340. Aerospace Medical Research Laboratory, Wright-Patterson Air Force Base, OH, 1970

OZONE
CAS: 10028-15-6

O_3

Synonym: Triatomic oxygen

Physical Form. Blue gas

Uses/Sources. Disinfectant for air and water; for bleaching textiles; in organic synthesis; produced in welding arcs, corona discharges by ultraviolet radiation, and around high-voltage electric equipment

Exposure. Inhalation

Toxicology. Ozone is an irritant of the mucous membranes and the lungs.

The primary target of ozone exposure is the respiratory tract.[1] Symptoms range from nose and throat irritation to cough, dyspnea, and chest pain. By analogy to animal studies, severe exposure may cause pulmonary edema and hemorrhage. Typically, the threshold for effects in humans is reported to be between 0.2 and 0.4 ppm. In one report, a single 2-hour exposure to 0.4 ppm resulted in increased levels of inflammatory cells and soluble factors capable of producing damage in the lower airways of 11 volunteers.[2] Exposure to 0.5 ppm 3 hr/day, 6 days/week, for 12 weeks caused a significant reduction in 1-second forced expiratory volume without subject symptoms.[3] Bronchial irritation, slight dry cough, and substernal soreness were reported in subjects exposed to 0.6 to 0.8 ppm ozone for 2 hours. Marked changes in lung function lasting up to 24 hours were also found.[4] A single 2-hour exposure to 1.5 to 2 ppm caused impaired lung function, chest pain, loss of coordinating ability, and difficulty in articulation. Cough and fatigue persisted for 2 weeks.[5] No deaths from exposure to ozone have been reported.[6]

Extrapulmonary toxic effects potentially attributable to ozone exposure include hematological changes, chromosomal effects in circulating lymphocytes, alterations in hepatic metabolism, reproductive effects, hormonal effects, central nervous system effects, changes in visual acuity, and altered susceptibility to infectious agents.[1,7] Some extrapulmonary effects may be secondary to respiratory system damage.

The toxic effects of ozone can be attributed to its strong oxidative capacity. Specifically, ozone may act by initiating peroxidation of polyunsaturated fatty acids present in the cell membrane and/or by direct oxidation of amino acids and proteins also found in the membranes.[8] If damage is severe, the cell dies; necrosis is commonly reported in the lungs of heavily exposed animals.[1] In animal studies, following acute ozone exposure, a characteristic ozone lesion occurs at the junction of the conducting airways and the gas-exchange region of the lung. This anatomical site is probably affected in humans as well.[1]

One of the principal modifiers of the magnitude of response to ozone is minute ventilation (Vt), which increases proportionally with increase in exercise workload.[9] Surprisingly, patients with mild to moderate respiratory disease do not appear to be more sensitive to threshold ozone concentrations than normal subjects are.[10]

The effects of ozone appear to be cumulative for initial exposures followed by adaptation. Of 6 subjects exposed to 0.5 ppm ozone 2 hr/day for 4 days, 5 showed cumulative effects of symptoms and lung-function tests for the first 3 days, followed by a return to near control values on day 4.[11] In animals exposure to 0.3 to 3 ppm for up to 1 hour permits the animals to withstand multilethal doses for months afterward.[12] However, repeated exposures impart protection from all forms of lung injury (eg, susceptibility to infectious agents, enzyme activities, inflammation). Initial ozone exposure may act to reduce cell sensitivity and/or increase the thickness of mucus, factors that may modify the accessibility and action of the gas.[8] It is not known how variations in the length, frequency, or magnitude of exposure modify the time course for tolerance.

A recent study found that ozone exposure produced alterations in vascular polymorphonuclear leukocytes. The increased cell motility and adherence of these cells may serve as biomarkers of ozone toxicity.[13]

The 1995 ACGIH threshold limit value (TLV) for ozone is 0.05 ppm (0.1 mg/m^3) with a STEL/ceiling of 0.2 ppm (0.4 mg/m^3).

REFERENCES

1. Menzel DB: Ozone: an overview of its toxicity in man and animals. *J Toxicol Environ Health* 13:183–204, 1984

2. Koren HS, Devlin RB, Graham DE, et al: Ozone-induced inflammation in the lower airways of human subjects. *Am Rev Respir Dis* 139:407–415, 1989

3. Bennett G: Ozone contamination of high altitude aircraft cabins. *Aerospace Med* 33:969–973, 1962

4. Young WA et al: Effect of low concentrations of ozone on pulmonary function. *J Appl Physiol* 19:765–768, 1964

5. Griswold SS et al: Report of a case exposure to high ozone concentrations for two hours. *Arch Ind Health* 15:108–118, 1957

6. Nasr ANM: Ozone poisoning in man: clinical manifestations and differential diagnosis—a review. *Clin Toxicol* 4:461–466, 1971

7. Lagerwerff JM: Prolonged ozone inhalation and its effects on visual parameters. *Aerospace Med* 34:479–486, 1963

8. Mehlman MA, Borek C: Toxicity and biochemical mechanisms of ozone. *Environ Res* 42:36–53, 1987

9. Kagawa J: Exposure-effect relationship of selected pulmonary function measurements in subjects exposed to ozone. *Int Arch Occup Environ Health* 53:345–358, 1984

10. Kulle TJ et al: Pulmonary function adaptation to ozone in subjects with chronic bronchitis. *Environ Res* 34:55–63, 1984

11. Hackney JD et al. Adaptation to short-term respiratory effects of ozone in men exposed repeatedly. *J Appl Physiol* 43:82–85, 1977

12. Stokinger HE, Scheel LD: Ozone toxicity: Immunochemical and tolerance-producing aspects. *Arch Environ Health* 4:327–334, 1962

13. Bhalla DK, Rasmussen RE, Daniels DS: Adhesion and motility of polymorphonuclear leukocytes isolated from the blood of rats exposed to ozone: potential biomarkers of toxicity. *Toxicol Appl Pharmacol* 123:177–186, 1993

PARAQUAT
CAS: 4685-14-7

$C_{12}H_{14}N_2$

Synonyms: 1,1'-Dimethyl-4,4'-bipyridinium dichloride; gramoxone; methylviologen

Physical Form. Yellow solid

Uses. Herbicide

Exposure. Inhalation; ingestion

Toxicology. Paraquat is an irritant of the eyes, mucous membranes, and skin; ingestion causes fibroblastic proliferation in the lungs.

In a study of 30 workers engaged in spraying paraquat over a 12-week period, approximately 50% of them had minor irritation of the eyes or nose; one worker had an episode of epistaxis.[1] Of 296 spray operators with skin exposure described as "gross and prolonged," 55 had damaged fingernails. The most common lesion was transverse white bands of discoloration, but other lesions were loss of nail surface, transverse ridging and gross deformity of the nail plate and, in some cases, loss of the nail.[2]

Paraquat is commonly combined in commercial herbicides with diquat, a related compound; in several instances, the commercial preparations splashed in the eyes have caused serious injury.[3,4] Effects have been loss of corneal and conjunctival epithelium, mild iritis, and residual corneal scarring. In contrast, in the eye of a rabbit, one drop of a 50% aqueous solution of pure paraquat caused slow development of mild conjunctival inflammation, and pure diquat proved even less irritating.[5] Presumably, the surfactants present in the commercial preparations are responsible for the severe eye injuries to humans.[4]

In a survey of 36 paraquat formulation workers, acute skin rashes and burns from a delayed caustic effect, eye injuries with conjunctivitis from splash injuries, nail damage, and minor epistaxis were common clinical com-

plaints.[6] Despite a mean exposure period of 5 years, there was no evidence of chronic effects on skin, mucous membranes, or general health. Comparison of a group of 27 Malaysian plantation spraymen, with a mean of 5.3 years of heavy paraquat use, to unexposed groups did not demonstrate any significant differences in pulmonary, renal, liver, and hematologic functions.[7] No abnormalities were attributable to paraquat exposure.[7]

In rats exposed to aerosols of paraquat, the LC_{50} for 6 hours was 1 mg/m³; death was delayed and resulted from pulmonary hemorrhage and edema.[8] In practice, the large particle size of agricultural sprays probably mitigates against this occurring in exposed workers.[9]

The results from ingestion by humans or injection in animals are in marked contrast to the irritant effects usually encountered in industrial exposure. There are numerous reports of fatal accidental and suicidal ingestion by humans.[9–11] In two cases, one person ingested about 114 ml of a 20% solution, and the other person was believed to have taken only a mouthful of the liquid, most of which was rejected immediately. The former died after 7 days; the latter person died after 15 days.[10] Initial symptoms included burning in the mouth and throat, nausea, vomiting, and abdominal pain with diarrhea. After 2 to 3 days, signs of liver and kidney toxicity developed, including jaundice, oliguria, and albuminuria; electrocardiogram changes were suggestive of toxic myocarditis with conduction defects. Shortly before death, respiratory distress occurred; at autopsy, findings in the lung included hemorrhage, edema, and massive solid areas that were airless owing to fibroblastic proliferation in the alveolar walls and elsewhere.[10] Early deaths from massive poisonings usually result from a combination of acute pulmonary edema, acute oliguric renal failure, and hepatic failure. Deaths from less massive poisonings typically result from pulmonary fibrosis, developing 1 to 3 weeks after ingestion.[9]

Intraperitoneal injection or oral administration to rats at doses that caused delayed death resulted in the same proliferative lesion in the lung; findings were alveolar, perivascular, and peribronchial edema, with cellular proliferation into the alveolar walls resulting in large solid areas of the lung with no air-containing cavities.[5]

There is no evidence that inhalation exposures in occupational settings cause the rapid progressive pulmonary fibrosis and injury to the heart, liver, and kidneys that occur from ingestion. Because of the low vapor pressure, there is little inhalation hazard. Spray droplets are usually too large to reach the alveoli. If exposure is excessive, droplets may be inhaled into the upper respiratory tract and cause nosebleed, sore throat, headache, and coughing from local irritant action. Rarely, dermal exposure to paraquat has resulted in systemic poisonings and deaths with renal and pulmonary damage.[9] Such episodes occurred with prolonged skin contact during spraying and exposure to concentrated solutions, or exposure to areas of pre-existing dermatitis; all could have been prevented with use of recommended work practices.

Workers involved in the manufacture of paraquat were found to have a high prevalence of hyperpigmented macules and hyperkeratosis, both of which may be premalignant skin lesions. Analysis of the data suggested that exposure to bipyridine precursors along with sunlight, rather than paraquat itself, was responsible.[12]

A mouse bioassay involving dietary exposure to 25, 50, and 75 mg/kg/day for 80 weeks yielded no evidence of carcinogenicity despite the occurrence of some deaths from respiratory disease.[9] A 2-year bioassay in rats exposed to paraquat in drinking water at 1.3 and 2.6 mg similarly resulted in lung pathology but no increased tumor incidence.[9] In general, paraquat was not genotoxic in a variety of in vitro and in vivo assays.[13]

Embryotoxicity has been observed only at doses that also cause significant maternal toxicity in rats, mice, and rabbits.[13]

The 1995 ACGIH threshold limit value–time-weighted average (TLV-TWA) for paraquat is 0.5 mg/m^3 for total dust and 0.1 mg/m^3 for the respirable fraction.

REFERENCES

1. Swan AAB: Exposure of spray operators to paraquat. *Br J Ind Med* 26:322–329, 1969
2. Hearn CED, Keir W: Nail damage in spray operators exposed to paraquat. *Br J Ind Med* 28:399–403, 1971
3. Cant JS, Lewis DRH: Ocular damage due to paraquat and diquat. *Br Med J* 2:224, 1968
4. Grant WM: *Toxicology of the Eye*, 3rd ed, pp 699–700. Springfield, IL, Charles C Thomas, 1986
5. Clark DG, McElligott TF, Hurst EW: The Toxicity of Paraquat. *Br J Ind Med* 23:126–132, 1966
6. Howard JK: A clinical survey of paraquat formulation workers. *Br J Ind Med* 36:220–223, 1979
7. Howard JK, Sabapathy NN, Whitehead PA: A study of the health of Malaysian plantation workers with particular reference to paraquat spraymen. *Br J Ind Med* 38:110–116, 1981
8. Gage JC: Toxicity of paraquat and diquat aerosols generated by a size-selective cyclone: effect of particle size distribution. *Br J Ind Med* 25:304–314, 1968
9. *International Programme on Chemical Safety, Environmental Health Criteria: 39-Paraquat and Diquat*, pp 13–128. Geneva, World Health Organization, 1984
10. Bullivant CM: Accidental poisoning by paraquat: report of two cases in man. *Br Med J* 1:1272–1273, 1966
11. Toner PG, Vetters JM, Spilg WGS, Harland WA: Fine structure of the lung lesion in a case of paraquat poisoning. *J Pathol* 102:182–185, 1970
12. Wang JD et al: Occupational risk and the development of premalignant skin lesions among paraquat manufacturers. *Br J Ind Med* 44:196–200, 1987
13. Trochimowicz HJ et al: Heterocyclic and miscellaneous nitrogen compounds. In Clayton GD, Clayton FE (eds): *Patty's Industrial Hygiene and Toxicology*, 4th ed, Vol II, Part E, *Toxicology*, pp 3390–3394. New York, John Wiley and Sons, Inc, 1994

PARATHION
CAS: 56-38-2

$C_{10}H_{14}NO_5PS$

Synonyms: *O,O*-Diethyl *O-p*-nitrophenyl phosphorothioate

Physical Form.   Brown or yellowish liquid

Uses.   Acaracide; insecticide

Exposure.   Inhalation; skin absorption; ingestion

Toxicology.   Parathion is an anticholinesterase agent.

Hundreds of deaths associated with parathion exposure have been reported. These deaths have resulted from accidental, suicidal, and homicidal poisonings. Parathion has been the cause of most poisonings of crop workers in the United States.[1] Fatal human poisonings have resulted from ingestion, skin exposure, and inhalation (with varying degrees of skin exposure).

Signs and symptoms of overexposure are caused by the inactivation of the enzyme cholinesterase, which results in the accumulation of acetylcholine at synapses in the nervous system, skeletal and smooth muscle, and secretory glands. The sequence of the development of systemic effects varies with the route of entry. The onset of signs and symptoms is usually prompt but may be delayed up to 12 hours.[1-4] After inhalation, respiratory and ocular effects are the first to appear, often within a few minutes after exposure. Respiratory effects include tightness in the chest and wheezing due to bronchoconstriction and excessive bronchial secretion; laryngeal spasms and excessive salivation may add to the respiratory distress; cyanosis may also occur. Ocular effects include miosis, blurring of distant vision, tearing, rhinorrhea, and frontal headache.

Following ingestion, gastrointestinal effects, such as anorexia, nausea, vomiting, abdominal cramps, and diarrhea, appear within 15 minutes to 2 hours. After skin absorption, localized sweating and muscular fasciculations in the immediate area occur usually within 15 minutes to 4 hours; skin absorption is somewhat greater at higher ambient temperatures and is increased by the presence of dermatitis.[3]

With severe intoxication by all routes, an excess of acetylcholine at the neuromuscular junctions of skeletal muscle causes weakness aggravated by exertion, involuntary twitchings, fasciculations and, eventually, paralysis. The most serious consequence is paralysis of the respiratory muscles. Effects on the central nervous system include giddiness, confusion, ataxia, slurred speech, Cheyne–Stokes respiration, convulsions, coma, and loss of reflexes. The blood pressure may fall to low levels, and cardiac irregularities, including complete heart block, may occur.[2]

Complete symptomatic recovery usually occurs within 1 week; increased susceptibility to the effects of anticholinesterase agents persists for up to several weeks after exposure. Daily exposure to concentrations that are insufficient to produce symptoms following a single exposure may result in the onset of symptoms. Continued daily exposure may be followed by increasingly severe effects.

The minimum lethal oral dose of parathion for humans has been estimated to range from less than 10 mg up to 120 mg.[6] In a study of 115 workers exposed to parathion under varying conditions, the majority excreted significant amounts of *p*-nitrophenol (a metabolite of parathion) in the urine, whereas only those with heavier exposures had a measurable decrease in blood cholinesterase.[7] Measurement of urinary *p*-nitrophenol can be useful in assessing parathion absorption in occupational or other settings.[1]

With dermal exposure in the occupational setting, onset of symptoms may be delayed from several hours to as long as 12 hours. This delay in onset, which is unusual for other organophosphate compounds, may occur even with poisonings that prove to be serious.[1]

Parathion itself is not a strong cholinesterase inhibitor, but one of its metabolites, paraoxon, is an active inhibitor. Paraoxon inactivates cholinesterase by phosphorylation of the active site of the enzyme to form the "diethylphosphoryl enzyme." Over the following 24 to 48 hours, there is a process, called aging, of conversion to the "monoethylphosphoryl enzyme." Aging is of clinical interest in the treatment of poisoning because cholinesterase reactivators such as pralidoxime (2-PAM, Protopam) chloride are ineffective after aging has occurred.

In the field, parathion is converted to varying degrees to paraoxon, which may persist on foliage and in soil. Exposure to paraoxon from weathered parathion residues by the dermal route upon re-entry by fieldworkers has resulted in anticholinesterase poisonings.[8]

In an animal bioassay, a dose-related increase in the incidence of adrenal cortical adenomas (with a few carcinomas at this site as well) has been observed in one strain of rats of both sexes. The significance of these lesions in aged rats is unclear. Other bioassays in mice and rats had sufficient limitations, such that the IARC deemed them inadequate for evaluation and concluded that there are insufficient data to evaluate the carcinogenicity of parathion for animals and no data for humans.[9] There are no documented cases of peripheral neuropathy from parathion exposure.[9]

The 1995 ACGIH threshold limit value–time-weighted average (TLV-TWA) for parathion is 0.1 mg/m^3, with a notation for skin absorption.

Diagnosis. *Signs and Symptoms:* Initial symptoms include headache, blurred vision, pallor, weakness, sweating, abdominal pain, nausea, vomiting, and diarrhea. Moderate to severe intoxication is typically characterized by miosis, lacrimation, excessive salivation, muscle fasciculations, dyspnea, cyanosis, convulsions, shock, cardiac arrhythmias, and coma.

Differential Diagnosis: Diagnosis is based primarily on a history of exposure and clinical evidence of diffuse parasympathetic stimulation. Careful observation of the effects of atropine and pralidoxime may be valuable. Patients with organophosphate poisoning are resistant to the action of atropine at moderate dosages: failure of 1 to 2 mg of atropine administered parenterally to produce signs of atropinization (flushing, mydriasis, tachycardia, or dry mouth) indicates anticholinesterase poisoning. Intravenous injection of 1 g pralidoxime generally causes some recovery from signs and symptoms.

Special Tests: Two types of cholinesterase are clinically significant: (1) true acetylcholinesterase, found principally in the nervous system and the red blood cells; and (2) pseudo- or butyrylcholinesterase, found in the plasma, the liver, and the nervous system. Whereas the action of both types is inhibited by organophosphates, the level of depression of red blood cell cholinesterase is a better indicator of clinically significant reduction of cholinesterase activity in the nervous system.

Laboratory evidence of depression of red blood cell cholinesterase to a level substantially below pre-exposure levels (at least 50% and usually much lower) is verification of poisoning. There is an imperfect correlation between the degree of depression of cholinesterase enzymes and the occurrence of symptoms. With a rapid drop in cholinesterase activity, generally reflecting an acute heavy exposure, there may be symptoms with only a 30% depression whereas, with slower drops to 70% depression, reflecting chronic low-level exposure, there may be no symptoms.[7]

If no pre-exposure baseline has been performed but symptoms are not sufficient to justify treatment with atropine, repeated testing during the recovery period demonstrating progressively increasing plasma and red blood cell cholinesterase levels over several days and weeks, respectively, suggests the diagnosis of anticholinesterase poisoning.

There are many different methods for estimation of cholinesterase content of blood, and associated with each method is a different set of normal values and a different set of reporting units.[10,11] The laboratory report of a cholinesterase determination should state the units involved, along with the appropriate normal range. Based on the Michel method, the normal range of red blood cell cholinesterase activity (delta pH/hr) is 0.39 to 1.02 for men and 0.34 to 1.10 for women.[11] The normal range of the enzyme activity (delta pH/hr) of plasma is 0.44 to 1.63 for men and 0.24 to 1.54 for women.

Treatment. Treatment of organophosphate poisoning ranges from simple removal from exposure in very mild cases to the provision of rigorous supportive and antidotal measures in severe cases.[2,3,12]

It is of great importance to decontaminate the patient without placing rescuers themselves at risk. The patient's contaminated clothing should be removed at once, preferably by medical staff who are protected against contamination, and the skin should be washed with generous amounts of soap or detergent and a flood of water, which is best accomplished under a shower or by submersion in a pond or other body of water if the exposure occurred in the field. Careful attention should be paid to cleansing of the skin and hair.

All severely poisoned patients should be treated in an intensive care unit if available.

In the moderate to severe case, because of pulmonary involvement, there may be need for artificial respiration using a positive-pressure method. Careful attention must be paid to removal of secretions and to maintenance of a patent airway. Maintenance of respiration is critical since death usually results from weakness of the muscles of respiration and accumulation of excessive secretions in the respiratory tract.[2] Heart rate, blood pressure, and ECG should be monitored routinely. Anticonvulsants such as diazepam may be necessary; in cases of repeated seizure activity, phenytoin should be considered as an alternative.[12]

As soon as cyanosis has been overcome, atropine should be given intravenously. (Atropine may induce ventricular fibrillation in the presence of cyanosis). If muscarinic symptoms reappear, the dose of atropine should be repeated; the dose should be titrated against signs of bronchospasm and bronchorrhea. A mild degree of atropinization should be maintained for at least 48 hours.[2]

Pralidoxime (2-PAM, Protopam) chloride is a cholinesterase reactivator that complements the action of atropine. It has its greatest effect in reversing the nicotinic action of anticholinesterase agents at skeletal neuromuscular junctions but has virtually no effect on central nervous system manifestations. Therapeutically effective oxime concentrations are reached with administration of 15 to 30 mg/kg body weight of 2-PAM. It has not been established whether it is more advantageous clinically to give 2-PAM by iv bolus injection or by continuous infusion following initial bolus doses. Treatment with 2-PAM in sufficient doses to produce and maintain adequate plasma oxime concentrations is recommended for as long as the patient exhibits signs of poisoning. Initial treatment with pralidoxime chloride will be most effective if given within 24 hours after poisoning.[2] Morphine, aminophylline, and phenothiazines are contraindicated because of documented experience of adverse reactions in cases of organophosphate poisoning.[10]

The patient should be attended and monitored continuously for not less than 24 hours, since serious and sometimes fatal relapses have occurred because of continuing absorption of the toxin, continuing production of the active metabolite(s), and/or dissipation of the effects of the antidote.

Regeneration of cholinesterase, primarily by synthesis of new enzyme, takes place at the rate of approximately 1%/day.[10] A patient who has recovered from the acute phase of poisoning remains hypersusceptible to anticholinesterases for up to several weeks.

Medical Control. Medical control involves preplacement and annual physical examination with determination of pre-exposure red blood cell and plasma cholinesterase activity. A person whose red blood cell cholinestersase falls to or below 40% of the pre-exposure baseline should be removed from further exposure until the activity returns to within 80% of the pre-exposure baseline.[13]

REFERENCES

1. Hayes WJ Jr, Laws ER Jr: *Handbook of Pesticide Toxicology*, Vol 2, *Classes of Pesticides*, pp 1040–1049. Academic Press, 1991
2. Taylor P: Anticholinesterase agents. In Gilman AG et al (eds): *Goodman and Gilman's The Pharmacological Basis of Therapeutics*, 7th ed, pp 110–129. New York, MacMillan, 1985
3. Koelle GB (ed): Cholinesterases and anticho-

linesterase agents. *Handbuch der Experimentellen Pharmakologie*, Vol 15, pp 989–1027. Berlin, Springer-Verlag, 1963

4. Namba T, Nolte CT, Jackrel J, Grob D: Poisoning due to organophosphate insecticides. *Am J Med* 50:475–492, 1971
5. Milby TH: Prevention and management of organophosphate poisoning. *JAMA* 216: 2131–2133, 1971
6. Hygienic Guide Series: Parathion. *Am Ind Hyg Assoc J* 30:308–312, 1969
7. Arterberry JD, Durham WF, Elliott JW, Wolfe HR: Exposure to parathion—measurement by blood cholinesterase level and urinary *p*-nitrophenol excretion. *Arch Environ Health* 3:476–485, 1961
8. Spear RC et al: Worker poisonings due to paraoxon residues. *J Occup Med* 19:411–414, 1977
9. *IARC Monographs on the Evaluation of the Carcinogenic Risk of Chemicals to Humans*, Vol 30, Miscellaneous pesticides, pp 153–181. Lyon, International Agency for Research on Cancer, 1983
10. Coye MJ, Lowe JA, Maddy KT: Biological monitoring of agricultural workers exposed to pesticides. I. Cholinesterase activity determinations. *J Occup Med* 28:619–627, 1986
11. Michel HO: Electrometric method for determination of red blood cell and plasma cholinesterase activity. *J Lab Clin Med* 34:1564–1568, 1949
12. Johnson MK, Vale JA: Clinical management of acute organophosphate poisoning: an overview. In Ballantyne B, Marrs TC (eds): *Clinical and Experimental Toxicology of Organophosphates and Carbamates*, pp 528–535, Butterworth-Heinemann, 1992
13. National Institute for Occupational Safety and Health: *Criteria for a Recommended Standard . . . Occupational Exposure to Parathion*, DHEW (NIOSH) Pub No 76-190, pp 13–36. Washington, DC, US Government Printing Office, 1976

PENTABORANE
CAS: 19624-22-7

B_5H_9

Synonyms: Dihydropentaborane (9); pentaboron undecahydride

Physical Form. Colorless liquid with a pungent odor

Uses. Reducing agent in propellant fuels

Exposure. Inhalation; skin absorption

Toxicology. Pentaborane is extremely toxic; it affects the nervous system and causes signs of narcosis and hyperexcitation.

In humans, the onset of symptoms may be delayed for up to 24 hours.[1] Minor intoxication causes lethargy, confusion, fatigue, inability to concentrate, headache, and feelings of constriction of the chest. With moderate intoxication, effects are more obvious and include thick speech; confused, sleepy appearance; transient nystagmus and drooping of the eyelids; and euphoria. With severe intoxication, there are signs of muscular incoordination; tremor and tonic spasms of the muscles of the face, neck, abdomen, and extremities; and convulsions and opisthotonos.[1–4]

In a fatal case involving extremely heavy accidental exposure with direct skin contact, there was rapid onset of seizures and opisthotonic spasms accompanied by severe metabolic acidosis without respiratory compensation.[5] The patient expired on day 8, and autopsy revealed severe bilateral necrotizing pneumonia, widespread fatty changes in the liver with centrilobular degeneration, widespread degeneration of the brain, and absence of mature spermatozoa in the testes. Another worker who was exposed while in an adjacent building survived but sustained severe neurologic damage. After 6 months, he demonstrated marked muscular weakness, incoordination, and spasticity. He could see only shapes and colors. A CT scan showed marked cortical atrophy and ventricular dilation. Institutionalization was required. Of 14 individuals with mild exposure to pentaborane from the same accident, 8 were judged to have mild cognitive deficits, as determined by various neuropsychological tests 2 months post-exposure.[6]

The concentrations of vapor and duration of exposures that cause mild, moderate, or severe intoxication are not documented. It has been estimated, on the basis of animal studies, that exposure for 60 minutes will cause slight signs of toxicity at 8 ppm, convulsions at 15 ppm, and death at 30 ppm. It was the clinical impression of one investigator that, in humans, a transient "wafting" odor did not produce symptoms (median odor threshold is 1 ppm) but that a "strong whiff, producing a penetrating feeling in the nose, usually produced symptoms."[1,7] Olfactory fatigue also occurs, and dangerous levels of pentaborane may not be readily detected.[1]

Severe irritation and corneal opacity of the eyes of test animals occurred from exposure to the vapor; the liquid on the skin of animals caused acute inflammation, with the formation of blisters, redness, and swelling.[8] Because the liquid may ignite spontaneously, fire and consequent burn damage may be a greater hazard than toxicity on contact with the liquid.[4]

The 1995 ACGIH threshold limit value-time-weighted average (TLV-TWA) is 0.005 ppm (0.013 mg/m³) with a STEL/ceiling of 0.015 ppm (0.039 mg/m³).

REFERENCES

1. Mindrum G: Pentaborane intoxication. *Arch Intern Med* 114:367–374, 1964
2. Lowe HJ, Freeman G: Boron hydride (borane) intoxication in man. *AMA Arch Ind Health* 16:523–533, 1957
3. Rozendaal HM: Clinical observations on the toxicology of boron hydrides. *AMA Arch Ind Hyg Occup Med* 4:257–260, 1951
4. Hygienic Guide Series: Pentaborane-9. *Am Ind Hyg Assoc J* 27:307–310, 1966
5. Yarbrough BE et al: Severe central nervous system damage and profound acidosis in persons exposed to pentaborane. *Clin Toxicol* 23:519–536, 1985–1986
6. Hart RP et al: Neuropsychological function following mild exposure to pentaborane. *Am J Ind Med* 6:37–44, 1984
7. Emergency Exposure Limits: Pentaborane-9. *Am Ind Hyg Assoc J* 27:193–195, 1966
8. Hughes RL, Smith IC, Lawless EW: III. Pentaborane (9) and derivatives. In Holzmann RT (ed): *Production of the Boranes and Related Research*, pp 294–301, 329–331, 433–489. New York, Academic Press, 1967

PENTACHLOROETHANE
CAS: 76-01-7

C_2HCl_5

Synonyms: Ethane pentachloride; pentalin

Physical Form. Dense, colorless liquid with a chloroformlike odor

Uses. May occur as an intermediate in the production of chlorinated ethylenes; previously used as a solvent for cellulose ethers, resins, and gums, for dry cleaning, in coal purification, as a soil-sterilizing agent, and as a chemical intermediate in the production of dichloroacetic acid

Exposure. Inhalation

Toxicology. Pentachloroethane is an irritant of the eyes and respiratory tract and may cause mild narcosis; chronic exposure causes hepatocellular carcinomas in mice and inflammation of the kidneys in rats.

No information is available on human exposures to pentachloroethane.

The lowest observed lethal concentrations observed for mice and rats were 35 g/m³ and 4238 ppm, respectively.[1]

Administration of 1000 mg/kg/day was lethal to 6 of 10 rats between the fourth and tenth days. The only observable clinical signs of toxicity were decreased motor activity and lethargy, and weight gain decrements in the survivors of the 500 mg/kg/day dose group. Neither gross nor histopathological examinations revealed lesions that could be attributed to the chemical. Mice survived 2 weeks at 1000 mg/kg/day but lost weight during the experiment.

Dose levels of 75 and 150 mg/kg/day were administered to rats by gavage during chronic (41–103 week) studies.[2,3] A significant dose-related increase in the incidence of chronic renal inflammation and a dose-related trend in the incidence of tubular-cell adenomas of the kidneys in males were noted although the survival of the high-dose group was significantly reduced compared to controls. The kidney lesions were distinguishable from the nephropathy normally seen in aging rats, and some of the dosed animals had both types of changes.

Mice were administered 250 or 500 mg/kg/day pentachloroethane by gavage for life. The hepatic carcinogenicity of pentachloroethane was clearly established despite reduced survival rates. The incidence of hepatocellular carcinomas was significantly increased in low-dose males and in treated females; there also was a significant dose-related increase in the incidence of hepatocellular adenomas in treated females.

It has been suggested that the mechanism of action of pentachloroethane may be similar to that of other chlorohydrocarbons that also induce a high incidence of hepatocellular carcinoma in mice and have little or no carcinogenic effect in rats. The carcinogenic potential may be mediated through the active metabolic intermediates, trichloroethylene and tetrachloroethylene, which are formed in some species but not in others.

It has also been noted that hexachloroethane is a major contaminant of the pentachloroethane used in the chronic studies. This compound also has been shown to induce hepatocellular carcinoma in mice but not rats although the low levels of hexachloroethane present in the pentachloroethane samples seem inadequate to produce the high incidence of tumors found in the pentachloroethane studies.

The IARC determined that there is limited evidence for the carcinogenicity of technical-grade pentachloroethane (containing hexachloroethane) to experimental animals and that no evaluation could be made of the carcinogenicity to humans.[1]

A 1995 ACGIH threshold limit value has not been established for pentachloroethane.

REFERENCES

1. *IARC Monographs on the Evaluation of Carcinogenic Risks to Humans*, Vol 41, Some halogenated hydrocarbons and pesticide exposures, pp 99–111. Lyon, International Agency for Research on Cancer, 1986
2. National Toxicology Program: *Carcinogenesis Bioassay of Pentachloroethane (CAS No 76-01-7) in F344/N Rats and B6C3F₁ Mice (Gavage Study)*. NTP-TR-232, NIH Pub No 83-1788, pp 1–149. Research Triangle Park, NC, US Department of Health and Human Services, 1983
3. Mennear JH, Haseman JK, Sullivan DJ, et al: Studies on the carcinogenicity of pentachloroethane in rats and mice. *Fund Appl Toxicol* 2:82–87, 1982

PENTACHLORONAPHTHALENE
CAS: 1321-64-8

$C_{10}H_3Cl_5$

Synonyms: Halowax 1013

Physical Form. Waxy white solid

Uses. In electric wire insulation; in lubricants

Exposure. Inhalation; skin absorption

Toxicology. Pentachloronaphthalene is toxic to the liver and skin.

The most striking human response to prolonged skin contact with the solid, or to shorter-term inhalation of the hot vapor, is chloracne.[1-3] This is an acneiform skin eruption characterized by papules, large comedones, and pustules, chiefly affecting the face, neck, arms, and legs. Pruritic erythematous and vasculoerythematous reactions have also been reported. The reaction is usually slow to appear and may take months to return to normal. Skin lesions are often accompanied by symptoms of systemic effects, including headache, vertigo, and anorexia. Liver damage characterized by toxic jaundice, which may progress to fatal hepatic necrosis, results from the repeated inhalation of higher concentrations of the hot fumes of the molten substance.[4]

Rats exposed to the vapor of a mixture of hexa- and pentachloronaphthalene at average concentrations of 1.16 mg/m³ for 16 hours daily for up to 4.5 months showed definite liver injury, whereas 8.8 mg/m³ produced some mortality and severe liver injury.[3]

Animal experiments have confirmed the greater toxicity of the more highly chlorinated members of the chloronaphthalene series up to and including hexachloronaphthalene.[3] Skin absorption has been demonstrated in animals and is suspected in humans.

The 1995 ACGIH threshold limit value-time weighted average (TLV-TWA) is 0.5 mg/m³, with a notation for skin absorption.

REFERENCES

1. Kleinfeld M, Messite J, Swencicki R: Clinical effects of chlorinated naphthalene exposure. *J Occup Med* 14:377–379, 1972
2. Greenburg L, Mayers MR, Smith AR: The systemic effects resulting from exposure to certain chlorinated hydrocarbons. *J Ind Hyg Toxicol* 21:29–38, 1939
3. Deichmann WB: Halogenated cyclic hydrocarbons. In Clayton GD, Clayton FE (eds): *Patty's Industrial Hygiene and Toxicology*, 3rd ed, Vol 2B, *Toxicology*, pp 3669–3675. New York, Wiley-Interscience, 1981
4. Cotter LH: Pentachlorinated naphthalenes in industry. *JAMA* 125:273–274, 1944

PENTACHLOROPHENOL

CAS: 87-86-5

C_6Cl_5OH

Synonyms: Penta; PCP, penchlorol; pentachlorophenate; Santophen 20; Dowicide 7

Physical Form. White to tan needlelike crystals

Uses. Wood preservative; insecticide for termite control; preharvest defoliant; general herbicide; fungicide

Exposure. Inhalation; skin absorption; ingestion

Toxicology. Pentachlorophenol has been reported to have adverse effects on the skin, eyes, respiratory system, nervous system, hematopoietic system, kidney, and liver; at high doses, it is fetotoxic to rats; it is carcinogenic to mice.

Deaths from pentachlorophenol have occurred following acute inhalation exposure, dermal exposure, and ingestion.[1,2]

Symptoms of poisoning can include rapid onset of profuse diaphoresis, hyperpyrexia, tachy cardia, tachypnea, generalized weakness, nausea, vomiting, abdominal pain, anorexia, headache, intense thirst, pain in the extremities, intermittent delirium, convulsions, progressive coma and, within hours of the onset of symptoms, death.[3] The acute toxicity results from the uncoupling of oxidative phosphorylation, causing stimulation of cell metabolism and accompanying hyperthermia.[1-4]

Postmortem examination in fatal cases has shown immediate onset of extreme rigor mortis in the muscles of the thighs and legs, edema and intra-alveolar hemorrhage in the lungs, cerebral edema and liver and kidney damage.[2,5] The risks of serious intoxication are increased during hot weather.[6]

Chronic exposure is associated with an increased prevalence of conjunctivitis, chronic sinusitis, bronchitis, polyneuritis, and dermatitis. Chloracne has been reported and is probably the result of dioxin contaminants in commercial-grade pentachlorophenol.[3] On the skin, solutions of pentachlorophenol as dilute as 1% may cause irritation if contact is repeated or prolonged.

Various hematological disorders, including aplastic anemia and hemolytic anemia, have been reported in humans following pentachlorophenol exposure. It has been suggested that pentachlorophenol causes blocking of the formation of adenosine triphosphate in red blood cells, leading to premature lysis.[7] In general, adverse hematological effects have been observed only in animals exposed to the technical-grade pentachlorophenol and not pure pentachlorophenol, suggesting that contaminants may play a role in hematotoxicity.[1]

Hepatic toxicity, as manifested by enlarged livers, fatty infiltration of the liver, centrilobular congestion and degeneration, and elevated serum enzyme levels, has been observed following fatal and nonfatal exposures. Again, animal studies have indicated that purified pentachlorophenol produces much less severe effects than those seen with the technical grade, imply a role for contaminants in hepatic toxicity.[1] Renal dysfunction, such as reduced glomerular filtration and tubular degeneration, appears to be mild and transient in cases of human exposures.[8]

Animal studies have shown that the immune system is sensitive to exposure.[9] Mice fed diets containing 50 or 500 ppm technical-grade pentachlorophenol showed greatly reduced immunocompetence in the form of increased susceptibility to the growth of transplanted tumors.

Given orally to pregnant rats at doses ranging from 5 to 50 mg/kg body weight per day, purified pentachlorophenol produced dose-related signs of fetotoxicity, including resorptions, subcutaneous edema, dilated ureters, and anomalies of the skull, ribs, and vertebrae.[10]

Although early animal studies found no sufficient evidence of carcinogenicity, a 2-year NTP report found clear evidence of carcinogenicity in mice for two technical-grade pentachlorophenol mixtures.[11,12] Male $B6C3F_1$ mice fed diets containing 100 or 200 ppm technical-grade pentachlorophenol had increased incidences of adrenal medullary and hepatocellular neoplasms; there was some evidence of carcinogenicity in female mice similarly exposed, as shown by increased incidences of hemangiosarcomas and hepatocellular neoplasms. For male and female mice given up to 600 ppm EC-7 pentachlorophenol, there were increased incidences of adrenal medullary and hepatocellular neoplasms (females also had hemangiosarcomas of the liver and spleen). It was concluded that pentachlorophenol was primarily responsible for the carcinogenic response in mice but that impurities may possibly play a role in the neoplastic process.[12]

Case reports and case-control studies in humans have suggested a possible association between cancer, including soft tissue sarcoma, acute leukemia, and Hodgkin's disease and occupational exposure to pentachlorophenol. However, in all cases, concomitant exposure to other toxic substances may have contributed to the effects seen.[1,11]

The IARC has determined that there is inadequate evidence for carcinogenicity in humans and sufficient evidence for carcinogenicity in experimental animals.[11]

Occupational exposure of 20 workers to pentachlorophenol at concentrations that ranged from 1.2 to 180 $\mu g/m^3$ for 3 to 34 years did not result in any increased incidence of sister chromatid exchanges or chromosomal aberrations.[13] In another report, significant increases in the incidence of dicentric chromosomes and acentric fragments were detected in the peripheral lymphocytes of exposed workers; the frequency of sister chromatid exchanges was not increased.[14]

The 1995 ACGIH threshold limit value–time-weighted average (TLV-TWA) for pentachlorophenol is 0.5 mg/m^3 with a notation for skin absorption.

REFERENCES

1. Agency for Toxic Substances and Disease Registry (ASTDR): *Toxicological Profile for Pentachlorophenol.* US Department of Health and Human Services, Public Health Service, pp 139, 1993
2. Jorens PG, Schepens PJC: Human pentachlorophenol poisoning. *Hum Exp Toxicol* 12:479–495, 1993
3. Wood S, Rom WN, White GL, et al: Pentachlorophenol poisoning. *J Occup Med* 25:527–530, 1983
4. Williams PL: Pentachlorophenol, an assessment of the occupational hazard. *Am Ind Hyg Assoc J* 43:799–810, 1982
5. Gray RE et al: Pentachlorophenol intoxication: report of a fatal case, with comments on the clinical course and pathologic anatomy. *Arch Environ Health* 40:161–164, 1985
6. Bergner H, Constantinidis P, Martin JH: Industrial pentachlorophenol poisoning in Winnipeg. *Can Med Assoc J* 92:448–451, 1965

7. Hassan AB, Hiseligmann H, Bassan HM: Intravascular hemolysis induced by pentachlorophenol. *Br Med J* 291:21–22, 1985

8. Begley J, Reichart AW, Seismen AW, et al: Association between renal function tests and pentachlorophenol exposure. *Clin Toxicol* 11:97–106, 1977

9. Kerkvliet NI et al: Immunotoxicity of pentachlorophenol (PCP): increased susceptibility to tumor growth in adult mice fed technical PCP-contaminated diets. *Toxicol Appl Pharmacol* 62:55–64, 1982

10. Schwetz BA et al: The effect of purified and commercial grade pentachlorophenol on rat embryonal and fetal development. *Toxicol Appl Pharmacol* 28:151–161, 1974

11. *IARC Monographs on the Evaluation of the Carcinogenic Risk of Chemicals to Humans*, Vol 53, Occupational exposures in insecticide application, and some pesticides, pp 380–394. Lyon, International Agency for Research on Cancer, 1991

12. National Toxicology Program: *Toxicology and Carcinogenesis Studies of Two Pentachlorophenol Technical-Grade Mixtures (CAS No 87-86-5) in B6C3F₁ Mice (Feed Studies)*, pp 1–265. NTP TR 349, NIH Pub No 89-2804. US Department of Health and Human Services, March 1989

13. Ziemsen B, Angerer J, Lehnart G: Sister chromatid exchange and chromosomal breakage in PCP exposed workers. *Int Arch Occup Environ Health* 59:413–417, 1987

14. Bauchinger M, Dresp J, Schmid E, et al: Chromosome exchanges in lymphocytes after occupational exposure to pentachlorophenol (PCP). *Mutat Res* 102:83–88, 1982

PENTAERYTHRITOL
CAS: 115-77-5

$C(CH_2OH)_4$

Synonyms: Tetramethylolmethane; 2,2-*bis*hydroxymethyl-1,3-propanediol

Physical Form. Solid

Uses. In the manufacture of pentaerythritol-tetranitrate; alkyd resins in surface-coating

compositions; pentaerythritol triacrylate and protective coatings; insecticides; pharmaceuticals

Exposure. Inhalation

Toxicology. Pentaerythritol is of very low toxicity.

Rats, dogs, and guinea pigs exposed for 6 hr/day for 90 days to 11,000 mg/m³ technical pentaerythritol showed no adverse effects.[1] The oral LD_{50} for guinea pigs was 11.3 g/kg; effects were tremors, ataxia, and loss of righting reflex.

Feeding studies in human subjects indicated that 85% of the pentaerythritol was eliminated unchanged in the urine.[2]

Skin application on rabbits of a saturated aqueous solution daily for 10 days caused no significant irritation; instillation of the same solution into the rabbit eye caused no irritation.[3]

The 1995 ACGIH threshold limit value–time-weighted average (TLV-TWA) for pentaerythritol is 10 mg/m³.

REFERENCES

1. Keplinger ML, Kay JH: Oral and inhalation studies on pentaerythritol. *J Appl Pharmacol* 6:351, 1964 (Abstract)

2. Berlow E, Barth RH, Snow JE: Physiological properties of the pentaerythritols. In: The Pentaerythritols. American Chemical Society Monograph 136, pp 39–40, New York, Reinhold, 1958

3. Pentaerythritol. *Documentation of the TLVs and BEIs*, 6th ed, pp 1183–1184. Cincinnati, OH, American Conference of Governmental Hygienists, 1991

PENTANE
CAS: 109-66-0

C_5H_{12}

Synonyms: Amyl hydride

Physical Form. Colorless liquid

Uses. Fuel; solvent; chemical synthesis

Exposure. Inhalation

Toxicology. In animals, pentane is a mucous membrane irritant and, at extremely high concentrations, causes narcosis; it is expected that severe exposure will produce the same effects in humans.

Human subjects exposed to 5000 ppm for 10 minutes did not experience mucous membrane irritation or other symptoms.[1]

Topical application of pentane to volunteers caused painful burning sensations accompanied by itching; after 5 hours, blisters formed on the exposed areas.[2]

A 5-minute exposure at 128,000 ppm produced deep anesthesia in mice; respiratory arrest occurred in 1 of the 4 animals during exposure.[3] Mice exposed to 32,000 or 64,000 ppm for 5 minutes showed signs of respiratory irritation and became lightly anesthetized during the recovery period.[3] No effects were observed for 5-minute exposures at 16,000 ppm or below.

The concentration of an alkane required for acute toxic effects decreases as the carbon number increases, and it is possible that this trend also applies to the effects of long-term exposure. Therefore, although there is no documentation of neurotoxic effects of pentane, higher exposure levels for longer periods may be necessary to demonstrate toxicity. Furthermore, most cases of neurotoxicity attributable to alkanes have involved mixed exposures; identification of a single causative chemical is difficult, and there may be an additive toxic effect from mixed exposures.[2]

The odor of pentane is readily detectable at 5000 ppm.[1]

The 1995 ACGIH threshold limit value-time-weighted average (TLV-TWA) is 600 ppm (1770 mg/m³) with a STEL/ceiling of 750 ppm (2210 mg/m³).

REFERENCES

1. Patty FA, Yant WP: *Report of Investigations—Odor Intensity and Symptoms Produced by Commercial Propane, Butane Pentane, Hexane, and Heptane Vapor, No 2979.* US Dept of Commerce, Bureau of Mines, 1929

2. National Institute for Occupational Safety and Health: *Criteria for a Recommended Standard . . . Occupational Exposure to Alkanes (C5–C8).* DHEW (NIOSH) Pub No 77-151. Washington, DC, US Government Printing Office, 1977

3. Swann HE Jr, Kwon BK, Hogan GK, Snellings WM: Acute inhalation toxicology of volatile hydrocarbons. *Am Ind Hyg Assoc J* 35:511–518, 1974

2,4-PENTANEDIONE
CAS: 123-54-6

$CH_3COCH_2COCH_3$

Synonyms: Acetylacetone, diacetyl methane, acetyl 2-propanone, 2,4-PD

Physical Form. Clear liquid with a rancid odor

Uses. Chemical intermediate; metal chelator; lubricant additive

Exposure. Inhalation; skin absorption

Toxicology. 2,4-Pentanedione is moderately irritating to the skin and eyes of animals; repeated exposure to high concentrations causes dyspnea, central nervous system damage, and death.

Information on human exposures is limited. Exposure to levels ranging from 2 to 14 ppm has been reported to produce nausea and headaches in several persons.[1]

For male and female rats, the combined LC_{50} for a 4-hour exposure was 1224 ppm.[2] Signs of toxicity before death included periocular, perinasal, and perioral wetness and encrustation, mouth and abdominal breathing, tremors, ataxia, and negative tail and toe pinch reflexes. Necropsy of animals that died showed dark red lungs, mottled livers, and gas-filled gastrointestinal tracts. Survivors did not show any gross pathology. At 919 ppm for 4 hr, there was decreased motor activity with recovery by the first postexposure day, and there were no signs of toxicity at 628 ppm.

Repeated exposure of rats to 650 ppm, 6 hr/day, 5 days/week, caused death in all females and 10 of 30 males between the second and sixth weeks of exposure.[3] Death was attributed to brain lesions consisting of degenerative changes in the deep cerebellar and vestibular nuclei and the corpora strata. Gliosis and malacia were observed in the same brain regions in approximately half of the surviving animals. Other changes included minimal squamous metaplasia in the nasal mucosa, decreased body and organ weights, lymphocytosis, and minor alterations in serum and urine chemistries. No deaths occurred in rats exposed at 307 ppm for 14 weeks, nor was there any significant brain pathology suggesting a well-defined threshold for effects.

Central neuropathologic lesions have also been described following gavage administration.[1] Rats developed weakness, ataxia, tremors, paresis, and rolling movements of the head following 100 to 150 mg/kg twice daily for periods from 3 to 61 days. Histologic changes induced by the shorter dosing schedules were perivascular edema, hemorrhage into the Virchow–Robin spaces, and endothelial swelling, all primarily localized in the cerebellum and brain stem. Chronic central nervous system lesions were bilateral, symmetrical areas of malacia and gliosis centered on the cerebellar peduncles, olivary nuclei, and lower brain stem.

The central neuropathological effects following inhalation exposure in rats appear to require a critical number of repeated exposures to high (>650 ppm) concentrations.[3] Thus, central neuropathological effects did not occur following acute exposure to potentially lethal concentrations or subchronic exposures at 307 ppm. The steep slope of the concentration-response relationship and the sharply defined exposure conditions for inducing central nervous system damage suggest that the mechanism of neurotoxicity may involve depletion of a biochemical pathway. Specifically, the similarity between the morphologic damage produced by 2,4-pentanedione and acute vitamin B deficiency and the ability of 2,4-pentanedione to inactivate lysyl residues suggest that the toxicity

of 2,4-pentanedione is due to its ability to produce deficiencies of thiamine, folic acid, and/or pyridoxine.[1]

Pregnant Fischer 344 rats were exposed to 2,4-pentanedione at 53, 202, or 398 ppm, 6 hr/day, on gestational days 6 to 15.[4] At the highest dose, there was maternal toxicity in the form of reduced body weight gain and fetotoxicity as reduced fetal body weight and a consistent pattern of reduced skeletal ossification; at 202 ppm, there was reduced fetal body weight gain. Embryotoxicity or teratogenicity were not observed at any concentration.

The percutaneous LD_{50} values for 24-hour occluded contact on the skin of rabbits were 1.41 ml/kg for males and 0.81 for females.[2] Times of death ranged from 45 min to 1 day. Signs of toxicity were dilated pupils within 15 to 30 minutes, convulsions in 1 animal, and excess blood-stained saliva. Local signs of inflammation were erythema, edema, and necrosis. In survivors, inflammation persisted for up to 7 days, with scab formation by 2 weeks. Instilled in the eye of rabbits, the liquid produced mild conjunctivitis without corneal injury.

Although irritant effects do not appear until high concentrations are reached, it is expected that the low odor threshold (0.01 ppm) could provide adequate warning of exposure to 2,4-pentanedione.[2]

2,4-Pentanedione caused sister chromatid exchange increases in CHO cells and an increase in the incidence of micronuclei in peripheral blood erythrocytes of mice.[5] It was not mutagenic in a *Salmonella typhimurium* assay.

A TLV has not been established for 2,4-pentanedione.

REFERENCES

1. Krasavage WJ, O'Donoghue JL, Divincenzo GD: 2,4-Pentanedione. In Clayton GD, Clayton FE (eds): *Patty's Industrial Hygiene and Toxicology*, 3rd ed, rev Vol 2C, pp 4773–4776. New York, Wiley-Interscience, 1982
2. Ballantyne B, Dodd DE, Myers RC, et al: The acute toxicity and primary irritancy of 2,4-pentanedione. *Drug Chem Toxicol* 9:133–146, 1986
3. Dodd DE, Garman RH, Pritts IM, et al: 2,4-Pentanedione: 9-day and 14-week vapor inha-

lation studies in Fischer-344 rats. *Fund Appl Toxicol* 7:329–339, 1986

4. Tyl RW, Ballantyne B, Pritts IM, et al: An evaluation of the developmental toxicity of 2,4-pentanedione in the Fischer 344 rat by vapour exposure. *Tox Ind Health* 6:461–474, 1990
5. Slesinski RS, Guzzie PJ, Ballantyne B: An in vitro and in vivo evaluation of the genotoxic potential of 2,4-pentanedione. *Environ Mutag* 9 (Suppl 8):44–49, 1987

PERCHLOROETHYLENE
CAS: 127-18-4

C_2Cl_4

Synonyms: Tetrachloroethylene; ethylene tetrachloride; 1,1,2,2-tetrachloroethylene; PCE

Physical Form. Colorless liquid

Uses. Solvent for dry cleaning and textile processing; chemical intermediate; in metal degreasing

Exposure. Inhalation

Toxicology. Perchloroethylene causes central nervous system depression and liver damage. Chronic exposure has caused peripheral neuropathy, and it is carcinogenic in experimental animals.

Occupational exposure has caused signs and symptoms of central nervous system depression, including dizziness, light-headedness, "inebriation," and difficulty in walking.[1]

Four human subjects exposed to 5000 ppm left a chamber after 6 minutes to avoid being overcome; they experienced vertigo, nausea, and mental confusion during the 10 minutes following cessation of exposure.[2] In an industrial exposure to an average concentration of 275 ppm for 3 hours, followed by 1100 ppm for 30 minutes, a worker lost consciousness; there was apparent clinical recovery 1 hour after exposure; the monitored concentration of perchloroethylene in the patient's expired air diminished slowly over a 2-week period.[3] During the second and third postexposure weeks, the results of liver-function tests became abnormal. Additional instances of liver injury following industrial exposure have been reported.[4]

Other effects on humans from inhalation of various concentrations are as follows: 2000 ppm, mild central nervous system depression within 5 minutes; 600 ppm, sensation of numbness around the mouth, dizziness, and some incoordination after 10 minutes.[5] In human experiments, 7-hour exposures at 100 ppm resulted in mild irritation of the eyes, nose, and throat; flushing of the face and neck; headache; somnolence; and slurred speech.[6] Prolonged exposure has caused impaired memory, numbness of extremities, and peripheral neuropathy, including impaired vision.[1]

Of 40 dry-cleaning workers, 16 showed signs of central nervous system depression and, in 21 cases, the autonomic nervous system was also affected.[7] Twenty dry-cleaning workers exposed for an average of 7.5 years to concentrations between 1 and 40 ppm had altered electrodiagnostic and neurologic rating scores.[7] Abnormal EEG recordings were found in 4 of 16 factory employees exposed to concentrations ranging from 60 ppm to 450 ppm for periods of 2 years to more than 20 years.[7]

The liquid on the skin for 40 minutes resulted in a progressively severe burning sensation beginning within 5 to 10 minutes and marked erythema, which subsided after 1 to 2 hours.[2]

Rats did not survive when exposed for longer than 12 to 18 minutes to 12,000 ppm. When exposed repeatedly to 470 ppm, they showed liver and kidney injury.[2] Cardiac arrhythmias owing to sensitization of the myocardium to epinephrine have been observed with certain other chlorinated hydrocarbons, but exposure of dogs to perchloroethylene concentrations of 5000 and 10,000 ppm did not produce this phenomenon.[8]

Rats exposed to 300 ppm, 7 hr/day, on days 6 to 15 of pregnancy showed reduced body weight and a slightly increased number of resorptions. Among litters of mice similarly exposed, the incidence of delayed ossification of skull bones, subcutaneous edema, and split sternebrae was significantly increased compared to those in controls.[9] In general, perchloroethylene is fetotoxic at concentrations that also produce maternal toxicity. Although a number of case-referent studies have suggested an effect of perchloroethylene on human reproductive parameters, including spontaneous abortions, sperm abnormalities, delayed conception, hormonal disturbances, and idiopathic infertility, the incompleteness of the studies precludes any significant conclusions.[10–13] However, exposure of pregnant women to perchloroethylene should be minimized.

Large gavage doses, approximately 500 and 1000 mg/kg per day for 78 weeks, caused a statistically significant increase in the incidence of hepatocellular carcinomas in mice.[14] Inhalation exposure by rats to 200 or 400 ppm for 2 years caused an increased incidence of mononuclear cell leukemia; a dose-related trend for a rare renal tubular neoplasm was observed in males.[15]

An increased incidence of hepatocellular adenomas and carcinomas was produced in mice with repeated exposure at 100 and 200 ppm.

The IARC has determined that there is sufficient evidence for carcinogenicity to animals.[16]

Following publication of the rodent carcinogenicity studies, a number of human epidemiologic surveys were conducted.

In a retrospective cohort mortality study of 1690 workers in the dry-cleaning industry, there was a statistically significant excess of observed deaths from urinary tract cancer. It was noted, however, that there was confounding exposure to petroleum solvents and that no data were given on exposure levels.[17]

In contrast, another large case-control study among persons employed as laundry workers, as dry cleaners, and in other occupations with similar chemical exposure did not show any clear association with bladder cancer.[18] Other cohort and proportionate mortality studies have variously reported excesses of lymphosarcomas, leukemias, and cancers of the skin, colon, lung, liver, and pancreas.[16]

A recent population-based case-control study found evidence for an association between perchloroethylene-contaminated drinking water and bladder cancer and leukemia.[19]

The large number of target sites, the small numbers of excess cancers, and the multiplicity of chemical exposures do not permit any definite conclusion to be drawn about perchloroethylene exposure and carcinogenicity to humans.[16] Although the IARC has determined that evidence for carcinogenicity to humans is inadequate, NIOSH recommends that this substance be handled as a potential human carcinogen.[16,20]

Genotoxic assays indicate that perchloroethylene itself is not mutagenic but that metabolites formed via the glutathione pathway are.[21]

Biological monitoring of perchloroethylene is possible by measuring levels of either the parent compound or metabolites in blood, urine, or exhaled air. Measurement of the parent compound in exhaled air is noninvasive and probably superior to measurement of metabolic products such as trichloroacetic acid that are not specific to perchloroethylene metabolism.[21]

The 1995 ACGIH threshold limit value–time-weighted average (TLV-TWA) for perchloroethylene is 25 ppm (170 mg/m^3) with a short-term exposure limit of 100 ppm (685 mg/m^3) and an A3 animal carcinogen designation.

REFERENCES

1. National Institute for Occupational Safety and Health: *Criteria for a Recommended Standard . . . Occupational Exposure to Tetrachloroethylene (Perchloroethylene).* DHEW (NIOSH) Pub No 76-185, pp 17–65. Washington, DC, US Government Printing Office, 1976
2. Hygienic Guide Series: Tetrachloroethylene (perchloroethylene). *Am Ind Hyg Assoc J* 26:640–643, 1965
3. Stewart RD, Erley DS, Schaffer AW, Gay HH: Accidental vapor exposure to anesthetic concentrations of a solvent containing tetrachloroethylene. *Ind Med Surg* 30:327–330, 1961
4. Stewart RD: Acute tetrachloroethylene intoxication. *JAMA* 208:1490–1492, 1969
5. von Oettingen WF: *The Halogenated Aliphatic, Olefinic, Cyclic, Aromatic, and Aliphatic-Aromatic Hydrocarbons Including the Halogenated Insecticides, Their Toxicity and Potential Dangers.* US Public Health Service Publication Pub No 414, pp 227–235. Washington DC, US Government Printing Office, 1955

6. Stewart RD, Baretta ED, Dodd HC, Torkelson TR: Experimental human exposure to tetrachloroethylene. *Arch Environ Health* 20:224–229, 1970

7. *Environmental Health Criteria 31*. Tetrachloroethylene, p 48. Geneva, World Health Organization, 1984

8. Reinhardt CF, Mullin LS, Maxfield ME: Epinephrine-induced cardiac arrhythmia potential of some common industrial solvents. *J Occup Med* 15:953–955, 1973

9. Schwetz BA, Leong BKJ, Gehring PJ: The effect of maternally inhaled trichloroethylene, perchloroethylene, methyl chloroform, and methylene chloride on embryonal and fetal development in mice and rats. *Toxicol Appl Pharmacol* 32:84–96, 1975

10. van der Gulden JWJ, Zielhuis GA: Reproductive hazards related to perchloroethylene. *Int Arch Occup Environ Health* 61:235–242, 1989

11. Kyyronen P, Taskinen H, Lindbohm M, et al: Spontaneous abortions and congenital malformations among women exposed to tetrachloroethylene in dry cleaning. *J Epidem Comm Health* 43:346–351, 1989

12. Ahlborg G Jr: Pregnancy outcome among women working in laundries and dry cleaning shops using tetrachloroethylene. *Am J Ind Med* 17:567–575, 1990

13. Olsen J, Hemminki K, Ahlborg G, et al: Low birthweight, congenital malformations, and spontaneous abortions among dry cleaning workers in Scandinavia. *Scan J Work Environ Health* 16:163–168, 1990

14. National Cancer Institute: *Bioassay of Tetrachloroethylene for Possible Carcinogenicity*. DHEW (NIH) Pub No 77-813. Washington DC, US Government Printing Office, 1977

15. National Toxicology Program: *Toxicology and Carcinogenesis Studies of Tetrachloroethylene (Perchloroethylene) (CAS No 127-18-4) in F344/N Rats and B6C3F₁ Mice (Inhalation Studies)*. DHHS (NTP) TR-311, pp 1–197. Washington, DC, US Government Printing Office, August 1986

16. *IARC Monographs on the Evaluation of the Carcinogenic Risks of Chemicals to Humans*, Suppl 4, pp 355–356. Lyon, International Agency for Research on Cancer, March 1987

17. Brown DP, Kaplan SD: Retrospective cohort mortality study of dry cleaner workers using perchloroethylene. *J Occup Med* 29:535–541, 1987

18. Smith EM, Miller ER, Woolson RF, et al: Bladder cancer risk among laundry workers, dry cleaners, and others in chemically related occupations. *J Occup Med* 27:295–297, 1985

19. Aschengrau A, Ozonoff D, Paulu C, et al: Cancer risk and tetrachloroethylene-contaminated drinking water in Massachusetts. *Arch Environ Health* 48:284–292, 1993

20. NIOSH: *Current Intelligence Bulletins: Summaries*, p 7. US Dept of Health and Human Services, September 1988

21. Agency for Toxic Substances and Disease Registry (ASTDR): *Toxicological Profile for Tetrachloroethylene*. US Department of Health and Human Services, Public Health Service, p 151, TP-92/18, 1993

PERCHLOROMETHYL MERCAPTAN
CAS 594-42-3

CCl₃SCl

Synonyms: PCM; perchloromethanethiol; trichloromethanesulfenyl chloride.

Physical Form. Yellow liquid

Uses. In the production of fungicides; vulcanizing accelerator in the rubber industry

Exposure. Inhalation

Toxicology. Perchloromethyl mercaptan is a severe pulmonary irritant and lacrimating agent; fatal exposure has also caused liver and kidney injury.

Humans can withstand exposures to 70 mg/m³ (8.8 ppm); eye irritation begins at 10 mg/m³ (1.3 ppm).[1] Acute exposure to higher concentrations may cause coughing, dyspnea, lacrimation, nausea, cyanosis, convulsions, and death due to lung edema.[2]

Of 3 chemical workers who were observed following accidental exposures to perchloromethyl mercaptan, 2 survived episodes of pulmonary edema, and the third died after 36 hours.[1] The fatality resulted from a spill of the liquid on the clothing and floor, with exposure to the vapor. At autopsy, there was necrotizing tracheitis, massive hemorrhagic pulmonary edema, marked toxic nephrosis, and vacuolization of centrilobular hepatic cells.

The liquid splashed on the skin may be expected to cause irritation.

Mice and cats exposed for 15 minutes at 45 ppm died within 1 to 2 days from pulmonary edema; the LC_{50} for mice was 9 ppm for 3 hours; repeated exposures over 3 months at 1 ppm resulted in the death of some of the mice tested.[1,2] Rats exposed at 1 ppm, 6 hr/day, 5 days/week, for 2 weeks had labored breathing, tremors, and nasal irritation; pulmonary edema was evident at autopsy.[3] No effects were seen at 0.13 ppm for the same exposure period.

The 1995 ACGIH threshold limit value-time-weighted average (TLV-TWA) is 0.1 ppm (0.76 mg/m^3).

REFERENCES

1. Althoff H: Todliche Perchlormethylmercaptan-Intoxikation, (fatal perchloromethyl mercaptan intoxication). *Arch für Toxikol* 31:121–135, 1973
2. Perchloromethyl mercaptan. *Documentation of the TLVs and BEIs*, 6th ed, pp 1195–1196. Cincinnati, OH, American Conference of Governmental Industrial Hygienists, 1991
3. Knapp HF, MacAskill SM, Axicker GM, et al: Effects in rats of repeated inhalation exposure to perchloromethyl mercaptan. *Toxicologist* 7: 191, 1987 (Abstract 762).

PERCHLORYL FLUORIDE
CAS: 7616-94-6

ClO_3F

Synonyms: Chlorine fluoride oxide; chlorine oxyfluoride

Physical Form. Gas

Uses. In organic synthesis to introduce F atoms into organic molecules; oxidizing agent in rocket fuels; insulator for high-voltage systems

Exposure. Inhalation

Toxicology. Perchloryl fluoride is an irritant of mucous membranes; in animals, it causes methemoglobinemia and pulmonary edema.

One report states that workers suffered symptoms of upper respiratory irritation from brief exposure to unspecified concentrations.[1] There are no reports of methemoglobinemia in humans from exposure to perchloryl fluoride. However, severe exposure may be expected to cause the formation of methemoglobin and resultant anoxia with cyanosis (especially evident in the lips, nose, and earlobes), severe headache, weakness, and dizziness.[2] The liquid is stated to produce moderately severe burns with prolonged contact.[3]

Dogs exposed to 220 ppm for 4 hours or 620 ppm for 2.5 hours developed hyperpnea, cyanosis, incoordination, and convulsions; methemoglobin levels were 29% and 71%, respectively.[4] In dogs that died from exposure, there was lung damage consisting of alveolar collapse and hemorrhage; pigment deposition in the liver, spleen, and bone marrow was observed.[5]

Repeated exposure of 3 species of animals to 185 ppm for 7 weeks caused the death of more than half of them, guinea pigs being the most susceptible; all the animals developed dyspnea, cyanosis, methomoglobinemia, alveolar edema, and pneumonitis.[4] With repeated exposure of animals to 104 ppm for 6 weeks, guinea pigs again succumbed, and other signs and symptoms were similar to those observed at 185 ppm; the normal fluoride levels were increased by a factor of 20 to 30 in the blood and 5 to 8 in the urine.[4]

The 1995 time-weighted average threshold limit value (TWA-TLV) is 3 ppm (13 mg/m^3) with a short-term excursion level (STEL/ceiling) of 6 ppm (25 mg/m^3).

Note: For a description of diagnostic signs, differential diagnosis, and medical control, including clinical laboratory treatment of overexposure to methemoglobin-forming agents, see "Aniline."

REFERENCES

1. Perchloryl fluoride. *Documentation of the TLVs and BEIs*, 5th ed, p 466. Cincinnati, OH, American Conference of Governmental Industrial Hygienists, 1986
2. Mangelsdorff AF: Treatment of methemoglobinemia. *AMA Arch Ind Health* 14:148–153, 1956
3. Boysen JE: Health hazards of selected rocket propellants. *Arch Environ Health* 7:77–81, 1963
4. Greene EA, Colbourn JL, Donati E, Weeks MH: The inhalation toxicity of perchloryl fluoride. US Army Chemical Research and Development Laboratories, Technical Report CRDLR 3010, pp 1–12. Army Chemical Center, Baltimore, MD, 1960
5. Greene EA, Brough R, Kunkel A, Rinehart W: Toxicity of perchloryl fluoride: an interim report. US Army Chemical Warfare Laboratories, CWL Technical Memorandum 26-5, pp 1–6. Army Chemical Center, Baltimore, MD 1958

PHENOL

CAS: 108-95-2

C_6H_5OH

Synonyms: Carbolic acid; phenic acid; phenylic acid; phenyl hydroxide; hydroxybenzene; oxybenzene

Physical Form.   White crystalline solid

Uses.   In the manufacture of phenolic resins, bisphenol A, alkyl phenols, and caprolactam; in disinfectants and antiseptics

Exposure.   Skin absorption; inhalation; ingestion

Toxicology.   Phenol is an irritant of the eyes, mucous membranes, and skin; systemic absorption causes nervous system toxicity as well as liver and kidney damage.

Phenol is not considered a serious respiratory hazard in the industry, largely because of its low volatility.[1] The skin is a primary route of entry for the vapor, liquid, and solid. The vapor readily penetrates the skin with an absorption efficiency equal to that for inhalation. Skin absorption can occur at low vapor concentrations, apparently without discomfort. Signs and symptoms can develop rapidly with serious consequences, including shock, collapse, coma, convulsions, cyanosis, respiratory arrest, and death.

A worker who accidentally fell into a shallow vat containing 40% phenol for a few seconds subsequently collapsed and suffered 50% body surface burns; he developed nausea and vomiting and was diagnosed as suffering from acute renal tubular necrosis.[2] After a number of days, respiratory distress also developed. Kidney function improved after 6 weeks, but the patient still showed marginal polyuria 1 year later.

A laboratory technician repeatedly exposed to the vapor (unknown concentration) and to the liquid spilled on the skin developed anorexia, weight loss, weakness, muscle pain, and dark urine.[3] During several months of nonexposure, there was gradual improvement in his condition but, after brief re-exposure, he suffered an immediate worsening of symptoms, with prompt darkening of the urine and tender enlargement of the liver.

Brief intermittent industrial exposures to vapor concentrations of 48 ppm of phenol (accompanied by 8 ppm of formaldehyde) caused marked irritation of eyes, nose, and throat.[1] Workers at the same plant who were continuously exposed to an average concentration of 4 ppm experienced no respiratory irritation.

Ingestion of lethal amounts causes severe burns of the mouth and throat, marked abdominal pain, cyanosis, muscular weakness, collapse, coma, and death. Tremor, convulsions, and muscle twitching have also occurred.[1,4] The minimal lethal oral dose in humans has been estimated to be approximately 140 mg/kg.[5,6]

Concentrated phenol solutions are severely irritating to the human eye and cause conjunctival swelling; the cornea becomes white and hypoesthetic. Loss of vision has occurred in some cases.[7]

In addition to systemic effects, contact with the solid or liquid can produce chemical burns.[1] Erythema, edema, tissue necrosis, and gangrene have been reported, and prolonged contact with dilute solutions may result in deposition of dark pigment in the skin (ochronosis).[1] Following severe chemical burns, progressive areas of depigmentation may also develop.

Rats and mice given doses of up to 120 mg/kg and 280 mg/kg, respectively, by gavage on days 6 to 15 of gestation showed dose-related signs of fetotoxicity with no evidence of teratogenic effects.[4] A Russian study demonstrated increased preimplantation loss and early postnatal death in the offspring of rats exposed to 0.13 and 1.3 ppm throughout pregnancy.[4]

Mice were treated twice weekly for 42 weeks by application of one drop of a 10% solution of phenol in benzene to the shaved dorsal skin; after 52 weeks, there were papillomas in 5 of 14 mice, and a single fibrosarcoma appeared at 72 weeks.[8] Phenol, as a nonspecific irritant, may promote development of tumors when applied in large amounts repeatedly to the skin.[1]

Phenol was not considered carcinogenic to rats or mice receiving 2500 to 5000 ppm in drinking water for 103 weeks, although an increased incidence of leukemia and lymphomas was detected in the low-dose male rats.[9]

In a case-control study of workers in various wood industries, an increased risk of tumors of the mouth and respiratory tract was associated with phenol exposure; however, the small number of cases and confounding exposures were inadequately controlled.[10,11] The IARC has determined that there is inadequate evidence for carcinogenicity of phenol in humans and experimental animals, and it is not classifiable as to its carcinogenicity to humans.[11]

In Bakelite factory workers, the urinary level of total phenol, free plus conjugated, was proportional to the air concentration of phenol up to 12.5 mg/m^3 of workroom air.[12]

Although phenol is a major metabolite of the leukemogen benzene, it does not exhibit any potential for myeloclastogenicity in animal tests.[13]

Phenol is detectable by odor at a threshold of 0.05 ppm.[14]

The 1995 ACGIH threshold limit value-time-weighted average (TLV-TWA) is 5 ppm (19mg/m^3) with a notation for skin absorption.

REFERENCES

1. National Institute for Occupational Safety and Health: *Criteria for a Recommended Standard ... Occupational Exposure to Phenol.* DHEW (NIOSH) Pub No 76-1967, pp 23–69. Washington, DC, US Government Printing Office, 1976

2. Foxall PJD, Bending MR, Gartland KPR, et al: Acute renal failure following accidental cutaneous absorption of phenol: application of NMR urinalysis to monitor the disease process. *Human Toxicol* 9:491–496, 1989

3. Merliss RR: Phenol marasmus. *J Occup Med* 14:55–56, 1972

4. US Environmental Protection Agency (EPA): *Summary Review of the Health Effects Associated with Phenol: Health Issue Assessment*, p 37. Washington, DC, US Government Printing Office, January 1986

5. Bruce RM, Santodonato J, Neal MW: Summary review of the health effects associated with phenol. *Toxicol Ind Health* 3:535–568, 1987

6. Agency for Toxic Substances and Disease Registry (ASTDR): *Toxicological Profile for Phenol.* US Public Health Service, pp 1–111. 1989

7. Grant WM: *Toxicology of the Eye*, 3rd ed, pp 720–721. Springfield, IL, Charles C Thomas, 1986

8. Boutwell RK, Bosch DK: The tumor-promoting action of phenol and related compounds for mouse skin. *Cancer Res* 19:413–424, 1959

9. National Cancer Institute: *Bioassay of Phenol for Possible Carcinogenicity.* CAS No 108-95-2, NCI-CG-TR-203, NTP-80-15. DHHS (NIH) Pub No 80-1759, p 123. Washington, DC, US Government Printing Office, August 1980

10. Kauppinen TP, Partanen TJ, Nurminen MM, et al: Respiratory cancers and chemical exposures in the wood industry: a nested case-control study. *Br J Ind Med* 43:84–90, 1986

11. *IARC Monographs on the Evaluation of Carcinogenic Risks to Humans*, Vol 47, pp 263–287. Lyon, International Agency for Research on Cancer, 1989

12. Ohtsuji H, Ideda M: Quantitative relationship between atmospheric phenol vapor and phenol in the urine of workers in Bakelite factories. *Br J Ind Med* 29:70–73, 1972
13. Gad-ed-Karim MM et al: Benzene myeloclastogenicity: a function of its metabolism. *Am J Ind Med* 7:475–484, 1985
14. Leonardos G et al: Odor threshold determination of 53 odorant chemicals. *J Air Pollutant Control Assoc* 19:91–95, 1969

---

p-PHENYLENEDIAMINE
CAS: 106-50-3

$C_6H_4(NH_2)_2$

Synonyms: p-Diaminobenzene; 1,4-benzenediamine

Physical Form. Colorless crystalline solid; with exposure to air, it turns red, brown and, finally, black

Uses. For dyeing furs; in hair-dye formulations; in photographic developers; in antioxidants

Exposure. Inhalation; skin absorption

Toxicology. p-Phenylenediamine is a sensitizer of the skin and respiratory tract and may produce bronchial asthma.

Frequent inflammation of the pharynx and larynx has been reported in exposed workers.[1] Very small quantities of the dust have caused asthmatic attacks in workers after periods of exposure ranging from 3 months to 10 years. Sensitization dermatitis has been reported from its use in the fur-dyeing industry. In this process, oxidation products of p-phenylenediamine are generated that are also strong skin sensitizers. Many instances of inflammation and damage of periocular and ocular tissue have been reported from contact with hair dyes containing p-phenylenediamine, presumably in sensitized individuals.[2–3]

p-Phenylenediamine did not cause developmental or teratogenic effects in rats, even at doses that were severely maternally toxic.[4]

Although p-phenylenediamine has been tested for carcinogenicity in mice by skin application and in rats by oral and subcutaneous administration, the IARC has determined that these studies are not adequate to evaluate carcinogenicity.[5] p-Phenylenediamine dihydrochloride was not carcinogenic in 2-year feeding studies with mice and rats.[6]

The 1995 ACGIH threshold limit value–time-weighted average (TLV-TWA) for p-phenylenediamine is 0.1 mg/m^3. In an analysis of structure and corresponding carcinogenicity, phenyldiamines appeared to be least active when the amine groups were para to one another and gained activity as they became ortho.[7] Accordingly, ACGIH has classified o-phenylenediamine as an A2 suspected human carcinogen.

REFERENCES

1. Goldblatt MW: Research in industrial health in the chemical industry. *Br J Ind Med* 12:1–20, 1955
2. Grant WM: *Toxicology of the Eye*, 2nd ed, pp 696–698. Springfield, IL, Charles C Thomas, 1974
3. Baer RL et al: The most common contact allergens. *Arch Dermatol* 108:74–78, 1973
4. Re TA, Loehr RF, Rodwell DE, et al: The absence of teratogenic hazard potential of p-phenylenediamine in Sprague–Dawley rats. *Fund Appl Toxicol* 1:421–425, 1981
5. *IARC Monographs on the Evaluation of the Carcinogenic Risk of Chemicals to Man*, Vol. 16, pp 125–142. Lyon, International Agency for Research on Cancer, 1978
6. National Cancer Institute: *Bioassay of p-Phenylenediamine Dihydrochloride for Possible Carcinogenicity.* TR-174, DHEW (NIH) 9-1730, 1979.
7. Milman HA, Peterson C: Apparent correlation between structure and carcinogenicity of phenylenediamines and related compounds. *Environ Health Perspect* 56:261–273, 1984

2-PHENYLETHANOL
CAS: 60-12-8

$C_6H_5CH_2CH_2OH$

Synonyms: Benzyl carbinol; PEA; phenyl-ethyl alcohol

Physical Form. Colorless viscous liquid

Uses. Fragrance; antimicrobial agent; in organic synthesis; preservative and food additive

Exposure. Inhalation

Toxicology. Phenylethanol is an irritant of the eyes and a teratogen in rats.

An 8-hour exposure of rats to an essentially saturated atmosphere failed to cause any deaths.[1]

A solution containing 0.5% phenylethanol and 0.9% sodium chloride caused a smarting sensation in human test subjects when dropped in the eye.[2] Application of a 1% solution to rabbit eyes caused irritation of the conjunctiva and transient clouding of the corneal epithelium.[3]

The liquid on the skin of human test subjects was not irritating or sensitizing.[4]

Daily oral doses of 4.3, 43, or 432 mg/kg to rats on days 6 to 15 of gestation caused abnormalities in 50%, 93%, and 100% of the animals.[5] Major malformations, including micromelia, vertebral openings, and skull defects were observed at the highest dose, whereas only skeletal variations occurred at 4.3 mg/kg.

A TLV has not been established for 2-phenylethanol.

REFERENCES

1. Carpenter CP et al: Range-finding toxicity data: List VIII. *Toxicol Appl Pharmacol* 28:313–319, 1974
2. Barkman R, Germanis M, Karpe G, Malmborg AS: Preservatives in drops. *Acta Ophthal* 47:461, 1969
3. Nakano M: Effect of various antifungal preparations on the conjunctiva and cornea of rabbits. *Yakuzaigku* 18:94, 1958
4. Greif N: Cutaneous safety of fragrance material as measured by the maximization test. *Am Perfum Cosmet* 82:54, 1967
5. Mankes RF et al: Effects of various exposure levels of 2-phenylethanol on fetal development and survival in Long–Evans rats. *J Toxicol Environ Health* 12:235–244, 1983

PHENYL ETHER
CAS: 101-84-4

$(C_6H_5)_2O$

Synonyms: 1,1'-Oxybisbenzene; diphenyl ether; diphenyl oxide

Physical Form. Colorless liquid

Uses. Heat-transfer medium; in perfuming soaps; in organic syntheses

Exposure. Inhalation

Toxicology. Phenyl ether appears to be of relatively low toxicity.

There are no reported effects in humans, although complaints about the disagreeable odor may occur.

Twenty exposures at 10 ppm, lasting 7 hr/day, caused eye and nose irritation in rats and rabbits but not in dogs.[1] At 4.9 ppm, there were no signs of irritation or toxicity. The acute lethal oral dose for rats and guinea pigs is 4.0 g/kg.[2] Rats receiving 2.0 g/kg and guinea pigs receiving 1.0 g/kg had liver and kidney injury at autopsy.[2] On the rabbit skin, the undiluted liquid is irritating if exposures are prolonged or repeated.[2]

The low vapor pressure of phenyl ether and its easily detectable odor should prevent exposure to hazardous concentrations.[2]

The 1995 ACGIH threshold limit value–time-weighted average (TLV-TWA) is 1 ppm (7 mg/m³) with a STEL/ceiling of 2 ppm (14 mg/m³).

REFERENCES

1. Hefner RE Jr et al: Repeated inhalation toxicity of diphenyl oxide in experimental animals. *Toxicol Appl Pharmacol* 33:78–86, 1975
2. Kirwin CJ Jr, Sandmeyer EE: Ethers. In Clayton GD, Clayton FE (eds): *Patty's Industrial Hygiene and Toxicology*, 3rd ed, Vol 2A, *Toxicology*, pp 2541–2543. New York, Wiley-Interscience, 1981

PHENYL GLYCIDYL ETHER
CAS: 122-60-1

$C_6H_5OCH_2CHOCH_2$

Synonyms: PGE; Phenoxypropenoxide; 2,3-epoxypropyl phenyl ether

Physical Form. Colorless liquid

Uses. Chemical intermediate with high solvency for halogenated materials

Exposure. Inhalation

Toxicology. Phenyl glycidyl ether (PGE) is an irritant of mucous membranes and skin and causes sensitization; it has caused nasal tumors in experimental animals.

Of 20 workers exposed to PGE, 13 had acute skin changes, including second-degree burns, vesicular rash, papules, and edema.[1] In another study of 15 workers with PGE-induced dermatitis, there was erythema with papules and vesicles.[2] Of these 15 workers, 8 reacted positively to patch tests.

During animal exposure studies, technicians experienced irritation of the eyes, nose, and respiratory tract.[1,2]

There are no reports describing systemic effects in humans, and the low vapor pressure should limit the risk of acute inhalation exposure.[2]

Intragastric LD_{50} values were 1.40 and 3.85 g/kg, respectively, for mice and rats.[1] The predominant effect was central nervous system depression, and death was due to paralysis of the respiratory muscles. Surviving animals exhibited a reversal of the depressant effect, with increased central nervous system activity manifested by hypersensitivity to sound, muscle twitching, and tremor.

No deaths were produced in mice exposed 4 hours, or rats exposed 8 hours, to saturated vapors at room temperature.

Rats exposed for 7 hr/day for 50 days to about 10 ppm showed no overt signs of toxicity and no deaths, although a few animals, when sacrificed, had mild pulmonary inflammation and nonspecific cellular changes in the liver.[1,2] Exposure to 5 and 12 ppm PGE for 30 hr/week for 13 weeks caused hair loss in rats attributed to direct irritation of the skin rather than to systemic toxicity.[3]

Chronic exposure of rats to 1 or 12 ppm, 6 hr/day, 5 days/week, for 2 years caused an increased incidence of rhinitis, squamous metaplasia and epidermal carcinomas of the nasal cavity.[4]

Exposure of pregnant rats to 11.5 ppm, 6 hr/day, on days 4 through 15 of gestation did not cause effects in mothers or their offspring.[2] Localized degeneration of the seminiferous tubules has been reported in some male rats exposed at this level.[2,5]

Direct application of PGE into rabbit eyes produced irritation ranging from mild to severe without permanent damage.[2]

PGE was mutagenic in *Salmonella typhimurium* assays.

The 1995 ACGIH threshold limit value–time-weighted average (TLV-TWA) for PGE is 0.1 ppm (0.6 mg/m³) with an A3 animal carcinogen designation.

REFERENCES

1. Hine CH, Kodama JK, Wellington JS, et al: The toxicology of glycidol and some glycidyl ethers. *AMA Arch Ind Health* 14:250–264, 1956
2. National Institute for Occupational Safety and Health: *Criteria for a Recommended Standard . . . Occupational Exposure to Glycidyl Ethers.* DHEW (NIOSH) Pub No 78-166, pp 1–196. Wash-

ington, DC, US Government Printing Office, 1978

3. Terrill JB, Lee KP: The inhalation toxicity of phenylglycidyl ether. I. 90-day inhalation study. *Toxicol Appl Pharmacol* 42:263–269, 1977
4. Lee KP, Schneider PW, Trochimowicz HJ: Morphological expression of glandular differentiation in the epidermoid nasal carcinomas induced by phenylglycidyl ether inhalation. *Am J Pathol* 111:140–148, 1983
5. Kodama JK, Guzman RJ, Dunlap NK, et al: Some effects of epoxy compounds on the blood. *Arch Environ Health* 2:50–61, 1961

PHENYLHYDRAZINE

CAS: 100-63-0

$C_6H_5N_2$

Synonyms: Hydrazinobenzine

Physical Form. Pale yellow crystal or oily liquid; becomes reddish-brown when exposed to air and light

Uses. Chemical intermediate; in the manufacture of dyes

Exposure. Inhalation; skin absorption

Toxicology. Phenylhydrazine causes hemolytic anemia and is a skin sensitizer; in animals, it has caused liver and kidney injury secondary to hemolytic anemia, and it is carcinogenic.

Historically, phenylhydrazine hydrochloride was used to induce hemolysis in the treatment of polycythemia vera (a disease marked by abnormally high erythrocyte counts).[1] Oral doses totaling 3 to 4 g were administered; in a few cases, thrombosis occurred during excessive hemolysis, but it apparently was not caused by phenylhydrazine hydrochloride alone.[1] Several mild cases of hemolytic anemia from occupational exposure have been reported.[2] Symptoms of intoxication have included fatigue, headache, dizziness, and vertigo.[3]

Phenylhydrazine is a potent skin sensitizer that causes eczematous dermatitis with swelling and vesiculation in a high proportion of individuals who have had repeated skin contact.[1,3] Based on results with other hydrazines, it is expected that phenylhydrazine could also be absorbed through the skin.[1]

The minimal lethal dose in mice by subcutaneous injection was 180 mg/kg; animals developed progressive cyanosis and dyspnea before death; at autopsy, there were degenerative lesions in the liver, kidneys, and other organs, with evidence of vascular damage.[4]

Hemoglobin concentration, hematocrit value, and erythrocyte count were significantly reduced in dogs receiving 20 mg/kg subcutaneously for 2 consecutive days.[5] At necropsy on day 5, the internal organs were dark brown and the spleen, liver, and kidneys were severely congested. Large amounts of blood pigments were found in these organs, and the spleen was 3 to 5 times the normal size.

A 1 mg dose of phenylhydrazine hydrochloride administered daily by gavage for 200 days to mice caused adenomas and adenocarcinomas of the lung in 53% of the animals, compared to 13% in the control group.[1] Consumption of 0.6 to 0.8 mg/day in drinking water for life resulted in an increased incidence of blood vessel tumors.[1] Although other studies have reported negative carcinogenicity results, NIOSH recommends that phenylhydrazine be regarded as a carcinogen.[1] The suspected carcinogenicity has been linked to its irritant effect.[6] The mechanism of action is postulated to be in direct alkylation of DNA, which is also closely connected to the toxic endpoint.

Rats injected intraperitoneally (10 or 20 mg/kg) during pregnancy had offspring with severe jaundice, anemia, and reduced performance in certain areas of learning.[1]

Phenylhydrazine has a faint aromatic odor that does not serve as an adequate warning property.[1]

The 1995 ACGIH threshold limit value–time-weighted average (TLV-TWA) is 0.1 ppm (0.44 mg/m³) with a notation for skin and an A2 suspected human carcinogen designation.

REFERENCES

1. National Institute for Occupational Safety and Health: *Criteria for a Recommended Standard . . . Occupational Exposure to Hydrazines,* DHEW (NIOSH) Pub No 78-172, p 279. Washington, DC, US Government Printing Office, 1978
2. Schuckmann Von F: Beobachtungen zur Frage verschiedener Formen der Phenylhydrazine Intoxikation (Observations on the question of different forms of phenylhydrazine poisoning). *Zbl Arbeitsmed* 11:338–341, 1969
3. von Oettingen WF: *The Aromatic Amino and Nitro Compounds, Their Toxicity and Potential Dangers.* US Public Health Service Pub No 271, pp 158–164. Washington, DC, US Government Printing Office, 1941
4. von Oettingen WF, Deichmann-Gruebler W: On the relation between the chemical constitution and pharmacological action of phenylhydrazine derivatives. *J Ind Hyg Toxicol* 18:1–16, 1936
5. Witchett CE: Exposure of dog erythrocytes in vivo to phenylhydrazine and monomethylhydrazine—a freeze-etch study of erythrocyte damage, p 33. Springfield, VA, US Department of Commerce, NTIS, 1975
6. Steinhoff D, Mohr U: The question of carcinogenic effects of hydrazine. *Exp Pathol* 33:133, 1988

PHENYL MERCAPTAN
CAS: 108-98-5

C_6H_6S

Synonyms: Benzenethiol; mercaptobenzene; thiophenol

Physical Form. Colorless liquid with a penetrating garliclike odor

Uses/Sources. As an intermediate in pesticide manufacture and in solvent formulations for the removal of polysulfide sealants

Exposure. Inhalation; skin absorption

Toxicology. Phenyl mercaptan is a central nervous system stimulant; the liquid is a severe eye and skin irritant.

In humans, phenyl mercaptan may cause headaches.[1] By analogy with effects in animals, it is expected that more severe exposures would cause central nervous system effects and other systemic injury.

In rats, the 4-hour inhalation LC_{50} was 33 ppm; it was 28 ppm for the mouse.[2] The oral LD_{50} for rats was 46 mg/kg, whereas the dermal LD_{50} was 300 mg/kg. In rabbits, the dermal LD_{50} was estimated to be 134 mg/kg. Effects by all exposure routes were consistent with central nervous system stimulation and included restlessness, increased respiration, muscular weakness, paralysis of the hind limbs, and cyanosis, followed by coma and death. Subacute inhalation in mice caused some kidney damage, necrosis of the liver, and occasional hemorrhages in the lungs. Effects were less severe in rats.

In rabbits, phenyl mercaptan, on direct contact, caused severe erythema and eye injury, with edema of the ocular conjunctiva and discharge, but with clearing in 16 days and complete reversal in 2 months.

An odor threshold of 0.00094 ppm has been reported.[3]

The 1995 ACGIH threshold limit value–time-weighted average (TLV-TWA) for phenyl mercaptan is 0.5 ppm (2.3 mg/m³).

REFERENCES

1. Sandmeyer EE: Organic sulfur compounds. In Clayton GD, Clayton FE (eds): *Patty's Industrial Hygiene and Toxicology,* 3rd ed, rev, Vol 2A, *Toxicology,* p 2080. New York, John Wiley & Sons, Inc, 1981
2. Fairchild EJ, Stokinger HE: Toxicological studies on organic sulfur compounds. I. Acute toxicity of some aliphatic and aromatic thiols (mercaptans). *Am Ind Hyg Asoc J* 19:171–189, 1958
3. Amoore JE, Hautala E: Odor as an aid to chemical safety: odor thresholds compared with threshold limit values and volatilities for 214 industrial chemicals in air and water dilution. *J Appl Toxicol* 3:272–290, 1983

N-PHENYL-β-NAPHTHYLAMINE
CAS: 135-88-6

$C_{10}H_7NHC_6H_5$

Synonyms: Anilinonaphthalene; 2-naphthyl-phenylamine; 2-phenylaminonaphthalene; N-phenyl-2-naphthylamine; PBNA

Physical Form. Gray to tan flakes or powder

Uses. Formerly, an antioxidant in rubber processing to impart heat, oxidation, and flex-cracking resistance in natural rubber, synthetic rubbers, and latexes; stabilizer in electrical-insulating silicone enamels

Exposure. Inhalation

Toxicology. N-Phenyl-β-naphthylamine (PBNA) is carcinogenic to experimental animals in some studies.

Leukoplakia, acne, and hypersensitivity to sunlight were observed in 36 workers exposed for prolonged periods to PBNA.[1] Dose levels or possible concurrent exposures to other substances were not reported.

The LD_{50} values are 8730 mg/kg for rats and 1450 mg/kg for mice. Acute vascular changes in the liver, lung, and brain as a result of venous congestion were observed. Daily inhalation by rats of 900 mg/m³ for 14 days caused weight loss, slight erythrocytopenia, and pulmonary emphysema.[1] Intragastric administration of 100 mg/kg/day to rats caused an increase in urinary protein after 1 month, whereas urinary hippuric acid and adrenal ascorbic acid decreased after 6 months, and a drop in urinary function occurred after 18 months. Lung and liver weights increased within 1 and 12 months, respectively, and changes were observed in the gastrointestinal tract after 6 months.

Reduced body weight, arched backs, rough coats, and diarrhea were observed in male rats receiving 12,500 ppm in the diet for 2 weeks and in females receiving 25,000 ppm; 50,000 ppm was associated with increased mortality for both sexes.[2] In 13-week studies, increased

mortality, lower weight gain, liver enlargement, and kidney lesions were found in rats and mice receiving up to 40,000 ppm. Other effects in rats included hematopoietic hypoplasia or atrophy of the femoral bone marrow, testicular hypospermatogenesis, lymphoid degeneration of the thymus, and lymphoid depletion of the spleen.

PBNA has been tested for carcinogenicity in a number of species without conclusive results. There was no evidence of carcinogenic activity in male or female rats or in male mice fed 2500 or 5000 ppm in the diet for 2 years.[2] The lack of carcinogenicity in rats may be related to an inability to metabolize PBNA to the known animal and human urinary bladder carcinogen β-naphthylamine. There was equivocal evidence for carcinogenicity in female mice, as indicated by the occurrence of two rare kidney tumors. Chemical-related nonneoplastic lesions, including nephropathy, karyomegaly, and hyperplasia, occurred in the kidneys of both species.

In a limited dog study, no bladder tumors were observed in three animals fed 540 mg, 5 days/week, for a period of 4.5 years.[3]

An increased incidence of carcinogenicity has been observed in other studies. In one strain of male mice given 464 mg/kg/day by stomach tube for 3 weeks, followed by a diet of 1206 mg/kg for 78 weeks, there was an increased frequency of tumor-bearing animals (7/17 vs. 0/16 for controls), with the increase due primarily to hepatomas (5/17 vs. 0/16 for controls).[4] A single subcutaneous injection of 464 mg/kg PBNA to one strain of female mice on the 28th day of life increased the total tumor incidence (5/18 vs. 9/154 for controls).[4] Repeated subcutaneous injection of 16 mg, 3 times/week for 9 weeks, caused an increased incidence of carcinomas of the lungs in treated mice (4/19 vs. 0/18 for controls).[5]

No excess of bladder tumors was found in men with known exposures to PBNA at a rubber tire factory.[6] During 1946 to 1970, there were 9 cases of bladder cancer among 4177 men versus 10.0 expected; of these 4177 workers, 3301 had known exposures to PBNA. These results contrast with those involving exposures before 1949, when workers at the factory also were exposed to β-naphthylamine (a known carcinogen); 23 bladder tumors were observed versus 10.3 expected, between 1946 and 1970, among 2081 men exposed to material containing β-naphthylamine.[1]

An increased risk of death from bladder cancer (33 vs. 22.7 expected) was reported in 40,000 rubber and cable workers who had mixed exposures to many rubber additives, including PBNA, but not to known carcinogens.[6]

Additional concern has been afforded PBNA because commercial PBNA contains 20 to 30 ppm of β-naphthylamine.[7] Furthermore, experimental evidence from human volunteers ingesting PBNA and workers inhaling PBNA dust indicates that β-naphthylamine is a metabolite of PBNA. Specifically, 3 to 4 μg β-naphthylamine was found in 24-hour samples of urine obtained from two volunteers who ingested 50 mg PBNA containing 0.7 μg β-naphthylamine.[8]

The IARC has concluded that there is limited evidence for carcinogenicity to animals and inadequate evidence for humans.[9] The ACGIH considers PBNA to be a suspected human carcinogen because β-naphthylamine is both an impurity and a human metabolite of PNBA.[7]

The threshold limit value (TLV) has no assigned numerical value.

REFERENCES

1. *IARC Monographs on the Evaluation of Carcinogenic Risk of Chemicals to Man*, Vol 16, Some aromatic amines and related nitro compounds, pp 325–341. Lyon, International Agency for Research on Cancer, 1978
2. National Toxicology Program: *NTP Technical Report on the Toxicology and Carcinogenesis Studies of N-phenyl-2-naphthylamine (CAS No 135-88-6) in F344/N Rats and B6C3F₁ Mice (Feed Studies)*, NTP TR 333, NIH Pub No 88-2589, pp 1–61. Research Triangle Park, NC, US Department of Health and Human Services, 1988
3. Gehrmann GH, Foulger JH, Fleming AJ: Occupational tumors of the bladder. In *Proceedings of the 9th International Congress of Industrial Medicine*, pp 472–475, London, Bristol Wright, 1949
4. Innes JRM, Ulland BM, Valerio MG, et al: Bioassay of pesticides and industrial chemicals for tumorigenicity in mice: a preliminary note. *J Natl Cancer Inst* 42:1101–1114, 1969
5. Wang H, Wang D, Dzeng R: Carcinogenicity of N-phenyl-1-naphthylamine and N-phenyl-2-naphthylamine in mice. *Cancer Res* 44:3098–3100, 1984
6. Fox AJ, Collier PF: A survey of occupational cancer in the rubber and cablemaking industries: analysis of deaths occurring in 1972–1974. *Br J Ind Med* 33:249–264, 1976
7. N-Phenyl-beta-naphthylamine. *Documentation of the TLVs and BEIs*, 6th ed, pp 1211–1213. Cincinnati, OH, American Conference of Governmental Industrial Hygienists (ACGIH), 1991
8. Moore RM Jr, Wolf BS, Stein HP, et al: Metabolic precursors of a known human carcinogen. *Science* 195:344, 1977
9. *IARC Monographs on the Evaluation of Carcinogenic Risks of Chemicals to Humans, Overall Evaluations of Carcinogenicity: An Updating of IARC Monographs Vols 1–42*, Suppl 7, pp 318–319. Lyon, International Agency for Research on Cancer, 1978

PHENYLPHOSPHINE
CAS: 638-21-1

$C_6H_5PH_2$

Synonym: PF

Physical Form. Colorless liquid with an objectionable odor

Uses/Sources. Exposure to phenylphosphine may occur when phenylphosphinates (used as catalysts and antioxidants) are heated above 200°C, yielding phenylphosphonic acid derivatives and phenylphosphine

Exposure. Inhalation

Toxicology. Phenylphosphine is a respiratory and skin irritant; multiple exposures in rodents cause hematologic changes and testicular degeneration in males.

Three volunteers exposed to 0.57 ppm reported an obnoxious odor after one shallow breath.[1]

The 4-hour LC_{50} in rats was 38 ppm.[1] Exposure caused clinical signs typical of respiratory irritation, including red ears, salivation, lacrimation, face pawing, and dyspnea. Exposure at 7.6 ppm, 4 hr/day, for 10 days caused transient dermatitis around the mouth and feet on conclusion of the exposures, in addition to signs of respiratory irritation. Weight loss was noted during the exposure period, but weight gain rate returned to normal during the recovery period. Histopathologic examination showed foci of red blood cell formation in the spleen that were still evident at the end of the recovery period.

In 90-day inhalation studies (6 hr/day, 5 days/week), rats became hypersensitive to sound and touch and had mild hyperemia at 0.6 ppm; at 2.2 ppm, there was greater increase in splenic red blood cell formation, mild hemolytic anemia, and dermatitis.[2] Severe testicular degeneration developed in the 2.2 ppm–exposed rats, which did not return to normal during a 10-week postexposure observation period. Dogs similarly exposed at 0.6 and 2.2 ppm had some loss of appetite, diarrhea, lacrimation, and hind-leg tremor. There was mild, reversible testicular degeneration in the males exposed at 2.2 ppm.

The 1995 ACGIH threshold limit value–ceiling (TLV-C) for phenylphosphine is 0.05 ppm (0.23 mg/m³).

REFERENCES

1. Waritz RS, Brown RM: Acute and subacute inhalation toxicities of phosphine, phenylphosphine and triphenylphosphine. *Am Ind Hyg Assc J* 36:452–458, 1975
2. Phenylphoshine. *Documentation of the TLVs and BEIs*, 6th ed, pp 1241–1242. Cincinnati, OH, American Conference of Governmental Industrial Hygenists (ACGIH), 1992

PHOSGENE
CAS: 75-44-5

Cl_2CO

Synonyms: Carbonyl chloride; carbon oxychloride

Physical Form. Gas at room temperature; yellowish liquid when compressed or refrigerated

Uses/Sources. Intermediate in organic synthesis, especially production of toluene diisocyanate and polymethylene polyphenylisocyanate; in metallurgy, to separate ores by chlorination of the oxides and volatilization; occurs as a product of combustion whenever a volatile chlorine compound comes in contact with a flame or very hot metal

Exposure. Inhalation

Toxicology. Phosgene gas is a severe respiratory irritant.

Immediate irritation of the human throat occurs at 3 ppm; 4 ppm causes immediate irritation of the eyes; 4.8 ppm causes cough; exposure above 50 ppm may be rapidly fatal.[1]

The LC_{50} in man is approximately 500 ppm/min.[2] Prolonged exposure to low concentrations (eg, 3 ppm for 170 minutes) is equally as fatal as acute exposure to higher concentrations (eg, 30 ppm for 17 minutes). Exposure to lower concentrations, however, may not lead to noteworthy initial symptoms, whereas higher concentrations cause heavy lacrimation, coughing, nausea, and dyspnea.[2]

The onset of severe respiratory distress may be delayed for up to 72 hours, the latent interval depending on the concentration and duration of exposure.[3] The delayed onset of pulmonary edema is characterized by cough, abundant quantities of foamy sputum, progressive dyspnea, and severe cyanosis. Pulmonary edema may progress to pneumonia, and cardiac failure may intervene. During the clinical latent period, phosgene reaches the terminal spaces of the lungs where hydrolysis occurs, yielding hydrochloric acid. Although hydrochloric acid may cause some of the toxic effects of phosgene, acylation of macromolecules may be the initiating event in phosgene toxicity. Membrane function breaks down, fluid leaks from the capillaries into the interstitial space, and gradually increasing pulmonary edema ensues.[2] In time, air spaces are diminished and the blood is thickened, leading to insufficient oxygen.[3,4] Death is due to asphyxiation or heart failure.[3,4]

Survivors of phosgene-induced pulmonary edema may expect a long recovery period.[5] Exertional dyspnea and reduced physical fitness may be apparent for several months to years after exposure. In exposures involving persons with pre-existing lung damage (eg, chronic bronchitis), there may be severe and progressive deterioration of lung function after toxic pulmonary edema due to phosgene, with no complete recovery.

Mortality experience among men occupationally exposed to phosgene in the years 1943 through 1945 was evaluated 30 years post-exposure.[6] No excess overall mortality, or mortality from diseases of the respiratory tract, was found in a group of chemical workers chronically exposed to levels with daily excursions above 1 ppm. Another group of this cohort, 106 workers acutely exposed at some time to a concentration probably greater than 50 ppm, included 1 death from pulmonary edema, which occurred within 24 hours of exposure, and 3 deaths versus 1.37 expected due to respiratory disease. No evidence of increased lung cancer mortality was found, but the small sample size was noted.

No chronic lung problems were found in 326 workers exposed to concentrations ranging from nondetectable to greater than 0.13 ppm.[4]

Of animals exposed to 0.2 ppm for 5 hours per day for 5 consecutive days, 41% developed pulmonary edema.[7] At 1 ppm, lung lesions that would be likely to cause serious clinical symptoms in humans were observed.[7] Splashes of liquefied phosgene in the eye may produce severe irritation.[3] Skin contact with the liquefied material may cause severe burns.[3]

The irritant properties of phosgene are not sufficient to give warning of hazardous concentrations. A trained observer can recognize 0.5 ppm as being "sweet" and, at about 1 ppm, the odor becomes typical of the musty or "new-mown hay" smell usually ascribed to phosgene. Workers exposed to phosgene can lose their ability to detect low concentrations through olfactory fatigue.

The 1995 ACGIH threshold limit value–time-weighted average (TLV-TWA) for phosgene is 0.1 ppm (0.40 mg/m³).

Diagnosis. Signs and symptoms include eye irritation; dryness or burning sensation of the throat; vomiting; cough, foamy sputum, dyspnea, pain in the chest, and cyanosis; and severe skin or eye burns from splashes of liquefied material.

Special Tests: Diagnostic studies should include an electrocardiogram, a sputum gram stain and culture, a differential white blood count, and an arterial blood gas analysis.

Treatment. Since the extent of phosgene exposure is usually not known, any exposed person should be treated as if the exposure is life-threatening.[2] Pulmonary edema may be forecast by X ray well before clinical symptoms appear. If intensive therapy is delayed, rapidly developing edema may be fatal. There is no specific therapy for phosgene-induced pulmonary injury. Glucocorticoids and positive pressure ventilation have been recommended, as well as supportive measures such as physical rest, antitussives, buffers, sedatives, antibiotics, antispasmodics and, possibly, diuretics.[8] Although methenamine (hexamethylene tetramin, Urotropin) has been recommended as a specific antidote, some reports suggest that it is effective only when administered prophylactically.[9] In animal studies, aminophylline and terbutaline have been effective on a postexposure basis.[10]

REFERENCES

1. Cucinell SA: Review of the toxicity of long-term phosgene exposure. *Arch Environ Health* 28:272–275, 1974
2. Diller WF: Medical phosgene problems and their possible solution. *J Occup Med* 20:189–193, 1978

3. Hygienic Guide Series: Phosgene. *Am Ind Hyg Assoc J* 29:308–311, 1968

4. National Institute for Occupational Safety and Health: *Criteria for a Recommended Standard ... Occupational Exposure to Phosgene.* DHEW (NIOSH) Pub No 76-137, pp 43, 55. Washington, DC, US Government Printing Office, 1976

5. Diller WF: Late sequelae after phosgene poisoning: a literature review. *Toxicol Ind Health* 1:129–136, 1985

6. Polednak AP: Mortality among men occupationally exposed to phosgene in 1943–1945. Environ Res 22:357–367, 1980

7. Cameron GR, Courtice FC, Foss GL, et al: First Report on Phosgene Poisoning: Part II. Ministry of Defense, UK Proton Report 2349, April 1942 (unclassified report)

8. Diller WF: Therapeutic strategy in phosgene poisoning. *Toxicol Ind Health* 1:93–99, 1985

9. Diller WF: The methenamine misunderstanding in the therapy of phosgene poisoning. *Arch Toxicol* 46:199–206, 1980

10. Kennedy TP, Michael JR, Hoidal JR, et al: Dibutyryl cAMP, aminophylline, and beta-adrenergic agonists protect against pulmonary edema caused by phosgene. *J Appl Physiol* 67:2542–2552, 1989

PHOSPHINE

CAS: 7803-51-2

PH_3

Synonyms: Hydrogen phosphide; phosphoretted hydrogen; phosphorus trihydride

Physical Form. Colorless gas

Uses. Insecticide used for fumigation; preparation of phosphonium halides; doping agent in semiconductor manufacture

Exposure. Inhalation

Toxicology. Phosphine is a severe pulmonary irritant.

Workers exposed intermittently to concentrations up to 35 ppm but averaging below 10 ppm complained of nausea, vomiting, diarrhea, tightness in the chest and cough, headache, and dizziness; no evidence of cumulative effects was noted.[1] Single severe exposures caused similar signs and symptoms, as well as excessive thirst, muscle pain, chills, a sensation of pressure in the chest, dyspnea; syncope, and stupor.[2] In a few cases of exposure, dizziness and staggering gait have also occurred.[1] From 1900 to 1958, 59 cases of phosphine poisoning were reported, with 26 deaths; the effect most frequently reported was marked pulmonary edema.[2] The acute lethal effects of phosphine are associated with its ability to inhibit electron transport and combine with heme iron in the presence of oxygen.[3]

Inhalation of phosphine released after fumigation using aluminum phosphide on a grain freighter resulted in acute illnesses among 29 of 31 crew members and 2 children, one of whom died.[4] Air concentrations measured 2 days after the onset of illness ranged from 0.5 ppm in some of the living quarters to 12 ppm at an air intake. The most common symptoms were headache, fatigue, nausea, vomiting, cough, and shortness of breath. Congestive heart failure with pulmonary edema and myocardial necrosis with inflammation were noted in the child who died. The other child had echocardiographic evidence of poor left ventricular function, an elevated MB (cardiac) isoenzyme fraction of creatine kinase, and an abnormal EKG, with resolution of abnormalities within 72 hours. No long-term clinical or laboratory abnormalities were observed in the survivors.

The long-term health sequelae of lower-level exposures have been examined. Fumigant applicators who were exposed to phosphine or to phosphine plus other pesticides 6 weeks to 3 months earlier had significantly increased stable chromosome rearrangement, primarily translocations in G-banded lymphocytes.[3] Less stable aberrations, including chromatid deletions and gaps, were significantly increased at the time of exposure but not at later time points. It is not known if these chromosome rearrangements have long-term biological significance.

Animals survived exposure to 5 ppm, 4 hr/day for 2 months, but 7 similar exposures at 10 ppm were fatal.[5]

Phosphine has a fishy or garliclike odor detectable at 2 ppm; the odor threshold does not provide sufficient warning of dangerous concentrations.

The 1995 ACGIH threshold limit value–time-weighted average (TLV-TWA) is 0.3 ppm (0.42 mg/m³) with a short-term excursion limit of 1 ppm (1.4 mg/m³).

REFERENCES

1. Jones AT, Jones RC, Longley EO: Environmental and clinical aspects of bulk wheat fumigation with aluminum phosphide. *Am Ind Hyg Assoc J* 25:376–379, 1964
2. Harger PN, Spolyar LW: Toxicity of phosphine, with a possible fatality from this poison. *AMA Arch Ind Health* 18:497–504, 1958
3. Garry VF, Griffith J, Danzl TJ, et al: Human genotoxicity: pesticide applicators and phosphine. *Science* 246:251–255, 1989
4. Wilson R et al: Acute phosphine poisoning aboard a grain freighter. *JAMA* 244:148–150, 1980
5. Hygienic Guide Series: Phosphine. *Am Ind Hyg Assoc J* 25:314–316, 1964

PHOSPHORIC ACID
CAS: 7664-38-2

H_3PO_4

Synonyms: Orthophosphoric acid; white phosphoric acid

Physical Form. Crystals or colorless liquid

Uses. In the manufacture of fertilizers, detergents, dental cements, pharmaceuticals, foods, and beverages; also used in electropolishing, engraving, sugar refining, and water-treatment industries

Exposure. Inhalation

Toxicology. Phosphoric acid mist is a mild irritant of the eyes, upper respiratory tract, and skin; the dust is especially irritating to skin in the presence of moisture.[1]

Unacclimated workers could not endure exposure to fumes of phosphorus pentoxide (the anhydride of phosphoric acid) at a concentration of 100 mg/m³; exposure to concentrations between 3.6 and 11.3 mg/m³ produced cough, whereas concentrations of 0.8 to 5.4 mg/m³ were noticeable but not uncomfortable.[2]

There is no evidence that phosphorus poisoning can result from contact with phosphoric acid.[3] The risk of pulmonary edema resulting from the inhalation of mist or spray is remote.[3] A subcohort of workers from 16 phosphate companies who were occupationally exposed to unspecified amounts of phosphoric acid had no significant increase in cause-specific mortality.[4]

Ingestion of concentrated solutions can produce nausea, vomiting, abdominal pain, hematemesis, bloody diarrhea, and burns of the mouth, esophagus, and stomach.[5] In one case, death occurred 19 days after ingestion as a result of recurrent internal hemorrhage; at autopsy, there was necrosis of the upper and lower digestive tract and of the pancreas.[2] In some cases, signs of obstruction and scarring may occur weeks to months after initial exposure.[5]

A dilute solution buffered to pH 2.5 caused a moderate brief stinging sensation but no injury when dropped in the human eye.[6] A 75% solution will cause severe skin burns.[1]

The 1995 ACGIH threshold limit value–time-weighted average (TLV-TWA) is 1 mg/m³, with a short term excursion limit of 3 mg/m³.

REFERENCES

1. Hygienic Guide Series: Phosphoric acid. *Am Ind Hyg Assoc Q* 18:175–176, 1957
2. Phosphoric acid. *Documentation of the TLVs and BEIs*, 6th ed, pp 1250–1251. Cincinnati, OH, American Conference of Governmental Industrial Hygienists (ACGIH), 1991
3. Chemical Safety Data Sheet SD-70, Phos-

phoric acid, pp 5–6, 12–13. Washington, DC, MCA, Inc, 1958

4. Checkoway H et al: Mortality among workers in the Florida phosphate industry. II. Cause-specific mortality relationships with work areas and exposures. *J Occup Med* 27:893–896, 1985
5. Von Burg R: Toxicology update: phosphoric acid/phosphates. *J Appl Toxicol* 12:301–303, 1992
6. Grant WM: *Toxicology of the Eye*, 3rd ed, pp 733–734. Springfield, IL, Charles C Thomas, 1986

PHOSPHORUS (Yellow)
CAS: 7723-14-0

P_4

Synonyms: Phosphorus (white)

Physical Form.   Yellowish or colorless transparent cystals that darken on exposure to light

Uses/Sources.   In the manufacture of rat poisons; for smoke screens; in gas analysis; in fireworks; the elemental material, which does not occur in nature, is a by-product in the production of phosphate fertilizer

Exposure.   Inhalation

Toxicology.   Yellow phosphorus fume is an irritant of the respiratory tract and eyes; the solid in contact with the skin produces deep thermal burns. Prolonged absorption of phosphorus causes necrosis of facial bones.

Yellow phosphorus burns spontaneously in air, and the vapor released is irritating to the respiratory tract. The early signs of systemic intoxication by phosphorus are abdominal pain, jaundice, and a garlic odor of the breath; prolonged intake may cause anemia, as well as cachexia and necrosis of bone, typically involving the maxilla and mandible (phossy jaw).[1-4] In chronic phosphorus intoxication, lowered potassium blood levels or increased chloride concentrations, along with leukopenia, have also been reported.[5]

The presenting complaints of overexposed workers may be toothache and excessive salivation. The oral mucosa may have a dull red appearance. One or more teeth may loosen, followed by pain and swelling of the jaw; healing may be delayed following dental procedures such as extractions; with necrosis of bone, a sequestrum may develop with sinus tract formation.[2] In a series of 10 cases, the shortest period of exposure to phosphorus fume (concentrations not measured) that lead to bone necrosis was 10 months (two cases), and the longest period of such exposure was 18 years.[2]

Although ingestion would not be expected in an occupational setting, the human lethal oral dose is about 1 mg/kg body weight.[5] Acute oral intoxication is characterized by an initial phase in which gastrointestinal effects such as nausea and vomiting predominate, followed by an apparent recovery period lasting up to 2 days which, in turn, is followed by the return of gastrointestinal symptoms plus signs of hepatic renal and cardiovascular involvement.

Yellow phosphorus fume causes severe eye irritation with blepharospasm, photophobia, and lacrimation; the solid produces severe injury in the eye.[6] Phosphorus burns on the skin are deep and painful; a firm eschar is produced and is surrounded by vesiculation.[7]

The 1995 ACGIH threshold limit value–time-weighted average (TLV-TWA) for yellow phosphorus is 0.02 ppm (0.1 mg/m^3).

REFERENCES

1. Rubitshy HJ, Myerson RM: Acute phosphorus poisoning. *Arch Int Med* 83:164–178, 1949
2. Hughes JPW et al: Phosphorus necrosis of the jaw: a present-day study. *Br J Ind Med* 19:83–99, 1962
3. Chemical Safety Data Sheet SD-16, Phosphorus, elemental, pp 3, 9–13. Washington, DC, MCA, Inc, 1947
4. Felton JS: Classical syndromes in occupational medicine: phosphorus necrosis—a classical occupational disease. *Am J Ind Med* 3:77–120, 1982
5. Beliles RP, Beliles EM: Phosphorus, selenium,

tellurium, and sulfur. In Clayton GD, Clayton FE (eds): *Patty's Industrial Hygiene and Toxicology*, 4th ed, Vol 2, Part A, *Toxicology*, pp 783–787. New York, Wiley-Interscience, 1993
6. Grant WM: *Toxicology of the Eye*, 3rd ed, pp 734–735. Springfield, IL, Charles C Thomas, 1986
7. Summerlin WT, Walder AI, Moncrief JA: White phosphorus burns and massive hemolysis. *J Trauma* 7:476–484, 1967

PHOSPHORUS OXYCHLORIDE
CAS: 10025-87-3

POCl₃

$POCl_3$

Synonyms: Phosphoryl chloride; phosphoryl trichloride

Physical Form. Colorless liquid

Uses/Sources. In the manufacture of pesticides, pharmaceuticals, plasticizers, gasoline additives, and hydraulic fluid

Exposure. Inhalation

Toxicology. Phosphorus oxychloride is strongly irritating to the skin, eyes, and respiratory tract.

Exposure to the vapors can cause cough, painful inflammation of the eyes, burning in the nose and throat, shortness of breath and, in severe exposures, death.[1] Both chronic and acute cases of occupational exposures have been recorded in the foreign literature.[2]

The 4-hour LC_{50} values for rats and guinea pigs were 48 and 53 ppm, respectively.[1] During exposure, the animals were restless and showed signs of irritation such as pawing and scratching the nose and head. Gasping and convulsions preceded death, which occurred within 48 hours of exposure. Signs of toxicity gradually abated in surviving animals and were not evident at the end of the 14-day observation period. Microscopic examination of tissues from

animals that died showed desquamation of the tracheal and bronchial epithelium, resulting in plugging of the lumen of the bronchioles. The alveolar spaces surrounding these lumen plugs became edematous and hemorrhagic. Surviving animals, when autopsied 14 days post-exposure, had no lesions attributable to phosphorus oxychloride.

The 1995 ACGIH threshold limit value–time-weighted average (TLV/TWA) is 0.1 ppm (0.63 mg/m³).

REFERENCES

1. Weeks MH, Musselman NP, Yevich PP, et al: Acute vapor toxicity of phosphorus oxychloride, phosphorus trichloride and methyl phosphonic dichloride. *Am Ind Hyg Assoc J* 25:470–475, 1964
2. Sassi C: Occupational poisoning: phosphorus-oxychloride. *Med Lav* 45:171–177, 1954

PHOSPHORUS PENTACHLORIDE
CAS: 10026-13-8

PCl₅

PCl_5

Synonyms: Phosphoric chloride; phosphorus perchloride

Physical Form. White to pale yellow, fuming crystalline mass with pungent, unpleasant odor

Uses. Catalyst in manufacture of acetylcellulose; chlorinating and dehydrating agent

Exposure. Inhalation

Toxicology. Phosphorus pentachloride fume is a severe irritant of the eyes and mucous membranes.

In humans, the fume causes irritation of the eyes and respiratory tract; cases of bronchitis have resulted from exposure.[1] Although not reported, delayed onset of pulmonary edema may occur. The material on the skin could be expected to cause dermatitis.

Exposure of mice to 120 ppm for 10 minutes was fatal.

The oral LD_{50} in rats is 660 mg/kg, and the inhalation LC_{50} for 4 hours is 205 mg/m³.[2]

Phosphorus pentachloride is expected to produce 67% more hydrogen chloride than an equimolar amount of phosphorus trichloride.[3] Accordingly, the ACGIH threshold limit value–time-weighted average (TLV-TWA) for phosphorus pentachloride is 0.1 ppm (0.85 mg/m³), half that of phosphorus trichloride.[3]

REFERENCES

1. Patty FA: As, P, Se, S, and Te. In Patty FA (eds): *Industrial Hygiene and Toxicology*, 2nd ed, Vol 2, *Toxicology*, p 885. New York, Wiley-Interscience, 1963
2. *Registry of Toxic Effects of Chemical Substances (RTECS)*, p 5477. Washington, DC, US Department of Health and Human Services, July 1986
3. Phosphorus pentachloride. *Documentation of the TLVs and BEIs*, 6th ed, pp 1257–1258. Cincinnati, OH, American Conference of Governmental Industrial Hygienists (ACGIH), 1991

Toxicology. Phosphorus pentasulfide is an irritant of the eyes, skin, and respiratory tract.

In the presence of moisture, phosphorus pentasulfide is readily hydrolyzed to hydrogen sulfide gas (qv) and phosphoric acid (qv).[1]

The primary effect of inhaled phosphorus pentasulfide is irritation of the respiratory tract; pulmonary irritation is expected at concentrations of 10 mg/m³.[2]

The oral LD_{50} in rats was 389 mg/kg; 500 mg applied to rabbit skin for 24 hours was moderately irritating, and 20 mg instilled in rabbit eyes for 24 hours was severely irritating.[3]

The 1995 ACGIH threshold limit value–time-weighted average (TLV-TWA) is 1 mg/m³, with a STEL/ceiling of 3 mg/m³.

REFERENCES

1. Phosphorus pentasulfide. *Documentation of the TLVs and BEIs*, 6th ed, pp 1259–1260. Cincinnati, OH, American Conference of Governmental Industrial Hygienists (ACGIH), 1991
2. Smyth HF Jr: Improved communication—hygienic standards for daily inhalation. *Am Ind Hyg Assoc J* 17:129–185, 1956
3. Lewis RJ Sr, Sweet DV (eds): *Registry of Toxic Effects of Chemical Substances (RTECS)*. DHHS (NIOSH) Pub No 84-101-6, p 5662. Washington, DC, US Department of Health and Human Services, July 1986

PHOSPHORUS PENTASULFIDE
CAS: 1314-80-3

P_2S_5

Synonyms: Phosphorus sulfide; phosphorus persulfide; thiophosphoric anhydride

Physical Form.   Light yellow crystals

Uses.   Intermediate in the manufacture of safety matches, ignition compounds, and lubricant additives

Exposure.   Inhalation

PHOSPHORUS TRICHLORIDE
CAS: 7719-12-2

PCl_3

Synonym: Phosphorus chloride

Physical Form.   Colorless, clear, fuming liquid

Uses.   Chlorinating agent; in manufacture of other phosphorus chloride compounds; in the manufacture of pesticides, surfactants, gasoline additives and dyestuffs

Exposure.   Inhalation

Toxicology.   Phosphorus trichloride vapor is a severe irritant of the eyes, mucous membranes, and skin.

The irritant effects of phosphorus trichloride result primarily from the action of the strong acids (hydrochloric acid and acids of phosphorus) formed on contact with water.[1]

Inhalation by humans could be expected to cause injury ranging from mild bronchial spasm to severe pulmonary edema; the onset of severe respiratory symptoms may be delayed for 2 to 6 hours and, after moderate exposure, the onset may not occur until 12 to 24 hours later.[1] Prolonged or repeated exposure to low concentrations may induce chronic cough and wheezing; pulmonary changes are nonfibrotic and nonprogressive.

Phosphorus trichloride causes severe burns in contact with the eyes, skin, or mucous membranes.[1] Although ingestion is unlikely to occur in industrial use, it will cause burns of the mouth, throat, esophagus, and stomach.[2]

Seventeen people exposed to phosphorus trichloride liquid and its hydration products following a tanker accident were evaluated.[3] Those closest to the spill experienced burning of the eyes, lacrimation, nausea, vomiting, dyspnea, and cough. Six patients had transient elevation of lactic dehydrogenase. Chest roentgenograms were normal. Pulmonary-function tests showed statistically significant decreases in vital capacity and FEV_1 in direct correlation with distance from the accident and duration of exposure. Of the 17 patients examined 1 month later, pulmonary-function tests showed improvement, suggesting that acute effects were due to phosphorus trichloride toxicity.[3]

In rats, the LC_{50} was 104 ppm for 4 hours; at autopsy, the chief finding was nephrosis; pulmonary damage was negligible.[2]

The 1995 ACGIH threshold limit value–time-weighted average (TLV-TWA) is 0.2 ppm (1.1 mg/m³) with a STEL/ceiling of 0.5 ppm (2.8 mg/m³).

REFERENCES

1. Chemical Safety Data Sheet SD-27, Phosphorus trichloride, pp 5, 15–19. Washington, DC, MCA, Inc, 1972
2. Weeks MH: Acute vapor toxicity of phosphorus oxychloride, phosphorus trichloride and methyl phosphonic dichloride. *Am Ind Hyg Assoc J* 25:470–475, 1964
3. Wason S et al: Phosphorus trichloride toxicity: preliminary report. *Am J Med* 77:1039–1042, 1984

PHTHALIC ANHYDRIDE
CAS: 85-44-9

$C_6H_4(CO)_2O$

Synonyms: Phthalic acid anhydride; phthalandione; 1,3-isobenzofurandione

Physical Form.   White crystalline solid

Uses.   In the production of plasticizers for vinyl, epoxy, and acetate resins; in alkyd resins; in the manufacture of dyes

Exposure.   Inhalation

Toxicology.   Phthalic anhydride is an irritant of the eyes, skin, and respiratory tract; it may also act as a sensitizer.

In workers, air concentrations of 30 mg/m³ (5 ppm) caused conjunctivitis; at 25 mg/m³ (4 ppm), there were signs of mucous membrane irritation.[1] Workers exposed to undetermined concentrations of mixed vapors of phthalic acid and phthalic anhydride developed, in addition to conjunctivitis, bloody nasal discharge, atrophy of the nasal mucosa, hoarseness, cough, occasional bloody sputum, bronchitis, and emphysema.[2] Several cases of bronchial asthma resulted; there was also skin sensitization, with occasional urticaria and eczematous response.

Phthalic anhydride is a direct but delayed

irritant of the skin; it is more severely irritating after contact with water because of the pronounced effects of the phthalic acid that is formed.[1] Prolonged or repeated exposure also may cause an allergic type of skin rash. Because phthalic anhydride is a known pulmonary and skin irritant, it is often difficult to differentiate between sensitization and irritation by clinical history.[3]

A group of 23 phthalic anhydride–exposed workers (air levels up to 17 mg/m³) had more work-related symptoms in their eyes and nose than 18 unexposed controls did.[4] The exposed workers also had significantly higher levels of total IgE than controls did, although values for specific IgE against phthalic anhydride did not differ. The investigators suggested that the irritant effect of phthalic anhydride on the mucous membranes facilitated the entry of other allergens, causing an increase in serum IgE levels. Lung-function tests did not reveal any significant impairment of large or small airways.

In another report, a worker who developed symptoms of rhinorrhea, lacrimation, and wheezing from exposure to phthalic anhydride over the period of a year was shown to have a positive patch test to the chemical and a high serum titer of specific IgE.[5]

A case of asthma was attributed to the release of phthalic anhydride during the grinding of cured moldings.[6] Unreacted phthalic anhydride may be trapped within cured resin and released during grinding or, alternatively, heat generated during grinding may lead to disruption of bonds between resin and hardener and cause release of phthalic anhydride vapor.[6]

In 2-year feeding studies, phthalic anhydride was not carcinogenic to rats or mice.[7]

The 1995 ACGIH threshold limit value–time-weighted average (TLV-TWA) is 1 ppm (6.1 mg/m³).

REFERENCES

1. Hygienic Guide Series: Phthalic anhydride. *Am Ind Hyg Assoc J* 28:395–398, 1967
2. Fassett DW: Organic acids and related compounds. In Patty FA (ed): *Industrial Hygiene and Toxicology*, 2nd ed, Vol 2, *Toxicology*, pp 1822–1823. New York, Wiley-Interscience, 1963
3. Flaherty DK, Gross CJ, Winzenburger P, et al: In vitro immunologic studies on a population of workers exposed to phthalic and tetrachlorophthalic anhydride. *JOM* 30:785–790, 1988
4. Nielsen J, Bensryd I, Almquist H, et al: Serum IgE and lung function in workers exposed to phthalic anhydride. *Intl Arch Occup Environ Health* 63:199–204, 1991
5. Maccia CA, Bernstein IL, Emmett EA, Brooks SM: In vitro demonstration of specific IgE in phthalic anhydride hypersensitivity. *Am Rev Respir Dis* 113:701–704, 1976
6. Ward MJ, Davies D: Asthma due to grinding epoxy resin cured with phthalic anhydride. *Clin Allerg* 12:165–168, 1982
7. National Cancer Institute: *Bioassay of Phthalic Anhydride for Possible Carcinogenicity*. NCI-CG-TR-159. DNEW (NIH) Pub No 79-1715. Washington, DC, US Government Printing Office, 1979

---

m-PHTHALODINITRILE
CAS: 626-17-5

$C_8H_4N_2$

Synonyms: 1,3-Benzenedicarbonitrile; 1,3-dicyanobenzene; isophthalonitrile

Physical Form.   Light-tan powder

Uses.   Intermediate in the manufacture of paints, varnishes, and agricultural chemicals

Exposure.   Inhalation

Toxicolgy.   *m*-Phthalodinitrile is a skin irritant in animals.

In humans, there have been no reports of adverse effects.

The oral LD_{50}s for rats, cats, and rabbits was 5000, 500, and 250 mg/kg, respectively.[1] Rats exposed to 190 or 1250 mg/m³, 6 hr/day for 2 weeks, had decreased food consumption and reduced body weight.[2] Alopecia was observed in the low-dose group, and rhinorrhea occurred in the high-dose animals. Pathological examination did not reveal any treatment-related effects. Slight skin reactions were observed following topical application to rabbits.

The 1995 ACGIH threshold limit value–time-weighted average (TLV-TWA) for *m*-phthalodinitrile is 5 mg/m³.

REFERENCES

1. Zeller H, Hofmann HTH, Thiess AM, et al: Toxicity of nitriles: results of experiments carried out on animals as well as occupational health experiences made during 15 years. *Zentralbl Arbeitsmed Arbeitschtz* 19:226–238, 1969
2. *m*-Phthalodinitrile. *Documentation of the TLVs and BEIs*, 6th ed, pp 1266–1267. Cincinnati, OH, American Conference of Governmental Industrial Hygienists (ACGIH), 1991

PICRIC ACID
CAS: 88-89-1

$C_6H_2(NO_2)_3OH$

Synonyms: 2,4,6-Trinitrophenol; carbazotic acid; picronitric acid

Physical Form. Yellow crystalline solid

Uses. High explosive; oxidant in rocket fuels; in the processing of leather; in metal etching

Exposure. Inhalation

Toxicology. Picric acid causes sensitization dermatitis; absorption of large amounts causes liver and kidney damage.

Dermatitis from skin contact with the chemical usually occurs on the face, especially around the mouth and the sides of the nose; the condition progresses from edema through the formation of papules and vesicles to ultimate desquamation.[1,2] The skin and hair of workers handling picric acid may be stained yellow.[1]

Inhalation of high concentrations of the dust by one worker caused temporary coma, followed by weakness, myalgia, anuria and, later, polyuria.[3] Following ingestion of 2 to 5 g of picric acid, which has a bitter taste, there may be headache, vertigo, nausea, vomiting, diarrhea, yellow coloration of the skin, hematuria, and albuminuria; high doses cause destruction of erythrocytes, hemorrhagic nephritis, and hepatitis.[3,4]

High doses, which cause systemic intoxication, will color all tissues yellow, including the conjunctiva and aqueous humor, and cause apparent yellow vision.[5] Corneal injury is stated to have resulted from a splash of a solution of picric acid in the eyes; dust or fume may cause eye irritation, which may be aggravated by sensitization.[5]

The LD_{50} for picric acid following oral dosing of male and female rats was 290 and 200 mg/kg, respectively.[6] Death was due to severe acidosis, with toxic doses of picric acid exceeding the buffering capacity of the blood. In rats, metabolism of picric acid is primarily limited to reduction of nitro groups of the aromatic ring and subsequent conjugation by acetate.

The 1995 ACGIH threshold limit value–time-weighted average (TLV-TWA) for picric acid is 0.1 mg/m³.

REFERENCES

1. Schwartz L: Dermatitis from explosives. *JAMA* 125:186–190, 1944
2. Sunderman FW, Weidman FD, Batson OV: Studies of the effects of ammonium picrate on man and certain experimental animals. *J Ind Hyg Toxicol* 27:241–248, 1945
3. von Oettingen WF: *The Halogenated Aliphatic, Olefinic, Cyclic, Aromatic, and Aliphatic-Aromatic Hydrocarbons Including the Halogenated Insecticides, Their Toxicity and Potential Dangers*, US Public Health Service Pub No 414, pp 150–154. Washington, DC, US Government Printing Office, 1941
4. Harris AH, Binkley OF, Chenoweth BM Jr: Hematuria due to picric acid poisoning at a

naval anchorage in Japan. *Am J Public Health*, 36:727–733, 1946
5. Grant WM: *Toxicology of the Eye*, 3rd ed. Springfield, IL, Charles C Thomas, 1986
6. Wyman JF, Serve MP, Hobson DW, et al: Acute toxicity, distribution, and metabolism of 2,4,6-trinitrophenol (picric acid) in Fischer 344 rats. *J Toxicol Environ Health* 37:313–327, 1992

PINDONE
CAS: 83-26-1

$C_{14}H_{14}O_3$

Synonyms: Pival; Pivalyl Valone; Tri-Ban; 2-pivaloyl-1,3-indanedione

Physical Form. Yellow powder

Uses. Rodenticide

Exposure. Inhalation; ingestion

Toxicology. Pindone is a vitamin K antagonist and causes inhibition of prothrombin formation, which results in hemorrhage.

There are no reports of effects in humans.

In rats, the ingestion of a single large dose of pindone causes rapid death from pulmonary and visceral congestion without hemorrhage and may not be related to vitamin K antagonism.[1] Death in animals from chronic exposure is due to multiple internal hemorrhage.

The 1995 ACGIH threshold limit value–time-weighted average (TLV-TWA) for pindone is 0.1 mg/m³.

REFERENCES

1. US Department of Health Service, US Department of Health, Education and Welfare: Operational memoranda on economic poisons, Public Health Service, pp 81–84. Atlanta, GA, Communicable Disease Center, 1956

PIPERAZINE DIHYDROCHLORIDE
CAS: 142-64-3

$C_4H_{12}N_2Cl_2$

Synonyms: Dihydrochloride salt of diethylene-diamine; piperazidine hydrochloride

Physical Form. White crystalline solid

Uses. In the manufacture of fibers, pharmaceuticals, and insecticides

Exposure. Inhalation

Toxicology. Piperazine dihydrochloride is an irritant and sensitizer.

Little information exists on the toxicology of piperazine dihydrochloride in humans or in animals. Acute human exposures to the dust have reportedly resulted in irritation to the eyes, mild to moderate skin burns, and sensitization.[1] Exposure levels and duration were not available. Occupational exposures have been associated with occasional cases of asthma.

The systemic toxicity appears to be low; the oral LD_{50} for rats was approximately 4.9 g/kg.

The 1995 ACGIH threshold limit value–time-weighted average (TLV-TWA) is 5 mg/m³.

REFERENCES

1. Piperazine dihydrochloride. *Documentaion of the TLVs and BEIs*, 6th ed, pp 1276–1277. Cincinnati, OH, American Conference of Governmental Industrial Hygienists (ACGIH), 1991

PLATINUM AND SOLUBLE SALTS
CAS: 7440-06-4

Pt

Compounds: Ammonium chloroplatinate; sodium chloroplatinate; platinic chloride; platinum chloride; sodium tertrachloroplatinate; potassium tetrachloroplatinate; ammonium tetrachloroplatinate; sodium hexachloroplatinate; potassium hexachloroplatinate; ammonium hexachloroplatinate

Physical Form. Crystalline solids

Uses. For jewelry; for oxygen sensors in internal combustion engines; in the chemical and electrical industries; in dentistry; for windings of high-temperature furnaces; in electroplating; in photography; chemotherapeutic anticancer agents

Exposure. Inhalation

Toxicology. Exposure to the complex salts of platinum, especially ammonium hexachloroplatinate and ammonium tetrachloroplatinate, but not elemental platinum, causes a progressive allergic reaction that may lead to pronounced asthmatic symptoms. Skin sensitization and eye irritation may also occur.

A syndrome characterized by runny nose, sneezing, tightness in the chest, shortness of breath, cyanosis, wheezing, and cough has been described following exposure to soluble complex platinum salts and is referred to variously as platinum allergy, platinum asthma, and platinosis.[1,2] Of 91 men employed in 4 platinum refineries and exposed to the dust or spray of the complex platinum salts, 52 experienced these symptoms.[1] The severity of response was greatest in workers crushing platinum salts, where airborne levels reached 1.7 mg/m³. Thirteen of the men also complained of dermatitis. Contact dermatitis has also been said to occur from exposure to platinum oxides and chlorides and, occasionally, from platinum itself.[3] Removal from platinum salt exposure results in almost immediate relief of asthma; the dermatitis usually clears in 1 to 2 days but may persist.[3]

Smokers have been found to be at increased risk of sensitization by platinum salts.[4] A historical perspective cohort study of 91 platinum refinery workers showed a four- to fivefold risk of developing a positive skin test to platinum salts in smokers. The risk of smokers developing symptoms was approximately twofold and, among recent employees, the rate of development of a positive skin test result was faster in smokers versus nonsmokers. Smoking is thought to act by increasing the serum levels of IgE.

The assumption that platinosis is due to an allergic response rather than to toxic or irritant effects is suggested by the following: (1) sensitivity appears following previous exposure that had no apparent effect; (2) only a fraction of exposed persons exhibit a response; and (3) affected subjects show increasingly high degrees of sensitivity to small amounts.[5] Of 306 platinum refinery workers exposed to unspecified levels, 38 had a positive skin-prick test to the platinum halide salts.[6]

The potent allergenicity of the divalent and tetravalent platinum compounds is thought to occur by conjugation with sulfhydryl-containing groups within proteins, thus forming immunogenic complexes.[6]

Soluble complex platinum salts have been used as anticancer agents, and atopic hypersensitivity has been provoked following repeated injections.[7] In the eyes, the dusts cause a burning sensation, lacrimation, and conjunctival hyperemia, sometimes associated with photophobia.[8]

The 1995 ACGIH threshold limit value-time-weighted average (TLV-TWA) is 1.0 mg/m³ for the metal dust and 0.002 mg/m³ for the soluble salts as Pt.

REFERENCES

1. Hunter D, Milton R, Perry KMA: Asthma caused by the complex salts of platinum. *Br J Ind Med* 2:92–98, 1945
2. Parrot JL, Herbert R, Saindelle A, Ruff F: Platinum and platinosis, allergy and histamine re-

lease due to some platinum salts. *Arch Environ Health* 19:685–691, 1969

3. Beliles RP: The metals. In Clayton GD, Clayton FE (eds): *Patty's Industrial Hygiene and Toxicology*, 4th ed, *Toxicology*, pp 2190–2194. New York, Wiley-Interscience, 1991
4. Venables KM, Dally MB, Nunn AF, et al: Smoking and occupational allergy in workers in a platinum refinery. *Br Med J* 299:939–942, 1989
5. National Research Council: Platinum group metals. *Medical and Biologic Effects of Environmental Pollutants*, p 232. Washington, DC, National Academy of Sciences, 1977
6. Murdoch RD et al: IgE antibody responses to platinum group metals: a large scale refinery survey. *Br J Ind Med* 43:37–43, 1986
7. Orbaek P: Allergy to the complex salts of platinum: a review of the literature and three case reports. *Scand J Work Environ Health* 8:141–145, 1982
8. Grant WM: *Toxicology of the Eye*, 3rd ed, p 748. Springfield, IL, Charles C Thomas, 1986

POLYBROMINATED BIPHENYLS

$C_{12}H_{(10-m-n)}Br_{(m+n)}$

Hexabromobiphenyl
Technical grades:
FireMaster BP-6 (CAS: 59536-65-1)
FireMaster FF-1 (CAS: 67774-32-7)

Octabromobiphenyl
Technical grade: Bromkal 80
(CAS: 61288-13-9)

Decabromobiphenyl
Technical grade: Flammex B 10
(CAS: 13654-09-6)

Physical Form. Solids

Uses. Polybrominated biphenyls (PBBs) are compounds that were used as flame retardants in electrical products and in business machines and motor housings. Only a small number of the 209 possible bromobiphenyl congeners have been synthesized and used. All the commercial products contained a mixture of several individual PBBs. Commercial production ceased in 1977

Exposure. Inhalation; skin

Toxicology. Polybrominated biphenyls (PBBs) are animal carcinogens; the liver is the main organ affected.

The majority of the human toxicity data of PBB stem from studies carried out after accidental addition of PBB to farm feed in Michigan in 1973, resulting in exposure of large numbers of the rural population of Michigan by ingestion of PBB-contaminated food. In a group of 933 Michigan farmers and residents, higher rates of dermatological, neurological, and musculoskeletal disorders were reported than in 229 unexposed Wisconsin farmers used as controls.[1] Disorders included rashes, acne, darkening or thickening of the skin, erythema, and hair loss.

A high prevalence of abnormal liver-function tests (SGOT and SGPT) was observed among 614 Michigan adults compared with 141 Wisconsin adults used as controls.[2]

A group of 55 workers who had been employed in the Michigan plant producing FireMaster BP-6 from 1970 to 1974 was examined, and all were found to have serum levels of PBB greater than 1 mg/l.[3] An increased prevalence of respiratory symptoms and skin disorders was seen in this group compared to the available data on PBB-exposed farmers in Michigan.

Animal studies in various species have shown that chronic PBB poisoning causes a wasting syndrome characterized by progressive decreased weight gain, with immediate moderate to severe body weight loss generally preceding death.[4]

The thyroid gland is a target organ in animals, although strong evidence for an effect in humans is lacking.[4] In rats exposed for acute and intermediate durations, effects have been decreases in serum levels of thyroxine and triiodothyronine, along with histological and ultrastructural changes in the thyroid.

When administered by oral gavage for 6 months at 10 mg/kg, technical-grade hexabromobiphenyl (FireMaster FF-1) induced hepatocellular carcinomas in mice and rats of both sexes and cholangiocarcinomas and neoplastic nodules of the liver in rats of both sexes.[5] The IARC considers that there is "sufficient evidence" that PBB is carcinogenic to experimental animals.[6]

A threshold limit value–time-weighted average (TLV-TWA) for PBB has not been assigned.

REFERENCES

1. Anderson HS, Lilis R, Selikoff IJ, et al: Unanticipated prevalence of symptoms among dairy farmers in Michigan and Wisconsin. *Environ Health Perspect* 23:217–226, 1978
2. Anderson HA, Holstein EC, Daum SM, et al: Liver function tests among Michigan and Wisconsin dairy farmers. *Environ Health Perspect* 23:333–339, 1978
3. Anderson HA, Wolff MS, Fischbein A, Selikoff IJ: Investigation of the health status of Michigan Chemical Corporation employees. *Environ Health Perspect* 23:187–191, 1978
4. Agency for Toxic Substances and Disease Registry: *Toxicological Profile for Polybrominated Biphenyls (PBBs)*, Public Health Service, Centers for Disease Control, Atlanta, GA, pp 78–88, 1993
5. National Toxicology Program: *Toxicology and Carcinogenesis Studies of a Polybrominated Biphenyl Mixture (Firemaster FF-1) in F344/N Rats and B6C3F₁ Mice (Gavage Studies)*, Technical Report Series No 244. Springfield, VA, National Technical Information Service, US Department of Commerce, 1983
6. *IARC Monographs on the Evaluation of the Carcinogenic Risk of Chemicals to Humans*, Vol 41, Some halogenated hydrocarbons and pesticide exposures, pp 261–292. Lyon, International Agency for Research on Cancer, December 1986

POLYTETRAFLUOROETHYLENE DECOMPOSITION PRODUCTS
CAS: 9002-84-0

Perfluoroisobutylene—382-21-8
Carbonyl fluoride—353-50-4
$(CF_2CF_2)_{n=ca\ 1000}$

Synonyms: Teflon; Algoflon; Fluon; Tetran; PTFE

Physical Form. Grayish white plastic

Uses. Coating on cooking utensils, reaction vessels, and other industrial applications to prevent sticking

Exposure. Inhalation

Toxicology. Fumes of heated polytetrafluoroethylene (PTFE) cause polymer fume fever, an influenzalike syndrome.

When PTFE is heated to between 315°C and 375°C, the fumes cause influenzalike effects, including chills, fever, and tightness in the chest, which last 24 to 48 hours.[1,2] Symptoms suggestive of pulmonary edema, including shortness of breath and chest discomfort, have been observed in a few instances.[3] Complete recovery usually has occurred within 12 to 48 hours after termination of exposure. The syndrome has been particularly associated with smoking of cigarettes contaminated with Teflon.

In rats, the LC_{50} dose for polytetrafluoroethylene heated at 595°C was 45 mg/m³ for a 30-minute exposure.[4] Conjunctival erythema and serous ocular and nasal discharge were observed immediately after exposure. Clinical signs included dyspnea, hunched posture, and lethargy. Pathologic findings included focal hemorrhages, edema, and fibrin deposition in the lungs. Disseminated intravascular coagulation developed in more than half the test animals, and its incidence and severity closely paralleled pulmonary damage.

The decomposition products, up to a temperature of 500°C, are principally the monomer tetrafluoroethylene but also include perfluoropropene, other perfluoro compounds containing 4 or 5 carbon atoms, and an unidentified particulate waxy fume.[5] From 500°C to 800°C, the pyrolysis product is carbonyl fluoride, which can hydrolyze to form HF and CO_2.

Experiments with rodents have shown that the PTFE pyrolysis particles rather than toxic gases are the toxic agent causing pulmonary edema and hemorrhage.[6] Mortality of rats was prevented by removal of the submicron particles by filtration, even though the concentration of the measured toxic gases was not significantly decreased.

Polytetrafluoroethylene implanted subcutaneously in animals has induced local sarcomas, suggesting a foreign body reaction rather than chemical carcinogenesis. The IARC has determined that there is insufficient evidence to assess the carcinogenic risk, especially with regard to occupational exposure in humans.[7]

There is no assigned threshold limit value for polytetrafluoroethylene, but air concentrations should be kept as low as possible.

REFERENCES

1. Zapp ZA Jr: Polyfluorines. *Encyclopaedia of Occupational Health and Safety*, Vol II, pp 1095–1097. New York, McGraw-Hill, 1972
2. Harris DK: Polymer-fume fever. *Lancet* 2: 1008–1011, 1951
3. Lewis CE, Kirby GR: An epidemic of polymer-fume fever. *JAMA* 191:375, 1965
4. Zook BC, Malek D, Kenney RA: Pathologic findings in rats following inhalation of combustion products of polytetrafluoroethylene (PTFE). *Toxicology* 26:25–36, 1983
5. National Institute for Occupational Safety and Health: *Criteria for a Recommended Standard . . . Occupational Exposure to Decomposition Products of Fluorocarbon Polymers*. DHEW (NIOSH) Pub No 77-193, pp 16, 63. Washington, DC, US Government Printing Office, 1977
6. Lee KP, Seidel WC: Pulmonary response to perfluoropolymer fume and particles generated under various exposure conditions. *Fund Appl Toxicol* 17:254–269, 1991
7. *IARC Monographs on the Evaluation of the Car-*

cinogenic Risk of Chemicals to Humans, Vol 19, Some monomers, plastics and synthetic elastomers, and acrolein, pp 285–297. Lyon, International Agency for Research on Cancer, 1979

PORTLAND CEMENT
CAS: 65997-15-1

Portland cement refers to a class of hydraulic cements in which the two essential constituents are tricalcium silicate ($3CaO \cdot SiO_2$), and dicalcium silicate ($2CaO \cdot SiO_2$), with varying amounts of alumina, tricalcium aluminate, and iron oxide. Portland cement is insoluble in water. The quartz content of most finished cements is below 1%; chromium may be present

Physical Form.   Solid

Uses.   In concrete where it constitutes approximately 15% of the finished concrete

Exposure.   Inhalation

Toxicology.   Portland cement is an irritant of the eyes and causes dermatitis.

Repeated and prolonged skin contact with cement can result in dermatitis of the hands, forearms, and feet; this is a primary irritant dermatitis and may be complicated in some instances by a secondary contact sensitivity to hexavalent chromium.[1] In a study of 95 cement workers, 15 had a mild dermatitis of the hands, which consisted of xerosis with erythema and mild scaling; of 20 workers who were patch-tested with 0.25% potassium dichromate, one person had a mild reaction and the others were negative.

In a survey of 2278 cement workers, it was concluded that exposure to the dust of finished Portland cement caused no significant findings on chest roentgenograms, even after heavy and prolonged exposures.[2] However, in a follow-up study of 195 of these workers after further exposure of 17 to 20 years, 13 showed increases in lung markings on roentgenograms; an additional 6 workers who had been exposed largely to raw dusts containing varying amounts of free silica had marked linear exaggeration with ill-defined micronodular shadows but no symptoms referable to the chest.[3]

In contrast, a study of 847 cement workers with at least 5 years of exposure to massive levels ranging up to 3020 mppcf in cement plants revealed that symptoms such as cough, expectoration, exertional dyspnea, wheezing, and chronic bronchitis syndromes were consistently more frequent than in a group of 460 control workers; a higher prevalence of these symptoms was also found in nonsmokers exposed to cement than in a control group of unexposed nonsmokers. It should be emphasized that these exposures were to cement dust not of Portland type.[3,4]

In a cross-sectional study of 2736 Portland cement workers and 755 controls, there were no significant differences in symptoms except that 5.4% of the cement workers had dyspnea compared to 2.7% of the controls.[5] The mean pulmonary-function indices were similar for the two groups. The mean exposure concentrations were 0.57 mg/m³ for respirable dust (ranging up to 46 mg/m³) and 2.90 mg/m³ for total dust (ranging up to 78.61 mg/m³). The authors concluded that a close relation between exposure to cement plant dust at levels existing in the United States and respiratory symptoms or ventilatory function is lacking.

The relation between exposure to Portland cement dust and cancer was examined in a population of 546 workers who had been exposed at some time before 1974 for 1 or more years. No increased risk of overall cancer, respiratory cancer, or stomach cancer was found among the cement workers compared to a referent population.[6]

Earlier studies suggested increases in various types of cancer, including lung and stomach cancer, with exposure to cement, but numerous limitations prevent any conclusions in this regard.[7-9]

The 1995 ACGIH threshold limit value-time weighted average (TLV-TWA) for Portland cement is 10 mg/m³ for total dust containing no asbestos and <1% crystalline silica.

REFERENCES

1. Perone VB et al: The chromium, cobalt, and nickel contents of American cement and their relationship to cement dermatitis. *Am Ind Hyg Assoc J* 35:301–306, 1974
2. Sander OA: Roentgen resurvey of cement workers. *AMA Arch Ind Health* 17:96–103, 1958
3. Kalacic I: Chronic nonspecific lung disease in cement workers. *Arch Environ Health* 26:78–83, 1973
4. Kalacic I: Ventilatory lung function in cement workers. *Arch Environ Health* 26:84–85, 1973
5. Abrons HL, Peterson MR, Sanderson WT, et al: Symptoms, ventilatory function, and environmental exposures in Portland cement workers. *Br J Ind Med* 45:368–375, 1988
6. Vestbo J, Knudsen KM, Raffn E, et al: Exposure to cement dust at a Portland cement factory and risk of cancer. *Br J Ind Med* 48:803–807, 1991
7. Rafnsson V, Johannesdottir SG: Mortality among masons in Iceland. *Br J Ind Med* 43:522–525, 1986
8. McDowall ME: A mortality study of cement workers. *Br J Ind Med* 41:179–182, 1984
9. McDowall M: Cement workers and cancer: Epidemiology at work? Editorial. *Br J Ind Med* 43:505–506, 1986

POTASSIUM HYDROXIDE
CAS: 1310-58-3

KOH

Synonyms: Caustic potash; KOH

Physical Form. White solid, usually as lumps, rods or pellets

Uses. Strong alkali; in manufacture of soft and liquid soaps; in manufacture of potassium carbonate for use in glass manufacture

Exposure. Inhalation

Toxicology. Potassium hydroxide is a severe

irritant of the eyes, mucous membranes, and skin.

The effects of potassium hydroxide are similar to those of other strong alkalies such as sodium hydroxide. The most significant industrial hazard is rapid tissue destruction of eyes or skin on contact with either the solid or with concentrated solutions.[1] Contact with the eyes causes disintegration and sloughing of conjunctival and corneal epithelium, corneal opacification, marked edema, and ulceration.[2] After 7 to 13 days, either gradual recovery begins or there is progression of ulceration and corneal opacification, which may become permanent. If KOH is not promptly removed from the skin, severe burns with deep ulceration will occur.

Although inhalation is usually of secondary importance, the effects from the dust or mist will vary from mild irritation to severe pneumonitis, depending on the severity of exposure.[1] Ingestion produces severe abdominal pain; corrosion of the lips, mouth, tongue, and pharynx; and the vomiting of large pieces of mucosa.

The 1995 ACGIH STEL/ceiling limit for potassium hydroxide is 2 mg/m³.

REFERENCES

1. National Institute for Occupational Safety and Health: *Criteria for a Recommended Standard . . . Occupational Exposure to Sodium Hydroxide.* DHEW (NIOSH) Pub No 76-105, pp 23–50. Washington, DC, US Government Printing Office, 1975
2. Grant WM: *Toxicology of the Eye*, 3rd ed, p 756. Springfield, IL, Charles C Thomas, 1986

PROPANE
CAS: 74-98-6

C_3H_8

Synonyms: Dimethylmethane; propyl hydride

Physical Form. Colorless, odorless gas

Uses. Fuel gas; refrigerant; in organic synthesis

Exposure. Inhalation

Toxicology. Propane is a simple asphyxiant. The determining factor in exposure is available oxygen. Minimal oxygen content of air in the workplace should be 18% by volume under normal atmospheric pressure, equivalent to Po_2 of 135 mm Hg.[1]

Exposure to 100,000 ppm propane for a few minutes produced slight dizziness in volunteers but was not noticeably irritating to the eyes, nose, or respiratory tract.[2] No adverse effects were reported in humans exposed to 10,000 ppm for 10 minutes or after exposure to 1000 ppm, 8 hr/day, for 9 days.

Intentional inhalation of 95% propane for approximately 1 minute produced feelings of euphoria, ataxia, and light-headedness; death, possibly due to hypoxemia secondary to propane inhalation, has been reported.[3,4]

Guinea pigs exposed at 47,000 to 55,000 ppm had tremors within 5 minutes, and nausea, retching, and stupor after 30 to 120 minutes. No effects were observed in monkeys exposed to approximately 750 ppm for 90 days.[5]

Direct contact with the liquefied product causes burns and frostbite.[6]

Propane is odorless, and atmospheres deficient in oxygen do not provide adequate warning.[1]

REFERENCES

1. *Threshold Limit Values and Biological Exposure Indices for 1994–1995*, p 7. Cincinnati, OH, American Conference of Governmental Industrial Hygienists (ACGIH), 1994
2. Gerarde HW: The aliphatic (open chain, acyclic) hydrocarbons. In Fassett DW, Irish DD (eds): *Patty's Industrial Hygiene and Toxicology*, 2nd ed, Vol 2, *Toxicology*, pp 1195–1198. New York, Interscience, 1963
3. Wheeler MG, Rozycki AA, Smith RP: Recreational propane inhalation in an adolescent male. *Clin Toxicol* 30:135–139, 1992
4. Siegel E, Wason S: Sudden death caused by inhalation of butane and propane. *N Engl J Med* 323:1638, 1990

5. Moore AF: Final report on the safety assessment of isobutane, isopentane, *n*-butane and protane. *J Am Coll Toxicol* 1:127–142, 1982
6. Sandmeyer EE: Aliphatic hydrocarbons. In Clayton GD, Clayton FE (eds): *Patty's Industrial Hygiene and Toxicology*, 3rd ed, Vol 2B, *Toxicology*, pp 3181–3182. New York, Wiley-Interscience, 1981

PROPANE SULTONE
CAS: 1120-71-4

$C_3H_6O_3S$

Synonyms: 1,3-Propane sultone; 3-hydroxy-1-propanesulfonic acid sultone; 1,2-oxathrolane 2,2-dioxide

Physical Form. White crystals or colorless liquid

Uses. Chemical intermediate to confer water solubility and anionic properties

Exposure. Inhalation

Toxicology. Propane sultone is a carcinogen in experimental animals and a suspected human carcinogen. No human data are available.[1]

It is a carcinogen in rats when administered orally, intravenously, or by prenatal exposure and a local carcinogen in mice and rats when given subcutaneously.[1]

In rats, twice-weekly oral doses by gavage of 56 mg/kg for 32 weeks, or 28 mg/kg for 60 weeks, resulted in several malignant manifestations, including tumors of the brain, ear duct, and small intestine, and leukemia.[2,3]

In mice, weekly subcutaneous injection of 0.3 mg caused tumors at the injection site in 21 of 30 mice, compared to no tumors in 30 controls.[4] Weekly subcutaneous injection of 15 and 30 mg/kg into rats resulted in the death of 7 of 12 and 11 of 11 animals, respectively, with local sarcomas.[2,5] A single subcutaneous dose of 100 mg/kg produced local sarcomas in all 18

treated rats. A single intravenous dose 150 mg/kg in 32 rats caused the death of 1 rat with a brain tumor after 235 days and the death of 9 others with malignant tumors of a variety of sites within 459 days. A single intravenous dose of 20 mg/kg given to pregnant rats on day 15 of gestation produced malignant neurogenic tumors in some of the offspring.[1]

Propane sultone is classified as an A2 suspected human carcinogen with no assigned threshold limit value.

REFERENCES

1. *IARC Monographs on the Evaluation of the Carcinogenic Risk of Chemicals to Man*, Vol 4, Some aromatic amines, hydrazine and related substances, *N*-nitroso compounds and miscellaneous alkylating agents, pp 253–258. Lyon, International Agency for Research on Cancer, 1974
2. Druckery H et al: Carcerogene alkylierende Substanzen. IV 1,3-Propanesultone und 1,4-Butansulton. *Z Krebsforsch* 75:69, 1970
3. Ulland B et al: Carcinogenicity of the industrial chemicals propylene imine and propane sultone. *Nature* 230:460, 1971
4. Van Durren BL et al: Carcinogenicity of isoesters of epoxides and lactones: Aziridine ethanol, propane sultone and related compounds. *J Natl Cancer Inst* 46:143, 1971
5. Druckery H, Kruse H, Preussman R: Propane sultone, a potent carcinogen. *Naturwissenschaften* 55:449, 1968

PROPARGYL ALCOHOL
CAS: 107-19-7

C_3H_4O

Synonyms: Ethynol carbinol; acetylene carbinol; propiolic alcohol; 2-propyn-1-ol; 2-propynyl alcohol

Physical Form. Clear to slightly straw-colored liquid

Uses. To prevent the hydrogen embrittlement of steel; corrosion inhibitor; solvent stabilizer; soil fumigant; chemical intermediate

Exposure. Inhalation; skin absorption

Toxicology. Propargyl alcohol is an irritant of the eyes and the skin; atmospheric concentrations of the chemical, readily attainable under room conditions, are dangerous to life, even with exposures of short duration; it is highly toxic when ingested and is easily absorbed through the skin in toxic amounts.

No reports of adverse effects in humans are available.

Following a 6-minute exposure to an essentially saturated atmosphere, 2 of 3 rats died, whereas a 12-minute exposure was fatal to all exposed animals.[1] A 2-hour LC_{50} of 850 ppm has been reported for both rats and mice.[2] Rats exposed to 80 ppm for 7 hours initially appeared to have eye irritation and to be lethargic.[1] Repeated exposures to this concentration for 5 days/week over a period of 3 months resulted in slight liver and kidney changes. Males had increased liver weights, and females had increases in both kidney and liver weights. Histopathological examination showed degenerative changes in these organs, with females showing the most injury.

Oral LD_{50} values of 50 mg/kg for the mouse, 60 mg/kg for the guinea pig, and 70 mg/kg for the rat have been reported.[1]

Applied to the skin of rabbits, propargyl alcohol causes hyperemia, edema, and some superficial necrosis.[1] It is rapidly absorbed through the skin of rabbits in lethal amounts, with the LD_{50} being approximately 16 mg/kg. A 10% solution is slightly irritating and may be lethal if exposure is extensive or prolonged. A 1% solution appears to be without adverse effects. Repeated dermal exposures of 10 mg/kg/day for 2 months or 20 mg/kg/day for 1 month caused no systemic effects, as evidenced by weight gain, hematology, and histopathological examination of the tissues.

Instilled in the eyes of rabbits, the undiluted material causes marked pain, irritation, and corneal injury; a 10% solution is slightly irritating, and a 1% solution has no effect.[1]

Propargyl alcohol induced chromosomal aberrations in CHO cells in vitro with and without metabolic activation; it was negative in the mouse bone marrow micronucleus test and in the *Salmonella*/mammalian microsome assay.[3]

Propargyl alcohol is reported to have a geraniumlike odor that is not adequate to provide warning of overexposure.

The 1995 ACGIH threshold limit value–time-weighted average (TLV-TWA) for propargyl alcohol is 1 ppm (2.3 mg/m³) with a notation for skin absorption.

REFERENCES

1. Rowe VK, McCollister SB: Alcohols. In Clayton GD, Clayton FE (eds): *Patty's Industrial Hygiene and Toxicology*, 3rd ed, Vol 2C, *Toxicology*, pp 4671–4673, New York, Wiley-Interscience, 1982
2. NIOSH: *Registry of Toxic Effects of Chemical Substances, 1985–86* ed, Vol 4, Sweet DV (ed). Washington, DC, US Department of Health and Human Services, Public Health Service, pp 3938–3939, 1987
3. Blakey DH, Maus KL, Bell R, et al: Mutagenic activity of 3 industrial chemicals in a battery of in vitro and in vivo tests. *Mut Res* 320:273–283, 1994

PROPENE
CAS: 115-07-01

C_3H_6

Synonyms: Propylene; methylethene; methylethylene

Physical Form. Colorless gas

Uses. In the production of polypropylene, acrylonitrile, isopropyl alcohol, and propene oxide, as well as gasoline and synthetic rubber; as an aerosol propellant or component

Exposure. Inhalation

Toxicology. Propene is of low toxicity. It is a simple asphyxiant and a mild anesthetic, with a physiological effect only at extremely high concentrations.

Concentrations of approximately 50% propene induced anesthesia in volunteers in 2 minutes, followed by complete recovery without adverse side effects.[1,2] In another report, exposure to 40%, 50%, and 75% for a few minutes caused reddening of the eyelids, flushing of the face, lacrimation, coughing, and sometimes flexing of the legs, without variation in respiratory or pulse rates or electrocardiograms. At 35% to 40%, two subjects vomited, and one complained of severe vertigo. Exposure to 13% for 1 minute or 6% for 2 minutes produced mild intoxication, paresthesias, and inability to concentrate. Propene also is considered to be a weak heart sensitization agent in humans.[3]

In rats, 40% propene caused light anesthesia with no other toxic symptoms within 6 hours; 55% propene for 3 to 6 minutes or 70% propene for 1 to 3 minutes produced deep anesthesia with no additional CNS disturbances.[1,2] Animal experiments with cats have shown no toxic signs when anesthesia was induced at concentrations of 20% to 31%; however, 70% propene resulted in a drop in blood pressure and an increased pulse rate, and an unusual ventricular ectopic beat occurred at concentrations ranging from 50% to 80%.

Limited information is available on the effects of chronic propene exposure. In mice, chronic exposure to minimal narcotic concentrations caused moderate to very slight fatty degeneration of the liver.[1,2]

No significant evidence for propene carcinogenicity was found in rats or mice exposed by inhalation to 5000 to 10,000 ppm for 6 hr/day for 103 weeks.[4] However, signs of nasal cavity pathology were observed in rats, including an increased incidence of nonneoplastic lesions consisting of epithelial hyperplasia and squamous metaplasia. In addition, inflammatory changes were noted, characterized by an influx of lymphocytes, macrophages, and granulocytes into the submucosa. A slight increase in the incidence of vascular tumors was observed in female mice. Other more limited animal studies also have failed to find a carcinogenic response by propene.[5]

The IARC has determined that there is inadequate evidence in humans and in experimental animals for the carcinogenicity of propene.[6] Overall, propene is not classifiable as to its carcinogenicity to humans.[6]

Propene has been reported to be nonmutagenic to both *Escherichia coli* and *Salmonella typhimurium*, either with or without metabolic activation; interestingly, the purported reactive metabolite of propene, propene oxide, is widely accepted as a mutagen.[2,6] Furthermore, propene oxide forms hemaglobin adducts in exposed animals.[6]

Propene gas is not an irritant to the skin, but direct contact with the liquid product may produce chemical burns.[1]

An important factor in the use of propene is that explosive concentrations of the gas are reached well before any physiological changes occur, and the gas or compressed liquid should be handled according to strict safety precautions.

According to the ACGIH, propene is classified as a simple asphyxiant.

REFERENCES

1. Sandmeyer EE: Aliphatic hydrocarbons. In Clayton GD, Clayton FE (eds): *Patty's Industrial Hygiene and Toxicology*, 3rd ed, Vol 2B, *Toxicology*, pp 3199–3201. New York, Wiley-Interscience, 1982
2. Gibson GG, Clarke SE, Farrar D, et al: Propene. In Snyder R (ed): *Ethel Browning's Toxicity and Metabolism of Industrial Solvents*, 2nd ed, Vol I, *Hydrocarbons*, pp 354–361. Amsterdam, Elsevier, 1987
3. Reinhardt CF, Azar A, Maxfield ME, et al: Cardiac arrhythmias and aerosol "sniffing." *Arch Environ Health* 22:265–279, 1971
4. Quest JA, Tomaszewski JE, Haseman JK, et al: Two year inhalation study of propylene in F344/N rats and B6C4F₁ mice. *Toxicol Appl Pharmacol* 76:288–295, 1984
5. Maltoni C, Ciliberti A, Cerretti D: Experimental contributions in identifying brain potential carcinogens in the petrochemical industry. *Ann NY Acad Sci* 381:216–249, 1982
6. *IARC Monographs on the Evaluation of Carcino-*

genic *Risks to Humans,* Vol 60, Some industrial chemicals, pp 161–180, Lyon, International Agency for Research on Cancer, 1994

β-PROPIOLACTONE
CAS: 57-57-8

$C_3H_4O_2$

Synonyms: BPL; 2-oxetanone; hydracrylic acid; β-lactone;

Physical Form. Colorless liquid

Uses. Vapor sterilant and disinfectant; intermediate in the production of acrylic acid and esters

Exposure. Inhalation; skin absorption

Toxicology. β-Propiolactone (BPL) is a skin irritant and a carcinogen in animals.

There are no reports of systemic effects in humans.

The 30-minute LC_{50} for rats was 250 ppm whereas, for 6 hours, the LC_{50} was 25 ppm.[1]

Although there is no epidemiologic evidence implicating BPL as a human carcinogen, the weight of experimental animal data suggests that BPL possesses a carcinogenic potential for humans.[2]

BPL applied to mouse skin 1 to 7 times (over a period of 2 weeks) as undiluted BPL or in solutions of corn oil or acetone at doses of 0.8 to 100 mg caused skin irritation; the effects ranged from erythema to hair loss and scarring.[3] Lifetime painting (3 times/week) using acetone and corn oil solutions showed that BPL produced both papillomas and cancer of the mouse skin; 0.25 mg in acetone caused papillomas in 12 of 30 animals and cancers in 3, whereas 5 mg produced tumors in 2 1 of 30 animals and cancers in 11. In corn oil, 0.8 mg caused tumors in 27 of 30 mice; 12 of the tumors were malignant.[3] Papillomas developed in 11 of 90 and

14 of 80 of the acetone and corn oil control groups, respectively.[3]

Following weekly subcutaneous injection of 0.73 BPL in tricaprylin for 503 days in 30 female mice, 89 mice developed fibrosarcomas, 3 adenocarcinomas, 7 squamous-cell carcinomas, and 3 squamous papillomas, all at the injection site. The number of months to first tumor was 7, and no local tumors developed in 110 controls treated with tricaprylin alone for up to 581 days.[4]

All of 10 rats injected biweekly for 44 weeks with 1 mg BPL in arachis oil developed injection-site sarcomas; no local sarcomas were observed in 7 controls given repeated injections of 0.5 mg arachis oil for 54 weeks.[5]

Repeated gastric administration of 10 mg BPL/0.5 ml tricaprylin/week for 70 weeks caused squamous-cell carcinomas of the forestomach in 3 of 5 rats; there were no tumors in controls treated with tricaprylin alone.[4]

The IARC has determined that β-propiolactone is carcinogenic in experimental animals and that it is possibly carcinogenic in humans.[6]

Because of high acute toxicity and demonstrated skin tumor production in animals, human contact by all routes should be avoided.[1]

The 1995 ACGIH threshold limit value–time-weighted average (TLV-TWA) is 0.5 ppm (1.5 mg/m³) with an A2 suspected human carcinogen designation.

REFERENCES

1. β-Propiolactone. *Documentation of TLVs and BEIs,* 6th ed, pp 1292–1293. Cincinnati, OH, American Conference of Governmental Industrial Hygienists (ACGIH), 1991
2. Department of Labor: Occupational safety and health standards—carcinogens. *Federal Register* 39:3757, 3786–3789, 1974
3. Palmes ED, Orris L, Nelson N: Skin irritation and skin tumor production by beta-propiolactone (BPL). *Am Ind Hyg J* 23:257–264, 1962
4. Van Duuren BL et al: Carcinogenicity of epoxides, lactones, and peroxy compounds. IV. Tumor response in epithelial and connective tissue in mice and rats. *J Natl Cancer Inst* 37:825–834, 1966

5. Dickens F, Jones HEH: Carcinogenic activity of a series of reactive lactones and related substances. *Br J Cancer* 15:85–100, 1961
6. *IARC Monographs on the Evaluation of the Carcinogenic Risk of Chemicals to Humans,* Suppl 7, *Overall Evaluation of Carcinogenicity: An Updating of IARC Monographs Vol 1—42,* p 70. Lyon, International Agency for Research on Cancer, 1974

PROPIONIC ACID
CAS: 79-09-4

CH₃CH₂COOH

CH3CH2COOH

Synonyms: Methylacetic acid; ethylformic acid; ethanecarboxylic acid; propanoic acid

Physical Form. Colorless, oily liquid with pungent odor

Uses/Sources. In synthesis of fungicides, herbicides, pharmaceuticals, flavorings, and perfumes; in production of propionates and cellulose propionate plastics; occurs naturally in dairy products

Exposure. Inhalation, ingestion, skin absorption

Toxicology. Propionic acid is an irritant to skin, eyes, and mucous membranes.

Propionic acid is a normal intermediary metabolite during the oxidation of fatty acids. It occurs ubiquitously in the gastrointestinal tract as an end product of microbial digestion of carbohydrates. It represents up to 4% of the normal total plasma fatty acids.[1]

Local damage may occur to skin, eye, or mucosal surfaces on contact with concentrated solutions.[1] No chronic or cumulative effects are known from industrial exposures.

In rats, the oral LD_{50} was 4.3 g/kg; the dermal LD_{50} in rabbits was 500 mg/kg.[1] Rats, mice, and hamsters fed diets containing 4% propionic acid for 7 days showed evidence of damage and cellular proliferation in the epithelium of the stomach.[2] Treatment-related histological changes in the epithelium of the forestomach, including acanthosis, hyperkeratosis, basal-cell hyperplasia, and intracellular vacuolation, were found in rats 14 days after treatment.[3] Persistent damage to cells of the forestomach and associated proliferative responses have been common factors in rodent forestomach tumorigenesis.[3] The relevance to humans, however, has not been determined.

Propionic acid does not appear to be genotoxic. In vitro mutagenicity assays with propionic acid, using *Salmonella typhimurium* or *Saccharomyces cerevisiae* were negative, with or without metabolic activation.[1]

No effect on maternal or fetal survival and no increase in the number of fetal abnormalities were seen following administration in the diet of pregnant mice and rats (up to 300 mg/kg/day for 10 days), hamsters (up to 400 mg/kg/day for 5 days), or rabbits (up to 400 mg/kg/day for 13 days).[1]

The 1995 ACGIH threshold limit value–time-weighted average (TLV-TWA) for propionic acid is 10 ppm (30 mg/m³).

REFERENCES

1. Guest D, Katz GV, Astill BD: Aliphatic carboxylic acids. In Clayton GD, and Clayton FE (eds): *Pattys's Industrial Hygiene and Toxicology,* 3rd ed, pp 4911–4913. New York, Wiley-Interscience, 1982
2. Harrison PTC, Grasso P, Badescu V: Early changes in the forestomach of rats, mice and hamsters exposed to dietary propionic and butyric acid. *Food Chem Toxicol* 29:367–371, 1991
3. Harrison PTC: Propionic acid and the phenomenon of rodent forestomach tumorigenesis: a review. *Food Chem Toxicol* 30:333–340, 1992

n-PROPYL ACETATE
CAS: 109-60-4

$C_3H_7COOCH_3$

Synonyms: Acetic acid, n-propyl ester

Physical Form. Colorless liquid

Uses. Solvent; in flavoring agents and perfumes

Exposure. Inhalation

Toxicology. In animals, propyl acetate is an irritant of the skin, eyes, and mucous membranes. At high concentrations, it causes narcosis, and it is expected that severe exposure will produce the same effect in humans.

No chronic or systemic effects have been reported in humans.

In cats, 24,000 ppm caused narcosis in 13 to 18 minutes; 30-minute exposures were lethal to some animals within 4 days post-exposure.[1] Exposure for 5 hours caused narcosis in cats at 7400 ppm and some deaths. Moderate irritation and salivation were observed at 5300 ppm for 6 hr/day.

Propyl acetate has a pearlike odor, but the odor threshold has not been determined.

The 1995 ACGIH threshold limit value-time-weighted average (TLV-TWA) is 200 ppm (835 mg/m³) with a STEL/ceiling of 250 ppm (1040 mg/m³).

REFERENCES

1. Sandmeyer EE, Kirwin CJ: Esters. In Clayton GD, Clayton FE (eds): *Patty's Industrial Hygiene and Toxicology*, 3rd ed, Vol 2A, *Toxicology*, pp 2273–2277. New York, Wiley-Interscience, 1981

n-PROPYL ALCOHOL
CAS: 71-23-8

$CH_3CH_2CH_2OH$

Synonyms: 1-Propanol; n-propanol; propyl alcohol

Physical Form. Clear liquid

Uses. Solvent; in organic syntheses

Exposure. Inhalation; minor skin absorption

Toxicology. n-Propyl alcohol is an irritant of the eyes and mucous membranes. At high concentrations, it causes narcosis in animals, and it is expected that severe exposure in humans will produce the same effect.

Based on acute animal studies, n-propyl alcohol appears to be slightly more toxic than isopropyl alcohol. No chronic effects have been reported in humans, although a human fatality has been ascribed to ingestion.[1] Exposure to 400 ppm for 3 to 5 minutes will reportedly produce mild irritation of the eyes, nose, and throat.[2]

Mice exposed to 3250 ppm developed ataxia in 90 to 120 minutes, and prostration was evident in 165 to 180 minutes; deep narcosis was manifest in 240 minutes at 4100 ppm.[2] Exposure to 13,120 ppm for 160 minutes or 19,680 ppm for 120 minutes was lethal to mice.[2] Exposure of rats to 20,000 ppm for 1 hour resulted in no mortalities during a 14-day postexposure observation period.[2] n-Propanol is not appreciably irritating to the skin of rabbits even after prolonged contact, but it can be absorbed in significant amounts if confined to the skin. Application of 38 ml/kg/day for 30 days resulted in the death of one-third of the rabbits.[2]

Instilled in rabbit eyes, 0.1 ml produced marked conjunctivitis, corneal opacities, and ulcerations.[2]

In a limited study, lifetime administration of n-propyl alcohol by intubation or subcutaneous injection caused severe liver injury, hematopoietic effects, and a number of malignant tumors not found in controls.[1]

The odor threshold (40 ppm) and irritant properties of *n*-propyl alcohol are expected to prevent inadvertent exposure to hazardous concentrations.[2]

The 1995 ACGIH threshold limit value-time-weighted average (TLV-TWA) is 200 ppm (492 mg/m³) with a STEL/ceiling of 250 ppm (614 mg/m³) and a notation for skin absorption.

REFERENCES

1. Gosselin RE et al: *Clinical Toxicology of Commercial Products*, 5th ed, p 218. Baltimore, MD, Williams and Wilkins, 1984
2. Rowe VK, McCollister SB: Alcohols. In Clayton GD, Clayton FE (eds): *Patty's Industrial Hygiene and Toxicology*, 3rd ed, Vol 2C, *Toxicology*, pp 4557–4561. New York, Wiley-Interscience, 1982

PROPYLENE DICHLORIDE
CAS: 78-87-5

$C_3H_6Cl_2$

Synonyms: 1,2-Dichloropropane; propylene chloride

Physical Form. Colorless liquid

Uses. Solvent; stain remover; chemical intermediate; fumigant

Exposure. Inhalation

Toxicology. Propylene dichloride is an eye and respiratory irritant; at very high concentrations, it is a central nervous system depressant and may cause liver injury.

Ingestion or inhalation of high levels caused severe liver damage, acute renal failure, hemolytic anemia, and disseminated intravascular coagulation in 3 reported cases.[1] Symptoms from inhalation included anorexia, abdominal pain, vomiting, ecchymoses, and hematuria. In all cases, more than 24 hours elapsed between exposure and onset of symptoms. Since 80% to 90% of propylene dichloride and its metabolites are eliminated within 24 hours, analysis of blood, urine, and feces for solvent is useless once symptoms appear.[1]

Workmen tolerated short-term exposures to 400 to 500 ppm without apparent adverse effects. Inhalation of a 98% solution over the course of an evening resulted in acute liver damage, as determined by laboratory tests (AST, ALT, total bilirubin, prothrombin); the patient recovered after 3 weeks of hospitalization.[1]

Guinea pigs repeatedly exposed to 2200 ppm for 7 hours developed severe conjunctival swelling, as well as signs of respiratory irritation and incoordination; 11 of 16 animals died after daily exposure and had severe liver injury and some kidney injury.[2] Rats dying from repeated inhalation of 1000 ppm showed weakness, general debility, and signs of respiratory irritation a few days prior to death; mice died after a few hours of exposure to 1000 ppm. In general, animals that survived 35 or more 7-hour exposures to 1000 to 2200 ppm showed no significant lesions at autopsy.

At 400 ppm, rats, guinea pigs, and dogs exposed for up to 140 daily 7-hour exposures showed no adverse effects.[3] There was a high percentage of mortality among mice repeatedly exposed to 400 ppm. In mice of a susceptible strain, hepatomas were found that were similar histologically to those induced by carbon tetrachloride. Oral administration of 100, 250, 500, or 1000 mg/kg to rats for up to 10 days caused body weight loss and central nervous system depression.[4] Morphological changes in the liver were apparent in the two highest-dose groups. Resistance to propylene dichloride hepatotoxicity over the 10 days of exposure was reflected by progressively lower serum enzyme levels and by decreases in the severity and incidence of toxic hepatitis and periportal vacuolization.

Female rats given 250 mg/kg/day by gavage for 103 weeks had a marginal, but statistically significant, increased incidence of adenomas of the mammary gland.[5] A dose-related increase in liver adenomas for both male and female mice was observed when treated with 125 or 250 mg/kg/day for 103 weeks. The NTP concluded that there was equivocal evidence of carcinogenicity in female rats and some evidence of carcinogenicity in male and female mice.[5]

Propylene dichloride was mutagenic in various strains of *Salmonella* and in mouse lymphoma cells, and it induced chromosomal aberrations in Chinese hamster cells.[6]

Some skin absorption may occur; the dermal LD_{50} for rabbits was 8.75 ml/kg.[1] The liquid is moderately irritating to the eye but does not cause serious or permanent injury.[7] Repeated or prolonged skin contact with propylene dichloride may result in skin irritation due to defatting.[7]

The liquid has a characteristic unpleasant chloroformlike odor; human subjects described the odor as "strong" at 130 to 190 ppm and "not noticeable" at 15 to 23 ppm.[7]

The 1995 ACGIH threshold limit value–time-weighted average (TLV-TWA) for propylene dichloride is 75 ppm (347 mg/m³) with a short-term excursion limit of 110 ppm (508 mg/m³).

REFERENCES

1. Pozzi C, Marai P, Ponti R, et al: Toxicity in man due to stain removers containing 1,2-dichloropropane. *Br J Ind Med* 42:770–772, 1985
2. Heppel LA, Neal PA, Highman B, Porterfield VT: Toxicology of 1,2-dichloropropane (propylene dichloride). I. Studies on effects of daily inhalations. *J Ind Hyg Toxicol* 28:1–8, 1946
3. Heppel LA, Highman B, Peak EG: Toxicology of 1,2-dichloropropane (propylene dichloride). IV. Effects of repeated exposures to a low concentration of the vapor. *J Ind Hyg Toxicol* 30:189–191, 1948
4. Bruckner JV, Mackenzie WF, Ramanathan R, et al: Oral toxicity of 1,2-dichloropropane: acute, short-term and long-term studies in rats. *Fund Appl Toxicol* 12:713–730, 1989
5. National Toxicology Program: *Toxicology and Carcinogenesis Studies of 1,2-Dichloropropane (Propylene Dichloride) (CAS No 78-87-5) in F344/N Rats and B6C3F₁ Mice (Gavage Studies)*. NTP-TR-263. NIH Pub No 86-2519, pp 1–182. DHEW, 1986
6. Agency for Toxic Substances and Disease Registry (ATSDR): *Toxicological Profile for 1,2-Dichloropropane*. US Public Health Service, p 119, 1989
7. Hygienic Guide Series: Propylene dichloride. *Am Ind Hyg Assoc J* 28:294–296, 1967

PROPYLENE GLYCOL DINITRATE
CAS: 6423-43-4

$C_3H_6N_2O_6$

Synonyms: Methylnitroglycol; propanediol dinitrate; dinitrate dipropylene glycol; PGDN

Physical Form.   Red-orange or colorless liquid with a disagreeable odor

Uses.   In the torpedo propellant Otto fuel II

Exposure.   Inhalation; skin absorption

Toxicology.   Propylene glycol dinitrate (PGDN) is a vasodilator and causes methemoglobin formation at high concentrations.

Male volunteers, 22 to 25 years of age, were exposed to PGDN at 0.03, 0.1, 0.2, 0.35, 0.5 and 1.5 ppm for single or daily exposures of various time periods.[1] At the 0.1 ppm exposure level, 2 of 9 subjects reported mild headaches; 7 of 9 had headaches at 0.2 ppm, which decreased dramatically with repeat exposures. At this level, most subjects could detect the odor for just 5 minutes. An alteration in visual evoked response was detected. Progressive throbbing headaches were noted in 7 of 9 volunteers exposed at 0.5 ppm and, after 6 hours, one subject was dizzy and nauseated. By 8 hours, 3 subjects had abnormal Romberg and heel-to-toe neurologic tests and narrowed pulse pressures with an increase in diastolic pressure. Alteration in visual evoked response was the only other effect noted. At 1.5 ppm, all 8 subjects could detect odor, had eye irritations within 40 minutes of exposure, and developed headaches. The headaches were so severe that exposure was stopped at 3 hours. All symptoms resolved within the subsequent 8 hours. There was no biochemical or hematological evidence of organ d amage in the studied exposure range.

A study of 87 naval employees chronically exposed to PGDN noted acute headaches and nasal congestion of presumed vascular origin but no chronic cardiovascular or neurotoxic disorders.[2] Twenty-nine subjects from this study group were tested before and immediately after PGDN exposure during torpedo maintenance procedure or turnaround. Significant changes in oculomotor function tests were observed although peak airborne concentrations were below 0.2 ppm. Although changes in these test scores were noted, there was no correlation between exposure levels and biologic effects. The authors concluded that PGDN could exert acute neurophysiologic effects but, at this exposure level, they were not functionally significant.

A cohort of 1352 Navy torpedomen munition workers exposed to PGDN between 1970 and 1979 had elevated rates and significantly elevated risks of angina pectoris and myocardial infarction.[3] The age-adjusted incidence rate for myocardial infarction was 18 of 10,000 in PGDN-exposed workers versus 8 of 10,000 for a group of nonexposed torpedomen; for angina pectoris, incidence rates were 9.8 of 10,000 in exposed workers versus 2.6 of 10,000 in nonexposed munitions workers. It was noted that 2 of the angina cases had no coronary atheromatous disease on angiography, a finding suggestive of vasoplastic etiology associated with PGDN exposure in these cases.

Pregnancy outcomes in women munitions workers were investigated between 1980 and 1983.[4] Spontaneous abortions among all female torpedo munitions workers were the same as or lower than those of hospital employees (enlisted female health-care workers) or all other Navy women. There were no spontaneous abortions among the few PGDN-exposed pregnant women.

Acute LD_{50} values have been reported in various animal species.[5] The oral LD_{50} in female rats was 1190 mg/kg; subcutaneous LD_{50}s (in mg/kg) were 463 for female rats, 524 for male rats, 1208 for female mice, and 200 to 300 for female cats. The LD_{50} dose in rats resulted in almost complete conversion of hemoglobin to methemoglobin, with lower conversion rates at lower doses. Death was due to anoxia. Methemoglobin levels were not measured in the mice or cats, but premorbid signs were consistent with methemoglobinemia in these species as well.

Blood pressure effects were recorded from cannulized femoral arteries in anesthetized rats following subcutaneous injection.[5] Maximal falls in blood pressure occurred within 30 minutes of injection. Small responses were seen at the 5 mg/kg level but, as the dose was increased, marked hypotension occurred.

Continuous 90-day exposure studies were conducted in rats, guinea pigs, dogs, and monkeys.[6] At 10 ppm, dogs had hemosiderin deposits in the liver, and similar pigment was found in the proximal convoluted tubules of the kidney. Guinea pigs showed foci of pulmonary hemorrhage at 15 ppm, whereas monkeys had increased serum urea nitrogen and decreased alkaline phosphatase, suggesting the possibility of renal damage at this level. At 35 ppm, hemosiderin deposits were found in the liver, spleen, and kidneys of dogs, female rats, and 4 of 9 monkeys. Methemoglobin values peaked at week 2, with values of 20% in dogs and monkeys. No changes in behavior patterns were observed in monkeys trained to perfor m a visual discrimination test and exposed continuously to 35 ppm for 90 days.

Rabbit skin applications were made daily for a 20-day subacute study.[6] At 1 g/kg, there was reversible erythema and no signs of systemic effect. At 2 g/kg, the rabbits appeared weak, were slightly cyanotic, and had rapid shallow breathing. At 4 g/kg, 13 of 14 animals were dead by the fifth application. Methemoglobin was measured at 35% at death. Autopsies showed overall weight loss and dark blue-gray internal organs, and the urinary bladder was markedly distended. The hemoglobin and hematocrit values were depressed, and urinary nitrates accounted for approximately 7% of the PGDN given at the 4 g/kg level.

Applied to rabbit eyes, 0.1 ml was only slightly irritating, and the irritation disappeared within 24 hours.[6]

Negative results were found in various mutagenic assays, including the Ames *Salmonella* assay (with or without microsomal activation), sister chromatid exchange assay in mouse lymphoma cells, mouse bone marrow cytogenic analysis, and mouse-dominant lethal assay.[4]

The 1995 ACGIH threshold limit value–time-weighted average (TLV-TWA) for propylene glycol dinitrate is 0.05 ppm (0.34 mg/m³) with a notation for skin absorption.

REFERENCES

1. Stewart RD, Peterson JE, Newton PE, et al: Experimental human exposure to propylene glycol dinitrate. *Toxicol Appl Pharmacol* 30:377–395, 1974
2. Horvath EP, Ilka RA, Boyd J, et al: Evaluation of the neurophysiologic effects of 1,2-propylene glycol dinitrate by quantitative ataxia and oculomotor function tests. *Am J Ind Med* 2:365–378, 1981
3. Forman SA, Helmkamp JC, and Bone CM: Cardiac morbidity and mortality associated with occupational exposure to 1,2-propylene glycol dinitrate. *J Occup Med* 29:445–450, 1987
4. Forman SA: A review of propylene glycol dinitrate toxicology and epidemiology. *Toxicol Lett* 43:51–65, 1988
5. Clark DG, Lichtfield MH: The toxicology, metabolism, and pharmacological properties of propylene glycol 1,2-dinitrate. *Toxicol Appl Pharmacol* 15:175–184, 1969
6. Jones RA, Strickland JA, Siegal J: Toxicity of propylene glycol 1,2-dinitrate in experimental animals. *Toxicol Appl Pharmacol* 22:128–137, 1972

PROPYLENE GLYCOL MONOMETHYL ETHER
CAS: 107-98-2

$C_4H_{10}O_2$

Synonyms: Propylene glycol methyl ether; PGME; 1-methoxy-2-propanol; Dowanol PM Glycol Ether; Propasol Solvent M; Poly-solv MPM Solvent

Physical Form. Colorless liquid

Uses. Solvent

Exposure. Inhalation

Toxicology. Propylene glycol monomethyl ether is low in systemic toxicity but causes irritation of the eyes, nose, and throat, with discomfort from the objectionable odor.

In human studies, 100 ppm was reported as having a transient objectionable odor. At 1000 ppm, there was irritation of the eyes, nose, and throat, along with signs of central nervous system impairment.[1]

The LC_{50} in rats was 10,000 ppm for 5 to 6 hours, with death caused by central nervous system depression.[2] Rats and monkeys exposed for 132 daily exposures to 800 ppm over a period of 186 days showed no evidence of adverse effects.

Exposure of rats to 3000 ppm, 6 hr/day, for a total of 9 days over an 11-day interval caused central nervous system depression that was reversible. No other effects were observed, including no adverse testicular effects.[3]

Exposure of pregnant rats and rabbits by inhalation to 500, 1500, or 3000 ppm for 6 hr/day on days 6 to 15 (rats) or 6 to 18 (rabbits) of gestation did not cause teratogenic or embryotoxic effects. Slight fetotoxicity in the form of delayed sternebral ossification was observed in the offspring of rats exposed at 3000 ppm—a dose that was also maternally toxic.[4] In a continuous breeding study, no change in reproductive parameters was observed in mice repeatedly administered 3333 mg/kg orally.[5]

The oral LD_{50} for rats was 6.6 g/kg and was on the order of 9.2 g/kg for dogs.[6,7] The dermal LD_{50} was in the range of 13 to 14 g/kg in rabbits, indicating minimal skin absorption.[2]

The liquid on the skin of rabbits caused only a very mild transient irritation after several weeks of constant application. In the rabbit eye, there was mild reversible irritation.

Propylene glycol monomethyl ether was not genotoxic in a variety of assays.[8]

The 1995 ACGIH threshold limit value–time-weighted average (TLV-TWA) for propylene glycol monomethyl ether is 100 ppm (369 mg/m³) with a short-term excursion limit (STEL) of 150 ppm (553 mg/m³).

REFERENCES

1. Stewart RD, Baretta ED, Dodd HC, Torkelson TR: Experimental human exposure to vapor of propylene glycol monomethyl ether. *Arch Environ Health* 20:218, 1970
2. Rowe VK, McCollister DD, Spencer HC, et al: Toxicology of mono-, di-, and tripropylene glycol methyl ethers. *Arch Ind Hyg Occup Med* 9:509, 1954
3. Miller RR, Ayres JA, Calhoun LL, et al: Comparative short-term inhalation toxicity of ethylene glycol monomethyl ether and propylene glycol monomethyl ether in rats and mice. *Toxicol Appl Pharmacol* 61:368, 1981
4. Hanley TR Jr, Calhoun LL, Yano BL, et al: Teratologic evaluation of inhaled propylene glycol monomethyl ether in rats and rabbits. *Fund Appl Toxicol* 4:784–794, 1984
5. Morrissey RE, Lamb JC, Morris RW, et al: Results and evaluations of 48 continuous breeding reproduction studies conducted in mice. Fund Appl Toxicol 13:747–777, 1989
6. Shideman FE, Procita L: Pharmacology of the monomethyl ethers of mono-, di-, and tripropylene glycol in the dog with observations of the auricular fibrillation produced by these compounds. *J Pharmacol Exp Therap* 102:79, 1951
7. Smyth HF Jr, Seaton J, Fisher L: Dose toxicity of some glycols and derivatives. *J Ind Hyg Toxicol* 23:259, 1941
8. McGregor DB: Genotoxicity of glycol ethers. *Environ Health Perspect* 57:97–104, 1984

PROPYLENE IMINE

CAS: 75-55-8

C_3H_7N

Synonyms: 2-Methylaziridine; 1,2-propyleneimine; 2-methylethylenimine

Physical Form. Flammable liquid

Uses. Intermediate in production of polymers, coatings, adhesives, textiles, and paper finishes

Exposure. Inhalation; skin absorption

Toxicology. Propylene imine vapor is an eye and respiratory tract irritant. It was carcinogenic to rats, the only species tested.

Inhalation may cause vomiting, breathing difficulty, and irritation of eyes, nose, and throat; on prolonged exposure, vapors tend to redden the whites of the eyes.[1]

Exposure of rats at 500 ppm for 4 hours was fatal, but inhalation for 2 hours resulted in no deaths.[2] Rats given 20 mg/kg by gavage twice weekly suffered from advanced flaccid paralysis after 18 weeks, and the mortality rate was high.[3] At 10 mg/kg, paralysis occurred to a lesser extent after 30 weeks. Granulocytic leukemia, squamous-cell carcinoma of the ear duct, and brain tumors (gliomas) were observed in the rats after 60 weeks at the 10 mg/kg dose; females showed mammary adenocarcinomas, a number of which metastasized to the lung.[3]

No information is available to assess the carcinogenic risk to humans.[4]

Instilled in the eye of a rabbit, a 5% aqueous solution produced corneal damage.[2] Contact with the liquid on the skin causes burns, and burns of the mouth and stomach would be expected from ingestion.[1]

The 1995 ACGIH threshold limit value–time-weighted average (TLV-TWA) is 2 ppm (4.7 mg/m³) with an A2-suspected human carcinogen classification.

REFERENCES

1. 2-Methylaziridine. *Dangerous Properties of Industrial Materials Report*, pp 85–90. July/August 1987
2. Carpenter CP et al: The acute toxicity of ethylene imine to small animals. *J Ind Hyg Toxicol* 30:2–6, 1948
3. Ulland B et al: Carcinogenicity of industrial chemicals propylene imine and propane sultone. *Nature* 230:460–461, 1971
4. *IARC Monographs on the Evaluation of the Carcinogenic Risk of Chemicals to Man*, Vol 9, pp 61–65, Lyon, International Agency for Research on Cancer, 1975

PROPYLENE OXIDE
CAS: 75-56-9

CH₃CHOCH₂

Synonyms: 1,2-Epoxypropane; propene oxide; methyloxirane; propylene epoxide

Physical Form. Colorless liquid

Uses. Primarily a chemical intermediate to produce polyether polyols, propylene glycols, and propylene glycol ethers; fumigant; preservative

Exposure. Inhalation

Toxicology. Propylene oxide is an irritant of the eyes, mucous membranes, and skin. At high concentrations, it causes narcosis in animals, and it is expected that severe exposure will produce the same effect in humans. It is carcinogenic in experimental animals.

No chronic or systemic effects have been reported in humans, although 3 cases of corneal burns from the vapor have been described.[1]

The LC₅₀ for rats exposed for 4 hours was 4000 ppm; for mice, it was 1740 ppm.[2] Rats and guinea pigs exhibited irritation, dyspnea, drowsiness, weakness, and some incoordination at concentrations of 2000 ppm or more.[3] Dogs exposed to 2030 ppm for 4 hours showed lacrimation, salivation, nasal discharge, and vomiting; there were some deaths.[2]

Rats, guinea pigs, rabbits, and a monkey were given repeated (79 or more) 7-hour exposures to 457 ppm. Irritation of the eyes and respiratory passages was noted in the rats and guinea pigs; rats had increased mortality owing to pneumonia.[3] There were no adverse effects on the monkey or the rabbits.[3] Rats exposed at 1500 ppm, 6 hr/day, 5 days/week, for 7 weeks developed ataxia in the hind legs. The main pathological change was axonal degeneration of the myelinated fibers in both the hind-leg nerve and the fasciculus gracilis.[4]

Rats exposed to 500 ppm, 7 hr/day, for 15 days 3 weeks prior to breeding and during gestation had a significant reduction in the numbers of corpora lutea, implants, and live fetuses.[5] For pregnant rats exposed on gestation days 6 to 15, there were no exposure-related effects, except for an increased frequency of seventh cervical ribs in fetuses at the maternally toxic exposure level of 500 ppm.[6] In another report, inhalation exposure at levels of up to 300 ppm over two generations did not produce any adverse effects on reproductive function.[7] Fetotoxicity was limited to minor skeletal abnormalities for exposed litters. Propylene oxide did not cause sperm abnormalities in mice treated 7 hr/day for 5 days by inhalation.[8]

Repeated subcutaneous administration of up to 2.5 mg/week for 95 weeks caused local sarcomas in mice.[9] Administered by oral gavage to rats twice a week for 2 years, propylene oxide caused a dose-dependent increase in forestomach tumors, which were mainly squamouscell carcinomas.[10]

In inhalation studies, there was some evidence of carcinogenicity in rats exposed at 400 ppm, as indicated by an increased incidence of papillary adenomas of the nasal turbinates.[11] In mice, there was clear evidence of carcinogenicity at this dose, as indicated by increased incidence of hemangiomas and hemangiosarcomas of the nasal turbinates. In the respiratory epithelium of the nasal turbinates, propylene oxide also caused suppurative inflammation, hyperplasia, and squamous metaplasia in rats, and inflammation in mice. The IARC has determined that there is sufficient evidence for carcinogenicity to animals.[8]

One case-control study in humans found no significant associations between exposure and various cancers; no information was given on exposure levels or possible confounding effects of other exposures.[8] The IARC has determined that there is inadequate evidence in humans for the carcinogenicity of propylene oxide but that it is possibly carcinogenic to humans.[8]

Propylene oxide was mutagenic to yeast, fungi, and bacteria. In mammalian cells in vitro, it also induced DNA damage and gene mutation, as well as sister chromatid exchange and chromosomal aberrations.[8] Propylene oxide forms adducts with proteins such as hemoglobin in a variety of species, including man. In mice, the concentration of the *N*-terminal valine adduct of propylene oxide in hemoglobin is linearly related to administered dose.

Aqueous solutions of 10% and 20% propylene oxide applied to the skin of rabbits caused hyperemia and edema when the duration of skin contact was 6 minutes or longer; severe exposures resulted in scar formation.[3]

The odor has been described as sweet, alcoholic, and like natural gas, ether, or benzene. The median detectable concentration is 200 ppm, which does not provide sufficient warning for prolonged or repeated exposures.[2]

The 1995 ACGIH threshold limit value–time-weighted average (TLV-TWA) for propylene oxide is 20 ppm (48 mg/m³).

REFERENCES

1. McLaughlin RS: Chemical burns of the human cornea. *Amer J Ophthal* 29:1355–1362, 1946
2. Jacobson KH, Hackley EB, Feinsilver L: The toxicity of inhaled ethylene oxide and propylene oxide vapors. *AMA Arch Ind Health* 13:237–244, 1956
3. Rowe VK, Hollingsworth RL, Oyen F, et al: Toxicity of propylene oxide determined on experimental animals. *AMA Arch Ind Health* 13:228–236, 1956
4. Ohnishi A, Yamamoto T, Murai Y, et al: Propylene oxide causes central-peripheral distal axonopathy in rats. *Arch Environ Health* 43:353–356, 1988
5. Hardin BD, Schuler RL, McGinnis PM, et al: Reproductive-toxicologic assessment of the epoxides ethylene oxide, propylene oxide, butylene oxide, and styrene oxide. *Scand J Work Environ Health* 9:94–102, 1983
6. Harris SB, Schardein JL, Ulrich CE, et al: Inhalation development toxicity study of propylene oxide in Fischer 344 rats. *Fund Appl Toxicol* 13:323–331, 1989
7. Hayes WC, Kirk HD, Gushow TS, et al: Effect of inhaled propylene oxide on reproductive parameters in Fischer 344 rats. *Fund Appl Toxicol* 10:82–88, 1988
8. *IARC Monographs on the Evaluation of Carcinogenic Risk to Humans*, Vol 60, Some industrial chemicals, pp 181–213. Lyon, International Agency for Research on Cancer, 1994
9. Dunkelberg H: Carcinogenic activity of ethylene oxide and its reaction products, 2-chloroethanol, 2-bromoethanol, ethylene glycol and diethylene glycol. I. Carcinogenicity of ethylene oxide in comparison with 1,2-propylene oxide after subcutaneous administration in mice. *Zbl Bakt Hyg I Abt Orig B* 174:383–404, 1981
10. Dunkelberg H: Carcinogenicity of ethylene oxide and 1,2-propylene oxide upon intragastric administration to rats. *Br J Cancer* 46:924–933, 1982
11. National Toxicology Program: *NTP Technical Report on the Toxicology and Carcinogenesis Studies of Propylene Oxide in F344/N Rats and B6C3F Mice (Inhalation Studies)*. NTP TR 26F, NIH Pub No 85-252F, pp 1–168, 1985

---

n-PROPYL NITRATE
CAS: 627-13-4

$C_3H_7NO_3$

Synonyms: Nitric acid *n*-propyl ester

Physical Form. Clear to yellow liquid

Uses. Fuel ignition promoter; rocket propellants; organic intermediate

Exposure. Inhalation

Toxicology. *n*-Propyl nitrate in animals causes anoxia from the formation of methemoglobin, as well as anemia and hypotension.

There have been no reports of human intoxication. It is speculated that, in humans, exposure severe enough to cause methemoglobin is unlikely because lower concentrations produce sufficient warning in the form of irritation, headache, and nausea.[1]

Exposure of rats to 10,000 ppm for 4 hours caused nasal irritation, dyspnea, methemoglobinemia, weakness, cyanosis, and death; dogs appeared to be more susceptible to *n*-propyl nitrate with a 4-hour LC$_{50}$ of 2500 ppm.[1] In dogs repeatedly exposed to 260 ppm for 26 weeks, hemoglobinuria and mild anemia appeared during the first 2 weeks of exposure but then subsided; at 900 ppm for 6 days, effects were cyanosis, methemoglobinemia, hemolytic anemia, hemoglobinuria, collapse, and death.[1]

Anesthetized dogs given 50 to 250 mg/kg intravenously immediately showed hypotension, arrest of gut activity, respiratory paralysis, hyperpnea, and moderate methemoglobinemia. Because death was produced with methemoglobin levels of only 4%, n-propyl nitrate intoxication may be caused in part by a direct action on vascular smooth muscle.[2] (It has been noted that the oral toxicity of n-propyl nitrate is very low compared to intravenously administered doses in which mg/kg doses were lethal vs. g/kg orally.[3])

The liquid instilled into the eyes of rabbits caused mild transient inflammation with no evidence of corneal damage.[3] The liquid applied to the skin of rabbits daily for 10 days caused staining, inflammation, and thickening of the skin but no evidence of systemic toxicity.[4]

The odor of n-propyl nitrate is detectable at 50 ppm and above.[1]

Note: For a description of diagnostic signs, differential diagnosis, and medical control, including clinical laboratory treatment of overexposure to methemoglobin-forming agents, see "Aniline."

The 1995 ACGIH threshold limit value-time-weighted average (TLV-TWA) is 25 ppm (107 mg/m³) with a STEL/ceiling of 40 ppm (172 mg/m³).

REFERENCES

1. Rinehart WE, Garbers RC, Greene EA, Stoufer RM: Studies on the toxicity of n-propyl nitrate vapor. *Am Ind Hyg Assoc J* 19:80–83, 1958
2. Murtha EF, Stabile DE, Wills JH: Some pharmacological effects of n-propyl nitrate. *J Pharmacol Exp Ther* 118:77–83, 1956
3. Sutton WL: Aliphatic nitro compounds, nitrates, nitrites. In Patty FA (ed): *Industrial Hygiene and Toxicology*, 2nd ed, Vol 2, *Toxicology*, pp 2090–2092. New York, Interscience, 1963
4. Hood DB: Toxicity of n-propyl nitrate and isopropyl nitrate, Haskell Laboratory for Toxicology and Industrial Medicine, Report No 21-53. Wilmington, El DuPont de Nemours and Company, 1953

PYRETHRUM
CAS: 8003-34-7

$C_{21}H_{28}O_3$
PI: $C_{22}H_{28}O_5$
PII: $C_{20}H_{28}O_3$

Synonyms: Pyrethrin I or II; Cinerin I or II; Jasmolin I or II. *Note.* Pyrethrum flowers yield "pyrethrum extract," of which the insecticidal constituents are collectively the "pyrethrins" or the "natural pyrethrins"

Physical Form.   Dust

Uses.   Insecticide

Exposure.   Inhalation

Toxicology.   Pyrethrum dust causes dermatitis and occasionally sensitization.

Under practical conditions, pyrethrum and its derivatives are probably some of the least toxic to mammals of all insecticides currently in use.[1] It was used for many years as an anthelminthic agent at a suggested oral dose of 20 mg/day for 3 days with no apparent ill effects. However, ingestion of 14 mg was lethal to a 2-year-old. Symptoms in an 11-month-old infant who ingested the powder included pallor, intermittent convulsions, vomiting, and bradycardia; there was extreme reddening of the lips and tongue and slight inflammation of the conjunctivas.[1]

Very young children are perhaps more susceptible to poisoning because they may not hydrolyze the pyrethrum esters efficiently.[1] Animal studies indicate that pyrethrum may undergo efficient destruction in the liver and/or be slowly absorbed from the gastrointestinal tract because oral LD_{50} values are several orders of magnitude higher than intravenous values.[1]

The primary effect in humans from exposure to pyrethrum is dermatitis.[2] The usual lesion is a mild erythematous dermatitis with vesicles, papules in moist areas, and intense pruritis; a bullous dermatitis may develop.[2]

In a study of workers engaged in processing pyrethrum powder, 30% had erythema, skin roughening, and pruritis, all of which subsided on cessation of exposure.[3] One of these workers had an anaphylactic-type reaction. Shortly after entering a dust-laden room, the worker's the facial skin turned red, and the worker felt a sensation of burning and itching. The cheeks and eyes rapidly became swollen, and pruritis became severe; the entire condition disappeared within 2 days after removal from exposure.[3]

Some persons exhibit sensitivity similar to pollinosis, with sneezing, nasal discharge, and nasal stuffiness.[2] A few cases of asthma due to pyrethrum mixtures have been reported; some of the people involved had a previous history of asthma, with allergy to a wide spectrum of substances.[2]

Dogs fed pyrethrins at a dietary level of 5000 ppm for 90 days showed tremor, ataxia, labored respiration, and salivation during the first month of exposure.[1] Rats given up to 5000 ppm in their diets for 2 years suffered no significant effects on growth or survival but had slight liver damage.[1] A daily gavage dose of 50, 100, or 150 mg/kg on days 6 to 15 of pregnancy caused an increased incidence of resorptions in rats compared to controls.[4]

The 1995 ACGIH threshold limit value–time-weighted average (TLV-TWA) is 5 mg/m³.

REFERENCES

1. Hayes WJ Jr: *Pesticides Studied in Man*, pp 75–80. Baltimore, MD, Williams and Wilkins, 1982
2. Hayes WJ Jr: *Clinical Handbook on Economic Poisons. Emergency Information for Treating Poisoning.* US Public Health Service Pub No 476, pp 74–76. Washington, DC, US Government Printing Office, 1963
3. Casida JE (ed): *Pyrethrum—The Natural Insecticide*, pp 123–142. New York, Academic Press, 1973
4. Khera KS et al: Teratogenicity study on pyrethrum and rotenone (natural origin) and ronnel in pregnant rats. *J Toxicol Environ Health* 10:111–119, 1982

PYRIDINE
CAS: 110-86-1

NC₅H₅

Synonyms: Azabenzene; azine

Physical Form. Colorless liquid

Uses. Solvent; in organic syntheses, especially agricultural chemicals

Exposure. Inhalation; skin absorption

Toxicology. Pyridine is an irritant and a central nervous system depressant; ingestion may cause liver and kidney damage.

Chemical plant workers chronically exposed to 6 to 12 ppm developed headache, vertigo, nervousness, sleeplessness, nausea, and vomiting.[1] Similar symptoms have occurred in workers repeatedly exposed to 125 ppm; in some cases, lower abdominal or back discomfort with urinary frequency was observed without associated evidence of liver or kidney damage.[2] Serious liver and kidney injury has been reported following oral administration of 1.8 to 2.5 ml of pyridine daily for 2 months in the treatment of epilepsy.[3] Skin irritation may result from prolonged or repeated contact with the chemical.

In animals, the major effects from administration of large doses by any route are local irritation and narcosis, whereas repeated feeding results in kidney and liver injury.[2] Exposure of rats to 23,000 ppm was lethal in 1.5 hours, and exposure to 3600 ppm for 6 hours was fatal to 2 of 3 rats tested.[2] The oral LD₅₀ for rats was 1.58 g/kg; the dermal LD₅₀ was 1 to 2 ml/kg in guinea pigs.[2] In the eye of a rabbit, a 40% solution caused corneal necrosis.

Carcinogenicity was not demonstrated in chronic subcutaneous studies in animals.[2]

Pyridine has an unpleasant odor detectable at 1 ppm; the odor is objectionable to unacclimatized individuals at 10 ppm but does not provide sufficient warning of hazardous concentrations because olfactory fatigue occurs quickly.[4]

The 1995 ACGIH threshold limit value–time-weighted average (TLV-TWA) for pyridine is 5 ppm (16 mg/m³).

REFERENCES

1. Teisinger J: Mild chronic intoxication with pyridine. *J Ind Hyg Toxicol* 30:58, 1948
2. Reinhardt CF, Brittelli MR: Heterocyclic and miscellaneous nitrogen compounds. In Clayton GD, Clayton FE (eds): *Patty's Industrial Hygiene and Toxicology*, 3rd ed, rev, Vol 2A, pp 2727–2731. New York, Wiley-Interscience, 1981
3. Pollack LJ, Finkelman I, Arieff AJ: Toxicity of pyridine in man. *Arch Intern Med* 71:95–106, 1943
4. Santodonato J et al: *Monograph on Human Exposure to Chemicals in the Workplace: Pyridine.* Washington, DC, National Cancer Institute, 1985

QUINONE
CAS: 106-51-4

$C_6H_4O_2$

Synonyms: *p*-Benzoquinone; 1,4-cyclohexadien-dione; *p*-quinone

Physical Form.   Yellow crystalline solid

Uses.   Oxidizing agent; in photography; for tanning hides; intermediate in the manufacture of dyes, fungicides, and hydroquionone

Exposure.   Inhalation

Toxicology.   Quinone affects the eyes.

Acute exposure causes conjunctival irritation and, in some cases, corneal edema, ulcer-ation, and scarring; transient eye irritation may be noted above 0.1 ppm and becomes marked at 1 to 2 ppm.[1] Chronic exposure causes the gradual development of changes characterized as: (1) brownish discoloration of the conjunctiva and cornea confined to the intrapalpebral fissure, (2) small opacities of the cornea, and (3) structural corneal changes that result in loss of visual acuity.[2,3] The pigmentary changes are reversible, but the more slowly developing structural changes in the cornea may progress. Although pigmentation may occur with less than 5 years of exposure, this is uncommon and usually is not associated with serious injury. Skin contact may cause discoloration, erythema, swelling, and the formation of papules and vesicles; prolonged contact may lead to necrosis. Systemic effects from industrial exposure have not been reported.

Administration of large doses of quinone to experimental animals caused local irritation, clonic convulsions, respiratory difficulties, drop in blood pressure, and death due to paralysis of the medullary centers. In chronic studies, quinone has been tested in mice by skin application and inhalation and in rats by subcutaneous injection. The IARC has determined that these studies are insufficient to evaluate the carcinogenicity of quinone.[4]

The odor and irritant properties do not provide adequate protection from levels capable of producing chronic eye injury.[1]

The 1995 ACGIH threshold limit value–time-weighted average (TLV-TWA) for quinone is 0.1 ppm (0.44 mg/m³).

REFERENCES

1. Hygienic Guide Series: Quinone. *Am Ind Hyg Assoc J* 24:194–195, 1963
2. Sterner JH, Oglesby F, Anderson B: Quinone vapors and their harmful effects to corneal and conjunctival injury. *J Ind Hyg Toxicol* 29:60–73, 1947
3. Anderson B, Oglesby F: Corneal changes from quinone hydroquinone exposure. *AMA Arch Ophthalmol* 59:495–501, 1958
4. *IARC Monographs on the Evaluation of the Carcinogenic Risk of Chemicals to Man*, Vol 15, Some fumigants, the herbicides 2,4-D and 2,4,5-T, chlorinated dibenzodioxins and miscellaneous

industrial chemicals, pp 255–261. Lyon, International Agency for Research on Cancer, August 1977

RESORCINOL
CAS: 108-46-3

$C_6H_4(OH)_2$

Synonyms: *m*-Dihydroxybenzene; resorcin; 1,3-benzenediol; 1,3-dihydroxybenzene

Physical Form. White crystals that turn pink on exposure to air

Uses. In manufacture of rubber products, wood adhesives, dyes, explosives, and cosmetics; in photography

Exposure. Inhalation; skin absorption

Toxicology. Resorcinol is an irritant of the eyes and the skin; in animals exposed at high concentrations, it affects the central nervous system.

Workers exposed to airborne levels of 10 ppm (45 mg/m³) for periods of 30 minutes or more reported no irritation or discomfort.[1] Application to the skin of solutions or ointments containing from 3% to 25% resulted in hyperemia, itching, dermatitis, edema, and corrosion.[2] Systemic effects from skin absorption have been restlessness, methemoglobinemia, convulsions, tachycardia, dyspnea, and death.[3] Ingestion of resorcinol induces similar signs and symptoms. Resorcinol has also been reported to cause sensitization and goiter.[3]

No toxic signs were observed in rats exposed by inhalation to 7800 mg/m³ (1733 ppm) for 1 hour or 2800 mg/m³ (625 ppm) for 8 hours.[1] When rats, rabbits, and guinea pigs were exposed to 34 mg/m³ for 6 hr/day for 2 weeks, no toxic effects were observed.[1]

Repeated gavage doses ranging from 55 mg/kg/day up to 450 mg/kg/day, 5 days/week, for 2 weeks caused tachypnea and hyperexcitability within 30 minutes of dosing to F344/N rats.[4] In 13-week studies, rats given 65 mg/kg/day or more had increased liver weights, whereas mice had significantly reduced adrenal weights when administered 28 mg/kg/day or more for the same period.[4] There was no evidence of carcinogenicity in rats or mice receiving up to 225 mg/kg/day, 5 days/week, for 2 years.[4] In a dermal oncogenicity study, 3 groups of female Swiss mice were treated with 0.02 ml of 5%, 25%, and 50% solutions of resorcinol in acetone twice weekly for 100 weeks.[5] The percentage of tumor-bearing animals was similar in the resorcinol-treated, untreated, and acetone-treated groups. Under the conditions of the test, resorcinol was considered noncarcinogenic.

A 10% solution in rabbit eyes has caused pain, conjunctivitis, and corneal vascularization.[6] Dry, powdered resorcinol applied to rabbit eyes has caused necrosis and corneal perforation.

No evidence of teratogenicity was found in rats receiving up to 250 mg/kg/day on days 6 through 15 of gestation.[7] This dose was maternally toxic, causing reduced body weight.

Although an epidemiologic study suggested that the occurrence of human thyroid disorder in certain geographic areas may be associated with the presence of resorcinol and related phenolic compounds in the food and water, no conclusions can be drawn from the data.[8]

The 1995 ACGIH threshold limit value–time-weighted average (TLV-TWA) is 10 ppm (45 mg/m³) with a short-term excursion limit of 20 ppm (90 mg/m³).

REFERENCES

1. Flickenger CW: The benzenediols: catechol, resorcinol and hydroquinone—review of the industrial toxicology and current industrial exposure limits. *Am Ind Hyg Assoc J* 37:596–606, 1976
2. Strakosch EA: Studies on ointments: ointments containing resorcinol. *Arch Dermatol Syph* 48:393, 1943
3. Deichmann WB, Keplinger ML: Phenols and phenolic compounds. In Clayton GD, Clayton FE (eds): *Patty's Industrial Hygiene and Toxicology*, 3rd ed, rev, Vol 2A, *Toxicology*, pp

2586–2589. New York, Wiley-Intersccience, 1981

4. National Toxicology Program: *Toxicology and Carcinogenesis Studies of Resorcinol (CAS No 108-46-3) in F344/N Rats and B6C3F₁ Mice (Gavage Studies).* NTP Technical Rept No 403, Pub No 92-2858. Research Triangle Park, NC, US Department of Health and Human Services, Public Health Service, National Institutes of Health, National Toxicology Program, 1992

5. Stenback F, Shubik P: Lack of toxicity and carcinogenicity of some commonly used cutaneous agents. *Toxicol Appl Pharmacol* 30:7–13, 1974

6. Estable JJ: The ocular effect of several irritant drugs applied directly to the conjunctiva. *Am J Ophthalmol* 31:837, 1948

7. DiNardo JC, Picciano JC, Schnetzinger RW, et al: Teratological assessment of five oxidative hair dyes in the rat. *Toxicol Appl Pharmacol* 78:163–166, 1985

8. Staff: Resorcinol linked to regional thyroid disorders. *Chem Engr News*, pp 66–67, April 28, 1986

RHODIUM AND COMPOUNDS
CAS: 7440-16-6

Rh

Principal Compounds: Rhodium trichloride; rhodium trioxide; rhodium (II) acetate; rhodium nitrate; rhodium potassium sulfate; rhodium sulfate; rhodium sulfite

Physical Form.   Silver-white metal

Uses.   Electroplating; in manufacture of rhodium-platinum alloys; in manufacture of high-reflectivity mirrors

Exposure.   Inhalation

Toxicology.   There are no data demonstrating acute or chronic rhodium-related diseases; irritation and sensitization have occasionally been reported in humans from exposure to the salts of rhodium. Solutions of insoluble salts splashed in the eye may cause mild irritation.

There are few reports of contact dermatitis from rhodium.[1-3] Of 12 workers in a precious-metal factory suffering from contact dermatitis, 7 were sensitized to rhodium, according to scratch-patch tests.[2] In a recent report, a woman working in a goldsmith's shop suffered occupational contact dermatitis from rhodium sulfate.[1] The investigators concluded that rhodium may be a potential sensitizer as a salt but not as a metal. Although metallic rhodium appears to have no sensitizing potential, when used as a coating for objects made of other metals, it may not prevent the sensitizing capacity of the underlying material (eg, nickel).[1]

The LD_{50} for rhodium trichloride in rabbits by intravenous injection was 215 mg/kg; the clinical signs presented shortly after injection were increasing lethargy and waning respiration.[4] There were no abnormal findings at autopsy, but the rapid onset of death suggested central nervous system effects.

A solution of rhodium trichloride in the eye of a rabbit gave a delayed injurious reaction; 0.1 mg solution adjusted to pH 7.2 with ammonium hydroxide was placed for 10 minutes in a rabbit eye after the corneal epithelium had been removed; an orange coloration of the cornea occurred which faded to faint yellow within 8 weeks.[5] During the first 2 to 3 weeks, the cornea was slightly hazy; in the third week, white opacities gradually developed and, finally, there were extensive opacification and vascularization.

Lifetime exposure to 5 ppm rhodium trichloride in the drinking water caused a minimally significant increase in malignant tumors in mice; lymphomas, leukemias, and adenocarcinomas were most prevalent.[6]

Chick embryos exposed to rhodium on the eighth day of incubation were stunted; mild reduction of limb size and feather growth inhibition were also observed.[7] A number of rhodium compounds have tested positive in bacterial assays for genetic-altering capability.[8]

The 1995 ACGIH threshold limit value–time-weighted average (TLV-TWA) are 1.0 mg/m³ for the metal, 1.0 mg/m³ as Rh—

insoluble compounds, and 0.01 mg/m^3 as Rh— soluble compounds.

REFERENCES

1. Cuadra J, Grau-Massanes M: Occupational contact dermatitis from rhodium and cobalt. *Cont Derm* 25:182–184, 1991
2. Bedello PG, Goitre M, Roncarolo G: Contact dermatis to rhodium. *Cont Derm* 17:111–112, 1987
3. Nakayama H, Imai T: Occupational contact uticaria, contact dermatitis and asthma caused by rhodium hypersensitivity. 6th International Symposium on Contact Dermatitis and Joint Meeting between ICDRG and JCDRG, Tokyo, 21 May 1982
4. Landholt RR, Berk HW, Russell HT: Studies on the toxicity of rhodium trichloride in rats and rabbits. *Toxicol Appl Pharmacol* 21:589–590, 1972
5. Grant WM: *Toxicology of the Eye*, 3rd ed, p 792. Springfield, IL, Charles C Thomas, 1986
6. Schroeder HA, Mitchener M: Scandium, chromium (VI) gallium, yttrium, rhodium, palladium, indium in mice: effects on growth and life span. *J Nutr* 101:1431–1438, 1971
7. Ridgway LP, Karnofsky DA: The effects of metals on chick embryo: toxicity and production of abnormalities in development. *Ann NY Acad Sci* 55:203–215, 1952
8. Warren G et al: Mutagenicity of a series of hexacoordinate rhodium III compounds. *Mut Res* 88:165–173, 1981

RONNEL
CAS: 299-84-3

$(CH_3O)_2P(S)OC_6H_2Cl_3$

Synonym: *O,O*-Dimethyl-*O*-(2,4,5-trichlorophenyl) phosphorothioate

Physical Form. White, crystalline powder

Uses. Systemic insecticide in livestock

Exposure. Inhalation; ingestion

Toxicology. Ronnel is a weak cholinesterase inhibitor and has low toxicity.

On both single and repeated doses, ronnel affects the pseudoesterase of the plasma rather than the true acetylcholinesterase of the red blood cells.[1]

In an experiment on humans to evaluate the primary skin-irritating and skin-sensitizing potential of ronnel, 50 subjects received 3 applications/week for 3 weeks of gauze saturated with a 10% suspension of ronnel in sesame oil; there were no significant effects on the skin.[1]

In male rats, the oral LD$_{50}$ was 1.7 g/kg; effects were salivation, tremor, diarrhea, miosis, and respiratory distress—all attributed to the anticholinesterase effect of ronnel.[1] Rats fed 50 mg/kg body weight in the diet for 105 days developed slight liver and kidney damage.

Dogs fed 10 mg/kg/day for 2 years showed no overt clinical signs, or evidence of any effect on urinalysis, hematologic analysis, organ weight measurement, or histological evaluation of the tissues; depression of plasma cholinesterase was the only significant finding.[2]

When a small amount of ronnel powder was placed in the eye of a rabbit, effects were slight discomfort and transient conjunctival irritation, which subsided within 48 hours.[1]

Daily oral administration of 600 or 800 mg/ kg ronnel to dams on days 6 through 15 of pregnancy caused a significant dose-related increase in fetuses with an extra rib.[3]

Ronnel has not been shown to potentiate the effect of other commonly used organophosphorus insecticides.

Note: For a description of diagnostic signs, differential diagnosis, and medical control, including clinical laboratory tests, as well as specific treatment of overexposure to anticholinesterase insecticides, see "Parathion".

The 1995 ACGIH threshold limit value–time-weighted average (TLV-TWA) is 10 mg/m^3.

REFERENCES

1. McCollister DD, Oyen F, Rowe VK: Toxicological studies of *O,O*-dimethyl-*O*-(2,4,5-tri-

chlorophenyl) phosphorothioate (ronnel) in laboratory animals. *J Agric Food Chem* 7:689–693, 1959

2. Worden AN et al: Effect of ronnel after chronic feeding to dogs. *Toxicol Appl Pharmacol* 23:1–9, 1972

3. Khera KS et al: Teratogenicity study on pyrethrum and rotenone (natural origin) and ronnel in pregnant rats. *J Toxic Environ Health* 10:111–119, 1982

ROTENONE
CAS: 83-79-4

$C_{23}H_{22}O_6$

Synonyms: Derrin; nicouline; tubatoxin

Physical Form.   Colorless crystals

Uses.   Insecticide; lotion for chiggers; emulsion for scabies

Exposure.   Inhalation

Toxicology.   Rotenone affects the nervous system and causes convulsions in animals.

The lethal oral dose in humans is estimated to be 0.3 to 0.5 g/kg.[1] Symptoms of absorption in humans (inferred mostly from animal studies) may include: numbness of oral mucous membranes, nausea, vomiting, abdominal pain, muscle tremor, incoordination, clonic convulsions, and stupor.[1] The dust is irritating to the eyes, skin, and respiratory tract.[2]

Animals repeatedly fed derris powder (a botanical source containing 9.6% rotenone) at levels from 312 to 5000 ppm developed focal liver necrosis and mild kidney damage.[2] The oral LD_{50} values vary greatly depending on particle size, manner of dispersion, activity of sample, and species tested. Values ranging from 25 mg/kg in rats to more than 3000 mg/kg in rabbits have been reported.[3]

At the cellular level, rotenone inhibits cellular respiration by blocking electron transport between flavoprotein and ubiquinone. It also inhibits spindle microtubule assembly.[3]

Rotenone has been reported to induce tumors in female Wistar rats. Of 40 female rats given daily intraperitoneal injections of 1.7 mg/kg/bw rotenone in sunflower oil for 42 days, over 60% developed mammary tumors 6 to 11 months after the end of treatment. Most of the tumors were mammary adenomas and 1 was a differentiated adenocarcinoma. None of the control animals had tumors when examined 19 months after treatment.[4]

Recent attempts to replicate these results have not been successful. Specifically, rotenone was not carcinogenic for the mammary gland in female Wistar rats when injected ip, 5 days/week for 8 weeks, at 1.0, 2.0 mg/kgbw in vehicles of sunflower oil or sunflower oil: chloroform.[5] Furthermore, tumors at other sites were not significantly different from those observed in control animals. Additional studies, including a 14-month oral gavage bioassay in Wistar rats, an 18-month ip injection bioassay in Sprague–Dawley rats, an 18-month feeding study in Syrian golden hamsters, and a 2-year feeding study in Fischer 344 rats and B6C3F$_1$ mice, have also shown no evidence of carcinogenicity for rotenone.[6,7]

Administered orally to rats on days 6 to 15 of pregnancy, 10 mg/kg was highly toxic to dams, killing 12 of 20; there was a significant decrease in the number of live fetuses per surviving dam and an increase in the proportion of resorptions.[8] In the 5 mg/kg group, there was an increased frequency of skeletal aberrations, such as an extra rib, delayed ossification of a sternebrum, and missing sternebrae.[8]

The 1995 ACGIH threshold limit value–time-weighted average (TLV-TWA) is 5 mg/m^3.

REFERENCES

1. Gosselin RE et al: *Clinical Toxicology of Commercial Products. Section III*, 5th ed, pp 366–368. Baltimore, MD, Williams and Wilkins, 1984
2. Negherbon WO (ed): *Handbook of Toxicology*, Vol III, p 665. Philadelphia, WB Saunders, 1957
3. Hayes WJ Jr: *Pesticides Studied in Man*, pp 82–

86. Baltimore, MD, Williams and Wilkins, 1982
4. Gosalvez M, Merchan J: Induction of rat mammary adenomas with the respiratory inhibitor rotenone. *Cancer Res* 33:3047–3050, 1973
5. Greenman DL, Allaben WT, Burger GT, et al: Bioassay for carcinogenicity of rotenone in female Wistar rats. *Fund Appl Toxicol* 20:383–390, 1993
6. Freundenthal RI, Thake DC, Baron RL: *Project Summary—Carcinogenic Potential of Rotenone:* Subchronic oral and peritoneal administration to rats and chronic dietary administration to Syrian golden hamsters. Health Effect Research Laboratory Report EPA-66/Si-81-03 7. Research Triangle Park, NC, US Environmental Protection Agency, 1981
7. Abdo KM, Eustis SL, Haseman J, et al: Toxicity and carcinogenicity of rotenone given in the feed to F344/N rats and B6C3F1 mice for up to two years. *Drug Chem Toxicol* 11:225–235, 1988
8. Khera KS et al: Teratogenicity study on pyrethrum and rotenone (natural origin) and ronnel in pregnant rats. *J Toxicol Environ Health* 10:111–119, 1982

SELENIUM AND COMPOUNDS
CAS: 7782-49-2

Se

Compounds: Selenium dioxide; selenium trioxide; selenium oxychloride; sodium selenite; sodium selenate; hydrogen selenide; selenic acid; selenium sulfide; selenium disulfide

Physical Form. Elemental selenium occurs as gray to black crystals; many compounds are solids although hydrogen selenide is a colorless gas

Uses/Sources. In electronics; selenium rectifiers and photocells; in xerography to coat the metal cylinders from which a photographic image is transferred; in glass and ceramics manufacture exposure also may occur during smelting and refining of ores containing selenium

Exposure. Inhalation

Toxicology. Selenium is an essential trace element that can be toxic in excessive amounts. Elemental selenium and selenium compounds as dusts, vapors, and fumes are irritants of the eyes, mucous membranes, and skin. Chronic exposure may cause central nervous system effects, gastrointestinal disturbances, and loss of hair and fingernails.

A group of workers briefly exposed to unmeasured but high concentrations of selenium fume developed severe irritation of the eyes, nose, and throat, followed by headaches. Transient dyspnea occurred in one case.[1] A case of accidental hydrogen selenide poisoning also resulted in irritation of the mucous membranes; following a brief recovery period, pulmonary edema, bronchitis, and bronchial pneumonia occurred.[2] Workers exposed to an undetermined concentration of selenium oxide developed bronchospasm and dyspnea, followed within 12 hours by metal fume fever (chills, fever, headache) and bronchitis, leading to pneumonitis in a few cases; all were asymptomatic within a week.[3]

In a study of workers in a selenium plant, workroom air levels ranged from 0.2 to 3.6 mg/m³, and urinary levels ranged from below 0.10 to 0.43 mg/l of urine. The chief complaints were garlic odor of the breath, metallic taste, gastrointestinal disturbances, and skin eruptions.[4]

An endemic disease in China, characterized by loss of hair and nails, skin lesions, and abnormalities of the nervous system, including some paralysis and hemiplegia, was attributed to chronic selenium poisoning.[5] The daily intake for 6 affected individuals averaged 5.0 mg versus 0.1 mg for people from an unaffected area. Changing the diet led to recoveries. There have been no reports of disabling chronic disease or death from industrial exposure.

An accidental spray of selenium dioxide into the eyes of a chemist caused superficial burns of the skin and immediate irritation of the eyes. Within 16 hours, the subject's vision was blurred, and the lower portions of both corneas appeared dulled. Sixteen days after the accident, the corneas were normal.[6]

The element selenium is not particularly irritating, but various compounds, such as selenium oxychloride and selenium dioxide, are strong vesicants.[7] Skin contact with the fume of heated selenium dioxide caused an acute, weeping dermatitis, with the development of hypersensitivity in some cases.[8] Selenium dioxide forms selenious acid when in contact with water; if allowed to penetrate beneath the fingernails, it causes an especially painful inflammatory reaction.[8] Compounds of selenium can be absorbed through the unbroken skin. Selenium sulfide, which is found in some shampoos, can penetrate the scalp and cause generalized toxic responses.[7]

In livestock, selenium has been found to be the cause of "blind staggers" and alkali disease. Blind staggers occurs as a result of acute ingestion of seleniferous plants and is characterized by impaired vision, depressed appetite, a tendency to wander in circles, paralysis, and death from respiratory failure.[9] A more chronic syndrome described in horses and livestock is alkali disease, which also is associated with consumption of grains or plants containing selenium. The disease is characterized by lack of vitality, loss of appetite, emaciation, deformed hoofs, loss of hair, erosion of the joints of long bones, anemia, cirrhosis, and cardiac atrophy.[9]

In a number of reproductive studies using a variety of selenium compounds, adverse effects have been seen only at doses that are associated with maternal toxicity[10,11]

Epidemiological studies in humans do not suggest an association between excess exposure to selenium and cancer.[11] Low levels of intake, however, have been associated with an increased risk of developing many kinds of cancers. With the exception of selenium sulfide, most animal studies have shown that selenium compounds inhibit tumorigenesis.[11] High doses of selenium sulfide administered by gavage caused liver tumors in rats and lung and liver tumors in female mice.[12] Mutagenic and antimutagenic effects of selenium have also been reported.[11,13]

(See separate entries for selenium hexafluoride and hydrogen selenide.)

The 1995 ACGIH threshold limit value–time-weighted average (TLV-TWA) for selenium and compounds is 0.2 mg/m^3, as Se.

REFERENCES

1. Clinton M Jr: Selenium fume exposure. *J Ind Hyg Toxicol* 29:225–226, 1947
2. Olson OE: Selenium toxicity in animals with emphasis on man. *J Am College Toxicol* 5:45–70, 1986
3. Wilson HM: Selenium oxide poisoning. *JAMA* 180(8):173–174, 1962
4. Glover JR: Selenium and its industrial toxicology. *Ind Med Surg* 39:50–54, 1970
5. Yang G, Wang S, Zhou R, et al: Endemic selenium intoxication of humans in China. *Am J Clin Nutr* 37:872–881, 1983
6. Middleton JM: Selenium burn of the eye. *AMA Arch Ophthalmol* 38:806–811, 1947
7. Wilber CG: Toxicology of selenium: a review. *Clin Toxicol* 17:171–230, 1980
8. Committee on Medical and Biological Effects of Environmental Pollutants, National Research Council: *Selenium*, pp 116–118. Washington, DC, National Academy of Sciences, 1976
9. Hogberg J, Alexander J: Selenium. In Friberg L et al (eds): *Handbook on the Toxicology of Metals*, 2nd ed, Vol II, *Specific Metals*, pp 482–520. Amsterdam, Elsevier, 1986
10. Domingo JL: Metal-induced developmental toxicity in mammals: a review. *J Toxicol Environ Health* 42:123–141, 1994
11. Agency for Toxic Substances and Disease Registry (ATSDR): *Toxicological Profile for Selenium*. US Public Health Service, p 185, 1989
12. National Cancer Institute: *Bioassay of Selenium Sulfide (Gavage) for Possible Carcinogenicity*. DHHS (NIH) Pub No 80-1750, p 130. Washington, DC, US Government Printing Office, 1980
13. Shamberger RJ: The genotoxicity of selenium. *Mutat Res* 154:29–48, 1985

SELENIUM HEXAFLUORIDE
CAS: 7783-79-1

SeF_6

Synonym: Selenium fluoride

Physical Form. Colorless gas

Uses. Gaseous electric insulator

Exposure. Inhalation

Toxicology. Selenium hexafluoride is a severe pulmonary irritant in animals; heavy exposure is expected to cause the same effect in humans.

There are no reports of human exposure to selenium hexafluoride.

Exposure of 4 animal species to 10 ppm for 4 hours was fatal; 5 ppm for 5 hours was not fatal but caused pulmonary edema, whereas 1 ppm produced no effects.[1] Animals exposed to 5 ppm for 1 hour daily for 5 days developed signs of pulmonary injury; 1 ppm for the same time period caused no effects.

The 1995 ACGIH threshold limit value–time-weighted average (TLV-TWA) for selenium hexafluoride is 0.05 ppm (0.16 mg/m³).

REFERENCES

1. Kimmerle G: Comparative investigation into the inhalation toxicity of the hexafluorides of sulfur, selenium, and tellurium. *Archiv Toxikol* 18:140–144, 1960

SILICA, AMORPHOUS—DIATOMACEOUS EARTH
CAS: 61790-53-2

SiO_2

Synonyms: Diatomite; diatomaceous silica; infusorial earth

Physical Form. Solid; soft, chalky powder

Uses. In the production of filters, polishes, absorbents, and insulators

Exposure. Inhalation

Toxicology. Amorphous silica, natural diatomaceous earth, is usually considered to be of low toxicity; however, pure amorphous silica is rarely found. Processing of amorphous silica by high-temperature calcining alters the silica from the benign amorphous to the pathogenic crystalline form (cristobalite), which causes fibrosis. Characteristically, natural diatomite contains no measurable cristobalite. Depending on the source, it may contain a low percentage of contaminating quartz, rarely over 2%. Non-flux-calcined diatomite may contain from 20% to 30% cristobalite, whereas flux-calcined diatomite may contain as much as 60% cristobalite.[1-3] Non-flux-calcined and flux-calcined diatomite can produce severe and disabling pneumoconiosis, which is attributed to their cristobalite content. Although a form of silicosis, it characteristically produces pathologic and radiographic changes, which are different from classical quartz silicosis. Diffuse rather than modular changes are more common.[2]

In a study of diatomaceous earth workers, those employed in the quarry for more than 5 years and exposed only to natural diatomaceous earth had no significant roentgenologic changes. Of others employed for more than 5 years in the milling process and exposed to calcined material, 17% had simple pneumoconiosis, and 23% had the confluent form, probably the result of fibrogenic action of the crystalline silica formed by calcination of the naturally occurring mineral.[3,4]

In humans, calcined diatomaceous earth pneumoconiosis is characterized roentgeno-graphically by fine linear and/or minute nodular shadows, either or both of which may be accompanied by conglomerate fibrosis. In the simple phase of the disease, the upper lobes are affected more than the lower lobes, and the condition progresses by an increase in the apparent number of the nodules, which rarely attain the density or size of nodules often seen in quartz silicosis.[4] In the early confluent stage of the disease, the linear and nodular changes in the upper lung fields become more circumscribed and homogeneous. Histologically, there is an absence of the focal, discrete, hyaline nodules or the whorled pattern of collagenous fibers of typical silicosis.[4,5]

Repeated exposure of guinea pigs to natural diatomaceous earth for periods of up to 50 weeks to average concentrations ranging from 60 to 124 mg/m^3 caused thickening of the alveolar septa by infiltration of macrophages, accumulation of large numbers of multinuclear cells containing dust particles, and lymphadenopathy but no proliferation of connective tissue.[6]

There is no evidence to associate any form of diatomaceous earth with human cancer, but studies directly addressing the question are limited.

Amorphous silica has been tested for carcinogenicity in a variety of animal studies by a number of routes.[7] Most of the tests were negative or were inadequate, primarily as a result of poorly defined physiochemical characteristics of the silica. The IARC concluded that evidence is inadequate to describe amorphous silica as carcinogenic in either experimental animals or humans. Crystalline silica, however, has been designated by the IARC as a probable human carcinogen (category 2A), based on "sufficient evidence" in experimental animals and "limited evidence" in man.[2,7,8] Therefore, although evidence for the carcinogenicity of crystalline silica in man is unconvincing, certainly from exposures insufficient to cause silicosis, appropriate hazard warnings are obligatory in the United States. These apply to all materials containing 0.1% or more of crystalline silica (quartz, cristobalite, and/or tridymite).[2]

The 1995 proposed threshold limit value–time-weighted average (TLV-TWA) for amorphous natural diatomaceous earth silica is 10 mg/m^3 for the inhalable particulate and 3 mg/m^3 for respirable dust containing no asbestos and <1% quartz.

REFERENCES

1. Cooper WC, Cralley LJ: *Pneumoconiosis in Diatomite Mining and Processing.* Public Health Services Pub No 601. Washington, DC, Government Printing Office, 1958
2. Cooper WC: *Effects of Diatomaceous Earth on Human Health—A Review of the Literature.* Long Beach, CA, International Diatomite Producers Assn, 1988
3. Dutra FR: Diatomaceous earth pneumoconiosis. *Arch Environ Health* 11:613–619, 1965
4. Oechsli WR, Jacobson G, Brodeur AE: Diatomite pneumoconiosis: roentgen characteristics and classification. *Am J Roentgenol Radium Ther Nucl Med* 85:263–270, 1961
5. Smart RH, Anderson WM: Pneumoconiosis due to diatomaceous earth—clinical and x-ray aspects. *Ind Med Surg* 21:509–518, 1952
6. Tebbens BD, Beard RR: Experiments on diatomaceous earth pneumoconiosis. I. Natural diatomaceous earth in guinea pigs. *AMA Arch Ind Health* 16:55–63, 1957
7. *IARC Monographs on the Evaluation of Carcinogenic Risk of Chemicals to Humans,* Vol 42, Silica and some silicates, pp 39–143. Lyon, International Agency for Research on Cancer, 1987
8. *IARC Monographs on the Evaluation of Carcinogenic Risks to Humans, Overall Evaluations of Carcinogenicity: An Updating of IARC Monographs, Vols 1–42,* Suppl 7. International Agency for Research on Cancer, 1987

SILICA, AMORPHOUS—FUME
CAS: 69012-64-2

SiO₂

SiO_2

Physical Form. Fine white powder with particle sizes generally below 1 μm. This is not the same as the commercial products fumes silica, silica gel, precipitated silica, or fused silica. It is formed during the electric arc production of elemental silicon from quartz, which is reduced to silicon monoxide, escaping from the furnace and oxidized by air to silicon dioxide. This condenses to form spherical particles. For the production of ferrosilicon, iron metal is added to the charge; during charging, there is also exposure to crystalline silica[1]

Uses. None; produced only as a by-product

Exposure. Inhalation

Toxicology. Amorphous silica fume exposure is associated with recurrent fever, similar to metal-fume fever, and nonprogressive pulmonary changes.

Adverse effects on the lungs of workers exposed to the fumes of ferrosilicon furnaces have been recognized since 1937. Subsequent clinical studies of workers exposed to amorphous silica fume in silicon and ferrosilicon plants reported pulmonary symptoms and X ray findings difficult to differentiate from classical silicosis due to crystalline silica, especially since there is often concurrent exposure to quartz dust during furnace operations.[1–4]

The disease process in workers exposed to silica fume was originally described as silicosis or acute silicosis, but it is now recognized that the X ray pattern and symptom complex are different from both, the severity of the symptoms is less, and there is apparently no progression. It has been postulated that heavy exposure to freshly formed silica fume causes an acute reaction similar to metal-fume fever. Continued or repeated exposure causes the "ferroalloy disease," which has been described.[1,2,5] This is characterized by recurrent fever over a period of 3 to 12 weeks, with the appearance of X ray markings similar to silicosis. The development of classical silicosis may be the result of long, continued exposure to amorphous silica fume or possibly concurrent exposure to crystalline silica.[1,5]

Of 900 African production workers in a ferroalloy plant, 35 cases of "ferroalloy worker disease" were identified over a 10-year period. These were either acute episodes of metal-fume fever or pulmonary fibrosis recognized by X ray. Over a period of 2 to 6 years after first diagnosis, 22 cases remained static, regressed, or returned to normal; 8 cases progressed, with increased fibrosis and nodulation by X ray.[4–6]

Autopsy of cases of alleged silicosis in Swedish ferrosilicon workers revealed no silicosis, and it was postulated that pulmonary conditions that had been recognized by X ray may have been due to unspecified infectious changes.[7] This study also concluded that "exposure to fumes and dust particles, for the most part amorphous SiO₂, in ferrosilicon alloy melting works, does not seem to give rise to a serious risk of silicosis although the additional handling of quartz in this industry certainly constitutes a grave risk of silicosis."

The 1995 ACGIH threshold limit value–time-weighted average (TLV-TWA) for amorphous silica fume is 2 mg/m³ for the respirable fraction of dust.

REFERENCES

1. Silica, amorphous-fume. *Documentation of TLVs and BEIs*, 6th ed, pp 1367–1370. Cincinnati, OH, American Conference of Governmental Industrial Hygienists (ACGIH), 1991
2. Princi F, Miller LH, Davier A, Cholak J: Pulmonary disease of ferroalloy workers. *J Occup Med* 4:301–310, 1962
3. Vitums VC, Edwards MJ, Niles NR, Borman JO: Pulmonary fibrosis from amorphous silica dust: a product of silica vapor. *Arch Environ Health* 32:62–68, 1977
4. Davis JCA: Inhalation hazards in the manufacture of silicon alloys. *Cent Afr J Med* 20(7):140, 1974
5. Taylor DM, Davies JCA: Ferro-alloy workers' disease: a report of a recent case against the

background of twelve years' experience. *Cent Afr J Med* 23:28, 1977
6. Bowie DS: Ferro-alloy workers' disease. *Cent Afr J Med* 24(5):81, 1978
7. Swenson A, Kvarnstrom K, Bruce T, et al: Pneumoconiosis in ferrosilicon workers—a follow-up study. *J Occup Med* 13:427–432, 1971

SILICA, CRYSTALLINE—QUARTZ
CAS: 14808-60-7

SiO₂

Synonyms: Silicon dioxide; silicic anhydride

Physical Form.   Colorless crystals

Uses.   In the manufacture of glass, porcelain, and pottery; metal casting; sandblasting; granite cutting; in the manufacture of refractory, grinding, and scouring compounds

Exposure.   Inhalation

Toxicology.   Crystalline silica causes silicosis, a form of disabling, progressive, and sometimes fatal pulmonary fibrosis characterized by the presence of typical nodulation in the lungs.[1]

The earliest lesions are seen in the region of the respiratory bronchioles. Lymphatics become obliterated by infiltration with dust-laden macrophages and granulation tissue. Morphologically, the typical lesion of silicosis is a firm nodule composed of concentrically arranged bundles of collagen; these nodules usually measure between 1 to 10 mm in diameter and appear around blood vessels and beneath the pleura, as well as in mediastinal lymph nodes. There may be conglomeration of nodules as the disease progresses, leading to massive fibrosis.[1] The pulmonary pleura is usually thickened due to fibrosis and is often adherent to the parietal pleura, especially over the upper lobes and in the vicinity of underlying conglomerate lesions.[2]

Histologically, the silicotic nodule consists of a relatively acellular, avascular core of hyalinized reticulin fibers arranged concentrically and blending with collagen fibers toward the periphery, which has well-defined borders.[3] The particles of silica responsible for the reaction are birefringent and can be visualized under polarized light if they exceed 1 μ in diameter. Silica in the lungs can be identified by X ray diffraction studies and by incinerating a portion of the lung, with subsequent analysis of the ash. The silica content of the normal lung should not exceed 0.2% dry weight.

The clinical signs and symptoms of silicosis tend to be progressive, with continued exposure to quantities of dust containing free silica, with advancing age, and with continued smoking habits.[1] Symptoms may also be exacerbated by pulmonary infections and cardiac decompensation. Symptoms include cough, dyspnea, wheezing, and repeated nonspecific chest illnesses. Impairment of pulmonary function may be progressive. In individual cases, there may be little or no decrement when simple discrete nodular silicosis is present but, when nodulations become larger or when conglomeration occurs, recognizable cardiopulmonary impairment tends to occur.

The progression of symptoms may continue after dust exposure ceases. Athough there may be a factor of individual susceptibility to a given exposure to silica dust, the risk of onset and the rate of progression of the pulmonary lesion are clearly related to the character of the exposure (dust concentration and duration).[1] The disease tends to occur after an exposure measured in years rather than in months. It is generally accepted that silicosis predisposes to active tuberculosis and that the combined disease tends to be more rapidly progressive than uncomplicated silicosis.

The earliest radiographic evidence of nodular silicosis consists of small discrete opacities of 1- to 3-mm diameter appearing in the upper lung fields. As the disease advances, discrete opacities increase in number and size and are seen in the lower as well as the other zones of the lung fields. Small conglomerations may then appear, subsequently developing into large, irregular, and sometimes massive opacities occupying the greater part of both lung fields. Bullae may be seen in the vicinity of conglomerations.[2]

A group of 972 granite shed workers was studied to relate exposure levels to incidence of silicosis.[4] The workers were grouped according to four average exposure levels: (1) 37 to 60, (2) 27 to 44, (3) 20, and (4) 3 to 9 mppcf. Those with the highest dust exposure showed development of early silicosis in 40% of the workers after 2 years, and 100% after 4 years of exposure. The development of silicosis in the remaining workers appeared to be proportional to the dust exposure. At the second-highest exposure level (27 to 44 mppcf), early stages of silicosis appeared after 4 years of exposure, and more advanced stages developed by the seventh year. In the group exposed at an average of 20 mppcf, there was little indication of severe effects on the health of the workers. In the lowest exposure group, where the average dust concentration was 6 mppcf (range 3 to 9 mppcf), there was no indication of any untoward effects of dust exposure on workers.

In some occupations, such as sandblasting and production of silica flour, exposure to high concentrations of silica over only a few years has produced a more rapidly progressive form of the disease termed accelerated silicosis. The symptoms are those of the more chronic disease, but clinical and radiologic progression is rapid.[5]

An acute form of silicosis has occurred in a few workers exposed to very high concentrations of silica over periods of as little as a few weeks. The history is one of progressive dyspnea, fever, cough, weight loss and, in severe cases, death within a year or two. In acute silicosis, the nodular pattern is absent, the lungs showing a diffuse ground-glass appearance, similar to pulmonary edema.[5]

Exposure of silica has also been related to chronic airflow limitation without radiographic changes and renal disease.[6,7]

Epidemiologic studies have been conducted in an effort to assess the role of silica exposure in the pathogenesis of lung cancer.[8–10]

Some studies of mining, quarry, tunnel, and foundry workers have shown moderately raised SMRs for lung cancer, ranging from 127 to 156.[11] However, the role of smoking or other contributing factors, such as radon exposure, cannot be excluded, and other large cohort studies have not found any increased risk for lung cancer.

In a number of cohort and case-control studies, lung cancer has been found to occur more frequently in persons diagnosed as having silicosis following occupational exposure.[8,9] For example, threefold risk of death from lung cancer was found in men who had received compensation for silicosis in Quebec between 1938 and 1985 compared to the general population.[12] In most investigations, information on cigarette smoking, confounding exposures, exposure levels, and the referent population have not been adequately determined. The association between the presence and severity of silicosis and lung cancer has not been established in other studies. A study based on necropsy records of South African miners in which dust levels were known and smoking histories were established showed no association between silica exposure and lung cancer.[13] A recent mortality cohort of female mine workers exposed to silica revealed increased deaths from lung cancer, but the excess was not related to the level or duration of exposure.[14]

In animal studies, significant increases in adenocarcinomas and squamous-cell carcinomas of the lung have occurred in rats following inhalation or intratracheal instillation in rats but not in hamsters.[9]

The IARC has determined that there is insufficient evidence for the carcinogenicity of crystalline silica to experimental animals and limited evidence for the carcinogenicity of crystalline silica to humans.[9]

The 1995 ACGIH threshold limit value–time-weighted average (TLV-TWA) for crystalline quartz silica is 0.1 mg/m^3 for the respirable fraction of dust.

REFERENCES

1. National Institute for Occupational Safety and Health: *Criteria for a Recommended Standard . . . Occupational Exposure to Crystalline Silica.* DHEW (NIOSH) Pub No 75-120. Washington, DC, US Government Printing Office, 1974

2. Parkes WR: *Occupational Lung Disorders*, 2nd ed, pp 142, 147–148. London, Butterworths, 1982

3. Levy SA: Occupational pulmonary diseases. In Zenz C (ed): *Occupational Medicine— Principles and Practical Applications*, pp 117, 129–134. Chicago, Year Book Medical Publishers, 1975

4. Russell AE, Britten RH, Thompson LR, Bloomfield JJ: *The Health of Workers in Dusty Trades—II. Exposure to Siliceous Dust (Granite Industry.* US Public Health Service Bull No 187. Washington, DC, US Government Printing Office, 1929

5. Seaton A. In Morgan WKC, Keith C: *Occupational Lung Diseases*, 2nd ed. p 686, Philadelphia, WB Saunders Co, 1984

6. Neukirch F, Cooreman J, Korobaeff M, et al: Silica exposure and chronic airflow limitation in pottery workers. *Arch Environ Health* 49:459–464, 1994

7. Goldsmith JR, Goldsmith DF: Fiberglass or silica exposure and increased nephritis or ESRD (end-stage renal disease). *Am J Ind Med* 23:873–881, 1993

8. Goldsmith DF, Winn DM, Shy CM: *Silica, Silicosis, and Cancer. Controversy in Occupational Medicine*. pp 1–536, Cancer research monographs V2. New York, Praeger, 1986

9. *IARC Monographs on the Evaluation of Carcinogenic Risk of Chemicals to Humans*, Vol 42, Silica and some silicates, pp 39–143. Lyon, International Agency for Research on Cancer, 1987

10. Holland LM: Crystalline silica and lung cancer: a review of recent experimental evidence. *Reg Toxicol Pharmacol* 12:224–237, 1990

11. McDonald JC: Silica, silicosis and lung cancer. *Br J Ind Med* 46:289–291, 1989

12. Infante-Rivard C, Armstrong B, Petitclerc M, et al: Lung cancer mortality and silicosis in Quebec, 1938–1985. *Lancet* 23/30:1504–1507, 1989

13. Hessel PA, Sluis-Cremer GK, Hnizdo E: Silica exposure, silicosis and lung cancer: a necropsy study. *Br J Ind Med* 47:4–9, 1990

14. Cocco PL, Carta P, Flore V, et al: Lung cancer mortality among female mine workers exposed to silica. *JOM* 36:894–898, 1994

SILICON
CAS: 7440-21-3

Si

Synonyms: None

Physical Form. Black to gray needlelike crystals

Uses. In manufacture of transistors, silicon diodes, and similar semiconductors; for making alloys such as ferrosilicon and silicon copper

Exposure. Inhalation

Toxicology. Silicon appears to be a biologically inert material.

Little information is available on the toxicology of pure elemental silicon, which is an inert material that appears to lack the property of causing fibrosis in lung tissue.[1] Silicon dust gave an inert response on intraperitoneal injection into guinea pigs and rats.[2] Another study, however, reported minimal pulmonary lesions in rabbits after the intratracheal injection of silicon dust at a high level of 25 mg.[3]

The 1995 threshold limit value–time-weighted average (TLV-TWA) is 10 mg/m^3 for total dust containing no asbestos and <1% crystalline silica.

REFERENCES

1. Silicon. *Documentation of the TLVs and BEIs*, 6th ed, pp 1387–1388. Cincinnati, OH, American Conference of Governmental Industrial Hygienists (ACGIH), 1991

2. McCord CP, Fredrick WG, Stolz S: The toxicity of silicon. *J Lab Clin Med* 23:278–279, 1937

3. Schepers GWH: Lung tumors of primates and rodents. *Ind Med Surg* 40:48–53, 1971

SILICON CARBIDE
CAS: 409-21-2

SiC

Synonyms: Carborundum; Crystolon; Carbonite; Carbofrax; Electrolon

Physical Form. Green to bluish-black iridescent crystals

Uses. In manufacture of abrasives and refractories, brake linings, heating elements, and thermistors

Exposure. Inhalation

Toxicology. Silicon carbide, in certain forms, may be a cause of pneumoconiosis in exposed workers.

Silicon carbide has generally been considered to be an inert dust with little adverse effect on the lungs.[1] Animal experiments have supported this view. In a recent study, rats injected intratracheally at 20 mg/day with silicon carbide dust for 50 exposures and observed for up to 12 months had no significant changes in the lungs.[2] Human studies, however, have reported abnormal chest radiographs compatible with pneumoconiosis and significant reductions in pulmonary functions among workers exposed to silicon carbide.[3,4] Pathological reports of silicon carbide pneumoconiosis identified silicon carbide but not significant amounts of other fibrogenic agents.[5,6]

In the silicon carbide manufacturing process, the major bioactive dusts identified are quartz particles and silicon carbide fibers generated in the process. In contrast to the silicon carbide fibers, silicon carbide particles were found in animal studies to be inert. The silicon carbide fibers have fibrogenic activities comparable to asbestos fibers of similar size and are likely to contribute to the pathogenesis of the interstitial lung disease of silicon carbide production workers.[3] Studies of exposure to silicon carbide whiskers (cylindrically shaped single crystals) in rats have also shown dose-related increases in the severity of alveolar, bronchiolar, and pleural wall thickening, and inflammatory lesions that did not reverse after a recovery period.[7]

These results have suggested that mineral dusts that are inert in a particulate form may have biological activity when they occur in a fibrous form.

The 1995 ACGIH threshold limit value–time-weighted average (TLV-TWA) is 10 mg/m³ for total dust containing <1% quartz.

REFERENCES

1. Parkes WR: *Occupational Lung Disorders,* 2nd ed, pp 130–131. London, Butterworths, 1982
2. Bruch J, Rehn B, Song W, et al: Toxicological investigations on silicon carbide. 2. In vitro cell tests and long term injection tests. *Br J Ind Med* 50:807–813, 1993
3. Begin R, Dufresne A, Cantin A, et al: Carborundum pneumoconiosis. *Chest* 95:842–849, 1989
4. Osterman JW, Greaves JA, Smith TJ, et al: Work related decrement in pulmonary function in silicon carbide production workers. *Br J Ind Med* 46:708–716, 1989
5. Funahashi A, Schueter D, Pintar KA, et al: Pneumoconiosis in workers exposed to silicon carbide. *Am Rev Respir Dis* 129:635–640, 1984
6. Hayashi H, Kajita A: Silicon carbide in lung tissue of a worker in the abrasives industry. *Am J Ind Med* 14:145–155, 1988
7. Lapin CA, Craig DK, Valerio MG, et al: A subchronic inhalation toxicity study in rats exposed to silicon carbide whiskers. *Fund Appl Toxicol* 16:128–146, 1991

SILICON TETRAHYDRIDE
CAS: 7803-62-5

SiH₄

Synonyms: Silane; monosilane

Physical Form. Colorless gas

Uses. In manufacture of solid-state devices; source of silicon for semiconductor manufacture

Exposure.   Inhalation

Toxicology.   Silicon tetrahydride is considered to be a skin, eye, and mucous membrane irritant.

There is no information regarding its toxicity to humans; by analogy with other tetrahydrides, it is thought to be an irritant.[1,2]

Silicon tetrahydride has a low acute toxicity in experimental animals. In rats, the 4-hour LC_{50} is 9600 ppm.[3] Rats exposed at 126 ppm for 1 hour were apparently unaffected.[2]

The potential for explosion, fire, and oxygen-deficient atmospheres constitutes the major hazards with silicon tetrahydride.

The 1995 ACGIH threshold limit value–time-weighted average (TLV-TWA) is 5 ppm (6.6 mg/m³).

REFERENCES

1. Wald PH, Becker CE: Toxic gases used in the microelectronics industry. In LaDou J (ed): *The Microelectronics Industry. State of the Art Reviews: Occupational Medicine*, Vol 1, pp 109–110, 1986
2. Silicon tetrahydride. *Documentation of the TLVs and BEIs*, 6th ed, pp 1394–1395. Cincinnati, OH, American Conference of Governmental Industrial Hygienists (ACGIH), 1991
3. Vernot EH, MacEwen JD, Haun CC, et al: Acute toxicity and skin corrosion data for some organic and inorganic compounds and aqueous solutions. *Toxicol Appl Pharmacol* 42:417–423, 1977

SILVER AND COMPOUNDS
CAS: 7440-22-4

Ag

Compounds: Silver nitrate; silver chloride; silver oxide; silver sulfide

Physical Form.   Elemental silver is a lustrous, white, solid metal.

Uses.   In photographic materials; in electrical and electronics products; in alloys and solders; in jewelry, mirrors, flatware, and coinage

Exposure.   Inhalation; oral; dermal

Toxicology.   The dust of silver and its soluble compounds causes local or generalized impregnation of the mucous membranes, skin and eyes with silver, a condition termed argyria.

Argyria may occur in an area of repeated or abrasive dermal contact with silver or silver compounds or, more extensively, over widespread areas of skin and the conjunctiva of the eyes following long-term oral or inhalation exposure.[1] Localized argyria occurs in the skin and eyes, where gray-blue patches of pigmentation are formed without evidence of tissue reaction.[2] Generalized argyria is recognized by the widespread pigmentation of the skin and may be seen first in the conjunctiva, with some localization in the inner canthus. Argyria of the respiratory tract has been described in two workers involved in the manufacture of silver nitrate. Their only symptom was mild chronic bronchitis. Bronchoscopy revealed tracheobronchial pigmentation. Biopsy of the nasal mucous membrane showed silver deposition in the subepithelial area.[2] It has been estimated that gradually accumulated intake of from 1 to 5 g of silver will lead to generalized argyria.[2]

Upper respiratory tract irritation has been observed in humans at estimated exposure levels of between 0.04 and 0.4 mg silver/m³ for less than 1 to greater than 10 years.[1] Irritant effects are considered to be related to the caustic properties of the various silver compounds rather than the silver itself.

Massive exposure to heated vapor of metallic silver for 4 hours by a workman caused lung damage with pulmonary edema.[3] Ingestion of 10 g silver nitrate is usually fatal. Large oral doses of the compound cause abdominal pain and rigidity, vomiting, convulsions, and shock.[4] Patients dying after intravenous administration of Collargol (silver plus silver oxide) showed necrosis and hemorrhage in the bone marrow, liver, and kidney.[4]

There is no historical information in humans to suggest that silver affects reproduction.[1] In an early animal study, no reduction in fertility or changes in spermatozoa were observed following 2 years of exposure to 89 mg silver/kg/day as silver nitrate or silver chloride in the drinking water.

Although fibrosarcomas have been reported in animals following subcutaneous embedding of silver foil, normal routes of exposure have not provided indications of carcinogenicity in animals or humans, and silver is not expected to be carcinogenic in humans.[1]

In genotoxic assays, the silver ion caused DNA strand breaks in vitro but was not mutagenic in *Salmonella typhymurium*.[1]

The 1995 ACGIH threshold limit value–time-weighted average (TLV-TWA) is 0.01 mg/m^3 for soluble compounds as Ag and 0.1 mg/m^3 for the metal dust and fume.

REFERENCES

1. Agency for Toxic Substances and Disease Registry (ATSDR): *Toxicological Profile for Silver.* US Department of Health and Human Services, Public Health Service, TP-90-24, p 145, 1990
2. Browning E: *Toxicity of Industrial Metals,* 2nd ed, pp 296–301. London, Butterworths, 1969
3. Forycki Z, Zegarski W, Bardzik J, et al: Acute silver poisoning through inhalation. *Bull Inst Maritime Trop Med Ingdynia* 34:199–203, 1983
4. US Environmental Protection Agency: *Ambient Water Quality, Criteria for Silver.* Springfield, VA, National Technical Information Service, PB81-117822, October 1980

SOAPSTONE

$3MgO\text{-}4SiO_2\text{-}H_2O$

Soapstone does not have a precise mineralogic definition but is of variable composition dependent on its source; talc is mined as soapstone, but some forms of soapstone have as little as 50% talc.

Synonyms: Steatite; massive talc

Physical Form. Talclike material of varying composition but generally grayish-white, fine, odorless powder. It is noncombustible and insoluble in water.

Uses. Pigment in paints and varnishes; filler for paper, rubber, and soap; for lubricating molds and machinery; heat insulator

Exposure. Inhalation

Toxicology. The fibrous talc in soapstone dust causes fibrotic pneumoconiosis; an increased incidence of cancer of the lungs and pleura has been reported.

In the development of talc pneumoconiosis, or talcosis, the subject initially is symptom-free, but cough and dyspnea develop as the disease progresses; cyanosis, digital clubbing, and cor pulmonale occur in advanced cases. The disease progresses slowly, even in the absence of continued exposure; occasionally, the disease may progress rapidly, with death occurring within a few years of a very heavy exposure.[1,2]

In an early report of 66 workers handling soapstone, no cases of pneumoconiosis were found in workers with an average dust exposure of 2.8 mg/m^3, but exposures ranging from 22 to 50 mg/m^3 caused severe cases.[3]

An epidemiologic study of 260 workers with 15 or more years of exposure to commercial talc dust, containing talc, tremolite, anthophyllite, carbonate dusts, and a small amount of free silica, revealed a four times greater than expected mortality rate from cancer of the lungs and pleura; in addition, a major cause of death among these workers was cor pulmonale—a result of the pneumoconiosis.[4,5]

The 1995 ACGIH threshold limit value–time-weighted average (TLV-TWA) for soapstone is 3 mg/m^3 as respirable dust containing no asbestos and <1% crystalline silica; and 6 mg/m^3 as inhalable dust containing no asbestos and <1% crystalline silica.

REFERENCES

1. Spiegel RM: Medical aspects of talc. In Goodwin A (ed): *Proceedings of the Symposium on Talc,* Bureau of Mines Report No 8639, pp 97–102.

Washington, DC, US Government Printing Office, 1973
2. Kleinfeld M, Messite J, Kooyman O, Zaki MH: Mortality among talc miners and millers in New York State. *Arch Environ Health* 14:663–667, 1967
3. Dreessen WC, DallaValle JM: The effects of exposure to dust in two Georgia talc mills and mines. *Pub Health Rept* 50:131–143, 1935
4. Blejer HP, Arlon R: Talc: a possible occupational and environmental carcinogen. *J Occup Med* 15:92–97, 1973
5. Kleinfeld M, Messite J, Zaki MH: Mortality experiences among talc workers: a follow-up study. *J Occup Med* 16:345–349, 1974

SODIUM FLUOROACETATE
CAS: 62-74-8

CH₂FCOONa

Synonyms: Compound 1080; fluoroacetic acid; sodium salt; Fratol; sodium monofluoroacetate

Physical Form. Fine white powder

Uses. Rodenticide (restricted use)

Exposure. Inhalation; ingestion

Toxicology. Sodium fluoroacetate is highly toxic and causes convulsions and ventricular fibrillation.

Fluoroacetate produces its toxic action by inhibiting the citric acid cycle.[1] The fluorine-substituted acetate is metabolized to fluorocitrate, which inhibits the conversion of citrate to isocitrate. There is an accumulation of large quantities of citrate in the tissue, and the cycle is blocked. The heart and central nervous system are the most critical tissues involved in poisoning by a general inhibition of oxidative energy metabolism.[1]

Onset of symptoms after ingestion is frequently delayed for 30 minutes to 2 hours; effects are vomiting, apprehension, auditory hallucinations, nystagmus, a tingling sensation of the nose, numbness of the face, facial twitching, and epileptiform convulsions.[2,3] After a period of several hours, there may be pulsus alterans, long sequences of ectopic heartbeats (often multifocal), tachycardia, ventricular fibrillation, and death.[3,4] The lethal oral dose in humans is estimated to be approximately 5.0 mg/kg.[4,5] In a fatal case of ingestion, autopsy findings included hemorrhagic pulmonary edema and degeneration of renal tubules.[5]

Applied as a 0.1% mixture in fish meal, and widely dispersed throughout a workplace as a rat poison, sodium fluoroacetate caused several employees to become seriously ill (details not given).[6] Exposure is thought to have occurred from airborne contamination although accidental ingestion cannot be ruled out.

In the only alleged case of chronic human poisoning, an exterminator repeatedly exposed over a period of 10 years presented with severe and progressive lesions of the renal tubular epithelium and with milder hepatic, neurologic, and thyroid dysfunctions.[7]

The 1995 ACGIH threshold limit value–time-weighted average (TLV-TWA) is 0.05 mg/m³, with a notation for skin absorption.

Treatment. Induce vomiting if convulsions are not imminent; administer sodium or magnesium sulfate in water (15 to 30 g).[4] Although the clinical efficacy of monoacetin (glyceral monoacetate) is not established, it probably should be administered.[4] Glyceral monoacetate appears to serve as an acetate donor to block fluoroacetate metabolism in a competitive manner. If monoacetin is not available, acetamide or ethyl alcohol may be given in the same doses.[4]

REFERENCES

1. Murphy SD: Toxic effects of pesticides. In Klaassen CD et al (eds): *Casarett and Doull's Toxicology: The Basic Science of Poisons*, 3rd ed, p 565. New York, Macmillan, 1986
2. Harrisson JWE et al: Acute poisoning with sodium fluoroacetate (compound 1080). *JAMA* 149:1520–1522, 1952
3. Hayes WJ Jr: *Clinical Handbook on Economic Poisons. Emergency Information for Treating Poisoning.* US Public Health Service Pub No 476,

pp 79–82. Washington, DC, US Government Printing Office, 1963

4. Gosselin RE et al: *Clinical Toxicology of Commercial Products*, Section III, 5th ed, pp 193–196. Baltimore, MD, Williams and Wilkins, 1984

5. Harrisson JWE, Ambrus JL, Ambrus CM: Fluoroacetate (1080) poisoning. *Ind Med Surg* 21:440–442, 1952

6. LaGoy PK, Bohrer RL, Halvorsen FH: The development of cleanup criteria for an acutely toxic pesticide at a contaminated industrial facility. *Am Ind Hyg Assoc J* 53:298–302, 1992

7. Parkin PJ: Chronic sodium monofluoroacetate (compound 1080) intoxication in a rabbiter. *NZ Med J* 85:93–99, 1977

SODIUM HYDROXIDE

CAS: 1310-73-2

NaOH

Synonyms: Caustic soda; caustic flake; lye; caustic; liquid caustic

Physical Form. White solid

Uses. In manufacture of rayon, mercerized cotton, soap, paper, aluminum, and petroleum products; metal cleaning; electrolytic extraction of zinc; tin plating; oxide coating

Exposure. Skin or eye contact; inhalation

Toxicology. Sodium hydroxide is highly corrosive; it is a severe irritant of the eyes, mucous membranes, and skin.

The most significant industrial hazard is rapid tissue destruction of eyes or skin on contact with either the solid or concentrated solutions.[1,2]

Contact with the eyes causes disintegration and sloughing of conjunctival and corneal epithelium, corneal opacification, marked edema, and ulceration; after 7 to 13 days, either gradual recovery begins, or there is progression of ulceration and corneal opacification.[3] Complications of severe eye burns are symblepharon with overgrowth of the cornea by a vascularized membrane, progressive or recurrent corneal ulceration, and permanent corneal opacification.[1]

On the skin, solutions of 25% to 50% cause the sensation of irritation within about 3 minutes; with solutions of 4%, this does not occur until after several hours.[1] If not promptly removed from the skin, severe burns with deep ulceration will occur. Exposure to the dust or mist may cause multiple small burns with temporary loss of hair.[2]

Although inhalation of sodium hydroxide is usually of secondary importance in industrial exposures, the effects from the dust or mist will vary from mild irritation of the nose at 2 mg/m^3 to severe pneumonitis, depending on the severity of exposure.[1,2]

In a recently reported case, severe obstructive airway disease was associated with chronic exposure to sodium hydroxide mists.[4] The worker, who for 20 years had daily exposure to boiling sodium hydroxide solutions, initially experienced tightness of the chest, dyspnea, cough, and eye irritation, which would resolve after he left the exposure area. Eventually, the worker began to suffer from mild exertional dyspnea and cough when not exposed. Physical examination, chest X ray, pulmonary-function tests, and arterial blood gases were all compatible with severe obstructive airway disease. It is probable that the massive and prolonged occupational exposure to the sodium hydroxide mists induced a bronchial inflammatory reaction leading to irreversible increased airway resistance.

Ingestion produces severe abdominal pain; corrosion of the lips, mouth, tongue, and pharynx; and vomiting of large pieces of mucosa. Cases of squamous-cell carcinoma of the esophagus have occurred with latent periods of 12 to 42 years after ingestion. These cancers were undoubtedly sequelae of tissue destruction and possibly scar formation rather than from a direct carcinogenic action of sodium hydroxide itself.[1]

The 1995 ACGIH short-term exposure limit (STEL/ceiling limit) is 2 mg/m^3.

REFERENCES

1. National Institute for Occupational Safety and Health: *Criteria for a Recommended Standard . . . Occupational Exposure to Sodium Hydroxide.* DHEW (NIOSH) Pub No 76-105, pp 23–50. Washington, DC, US Government Printing Office, 1975
2. Chemical Safety Data Sheet SD-9, Caustic soda, pp 5, 16–17. Washington, DC, MCA, Inc, 1968
3. Patty FA: Alkaline materials. In Fassett DW, Irish DD (eds): *Patty's Industrial Hygiene and Toxicology,* 2nd ed, Vol 2, *Toxicology,* pp 867–868. New York, Interscience, 1963
4. Rubin AE, Bentur L, Bentur Y: Obstructive airway disease associated with occupational sodium hydroxide inhalation. *Br J Ind Med* 49:213–214, 1992

SODIUM METABISULFITE
CAS: 7681-57-4

$Na_2S_2O_5$

Synonyms: Disodium disulfite; sodium pyrosulfite

Physical Form. White powder or crystal with the odor of sulfur dioxide

Uses. Preservative in food and wine; antioxidant in pharmaceuticals

Exposure. Ingestion; inhalation

Toxicology. Sodium metabisulfite may cause bronchospasm, oculonasal symptoms, and urticaria in sulfite-sensitive individuals; irritation of mucous membranes may occur from inhalation of the dust.

Two workers died while applying dry sodium metabisulfite in a ship hold.[1] Postmortem examination showed diffuse pulmonary edema consistent with death secondary to asphyxia and visceral congestion.

Sodium metabisulfite can trigger broncho-constriction in asthmatic subjects. In one study, 30 asthmatic subjects inhaled sodium metabisulfite in concentrations of 6.2, 12.5, 50, and 100 mg/ml.[2] All the asthmatic subjects responded as determined by decline in FEV_1. The response occurred within 1 minute, and most subjects recovered to baseline after 30 or 40 minutes. Neither enhanced sensitivity to subsequent histamine inhalation nor refractoriness to subsequent sodium metabisulfite inhalation was found. None of the nonasthmatic, nonatopic subjects responded to sodium metabisulfite, but inhalation of high doses may cause mild bronchoconstriction in these individuals.

The mechanism of action of bronchoconstriction may involve the liberation of sulfur dioxide gas from sodium metabisulfite which, in turn, acts on the parasympathetic nerves in the lung.[2]

In mice exposed to aerosols of sodium metabisulfite, there was sensory irritation of the upper respiratory tract.[3]

A low order of systemic toxicity was found in chronic feeding studies with rats. Administered in the diet for 2 years, 0.215% sodium metabisulfite caused no adverse effects.[4] Reproductive parameters were not affected in three generation feeding studies in rats at concentrations up to 13 mM/kg/day.[4]

The 1995 ACGIH threshold limit value–time-weighted average (TLV-TWA) for sodium metabisulfite is 5 mg/m^3.

REFERENCES

1. Atkinson DA, Sim TC, Grant JA: Sodium metabisulfite and SO_2 release: an unrecognized hazard among shrimp fishermen. *Ann Allergy* 71:563–566, 1993
2. Wright W, Zhang YG, Salome CM, et al: Effect of inhaled preservatives on asthmatic subjects 1. Sodium metabisulfite. *Am Rev Respir Dis* 141:1400–1404, 1990
3. Alarie Y, Wakisaka I, Oka S: Sensory irritation by sulfite aerosols. *Environ Physiol Biochem* 3:182–184, 1973
4. Til HP, Feron VJ, De-Goot AP: The toxicity of sulphite. 1. Long-term feeding and multigeneration studies in rats. *Food Cosmet Toxicol* 10:291–310, 1972

STIBINE
CAS: 7803-52-3

SbH₃

Synonyms: Antimony hydride; hydrogen antimonide

Physical Form. Colorless gas

Sources. Produced accidentally as a result of the generation of nascent hydrogen in the presence of antimony; formed when acid solutions of antimony compounds are treated with reducing agents

Exposure. Inhalation

Toxicology. Stibine is a hemolytic agent in animals; it is expected that the same effect will occur in humans.

No clear-cut case of fatal stibine poisoning in humans has been reported.[1-4] Workers exposed to a mixture of gases (concentrations unmeasured) of stibine, arsine, and hydrogen sulfide developed headache, weakness, nausea, abdominal and lumbar pain, hemoglobinuria, hematuria, and anemia.[1] Although these signs and symptoms are clearly manifestations of acute hemolytic anemia, it is not possible to determine the relative contribution of arsine, which is also a hemolytic agent. By analogy to other effects caused by arsine, additional signs of stibine poisoning may be leukocytosis and jaundice.

Guinea pigs exposed to 65 ppm of stibine for 1 hour developed hemoglobinuria, followed within a few days by profound anemia.[5] Stibine is also a pulmonary irritant in animals, causing pulmonary congestion and edema and, ultimately, death in cats and dogs following a 1-hour exposure at 40 to 45 ppm.[5]

The 1995 ACGIH threshold limit value–time-weighted average (TLV-TWA) is 0.1 ppm (0.51 mg/m³).

REFERENCES

1. Dernehl C, Stead FM, Nau CA: Arsine, stibine and H₂S: accidental generation in a metal refinery. *Ind Med Surg* 13:361, 1944
2. Stokinger HE: The metals. In Clayton GD, Clayton FE (eds): *Patty's Industrial Hygiene and Toxicology*, 3rd ed, Vol 2A, *Toxicology*, p 1511. New York, Wiley-Interscience, 1981
3. Hygienic Guide Series: Stibine. *Am Ind Hyg Assoc J* 21:529–530, 1960
4. Pinto SS: Arsine poisoning: Evaluation of the acute phase. *J Occup Med* 18:633–635, 1976
5. Webster SH: Volatile hydrides of toxicological importance. *J Ind Hyg Toxicol* 28:167–182, 1946

STODDARD SOLVENT
CAS: 8052-41-3
15% to 20% aromatic hydrocarbons
80% to 85% paraffin & naphthenic hydrocarbons

Synonyms: White spirits; safety solvent; varnoline

Physical Form. Colorless liquid

Uses. Dry cleaning; degreasing; paint thinner

Exposure. Inhalation

Toxicology. Stoddard solvent is a mild central nervous system depressant and a mucous membrane irritant.

Stoddard solvent is a mixture of predominantly C₉ to C₁₁ hydrocarbons, of which 30% to 50% are straight- and branched-chain paraffins, 30% to 40% naphthenes, and 10% to 20% aromatic hydrocarbons.[1] Although uses may differ, Stoddard solvent is chemically similar to mineral spirits, and the terms have been used interchangeably.[1]

Of 6 volunteers exposed to 150 ppm of Stoddard solvent for 15 minutes, 1 had transitory eye irritation; at 470 ppm (2700 mg/m³), all subjects had eye irritation, and 2 had slight dizziness.[2] In another study, 8 volunteers exposed at 4000 mg/m³ for 50 minutes had some changes in simple reaction time tests but not in perceptual speed, short-term memory, or manual dexterity compared to pre- and postexposure self-controls.[3]

Reports of Stoddard solvent as an etiologic agent in the development of aplastic anemia are of questionable validity.[1] Skin exposure may cause dermatitis and sensitization.[1]

Rats exposed to Stoddard solvent at a level of 1400 ppm for 8 hours exhibited eye irritation, bloody exudate around the nostrils, and slight loss of coordination. Exposure of a dog resulted in increased salivation at 3 hours, tremor at 4 hours, and clonic spasms after 5 hours; in cats, 1700 ppm caused tremors, convulsions and, finally, death after 2.5 to 7.5 hours.[2] No significant effects were observed in dogs exposed 6 hr/day for 65 days to 330 ppm; there were elevated blood urea nitrogen levels and marked tubular regeneration in the kidneys of rats similarly exposed.[2] Renal effects noted in rats are consistent with a mechanism that appears to be unique to male rats (ie, interactions with α_{2u}-globulin), and it is unlikely that Stoddard solvent would cause similar effects in humans.[4]

Stoddard solvent was not genotoxic in most standard animal assays.[4]

The odor threshold is 0.9 ppm; the odor and irritative properties probably do not provide adequate warning of dangerous concentrations.[2]

The 1995 ACGIH threshold limit value–time-weighted average (TLV-TWA) is 100 ppm (525 mg/m³).

REFERENCES

1. National Institute for Occupational Safety and Health: *Criteria for a Recommended Standard . . . Occupational Exposure to Refined Petroleum Solvents*. DHEW (NIOSH) Pub No 77-192. Washington, DC, US Government Printing Office, 1977
2. Carpenter CP et al: Petroleum hydrocarbon toxicity studies. III. Animal and human response to vapors of Stoddard solvent. *Toxicol Appl Pharmacol* 32:282–297, 1975
3. Gamberale F, Annwall G, Hultengren M: Exposure to white spirit: II. Psychological func-

tions. *Scand J Work Environ Health* 1:31–39, 1975
4. Agency for Toxic Substances and Disease Registry (ASTDR): *Toxicological Profile for Stoddard Solvent*. US Department of Health and Human Services, Public Health Service, p 87, 1993

STRYCHNINE
CAS: 57-24-9

$C_{21}H_{22}N_2O_2$

Synonym: Stricnina

Physical Form. White crystalline powder

Uses. Rodenticide

Exposure. Inhalation; ingestion

Toxicology. Strychnine is a potent convulsant.

Strychnine poisoning occurs from accidental and intentional ingestion, and from misuse as a therapeutic agent.[1] Doses of 5 to 7 mg cause muscle tightness, especially in the neck and jaws, and twitching of individual muscles, especially in the little fingers.[1]

The lethal oral dose in man is probably around 100 mg, but doses as low as 16 mg have reportedly been fatal, whereas doses of 2000 mg have been survived.[1,2] After ingestion, effects usually occur within 10 to 30 minutes and include stiffness of the face and neck muscles and increased reflex excitability.[3] Strychnine acts by altering nerve impulses in the spinal cord, resulting in a decreased threshold for stimulation and, hence, a hyperexcitable state. Any sensory stimulus may produce a violent motor response which, in the early stages of intoxication, tends to be a coordinated extensor thrust and, in later stages, may be a tetanic convulsion with opisthotonos; anoxia and cyanosis develop rapidly. Between convulsions, muscular relaxation is complete, breathing is resumed, and cyanosis lessens.[1] Because sensation is unaffected, the convulsions are painful and lead to overwhelming fear. As many as 10 convulsions, separated by intervals of 10 to 15 minutes, may be experienced, but death often occurs after the second to fifth convulsion, and even the first convulsion may be fatal if sustained; death is commonly due to asphyxia.[2,3] If recovery occurs, it is remarkably prompt and complete in spite of the violence of the illness; muscle soreness may persist for a number of days.[1]

In fatal cases, the pathological findings are entirely nonspecific. They usually consist of petechial hemorrhages and congestion of the organs, indicating combined action of severe convulsions and anoxia.[1] Compression fractures and related injury may be found in cases with violent tetany.[1]

The 1995 ACGIH threshold limit value–time-weighted average (TLV-TWA) for strychnine is 0.15 mg/m³.

Diagnosis. Signs and symptoms include stiffness of neck and facial muscles, increased reflex excitability, tetanic convulsions with opisthotonos, and cyanosis.

Differential Diagnosis: Other causes of convulsions must be differentiated from strychnine exposure. These include idiopathic epilepsy; hypertensive encephalopathy; metabolic disturbances such as hypoglycemia, hypocalcemia, uremia, and porphyria; hypoxic encephalopathy; infections such as viral encephalitis; and exposure to other convulsants. Fully developed tetanus closely resembles strychnine poisoning with generalized increased rigidity and convulsive spasms of skeletal muscles.[4]

Special Tests: Except for the demonstration of the poison itself, laboratory findings are not helpful in diagnosis or treatment.[1] In general, blood levels of strychnine do not correlate with severity of toxic effect.

Treatment. Because of the very rapid absorption of strychnine and the danger of precipitating convulsions, emptying the stomach should not be attempted unless it can be done immediately after ingestion or after the patient is completely protected against convulsions.[1] Potassium permanganate or charcoal may be given to reduce and delay absorption if symptoms are minimal or absent. Tea and coffee should be avoided because of the synergistic stimulation of caffeine. If symptoms develop, a rapid-acting sedative such as diazepam should be administered; slow, intravenous injection of mephenesin has also been successful in treatment.[1] Endotracheal intubation is an important safeguard.[5] If convulsions develop, administration of succinylcholine or curare and artificial ventilation may be required.[1] Strychnine is so rapidly metabolized that little is gained from accelerated drug removal by dialysis or hemoperfusion. Morphine is contraindicated because of respiratory depression. In the past, it was recommended that the patient should be placed in a quiet, darkened room and protected from sudden, unexpected stimuli.[1] The patient often may tolerate manipulation without convulsions, provided all motions are expected and gentle. Although avoidance of unexpected stimuli remains a useful element in management of the patient, the availability of effective drugs has caused greater attention to be given to treating the patient in such a way that he reacts normally to stimuli.[1]

REFERENCES

1. Hayes WJ Jr, Laws ER Jr: *Handbook of Pesticide Toxicology*, Vol 2, *Classes of Pesticides*, pp 615–619. New York, Academic Press, 1991
2. Gosselin RE et al: *Clinical Toxicology of Commercial Products*, Section III, 5th ed, pp 375–379. Baltimore, MD, Williams and Wilkins, 1984
3. Franz DN: Central nervous system stimulants. In Goodman LS, Gilman A (eds): *The Pharmacological Basis of Therapeutics*, 7th ed, pp 582–584. New York, Macmillan, 1985
4. Beaty HN: Tetanus. In Petersdorf RG et al (eds): *Harrison's Principles of Internal Medicine*, 10th ed, p 1004. New York, McGraw-Hill, 1983
5. Victor M, Adams R: Sedatives, stimulants, and psychotropic drugs. In Petersdorf RG et al (eds): *Harrison's Principles of Internal Medicine*, 10th ed, p 1301. New York, McGraw-Hill, 1983

STYRENE, MONOMER
CAS: 100-42-5

$C_6H_5CHCH_2$

Synonyms: Vinylbenzene;   phenylethylene; cinnamene

Physical Form.   Colorless to yellowish oily liquid

Uses.   Solvent for synthetic rubber and resins; intermediate in chemical synthesis; in manufacture of polymerized synthetic materials

Exposure.   Inhalation; skin absorption

Toxicology.   Styrene is an irritant of the skin, eyes, and mucous membranes and is neurotoxic.

Humans exposed to 376 ppm experienced eye and nasal irritation within 15 minutes; after 1 hour at 376 ppm, effects were headache, nausea, decreased dexterity and coordination, and other signs of transient neurologic impairment.[1] Subjective complaints, including headache, fatigue, and difficulty concentrating, have been reported following 90-minute experimental exposures at concentrations as low as 50 ppm.[2] Some acute effects on neuropsychological tests of verbal learning skills and visuoconstructive abilities have been demonstrated among workers exposed to mean concentrations of about 50 ppm.[3] Alterations in electroencephalographs and nerve conduction velocities have also been reported.[4]

The rate of absorption of the liquid through the skin of the hand and the forearm in man was 9 to 15 mg/cm^2/hr.[5] Prolonged or repeated exposure may lead to dermatitis due to defatting action on the skin.[6]

Rats and guinea pigs exposed to 10,000 ppm became comatose in a few minutes and died after 30 to 60 minutes of exposure.[7] Animals exposed to 2500 ppm showed weakness and stupor, followed by incoordination, tremor, coma, and death in 8 hours.[7] Rats and guinea pigs showed signs of eye and nasal irritation after exposure to 1300 ppm for 8 hr/day, 5 days/week for 6 months.[6]

Although high-level experimental exposure to animals has resulted in evidence of liver damage, there is no clear-cut evidence of human liver toxicity from industrial exposures.[2] Liver enzymes and serum bile acid concentrations among 34 workers with average 30 to 40 ppm styrene exposures for a mean of 5.1 years did not differ significantly from a control group of unexposed workers.[8]

Epidemiologic studies of styrene-exposed workers have not revealed an excess in overall cancer incidence. Although some mortality studies have identified several cases of lymphoma or leukemia among styrene workers that were greater than the expected occurrence, the numbers are small, and there is often concomitant exposure to other agents.[9-12]

The IARC has determined that there is inadequate evidence for carcinogenicity to humans; experimental animal studies have shown limited evidence of carcinogenicity (pulmonary tumors in male mice and hepatocellular adenomas in females following gastric intubation of styrene).[12] However, there is sufficient evidence for the carcinogenicity of styrene oxide, a metabolite of styrene.[12]

An increased incidence of chromosome aberrations and micronuclei in peripheral lymphocytes has been reported in occupationally exposed workers. Additional studies have found a slight increase in the incidence of sister chromatid exchanges, whereas no increase has been found in several others.[12] Both DNA and protein adducts are formed in man following styrene exposure.

Limited studies of the effects of styrene on reproduction are available, including conflicting reports of association between exposure and birth defects and fetal loss. In one of the more recent reports, women who worked at the most highly exposed jobs had offspring with adjusted birth weights of 4% less than the offspring of unexposed women.[13] Styrene does not appear to be teratogenic in experimental animals, nor does it appear to cause any specific developmental or reproductive toxicity.[6,12]

The odor threshold is 0.1 ppm; the disagreeable odor and the eye and nose irritation make the inhalation of seriously acute toxic quantities unlikely, although the warning properties may not be sufficient for prolonged exposures.

The 1995 threshold limit value–time-weighted average (TLV-TWA) is 50 ppm (213 mg/m^3) with a short-term excursion limit of 100 ppm (426 mg/m^3) and a notation for skin absorption.

REFERENCES

1. Stewart RD, Dodd HC, Baretta ED, Schaffer AW: Human exposure to styrene vapor. *Arch Environ Health* 16:656–662, 1968
2. National Institute for Occupational Safety and Health: *Criteria for a Recommended Standard ... Occupational Exposure to Styrene.* DHHS (NIOSH) Pub No 83-119, pp 121–130. Washington, DC, US Government Printing Office, 1983
3. Mutti A et al: Exposure-effect and exposure-response relationships between occupational exposure to styrene and neuropsychological functions. *Am J Ind Med* 5:275–286, 1984
4. Pahwa R, Kalra J: A critical review of the neurotoxicity of styrene in humans. *Vet Human Toxicol* 35:516–519, 1993
5. Dutkiewicz T, Tyras H: Skin absorption of toluene, styrene, and xylene by man. *Br J Ind Med* 25:243, 1968
6. Bond JA: Review of the toxicology of styrene. *CRC Crit Rev Toxicol* 19:227–249, 1989
7. Sandmeyer EE: The aromatic hydrocarbons. In Clayton GD, Clayton FE (eds): *Patty's Industrial Hygiene and Toxicology*, 3rd ed, Vol 2A, *Toxicology*, pp 3312–3319. New York, Wiley-Interscience, 1981
8. Harkonen H, Lehtniewi A, Aitio A: Styrene exposure and the liver. *Scand J Work Environ Health* 10:59–61,1984
9. Hodgson J, Jones R: Mortality of styrene production, polymerization and processing workers of a site in northwest England. *Scand J Work Environ Health* 11:347–352, 1985
10. Okun A et al: Mortality patterns among styrene-exposed boatbuilders. *Am J Ind Med* 8:193–205, 1985
11. Ott MG, Kolsear RC, Scharnweber HC, et al: A mortality survey of employees engaged in the development or manufacture of styrene-based products. *JOM* 22:445–460, 1980
12. *IARC Monographs on the Evaluation of Carcinogenic Risks to Humans*, Vol 60, Some industrial chemicals, pp 233–305. Lyon, International Agency for Research on Cancer, 1994
13. Lemasters GK, Samuels SJ, Morrison JA, et al: Reproductive outcomes of pregnant workers employed at 36 reinforced plastics companies. II. Lowered birth weight. *JOM* 31:115–120, 1989

STYRENE OXIDE
CAS: 96-09-3

C_8H_8O

Synonyms: Epoxyethylbenzene; epoxystyrene; phenylethylene oxide; phenyloxirane

Physical Form. Colorless to pale straw-colored liquid

Uses. Intermediate in the production of styrene glycol and its derivatives; reactive diluent in the epoxy-resin industry; chemical intermediate for making β-phenethyl alcohol, fragrance material

Exposure. Inhalation; skin

Toxicology. Styrene oxide is a skin and eye irritant and may produce skin sensitization; it is carcinogenic in experimental animals.

Tests with human subjects indicate that styrene oxide is capable of causing moderate skin irritation and skin sensitization.[1] These effects may result from single or repeated contact with the undiluted liquid and with solutions as dilute as 1%. Experience indicates that persons who have become sensitized may react severely to contact with the vapor as well as with the liquid material.

In rats, exposure to 1000 ppm was lethal to 2 of 6 animals within 4 hours.[2] Repeated 7-hour exposures at 300 ppm were rapidly fatal to 40% of female rats, and extensive mortality occurred in rats receiving prolonged exposure to 100 ppm.[3] Toxicity also was marked in the rabbit, with prolonged and repeated exposures at 15 to 50 ppm producing mortality.[3] Histopathological changes in rats and rabbits included metaplasia and hyperplasia of the lungs.

Inhalation exposure during gestation by rats and rabbits produced reproduction and developmental toxicity as well as maternal toxicity.[3] Exposure to 15 or 50 ppm for 7 hr/day on days 1 to 24 of gestation resulted in maternal toxicity (increased mortality, decreased food consumption, and weight gain) and increased the frequency of resorptions in rabbits. Exposure of rats to 100 ppm on days 1 to 19 of gestation decreased fecundity by significantly increasing preimplantation loss. Fetal size, including crown-rump length and weight, also tended to be decreased by exposure in both species. It has not been established whether the developmental effects are direct effects or are the result of maternal toxicity.

The liquid is slowly absorbed by the skin and may reach toxic levels in rabbits over a 24-hour period with an LD_{50} of 2.8 g/kg.[1]

Intraperitoneal injection has been associated with hepatic damage in rats, causing a decrease in the activities of mixed-function oxidases and in cytochrome P-450 content.[4]

In a long-term bioassay, styrene oxide administered to rats by gavage (250 or 50 mg/kg daily for 1 year) produced a high incidence of tumors in the forestomach (papillomas, acanthomas, and in situ and invasive squamous-cell carcinomas).[5] Styrene oxide also increased the incidence of squamous-cell papillomas and carcinomas of the forestomach in mice when administered by gavage at doses of 375 or 750 mg/kg for 2 years.[6]

Prenatal exposure followed by postnatal oral administration of 96 weekly doses of 100 to 150 mg/kg also produced a significantly increased incidence of forestomach tumors, including papillomas and carcinomas in rats.[7]

No increase in the incidence of skin tumors was observed in two mice studies following topical application.[2,8]

Both positive and negative findings have been reported in genotoxic assays of styrene oxide. It has induced gene mutations in bacteria and rodent cells in vitro and caused chromosomal aberrations and sister chromatid exchange both in vivo and in vitro.[9] Styrene oxide

forms covalent adducts with DNA in humans, rats and mice.

The IARC has determined that there is sufficient evidence for the carcinogenicity of styrene oxide to experimental animals and that, although there is inadequate evidence for the carcinogenicity to humans, it should be regarded as probably carcinogenic to humans.[9]

The ACGIH has not determined a threshold limit value for styrene oxide.

REFERENCES

1. Hine CH, Rowe, VK, White ER, et al: Epoxy compounds. In Clayton GD, Clayton FE (eds): *Patty's Industrial Hygiene and Toxicology*, 3rd ed, rev, Vol 2B, *Toxicology*, pp 2192–2194. New York, Wiley-Interscience, 1981
2. Weil CS, Condra N, Haun C, et al: Experimental carcinogenicity and acute toxicity of representative epoxides. *Am Ind Hyg Assoc J* 24:305–325, 1963
3. Sikov MR, Cannon WC, Carr DB, et al: Reproductive toxicology of inhaled styrene oxide in rats and rabbits. *J Appl Toxicol* 6:155–164, 1986
4. Parkki MG, Marniemi J, Vainio H: Action of styrene and its metabolites styrene oxide and styrene glycol on activities of xenobiotic biotransformation enzymes in rat liver in vivo. *Toxicol Appl Pharmacol* 38:59–70, 1976
5. Conti B, Maltoni C, Perion G, et al: Long-term carcinogenicity bioassays on styrene administered by inhalation, ingestion and injection and styrene oxide administered by ingestion on Sprague–Dawley rats, and para-methylstyrene administered by ingestion in Sprague–Dawley rats and Swiss mice. *Ann NY Acad Sci* 534:203–234, 1988
6. Lijinsky W: Rat and mouse forestomach tumors induced by chronic oral administration of styrene oxide. *J Natl Cancer Inst* 77:471–476, 1986
7. Ponomarkov V, Cabral JRP, Wahrendorg J, et al: A carcinogenicity study of styrene-7,8-oxide in rats. *Cancer Lett* 24:95–101, 1984
8. Van Duuren BL, Nelson H, Orris L, et al: Carcinogenicity of epoxides, lactones, and peroxy compounds. *J Natl Cancer Inst* 31:41–55, 1963
9. *IARC Monographs on the Evaluation of Carcinogenic Risks to Humans*, Vol 60, Some industrial

chemicals, pp 321–346. Lyon, International Agency for Research on Cancer, 1994

SULFOLANE
CAS: 126-33-0

$C_4H_8SO_2$

Synonyms: 1,1-Dioxidetetrahydrothiofuran; 1,1-dioxothiolan; cyclotetramethylene sulfone; dioxothiolan; sulfoxaline; tetrahydrothiophene 1,1-dioxide; tetramethylene sulfone; thiocyclopentane dioxide; thiophane dioxide

Physical Form. Colorless, oily liquid (solid at 15°C)

Uses. Process solvent for extractions of aromatics and for purification of acid gases

Toxicology. Sulfolane is a convulsant in animals.

Sulfolane is not highly toxic. Oral LD_{50} values in the rat range from 1846 to 2500 mg/kg.[1] Symptoms of neurotoxicity have been observed in rats, dogs, and monkeys after ingestion, injection, inhalation, or dermal application. Effects included convulsions, hyperactivity, tremor, and ataxia.[2] In acute inhalation studies, no rats died in the 2 weeks following 4-hour exposures to levels as high as 12,000 mg/m³.[2] Dogs exposed continuously to 200 mg/m³ for 7 days experienced convulsions.

The liquid is not irritating to the skin and is mildly irritating to the eyes.[3] It was not a sensitizer in the guinea pig.[4,5]

A threshold limit value has not been established for sulfolane.

REFERENCES

1. Anderson ME et al: Sulfolane-induced convulsions in rodents. *Res Commun Chem Pathol Pharmacol* 15:571, 1976
2. Anderson ME et al: The inhalation toxicity of sulfolane (tetrahydrothiopene-1,1-dioxide). *Toxicol Appl Pharmacol* 40:463, 1977
3. Weiss G (ed): *Hazardous Chemicals Data Book*, p 840. Park Ridge, NJ, Noyes Data Corporation, 1980
4. Brown VKH, Ferrigan LW, Stevenson DE: Acute toxicity and skin irritation properties of sulfolane. *Br J Ind Med* 23:302, 1966
5. Phillips Petroleum Co: FYI-OTS-0484-034 Supplement Sequence D. *Summary of Toxicity of Sulfolane*. Washington, DC, Office of Toxic Substances, US Environmental Protection Agency, June 6, 1983

SULFUR DIOXIDE
CAS: 7446-09-5

SO_2

Synonyms: Sulfurous anhydride; sulfurous oxide

Physical Form. Colorless gas

Uses. Intermediate in the manufacture of sulfuric acid and sulfite pulp; casting of nonferrous metal; in the food industry as a biocide and a preservative

Exposure. Inhalation

Toxicology. Sulfur dioxide is a severe irritant of the eyes, skin, and upper airways of the respiratory tract.

The irritant effects of sulfur dioxide are due to the rapidity with which it forms sulfurous acid on contact with moist membranes.[1,2] Approximately 90% of all sulfur dioxide inhaled is absorbed in the upper respiratory passages, where most effects occur; however, it may produce respiratory paralysis and may also cause pulmonary edema.[2] In fatal cases, histopathological examination of the lungs has revealed pulmonary edema and alveolar hemorrhage.[3]

Exposure to concentrations of 10 to 50 ppm for 5 to 15 minutes causes irritation of the eyes, nose, and throat; rhinorrhea, choking; cough and, in some instances, reflex bronchoconstriction with increased pulmonary resistance.[2]

The phenomenon of adaptation to irritating concentrations is a recognized occurrence in experienced workers.[2] Workers repeatedly exposed to 10 ppm experienced upper respiratory irritation and some nosebleeds, but the symptoms did not occur at 5 ppm. In another study, initial cough and irritation did occur at 5 ppm and 13 ppm but subsided after 5 minutes of exposure.[4]

In a human experimental study with the subjects breathing through the mouth, brief exposure to 13 ppm caused a 73% increase in pulmonary flow resistance, 5 ppm resulted in a 40% increase, and 1 ppm produced no effects.[4]

Studies of individuals with mild asthma have demonstrated much greater sensitivity to low levels of sulfur dioxide exposure, particularly during exercise. Exposures to concentrations of 0.5 to 0.1 ppm during exercise resulted in significant increases in airway resistance in these subjects.[5] At rest, exposures to 1 ppm resulted in significant increases in airway resistance in mild asthmatics.[6]

Epidemiologic studies of workers chronically exposed to sulfur dioxide, such as copper smelters, have yielded conflicting results regarding excessive occurrence of chronic respiratory disease, chronic bronchitis, or decrements in pulmonary function. Such studies are plagued by the confounding effect of smoking and difficulties in exposure assessment. Overall, the evidence for chronic effects in man, including carcinogenicity, is quite limited.[3,7]

In one animal study, a significant increase in lung tumors was observed in female mice exposed by inhalation.[3] In genotoxic assays, sulfur dioxide did not induce sister chromatid exchange, chromosomal aberrations, or micronucleus formation in the bone marrow of mice or Chinese hamsters.[3]

The IARC has determined that there is limited evidence for the carcinogenicity of sulfur dioxide in experimental animals and inadequate evidence in humans.

Although a variety of environmental exposures involving sulfur dioxide have been linked to human reproductive effects, there is no clear relationship between sulfur dioxide concentrations and adverse reproductive outcomes.[3]

Exposure of the eyes to liquid sulfur dioxide from pressurized containers causes corneal burns and opacification resulting in a loss of vision.[2] The liquid on the skin produces skin burns from the freezing effect of rapid evaporation.[2]

The 1995 ACGIH threshold limit value–time-weighted average (TLV-TWA) is 2 ppm (5.2 mg/m^3) with a short-term excursion limit of 5 ppm (13 mg/m^3).

REFERENCES

1. National Institute for Occupational Safety and Health: *Criteria for a Recommended Standard . . . Occupational Exposure to Sulfur Dioxide*, pp 16–54. Washington, DC, US Government Printing Office, 1974
2. Department of Labor: Occupational exposure to sulfur dioxide. *Federal Register* 40:54520–54534, 1975
3. *IARC Monographs on the Evaluation of Carcinogenic Risks to Humans*. Vol 54, Occupational exposures to mists and vapours from strong inorganic acids; and other industrial chemicals. Lyon, International Agency for Research on Cancer, 1992
4. Whittenberger JL, Frank RN: Human exposures to sulfur dioxide. *Arch Environ Health* 7:244–245, 1963
5. Sheppard D et al: Exercise increases sulfur dioxide–induced bronchoconstriction in asthmatic subjects. *Am Rev Resp Disease* 123:486–491, 1981
6. Sheppard D et al: Lower threshold and greater bronchomotor responsiveness of asthmatic subjects to sulfur dioxide. *Am Rev Resp Disease* 122:873–878, 1980
7. Federspiel C et al: Lung function among employees of a copper mine smelter: lack of effect of chronic sulfur dioxide exposure. *J Occup Med* 22:438–444, 1980

SULFUR HEXAFLUORIDE
CAS: 2551-62-4

SF_6

Synonyms: Sulfur fluoride

Physical Form. Colorless, odorless gas

Uses. Dielectric for high-voltage equipment

Exposure. Inhalation

Toxicology. Sulfur hexafluoride is an agent of low toxicity; at extremely high levels, it has a mild effect on the nervous system.

In humans, inhalation of 80% sulfur hexafluoride and 20% oxygen for 5 minutes produced peripheral tingling and a mild excitement stage, with some altered hearing in most subjects.[1] According to the ACGIH, the chief hazard, as with other inert gases, would be asphyxiation as a result of the displacement of air by this heavy gas.[2]

Rats exposed for many hours to an atmosphere containing 80% sulfur hexafluoride and 20% oxygen gave no perceptible indications of intoxication, irritation, or other toxicologic effects.[3] Electrical discharges and high temperatures will cause sulfur hexafluoride decomposition.[2,4] Although some decomposition products are highly toxic, the concentrations produced and their practical significance under usual working conditions are undetermined.

The 1995 ACGIH threshold limit value–time-weighted average (TLV-TWA) for sulpher hexafluoride is 1000 ppm (5970 mg/m³).

REFERENCES

1. Glauser SC, Glauser EM: Sulfur hexafluoride—a gas not certified for human use. *Arch Environ Health* 13:467, 1966
2. Sulfur hexafluoride. *Documentation of the TLVs and BEIs*, 6th ed, pp 1459–1460. Cincinnati, OH, American Conference of Governmental Industrial Hygienists (ACGIH), 1991
3. Lester D, Greenberg LA: The toxicity of sulfur hexafluoride. *Arch Ind Hyg Occup Med* 2:348–349, 1950
4. Griffin GD et al: On the toxicity of sparked SF6. *IEEE Transactions on Electrical Insulation*, Vol EI-18, No 5, pp 551–552. Washington, DC, Institute of Electrical and Electronic Engineers, 1983

SULFURIC ACID
CAS: 7664-93-9

H_2SO_4

Synonyms: Oil of vitriol; sulphuric acid

Physical Form. Colorless liquid

Uses. In fertilizer manufacturing; in metal cleaning; in manufacture of chemicals, plastics, and explosives; in petroleum refining; in pickling of metal

Exposure. Inhalation

Toxicology. Sulfuric acid is a severe irritant of the respiratory tract, eyes, and skin; contact with the teeth causes dental erosion; cancer of the respiratory tract has been associated with chronic exposure.

Concentrated sulfuric acid destroys tissue as a result of its severe dehydrating action, whereas the dilute form acts as a milder irritant because of its acidic properties.

In human subjects, concentrations of about 5 mg/m³ were objectionable, usually causing cough, with an increase in respiratory rate and impairment of ventilatory capacity.[1]

In a study of 248 workers, no significant association was found between exposure to vapor concentrations of up to 0.42 mg/m³ (2.6 to 10 μgm mass median diameter) and symptoms of cough, phlegm, dyspnea, and wheezing.[2] However, the FVC in the highest-exposure group was reduced compared to that of a low-exposure group. Repeated exposure of workers to unspecified sulfuric acid concentrations reportedly has caused chronic conjunctivitis, tracheobronchitis, stomatitis, and dermatitis.[3]

The dose-effect relationship for chronic exposure is difficult to determine because of the number of factors that influence toxicity, including the particle size of the mist, presence of particulates, synergistic and protective agents, and humidity.[4] In regard to particle size, the smallest aerosol particles appear to cause the greatest alteration in pulmonary function and more microscopic lesions because of their ability to penetrate deeply into the lungs.[3] Larger particles that deposit in the upper lung may be more acutely harmful because reflexive bronchoconstriction occurs. Very large particles that penetrate only the nasal passages and upper respiratory tract would not lead to either effect. Adsorbed onto other particulates, sulfuric acid may be carried farther into the respiratory tract.[4] Synergism has been demonstrated between sulfuric acid and ozone, sulfur dioxide, and metallic aerosols.[4] Increased ammonia concentration in expired air affords protection. Because of the hygroscopic nature of sulfuric acid, humidity directly affects particle size and, hence, toxicity.[4]

In guinea pigs, aerosols of larger, but still respirable, size were more lethal than those of smaller size.[5] For 8-hour exposures, the LC_{50} was 30 mg/m^3 for mists of 0.8 μm and was greater than 109 mg/m^3 for 0.4 μm mists.[5] Animals that died from exposure to the larger mists had hyperinflated lungs, whereas those that died from the smaller mists also had hemorrhage and transudation. Changes in pulmonary function, however, were more severe for aerosols of smaller diameter.[6] The concentration producing a 50% increase in pulmonary flow resistance was 0.3 mg/m^3 for 0.3 μm particles, 0.7 mg/m^3 for 1 μm, 6 mg/m^3 for 3.4 μm, and 30 mg/m^3 for 7 μm. Long-term exposure of monkeys at concentrations between 0.1 and 1 mg/m^3, regardless of particle size, produced slight but increasingly severe microscopic pulmonary lesions.[7] Impairment of pulmonary ventilation occurred above 2.5 mg/m^3.

The corrosive effects of sulfuric acid on teeth with chronic exposure are well established.[3] The damage, etching of dental enamel followed by erosion of enamel and dentine with loss of tooth substance, is limited to the parts of the teeth that are exposed to direct impingement of acid mist on the surface. Although etching typically occurs after years of occupational exposure, in one case, exposure to an average of 0.23 mg/m^3 for 4 months was sufficient to initiate erosion.[2]

Splashed in the eye, the concentrated acid causes extremely severe damage, often leading to blindness, whereas dilute acid produces more transient effects, from which recovery may be complete.[8] Chemical burns are the most commonly encountered occupational hazard. Initially, the zone of contact is bleached and turns brown prior to the formation of a clearly defined ulcer on a light red background. The wounds are slow to heal, and scarring may result in functional inhibition. Severe burns have been fatal. A worker sprayed in the face with liquid fuming sulfuric acid suffered skin burns of the face and body, as well as pulmonary edema from inhalation.[3] Sequelae were pulmonary fibrosis, residual bronchitis, and pulmonary emphysema; in addition, necrosis of the skin resulted in marked scarring.[3]

Although ingestion of the liquid is unlikely in ordinary industrial use, the highly corrosive nature of the substance will produce serious burns of the mouth and the esophagus.[3]

A number of studies have indicated that exposures to sulfuric acid or to acid mist in general are associated with laryngeal cancer.[9-12] In a nested case-referent study, a 13-fold excess risk of laryngeal cancer was found among chemical refinery workers with the highest levels of sulfuric acid exposure compared to those least exposed; a fourfold risk for moderately exposed workers versus those least exposed was also found.[9] Nine cases of laryngeal cancer were identified (vs. 3.92 expected) in steelworkers exposed to sulfuric acid mist for a minimum of 6 months prior to 1965.[10] Exposure levels averaged about 0.2 mg/m^3, and the average duration of exposure was 9.5 years. Excess risks for laryngeal cancer were also found in a Swedish study of a cohort of workers in steel pickling.[11] And, finally, in a recent population-based case-referent report from Canada, there was an association between exposure to sulfuric acid in the workplace, particularly at higher concentrations and over longer periods of time, and the development of laryngeal cancer.[12]

The IARC has determined that there is sufficient evidence that occupational exposure to strong inorganic acid mists containing sulfuric acid is carcinogenic to humans.[13]

Significant increases in the incidences of sister chromatid exchange, micronucleus formation, and chromosomal aberrations in peripheral lymphocytes were observed in a single study of workers engaged in the manufacture of sulfuric acid.[13]

Sulfuric acid mist was not teratogenic in mice or rabbits exposed 7 hours a day to 20 mg/m^3 during the period of major organogenesis.[14]

The 1995 threshold limit value (TLV-TWA) is 1 mg/m^3, with a short-term excursion level (STEL/ceiling) of 3 mg/m^3.

REFERENCES

1. Amdur MO, Silverman L, Drinker P: Inhalation of sulfuric acid mist by human subjects. *AMA Arch Ind Hyg Occup Med* 6:305–313, 1952
2. Gamble J et al: Epidemiological–environmental study of lead acid battery workers. III. Chronic effects of sulfuric acid on the respiratory system and teeth. *Environ Res* 35:30–52, 1984
3. National Institute for Occupational Safety and Health: *Criteria for a Recommended Standard . . . Occupational Exposure to Sulfuric Acid.* DHEW (NIOSH) Pub No 74-128, pp 19–49. Washington, DC, US Government Printing Office, 1974
4. Health effects assessment for sulfuric acid. Report No EPA/540/1-86/031, p 33. Washington, DC, US Environmental Protection Agency: Environmental Criteria and Assessment Office, 1984
5. Wolff RK et al: Toxicity of 0.4- and 0.8-μm sulfuric acid aerosols in the guinea pig. *J Toxicol Environ Health* 5:1037–1047, 1979
6. Amdur MO et al: Respiratory response of guinea pigs to low levels of sulfuric acid. *Environ Res* 5:418–423, 1978
7. Alarie YC et al: Long-term exposure to sulfur dioxide, sulfuric acid mist, fly ash, and their mixtures—results of studies in monkeys and guinea pigs. *Arch Environ Health* 30:254–263, 1975
8. Grant WM: *Toxicology of the Eye*, 3rd ed, pp 866–868. Springfield, IL, Charles C Thomas, 1986
9. Soskolne CL et al: Laryngeal cancer and occupational exposure to sulfuric acid. *Am J Epidemiol* 120:358–369, 1984
10. Steenland K, Schnorr T, Beaumont J, et al: Incidence of laryngeal cancer and exposure to acid mists. *Br J Ind Med* 45:766–776, 1988
11. Ahlborg G, Hogstedt C, Sundell L, et al: Laryngeal cancer and pickling house vapors. *Scand J Work Environ Health* 7:239–240, 1981
12. Soskolne CL, Jhangri GS, Siemiatycki J, et al: Occupational exposure to sulfuric acid in southern Ontario, Canada, in association with laryngeal cancer. *Scand J Work Environ Health* 18:225–232, 1992
13. *IARC Monographs on the Evaluation of Carcinogenic Risks to Humans*, Vol 54, Occupational Exposures to mists and vapours from strong inorganic acids; and other industrial chemicals. pp 41–130. Lyon, International Agency for Research on Cancer, 1992
14. Murray FJ et al: Embryotoxicity of inhaled sulfuric acid aerosol in mice and rabbits. *J Environ Sci Health* 13:251–266, 1979

SULFUR MONOCHLORIDE
CAS: 10025-67-9

S_2Cl_2

Synonyms: Sulfur chloride; sulfur subchloride; disulfur dichloride

Physical Form. Nonflammable, light-amber to yellowish-red, fuming, oily liquid

Uses. Intermediate and chlorinating agent in manufacture of organics, sulfur dyes, insecticides, and synthetic rubber

Exposure. Inhalation

Toxicology. Sulfur monochloride is an irritant of the eyes, mucous membranes, and skin.

On contact with water, it decomposes to form hydrogen chloride and sulfur dioxide; because this occurs rapidly, it acts primarily as an upper respiratory irritant and does not ordinarily reach the lungs.[1] However, exposure to high concentrations may cause pulmonary edema.[2]

Concentrations of 2 to 9 ppm are reported to be mildly irritating to exposed workers.[3] Splashes of the liquid in the eyes produce severe immediate damage, which may result in permanent scarring.[2] The liquid on the skin will produce irritation and burns if not promptly removed.[2]

Exposure of mice to 150 ppm for 1 minute is fatal.[1] In cats, some deaths occurred following a 15-minute exposure to 60 ppm.

The 1995 ACGIH STEL/ceiling limit for sulfur monochloride is 1 ppm (5.5 mg/m³).

REFERENCES

1. Patty FA: As, P, Se, S, and Te. In Patty FA (ed): *Industrial Hygiene and Toxicology*, 2nd ed, Vol 2, *Toxicology*, pp 905–906. New York, Interscience, 1963
2. Chemical Safety Data Sheet SD-77, Sulfur chlorides, pp 5, 11–13. Washington, DC, MCA, Inc, 1960
3. Elkins HB: *The Chemistry of Industrial Toxicology*, 2nd ed, p 81. New York, John Wiley and Sons, Inc, 1959

SULFUR PENTAFLUORIDE
CAS: 5714-22-77

S_2F_{10}

Synonym: Disulfur decafluoride

Physical Form. Colorless liquid

Source. Production by-product of synthesis of sulfur hexafluoride

Exposure. Inhalation

Toxicology. Sulfur pentafluoride is a severe pulmonary irritant in animals; severe exposure is expected to cause the same effect in humans.

Exposure of rats to 1 ppm for 16 to 18 hours was fatal; 0.5 ppm caused pulmonary edema and hemorrhage; 0.1 ppm caused irritation of the lungs; 0.01 ppm had no effect.[1] Nonfatal exposure of rats to 10 ppm for 1 hour caused pulmonary hemorrhage.[1]

The 1995 ACGIH STEL/ceiling limit for sulfur pentafluoride is 0.01 ppm (0.10 mg/m³).

REFERENCES

1. Greenberg LA, Lester D: The toxicity of sulfur pentafluoride. *AMA Arch Ind Hyg Occup Med* 2:350–352, 1950

SULFUR TETRAFLUORIDE
CAS: 7783-60-0

SF₄

Synonyms: None

Physical Form. Colorless gas with odor similar to that of sulfur dioxide

Uses/Sources. Fluorinating agent in the production of water and oil-repellent materials and lubricity improvers; found as a degradation product of sulfur hexafluoride

Exposure. Inhalation

Toxicology. Sulfur tetrafluoride is extremely irritating and corrosive to the respiratory tract, skin, and eyes.

Six workers were exposed to degradation products of sulfur hexafluoride during electrical repair work.[1] One degradation product, sulfur tetrafluoride, was identified from worksite measurements. Unprotected exposure totaling approximately 6 hours occurred over a 12-hour period in an enclosed underground space. Workers initially noticed an odor like that of a burning battery and experienced eye irritation with tears, dry and burning throat, and tightness in the chest. The workers went above ground, and symptoms abated after approximately 15 minutes. Subsequent underground visits resulted in a recurrence of symptoms. Repair work was stopped after workers experienced tightness in the chest, shortness of breath, headache, fatigue, nosebleed, nausea, and vomiting. Most symptoms resolved within a week of exposure, with some intermittent epistaxis persisting for up to a month. Radiographic evidence of multilobar atelectasis was present in one worker, and a second worker developed chest tightness on exposure to cold air and transitory changes on pulmonary-function tests. Examination of the workers 1 year later did not reveal any persistent adverse consequences.

At the time of the incident, worksite measurements qualitatively identified sulfur tetrafluoride in the air samples. It was suggested that intense heat caused sulfur hexafluoride to decompose to sulfur tetrafluoride, which escaped as a pipe was opened at the worksite. Subsequent to this incident, it has been noted that, because sulfur hexafluoride is an odorless gas, any odors present in areas containing heated sulfur hexafluoride must be considered to be coming from decomposition products, which are significant health hazards.

In animal studies, a 4-hour exposure at 19 ppm was lethal to 1 of 2 rats.[2] Eye irritation and irregular breathing were observed, and there was evidence of pulmonary edema at necropsy. Rats exposed at 4 ppm for 4 hr/day for 10 days had dyspnea, weakness, and nasal discharge.

The 1995 ACGIH ceiling–threshold limit value (C-TLV) is 0.1 ppm (0.44 mg/m^3).

REFERENCES

1. Kraut A, Lilis R: Pulmonary effects of acute exposure to degradation products of sulphur hexafluoride during electrical cable repair work. Br J Ind Med 47:829–832, 1990
2. Clayton JW Jr: The toxicity of fluorocarbons

with special reference to chemical constitution. J Occup Med 4:262–273, 1962

SULFURYL FLUORIDE
CAS: 2699-79-8

SO$_2$F$_2$

Synonyms: Sulfuric oxyfluoride; sulfuryl difluoride; Vikane

Physical Form. Colorless gas

Uses. Insect fumigant

Exposure. Inhalation

Toxicology. Sulfuryl fluoride is a central nervous system depressant and a pulmonary irritant in animals.

A worker exposed to an undetermined concentration of a mixture of sulfuryl fluoride and 1% chloropicrin for 4 hours developed nausea, vomiting, abdominal pain, and pruritis; physical examination revealed conjunctivitis, rhinitis, pharyngitis, and paresthesia of the right leg, all of which rapidly subsided.[1] The role of sulfuryl fluoride in this case is not known, but the signs and symptoms are those expected of chloropicrin overexposure.

Two fatalities occurred after re-entry into a home fumigated with sulfuryl fluoride.[2] The male experienced severe dyspnea and cough, followed by generalized seizure and cardiopulmonary arrest within 24 hours. The female initially had weakness, nausea, and repeated vomiting; within 4 days, there was severe hypoxemia and diffuse pulmonary infiltrates. Ventricular fibrillation occurred and death on day 6. The concentration of sulfuryl fluoride gas was not available, and the difference in time of death for the two individuals was not explainable.

Evaluation of workers occupationally exposed to sulfuryl fluoride found no effects attributable to exposure in a series of psychological and neurological tests compared to individuals with no history of exposure.[3]

Acute exposure of rats to high concentrations (up to 40,000 ppm) has resulted in convulsions, pulmonary edema, respiratory arrest, and death.[4] In rats repeatedly exposed at 600 ppm, death was attributed to renal papillary necrosis; renal toxicity was not present in rabbits similarly exposed. Exposure of rabbits to 300 or 600 ppm resulted in convulsions and hyperactivity, moderate inflammation of nasal tissues, and some inflammation of the trachea or bronchi. Subchronic studies found that rats exposed at 300 ppm had mottled incisor teeth, minimal renal effects, pulmonary histiocytosis, inflammation of nasal tissues, and cerebral vacuolation.

Exposure to sulfuryl fluoride was not teratogenic in either rats or rabbits exposed to levels of up to 225 ppm during periods of major organogenesis; fetotoxic effects in the form of reduced body weights were observed in rabbits only at levels that produced maternal weight loss.[5]

Because the gas is odorless and colorless, there are no warning properties of overexposure.

The 1995 ACGIH threshold limit value–time-weighted average (TLV-TWA) is 5 ppm (21 mg/m³) with a STEL/ceiling of 10 ppm (42 mg/m³).

REFERENCES

1. Taxay EP: Vikane inhalation. *J Occup Med* 8:425–426, 1966
2. Nuckolls JG, Smith DC, Walls WE, et al: Fatalities resulting from sulfuryl fluoride exposure after home fumigation—Virginia. *Morbid Mortal Week Rept* 36:602–611, 1987
3. Anger WK, Moody L, Burg J, et al: Neurobehavioral evaluation of soil and structural fumigators using methyl bromide and sulfuryl fluoride. *Neurotoxicology* 7:137–156, 1986
4. Eisenbrandt DL, Nitschke KD: Inhalation toxicity of sulfuryl fluoride in rats and rabbits. *Fund Appl Toxicol* 12:540–557, 1989
5. Hanley TR Jr, Calhoun LL, Kociba RJ, et al: The effects of inhalation exposure to sulfuryl fluoride on fetal development in rats and rabbits. *Fund Appl Toxicol* 13:79–86, 1989

TALC (NONASBESTOS FORM)
CAS: 14807-96-6

$Mg_3Si_4O_{10}(OH)_2$

Synonym: Nonfibrous talc

Physical Form.   Talc as a pure chemical compound is hydrous magnesium silicate. Talc is usually crystalline, flexible, and soft, with the purity and physical form of any sample depending on the source of the talc and on the minerals found in the ore body from which it is refined

Uses.   For clarifying liquid by filtration; pigment; for lubricating molds and machinery; electric and heat insulators; cosmetics

Exposure.   Inhalation

Toxicology.   The nonasbestos form of talc, also termed nonfibrous or pure talc, has not been proven to cause the effects produced by exposure to fibrous talc, namely, fibrotic pneumoconiosis and an increased incidence of cancer of the lungs and pleura.

Although there are a number of contradictory reports regarding the effects of talc, the contradiction has been ascribed to the differences in the mineral composition of the various talcs, which include pure talc, talc associated with silica and other nonasbestiform minerals, and talc containing asbestiform fibers such as tremolite and anthophyllite.[1]

In a study of 20 workers exposed for 10 to 36 years to talc described as "pure," at levels ranging from 15 to 35 mppcf, no evidence of pneumoconiosis was found.[2,3] In another study that compared the pulmonary function of workers exposed to either fibrous or nonfibrous talc, it was concluded that, although the fibrous form was the more pathogenic type, both talcs produced pulmonary fibrosis; no data were presented to document the types of talc involved.[4]

A study of 260 workers with 15 or more years of exposure to commercial talc dust (containing not only talc but also tremolite, anthophyllite, carbonate dusts, and a small amount of free silica) revealed a 40-fold greater than expected proportional mortality from cancer of the lungs and pleura. In addition, a major cause of death was cor pulmonale, a result of the pneumoconiosis; the effects were probably due to the asbestos-form contaminants.[5,6] The role of nonfibrous talc in these disease states could not be assessed.

In a study of 80 talc workers, there was an excess prevalence of productive cough and of criteria of chronic obstructive lung disease (COLD) when compared with 189 nonexposed workers.[7] The increase in COLD and wheezing occurred only among smokers. Those talc workers with more than 10 years of exposure had significantly decreased FEV_1; none of the talc workers had chest X rays definitely consistent with classical talc pneumoconiosis.[7] Exposure had been to talc of industrial grade with less than 1% silica and less than 2 fibers of asbestos/ml at levels of 0.51 to 3.55 mg/m^3, with most of the workers exposed to less than 1 mg/m^3 (or 2 mppcf).

A mortality study of 392 miners and millers of nonasbestos talc in Vermont showed an excess of deaths due to nonmalignant respiratory disease among millers and an excess of lung cancer mortality among miners.[8] The fact that the excess lung cancer mortality was observed for miners and not millers, despite probable higher dust exposure, led the investigators to conclude that other etiologic agents, either alone or in combination with talc dust, affected the miners.[8]

Another historical cohort study of 655 workers in a New York talc mine and mill revealed no significant differences in death rates from all causes, from cancer of the respiratory system, and from nonmalignant respiratory disease for the period ranging from 1948 to 1978.[9] However, workers with previous occupational histories were found to have excessive mortality from lung cancer and from nonmalignant respiratory tract disease, again suggesting another etiologic agent. No excess cancer risk or cause-specific mortality was found in a cohort mortality study of 94 talc miners and 295 talc millers from Norway who were exposed to a nonasbestiform talc with low quartz content.[10]

A 1-year follow-up of 103 miners and millers of talc ore free from asbestos and silica showed an association between exposure and small opacities on chest radiographs; the annual loss in FEV_1 and FVC was greater than expected and could not be wholly attributed to cigarette smoking.[11] However, effects on pulmonary function in nonsmokers was not associated with lifetime or current talc exposure.[11]

In inhalation studies with hamsters exposed to 8 mg/m^3 at a cumulative dust dose ranging from 15 to 6000 mg/m^3, no talc-induced lung lesions were found.[12] However, Italian talc, containing some quartz, was fibrogenic in specified-pathogen-free rats exposed to a respirable dust concentration of 10.8 mg/m^3—the cumulative dust doses being approximately 4100, 8200, and 16,400 mg/m^3 for 3-, 6-, and 12-month exposures.[13] There was some evidence of progression of the fibrosis after exposure to talc had been discontinued in the animals exposed for the longest period of time. The IARC has determined that there is inadequate evidence for carcinogenicity of talc to experimental animals; there is inadequate evidence for the carcinogenicity to humans of talc not containing asbestiform fibers, whereas there is carcinogenicity to humans of talc containing asbestiform fibers.[14]

Recently, in a National Toxicology Program study, rats exposed to dust levels of 18 mg/m^3 (with occasional higher excursions) were found to have impaired respiratory function, increased lung weights, inflammatory and proliferative processes in the lungs, interstitial fibrosis, hyperplasia of the alveolar epithelium and, occasionally, squamous metaplasia.[15] Incidences of alveolar/bronchiolar adenomas and carcinomas were significantly higher in females but not males, and pheochromocytomas of the adrenal medulla occurred in both sexes. Mice similarly exposed had inflammation in the lungs but no hyperplasia, fibrosis, or pulmonary neoplasms. It has been suggested that the high doses used in this study may have overwhelmed the bronchopulmonary clearance mechanism leading to the fibrotic tissue response.[1] Under conditions that don't overload the lung, natural defense mechanisms such as macrophages and mucociliary clearance can ordinarily cope with the lung burden without lesion development.

In vitro assay of a number of respirable talc specimens of high purity demonstrated a modest but consistent cytotoxicity to macrophages; the investigators conclude that the talcs would be expected to be slightly fibrogenic in vivo.[16]

The 1995 ACGIH threshold limit value–time-weighted average (TLV-TWA) for talc (containing no asbestos fibers) is 2 mg/m³.

REFERENCES

1. Wehner AP: Biological effects of cosmetic talc. *Food Chem Toxic* 32:1173–1184, 1994
2. Spiegel RM: Medical aspects of talc. In Goodwin A (ed): *Proceedings of the Symposium on Talc*, Bureau of Mines Report No 8639, pp 97–102. Washington, DC, US Government Printing Office, 1973
3. Hogue WL, Mallette FS: A study of workers exposed to talc and other dusting compounds in the rubber industry. *J Ind Hyg Toxicol* 31:359–364, 1949
4. Kleinfeld M et al: Lung function in talc workers—a comparative physiologic study of workers exposed to fibrous and granular talc dusts. *Arch Environ Health* 9:559–566, 1964
5. Kleinfeld M, Messite J, Kooyman O, Zaki MH: Mortality among talc miners and millers in New York State. *Arch Environ Health* 14:663–667, 1967
6. Kleinfeld M, Messitte J, Zaki MH: Mortality experiences among talc workers: a follow-up study. *J Occup Med* 16:345–349, 1974
7. Fine LJ, Peters JM, Burgess WA, Di Berardinis LJ: Studies of respiratory morbidity in rubber workers. Part IV. Respiratory morbidity in talc workers. *Arch Environ Health* 31:195–200, 1976
8. Selevan SG et al: Mortality patterns among miners and millers of non-asbestiform talc: preliminary report. *J Environ Pathol Toxicol* 2:273–284, 1979
9. Stille WT, Tabershaw IR: The mortality experience of upstate New York talc workers. *J Occup Med* 24:480–484, 1982
10. Wergeland E, Andersen A, Baerheim A: Morbidity and mortality in talc-exposed workers. *Am J Ind Med* 17:505–513, 1990
11. Wegman DH et al: Evaluation of respiratory effects in miners and millers exposed to talc free of asbestos and silica. *Br J Ind Med* 39:233–238, 1982
12. Wehner AP, Zwicker GM, Cannon WC, et al: Inhalation of talc baby powder by hamsters. *Food Cosmet Toxicol* 15:121–129, 1977
13. Wagner JC et al: An animal model for inhalation exposure to talc. In Lemen R, Dement JM (eds): *Dusts and Disease*, p 389. Proceedings of the Conference on Occupational Exposure to Fibrous and Particulate Dust and Their Extension into the Environment. Park Forest South, IL, Pathotox Publishers, 1979
14. *IARC Monographs on the Evaluation of the Carcinogenic Risk of Chemical to Humans*, Vol 42, Silica and some silicates, pp 185–224. Lyon, International Agency for Research on Cancer, 1987
15. National Toxicology Program: *NTP Technical Report on the Toxicology and Carcinogenesis Studies of Talc in F344/N Rats and B6C3F₁ Mice.* NIH Publication No 92-3152. National Institutes of Health, Bethesda, MD, 1992
16. Davies R, Skidmore JW, Griffiths DM, et al: Cytotoxicity of talc for macrophages in vitro. *Food Chem Toxicol* 21:201–207, 1983

TANTALUM
CAS: 7440-25-7

Ta

Synonyms: None

Physical Form. Solid (powder)

Uses. In manufacture of capacitors and other electronic components; in chemical equipment and corrosion-resistant tools

Exposure. Inhalation

Toxicology. Tantalum has a low order of toxicity but has produced transient inflammatory lesions in the lungs of animals.

Surgical implantation of tantalum metal products as plates and screws has not shown any adverse tissue reaction, thus demonstrating its physiological inertness.[1]

Intratracheal administration to guinea pigs of 100 mg tantalum oxide produced transient bronchitis, interstitial pneumonitis, and hyperemia, but it was not fibrogenic.[2] There were some slight residual sequelae in the form of focal hypertrophic emphysema and organizing pneumonitis around metallic deposits; and there was slight epithelial hyperplasia in the bronchi and bronchioles. Doses as high as 8000 mg/kg given orally produced no untoward effects in rats.[3]

The 1995 threshold limit value–time-weighted average (TLV-TWA) for tantalum metal and oxide dusts as Ta is 5 mg/m³.

REFERENCES

1. Stokinger HE: The metals. In Clayton GD, Clayton FE (eds): *Patty's Industrial Hygiene and Toxicology*, 3rd ed, rev, Vol 2A, *Toxicology*, pp 1492–2060. New York, Wiley-Interscience, 1981
2. Schepers GWH: The biologic action of tantalum oxide. *AMA Arch Ind Health* 12:121–123, 1955
3. Cochran KW et al: Acute toxicity of zirconium, columbium, strontium, lanthanum, cesium, tantalum and yttrium. *Arch Ind Hyg Occup Med* 1:637–650, 1950

TELLURIUM
CAS: 13494-80-9

Te

Synonyms: None

Physical Form. Grayish-white, lustrous, brittle crystalline solid; or dark-gray to brown amorphous powder

Uses. Coloring agent in chinaware, porcelains, and glass; reagent in producing black finish on silverware; rubber manufacturing; component of many alloys

Exposure. Inhalation

Toxicology. Tellurium causes garlic odor of the breath and malaise in humans.

Serious cases of tellurium intoxication have not been reported from industrial exposure. Iron foundry workers exposed to concentrations between 0.01 and 0.1 mg/m³ complained of garlic odor of the breath and sweat, dryness of the mouth and metallic taste, somnolence, anorexia, and occasional nausea; urinary concentrations ranged from 0 to 0.06 mg/l. Somnolence and metallic taste in the mouth did not appear with regularity until the level of tellurium in the urine was at least 0.01 mg/l.[1] Skin lesions in the form of scaly, itching patches and loss of sweat function occurred in workers exposed to tellurium dioxide in an electrolytic lead refinery.[2]

Hydrogen telluride has caused pulmonary irritation and hemolysis of red blood cells in animals; this gas is very unstable, however, and its occurrence as an actual industrial hazard is unlikely.[1,3]

In animals, acute tellurium intoxication results in restlessness, tremor, diminished reflexes, paralysis, convulsions, somnolence, coma, and death.[4] Administration to pregnant rats of 500 to 3000 ppm tellurium in the diet resulted in a high incidence of hydrocephalic offspring.[5] Weanling rats fed elemental tellurium at a level of 1% (10,000 ppm) in the diet developed a neuropathy characterized by segment demyelination; remyelination and functional recovery occurred despite continued administration of tellurium.[6] Both skeletal and soft tissue malformations (primarily hydrocephalus) were noted in the offspring of rats exposed to 3000 or 15,000 ppm in the diet on days 6 to 15 of gestation, but significant maternal toxicity was also noted.[7] Similarly, skeletal delays and nonspecific abnormalities occurred in the offspring of rabbits only at dosages (5250 ppm in the diet) well in excess of levels that produced significant maternal toxicity.

The 1995 ACGIH threshold limit value

(TLV-TWA) for tellurium and compounds as Te is 0.1 mg/m³.

See separate monograph on tellurium hexafluoride.

REFERENCES

1. Hygienic Guide Series: Tellurium. *Am Ind Hyg Assoc J* 25:198–201, 1964
2. Browning E: *Toxicity of Industrial Metals.* 2nd ed, pp 310–316. London, Butterworths, 1969
3. Cerwenka EA, Cooper WC: Toxicology of selenium and tellurium and their compounds. *Arch Environ Health* 3:189–200, 1961
4. Cooper WC (ed): *Tellurium,* pp 313–321. New York, Van Nostrand Reinhold Co, 1971
5. Duckett S: Fetal encephalopathy following ingestion of tellurium. *Experientia* 26:1239–1241, 1970
6. Lampert P, Garro F, Pentschew A: Tellurium neuropathy. *Acta Neuropathol* 15:308–317, 1970

TELLURIUM HEXAFLUORIDE
CAS: 7783-80-4

TeF_6

Synonyms: None

Physical Form. Gas

Source. By-product of ore refining

Exposure. Inhalation

Toxicology. Tellurium hexafluoride is a pulmonary irritant in animals; severe exposure is expected to cause the same effect in humans.

Human exposure has caused headache and dyspnea.[1,2] Two subjects accidentally exposed to tellurium hexafluoride following leakage of 50 g into a small laboratory experienced garlic breath, fatigue, a bluish-black discoloration of the webs of the fingers, and streaks on the neck and face.[3] Complete recovery occurred without treatment.

Rodents exposed to 1 ppm for 1 hour had increased respiratory rates, whereas a 4-hour exposure at this concentration caused pulmonary edema.[4] However, repeated exposure at 1 ppm for 1 hr/day for 5 days produced no effect; 5 ppm for 4 hours was fatal.

The 1995 ACGIH threshold limit value–time-weighted average (TLV-TWA) is 0.02 ppm (0.10 mg/m³) as Te.

REFERENCES

1. Cooper WC: *Tellurium,* pp 317, 320–321. New York, Van Nostrand Reinhold Co, 1971
2. Cerwenka EA Jr, Cooper WC: Toxicology of selenium and tellurium and their compounds. *Arch Environ Health* 3:71–82, 1961
3. Blackadder ES, Manderson WG: Occupational absorption of tellurium: a report of two cases. *Br J Ind Med* 32:59–61, 1975
4. Kimmerle G: Comparative research on the inhalation toxicity of selenium sulfide and tellurium hexafluoride. *Arch Toxicol* 18:140–144, 1960

TERPHENYLS
CAS: 26140-60-3

$C_{18}H_{14}$

Synonyms: Phenylbiphenyls; diphenylbenzenes; triphenyls; *o*-terphenyl; *m*-terphenyl; *p*-terphenyl

Physical Form. Colorless or light-yellow solids

Uses. Coolant for heat exchange in nuclear reactors

Exposure. Inhalation

Toxicology. Terphenyls are irritants of the eyes, mucous membranes, and skin.

There are no well-documented studies showing the effects of terphenyls on humans. Clinical studies of an exposed group of workers showed no ill effects from prolonged exposure to 0.1 to 0.9 mg/m³.[1] Workers have experienced eye and respiratory irritation at levels above 10 mg/m³.[2] As a class of compounds, organic coolants (including terphenyls) have caused transient headache and sore throat.[1] In addition, cases of dermatitis have been attributed to skin contact with organic coolant compounds.[1]

Inhalation by rats of relatively high concentrations (660–3390 mg/m³) of mixed and single isomers for periods of 1 hour for up to 14 days caused tracheobronchitis, pulmonary edema, and death at the higher concentrations.[3] In rats, the oral LD_{50} for *o*-terphenyl was 1.9 g/kg; for *m*-terphenyl, it was 2.4 g/kg; and, for *p*-terphenyl, it was greater than 10 g/kg.[4]

Transient morphological changes in mitochondria of pulmonary cells were found in rats exposed to 50 mg/m³ terphenyls, 7 hr/day, for up to 8 days.[5] The number of vacuolated mitochondria increased with days of exposure.[5]

The 1995 ACGIH ceiling–threshold limit value (C-TLV) is 0.53 ppm (5 mg/m³).

REFERENCES

1. Weeks JL, Lentle BC: Health considerations in the use of organic reactor coolants. *J Occup Med* 12:246–252, 1970
2. Testa C, Masi G: Determination of polyphenyls in working environments of organic reactors by spectrophotometric methods. *Analyt Chem* 36:2284–2287, 1964
3. Haley TJ et al: Toxicological studies on polyphenyl compounds used in atomic reactor moderator-coolants. *Toxicol Appl Pharmacol* 1:515–523, 1959
4. Cornish HH, Bahor RE, Ryan RC: Toxicity and metabolism of ortho-, meta-, and para-terphenyls. *Am Ind Hyg Assoc J* 23:372–378, 1962
5. Adamson IYR et al: The acute toxicity of reactor polyphenyls on the lung. *Arch Environ Health* 19:499–504, 1969

1,1,2,2-TETRACHLORO-1,2-DIFLUOROETHANE
CAS: 76-12-0

$C_2Cl_4F_2$

Synonyms: TCDF; Refrigerant 112

Physical Form. Colorless solid or liquid

Uses. Refrigerant; solvent extractant; blowing or foaming agent

Exposure. Inhalation

Toxicology. At high concentrations, 1,1,2,2-tetrachloro-1,2-difluoroethane affects the nervous system and causes pulmonary edema in animals; it is expected that severe exposure in humans will produce the same effects.

There are no reports of adverse effects in humans.

Rats exposed to 30,000 ppm died within 1 hour after onset of exposure with severe pulmonary hemorrhage.[1] At 15,000 ppm, rats exhibited excitability, incoordination, coma, rapid respiration, tremor, and convulsions; 3 of 4 died in 3 hours with pulmonary edema and hyperemia of the lungs and liver.[2] Exposure at 5000 ppm for 18 hours caused coma, pulmonary damage, and death.[1] Rats survived 10 exposures of 4 hours each at 3000 ppm with rapid, shallow respiration, hyperresponsiveness, and slight incoordination; recovery was immediate after exposure.[2] Decreased leukocyte count occurred in female rats exposed to 1000 ppm, 6 hr/day, for 31 days.[2]

1,1,2,2-Tetrachloro-1,2-difluoroethane was mildly irritating to rabbit eyes and guinea pig skin.

The 1995 ACGIH threshold limit value–time-weighted average (TLV-TWA) is 500 ppm (4170 mg/m³).

REFERENCES

1. Greenberg LA, Lester D: Toxicity of the tetrachlorodifluoroethanes. *Arch Ind Hyg Occup Med* 2:345–347, 1950

2. Clayton JW Jr, Sherman H, Morrison SD, et al: Toxicity studies on 1,1,2,2-tetrachloro-1,2-difluoroethane and 1,1,1,2-tetrachloro-2,2-difluoroethane. *Am Ind Hyg Assoc J* 27:332, 1966

1,1,2,2-TETRACHLOROETHANE
CAS: 79-34-5

CHCl₂CHCl₂

Synonyms: Acetylene tetrachloride; sym-tetrachloroethane; 1,1-dichloro-2,2-dichloroethane

Physical Form. Heavy, clear liquid

Uses. Intermediate in the production of trichloroethylene, tetrachloroethylene, and 1,2-dichloroethylene; formerly used as a solvent, insecticide, and fumigant

Exposure. Inhalation; skin absorption

Toxicology. Tetrachloroethane is toxic to the liver and causes central nervous system depression and gastrointestinal effects.

Reports of industrial experience indicate that cases of intoxication most commonly have presented symptoms of gastrointestinal irritation (nausea, vomiting, abdominal pain, anorexia) and liver involvement (liver enlarged and tender, jaundice, bilirubinuria).[1,2] Jaundice sometimes progressed to cirrhosis and was often accompanied by delirium, convulsions, coma, and death. Other cases have been characterized primarily by central nervous system effects (dizziness, headache, irritability, nervousness, insomnia, paresthesia, and tremors).[1,2]

In one study, exposure of two men at 116 ppm for 20 minutes caused dizziness and mild vomiting; at 146 ppm, dizziness occurred after 10 minutes, mucosal irritation at 12 minutes, and fatigue within 20 minutes.[2] Concentrations of up to 335 ppm produced the same symptoms with shorter exposure times. Occupational ex-posure to concentrations ranging from 1.5 to 247 ppm caused signs of liver injury such as hepatomegaly and increased serum bilirubin. These signs were still found after air concentrations had been reduced below 36 ppm.[2] Among a group of workers in India exposed to 20 to 65 ppm, effects were nausea, vomiting, and abdominal pain, along with a high incidence of tremor of the fingers.[3]

Oral ingestion of 3 ml caused coma or impaired consciousness in 8 adult patients mistakenly administered tetrachloroethane.[2] Dermal absorption has been suspected in some poisoning cases.[2] Skin exposure may also produce dermatitis as a result of defatting action; in rare cases, the dermatitis may be caused by hypersensitivity to the substance.[4]

Treatment of mice during gestation caused embryotoxic effects and a low incidence of malformations.[5] Administration of 3.2 mg/kg/day to rats for 27 weeks caused irreversible histopathological changes in the testes.[6]

1,1,2,2-Tetrachloroethane administered by gavage produced an increased incidence of hepatocellular carcinomas in mice but not in rats.[7] In one epidemiological study of exposed army workers, there was a slight increase in deaths due to genital cancer and leukemia.[8] Exposure levels were not available, and other confounding factors may have been present, so that no definite conclusions could be drawn from the study.[8] There is limited evidence of carcinogenicity in experimental animals according to the IARC, and inadequate evidence for humans.

In in vitro genotoxic assays, 1,1,2,2-tetrachloroethane gave negative results for gene mutation, chromosomal aberration, DNA repair and synthesis, and cell transformation.[6] It did induce sister chromatid exchange in Chinese hamster ovary cells with and without metabolic activation.

Tetrachloroethane has a mild, sweetish odor detectable at 3 ppm that may not provide sufficient warning of dangerous levels because of olfactory fatigue.

The 1995 ACGIH threshold limit value–time-weighted average (TLV-TWA) for 1,1,2,2-tetrachloroethane is 1 ppm (6.9 mg/m³) with a notation for skin absorption.

REFERENCES

1. von Oettingen WF: *The Halogenated Aliphatic, Olefinic, Aromatic, and Aliphatic-Aromatic Hydrocarbons Including the Halogenated Insecticides, Their Toxicity and Potential Dangers.* US Public Health Service Publication Pub No 414, pp 158–164. Washington, DC, US Government Printing Office, 1955
2. National Institute for Occupational Safety and Health: *Criteria for a Recommended Standard . . . Occupational Exposure to 1,1,2,2-Tetrachloroethane.* DHEW (NIOSH) Pub No 77-121, p 143. Washington, DC, US Government Printing Office, 1976
3. Lobo-Mendonca R: Tetrachloroethane—a survey. *Br J Ind Med* 20:50–56, 1963
4. Chemical Safety Data Sheet SD-34, Tetrachloroethane, p 12. Washington, DC, MCA, Inc, 1949
5. *IARC Monographs on the Evaluation of the Carcinogenic Risk of Chemicals to Humans,* Vol 20, Some halogenated hydrocarbons, pp 477–489. Lyon, International Agency for Research on Cancer, 1979
6. Agency for Toxic Substances and Disease Registry (ATSDR): *Toxicological Profile for 1,1,2,2-Tetrachloroethane.* US Public Health Service, p 87, 1989
7. National Cancer Institute: *Bioassay of 1,1,2,2-Tetrachloroethane for Possible Carcinogenicity.* DHEW (NIOSH) Pub No 78-827. Washington, DC, US Department of Health, Education and Welfare, 1978
8. Norman JE Jr, Robinette CD, Fraumeni JF Jr: The mortality experience of Army World War II chemical processing companies. *J Occup Med* 23:818–822, 1981

TETRACHLORONAPHTHALENE

CAS: 1335-88-2

$C_{10}H_4Cl_4$

Synonym: Halowax

Physical Form. Solid

Uses. In synthetic wax; in dielectrics in capacitors; in wire insulation

Exposure. Inhalation; skin absorption

Toxicology. Tetrachloronaphthalene may cause liver injury.

Experiments on human volunteers showed tetrachloronaphthalene to be nonacneigenic as opposed to the penta- and hexachloro- derivatives, which produce very severe chloracne.[1]

Rats exposed 16 hr/day to 10.97 mg/m³ of tri- and tetrachloronaphthalene vapor for up to 4.5 months had slight liver injury.[2] When a mixture of tetra- and pentachloronaphthalene was fed to rats at a dose of 0.5 mg/day for 2 months, definite liver injury and some mortality occurred.[2]

The 1995 ACGIH threshold limit value–time-weighted average (TLV-TWA) for tetrachloronaphthalene is 2 mg/m³.

REFERENCES

1. Shelley WB, Kligman AM: The experimental production of acne by penta- and hexachloronaphthalenes. *Arch Dermatol* 75:689–695, 1957
2. Deichmann WB: Halogenated cyclic hydrocarbons. In Clayton GD, Clayton FE (eds): *Patty's Industrial Hygiene and Toxicology,* 3rd ed, rev, Vol 2B, *Toxicology,* pp 3669–3675. Wiley-Interscience, 1981

TETRAETHYL LEAD
CAS: 78-00-2

$Pb(C_2H_5)_4$

Synonyms: Lead tetraethyl; TEL; tetraethylplumbane

Physical Form. Colorless liquid

Uses. Gasoline additive to prevent "knocking" in motors

Exposure. Inhalation; skin absorption; ingestion

Toxicology. Tetraethyl lead (TEL) affects the nervous system and causes mental aberrations—including psychosis, mania, and convulsions—and death.

Of approximately 150 reported fatal cases of TEL poisoning, most have been related to early production methods, to cleaning of leaded gasoline storage tanks without protective equipment, and to suicidal or accidental ingestion.[1] Milder cases of intoxication have been caused by exposures to leaded gasoline in the workplace.[1]

The absorption by humans of a sufficient quantity of tetraethyl lead, either briefly at a high rate (100 mg/m^3 for 1 hour) or for prolonged periods at a lower rate, causes intoxication.[2] The interval between exposure and the onset of symptoms varies inversely with dose and may last 1 hour to several days.[1] This clinical latency is related to the time it takes for TEL to be absorbed, distributed, and metabolized to triethyl lead before toxic action develops.[1]

The signs and symptoms of TEL intoxication differ in many respects from those of inorganic lead intoxication and are often vague and easily missed. The initial or prodromal symptoms are nonspecific and include asthenia, weakness, fatigue, headache, nausea, vomiting, diarrhea, and anorexia.[1] Insomnia is usually present, and any sleep is light, usually with nightmares. Signs of nervous system involvement may then develop (ataxia, tremor, hypotonia) as well as bradycardia and hypothermia, referred to as the TEL triad.[1]

More severe intoxication causes recurrent or nearly continuous episodes of disorientation, hallucinations, facial contortions, and intense hyperactivity, which require that the individual be restrained. Such episodes may convert abruptly into maniacal or violent convulsive seizures, which may terminate in coma and death.[2] Autopsy reports from humans who succumbed to TEL poisoning confirm that the brain is the critical target organ, and both focal and generalized damage have been described. For survivors of TEL poisoning, recovery may take many weeks or months.[1] There is some question as to whether all changes are reversible following heavy or long-term exposures.[1]

During intoxication, there is a striking elevation in the rate of excretion of lead in the urine but only a negligible or slight elevation in the concentration of lead in the blood.[2,3] In severe intoxication, the urine lead is rarely less than 350 μg/l of urine, whereas the blood lead is rarely more than 50 μg/100 g of blood.[2,3] There is also a total absence of morphological or chemical abnormalities in the erythrocytes—in sharp contrast to intoxication caused by inorganic lead.[2]

In a mortality study of 592 workers, the mean exposure time to TEL was 17.9 years, and urinary lead levels during this period did not exceed 180 μg/l. The incidence of death in this group and in a control group of employees was less than that expected in the general population, and there were no peculiarities in the specific causes of death in either group.[4] In a similar study of a different cohort of these exposed workers, there were no significant health differences compared to a control group.[5]

A cohort of gasoline depot workers exposed to a mean external tetraethyl lead concentration of 84.8 μg/m^3 (as Pb) had a statistically increased frequency of appearance of tremor and sinus bradycardia (vs. controls).[6] No clinical neurological or neurobehavioral findings were found following long-term exposure at a chemical manufacturing plant where TEL exposures ranged from 0.6 to 43.1 μg/m^3 (as Pb).[7]

Tetraethyl lead is not an irritant, and no unpleasant sensations are related to skin contact or inhalation.[1] The ability to penetrate skin makes reliance on airborne concentrations impractical.[2]

Of 41 female Swiss mice that survived for 36 weeks after a single subcutaneous injection of 0.6 mg, 5 developed malignant lymphomas during the next 15 weeks; the significance of the study cannot be evaluated because this tumor occurs spontaneously with a variable incidence in the mouse strain used.[8,9]

Teratogenic effects have not been observed after exposure to maximally tolerated doses in mice or rats.[1] Rodent embryos may serve as a poor model for human fetuses because the hepatic microsomal metabolizing enzymes do not develop until after birth in rodents while these enzymes develop early in humans.

The 1995 ACGIH threshold limit value–time-weighted average (TLV-TWA) is 0.1 mg/m³ as Pb, with a notation for skin absorption.

Diagnosis. Signs and symptoms include insomnia, lassitude, nightmares, anxiety; tremor, hyperreflexia, spasmodic muscular contractions; bradycardia, hypotension, hypothermia, pallor; nausea, anorexia, weight loss; disorientation, hallucinations, psychosis, mania, convulsions, coma; and eye irritation.

Differential Diagnosis: Other causes of convulsions must be differentiated from tetraethyl lead exposure. These include idiopathic epilepsy; hypertensive encephalopathy; metabolic disturbances such as hypoglycemia, hypocalcemia, uremia, and porphyria; hypoxic encephalopathy; infections such as viral encephalitis; and exposure to other convulsants.

A history of exposure to TEL and the presence of the aforementioned signs and symptoms should confirm the diagnosis.

Special Tests: An analysis of the urinary concentration of lead is useful in evaluating the amount of TEL absorption. Blood lead concentration is an unreliable index of TEL absorption. A urine concentration of lead of 150 μg/l, corrected to a specific gravity of 1.024, indicates a dangerous degree of absorption; if the level rises to 180 μg/l, the worker should be removed from exposure. TEL poisoning is typically associated with levels of 300 μg/l or more; if the level is less than 100 μg/l at the time of symptoms, TEL absorption is not the cause.[10] (Recent studies suggest that levels of urinary diethyl lead may be the most sensitive and specific indicator of tetraethyl lead exposure.[6])

Treatment. If TEL gets on the skin, immediately remove it by rinsing first with kerosene or a similar petroleum distillate product, and then wash with soap and water. Chelating agents are not useful for organolead poisoning.[11] Heavy and prolonged sedation with short-acting barbiturates in a hospital provides the most effective therapy available. Drugs with a cortical effect, such as morphine, are contraindicated because they may worsen symptoms.[1] Fluid and electrolyte balance must be carefully maintained; this may be difficult because of the patient's hyperactivity.

Persons suspected of having TEL intoxication should be kept under close surveillance because personality changes may occur and be manifested in a suicide attempt. Relapses during recovery are common.

REFERENCES

1. Grandjean P: Organolead exposures and intoxications. In Grandjean P (ed): *Biological Effects of Organolead Compounds*, pp 1–278. Boca Raton, FL, CRC Press, 1984
2. Kehoe R: Lead, alkyl compounds. In International Labour Office: *Encyclopaedia of Occupational Health and Safety*, Vol II, pp 1197–1199. Geneva, 1983
3. Fleming AJ: Industrial hygiene and medical control procedures—manufacture and handling of organic lead compounds. *Arch Environ Health* 8:266–270, 1964
4. Robinson TR: 20-year mortality of tetraethyl lead workers. *J Occup Med* 17:601–605, 1974
5. Robinson TR: The health of long service tetraethyl lead workers. *J Occup Med* 18:31–40, 1976
6. Zhang W, Zhang G, He H, et al: Early health effects and biological monitoring in persons occupationally exposed to tetraethyl lead. *Int Arch Occup Environ Health* 65:395–399, 1994
7. Seeber A, Kiesswetter E, Neidhart B, et al: Neurobehavioral effects of long term exposure to tetraalkyllead. *Neurotoxicol Teratol* 12:653–655, 1990
8. *IARC Monographs on the Evaluation of the Carcinogenic Risk of Chemicals to Man*, Vol 2, Some inorganic and organometallic compounds, pp 150–160. Lyon, International Agency for Research on Cancer, 1973
9. Epstein SS, Mantel N: Carcinogenicity of tetraethyl lead. *Experientia* 24:580–581, 1968
10. Tsuchiya K: Lead. In Friberg L, Norsberg G, Vouk V (eds): *Handbook on the Toxicology of Metals*, 2nd ed, Vol II, *Specific Metals*, pp 340–342. Amsterdam, Elsevier, 1986
11. Chisholm JJ Jr: Treatment of lead poisoning. *Mod Treatment* 8:593–611, 1971

TETRAETHYL PYROPHOSPHATE
CAS: 107-49-3

$C_8H_{20}O_7P_2$

Synonyms: Ethyl pyrophosphate; phosphoric acid tetraethyl ester; TEPP; Tetron; NIFOS; TEP

Physical Form. Colorless, odorless liquid (pure); amber liquid (crude)

Uses. Insecticide

Exposure. Inhalation; skin absorption; ingestion

Toxicology. Tetraethyl pyrophosphate is a highly toxic anticholinesterase agent.

Signs and symptoms of overexposure are caused by the inactivation of the enzyme cholinesterase, which results in the accumulation of acetylcholine at synapses in the nervous system, skeletal and smooth muscle, and secretory glands.[1-3] The sequence of the development of systemic effects varies with the route of entry.[2] The onset of signs and symptoms is usually prompt but may be delayed for up to 12 hours.[1,2] After inhalation, respiratory and ocular effects are the first to appear, often within a few minutes of exposure. Respiratory effects include tightness in the chest and wheezing due to bronchoconstriction and excessive bronchial secretion; laryngeal spasms and excessive salivation may add to the respiratory distress; cyanosis may also occur. Ocular effects include miosis, blurring of distant vision, tearing, rhinorrhea, and frontal headache.

After ingestion, gastrointestinal effects such as anorexia, nausea, vomiting, abdominal cramps, and diarrhea appear within 15 minutes, and muscular fasciculations in the immediate area occur, usually within 15 minutes to 4 hours. The lowest lethal oral dose in man was approximately 1.4 g/kg; oral doses of 0.3 mg/kg have caused abnormal muscle contractions, gastrointestinal upset, and wakefulness.[4]

Skin absorption is somewhat greater at higher ambient temperatures and is increased by the presence of dermatitis.[1,2]

With severe intoxication by all routes, an excess of acetylcholine at the neuromuscular junctions of skeletal muscle causes weakness aggravated by exertion, involuntary twitchings, fasciculations and, eventually, paralysis.[2] The most serious consequence is paralysis of the respiratory muscles; in fatal cases, death usually occurs within 24 hours.[2] Effects on the central nervous system include giddiness, confusion, ataxia, slurred speech, Cheyne–Stokes respiration, convulsions, coma, and loss of reflexes. The blood pressure may fall to low levels, and cardiac irregularities, including complete heart block, may occur.[1]

In nonfatal cases, recovery usually occurs within 1 week, but increased susceptibility to anticholinesterase agents persists for up to several weeks after exposure.[2] Daily exposure to concentrations that are insufficient to produce symptoms following a single exposure may result in the onset of symptoms. Continued daily exposure may be followed by increasingly severe effects.[1]

Mild intoxication was reported in 15 people exposed to a dust of 1% TEPP; the predominant symptom was shortness of breath, which occurred after breathing the dust-laden air for 30 minutes. Symptoms rapidly abated after exposure was terminated.[5]

Eye exposure can produce visual disturbances without affecting blood cholinesterase levels. Exposed crop duster pilots, unable to judge distances, have been involved in accidents. Volunteers instilled with 2 drops of 0.1% TEPP 30 minutes apart experienced maximal miosis without any change in blood cholinesterase.[6]

TEPP inactivates cholinesterase by phosphorylation of the active site of the enzyme to form the diethylphosphoryl enzyme. Over the following 24 to 48 hours, there is a process, termed aging, of conversion to the monoethylphosphoryl enzyme. Aging is of clinical interest in the treatment of poisoning because cholinesterase reactivators such as pralidoxime (2-PAM, Protopam) chloride are ineffective after aging has occurred.

The 1995 threshold limit value–time-weighted average for TEPP is 0.004 ppm (0.047 mg/m³) with a notation for skin absorption.

Note: For a description of diagnostic signs, differential diagnosis, and medical control, including clinical laboratory tests, as well as specific treatment of overexposure to anticholinesterase insecticides, see "Parathion."

REFERENCES

1. Koelle GB (ed): Cholinesterases and Anticholinesterase Agents. *Handbuch der Experimentellen Pharmakologie*, Vol 15, pp 989–1027. Berlin, Springer-Verlag, 1963
2. Taylor P: Anticholinesterase agents. In Gilman AG et al (eds): *Goodman and Gilman's The Pharmacological Basis of Therapeutics*, 7th ed, pp 110–129. New York, Macmillan, 1985
3. Hayes WJ Jr: Clinical Handbook on Economic Poisons, *Emergency Information for Treating Poisoning*. US Public Health Service Pub No 476, pp 12–23, 40–42. Washington, DC, US Government Printing Office, 1963
4. Communicable Disease Center, US Public Health Service: *Clinical Memoranda on Economic Poisons*, Pub No. 476. US Dept of Health Education, and Welfare, Public Health Service, Atlanta, GA, 1956
5. Quinby GE, Doornink GM: Tetraethyl pyrophosphate poisoning following airplane dusting. *JAMA* 191:95–100, 1965
6. Hayes WJ Jr: *Pesticides Studied in Man*, pp 391–394. Baltimore, MD, Williams and Wilkins, 1982

TETRAHYDROFURAN

CAS: 109-99-9

$(C_2H_4)_2O$

Synonyms: Cyclotetramethylene oxide; diethylene oxide; THF; tetramethylene oxide

Physical Form. Colorless liquid

Uses. Widely used as an industrial solvent, especially for plastic resins; reaction medium; in a coating agent used in the production of audio- and videotapes

Exposure. Inhalation

Toxicology. Tetrahydrofuran (THF) is an upper respiratory tract irritant; at high concentrations, it is a central nervous system depressant.

Two workers who had been exposed to glue containing THF for up to 8 hours in a confined space had nausea, headache, dizziness, dyspnea, and chest pain.[1] Clinical examination disclosed conjunctival irritation and alteration in liver enzymes. Symptoms disappeared within a few hours after exposure ceased, and liver enzymes returned to normal within 2 weeks.

Administered to mice, 49,000 ppm for 51 minutes resulted in narcosis, muscular hypotonia, disappearance of corneal reflexes, then coma, followed by death.[2] The LC_{50} was estimated to be 21,000 ppm in rats exposed for 3 hours.[3] Repeated exposure of rats to concentrations ranging from 100 ppm to 5000 ppm for 12 weeks caused a dose-related increase in irritation of the mucous membranes. At the 5000 ppm level, there were marked edema or opacity of the cornea, salivation, and discharge or bleeding in the nasal mucosa.

In subchronic studies, mice and rats were exposed at 0, 66, 200, 600, 1800, and 5000 ppm, 6 hr/day, 5 days/week, for 13 weeks.[4] Rats were ataxic at the high dose, and mice exposed to 1800 or 5000 ppm appeared to be in a state of narcosis. A minimal to mild centrilobular hepatocytomegaly also occurred in the mice exposed at 5000 ppm. Stomach lesions, limited to the rats, were thought to occur from direct contact of THF ingested during the inhalation exposure period.

Exposure of pregnant CD-1 mice for 6 hr/day on days 6 through 17 of gestation was embryotoxic at 1800 ppm and 5000 ppm, as indicated by a reduction in the number of live fetuses per litter (95% resorptions in the 5000 ppm group).[5] These doses were also maternally toxic, producing narcosis in dams at 1800 ppm and significant lethality at 5000 ppm. There were no statistically significant differences in the incidences of malformations or variations.

In rats similarly exposed, maternal and fetal body weights were significantly reduced at the 5000 ppm exposure level.

THF was nonmutagenic in the *Salmonella* microsomal assay.[6]

Recent studies have suggested that measurement of THF concentration in the urine may be a useful biological indicator of occupational exposure to THF, whereas exhaled breath and blood analyses may be less suitable.[7]

The liquid has an ethereal odor similar to that of acetone and a pungent taste.

The 1995 threshold limit value–time-weighted average (TLV-TWA) is 200 ppm (590 mg/m³) with a short-term exposure limit (STEL) of 250 ppm (737 mg/m³).

REFERENCES

1. Garnier R, Rosenberg N, Puissant JM, et al: Tetrahydrofuran poisoning after occupational exposure. *Br J Ind Med* 46:677–678, 1989
2. Sax NI: *Hazardous Chemicals Information Annual.* Vol 1, pp 640–644. New York, Van Nostrand Reinhold Information Services, 1986
3. Katahira T, Teramoto K, Horiguchi S: Experimental studies on the acute toxicity of tetrahydrofuran in animals. *Japan J Indust Health* 24:373–378, 1982
4. Chhabra RS, Elwell MR, Chou B, et al: Subchronic toxicity of tetrahydrofuran vapors in rats and mice. *Fund Appl Toxicol* 14:338–345, 1990
5. Mast TJ, Weigel RJ, Westerberg RB, et al: Evaluation of the potential for developmental toxicity in rats and mice following inhalation exposure to tetrahydrofuran. *Fund Appl Toxicol* 18:255–265, 1992
6. Maron D, Katzenellenbogen J, Ames B: Compatibility of organic solvents with the *Salmonella* microsome test. *Mutat Res* 88:343–350, 1981
7. Ong CN, Chia SE, Phoon WH, et al: Biological monitoring of occupational exposure to tetrahydrofuran. *Br J Ind Med* 48:616–621, 1991

TETRALIN
CAS: 119-64-2

$C_{10}H_{12}$

Synonyms: 1,2,3,4-tetrahydronaphthalene; tetraline; tetranap; benzocyclohexane

Physical Form.   Colorless liquid

Uses.   Solvent for fats and oils; alternative for turpentine in polishes and paint

Exposure.   Inhalation

Toxicology.   Tetralin is an irritant to the skin, eyes, and mucous membranes and may cause neurological disturbances at high concentrations.

The hallmark for tetralin exposure in man is the production of green-colored urine.[1] Two painters who used varnishes containing tetralin in a poorly ventilated area had intense irritation of the mucous membranes, profuse lacrimation, headache, stupor, and the characteristic green urine. Hospital patients on a ward whose floor had been waxed with a tetralin-based polish experienced similar symptoms, including eye irritaion, headache, nausea, diarrhea, and green urine. Asthenia also was observed in subjects who had slept in rooms waxed with a tetralin-based polish.

In a human case involving ingestion of approximately 1 to 1.5 mg/kg, effects consisted of nausea, vomiting, and green-gray urine.[2] Clinical changes included proteinuria and elevated serum levels of bilirubin, creatine, alkaline phosphatase, lactic dehydrogenase, and glutamic oxaloacetic transaminase. All signs and symptoms returned to normal within 2 weeks.

Exposure to tetralin-saturated vapor for 8 hours was not lethal to rats.[3] Acute intoxication of guinea pigs, exposed either orally (0.25 ml/day), percutaneously by application to 6 cm² of shaved skin), or by inhalation (1.42 mg/l for 8 hr/day), produced loss of weight, tremors, paralysis of the hindquarters, and difficult respiration. Applied to the skin of guinea pigs, the liquid caused erythema, drying, and defatting. In rabbits, the irritant dermal and ocular dose was 500 mg, and the dermal LD_{50} was 17.3 g/kg.

A threshold limit value has not been established for tetralin.

REFERENCES

1. Longacre SL: Tetralin. In Snyder R (ed): *Ethel Browning's Toxicity and Metabolism of Industrial Solvents*, 2nd ed, Vol I, *Hydrocarbons*, pp 143–152. Amsterdam, Elsevier, 1987
2. Drayer DE, Reidenberg MM: Metabolism of tetralin and toxicity of Cuprex in man. *Drug Metab Dispos* 1:577–579, 1973
3. Smyth HF, Carpenter CP, Weil CS: Range-finding toxicity data: List IV. *AMA Arch Ind Hyg Occup Med* 4:119–122, 1951

TETRAMETHYL LEAD
CAS: 75-74-1

$Pb(CH_3)_4$

Synonyms: Lead tetramethyl; TML

Physical Form. Colorless liquid

Uses. Gasoline additive, especially to aviation and premium grades with high aromatic content

Exposure. Inhalation; skin absorption; ingestion

Toxicology. Tetramethyl lead (TML) affects the nervous system in animals.

Accidental human exposure to a high level of TML liquid for approximately 5 minutes caused no signs or symptoms of lead poisoning. Significant exposure was corroborated by high levels of urinary lead, averaging almost 1000 µg/24 hour for the first 4 days post-exposure.[1] By comparison, urinary lead levels of less than 750 µg/24 hour following tetraethyl lead (TEL) exposure have been associated with confusion, agitation, and acute toxic delirium.[1]

In a plant, 21 workers were exposed at different times to TEL and then to TML under similar conditions for similar periods. TML had 3 times the airborne level found during TEL production, yet the urinary lead levels were nearly the same in both cases; this suggests that TML is absorbed more slowly than TEL.[2] No signs or symptoms of toxicity were noted.

In rats, the approximate oral LD_{50} for TML is 108 mg/kg versus 17 mg/kg for TEL. Effects were tremor, hyperactivity, and convulsions.[3] Inhalation studies on rats showed TML to have less than one-tenth the toxicity of TEL.[4] In dogs and mice, however, the reverse is true, with TML being more potent than TEL.[5]

Prudent practice suggests that TML be treated as if it were TEL.[5] Further caution is indicated by recent reports that the degradation product of TML, trimethyllead, acts differently from higher trialkyllated compounds, inducing lipid peroxidation.[6] This difference indicates a potential for more severe chronic toxicity from TML exposure.

The 1995 ACGIH threshold limit value–time-weighted average (TLV-TWA) for TML is 0.15 mg/m³ as lead, with a notation for skin absorption.

REFERENCES

1. Gething J: Tetramethyl lead absorption: a report of human exposure to a high level of tetramethyl lead. *Br J Ind Med* 32:329–333, 1975
2. deTreville RTP, Wheeler HW, Sterling T: Occupational exposure to organic lead compounds—the relative degree of hazard in occupational exposure to air-borne tetraethyllead and tetramethyllead. *Arch Environ Health* 5:532–536, 1962
3. Schepers GWH: Tetraethyl lead and tetramethyl lead—comparative experimental pathology. Part I. Lead absorption and pathology. *Arch Environ Health* 8:277–295, 1964
4. Cremer JE, Callaway S: Further studies on the toxicity of some tetra and trialkyl lead compounds. *Br J Ind Med* 18:277–282, 1961
5. Grandjean P: *Biological Effects of Organolead Compounds*, p 278. Boca Raton, FL, CRC Press, Inc, 1983

6. Kamstock ER et al. Trialkyl lead metabolism and lipid peroxidation *in vivo* in vitamin E— and selenium deficient rats as measured by ethane production. *Toxicol Appl Pharmacol* 54:251–257, 1980

TETRAMETHYL SUCCINONITRILE
CAS: 3333-52-6

$C_8H_{12}N_2$

Synonyms: TMSN; tetramethylsuccinic acid dinitrile

Physical Form.   Crystalline solid

Source.   Breakdown product of azobisisobutyronitrile used as a blowing agent for the production of vinyl foam; by-product of a polymerization catalyst in photocopier toner

Exposure.   Inhalation; skin absorption

Toxicology.   Tetramethyl succinonitrile (TMSN) is a convulsant.

Exposure involving TMSN occurred in a group of 16 workers using azoisobutyronitrile over an 18-month period in the production of polyvinyl chloride foam.[1] There were five cases of convulsions and unconsciousness. Other reported symptoms included headache, dizziness, nausea, and vomiting. Although an unknown concentration of tetramethylsuccinonitrile was the suspected etiologic agent, it was noted that exposure to a number of other substances also occurred. All symptoms subsided following installation of improved ventilation in the work area.

Exposure of rats to the vapor at 60 ppm for 2 to 3 hours, or to 6 ppm for 30 hours, caused death.[1] Mice exposed to 22 ppm had muscle spasms and died within 2 to 3 hours.[1,2] Rats, guinea pigs, rabbits, and dogs administered tetramethylsuccinonitrile by a variety of routes developed violent convulsions and asphyxia, which eventually led to death from 1

minute to 5 hours following convulsions.[1] In a variety of species, LD_{50} values for intravenous, intraperitoneal, subcutaneous, and oral administration ranged from 17.5 mg/kg to 30 mg/kg.[1] Administration of a quick-acting barbiturate followed by phenobarbital reduced the toxicity of tetramethylsuccinonitrile given in doses of up to 50 mg/kg.[1,2]

Parental injection of TMSN caused some fetal malformation and embryonic death but only at doses that caused severe maternal toxicity.[3]

The 1995 ACGIH threshold limit value–time-weighted average (TLV-TWA) for TMSN is 0.5 ppm (2.8 mg/m³) with a notation for skin absorption.

REFERENCES

1. National Institute for Occupational Safety and Health: *Criteria for a Recommended Standard . . . Occupational Exposure to Nitriles.* DHEW (NIOSH) Pub No 78-212, p 155. Washington, DC, US Government Printing Office, 1978
2. Harger RN, Hulpieu HR: Toxicity of tetramethyl succinonitrile and the antidotal effects of thiosulphate, nitrile and barbiturates. *Fed Proc* 8 (Abstract):205, 1949
3. Doherty PA, Smith RP, Ferm VH: Comparison of the teratogenic potential of two aliphatic nitriles in hamsters: succinonitrile and tetramethyl succinonitrile. *Fund Appl Toxicol* 3:41–48, 1983

TETRANITROMETHANE
CAS: 509-14-8

$C(NO_2)_4$

Synonym: TNM

Physical Form.   Pale yellow liquid

Uses.   Oxidizer in rocket propellants; explosive in admixture with toluene; reagent for detecting presence of double bonds in organic compounds

Exposure. Inhalation

Toxicology. Tetranitromethane vapor is a severe irritant of the eyes and respiratory tract; it can cause mild methemoglobinemia.

In workers, various studies showed that exposure caused irritation of the eyes, nose, and throat; dizziness; headache; chest pain; dyspnea; and, rarely, skin irritation.[1]

Severe exposure may be expected to cause the formation of methemoglobin and resultant anoxia with cyanosis (especially evident in the lips, nose, and earlobes); other effects are weakness, dizziness, and severe headache.[2-4] Concentrations in excess of 1 ppm cause lacrimation and upper respiratory irritation, whereas 0.4 ppm may cause mild irritation.[2] The liquid on the skin may cause mild burns.[2]

The LC_{50} for rats was 1230 ppm for 36 minutes; effects included lacrimation, rhinorrhea, gasping, and cyanosis. Pulmonary edema was present at autopsy.[1] At 300 ppm, all rats died within 40 to 90 minutes, whereas exposure to 33 ppm caused deaths in 3 to 10 hours.[1] Exposure to 6.35 ppm for 6 hr/day, 5 days/week, for 6 months resulted in the death of 11 of 19 rats; similar exposure to dogs caused mild symptoms the first 2 days, followed by complete recovery.[1] In three species of animals, intravenous injection caused methemoglobinemia, anemia, damage to the central nervous system, and pulmonary edema.[1]

Rats and mice were exposed 6 hr/day, 5 days/week, for 2 years at 2 or 5 ppm or 0.5 or 2 ppm, respectively.[5,6] Tetranitromethane was found to cause mild irritation and hyperplastic lesions in the nasal passages. Nearly all animals exposed at the higher dose levels developed alveolar/bronchiolar adenoma or carcinoma; squamous-cell neoplasms of the lung also occurred in exposed rats. The carcinogenic activity of tetranitromethane appears to be the result of chronic epithelial irritation, mitotic stimulation, and ensuing hyperplastic response.[6]

Tetranitromethane was genotoxic in a number of assays inducing chromosomal aberrations and sister chromatid exchanges in Chinese hamster ovary cells.[6]

The 1995 ACGIH threshold limit value–time-weighted average (TLV-TWA) is 0.005 ppm (0.04 mg/m³) with an A2 suspected human carcinogen designation.

REFERENCES

1. Horn HJ: Inhalation toxicology of tetranitromethane. *AMA Arch Ind Hyg Occup Med* 10:213–222, 1954
2. Hygienic Guide Series: Tetranitromethane. *Am Ind Hyg Assoc J* 25:513–515, 1964
3. Hager KF: Tetranitromethane. *Ind Engr Chem* 41:2168–2172, 1949
4. Rieder RF: Methemoglobinemia and sulmethemoglobinemia. In Wyngaarden JB, Smith LH (eds): *Cecil Textbook of Medicine*, 16th ed, p 896. Philadelphia, WB Saunders Co, 1982
5. Bucher JR, Huff JE, Jokinen MP, et al: Inhalation of tetranitromethane causes nasal passage irritation and pulmonary carcinogenesis in rodents. *Cancer Letts* 57:95–101, 1991
6. National Toxicology Program: *Toxicology and Carcinogenesis Studies of Tetranitromethane (CAS No 509-14-8) in F344/N Rats and B6C3F₁ Mice (Inhalation Studies)*, NTP No 386. NTIS Pub No PB-91-113-373. National Technical Information Service, Springfield, VA, 1990

TETRASODIUM PYROPHOSPHATE
CAS: 7722-88-5

$Na_4P_2O_7$

Synonyms: Sodium pyrophosphate; tetrasodium diphosphate; TSPP

Physical Form. White crystalline powder

Uses. Water softener; metal cleaner; dispersing and emulsifying agent

Exposure. Inhalation

Toxicology. Tetrasodium pyrophosphate (TSPP) is of low toxicity, but the dust may be irritating to the eyes and upper respiratory tract and skin.

Mild to moderate skin and eye irritation has occurred with acute exposure to the dust.

In rats, the oral LD_{50} ranges between 1 and 3 g/kg.[1] Applied to rabbit eyes, TSPP can cause severe irritation and corneal injury. There were no adverse effects in rats fed 50 mg/kg/day for 1 year.

Injected into chick embryos, TSPP produced terata.[2]

The 1995 ACGIH threshold limit value–time-weighted average (TLV-TWA) is 5 mg/m³.

REFERENCES

1. Tetrasodium pyrophosphate. *Documentation of TLVs and BEIs*, 6th ed, pp 1529–1530. Cincinnati, OH, American Conference of Governmental Industrial Hygienists (ACGIH), 1991
2. Verrett MJ, Scott WF, Reynaldo EF, et al: Toxicity and teratogenicity of food additive chemicals in the developing chicken embryo. *Toxicol Appl Pharmacol* 56:265–273, 1980

TETRYL
CAS: 479-45-8

$C_7H_5N_5O_8$

Synonyms: 2,4,6-Trinitrophenylmethyl-nitramine; tetralite; nitramine; *n*-methyl-*n*,2,4,6-tetranitroaniline

Physical Form. Yellow crystals

Uses. Once widely used as a military explosive but no longer manufactured or used in the United States

Exposure. Inhalation; skin absorption

Toxicology. Tetryl causes contact and sensitization dermatitis and irritation of the upper respiratory tract.

Contact with tetryl causes a bright yellow staining, most often seen on the palms, face, and neck and in the hair.[1] The irritant effects on the upper respiratory tract are variously localized from the nostrils to the bronchi and cause burning, itching, sneezing, nasal discharge, epistaxis, and cough. The symptoms may begin the first day of exposure or as late as the third month; on removal from exposure, the symptoms typically regress over 2 to 4 weeks.[1]

Dermatitis in workers may appear as early as the first week of exposure to the dust with itching of, and around, the eyes; there is a progression to erythema and edema occurring most often on the nasal folds, cheeks, and neck; papules and vesicles may develop; the remainder of the body is rarely affected.[1] The severest forms show massive generalized edema, with partial obstruction of the trachea due to swelling of the tongue, and require hospitalization; exfoliation usually occurs after the edema subsides.[1] The majority of these effects occur between the 10th and 20th days of exposure; on cessation of exposure, there is rapid abatement of the mild symptoms and, after 3 to 10 days, disappearance of physical signs.[1] Some individuals have become sensitized to tetryl and developed a rash in response to recontact with even small amounts of the substance.[2]

Other effects reported in tetryl workers are irritability, fatigue, malaise, headache, lassitude, insomnia, nausea, and vomiting.[1] Anemia, of either the marrow depression or deficiency type, has been observed among tetryl workers.[1] Conjunctivitis may be caused by rubbing the eyes with contaminated hands or by airborne dust; keratitis and iridocyclitis have occurred.[3]

Tetryl has been reported to cause irreversible liver damage and death following chronic heavy exposure.[4]

A number of in vitro genotoxic assays in bacteria and fungi suggest that tetryl is a direct-acting genotoxin.[2]

The 1995 ACGIH threshold limit value–time-weighted average (TLV-TWA) is 1.5 mg/m³.

REFERENCES

1. Bergman BB: Tetryl toxicity: a summary of ten years' experience. *AMA Arch Ind Hyg Occup Med* 5:10–20, 1952
2. Agency for Toxic Substances and Disease Registry (ATSDR): *Toxicological Profile for Tetryl.* US Department of Health and Human Services, Public Health Service, p 77, 1993
3. Troup HE: Clinical effects of tetryl (CE powder). *Br J Ind Med* 3:20–23, 1946
4. Hardy HL, Maloof CC: Evidence of systemic effect of tetryl, with summary of available literature. *AMA Arch Ind Hyg Occup Med* 1:545–555, 1950

THALLIUM
CAS: 7440-28-0

Tl

Compounds. Thallium acetate; thallium chloride

Physical Form. Bluish-white, very soft, inelastic, easily fusible heavy metal

Uses. In the semiconductor industry for the production of switches and closures; in the pharmaceutical industry for cardiac imaging; in the manufacture of optical glass; formerly used as a rodenticide and insecticide until banned in the United States in 1972

Exposure. Inhalation; skin absorption; ingestion

Toxicology. Thallium is one of the most toxic of the heavy metals; it affects primarily the nervous system, gastrointestinal tract, and hair.

The lethal oral dose of thallium acetate for humans is estimated to be about 12 mg/kg body weight.[1] Although symptoms may be nonspecific as a result of multiorgan toxicity, gastroenteritis, polyneuropathy, and hair loss are the dominant clinical features of thallium poisoning.[2] In fatal cases, however, death has been regularly attributed to cardiac or respiratory failure, which may overshadow the characteristic manifestation of neuropathy.[3] A latent period of hours to 1 to 2 days may follow acute exposure.[2] Nausea, vomiting, diarrhea, abdominal pain, and gastrointestinal hemorrhage are common initial complaints. These symptoms are followed, or accompanied, by ptosis and strabismus; peripheral neuritis; pain, weakness, and paresthesias in the legs; tremor; and retrosternal tightness and chest pain.[1,4] Severe and abrupt alopecia is pathognomonic of the toxic effects of thallium and usually, but not always, occurs after 2 to 3 weeks.[4,5]

Severe intoxication has resulted in prostration, tachycardia, blood pressure fluctuations, convulsive seizures, choreiform movements, and psychosis. Recovery may be complete, but permanent residual effects such as ataxia, optic atrophy, tremor, mental abnormalities, and foot drop have been reported.[4] In cases of fatal intoxication, typical autopsy findings include pulmonary edema, necrosis of the liver, nephritis, and degenerative changes in peripheral axons.[1]

Prolonged ingestion of thallium produces a variable clinical picture, which includes stomatitis, tremor, cachexia, polyneuropathy, alopecia, and emotional disturbance.[4] Alopecia may be the best-known effect of chronic poisoning, with epilation beginning about 10 days after ingestion and complete hair loss occurring in about 1 month.[2]

In a study of 15 workers who had handled solutions of organic thallium salts over a 7.5-year period, 6 workers suffered thallium intoxication. Chief complaints were abdominal pain, fatigue, weight loss, pain in the legs, and nervous irritability; 3 of the workers had albuminuria, and 1 had hematuria.[6]

In another cohort study, no statistically significant clinical effects were found, even though urinary concentrations ranging up to 236 μg/l indicated exposures above the TLV of 0.1 mg/m³.[7] A urine thallium concentration of 100 μg/l corresponds approximately to a 40-hr/week exposure at 0.1 mg/m³, and normal values range between 0.6 and 2.0 μg/l.[7]

Several mechanisms have been postulated to account for thallium's toxicity, including ligand formation with sulfhydryl groups of enzymes and transport proteins, inhibition of cellular respiration, interaction with riboflavin and riboflavin-based cofactors, alteration of the activity of K^+-dependent proteins, and disruption of intracellular calcium homeostasis.[2]

In 6 cases of thallium intoxication of pregnant women during their first trimester, no congenital abnormalities were observed.[7] Exposure of pregnant mice, rabbits, or rats produced slight embryotoxic effects at maternally toxic doses.[8]

Administered in the drinking water to male rats for 60 days, 0.7 mg thallium/day, as thallium sulfate, caused abnormalities in testicular morphology, function, and biochemistry.[9] Effects included increased epididymal sperm with increased numbers of immature cells, decreased sperm motility, and reduced testicular β-glucuronidase.

Thallium was genotoxic in a variety of assays, inducing single strand breaks in mouse cell cultures, dominant lethals in male rats in vivo, and DNA damage in bacterial systems.[3]

The 1995 ACGIH threshold limit value–time-weighted average (TLV-TWA) for thallium and soluble compounds is 0.1 mg/m³ as Tl, with a notation for skin absorption.

Special Tests: Thallium in the urine indicates that systemic absorption has occurred. Urinary thallium in normal unexposed subjects is less than 1.5 µg/l.[10] No relationship between excretion rate of thallium and either exposure or appearance of signs and symptoms has been established.

REFERENCES

1. Browning E: *Toxicity of Industrial Metals*, 2nd ed, pp 317–322. London, Butterworths, 1969
2. Mulkey JP, Oehme FW: A review of thallium toxicity. *Vet Hum Toxicol* 35:445–453, 1993
3. Agency for Toxic Substances and Disease Registry (ATSDR): *Toxicological Profile for Thallium*. US Department of Health and Human Services, Public Health Service, TP-91/26, p 90, 1992
4. Paulson G, Vergara G, Young J, Bird M: Thallium intoxication treated with dithizone and hemodialysis. *Arch Intern Med* 129:100–103, 1972
5. Bank WJ, Pleasure DE, Suzuki K, et al: Thallium poisoning. *Arch Neurol* 26:456–464, 1972
6. Richeson EM: Industrial thallium intoxication. *Ind Med Surg* 27:607–619, 1958
7. Marcus RL: Investigation of a working population exposed to thallium. *J Soc Occup Med* 35:4–9, 1985
8. Dolgner R et al: Repeated surveillance of exposure to thallium in a population living in the vicinity of a cement plant emitting dust containing thallium. *Int Arch Occup Environ Health* 52:69–94, 1983
9. Formigli L, Scelsi R, Pogg P, et al: Thallium-induced testicular toxicity in the rat. *Environ Res* 40:531–539, 1986
10. Lauwerys RR: *Industrial Chemical Exposure: Guidelines for Biological Monitoring*, pp 48–49. Davis, CA, Biomedical Publications, 1983

THIOACETAMIDE
CAS: 62-55-5

CH_3CSNH_2

Synonyms: Acetothioamide; ethanethioamide; TAA

Physical Form. Colorless crystals

Uses. Laboratory reagent used as a substitute for hydrogen sulfide

Exposure. Inhalation; skin absorption

Toxicology. Thioacetamide can cause liver and pulmonary damage; it is carcinogenic to experimental animals.

Exposure to high concentrations may cause irritation of the nose, throat, and lungs; even higher exposure may result in pulmonary edema. High exposure can also cause liver injury severe enough to result in death.[1]

In female Wistar rats, administration of 50 mg/kg twice weekly for 30 weeks resulted in hepatic necrosis.[2] Slight to moderate cirrhosis was observed in male albino rats fed 0.005% or 0.01% in the diet for 18 months; 1 of 6 survivors developed hepatic-cell adenoma.[3] Thioacetamide induced liver-cell tumors in mice and liver and bile duct tumors in rats following chronic administration of 0.03% in the diet. Cirrhosis, neoplastic nodules, cholangiofibromas, hepatocarcinomas, and cholangiocarcinomas occurred in male ACI rats fed 0.035% in the diet for 1 year.[4] In hamsters, 2.5 mg given by gavage once a week for 30 weeks was not carcinogenic.

The IARC has determined that there is sufficient evidence for the carcinogenicity of thioacetamide to animals; no data are available for humans.[5]

The ACGIH has not established a threshold limit value for thioacetamide.

REFERENCES

1. New Jersey Department of Health: *Right to Know Project.* CN 368, Trenton, NJ
2. Munoz Torres E, Paz Bouza JI, Lopez Bravo A, et al: Experimental thioacetamide-induced cirrhosis of the liver. *Histol Histopathol* 6:95–100, 1991
3. *IARC Monographs on the Evaluation of the Carcinogenic Risk of Chemicals to Man*, Vol 7, Some antithyroid and related substances, nitrofurans and industrial chemicals, pp 77–81. Lyon, International Agency for Research on Cancer, 1974
4. Becker FD: Thioacetamide hepatocarcinogenesis. *J Natl Cancer Inst* 71:553–556, 1983
5. *IARC Monographs on the Evaluation of the Carcinogenic Risks to Humans, Overall Evaluations of Carcinogenicity: An Updating of IARC Monographs Vol 1–41*, Suppl 7, p 72. Lyon, International Agency for Research on Cancer, 1987

4,4'-THIOBIS(6-*tert*-BUTYL-*m*-CRESOL)
CAS: 96-69-5

$C_{22}H_3O_2S$

Synonyms: 4,4'-thio-bis(3-methyl-6-*tert*-butylphenol); bis(3-*tert*-butyl-4-hydroxy-6-methylphenyl)sufide; TBBC

Physical Form. Fine, white to gray powder

Uses. Antioxidant in the rubber and plastics industry; stabilizer in polyethylene and polyolefin packaging materials for foodstuffs

Exposure. Ingestion; inhalation

Toxicology. 4,4'-Thiobis(6-*tert*-butyl-*m*-cresol) or TBBC is of low systemic toxicity in animals; allergic contact dermatitis has been reported in humans.

Allergic contact dermatitis developed on the hands and face of 2 patients after exposure to latex examination gloves.[1] Both patients were patch-test negative to the usual rubber allergens, but both had a positive test reaction to TBBC.

In 15-day feeding studies, groups of rats and mice were fed diets containing 1000, 2500, 5000, 10,000, or 25,000 ppm TBBC.[2] All 25,000 and some 10,000 ppm rats died; rats in the 5000 and 10,000 ppm group consumed less food than controls and had significant weight loss and diarrhea. Renal papillary and tubule necroses were the principal lesions attributed to TBBC exposure in the 10,000 ppm group. Focal necrosis of the glandular stomach also occurred in some 10,000 ppm rats. Some mice did not survive exposure at 5000 and 10,000 ppm, and 25,000 ppm was lethal to all. Weight loss, diarrhea, and renal tubule necrosis was similar to those observed in rats.

Histopathological findings in rats fed 2500 or 5000 ppm for 13 weeks included hypertrophy of Kupffer's cells, bile duct hyperplasia, and individual cell necrosis of hepatocytes; pigmentation and degeneration of the renal cortical tubule epithelial cells were also present.[2] In male rats exposed at 1000 ppm and above, hematocrit and hemoglobin concentrations and mean erythrocyte volume were significantly lower than those in controls. Mice survived exposures of up to 2500 ppm in their diets for 13 weeks. Body weights were significantly lower in the high-dose groups and corresponded with reduced feed consumption. Kupffer's-cell hypertrophy, bile duct hyperplasia, increased spleen weights, and an increase in size and number of macrophages in mesenteric lymph nodes were present in the 2500 ppm–treated mice.

Two-year studies of rats administered up to 2500 ppm and mice administered up to 1000 ppm in the diets were associated with Kupffer's-cell hypertrophy, cytoplasmic vacuolization, and mixed cell foci in the liver of rats, but there were no significant pathological findings in mice. There was no evidence of carcinogenic activity in either species.

TBBC was not mutagenic in *Salmonella typhimurium* strains with or without metabolic activation. In Chinese hamster ovary cells, TBBC induced an increase in sister chromatid exchanges, but there were no increases in chromosomal aberrations.

The 1995 ACGIH threshold limit value–time-weighted average (TLV-TWA) for 4,4′-thiobis(6-*tert*-butyl-*m*-cresol) is 10 mg/m³.

REFERENCES

1. Rich P, Belozer ML, Norris P, et al: Allergic contact dermatitis to two antioxidants in latex gloves: 4,4-thiobis(6-*tert*-butyl-meta-cresol) (Lowinox 44536) and butylhydroxyanisole: allergen alternatives for glove allergic patients. *J Am Acad Dermatol* 24:37–43, 1991
2. National Toxicology Program: *NTP Technical Report on the Toxicology and Carcinogenesis Studies of 4,4′-Thiobis(6-t-butyl-m-cresol) (CAS No 96-69-5) in F344 rats and B6C3F₁ Mice (Feed Studies)*, NTP TR 435. NIH Pub No 93-3166, US Department of Health and Human Services, Public Health Service, National Institutes of Health, 1993

THIOGLYCOLIC ACID
CAS: 68-11-1

$C_2H_4O_2S$

Synonyms: Mercaptoacetic acid; thiovanic acid

Physical Form. Colorless liquid

Uses. In the formulations of permanent wave solutions and depilatories; in pharmaceutical manufacture; stabilizer in vinyl plastics

Exposure. Inhalation; skin absorption

Toxicology. Thioglycolic acid is corrosive to the skin, eyes, and mucous membranes on contact.

In one reported case, thioglycolic acid accidentally splashed onto the eyes, face, legs, and arms caused second-degree burns of the skin.[1] Within 2 hours, the corneas became clouded, and the conjunctivas were edematous. Over the course of several months, the cornea cleared, and necrotic conjunctivas regenerated and vascularized, leaving slightly impaired vision.

In rats, the oral LD_{50} is less than 50 mg/kg.[2] Applied to the skin of guinea pigs, 5 ml/kg of a 10% solution caused weakness, gasping, convulsions, and death. Two drops of a 10% solution instilled in rabbit eyes caused immediate pain, and the epithelium turned gray within seconds; the conjunctivas were hyperemic, with moderate discharge, and corneas were opaque at 2 days.[2] Corneal clouding cleared gradually, but not completely, within 6 weeks.

The 1995 ACGIH threshold limit value–time-weighted average (TLV-TWA) is 1 ppm (3.8 mg/m³) with a notation for skin absorption.

REFERENCES

1. Grant WM: *Toxicology of the Eye*, 3rd ed, p 905. Springfield, IL, Charles C Thomas, 1986
2. Fassett DW: Organic acids and related compounds. In Fassett DW, Irish DD (eds): *Industrial Hygiene and Toxicology*, 2nd ed, rev, Vol 2, pp 1807–1808. New York, John Wiley & Sons, 1963

THIONYL CHLORIDE
CAS: 7719-09-7

$SOCl_2$

Synonyms: Sulfurous oxychloride; thionyl dichloride

Physical Form. Colorless to pale yellow liquid with a suffocating odor

Uses/Sources. In the manufacture of lithium batteries; in the synthesis of herbicides, surfactants, drugs, vitamins, and dyestuffs

Exposure. Inhalation

Toxicology. Thionyl chloride may cause severe irritation of the skin, eyes, and mucous membranes, as well as potentially serious lung injury.

Two cases of accidental thionyl chloride exposure resulting in lung injury that varied from relatively mild and reversible interstitial lung disease to a severe form of bronchiolitis obliterans have recently been reported.[1] In the first case, a 30-year-old worker was exposed when a thionyl chloride tank burst in an open space. The worker was asymptomatic until dyspnea gradually developed 2 weeks after his exposure. The patient was mildly dyspneic, with 22 respirations per minute, and lung-function tests showed moderate restrictive dysfunction. Following treatment by salbutamol inhalations, oral aminophylline, and prednisone, 60 mg, the patient's condition improved within 2 weeks, and prednisone dosage was tapered to 20 mg/day. This, however, was followed by a relapse, which was treated successfully by doubling the prednisone dose and slowly tapering off over a total period of 6 months.

In the second case, a 23-year-old worker suffered short-term exposure to thionyl chloride fumes in an enclosed space. Acute effects included second-degree chemical burns on the ankle, wrist, tongue, nasal septum, and corneas. The patient was not dyspneic, and chest radiographs were normal. Arterial blood gases showed mild, partially compensated metabolic acidosis with a lower partial pressure of oxygen, and lung-function tests showed mild restrictive change. Hydrocortisone treatment (300 mg/day intravenously) was initiated to prevent or minimize the risk of lung injury. The dose was reduced to 50 mg/day after 3 days and, on discharge, a regimen of 10 mg/day of prednisone was prescribed. Following a latent, clinically asymptomatic phase of over 2 weeks, the patient was readmitted with acute respiratory failure. Chest radiographs showed bilateral hyperinflated lungs with verticalization of the heart.

Lung-function tests showed a severe mixed restrictive and obstructive pattern that was unresponsive to bronchodilators. A clinical diagnosis of bronchiolitis obliterans secondary to the inhalation of thionyl chloride fumes was made. Other complications included spontaneous pneumothorax and bronchopleural fistula. The patient ultimately survived but was left permanently disabled.

The clinicians noted that, although the first patient responded well to steroid therapy, steroids may be less useful in more severe cases of bronchiolitis obliterans. Specifically, steroid treatment should be stopped if no improvement is seen during the first days because this treatment may increase the risk of lung infection in the presence of a denuded lung epithelium.

In an earlier report, a worker exposed to an unknown concentration of thionyl chloride for approximately 6 minutes after a battery cell exploded succumbed to fulminant pulmonary edema 3 hours after the accident.[2]

The toxicity of thionyl chloride is attributed to the formation of sulfur dioxide and hydrogen chloride in contact with water. The reaction of 1 molecule of thionyl chloride with 1 molecule of water yields 2 molecules of hydrogen chloride and 1 of sulfur dioxide. Therefore, 1 ppm of thionyl chloride produces a total irritant gas concentration equivalent to 3 ppm.[3]

The 1995 ACGIH threshold limit value–ceiling (TLV-C) for thionyl chloride is 1 ppm (4.9 mg/m^3).

REFERENCES

1. Konichezky S, Schattner A, Ezri T, et al: Thionyl-chloride-induced lung injury and bronchiolitis obliterans. *Chest* 104:971–973, 1993
2. Ducatman AM, Ducatman BS, Barnes JA: Lithium battery hazard: old fashioned planning implications of new technology. *J Occup Med* 30:309–311, 1988
3. Thionyl chloride. *Documentation of the TLVs and BEIs*, 6th ed, pp 1543–1544. Cincinnati, OH, American Conference of Governmental Industrial Hygienists (ACGIH), 1991

THIRAM
CAS:137-26-8

$C_6H_{12}N_2S_4$

Synonyms: Tetramethylthiuram disulfide; TMTD; tetramethylthioperoxydicarbonic diamide; bis(dimethylthiocarbamyl)disulfide

Physical Form. White or yellow crystals

Uses. Agricultural fungicide; rubber accelerator

Exposure. Inhalation

Toxicology. Thiram is an irritant of the eyes, mucous membranes, and skin and causes sensitization dermatitis; adverse reproductive effects have been reported in experimental animals.

Thiram is the methyl analog of disulfiram or Antabuse, a drug used to establish a conditioned reflex of fear of alcohol in the treatment of alcoholism.[1] Ingestion of even a small amount of alcohol by a person undergoing Antabuse therapy is followed by distressing and occasionally dangerous symptoms, including flushing, palpitations, headache, nausea, vomiting, and dyspnea. The systemic "Antabuse-alcohol" syndrome is apparently rare in thiram-exposed workers, but it has been reported.[2] In one case, a man became ill and died 4 days after treating seed with thiram. Although he received substantial exposure over 10 hours, it is unclear whether he received enough thiram to produce death without associated alcohol ingestion.[2] A skin reaction, without other systemic effects, is said to occur in chronically exposed workers after ingestion of alcohol. The response of the skin is rapid and takes the form of flushing, erythema, pruritis, and urticaria.[1] Thiram without alcohol can produce dermatitis but only in a few susceptible people. Sensitization dermatitis in the form of eczema has occurred on the hands, forearms, and feet.[1,3]

In mice and male rats, the oral LD$_{50}$ was approximately 4g/kg; symptoms of toxicity were ataxia and hyperactivity followed by inac-

tivity, loss of muscular tone, labored breathing, clonic convulsions, and death within 2 to 7 days.[4]

Daily administration of 132 mg/kg bw in the diet for 13 weeks decreased the fertility of CD rats; 14 days at 96 mg/kg altered the estrus cycle of females.[5] Gavage doses in rats of 25 mg/kg/day for 90 days produced a significant increase in relative testes weight and mild pathomorphological changes indicative of testicular dysfunction.[6] A significant increase in the frequency of abnormal sperm was found in mice following a single subcutaneous dose of 1000 mg/kg or 5 repeated doses at 250 mg/kg bw.[7]

Thiram was teratogenic at maternally toxic doses, causing, primarily skeletal malformations in hamsters given a single oral dose of 250 mg/kg during the period of organogenesis and in mice given oral doses of 5 to 30 mg per animal daily between days 6 and 17 of pregnancy.[8,9]

A dietary level of 1000 ppm for 2 years produced weakness, ataxia, and varying degrees of paralysis of the hind legs of rats.[2] In a recent study, rats were administered 3, 30, or 300 ppm in the diets for up to 2 years, and dogs were given 0.4, 4, or 40 mg/day for up to 2 years.[10] Rats of the high-dose group had retarded growth, and females had anemia, regressive changes of the sciatic nerve, and atrophy of the calf muscle. Dogs in the high-dose group had severe toxic signs, including vomiting, salivation, and clonic convulsions, and did not survive the first year of treatment. Ophthalmological changes included fundal hemorrhage, miosis, and desquamation of the retina. At the middose range, dogs had liver failure and females also had kidney damage. There were no increased incidences of any tumors.

Thiram also was not carcinogenic in rats in chronic feeding studies or by gavage, or in mice by single subcutaneous injection.[2,5] The IARC has noted, however, that thiram can react with nitrite under mildly acidic conditions, simulating those in the human stomach, to form *N*-nitrosodimethylamine, which is carcinogenic in a number of species.[5] Dietary administration of 500 ppm thiram plus 2000 ppm sodium nitrite for 2 years caused a high incidence of nasal cavity tumors in rats versus no tumors in controls or in animals given only one compound.[11]

The IARC has determined that there is inadequate evidence in experimental animals and in humans for the carcinogenicity of thiram.

Thiram was genotoxic to insects, plants, fungi, and bacteria: it induced sister chromatid exchange and unscheduled DNA synthesis in cultured human cells.

The 1995 ACGIH threshold limit value–time-weighted average (TLV-TWA) for thiram is 1 mg/m³.

REFERENCES

1. Shelley WB: Golf-course dermatitis due to thiram fungicide. *JAMA* 188:415–417, 1964
2. Hayes WJ Jr: *Pesticides Studied in Man*, pp 603–606. Baltimore, MD, Williams and Wilkins, 1982
3. Fogh A, Pock-Steen B: Contact sensitivity to thiram in wooden shoes. *Contact Derm* 27:348, 1992
4. Lee CC et al: Oral toxicity of ferric dimethyl-dithiocarbamate (ferbam) and tetramethyl-thiram disulfide (thiram) in rodents. *J Toxicol Environ Health* 4:93–106, 1978
5. *IARC Monographs on the Evaluation of Carcinogenic Risk to Humans.* Vol 53, pp 403–420. Lyon, International Agency for Research on Cancer, 1991
6. Mishra VK, Srivastava MK, Raizada RB: Testicular toxicity of thiram in rat: morphological and biochemical evaluations. *Ind Health* 31:59–67, 1993
7. Hemavathi E, Rahiman MA: Toxicological effects of ziram, thiram, and dithane m-45 assessed by sperm shaped abnormalities in mice. *J Toxicol Environ Health* 38:393–398, 1993
8. Robens JF: Teratologic studies of carbaryl, diazinon, norea, disulfiram, and thiram in small laboratory animals. *Toxicol Appl Pharmacol* 15:152–173, 1969
9. Roll R: Teratologische Untersuchungen mit Thiram (TMTD) an zwei Mausestammen. *Arch Toxicol* 27:163–186, 1971
10. Maita K, Tsuda S, Shirasu Y: Chronic studies with thiram in Wistar rats and beagle dogs. *Fund Appl Toxicol* 16:667–686, 1991
11. Lijinsky W: Induction of tumors of the nasal

cavity in rats by concurrent feeding of thiram and sodium nitrite. *J Toxicol Environ Health* 13:609–614, 1984

TIN (INORGANIC COMPOUNDS)
CAS: 7440-31-5

Sn

Compounds: Stannic oxide; tin tetrachloride; stannic chloride; stannous chloride; stannous sulfate; sodium stannate; potassium stannate

Physical Form. Solid

Uses. In protective coatings and alloys; in glass bottle manufacture

Exposure. Inhalation

Toxicology. Inorganic tin salts are irritants of the eyes and skin.

No systemic effects have been reported from industrial exposure. Some inorganic tin compounds can cause skin or eye irritation because of acid or alkaline reaction produced with water. Tin tetrachloride, stannous chloride, and stannous sulfate are strong acids; sodium and potassium stannate are strong alkalies.[1] Makers of glass bottles who were exposed to a hot mist of stannic chloride (0.10 to 0.18 mg/m³) and hydrogen chloride (5 ppm) had an excess of symptoms of respiratory irritation over workers exposed predominantly to hydrogen chloride in the same plant.[2] Exposure to dust and fume of tin oxide results in stannosis, a rare benign pneumoconiosis.[3]

Ingested inorganic tin exhibits only moderate toxicity, probably because of poor absorption and rapid tissue turnover. However, consumption of food and fruit juices heavily contaminated with tin compounds in the range of 1400 ppm or more results in symptoms of gastrointestinal irritation, including nausea, abdominal cramps, vomiting, and diarrhea.[4]

In animals, high doses of soluble tin salts

induce neurologic disturbances.[4] Subcutaneous injection of animals with sodium stannous tartrate at a daily dose of 12.5 mg/kg was fatal. Death was preceded by vomiting, diarrhea, and paralysis with twitching of the limbs.[5] Daily administration to a dog of stannous chloride in milk at a level of 500 mg/kg produced paralysis after 14 months.[1]

Administration of 1 and 3 mg Sn/kg bw to rats resulted in inhibition of various enzymes, including hepatic succinate dehydrogenase and the acid phosphatase of the femoral epiphysis. Tin also appears to interact with the absorption and metabolism of biological essential metals such as copper, zinc, and iron and to influence heme metabolism.[4]

Limited animal testing with stannous chloride has not revealed evidence of carcinogenic potential.[6] Mixed results have been observed in genotoxic assays.

The 1995 ACGIH threshold limit value–time-weighted average (TLV-TWA) for tin (metal, oxide, and inorganic compounds except SnH_4) is 2 mg/m³.

REFERENCES

1. Stauden A (ed): *Kirk-Othmer Encyclopedia of Chemical Technology*, 2nd ed, Vol 20, pp 323–325. New York, Interscience, 1972
2. Levy BS, Davis F, Johnson B: Respiratory symptoms among glass bottle makers exposed to stannic chloride solution and other potentially hazardous substances. *J Occup Med* 27:277–282, 1985
3. Dundon CE, Hughes JP: Stannic oxide pneumoconiosis. *Am J Roentgen* 63:797–812, 1950
4. Schafer SG, Femfurt U: Tin—a toxic heavy metal? A review of the literature. *Reg Toxicol Pharmacol* 4:57–69, 1984
5. Barnes JM, Stoner HB: The toxicology of tin compounds. *Pharmacol Rev* 11:214–216, 1959
6. Agency for Toxic Substances and Disease Registry (ATSDR): *Toxicological Profile for Tin*. US Department of Health and Human Services, Public Health Service, p 148, TP-91/27, 1992

TIN (ORGANIC COMPOUNDS)
CAS: 7440-31-5

Sn

Compounds: Triethyltin iodide; dibutyltin chloride; tributyltin chloride; triphenyltin acetate; bis(tributyltin)oxide; triphenyltin chloride

Physical Form. Solids and liquids

Uses. Stabilizers in polymers; biocides; catalysts

Exposure. Inhalation; skin absorption

Toxicology. Organotin compounds cause irritation of the eyes, mucous membranes, and skin; some produce cerebral edema, and others cause hepatic necrosis.

The most toxic of the organotin compounds are the trialkyltins, followed by the dialkyltins and monoalkyltins.[1] The tetra-alkyltins are metabolized to their trialkyltin homologues; their effects are those of the trialkyltins, with severity of effects dependent on the rate of metabolic conversion. In each major organotin group, the ethyl derivative is the most toxic.[1]

Triethyltin: Oral administration of a French medication (Stalinon, containing diethyltin diiodide and isolinoleic esters) for treatment of human furunculosis resulted in 217 cases of poisoning, of which 102 were fatal.[1,2] The capsules were found to be contaminated with triethyltin and other organotin compounds. After a latent period of 4 days, effects were severe, persistent headache; vertigo; visual disturbances (including photophobia); abdominal pain; vomiting; and urinary retention. The more severe cases showed transient or permanent paralysis and psychic disturbances. Residual symptoms in those who recovered included persistent headache, diminished visual acuity, paresis, focal anesthesia and, in four severe cases, flaccid paraplegia with incontinence. The most significant lesion found at autopsy was cerebral edema.

Tributyltin: Workers exposed to the vapor or fume of tributyltin compounds developed sore throat and cough several hours after exposure.[3] When a worker was splashed in the face with a tributyltin compound, lacrimation and severe conjunctivitis appeared within minutes, despite immediate lavage, and persisted for 4 days. At the end of 7 days, the eyes appeared normal.[3] Chemical burns may result after only brief contact with the skin. Pain is usually moderate, and itching is the chief symptom. Healing is usually complete within 7 to 10 days.[3]

Triphenyltin Acetate: Liver damage has occurred from occupational exposure to triphenyltin acetate.[1] In two cases, both developed hepatomegaly; one had slightly elevated SGPT and SGOT activity. Occupational exposure to a 20% solution produced skin irritation 2 to 3 days after prolonged contact with contaminated clothing. Other nonspecific effects of exposure have included headache, nausea, vomiting, diarrhea, and blurred vision.[1]

Trimethyltin: Induction of overt neurological and behavioral changes in rodents, including aggression, hyperexcitability, tremor, spontaneous seizures, and hyperreactivity by trimethyltin compounds, are well documented.[4]

Bis(tributyltin) Oxide (TBTO): TBTO is an irritant of the eyes and respiratory tract.[1]

In chronic rodent studies, no evidence of carcinogenicity was found in studies with triphenyltin acetate, triphenyltin hydroxide or dibutlytin acetate.[5] Tributyltin oxide was associated with an increased incidence of benign pituitary and phenochromocytomas in rats, which was attributed to a direct action on the endocrine glands rather than a carcinogenic effect.[5] Results from most genotoxic assays for organic tin have been negative.

Developmental effects including decreased fetal weights and increased incidences of cleft palate have occurred in mice at doses of TBTO that also produce maternal toxicity.[5]

The 1995 ACGIH threshold limit value–time-weighted average (TLV-TWA) is 0.1 mg/m^3 as Sn, with a short-term excursion limit of 0.2 mg/m^3 and a notation for skin absorption.

REFERENCES

1. National Institute for Occupational Safety and Health: *Criteria for a Recommended Standard . . . Occupation Exposure to Organotin Compounds.* DHEW (NIOSH) Pub No 77-115, pp 26–105. Washington, DC, US Government Printing Office, 1976
2. Barnes JM, Stoner HB: The toxicology of tin compounds. *Pharmacol Rev* 11:211–231, 1959
3. Lyle WH: Lesions of the skin in process workers caused by contact with butyl tin compounds. *Br J Ind Med* 15:193–196, 1958
4. Chang LW: Neuropathology of trimethyltin: A proposed pathogenic mechanism. *Fund Appl Toxicol* 6:217–232, 1986
5. Agency for Toxic Substances and Disease Registry (ASTDR): *Toxicological Profile for Tin.* US Department of Health and Human Services, Public Health Service, p 147, TP-91/27, 1992

TITANIUM DIOXIDE
CAS: 13463-67-7

TiO$_2$

Synonyms: Unitane; rutile; anatase; octahedrite; brookite

Physical Form. White powder

Uses. In welding rod coating; in acid-resistant vitreous enamels; white pigment for paints; in acetate rayon; in ceramics

Exposure. Inhalation

Toxicology. Titanium dioxide is a mild pulmonary irritant and is generally regarded as a nuisance dust.

Of 15 workers who had been exposed to titanium dioxide dust, 3 showed radiographic signs in the lungs resembling "slight fibrosis," but disabling injury did not occur. The magnitude and duration of exposure were not specified.[1,2] In the lungs of 3 workers involved in processing titanium dioxide pigments, deposits of the dust in the pulmonary interstitium were associated with cell destruction and slight fibrosis; the findings indicated that titanium dioxide is a mild pulmonary irritant.[3]

Rats exposed 6 hr/day for 5 days to 50 mg/m[3] and examined at various intervals after exposure showed no pulmonary response to titanium dioxide, as determined by bronchoalveolar lavage fluid parameters or histopathology.[4] Repeated exposure of rats to concentrations of 10 to 328 mppcf of air for as long as 13 months showed small focal areas of emphysema, which were attributed to large deposits of dust. There was no evidence of any specific lesion being produced by titanium dioxide.[5]

In a 2-year inhalation bioassay, exposure to 250 mg/m[3] titanium dioxide resulted in the development of squamous-cell carcinomas in 13 of 74 female rats and in 1 of 77 male rats, as well as an increase in bronchioloalveolar adenomas. No excess tumor incidence was observed at 50 mg/m[3].[6] Given the extremely high concentration exposures, the unusual histology and location of the tumors, and the absence of metastases, the authors questioned the biological relevance of these tumors to man.[6] There was no evidence that titanium dioxide–coated mica produced either toxicological or carcinogenic results when administered in the diet of F344 rats for 130 weeks at concentrations as high as 5%.[7]

Genotoxic studies of titanium dioxide have yielded negative results.[7]

The ACGIH threshold limit value–time-weighted average (TLV-TWA) for titanium dioxide is 10 mg/m[3].

REFERENCES

1. Browning E: *Toxicity of Industrial Metals*, 2nd ed, pp 331–335. London, Butterworths, 1969
2. AIHA Hygienic Guide Series: *Titanium Dioxide*. Akron, OH, American Industrial Hygiene Association, 1978
3. Elo R, Maatta K, Uksila E, Arstila AU: Pulmonary deposits of titanium dioxide in man. *Arch Pathol* 94:417–424, 1972
4. Driscoll KE, Lindenschmidt RC, Maurer JK, et al: Pulmonary response to inhaled silica or titanium dioxide. *Toxicol Appl Pharmacol* 111:201–210, 1991
5. Christie H, Mackay RJ, Fisher AM: Pulmonary effects of inhalation of titanium dioxide by rats. *Am Ind Hyg Assoc J* 24:42–46, 1963
6. Lee K et al: Pulmonary response of rats exposed to titanium dioxide (TiO_2) by inhalation for two years. *Toxic Appl Pharmacol* 79:179–192, 1985
7. Bernard BK, Osheroff MR, Hofmann A, et al: Toxicology and carcinogenesis studies of dietary titanium dioxide–coated mica in male and female Fischer 344 rats. *J Toxicol Environ Health* 29:417–429, 1990

TOLUENE
CAS: 108-88-3

$C_6H_5CH_3$

Synonyms: Toluol; methylbenzene; phenylmethane

Physical Form. Colorless liquid

Uses. In manufacturing of benzene and other chemicals; solvent for pains and coatings; component of gasoline

Exposure. Inhalation; skin absorption

Toxicology. Toluene causes central nervous system depression.

Exposure to extremely high concentrations of toluene (5000 to 30,000 ppm) may cause mental confusion, loss of coordination, and unconsciousness within a few minutes. Controlled exposure of human subjects to 200 ppm for 8 hours produced mild fatigue, weakness, confusion, lacrimation, and paresthesias of the skin. At 600 ppm for 8 hours, other effects included euphoria, headache, dizziness, dilated pupils, and nausea. At 800 ppm for 8 hours, symptoms were more pronounced, and aftereffects included nervousness, muscular fatigue, and insomnia persisting for several days.[1-4]

Subjects exposed to 100 ppm of toluene for 6 hours complained of eye and nose irritation and, in some cases, headache, dizziness, and a feeling of intoxication. However, no significant difference in performance on a variety of neurobehavioral tests was noted. No symptoms were noted at 10 or 40 ppm.[5]

Chronic organic brain dysfunction, associated with cerebral and cerebellar atrophy, has been described following long-term inhalational abuse of toluene among glue-sniffers exposed to very high concentrations. Several studies of workers chronically exposed to toluene or mixtures of toluene and other solvents have suggested minor abnormalities on neuropsychological testing or differences in performance on such testing compared to unexposed controls.[6] However, a recent study of 43 rotogravure printers exposed to estimated mean levels of 117 pm for a mean of 22 years failed to demonstrate significant clinical neuroradiological, neurophysiological, or neuropsychological differences compared to a control group of 31 unexposed printers.[7]

Severe but reversible liver and kidney injury occurred in a person who was a glue-sniffer for 3 years. The chief component of the inhaled solvent was toluene (80% v/v); other ingredients were not listed.[3] In workers exposed for many years to concentrations in the range of 80 to 300 ppm, there was no clinical or laboratory evidence of altered liver function.[3]

Toluene exposure does not result in the hematopoietic effects caused by benzene. The myelotoxic effects previously attributed to toluene are judged by more recent investigations to be the result of concurrent exposure to benzene present as a contaminant in toluene solutions.[3] Most of the toluene absorbed from inhalation is metabolized to benzoic acid, conjugated with glycine in the liver to form hippuric acid, and excreted in the urine. The average amount of hippuric acid excreted in the urine by persons not exposed to toluene is approximately 0.7 to 1.0 g/l of urine.[3]

The liquid splashed in the eyes of two workers caused transient corneal damage and conjunctival irritation; complete recovery occurred within 48 hours.[3] Repeated or prolonged skin contact with liquid toluene has a defatting action, causing drying, fissuring, and dermatitis.

In humans, chronic maternal inhalation abuse of toluene during pregnancy has been associated with teratogenic effects in a number of case reports. Manifestations include microcephaly, central nervous system dysfunction, attentional deficits, and developmental delay, with language impairment and growth retardation.[8] Phenotypic abnormalities may include a small midface, short palpebral fissures with deep-set eyes, low-set ears, flat nasal bridge with a small nose, micrognathia, and blunt fingertips. Interpretation of these human results may be confounded by the contribution of multiple chemical exposures.[9] Furthermore, it has been noted that only excessively high doses, possibly on the order of 30,000 ppm, which produce maternal toxicity, have been associated with developmental effects.

A number of animal studies have found evidence of developmental toxicity. Embryo and fetotoxicity consisting of reduced fetal weight and retarded skeletal development but without teratogenicity have been observed in mice and rats exposed to 133 ppm and 266 to 399 ppm, respectively.[6] Exposure to 2000 ppm, 6 hr/day, for 80 days prior to mating and through lactation produced no significant maternal toxicity but caused retardation of both fetal and postnatal development in rats.[9]

A chronic inhalation study found no evidence of carcinogenic activity in rats exposed at concentrations of 600 or 1200 ppm for 2 years or in mice exposed at 120, 600, or 1200 ppm for the same duration.[10]

The IARC has determined that there is inadequate evidence for the carcinogenicity of toluene in experimental animals and in humans.[11] Results of in vitro and in vivo assays generally indicate that toluene is not genotoxic.[12] Reports of increased incidences of sister chromatid exchanges and chromatid breaks in exposed workers are confounded by concurrent exposure to other organic chemicals.[12]

The 1995 ACGIH threshold limit value–time-weighted average (TLV-TWA) for toluene is 50 ppm (188 mg/m³) with a notation for skin absorption.

REFERENCES

1. Department of Labor: Occupational exposure to toluene. *Federal Register* 40:46206–46219, 1975
2. von Oettingen WF, Neal PA, Donahue DD: The toxicity and potential dangers of toluene—preliminary report. *JAMA* 113:578–584, 1942
3. National Institute for Occupational Safety and Health: *Criteria for a Recommended Standard . . . Occupational Exposure to Toluene.* DHEW (NIOSH) Pub No (HSM) 7311023, pp 14–45. Washington, DC, US Government Printing Office, 1973
4. Low LK, Meeks JR, Mackerer CR: Health effects of the alkylbenzenes. I. Toluene. *Toxicol Ind Health* 4:49–75, 1988
5. Anderson I et al: Human response to controlled levels of toluene in six-hour exposures. *Scand J Work Environ Health* 9:405–418, 1983
6. Environmental Protection Agency: Health Assessment Document for Toluene, Research Park Triangle, NC, NTIS, 1983.
7. Juntunen J et al: Nervous system effects of long-term occupational exposure to toluene. *Acta Neurol Scand* 75:512–517, 1985
8. Hersh JH: Toluene embryopathy: two new cases. *J Med Genetics* 26:333–337, 1987
9. Donald JM, Hooper K, Hopenhayn-Rich C: Reproductive and developmental toxicity of toluene: a review. *Environ Health Perspect* 94:237–244, 1991
10. National Toxicology Program: *NTP Technical Report on the Toxicology and Carcinogenesis Studies of Toluene (CAS No 108-88-3) in F344/N Rats and B6C3F₁ Mice*, NTR TR 371, NIH Pub No 89-2826, 1989
11. *IARC Monographs on the Evaluation of Carcinogenic Risks to Humans*, Vol 47, Some organic solvents, resin monomers and related compounds, pigments and occupational exposures in paint manufacture and painting, International Agency for Research on Cancer, 1989
12. Agency for Toxic Substances and Disease Registry (ATSDR): *Toxicological Profile for Toluene.* US Department of Health and Human Services, Public Health Service, p 139, 1992

TOLUENE-2,4-DIISOCYANATE
CAS: 854-84-9

$CH_3C_6H_3(NCO)_2$

Synonyms: TDI; toluene diisocyanate

Physical Form.   Colorless liquid; aerosol

Uses.   In production of polyurethane foams and plastics; in polyurethane paints and wire coatings; the most commonly used material is a mixture of 80% 2,4 isomer and 20% 2,6 isomer

Exposure.   Inhalation

Toxicology.   Toluene-2,4-diisocyanate (TDI) is a strong irritant of the eyes, mucous membranes, and skin and is a potent sensitizer of the respiratory tract.

Exposure of humans to sufficient concentrations causes irritation of the eyes, nose, and throat; a choking sensation; and a productive cough of paroxysmal type, often with retrosternal soreness and chest pain.[1,2] If the breathing zone concentration reaches 0.5 ppm, the possibility of respiratory response is imminent.[3] Depending on length of exposure and level of concentration above 0.5 ppm, respiratory symptoms will develop with a latent period of 4 to 8 hours.[3] Higher concentrations produce a sensation of oppression or constriction of the chest. There may be bronchitis and severe bronchospasm; pulmonary edema may also occur. Nausea, vomiting, and abdominal pain may complicate the presenting symptoms. On removal from exposure, the symptoms may persist for 3 to 7 days.

Although the acute effects of TDI may be severe, their importance is overshadowed by respiratory sensitization in susceptible persons; this has occurred after repeated exposure to levels of 0.02 ppm TDI and below.[2] The onset of symptoms of sensitization may be insidious, becoming progressively more pronounced with continued exposure over a period of days to months. Initial symptoms are often nocturnal dyspnea and/or nocturnal cough with progression to asthmatic bronchitis.[1] Immediate, late, and dual patterns of bronchospastic response to laboratory exposure to TDI in sensitized individuals have been observed, confirming the clinical findings of nocturnal symptoms in some exposed workers. The time from initial employment to the development of symptoms suggestive of asthma has been reported to vary from 6 months to 20 years.[4,5]

In another pattern of sensitization response, a worker who has had only minimal upper respiratory symptoms or no apparent effects from several weeks of low-level exposure may suddenly develop an acute asthmatic reaction to the same or slightly higher level. The asthmatic reaction may be severe, sometimes resulting in status asthmaticus, which may be fatal if exposure continues.[1]

Susceptibility to TDI-induced asthma does not require a prior history of atopy or allergic conditions, and sensitization may not be any more common in atopics.[6] Given sufficient exposure, it appears that virtually any person may become sensitized. The proportion of individuals with TDI asthma in working populations has varied from 4.3% to 25%.[7] There is some evidence that this percentage decreases with decreasing air concentrations. Exposure to spills of TDI appear to increase the risks of sensitization. The pathophysiology of TDI-induced asthma is unknown; both immunologic and nonimmunologic pharmacologic mechanisms have been postulated. Amines may play a causative role in TDI-induced asthma.[8] It is clear, however, that TDI-induced asthma is not mediated solely by a type I hypersensitivity response associated with IgE antibody.[6]

Several studies have provided evidence of cross-shift and progressive annual declines in FEV_1 and FEF 25–75% among asymptomatic workers without evidence of TDI asthma, when exposed to low levels of TDI (below 0.02 ppm and as low as 0.003 ppm). The annual declines were two- to threefold greater than expected, appeared dose-related, and correlated with observed cross-shift declines. Workers, in general, exhibited no acute or chronic symptoms related to these exposures or pulmonary-function decrements.[9,10]

The diagnosis of TDI-induced asthma relies primarily on the clinical history in a worker with known exposure; it is recognized that symptoms (wheezing, dyspnea, cough) may develop at night long after the end of the shift. Serial measurement of peak flow rates by the worker may aid in making the diagnosis.[11] Nonspecific bronchial hyperreactivity to histamine or methacholine is frequently, but not invariably, present in patients with TDI-induced asthma. Its absence may reflect that the asthma is quiescent because there has been no recent exposure, and re-exposure may lead to hyperreactivity. Failure to demonstrate nonspecific hyperreactivity on a single test does not exclude the diagnosis of TDI-induced asthma.[12] RAST testing for IgE antibodies against *p*-tolyl monoisocyanate antigens is probably not useful because of the occurrence of false-positive (in exposed but asymptomatic workers) and false-negative results.[13] Specific broncho provocation challenge with TDI is a definitive way to make the diagnosis but is often not practical because of the need for prolonged observation for late reactions and because of the risk of severe reactions.

Following removal from exposure, some patients have had resolution of symptoms. However, there is evidence from several studies that individuals with TDI-induced asthma may continue to have symptoms of dyspnea, wheezing, and bronchial hyperreactivity for 2 or more years following cessation of exposure.[14-16] In one study, patients with TDI-induced asthma who continued to have exposure to TDI for 2 more years had, as a rule, marked abnormal decreases in spirometric parameters and increases in nonspecific hyperreactivity.[14] In another study, 6 of 12 workers with a convincing history of TDI-induced asthma had positive responses to specific bronchial provocation testing with low concentrations of TDI (up to 0.02 ppm) at a mean of 4.5 years after cessation of exposure. These persons had persistent respiratory symptoms requiring daily treatment for asthma and persistent airway hyperreactivity.[15] Once sensitized, it is clear that patients can react to concentrations of 0.005 ppm or less.[7]

Bronchial biopsies of subjects with occupational asthma induced by TDI revealed pathological features such as increased number of inflammatory cells in the airway mucosa and thickening of subepithelial collagen.[17]

Splashes of TDI liquid in the eye cause severe conjunctival irritation and lacrimation. On the skin, the liquid produces a marked inflammatory reaction. Sensitization of the skin occurs but is uncommon as a result of proper work practices. There seems to be little relation between skin sensitivity and respiratory sensitivity to TDI.[1]

Commercial-grade TDI, consisting of 80% 2,4-TDI and 20% 2,6-TDI, was administered by gavage to female rats and mice at doses of 60 or 120 mg/kg, while male rats received 30 or 60 mg/kg and male mice received 120 or 240 mg/kg.[18] The major nonneoplastic lesions observed in rats were dose-related increases in acute bronchopneumonia and, in male mice, there was cytomegaly of the renal tubular epithelium. Despite early mortality in all groups, TDI was carcinogenic to both species, causing pancreatic acinar cell adenomas in male rats; pancreatic islet cell adenomas, neoplastic nodules of the liver, and mammary gland tumors in female rats; and subcutaneous fibromas and fibrosarcomas in both sexes. In female mice, there was an increase in hemangiomas and hepatocellular adenomas. The pattern of multiple tumor sites was similar to that found with 2,4-diaminotoluene. Metabolic studies have shown that common metabolites are produced from the 2,4-TDI isomer and from 2,4-diaminotoluene, suggesting that the 2,4-isomer in the commercial-grade TDI was responsible for the carcinogenic activity.

In genotoxic assays, TDI has produced chromosomal aberrations, base-pair substitution, frameshift mutations, and DNA strand breaks of human white blood cells in vitro.[19]

Biological monitoring of TDI exposure levels has been accomplished with postshift analysis of urinary toluene.[20]

The 1995 threshold limit value–time-weighted average (TLV-TWA) for toluene-2,4-diisocyanate is 0.005 ppm (0.036 mg/m³) with a short-term excursion limit of 0.02 ppm (0.14 mg/m³).

REFERENCES

1. National Institute for Occupational Safety and Health: *Criteria for a Recommended Standard . . . Occupational Exposure to Toluene Diisocyanate.* DHEW (NIOSH) Pub No (HSM) 73-11022. Washington, DC, US Government Printing Office, 1973

2. Elkins HB, McCarl GW, Brugsch HG, Fahy JP: Massachusetts experience with toluene diisocyanate. *Am Ind Hyg Assoc J* 23:265–272, 1962

3. Rye WA: Human responses to isocyanate exposure. *J Occup Med* 15:306–307, 1973

4. O'Brien I, Harris M, Burge P, Pepys J: Toluene diisocyanate–induced asthma. *Clin Allerg* 9:1, 1979

5. Chester E et al: Patterns of airway reactivity to asthma produced by exposure to toluene diisocyanate. *Chest* 75:229, 1979

6. Bernstein I: Isocyanate-induced pulmonary diseases: a current perspective. *J Allerg Clin Immun* 70:24–31, 1982

7. Toluene-2,4-Diisocyanate. *Documentation of the TLVs and BEIs*, 5th ed, pp 580–585. Cincinnati, OH, American Conference of Governmental Industrial Hygienists, 1986

8. Belin L et al: Amines: possible causative agents in the development of bronchial hyperreactivity in workers manufacturing polyurethanes from isocyanates. *Br J Ind Med* 40:251–257, 1983

9. Diem JE et al: Five-year longitudinal study of workers employed in a new toluene diisocyanate manufacturing plant. *Am Rev Resp Dis* 126:420–428, 1982

10. Wegman D et al: Accelerated loss of FEV-1 in polyurethane production workers: a four-year prospective study. *Am J Ind Med* 3:209–215, 1982

11. Burge P, O'Brien I, Harries M: Peak flow rate record in the diagnosis of occupational asthma due to isocynates. *Thorax* 34:317, 1979

12. Burge P: Nonspecific bronchial hyperreactivity in workers exposed to toluene diisocyanate diphenylmethane diisocyanate and colophony. *Europ J Resp Dis* 63(Suppl 123):91–96, 1982

13. Butcher B et al: Radioallergosorbent testing with p-tolyl monoisocyanate in toluene diisocyanate workers. *Clin Allerg* 13:31–34, 1983

14. Paggiaro P et al: Follow-up study of patient

with respiratory disease due to toluene diiso-cyanate (TDI). *Clin Allerg* 14:463–469, 1984
15. Moller D et al: Chronic asthma due to toluene diisocyanate. *Chest* 90:494–499, 1986
16. Luo CJ, Nelson KG, Fishbein A: Persistent reactive airway dysfunction syndrome after exposure to toluene diisocyanate. *Br J Ind Med* 47:239–241, 1990
17. Saetta M, Di Stefano A, Maestrelli P, et al: Airway mucosal inflammation in occupational asthma induced by toluene diisocyanate. *Am Rev Respir Dis* 145:160–168, 1992
18. Dieter MP, Boorman GA, Jameson CW, et al: The carcinogenic activity of commercial grade toluene diisocyanate in rats and mice in relation to the metabolism of the 2,4- and 2,6-TDI isomers. *Toxicol Ind Health* 6:599–621, 1990
19. Marczynski B, Czuppon AB, Marek W, et al: Indication of DNA strand breaks in human white blood cells after in vitro exposure to toluene diisocyante. *Tox Ind Health* 8:157–169, 1992
20. Maitre A, Berode M, Perdix A, et al: Biological monitoring of occupational exposure to toluene diisocyanate. *Int Arch Occup Environ Health* 65:97–100, 1993

TOLUIDINE

o-   *CAS: 95-53-4*
m-   *CAS: 108-44-1*
p-   *CAS: 106-49-0*

$CH_3C_6H_4NH_2$

Synonyms: ortho-Aminotoluene; 1 methyl-2-aminobenzene; 2-methyl-aniline

Physical Form. Clear to light-yellow liquid

Uses. Dye intermediate

Exposure. Inhalation; skin absorption

Toxicology. *o*-Toluidine causes anoxia, as a result of the formation of methemoglobin, and hematuria; ortho-toluidine hydrochloride is carcinogenic in experimental animals. Although

the meta- and para- isomers are assumed to produce comparable toxic effects, meta-toluidine seems to have no carcinogenic activity.

Signs and symptoms of overexposure are due to the loss of oxygen-carrying capacity of the blood. The earliest manifestations of poisoning in humans are headache, along with cyanosis of the lips, the mucous membranes, the fingernail beds, and the tongue.[1] Minor degrees of hypoxia may lead to a temporary sense of well-being and exhilaration. As the lack of oxygen increases, however, there is growing weakness, dizziness, and drowsiness, leading to stupor, unconsciousness, and even death if treatment is not prompt. Exposure to 10 ppm for more than a short time may lead to symptoms of illness, and 40 ppm for 60 minutes may cause severe toxic effects.[2] Transient microscopic hematuria has been observed in *o*-toluidine workers, presumably of renal origin given that no alterations in the bladder mucosa were observed by cystoscopy.[3]

In general, higher ambient temperatures increase susceptibility to cyanosis from exposure to methemoglobin-forming agents.[4]

Rats survived an 8-hour exposure to concentrated vapor.[5] Animals exposed from 6 to 23 ppm for several hours developed mild methemoglobinemia.[6] In the eye of a rabbit, the liquid caused a severe burn.[5] Excessive drying of the skin may result from repeated or prolonged contact.[1] The meta- and para- isomers of toluidine show the same toxicity profile and dose range as ortho-toluidine; similar effects from exposure are expected, although these isomers have not been tested as extensively as *o*-toluidine.[7]

o-Toluidine hydrochloride was carcinogenic in mice fed diets containing 1000 or 3000 mg/kg for 2 years, producing hepatocellular carcinomas or adenomas in females and hemangiosarcomas at multiple sites in males.[8] In another strain of mice fed diets of 16,000 ppm for 3 months and then 8000 ppm for an additional 15 months or 32,000 ppm for 3 months followed by 16,000 ppm for 15 months, there were significant dose dependent increases in the incidences of vascular tumors.[9] *o*-Toluidine hydrochloride was also carcinogenic in rats fed 0.028 M/kg diet for 72 weeks, producing tumors of multiple organs.[10] *p*-Toluidine was carcinogenic to mice after oral administration, producing liver tumors; *m*-toluidine was not carcinogenic in any reports.[7]

Epidemiological studies have dealt with workers exposed only to *o*-toluidine in combination with other chemicals. In a recent report, chemical workers exposed to *o*-toluidine and aniline had increased incidences of bladder cancer.[11] Among 1749 workers, 13 cases of bladder cancer were observed versus 3.61 expected. Increased risk of bladder cancer was strongly associated with duration of employment in the department where *o*-toluidine and aniline were used. The investigators suggested that, because *o*-toluidine was a more potent animal bladder carcinogen than aniline, it was more likely to be the etiologic agent responsible for the bladder cancer excesses in this plant although aniline may have played a role.

The IARC has determined that there is sufficient evidence for carcinogenicity of *o*-toluidine hydrochloride in animals and that it should be regarded as if it presented a carcinogenic risk to humans.[12,13]

Skin absorption of toluidines is considered to be a potential hazard. Recent estimations of workplace exposures have included individual dermal badges and surface wipes in addition to airborne monitoring.[14]

The 1995 ACGIH threshold limit value–time-weighted average (TLV-TWA) for the toluidines is 2 ppm (8.8 mg/m^3) with a notation for skin absorption; the ortho- and para- isomers have an A2 suspected human carcinogen designation.

Note: For a description of diagnostic signs, differential diagnosis, and medical control, including clinical laboratory treatment of overexposure to methemoglobin-forming agents, see "Aniline."

REFERENCES

1. Chemical Safety Data Sheet SD-82, Toluidine, pp 13–14. Washington, DC, MCA, Inc, 1961
2. Goldblatt MW: Research in industrial health in the chemical industry. *Br J Ind Med* 12:1–20, 1955
3. Hamblin DO: Aromatic nitro and amino compounds. In Patty FA (ed): *Industrial Hygiene and Toxicology.* 2nd ed, Vol 2, *Toxicology,* pp 2123, 2155. New York, Interscience, 1963
4. Linch AL: Biological monitoring for industrial exposure to cyanogenic aromatic nitro and aminocompounds. *Am Ind Hyg Assoc J* 35:426–432, 1974
5. Smyth HF Jr et al: Range-finding toxicity data: List VI. *Am Ind Hyg Assoc J* 23:95–96, 103, 1962
6. Henderson Y, Haggard HW: *Noxious Gases,* 2nd ed, p 288. New York, Reinhold, 1943
7. Beard RR, Noe JT: Aromatic and nitro compounds. In Clayton GD, Clayton FE (eds): *Patty's Industrial Hygiene and Toxicology* 3rd ed, rev, Vol 2A, *Toxicology,* pp 2483–2484, New York, Wiley-Interscience, 1981
8. National Cancer Institute: *Bioassay of o-Toluidine Hydrochloride for Possible Carcinogenicity,* TR-153. DHEW (NIH) Pub No 79-1709. Washington, DC, US Government Printing Office, 1979
9. Weisburger EK et al: Testing of twenty-one environmental aromatic amines or derivatives for long term toxicity or carcinogenicity. *J Environ Pathol Toxicol* 2:325–356, 1978
10. Hecht SS et al: Comparative carcinogenicity of *o*-toluidine hydrochloride and o-nitrosotoluene in F-344 rats. *Cancer Lett* 16:103–108, 1982
11. Ward E, Carpenter A, Markowitz S, et al: Excess number of bladder cancers in workers exposed to ortho-toluidine and aniline. *J Natl Cancer Inst* 83:501–506, 1991
12. *IARC Monographs on the Evaluation of the Carcinogenic Risk of Chemicals to Humans,* Vol 27, pp 155–175. Lyon, International Agency for Research on Cancer, 1982
13. *IARC Monographs on the Evaluation of Carcinogenic Risks to Humans,* Suppl 7, *Overall Evaluations of Carcinogenicity: An Updating of IARC Monographs Vols 1–to 42,* pp 362-363. Lyon, International Agency for Research on Cancer, 1987
14. Pendergrass SM: An approach for estimating workplace exposure to *o*-toluidine, aniline, and nitrobenzene. *Am Ind Hyg Assoc J* 55:733–737, 1994

TOXAPHENE
CAS: 8001-35-2

$C_{10}H_{10}Cl_8$ *(approximate)*

Synonyms: Chlorinated camphene; polychlorocamphene; octachlorocamphene

Physical Form. Yellow, waxy solid

Uses. Insecticide

Exposure. Inhalation; skin absorption; ingestion

Toxicology. Toxaphene is a central nervous system stimulant; it is carcinogenic in experimental animals.

Toxaphene is a mixture of at least 670 chlorinated camphenes; differences in toxicity have been observed for various toxaphene fractions or components.[1]

Most fatal cases of poisoning have been due to accidental ingestion, resulting in convulsions and death due to respiratory arrest.[1-4] The lethal oral dose for humans is estimated to be 2 to 7 g.[2]

Symptoms of acute intoxication are salivation, hyperexcitability, behavioral changes and, in severe cases, convulsions and death.[1] Convulsions may be preceded by nausea, vomiting, and muscle spasms or may begin without antecedent symptoms.[2] Onset of symptoms occurs within 4 hours, with death occurring from 4 to 24 hours post-exposure. Nonfatal poisoning has been characterized by nausea, mental confusion, jerking of the arms and legs, and convulsions.[3,4]

One proposed mechanism for toxaphene-induced neurotoxicity is that it acts as a non-competitive γ-aminobutyric acid antagonist at the chloride channel in brain synaptosomes. Substances that bind to the GABA-regulated chloride channel induce convulsions by inhibiting chloride flux, thus allowing brain cells to depolarize and fire spontaneously.[1,5]

Few cases of intoxication due to occupational exposure have been reported and, of these, two cases of pneumonitis in insecticide sprayers are of dubious validity.[6] In one acute study, 25 volunteers were exposed to 500 mg/m³ for 30 minutes for 10 days.[7] Following a 3-week respite, the exposure was repeated for 3 days. Each subject was thought to have absorbed 1 mg/kg/day. Physical examination and blood and urine tests revealed no toxic manifestations.

In subchronic animal studies, rats fed diets containing 4, 20, 100, or 500 ppm of the compound showed no clinical signs of toxicity; dose-dependent histological changes were observed in the kidney, thyroid, and liver.[8] For dogs administered 0.2, 2.0, and 5.0 mg/kg/day for 13 weeks by capsule, there were mild to moderate dose-dependent histological changes in the liver and thyroid, but no clinical signs of toxicity were observed.[8]

Toxaphene is less toxic when applied to the skin compared to oral administration.[1] Dermal LD_{50}s ranging from 7.8 to 45 g/kg have been obtained in laboratory animals. Applied to rabbit skin for 4 hours toxaphene was mildly irritating; a 0.5% solution was nonirritating to the forearms and faces of volunteers.

No fetal anatomical defects were observed in rats and mice at doses ranging from 0.05 to 75 mg/kg/day.[1] Adverse developmental effects, such as impaired righting reflexes, have been observed in rats at doses below those required to produce maternal toxicity.[9] There was no evidence that toxaphene interfered with fertility or pup survival and growth when male and female rats were fed toxaphene in their diet at concentrations as high as 25 mg/kg/day and then mated.[10]

Toxaphene caused a dose-related increase in hepatocellular carcinomas in mice fed 98 or 198 ppm for 80 weeks. In rats, there was a significantly increased incidence of neoplastic thyroid lesions at the high dose.[11] The IARC has determined that sufficient evidence exists in rodents to regard toxaphene as if it presented a carcinogenic risk to humans.[12]

Toxaphene has been found to be genotoxic in a number of assays.[1] It is mutagenic in *Salmonella typhimurium*, and increases the frequency of sister chromatid exchanges in cell culture; toxaphene-exposed individuals have a higher incidence of chromosomal aberrations in lymphocytes than controls.

The ACGIH threshold limit value–time-weighted average (TLV-TWA) is 0.5 mg/m³, with a short-term excursion limit (STEL) of 1 mg/m^3 and a notation for skin absorption.

REFERENCES

1. Agency for Toxic Substances and Disease Registry (ATSDR): *Toxicological Profile for Toxaphene*. US Department of Health and Human Services, Public Health Service, TP-90-26, p 161, 1990
2. Starmont RT, Conley BE: Pharmacologic properties of toxaphene, a chlorinated hydrocarbon insecticide. *JAMA* 149:1135–1137, 1952
3. McGee LC, Reed HL, Fleming JP: Accidental poisoning by toxaphene. *JAMA* 149:1124–1126, 1952
4. Hayes WJ Jr: *Clinical Handbook on Economic Poisons, Emergency Information for Treating Poisoning*. US Public Health Service Pub No 476, pp 47–50, 71–73. Washington, DC, US Government Printing Office, 1963
5. Matsumura F, Tanaka K: Molecular basis of neuroexcitatory actions of cyclodiene-type insecticides. In Narahashi T (ed): *Cellular and Molecular Neurotoxicology*. New York, Raven Press, pp 225–240, 1984
6. Warraki S: Respiratory hazards of chlorinated camphene. Arch Environ Health 7:137–140, 1963
7. Keplinger ML: Use of humans to evaluate safety of chemicals. *Arch Environ Health* 6:342–349, 1963
8. Chu I, Villeneuve DC, Sun C, et al: Toxicity of toxaphene in the rat and beagle dog. *Fund Appl Toxicol* 7:406–418, 1986
9. Olson KL, Matsumura F, Boush GM: Behavioral effects on juvenile rats from perinatal exposure to low levels of toxaphene, and its toxic components, toxicant A, and toxicant B. *Arch Environ Contam Toxicol* 9:247–257, 1980
10. Chu I, Secours V, Villeneuve DC, et al: Reproduction study of toxaphene in the rat. *J Environ Sci Health* 23:101–126, 1988
11. National Cancer Institute. *Bioassay of Toxaphene for Possible Carcinogenicity*. DHEW (NIH) Pub No 79-837. Bethesda, MD, Carcinogenesis Testing Program, Division of Cancer Cause and Prevention, 1979
12. *IARC Monographs on the Evaluation of the Carcinogenic Risk of Chemicals to Man*, Vol 20, Some halogenated hydrocarbons, pp 327–348. Lyon, International Agency for Research on Cancer, 1978

TRIBUTYL PHOSPHATE
CAS: 126-73-8

$(C_4H_9)_3PO_4$

Synonyms: TBP; phosphoric acid tributyl ester

Physical Form. Colorless liquid

Uses. Antifoaming agent; plasticizer for cellulose esters, lacquers, plastic, and vinyl resins

Exposure. Inhalation

Toxicology. Tributyl phosphate (TBP) is an irritant of the eyes, mucous membranes, and skin; it causes pulmonary edema in animals, and severe exposure is expected to cause the same effect in humans.

Workers exposed to unspecified concentrations of vapor complained of headache and nausea; hot vapor was severely irritating to the eyes and throat.[1] The liquid on the skin is said to be irritating.[2]

In rats, 123 ppm for 6 hours caused respiratory irritation.[2] The oral LD$_{50}$ for rats was 3 g/kg; effects included weakness, dyspnea, pulmonary edema, and muscle twitching.[2] Administered by gavage to rats for 5 days/week for 18 weeks, doses of 0.20 g and above caused diffuse hyperplasia of the urinary bladder epithelium.[3] The liquid dropped on the eye of a rabbit caused temporary epithelial injury and discomfort.[4]

Recent reports suggest that TBP has negligible risk of causing organophosphorus compound–induced delayed neurotoxicity.[5] Two oral doses of 1500 mg/kg TBP, separated by a 21-day interval, did not produce delayed neurotoxicity in hens; neither neurological deficits nor histopathological changes characteristic of organophosphorus compound–induced delayed neurotoxicity were observed. Although some electrophysiological and histopathological changes have been reported in rat peripheral nerve following doses of 6000 mg/kg over 2 weeks, the damage is not considered characteristic of delayed neuropathy.

In vitro, tributyl phosphate caused weak inhibition of cholinesterase in human erythrocytes and plasma.[6] In contrast, TBP resulted in a two- to threefold increase in plasma butyryl-cholinesterase activity in the hen.[5] The significance of this result is not known.

The 1995 ACGIH threshold limit value–time-weighted average (TLV-TWA) is 0.2 ppm (2.2 mg/m³).

REFERENCES

1. Tributyl phosphate. *Documentation of the TLVs and BEIs*, 6th ed, pp 1600–1601. Cincinnati, OH, American Conference of Governmental Industrial Hygienists (ACGIH), 1991
2. Sandmeyer EE, Kirwin CJ Jr: Ethers. In Clayton, GD, Clayton FE (ed): *Patty's Industrial Hygiene and Toxicology*, 3rd ed, rev, Vol 2A, *Toxicology*, pp 2370, 2379. New York, Wiley-Interscience, 1981
3. Latham L, Long G, Broxup B: Induction of urinary bladder hyperplasia in Sprague–Dawley rats orally administered tri-*n*-butyl phosphate. *Arch Environ Health* 40:306–310, 1985
4. Grant WM: *Toxicology of the Eye*, 2nd ed, p 1032. Springfield, IL, Charles C Thomas, 1974
5. Carrington CD, Lapadula DM, Othman M, et al: Assessment of the delayed neurotoxicity of tributyl phosphate, tributoxyethyl phosphate, and dibutylphenyl phosphate. *Tox Ind Health* 6:415–423, 1989
6. Sabine JC, Hayes FN: Anticholinesterase activity of tributyl phosphate. *AMA Arch Ind Hyg Occup Med* 6:174–177, 1952

TRICHLOROACETIC ACID
CAS: 76-03-9

CCl₃COOH

Synonyms: TCA; trichloroethanoic acid

Physical Form. Crystals

Uses/Source. Reagent for albumin detection; in making herbicides; found as a by-product following chlorination of water containing humic materials

Exposure. Ingestion; skin contact

Toxicology. Trichloroacetic acid (TCA) is corrosive to the skin and eyes.

There is little information available concerning the general toxicity of TCA. It is a relatively strong acid; the medical reports of acute exposure show mild to moderate skin and eye burns.

In animal studies, 500 mg/kg was fatal to mice by intraperitoneal administration, and the reported oral LD_{50}s were 3.3 g/kg for rats and 5.0 g/kg for mice.[1]

Current concern regarding TCA arises from chronic low-level exposure via chlorinated drinking water. In 90-day subchronic studies, 5000 ppm in the drinking water caused increased liver and kidney to body weight ratios in rats.[2] Increased hepatic peroxisome activity and histopathological changes in the liver and kidneys were also observed.

Administered in the drinking water of mice for 61 weeks, 2 or 5 g/l TCA caused hepatocellular carcinomas and adenomas. Following a single intraperitoneal injection of ethylnitrosourea, TCA (2 or 5 g/l in the drinking water for 61 weeks) increased the tumor incidence from 5% to 48%.[3] TCA was not carcinogenic in rats.[4]

Developmental studies have evaluated the effects of TCA in the rat; animals were dosed by oral intubation on gestation days 6 to 15 with 330, 800, 1200, or 1800 mg/kg/day.[5] There were no maternal deaths associated with toxicity, but weight gain during treatment was reduced at levels of 800 mg/kg and above. Maternal spleen and kidney weights also increased in a dose-dependent manner. The mean percent of resorbed implants per litter was 34, 62, and 90 at 800, 1200, and 1800 mg/kg, respectively. Live fetuses showed dose-dependent reductions in weight and length. The mean frequency of soft tissue malformations (primarily in the cardiovascular system) ranged from 9% at the low dose to 97% at the high dose. Skeletal malformations were found only at the two highest doses and were principally in the orbit. The authors considered TCA to be developmentally toxic in the rat at doses of 330 mg/kg and above, which also caused slight maternal toxicity.

The 1995 ACGIH threshold limit value–

time-weighted average (TLV-TWA) is 1 ppm (6.7 mg/m³).

REFERENCES

1. Trichloroacetic acid. *Documentation of the TLVs and BEIs*, 6th ed, pp 1602–1604. Cincinnati, OH, American Conference of Governmental Industrial Hygienists (ACGIH), 1991
2. Mather GG, Exon JH, Koller LD: Subchronic 90 day toxicity of dichloroacetic and trichloroacetic acid in rats. *Toxicology* 64:71–80, 1990
3. Herren-Freund SL, Pereira MA, Khoury MD, et al: The carcinogenicity of trichloroethylene and its metabolites, trichloroacetic acid and dichloroacetic acid, in mouse liver. *Toxicol Appl Pharmacol* 90:183–189, 1987
4. DeAngelo AB, Daniel FB: An evaluation of carcinogenicity of the chloroacetic acids in the male F344 rat. *Toxicologist* 12:206, 1992
5. Smith MK, Randall JL, Read EJ, et al: Teratogenic activity of trichloroacetic acid in the rat. *Teratology* 40:445–451, 1989

1,2,4-TRICHLOROBENZENE
CAS: 120-82-1

$C_6H_3Cl_3$

Synonyms: Unsymmetrical trichlorobenzene

Physical Form. Colorless liquid

Uses. Dye carrier; herbicide intermediate; heat-transfer medium; dielectric fluid in transformers; lubricant

Exposure. Inhalation; skin absorption

Toxicology. 1,2,4-Trichlorobenzene may cause eye and throat irritation; at high concentrations, it may produce hepatic toxicity.

In certain individuals, eye and throat irritation may occur at 3 to 5 ppm.[1]

The single oral LD_{50} value was 756 mg/kg in rats and 766 mg/kg in mice.[2] The dermal LD_{50} was 11 g/kg in rats.[2] Repeated exposures at 70 and 200 ppm, 6 hr/day, for 15 days caused lethargy and reduced body weight gain in animals.[3] Male rats, rabbits, and monkeys were exposed at 0, 25, 50, or 100 ppm, 7 hr/day, 5 days/week, for 26 weeks.[4] No differences were seen in body weight measurements, hematology, serum biochemistry, pulmonary function, or eye examination between any of the animals and their controls. Microscopic changes were observed in the rat liver and kidney parenchyma following 4 or 13 weeks of exposure but not after 26 weeks.

Topical application to rabbit ears 3 times/week for 13 weeks caused some local dermal irritation due to defatting action.[5]

Embryonic effects were observed only at treatment levels associated with severe maternal toxicity.[6] Administered to rats on days 9 to 13 of gestation, 360 mg/kg/day caused retarded embryonic development in the form of reduced head length, crown-rump length, somite number, and protein content; maternal deaths (2/9 rats) and significantly decreased body weight gain were also seen.

The 1995 ACGIH ceiling–threshold limit value is 5 ppm (37 mg/m³).

REFERENCES

1. 1,2,4-Trichlorobenzene. *Documentation of the TLVs and BEIs*, 6th ed, pp 1605–1606. Cincinnati, OH, American Conference of Governmental Industrial Hygienists (ACGIH), 1991
2. Brown VKH, Muir C, Thorpe J: The acute toxicity and skin irritant properties of 1,2,4-trichlorobenzene. *Ann Occup Med* 12:209–212, 1969
3. Gage JC: The subacute inhalation toxicity of 109 industrial chemicals. *Br J Ind Med* 27:1–18, 1970
4. Coate WB, Schoenfisch WH, Lewis TR, et al: Chronic inhalation exposure of rats, rabbits, and monkeys to 1,2,4-trichlorobenzene. *Arch Environ Health* 32:249–255, 1977
5. Powers MB, Coate WB, Lewis TR: Repeated topical applications of 1,2,4-trichlorobenzene. Effects on rabbit ears. *Arch Environ Health* 30:165–167, 1975
6. Kitchin KT, Ebron MT: Maternal hepatic and

embryonic effects of 1,2,4-trichlorobenzene in the rat. *Environ Res* 31:362–373, 1983

1,1,1-TRICHLOROETHANE
CAS: 71-55-06

CH₃CCl₃

Synonyms: Methylchloroform; methyltrichloromethane; trichloromethylmethane; α-trichloroethane

Physical Form. Colorless liquid

Uses. Solvent to clean metals, plastic molds, motors, electronic gear, and semiconductors; extraction solvent; aerosol propellant; dry-cleaning solvent

Exposure. Inhalation; skin absorption

Toxicology. 1,1,1-Trichloroethane causes central nervous system depression.

Human deaths after inhalation exposure have been attributed to respiratory failure secondary to central nervous system depression and to cardiac arrhythmias.[1,2] Lethal arrhythmias may result from sensitization of the heart to epinephrine.

Based on effects caused in monkeys and rats, the following effects are expected in humans: 20,000 ppm for 60 minutes, coma and possibly death; 10,000 ppm for 30 minutes, marked incoordination; 2000 ppm for 5 minutes, disturbance of equilibrium.[3] Human subjects exposed to 900 to 1000 ppm for 20 minutes experienced light-headedness, incoordination, and impaired equilibrium; transient eye irritation has also been reported at similar concentrations.[1] Impairments in psychomotor task performance, such as reaction time, perceptual speed, and manual dexterity, have been demonstrated at levels around 350 ppm.[4,5] Other studies at similar exposure levels have failed to show any impairment, but the type of task chosen to test behavioral effects and the times at which behavioral measures were sampled during the course of exposure may explain the variations from study to study.[4]

Recent case reports have associated chronic long-term exposure with peripheral sensory neuropathy and toxic encephalitis.[6,7] In one instance, a woman with daily exposure to 1,1,1-trichloroethane and considerable potential for dermal exposure, developed perioral tingling accompanied by discomfort in her hands and feet; the oral and hand symptoms disappeared after removal from exposure.[6] In another report, a group of 28 workers with long-term repetitive high exposures to 1,1,1-trichloroethane had significant deficits in memory, intermediate memory, rhythm, and speed, as determined by a neuropsychological battery of tests.[7] Evidence of long-term central nervous system damage has also been suggested from animal studies. Gerbils exposed at 210 ppm and 1000 ppm for 3 months had increased glial fibrillary acid protein, which is considered to be a marker for astrogliosis, which is associated with brain injury.[8]

An epidemiologic study of 151 matched pairs of exposed textile workers revealed no evidence of cardiovascular, hepatic, renal, or other effects as a function of exposure; for some workers, exposures exceeded 200 ppm, and duration of exposure ranged from several months to 6 years.[9]

A few scattered reports have indicated mild kidney and liver injury in humans from severe exposure; animal experiments have confirmed the potential for liver but not for kidney injury.[1,10]

The liquid is mildly irritating when applied to the skin or instilled directly into the eyes.[2]

In a carcinogenicity study, rats and mice were given the liquid orally at two different dose levels, 5 days a week for 78 weeks.[11] Both female and male test animals exhibited early mortality compared to untreated controls, and a variety of neoplasms were found in both treated animals and controls. Although rats of both sexes demonstrated a positive dose-related trend, no relationship was established between the dosage groups and the species, sex, type of neoplasm, or site of occurrence. The IARC concluded that an evaluation of the carcinogenicity of 1,1,1-trichloroethane could not be made.[12] In a more recent study, rats exposed at 1500 ppm for 6 hr/day, 5 days/week, for 2 years showed no oncogenic effects.[13]

Inhalation exposure of female rats before mating and during pregnancy to 2100 ppm caused an increased incidence of skeletal and soft tissue variation in the offspring, indicative of developmental delay; no persistent detrimental effects were found in the offspring at 12 months of age.[14]

The genotoxic data are largely negative although 1,1,1-trichloroethane was mutagenic in some *Salmonella* assays and induced chromosomal aberrations in Chinese hamster ovary cells and cell transformation in mammalian systems.[2]

The odor threshold has been described by various investigators as ranging from 16 to 400 ppm.[1]

The 1995 threshold limit value–time-weighted average (TLV-TWA) is 350 ppm (1910 mg/m^3) with a short-term excursion level (STEL) of 450 ppm (2460 mg/m^3)

REFERENCES

1. National Institute for Occupational Safety and Health: *Criteria for a Recommended Standard ... Occupational Exposure to 1,1,1-Trichloroethane (Methyl Chloroform)*. DHEW (NIOSH) Pub No 76-184, pp 16–96. Washington, DC, US Government Printing Office, 1976
2. Agency for Toxic Substances and Disease Registry (ATSDR): *Toxicological Profile for 1,1,1-Trichloroethane*. US Department of Health and Human Services, Public Health Service, p 213, 1993
3. 1,1,1-Trichloroethane—emergency exposure limits. *Am Ind Hyg Assoc J* 25:585, 1964
4. Mackay CJ et al: Behavioral changes during exposure to 1,1,1-trichloroethane: time-course and relationship to blood solvent levels. *Am J Ind Med* 11:223–239, 1987
5. Gamberale F, Hultengren M: Methylchloroform exposure. II. Psychophysiological functions. *Work Environ Health* 10:82–92, 1973
6. House RA, Liss GM, Wills MC: Peripheral sensory neuropathy associated with 1,1,1-trichloroethane. *Arch Environ Health* 49:196–199, 1994
7. Kelafant GA, Berg RA, Schleenbaker R: Toxic encephalopathy due to 1,1,1-trichloroethane exposure. *Am J Ind Med* 25:439–446, 1994
8. Rosengren LE, Aurell A, Kjellstrand P, et al: Astrogliosis in the cerebral cortex of gerbils after long-term exposure to 1,1,1-trichloroethane. *Scand J Work Environ Health* 11:447–455, 1985
9. Kramer C et al: Health of workers exposed to 1,1,1-trichloroethane: a matched-pair study. *Arch Environ Health* 33:331–342, 1978
10. Cohen C, Frank AL: Liver disease following occupational exposure to 1,1,1-trichloroethane: a case report. *Am J Ind Med* 26:237–241, 1994
11. National Cancer Institute: *Bioassay of 1,1,1-Trichloroethane for Possible Carcinogenicity*, Technical Report Series No 3. DHEW (NIOSH) Pub No 77-803. Washington DC, US Government Printing Office, 1977
12. *IARC Monographs on the Evaluation of the Carcinogenic Risk of Chemicals to Humans*, Vol 20, Some halogenated hydrocarbons, pp 515–531. Lyon, International Agency for Research on Cancer, 1979
13. Quast JF, Calhoun LL: Chlorothene VG: a chronic inhalation toxicity and oncogenicity study in rats and mice. Part II. Results in rats. Final report, February 5, 1986, pp 1–165. Midland, MI, Mammalian and Environmental Toxicology Research Laboratory, Dow Chemical, 1986
14. York RG, Sowry BM, Hastings L, et al: Evaluation of teratogenicity and neurotoxicity with maternal inhalation exposures to methyl chloroform. *J Toxicol Environ Health* 9:251–266, 1982

1,1,2-TRICHLOROETHANE
CAS: 79-00-5

$C_2H_3Cl_3$

Synonyms: Vinyl trichloride; ethane trichloride; β-trichloroethane; TCE

Physical Form. Colorless liquid

Uses. Intermediate in the production of vinylidene chloride; solvent

Exposure. Inhalation; skin absorption

Toxicology. In animals, 1,1,2-trichloroethane is a central nervous system depressant and causes liver and kidney damage; it is expected that severe exposure will produce the same effects in humans.

No cases of human intoxication or systemic effects from industrial exposure have been reported.[1]

The lethal concentration for rats was 2000 ppm for 4 hours, with the deaths occurring during a 14-day observation period.[2] An 8-hour exposure to 500 ppm was also lethal to about half of the exposed rats.[3] Rats exposed to 250 ppm for 4 hours survived but showed liver and kidney necrosis.[4] Repeated exposure to 30 ppm resulted in minor liver changes in female rats.

Application of 0.5 ml to the skin of guinea pigs was lethal to all animals within 3 days, whereas 0.25 ml was fatal to 5 of 20 animals.[5] No effects were observed with repeated application of 0.1 ml to the forearm of a volunteer. However, the liquid caused stinging, burning, and whitening of the skin when placed under occlusion for 5 minutes[6] The liquid is considered a slight eye irritant when instilled in rabbit eyes.

Mice treated by intraperitoneal injection with anesthetic doses showed moderate hepatic dysfunction and renal dysfunction. At autopsy, findings were centrilobular necrosis of the liver and tubular necrosis of the kidneys; the 24-hour LD_{50} for intraperitoneal injection was 0.35 mg/kg.[7] The LC_{50}s for 1,1,2-trichloroethane administered by a single gavage dose to male and female mice were 378 and 491 mg/kg, respectively.[8] Above 450 mg/kg, animals became sedated within an hour, and deaths from central nervous system depression occurred within 24 hours. Necropsies showed irritation of the upper gastrointestinal tract, pale livers, and some lung damage. Dose-dependent altera tions in hepatic microsomal enzyme activities and serum enzyme levels were found in mice given 1,1,2-trichloroethane in their drinking water for 90 days.[8]

Administered orally to pregnant mice, 1,1,2-trichloroethane caused no reduction in neonate survival or in neonatal weight at doses that were maternally toxic.[9]

A significant increase in hepatocellular carcinomas occurred in mice given 195 or 390 mg/kg/day by gavage for 78 weeks.[10] Adrenal pheochromocytomas were also increased for the high-dose female mice. No neoplasms were observed at statistically significant incidences in rats given up to 92 mg/kg/day.

The IARC has determined that there is limited evidence that 1,1,2-trichloroethane is carcinogenic in experimental animals and that 1,1,2-trichloroethane is not classifiable as to its carcinogenicity to humans.[11]

1,1,2-Trichloroethane was not mutagenic to bacteria, but it did induce morphological transformation in cultured mammalian cells and chromosomal malsegregation in fungus.[11]

The 1995 threshold limit value–time-weighted average (TLV-TWA) is 10 ppm (55 mg/m³) with a notation for skin absorption.

REFERENCES

1. National Institute for Occupational Safety and Health: *Current Intelligence Bulletin 27, Chloroethanes: Review of Toxicity.* DHEW (NIOSH) Pub No 78-181, p 22, 1978
2. Carpenter CP, Smyth HF Jr, Pozzani UC: The assay of acute vapor toxicity, and the grading and interpretation of results on 96 chemical compounds. *J Ind Hyg Toxicol* 31:343–346, 1949
3. Smyth HF Jr, Carpenter CP, Weil CS, et al: Range-finding toxicity data: List VII. *Am Ind Hyg Assoc J* 30:470–476, 1969
4. Torkelson TR, Rowe VK: Halogenated aliphatic hydrocarbons. In Clayton GD, Clayton FE (eds): *Patty's Industrial Hygiene and Toxicology*, 3rd ed, rev, Vol 2B, *Toxicology*, pp 3510–3513. New York, Wiley-Interscience, 1981
5. Wahlberg JE: Percutaneous toxicity of solvents. A comparative investigation in the guinea pig with benzene, toluene, and 1,1,2-trichloroethane. *Ann Occup Hyg* 19:226–229, 1976
6. Agency for Toxic Substances and Disease Registry (ATSDR): *Toxicological Profile for*

1,1,2-Trichloroethane. US Public Health Service,
   p 109, 1989
7. Klassen CD, Plaa GL: Relative effects of various chlorinated hydrocarbons on liver and kidney function in mice. *Toxicol Appl Pharmacol* 9:139–151, 1966
8. White KL Jr, Sanders VM, Barnes DW, et al: Toxicology of 1,1,2-trichloroethane in the mouse. *Drug Chem Toxicol* 8:333–335, 1985
9. Seidenberg JM, Anderson DG, Becker RA: Validation of an in vivo developmental toxicity screen in the mouse. *Teratogen Carcinogen Mutagen* 6:361–374, 1986
10. National Cancer Institute: *Bioassay of 1,1,2-Trichloroethane for Possible Carcinogenicity,* Carcinogenesis Technical Report Series No 74. NCI-CG-TR-74. Washington, DC, US Department of Health, Education and Welfare, 1978
11. *IARC Monographs on the Evaluation of Carcinogenic Risks to Humans,* Vol 52, Chlorinated drinking-water; chlorination by-products; some other halogenated compounds; cobalt and cobalt compounds, pp 337–355. Lyon, International Agency for Research on Cancer, 1991

TRICHLOROETHYLENE
CAS: 79-01-6

C_2HCl_3

Synonyms: TCE; 1,1,2-trichloroethylene; trichloroethene, 1,1-dichloro-2-chloroethylene; acetylene trichloride; ethylene trichloride

Physical Form. Colorless liquid

Uses. Degreasing solvent; dry cleaning and extraction; chemical intermediate; limited use as an anesthetic and analgesic

Exposure. Inhalation

Toxicology. Trichloroethylene (TCE) is primarily a central nervous system depressant. Although it is carcinogenic at high doses in experimental animals, it is not considered a human carcinogen at low exposure levels.

Inhalation of concentrations in the range of 5000 to 20,000 ppm has been used to produce light anesthesia.[1] Recovery from unconsciousness is usually uneventful, but ventricular arrhythmias and death from cardiac arrest have occurred rarely. Exposure of volunteers to 500 to 1000 ppm has resulted in some symptoms of CNS disturbance, such as dizziness, light-headedness, lethargy, and impairment in visual-motor response tests. In general, no significant signs of toxicity or impaired performance have been noted in subjects acutely exposed to 300 ppm or less.

Prenarcotic symptoms, including visual disturbances and feelings of inebriation, occurred in workers exposed to mean levels of 200 to 300 ppm. Some evidence of mild liver dysfunction has occurred in workers exposed to levels sufficient to produce marked CNS effects. Prolonged exposure at toxic levels may also result in hearing defects.

Workers exposed to average levels of TCE estimated to be 100 to 200 ppm have reported increased incidence of fatigue, vertigo, dizziness, headaches, memory loss, and impaired ability to concentrate. Other effects noted at about 100 ppm and above include paresthesia, muscular pains, and gastrointestinal disturbances.

Intolerance to alcohol, presenting as a transient redness affecting mainly the face and neck (trichloroethylene flush) has frequently been observed following repeated exposure to TCE and alcohol ingestion. It has been suggested that ingestion of alcohol may potentiate the effect of TCE intoxication.[2]

TCE is mildly irritating to the skin; repeated contact may cause chapping and erythema because of defatting.[1] Direct eye contact produces injury to the corneal epithelium; recovery usually occurs within a few days.[1]

Breath analysis for TCE has provided a more accurate index of exposure than the measurement of metabolites (trichloroethanol and trichloroacetic acid) in the urine.[3]

Technical-grade TCE (later shown to be contaminated with other chemicals) has been found to cause liver cancer in B6C3F$_1$ mice but not in Osborne–Mendel rats in an NCI study.[4] Intragastric administration of 2.4 g/kg, 5 times/week for 78 weeks, resulted in hepatocellular carcinomas in 31 of 48 male mice. At 1.2 g/kg, 26 of 50 males were affected, whereas male controls had a 5% liver cancer rate. Among female mice, 11 of 47 developed liver hepatocellular carcinomas, versus only 1 of 80 control animals.[4] In a second gavage bioassay using epichlorohydrin-free reagent grade TCE, results paralleled the NCI study; significantly elevated incidences of hepatocellular adenomas and carcinomas occurred in mice administered 1.0 g/kg for 2 years.[5] An increase in renal adenocarcinomas was also found in male rats.[5]

Mice, rats, and hamsters inhaling up to 500 ppm, 6 hr/day, 5 days/week for 18 months, showed no increase in tumor formation except for an increased incidence of malignant lymphomas in female MRI mice.[6] This strain normally has a high spontaneous incidence of lymphomas, and the significance of TCE exposure is unclear. ICR mice exposed at 150 and 450 ppm for 107 weeks developed a 16% and 15% incidence of adenocarcinomas of the lungs versus 2% for controls.[7] Rats did not show a higher incidence at any site.

Epidemiologic studies have not shown a potent carcinogenic effect in TCE-exposed populations.[8-10] A mortality study of 2117 workers exposed at some time between 1963 and 1976 showed no increase in overall mortality or cancer deaths.[10] Limitations of the study include short latency period, young age of cohort, no direct data on exposure levels, exposure to other chemicals, and possible inclusion of unexposed workers. A recent update of a Swedish cohort study of 1670 workers found no evidence of a carcinogenic effect correlating with trichloroethylene exposure.[11] It was noted that exposures were relatively low (<20 ppm) for most of these workers

TCE carcinogenesis may require exposure to high doses sufficient to cause cellular necrosis.[12] Repeated cycles of necrosis and regeneration would occur with the emergence of hyperplasia and then neoplasia. Low exposures commonly encountered in human studies are not sufficient to initiate the carcinogenic process.

Results from genotoxic studies suggest that TCE is a very weak indirect mutagen.[13]

No evidence of teratogenic effects has been seen in rodent assays.[1] At 1800 ppm, 6 hr/day, on days 0 to 20 of gestation, there were some fetotoxic effects, including incomplete ossification of the sternum in rats.[14] At 300 ppm, on days 6 to 15 of gestation, there were slight fetotoxic effects in mice but not rats.[15] In humans, there is no evidence of an increased incidence of adverse effects in the offspring of female TCE-exposed workers. Increased incidences of menstrual disorders in women workers and of decreased libido in males have been reported in workers exposed to levels sufficient to produce marked CNS disturbances.[1]

The 1995 threshold limit value–time-weighted average (TLV-TWA) is 50 ppm (269 mg/m³) with a short-term excursion level of 100 ppm (537 mg/m³) and an A5 designation (not suspected as a human carcinogen).

REFERENCES

1. Fielder RJ et al: Toxicity Review 6. Trichloroethylene. *Health and Safety Executive*, pp 1–70. London, Her Majesty's Stationery Office, 1982
2. National Institute for Occupational Safety and Health: *Criteria for a Recommended Standard . . . Occupational Exposure to Trichloroethylene*. DHEW (NIOSH) Pub No (HSM) 73-11025, pp 15–40. Washington, DC, US Government Printing Office, 1976
3. Stewart RD, Hake CL, Peterson JE: Use of breath analysis to monitor trichloroethylene exposures. *Arch Environ Health* 29:6–13, 1974
4. National Cancer Institute: Carcinogenesis Bioassay of Trichloroethylene, TR-2. DHEW (NIH) Pub No 76-802. Washington, DC, US Department of Health, Education and Welfare, 1976
5. Kimbrough RD et al: Trichlorethylene: an update. *J Toxicol Environ Health* 15:369–383, 1985
6. Henschler D, Romer W, Elasser HM, et al: Carcinogenicity study of trichloroethylene by long-term inhalation in three animal species. *Arch Toxicol* 43:237–248, 1980
7. Fukuda K, Takemoto K, Tsuruta H: Inhala-

tion carcinogenicity of trichloroethylene in mice and rats. *Ind Health* 21:243–254, 1983

8. Axelson O, Andersson K, Hogstedt C, et al: A cohort study on trichloroethylene exposure and cancer mortality. *J Occup Med* 20:194–196, 1978

9. Shindell S, Ulrich S: A cohort study of employees of a manufacturing plant using trichloroethylene. *J Occup Med* 27:577–579, 1985

10. Tola S, Vilhunen R, Jarvinen E, et al: A cohort study on workers exposed to trichloroethylene. *J Occup Med* 22:737–740, 1980

11. Axelson O, Selden A, Andersson K, et al: Updated and expanded Swedish cohort study on trichloroethylene and cancer risk. *JOM* 36:556–562, 1994

12. Steinberg AD, DeSesso JM: Have animal data been used inappropriately to estimate risks to humans from environmental trichloroethylene? *Reg Toxicol Pharmacol* 18:137–153, 1993

13. Agency for Toxic Substances and Disease Registry (ATSDR): *Toxicological Profile for Trichloroethylene*. US Department of Health and Human Services, TP-92/19, p 155, 1993

14. Dorfmueller MA, Henne SP, York RG, et al: Evaluation of teratogenicity and behavioural toxicity with inhalation exposure of maternal rats to trichloroethylene. *Toxicology* 14:153–166, 1979

15. Schwetz BA, Leong KJ, Gehring PJ: The effect of maternally inhaled trichloroethylene, perchloroethylene, methyl chloroform and methylene chloride on embryonal and fetal development in mice and rats. *Toxicol Appl Pharmacol* 32:84–96, 1975

TRICHLOROFLUOROMETHANE
CAS: 75-69-4

FCCl$_3$

Synonyms: Freon 11; fluorotrichloromethane; fluorocarbon 11

Physical Form. Colorless liquid

Uses. Aerosol propellant; refrigerant and blowing agent; solvent for cleaning and degreasing

Exposure. Inhalation

Toxicology. Trichlorofluoromethane causes narcosis and death from respiratory depression at extremely high concentrations. Death can also occur following cardiac sensitization.

Exposure of volunteers to 250, 500, or 1000 ppm for up to 8 hours did not produce adverse effects.[1] Chronic exposure 6 hr/day for 20 days to 1000 ppm caused a slight but insignificant decrement in cognitive tests; there were no changes in pulmonary function or cardiac rhythm.[1] Workmen near a large area of spilled trichlorofluoromethane experienced narcotic effects, including loss of consciousness; prolonged tachycardia was also observed in one of those exposed.[2] Accidental ingestion caused necrosis and multiple perforations of the stomach.[2]

Sudden deaths from "sniffing" aerosols have been associated with a number of chlorofluorocarbons. The deaths are thought to be due to ventricular fibrillation following cardiac sensitization.[3]

Exposure of rats to 500,000 ppm for 1 minute, 150,000 ppm for 8 minutes, or 100,000 ppm for 30 minutes was always fatal.[4] At 66,000 ppm, 1 of 4 rats died within 2 hours, but all survived 4 hours at 36,000 ppm.[2] Symptoms at the higher dose levels included rapid or labored breathing, twitching, unresponsiveness, or unconsciousness.

No symptoms were observed in rats, guinea pigs, monkeys, or dogs continuously exposed to 1000 ppm for 90 days or exposed to 10,250 ppm, 8 hr/day, for 6 weeks.[5]

Cardiac arrhythmias have been provoked in a number of species. Inhalation of 3500 to 6100 ppm by dogs for 5 minutes caused ventricular fibrillation and cardiac arrest following injection of epinephrine.[3] The minimal concentration that elicited cardiac arrhythmias in the anesthetized monkey was 50,000.[6]

Cardiac sensitization is unlikely to occur in man in the absence of any effects on the CNS, and dizziness should act as an early warning that a dangerous concentration is being reached.[7]

Administered by gavage at 3925 mg/kg/day for 78 weeks, trichlorofluoromethane was not carcinogenic to mice; results from rats are inconclusive because of poor survival rates.[8]

The 1995 ACGIH short term excursion ceiling limit (STEL/C) for trichlorofluoromethane is 1000 ppm (5620 mg/m³).

REFERENCES

1. Stewart RD et al: Physiological response to aerosol propellants. *J Environ Health Perspect* 26:275–285, 1978
2. EI DuPont Co, Haskell Laboratory: Toxicity Review: Freon 11 (unpublished), p 25, Wilmington, DE, 1982
3. Reinhardt CF et al: Cardiac arrhythmias and aerosol "sniffing." *Arch Environ Health* 22:265–279, 1971
4. Lester D, Greenburg LA: Acute and chronic toxicity of some halogenated derivatives of methane and ethane. *Arch Ind Hyg Occup Med* 2:335–344, 1950
5. Jenkins LJ et al: Repeated and continuous exposures of laboratory animals to trichlorofluoromethane. *Toxic Appl Pharmacol* 16:133–142, 1970
6. Belej MA et al: Toxicity of aerosol propellants in the respiratory and circulatory systems. IV. Cardiotoxicity in the monkey. *Toxicology* 2:381–395, 1974
7. Clark DG, Tinston DJ: Acute inhalation toxicity of some halogenated and non-halogenated hydrocarbons. *Human Toxicol* 1:239–247, 1982
8. National Cancer Institute: *Bioassay of Trichlorofluoromethane for Possible Carcinogenicity.* CAS No 75-69-4. MCI-CGTR-106, p 46. US Department of Health, Education and Welfare, 1978

TRICHLORONAPHTHALENE
CAS: 1321-65-9

$C_{10}H_5Cl_3$

Synonyms: 1,4,5-Trichloronaphthalene; 1,4,6-trichloronaphthalene

Physical Form. White solid

Uses. Electric wire insulation; lubricants

Exposure. Inhalation; skin absorption

Toxicology. Trichloronaphthalene is moderately toxic to the liver.

Industrial exposure to trichloronaphthalene (usually mixed with tetrachloronaphthalene) has been relatively free of untoward effects compared to the more highly chlorinated naphthalenes.[1] No fatal cases of liver injury have been reported, but one instance of toxic hepatitis supposedly resulted from exposure to 3 mg/m³.[2] Although there are several reports of chloracne from exposure to trichloronaphthalene, they do not stand up well to critical analysis.[1] Experiments on human volunteers showed that the mist was entirely nonacneigenic as opposed to the penta- and hexachloro- derivatives, which produce severe chloracne.[3]

Rats exposed 16 hr/day for 2.5 months to 11 mg/m³ of trichloronaphthalene containing some tetrachloronaphthalene showed slightly swollen liver cells with granular cytoplasm.[4]

The more highly chlorinated naphthalenes show a much greater toxicity.[1]

The 1995 ACGIH threshold limit value–time-weighted average (TLV-TWA) is 5 mg/m³ with a notation for skin absorption.

REFERENCES

1. Deichmann WB: Halogenated cyclic hydrocarbons. In Clayton GD, Clayton FE (eds). *Patty's Industrial Hygiene and Toxicology*, pp 3669–3675. New York, Wiley-Interscience, 1981
2. Mayers MR, Smith AR: Systemic effects from exposure to certain of the chlorinated naphthalenes. *NY Ind Bull* 21:30–33, 1942
3. Shelley WB, Klingman AM: The experimental production of acne by penta- and hexachloronaphthalenes. *Arch Dermatol* 75:689–695, 1957
4. Drinker CK, Warren MF, Bennett GA: The problem of possible systemic effects from certain chlorinated hydrocarbons. *J Ind Hyg Toxicol* 19:283–311, 1937

2,4,6-TRICHLOROPHENOL
CAS: 88-06-2

$C_6H_3Cl_3O$

Synonyms: Dowicide 2S, Omal, Phenachlor

Physical Form. Yellow flakes

Uses. Wood preservative; disinfectant; fungicide, herbicide, defoliant

Exposure. Inhalation, skin absorption

Toxicology. In experimental animals, 2,4,6-trichlorophenol causes cancer and toxic effects in the liver and hematological system. There is no reliable information regarding exposure and toxic effects in humans.

The acute intraperitoneal LD_{50} in rats is 276 mg/kg.[1] Signs of toxicity prior to death included sluggishness, hypotonia, elevated body temperature, labored breathing, altered respiratory rate, and central nervous system effects, including convulsions, tremors, coma, excited behavior, and incoordination.[2] It has been suggested that 2,4,6-trichlorophenol acts by interfering with mitochondrial oxidative phosphorylation and inhibition of cytochrome P450–dependent mixed-function oxidases.[1]

Hepatic and splenic lesions were observed following subchronic oral studies in rodents.[3] Rats exposed to 2300 mg/kg/day in the diet for 7 weeks experienced a "moderate to marked" increase in splenic hematopoiesis.[3] A high incidence of bone marrow hyperplasia and leukocytosis occurred in rats following chronic exposure to about 1300 mg/kg/day in the diet.

No developmental effects were noted in offspring of female rats exposed to 2,4,6-trichlorophenol throughout gestation or in the offspring of treated males and untreated females.[4,5] Reduced mean litter size was observed in rats following exposure to 42 mg/kg/day in drinking water, but not at 4.2 mg/kg/day.[5] Reproductive function and litter size were not affected in rats administered as much as 1000 mg/kg/day by gavage.[4]

A statistically significant increase in monocytic leukemia was observed in male rats chronically administered either 250 or 650 mg/kg/day.[3] In addition, there was a statistically significant increase in hepatocellular tumors in male (both dose levels) and female (high dose only) mice. Though there is limited evidence supporting the carcinogenicity of chlorophenols as a general class of chemicals to humans, there are no data from which to evaluate the possible carcinogenicity of 2,4,6-trichlorophenol, specifically, in humans.[6] The International Agency for Research on Cancer has indicated that there is sufficient evidence for carcinogenicity to experimental animals, and the EPA has classified 2,4,6-trichlorophenol as a probable human carcinogen based on the animal data.[6,7]

2,4,6-Trichlorophenol has been evaluated for genotoxicity in a variety of in vivo and in vitro assays, and results are inconclusive.[7] Although a majority of the studies reported negative results, some positive results in bacteria, yeast, and mammalian cells suggest that 2,4,6-trichlorophenol may have some genotoxic potential.[7]

A TLV-TWA has not been established for 2,4,6-trichlorophenol.

REFERENCES

1. *IARC Monographs on the Evaluation of the Carcinogenic Risk of Chemicals to Humans*, Vol. 20, Some halogenated hydrocarbons, p 360. Lyon, International Agency for Research on Cancer, 1979.
2. Farquaharson ME, Gage JC, Northover, J: The biological action of chlorophenols. *Br J Pharmacol* 13:20–24, 1958
3. National Cancer Institute: *Bioassay of 2,4,6-Trichlorophenol for Possible Carcinogenicity.* Bethesda, MD, National Institutes of Health, DHEW (NIH) Publ No 79-1711, 1979
4. Blackburn K et al: Evaluation of the reproductive toxicology of 2,4,6-trichlorophenol in male and female rats. *Fund Appl Toxicol* 6:233–239, 1986
5. Exon JH, Koller LD: Toxicity of 2-chlorophenol, 2,4-dichlorophenol and 2,4,6-trichlorophenol. In Jolley RL et al (eds): *Water Chlorination*, Vol. 5, *Chemistry, Environmental Impact*

and Health Effects. Chelsea, MI, Lewis Publishers, 1985

6. *IARC Monographs on the Evaluation of Carcinogenic Risks to Humans, Overall Evaluations of Carcinogenicity: An Updating of Vol 1–42*, Suppl 7, pp 154–156. Lyon, International Agency for Research on Cancer, 1987

7. Agency for Toxic Substances and Disease Registry (ATSDR): *Toxicological Profile for 2,4,6-Trichlorophenol*. US Department of Health and Human Services, Public Health Service, TP-90-28, p 119, 1990

2,4,5-TRICHLOROPHENOXYACETIC ACID

CAS: 93-76-5

$C_8H_5Cl_3O$

Synonym: 2,4,5-T

Physical Form. Solid

Uses. Formerly used as a herbicide in brush control. Production was terminated in the United States in 1979 when the Environmental Protection Agency, in an emergency action, suspended all uses because of contamination with 2,3,7,8-tetrachlorodibenzo-*p*-dioxin (TCDD). In October 1983, all registrations for use of 2,4,5-T were canceled by the US Department of Agriculture because of concerns over the potential of dioxin contamination to produce birth defects and cancer, despite no firm evidence that 2,4,5-T alone had contributed to teratogenesis or carcinogenesis in humans[1]

Exposure. Inhalation

Toxicology. 2,4,5-Trichlorophenoxyacetic acid (2,4,5-T) is of low-order acute toxicity; at high doses, it is teratogenic in experimental animals.

Eleven men in two separate experiments experienced no clinical effects after ingestion of 5 mg/kg 2,4,5-T. Most did report a metallic taste lasting 1 to 2 hours after ingestion.[1]

Most, if not all, occupational illness associated with 2,4,5-T (such as chloracne) has been found to be the result of product contamination with 2,3,7,8-tetrachlorodibenzo-*p*-dioxin (TCDD).[2] TCDD is extremely toxic to animals, and exposure has also been associated with liver-function impairment, peripheral neuropathy, personality changes, porphyria cutanea, hypertrichosis, and hyperpigmentation in humans.[3] TCDD is a chlorinated dioxin, one of a large number of related compounds referred to as "dioxins"; it has no functional use and is not intentionally produced. It has been identified as the responsible toxic agent in several industrial disasters, such as accidental releases at Nitro, WV, in 1949 and at Seveso, Italy, in 1976.[3,4] The role of dioxin contaminants must also be considered in the discussion of 2,4,5-T toxicology.

A study of 204 workers exposed from 1 month to 20 years to 2,4,5-T and its contaminants (concentrations unspecified) showed no evide nce of increased risk for cardiovascular disease, hepatic disease, renal damage, central or peripheral nervous system effects, reproductive problems, or birth defects.[3] Clinical evidence of chloracne persisted in 55.7%, and an association between exposure and history of upper GI tract ulcer was found.

The oral LD_{50} for dogs was in the range of 100 mg/kg; effects were limited to a slight or moderate stiffness in the hind legs with development of ataxia.[5] Dogs survived 10 mg/kg/day for 90 days without illness. In rats fed diets containing 2000 ppm 2,4,5-T (<0.05 TCDD), the minimal cumulative fatal dose was approximately 900 mg/kg.[6]

Concern about the toxicology of 2,4,5-T has centered on its teratogenic action in experimental animals.[2] Although the first studies were carried out with 2,4,5-T contaminated by 30 ppm TCDD, subsequent experiments using analytical-grade 2,4,5-T (<0.05% TCDD) showed that 100 mg/kg/day administered subcutaneously to mice on days 6 through 15 of gestation caused an increased incidence of cleft palates.[7] In a recent report, 2,4,5-T administered by gavage on gestational days 6 through 14 to various stocks and strains of mice caused developmental toxicity at doses below those producing discernible maternal toxicity.[8] The most significant prenatal effects were cleft palate, embryolethality, and intrauterine growth retardation. The number of viable fetuses per litter and mean fetal weight decreased with increasing dose and embryolethality increased.[8] 2,4,5-T containing no detectable TCDD was feticidal and teratogenic to hamsters when administered orally on days 6 to 10 of gestation at a dosage of 100 mg/kg/day.[9] At 80 mg/kg/day, there was a reduction in the number of pups per litter, in fetal weight, and in survival.[9] Rats, rabbits, and monkeys have appeared relatively resistant to teratogenic effects in a number of studies.[1,2]

An epidemiologic investigation of New Zealand chemical applicators using 2,4,5-T found no significant differences in the rate of congenital defects, stillbirths, or miscarriages compared to control.[10]

Several epidemiological studies in Sweden suggested an association between exposure to phenoxyherbicides (and/or their contaminants) and soft tissue sarcomas.[11] There has also been widespread concern among Vietnam veterans that exposure to the defoliant Agent Orange, which contains equal quantities of 2,4-D and 2,4,5-T (with its contaminant TCDD), might increase their risk of adverse health effects, particularly various forms of cancer.[2] Information to support or refute these claims is fragmentary, and no conclusions as to the carcinogenicity of 2,4,5-T can be made at this time. Animal studies do not support the notion that 2,4,5-T itself is carcinogenic.[12] Chronic feeding studies in rats did not produce an increased tumor incidence, even at doses of 30 mg/kg/day, which produced toxic effects.[12] The IARC has determined that there is inadequate evidence for carcinogenicity in both animals and humans.[13]

The 1995 ACGIH threshold limit value–time-weighted average (TLV-TWA) is 10 mg/m³.

REFERENCES

1. Hayes WJ Jr: *Pesticides Studied in Man*, pp 526–533. Baltimore, MD, Williams and Wilkins, 1982

2. Murphy SD: Toxic effects of pesticides. In Klaasen CD et al (eds): *Casarett and Doull's Toxicology. The Basic Science of Poisons*, 3d ed, pp 554–555. New York, Macmillan, 1986

3. Suskind RR, Hertzberg VS: Human health effects of 2,4,5-T and its toxic contaminants. *JAMA* 251:2372–2380, 1984

4. Coulston F, Pocchiari F: Accidental Exposure to Dishins: Human Health Aspects, pp 5–67. New York, Academic Press, 1983

5. Drill VA, Hiratzka T: Toxicity of 2,4-dichlorophenoxyacetic acid and 2,4,5-trichlorophenoxyacetic acid. *AMA Arch Ind Hyg Occup Med* 7:61–67, 1953

6. Chang H et al: Effects of phenoxyacetic acids on rat liver tissues. *J Agric Food Chem* 22:62–65, 1974

7. Moore JA, Courtney KD: Teratology studies with the trichlorophenoxy acid herbicides, 2,4,5-T and silvex. *Teratology* 4 (Abstract):36, 1971

8. Holson JF, Gaines TB, Nelson CJ, et al: Developmental toxicity of 2,4,5-trichlorophenoxyacetic acid (2,4,5-T). *Fund Appl Toxicol* 19:286–297, 1992

9. Collins TFX, Williams CH: Teratogenic studies with 2,4,5-T and 2,4-D in the hamster. *Bull Environ Contam Toxicol* 6:559–567, 1971

10. Smith AH et al: Preliminary report of reproductive outcomes among pesticide applicators using 2,4,5-T. *NZ Med J* 93:177–179, 1981

11. Johnson ES: Review. Association between soft tissue sarcomas, malignant lymphomas, and phenoxy herbicides/chlorophenols: evidence from occupational cohort studies. *Fund Appl Toxicol* 14:219–234, 1990

12. Kociba RJ et al: Results of a two-year chronic toxicity and oncogenic study of rats ingesting diets containing 2,4,5-trichlorophenoxyacetic acid (2,4,5-T). Food Cosmet Toxicol 17:205–221, 1979

13. *IARC Monographs on the Evaluation of the Carcinogenic Risk of Chemicals to Humans.* Suppl 4, pp 235–238. Lyon, International Agency for Research on Cancer, 1982

1,2,3-TRICHLOROPROPANE
CAS: 96-18-4

$C_3H_5Cl_3$

Synonyms: Glycerol trichlorohydrin; allyl trichloride; trichlorohydrine

Physical Form. Colorless liquid

Uses. Intermediate in the manufacture of pesticides and polysulfide rubbers; formerly used as a solvent and extractive agent

Exposure. Inhalation; skin absorption

Toxicology. 1,2,3-Trichloropropane is an irritant of the eyes and mucous membranes; in experimental animals, it has caused hepatic, renal, hematological, and central nervous system effects; it is carcinogenic to rodents exposed orally.

Human subjects exposed to 100 ppm for 15 minutes noted eye and throat irritation and objected to the unpleasant odor.[1] Ingestion of 3 g caused drowsiness, headache, unsteady gait, and lumbar pain.[2]

In rats, 1000 ppm caused death in 5 of 6 animals after 4 hours of exposure.[3] Of 15 mice, 8 did not survive exposure to 5000 ppm for 20 minutes; liver damage accounted for 4 additional deaths after 7 to 10 days.[2] Daily 10-minute exposures to 2500 ppm for 10 days resulted in the death of 7 of 10 mice tested.[2]

Oral LD$_{50}$s ranging from 150 mg/kg to 450 mg/kg have been determined in rats.[4] Prior to death, signs suggestive of central nervous system damage have included piloerection, salivation, ataxia, and coma; hemorrhagic damage to the liver and kidneys was also observed. Repeated gavage administration of 250 mg/kg caused hepatic and renal necrosis severe enough to cause death within 2 weeks in both mice and rats.[4-6] Increased liver weights and altered enzyme levels were found in rats at doses as low as 16 mg/kg/day for 17 weeks, whereas 32 mg/kg/day for the same period had increased kidney weights and slight inflammation. In another report, subacute gavage exposure of rats with 0.80 mM/kg/

day for 10 days caused myocardial degeneration and necrosis in addition to mild hepatotoxicity.[7]

Oral exposures in the near-lethal range also produced pathological changes in the nasal turbinates of both mice and rats.[4-6] Effects included inflammation and necrotic alterations in the dorsal posterior of the nasal passages. Other effects in rats following repeated gavage administration were hyperkeratosis and/or acanthosis of the esophagus and stomach (doses greater than 63 mg/kg/day) and nonregenerative anemia, as indicated by decreased hematocrit, hemoglobin, and erythrocyte counts (doses of 16 mg/kg day).

1,2,3-Trichloropropane was carcinogenic in Fischer-344 rats and B6C3F$_1$ mice when administered for 2 years by gavage.[8] Rats given 3 mg/kg/day or more and mice given 6 mg/kg/day or more had increased incidences of squamous-cell papillomas and/or carcinomas in the oral mucosa and/or the forestomach. Increased incidences of other tumors included: pancreatic acinar adenoma, renal tubule adenoma, and adenoma and carcinoma of the preputial gland in male rats; clitoral gland adenoma and carcinoma and mammary gland adenocarcinoma in female rats; hepatocellular adenoma and carcinoma and Harderian gland adenoma in male and female mice; and uterine neoplasms in female mice.

The carcinogenicity of 1,2,3-trichloropropane is consistent with positive genotoxic findings, which have included mutagenicity in *Salmonella typhimurium* and induction of sister chromatid exchanges in cultured hamster cells.[4]

Intraperitoneal doses causing maternal toxicity in rats were not fetotoxic or teratogenic.[9] Male rats administered 80 mg/kg/day by gavage for 5 days and then mated with an untreated female did not produce any meaningful changes in indices such as numbers of implants and number of live embryos compared to controls.[10] Oral administration for up to 4 months at near-lethal levels caused decreased testes and epididymis weights in rats and mice but no effects on testicular histology, sperm counts, or sperm morphology.[4-6]

The liquid was irritating to the skin of rabbits with prolonged or repeated exposure and was also extremely irritating when instilled in rabbit eyes.[4] The dermal LD_{50} absorption was 2.5 g/kg.[3]

The ACGIH threshold limit value–time-weighted average (TLV-TWA) for 1,2,3-trichloropropane is 10 ppm (60 mg/m^3) with a notation for skin absorption.

REFERENCES

1. Silverman L, Schulte HF, First MW: Further studies on sensory response to certain industrial solvent vapors. *J Ind Hyg Toxicol* 28:262–266, 1946
2. McOmie WA, Barnes TR: Acute and subacute toxicity of 1,2,3-trichloropropane in mice and rabbits. *Fed Proc* 8:319, 1948
3. Smyth HF Jr, Carpenter CP, Weil CS, et al: Range-finding toxicity data: List VI. *Am Ind Hyg Assoc J* 23:95–107, 1962
4. Agency for Toxic Substances and Disease Registry (ATSDR): *Toxicological Profile for 1,2,3-Trichloropropane.* US Department of Health and Human Services, Public Health Service, p 93. TP-91/28, 1992
5. National Toxicology Program (NTP): Final report. 120-Day toxicity gavage study of 1,2,3-trichloropropane in Fischer 344 rats. Hazelton Laboratories New York, NY (unpublished), 1983
6. National Toxicology Program (NTP): Final report. 120-Day toxicity gavage study of 1,2,3-trichloropropane in B6C3F$_1$ mice. Hazelton Laboratories New York, NY (unpublished), 1983
7. Merrick BA, Robinson M, Condie LW: Cardiopathic effect of 1,2,3-trichloropropane after subacute and subchronic exposure in rats. *J Appl Toxicol* 11:179–187, 1991
8. National Toxicology Program (NTP): *Toxicology and Carcinogenesis Studies of 1,2,3-Trichloropropane (CAS No 96-18-4) in F344/N Rats and B6C3F$_1$ Mice (Gavage Studies).* Research Triangle Park, NC, Department of Health and Human Services, National Institutes of Health, Technical Report Series No 384, NIH Pub 91-2839
9. Hardin BD, Bond GP, Sikov MR, et al: Testing of selected workplace chemicals for teratogenic potential. *Scand Work Environ Health* 7(Suppl 4):66–75, 1981
10. Saito-Suzuki R, Teramoto S, Shirasu Y: Dominant lethal studies in rats with 1,2-dibromo-3-chloropropane and its structurally related compounds. *Mut Res* 191:321–327, 1982

1,1,2-TRICHLORO-1,2,2-TRIFLUOROETHANE

CAS: 76-13-1

CCl_2CF_3

Synonyms: Refrigerant 113; fluorocarbon 113; Freon 113; F-113; TCTFE

Physical Form. Colorless gas; volatile liquid

Uses. Solvent for cleaning electronic equipment and degreasing machinery; refrigerant; dry-cleaning agent

Exposure. Inhalation

Toxicology. 1,1,2-Trichloro-1,2,2-trifluoroethane (TCTFE) is a central nervous system depressant, a cardiac sensitizer, and a mild mucous membrane irritant.

Although 1,1,2-trichloro-1,2,2-trifluoroethane is not considered extremely toxic, several deaths have occurred when the chemical was used as a cleaning agent in small, closed, unventilated areas.[1] The vapor acts by displacing oxygen in the victim's immediate breathing zone, resulting in asphyxia, followed by pulmonary edema and death. Symptoms such as headache, light-headedness, dizziness, or drowsiness may or may not precede collapse.

In experimental human studies, exposure to 4500 ppm for 30 to 100 minutes resulted in significant impairment in tests of manual dexterity and vigilance. Subjects reported loss of concentration and a tendency to somnolence, which disappeared 15 minutes after the exposure ended; at 1500 ppm, no effects were observed.[2] More prolonged human exposures of 6 hours daily, 5 days/week, for 2 weeks at concentrations of approximately 500 and 1000 ppm caused mild throat irritation on the first day; there was no decrement in performance of complex mental tasks.[3] No signs or symptoms of adverse effects were found among 50 workers exposed to levels ranging from 46 to 4700 ppm for an average duration of 2.8 years.[4]

The liquid dissolves the natural oils of the skin, and dermatitis may occur as a result of repeated contact; one worker experienced drying of the skin attributed to contact with TCTFE.[4,5]

Pharmacokinetic studies have not determined whether TCTFE is metabolized by man or eliminated unchanged.[6]

Animal studies have indicated low acute toxicity from inhaled TCTFE: the LC_{50} for 2-hour exposures ranged from 50,000 to 120,000 ppm for a number of species.[7] Dogs exposed at 11,000 to 13,000 ppm for 6 hours showed lethargy, nervousness, vomiting, and tremors—all reversible within 15 minutes after exposure. Chronic exposure of rats and rabbits to 12,000 ppm for up to 2 years caused no adverse effects. Rats exposed by whole-body inhalation to 2000, 10,000, or 20,000 ppm, 6 hr/day, 5 days/week, for 24 months, showed no microscopic evidence of compound-related toxicity or carcinogenicity.[8] Observations of appearance, behavior, mortality, and clinical laboratory measurements were unremarkable, except for a 5% to 10% decrease in body weight gains at the 10,000 and 20,000 ppm exposure levels.

In dogs, cardiac sensitization to intravenously administered epinephrine occurred at concentrations of 5000 to 10,000 ppm.[9] Concentrations greater than 25,000 ppm were necessary to produce arrhythmias in animals under anesthesia.[10]

Occluded contact with rabbit skin of 5 g/kg/day for 5 days caused local necrosis of skin and enlargement of liver cells; no effects were observed after 20 weeks of applications to uncovered skin.[7] The liquid produced no significant irritation in a rabbit eye test.[7]

TCTFE is odorless, tasteless, and colorless and provides no warning of overexposure.[1]

The 1995 ACGIH threshold limit value–time-weighted average (TLV-TWA) is 1000 ppm (7670 mg/m³) with a ceiling exposure of 1250 ppm (9590 mg/m³).

REFERENCES

1. Voge VM: Freon: an unsuspected problem. *Aviat Space Environ Med* 60(10, Suppl):B27–B28, 1989
2. Stopps GJ, McLaughlin M: Psychophysiological testing of human subjects exposed to solvent vapors. *Am Ind Hyg Assoc J* 28:43–50, 1967
3. Reinhardt CF et al: Human exposures to fluorocarbon 113. *Am Ind Hyg Assoc J* 32:143–152, 1971
4. Imbus HR, Adkins C: Physical examination of workers exposed to trichlorotrifluoroethane. *Arch Environ Health* 24:257–261, 1972
5. Hygienic Guide Series: 1,1,2-Trichloro-1,2,2-trifluoroethane. *Am Ind Hyg Assoc J* 29:521–525, 1968
6. Auton TR, Woollen BH: A physiologically based mathematical model for the human inhalation pharmacokinetics of 1,1,2-trichloro-1,2,2-trifluoroethane. *Int Arch Occup Environ Health* 63:133–138, 1991
7. 1,1,2-Trichloro-1,2,2-trifluoroethane. *Documentation of the TLVs and BEIs*, 6th ed, pp 1631–1634. Cincinnati, OH, American Conference of Governmental Industrial Hygienists (ACGIH), 1991
8. Trochimowicz HJ, Rusch GM, Chiu T, et al: Chronic inhalation toxicity/carcinogenicity study in rats exposed to fluorocarbon 113 (FC-113). *Fund Appl Toxicol* 11:68–75, 1988
9. Reinhardt CF, Mullin LS, Maxfield ME: Cardiac sensitization potential of some common industrial solvents. *Ind Hyg News Rept* 15 3–4, 1972
10. Aviado DM: Toxicity of aerosol propellants in the respiratory and circulatory systems. X. Proposed classification. *Toxicology* 3:321–332, 1975

TRIETHANOLAMINE
CAS: 102-71-6

N(CH₂CH₂OH)₃

$N(CH_2CH_2OH)_3$

Synonyms: 2,2',2'',-Nitrilotriethanol; tri(hydroxyethyl)amine

Physical Form. Clear, colorless, viscous liquid with ammonia odor

Uses. In manufacture of emulsifiers and dispersing agents; in cosmetic formulations; in household and commercial cleaners and detergents

Exposure. Inhalation

Toxicology. Triethanolamine is a moderate irritant to the eyes and skin. There have been no reports of human injury.

The acute toxicity of triethanolamine is low, as reflected in the high values for the oral LD$_{50}$ in rats of 4.2 to 11.3 g/kg.[1-3]

In rats fed 0.73 g/kg daily for 90 days, the only major effect was fatty degeneration of the liver.[2,4] There were no effects at 0.08 g/kg.

When triethanolamine was applied to the skin of rabbits for 72 hours, there was moderate hyperemia, edema, and necrosis.[5] In a guinea pig sensitization test, there was no evidence of sensitization.[6]

In the eyes of rabbits, one drop caused moderate, transient injury at 24 hours.[7]

Triethanolamine in the diet of ICR mice at levels of 0.03% or 0.3% caused a significant increase in the occurrence of thymic and non-thymic tumors in lymphoid tissues of females.[8] It has recently been suggested that this increase in lymphomas in female ICR mice may be attributable to an unusually low incidence of lymphomas reported in the control animals.[9] (The lymphoma incidence reported in the treated groups is similar to that usually found in controls.) In a follow-up study, B6C3F$_1$ mice administered 1% or 2% triethanolamine in drinking water for 82 weeks showed no dose-related increase in the incidence of any tumor.[9]

Administered continuously to rats as 1% or 2% of the drinking water for up to 2 years, triethanolamine was not carcinogenic, but it was toxic to the kidneys, especially in female animals.[10] Triethanol amine had no carcinogenic or cocarcinogenic activity when dermally applied to mice for 18 months.[11]

Triethanolamine was not mutagenic in bacterial screening assays, nor did it induce unscheduled DNA synthesis in another test.[9]

The 1995 ACGIH threshold limit value–time-weighted average (TLV-TWA) for triethanolamine is 5 mg/m³.

REFERENCES

1. Cosmetic, Toiletry and Fragrance Association (CTFA). Submission of data by CTFA (2-9-59). Acute oral toxicity of triethanolamine, 1973
2. Mellon Institute. Submission of data by FDA. Mellon Institute of Industrial Research, University of Pittsburgh, Special report on the acute and subacute toxicity of mono-, di-, and triethanolamine, Carbode and Carbon Chem Div, UCC Industrial Fellowship No 274-13 (Report 13-67), August 18, 1950
3. Mellon Institute. Submission of data to FDA. Letter from HF Smyth Jr to ER Weidlein Jr, Union Carbide Chemicals Co NY, Acute oral toxicity of triethanolamine, June 16, 1961
4. Smyth HF Jr et al: Range-finding toxicity data: List IV. *AMA Arch Ind Hyg Occup Med* 4:119, 1951
5. Cosmetic, Toiletry and Fragrance Association (CTFA). Submission of data by CTFA. CIR safety data test summary, primary skin irritation and eye irritation of triethanolamine, 1959
6. Life Science Research: Submission of data by CTFA (2-5-50). Dermal sensitization test in guinea pigs (TEA), 1975
7. Carpenter CP, Smyth HF Jr: Chemical burns of the rabbit cornea. *Am J Ophthalmol* 29:1363, 1946
8. Hoshino H, Tanooka H: Carcinogenicity of triethanolamine in mice and its mutagenicity after reaction with sodium nitrite in bacteria. *Cancer Res* 38:3918, 1978
9. Konishi Y, Denda A, Uchida K, et al: Chronic toxicity carcinogenicity studies of triethanol-

amine in B6C3F$_1$ mice. *Fund Appl Toxicol* 18:25–29, 1992
10. Maekawa A, Onodera H, Tanigawa H, et al: Lack of carcinogenicity of triethanolamine in F344 rats. *J Toxicol Environ Health* 19:345–357, 1986
11. Beyer KH Jr et al: Final report on the safety assessment of triethanolamine, diethanolamine, and monoethanolamine. *J Am College Toxicol* 1:183–235, 1983

TRIETHYLAMINE
CAS: 121-44-8

$(C_2H_5)_3N$

Synonyms: TEA; *N,N*-diethylethanamine

Physical Form. Colorless liquid

Uses. In the manufacture of waterproofing agents; corrosion inhibitor; propellant

Exposure. Inhalation

Toxicology. Triethylamine is an irritant of the eyes and mucous membranes.

Two volunteers exposed to approximately 4.5 ppm for 8 hours experienced slight subjective visual disturbances.[1] At 12 ppm for 1 hour, subjects experienced heavy hazing of the visual field, an inability to distinguish outlines of objects 100 m or more away, and bluish halos around lights. There was pronounced increase in corneal thickness. The investigators suggest that the decrease in visual acuity may be severe enough to cause accidents in the workplace or in traffic at the end of work. Effects are reversible, and it appears that even repeated bouts of edema do not cause permanent damage to the cornea.

Among 19 workers repeatedly exposed to time-weighted average levels of 3 ppm with brief excursions to higher levels, 5 workers reported foggy vision, blue haze, and halo phe-nomena on 47 occasions over an 11-week period.[2]

Exposure of 6 rats to 1000 ppm for 4 hours was lethal to 1.[3] Rabbits survived exposures to 100 ppm daily for 6 weeks, but showed pulmonary irritation, myocardial degeneration, and cellular necrosis of liver and kidneys; at 50 ppm, the effects on lung, liver, and kidneys were less severe, but there was also damage to the cornea.[4] Rats appeared to be less sensitive to the effects of triethylamine than rabbits.[5] Exposure to 25 or 247 ppm, 6 hr/day, 5 days/week, for up to 28 weeks caused no statistically significant treatment-related effects on body weight gain, organ weights, hematology, clinical chemistry, or electrocardiographic indices.[5] No gross or histopathological lesions attributable to exposure were noted at autopsy.

In rabbits, skin contact caused irritation.[3]

The 1995 ACGIH threshold limit value–time-weighted average (TLV-TWA) is 1 ppm (4.1 mg/m^3) with a STEL/ceiling of 5 ppm (20.7 mg/m^3) and a notation for skin.

REFERENCES

1. Akesson B et al: Visual disturbances after experimental human exposure to triethylamine. *Br J Ind Med* 42:848–850, 1985
2. Akesson B et al: Visual disturbances after industrial triethylamine exposure. *Int Arch Occup Environ Health* 47:297–302, 1986
3. Smyth HF Jr et al: Range-finding toxicity data: List IV. *AMA Arch Ind Hyg Occup Med* 4:109–122, 1951
4. Brieger H, Hodes WA: Toxic effects of exposure to vapors of aliphatic amines. *AMA Arch Ind Hyg Occup Med* 3:287–291, 1951
5. Lynch DW, Moorman WF, Lewis TR, et al: Subchronic inhalation of triethylamine vapor in Fischer 344 rats: organ system toxicity. *Toxicol Ind Health* 6:403–414, 1990

TRIETHYLENE TETRAMINE
CAS: 112-24-3

$H_2N(CH_2)_2-N(CH_2)_2NH-(CH_2)_2NH_2$

Synonyms: TETA; Araldite hardener HY 951; DEH 24; TECZA; 1,3,7,10-Tetraazadecane; Trien

Physical Form. Slightly viscous, yellow liquid; commercially available form is 95% to 98% pure (impurities include linear, branched, and cyclic isomers)

Uses. Hardener/cross-linker for epoxy resins; metal chelator; constituent of wet-strength paper resins; copolymer with fatty acids in metal spray coatings; constituent of synthetic elastomer formulations

Exposure. Inhalation

Toxicology. Triethylene tetramine (TETA) is a strong irritant of the eyes, mucous membranes, and skin and is a sensitizer of the respiratory tract and skin.

Exposure to the vapor causes irritation of the eyes, nose, throat, and respiratory tract.[1] Exposure to hot vapor causes itching of the face with erythema and edema.[2]

Sensitization of the respiratory tract has followed chronic exposure to fumes or dust of TETA, manifested by bronchial asthma.[3-5] One worker developed asthma after laminating aircraft windows using an epoxy resin/TETA formulation for 6 months.[5] In an environmental chamber with simulated job conditions, the worker developed flulike symptoms and asthmatic breathing after 2 hours with the resin/TETA mixture. Similar exposure to the resin alone did not produce the symptoms.

TETA on the skin causes irritation and dermatitis, and continued exposure can induce allergic contact dermatitis.[3] Cross-sensitization to other amines has occurred.

TETA was teratogenic when fed to rats at 1.67% in the diet.[6] The dihydrochloride salt administered to mice at 3000, 6000 or 12,000 mg/l of drinking water caused a dose-related increased frequency of gross brain abnormalities such as hemorrhages, delayed ossification of the cranium, hydrocephaly, exencephaly, and microcephaly.[7] Microscopically, disorganization of neuronal cell layers, spongiform changes in white matter, and reduced myelin development were noted in the cerebrums of treated animals.

ACGIH has not established a threshold limit value for triethylene tetramine.

REFERENCES

1. Spitz RD: Diamines and higher amines, aliphatic. In Grayson M, Eckroth D (eds): *Kirk-Othmer Encyclopedia of Chemical Technology*, 3rd ed. New York, Wiley-Interscience, 1979
2. Beard RR, Noe JT: Aliphatic and alicyclic amines. In *Patty's Industrial Hygiene and Toxicology*, Vol 2B, *Toxicology*, pp 3235–3273. New York, John Wiley & Sons, Inc, 1971
3. Eckardt RE, Hindin R: The health hazards of plastics. *J Occup Med* 15:808–819, 1973
4. Eckardt RE: Occupational and environmental health hazards in the plastics industry. *Environ Health Perspect* 17:103–196, 1976
5. Fawcett IW, Taylor AJN, Pepys J: Asthma due to inhaled chemical agents—epoxy resin systems containing phthalic acid anhydride, trimellitic acid anhydride and triethylene tetramine. *Clin Allergy* 7:1–14, 1977
6. Cohen NL, Keen CL, Lonnerdal B, Hurley LS: Low tissue copper and teratogenesis in triethylenetetramine-treated rats. *Fed Proc* 41:944, 1982
7. Tanaka H, Inomata K, Arima M: Teratogenic effects of triethylene tetramine dihydrochloride on the mouse brain. *J Nutri Sci Vitamin* 39:177–188, 1993

TRIFLUOROBROMOMETHANE
CAS: 75-63-8

CF₃Br

Synonyms: Bromotrifluoromethane; trifluoro-monobromomethane; Freon 1301; Halon 1301

Physical Form. Colorless gas

Uses. Fire-extinguishing agent; refrigerant

Exposure. Inhalation

Toxicology. Trifluorobromomethane in animals causes sensitization of the myocardium to epinephrine, as well as central nervous system effects.

Human exposure to 70,000 ppm for 3 minutes caused no adverse effects.[1] Light-headedness, paresthesia, and diminished performance were reported during exposures of up to 100,000 ppm; at 150,000, a feeling of impending unconsciousness developed.[2]

In dogs and rats repeatedly exposed to 23,000 ppm, there were no toxic signs or pathologic changes.[2] Monkeys exposed to concentrations of 200,000 ppm were lethargic and suffered spontaneous cardiac arrhythmias within 5 to 40 seconds of exposure.[3] Dogs exposed to concentrations of 200,000 ppm or greater became agitated within 1 to 2 minutes, and tremor occurred within 3 minutes.[3] Epileptiform convulsions characterized by generalized rigidity, apnea, and cyanosis of the tongue were observed in about half of the dogs exposed to 500,000 to 800,000 ppm. Intravenous injection of a pressor dose of epinephrine produced arrhythmias in all animals exposed to 400,000 ppm; larger doses of epinephrine (5–10 μg/kg) caused ventricular fibrillation with cardiac arrest in dogs and spontaneous defibrillation in monkeys.

The 1995 threshold limit value–time-weighted average (TLV-TWA) for trifluorobromomethane is 1000 ppm (6090 mg/m³).

REFERENCES

1. Smith DG, Harris DJ: Human exposure to Halon 1301 (CBrF₃) during simulated aircraft cabin fires. *Aerosp Med* 44:198–201, 1973
2. Trifluorobromomethane. *Documentation of the TLVs and BEIs*, 6th ed, pp 1640–1641. Cincinnati, OH, American Conference of Governmental Industrial Hygienists, 1986
3. Van Stee EW, Back KC: Short-term inhalation of bromotrifluoromethane. *Toxicol Appl Pharmacol* 15:164–174, 1969

TRIMELLITIC ANHYDRIDE
CAS: 552-30-7

C₉H₄O₅

Synonyms: Anhydrotrimellitic acid; 1,2,4-benzenetricarboxylic acid anhydride; 1,3-dihydro-1,3-dioxo-5-isobenzofurancarboxylic acid; TMA; TMAN (preferred)

Physical Form. White crystalline solid

Uses. Curing agent for epoxy and other resins; vinyl plasticizer; in anticorrosive surface coatings, polymers, paints, dyes, and pharmaceuticals

Exposure. Inhalation

Toxicology. Trimellitic anhydride (TMAN) causes both respiratory irritation and immunologic respiratory disease.

In humans, four clinical syndromes induced by inhalation of TMAN dust and fume have been described.[1-3] The first is a direct irritant syndrome characterized by cough and upper airway irritation related to the irritant properties of the anhydride at high-dose exposures.

The second syndrome, characterized by rhinitis, asthma, or both, is an immediate-type airway response mediated by IgE antibodies directed against trimellityl–human protein conjugates. A latent period, ranging from weeks to years, is required between the sensitizing exposure and the onset of symptoms but, once sensitization has occurred, symptoms appear almost immediately on re-exposure.

The third condition, late respiratory systemic syndrome, is characterized by cough, mucus production, occasional wheezing, and systemic symptoms of malaise, chills, fever, and aching muscles and joints, occurring 4 to 12 hours after exposure. This syndrome has been termed TMA flu and clinically resembles hypersensitivity pneumonitis with visible chest X-ray infiltrates. High levels of IgG serum antibody and total serum antibody directed against trimellityl–human protein conjugates accompany the syndrome, and a latent period of exposure before the onset of symptoms is typical.

The fourth condition, termed pulmonary disease–anemia syndrome, is characterized by dyspnea, hemoptysis, pulmonary infiltrates, restrictive lung disease, and anemia. It occurs with high-dose exposure to fumes when heated metal surfaces are sprayed with TMAN-containing materials. High titers of antibody to trimellityl human proteins and erythrocytes have been found in affected workers.

It is thought that low-molecular-weight compounds such as TMAN cannot directly elicit immunologic sensitization; however, they can act as haptens.[4] Thus, TMAN combines with human serum albumin or with human erythrocytes to form antigens, against which numerous types of antibodies can be found. During TMAN conjugation, the anhydride group is lost, and trimellityl-protein complexes, such as TMAN–human serum albumin (TMAN-HSA), are formed.

The TMAN levels associated with various respiratory effects have not been clearly defined.[5] However, workers who may have been exposed to up to 7.5 mg/m^3 of TMAN during the manufacture of epoxy paint complained of irritation of the eyes, nose, and throat; shortness of breath; cough; nausea; headache; and skin irritation. Symptoms of chest pain and respiratory tract irritation have been reported in workers exposed to levels in the range of 0.1 to 10 mg/m^3. In one study, intermittent exposure to levels ranging up to 2.1 mg/m^3 caused late respiratory systemic syndrome and allergic rhinitis in a portion of the exposed workers.[5] Reduction

of the work levels to 0.03 mg/m^3 coincided with symptomatic improvement in the 3 workers with late respiratory systemic syndrome and with a fall in total antibody binding to trimellityl–human serum albumin. However, the continued low-level exposure was sufficient to elicit and maintain a specific IgE immune response in a worker who eventually developed lacrimation and rhinorrhea. Further study of this same population showed that workers exposed only after the low levels were in place developed no immunologic syndromes and had insignificant antibody responses.[6]

A follow-up study of 29 workers with TMAN-induced immunological lung disease who had been moved to low-exposure jobs for more than 1 year revealed that workers with late asthma or late respiratory systemic syndrome had improved symptoms, improved pulmonary functions, and lower total antibody against TMAN-HSA.[7] In contrast, 7 of 12 workers with asthma rhinitis continued to have moderate to severe symptoms, abnormal pulmonary functions, and elevated IgE against TMAN-HSA. Elevated IgE against TMAN-HSA appears to be a marker for the subpopulation of workers with asthma rhinitis that does not improve.

In animal studies, TMAN had a low acute toxicity when administered by the oral or the percutaneous route.[8] Rats given 10,000 ppm in the diet for 90 days showed an increase in the number of white blood cells. Inhalation of 0.2 mg/m^3 and above for 14 days was associated with hemorrhagic foci in the lungs.

There are no reports of carcinogenicity associated with TMAN exposure. No teratogenic effects or developmental toxicity was seen in rats or guinea pigs exposed to 500 mg/m^3 for 6 hr/day during their period of major organogenesis.[9]

The 1995 ACGIH ceiling–threshold limit value (C-TLV) for trimellitic anhydride is 0.04 mg/m^3.

References

1. McGrath KG, Zeiss R, Patterson R: Allergic reactions to industrial chemicals. *Clin Immunol Rev* 2:1–58, 1983

2. Zeiss, CR, Wolkonsky P, Pruzansky JJ, Patterson R: Clinical and immunologic evaluation of trimellitic anhydride workers in multiple industrial settings. *J Allergy Clin Immunol* 70:15–18, 1982

3. Letz G, Wygofski L, Cone JE, et al: Trimellitic anhydride exposure in a 55-gallon drum manufacturing plant: clinical, immunologic, and industrial hygiene evaluation. *Am J Ind Med* 12:407–417, 1987

4. National Institutes of Health: Occupational asthma from a low-molecular-weight organic chemical. *JAMA* 244:1667–1668, 1980

5. Bernstein DI, Roach DE, McGrath K, et al: The relationship of airborne trimellitic anhydride–induced symptoms and immune responses. *J Allergy Clin Immunol* 72:709–713, 1983

6. McGrath K, Roach D, Zeiss R, Patterson R: Four-year evaluation of workers exposed to trimellitic anhydride. *J Occup Med* 26:671–675, 1984

7. Grammer LC, Shaughnessy MA, Henderson J, et al: A clinical and immunological study of workers with trimellitic–anhydride-induced immunologic lung disease after transfer to low exposure jobs. *Am Rev Resp Dis* 148:54–57, 1993

8. Toxicity Review 8. Trimellitic anhydride. *Health and Safety Executive.* London, Her Majesty's Stationery Office, 1983

9. Ryan BM, Hatoum NS, Zeiss CR, Garvin PJ: Immunoteratologic investigation of trimellitic anhydride (TMA) in the rat and guinea pig. *Teratology* 39:477–478, 1989 (Abstract)

TRIMETHYLAMINE

CAS: 75-50-3

$(CH_3)_3N$

Synonyms: N,N-Dimethylmethanamine; TMA

Physical Form.   Colorless gas

Uses.   Insect attractant; warning agent for natural gas; in organic synthesis

Exposure.   Inhalation

Toxicology.   Trimethylamine is a skin, eye, and respiratory irritant.

In an accidental exposure, a blast of vapor that struck the eye, of a student caused the epithelium to be lost from the cornea. There was no edema of the corneal stroma, and the eye was completely normal within 4 or 5 days; the exposure was thought to be minimal.

Tests of single drops of aqueous solutions applied to the eyes of animals have shown that 1% solution causes severe irritation, 5% causes hemorrhagic conjunctivitis, and 16% causes severe reaction with conjunctival hemorrhages, corneal edema, and opacities, followed by some clearing but much vascularization.[1]

In rats, 3500 ppm for 4 hours was considered an approximate lethal concentration. Rats exposed 6 hr/day for 10 days to 0, 75, 250, or 750 ppm had dose-dependent degenerative changes in the olfactory and respiratory epithelium.[2] Degeneration of the tracheal mucosa was also observed at the two higher doses.

Administered to mouse embryo cultures in vitro, trimethylamine was teratogenic, causing neural tube defects and inhibiting embryonic growth.[3] Trimethylamine may exert these effects by reducing macromolecular synthesis.

The 1995 ACGIH threshold limit value–time-weighted average (TLV-TWA) for trimethylamine is 5 ppm (12 mg/m³) with a short-term excursion limit of 15 ppm (36 mg/m³).

REFERENCES

1. Grant WM: *Toxicology of the Eye,* 3rd ed, p 952. Springfield, IL, Charles C Thomas, 1986

2. Kinney LA, Burgess BA, Chen HC, et al: Inhalation toxicology of trimethylamine (TMA). *Inhal Toxicol* 2:41–51, 1990

3. Guest I, Varma DR: Teratogenic and macromolecular synthesis inhibitory effects of trimethylamine on mouse embryos in culture. *J Toxicol Environ Health* 36:27–41, 1992

TRIMETHYL BENZENE
CAS: 25551-13-7

Mesitylene—CAS: 108-67-8
Pseudocumene—CAS: 95-63-6
Hemimellitene—CAS: 526-73-8

$(CH_3)_3C_6H_3$

Synonyms: The three isomers of trimethyl benzene are mesitylene (1,3,5-trimethylbenzene, *sym*-trimethylbenzene, 1,3,5-TMB); pseudocumene (1,2,4-trimethylbenzene, pseudocumol, 1,2,4-TMB); and hemimellitene (1,2,3-trimethylbenzene, 1,2,3-TMB)

Physical Form.   Colorless liquid

Uses.   Raw material in chemical syntheses; ultraviolet stabilizer in plastics; in solvents; constituent of gasoline

Exposure.   Inhalation

Toxicology.   Trimethyl benzene is an eye, nose, and respiratory irritant; at high concentrations, it causes central nervous system depression.

In one of the few reports of human exposure, 27 workers exposed for a number of years to a paint thinner containing primarily pseudocumene (50%) and mesitylene (30%), plus other alkylbenzenes in unspecified amounts, had signs and symptoms of impairment of the respiratory, nervous, and hematopoietic systems.[1] Approximately 70% of the workers complained of headaches and drowsiness, with 51% suffering from anemia, 30% displaying signs of asthmalike bronchitis, and 30% showing a tendency to hemorrhage. The presence of other hydrocarbons and unidentified additives, which accounted for 20% of the thinner, suggested that the blood disturbances were probably due to a contaminant.

Mice exposed at 5100 to 7140 ppm mesitylene for 2 hours suffered from a loss of righting reflex and, at 7140 to 9180 ppm for 2 hours, mice showed depression of the central nervous system.[1,2] Similar exposures to pseudocumene produced similar results with 8130 ppm for 2 hours causing a loss of righting reflex and 8130 to 9140 causing a loss of reflexes and depression of the central nervous system.

In rats, exposure to 2400 ppm mesitylene for 24 hours, caused death due to respiratory failure and depression of the central nervous system in 4 of 10 animals; at 612 ppm for 24 hours, there were no adverse effects.[1] Rats exposed at 600 ppm, 6 hr/day, 6 days/week for 5 weeks, showed no hematological or biochemical changes. Experiments with pseudocumene showed nose and eye irritation, respiratory difficulty, lethargy, tremors, and reduced weight gain with 12 exposures to 2000 ppm for 16 hours each; at 1000 ppm, there was only slight eye and nose irritation.[3]

Inhalation of mixed trimethylbenzene by rats for 4 hr/day for 6 months at 200 ppm caused inhibition of phagocytic activity of the leukocytes.

It is expected that repeated skin exposure to the liquid will cause drying and cracking of the skin.

The 1995 ACGIH threshold limit value–time-weighted average (TLV-TWA) for trimethyl benzene is 25 ppm (123 mg/m³).

REFERENCES

1. Snyder R (ed): *Ethel Browning's Toxicity and Metabolism of Industrial Solvents*, 2nd ed, Vol I, *Hydrocarbons*. pp 121–142, New York, Elsevier, 1987
2. Sandmeyer EE: Aromatic hydrocarbons. In Clayton GD, Clayton FE (eds): *Patty's Industrial Hygiene and Toxicology*, 3rd ed, rev, Vol 2B, *Toxicology*, pp 3300–3302. New York, Wiley-Interscience, 1981
3. Gage JC: The subacute inhalation toxicity of 109 industrial chemicals. *Br J Ind Med* 27:1–18, 1970

TRIMETHYL PHOSPHITE
CAS: 121-45-9

$C_3H_9O_3P$

Synonyms: Methyl phosphite; phosphorus acid trimethyl ester; TMP; trimethoxyphosphine

Physical Form. Colorless liquid

Uses. Primarily in the synthesis of organophosphate insecticides; also in the production of flame-retardant polymers and textiles.

Exposures. Inhalation; skin absorption

Toxicology. Trimethyl phosphite is a skin irritant, and high levels may cause ocular damage.

The adverse effects of trimethyl phosphite on man are largely unknown. At one plant with average exposures between 0.3 and 4 ppm, and excursions as high as 15 ppm, examination of 179 workers showed no adverse effects associated with occupational exposure.[1] Odors approaching 20 ppm are considered to be objectionable.

In rats, the oral LD_{50} was 2.5 g/kg.[2] A 4-hour LC_{50} of greater than 10,000 ppm was also found in rats, with the animals indicating respiratory distress, irritation, and discomfort. Applied to the skin of rabbits, trimethyl phosphite caused moderately severe irritation, and the dermal LD_{50} was 2.6 g/kg.

Rats were exposed at 600, 300, or 100 ppm for 6 hr/day, 5 days/week, for 4 weeks.[1] At the highest dose, severe cataracts developed, and 70% of the animals died; there was histological evidence of lung inflammation. The middose was lethal to 10% of the exposed group and caused mild cataracts. At the low dose, there were mild, reversible striate opacities. In further studies, no effects were observed in animals exposed at 10 ppm.

Trimethyl phosphite was administered by gavage to pregnant rats at rates of 16, 49, or 164 mg/kg/day on gestation days 6 through 15.[3] Teratologic evaluation revealed gross fetal abnormalities, skeletal defects, and soft tissue defects, and an increased frequency of fetal resorption rates was observed at 164 mg/kg/day. No changes were observed at the lower dose levels. The teratogenic effect of trimethyl phosphite may be associated with inhibition of cholinesterase activity.

Trimethyl phosphite was genotoxic in mouse lymphoma assays but was not mutagenic in various bacterial assays.[1]

The 1995 ACGIH threshold limit value–time-weighted average (TLV-TWA) for trimethyl phosphite is 2 ppm (10 mg/m³).

REFERENCES

1. Trimethyl phosphite. *Documentation of the TLVs and BEIs*, 6th ed, pp 1650–1651. Cincinnati, OH, American Conference of Governmental Industrial Hygienists (ACGIH), 1991
2. Levin L, Gabriel KL: The vapor toxicity of trimethyl phosphite. *Am Ind Hyg Assoc J* 34:286–291, 1973
3. Mehlman MA, Craig PH, Gallo MA: Teratological evaluation of trimethyl phosphite in the rat. *Toxicol Appl Pharmacol* 72:119–123, 1984

2,4,6-TRINITROTOLUENE
CAS:118-96-7

$C_7H_5N_3O_6$

Synonyms: TNT; trinitrotoluol; *sym*-trinitrotoluene; 1-methyl- 2,4,6-trinitrobenzene

Physical Form. Colorless, monoclinic prisms or crystals; commercial crystals are yellow

Uses. Explosives

Exposure. Inhalation; skin absorption

Toxicology. 2,4,6-Trinitrotoluene (TNT) causes liver damage and aplastic anemia.

Deaths from aplastic anemia and toxic hepatitis were reported in TNT workers prior to the 1950s; with improved industrial practices, there have been few reports of fatalities or serious health problems related to its use.[1]

Exposures exceeding 0.5 mg/m³ cause destruction of red blood cells.[2] Hemolysis is partially compensated for by enhanced regeneration of red blood cells in the bone marrow, which is manifested as an increased percentage of reticulocytes in peripheral blood.[2] Among some groups of workers, there is a reduction in average hemoglobin and hematocrit values.[2] Workers deficient in glucose-6-phosphate dehydrogenase may be particularly at risk of acute hemolytic disease.[3] Three such cases occurred after a latent period of 2 to 4 days and were characterized by weakness, vertigo, headache, nausea, pallor, enlarged liver and spleen, dark urine, decreased hemoglobin levels, and reticulocytosis.[3] Although no simultaneous measurements of atmospheric levels were available, measurement on other occasions showed levels up to 3.0 mg/m³.[3]

Above 1.0 mg/m³, the liver is unable to handle the increased amounts of red blood cell breakdown products, and indirect bilirubin levels rise.[2] Elevations of liver-function enzymes may occur, particularly in new employees or those recently exposed to higher levels. There are suggestions of marked individual susceptibility to liver damage, with most not showing effects unless exposures considerably exceed 1.0 mg/m³.[2]

A characteristic TNT cataract is reportedly produced with exposures regularly exceeding 1.0 mg/m³ for more than 5 years.[2] In one study, 6 of 12 workers had bilateral peripheral cataracts, visible only with maximum dilation.[4] The opacities did not interfere with visual acuity or visual fields. The induced cataracts may not regress once exposure ceases although progression is arrested.

The vapor or dust can cause irritation of mucous membranes, resulting in sneezing, cough, and sore throat.[5] Although intense or prolonged exposure to TNT may cause some cyanosis, it is not regarded as a strong producer of methemoglobin.[6] Other occasional effects include leukocytosis or leukopenia, peripheral neuritis, muscular pains, cardiac irregularities, and renal irritation.[2] The skin, hair, and nails of exposed workers may be stained yellow.[5]

Trinitrotoluene is absorbed through skin fairly rapidly, and reference to airborne levels of vapor or dust may underestimate total systemic exposure if skin exposure also occurs.[2] Apparent differences in dose-response relationships based only on airborne levels may be explained by differences in dermal absorption.[2] TNT causes sensitization dermatitis; the hands, wrist, and forearms are most commonly affected, but skin at friction points such as the collar line, belt line, and ankles is also often involved. Erythema, papules, and an itchy eczema can be severe.[7]

Rats administered 50 mg/kg/day in their diets had anemia, splenic lesions, and liver and kidney damage.[8] Testicular atrophy and atrophic seminiferous tubules have been reported in rats following 13 weeks of treatment at high doses.[9]

Hyperplasia and carcinoma of the urinary bladder were also observed in females exposed for 24 months. A statistically significant incidence of leukemia and/or malignant lymphoma of the spleen was present in female mice receiving 70 mg/kg/day for 24 months.[9]

In bacterial and mammalian in vitro cell systems, TNT is a direct-acting mutagen.[9] However, inclusion of exogenous metabolic activation appears to abolish the genotoxicity. In vivo assays of TNT have not shown TNT to be genotoxic, suggesting that TNT may be reduced to nonmutagenic metabolic products in the whole animal.

The 1995 ACGIH threshold limit value–time-weighted average (TLV-TWA) for 2,4,6-trinitrotoluene is 0.5 mg/m³, with a notation for skin absorption.

REFERENCES

1. Woollen BH, Hall MG, Craig R, et al: Trinitrotoluene: Assessment of occupational absorption during manufacture of explosives. *Br J Ind Med* 43:465–473, 1986
2. Hathaway JA: Subclinical effects of trinitrotoluene: a review of epidemiology studies. In Richert DE (ed): *Toxicity of Nitroaromatic Compounds*, pp 255–274. New York, Hemisphere Publishing Co, 1985
3. Djerassi LS, Vitany L: Haemolytic episode in

G6-PD deficient workers exposed to TNT. *Br J Ind Med* 32:54–58, 1975

4. Harkonen H, Karki M, Lahti A, et al: Early equatorial cataracts in workers exposed to trinitrotoluene. *Am J Ophthal* 95:807–810, 1983
5. Hygienic Guide Series: 2,4,6-Trinitrotoluene (TNT). *Am Ind Hyg Assoc J* 25:516–519, 1964
6. Goodwin JW: Twenty years handling TNT in a shell loading plant. *Am Ind Hyg Assoc J* 33:41–44, 1972
7. Schwartz L: Dermatitis from explosives. *JAMA* 125:186–190, 1944
8. Levine BS et al: Two-year chronic oral toxicity/carcinogenicity study on the munitions compound trinitrotoluene (TNT) in rats. *Toxicologist* 5:(Abstract 697)175, 1985
9. Agency for Toxic Substances and Disease Registry (ATSDR): *Toxicological Profile for 2,4,6-Trinitrotoluene.* US Department of Health and Human Services, Public Health Service, p 125, 1993

TRIORTHOCRESYL PHOSPHATE
CAS: 78-30-8

$(CH_3C_6H_4)_3PO_4$

Synonyms: TOCP; tri-*o*-tolyl phosphate; phosphoric acid, tri-*o*-tolyl ester

Physical Form. Colorless or pale yellow liquid

Uses. Plasticizer in vinyl plastics, lacquers, and varnishes; flame retardant

Exposure. Inhalation; skin absorption; ingestion

Toxicology. Tri-*o*-cresyl phosphate (TOCP) causes peripheral neuropathy with flaccid paralysis of the distal muscles of the upper and lower extremities, followed in some cases by spastic paralysis.

Reports of intoxication from occupational exposure are rare.[1] However, thousands of people have been poisoned by the accidental ingestion of TOCP in contaminated foods and beverages.[1] The most notable example was the consumption of an adulterated Jamaica ginger extract ("Jake") in 1930.[2] Shortly after ingestion, there may be nausea, vomiting, diarrhea, and abdominal pain.[1] After a symptom-free interval of 3 to 28 days, most patients complain of sharp, cramplike pains in the calf muscles; some patients complain of numbness and tingling in the feet and sometimes the hands.[1,3] Within a few hours, there is increasing weakness of the legs and feet, progressing to bilateral foot drop.[3] After an interval of another 10 days, weakness of the fingers and wrist drop develop, but the paralysis is not usually as severe as that which occurs in the feet and legs. This process does not extend above the elbows; the thigh muscles are infrequently involved. Sensory changes, if they occur, are minor.[4,5]

With severe intoxication, lesions of the anterior horn cells and the pyramidal tracts may also occur.[5,6] Muscular weakness may increase over a period of several weeks or months; recovery may take months or years and, in 25% to 30% of cases, permanent residual effects remain, usually confined to the lower limbs.[3,5] Gait impairment, permanent in some, was called "Jake Walk."[2]

Fatalities are rare and occur principally in those who have taken large quantities in a short period of time; autopsy of 6 human cases revealed involvement of anterior horn cells and demyelination of nerve cells.[4] The lethal dose for humans by ingestion is about 1.0 g/kg; severe paralysis has been produced by ingestion of 6 to 7 mg/kg.[4]

In workers engaged in the manufacture of aryl phosphates (including up to 20% TOCP) and exposed to concentrations of aryl phosphates at 0.2 to 3.4 mg/m³, there was some inhibition of plasma cholinesterase, but no correlation of this effect with degree of exposure or with minor gastrointestinal or neuromuscular symptoms.[7,8] No effects on the eyes or skin have been reported; TOCP is readily absorbed through the skin without local irritant effects.

In affected cats and hens, extensive damage is observed in the spinal cord and sciatic nerves; damage to the myelin sheath and Schwann cells is secondary to the destructive lesion in the axon, which starts at the distal end of the longer axons.[9]

No evidence of teratogenic effects was observed in the offspring of rats orally dosed with 90, 175, or 350 mg/kg/day from gestation days 6 through 18.[10] TOCP is, however, a reproductive toxin in male rats, causing testicular toxicity and decreased fertility.[11] With the administration of 150 mg TOCP/kg/day by gavage, there was an increase in the number of necrotic spermatids and, by day 14, 90% of the seminiferous tubules were devoid of sperm.[12] Atrophy of the seminiferous tubules also occurred in male rats fed 6600 ppm or 13,000 ppm of a mixed isomer of tricresyl phosphate for 13 weeks.[13]

In chronic two-year feeding studies of the mixed isomer, there was no evidence of carcinogenicity in rats given up to 300 ppm or in mice given up to 250 ppm in the diet.[13] Tricresyl phosphate was not mutagenic in *Salmonella typhimurium*, nor did it induce chromosomal aberrations or sister chromatid exchange in Chinese hamster ovary cells.[13]

The 1995 ACGIH threshold limit value–time-weighted average (TLV-TWA) for triorthocresyl phosphate is 0.1 mg/m^3 with a notation for skin absorption.

REFERENCES

1. Hygienic Guide Series: Triorthocresylphosphate. *Am Ind Hyg Assoc J* 24:534–536, 1963
2. Morgan JP, Tulloss TC: The Jake Walk blues—a toxicologic tragedy mirrored in American popular music. *Ann Intern Med* 85:804–808, 1976
3. Susser M, Stein Z: An outbreak of tri-ortho-cresyl phosphate (TOCP) poisoning in Durban. *Br J Ind Med* 14:11–120, 1957
4. Fassett DW: Esters. In Patty FA (ed): *Industrial Hygiene and Toxicology*, 2nd ed, Vol 2, *Toxicology*, pp 1853, 1914–1925, 1935–1937. New York, Interscience, 1963
5. Hunter D, Perry KMA, Evans RB: Toxic polyneuritis arising during the manufacture of tricresyl phosphate. *Br J Ind Med* 1:227–231, 1944
6. Vora DD: Toxic polyneuritis in Bombay due to ortho-cresyl- phosphate poisoning. *J Neurol Neurosurg Psychiatry* 25:234–242, 1962
7. Tabershaw IR, Kleinfeld M, Feiner B: Manufacture of tricresyl phosphate and other alkyl phenyl phosphates: an industrial hygiene study. I. Environmental factors. *AMA Arch Ind Health* 15:537–540, 1957
8. Tabershaw IR, Kleinfeld M: Manufacture of tricresyl phosphate and other alkyl phenyl phosphates: an industrial hygiene study. II. Clinical effects of tricresyl phosphate. *AMA Arch Ind Health* 14:541–544, 1957
9. Johnson MK: Delayed neurotoxic action of some organophosphorus compounds. *Br Med Bull* 25:231–235, 1969
10. Tocco DR, Randall JL, York RG, et al: Evaluation of the teratogenic effects of tri-ortho-cresyl phosphate in the Long–Evans hooded rat. *Fund Appl Toxicol* 8:291–297, 1987
11. Chapin RE, Phelps JL, Burka, et al: The effects of tri-*o*-cresyl phosphate and metabolites on rat sertoli cell function in primary culture. *Toxicol Appl Pharmacol* 108:194–204, 1991
12. Somkuti SG, Lapadula DM, Chapin RE, et al: Light and electron microscopic evidence of tri-*o*-cresyl phosphate (TOCP)-mediated testicular toxicity in Fischer 344 rats. *Toxicol Appl Pharmacol* 107:35–46, 1991
13. National Toxicology Program: *NTP Technical Report on the Toxicology and Carcinogenesis Studies of Tricresyl Phosphate (CAS No 1330-78-5) in F344 Rats and B6C3F$_1$ Mice (Gavage and Feed Studies)*. NTP TR 433, NIH Pub No 93-3164. US Department of Health and Human Services, Public Health Service, National Institutes of Health, 1994

TRIPHENYL AMINE
CAS: 603-34-9

$(C_6H_5)_3N$

Synonyms. N,N-Diphenylaniline; N,N-diphenylbenzenamine; triphenylamine

Physical Form. Colorless crystalline solid

Uses. Coating on photographic film as a photoconductor

Exposure. Inhalation

Toxicology. Triphenyl amine is considered

to have low systemic toxicity, but it may act as a slight skin irritant.

Adverse effects have not been reported in humans.[1]

In rats, the oral LD_{50} was between 3200 and 6400 mg/kg. Clinical signs were unremarkable, with death delayed up to 11 days. In mice, the LD_{50} ranged between 1600 and 3200 mg/kg, with deaths delayed up to 2 days.

Applied to the skin of guinea pigs for 24 hours, a 10% solution caused only slight erythema at 10 to 20 ml/kg, whereas 5 ml/kg caused no effect. There was no evidence of systemic toxicity following topical application. Triphenyl amine was not a skin sensitizer, as determined by repeated application of a 0.1 M solution to guinea pigs.

The 1995 ACGIH threshold limit value–time-weighted average (TLV-TWA) is 5 mg/m³.

REFERENCES

1. Triphenyl amine. *Documentation of the TLVs and BEIs*, 6th ed, pp 1658–1659. Cincinnati, OH, American Conference of Governmental Hygienists (ACGIH), 1991

TRIPHENYL PHOSPHATE
CAS: 115-86-6

$(C_6H_5O)_3PO_4$

Synonyms: TPP; Celluflex TP

Physical Form. Solid

Uses. Noncombustible substitute for camphor in celluloid; for impregnating roofing paper; plasticizer in lacquers and varnishes

Exposure. Inhalation

Toxicology. Triphenyl phosphate is of low toxicity in humans.

A group of 16 workers exposed to vapor, mist, or dust at an average concentration of 3.5 mg/m³, and occasionally as high as 40 mg/m³, for 8 to 10 years exhibited no signs of illness; the only positive finding was a slight but statistically significant reduction in erythrocyte cholinesterase activity.[1] In workers engaged in the manufacture of aryl phosphates (including tri phenyl phosphates and up to 20% tri-o-cresyl phosphate) and exposed to concentrations of aryl phosphates of 0.2 to 3.4 mg/m³, there was some inhibition of plasma cholinesterase but no correlation of this effect with degree of exposure or with minor gastrointestinal or neuromuscular symptoms.[2,3]

Anecdotal cases of contact dermatitis from triphenyl phosphate have been reported.[4] A positive patch test to 5% triphenyl phosphate occurred in a hobbyist who worked with a plastic glue and had symptoms of psoriasiform dermatitis of both palms.

Of 6 cats given a single intraperitoneal injection of triphenyl phosphate at 0.1 to 0.5 g/kg, 2 developed paralysis after 16 to 18 days.[1] The effects of triphenyl phosphate on the eye have not been reported; application in ethanol to the skin of mice produced no more irritation than was expected from the solvent.[1]

Triphenyl phosphate was not teratogenic or maternally toxic when fed to rats from 4 weeks post-weaning for 91 days through mating and gestation at levels of up to 1% of the diet.[5]

The 1995 ACGIH threshold limit value–time-weighted average (TLV-TWA) for triphenyl phosphate is 3 mg/m³.

REFERENCES

1. Sutton WL et al: Studies on the industrial hygiene and toxicology of triphenyl phosphate. *Arch Environ Health* 1:45–48, 1960
2. Tabershaw IR, Kleinfeld M, Feiner B: Manufacture of tricresyl phosphate and other alkyl phenyl phosphates: an industrial hygiene study. I. Environmental factors. *AMA Arch Ind Health* 15:537–540, 1957
3. Tabershaw IR, Kleinfeld M, Feiner B: Manufacture of tricresyl phosphate and other alkyl phenyl phosphates: an industrial hygiene study. II. Clinical effects of tricresyl phosphate. *AMA Arch Ind Health* 15:541–544, 1957
4. Camarasa JG, Serra-Baldrich E: Allergic con-

tact dermatitis from triphenyl phosphate. *Contact Derm* 26:264–265, 1992
5. Welsh JJ, Collins TFX, Whitby KE, et al: Teratogenic potential of triphenyl phosphate in Sprague–Dawley (Spartan) rats. *Toxicol Ind Health* 3:357–369, 1987

TRIPHENYL PHOSPHITE
CAS: 101-02-0

$P(OC_6H_5)_3$

Synonyms: Phenyl phosphite; triphenoxyphosphine; TPP

Physical Form. Water-white to pale yellow solid (below 22°C) or oily liquid

Uses. Stabilizer/antioxidant for vinyl plastics and polyethylene, polypropylene, styrene copolymers, and rubber

Exposure. Inhalation

Toxicology. Triphenyl phosphite is a skin irritant and sensitizer in humans and is neurotoxic in laboratory animals.

Systemic effects have not been reported in humans.

In an early study in rats, subcutaneous injections of triphenyl phosphite caused two distinct stages of neurotoxic action.[1] The early, rapidly developing stage was characterized by fine or coarse tremor, usually involving the large muscle groups. The tremor disappeared in surviving animals within a few hours. The later stage occurred several days after treatment and was characterized by hyperexcitability, some spasticity, and incoordination, followed by partial flaccid paralysis of the extremities. The posterior extremities were usually more affected.

In the same study, cats received a one-time subcutaneous injection of 0.1, 0.2, 0.3, or 0.5 mg/kg. At the lower doses, the compound produced ataxia and paresis of the extremities after several days. The intermediate dose (0.3

ml/kg) was eventually lethal in two animals, and produced rapidly progressing ataxia on day 6, followed in 1 to 2 days by extensor rigidity. The highest dose (0.5 ml/kg) produced death within 30 hours.

In a later report, rats were injected with two 1.0 ml/kg (1184 mg/kg) subcutaneous injections spaced 1 week apart, and were sacrificed after the second injection.[2] Dysfunctional changes, including tail rigidity, circling, and hind-limb paralysis were noted. However, the pattern of triphenyl phosphite–induced spinal cord damage in conjunction with marginal neurotoxic esterase inhibition suggested that this toxic neuropathy differed from those previously described for organophosphorus-induced delayed neuropathy. Follow-up studies of the central nervous system of rats found widespread axonal and terminal degeneration involving not only the spinal cord and brainstem, as is the case with other organophosphorus compounds, but also the midbrain, thalamus, and cerebral cortex.[3] Subcutaneous injection of triphenyl phosphite in the hen also produced patterns of severe and widespread central nervous system neuropathology, including damage to the spinal cord, brainstem, and other higher-order centers responsible for sensorimotor, visual, and auditory information.[4]

Applied to human skin in patch tests, triphenyl phosphite diluted 1:3 with cold cream produced slight irritation in two-thirds of volunteers tested after a 48-hour contact time. A challenge with the compound 14 days later produced a moderate sensitization reaction.[5] When applied to the skin of laboratory animals, the undiluted chemical was severely irritating and produced moderate sensitization. Instillation of 0.1 ml of triphenyl phosphite into the eyes of rabbits did not produce primary eye irritation.[6] However, another report lists triphenyl phosphite as an eye irritant.[7]

The ACGIH has not established a threshold limit value for triphenyl phosphite.

REFERENCES

1. Smith MI, Lillie RD, Elvove E, Stohlman EF: The pharmacologic action of phosphorous acid

esters of the phenols. *J Pharmacol Exp Therap* 49:78–79, 1933

2. Veronesi B, Padilla S, Newland D: Biochemical and neuropathological assessment of triphenyl phosphite in rats. *Toxicol Appl Pharmacol* 83:203–210, 1986
3. Lehning E, Tanaka D Jr, Bursian SJ: Widespread axonal and terminal degeneration in the forebrain of the rat after exposure to triphenyl phosphite (TPP). *Toxicologist* 10:341–345. 1990
4. Tanaka D Jr, Bursian SJ, Lehning EJ: Neuropathological effects of triphenyl phosphite on the central nervous system of the hen (*Gallus domesticus*). *Fund Appl Toxicol* 18:72–78, 1992
5. Mallette FS, VonHaam E: Studies on the toxicity and skin effects of compounds used in the rubber and plastics industries. *Arch Ind Hyg Occup Med* 5:311–317, 1952
6. Borg-Warner Chemicals: FYI-OTS-0785-0422 FLWP. Sequence D. Primary eye irritation tests of triphenyl phosphite in rabbits. Washington, DC, US Environmental Protection Agency, Office of Toxic Substances, 1980
7. Sandmeyer EE, Kirwin CJ: Esters. In Clayton GD, Clayton FE (eds): *Patty's Industrial Hygiene and Toxicology*, 3rd ed, rev, Vol 2A, *Toxicology*, pp 2362–2377. Wiley-Interscience, 1981

TUNGSTEN AND COMPOUNDS
CAS: 7440-33-7

W

Synonyms: Wolfram

Compounds: Tungsten carbide; tungsten sulfide; tungsten carbonyl; tungsten chloride; tungsten fluoride; tungsten oxychloride; tungsten silicide; tungsten oxide; tungstic acid; various tungstates

Physical Form. Gray, hard, brittle metal

Uses. In ferrous and nonferrous alloys; filaments in incandescent lamps; heating elements; welding electrodes; in manufacture of abrasives and tools; in manufacture of textiles and ceramics

Exposure. Inhalation

Toxicology. The soluble compounds of tungsten are distinctly more toxic than the insoluble forms.

Tungsten and tungsten carbide are considered inert dusts. Tungsten metal and tungsten carbide have not appeared to exert a significant effect on the respiratory system. Studies in a number of factories producing tungsten carbide products have found increased incidence of pulmonary fibrosis. This "hard-metal disease," however, is thought to be related to cobalt exposure, with which tungsten carbide is fused. It has been suggested that tungsten carbide might enhance the solubility of cobalt in protein-containing fluid.[1] A study by Russian investigators reportedly indicated an incidence of pulmonary fibrosis of 9% to 11% among employees exposed to tungsten and not cobalt, but no details are available.[2]

No allergic reactions to tungstate were observed in patch testing of 853 individuals who were working, or had worked, with tungsten carbide in hard-metal manufacture. Irritant pustular reactions appeared in 2% of the patch tests.[3]

No acute effects were produced in rats following intratracheal injection with 5% suspensions of metallic tungsten powder and of tungsten carbide. Following intratracheal instillation of tungsten carbide, no cellular reaction (other than that expected from an inert dust) was observed in rats during an 18-week follow-up period.[1] Focal interstitial pneumonitis and bronchiolitis were observed in guinea pigs following intratracheal injection of tungsten metal in three 50 mg doses. Nearly complete recovery was observed after 1 year. Only negligible reactions were observed following the same treatment with tungsten carbide dust.[1] Inserted into rabbit eyes for 1 year, tungsten caused no reaction and was classified as completely inert.[4]

There are no reports of occupational exposure to soluble compounds of tungsten, but they show considerable systemic toxicity in animal experiments.[2] The LD_{50} of sodium tungstate by subcutaneous injection in rats is 140 to 160 mg/kg as tungsten. Both sodium tungstate and tungsten oxide were lethal to rats when fed a diet containing 0.5% as tungsten. Guinea pigs treated orally or intravenously with tungsten or sodium tungstate experienced anorexia, colic, weight loss, incoordination, trembling, and dyspnea prior to a delayed death.

The 1995 ACGIH threshold limit value–time-weighted average (TLV-TWA) is 1 mg/m^3, as W for soluble compounds, with a short-term excursion limit of 3 mg/m^3, and 5 mg/m^3 as W for insoluble compounds, with a short-term excursion limit of 10 mg/m^3.

REFERENCES

1. Kazantzis G: Tungsten. In Friberg L, Nordberg GF, Vouk VB (eds): *Handbook on the Toxicology of Metals*, 2nd ed, pp 610–621. Amsterdam, Elsevier, 1986
2. National Institute for Occupational Safety and Health: *Criteria for a Recommended Standard . . . Occupational Exposure to Tungsten and Cemented Tungsten Carbide*. Department of Health, Education and Welfare, Pub No 77-227, NTIS Pub No PB-275-594. Springfield, VA, National Technical Information Service (NIOSH), 1977
3. Rystedt I, Fischer T, Lagerholm B: Patch testing with sodium tungstate. *Contact Derm* 9:69–73, 1983
4. Beliles R: The metals. In Clayton GD, Clayton FE (eds): *Patty's Industrial Hygiene and Toxicology*, 4th ed, Vol 2C, *Toxicology*, pp 2289–2297, 1994

TURPENTINE
CAS: 8006-64-2

$C_{10}H_{16}$

Synonyms: Spirit of turpentine; oil of turpentine; wood turpentine

Physical Form. Volatile liquid, colorless or yellow, which varies in composition according to its source and method of production. A typical analysis of turpentine is: α-pinene, 82.4%; camphene, 8.7%; β-pinene, 2.1%; unidentified natural turpenes, 6.8%

Uses. Solvent

Exposure. Inhalation; skin absorption; ingestion

Toxicology. Turpentine is a skin and mucous membrane irritant and a central nervous system depressant.

Several human subjects had nose and throat irritation at exposures of 75 ppm for 3 to 5 minutes; 175 ppm was intolerable to the majority.[1] Although often reported in the older literature, toxic nephrosis, characterized by albuminuria, dysuria, hematuria, and glycosuria, is seldom seen today with turpentine overexposure.[2] The apparent rarity of renal lesions in current poisonings may be related to the change in composition of domestic turpentine; turpentine is now more "pure" because of the removal of a hydroperoxide of δ-3-carene.[2]

The mean lethal dose for humans by ingestion probably lies between 120 and 180 ml.[2] Symptoms include burning pain in the mouth and throat, abdominal pain, nausea, vomiting and, occasionally, diarrhea.[2] Central nervous system effects are excitement, ataxia, confusion, and stupor. Convulsions may occur several hours after ingestion. Fever and tachycardia are common, and death is usually attributed to respiratory failure.[2]

The liquid may cause conjunctivitis and corneal burns.[3] Turpentine from any source is a skin irritant if allowed to remain in contact for a sufficient length of time; hypersensitivity occurs in some persons.[3] A study of nearly 85,000 patients between 1979 and 1988 from 5 different countries found that less than 1.8% had positive patch tests to 10% turpentine in oil.[4] The liquid can be absorbed by the skin and mucous membranes, and intoxication by this route has been reported.[2]

The LC_{50} for rats was 3590 ppm for 1 hour and 2150 ppm for 6 hours; hyperpnea, ataxia, tremor, and convulsions were noted.[5] Mucous membrane irritation, particularly of the eyes, and mild convulsions were observed in cats exposed to 540 to 720 ppm for a few hours.[2]

The 1995 ACGIH threshold limit value–time-weighted average (TLV-TWA) is 100 ppm (556 mg/m^3).

REFERENCES

1. Nelson KW et al: Sensory response to certain industrial solvent vapors. *J Ind Hyg Toxicol* 25:282–285, 1943
2. Gosselin RE, Smith RP, Hodge HC: *Clinical Toxicology of Commercial Products, Section III*, 5th ed, pp 393–394. Baltimore, MD, Williams and Wilkins, 1984
3. Hygienic Guide Series: Turpentine. *Am Ind Hyg Assoc J* 28:297–300, 1967
4. Rudzki E, Berova N, Czernielewski A, et al: Contact allergy to oil of turpentine: a 10-year retrospective view. *Contact Derm* 24:317–318, 1991
5. Sperling F, Marcus WL, Collins C: Acute effects of turpentine vapors on rats and mice. *Toxicol Appl Pharmacol* 10:8–20, 1967

URANIUM

CAS: 7440-61-1

U

Synonyms: Soluble: Uranyl nitrate; uranyl fluoride; uranium hexafluoride. *Insoluble:* Uranium dioxide; uranium tetrafluoride

Physical Form. Solids

Uses. Nuclear fuel; in weapons systems; in photography; cayalyst

Exposure. Inhalation

Toxicology. Insoluble compounds of uranium are respiratory irritants, whereas soluble compounds are also toxic to the kidneys.

Soluble Compounds: Animals repeatedly exposed to dusts of soluble uranium compounds in concentrations from 3 to 20 mg/m^3 died of pulmonary and renal damage; both feeding and percutaneous toxicity studies on animals indicated that the more soluble compounds are the more toxic.[1] In animals, effects on the liver are a consequence of the acidosis and azotemia induced by renal dysfunction.[1]

Animal studies indicate that the primary toxic effect of uranium exposure is on the kidney, with particular damage to the proximal tubules. Functionally, this may result in increased excretion of glucose and amino acids. Structurally, the necrosis of the tubular epithelium leads to the formation of cellular casts in the urine. If exposure is insufficient to cause death from renal failure, the tubular lesion is reversible with epithelial regeneration. Although bone is the other major site of deposition, there is no evidence of toxic or radiocarcinogenic effects to bone or bone marrow from experimental studies.[2]

Insoluble Compounds: These compounds are generally considered to be less toxic than the soluble compounds.[3] Repeated exposures of three animal species to uranium dioxide dust at a concentration of 5 mg uranium/m^3 for periods of up to 5 years resulted in no kidney injury. More than 90% of the uranium found in the body was in the lungs and tracheobronchial lymph nodes (TLN).[2] Fibrotic changes suggestive of radiation injury were seen occasionally in the TLN of dogs and monkeys and in the lungs of monkeys after exposure periods longer than 3 years; the estimated alpha dose to tissues was greater than 500 rads for lungs and 7000 rads for TLN.[4]

Uranium: Rats injected with metallic uranium in the femoral marrow and in the chest wall developed sarcomas; whether this was due to metallocarcinogenic or radiocarcinogenic action could not be determined.[1] The increased incidence of lung cancer reported among uranium miners is probably the result of exposure to radon gas and its particulate daughters rather than to uranium dust.[5] In some mining cohorts, part of the lung cancer excess has also been attributed to arsenic exposure, which occurred in men who worked in both uranium and gold mines.[6]

In a group of uranium mill workers, there was an excess of deaths from malignant disease of lymphatic and hematopoietic tissue; data from animal experiments suggested that this excess may have resulted from irradiation of lymph nodes by thorium-230, a disintegration product of uranium.[5] Some absorbed uranium is deposited in bone. A potential risk of radiation effects on bone marrow has been postulated, but extensive clinical studies on exposed workers have disclosed no hematologic abnormalities.[7,8]

Accidental exposure of workers to a mixture of uranium hexafluoride, uranyl fluoride, hydrofluoric acid, and live steam caused lacrimation, conjunctivitis, shortness of breath, paroxysmal cough, rales in the chest, nausea, vomiting, skin burns, transitory albuminuria, and elevation of blood urea nitrogen.[7] Two deaths occurred among the most heavily exposed workers shortly after exposure. The persons having the greatest exposure showed the highest urinary uranium levels. In addition, their urinary abnormalities were the most severe, including albuminuria plus red cells and casts in the urinary sediment, and blood urea nitrogen remained elevated for several weeks. The injurious effects observed on the skin, eyes, and respiratory tract were apparently caused by the irritant action of the hydrofluoric acid, whereas the uranium was believed to be responsible for the transient renal changes.

No evidence of chronic toxicity, either chemical or radiation, was observed for any uranium compound during the first 6 years of the US atomic energy program; all exposed workers were under very close medical surveillance.[1] Several uranium compounds tested on the eyes of animals caused severe eye damage as well as systemic poisoning. The anion and its hydrolysis products determine the degree of injury.[9,10] A hot nitric acid solution of uranyl nitrate spilled on the skin caused skin burns, nephritis, and heavy-metal encephalopathy.[9] Prolonged skin contact with uranium compounds should be avoided because of potential radiation damage to basal cells. Dermatitis has occurred as a result of handling uranium hexafluoride.[9]

In genotoxic assays, significant increases in frequencies of chromosomal aberrations in peripheral lymphocytes have been reported in uranium miners.[3] This effect has been attributed to radon daughter products and, more recently, to mutagenic mycotoxins produced by molds and present in the uranium mines.[11]

Oral administration of 3 mg uranium/kg/ day as uranyl acetate dihydrate to pregnant mice on gestation days 6 through 15 caused an increase in fetotoxicity (stunted fetuses, external and skeletal malformations, and developmental variations) and maternal toxicity.[12] In reproductive studies, no adverse effects were observed in testicular function or spermatogenesis in male mice treated with up to 80 mg/kg/day uranyl acetate dihydrate for 64 day.[13]

The 1995 ACGIH threshold limit value–time-weighted average (TLV-TWA) for uranium (soluble and insoluble compounds as U) is 0.2 mg/m^3, with a short-term excursion limit of 0.6 mg/m^3.

REFERENCES

1. Beliles RP: The metals. In Clayton GD, Clayton FE (eds): *Patty's Industrial Hygiene and Toxicology*, 4th ed, Vol IIC, *Toxicology*, pp 2300–2317. New York, Wiley-Interscience, 1994

2. US Environmental Protection Agency: Drinking Water Criteria Document for Uranium. Washington, DC, Office of Drinking Water, F-198, 1985

3. Agency for Toxic Substances and Disease Registry (ATSDR): *Toxicological Profile for Uranium.* US Department of Health and Human Services, Public Health Service, p 201, TP-90-29, 1990

4. Leach LJ, Maynard EA, Hodge CH, et al: A five-year inhalation study with natural uranium dixoide (UO$_2$) dust. I. Retention and biologic effect in the monkey, dog and rat. *Health Phys* 18:599–612, 1970

5. Archer VE, Wagoner JK, Lundin FE Jr: Cancer mortality among uranium mill workers. *J Occup Med* 15:11–14, 1973

6. Kusiak RA, Ritchie AC, Muller J, et al: Mortality from lung cancer in Ontario uranium miners. *Br J Ind Med* 50:920–928, 1993

7. Voegtlin C, Hodge HC: *Pharmacology and Toxicology of Uranium Compounds*, Vol 1, pp 413–414. New York, McGraw-Hill, 1949

8. Voegtlin C, Hodge HC: *Pharmacology and Toxicology of Uranium Compounds*, Vol. 2, pp 687–689, 993–1017. New York, McGraw-Hill, 1949

9. Hygienic Guide Series: Uranium (natural) and its compounds. *Am Ind Hyg Assoc J* 30:313–317, 1969

10. Grant WM: *Toxicology of the Eye*, 2nd ed, p 1073. Springfield, IL, Charles C Thomas, 1974

11. Sram RJ, Dobias L, Rossner P, et al: Monitoring genotoxic exposure in uranium mines. *Environ Health Perspect* 101(Suppl 3):155–158, 1993
12. Domingo JL, Paternain JL, Llobet JM, et al: The developmental toxicity of uranium in mice. *Arch Environ Health* 44:395–398, 1989
13. Llobet JM, Sirvent JJ, Ortega A, et al: Influence of chronic exposure to uranium on male reproduction in mice. *Fund Appl Toxicol* 16:821–829, 1991

---

n-VALERALDEHYDE
CAS: 110-62-3

$C_5H_{10}O$

Synonyms: Amyl aldehyde; butyl formal; pentanal; valeric aldehyde

Physical Form. Colorless liquid

Uses. In food flavorings; in the acceleration of rubber vulcanization

Exposure. Inhalation; ingestion

Toxicology. *n*-Valeraldehyde has low systemic toxicity but is considered an eye and skin irritant.

No effects from exposure have been reported in humans.

The oral LD_{50} for rats was 4.6 g/kg, and the rabbit dermal LD_{50} was 4.9 g/kg.[1] Three of 6 rats succumbed to 4 hours of exposure at 4000 ppm. Ten-hour exposure at 670 ppm caused some deaths in mice and guinea pigs but not in rabbits.[2] Mice exposed in a head-only exposure chamber at 1100 ppm for 10 minutes had a 50% decrease in respiratory rate.[3]

Applied to the rabbit eye or guinea pig skin, the liquid was severely irritating. An odor threshold of 0.028 ppm has been reported.[4]

The 1995 ACGIH threshold limit value–time-weighted average (TLV-TWA) for *n*-valeraldehyde is 50 ppm (176 mg/m³).

REFERENCES

1. Smyth HF Jr, Carpenter CP, Weil CS, et al: Range-finding toxicity data: List VII. *Am Ind Hyg Assoc J* 30:470–476, 1969
2. Salem H, Cullumbine H: Inhalation toxicities of some aldehydes. *Toxicol Appl Pharmacol* 2:183–187, 1960
3. Steinhagen WH, Barrow CS: Sensory irritation structure-activity study of inhaled aldehydes in B6C3F₁ and Swiss–Webster Mice. *Toxicol Appl Pharmacol* 72:495–503, 1984
4. Amoore JE, Hautala E: Odor as an aid to chemical safety: odor thresholds compared with threshold limit values and volatilities for 214 industrial chemicals in air and water dilution. *J Appl Toxicol* 3:272–290, 1983

VANADIUM PENTOXIDE
CAS: 1314-62-1

V_2O_5

Synonyms: Vanadic anhydride; divanadium pentoxide; vanadium oxide; vanadic acid

Physical Form. Yellow to rust-brown crystals

Uses. In the production of high-strength steel alloys; catalyst in oxidation reactions; in pesticides; in dyes and inks

Exposure. Inhalation

Toxicology. Vanadium pentoxide primarily affects the respiratory system.

Fume is recognized as being generally more toxic than dust because of the smaller particle size of fume, which allows more complete penetration to the small airways of the lungs.

Sixteen workers exposed to concentrations of dust (and possibly some fume) in excess of 0.5 mg/m³, with particle sizes ranging from 0.1 to 10 µ, developed conjunctivitis, nasopharyngitis, hacking cough, fine rales, and wheezing. In 3 workers exposed to the highest concentrations, the onset of symptoms occurred at the end of the first workday.[1] The bronchospastic element in the more seriously ill persisted for 48 hours after removal from exposure; rales lasted for 3 to 7 days and, in several cases, the cough lasted for up to 14 days.[1] Among those with acute intoxication, there was dramatically increased severity of symptoms from repeated exposures of less time and intensity.

Absorbed vanadium is primarily excreted in the urine, and it was detectable in 12 of the workers for periods of up to 2 weeks. Urinary vanadium concentrations were elevated in workers exposed to mean air concentrations of 0.1 to 0.28 mg/m³, but there was no correlation between the air and urinary concentrations. Although most absorbed vanadium was excreted within 1 day after cessation of exposure, increased excretion relative to unexposed controls continued for more than 2 weeks among chronically exposed workers.[2]

Workers exposed to a mixture of ammonium metavanadate and vanadium pentoxide at concentrations near 0.25 mg/m³ developed green tongue, metallic taste, throat irritation, and cough.[3] Of 36 workers examined 8 years after their original exposure to vanadium pentoxide, there was no evidence of either pneumoconiosis or emphysema, although 6 of the workers still had bronchitis with rhonchi resembling asthma and bouts of dyspnea.[4]

Two volunteers exposed to a concentration of 1 mg/m³ for 8 hours developed a persistent cough, which lasted for 8 days; 21 days after the original exposure, re-exposure for 5 minutes to a heavy cloud of vanadium pentoxide dust occurred and, within 16 hours, marked cough developed; the following day, rales and expiratory wheezes were present throughout the entire lung field, but pulmonary function was normal.[5] Subjects exposed to a concentration of 0.2 mg/m³ for 8 hours developed a loose cough the following morning; other subjects exposed for 8 hours to 0.1 mg/m³ developed a slight cough with increased mucus, which lasted 3 to 4 days.

Although there have been some cases of emphysema observed among workers with exposure to vanadium pentoxide, other possible causes, such as smoking, were not excluded.

Cases of asthma have occurred more frequently, suggesting that this may be an effect of chronic exposure.[3] Recent animal studies in cynomolgus monkeys have not found evidence of increased pulmonary reactivity to vanadium pentoxide following repeated exposures; cytological/immunological and skin test results also indicated the absence of allergic sensitization.[6]

Exposure to the dust can cause eye irritation, and skin rashes have been reported. Green discoloration of the tongue may occur as a result of direct deposition of vanadium.[7]

No adverse effects on fertility, reproduction, or parturition were found when male and female rats were treated with sodium metavanadate by gavage and then mated.[8]

Vanadium pentoxide was genotoxic in a number of assays, inducing gene mutations in bacteria.[7]

The 1995 ACGIH threshold limit value–time-weighted average (TLV-TWA) for vanadium pentoxide is 0.05 mg/m³ as respirable dust or fume.

REFERENCES

1. Zenz C, Bartlett JP, Thiede WH: Acute vanadium pentoxide intoxication. *Arch Environ Health* 5:542–546, 1962
2. Kiviluoto M: Serum and urinary vanadium of workers processing vanadium pentoxide. *Int Arch Occup Environ Health* 48:251–256, 1981
3. National Institute for Occupational Safety and Health: *Criteria for a Recommended Standard . . . Occupational Exposure to Vanadium.* DHEW (NIOSH) Pub No 77-222, p 142. Washington, DC, US Government Printing Office, 1977
4. Sjoberg SG: Follow-up investigation of workers at a vanadium factory. *Acta Med Scand* 154:381, 1956
5. Zenz C, Berg BA: Human responses to controlled vanadium pentoxide exposure. *Arch Environ Health* 14:709–712, 1967
6. Knecht EA, Moorman WJ, Clark JC, et al: Pulmonary reactivity to vanadium pentoxide following subchronic inhalation exposure in a non-human primate animal model. *J Appl Toxicol* 12:427–434, 1992
7. Agency for Toxic Substances and Disease Registry (ATSDR): *Toxicological Profile for Vanadium.* US Department of Health and Human

Services, Public Health Service, TP-91/29, p 106, 1992

8. Domingo JL, Paternain JL, Llobet JM, et al: Effects of vanadium reproduction, gestation, parturition, and lactation in rats upon oral administration. *Life Sci* 39:819–824, 1986

VINYL ACETATE
CAS: 108-05-4

$C_4H_6O_2$

Synonyms: 1-Acetoxyethylene; acetic acid ethenyl ester; ethanoic acid; ethenyl ethanoate; vinyl ethanoate

Physical Form. Colorless liquid that polymerizes to a transparent solid on exposure to light

Uses. In the production of vinyl acetate polymers

Exposure. Inhalation

Toxicology. Vinyl acetate is an irritant of the eyes, nose, and throat; it is carcinogenic in experimental animals at high doses. Volunteers exposed to vinyl acetate showed a wide variation in individual sensitivity to its irritant effects; 1 of 3 had throat irritation at 20 ppm for 4 hours, whereas 72 ppm for 30 minutes produced eye irritation in 3 of 4 participants.[1] All subjects agreed that they could not work at 72 ppm for 8 hours.

From a study of 21 workers exposed for an average of 15 years at concentrations of 5 to 10 ppm (with occasional excursions above 300 ppm), vinyl acetate produced no serious chronic effects.[2] Some subjects were sensitive at concentrations of about 6 ppm, and concentrations above 20 ppm produced irritation in most persons.

Prolonged dermal contact, such as that afforded by clothing wet with vinyl acetate, may result in severe irritation or blistering of the skin in some persons.[1] Direct eye contact with the liquid or vapor can cause irritation of the eyes.[3]

The LC_{50} for 4 hours in rats was 14,000 mg/m^3 (about 4667 ppm).[4] Dogs exposed 6 hours daily for several weeks, starting at 91 ppm and ending after 11 weeks at 186 ppm, exhibited eye irritation and lacrimation.[5] Rats exposed repeatedly to 100 ppm showed no effects.[6]

Rats administered 0, 200, 1000, or 5000 ppm in the drinking water from the time of gestation up to 104 weeks showed no evidence of systemic organ toxicity and/or carcinogenicity.[7] Decreased food consumption and concurrent body weight decrement were observed in the high-dose group. In rats and mice exposed at 0, 50, 200, or 600 ppm by inhalation for 2 years, significant histopathological changes were noted in the nasal cavity at the two highest dose levels.[8] Epithelial atrophy, basal-cell hyperplasia, and regenerative effects (squamous metaplasia and respiratory metaplasia of the olfactory epithelium) were observed in both species. In rats, the total tumor incidence in the high-exposure group was 9% and included papillomas, squamous-cell carcinoma and carcinoma in situ in olfactory regions, and papillomas of the respiratory region. It has been noted that the unique nature, both structurally and functionally, of the rodent nasal cavity may make it an unsuitable model for assessing human risk.

One human study of workers in a US synthetic chemical plant failed to find any specific association between exposure to vinyl acetate and excess lung cancer.[9]

Vinyl acetate has been tested for teratogenicity in inhalation and oral assays.[3] Pregnant rats exposed to levels as high as 1000 ppm by inhalation or 5000 ppm in drinking water on gestation days 6 to 15 had significantly reduced weight gain during exposure. The fetuses of the rats exposed via inhalation were also significantly smaller than control fetuses and had an increased incidence of minor skeletal defects. However, investigators thought that the fetal effects were a consequence of the maternal growth retardation and not of vinyl acetate treatment. In the drinking water study, there were no significant effects on the fetuses, and the investigators concluded that vinyl acetate did not elicit embryolethality, embryotoxicity, or teratogenicity.

Vinyl acetate was genotoxic in a number of mammalian system assays, inducing micronuclei, chromosomal aberrations, sister chromatid exchange, and DNA cross-links.[3]

The 1995 ACGIH threshold limit value–time-weighted average for vinyl acetate is 10 ppm (35 mg/m^3) with a short-term excursion limit of 15 ppm (53 mg/m^3) and an A3 animal carcinogen designation.

REFERENCES

1. National Institute for Occupational Safety and Health: *Criteria for a Recommended Standard . . . Occupational Exposure to Vinyl Acetate.* DHEW (NIOSH) Pub No 78-205, p 78. Washington, DC, US Government Printing Office, 1978
2. Deese DE, Joyner RE: Vinyl acetate—a study of chronic human exposure. *Am Ind Hyg Assoc J* 30:449, 1969
3. Agency for Toxic Substances and Disease Registry (ATSDR): *Toxicological Profile for Vinyl Acetate.* US Department of Health and Human Services, Public Health Service, TP-91-30, p 140, 1992
4. Carpenter CP et al: The assay of acute vapor toxicity, and the grading and interpretation of results on 96 chemical compounds. *J Ind Hyg Toxicol* 31:343–346, 1949
5. Haskell Laboratory: Report of toxicity of vinyl acetate. Wilmington, DE, EI DuPont de Nemours & Co, January 1967
6. Gage JC: The subacute inhalation toxicity of 109 industrial chemicals. *Br J Ind Med* 27:1–18, 1970
7. Bogdanffy MS, Tyler TR, Vinegar MB, et al: Chronic toxicity and oncogenicity study with vinyl acetate in the rat: in utero exposure in drinking water. *Fund Appl Toxicol* 23:206–214, 1994
8. Bogdanffy MS, Dreef-Van Der Meulen HC, Beems RB, et al: Chronic toxicity and oncogenicity inhalation study with vinyl acetate in the rat and mouse. *Fund Appl Toxicol* 23:215–229, 1994
9. Waxweiler RJ, Smith AH, Falk H, et al: Excess lung cancer risk in a synthetic chemicals plant. *Environ Health Perspect* 41:159–165, 1981

VINYL BROMIDE
CAS: 593-60-2

C$_2$H$_3$Br

Synonyms: Bromoethene; bromoethylene

Physical Form.   Gas

Uses.   In the production of flame-resistant plastics and thermoplastic resins

Exposure.   Inhalation

Toxicology.   Vinyl bromide causes central nervous system (CNS) depression in animals at high levels and has a carcinogenic action in rats.

There are no data on human exposures.

Exposure of rats to 100,000 ppm for 15 minutes resulted in deep anesthesia and death.[1] Exposure to 50,000 ppm caused anesthesia in 25 minutes and was lethal after exposure for 7 hours.

A significant decline in animal body weights was the only treatment-related effect following exposure at 10,000 ppm, 7 hr/day, for 4 weeks. In a 6-month inhalation study in a number of species, serum bromide levels increased following exposure to 250 and 500 ppm.

In male and female rats exposed to 10, 50, 250, or 1250 ppm vinyl bromide in a lifetime inhalation study, there was a dose-related increase in angiosarcomas of the liver in both sexes.[2] A significant increase in hepatocellular neoplasms was also seen in male rats exposed at 250 ppm and in female rats exposed at 10, 50, and 250 ppm. The lack of increase in hepatocellular neoplasms in rats at the 1250 level was probably due to their early mortality and termination at 72 weeks. In limited mice studies, no local tumors were produced by skin application or subcutaneous administration.[3] The IARC has determined that there is sufficient evidence for the carcinogenicity of vinyl bromide to experimental animals, but no data are available for humans.[3]

The liquid was moderately irritating to the rabbit eye but essentially nonirritating to the skin.

The 1995 threshold limit value–time-weighted average (TLV-TWA) for vinyl bromide is 5 ppm (22 mg/m^3) with an A2 suspected human carcinogen designation.

REFERENCES

1. Leong BKJ, Torkelson TR: Effects of repeated inhalation of vinyl bromide in laboratory animals with recommendations for industrial handling. *Am Ind Hyg Assoc J* 31:1–11, 1970
2. Benya TJ et al: Inhalation carcinogenicity bioassay of vinyl bromide in rats. *Toxicol Appl Pharmacol* 64:367–379, 1982
3. *IARC Monographs on the Evaluation of the Carcinogenic Risk of Chemicals to Humans*, Vol 39, Some chemicals used in plastics and elastomers, pp 133–145. Lyon, International Agency for Research on Cancer, 1986

VINYL CHLORIDE
CAS: 75-01-4

C$_2$H$_3$Cl

Synonyms: Chlorethene; chloroethylene; ethylene monochloride

Physical Form. A colorless gas, but usually handled as a liquid under pressure

Uses. In the production of polyvinyl chloride resins; in organic synthesis

Exposure. Inhalation

Toxicology. Occupational exposure to vinyl chloride is associated with an increased incidence of angiosarcoma of the liver and other malignant tumors, acro-osteolysis, Raynaud's syndrome, scleroderma, thrombocytopenia, circulatory disturbances, and impaired liver function. Very high concentrations cause central nervous system (CNS) depression.

Humans exposed to 20,000 ppm for 5 minutes experienced dizziness, light-headedness, nausea, and dulling of vision and auditory cues.[1] For 5-minute exposures, 8000 ppm caused some dizziness although 4000 ppm was without effect. Longer exposures at 1000 pm may cause drowsiness, faltering gait, visual disturbances, and numbness and tingling in the extremities.[2]

Chronic exposure to high levels of vinyl chloride vapor has resulted in a syndrome termed vinyl chloride disease, which includes the following symptoms: enhanced collagen deposition and thickening of the subepidermal layer of the skin; Raynaud's phenomenon (arteriole constriction causing whitening of the fingers and numbness); and, in some cases, acro-osteolysis (resorption of the terminal phalanges).[3] Raynaud's phenomenon was often the first manifestation noted by a majority of subjects with vinyl chloride disease, suggesting that the vascular lesion anteceded the bone changes in most cases.[4] Radiologic findings in patients with acro-osteolysis included lytic lesions in the distal phalanges of the hands, in the styloid processes of the ulna and radius, and in the sacroiliac joints.[4] Vinyl chloride disease has been associated with exposure to several hundred parts per million for periods ranging from months to years; no new cases have been reported in the United States since 1974, when occupational exposure levels were reduced to 1 ppm.[3] Other effects in exposed workers include thombocytopenia; hepatic changes, including hypertrophy and hyperplasia of hepatocytes and fibrosis; and increased levels of circulating immune complexes.[3] Of 20 autoclave cleaners with exposure to vinyl chloride, 16 had thrombocytopenia, 7 had splenomegaly, 6 had hepatomegaly, 4 had fibrosis of the liver capsule, and 4 had signs of acro-osteolysis.[5]

Vinyl chloride has been associated with cancer in humans in a number of epidemiological studies. In four facilities engaged in the polymerization of vinyl chloride for at least 15 years, workers exposed for at least 5 years had a significant number of excess deaths due to malignant neoplasms (35 deaths observed vs. 23.5 expected).[6] The excesses were found for four organ systems: CNS (3 observed vs. 0.9 expected), respiratory system (12 observed vs. 7.7 expected); hepatic system (7 observed vs. 0.6 expected), and lymphatic and hematopoietic systems (4 observed vs. 2.5 expected).

By 1975, over 30 cases of angiosarcoma of the liver had been reported among vinyl chloride polymerization workers in the United States and nine other nations.[7] Because this tumor is extremely rare, the occurrence of these cases under similar occupational conditions strongly suggests a causal relationship to some phase of vinyl chloride production.[8] Clinical features of 7 patients with the malignancy varied from no signs or symptoms to weakness, pleuritic pain, abdominal pain, weight loss, gastrointestinal bleeding, and hepatosplenomegaly. Liver-function abnormalities were present in all subjects but without a consistent pattern.[8] In addition to the malignant tumors, four cases of nonmalignant hepatic disease characterized by portal fibrosis and portal hypertension have been attributed to vinyl chloride exposure.[8]

A large multicentric cohort study of European vinyl chloride workers revealed a nearly threefold increase in liver cancer based on 24 observed deaths versus 8.4 expected. The excess was clearly related to time since first exposure, duration of employment, and estimated ranked and quantitative exposures.[9] A recent cohort study of 10,173 US men who had worked at least 1 year in jobs involving exposure to vinyl chloride confirmed a significant mortality excess in angiosarcoma (15 deaths), cancer of the liver and bilary tract (SMR=641), and cancer of the central nervous system (SMR=180).[10]

The tumorigenic potential of vinyl chloride has been confirmed in a number of animal studies. Zymbal gland carcinomas, nephroblastomas, and angiosarcomas were the prevailing tumors in treated rats. Results ranged from a 16% tumor incidence at an exposure level of 250 ppm to a 39% incidence at 10,000 ppm. In mice, liver angiosarcomas, pulmonary adenomas, and mammary carcinomas were observed after exposures ranging from 50 to 10,000 ppm.[11] The development of some tumors depended more on duration of exposure than on concentration of vinyl chloride.[11]

Vinyl chloride was genotoxic in a variety of assays, both in vivo and in vitro.[3] A number of studies have found an increase in chromosomal aberrations in cultured lymphocytes of exposed workers. The chromatid breaks appear to be nonrandom, suggesting that vinyl chloride or its metabolites interact with specific areas of the genome.

The IARC has determined that there is sufficient evidence for carcinogenicity to humans and to animals.[12]

Developmental effects in animals, consisting of resorptions, decreased litter size, and reduced fetal weight, have generally been observed at doses that produce some maternal toxicity.[3] In humans, prenatal loss and congenital abnormalities have been reported in some communities located near vinyl chloride plants, but no association between effects and exposure have been established.

The 1995 ACGIH threshold limit value–time-weighted average (TLV-TWA) for vinyl chloride is 5 ppm (13 mg/m^3) with an A1 confirmed human carcinogen designation.

REFERENCES

1. Lester D, Greenberg LA, Adams WR: Effects of single and repeated exposures of humans and rats to vinyl chloride. *Am Ind Hyg Assoc J* 24:265–275, 1963
2. Easter MD, Von Burg R, et al: Toxicology update: vinyl chloride. *J Appl Toxicol* 14:301–307, 1994
3. Agency for Toxic Substances and Disease Registry (ATSDR): Toxicological Profile for Vinyl Chloride. US Department of Health and Human Services, Public Health Service, p 157, TP-92/20, 1993
4. Dodson VN et al: Occupational acroosteolysis. III. A clinical study. *Arch Environ Health* 22:83–91, 1971
5. Marstellar HJ, Lelbach WK, Muller R, et al: [Chronic toxic liver lesions in the PVC (polyvinyl chloride)-producing workers]. Deutsch Med Wochenschrift 98:2311–2314, 1973
6. Waxweiler RJ, Stringer W, Wagner JK, et al: Neoplastic risk among workers exposed to vinyl chloride. *Ann NY Acad Sci* 271:40–48, 1976
7. Lloyd JW: Angiosarcoma of the liver in vinyl chloride/polyvinyl chloride workers. *J Occup Med* 17:333–334, 1975
8. Falk H, Creech JL Jr, Heath CW Jr, et al:

Hepatic disease among workers at a vinyl chloride polymerization plant. *JAMA* 230: 59–63, 1974

9. Simonato L, L'Abbe KA, Anderson A, et al: A collaborative study of cancer incidence and mortality among vinyl chloride workers. *Scand J Work Environ Health* 17:159–169, 1991

10. Wong O, Whorton MD, Foliart DE, et al: An industry-wide epidemiologic study of vinyl chloride workers, 1942–1982. *Am J Ind Med* 20:317–334, 1991

11. Maltoni C, Lefemine G: Carcinogenicity bioassays of vinyl chloride. I. Research plan and early results. *Environ Res* 7:387–405, 1974

12. *IARC Monographs on the Evaluation of Carcinogenic Risk to Humans, Overall Evaluations of Carcinogenicity: An Updating of IARC Monographs Vols 1–42*, Suppl 7, pp 373–376. Lyon, International Agency for Research on Cancer, 1987

4-VINYLCYCLOHEXENE
CAS: 100-40-3

C_8H_{12}

Synonyms: 4-Ethenylcyclohexene; 1-vinyl-3-cyclohexene; VCH

Physical Form. Colorless liquid

Uses/Sources. Intermediate in the production of flame retardants, flavors, fragrances, and vinyl cyclohexene dioxide (which is used in the manufacture of epoxy resins); found in gases discharged during the process of curing rubber in tire manufacturing

Exposure. Inhalation; skin absorption

Toxicology. 4-Vinylcyclohexene (VCH) is a moderate skin irritant and causes ovarian toxicity in mice. It is carcinogenic in some animal species when metabolically activated.

In one isolated report, Russian rubber workers exposed to concentrations averaging 271 to 542 ppm, with excursions to 677 ppm, were reported to suffer keratitis, rhinitis, headache, leukopenia, neutrophilia, lymphocytosis, and impairment of pigment and carbohydrate metabolism.[1] It is not clear what other confounding chemical exposures may have been concurrent. It has also been noted that these exposure levels contrast significantly with those found in the discharged off-gases to which domestic rubber workers were exposed in the tire-curing process and which measured 118 ppb.[2]

In rats, the oral LD_{50} was 2.6 g/kg.[3] Inhalation of 8000 ppm for 4 hours was lethal to 4 of 6 rats.[3] The liquid produced moderate irritation when applied to the skin of rabbits and caused a small necrotic area on the cornea when instilled into a rabbit eye.

In 14-day studies, all rats and most mice died when administered VCH by gavage in corn oil at doses greater than or equal to 1250 mg/kg/day.[1] Prior to death, tremors and inactivity were observed in mice, whereas rats showed central nervous system depression, tremors, and gastrointestinal distress. No compound-related gross or histopathologic effects were observed.

Final body weights were reduced in 13-week studies in male rats receiving doses of 400 mg/kg/day or more, in female rats receiving 800 mg/kg/day, and in female mice receiving 600 mg/kg/day.[1] Compound-related histopathologic effects included hyaline droplet degeneration of the proximal convoluted tubules of the kidney in male rats, and a reduction in the number of primary follicles and mature graafian follicles in the ovaries of female mice dosed at the 1200 mg/kg/day level. No compound-related gross or histopathologic effects were observed in female rats or male mice in this study.

Administration of VCH by oral intubation to rats, 5 days/week for 2 years, at 0, 200, or 400 mg/kg/day was associated with slightly increased incidence of epithelial hyperplasia of the forestomach and squamous-cell papillomas or carcinomas of the skin in high-dose males.[1] Poor survival of the dosed animals may have compromised the study. In mice similarly dosed, the number of uncommon ovarian neoplasms and ovarian pathologies was significantly increased in the females, and there was some suggestion of increased numbers of adrenal gland adenomas in the high-dose females. Among high-dose male mice, there were scattered instances of lymphomas and cancers of the lung but, again, poor survival confounded results.

Toxicological studies have suggested that the species specificity for induction of ovarian tumors (produced in mice but not rats) occurs because the blood level of the ovotoxic VCH metabolite, VCH-1,2-epoxide, is dramatically higher in VCH-treated female mice compared to rats.[4] VCH has been shown to be metabolized by the liver of mice to the ovotoxic metabolite (VCH-1,2-epoxide), which circulates in blood and is delivered to the ovary, where it destroys small oocytes. This destruction of small oocytes is considered to be an early event in carcinogenesis. Species difference in epoxidation of VCH by hepatic microsomes correlates well with the differences observed in the blood concentration of VCH-1,2-epoxide and VCH ovarian toxicity. Further in vitro studies have found that the rate of VCH epoxidation in humans by human hepatic microsomes was 13 times and 2 times lower than epoxidation by mouse and rat systems, respectively.[5] Therefore, if the rate of hepatic VCH epoxidation is the main factor that determines the ovotoxicity of VCH, then rats may be a more appropriate animal model for humans.

In a reproductive study using the continuous breeding protocol, 500 mg/kg/day administered by gavage to mice caused slight generalized toxicity and reduced the number of oocytes by 33% in females and testicular sperm count by 17% in males but did not adversely affect the reproductive competence of F_0 and F_1 generations.[6]

4-Vinylcyclohexene was not mutagenic in *Salmonella typhimurium*; however, several of its metabolites, including 4-vinylcyclohexene dioxide, are genotoxic.[1,7]

The IARC has determined that there is inadequate evidence in humans, but sufficient evidence in animals, for the carcinogenicity of 4-vinylcyclohexene.[7]

The 1995 ACGIH threshold limit value–time-weighted average (TLV-TWA) for 4-vinylcyclohexene is 0.1 ppm (0.4 mg/m³) with an A2 suspected human carcinogen designation.

REFERENCES

1. National Toxicology Program: *Toxicology and Carcinogenesis Studies of 4-Vinylcyclohexene (CAS No 100-40-3) in F344/N Rats and B6C3F₁ Mice (Gavage Studies)*. NTP TR 303, NIH Pub No 86-2559, Research Triangle Park, NC, US Department of Health and Human Services, 1986
2. Rappaport SM, Fraser DA: Air sampling and analysis in a rubber vulcanization area. *Am Ind Hyg Assoc J* 38:205–210, 1977
3. Smyth HF Jr, Carpenter CP, Weil CS, et al: Range-finding toxicity data: List VII. *Am Ind Hyg Assoc J* 30:470–476, 1969
4. Smith BJ, Mattison DR, Sipes IG: The role of epoxidation in 4-vinylcyclohexene-induced ovarian toxicity. *Toxicol Appl Pharmacol* 105:372–381, 1990
5. Smith BJ, Sipes IG: Epoxidation of 4-vinylcyclohexene by human hepatic microsomes. *Toxicol Appl Pharmacol* 109:367–371, 1991
6. Grizzle TB, George JD, Fail PA, et al: Reproductive effects of 4-vinylcyclohexene in Swiss mice assessed by a continuous breeding protocol. *Fund Appl Toxicol* 22:122–129, 1994
7. *IARC Monographs on the Evaluation of Carcinogenic Risks to Humans*. Vol 60, Some industrial chemicals, pp 347–359. Lyon, International Agency for Research on Cancer, World Health Organization, 1994

VINYL CYCLOHEXENE DIOXIDE
CAS: 106-87-6

C8H12O2

Synonyms: 4-Vinylcyclohexene diepoxide; 1,2-epoxy-4-(epoxyethyl)cyclohexane; 1-epoxyethyl-3,4-epoxycyclohexane; VCD

Physical Form. Colorless or pale yellow liquid

Uses. Chemical intermediate; reactive diluent for diepoxides and epoxy resins.

Exposure. Inhalation; skin absorption

Toxicology. Vinyl cyclohexene dioxide (VCD) is an irritant to the skin, eyes, and respiratory system. It is carcinogenic in experimental animals.

In humans, VCD is considered to be a mild to moderate skin irritant although occasional instances of marked irritation have been reported. In one case, severe vesiculation of the skin of both feet occurred when a worker wore shoes previously contaminated with VCD.[1] A single case of allergic contact dermatitis has also been reported in a worker whose gloves were permeable to the VCD.[2] Systemic illness in humans has not been reported in association with exposure.[1]

In rats, the inhalation LC_{50} is 800 ppm for 4 hours, and the oral LD_{50} is 2.1 g/kg.[3] Dermal application of the undiluted material to rabbits caused edema and redness equivalent to a moderate first-degree burn. The liquid can penetrate the skin and is more toxic when applied dermally than by other routes. The dermal LD_{50} in rabbits is 0.62 ml/kg body weight.

VCD is an alkylating agent and is selectively active against rapidly dividing cells, such as the blood-forming elements in the bone marrow, lymphoid tissues, and reproductive organs.[4] Immunotoxic effects were observed in mice after 5-day dermal exposures at 10 mg/day.[3] Hematologic studies indicated a significant decrease in the leukocyte count that was related to the decreased numbers of circulating lymphocytes at this same dose. A decrease in the lymphoproliferative response to phytohemagglutinin and concanavalin A and suppression of the antibody plaque-forming cell response indicated that VCD was immunosuppressive.

Repeated intramuscular injections of 400 mg/kg VCD to male rats for 7 days decreased the size of the spleen, thymus, and testis and resulted in enlarged adrenal glands.[4] The leukocyte count fell more than 60% and the myeloid to erythroid ratio was increased.

In 14-day dermal studies, rats receiving 139 mg/rat or higher for males and 112 mg/rat or higher for females died before the end of the treatment period.[3] Necropsy revealed congestion and/or hypoplasia of the bone marrow, and most of the rats had acute nephrosis. Skin lesions included epidermal necrosis and ulcer-ation, epidermal hyperplasia, and hyperkeratosis. Male rats receiving 68 mg/rat and females receiving 57 mg/rat had final mean body weights lower than those of control animals; skin lesions were similar to those seen at the higher dose levels but of less severity.

All rats survived dermal doses of up to 60 mg/rat administered over 13 weeks.[3] Mean body weights were up to 14% lower than controls, and redness, scabs, and ulceration occurred at the application site. In mice, applications of up to 10 mg/mouse produced increased liver and kidney weights. Compound-related skin lesions included sebaceous gland hyperplasia and hyperplasia and hyperkeratosis of the stratified squamous epithelium at the site of application; ovarian atrophy was also considered to be compound-related.

Two-year studies were conducted by administering VCD in acetone by dermal application, 5 days per week for over 100 weeks, to groups of rats of each sex at 0, 15, or 30 mg/animal and to groups of mice at 0, 2.5, 5, or 10 mg/animal. Acanthosis and sebaceous gland hypertrophy of skin from the scapula were observed at increased incidences in both species. Squamous-cell papillomas in male rats and squamous-cell carcinomas in males and females were observed in exposed rats at an increased incidence. The combined incidence of basal-cell adenomas or carcinomas were also increased in both sexes. Squamous-cell carcinomas were found in the exposed mice. Follicular atrophy and tubular hyperplasia of the ovary in female mice were increased with increasing dose, and the combined incidences of luteomas, granulosa-cell tumors, benign mixed tumors, or malignant granulosa-cell tumors in mid- and high-dose female mice were increased. Increased incidences of lung neoplasms in females may also have been related to chemical exposure.

Under the conditions of the study, there was clear evidence of carcinogenicity in rats, as shown by squamous-cell and basal-cell neoplasms of the skin, and in mice, as shown by squamous-cell carcinomas of the skin and ovarian neoplasms in females. No information has been reported on the carcinogenicity of VCD in humans.

A number of early studies also demonstrated the carcinogenicity of VCD in rodents. Dermal application of 16 mg, 5 days per week for 12 months, resulted in squamous-cell carcinomas or sarcomas in 9 of 20 exposed male mice.[5] One skin neoplasm and 4 malignant lymphomas occurred in 16 of 20 mice surviving a total dermal dose of 70 mg over 14 months.[6]

IARC has determined that there is sufficient evidence in experimental animals and inadequate evidence in humans for the carcinogenicity of VCD.[7]

VCD has been found to be mutagenic in a number of bacterial tester strains in the presence and absence of mammalian microsomal metabolic activation. It induced direct sister chromatid exchange and chromosomal aberrations in cultured Chinese hamster ovary cells.[7]

The 1995 ACGIH threshold limit value–time-weighted average (TLV-TWA) is 10 ppm (57 mg/m³) with an A2 suspected human carcinogen designation.

REFERENCES

1. Vinyl cyclohexene dioxide. *Documentation of the TLVs and BEIs*, 6th ed, pp 1708–1710. Cincinnati, OH, American Conference of Governmental Industrial Hygienists (ACGIH), 1991
2. Dannaker CJ: Allergic sensitization to a non-bispenol A epoxy of the cycloaliphatic class. *JOM* 30:641–643, 1988
3. US National Toxicology Program: *Toxicology and Carcinogenesis Studies of 4-Vinyl-1-Cyclohexene Diepoxide (CAS No 106-87-6) in F344/N Rats and B6C3F₁ Mice (Dermal Studies).* Technical Report No 362, NIH Pub 90-2817, Research Triangle Park, NC, US Department of Health and Human Services, 1989
4. Kodama JK, Guyman RJ, Dunlap MK, et al: Some effects of epoxy compounds on the blood. *Arch Environ Health* 2:56–57, 1961
5. Hendry JA, Homer RF, Rose FL, et al: Cytotoxic agents. II. Bisepoxides and related compounds. *Br J Pharmacol* 6:235–255, 1951
6. Kotin P, Falk HL: Organic peroxides, hydrogen peroxide, epoxides and neoplasia. *Radiat Res* 3(Suppl):193–211, 1963
7. *IARC Monographs on the Evaluation of Carcinogenic Risks to Humans*, Vol 60, Some industrial chemicals, pp 347–359. Lyon, International Agency for Research on Cancer, 1994

VINYLIDENE CHLORIDE
CAS: 75-35-4

$C_2H_2Cl_2$

Synonyms: 1,1-Dichloroethylene; VDC; *asym*-dichloroethylene; 1,1-dichloroethene; 1,1-DCE; vinylidene dichloride

Physical Form. Clear liquid that is highly flammable and reactive and, in the presence of air, can form complex peroxides in the absence of chemical inhibitors

Uses. In the production of copolymers of high vinylidene chloride content, the other major monomer usually being vinyl chloride, such as in Saran and Velon for films and coatings

Exposure. Inhalation

Toxicology. Vinylidene chloride (VDC) causes CNS depression at high levels, and repeated exposure to lower concentrations results in liver and kidney damage in experimental animals.

Limited information is available on the human health effects following exposure to VDC.[1] Upper airway irritation consisting of inflammation of mucous membranes has been reported following acute exposure, whereas central nervous system toxicity has been associated with levels of 4000 ppm.[1]

In male rats, the 4-hour LC_{50} was 6350 ppm.[2] The oral LD_{50} of VCD in corn oil was 1500 mg/kg in male rats.[3] Rats exposed 6 hr/day for 20 days to 200 ppm exhibited only slight nasal irritation.[4]

Results from animal studies indicate that the liver is a primary target for VCD-induced toxicity.[1] Hepatotoxicity following both inhalation and oral exposures has ranged from biochemical changes, including increases in serum enzyme markers of liver dysfunction and induction of hepatic enzymes, to marked histological changes, including centrilobular vacuolization, swelling, degeneration, and necrosis. The effects appear to follow a dose-response relationship and may also be influenced by duration of exposure. Mice exposed to 50 ppm for 6 hours exhibited slight centrilobular swelling, whereas hepatic degeneration was observed in mice exposed to concentrations of up to 200 ppm, 6hr/day, 5 days/week for 10 days. Vinylidene chloride also affects several liver enzymes: it decreases the activity of hepatic glucose-6-phosphatase and the content of glutathione, and it increases serum alanine α-ketoglutanate transaminase activity and liver content of triglycerides.[5]

Renal toxicity—including enzyme changes, hemoglobinuria, and tubular swelling—degeneration, and necrosis have been observed in experimental animals following VDC exposure.[1] Severe histological lesions of the kidney were observed in mice following acute exposure to 10 to 50 ppm of VCD, whereas exposures of up to 300 ppm were necessary to produce the same effects in rats.

Studies in mice have shown that selective covalent binding of VDC occurs in the proximal tubules, the liver lobules, and the mucosa of the upper respiratory tract and corresponds to sites of potential toxicity.[6] Additional events, such as depletion of glutathione, appear to be necessary for VDC-induced cell death to occur.

In rats, ingestion of drinking water containing up to 200 ppm vinylidene chloride caused mild, dose-related changes in the liver but did not affect the reproductive capacity through 3 generations that produced 6 sets of litters.[7] Prenatal exposure to doses ranging from 15 to 450 ppm resulted in skeletal defects in rats, rabbits, and mice and also caused maternal toxicity in the form of decreased body weight and death.[1]

In a carcinogenicity study, Swiss mice were exposed to 10 or 25 ppm for 4 hr/day, 5 days/week for 52 weeks.[8,9] After 98 weeks, 25 ppm had caused kidney adenocarcinomas in 24 of 150 males and 1 of 150 females, whereas none was seen in the control group. Rats exposed to 75 ppm, 6 hr/day, 5/days week for 18 months,

and then held until 24 months, showed a reversible hepatocellular fatty change but no increase in tumor incidence that could be attributed to vinylidene chloride exposure.[10] Several other studies in other strains of mice, rats, and hamsters did not produce carcinogenic effects.[5]

In one epidemiological study of 138 exposed workers, no excess of cancer cases was found, but follow-up was incomplete; nearly 40% of the workers had less than 15 years' latency since first exposure, and only 5 deaths were observed.[11] The IARC has determined that there is inadequate evidence for carcinogenicity to humans and limited evidence for carcinogenicity to animals.[5]

VCD was genotoxic in a number of test systems; it induced chromosome aberrations and sister chromatid exchanges in cultured mammalian cells and DNA damage in mice in vivo; gene mutations were observed in vitro for bacteria, yeast, and plant cells following metabolic activation.[1]

The liquid is moderately irritating to the eyes and irritating to the skin of rabbits.

The 1995 ACGIH threshold limit value–time-weighted average (TLV-TWA) is 5 ppm (20 mg/m³) with a short-term excursion level of 20 ppm (79 mg/m³).

REFERENCES

1. Agency for Toxic Substances and Disease Registry (ASTDR): *Toxicological Profile for 1,1-Dichloroethene.* US Department of Health and Human Services, Public Health Service, p 123, 1992
2. Siegel J, Jones RA, Coon RA, Lyon JP: Effects on experimental animals of acute, repeated and continuous inhalation exposures to dichloroactylene mixture. *Toxicol Appl Pharmacol* 18:168, 1971
3. Jenkins LF Jr, Trabulus MJ, Murphy SD: Biochemical effects of 1,1-dichloroethylene. *Toxicol Appl Pharmacol* 23:501, 1972
4. Gage JC: The subacute inhalation toxicity of 109 industrial chemicals. *Br J Ind Med* 27:1, 1970
5. *IARC Monographs on the Evaluation of Carcinogenic Risks to Humans,* Vol 39, Some chemicals used in plastics and elastomers, pp 195–226.

Lyon, International Agency for Research on Cancer, 1986

6. Brittebo EB, Darnerud PO, Eriksson C, et al: Nephrotoxicity and covalent binding of 1,1-dichloroethylene in buthionine sulphoximine-treated mice. *Arch Toxicol* 67:605–612, 1993
7. Nitschke KD, Smith FA, Quast JF, et al: A three-generation rat reproductive toxicity study of vinylidene chloride in the drinking water. *Fund Appl Toxicol* 3:75–79, 1983
8. Maltoni C: Recent findings on the carcinogenicity of chlorinated olefins. *Environ Health Persp* 21:1–5, 1977
9. Maltoni C, Cotti G, Morisi L, Chieco P: Carcinogenicity bioassays of vinylidene chloride: research plans and early results. *Med Lav* 58:241–262, 1977
10. Quast JF, McKenna MJ, Rampy LW, et al: Chronic toxicity and oncogenicity study on inhaled vinylidene chloride in rats. *Fund Appl Toxicol* 6:105–144, 1986
11. Ott MG, Fishbeck WA, Townsend JC, et al: A health study of employees exposed to vinylidene chloride. *J Occup Med* 18:735–738, 1976

VINYLTOLUENE
CAS: 25013-15-4

C_9H_{10}

Synonyms: Ethenylmethylbenzene; methylstyrene; tolyethylene; methylvinylbenzene

Physical Form. Colorless liquid

Uses. Reactive monomer in the production of polymers and coatings

Exposure. Inhalation

Toxicology. Vinyltoluene is an irritant of the eyes and mucous membranes; at high concentrations, it causes narcosis in animals, and it is expected that severe exposure will produce the same effect in humans.

Commercial vinyltoluene is a mixture of meta and para isomers with small amounts of ortho isomer.[1]

Human subjects exposed to 200 ppm detected a strong odor, but excessive discomfort was not experienced; at 400 ppm, there was strong eye and nasal irritation.[2] Central nervous system effects, such as depression, poor memory, and slow visuomotor performance, have been associated with heavy exposures.[1]

Exposure of rats and guinea pigs to 1350 ppm for 7 hr/day for 100 days caused the death of some of the rats and slight liver damage in guinea pigs; there were no effects in female monkeys at this concentration.[2]

Rats tolerated exposure to 300 ppm for 60 hours without clinical symptoms although they appeared relatively inactive.[3] At this concentration, vinyltoluene was found to accumulate in perirenal fat and was more effective than styrene, xylene, or toluene in producing neurochemical effects as determined by enzyme assays.

Mice administered 0, 10, 50, or 250 mg/kg and rats given 0, 10, 50, 250, or 500 mg/kg by gastric intubation once a day for 83 and 107 weeks, respectively, showed no treatment-related increases in malignant or benign tumors.[4] Chronic inhalation experiments in B6C3F$_1$ mice (10 or 25 ppm) and Fischer 344/N rats (100 or 300 ppm) caused hyperplasia of the respiratory epithelium of the nasal passages in both species, but there was no evidence of treatment-related increases in the incidences of any tumor.[5]

The IARC has determined that there is evidence suggesting the lack of carcinogenicity of vinyltoluene in experimental animals and inadequate evidence in humans.[1]

Vinyltoluene induces sister chromatid exchange and chromosomal aberrations in cultured human lymphocytes and micronuclei in mouse bone marrow cells in vivo.

The liquid dropped in the eyes of rabbits caused slight conjunctival irritation.[2] Applied to rabbit skin, vinyltoluene produced erythema with the development of edema and superficial necrosis.[2]

Vinyltoluene has a disagreeable odor detectable at 50 ppm.[2]

The ACGIH threshold limit value–time-weighted average (TLV-TWA) for vinyltoluene is 50 ppm (242 mg/m³) with a short-term excursion limit of 100 ppm (483 mg/m³).

REFERENCES

1. *IARC Monographs on the Evaluation of Carcinogenic Risks to Humans*, Vol 60, Some industrial chemicals, pp 373–388. Lyon, International Agency for Research on Cancer, 1994
2. Wolf MA, Rowe VK, McCollister DD, et al: Toxicological studies of certain alkylated benzenes and benzene. *AMA Arch Ind Health* 14:387–398, 1956
3. Savolainen H, Pfaffli P: Neurochemical effects of short-term inhalation exposure to vinyltoluene vapor. *Arch Environ Contam Toxicol* 10:511–517, 1981
4. Conti B, Maltoni C, Perino G, et al: Long-term carcinogenicity bioassays on styrene administered by inhalation, ingestion and injection and styrene oxide administered by ingestion in Sprague–Dawley rats, and para-methylstyrene administered by ingestion in Sprague–Dawley rats and Swiss mice. *Ann NY Acad Sci* 534:203–234, 1988
5. US National Toxicology Program: *Toxicology and Carcinogenesis Studies of Vinyl Toluene (Mixed Isomers) (65%–71% Meta-Isomer and 32%–35% Para-Isomer) (CAS No 25013-15-4) in F344/N Rats and B6C3F₁ Mice (Inhalation Studies)*, Technical Report Series No 375. NIH Publ No 90-2830, Research Triangle Park, NC, US Department of Health and Human Services, 1990

VM & P NAPHTHA
CAS: 8032-32-4

Synonyms: Varnish makers' and printers' naphtha; light naphtha, dry cleaners' naphtha; spotting naphtha

Physical Form. Clear, colorless to yellow liquid; petroleum distillate containing C_5 to C_{11} hydrocarbons; a typical composition is paraffins 55.5%, naphthenes 30.3%, alkyl benzene 11.7%, dichloroparaffins 2.4%, benzene less than 0.1%

Uses. Diluent for paints, coatings, resins, printing inks, rubbers, and cements; solvent

Exposure. Inhalation

Toxicology. VM & P naphtha vapor is a central nervous system depressant and a mild irritant of the eyes and upper respiratory tract.

In human tests, exposure to 880 ppm (4100 mg/m³) for 15 minutes resulted in eye and throat irritation with olfactory fatigue.[1] The chief effect of exposure to high levels of the vapor is reported to be CNS depression.[2,3] However, in an accidental brief exposure of 19 workers from an overheated solvent tank, the chief effect was dyspnea, which lasted for several minutes after the exposure.[2] Two of the workers were cyanotic, with tremor and nausea, but these were of brief duration. The absence of CNS depression was noteworthy.

The LC_{50} in rats was 3400 ppm for 4 hours; incoordination was observed.[4] In rats and beagle dogs exposed to 500 ppm for 30 hours weekly for 13 weeks, there was no evidence of latent or chronic effects.

The 1995 ACGIH threshold limit value–time-weighted average (TLV-TWA) for VM & P naphtha is 300 ppm (1370 mg/m³).

REFERENCES

1. Carpenter CP et al: Petroleum hydrocarbon toxicity series. IV. Animal and human response to vapors of rubber solvent. *Toxicol Appl Pharmacol* 33:526, 1975
2. Wilson WF: Toxicology of petroleum naphtha distillate vapors. *J Occup Med* 18:821, 1976
3. National Institute for Occupational Safety and Health: *Criteria for a Recommended Standard . . . Occupational Exposure to Refined Petroleum Solvents.* DHEW (NIOSH) Pub No 77-192. Washington, DC, US Government Printing Office, 1977
4. Carpenter CP et al: Petroleum hydrocarbon toxicity series. II. Animal and human response to vapors of varnish makers' and printers' naphtha. *Toxicol Appl Pharmacol* 32:263, 1975

WARFARIN

CAS: 81-81-2

$C_{19}H_{16}O_4$

Synonyms: 3-(α-Acetonylbenzyl)-4,hydroxy-coumarin; coumadin; compound 42

Physical Form.   Colorless crystals

Uses.   Rodenticide; used clinically as an anticoagulant

Exposure.   Inhalation; skin absorption

Toxicology.   Warfarin causes hypoprothrombinemia and vascular injury, which results in hemorrhage.

Warfarin acts as a vitamin K antagonist and suppresses the hepatic formation of prothrombin and of factors VII, IX, and X, causing a markedly reduced prothrombin activity of the blood.[1,2] Warfarin also causes dilatation and engorgement of blood vessels and an increase in capillary fragility.[1] The two effects can combine to produce hematomas, severe blood loss, and death from shock or hemorrhage.[2] The inhibition of prothrombin formation does not become apparent until the prothrombin reserves are depleted, which usually requires exposure for a number of days.[1] Accidental ingestion of approximately 2 mg/kg/day for 15 days by 14 family members caused massive bruising and hematomas on the buttocks, and at the knee and elbow joints after 7 to 10 days; gum and nasal bleeding subsequently appeared, and blood was noted in the urine and feces.[2] Two individuals succumbed to the poisoning; the other 12 recovered following treatment.

A farmer whose hands were intermittently wetted with a 0.5% solution of warfarin over a period of 24 days developed gross hematuria 2 days after the last contact with the solution; the following day, spontaneous hematomas appeared on the arms and legs.[3] Within 4 days, other effects included epistaxis, punctate hemorrhages of the palate and mouth, and bleeding from the lower lip. Four days later, after treatment for 2 days with phytonadione, hematologic indices had returned to the normal range. Other effects of warfarin intoxication have included back pain, abdominal pain, vomiting, and petechiae of the skin.[1,4]

Teratogenic effects have been observed in humans following maternal warfarin exposure.[2] The effects are primarily seen in the nasal region of the fetus and include nasal hypoplasia, bone stippling, and mental retardation. Central nervous system abnormalities due to localized hemorrhaging and scarring have occurred following second or third trimester exposures, whereas exposure during early pregnancy may result in dysmorphism.[5,6]

Treatment for warfarin poisoning includes vitamin K administration and, in severe cases, transfusions of whole blood.[2]

The 1995 ACGIH threshold limit value–time-weighted average (TLV-TWA) for warfarin is 0.1 mg/m^3.

REFERENCES

1. Gosselin RE, Hodge HC, Smith RP: *Clinical Toxicology of Commercial Products, Section III*, 5th ed, pp 395–397. Baltimore, MD, Williams and Wilkins, 1984
2. Hayes AW (ed): *Principles and Methods of Toxicology*, 2nd ed, p 155. New York, Raven Press, 1989
3. Fristedt B, Sterner N: Warfarin intoxication from percutaneous absorption. *Arch Environ Health* 11:205, 1965
4. Hayes WJ Jr, Laws ER Jr: *Handbook of Pesticide Toxicology.* Vol 3, *Classes of Pesticides*, pp 1291–1297. Academic Press, Inc., New York, 1991
5. Hall JG, Pauli RM, Wilson KM: Maternal and fetal sequelae of anticoagulation during pregnancy. *Am J Med* 68:122–140, 1980
6. Ruthnum P, Tolmie JL: Atypical malformations in an infant exposed to Warfarin during the first trimester of pregnancy. *Teratol* 36:299–301, 1987

WOOD DUST

Physical Form. Wood is a complex biologic and chemical material consisting primarily of cellulose, hemicellulose, and lignin. The two general classes are hardwood and soft wood, each with its own structure and composition. Woods also may contain a variety of organic compounds, including glycosides, quinones, tannins, stilbenes, terpenes, aldehydes, and coumarins.[1] Not only is the composition of wood extremely variable from species to species, but different parts of the same tree may have different compositions. Various solvents, adhesives, fungicides, insecticides, and microorganisms also may be associated with wood. As a result of this variability, wood dust cannot be treated as a single agent.[2]

Exposure. Inhalation and skin contact

Toxicology. Wood dust exposure may cause eye and skin irritation, respiratory effects, and hardwood nasal cancer. Irritation of the skin and eyes resulting from contact with wood dust is relatively common and may result from mechanical action (eg, irritation caused by bristles and splinters), chemical irritation, sensitization, or a combination of these factors.[1]

Primary irritant dermatitis caused by wood contact consists of erythema and blistering, which may be accompanied by erosions and secondary infections. Irritant chemicals typically are found in the bark or the sap of the outer part of the tree. Therefore, loggers and persons involved in initial wood processing are most affected. In most reports of contact dermatitis, hardwoods of tropical origin have been implicated although other woods, including pine, spruce, Western red cedar, elm, and alder, have been cited.

Allergic dermatitis arising from exposure to wood substances is characterized by redness, scaling, and itching, which may progress to vesicular dermatitis after repeated exposures. The hands, forearms, eyelids, face, neck, and genitals generally are first affected. Allergic dermatitis may appear after several years' contact but typically ensues after a few days or a few weeks of contact. Chemicals causing sensitization generally are found in the heartwood; therefore, workers involved in secondary wood processing (carpenters, sawyers, furniture makers) are more often affected than persons involved in initial processing. Numerous sensitizing agents in wood have been identified, including lapachol (teak), usnic acid (Western red cedar), quinones (rosewood), and anthothecol (African mahogany).[3]

Another type of wood-related dermatitis is woodcutters' eczema, which is not caused by contact with wood or wood dust, but rather by contact with epiphytes, lichens, and liverworts growing on bark or shrubs.

Respiratory ailments associated with wood dust exposure include irritation, bronchitis, nasal mucociliary stasis, impairment of ventilatory function, and asthma.

A correlation between the incidence of sinusitis, sneezing, watery nasal discharge, nasal mucosal irritation, and cough and wood dust concentration was found in German furniture workers. Fourteen persons were exposed to a dust concentration below 5 mg/m^3, 15 to between 5 and 9 mg/m^3, 26 to between 10 and 19 mg/m^3, and 36 to 20 mg/m^3 or more.[3]

Middle ear symptoms occurred significantly more frequently among Danish furniture workers exposed to dust levels above 5 mg/m^3.[4] Other symptoms, such as sinus inflammation, long-lasting colds, asthma, nosebleed, and sneezing, also occurred more frequently in the higher-exposure group.

Impairment of mucociliary clearance, the rate at which mucus is transported from the nose to the pharynx, was found in a study of 68 Danish hardwood furniture workers.[5] Mucostasis (defined as a nasal transit time of 40 or more minutes) increased in direct proportion to the dust concentration; at 25.5 mg/m^3, 63% had mucotasis versus 11% at 2.2 mg/m^3.

Obstructive lung disease, as measured by pulmonary-function tests, has been associated with wood dust exposure. Vermont woodworkers with hardwood or pine dust exposures greater than 10 mg-years/m^3 generally had lower pulmonary function, as determined by FEV_1/FVC, than those with exposure indices of 0 to 2 mg-years/m^3.[6] Higher exposures also significantly lowered values of the maximal midexpiratory flow rate (MMEFR) compared to theoretical values.

Although no dose-response relationship was established, a study of employees from five plants with dust levels ranging from 0.46 to 8.3 mg/m^3 found decreases in FEV$_1$ and FVC of up to 0.19 liter during the workshift for workers employed at the dustier furniture plant.[7]

A hypersensitivity reaction leading to asthma (defined as reversible airway obstruction) has been reported from exposure to a number of wood dusts, including oak, mahogany, and redwood, as well as more exotic woods, such as iroko cocobolo, zebrawood, and abiruana.[1,2] Connecting asthma to wood dust exposure has been difficult because, frequently, the subject has worked with wood for years with no reaction.[2] Sensitization typically begins as eye and nose irritation, followed by nonproductive cough and difficulty in breathing. In a sensitized individual, exposure may produce an immediate onset of symptoms and rapid reversibility, or a delayed onset of 5 to 8 hours with a more gradual reversibility.

Immunologic findings in individuals with wood dust–induced asthma also vary.[1] In some cases, a Type 1 allergic reaction is confirmed by the presence of IgE antibodies. Positive skin reactions and the presence of precipitating antibodies to wood dust or extracts may or may not occur.

Extensive studies have been done on a clearly defined asthma syndrome produced by exposure to Western red cedar.[8–10] Plicatic acid has been identified as the etiologic agent. The Western red cedar asthma syndrome includes rhinitis, conjunctivitis, wheezing, cough, and nocturnal attacks of breathlessness characterized by a precipitous decline in FEV$_1$. There is no apparent relation between skin sensitivity and respiratory changes. No precipitating IgG antibodies are found in the serum of sensitized individuals, and circulating IgE antibodies are present in about one-third of affected individuals.

It has been estimated that approximately 5% of exposed workers are affected.[1] The asthmatic reaction is species-specific; subjects who exhibit asthma with one type of wood dust show no reaction when challenged with another type.[2]

Other syndromes are associated with exposure to fungi present on wood.[11] Organic dust toxic syndrome is characterized by generalized feelings of feverishness, often accompanied by dry cough, fatigue, and shaking chills; it appears to be caused by high-dose exposures to fungal spores in moldy materials. Extrinsic allergic alveolitis has been associated with fungi found in bark and wood flour. Findings include abnormal X ray, reduced lung volume, and serum precipitating antibodies to fungal antigens. Lung biopsy studies have reported interstitial infiltration with granuloma formation.

The association between nasal cancer and occupations involving exposure to wood dust has been established from case reports and epidemiologic studies.[12] This relationship first was noted in the late 1960s in Great Britain, where the incidence of nasal adenocarcinoma, a rare type of nasal cancer, among woodworkers in the furniture industry was found to be 10 to 20 times greater than among other woodworkers and 100 times greater than in the general population.[1] In a 19-year follow-up study of 8141 Swedish furniture workers, nasal adenocarcinoma was 62 times higher than expected, whereas sinonasal adenocarcinoma and sinonasal carcinoma were 44 and 7 times higher than expected, respectively.[13]

A study of deaths in furniture-making counties of North Carolina found that 8 of 37 (21.6%) people dead from nasal cancer had been employed in the furniture industry, whereas only 5 of the 73 (6.8%) controls had been so employed.[14] Of 215 patients with nasal cancer in Connecticut, 2.8% probably had been occupationally exposed to wood dust versus only 0.8% of 741 persons dying of other cancers with similar exposures.[15]

Although estimates of the relative risk of nasal adenocarcinoma vary considerably because of differences in exposure levels, types of wood dust, latency periods, selection of controls, and other confounding factors, the IARC has concluded that there is sufficient evidence that nasal adenocarcinomas have been caused by exposure in the furniture-making industry.[12] The carcinogenic agent(s) in wood dust has not been identified, nor has the importance of particle size and shape been investigated.[16]

It has been postulated that wood dust carcinoma results from a multistep process: exposure causes loss of cilia and hyperplasia of the goblet cells and initiation of cuboid-cell metaplasia, followed (after a quiescent period) by squamous-cell metaplasia.[8] Decades later, cellular aplasia leads to nasal adenocarcinoma. The time between first occupational exposure to wood dust and the development of nasal cavity adenocarcinoma averages 40 years.[16]

Other cancers, including lung, Hodgkin's disease, multiple myeloma, stomach, and colorectal cancer and lymphosarcoma, have been mentioned in relation to wood dust exposure. Data are insufficient and inconclusive regarding the relationship between occupational exposure to wood dust and cancers other than nasal adenocarcinoma.[1]

The 1995 ACGIH threshold limit value–time-weighted average (TLV-TWA) for hardwoods is 1 mg/m³, and 5 mg/m³ for soft woods, with a short-term excursion limit of 10 mg/m³.

REFERENCES

1. Tatken RL et al: *Health Effects of Exposure to Wood Dust: A Summary of the Literature*, pp 1–157. Cincinnati, OH, US Department of Health and Human Services, Public Health Service, Centers for Disease Control, National Institute for Occupational Safety and Health, 1977

2. Meola A: Toxic effects of wood dust exposure. *Prof Saf*, 26–29, March 1984

3. *IARC Monographs on the Evaluation of the Carcinogenic Risk of Chemicals to Humans*, Vol 25, Wood, leather and some associated industries, pp 99–138. Lyon, International Agency for Research on Cancer, 1981

4. Solgaard J, Anderson I: Airway function and symptoms in woodworkers. *Ugeskr Laeg* 137:2593–2599, 1975

5. Anderson HC et al: Nasal cancers, symptoms and upper airway function in woodworkers. *Br Med J* 34:201–207, 1977

6. Whitehead LW et al: Pulmonary function status of workers exposed to hardwood or pine dust. *Am Ind Hyg Assoc* 42:178–186, 1981

7. Zuhair YS et al: Ventilatory function in workers exposed to tea and wood dust. *Br J Ind Med* 38:339–345, 1981

8. Goldsmith DF, Shy CM: Respiratory health effects from occupational exposure to wood dusts. *Scand J Work Environ Health* 14:1–15, 1988

9. Chan-Yeung M et al: Occupational asthma and rhinitis due to Western red cedar *(Thuja plicata)*. *Am Rev Respir Dis* 108:1094–1102, 1973

10. Chan-Yeung M et al: Follow-up study of 232 patients with occupational asthma caused by Western red cedar *(Thuja plicata)*. *J Allergy Clin Immunol* 79:792–796, 1987

11. Enarson DA, Chan Yeung, M: Characterization of health effects of wood dust exposures. *Am J Ind Med* 17:33–38, 1990

12. *IARC Monographs on the Evaluation of the Carcinogenic Risks to Humans, Overall Evaluations of Carcinogenicity: An Updating of IARC Monographs Vols 1–42*, Suppl 7, pp 378–380. Lyon, International Agency for Research on Cancer, 1987

13. Gerhardsson MR et al: Respiratory cancer in furniture workers. *Br J Ind Med* 42:403–405, 1985

14. Brinton LA et al: A death certificate analysis of nasal cancer among furniture workers in North Carolina. *Cancer Res* 37:3473–3474, 1977

15. Wills JH: Nasal carcinoma in woodworkers: a review. *J Occup Med* 24:526–530, 1982

16. Nylander LA, Dement JM: Carcinogenic effects of wood dust: review and discussion. *Am J Ind Med* 24:619–647, 1993

XYLENE
CAS: 133-20-7

o-Xylene—CAS: 95-47-6
m-Xylene—CAS: 108-38-3
p-Xylene—CAS: 106-42-3

$C6H_4(CH_3)_2$

Synonyms: Xylol; dimethylbenzene

Physical Form. Colorless liquid

Uses. Solvent; in manufacture of certain organic compounds; cleaning agent; component of fuels

Exposure. Inhalation; skin absorption

Toxicology. Xylene vapor is an irritant of the eyes, mucous membranes, and skin; at high concentrations, it causes narcosis.

Three painters working in the confined space of a fuel tank were overcome by xylene vapor estimated to be 10,000 ppm; they were not found until 18.5 hours after entering the tank, and one died from pulmonary edema shortly thereafter. The other two workers recovered completely in 2 days; both had temporary hepatic impairment (inferred from elevated serum transaminase levels), and one had evidence of temporary renal impairment (increased blood urea and reduced creatinine clearance).[1]

Giddiness, anorexia, and vomiting occurred in a worker exposed to a solvent containing 75% xylene at levels of 60 to 350 ppm, with possible higher excursions.[2] In another report, 8 painters exposed to a solvent consisting of 80% xylene and 20% methylglycolacetate experienced headache, vertigo, gastric discomfort, dryness of the throat, and signs of slight drunkenness.[3]

Volunteers exposed to 460 ppm for 15 minutes had slight tearing and light-headedness.[4] A level of 230 ppm was not considered objectionable by most of these subjects. However, in an earlier study, the majority of subjects found 200 ppm irritating to the eyes, nose, and throat and judged 100 ppm to be the highest concentration subjectively satisfactory for an 8-hour exposure.[5]

Prior to 1940, most reports on the possible chronic toxicity of xylene also involved exposure to solvents that contained high percentages of benzene or toluene as well as other compounds. Consequently, the effects attributed to xylene in these are questionable.[6] Blood dyscrasias, such as those reportedly caused by benzene exposure, have not been associated with the xylenes.[6]

Both human and animal data suggest that mixed xylene, *m*-xylene, *o*-xylene, and *p*-xylene, all produce similar effects although the potency with regard to a given effect may vary with individual isomers.[7] In mice, the 6-hour LC_{50} values for *m*-, *o*-, and *p*-xylene were determined to be 5267, 4595, and 3907 ppm, respectively.[7] The 4-hour LC_{50} value for mixed xylene in rats ranged from 6350 to 6700 pm.

Exposure of rats to 1600 ppm for 2 or 4 days produced mucous membrane irritation, incoordination, narcosis, weight loss, increased erythrocyte count, and death. Exposure to 980 ppm for 7 days caused leukopenia, kidney congestion, and hyperplasia of the bone and spleen.[5]

Repeated exposure of rabbits to 1150 ppm of a mixture of isomers of xylene for 40 to 55 days caused a reversible decreased 1480 in red and white cell count and an increase in thrombocytes; exposure to 690 ppm for the same time period caused only a slight decrease in the white cell count.[8]

Fetotoxic effects, including altered enzyme activities in rat pups, have been reported following inhalation exposure to xylenes.[9] Oral treatment has resulted in prenatal mortality, growth inhibition, and malformations, primarily cleft palate, but only at maternally toxic doses. No reproductive effects were found in rats following inhalation of 500 ppm xylene before mating and during gestation and lactation.[7] Following intermediate and chronic exposures, there was no histological evidence of reproductive organ damage in mice administered 1000 mg/kg/day or rats given 800 mg/kg/day.[10]

In 2-year gavage studies, there was no evidence of carcinogenicity of mixed xylenes for male or female rats given 250 or 500 mg/kg/day, or for male or female mice given 500 or 1000 mg/kg/day.[10]

Mixed xylene and the individual xylene isomers have tested negative in a wide variety of genotoxic assays; they are considered to be nonmutagenic.[7]

Repeated application of 95% xylene to rabbit skin caused erythema and slight necrosis. Instilled in rabbit eyes, it produced conjunctival irritation and temporary corneal injury. Exposure to the vapors produced reversible vacuoles in the corneas of cats.

The odor threshold has been reported as 1 ppm.[6]

The 1995 ACGIH threshold limit value–time-weighted average (TLV-TWA) for xylene (*o*-, *m*-, *p*-isomers) is 100 ppm (434 mg/m³) with a short-term excursion limit of 150 ppm (651 mg/m³).

REFERENCES

1. Morley R, Eccleston DW, Douglas CP, et al: Xylene poisoning—a report on one fatal case and two cases of recovery after prolonged unconsciousness. *Br Med J* 3:442–443, 1970
2. Glass WI: A case of suspected xylol poisoning. *NZ Med J* 60:113, 1961
3. Goldie I: Can xylene (xylol) provoke convulsive seizures? *Ind Med Surg* 29:33–35, 1960
4. Carpenter CP, Kinkead ER, Geary DJ, et al: Petroleum hydrocarbon toxicity studies. V. Animal and human responses to vapors of mixed xylenes. *Toxicol Appl Pharmacol* 33:543–558, 1975
5. National Institute for Occupational Safety and Health: *Criteria for a Recommended Standard ... Occupational Exposure to Xylene.* DHEW (NIOSH) Pub No 75-168. Washington, DC, US Government Printing Office, 1975
6. Von Burg R: Toxicology updates: xylene. *J Appl Toxicol* 2:269–271, 1982
7. Agency for Toxic Substances and Disease Registry (ATSDR); *Toxicological Profile for Xy-lenes.* US Department of Health and Human Services, Public Health Service, Atlanta, GA, pp 1–209, 1993
8. Fabre R et al: Toxicological research on replacement solvents for benzene. IV. Study of xylenes. *Arch Mal Prof Med Trav Secur Soc* 21:301, 1960
9. Hood RD, Ottley MS: Developmental effects associated with exposure to xylene: a review. *Drug Chem Toxicol* 8:281–297, 1985
10. National Toxicology Program: *Toxicology and Carcinogenesis Studies of Xylenes (Mixed) in F344/N Rats and B6C3F Mice (Gavage Studies).* NTP, TR 327, NIH Pub No 87-2583. Atlanta, GA, US Department of Health and Human Services, 1986

XYLIDINE (MIXED ISOMERS)
CAS: 1300-73-8

$C_8H_{11}N$

Synonyms: Aminodimethylbenzene; dimethylaniline

Physical Form. Liquid, except that *o*-4-xylidine is a solid

Uses. Chemical intermediate in the manufacture of pesticides, dyes, antioxidants, pharmaceuticals, synthetic resins, and fragrances

Exposure. Inhalation

Toxicology. Xylidine causes liver damage in experimental animals and is a mild methemoglobin former; it has caused tumors of the nasal cavity in rats.

There are six isomeric forms of xylidenes, with the commercial product consisting primarily of the 2,4- and 2,6-isomers.[1]

The oral LD$_{50}$ in rats ranged from 470 mg/kg for 2,4-xylidine to 1300 mg/kg for 2,5-xylidine.[2] Although cyanosis has been observed in severely intoxicated animals, methemoglobin-induced hypoxia did not appear to be severe enough to be the cause of death.

The extent of methemoglobin formation from xylidines appears to be species-dependent, with cats more susceptible than humans and dogs less susceptible.[3] Administered intravenously to cats, 0.28 mM/lg produced 10% methemoglobin in cats, whereas similar exposure in dogs did not produce methemoglobin.[3] Rats given 20 mg of the same isomer produced less than 3% methemoglobin.[4]

Oral doses of 2,4-, 2,5-, and 2,6-xylidine administered to dogs for 4 weeks caused hepatotoxicity at doses of 2, 20, and 50 mg/kg/day; all three isomers induced fatty degeneration of the liver, with the 2,6-isomer the most toxic.[5] In rats, doses up to 700 mg/kg for 4 weeks caused hepatomegaly, but liver histology was normal.

Chronic 2-year studies showed a significant increase in the incidences of adenomas and carcinomas of the nasal cavity in high-dose rats fed diets containing 3000 ppm of 2,6-xylidine.[1] The carcinomas were highly invasive and frequently destroyed the nasal turbinates and nasal septum. Rhabdomyosarcomas, a rare tumor of the nasal cavity, were also observed in the high-dose males and females. The nonneoplastic lesions observed in the nasal cavity included acute inflammation, epithelial hyperplasia, and squamous metaplasia. In addition, subcutaneous fibromas and fibrosarcomas occurred in both males and females, and there was an increased incidence of neoplastic nodules in the livers of female rats.

The IARC has determined that there is sufficient evidence for the carcinogenicity of 2,6-xylidine in experimental animals and inadequate evidence in humans.[6] Overall, 2,6-xylidine is considered possibly carcinogenic to humans.

In genotoxic assays, 2,6-xylidine induced sister chromatid exchanges and chromosomal aberrations in cultured mammalian cells but did not induce micronuclei in the bone marrow of mice; conflicting results have been reported in the *Salmonella typhimurium* assay.[6]

The 1995 ACGIH threshold limit value–time-weighted average (TLV-TWA) for xylidine (mixed isomers) is 0.5 ppm (2.5 mg/m^3)

with an A2 suspected human carcinogen classification.

REFERENCES

1. National Toxicology Program: NTP technical report on the toxicology and carcinogenesis studies of 2,6-xylidine (2,6-dimethylaniline) (CAS No 87-62-7). In *Charles River CD Rats (Feed Studies)*. NTP TR 278, NIH Pub No 90-2534, pp 1–138. Research Triangle Park, NC, US Department of Health and Human Services, 1990
2. Vernot, EH, MacEwen ID, Haun GG, Kinkead ER: Acute toxicity and skin corrosion data for some organic and inorganic compounds and aqueous solutions. *Toxicol Appl Pharmacol* 42:417, 1977
3. McLean S, Starmer GA, Thomas J: Methaemoglobin formation by aromatic amines. *J Pharmacol* 21:441, 1969
4. Lindstrom HV, Bowie WG, Wallace WG, et al: The toxicity and metabolism of mesidine and pseudocumidine in rats. *J Pharmacol Exper Ther* 167:223, 1969
5. Magnusson G, Majeed SK, Down WH, et al: Hepatic effects of xylidine isomers in rats. *Toxicology* 12:63–74, 1979
6. *IARC Monographs on the Evaluation of Carcinogenic Risks to Humans*, Vol 57, Occupational exposures of hairdressers and barbers and personal use of hair colourants; some hair dyes, cosmetic colourants, industrial dyestuffs and aromatic amines, pp 323–335. Lyon, International Agency for Research on Cancer, 1993

YTTRIUM
CAS: 7440-65-5

Y

Compounds: Yttrium chloride; yttrium nitrate; yttrium oxide; yttrium phosphate

Physical Form. White powder

Uses. Yttrium is mixed with rare earths as

phosphors for color television receivers; the oxide is used for mantles in gas and acetylene lights; compounds are used in ceramics and superconductors

Exposure. Inhalation

Toxicology. Yttrium compounds cause pulmonary irritation in animals.

No effects in humans have been reported.

Intratracheal administration of 50 mg yttrium oxide in rats caused granulomatous nodules to develop in the lungs by 8 months.[1] Nodules in the peribronchial tissue compressed and deformed several bronchi; the surrounding lung areas were emphysematous, the interalveolar walls thin and sclerotic, and the alveolar cavities dilated. Intraperitoneal injection of yttrium chloride in animals caused peritonitis with serous or hemorrhagic ascites.[2] It was speculated that the development of ascites may have been related to the acidity of the administered solution rather than to the yttrium.[2] In a more recent report, the toxicity of intratracheally administered yttrium chloride, as determined by lactate dehydrogenase activity in bronchoalveolar lavage fluid, was judged to be higher than zinc oxide but lower than cadmium compounds.[3]

Intravenous administration of 1 mg yttrium chloride to rats caused formation of colloidal material in blood plasma, which accumulated primarily in the liver and spleen, causing injury to these organs.[4]

Application of a 0.1 M solution of yttrium chloride to the eyes of rabbits caused no injury; similar exposure of eyes from which the corneal epithelium had been removed resulted in immediate slight haziness of the cornea, which subsequently be came opaque and vascularized.[5]

The ACGIH threshold limit value–time-weighted average (TLV-TWA) for yttrium and compounds is 1 mg/m³ as yttrium.

REFERENCES

1. Stokinger HE: The metals. In Clayton GD, Clayton FE (eds): *Patty's Industrial Hygiene and Toxicology*, 3rd ed, rev, Vol 2A, *Toxicology*, p 1682. New York, Wiley-Interscience, 1981

2. Steffee CH: Histopathologic effects of rare earths administered intraperitoneally to rats. *AMA Arch Ind Health* 20:414–419, 1959
3. Hirano S, Kodama N, Shibata K, et al: Distribution, localization, and pulmonary effects of yttrium chloride following intratracheal instillation into the rat. *Toxicol Appl Pharmacol* 104:301–311, 1990
4. Hirano S, Kodama N, Shibata K, et al: Metabolism and toxicity of intravenously injected yttrium chloride in rats. *Toxicol Appl Pharmacol* 121:224–232, 1993
5. Grant WM: *Toxicology of the Eye*, 3rd ed, p 986. Springfield, IL, Charles C Thomas, 1986

ZINC CHLORIDE
CAS: 7646-85-7

ZnCl₂

Synonym: Zinc dichloride fume

Physical Form. White fume

Uses. Smoke generators; flux in soldering

Exposure. Inhalation

Toxicology. Zinc chloride fume is an irritant of the eyes, mucous membranes, and skin and, at very high concentrations, causes pulmonary edema.

Ten deaths and 25 cases of nonfatal injury occurred among 70 persons exposed to a high concentration of zinc chloride released from smoke generators.[1] Presenting symptoms were conjunctivitis (two cases with burns of the corneas), irritation of nose and throat, cough with copious sputum, dyspnea, constrictive sensation in the chest, stridor, retrosternal pain, nausea, epigastric pain, and cyanosis. Of the 10 fatalities, a few died immediately or within a few hours with pulmonary edema, whereas those who survived longer developed bronchopneumonia. Between the second and fourth days after exposure, almost all cases developed moist, adventitious sounds in the lungs, and the majority continued to present a pale, cyanotic color. A prominent feature was the disparity between the severe symptoms and the paucity of physical signs in the lungs. Recovery in survivors occurred within 1 to 6 weeks after the incident.

In a firefighter who was fatally exposed to a high but undetermined concentration of zinc chloride fume from a smoke generator, presenting symptoms were nausea, sore throat, and tightness in the chest, aggravated by deep inspiration.[2] The patient improved initially but developed tachypnea, substernal soreness, fever, cyanosis, and coma. The lung fields were clear on auscultation despite diffuse pulmonary infiltrations seen on the chest roentgenogram. Death occurred 18 days after exposure, and autopsy revealed active fibroblastic proliferation of lung tissue and cor pulmonale.

In an airport disaster drill, an outdoor exposure to zinc chloride aerosol following the detonation of a smoke bomb resulted in upper respiratory tract irritative symptoms in the victims, correlating with the presumed intensity and duration of exposure.[3] Questionnaire responses by 81 exposed individuals most commonly reported cough, hoarseness, and sore throat, with onset primarily at the time of exposure. Other symptoms among individuals with self-reported moderate and heavy exposures included listlessness, metallic taste, light-headedness, chest tightness, and soreness in the chest. Wheezing was relatively uncommon and, by spirometry, 1 to 2 days after exposure, the mean results for FEV_1 and FVC as a percentage of predicted values were normal. The predominance of upper respiratory symptoms was attributed to the solubility and hygroscopic tendency of zinc chloride, resulting in upper respiratory tract deposition. On dissolution of zinc chloride, both hydrochloric acid and zinc oxychloride are formed, contributing to the corrosive action. Most of the exposed victims became asymptomatic within 48 hours, but symptoms persisted in a few patients for up to several weeks.[3]

Accidental instillation in a human eye of one drop of a 50% zinc chloride solution caused immediate severe pain, which persisted despite immediate irrigation with water. The corneal epithelium was burned, and corneal vascularization followed. After many weeks, areas of opacification and vascularization remained in the cornea.[4] Zinc chloride has caused ulceration of the fingers, hands, and forearms of workers who used it as flux in soldering.[5]

Mice and rats exposed to 122 mg zinc/m³ as zinc chloride for 1 hr/day, 5 days/week, survived 20 weeks of exposure but showed increased macrophages in lungs when sacrificed 13 months following exposure.[6] In guinea pigs, 120 mg zinc/m³ as zinc chloride for 1 hr/day, 5 days/week, for up to 3 weeks was lethal; at necropsy, focal alveolitis, consolidation, emphysema, infiltration with macrophages, and fibrosis were observed.

Injection of zinc chloride solution into the testes of 49 Syrian hamsters resulted in areas of necrosis occupying about 25% of each testis; two embryonal carcinomas of the testis were found 10 weeks later at necropsy.[7] There is no evidence that zinc compounds are carcinogenic after administration by any other route.[8]

In general, exposure to zinc chloride does not increase mutation frequencies in bacterial or mammalian test systems.[9]

The 1995 ACGIH threshold limit value–time-weighted average (TLV-TWA) for zinc chloride fume is 1 mg/m³, with a short-term excursion level of 2 mg/m³.

REFERENCES

1. Evans EH: Casualties following exposure to zinc chloride smoke. *Lancet* 2:368–370, 1945
2. Milliken JA, Waugh D, Kadish ME: Acute interstitial pulmonary fibrosis caused by a smoke bomb. *Can Med Assoc J* 88:36–39, 1963
3. Schenker MB, Speizer FE, Taylor JO: Acute upper respiratory symptoms resulting from exposure to zinc chloride aerosol. *Environ Res* 25:317–324, 1981
4. Grant WM: *Toxicology of the Eye*, 3rd ed, pp 986–987. Springfield, IL, Charles C Thomas, 1986
5. Stokinger HE: The metals (excluding lead). In Fassett DW, Irish DD (eds): *Patty's Industrial Hygiene and Toxicology*, 2nd ed, Vol 2, *Toxicology*, pp 1182–1188. New York, Interscience, 1963
6. Marrs TC, Colgrave HF, Edginton JAG, et al: The repeated dose toxicity of a zinc oxide/

hexachloroethane smoke. *Arch Toxicol* 62:123–132, 1988

7. Guthrie J, Guthrie OA: Embryonal carcinomas in Syrian hamsters after intratesticular inoculation of zinc chloride during seasonal testicular growth. *Cancer Res* 34:2612–2613, 1974
8. Elinder CG: Zinc. In Friberg L, Nordberg GF, Vouk VB (eds): *Handbook on the Toxicology of Metals*. Vol II, *Specific Metals*, pp 664–679. New York, Elsevier, 1986
9. Agency for Toxic Substances and Disease Registry (ATSDR): *Toxicological Profile for Zinc*. US Department of Health and Human Services, Public Health Service, pp 1–133, 1992

ZINC OXIDE
CAS: 1314-13-2

ZnO

Synonyms: Calamine; zincite

Physical Form. White to yellowish powder that may exist as a fume or dust

Uses. Metallic zinc in galvanizing, electroplating, dry cells, and alloying; zinc oxide in pigments

Exposure. Inhalation

Toxicology. Inhalation of zinc oxide fume causes an influenzalike illness termed metal-fume fever.

During human exposure to zinc oxide fume, effects are dryness and irritation of the throat, a sweet or metallic taste, substernal tightness and constriction in the chest, and a dry cough.[1-4] Several hours following exposure, the subject develops chills, lassitude, malaise, fatigue, frontal headache, low back pain, muscle cramps and, occasionally, blurred vision, nausea, and vomiting. Physical signs include fever, perspiration, dyspnea, rales through the chest, and tachycardia; in some instances, there has been a reversible reduction in pulmonary vital capacity. There is usually leukocytosis, which may reach 12,000 to 16,000/cm.[3] The pathogenesis of the syndrome is not clear, but an allergic response has been suggested, with zinc entering the blood circulation and forming a sensitizing complex with plasma proteins.[5]

An attack usually subsides after 6 to 12 hours but may last for up to 24 hours; recovery is usually complete.[3] Most workers develop an immunity to these attacks, but it is quickly lost; attacks tend to be more severe on the first day of the workweek.[3] Despite the severity of the acute subjective symptoms, there appear to be no consistent functional or pathological respiratory effects attributable to chronic exposure.[1]

The critical factor in the development of the syndrome is the size of the ultrafine zinc oxide particles produced when zinc is heated to temperatures approaching its boiling point of 907°C in an oxidizing atmosphere.[5] The particles must be small enough (<1 μm) to reach the alveoli when inhaled. The syndrome is not produced when normal zinc oxide powder is either inhaled or taken orally.[2] Only freshly formed fume causes the illness, presumably because flocculation occurs in the air, with formation of larger particles that are deposited in the upper respiratory tract and do not penetrate deeply into the lungs.[6]

Data on exposure concentrations and durations associated with metal-fume fever are insufficient.[7] Early reports found moderate symptoms following 12 minutes at 600 mg/m³; other investigators found no signs of chronic toxicity with occupational exposures of 3 to 15 mg/m³ for periods of up to 35 years. In a recent report, each of 4 volunteers reported one or more of the symptoms of metal-fume fever (sore throat, chest tightness, and/or headache) approximately 6 to 10 hours after a 2-hour exposure at 5 mg/m³ (particle size 0.06 μm).[8] The symptoms were not accompanied by changes in pulmonary function. Other investigators have found that exposure to 23 to 171 mg/m³ for up to 30 minutes results in the increase of several pulmonary lavage parameters, including neutrophils, macrophages, and activated T lymphocytes.[9]

A short-term study of guinea pigs exposed to zinc oxide fume for 3 hr/day for 6 days at the TLV of 5 mg/m³ revealed pulmonary function changes and morphologic evidence of small airway inflammation and edema. Pulmonary flow resistance increased, compliance decreased, and lung volumes and carbon monoxide diffusing capacity decreased. Some of these changes persisted for the 72-hour duration of postexposure follow-up.[10]

Zinc oxide is not considered to be a skin irritant.[4] Early reports of workers whose skin was frequently covered with zinc oxide and who developed pustules on the axilla and inner thighs attributed this condition to clogging of sebaceous glands by sweat, bacteria, and dust, with subsequent infection.

In general, genotoxic studies have not found evidence for mutagenicity of zinc.[4]

The larger particle sizes of zinc oxide dusts are considered nuisance dusts that have little adverse effect on the lung and do not produce significant organic disease when exposures are kept under reasonable control.[1]

The ACGIH threshold limit value–time-weighted average (TLV-TWA) for zinc oxide fume is 5 mg/m³, with a short-term exposure limit of 10 mg/m³; the dust has a TLV-TWA of 10 mg/m³.

REFERENCES

1. National Institute for Occupational Safety and Health: *Criteria for a Recommended Standard . . . Occupational Exposure to Zinc Oxide.* DHEW (NIOSH) Pub No 76-104, pp 36–38. Washington, DC, US Government Printing Office, 1975
2. McCord CP: Metal-fume fever as an immunological disease. *Ind Med Surg* 29:101–107, 1960
3. Rohrs LC: Metal-fume fever from inhaling zinc oxide. *AMA Arch Ind Health* 16:42–47, 1957
4. Agency for Toxic Substances and Disease Registry (ATSDR): *Toxicological Profile for Zinc.* US Department of Health and Human Services, Public Health Service, p 133, 1992
5. Brown JJL: Zinc fume fever. *Br J Radiol* 61:327–329, 1988
6. Hygienic Guide Series: Zinc oxide. *Ann Ind Hyg Assoc J* 30:422–424, 1969
7. Elinder CG: Zinc. In Friberg L, Nordbert GF, Vouk VB (eds): *Handbook on the Toxicology of Metals*, Vol II, *Specific Metals*, pp 664–679. New York, Elsevier, 1986
8. Gordon T, Chen LC, Fine JM, et al: Pulmonary effects of inhaled zinc oxide in human subjects, guinea pigs, rats, and rabbits. *Am Ind Hyg Assoc J* 53:503–509, 1992
9. Blanc PD, Wong H, Bernstein MS, et al: An experimental human model of metal fume fever. *Ann Int Med* 114:930–936, 1991
10. Lam HF, Conner MW, Rogers AE, et al: Functional and morphologic changes in the lungs of guinea pigs exposed to freshly generated ultrafine zinc oxide. *Toxicol Appl Pharmacol* 78:29–38, 1985

ZIRCONIUM COMPOUNDS
CAS: 7440-67-7

Zr

Synonyms: Zirconium dioxide; zirconium silicate; zirconium tetrachloride

Physical Form. Solids

Uses. Structural material for atomic reactors; ingredient in priming and explosive mixtures; reducing agent; pigment; textile water repellent

Exposure. Inhalation

Toxicology. Zirconium compounds are of low toxicity although granulomas have been produced by repeated topical applications of zirconium salts to human skin.

A study of 22 workers exposed to fume from a zirconium reduction process for 1 to 5 years revealed no abnormalities related to the exposure.[1] There are no well-documented cases of toxic effects from industrial exposure. Two persons given zirconium malate in 50 mg intravenous injections developed vertigo.[2] Granulomas of the human axillary skin have occurred from use of deodorants or poison ivy remedies containing zirconium.[3]

In rats, the oral LD$_{50}$ of several zirconium

compounds ranged from 1.7 to 10 g/kg.[4] Animals acutely poisoned by zirconium compounds show progressive depression and decrease in activity until death.[2] Repeated inhalation of zirconium tetrachloride mist by dogs for 2 months at 6 mg zirconium/m^3 caused slight decreases in hemoglobin and in erythrocyte counts, with some increased mortality over that of controls. These effects may have been due to the liberation of hydrogen chloride.[4] Animals exposed to zirconium dioxide dust for 1 month at 75 mg zirconium/m^3 showed no detectable effects. Rats exposed to high concentrations of zirconium silicate dust for 7 months developed radiographic shadows in the lungs; these were attributed solely to the deposition of the radiopaque particles since histologic examination showed no cellular reaction. The addition of 5 ppm of zirconium as the sulfate to the drinking water of mice for their lifetime did not increase the incidence of tumors.[5]

The 1995 ACGIH threshold limit value–time-weighted average (TLV-TWA) for Zr is 5 mg/m^3, with a short-term excursion level of 10 mg/m^3.

REFERENCES

1. Reed CE: A study of the effects on the lung of industrial exposure to zirconium dusts. *AMA Arch Ind Health* 13:578–580, 1956
2. Smith IC, Carson BL: *Trace Metals in the Environment*, Vol 3, *Zirconium*, p 405. Ann Arbor, MI, Ann Arbor Science Publishers, 1978
3. Shelley WB, Hurley HJ: The allergic origin of zirconium deodorant granulomas. *Br J Dermatol* 70:75–101, 1958
4. Stokinger HE: The metals. In Clayton GD, Clayton FE (eds): *Patty's Industrial Hygiene and Toxicology*, 3rd ed, rev, Vol 2A, *Toxicology*, pp 2049–2059. New York, Wiley-Interscience, 1981
5. Kanisawa M, Schroeder HA: Life term studies on the effect of trace elements on spontaneous tumors in mice and rats. *Cancer Res* 29:892–895, 1969

CAS NUMBER INDEX

INDEX OF COMPOUNDS AND SYNONYMS